Teacher's Edition

Integrated 1
Mathematics

ALGEBRA

discrete

x^2

Authors

Senior Authors

Rheta N. Rubenstein

Timothy V. Craine

Thomas R. Butts

Kerry Cantrell

Linda Dritsas

Valarie A. Elswick

Joseph Kavanaugh

Sara N. Munshin

Stuart J. Murphy

Anthony Piccolino

Salvador Quezada

Jocelyn Coleman Walton

McDougal Littell
A Houghton Mifflin Company
Evanston, Illinois   Boston Dallas

Authors

Senior Authors

Rheta N. Rubenstein Professor of Education, University of Windsor, Windsor, Ontario

Timothy V. Craine Assistant Professor of Mathematical Sciences, Central Connecticut State University, New Britain, Connecticut

Thomas R. Butts Professor of Mathematics Education, University of Texas at Dallas, Dallas, Texas

Kerry Cantrell Mathematics Department Head, Marshfield High School, Marshfield, Missouri

Linda Dritsas Mathematics Coordinator, Fresno Unified School District, Fresno, California

Valarie A. Elswick Mathematics Teacher, Roy C. Ketcham Senior High School, Wappingers Falls, New York

Joseph Kavanaugh Academic Head of Mathematics, Scotia-Glenville Central School District, Scotia, New York

Sara N. Munshin Mathematics Teacher, Los Angeles Unified School District, Los Angeles, California

Stuart J. Murphy Visual Learning Specialist, Evanston, Illinois

Anthony Piccolino Assistant Professor of Mathematics and Computer Science, Montclair State University, Upper Montclair, New Jersey

Salavador Quezada Mathematics Teacher, Theodore Roosevelt High School, Los Angeles, California

Jocelyn Coleman Walton Educational Consultant, Mathematics K-12, and former Mathematics Supervisor, Plainfield High School, Plainfield, New Jersey

The authors wish to thank **Jane Pflughaupt**, Mathematics Teacher, Pioneer High School, San Jose, California, for her contribution to this Teacher's Edition.

ISBN: 0-395-85503-9

1 2 3 4 5 6 7 8 9 10–VJM–03 02 01 00 99 98 97

Contents of the Teacher's Edition

PHILOSOPHY
of Integrated Mathematics

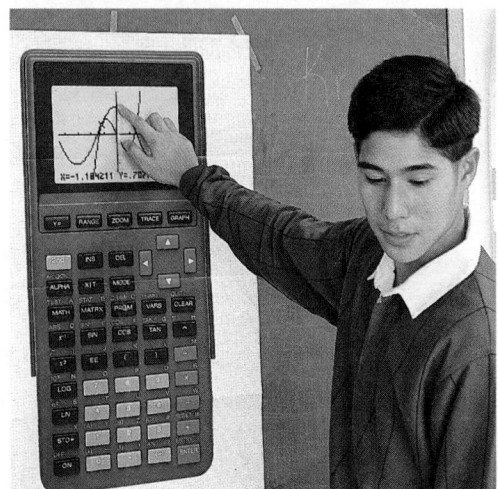

Goals of the Course

Integrated Mathematics has been written to prepare your students for success in college, and in their careers and daily lives in the 21st century, by helping them develop their abilities to:

- Explore and solve mathematical problems
- Think critically
- Work cooperatively with others
- Communicate ideas clearly

Underlying Concept

This program is built on the idea that students develop better conceptual understanding of mathematics and stronger problem solving skills when they:

- See the connections among different branches of mathematics
- Are actively involved in the learning process
- Study mathematics that is meaningful

Accessible and Inviting Mathematics

Integrated Mathematics was designed to make mathematics accessible and inviting. It opens the door to mathematics, and to college and career opportunities, for more students by incorporating a variety of different teaching strategies, including:

- Visual and hands-on approaches
- Real-life applications
- Exploratory activities and projects
- Use of technology
- Group work
- Open-ended problem solving

A Manageable Program

Integrated Mathematics makes it easy for you to manage these teaching strategies by incorporating them directly into the textbook — at the places where you would use them in teaching. In addition, ongoing assessment that matches the instruction is included throughout the course.

Basis of the Curriculum

This program is based on the recommendations of the National Council of Teachers of Mathematics and other curriculum groups that emphasize problem solving, critical thinking, communication, and connections among mathematical topics and connections between mathematics and other subject areas.

Mathematical Content

Over a three-year period, *Integrated Mathematics* teaches the same mathematical topics as a contemporary Algebra 1/Geometry/Algebra 2 sequence. The difference is in the organization of the content.

Instead of being divided into separate courses, algebra and geometry are taught in each of the three years. In addition, topics from logical reasoning, measurement, probability, statistics, discrete mathematics, and functions are interwoven throughout each year.

Advantages of an Integrated Approach

With an integrated approach, your students can:

➤ **Learn more mathematics**
➤ **Solve problems that are more realistic and more interesting**
➤ **Have better retention of what they have learned**

Field Testing

Preliminary versions of this book were tried out by hundreds of teachers and thousands of students in many different types of classrooms nationwide. Their comments and suggestions have guided the development of this book.

Here is what some teachers who piloted the book have said:

"The kids get involved and they enjoy it. I think that helps them tremendously in learning."

"Very often there's an 'ah-hah!' at the end, and 'oh, that's why this works.' I think they're feeling much more confident in what mathematics is, and they seem to be remembering it for a longer time."

"I really think students see mathematics as a whole. In problem solving they're a lot more willing to try things in different ways and pull in things from different areas."

Contents

Unit 1   This unit features the integration of topics and the teaching/learning strategies that continue throughout the book. Concepts from algebra, geometry, and data analysis are introduced, as are activities involving exploration, cooperative learning, communication, and use of technology.

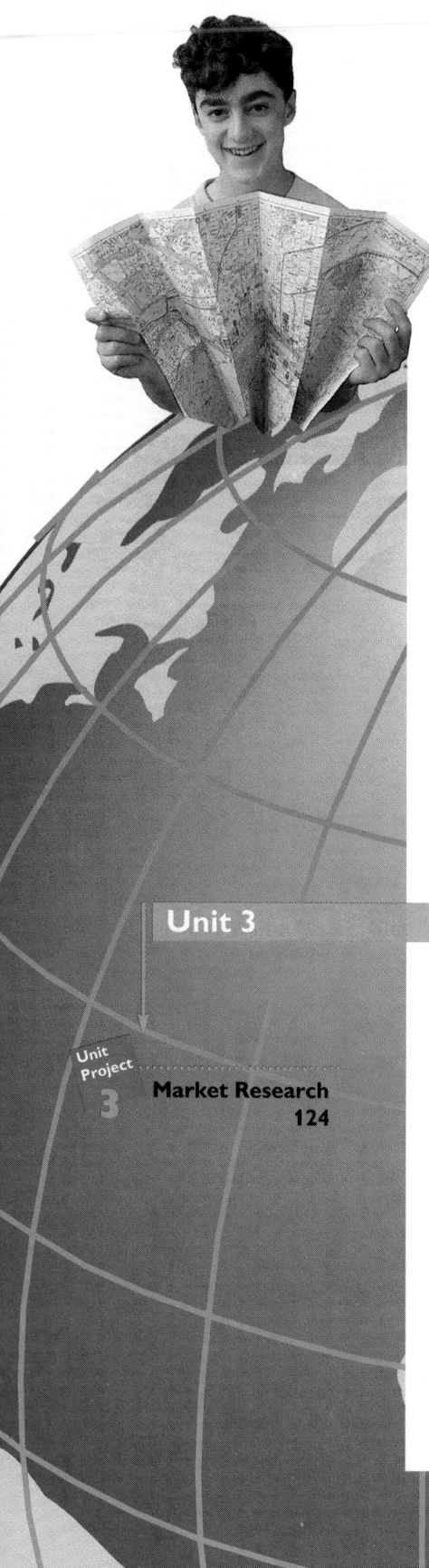

Unit 3   Representing Data

Table of Contents

Unit 4   Coordinates and Functions

Animation 180

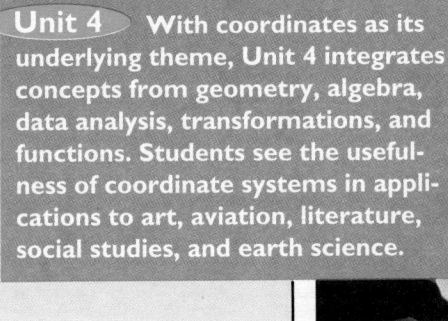

Unit 4   With coordinates as its underlying theme, Unit 4 integrates concepts from geometry, algebra, data analysis, transformations, and functions. Students see the usefulness of coordinate systems in applications to art, aviation, literature, social studies, and earth science.

viii

Table of Contents

Unit 5   Modeling everyday situations using linear equations and geometric formulas is the focus of this unit. Technology, manipulatives, and visual displays are used throughout to facilitate learning.

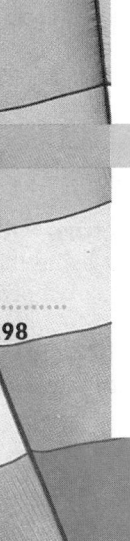

Table of Contents

Unit 6   In this unit, ratio is the key idea that unifies ideas from algebra, geometry, probability, statistics, and trigonometry. Explorations occur throughout the unit, and numerous research, hands-on, cooperative, and communication activities help reinforce learning.

Unit 7   Direct Variation

Unit Project

Unit 7 Using direct variation as the
underlying idea, Unit 7 ties together
concepts from algebra, geometry, and
trigonometry. Through the use of
explorations and numerous applications,
students apply the mathematical concepts of the unit to everyday situations.

x

Table of Contents

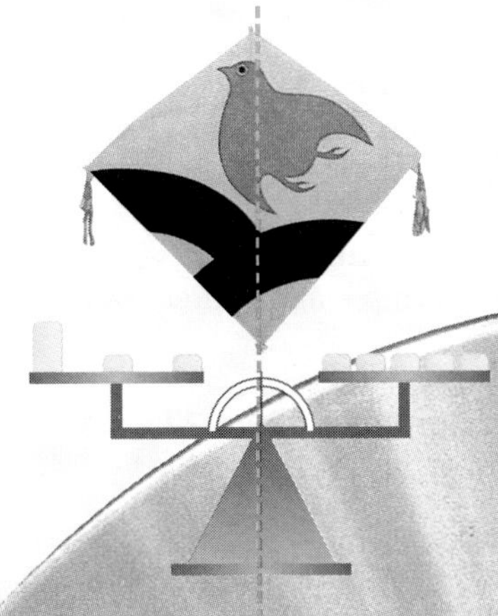

Unit 8   Building on the discussion of direct variation in Unit 7, this unit demonstrates how linear equations can be used to model a variety of different everyday situations. To aid understanding, many concepts are presented both visually and verbally.

Table of Contents

Unit 9   This unit focuses on logical reasoning, a topic that underlies all of mathematics. In explorations and problems, students use logical reasoning to discover and explain concepts relating to algebra, geometry, and probability.

Unit 10   **Quadratic Equations as Models**

Unit 10 Like Unit 5, this unit uses equations to model problem situations, this time concentrating on quadratic equations. Students use both algebra tiles and a graphics calculator to explore quadratic functions. Connections are made to business, art, urban planning, history, and biology.

xii   **Table of Contents**

Integrated Mathematics   Topic Spiraling

This chart shows how mathematical strands are spiraled over the three years of the *Integrated Mathematics* program.

	Course 1	Course 2	Course 3
Algebra	Linear equations Linear inequalities Multiplying binomials Factoring expressions	Quadratic equations Linear systems Rational equations Complex numbers	Polynomial functions Exponential functions Logarithmic functions Parametric equations
Geometry	Angles, polygons, circles Perimeter, circumference Area, surface area Volume Trigonometric ratios	Similar and congruent figures Geometric proofs Coordinate geometry Transformational geometry Special right triangles	Inscribed figures Transforming graphs Vectors Triangle trigonometry Circular trigonometry
Statistics, Probability	Analyzing data and displaying data Experimental and theoretical probability Geometric probability	Sampling methods Simulation Binomial distributions	Variability Standard deviation z-scores
Logical Reasoning	Conjectures Counterexamples If-then statements	Inductive and deductive reasoning Valid and invalid reasoning Postulates and proof	Identities Contrapositive and inverse Comparing proof methods
Discrete Math	Discrete quantities Matrices to display data Lattices	Matrix operations Transformation matrices Counting techniques	Sequences and series Recursion Limits

Teachers today are being asked to teach more mathematics, and better mathematics, to more students.

Integrated *Approach*

Integrated Mathematics interweaves mathematical topics and contemporary teaching strategies throughout the course. Key mathematical strands are spiraled through the units — and integrated within individual sections.

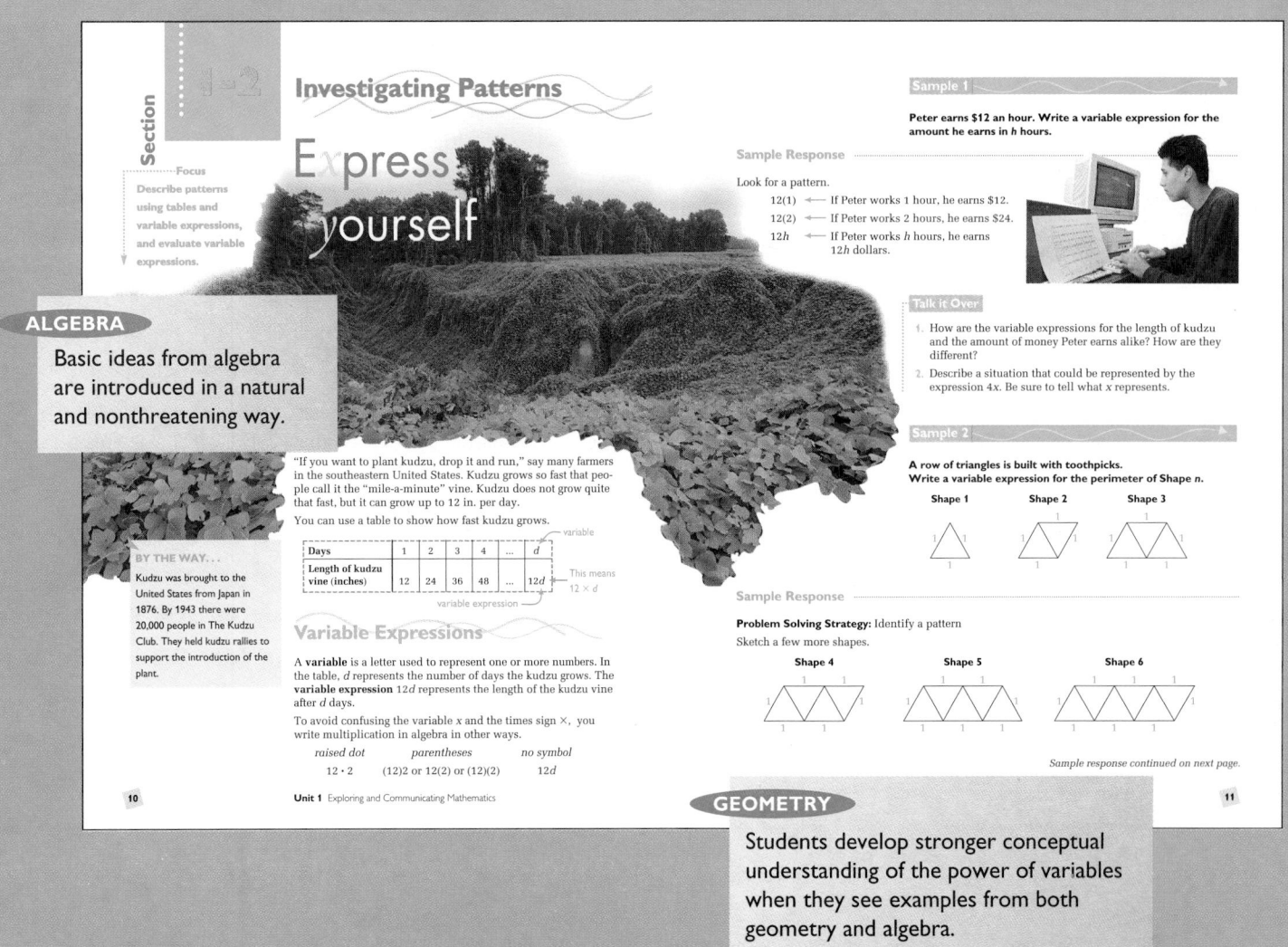

ALGEBRA

Basic ideas from algebra are introduced in a natural and nonthreatening way.

GEOMETRY

Students develop stronger conceptual understanding of the power of variables when they see examples from both geometry and algebra.

ALGEBRA

Solving linear equations 2-7, 2-8, 3-2, 5-1, 5-2, 5-3, 5-6, 6-3, 7-4

Writing equations to model problem situations 2-8, 5-1, 5-3, 5-8, 7-2, 8-1, 8-2

Coordinate graphing 4-2, 4-6, 4-7, 5-1, 7-2, 7-4, 8-1, 8-5, 8-6, 8-7, 10-7

Solving inequalities 5-4, 8-6

Solving systems of equations 5-8, 8-5

Multiplying binomials 10-6

Factoring trinomials 10-7

Solving quadratic equations 10-8

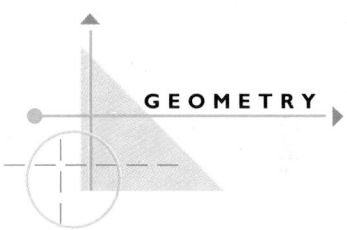

GEOMETRY

Angles 1-7, 2-5, 5-8, 6-7, 7-1, 7-3

Polygons 1-6, 1-7, 2-5, 4-2, 5-7, 6-5, 6-6, 6-7, 7-1

Circles 7-3, 7-6

Length, perimeter, and circumference 1-7, 2-4, 7-3, 9-1

Area and surface area 2-4, 5-7, 7-6, 9-5, 9-8

Volume 2-9, 9-6, 9-7, 9-8

Congruence and similarity 1-6, 2-4, 2-5, 6-5, 6-6, 9-8

STATISTICS & PROBABILITY

Analyzing and displaying data 1-1, Unit 3, 4-5, 6-4, 7-2, 8-1, 9-8

Probability 2-1, 6-2, 9-4

DISCRETE MATH

$_5P_5$

Discrete (countable) quantities 2-1, 3-3

Matrices 3-1

Lattices 8-2, 8-6

SPIRAL LEARNING

Section 1-2 introduces basic ideas from algebra and geometry that form a foundation for later work. Some sections where key topics that build on these ideas are integrated into the course are listed here.

DATA ANALYSIS

Organizing data in a table and looking for patterns in data are important problem solving strategies in many areas of mathematics.

Make a table of the perimeters.

Shape number	1	2	3	4	5	6	n
Perimeter	3	4	5	6	7	8	?

Look for a pattern. The perimeter is 2 more than the shape number.

An expression for the perimeter of Shape n is $n + 2$.

Evaluating Variable Expressions

When you evaluate a variable expression, you substitute numbers for the variables and find the result.

Sample 3

Suppose a kudzu vine grows 12 in. a day. How long is the vine after each number of days?

a. 7   b. 30   c. 365

Sample Response

Write the variable expression first.

a. $12d$
$12(7)$ ← Substitute 7 for d.
84
After a week, the kudzu is 84 in. long.

b. $12d$
$12(30)$ ← Substitute 30 for d.
360
After a month, the kudzu is 360 in. long.

c. $12d$
$12(365)$ ← Substitute 365 for d.
4380
After a year, the kudzu is 4380 in. long.

Talk it Over

3. Sample 2 shows triangles made from toothpicks. How many toothpicks would you need to build a row of a million triangles?

4. Evaluate each expression when $a = 3$ and $b = 7$.

a. $5a$ b. $\frac{b}{14}$

c. ab d. $6a + b$

⋯▸ Now you are ready for:
Exs. 1–25 on pp. 14–16

I want *all* my students to have the mathematical and problem solving skills they need. But some students just don't seem interested in learning math.

How can I get them involved?

Active
Learning

Integrated Mathematics makes it easy for you to get your students actively involved because projects, explorations, activities, and discussion questions are built right into the book.

EXPLORATIONS

The Explorations get students involved in investigating math. They build strong conceptual understanding by helping students move from the concrete to the abstract.

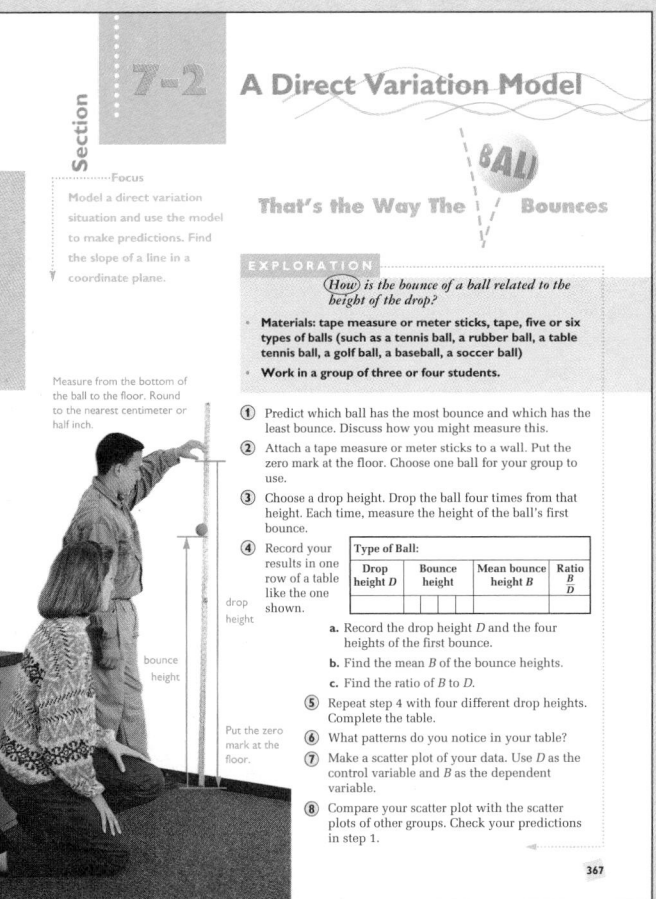

Section 7-2

A Direct Variation Model

Focus
Model a direct variation situation and use the model to make predictions. Find the slope of a line in a coordinate plane.

That's the Way The **BALL** Bounces

EXPLORATION

How is the bounce of a ball related to the height of the drop?

- **Materials:** tape measure or meter sticks, tape, five or six types of balls (such as a tennis ball, a rubber ball, a table tennis ball, a golf ball, a baseball, a soccer ball)
- **Work in a group of three or four students.**

1. Predict which ball has the most bounce and which has the least bounce. Discuss how you might measure this.
2. Attach a tape measure or meter sticks to a wall. Put the zero mark at the floor. Choose one ball for your group to use.
3. Choose a drop height. Drop the ball four times from that height. Each time, measure the height of the ball's first bounce.
4. Record your results in one row of a table like the one shown.

Type of Ball:			
Drop height D	Bounce height	Mean bounce height B	Ratio $\frac{B}{D}$

   a. Record the drop height D and the four heights of the first bounce.
   b. Find the mean B of the bounce heights.
   c. Find the ratio of B to D.
5. Repeat step 4 with four different drop heights. Complete the table.
6. What patterns do you notice in your table?
7. Make a scatter plot of your data. Use D as the control variable and B as the dependent variable.
8. Compare your scatter plot with the scatter plots of other groups. Check your predictions in step 1.

Measure from the bottom of the ball to the floor. Round to the nearest centimeter or half inch.

drop height

bounce height

Put the zero mark at the floor.

367

Each unit begins with a Unit Project. These projects give your students a chance to work on the types of open-ended, long-range problems that prepare them for future careers.

The Unit Projects
➤ put the mathematics in context
➤ unify related mathematical topics
➤ give all students an opportunity to participate and contribute

A Mathematical Model: Direct Variation

To model the behavior of a bouncing ball with an equation, look for a relationship between the two variables, the drop height D and the mean bounce height B.

Sample 1

In doing the Exploration, Julio, Tara, and Linda recorded the values shown in the table.

a. **Writing** Julio thinks the relationship between D and B can be modeled by direct variation. Explain his reasoning.

b. Model the behavior of the ball with a direct variation equation relating D and B. What is the variation constant for the ball?

Drop height D (in.)	Mean bounce height B (in.)	Ratio $\frac{B}{D}$
20	15.0	0.75
30	24	0.8
40	32.5	0.8125
50	41	0.82
60	48.5	$0.80\overline{83}$

Sample Response

a. When two variables have a constant ratio, you can say that one variable varies directly with the other.

State the meaning of direct variation.

Although the ratios $\frac{B}{D}$ are not all exactly the same, to the nearest tenth each ratio is 0.8. Direct variation is a good model for the relationship between D and B.

Explain why the data fit a direct variation model.

b. The direct variation equation $\frac{B}{D} = 0.8$ models the behavior of the bouncing ball. The variation constant for the ball is about 0.8.

Unit Project 7: Design a Sports Arena

Your project is to make a scale drawing or a model of an arena for skateboarding or in-line skating competitions.

Your group's design should include at least one ramp for launching jumps. The ramp may be straight or curved. You may want to create several "stations," each with a different kind of ramp.

Along with your design, you should include an estimate of the cost of building your arena.

Construction of the skateboard bowl arena at The Hanger in Charleston, N.C.

One arena can be the site of a concert on Friday, a basketball game Saturday afternoon, a hockey game Saturday night, and a trip to the circus on Sunday! With a good design, a multi-event arena can change from a floor of ice to hardwood within a few hours. The design of other arenas, like those for skateboarding events, affects the difficulty level of the competition.

357

MULTIPLE REPRESENTATIONS

Seeing different representations of the same problem situation (table, equation, graph) helps students become creative problem solvers.

Sample 2

To model the data from the table in Sample 1, Tara made a scatter plot of the five data points (D, B). Then she drew a fitted line.

Use Tara's graph to predict how high the ball will bounce if she drops it from a height of 42 in.

Sample Response

1 From the 42 in. mark on the horizontal axis, draw a line up until it reaches the fitted line.

The drop height is the control variable.

2 Then draw a line to the left until it meets the vertical axis at about 34 in.

The bounce height depends on the drop height.

If Tara drops the ball from a height of 42 in., it will bounce to about 34 in.

BY THE WAY...

The fuzzy felt on tennis balls slows them down and gives them a better "grip" on different surfaces. The dimples on a golf ball lift them farther through the air.

Talk it Over

1. Use the equation in part (b) of Sample 1 to predict how high the ball will bounce if the group drops it from a height of 42 in. Compare your answer to the prediction in Sample 2.

2. Instead of using the equation $\frac{B}{D} = 0.8$ in question 1, Linda used the equation $B = 0.8D$. Do you think this equation is correct? Explain.

3. Julio wants the ball to bounce to a height of 12 in. How can Tara's graph help Julio estimate the drop height he should use?

4. What algebraic equation could Julio solve to find the drop height that will give a bounce height of 12 in.? Show how to solve it.

Finding Slope Using Coordinates

In Section 7-1, you learned about the slope of a line. You can use the coordinates of two points on a line to find the line's slope.

7-2 A Direct Variation Model   369

DISCUSSION

The Talk it Over questions encourage students to think and talk about what they are learning.

PROJECT EXERCISES

The Working on the Unit Project exercises in each section help students build the knowledge and the skills they need to complete the project.

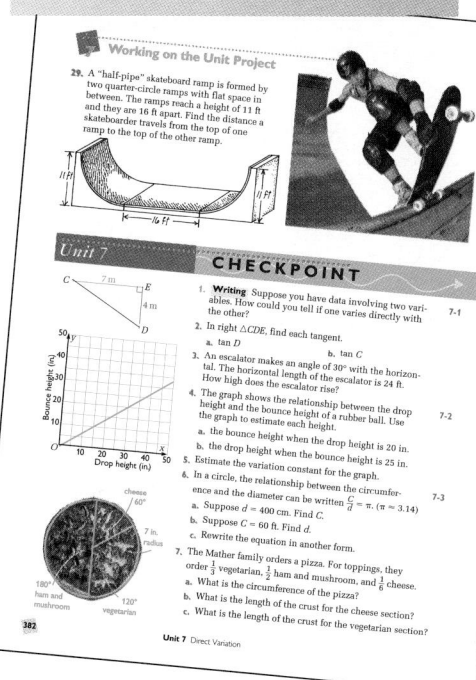

Working on the Unit Project

29. A "half-pipe" skateboard ramp is formed by two quarter-circle ramps with flat space in between. The ramps reach a height of 11 ft and they are 16 ft apart. Find the distance a skateboarder travels from the top of one ramp to the top of the other ramp.

Unit 7   CHECKPOINT

1. **Writing** Suppose you have data involving two variables. How could you tell if one varies directly with the other?   7-1

2. In right $\triangle CDE$, find each tangent.
   a. tan D b. tan C

3. An escalator makes an angle of 30° with the horizontal. The horizontal length of the escalator is 24 ft. How high does the escalator rise?

4. The graph shows the relationship between the drop height and the bounce height of a rubber ball. Use the graph to estimate each height.   7-2
   a. the bounce height when the drop height is 20 in.
   b. the drop height when the bounce height is 25 in.

5. Estimate the variation constant for the graph.

6. In a circle, the relationship between the circumference and the diameter can be written $\frac{C}{d} = \pi$. ($\pi \approx 3.14$)   7-3
   a. Suppose $d = 400$ cm. Find C.
   b. Suppose $C = 60$ ft. Find d.
   c. Rewrite the equation in another form.

7. The Mather family orders a pizza. For toppings, they order $\frac{1}{3}$ vegetarian, $\frac{1}{2}$ ham and mushroom, and $\frac{1}{6}$ cheese.
   a. What is the circumference of the pizza?
   b. What is the length of the crust for the cheese section?
   c. What is the length of the crust for the vegetarian section?

382   Unit 7 Direct Variation

T17

I'm tired of my students asking, "When am I ever going to use this?" I want them to realize that mathematics is useful and powerful.

How can I convince them?

Meaningful *Mathematics*

Integrated Mathematics focuses on important concepts in mathematics and shows how they can be applied to solve a wide variety of types of problems in daily life and in careers.

MATHEMATICAL MODELING

Modeling problem situations is an essential topic that is emphasized throughout the course. Students see many examples of linear and quadratic modeling.

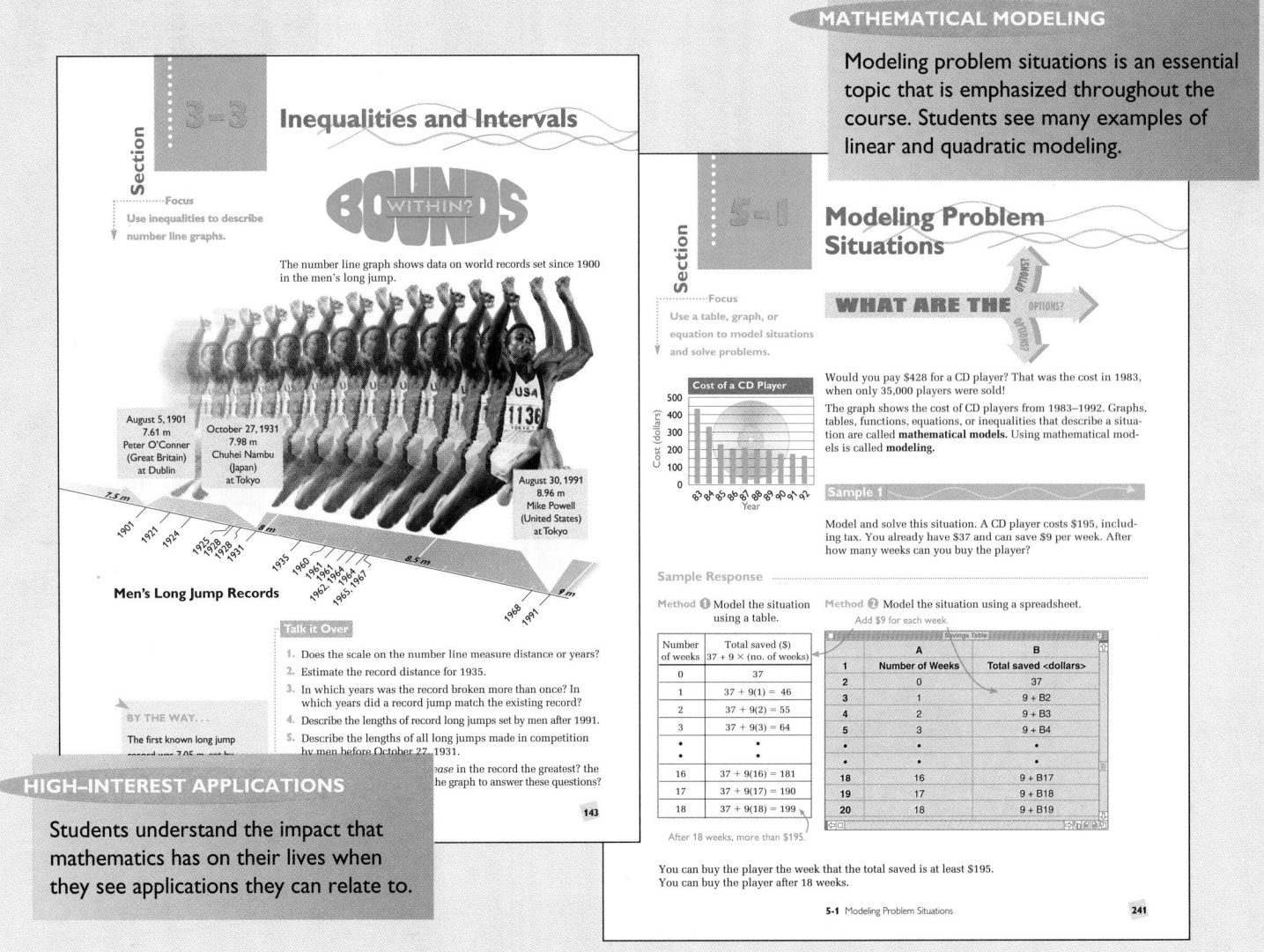

Section 3-3

Inequalities and Intervals

BOUNDS WITHIN?

— Focus
Use inequalities to describe number line graphs.

The number line graph shows data on world records set since 1900 in the men's long jump.

August 5, 1901
7.61 m
Peter O'Conner
(Great Britain)
at Dublin

October 27, 1931
7.98 m
Chuhei Nambu
(Japan)
at Tokyo

August 30, 1991
8.96 m
Mike Powell
(United States)
at Tokyo

Men's Long Jump Records

Talk it Over

1. Does the scale on the number line measure distance or years?
2. Estimate the record distance for 1935.
3. In which years was the record broken more than once? In which years did a record jump match the existing record?
4. Describe the lengths of record long jumps set by men after 1991.
5. Describe the lengths of all long jumps made in competition by men before October 27, 1931.

BY THE WAY...
The first known long jump record was 7.05 m set by ...

...ease in the record the greatest? the ... he graph to answer these questions?

143

HIGH–INTEREST APPLICATIONS

Students understand the impact that mathematics has on their lives when they see applications they can relate to.

Section 5-1

Modeling Problem Situations

WHAT ARE THE OPTIONS?

— Focus
Use a table, graph, or equation to model situations and solve problems.

Would you pay $428 for a CD player? That was the cost in 1983, when only 35,000 players were sold!

The graph shows the cost of CD players from 1983–1992. Graphs, tables, functions, equations, or inequalities that describe a situation are called **mathematical models.** Using mathematical models is called **modeling.**

Sample 1

Model and solve this situation. A CD player costs $195, including tax. You already have $37 and can save $9 per week. After how many weeks can you buy the player?

Sample Response

Method ❶ Model the situation using a table.

Number of weeks	Total saved ($) 37 + 9 × (no. of weeks)
0	37
1	37 + 9(1) = 46
2	37 + 9(2) = 55
3	37 + 9(3) = 64
•	•
•	•
16	37 + 9(16) = 181
17	37 + 9(17) = 190
18	37 + 9(18) = 199

After 18 weeks, more than $195.

Method ❷ Model the situation using a spreadsheet.

Add $9 for each week.

	A	B
	Number of Weeks	**Total saved <dollars>**
1		
2	0	37
3	1	9 + B2
4	2	9 + B3
5	3	9 + B4
•	•	•
•	•	•
18	16	9 + B17
19	17	9 + B18
20	18	9 + B19

You can buy the player the week that the total saved is at least $195.
You can buy the player after 18 weeks.

5-1 Modeling Problem Situations   241

Section 10-2 Transforming Parabolas

Focus
Translate and reflect the graph of $y = x^2$.

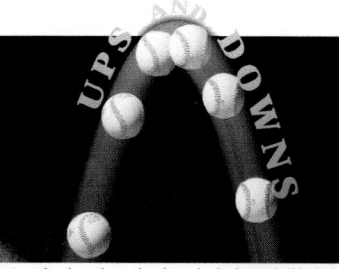

A strobe photo shows that the path of a thrown ball looks like an upside-down U.

A thrown or batted ball, water rising from a fountain, a whale's spout, and the light trails of fireworks all follow a curve called a **parabola**. This curve has the same general shape as the graph of the *squaring function* $y = x^2$.

Talk it Over

1. Look at the graph of $y = x^2$.
   a. What is the line of symmetry of the graph?
   b. The point of a parabola that lies on the line of symmetry is called the **vertex** of the parabola. What are the coordinates of the vertex of the graph?

2. Look at the graph of $y = -x^2$ (read $-x^2$ as "the opposite of x^2").
   a. What is the line of symmetry of the graph?
   b. What are the coordinates of the vertex of the graph?
   c. Suppose you want to find the value of y when $x = 2$. Which do you write, $(-2)^2$ or -2^2? Does it matter?

3. How do the values of y compare for $y = -x^2$ and $y = x^2$ when $x = 1$? when $x = 0$? when $x = -2$? in general?

4. The graph of $y = -x^2$ is a reflection of the graph of $y = x^2$. What is the line of reflection?

10-2 Transforming Parabolas   **555**

USING TECHNOLOGY

There are many opportunities in the concept developments and in the exercises and problems for you to incorporate technology into your course.

---

connection to **GEOGRAPHY**

The map shows Australia.

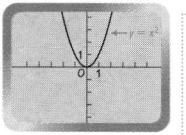

---

connection to **SCIENCE**

21. The temperature of lightning is about 50,000°F. This is more than 4.5 times the temperature of the sun's surface. Write and solve an inequality to find a maximum possible temperature for the sun's surface.

22. Cirrus clouds are found above 7 km. This is 3.5 times as high as cumulus clouds. Write and solve an inequality to determine the heights of cumulus clouds.

Weather Clouds

U.S. WEATHER SERVICE—Cumulus clouds that grow higher than 50,000 ft are called Cumulonimbus Towers. When these clouds reach a temperature between 32°F and 39°F, ice crystals start to form and fall to Earth as snow or rain. Cumulonimbus Towers are responsible for up to three-fourths of the world's rain and some of its most dangerous weather.

23. **Reading** Write two inequalities about the information in this article. Be sure to tell what your variables represent.

24. **Career** To study clouds, Joanne Simpson, a meteorologist and chief scientist at NASA, once flew directly into them in a single engine plane. Today she uses satellites, airplanes, and computers to study weather patterns. Simpson was the first person to write a paper about using math models to study cloud growth.

a. **Research** Find the names and height ranges for two types of clouds other than cumulus, cirrus, or cumulonimbus towers. Write an inequality to show the relationship between their heights. Draw a picture of each type of cloud.

b. **Writing** Joanne Simpson says that "science isn't just cold, hard facts. The crux is fitting unfitted things together and making them hang together. It's similar to composing music or art or poetry." Do you think this statement is true for mathematics? Explain why or why not.

Joanne Simpson, Meteorologist and NASA scientist.

5-4 Inequalities with One Variable   **267**

Use the double bar graph. Name each country described. Use the age scale to estimate the measure of each bar for that country.

11. the country where the average age of brides and grooms is the same

12. the country where the average age of grooms is less than the average age of brides

13. the country with the greatest difference between the average age of grooms and the average age of brides

om walls with posters.
plans to use only

ster? Can Midori
overlapping

ing Measures: Length and Area   **83**

INTERDISCIPLINARY·PROBLEMS

These theme exercises connect mathematics to other subjects areas. They illustrate the power of mathematics as a problem solving tool.

T19

I've heard that math educators need to make some changes. I'd like to try new teaching strategies, but I'm not sure how to get started.

How can I make change work for me?

Teaching

Flexibility

The Integrated Mathematics program provides teaching flexibility that enables you to incorporate the changes that are right for you and your students — at a pace that is comfortable for you. This program makes it easy for you to try new approaches and accommodate different learning styles.

INTERACTIVE TEACHING

Teaching questions and exploratory activities are easy to incorporate into your teaching because they are right in the textbook.

LEARNING STYLES

Integrated Mathematics helps you address different learning styles by presenting concepts visually, verbally, and kinesthetically.

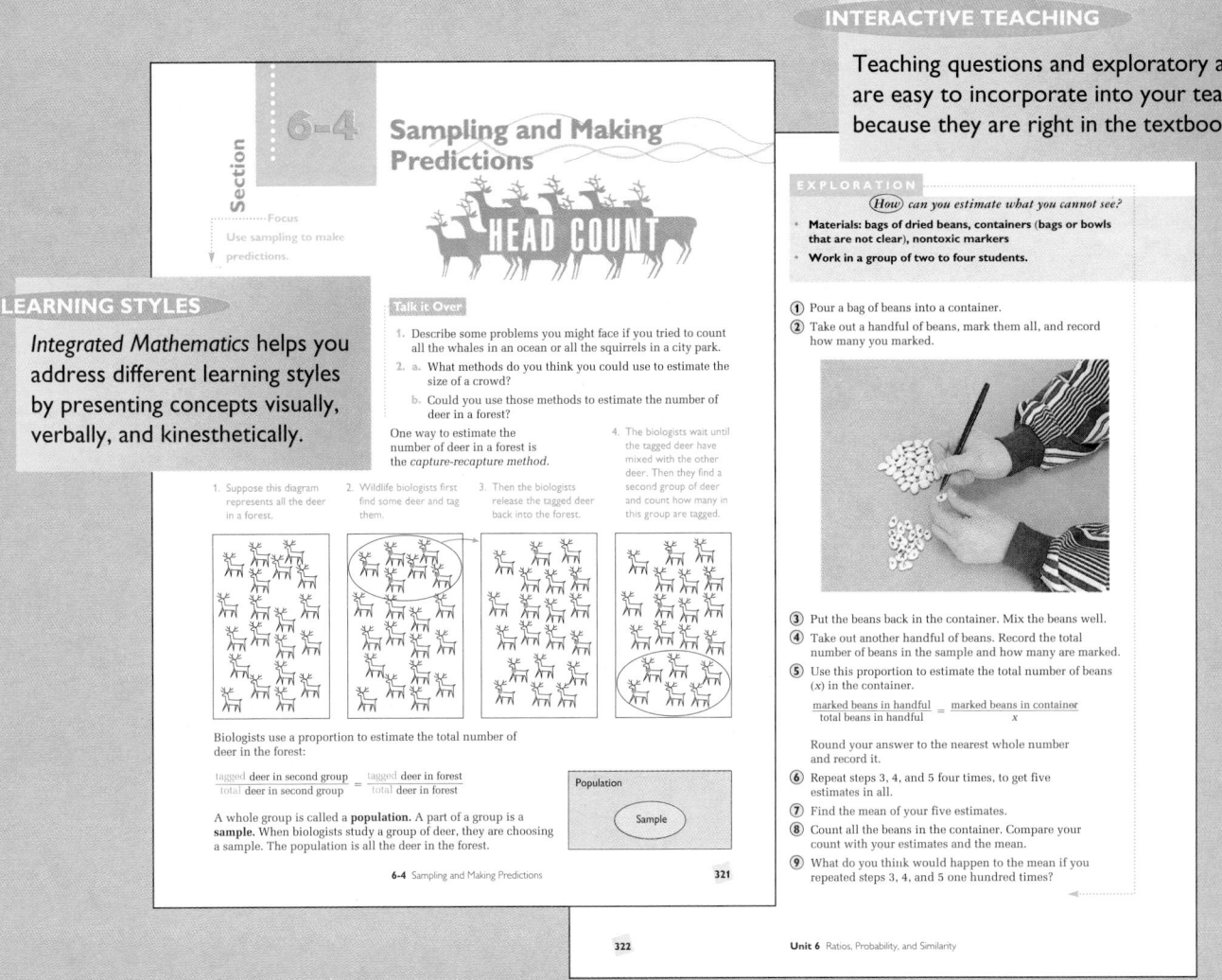

COMPLETE TEACHING SUPPORT

In the Teacher's Edition, each section has a planning list that tells you the materials you will need and the other resources that are available to help you present, extend, or reinforce the section.

VARIED PRACTICE AND EXTENSIONS

The wide variety of exercises, problems, and activities in the textbook and in the support materials allows you to tailor the course to the needs of your students and your teaching preferences.

red	⊬⊦⊤ I
green	IIII
blue	⊬⊦⊤

connection to SOCIAL STUDIES

Exit polling is used to predict who will win an election. In an exit poll, people leaving a voting place are asked how they voted.

7. In one exit poll, a sample of 400 voters was polled out of about 483,000 people voting in the election. The table shows the results. The margin of error is 6%.

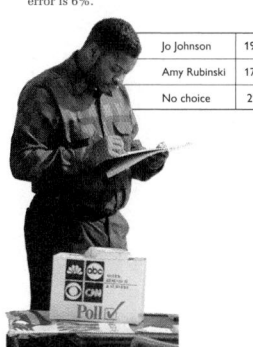

Jo Johnson	196
Amy Rubinski	175
No choice	29

a. Estimate the interval for the number of votes expected to go to Jo Johnson.

b. Whom do you expect to win the election? Explain your choice.

c. In a poll taken before the election, a sample of 400 out of about 1,295,000 registered voters were asked which candidate they supported. This poll predicted that Amy Rubinski would win by a wide margin. Give as many reasons as you can why the result does not agree with the exit poll.

8. **Research** Find out what percent of the people in your state who are eligible to vote are registered voters, and what percent of the registered voters voted in the last national election. Do the results surprise you?

9. **Career** A quality tester tests a random sample of 20 items. The graph shows the results. Item numbers are on the horizontal axis and item weights in grams are on the vertical axis. Each item should weigh 10 g, but any weight in the interval 10 ± 0.1 g is acceptable. There are 4000 items in the lot. Estimate the number of unacceptable items in the lot.

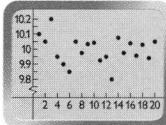

► Now you are ready for Exs. 5–17 on pp. 325–327

326

MAKING ASSIGNMENTS

The assignment guide in the Teacher's Edition helps you choose the exercises and problems that are best for your students.

6-4 Exercises and Problems

1. a. In the Exploration, what is the sample?
   b. In the Exploration, what is the population?

2. **Writing** Explain how the Exploration is like the method used by biologists to estimate the number of deer in a forest.

3. **Reading** Compare how the word *population* is used in this section with the way it is used in everyday language.

4. Biologists captured 400 penguins, tagged them, and released them back into the same region. Later, the penguin population of the region was sampled once a month for four months.

	Month 1	Month 2	Month 3	Month 4
Size of sample	200	100	150	100
Tagged animals in sample	50	30	45	24

a. Estimate the size of the population of penguins in the region for each month.

b. Find the mean of the four estimates you made in part (a).

c. Suppose the biologists combine the four samples into one large sample. Estimate the size of the penguin population in the region.

d. Compare your answers in parts (b) and (c).

5. A town's high school has 627 students. A sample of 200 people living in the town is picked at random. Of the 200, 40 are students at the high school.

a. About how many people live in the town?

b. Estimate an interval for the number of people living in the town. Assume a 4% margin of error.

6-4 Sampling and Making Predictions   **325**

Objectives and Strands
See pages 298A and 298B.

Spiral Learning
See page 298B.

Materials List
➤ Bags of dried beans
➤ Containers (opaque)
➤ Nontoxic markers
➤ Graph paper

Recommended Pacing
Section 6–4 is a two-day lesson.

Day 1

Pages 321–322: Talk it Over through Exploration, *Exercises 1–4*

Day 2

Pages 323–325: Sampling through Look Back, *Exercises 5–17*

Extra Practice
See pages 629–631.

Warm-Up Exercises
Warm-Up Transparency 6-4

Support Materials
➤ Practice 47
➤ Enrichment 42
➤ Study Guide 6-4
➤ Problem Set 13
➤ Diagram Masters 2, 16
➤ Using IBM Plotter Plus Disk: Sampling Experiment
➤ Quiz 6-4
➤ Alternative Assessment 8

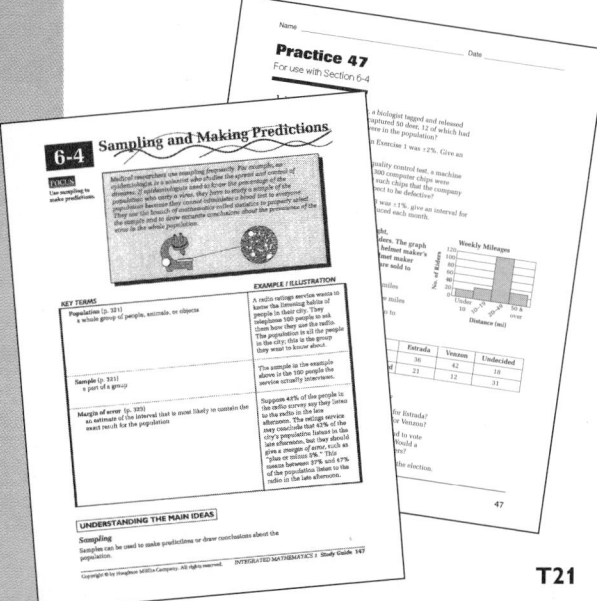

T21

A Complete Program

Teaching Support

The *Integrated Mathematics* program supports the full range of teaching and learning needs.

TEACHER'S EDITION

➤ Student pages with answers and teaching commentary, planning information, assignment guide

➤ **Solution Key** with solutions to all problems also available

ASSESSMENT BOOK

➤ Short quizzes
➤ Unit tests, Forms A and B
➤ Spanish unit tests
➤ Alternative assessment questions

OVERHEAD TRANSPARENCIES

➤ **Warm-Up** exercises for each section
➤ Multi-color **Overhead Visuals** with teaching suggestions

TEACHER'S RESOURCES

➤ Assessment Book
➤ Warm-Up Exercises transparencies
➤ Using TI-81 and TI-82 Calculators
➤ Using Plotter Plus
➤ Graphing Software

➤ Multi-Language Glossary
➤ Study Guides
➤ Project Book
➤ Explorations Lab Manual
➤ Problem Bank
➤ Activity Bank
➤ Practice Bank

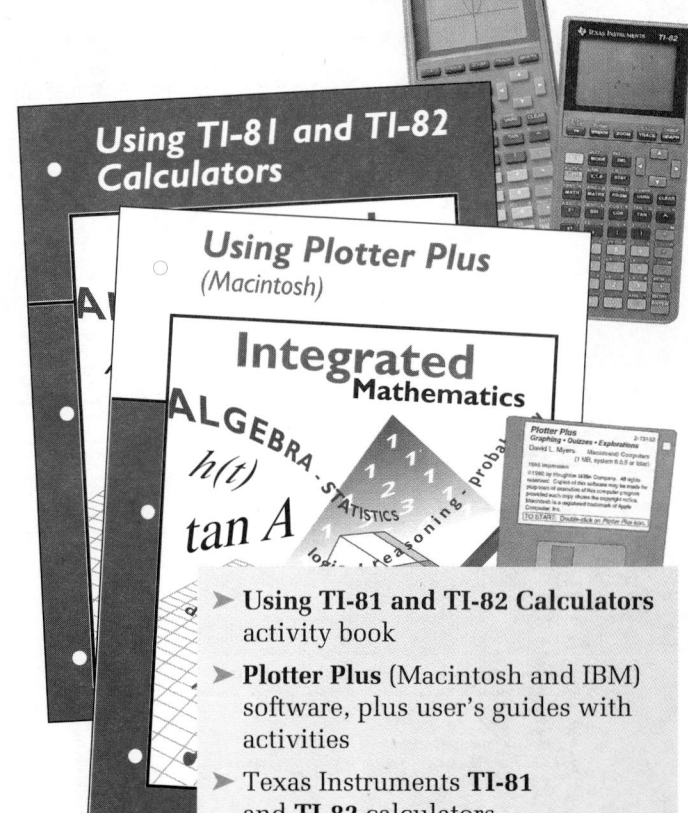

➤ **Using TI-81 and TI-82 Calculators** activity book

➤ **Plotter Plus** (Macintosh and IBM) software, plus user's guides with activities

➤ Texas Instruments **TI-81** and **TI-82** calculators

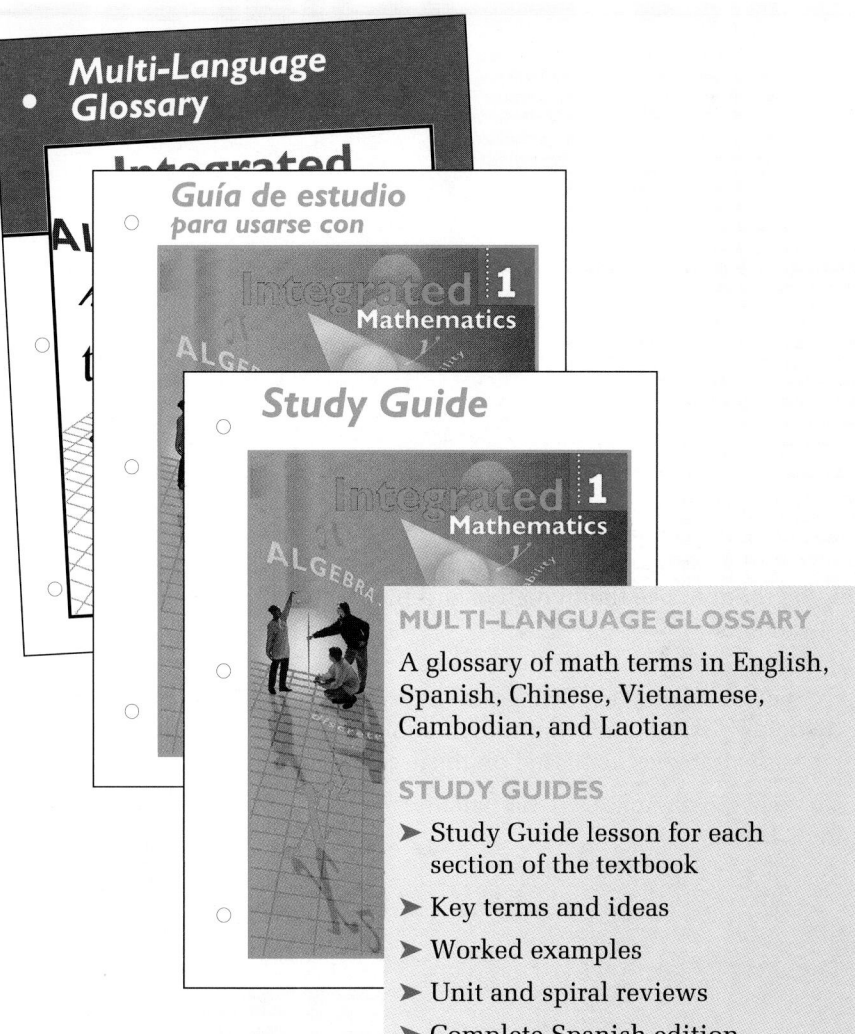

MULTI-LANGUAGE GLOSSARY

A glossary of math terms in English, Spanish, Chinese, Vietnamese, Cambodian, and Laotian

STUDY GUIDES

➤ Study Guide lesson for each section of the textbook

➤ Key terms and ideas

➤ Worked examples

➤ Unit and spiral reviews

➤ Complete Spanish edition

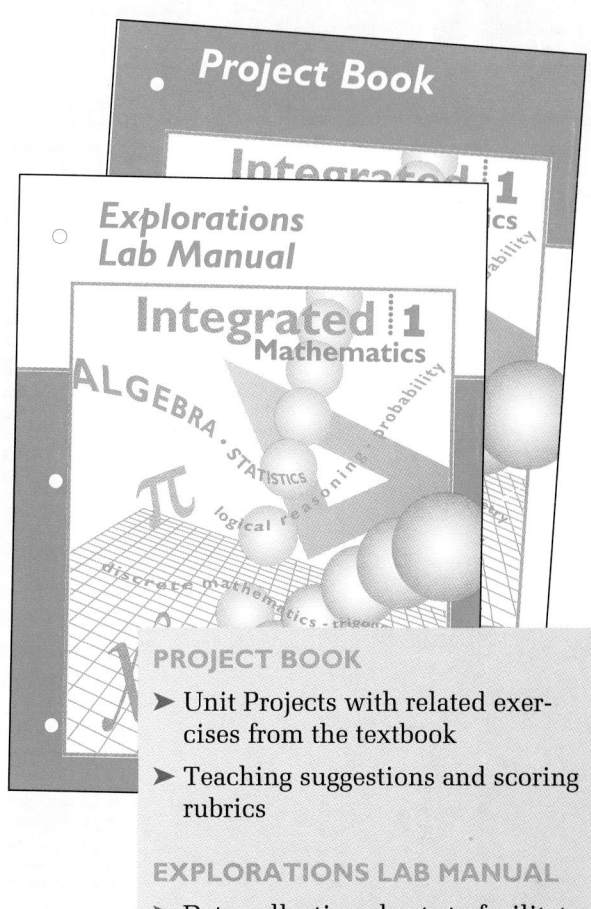

PROJECT BOOK

➤ Unit Projects with related exercises from the textbook

➤ Teaching suggestions and scoring rubrics

EXPLORATIONS LAB MANUAL

➤ Data collection sheets to facilitate textbook Explorations

➤ Additional Explorations

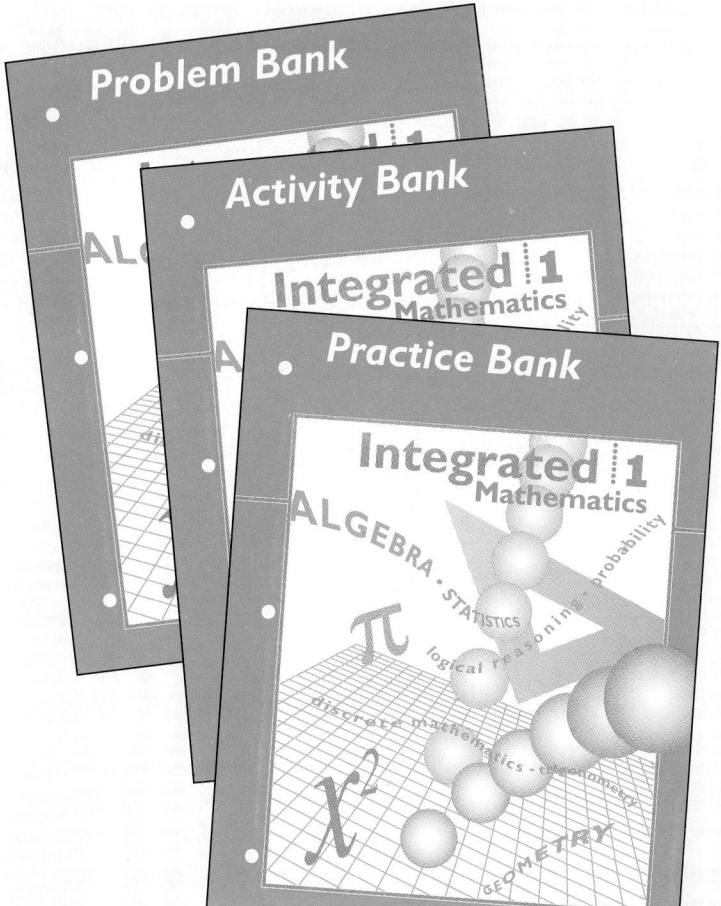

PROBLEM BANK

➤ Additional problems for each section

➤ Unifying Problems for each unit

ACTIVITY BANK

➤ Family involvement activities for each unit

➤ Enrichment activities for each section

PRACTICE BANK

➤ Practice exercises and problems for each textbook section

➤ Cumulative reviews for each unit

Effective Learning and Teaching

by **Gerlena R. Clark**
Mathematics Consultant
Los Angeles County Office of Education; Los Angeles, California

*"All students
can learn
mathematics."*

The Challenge

American businesses want their future employees to be able to work with others, to solve problems, to read and understand the principles of mathematics, and to communicate ideas. The primary question for teachers today is: How do I help students build a foundation of skills and information while simultaneously encouraging them to use their creative and intellectual abilities to solve real-world problems?

Recent research on how students learn suggests some strategies for accomplishing the challenging goal of preparing students for their lives as adults in the 21st century. These strategies are based on the assumption that:

> All students can learn mathematics if mathematics is taught in the way that students learn.

How Do Students Learn Mathematics?

Learning in the traditional manner is not sufficient in a world that demands attitudes that are conducive to creativity, as well as specific knowledge and skills. In the past students were expected to acquire facts through drill, practice, and memorization. The teacher was the giver of knowledge through lecturing and demonstrating.

Now we know that:

- Students must be actively involved in the learning process.
- Students learn best through dialogue, discussion, and interaction with others.
- Students benefit from reviewing, critiquing, and revising another's work as well as their own.
- What students learn is connected to how they learn.
- Students learn by experiencing tasks that are as closely aligned to real life as possible.
- Students learn by making connections to what they already know about the task and the real world.

We may summarize these statements by defining:

- **knowledge** as the result of individuals constructing meaning for themselves, by creating rules and hypotheses to explain what they've experienced.
- **intelligence** as a function of experiences. The brain learns best through first-hand experiences.

It is difficult to learn, understand, and apply experiences that are second hand (such as pictures and models) or last hand (such as the symbols found in language and in mathematics). Therefore, the more hands-on experiences students have, the more they can make mathematical concepts "come alive" for themselves.

We must remember that the human brain is a pattern-seeking device that is constantly trying to use patterns to understand the world about us. Since we acquire knowledge through many avenues, no one avenue should be considered the "right" way.

An Example

The following example illustrates how different students, or groups of students working together, can use different approaches to solve the same problem.

The Triple Twelve Birthday Problem

Sarah Witkowsky will be twelve years old on the twelfth day of the twelfth month. To commemorate this once-in-a-lifetime numerical event, her grandparents decide to give Sarah a number of gifts beginning on December 1 and continuing through December 12, her birthday.

The number of gifts Sarah gets each day will be equal to the number of the day of the month. Thus she gets

> 1 gift on December 1
> 2 gifts on December 2
> 3 gifts on December 3, and so on.

How many gifts in all does Sarah get for her birthday?

Some students approach this task from a very concrete level and use manipulatives to help them find a solution.

Solution 1: Concrete Approach

➤ Use an egg carton that holds 12 eggs.

➤ Number each cell of the carton from 1 to 12.

➤ Place beans (or some other type of counter) in each of the twelve cells to represent the number of gifts on the day corresponding to the cell number.

    Cell 1 gets 1 bean.
    Cell 2 gets 2 beans.
    Cell 3 gets 3 beans.
          •
          •
          •
    Cell 12 gets 12 beans.

➤ Count all the beans in the 12 cells to find the total number of gifts (78) that she receives during the 12 days.

Other students may understand the problem in terms of arithmetic operations.

Solution 2: Numerical Approach

➤ Add the numbers 1 through 12 sequentially.

$$1 + 2 + 3 + 4 + 5 + 6 + 7 + 8 + 9 + 10 + 11 + 12 = 78$$

Still other students may view the problem geometrically. They may also use the strategy of solving a smaller problem and then generalizing their finding to solve the larger problem.

Solution 3: Geometric Approach

➤ Use square tiles to represent the number of gifts.
➤ Start by finding the number of gifts for 4 days.

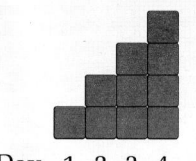

➤ Form a staircase figure.

Day 1 2 3 4

➤ Now make an identical staircase.

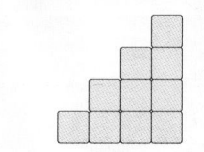

➤ Put the two staircases together to form a rectangle.

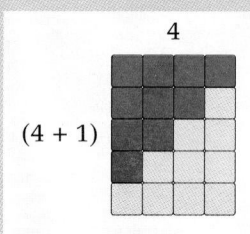

4

(4 + 1)

➤ Now find the number of tiles in the rectangle.
$4 \times (4 + 1) = 20$ tiles.

➤ Since we want the number of tiles in just one staircase, we need to divide by 2. Thus the number of tiles in one staircase is
$$\frac{4 \times (4 + 1)}{2} = 10 \text{ tiles.}$$

➤ The problem is to find the number of gifts received over 12 days. We can generalize from the result for 4 days that the result for n days will be
$$\frac{n \times (n + 1)}{2} \text{ tiles.}$$

➤ Substituting 12 for n gives
$$\frac{12 \times (12 + 1)}{2} = \frac{12 \times 13}{2} = 78.$$

Sarah receives 78 gifts in all.

How Does This Learning Take Place in the Classroom?

If we put what we know about learning into action, what will be happening in the classroom?

In the classroom, students should:

➤ work with objects to represent mathematical models

➤ work in cooperative groups, or in pairs, as the task dictates

➤ write results, or outline strategies

➤ discuss mathematical ideas

➤ ask and answer each other's questions

In the classroom, the teacher should:

➤ allow time for students to think through problem formation and solution

➤ maintain an atmosphere of freedom for students' expressions

➤ encourage mathematical arguments with questions such as, Do you agree or disagree? Why?

➤ not focus on the "correct" answer, but allow discussion on alternative answers and solution procedures

➤ avoid paraphrasing what students say; ask students to clarify their own thinking

➤ model expected behavior for working in a group and for solving problems

➤ ask questions that will allow students to go beyond one-dimensional responses

➤ encourage students to go to each other for assistance

➤ encourage students to revise their written responses

➤ allow students to self-assess as well as assist in the assessment of others

Meeting the Challenge

All of this implies a change in the way students learn and in the way teachers teach. Making this change may not be simple or easy at first, but it will become simpler and easier with time. And the result—students who know and use mathematics, who are mathematically empowered—will be of benefit to all.

In *Integrated Mathematics 1*
See student pages 88, 100, 162, 170, and 447.

Cooperative Learning

by *Judith Collison*
Research Associate
Technical Education Research Center; Cambridge, Massachusetts

*"Group work
can decrease or
eliminate math
anxiety."*

Benefits of Cooperative Learning

The NCTM *Curriculum and Evaluation Standards* stress the importance of developing skills of collaboration in mathematics teaching and learning. Developing skills of group participation is an important goal of all education, but it is especially useful in mathematics. Research has shown that group work in math classes has decreased or eliminated math anxiety; increased motivation, flexibility, confidence, self-esteem, curiosity, and perseverance; improved ability to solve problems and to communicate mathematically; and resulted in more positive attitudes towards mathematics.

When and how should cooperative strategies be used?

Teachers should use collaboration to create a sense of community and trust among students, as well as to create a deeper and more personal understanding of mathematics.

Cooperative forms are natural for problems that seem too big, too time-consuming, or too complex to be tackled by one person, and problems that require multiple perspectives, ability levels, or discussions.

Adequate preparations are key to the success of cooperative group work. The teacher needs to decide how to configure the groups, what type of group work will be used, how the problem will be divided among or within the groups, and how the work of the groups and of individual group members will be assessed.

Before students embark on their assignment, the teacher should verify that all groups and group members understand all instructions and expectations.

How should groups be formed?

The teacher selects the method by which group membership is determined. The composition of the group may be decided by the teacher or the students, according to some criteria, or randomly.

Types of group structure

Most typical cooperative structures include students working in *pairs,* in *small groups* made up of three or more members, or as a *whole class.* The teacher needs to choose the structure most useful for the activity or problem at hand. The following descriptions of the various types of cooperative structures are adapted from Neil Davidson's *Cooperative Learning in Mathematics: A Handbook for Teachers* (Addison-Wesley, 1990).

INTERVIEW (2-4 participants in each group) Most useful for getting students acquainted with each other in order to begin forming a sense of community. Members of the group ask each other questions dealing with either personal information or with applications of mathematics to their lives. They then share the information with the larger group or with the whole class.

Teacher's Choice
(by ability, social, psychological, or random grouping)

HOMOGENEOUS	HETEROGENEOUS	RANDOM
ability level	mixed levels	counting off
talents/ interests	complementary talents/ interests	according to height
learning style	combination of learning styles	arbitrary numbering e.g., phone or social
social group	diverse ethnic represen- tation	security numbers
psychological group		

Student's Choice

SELF-SELECTION
Students choose their own working partners.

MODIFIED SELF-SELECTION
Students list first, second, and third choices, and the teacher constructs groups based on these preferences.

THINK-PAIR-SHARE Useful for developing communication about concepts and procedures, and practice in problem solving. Students think about the problem alone, then discuss possible solutions with their partners and agree on the correct solution. They share their conclusions with the rest of the class. Having students work in pairs is probably the best cooperative format when using computers in problem solving.

PEER PRACTICE AND DRILL There are several versions of this format. It provides opportunities to practice the mastery of skills or concepts.

a. **Partner drill** Students take turns asking their partners questions.

b. **Flashcards** Each student makes up flash cards with questions for a partner. The student presenting the cards should have the correct answer(s) for each question asked.

c. **Teams-Games-Tournaments** In every round, individual students from each team compete with each other. The team's score is determined by members of each team.

d. **Peer-pair-problem solving** (also called "pairs check") Each of the students in the pair has a unique role. One is the "solver" or "performer," the other is the "checker" or "coach." The "solver" works on solving the problem. The "checker" observes, gives hints, points out errors, and gives positive feedback and encouragement to the partner. The "checker" can only give suggestions, not part or all of the solution. Roles are reversed for the next problem to be solved.

JIGSAW Most useful for solving large or complex problems or doing an extended class project. The task is divided into several component parts, approaches, or topics.

a. Each group member is assigned to work on a different aspect of the problem. For example, if a problem requires students to make measurements, check and tabulate data, and report on results, a "measurer," "checker," "tabulator," and "reporter" may be designated.

b. Each group works on a different part of the problem or project. Each group's effort becomes part of the unified effort of the whole class, as pieces of a jigsaw puzzle fit together to form a larger, complete picture.

ROUNDTABLE This method is not highly interactive, but is excellent for brainstorming, for generating enthusiasm, and for generating a large number of answers to problems with multiple solutions. Each group has one piece of paper. After the teacher poses the question or problem, students write a solution or suggestion on the paper, then pass the paper to the next student. Within a given time frame, the paper continues to be circulated among the group (or the whole class). A student may choose to pass a round without penalty.

WHOLE CLASS AS A GROUP The whole class as a group may decide on classroom procedures, divide work among the groups, decide on topics for class projects, brainstorm, present the results of their project to the school or the community, go over homework, review materials, or play mathematical games.

Is cooperative learning always an appropriate methodology?

Clearly, it is not always appropriate to use cooperative learning strategies in a mathematics class, nor should group work be the only strategy used. Students need to be engaged in a combination of individual and collaborative efforts. In addition to developing a common vocabulary and shared understanding of concepts, students need to develop a personal voice and a personal understanding of the ideas.

In ***Integrated Mathematics 1*** See student pages 127, 367, and 485 and side-column commentary on pages 62 and 436.

Enhancing Mathematics Learning Using Graphing Technology

by **Bill Leonard**
Mathematics Instructor
Shawnee Mission West High School; Shawnee Mission, Kansas

"Graphing technology helps students make connections between logic, symbols, and visualization."

Benefits of Graphing Technology

Developments in graphing software and graphics calculators have had a great impact on the teaching and learning of mathematics and have allowed teachers and students opportunities that were unavailable before. Teachers can present the same problem situation algebraically and geometrically within the same lesson. Students can visualize problems much more easily than in the past, and can attempt to solve problems that in the past might have been too complicated to deal with using only paper and pencil and algorithms.

Graphing technology can be used to facilitate students' many different individual learning styles, including verbal/linguistic, visual/spatial, logical, and kinesthetic. Graphing technology works as well with students whose learning style is interpersonal as with those who learn best working on their own.

Perhaps the greatest strength of graphing technology is its facility in helping students to make connections between logic, symbols, and visualization.

Features that Enhance Different Learning Styles

The relationship between two lines that have equal slope but different y-intercepts can quickly be seen by graphing the equations of two or more such lines using graphing software or a graphics calculator. The image on the screen helps all students, and especially visual learners, conclude that lines with equal slope but different y-intercepts are parallel. The effect of changes in the y-intercept for a line with given slope is also apparent, which helps make the slope-intercept form of an equation, $y = mx + b$, more meaningful. While an example such as this is particulary beneficial to the visual learner, the hands-on approach is also helpful to the kinesthetic learner.

Lines with the Same Slope

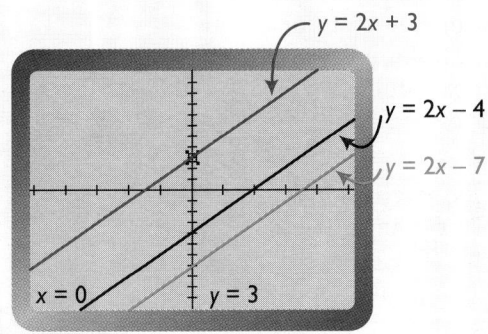

$y = 2x + 3$

$y = 2x - 4$

$y = 2x - 7$

$x = 0$

$y = 3$

Reflections

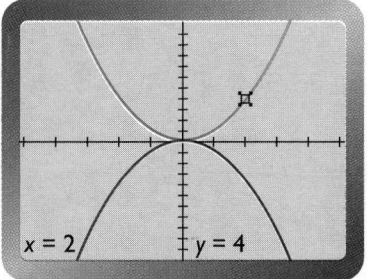

The relationship between changes in a quadratic equation and resulting translations of the graph can also be viewed quickly using graphing technology. The examples at the left and below demonstrate how graphing technology can be used to show how changes in an equation affect its graph. For example, when the graphs of $y_1 = x^2$ and $y_2 = -x^2$ are displayed on the same screen, students can see that the effect of multiplying the coefficient of x^2 by -1 is to reflect the graph of the original equation in the x-axis. When 3 is added to the equation, the graph is translated up 3 units. When 3 is added to the x-term before squaring, the graph is shifted 3 units left.

Vertical Shift

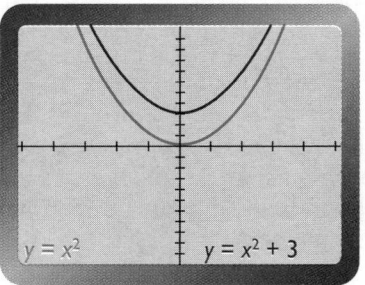

Horizontal Shift

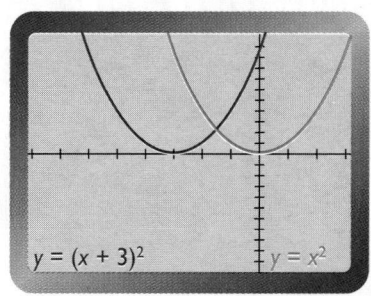

Solving a quadratic equation graphically by hand may be time-consuming, but, worse, may produce a graph from which it is extremely difficult to estimate solutions. However, using a graphics calculator or software produces the graph quickly. The student can use the TRACE and ZOOM features to estimate solutions quickly and accurately.

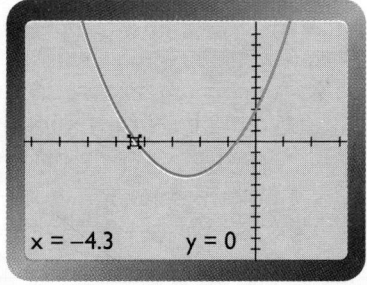

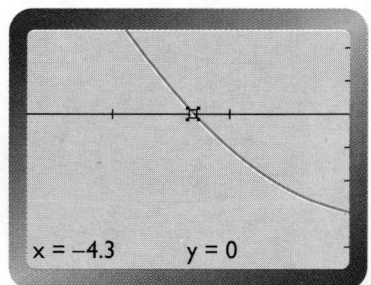

Meeting the Needs of Students for the Year 2000 and Beyond

As technology continues to become more and more a part of everyday life, students need to develop a familiarity and ease with using technology. They also need to recognize that math can be used to model and solve real-life problems. Graphics software and graphics calculators are successful tools for enabling students to achieve both of these goals. What may be just as important, they are fun to use!

In *Integrated Mathematics 1*
See student pages 277, 556, and 611 and side-column commentary on pages 20, 214, and 292.

Using Manipulatives to Develop Understanding

by **Valarie A. Elswick**
Mathematics Teacher
Roy C. Ketcham Senior High School; Wappingers Falls, New York

*"Using manipula-
tives encourages
active participa-
tion."*

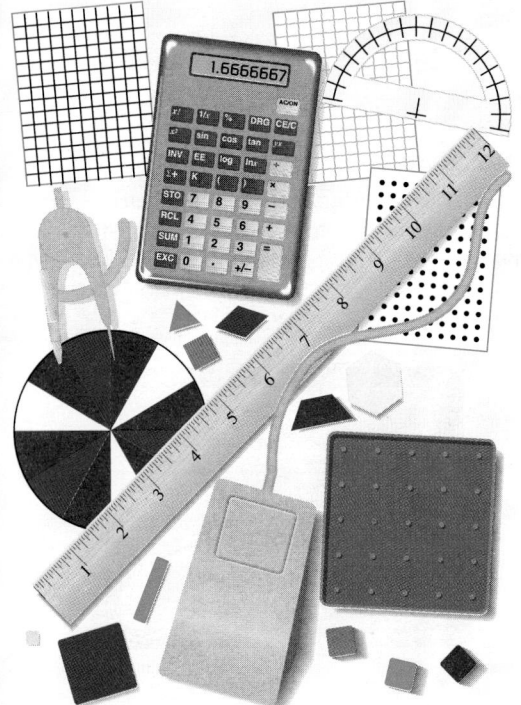

Incorporating Manipulatives

Are the materials shown at the left available to you and your students?

These are some of the many different types of manipulatives that can be found in mathematics classes around the country. Many mathematics teachers today are learning to incorporate manipulatives into their lessons to improve student comprehension. At all levels of mathematics instruction, the use of manipulatives can help students build conceptual understanding and nurture their learning of abstract ideas.

Student Goals

Manipulatives can help students understand specific topics and procedures. Over a period of time manipulatives can also help students achieve some broader goals. For example, manipulative activities help instill in students a sense of confidence in their ability to think and communicate mathematically, especially when group work is involved. While working with manipulatives, students have the opportunity to take chances, make several tries, and reach appropriate decisions in selecting strategies and techniques. The potential exists for students to take intellectual risks by raising questions, formulating conjectures, and presenting solutions.

Using manipulatives encourages active participation by all students. Students can be asked to investigate, explore, predict, test, develop, describe orally and in a written format, discuss in a group or with the whole class, justify, solve, and use and apply ideas.

The Teacher's Role

Preparing the Activity

Using manipulatives in a lesson can be very exciting and motivating. It can also challenge teachers to develop lesson plans that keep students actively involved and prevent inappropriate use of the manipulatives. Advance planning and preparation are essential for successful use of manipulatives.

➤ Thought should be given to selecting the best manipulative for the learning objective.

➤ Materials should be organized for easy use and distribution to the class.

➤ An evaluation process appropriate to the activity should be selected.

Facilitating the Activity

Before beginning a manipulative activity, students need to have an understanding of the procedure and directions for getting the materials and beginning their work. They might be asked to reflect on the mathematics involved and directed to communicate their thoughts in a verbal or written format at the end of the activity. Once the activity begins, the teacher should move among the students, listening to the discussions and exchanges of ideas. As the activity continues, the teacher may need to answer questions or redirect the focus.

Making Connections

Using manipulatives enables students to make connections between mathematical topics, such as algebra and geometry. Algebra tiles and geoboards are two types of manipulatives that can be used to connect algebra and geometry, both symbolically and visually, as these examples illustrate.

Adding Polynomials Using Algebra Tiles

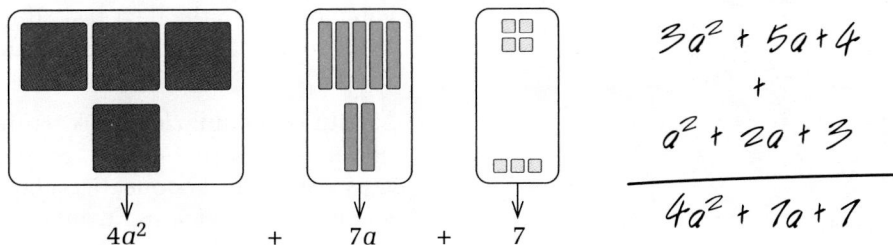

$4a^2$ + $7a$ + 7

$$3a^2 + 5a + 4$$
$$+$$
$$a^2 + 2a + 3$$
$$\overline{}$$
$$4a^2 + 7a + 7$$

Exploring Triangles on the Geoboard

How many different triangles can you form on a 3-by-3 geoboard (or on 3-by-3 dot paper)?

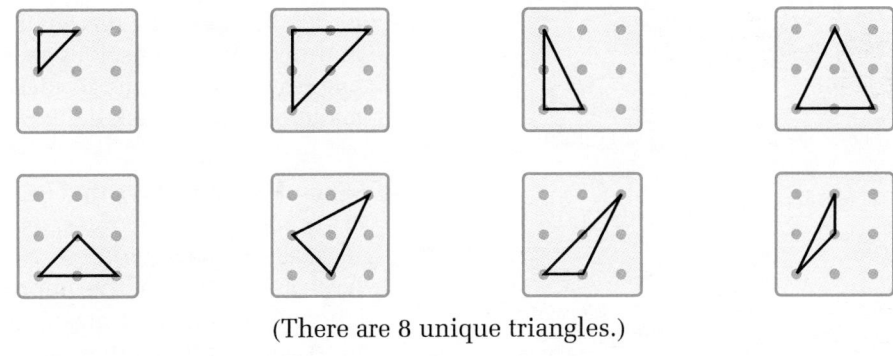

(There are 8 unique triangles.)

Summary of Benefits

Working with manipulatives can help students overcome language barriers, improve their listening and speaking skills, and increase their receptivity to a variety of concepts and approaches. Manipulatives motivate students by involving them in the learning process. Manipulatives also develop students' confidence in their ability to solve problems and to reason and communicate mathematically.

In *Integrated Mathematics 1*
See student pages 13, 39, and 436 and side-column commentary on pages 88, 155, and 206.

Writing in Mathematics

by **Joan C. Countryman**
Head of School
Lincoln School; Providence, Rhode Island

"When students write they learn that mathematics is a human endeavor."

"Does your answer make sense?"

The student had come in for extra help and we were going over the homework problems. In the silence, I looked up, expecting some sort of defense for his solution to the exercise, but as our eyes met I realized that my question had startled him.

"Is it supposed to make sense?"

For too many students, studying mathematics has nothing to do with making sense. Practicing the steps, learning the rules, passing tests, adding, subtracting, multiplying, dividing—these are the activities of math class. If you get the right answer, the one in the back of the book, or the one your friend got, you can move on to the next task. If not, you must retrace your steps, find the mistake, do the calculations once more.

My student's quizzical look reminded me that, for some math students, none of it makes sense. I needed to find a way to help the young people enrolled in my classes build connections between what they already knew and what I wanted them to learn. I wanted to help them learn to stop and think about what we were doing in class. Why does the graph rise here and fall there? What does x represent in this example? Are squares of numbers always bigger than the numbers? Why not multiply and divide in this case?

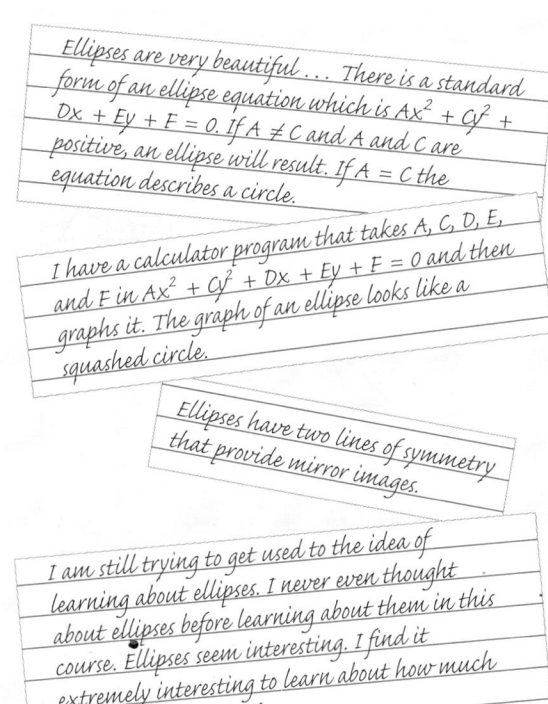

Ellipses are very beautiful ... There is a standard form of an ellipse equation which is $Ax^2 + Cy^2 + Dx + Ey + F = 0$. If $A \neq C$ and A and C are positive, an ellipse will result. If $A = C$ the equation describes a circle.

I have a calculator program that takes A, C, D, E, and F in $Ax^2 + Cy^2 + Dx + Ey + F = 0$ and then graphs it. The graph of an ellipse looks like a squashed circle.

Ellipses have two lines of symmetry that provide mirror images.

I am still trying to get used to the idea of learning about ellipses. I never even thought about ellipses before learning about them in this course. Ellipses seem interesting. I find it extremely interesting to learn about how much one equation can do for you.

I started asking my students to write about their work in mathematics because I thought that writing might help them move beyond a mechanical approach to learning, an approach that they found annoying but familiar. I wanted to help them discover the questions that are central to the discipline, the questions that mathematicians might pose. I also wanted them to think about themselves as learners. Over the years, I have found that one way to help students clarify, express, and reflect on their work in mathematics is to ask them to write to learn.

When I talk about writing in math I mean writing in its broadest sense. I ask students to take notes, make lists, record their observations and feelings, as well as to write essays, term papers, and stories. Having students write supports an active approach to teaching and learning. I expect my students to construct meaning. In order to make sense of the material, they must connect new information to what they already know. Writing helps them to learn to ask their own questions, and to explore some of the questions that I pose.

Many teachers use the writing process across the grades to help students construct mathematical knowledge. The examples at the left are from the math journals of tenth graders. The students recorded their observations about ellipses during a unit on analytic geometry. These brief comments, written at the end of class, provided insight for the teacher about how the students understood their work in conics.

Keeping journals and writing word problems are just two of many activities that math teachers might require of their students. Autobiographies, the stories of their growth as math students, written in the first weeks of a math course, can serve to inform teachers about students' initial perceptions of mathematics and their own learning styles. Letters to parents or friends provide current accounts of coursework. Study guides, test questions, and lesson summaries can serve as excellent reviews. One way to get started is simply to ask students to write a comment, on the back of the homework, about the exercises they have completed. Which was the most difficult? Which provided the most insight on the material? Which ones were easy to complete?

Advocates of writing across the curriculum are not suggesting that all teachers assign essays and correct them as an English teacher might. Instead we imagine that teachers might ask students to think and write as essayists, scientists, historians, and mathematicians do, posing questions, and solving problems by writing and reflecting on the material of the discipline.

Most useful to us as teachers is the writing that provides insights on how our students think about their work. The following example is from a mathematics student who was also studying physics. For the teacher, the student's comments revealed the depth of his thinking about mathematical concepts of real phenomena.

> Since I have been very interested and involved in physics, I wanted to find a topic for my final paper that would in some way investigate some of the principles we were studying in physics. I also wanted to study empirically one of these topics, that is, to take my own data, and develop equations based not on a textbook but on my own data-taking and analysis. For these reasons, I decided to examine the behavior of different balls as they bounce, specifically to establish a relationship between the height from which a ball is dropped and the height that it then bounces. While I worked on this project, I also became interested in the time over which a ball continues to bounce, and I added my investigations in this area to my paper.

My hope is that one day students and teachers will write to learn freely. Pages of notes, stories, plays, lists, poems, sketches, and journal entries about math, language, literature, science, and history will help students make sense of the world in which they live. Teachers of all subjects will serve as coaches and experts about the learning process, knowledgeable about students—who they are, what they can do, what they know, and what they need to know.

What might students learn when they write about math? What can we as teachers learn from their writing? First, when students write they learn that mathematics is a human endeavor, one that comes not from the sky, but from the work of human minds. In fact, when students write some of the mathematics comes from their own minds. Second, when reading students' writing, teachers discover that learners, like mathematicians, must construct the mathematics for themselves. If we give them time to do this, they will succeed in constructing their understanding of the material. Finally, students and teachers will learn that meaning lies not in the words and symbols themselves, but in the ways that we use those words and symbols to make sense of information.

In *Integrated Mathematics 1*
See student pages 46, 286, and 480 and side-column commentary on pages 6, 231, and 431.

Developing Good Problem Solvers

by **Martha E. Wilson**
Preparatory Mathematics Specialist
Mathematical Sciences Teaching and Learning Center
University of Delaware; Newark, Delaware

"Teaching itself is a complex problem solving process."

The Goal of Mathematics Instruction

Mathematics educators have agreed for some time that problem solving is a very important, if not the most important, goal of mathematics instruction. Students who learn mathematics through drill in routine operations lose interest in the subject and miss the opportunity for intellectual development. Stimulating them to solve appropriately challenging problems, and helping them to solve those problems, provides students with interest in and tools for independent thinking.

Research in mathematics education has not been able to identify any one single way of teaching that is "best" for developing problem solving skills in mathematics for all students in all situations. There are, however, examples of good problem solving that point to actions that teachers might take to help students develop those skills.

Cooperative Learning

The National Council of Teachers of Mathematics *Curriculum and Evaluation Standards* and *Professional Standards for Teaching Mathematics* both call for a classroom where students have an opportunity to explore and investigate ideas, develop conjectures, and verify hypotheses. Cooperative learning in mathematics is often recommended as one way to help students because:

➤ they become active participants in the classroom

➤ they may be encouraged to discuss and communicate their ideas about mathematics in an environment that is less threatening than in whole-class discussions

➤ when groups of students struggle with an interesting problem, the outcome for many students is a new and refreshing view of doing mathematics

Whole-Group Instruction

Research also suggests that an active, problem solving approach to learning can be accomplished in whole-group instruction, where the teacher directs the activities of the entire class. Students can become actively involved when the teacher selects and presents an interesting idea or problem and then leads students to a discovery of concepts and connections through a series of questions. Questions should develop a train of thought in logical sequence and should include these types:

➤ some moderately challenging, to stimulate thinking

➤ some factual, to bring out important facts or information

➤ some requiring considerable thought and formulation of a conclusion

The Teacher's Role

The role of the teacher in preparing for either model of instruction becomes one of organizing for learning by selecting an engaging task or problem. An understanding of the background knowledge and interests of the students and thoughtful planning of future lessons is necessary for the design of this task. Good tasks prompt an interest in investigation whether they are presented for group work or for whole-class discussion.

Selecting Good Tasks

A task should be:

➤ set in a context that will engage the interest of students

➤ complex and difficult enough to challenge students' thinking, but not so difficult that they will give up quickly

➤ solvable by more than one method, so that a subsequent discussion can point out connections and the possibility of multiple approaches

In discussions, the teacher must prepare to ask the questions that lead to clear and concise mathematical conclusions and emphasize the connections that can be made.

Teachers as Problem Solvers

Research on teaching and learning mathematics is still incomplete, and there is no indication that teaching mathematics must be done in a single prescribed way in order for students to become good mathematical problem solvers. Teachers may choose among a variety of styles, including teacher presentation, large-group activities, small-group cooperative learning, and combinations of these. As researchers have investigated several modes of teaching in search for the one that produces good problem solvers, they seem to have found that teaching itself is a complex problem solving process.

Example
Not this: What is the area of the polygon on the geoboard?
But this: How many other polygons with area 12 square units can you make on the geoboard?

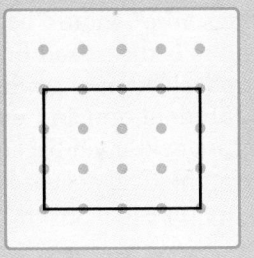

In _Integrated Mathematics 1_
See student pages 102, 138, and 425 and side-column commentary on pages 17, 193, and 317.

14. During a golf tournament, Chet Washington hits a ball that lands 200 yd from the tee. The path of the ball is a parabola.

   a. Sketch a possible path for the ball, using the *x*-axis as the ground and the origin as the tee.

   b. What are the *x*-intercepts of the graph?

   c. What do the *x*-intercepts mean in terms of the situation?

   d. **Writing** Amalia wrote the equation $y = -2x(4x - 800)$ for a possible path of the ball. Elizabeth wrote the equation $y = -2x(2x - 100)$. Which student's equation *cannot* be correct? Why?

From **Integrated Mathematics,** page 565

Teaching Students with Limited English Proficiency

by **Cesar Larriva**
Doctoral Student, Department of Education
Stanford University; Stanford, California

> *"The strategies for teaching students with limited English proficiency work well for all students."*

Challenge

The educational backgrounds of LEP (limited English proficient) students in general are as varied as the nationalities they represent. Some students have received top-quality public or private education in their native countries while others have barely attended school. The students' first (native) language proficiency can also vary considerably and has a strong influence on student academic success in the second language setting. The LEP student generally comes from a culture where education is valued, and the school

and the teacher are highly respected. These students are generally very motivated by their desire to succeed. They are also motivated by their desire to please the teacher and by peer pressure from other LEP students, which is largely achievement oriented.

The wide range of mathematics skills, English proficiency, and first language proficiency of LEP students within a classroom makes it challenging to meet the needs of all students. Meeting this challenge therefore requires special strategies.

Sheltering Techniques

Sheltering techniques are an important tool. Sheltered English instruction is a method of delivering subject matter (e.g., mathematics) instruction to LEP students using English as the medium for instruction. A language can be learned only if it is presented as comprehensible input. Information is comprehensible when the vocabulary and language used are familiar to the learner and the information is presented in a meaningful context.

In sheltered English instruction, the delivery of the message is simplified (sheltered), but subject matter remains challenging; material is not "watered down." Instruction in the sheltered English classroom is adjusted to ensure student comprehension. Input can be made more comprehensible by the following techniques.

Teacher's Speech

Language is simplified by avoiding compound sentences, by favoring simple grammatical structures such as the present tense, by limiting the vocabulary, and by avoiding use of idioms. A phrase such as "I want you to stay on top of things here" should be used only if its meaning is discussed first. Content is emphasized over grammatical accuracy. Student oral and written responses are evaluated based on content not grammatical accuracy. Important ideas are repeated several times for emphasis.

Providing Clues

Effort should constantly be made to provide contextual clues. These may be graphical representations such as photographs or graphs. The clues may be in written form, such as the posting of important vocabulary on a chalk or bulletin board; a written vocabulary word can be pointed to as the teacher uses it in the sentence during lecture. Clues may be physical, such as real objects and scaled models.

Acceptance

The classroom culture is supportive, motivational, and non-threatening so that the student's defense mechanism (which hinders participation) is low. LEP students experience a pre-speech stage or silent period during which active listening and learning occur without language production. Students will produce language when ready and should be encouraged but not pushed.

Manipulatives

Concepts are contextualized and communication is facilitated through the use of hands-on activities and manipulatives, such as algebra tiles. A great deal of emphasis is placed on making the abstract more concrete. For example, students can physically act out or model a problem. The diagram shows students physically plotting a linear equation on a tennis court or a school yard on which coordinate axes have been drawn using chalk.

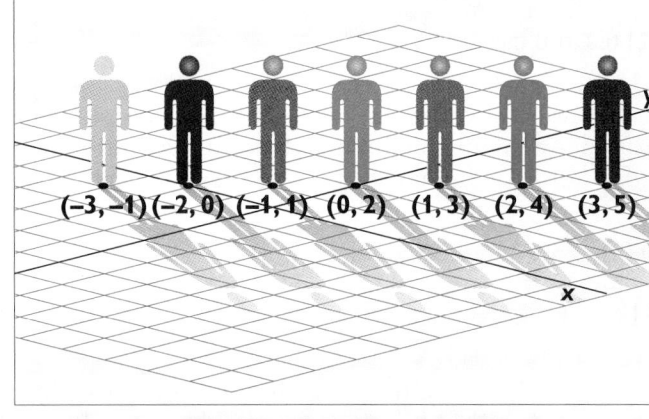

Students physically model the function
$y = x + 2$.

(−3, −1) (−2, 0) (−1, 1) (0, 2) (1, 3) (2, 4) (3, 5)

Any technique like this that facilitates communication by decreasing the reliance on language is beneficial.

Prior Experience

Provide sufficient preparation and background when introducing a new topic. A lesson on probability using playing cards, for example, will definitely require an explanation of the playing cards themselves, since many LEP students have never seen these cards that are so familiar to us. New concepts should be presented in a context that is meaningful to the student.

It is important to recognize that LEP students have a wealth of prior skills and experience to draw upon. Many also have highly developed informal math skills. It is our job to help students utilize their prior knowledge. Since thinking skills transfer from one language to another, problems involving experiences such as travel, money, and school life can trigger students' interest and motivation. Classroom activities should encourage students to draw upon these.

Cooperative Learning

Related to the goal of encouraging LEP student participation is the goal of creating a positive feeling toward mathematics and the mathematics classroom. Maintaining a friendly non-threatening, supportive environment is an essential aspect of encouraging participation and thus learning. Additionally, fostering a feeling of community is essential in a classroom for students to acquire lasting knowledge. The community should generate and sustain a mathematics culture.

Cooperative learning groups play an important role in promoting the exchange of values and providing a forum for participation in the mathematics culture. The culture values and rewards inquiry, effort, and risk taking. It is the responsibility of the teacher to facilitate the development of such a culture within the classroom. Student conversational interaction is encouraged through cooperative learning activities. I am careful to seat LEP students next to others who can translate and/or provide support. Small groups provide a safer environment since many students are reluctant to ask questions or offer answers in a large class setting. Students pool their talents and strengths (e.g. language, computational, and problem solving abilities) to piece together solutions to problems.

The teacher should allow for a reasonable noise level since a class of 30 or 40 engaged in cooperative learning can generate considerable noise. Classroom management skills become important in maintaining the balance between the organized group debates which develop, and anarchy.

Language Development

Vocabulary (mathematical, technical, and general) is taught as part of subject matter instruction rather than in the traditional method that relied on vocabulary lists without connection to meaningful contexts. Therefore vocabulary is always presented as an incidental part of a lesson and is connected to real ideas and objects. In this way vocabulary learning takes on a new meaning for students; it has a purpose.

Students' language development is central to success for LEP students in the mathematics classroom and thus deserves added attention. Collaborative problem solving is an effective way to foster written and verbal communication between students. The student discussions create opportunities for students to hear themselves and other students using the language of mathematics. Students must be given opportunities to write and speak mathematically with each other as well as the teacher. Additional methods of promoting language development include the use of investigations and projects in collaborative group settings. Assigning portfolios and journal writing is also useful. Do not avoid assigning problems that require writing. Instead, use these problems as vehicles for students to develop oral and written language skills in cooperative group settings.

Effective Teaching

The instructional approaches recommended thus far for teaching mathematics to limited English proficient students are in fact nothing more than good teaching techniques that work well for all students. It is therefore possible to accommodate the needs of limited English proficient students and native English speakers concurrently without compromising the learning of either.

In *Integrated Mathematics 1*
See student pages 137, 272, and 455 and side-column commentary on pages 12, 145, and 198.

Visual Learning Strategies

by **Stuart J. Murphy**
Visual Learning Specialist
Evanston, Illinois

"Linking the visual and the verbal is a powerful teaching tool."

Our Visual Environment

There is no question that we are living in an intensely visual environment. Whether from television and videos, or magazines and books, information regularly comes to us in a variety of formats. In addition to text, these formats include charts and graphs, maps and diagrams, photographs and illustrations, symbols and cartoons.

Even the way in which text is presented has become more varied to include a greater use of highlighted phrases, headlines, call-outs, and captions. The need to absorb more information—in more formats—has never been greater.

With this need come many learning opportunities. There is growing evidence that comprehension increases when verbal information is augmented by high-quality visual displays.

Linking the visual and the verbal is a powerful teaching tool. Such a link interests and motivates learners, provides more information, and reaches a broader audience than either method alone.

Visual Learning in Mathematics

In the study of mathematics, visual learning strategies play an especially important role. Understanding symbols—and the use of symbols within the concise language of mathematics—is critical to the ability to comprehend and express mathematical ideas.

Icons and symbols also play an important role in the use of technology in mathematical instruction. Calculators and computers use a carefully constructed symbolic language to provide direction and convey meaning to users.

Visual presentations help us to model mathematical ideas, to see patterns, and to understand relationships. Indeed, a basic understanding of many important mathematical concepts—concepts such as comparison, scale, dimension, translation, and perspective—depends upon the ability of the student to visualize.

A better understanding of how information is conveyed visually can also help us as we work with students who are visual/spatial learners, students who have limited English proficiency, and students who come from a variety of socio-economic and cultural backgrounds. In fact, using visual learning strategies can help us increase the learning potential of all students.

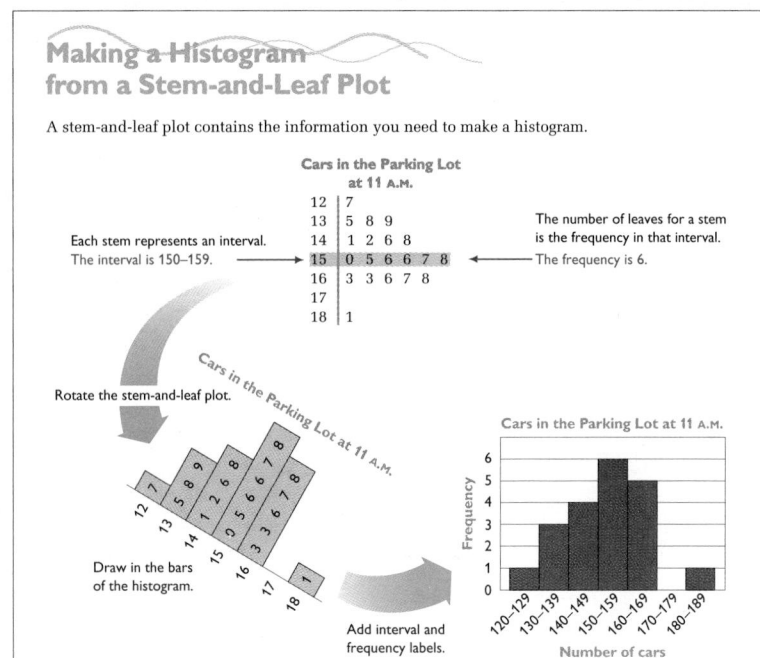

Making a Histogram from a Stem-and-Leaf Plot

A stem-and-leaf plot contains the information you need to make a histogram.

Cars in the Parking Lot at 11 A.M.

12	7
13	5 8 9
14	1 2 6 8
15	0 5 6 6 7 8
16	3 3 6 7 8
17	
18	1

Each stem represents an interval. The interval is 150–159.

The number of leaves for a stem is the frequency in that interval. The frequency is 6.

Rotate the stem-and-leaf plot.

Cars in the Parking Lot at 11 A.M.

Draw in the bars of the histogram.

Add interval and frequency labels.

Cars in the Parking Lot at 11 A.M.

Number of cars

Frequency

Photographs illustrate that the information students are learning is relevant to their lives.

Integrated Mathematics includes a carefully planned visual learning strand to help students develop visual learning skills. Pages are designed to allow easy access to the material being presented and to provide multiple points of entry, including images, titles, diagrams, and call-outs.

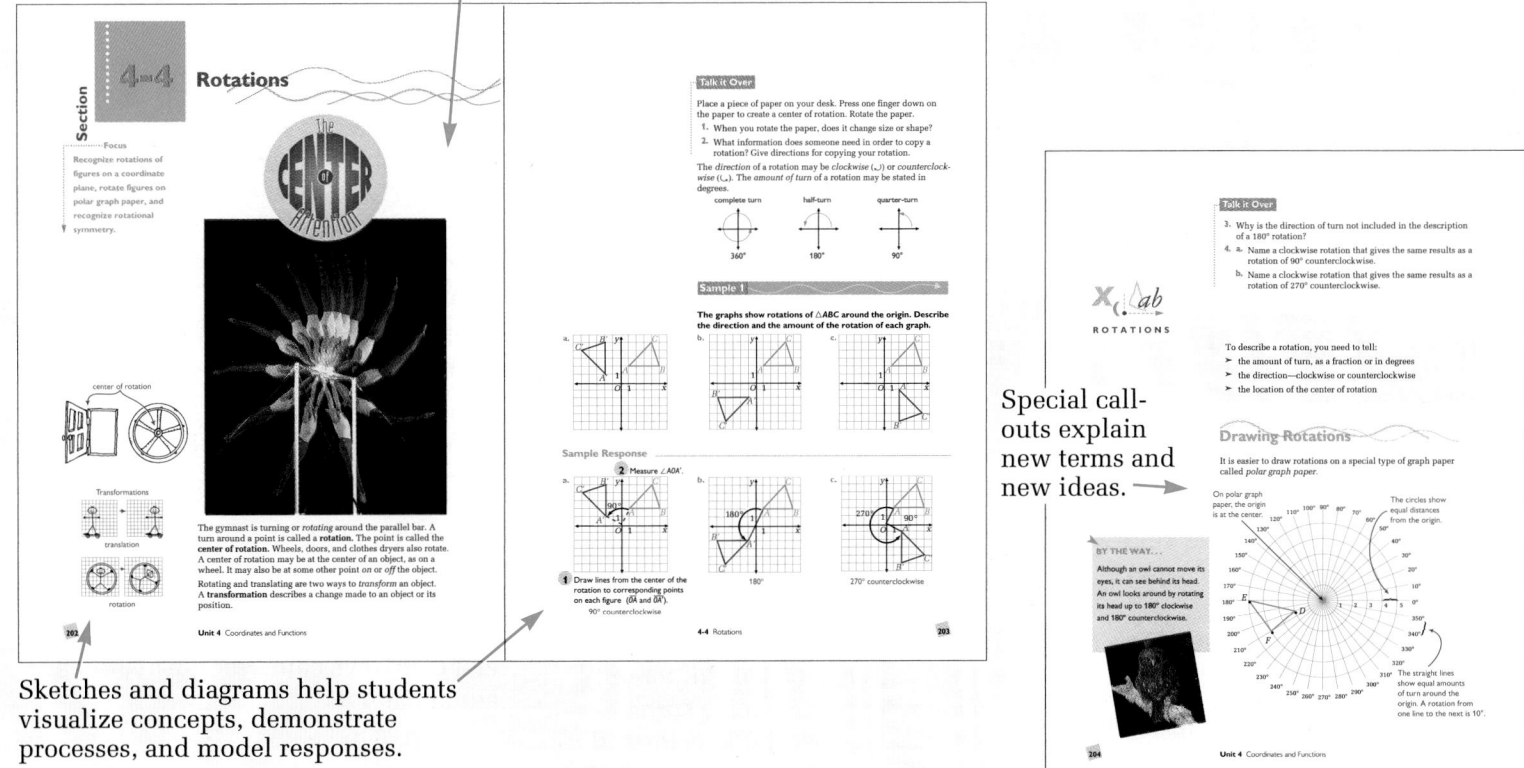

Special call-outs explain new terms and new ideas.

Sketches and diagrams help students visualize concepts, demonstrate processes, and model responses.

Strategies for Developing Visual Skills

Here are some visual learning strategies that you can use on an ongoing basis.

➤ Display visual materials to interest, excite, and motivate students.

➤ Emphasize photos within the text that demonstrate real applications of mathematical concepts and ask students to consider and discuss other examples.

➤ Explain—or have students explain—the diagrams within a lesson.

➤ Develop—or have students develop—ways to visualize abstract concepts.

➤ When students are having trouble understanding a concept, try to explain the concept without using words.

➤ Encourage students to:

• draw and sketch as part of their note-taking and journal practice

• take photographs or clip photos from magazines to connect related ideas

• demonstrate their thinking by mapping out the steps or acting out the process

• show their understanding by drawing a concept map

• construct charts, graphs, and diagrams to explain concepts

Using these strategies and the images that have been provided in *Integrated Mathematics* will help your students develop their visual learning skills—help them to link visual to verbal, process to concept, and learning to life.

In **Integrated Mathematics 1**
See student pages 158, 208, and 433 and side-column commentary on pages 7, 80, and 364.

Assessment Methods

by **Karen S. Norwood**
Assistant Professor, Department of Mathematics and Science Education
North Carolina State University; Raleigh, North Carolina

> ## "Assessment needs to be embedded in the instructional process."

Investigate the shape and the dimensions of the pen with the largest area that can be constructed with 36 feet of fencing.

Explain which is larger:

π^6 or 6π

Do not use your calculator.

Write a paragraph explaining how the sine and the cosine ratios are alike and how they are different.

Demonstration
Tell the class everything you know about the graphs of

A $y = x^2 + 2x + 1$

and

B $y = -x^2 + 2$

Use a graphics calculator or computer software if you wish.

Assessment Goals

The purpose of assessment in mathematics is to improve and evaluate learning and teaching. In the teaching-learning process, it is imperative that assessment be used to broaden and inform, rather than restrict, the process. Assessment needs to be embedded in the instructional process, instead of being apart from it. This view was well stated in the NCTM's *Curriculum and Evaluation Standards for School Mathematics.* "In an instructional environment that demands a deeper understanding of mathematics, testing instruments that call for only the identification of single correct responses no longer suffice. Instead, our instruments must reflect the scope and intent of our instructional program to have students solve problems, reason, and communicate."

Traditional paper-and-pencil tests are incomplete measures of achievement. In fact, no single type of assessment can serve all the information needs of an educational institution. Using alternative assessment methods provides a more equitable measure of a student's mathematical progress, has less potential for bias, and encourages respect for diversity by modeling appreciation for varied approaches to a problem. The goals of alternative assessment are to:

➤ find out what the students already know

➤ evaluate the depth of the students' conceptual understanding and their ability to transfer this understanding to new and different situations

➤ evaluate the students' ability to communicate their understanding mathematically, make mathematical connections, and reason mathematically

➤ plan the mathematics instruction in order to achieve the objectives

➤ report individual student progress and show growth in mathematical maturity

➤ analyze the overall effectiveness of the mathematical instruction

When using alternative assessment, it is important to start slowly, so as not to become overwhelmed. Journal writing is a good place to start. Once you become comfortable with this technique, try to add another alternative assessment strategy to your repertoire. Don't try to use alternative assessment alone; involve colleagues, parents, and administrators.

Scoring

There are several ways to score alternative assessment assignments. One of the most simple methods is to divide papers into piles labeled "satisfactory" and "unsatisfactory." Then assign a grade from 0 to 3 based on the following criteria. Satisfactory papers are given a grade of 3 if the student gives a clear explanation with appropriate diagrams or graphs, and a score of 2 if the student's work is complete and shows understanding but contains computational errors or minor flaws in explanation. Unsatisfactory papers are given a 1 if the work is incomplete and contains serious conceptual errors along with flagrant computational errors. A score of 0 is given if little or no effort was made to complete the assignment.

Some people prefer to use a scale of 1 to 4, where 4 indicates excellent and 1 indicates unacceptable work. Five- and six-point scales are also used.

Alternative Assessment Formats

Several types of alternative assessment items are appropriate to the mathematics classroom.

JOURNALS Regular use of a journal encourages students to express complex mathematical concepts in words. Writing helps to make students aware of what they do and don't understand, what they can and cannot do. Reading a journal gives the teacher insight into the student's understanding.

RESEARCH PROJECTS Group or individual research projects allow students to investigate topics that encompass many mathematical concepts and their real-world applications. Examples of such projects are the Unit Projects in this book.

DEMONSTRATION/PERFORMANCE ASSESSMENT Teachers can assess their students' comprehension of a mathematical concept by asking them to explain the concept in their own words using such items as compasses, graph paper, calculators, and computers.

PROBLEM SOLVING Problem solving is considered to be the link between facts and algorithms and the real-life problem situations that we all face. Problem-solving activities include nonroutine problems where the strategy necessary to solve the problem is not immediately apparent, and analysis and synthesis of previously learned knowledge are required.

PORTFOLIOS As artists and writers use portfolios to show off their best work, mathematics students can use portfolios to document their growth and the development of their mathematical power. Portfolios can be used to assess a student's mathematical reasoning, understanding, attitudes, and ability to communicate mathematically.

Both the teacher and the student should have input into selecting what will be included in a portfolio. For example, the teacher might determine how many pieces are to be included and the categories from which they will come. The student might be allowed to choose the pieces. The portfolio should include a table of contents and a cover letter. Each included work should be labeled with the date, a description of the task or problem, and the identity of the person who selected the work. A self-assessment should also be included.

The contents of a portfolio might include:

- open-ended questions, problems, and tasks, in which the student is asked to formulate hypotheses, explain a mathematical situation, make a generalization, and so on, either orally or in writing

- research projects

- presentations, discussions, and debates

- journal entries

- cooperative learning activities

- math logs: problems assigned by the teacher which require that the student not only show computations, but validate the solution

- problem solving

- investigations

- models and simulations

- interviews: students talk, individually or in groups, while the teacher listens and asks questions. The teacher may encourage students to further elaborate in an interview by using phrases such as, "I am interested in your thinking," or "I understand it better now, but..."

- photographs of items the student may have produced that are too bulky to fit in a portfolio

- work dealing with the same mathematical idea sampled at different times

- copies of awards or prizes

In *Integrated Mathematics 1*
See student pages 174, 260, and 521 and side-column commentary on pages 25, 163, and 313.

Special Planning Pages for Every Unit

4 Coordinates and Functions

OVERVIEW

➤ **Unit 4** introduces coordinates as a means of describing locations in the plane. Students learn how the coordinates of the vertices of polygons are affected by translations on the coordinate plane. Some prior experience with locating points on the coordinate plane is assumed. Students may refresh their knowledge of a coordinate system by using the **Student Resources Toolbox** on page 660.

➤ Students apply coordinate geometry to the statistical topic of scatter plots and fitted lines for making correlations and predictions.

➤ In Unit 4, students begin their study of functions. They learn to recognize functions from the shapes of their graphs, to represent functions with tables of ordered pairs, and to graph functions from equations.

➤ Animation is the theme of the Unit Project. Students employ transformations on the coordinate plane to create a flip book illustrating animated motion.

➤ Connections to social studies, art, literature, earth science, and computer science are integrated into the teaching materials and exercises.

➤ Problem-solving strategies used in Unit 4 include *Use a Diagram* (Section 4-2), *Break the Problem into Parts* (Section 4-2), and *Identify a Pattern* (Section 4-7).

Unit Objectives

Section	Objectives	NCTM Standards
4-1	• Use coordinate systems to solve problems about locations in different settings.	1, 2, 4, 5, 8
4-2	• Learn about coordinate geometry.	1, 2, 4, 5, 8
	• Identify polygons on a coordinate plane.	
	• Find the areas of polygons drawn on a coordinate plane.	
4-3	• Translate figures on a coordinate plane.	1, 2, 4, 5, 8
	• Recognize translational symmetry.	
4-4	• Recognize rotations of figures on a coordinate plane.	1, 2, 4, 5, 8
	• Rotate figures on polar graph paper.	
	• Recognize rotational symmetry.	
4-5	• Make and interpret a scatter plot of data.	1, 2, 3, 4, 5, 6, 8, 10
	• Use scatter plots to make predictions.	
4-6	• Understand what a function is.	1, 2, 4, 5, 6, 8
	• Identify control variables and dependent variables.	
	• Draw graphs of functions.	
	• Recognize functions.	
4-7	• Change from one representation of a function to another.	1, 2, 4, 5, 6, 8
	• Learn about some basic functions.	

Unit Overview

➤ **Overview** provides a summary of the mathematical topics, applications, and problem solving strategies for each unit.

➤ **Unit Objectives** gives objectives and NCTM Standards for each section.

Section	Connections to Prior and Future Concepts
4-1	**Section 4-1** reviews the coordinate system, which many students have studied in previous courses. The coordinate system is developed and used throughout Unit 4. Coordinate graphing is also important in Units 6, 7, 8, and 10 of Book 1. Coordinate geometry is studied in Units 2, 3, 4, 5, 9, and 10 of Book 2 and in Units 1–3, 5, 7, and 9 of Book 3.
4-2	**Section 4-2** uses the coordinate system presented in Section 4-1 to introduce area concepts. Students identify polygons and find the areas of polygons drawn on the coordinate plane. Area is further studied in Sections 5-7, 7-6, 9-5, and 9-8 of Book 1 and in Unit 6 of Book 2.
4-3	**Section 4-3** introduces translations on the coordinate plane and translational symmetry. Transformation concepts are extended in Sections 4-4, 6-6, 10-1, and 10-2 of Book 1; in Units 3–5 and 8 of Book 2; and in Units 3, 9, and 10 of Book 3.
4-4	**Section 4-4** continues the study of transformations begun in Sections 1-6 and 4-3 by introducing rotations and rotational symmetry. Transformations are studied throughout all three books of the series.
4-5	**Section 4-5** extends the study of statistical graphing begun in Unit 3. Statistical graphing and the coordinate system are combined in the exploration of scatter plots. The examination of data analysis and graphing continues throughout Book 1 and is further developed in Books 2 and 3, as already indicated.
4-6	**Section 4-6** combines the concepts presented in Sections 4-1 and 4-5 to introduce functions and graphs of functions. The study of functions and their graphs, which is a major focus of mathematics, continues throughout the remainder of Book 1 and is extended in most of the units of Book 2 and Book 3.
4-7	**Section 4-7** continues the investigation of functions begun in Section 4-6 by relating functions, graphs, and equations. Functions, graphs, and their equations are given special emphasis in Units 5, 7, 8, and 10 of Book 1; in Units 2, 4, and 9 of Book 2; and in Units 1, 2, 4, 5, 7, and 9 of Book 3.

Integrating the Strands

Strands	Sections
Number	4-2, 4-5, 4-6
Algebra	4-2, 4-3, 4-5, 4-6, 4-7
Functions	4-6, 4-7
Measurement	4-1, 4-2, 4-3, 4-5
Geometry	4-1, 4-2, 4-3, 4-4, 4-6, 4-7
Statistics and Probability	4-5, 4-6
Discrete Mathematics	4-1
Logic and Language	4-1, 4-2, 4-3, 4-4, 4-5, 4-7

Topic Integration

➤ **Topic Spiraling** gives connections to past and future learning.

➤ **Integrating the Strands** shows the integration of mathematical strands throughout the unit.

4 Coordinates and Functions

Section Planning Guide

➤ Essential exercises and problems are indicated in boldface.
➤ Ongoing work on the Unit Project is indicated in color.
➤ Exercises and problems that require student research, group work, manipulatives, or graphing technology are indicated in the column headed "Other."

Section	Materials	Pacing	Standard Assignment	Extended Assignment	Other
4-1		Day 1	**1–15**, 16–23, 29–35, 36	**1–15**, 16–23, 25–35, 36	24
4-2	long pieces of string	Day 1	**1–3**, 4, 5, **6–11**	**1–3**, 4, 5, **6–11**	
		Day 2	**12–15**, **17**, **18**, 20–26, 27	**12–15**, **17**, **18**, 20–26, 27	16, 19
4-3		Day 1	1, **2–10**, 11–15, **16–22**, 24–31, 32–34	1, **2–10**, 11–15, **16–22**, 24–31, 32–34	23
4-4	polar graph paper, protractor, scissors	Day 1	1, 2, **3–7**, 8–11, **12–15**, 23, 24, 25, 26	1, 2, **3–7**, 8–11, **12–15**, 16–19, 23, 24, 25, 26	20–22
4-5	tape measure or meter stick, graphing technology, dry spaghetti	Day 1	**1**, **3–8**, 9	**1**, **3–8**, 9	2
		Day 2	**12–15**, **17–21**, 24–30, 31	10, 11, **12–15**, **17–21**, 24–30, 31	16, 22, 23, 31
4-6	geometric drawing software	Day 1	**1–11**, 13–25, 26	**1–11**, 13–25, 26	12
4-7	graphing technology	Day 1	**1–10**, 11	**1–10**, 11	
		Day 2	12, **16–20**	12, **16–20**	
		Day 3	**21–38**, 39–44, 45	**21–38**, 39–44, 45	13–15
Review		Day 1	Unit Review	Unit Review	
Test		Day 2	Unit Test	Unit Test	

Yearly Pacing	Unit 4 Total	Units 1–4 Total	Remaining	Total
	15 days (2 for Unit Project)	63 days	101 days	164 days

Support Materials

➤ See **Project Book** for notes on Unit 4 Project: Making a Flip Book.
➤ UPP and disk refer to **Using Plotter Plus** booklet and **Plotter Plus** disk.
➤ Warm-up exercises for each section are available on **Warm-Up Transparencies**.

Section	Study Guide	Practice Bank	Problem Bank	Activity Bank	Explorations Lab Manual	Assessment Book	Visuals	Technology
4-1	4-1	Practice 27	Set 8	Enrich 24		Quiz 4-1		
4-2	4-2	Practice 28	Set 8	Enrich 25	Masters 2, 3	Quiz 4-2		
4-3	4-3	Practice 29	Set 8	Enrich 26	Master 2	Quiz 4-3		
4-4	4-4	Practice 30	Set 8	Enrich 27	Masters 2, 4	Quiz 4-4 Test 14		Folder 5
4-5	4-5	Practice 31	Set 9	Enrich 28	Masters 2, 12	Quiz 4-5		
4-6	4-6	Practice 32	Set 9	Enrich 29	Master 2	Quiz 4-6		
4-7	4-7	Practice 33	Set 9	Enrich 30	Masters 2, 3, 13	Quiz 4-7 Test 15		
Unit 4	Unit Review	Practice 34	Unifying Problem 4	Family Involve 4		Tests 16, 17		

180C

Teaching Information

➤ **Section Planning Guide** gives materials, pacing, and suggested assignments for each section.

➤ **Support Materials** lists all support materials for each section.

UNIT TESTS

OUTSIDE RESOURCES

Form A

Spanish versions of these tests are on pages 134–137 of the **Assessment Book**.

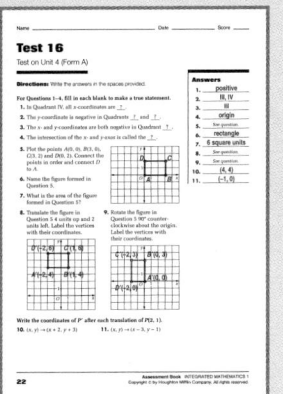

Form B

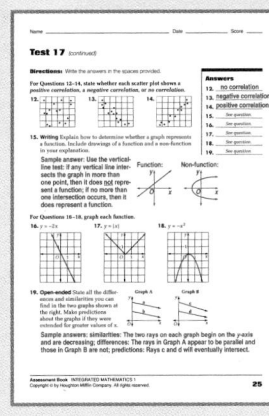

22   Assessment Book INTEGRATED MATHEMATICS 1
Copyright © by Houghton Mifflin Company. All rights reserved.

23

24   Assessment Book INTEGRATED MATHEMATICS 1
Copyright © by Houghton Mifflin Company. All rights reserved.

25

180D

Books/Periodicals

Borlaug, Victoria. "From Algebra to Calculus." *Mathematics Teacher* (December 1992): pp. 282–287.

Geometry from Multiple Perspectives. Grades 9–12 Addenda. Reston, VA: NCTM, 1993.

Manipulatives

Algeblocks. Cincinnati, OH: Southwestern Publications.

Software

Schwartz, Judah and Michal Yerushalmy. *Geometric Supposer.* Scotts Valley, CA: Sunburst.

Hofer, Alan. *Graph Whiz.* Macintosh. Acton, MA: William K. Bradford Publishing.

Harvey, Wayne, Judah Schwartz, and Michal Yerushalmy. *Visualizing Algebra: The Function Analyzer.* Scotts Valley, CA: Sunburst.

Videos

Visualizing Algebra: A New Vision for Learning and Teaching Algebra. Reston, VA: NCTM, 1993.

Teacher's Resources

➤ **Unit Tests** shows reduced facsimiles of Unit Tests, Forms A and B.

➤ **Outside Resources** lists books, periodicals, manipulatives, software, and videos.

Facsimiles of the *Practice* masters appear in the side columns next to the section that they accompany.

Using the TE

A Teaching Plan for Every Section

Planning

A column referencing

➤ **Objectives and Strands**
➤ **Spiral Learning**
➤ **Materials List**
➤ **Recommended Pacing**
➤ **Toolbox References**
➤ **Extra Practice**
➤ **Warm-Up Exercises**
➤ **Support Materials**

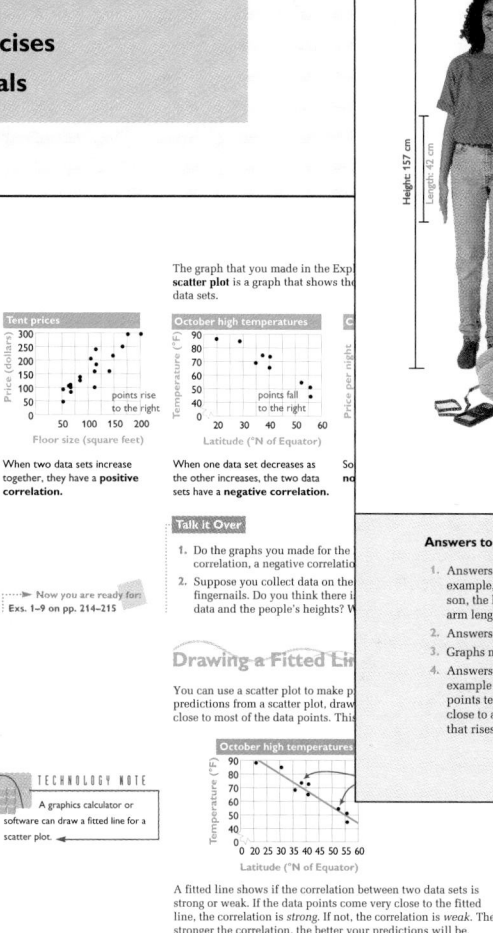

Section 4-5

Scatter Plots

Focus
Make and interpret a scatter plot of data, and use scatter plots to make predictions.

GET TO THE POINT

EXPLORATION

(*Do*) *taller people have longer arms?*

- Materials: tape measure or meter stick, graph paper, graphics calculator or software (optional)
- Work in a group of at least five students.

① Predict what the relationship between height and lower-arm length might be.

② Measure and record the height of each person in your group to the nearest centimeter.   Measure and record each person's lower-arm length.

Name	Height (cm)	Lower-arm length (cm)
Michelle	157	42
?	?	?

③ Use graph paper or a graphics calculator to plot your data.

④ Describe any trends you see in your graph. Does your graph support the prediction you made in step 1?

⑤ Display your table and graph along with those of the other groups. How are the data and the graphs alike or different?

⑥ Plot the data from all the groups on one graph.

⑦ Compare this graph with the first graph your group made. Does the new graph support the prediction you made in step 1?

⑧ Suppose a student is 171 cm tall. Use the graph to predict the student's lower-arm length.

⑨ Measure the wrist circumference and back length of each person in your group. Save these data for use in Exercise 3.

4-5 Scatter Plots   211

Answers to Exploration

1. Answers may vary. For example, the taller the person, the longer the lower-arm length.   height increases, lower-arm length increases, supporting the prediction in step 1.

2. Answers may vary.

3. Graphs may vary.

4. Answers may vary. An example is given. The data points tend to lie on or are close to an imaginary line that rises to the right. As

5–6. Answers and graphs may vary.

7. Answers may vary.

8. Answers may vary.

9. Answers may vary.

211

PLANNING

Objectives and Strands
See pages 180A and 180B.

Spiral Learning
See page 180B.

Materials List
➤ Tape measure or meter stick (one for each group of 5)
➤ Graph paper
➤ Graphics calculator or graphing software (optional)
➤ Pieces of dry spaghetti

Recommended Pacing
Section 4-5 is a two-day lesson.
Day 1
Pages 211–212: Exploration through Talk it Over, *Exercises 1–9*
Day 2
Pages 212–214: Drawing a Fitted Line through Look Back, *Exercises 10–31*

Extra Practice
See pages 625–627.

Warm-Up Exercises
Warm-Up Transparency 4-5

Support Materials
➤ Practice 31
➤ Enrichment 28
➤ Study Guide 4-5
➤ Problem Set 9
➤ Diagram Masters 2, 12
➤ Quiz 4-5

TEACHING

Exploration
The goal of the Exploration is to give students experience in gathering data, organizing it into a table, and graphing it as a scatter plot.

Communication: Listening
Before students get under way on the Exploration activity, explain exactly what is meant in step 1 by the term *lower-arm length*. As used here, the term means the number of centimeters from the elbow to the tip of the middle finger.

Error Analysis
It is common for students to make misleading graphs when there are repeated data points. What would you do in step 3 of the Exploration if both Jeanne and Sue have a height of 155 cm and a lower-arm length of 37 cm? How would you plot the point (155, 37) *twice* and *show* that you have done so?

Discuss this difficulty with the class. Point out the desirability of being able to count data points in the graph to be sure there are as many data points as there are students in the class (step 6). Then agree on a scheme for showing repeated points. One possibility is illustrated below.

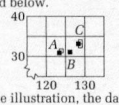

In the illustration, the data point at *A* occurs twice, at *B* once, and at *C* three times. These considerations can be important in drawing scatter plots of real-world data and in drawing fitted lines.

Talk it Over
Use questions 1 and 2 to take a closer look at the idea of correlation. The notion of correlation has to do with how strongly one set of data is related to another.

212

The graph that you made in the Expl[...] **scatter plot** is a graph that shows th[...] data sets.

Tent prices
Price (dollars): 300, 250, 200, 150, 100, 50, 0
Floor size (square feet): 50 100 150 200
points rise to the right

When two data sets increase together, they have a **positive correlation**.

October high temperatures
Temperature (°F): 90, 80, 70, 60, 50, 40, 0
Latitude (°N of Equator): 20 30 40 50 60
points fall to the right

When one data set decreases as the other increases, the two data sets have a **negative correlation**.

So[...]
n[...]

➤ Now you are ready for Exs. 1–9 on pp. 214–215

Drawing a Fitted Lin[...]

You can use a scatter plot to make p[...] predictions from a scatter plot, draw[...] close to most of the data points. This[...]

TECHNOLOGY NOTE
A graphics calculator or software can draw a fitted line for a scatter plot.

October high temperatures
Temperature (°F): 90, 80, 70, 60, 50, 40, 0
Latitude (°N of Equator): 0 20 25 30 35 40 45 50 55 60

A fitted line shows if the correlation between two data sets is strong or weak. If the data points come very close to the fitted line, the correlation is *strong*. If not, the correlation is *weak*. The stronger the correlation, the better your predictions will be.

Unit 4 Coordinates and Functions

Answers to Talk it Over

1. Answers may vary. A positive correlation is expected.

2. Answers may vary. One might reason there is no correlation or that there is a negative correlation—girls, who tend to be shorter, tend to have longer fingernails than boys.

Answers

Answers to Explorations, Talk it Over questions, Look Back questions, and Exercises and Problems are conveniently located at the bottom of each page.

In addition to the section side-column notes, a **Quick Quiz** is provided at each Checkpoint in the student book, as well as at the end of each unit.

Teaching

Notes on

➤ **Explorations**

➤ **Talk it Over questions**

➤ **Additional Samples**

➤ **Error Analysis**

➤ **Problem Solving**

➤ **Using Technology**

➤ **Using Manipulatives**

➤ **Communication**

➤ **Cooperative Learning**

➤ **Reasoning**

➤ **Mathematical Procedures**

➤ **Limited English Proficiency**

➤ **Multicultural Information**

➤ **Visual Thinking**

➤ **Look Back questions**

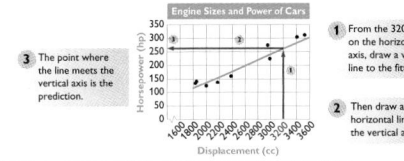

Sample

The table shows the piston displacement and horsepower (hp) of nine similarly powered cars. What horsepower do you predict for a 3200 cc engine?

Car	A	B	C	D	E	F	G	H	I
Displacement (cc)	2389	1836	3405	3535	1840	2977	2986	1998	2164
Horsepower	155	134	300	310	140	270	220	119	130

Sample Response

Make a scatter plot of the data. Draw a fitted line. Then predict.

3 The point where the line meets the vertical axis is the prediction.

1 From the 3200 mark on the horizontal axis, draw a vertical line to the fitted line.

2 Then draw a horizontal line to the vertical axis.

A 3200 cc engine in a similarly powered car should have about 270 hp.

Talk it Over

3. In the Sample, can you be certain that a 3200 cc engine will have 270 hp? Why or why not?

4. Use the scatter plot in the Sample to predict the displacement of an engine in a similarly powered car that has 200 hp.

5. A fitted line must lie close to the points on a scatter plot, but there is not one correct placement. Decide whether each line is appropriate as a fitted line for the scatter plot. Explain your decisions.

a. **Ratings of ice chests**
b. **Ratings of ice chests**

4-5 Scatter Plots

213

Communication: Discussion

Before discussing the Sample, ask students to give some examples from everyday life of quantities they think will have a strong or weak correlation. Ask them to explain their thinking.

Ask what a fitted line would look like for a scatter plot that indicates no correlation between the sets of data.

Additional Sample

The table shows the list price and quality rating (from 0 to 100) for seven brands of miniature TV sets. Predict the quality rating for a TV set that lists for $500.

TV Brand	List Price ($)	Rating
A	600	86
B	550	75
C	300	50
D	650	70
E	450	51
F	369	48
G	300	40

Ratings of TV sets

Using the scatter plot and fitted line shown above, a rating of 64 seems a reasonable prediction.

Talk it Over

Use questions 3–5 to be sure students understand what a fitted line is and how it can be used to make reasonable predictions.

213

Answers to Talk it Over

3. No. 270 horsepower is a good prediction based on the data, but the actual car may have a different horsepower.

4. about 2600 cc

5. Answers may vary. Examples are given.
a. No. All of the points lie below the line, so the line does not lie as close as it could to many of the points.

b. Yes. About as many points lie above the line as lie below it and the line appears to be close to as many points as possible.

Reasoning

Be sure to point out the Watch Out bubble at the top of the page. Ask students to give some examples of cause-and-effect relationships to check their understanding of what cause-and-effect means.

Look Back

You can have each group that worked on the Exploration reconvene to discuss the Look Back questions. Each group can then select a spokesperson who will present the group's thinking to the class.

APPLYING

Suggested Assignment

Standard 1, 3–9, 12–15, 17–21, 24–31

Extended 1, 3–15, 17–21, 24–31

Integrating the Strands

Number Ex. 30

Algebra Exs. 27–29

Measurement Exs. 24–26

Statistics and Probability Exs. 2, 3, 5–23

Logic and Language Exs. 1, 4, 8–11, 20–22, 31

Using Technology

In Ex. 2(b), students are asked to trace along a calculator-generated fitted line to the point whose x-coordinate is 171. Unless the range settings are carefully chosen, it will not be possible to hit the x-coordinate 171 exactly. For the data students gathered in the Exploration, the following range settings will make it possible to land exactly on this point.

	TI-81	TI-82
Xmin:	100	100
Xmax:	195	194
Ymin:	0	0
Ymax:	63	63

A good scale setting for each axis is 10.

214

Watch Out!

A correlation between two sets of data does not necessarily mean there is a cause-and-effect relationship.

➤ Suppose there is a positive correlati[on]... students who go to their school's bas[ket]... ber of games won. Here are some co[...]

➤ Stronger student support causes [...]

➤ The team's success causes more s[...]

➤ Something else, such as a new gy[m]... dance to rise and the team to win[...]

Look Back ◀

Do you think that the scatter plots y[...] Exploration show a strong or a weak [...] height and lower-arm length? How d[...]

➤ Now you are ready for: Exs. 10–31 on pp. 215–217

4-5 **Exercises and Problems**

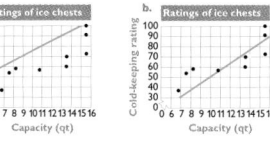

1. Use the graph you made in the Exploration.

  a. Suppose two people have the same height and different lower-arm lengths. How would this look on the graph?

  b. Suppose two people have different heights and the same lower-arm length. How would this look on the graph?

2 TECHNOLOGY Use a graphics calculator or graphing software.

  a. Plot the data that you collected in the Exploration using the statistics menu of the calculator. Does it look like the graph that you drew in Step 3 of the Exploration? If not, change the range until it does. What is the range that you need to use?

  b. Use the *linear regression* feature to find a fitted line for the data. *Trace* along the line until you find the point with x-coordinate 171. What is the y-coordinate of the [...] point?

  c. Explain what the y-coordinate that you found in part (b) represents. How is it related to the prediction that you made in Step 8 of the Exploration?

3. Use one of the data sets that you collected in Step 9 of the Exploration.

  a. What kind of correlation, if any, do you think that there is between the data set and height? Why?

  b. Test your prediction by making a scatter plot. (Put height on the horizontal axis.)

  c. Describe the correlation shown by the scatter plot. Does your scatter plot support your prediction? Explain.

BY THE WAY...

In his first year as a professional basketball player, Shawn Bradley wore number 76 for the Philadelphia 76ers. He is 7 ft 6 in. tall. Can you predict his lower-arm length?

214   **Unit 4** Coordinates and Functions

Answers to Look Back

Answers may vary depending on results.

Answers to Exercises and Problems

1. a. One point would be directly above the other.

  b. One point would lie to the left of the other.

2. Answers may vary depending on data used.

3. a. Predictions may vary.

  b. Graphs may vary.

c. Answers may vary, based on graphs in part (b).

4. Answers may vary. Examples are given. Car manufacturers might need to know the relationship between height and other measurements because they need to determine where to place the steering wheel and front seat of a

---

Applying

Notes on

➤ **Suggested Assignments**

➤ **Integrating the Strands**

➤ **Problem Solving**

➤ **Using Technology**

➤ **Using Manipulatives**

➤ **Cooperative Learning**

➤ **Reasoning**

➤ **Multicultural Information**

➤ **Unit Projects**

➤ **Careers**

➤ **Applications**

➤ **Research**

➤ **Visual Thinking**

➤ **Interdisciplinary Problems**

➤ **Assessment**

Using the Section Planning Guide

Pacing Chart

A yearly Pacing Chart and daily assignments are provided for two levels of courses—a standard course and an extended course. Both levels provide for 164 days, including days for using the Unit Openers, completing the Unit Project, and review and testing. The Pacing Chart below shows the number of days allotted for each unit of both courses. Semester and trimester divisions are indicated by a red rule and blue rules, respectively.

Unit	1	2	3	4	5	6	7	8	9	10
Standard Course	16	18	14	15	15	16	16	16	19	19
Extended Course	16	18	14	15	15	16	16	16	19	19

trimester   semester   trimester

Standard Course

The standard course is intended for students who enter with typical mathematical and problem solving skills. The course covers all ten units. The daily assignments include all the essential exercises and problems plus a number of other exercises that focus on higher-order thinking skills.

Extended Course

The extended course is intended for students who enter with strong mathematical and problem solving skills and who are able to understand new concepts quickly. The course covers all ten units. The daily assignments include all the essential exercises plus many other exercises that focus on higher-order thinking skills. It is recommended that these students be assigned some of the exercises that are listed in the Other column of the Section Planning Guide.

Section Planning Guide

The Section Planning Guide for each unit is located on the interleaved pages preceding the unit. A part of the Section Planning Guide for Unit 4 is shown here. A key describing the exercises and problems for the assignments is given in each Section Planning Guide.

Section Planning Guide

➤ Essential exercises and problems are indicated in boldface.
➤ Ongoing work on the Unit Project is indicated in color.
➤ Exercises and problems that require student research, group work, manipulatives, or graphing technology are indicated in the column headed "Other."

Section	Materials	Pacing	Standard Assignment	Extended Assignment	Other
4-1		Day 1	**1–15**, 16–23, 29–35, 36	**1–15**, 16–23, 25–35, 36	24
4-2	long pieces of string	Day 1	**1–3**, 4, 5, **6–11**	**1–3**, 4, 5, **6–11**	
		Day 2	**12–15, 17, 18**, 20–26, 27	**12–15, 17, 18**, 20–26, 27	16, 19
4-3		Day 1	1, **2–10**, 11–15, **16–22**, 24–31, 32–34	1, **2–10**, 11–15, **16–22**, 24–31, 32–34	23
4-4	polar graph paper, protractor, scissors	Day 1	**1**, 2, **3–7**, 8–11, **12–15**, 23, 24, 25, 26	1, 2, **3–7**, 8–11, **12–15**, 16–19, 23, 24, 25, 26	20–22

Integrated Mathematics 1

Authors

Senior Authors

Rheta N. Rubenstein

Timothy V. Craine

Thomas R. Butts

Kerry Cantrell

Linda Dritsas

Valarie A. Elswick

Joseph Kavanaugh

Sara N. Munshin

Stuart J. Murphy

Anthony Piccolino

Salvador Quezada

Jocelyn Coleman Walton

McDougal Littell
A Houghton Mifflin Company
Evanston, Illinois   Boston Dallas

i

Authors

Senior Authors

Rheta N. Rubenstein   Professor of Education, University of Windsor, Windsor, Ontario

Timothy V. Craine   Assistant Professor of Mathematical Sciences, Central Connecticut State University, New Britain, Connecticut

Thomas R. Butts   Professor of Mathematics Education, University of Texas at Dallas, Dallas, Texas

Kerry Cantrell   Mathematics Department Head, Marshfield High School, Marshfield, Missouri

Linda Dritsas   Mathematics Coordinator, Fresno Unified School District, Fresno, California

Valarie A. Elswick   Mathematics Teacher, Roy C. Ketcham Senior High School, Wappingers Falls, New York

Joseph Kavanaugh   Academic Head of Mathematics, Scotia-Glenville Central School District, Scotia, New York

Sara N. Munshin   Mathematics Teacher, Los Angeles Unified School District, Los Angeles, California

Stuart J. Murphy   Visual Learning Specialist, Evanston, Illinois

Anthony Piccolino   Assistant Professor of Mathematics and Computer Science, Montclair State University, Upper Montclair, New Jersey

Salvador Quezada   Mathematics Teacher, Theodore Roosevelt High School, Los Angeles, California

Jocelyn Coleman Walton   Educational Consultant, Mathematics K-12, and former Mathematics Supervisor, Plainfield High School, Plainfield, New Jersey

All authors contributed to the planning and writing of the series. In addition to writing, the Senior Authors played a special role in establishing the philosophy of the program, planning the content and organization of topics, and guiding the work of the other authors.

Field Testing The authors give special thanks to the teachers and students in classrooms nationwide who used a preliminary version of this book. Their suggestions made an important contribution to its development.

ISBN: 0-395-85502-0   1 2 3 4 5 6 7 8 9 10–VJM–03 02 01 00 99 98 97

Welcome  to Integrated Mathematics!

Course Goals

This new program has been written to prepare you for success in college, in careers, and in daily life in the 21st century.

It helps you develop the ability to:

➤ **Explore and solve mathematical problems**

➤ **Think critically**

➤ **Work cooperatively with others**

➤ **Communicate ideas clearly**

Mathematical Topics

You can learn more with *Integrated Mathematics* because the mathematical topics are integrated. Over a three-year period, this program teaches all the essential topics in the *Algebra 1/Geometry/Algebra 2* sequence, plus many other interesting, contemporary topics.

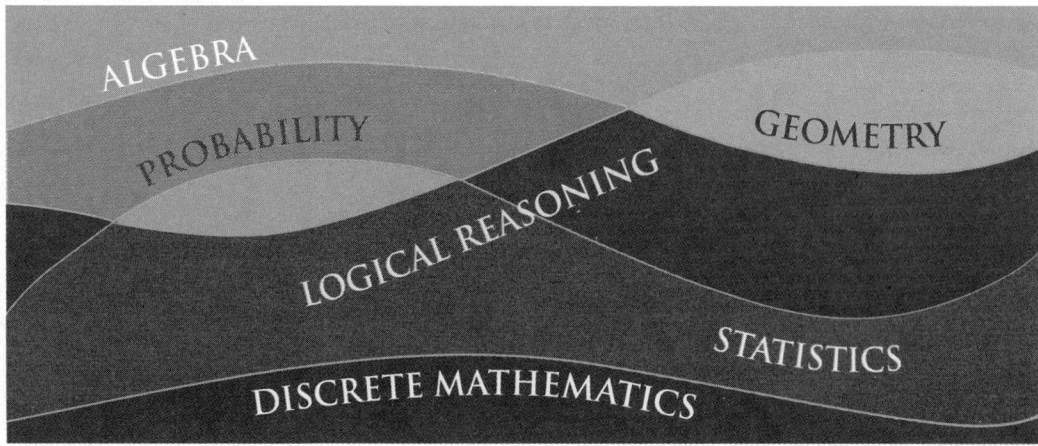

ALGEBRA
PROBABILITY
GEOMETRY
LOGICAL REASONING
STATISTICS
DISCRETE MATHEMATICS

Course Structure

- Algebra and Geometry are taught in each of the three years.
- Topics from Logical Reasoning, Measurement, Probability, Statistics, Discrete Mathematics, and Functions are interwoven throughout.
- Topics are spiraled throughout the course, so that you continually build on what you have learned.

Advantages of this Program

Integrated Mathematics develops clear understanding of topics and strong problem solving skills by giving you opportunities to:

- **Get actively involved in learning**
- **Study meaningful mathematics**
- **See connections among different branches of mathematics**
- **Try a wide variety of types of problems, including real-world applications and long-term projects**
- **Use calculators and computers**

Contents

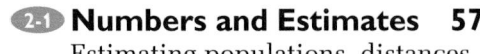

Unit 4	Coordinates and Functions

Unit Project 4

Table of Contents

Unit 7   Direct Variation

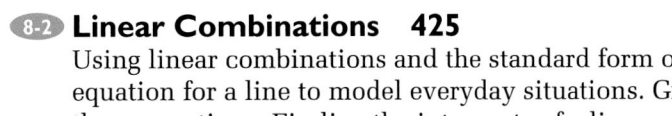

Unit 10   Quadratic Equations as Models

 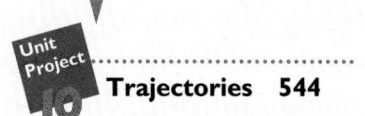

Integrated Mathematics   Topic Spiraling

This chart shows how mathematical strands are spiraled over the three years of the *Integrated Mathematics* program.

	Course 1	Course 2	Course 3
Algebra	Linear equations Linear inequalities Multiplying binomials Factoring expressions	Quadratic equations Linear systems Rational equations Complex numbers	Polynomial functions Exponential functions Logarithmic functions Parametric equations
Geometry	Angles, polygons, circles Perimeter, circumference Area, surface area Volume Trigonometric ratios	Similar and congruent figures Geometric proofs Coordinate geometry Transformational geometry Special right triangles	Inscribed figures Transforming graphs Vectors Triangle trigonometry Circular trigonometry
Statistics, Probability	Analyzing data and displaying data Experimental and theoretical probability Geometric probability	Sampling methods Simulation Binomial distributions	Variability Standard deviation z-scores
Logical Reasoning	Conjectures Counterexamples If-then statements	Inductive and deductive reasoning Valid and invalid reasoning Postulates and proof	Identities Contrapositive and inverse Comparing proof methods
Discrete Math	Discrete quantities Matrices to display data Lattices	Matrix operations Transformation matrices Counting techniques	Sequences and series Recursion Limits

Table of Contents

xiii

What Students are Saying...

Who has the **BEST IDEAS** about how mathematics should be taught? **Students** and **teachers**, of course! That is why preliminary versions of this book were tried out by thousands of students in hundreds of different CLASSROOMS NATIONWIDE. The suggestions from these students and their teachers have been incorporated into this book.

Here is what some of the students who have already studied this course have said.

$\sqrt{a}$

π

> I like math much more now. I can see that it's not just finding a right answer. The thought process is really important.

> I liked working with other students on the projects and explorations. It's good practice for working on project teams in our future jobs.

xiv

Get Involved

This course may be different from ones you have taken before. This is not a year to sit back and just listen. In this course you will be

- ➤ TALKING about mathematics
- ➤ working **together** to explore ideas
- ➤ gathering **DATA**
- ➤ looking for **patterns**
- ➤ making and testing **predictions**.

Your ideas and viewpoints are important. Sharing them with others will help everyone learn more. So don't hold back. Jump right in and get involved.

Guide to Your Course

The next ten pages will give you an overview of the organization of your book and a preview of what you will be learning in the course. They will help you get off to a good start.

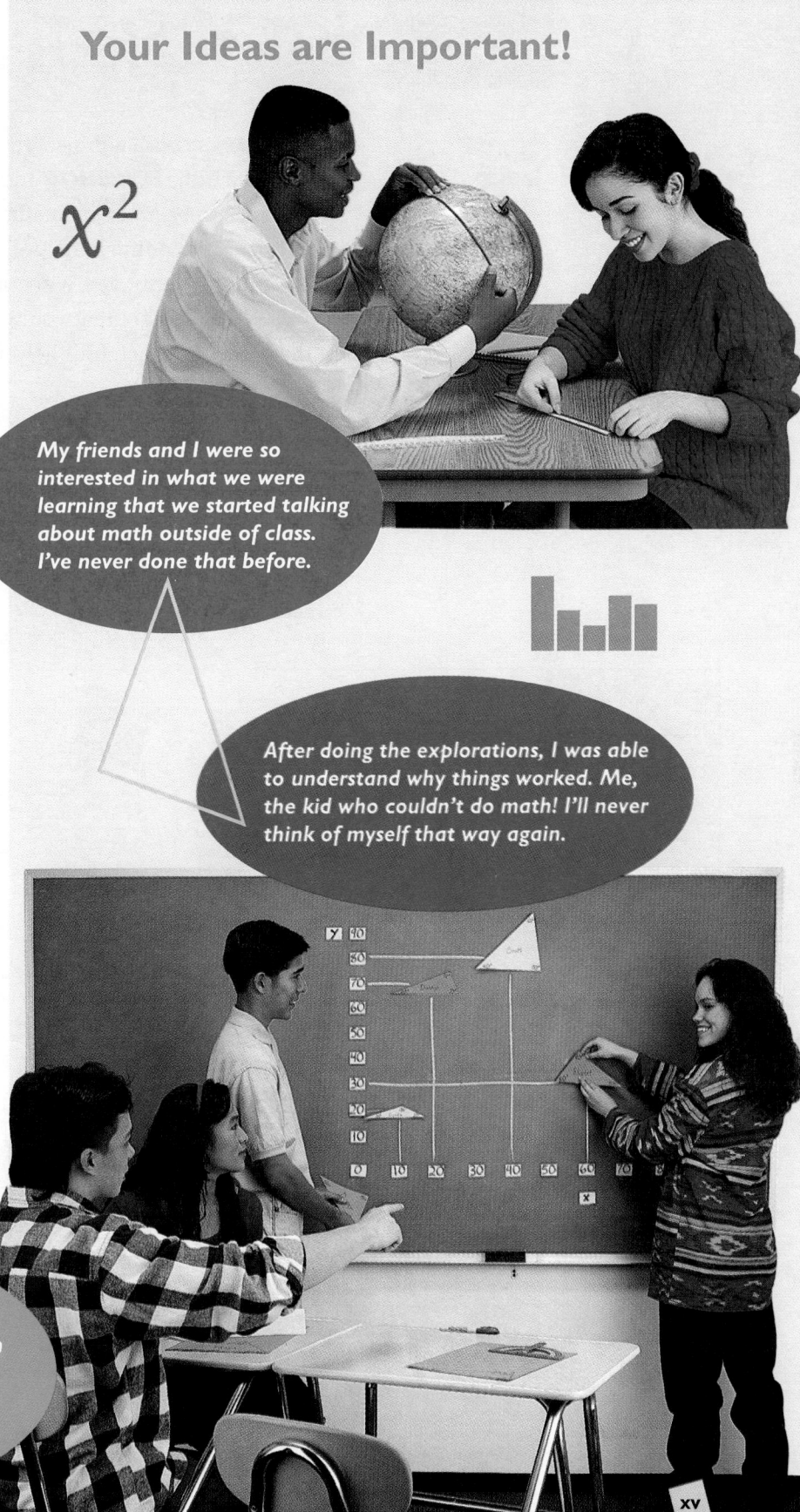

x^2

My friends and I were so interested in what we were learning that we started talking about math outside of class. I've never done that before.

After doing the explorations, I was able to understand why things worked. Me, the kid who couldn't do math! I'll never think of myself that way again.

I was used to working alone. Now I realize that I can learn more by working with other students. I have much more respect for their ideas.

Unit Projects

Each unit begins with a project that sets the stage for the mathematics you will be *learning* in the unit. The project gives you a chance to begin DOING mathematics right away. Don't expect to be able to finish the project immediately, though. As you study the unit, you will gradually develop the skills and the **INFORMATION** that you will need to complete it. The first three pages of each unit help you get started.

> **Project Theme**
>
> Each project has a theme, like animation, that relates the mathematics of the unit to careers and to daily life.

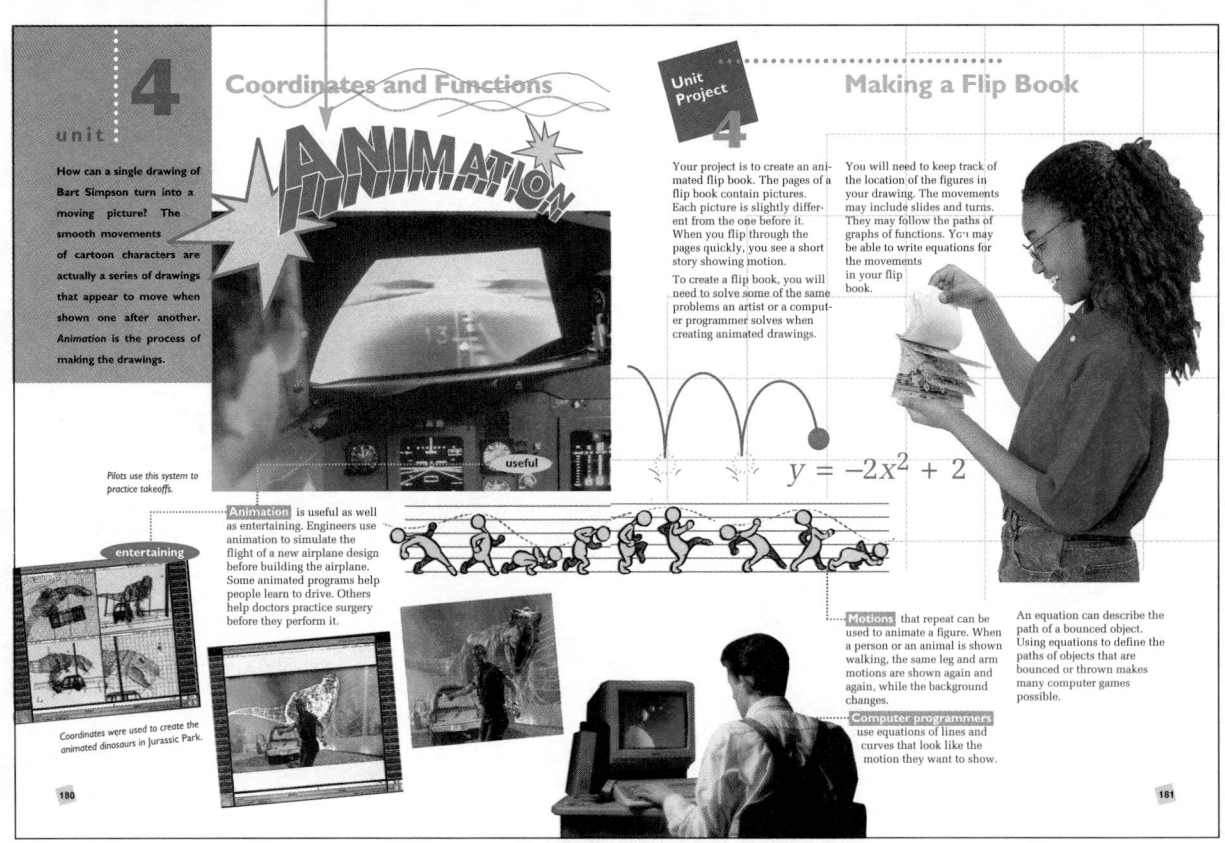

unit 4

Coordinates and Functions

ANIMATION

How can a single drawing of Bart Simpson turn into a moving picture? The smooth movements of cartoon characters are actually a series of drawings that appear to move when shown one after another. *Animation is the process of making the drawings.*

Pilots use this system to practice takeoffs.

Animation is useful as well as entertaining. Engineers use animation to simulate the flight of a new airplane design before building the airplane. Some animated programs help people learn to drive. Others help doctors practice surgery before they perform it.

Coordinates were used to create the animated dinosaurs in Jurassic Park.

useful

entertaining

Unit Project 4

Making a Flip Book

Your project is to create an animated flip book. The pages of a flip book contain pictures. Each picture is slightly different from the one before it. When you flip through the pages quickly, you see a short story showing motion.

To create a flip book, you will need to solve some of the same problems an artist or a computer programmer solves when creating animated drawings.

You will need to keep track of the location of the figures in your drawing. The movements may include slides and turns. They may follow the paths of graphs of functions. You may be able to write equations for the movements in your flip book.

$$y = -2x^2 + 2$$

Motions that repeat can be used to animate a figure. When a person or an animal is shown walking, the same leg and arm motions are shown again and again, while the background changes.

An equation can describe the path of a bounced object. Using equations to define the paths of objects that are bounced or thrown makes many computer games possible.

Computer programmers use equations of lines and curves that look like the motion they want to show.

180

181

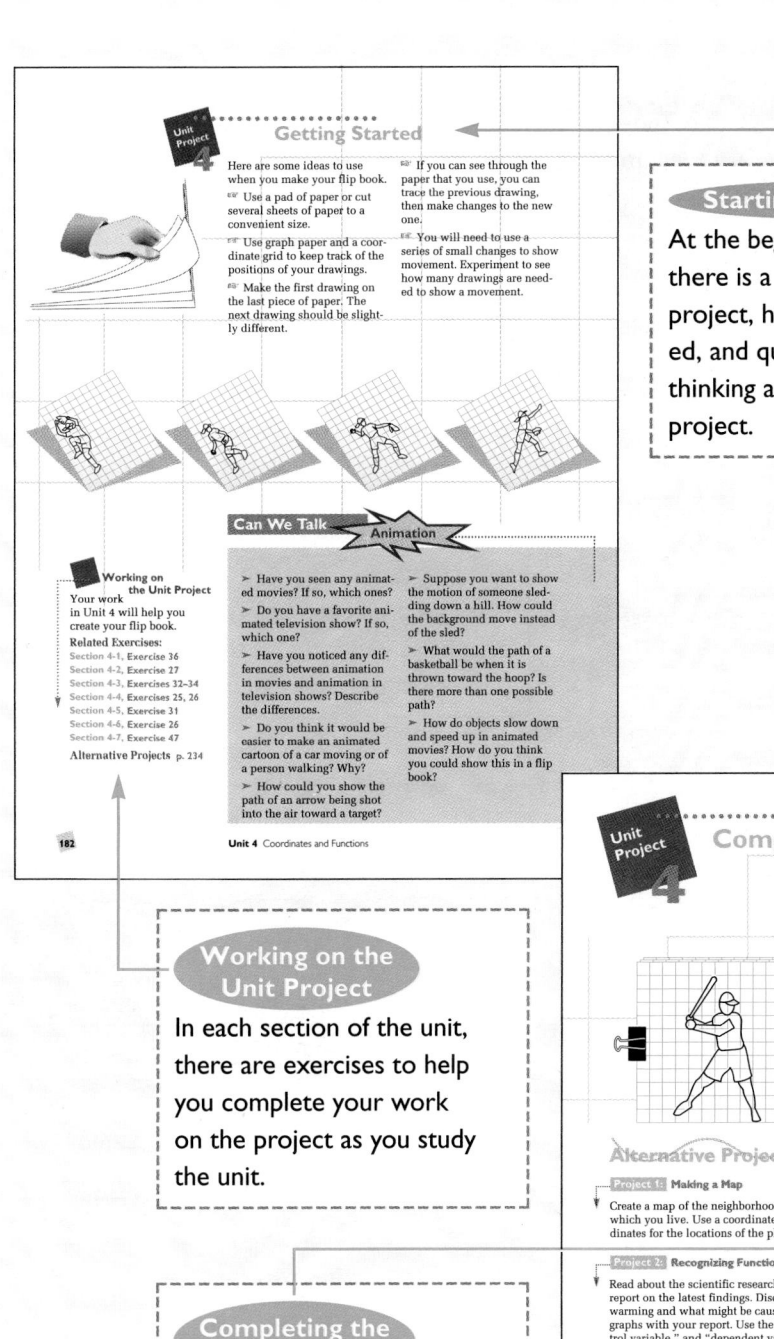

Getting Started

Here are some ideas to use when you make your flip book.

☞ Use a pad of paper or cut several sheets of paper to a convenient size.

☞ Use graph paper and a coordinate grid to keep track of the positions of your drawings.

☞ Make the first drawing on the last piece of paper. The next drawing should be slightly different.

☞ If you can see through the paper that you use, you can trace the previous drawing, then make changes to the new one.

☞ You will need to use a series of small changes to show movement. Experiment to see how many drawings are needed to show a movement.

Can We Talk

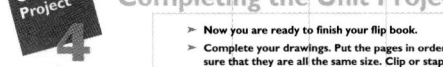
Animation

➤ Have you seen any animated movies? If so, which ones?

➤ Do you have a favorite animated television show? If so, which one?

➤ Have you noticed any differences between animation in movies and animation in television shows? Describe the differences.

➤ Do you think it would be easier to make an animated cartoon of a car moving or of a person walking? Why?

➤ How could you show the path of an arrow being shot into the air toward a target?

➤ Suppose you want to show the motion of someone sledding down a hill. How could the background move instead of the sled?

➤ What would the path of a basketball be when it is thrown toward the hoop? Is there more than one possible path?

➤ How do objects slow down and speed up in animated movies? How do you think you could show this in a flip book?

Working on the Unit Project

Your work in Unit 4 will help you create your flip book.

Related Exercises:
Section 4-1, Exercise 36
Section 4-2, Exercise 27
Section 4-3, Exercises 32–34
Section 4-4, Exercises 25, 26
Section 4-5, Exercise 31
Section 4-6, Exercise 26
Section 4-7, Exercise 47

Alternative Projects p. 234

182   **Unit 4** Coordinates and Functions

Starting the Project

At the beginning of each unit, there is a description of the project, hints for getting started, and questions to get you thinking and talking about the project.

Working on the Unit Project

In each section of the unit, there are exercises to help you complete your work on the project as you study the unit.

Completing the Unit Project

Now you've completed the unit. You can finish the project and present your results. You are ready to look back over what you have learned.

Completing the Unit Project

➤ Now you are ready to finish your flip book.

➤ Complete your drawings. Put the pages in order and make sure that they are all the same size. Clip or staple the pages together.

➤ Write a script to narrate while the pages are being flipped.

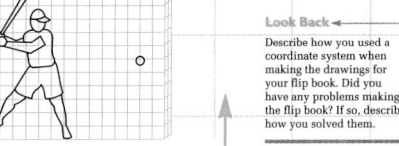

Look Back

Describe how you used a coordinate system when making the drawings for your flip book. Did you have any problems making the flip book? If so, describe how you solved them.

Alternative Projects

Project 1: Making a Map

Create a map of the neighborhood around your school or of the neighborhood in which you live. Use a coordinate system and make a map key by listing the coordinates for the locations of the places on the map. Include a scale on your map.

Project 2: Recognizing Functions in Science

Read about the scientific research on global warming. Write a report on the latest findings. Discuss the evidence for global warming and what might be causing it. Include data tables and graphs with your report. Use the phrases "is a function of," "control variable," and "dependent variable" to describe the data.

Project 3: Collecting and Graphing Data

Collect data that can be displayed in a scatter plot. Write a report and include:
➤ the method(s) you used to gather the data
➤ a summary of the data
➤ a scatter plot of the data
➤ the type of correlation the scatter plot shows and the reasons for it
➤ several questions that can be answered by using the scatter plot

⬤ Tropical forest
◯ Former tropical forest area

234   **Unit 4** Completing the Unit Project

Unit Projects

Explorations

Explorations are an important part of this course. They will help you DISCOVER, UNDERSTAND, and CONNECT mathematical ideas. In the Explorations, you will be measuring, gathering **data**, looking for **patterns**, and making generalizations. You will be working with others and **sharing your ideas**.

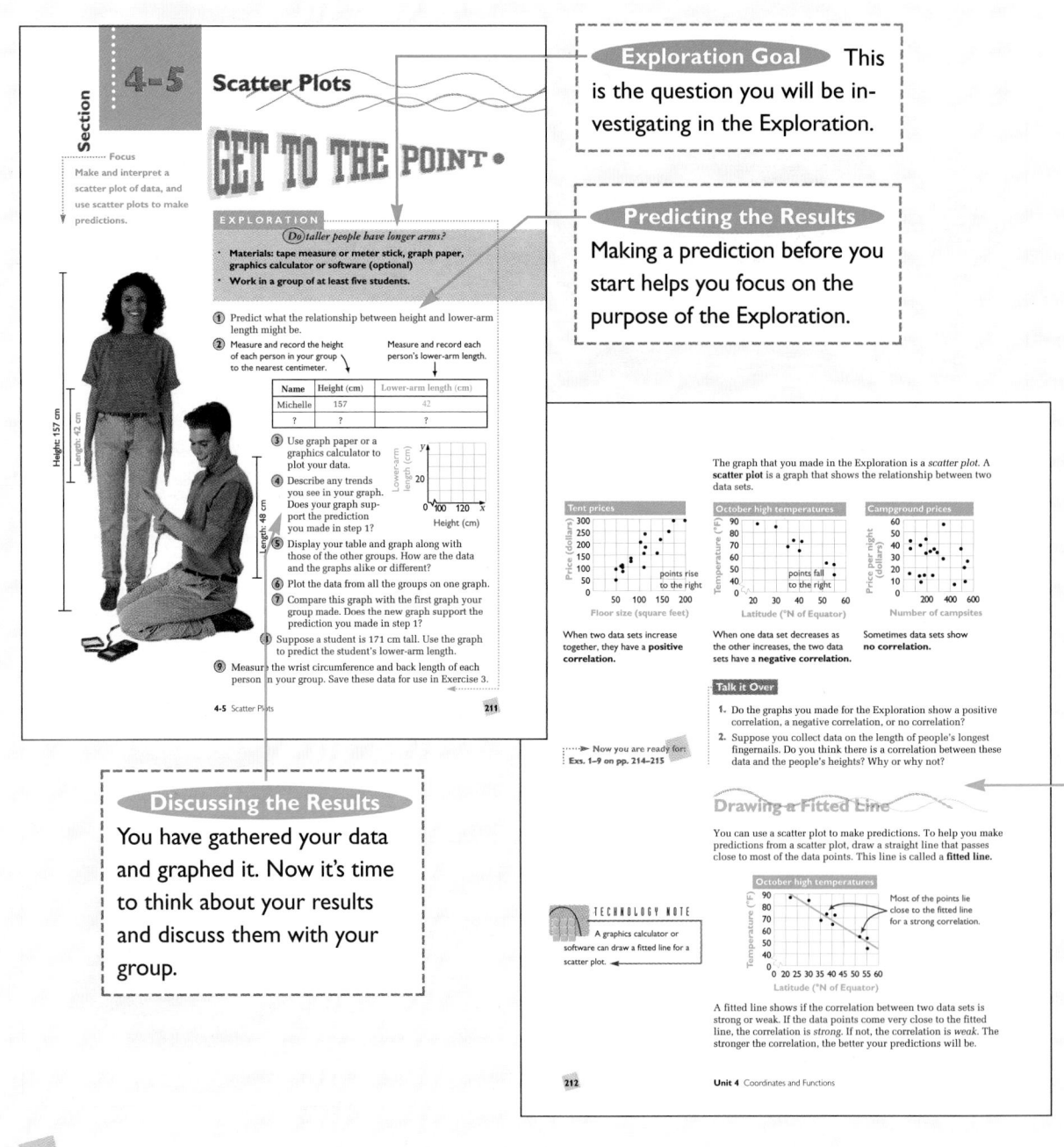

Exploration Goal This is the question you will be investigating in the Exploration.

Predicting the Results
Making a prediction before you start helps you focus on the purpose of the Exploration.

Discussing the Results
You have gathered your data and graphed it. Now it's time to think about your results and discuss them with your group.

Explorations

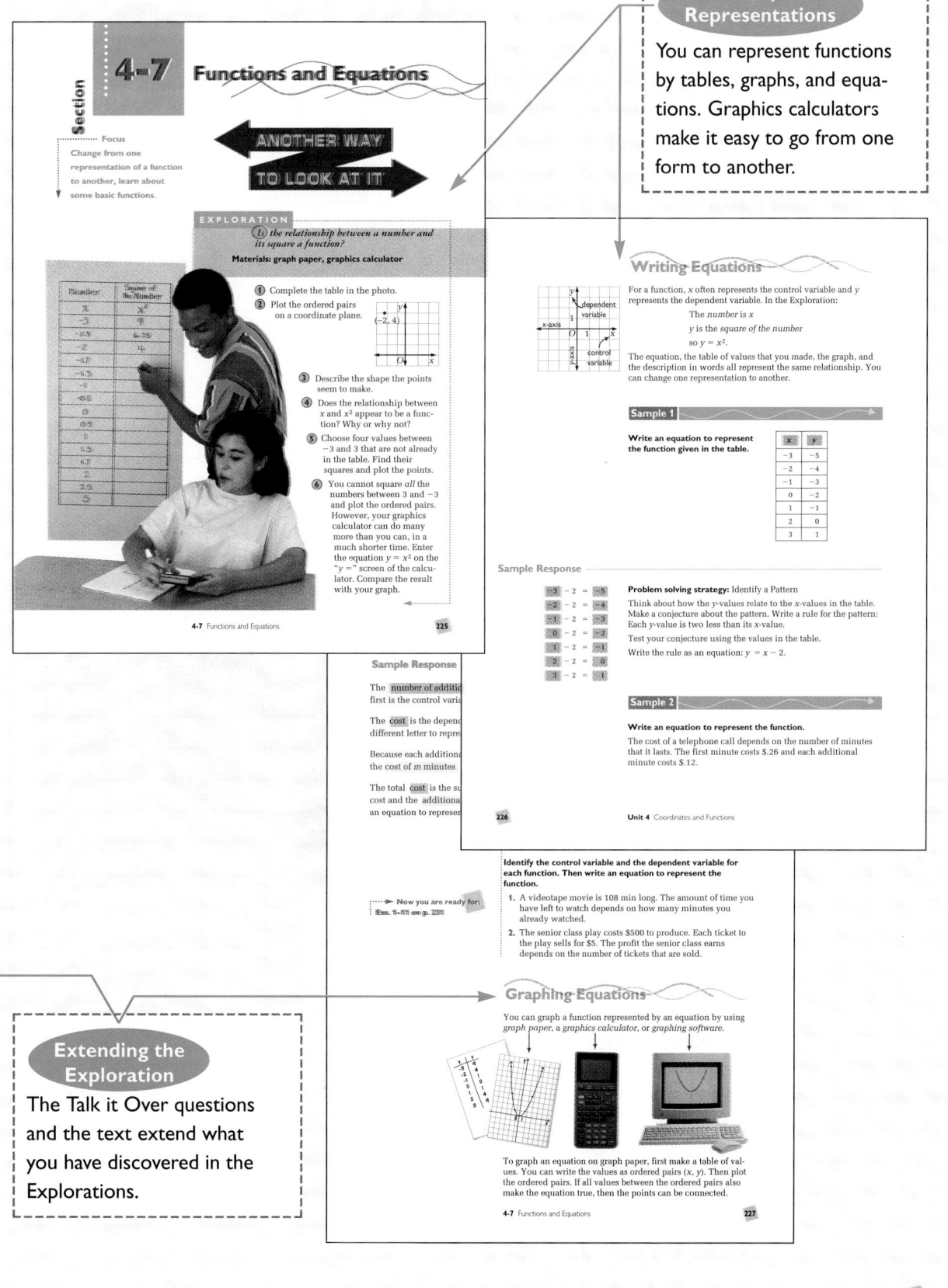

Multiple Representations

You can represent functions by tables, graphs, and equations. Graphics calculators make it easy to go from one form to another.

Extending the Exploration

The Talk it Over questions and the text extend what you have discovered in the Explorations.

Section 4-7 Functions and Equations

ANOTHER WAY TO LOOK AT IT

Focus
Change from one representation of a function to another, learn about some basic functions.

EXPLORATION

Is the relationship between a number and its square a function?

Materials: graph paper, graphics calculator

1. Complete the table in the photo.
2. Plot the ordered pairs on a coordinate plane.
3. Describe the shape the points seem to make.
4. Does the relationship between x and x^2 appear to be a function? Why or why not?
5. Choose four values between -3 and 3 that are not already in the table. Find their squares and plot the points.
6. You cannot square *all* the numbers between 3 and -3 and plot the ordered pairs. However, your graphics calculator can do many more than you can, in a much shorter time. Enter the equation $y = x^2$ on the "$y =$" screen of the calculator. Compare the result with your graph.

4-7 Functions and Equations   225

Writing Equations

For a function, x often represents the control variable and y represents the dependent variable. In the Exploration:

The *number* is x

y is the *square of the number*

so $y = x^2$.

The equation, the table of values that you made, the graph, and the description in words all represent the same relationship. You can change one representation to another.

Sample 1

Write an equation to represent the function given in the table.

x	y
-3	-5
-2	-4
-1	-3
0	-2
1	-1
2	0
3	1

Sample Response

-3 - 2 = -5
-2 - 2 = -4
-1 - 2 = -3
0 - 2 = -2
1 - 2 = -1
2 - 2 = 0
3 - 2 = 1

Problem solving strategy: Identify a Pattern

Think about how the y-values relate to the x-values in the table. Make a conjecture about the pattern. Write a rule for the pattern: Each y-value is two less than its x-value.

Test your conjecture using the values in the table.

Write the rule as an equation: $y = x - 2$.

Sample 2

Write an equation to represent the function.

The cost of a telephone call depends on the number of minutes that it lasts. The first minute costs \$.26 and each additional minute costs \$.12.

226   Unit 4 Coordinates and Functions

Sample Response

The number of addition... first is the control varia...

The cost is the depend... different letter to repre...

Because each additiona... the cost of m minutes

The total cost is the su... cost and the additiona... an equation to represen...

► Now you are ready for: Exs. 5–11 on p. 228

Identify the control variable and the dependent variable for each function. Then write an equation to represent the function.

1. A videotape movie is 108 min long. The amount of time you have left to watch depends on how many minutes you already watched.
2. The senior class play costs \$500 to produce. Each ticket to the play sells for \$5. The profit the senior class earns depends on the number of tickets that are sold.

Graphing Equations

You can graph a function represented by an equation by using *graph paper*, a *graphics calculator*, or *graphing software*.

To graph an equation on graph paper, first make a table of values. You can write the values as ordered pairs (x, y). Then plot the ordered pairs. If all values between the ordered pairs also make the equation true, then the points can be connected.

4-7 Functions and Equations   227

Section Organization

This book has been written to help you **UNDERSTAND** mathematics and to see the many different ways it can be applied. The organization of the material within sections is patterned after the **way that you learn**: Ideas are introduced. You **EXPLORE** them, **think** about them, and TALK about them with other students. You check that you understand them by working through some SAMPLE PROBLEMS. Before going on, you pause and **look back** at what you have learned.

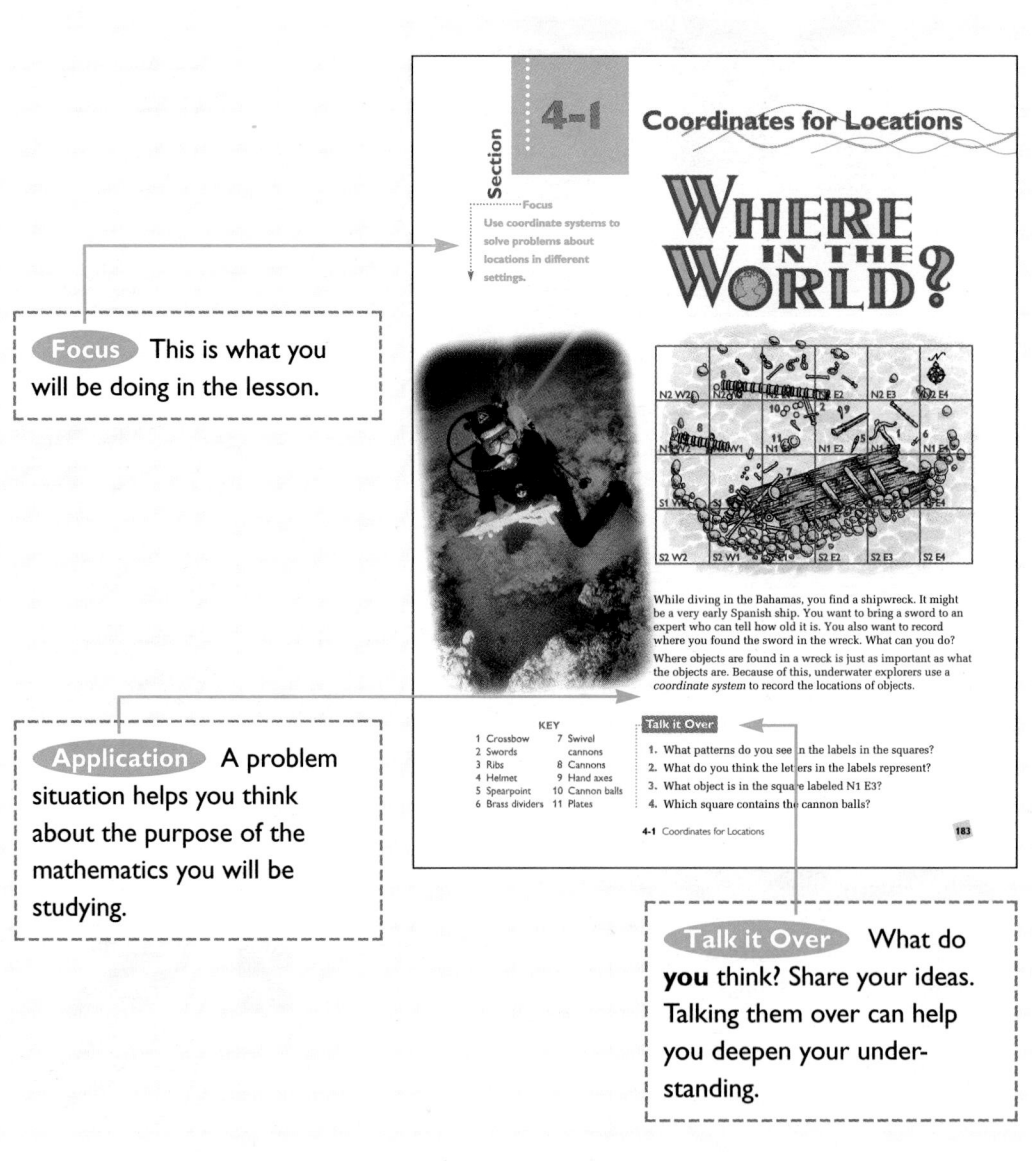

Focus This is what you will be doing in the lesson.

Application A problem situation helps you think about the purpose of the mathematics you will be studying.

Talk it Over What do **you** think? Share your ideas. Talking them over can help you deepen your understanding.

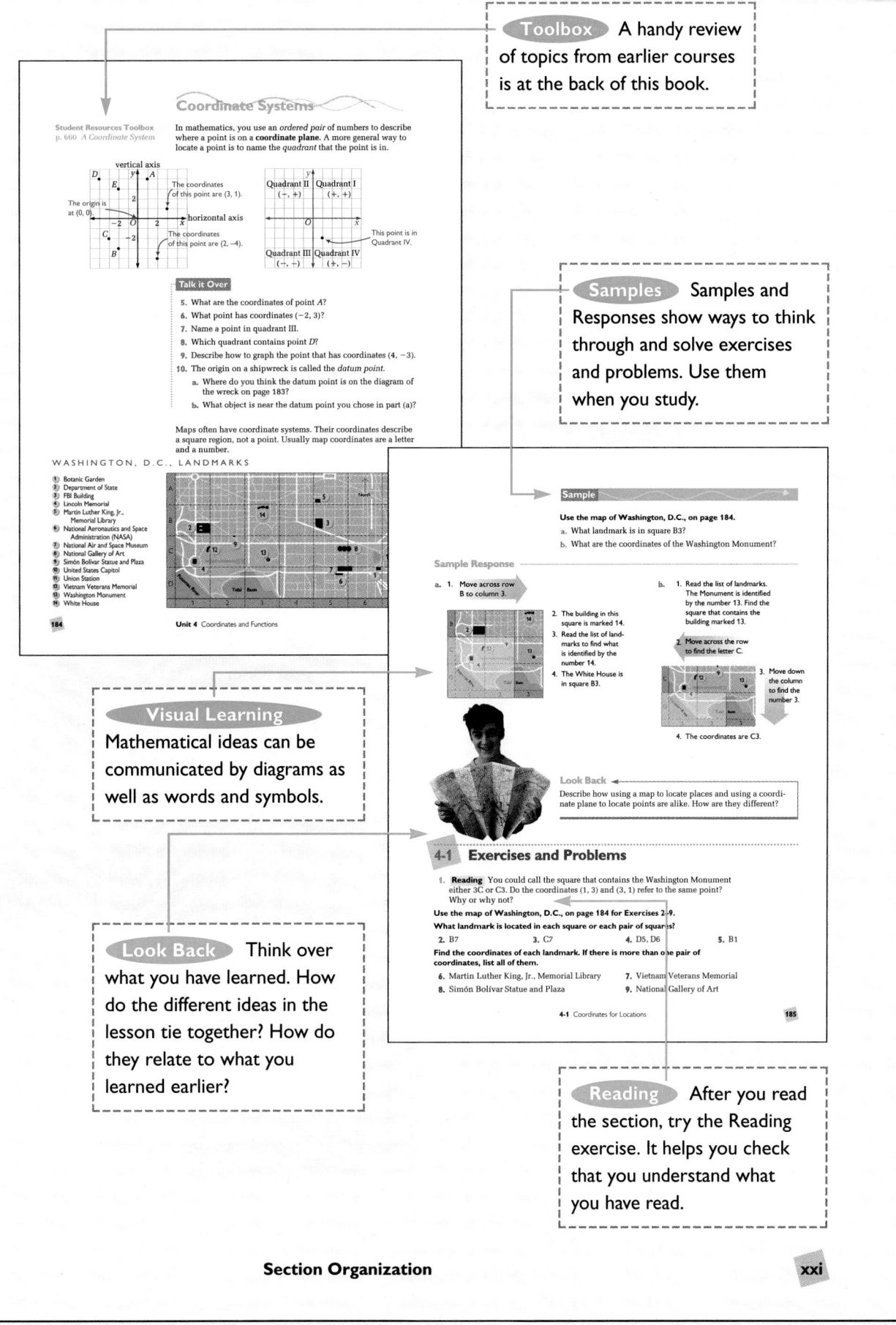

Toolbox A handy review of topics from earlier courses is at the back of this book.

Samples Samples and Responses show ways to think through and solve exercises and problems. Use them when you study.

Visual Learning Mathematical ideas can be communicated by diagrams as well as words and symbols.

Look Back Think over what you have learned. How do the different ideas in the lesson tie together? How do they relate to what you learned earlier?

Reading After you read the section, try the Reading exercise. It helps you check that you understand what you have read.

Coordinate Systems

Student Resources Toolbox
p. 660 *A Coordinate System*

In mathematics, you use an *ordered pair* of numbers to describe where a point is on a **coordinate plane**. A more general way to locate a point is to name the *quadrant* that the point is in.

The coordinates of this point are (3, 1).
The origin is at (0, 0).
The coordinates of this point are (2, −4).

| Quadrant II (−, +) | Quadrant I (+, +) |
| Quadrant III (−, −) | Quadrant IV (+, −) |

This point is in Quadrant IV.

Talk it Over

5. What are the coordinates of point *A*?
6. What point has coordinates (−2, 3)?
7. Name a point in quadrant III.
8. Which quadrant contains point *D*?
9. Describe how to graph the point that has coordinates (4, −3).
10. The origin on a shipwreck is called the *datum point*.
   a. Where do you think the datum point is on the diagram of the wreck on page 183?
   b. What object is near the datum point you chose in part (a)?

Maps often have coordinate systems. Their coordinates describe a square region, not a point. Usually map coordinates are a letter and a number.

WASHINGTON, D.C., LANDMARKS

1. Botanic Garden
2. Department of State
3. FBI Building
4. Lincoln Memorial
5. Martín Luther King, Jr., Memorial Library
6. National Aeronautics and Space Administration (NASA)
7. National Air and Space Museum
8. National Gallery of Art
9. Simón Bolívar Statue and Plaza
10. United States Capitol
11. Union Station
12. Vietnam Veterans Memorial
13. Washington Monument
14. White House

184   Unit 4 Coordinates and Functions

Sample

Use the map of Washington, D.C., on page 184.
a. What landmark is in square B3?
b. What are the coordinates of the Washington Monument?

Sample Response

a. 1. Move across row B to column 3.
   2. The building in this square is marked 14.
   3. Read the list of landmarks to find what is identified by the number 14.
   4. The White House is in square B3.

b. 1. Read the list of landmarks. The Monument is identified by the number 13. Find the square that contains the building marked 13.
   2. Move across the row to find the letter C.
   3. Move down the column to find the number 3.
   4. The coordinates are C3.

Look Back

Describe how using a map to locate places and using a coordinate plane to locate points are alike. How are they different?

4-1 Exercises and Problems

1. **Reading** You could call the square that contains the Washington Monument either 3C or C3. Do the coordinates (1, 3) and (3, 1) refer to the same point? Why or why not?

Use the map of Washington, D.C., on page 184 for Exercises 2–9.
What landmark is located in each square or each pair of squares?

2. B7   3. C7   4. D5, D6   5. B1

Find the coordinates of each landmark. If there is more than one pair of coordinates, list all of them.

6. Martín Luther King, Jr., Memorial Library   7. Vietnam Veterans Memorial
8. Simón Bolívar Statue and Plaza   9. National Gallery of Art

4-1 Coordinates for Locations   185

Section Organization

Exercises and Problems

Each section has a wide variety of exercises and problems. Some **PRACTICE** and **EXTEND** the concepts and skills you have learned. Others apply the concepts to everyday situations and **explore connections** to other subject areas and to careers. The problems help you sharpen your **THINKING** and **problem solving skills**.

Here are some of the types of exercises and problems you will see in the course.

Practicing and Extending

These exercises build on the ideas in the Talk it Over questions and the Sample on the preceding pages.

Connections to...

Sometimes math shows up where you don't expect it — in social studies, science, art, literature, music, sports.

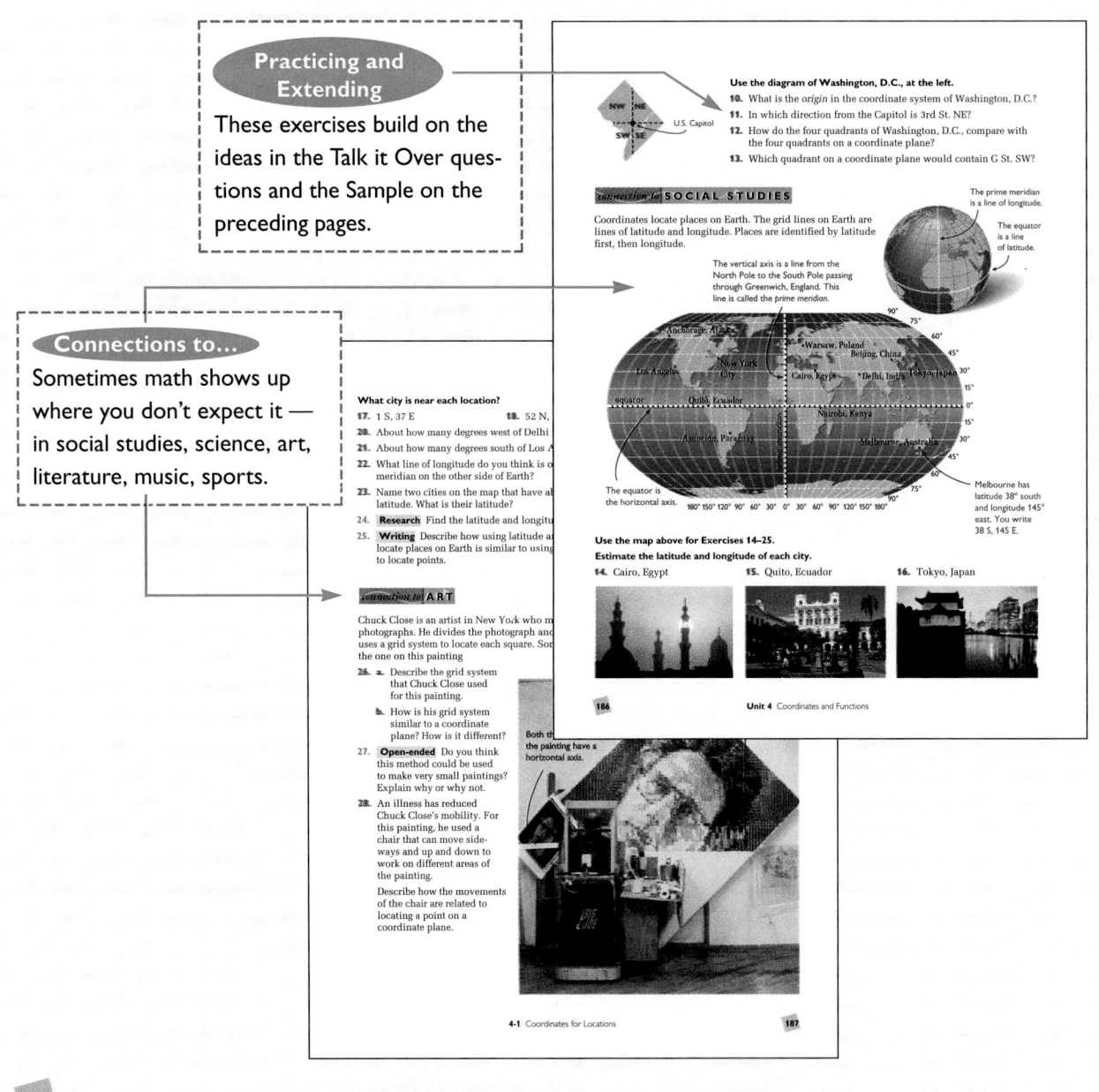

Use the diagram of Washington, D.C., at the left.

10. What is the *origin* in the coordinate system of Washington, D.C.?

11. In which direction from the Capitol is 3rd St. NE?

12. How do the four quadrants of Washington, D.C., compare with the four quadrants on a coordinate plane?

13. Which quadrant on a coordinate plane would contain G St. SW?

connection to SOCIAL STUDIES

Coordinates locate places on Earth. The grid lines on Earth are lines of latitude and longitude. Places are identified by latitude first, then longitude.

The prime meridian is a line of longitude.

The equator is a line of latitude.

The vertical axis is a line from the North Pole to the South Pole passing through Greenwich, England. This line is called the *prime meridian*.

The equator is the horizontal axis.

Melbourne has latitude 38° south and longitude 145° east. You write 38 S, 145 E.

Use the map above for Exercises 14–25.
Estimate the latitude and longitude of each city.

14. Cairo, Egypt   **15.** Quito, Ecuador   **16.** Tokyo, Japan

186   **Unit 4** Coordinates and Functions

What city is near each location?

17. 1 S, 37 E   **18.** 52 N,

20. About how many degrees west of Delhi

21. About how many degrees south of Los A

22. What line of longitude do you think is o
meridian on the other side of Earth?

23. Name two cities on the map that have a
latitude. What is their latitude?

24. **Research** Find the latitude and longitu

25. **Writing** Describe how using latitude a
locate places on Earth is similar to using
to locate points.

connection to ART

Chuck Close is an artist in New York who m
photographs. He divides the photograph and
uses a grid system to locate each square. Sor
the one on this painting

26. a. Describe the grid system
that Chuck Close used
for this painting.

b. How is his grid system
similar to a coordinate
plane? How is it different?

Both th
the painting have a
horizontal axis.

27. **Open-ended** Do you think
this method could be used
to make very small paintings?
Explain why or why not.

28. An illness has reduced
Chuck Close's mobility. For
this painting, he used a
chair that can move side-
ways and up and down to
work on different areas of
the painting.

Describe how the movements
of the chair are related to
locating a point on a
coordinate plane.

4-1 Coordinates for Locations   187

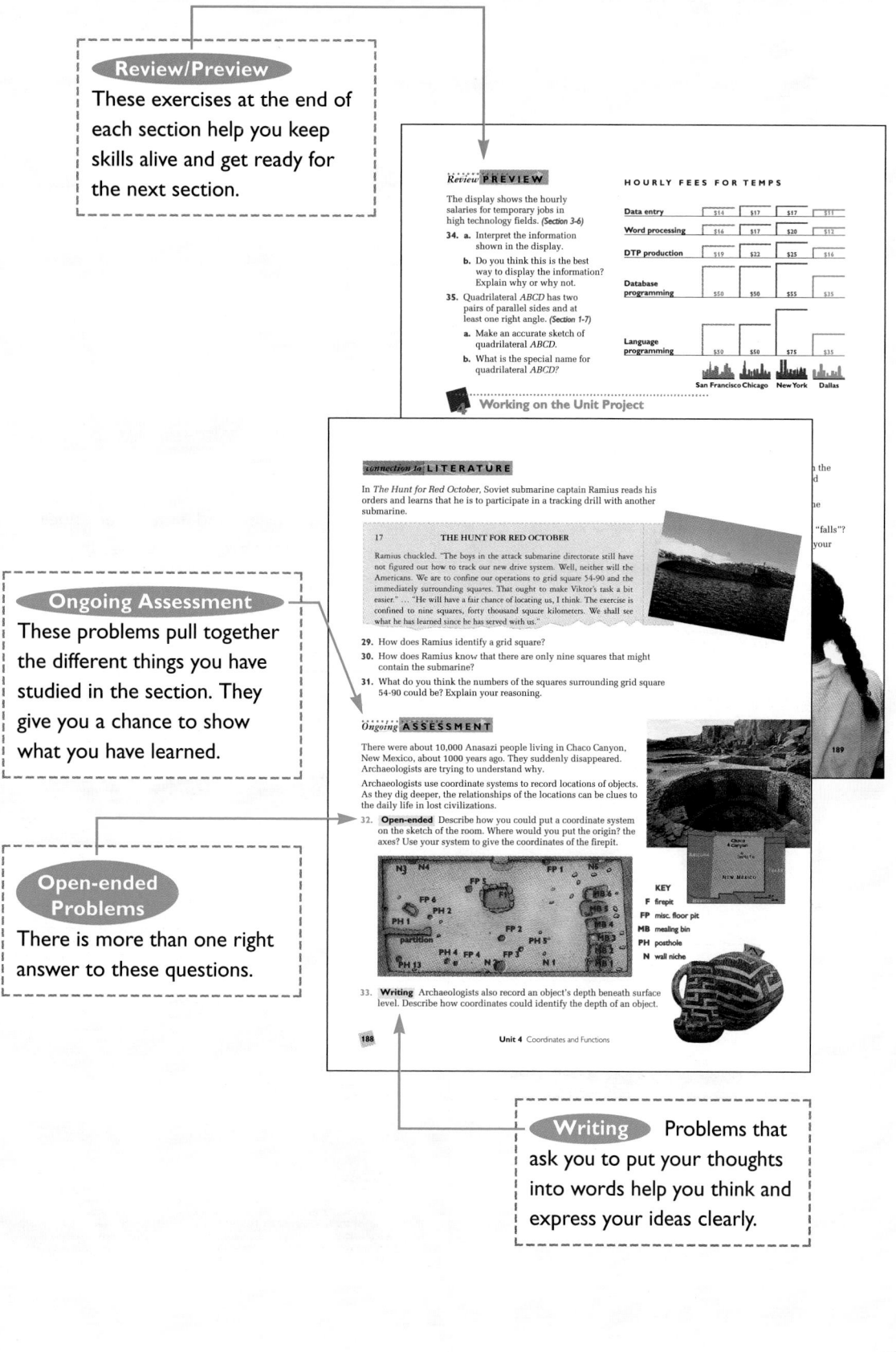

Review PREVIEW

The display shows the hourly salaries for temporary jobs in high technology fields. *(Section 3-6)*

34. a. Interpret the information shown in the display.

b. Do you think this is the best way to display the information? Explain why or why not.

35. Quadrilateral *ABCD* has two pairs of parallel sides and at least one right angle. *(Section 1-7)*

a. Make an accurate sketch of quadrilateral *ABCD*.

b. What is the special name for quadrilateral *ABCD*?

HOURLY FEES FOR TEMPS

Data entry — $14, $17, $17, $11
Word processing — $16, $17, $20, $13
DTP production — $19, $22, $25, $16
Database programming — $50, $50, $55, $35
Language programming — $30, $50, $75, $35

San Francisco Chicago New York Dallas

Working on the Unit Project

connection to LITERATURE

In *The Hunt for Red October*, Soviet submarine captain Ramius reads his orders and learns that he is to participate in a tracking drill with another submarine.

17   THE HUNT FOR RED OCTOBER

Ramius chuckled. "The boys in the attack submarine directorate still have not figured out how to track our new drive system. Well, neither will the Americans. We are to confine our operations to grid square 54-90 and the immediately surrounding squares. That ought to make Viktor's task a bit easier." ... "He will have a fair chance of locating us, I think. The exercise is confined to nine squares, forty thousand square kilometers. We shall see what he has learned since he has served with us."

29. How does Ramius identify a grid square?

30. How does Ramius know that there are only nine squares that might contain the submarine?

31. What do you think the numbers of the squares surrounding grid square 54-90 could be? Explain your reasoning.

Ongoing ASSESSMENT

There were about 10,000 Anasazi people living in Chaco Canyon, New Mexico, about 1000 years ago. They suddenly disappeared. Archaeologists are trying to understand why.

Archaeologists use coordinate systems to record locations of objects. As they dig deeper, the relationships of the locations can be clues to the daily life in lost civilizations.

32. Open-ended Describe how you could put a coordinate system on the sketch of the room. Where would you put the origin? the axes? Use your system to give the coordinates of the firepit.

KEY
F   firepit
FP misc. floor pit
MB mealing bin
PH posthole
N   wall niche

33. Writing Archaeologists also record an object's depth beneath surface level. Describe how coordinates could identify the depth of an object.

188   **Unit 4** Coordinates and Functions

Review and Assessment

With this book you **REVIEW** and **ASSESS** your progress as you go along. In each unit there are one or two Checkpoints for `self-assessment`, plus a thorough Unit Review and Assessment at the end.

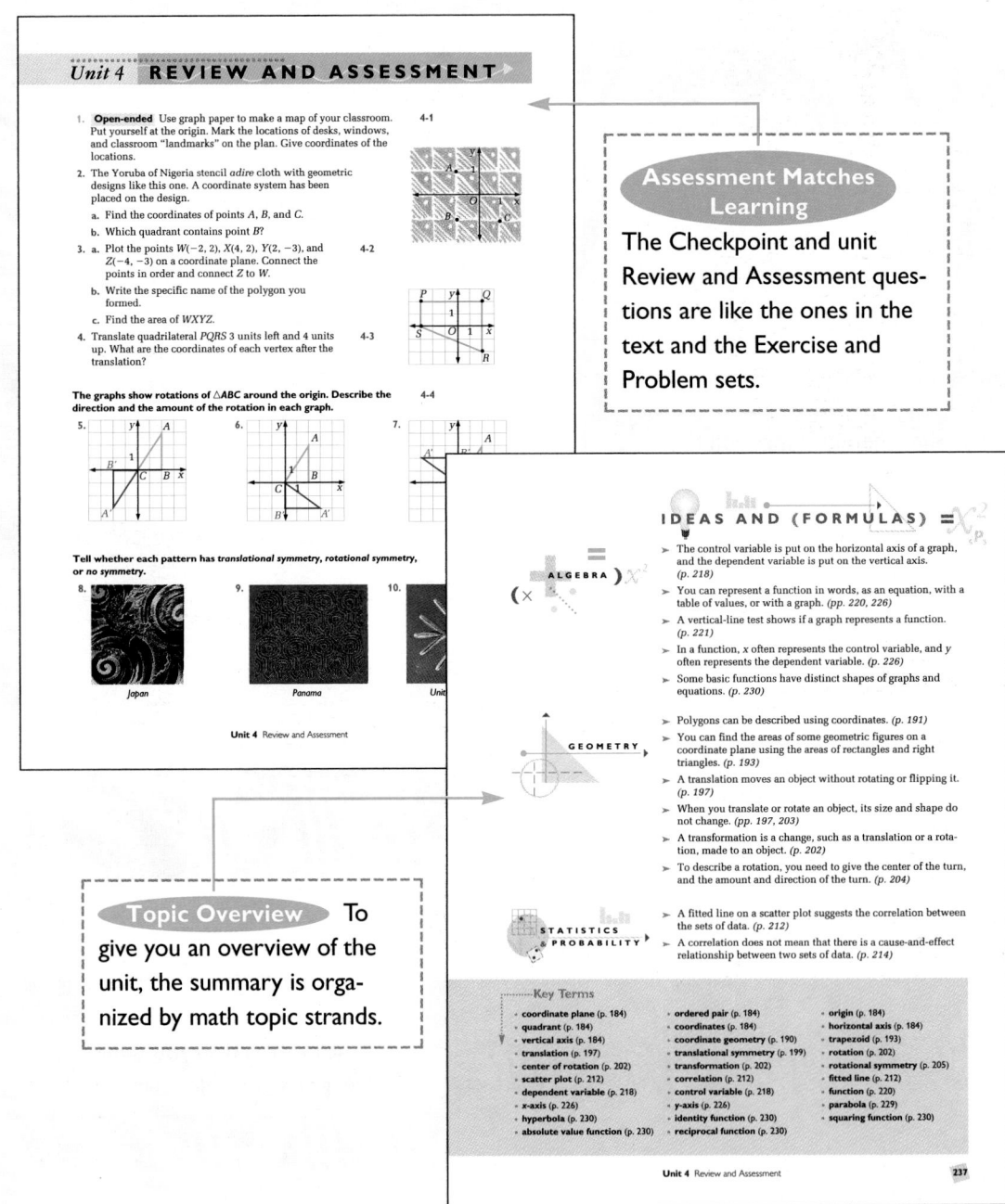

Assessment Matches Learning

The Checkpoint and unit Review and Assessment questions are like the ones in the text and the Exercise and Problem sets.

Topic Overview To give you an overview of the unit, the summary is organized by math topic strands.

Review and Assessment

Technology

In this course you will see many different ways that CALCULATORS and COMPUTERS can make exploring ideas and solving problems easier.

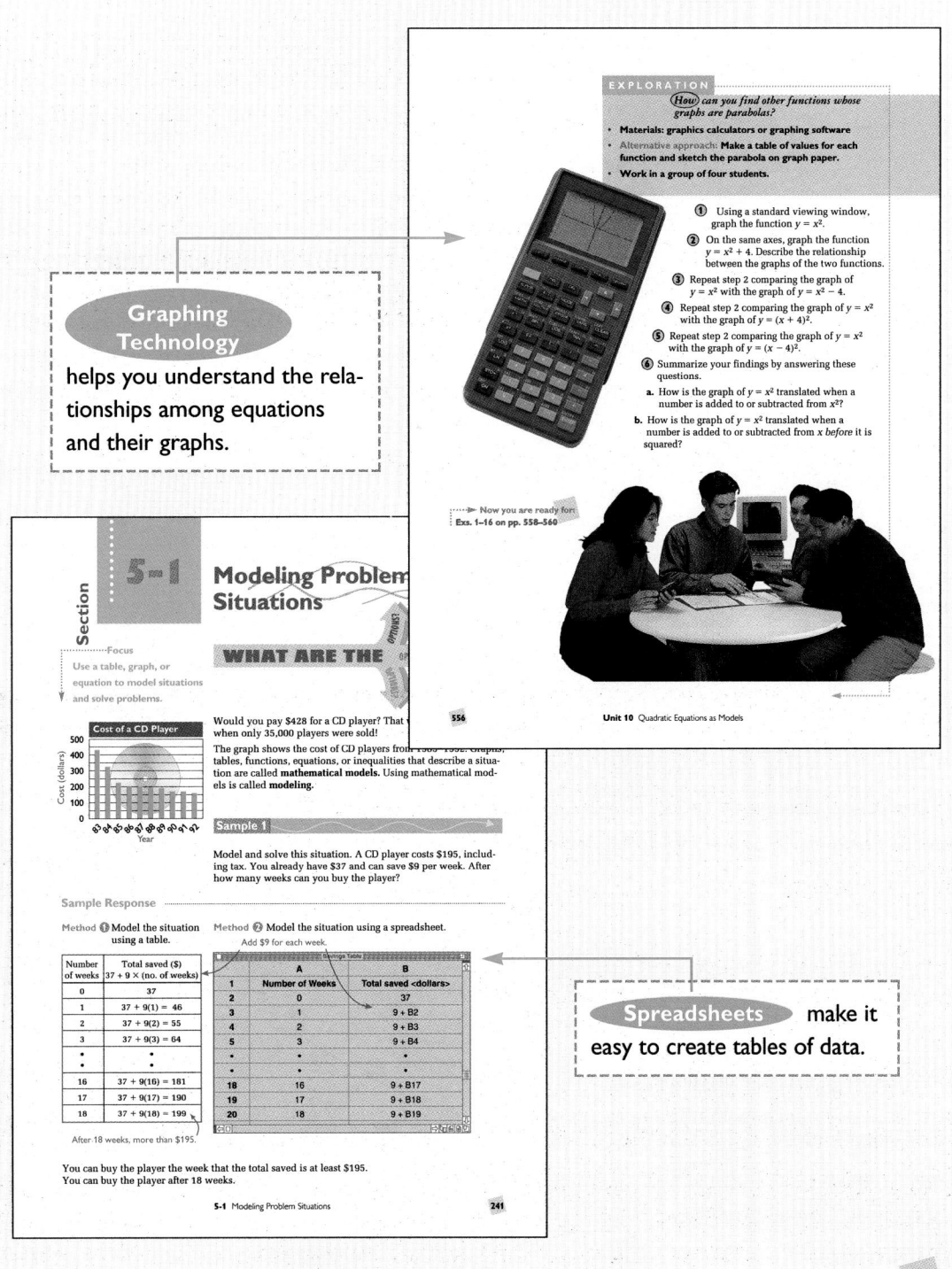

Graphing Technology helps you understand the relationships among equations and their graphs.

EXPLORATION

How can you find other functions whose graphs are parabolas?

- Materials: graphics calculators or graphing software
- Alternative approach: Make a table of values for each function and sketch the parabola on graph paper.
- Work in a group of four students.

① Using a standard viewing window, graph the function $y = x^2$.

② On the same axes, graph the function $y = x^2 + 4$. Describe the relationship between the graphs of the two functions.

③ Repeat step 2 comparing the graph of $y = x^2$ with the graph of $y = x^2 - 4$.

④ Repeat step 2 comparing the graph of $y = x^2$ with the graph of $y = (x + 4)^2$.

⑤ Repeat step 2 comparing the graph of $y = x^2$ with the graph of $y = (x - 4)^2$.

⑥ Summarize your findings by answering these questions.

a. How is the graph of $y = x^2$ translated when a number is added to or subtracted from x^2?

b. How is the graph of $y = x^2$ translated when a number is added to or subtracted from x *before* it is squared?

Now you are ready for: Exs. 1–16 on pp. 558–560

556   **Unit 10** Quadratic Equations as Models

Section 5-1   Modeling Problem Situations

Focus
Use a table, graph, or equation to model situations and solve problems.

WHAT ARE THE

Would you pay $428 for a CD player? That when only 35,000 players were sold!

The graph shows the cost of CD players from tables, functions, equations, or inequalities that describe a situation are called **mathematical models**. Using mathematical models is called **modeling**.

Cost of a CD Player

Sample 1

Model and solve this situation. A CD player costs $195, including tax. You already have $37 and can save $9 per week. After how many weeks can you buy the player?

Sample Response

Method ① Model the situation using a table.

Number of weeks	Total saved ($) 37 + 9 × (no. of weeks)
0	37
1	37 + 9(1) = 46
2	37 + 9(2) = 55
3	37 + 9(3) = 64
•	•
•	•
16	37 + 9(16) = 181
17	37 + 9(17) = 190
18	37 + 9(18) = 199

After 18 weeks, more than $195.

Method ② Model the situation using a spreadsheet.

Add $9 for each week.

	A	B
	Number of Weeks	Total saved <dollars>
1		37
2	0	9 + B2
3	1	9 + B3
4	2	9 + B4
5	3	•
•	•	•
•	•	•
18	16	9 + B17
19	17	9 + B18
20	18	9 + B19

You can buy the player the week that the total saved is at least $195.
You can buy the player after 18 weeks.

5-1 Modeling Problem Situations   241

Spreadsheets make it easy to create tables of data.

Technology   xxv

Exploring and Communicating Mathematics

OVERVIEW

➤ In **Unit 1,** students reinforce their previous mathematical experience with variables, algebraic expressions, and polygons while they begin to explore and develop their mathematical communication skills. The **Student Resources Toolbox** on pages 640–665 provides a convenient review of key topics from previous courses.

➤ Students are given opportunities to use tables, diagrams, and concept maps; recognize and describe patterns; further their understanding of evaluating expressions and the order of operations; and investigate symmetry and congruence.

➤ The Unit Project is based on the creation of logos as a geometric form of graphic design and as a form of communication. The goal of this project is for students to create a special logo for their mathematics class.

➤ Connections to language arts, agriculture, literature, the U. S. Census, sign language, graphic design, and quilting patterns are some of the topics included in the teaching materials and the exercises.

➤ A problem-solving strategy used in Unit 1 is *Identify a Pattern* (Section 1-2).

Unit Objectives

Section	Objectives	NCTM Standards
1-1	• Make statements about information presented in tables, graphs, and concept maps.	1, 2, 3, 4, 10
1-2	• Describe patterns using tables and variable expressions.	1, 2, 3, 4, 5, 6
	• Evaluate variable expressions.	
1-3	• Use exponents to express repeated multiplication.	1, 2, 3, 4, 5
	• Evaluate powers of variables.	
	• Make conjectures.	
1-4	• Use the order of operations to evaluate expressions.	4, 5, 14
1-5	• Use the distributive property to simplify calculations, rewrite expressions, and combine like terms.	1, 4, 5, 14
1-6	• Create and identify congruent polygons.	1, 4, 7
	• Name corresponding parts of congruent polygons.	
1-7	• Build quadrilaterals from congruent triangles and write expressions for their perimeters.	1, 2, 4, 7
	• Find lines of symmetry.	
	• Communicate ideas through writing.	

Section	Connections to Prior and Future Concepts
1-1	**Section 1-1** stresses the many ways in which communication takes place in our everyday lives. Students have a variety of opportunities to interpret and communicate information using diagrams, pictures, signs, logos, tables, graphs, and concept maps. Opportunities to practice and develop these skills are integrated in the lessons throughout Books 1, 2, and 3 of this series; for example, in Unit 1 of Book 2 and Units 1–3 of Book 3.
1-2	**Section 1-2** introduces variables and the uses of variable expressions. Students use algebra tiles to model variable expressions. Variable expressions are used to help establish patterns and solve problems. Writing and evaluating algebraic expressions is an important skill throughout Books 1, 2, and 3; for example, Unit 2 of Book 2 and Unit 2 of Book 3.
1-3	**Section 1-3** develops the skill of writing and evaluating powers of numbers and variables. Some students may have had experience working with exponents. However, this may be the first time that students have been asked to make conjectures and find counterexamples. Both activities will be developed further in Units 2, 5 and 9 of Book 2 and Units 2 and 5 of Book 3.
1-4	**Section 1-4** stresses the use of the order of operations that is important for evaluating and simplifying arithmetic and algebraic expressions. This may be a review for many students, but students usually need to refresh their memory of the rules as they begin a new mathematics class.
1-5	**Section 1-5** covers the use of the distributive property as it applies to simplifying expressions and combining like terms. Mastery of the distributive property and its applications is critical in Unit 10 of Book 1, Units 3 and 4 of Book 2, and Unit 2 of Book 3. (Other properties of addition and multiplication are reviewed on page 640 of the Toolbox.)
1-6	**Section 1-6** provides exploration activities to help reacquaint students with various polygons and their properties while developing the concept of congruency. The motions of slides, turns, and flips, introduced here, are covered more thoroughly in Sections 4-3, 4-4, 6-6, 10-1, and 10-2 of Book 1. A transformation strand continues in Units 3–5 of Book 2 and in Units 5, 9, and 10 of Book 3.
1-7	**Section 1-7** explores quadrilaterals formed by congruent triangles. Students draw diagrams, record perimeters, and summarize their findings. They review properties of quadrilaterals and the concept of symmetry. Properties of geometric figures and symmetry are covered further in Unit 10 of Book 1; Units 1, 4, 5, 7, and 8 of Book 2; and Unit 3 of Book 3. Students practice effective communication throughout Books 1, 2, and 3 in Talk it Over exercises, Writing exercises, Open-ended exercises, Research exercises, Critical Thinking exercises, unit projects, and so on.

Integrating the Strands

Strands	Sections
Number	1-1, 1-2, 1-3, 1-5, 1-7
Algebra	1-2, 1-3, 1-4, 1-5, 1-6, 1-7
Geometry	1-1, 1-2, 1-3, 1-4, 1-5, 1-6, 1-7
Statistics and Probability	1-3, 1-7
Discrete Mathematics	1-2, 1-3
Logic and Language	1-1, 1-2, 1-3, 1-4, 1-5, 1-6, 1-7

Exploring and Communicating Mathematics

Section Planning Guide

➤ Essential exercises and problems are indicated in boldface.
➤ Ongoing work on the Unit Project is indicated in color.
➤ Exercises and problems that require student research, group work, manipulatives, or graphing technology are indicated in the column headed "Other."

Section	Materials	Pacing	Standard Assignment	Extended Assignment	Other
1-1		Day 1	**9–27**, 30–36, 37	1–8, **9–27**, 30–36, 37	29
1-2	algebra tiles	Day 1	**1–15**, 19–25	**1–15**, 16–25	
		Day 2	**26–28**, **30–32**, 41–46, 47	**26–28**, 29, **30–32**, 34, 36–46, 47	33, 35
1-3	calculator, algebra tiles	Day 1	**1–13**, **15–19**, **21–26**	**1–13**, 14, **15–19**, 20, **21–26**, 27	
		Day 2	**28–40**, 41–44, 45	**28–40**, 41–44, 45	
1-4	graphics calculator, algebra tiles	Day 1	**3–20**, **23–25**, 26–29, 35–41, 42	1, 2, **3–20**, **23–25**, 26–31, 35–41, 42	21, 22, 32–34
1-5	algebra tiles	Day 1	**1–20**	**1–20**, 21	
		Day 2	**24–30**, 32–40, 41	22, 23, **24–30**, 32–40, 41	31
1-6	dot paper	Day 1	**1–9**, 11, 12	**1–9**, 11, 12	10
		Day 2	13, 14, 15, **16–24**, 26–31, 32	13, 14, 15, **16–24**, 26–31, 32	25
1-7	scissors, toothpicks	Day 1	**1–16**, **18–20**	**1–16**, 17, **18–20**	
		Day 2	**24**, 25, 26, **27**, 28–35, 36	21–23, **24**, 25, 26, **27**, 28–35, 36	
Review		**Day 1**	**Unit Review**	**Unit Review**	
Test		**Day 2**	**Unit Test**	**Unit Test**	

Yearly Pacing	Unit 1 Total	Remaining	Total
	16 days (2 for Unit Project)	148 days	164 days

Support Materials

➤ See **Project Book** for notes on Unit 1 Project: Design a Logo.
➤ "UPP" and "disk" refer to **Using Plotter Plus** booklet and **Plotter Plus** disk.
➤ Warm-up exercises for each section are available on **Warm-Up Transparencies**.
➤ "FI," "PC," "GI," "MA," and "Stats!" refer, respectively, to the McDougal Littell Mathpack software Activity Books for **Function Investigator, Probability Constructor, Geometry Inventor, Matrix Analyzer,** and **Stats!**.

Section	Study Guide	Practice Bank	Problem Bank	Activity Bank	Explorations Lab Manual	Assessment Book	Visuals	Technology
1-1	1-1	Practice 1	Set 1	Enrich 1		Quiz 1-1		
1-2	1-2	Practice 2	Set 1	Enrich 2	Masters 6, 8	Quiz 1-2	Folder 1	
1-3	1-3	Practice 3	Set 1	Enrich 3	Masters 6, 7	Quiz 1-3	Folder 1	
1-4	1-4	Practice 4	Set 1	Enrich 4	Masters 6, 7	Quiz 1-4 Test 1	Folder 1	
1-5	1-5	Practice 5	Set 2	Enrich 5	Add. Expl. 1 Masters 6, 7	Quiz 1-5	Folder 1	FI Act. 1 UPP, page 25
1-6	1-6	Practice 6	Set 2	Enrich 6	Masters 1, 9	Quiz 1-6		
1-7	1-7	Practice 7	Set 2	Enrich 7		Quiz 1-7 Test 2		GI Act. 29
Unit 1	Unit Review	Practice 8	Unifying Problem 1	Family Involve 1		Tests 3, 4		

Form A

Spanish versions of these tests are on pages 122–125 of the **Assessment Book**.

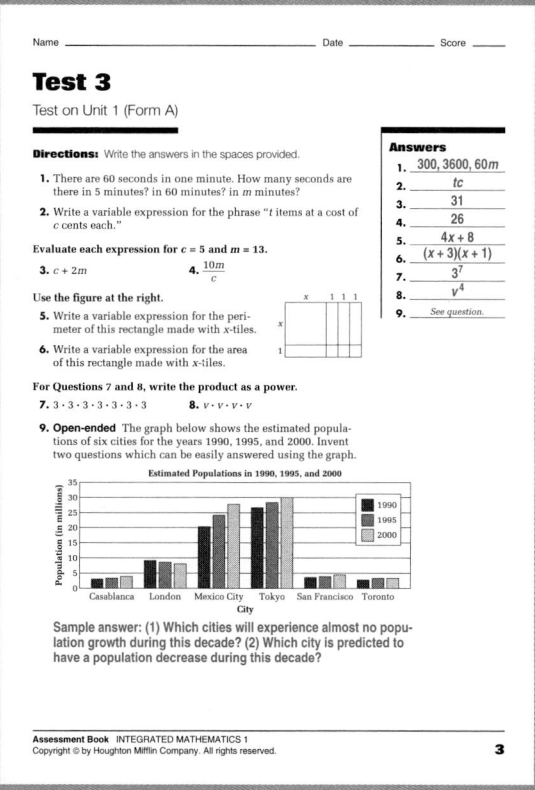

Name _____ Date _____ Score _____

Test 3

Test on Unit 1 (Form A)

Directions: Write the answers in the spaces provided.

1. There are 60 seconds in one minute. How many seconds are there in 5 minutes? in 60 minutes? in m minutes?

2. Write a variable expression for the phrase "t items at a cost of c cents each."

Evaluate each expression for $c = 5$ and $m = 13$.

3. $c + 2m$ 4. $\dfrac{10m}{c}$

Use the figure at the right.

5. Write a variable expression for the perimeter of this rectangle made with x-tiles.

6. Write a variable expression for the area of this rectangle made with x-tiles.

For Questions 7 and 8, write the product as a power.

7. $3 \cdot 3 \cdot 3 \cdot 3 \cdot 3 \cdot 3 \cdot 3$ 8. $v \cdot v \cdot v \cdot v$

9. **Open-ended** The graph below shows the estimated populations of six cities for the years 1990, 1995, and 2000. Invent two questions which can be easily answered using the graph.

Estimated Populations in 1990, 1995, and 2000

Sample answer: (1) Which cities will experience almost no population growth during this decade? (2) Which city is predicted to have a population decrease during this decade?

Answers
1. $300, 3600, 60m$
2. tc
3. 31
4. 26
5. $4x + 8$
6. $(x + 3)(x + 1)$
7. 3^7
8. v^4
9. *See question.*

3

Name _____ Date _____ Score _____

Test 3 (continued)

Directions: Write the answers in the spaces provided.

Write as a power of ten.

10. $10^3 \cdot 10^8$ 11. $\dfrac{10^{19}}{10^4}$

Calculate according to the order of operations.

12. $20 \div 2^2$ 13. $9 \div 3 \cdot 4$ 14. $24 - (8 - 6)^3$

Use the distributive property to rewrite each expression without parentheses.

15. $7(a + b)$ 16. $6(2m - n)$ 17. $\frac{1}{4}(12x + 8)$

For Questions 18 and 19, combine like terms.

18. $7r + 18r$ 19. $4(x + 2) + 3x^2 + 5(x + 1)$

20. **Writing** Geoffrey wrote this mathematical sentence: $7x^3 + 5x^3 = 12x^6$. Explain his mistake and how to correct it.
Sample answer: When Geoffrey combined like terms, he should not have added the exponents. When adding like terms, the sum is also a like term; the correct answer is $12x^3$.

Use the figure at the right.

21. Mark the congruent lengths using tick marks.

22. Draw all lines of symmetry for the polygon.

Mark congruent lengths by using letters. Then write a variable expression for the perimeter of the polygon.   Choice of letters may vary.

23. perimeter: $2a + 4b$ 24. perimeter: $x + y + 2z$

25. How many lines of symmetry does a square have?

Answers
10. 10^{11}
11. 10^{15}
12. 5
13. 12
14. 16
15. $7a + 7b$
16. $12m - 6n$
17. $3x + 2$
18. $25r$
19. $3x^2 + 9x + 13$
20. *See question.*
21. *See question.*
22. *See question.*
23. *See question.*
24. *See question.*
25. 4

4

Form B

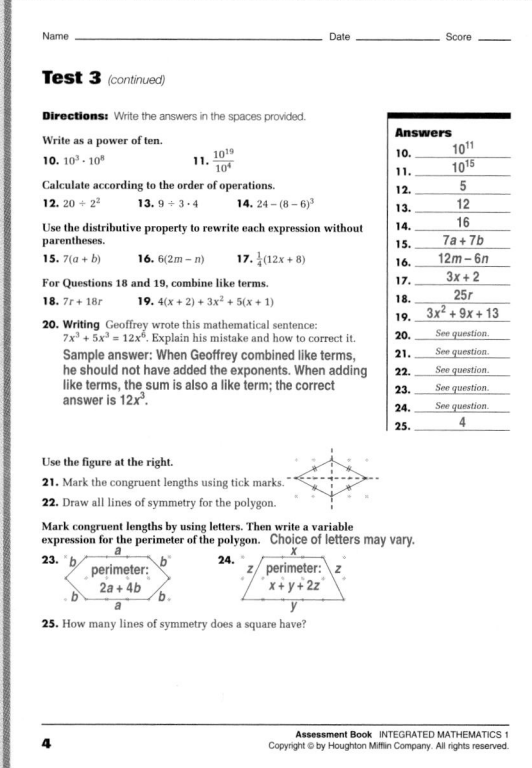

Name _____ Date _____ Score _____

Test 4

Test on Unit 1 (Form B)

Directions: Write the answers in the spaces provided.

1. There are 12 inches in one foot. How many inches are there in 5 feet? in 10 feet? in f feet?

2. Write a variable expression for the phrase "d dollars saved each week for w weeks."

Evaluate each variable expression for $c = 6$ and $m = 15$.

3. $c + 3m$ 4. $\dfrac{10m}{5c}$

Use the figure at the right.

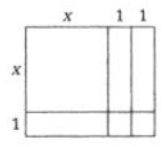

5. Write a variable expression for the perimeter of this shape made with x-tiles.

6. Write a variable expression for the area of this shape made with x-tiles.

For Questions 7 and 8, write the product as a power.

7. $4 \cdot 4 \cdot 4 \cdot 4 \cdot 4 \cdot 4$ 8. $k \cdot k \cdot k$

9. **Open-ended** The graph below shows the 1990 energy use and the predicted future energy use for various modes of transportation and utilities. Invent two questions which can be easily answered using the graph.

1990 Energy Use and Potential Use

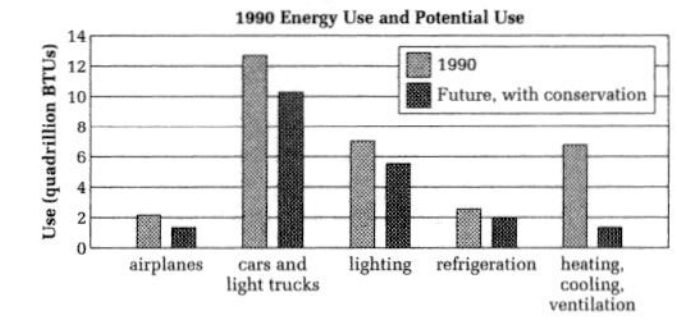

Sample answer: (1) For which item is the greatest decrease in energy use predicted? (2) For which item is little or no change in energy use predicted?

Answers
1. $60, 120, 12f$
2. dw
3. 51
4. 5
5. $4x + 6$
6. $(x + 2)(x + 1)$
7. 4^6
8. k^3
9. *See question.*

5

Name _____ Date _____ Score _____

Test 4 (continued)

Directions: Write the answers in the spaces provided.

Write as a power of ten.

10. $10^7 \cdot 10^5$ 11. $\dfrac{10^{19}}{10^6}$

Calculate according to the order of operations.

12. $24 \div 2^3$ 13. $8 \div 4 \cdot 3$ 14. $32 - (8 - 6)^3$

Use the distributive property to rewrite each expression without parentheses.

15. $5(h + k)$ 16. $3(7r - t)$ 17. $\frac{1}{3}(12x + 9)$

For Questions 18 and 19, combine like terms.

18. $17v + 11v$ 19. $4(m + 5) + 3m^2 + 6(m + 1)$

20. **Writing** Suzanna wrote this mathematical sentence: $9x^4 + 5x^4 = 14x^8$. Explain her mistake and how to correct it.
Sample answer: When Suzanna combined like terms, she should not have added the exponents. When adding like terms, the sum is also a like term; the correct answer is $14x^4$.

Use the figure at the right.

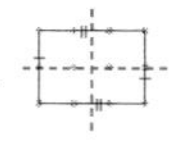

21. Mark the congruent lengths using tick marks.

22. Draw all lines of symmetry for the polygon.

Mark congruent lengths by using letters. Then write a variable expression for the perimeter of the polygon.   Choice of letters may vary.

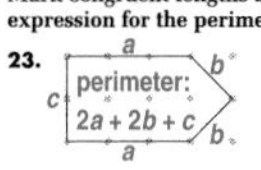

23. perimeter: $2a + 2b + c$ 24. perimeter: $2a + 2b$

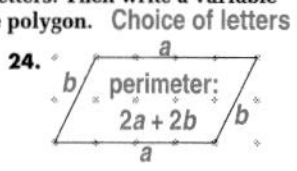

25. How many lines of symmetry does an ellipse have?

Answers
10. 10^{12}
11. 10^{13}
12. 3
13. 6
14. 24
15. $5h + 5k$
16. $21r - 3t$
17. $4x + 3$
18. $28v$
19. $3m^2 + 10m + 26$
20. *See question.*
21. *See question.*
22. *See question.*
23. *See question.*
24. *See question.*
25. 2

6

Software Support

McDougal Littell Mathpack

Function Investigator
Geometry Inventor

Outside Resources

Books/Periodicals

Countryman, Joan. *Writing to Learn Mathematics.* Portsmouth, NH: Heinemann Educational Books, 1992.

Driscoll, Mark and Jere Confrey, eds. *Teaching Mathematics: Strategies that Work.* Chelmsford, MA: Northeast Regional Exchange, Inc., 1985.

Novak, Joseph D. and Bob Gowin. *Learning How to Learn.* New York: Cambridge University Press, 1984.

Kennedy, Jane B. "Area and Perimeter Connections." *Mathematics Teacher* (March 1993): pp. 218–221.

Edwards, Ronald. *AlgeCadabra! Algebra Magic Tricks.* Pacific Grove, CA: Thinking Press and Software, 1992.

Manipulatives

Handiart Grids and Charts. Mt. Pleasant, MI: Tricon Publishing, 1993.

Software

Hoffer, Alan. *Algebra Expresser.* Macintosh. Acton, MA: William K. Bradford Publishing.

IBM Mathematics Exploration Toolkit. IBM, Armonk, NY.

Videos

Algebra for Everyone: Videotape and Discussion Guide. A project of the NCTM Mathematics Education Trust. Reston, VA: NCTM, 1991.

PROJECT GOALS

➤ Students learn that various images from this unit can be used to design a team logo for use as a book cover to identify their math class.

➤ When designing a logo, students choose various shapes, patterns, symbols, numbers or terms and describe the reasons why each of these elements is part of the logo.

➤ Students work together in a cooperative group and appreciate each other's strong points.

PROJECT PLANNING

Materials List

➤ Ruler
➤ Blank book cover
➤ Markers
➤ Poster board or blank bookmarks

Project Teams

Before students begin working as part of a team, have individual students make a list of logos they have seen in television commercials, in professional sports, and in places in their community.

Have students work on the project in groups of three. One way for the group to distribute the work that takes advantage of special talents students may have is as follows:

1. Coordinator: collects, analyzes, and summarizes all possible ideas from the group.

2. Writer: describes the elements of the logo.

3. Illustrator: draws and colors the logo on the book cover.

Individual students will need a few days to examine the unit and find appropriate images to use in the team logo. The team needs to discuss the individual choices and select one for the team.

unit 1

If you were driving in Morocco and you saw this sign, you would probably know what to do even without knowing how to read the Arabic word. The image of a red octagon means **STOP** to many people around the world. When you travel, international symbols and signs can help you find your way around and follow local rules.

Exploring and Communicating Mathematics

The radiating lines represent a type of headdress and show the importance of the sun in Native American culture.

The base of the symbol represents Mother Earth and emphasizes the relationship between nature and Native American people.

symmetry

A graphic artist may create many designs before one is chosen as a logo for a team or an organization.

Larry DesJarlais, Jr., a North Dakota Chippewa and faculty member at the Institute of American Indian Arts, designed this logo for the Smithsonian Institution's National Museum of the American Indian National Campaign. The geometric shapes and symmetry in his design form the image and have special meaning.

General Rubric for Unit Projects

Each unit project can be evaluated in many possible ways. The following rubric is just one way to evaluate these open-ended projects. It is based on a 4-point scale.

4 The student fully achieves all mathematical and project goals. The presentation demonstrates clear thinking and explanation.

3 The student substantially achieves the mathematical and project goals. The main thrust of the project and the mathematics behind it is understood, but there may be some minor misunderstanding of content, errors in computation, or weakness in presentation.

2 The student partially achieves the mathematical and project goals. A limited grasp of the main mathematical ideas or project requirements is demonstrated. Some of the work may be incomplete, misdirected, or unclear.

1 The student makes little progress toward accomplishing the goals of the project because of lack of understanding or lack of effort.

Design a Logo

Your group project is to create a team logo to identify your math class. Your work in this unit will help you choose various shapes, patterns, math symbols, numbers, and words to use in your logo.

You will put your logo on a book cover to fit your math book. On the inside flap of the front cover, you will describe your group's design and explain why you included each design element.

The **images and colors** on flags often tell a story about the history or geography of a country.

symbols

The motto of Brazil is "Order and Progress."

Blue and white are the colors of Portugal, an important country in Brazil's history.

There is one star for every state in Brazil, and one to represent the territories.

The green background represents the land of Brazil.

The constellation is one seen over the capital city of Rio de Janiero.

The gold represents the minerals found in the soil.

BRAZIL

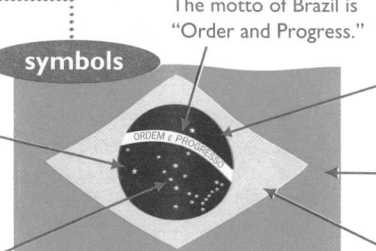

sports

Team recognition is important on the playing field. Sports organizations use logos and colors to identify their team and sport. How can you use these logos to tell which sport each team plays? What other information do the logos show?

1

Suggested Rubric for Unit Project

4 The logo design on the book cover uses patterns, symbols, numbers, and shapes in a way that is mathematically accurate. The written description and explanation for each element of the design are accurate and thorough. Reasons why this design identifies the math class are insightful. The logo is visually clear and appealing and may also be enlarged for use on a poster or reduced for use on a bookmark.

3 The logo design on the book cover uses the mathematics adequately. The written description and explanation of the design elements is not as thorough as it could be. The visual effect of the logo is not entirely clear.

2 The logo design on the book cover uses the mathematical elements in an incomplete or somewhat confusing manner. The description of each element of the design is incom-

plete. The reasons describing why this design identifies the math class are lacking or need more elaboration. This project should be returned with suggestions for improvements and a new deadline.

1 This project cannot be evaluated. It is illegible, incomplete, or not understandable. The design should be returned with a new deadline for completion. The group should be encouraged to speak with the teacher as soon as possible.

Support Materials

The *Project Book* contains information about the following topics for use with this Unit Project.

➤ Project Description
➤ Teaching Commentary
➤ Working on the Unit Project Exercises
➤ Completing the Unit Project
➤ Assessing the Unit Project
➤ Alternative Projects
➤ Outside Resources

Limited English Proficiency

Creating a logo is a valuable project for students acquiring English. It reinforces the visual elements of mathematical concepts and requires written descriptions of these concepts. Have a student fluent in English be a partner to a student acquiring English so that the important concepts and terms of the Unit Project are understood.

ADDITIONAL BACKGROUND

Multicultural Note

Symbols were used before written language. Today, symbols communicate without regard to literacy or language barriers. International symbols are based on simple images; for example, a picture of an airplane represents an airport.

Since the 1930s, many nations have been working together to create a system of symbols that can be used on highways and in public places throughout the world. The International Organization for Standardization (ISO), based in Geneva, Switzerland, promotes and coordinates this work worldwide. Today, there are thousands of internationally approved symbols for use in such fields as transportation, computer technology, agriculture, and finance.

Colophon

A *colophon* is an inscription placed at the end of a book or manuscript, usually with facts relative to its production. Beginning in the 6th century, scribes wrote a final paragraph at the end of a manuscript. This colophon, or "finishing touch," included the title of the work, the scribe's name, and the date and place of completion. In the 15th century, some printers included emblems at the end of the colophon. Today, some publishers use an emblem on the spine or title page of their books. Point out Houghton Mifflin's emblem (the boy on a dolphin) on the spine of the student book.

ALTERNATIVE PROJECTS

Project 1, page 50

The Fibonacci Sequence

Describe the number pattern exhibited by a Fibonacci sequence. Research the sequence, write a report about it, and describe how it is found in nature.

Project 2, page 50

Create a Polygon Pattern

Create patterns using regular polygons. Describe the kinds of patterns that can be built by using one or more types of polygons. Sketch the patterns, describe the symmetry of each design, and use color to show how it can affect the symmetry of a design.

Getting Started

For this project you should work in a group of three students. Here are some ideas to help you get started.

☞ As a group, describe what materials to use and where you will get them. Can you use any recycled materials?

☞ Measure the cover of your math book. Think about the size and position of your logo.

☞ It may be a good idea to fit a blank cover on your book and then mark the area where you want to put your logo.

☞ Book covers can be made in many ways. Which methods do you know? Which method protects the book the best?

Will you center your design?

Will you have one large logo or a repeating pattern of many small logos?

Other than on the front, where else on your cover could you place the logo?

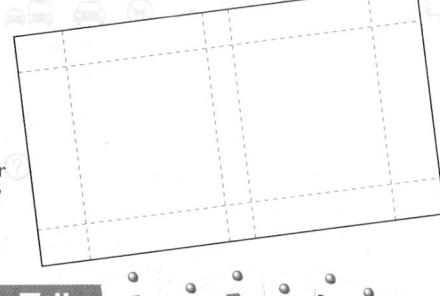

Can We Talk

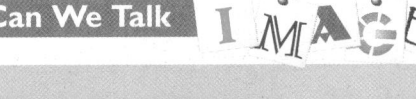

Working on the Unit Project

Your work in Unit 1 will help you design your logo.

Related Exercises:
Section 1-1, Exercise 37
Section 1-2, Exercise 47
Section 1-3, Exercise 45
Section 1-4, Exercise 42
Section 1-5, Exercise 41
Section 1-6, Exercise 32
Section 1-7, Exercise 36

Alternative Projects p. 50

➤ Describe some famous logos. What shapes or images did the designers use? Why do you think those images were used?

➤ What logos can you recognize just by their shape, color, or style of print? Why do you think you remember these images?

➤ What feelings do lighthouses, rocks, and doves communicate? What organizations might use these symbols?

➤ Does your school have a symbol that is used on stationery or official documents? If so, how do the parts of that logo represent your school?

➤ What do you think these international symbols mean? Why do you think these symbols were created?

➤ What do the symbols and colors on the United States flag represent? Why do you think these images were chosen?

Unit 1 Exploring and Communicating Mathematics

Answers to Can We Talk?

➤ Examples include airline logos, designer jeans logos, organization logos, union logos, and so on. Lettering, color, and design elements can be chosen because they are symbolic of the company or eye-catching.

➤ Answers may vary.

➤ Lighthouses could suggest guidance, rocks could suggest strength, doves could represent peace.

➤ Answers may vary.

➤ no smoking; telephone; information; taxi; These logos were created to enhance international communication.

➤ The stripes represent the 13 original colonies. The stars represent the number of states. One thought is that the stars were chosen because they resemble heaven and the stripes resemble rays of light.

It is not known for sure why red, white, and blue were chosen as the colors of the flag. Some believe it is because those are the colors of England, the "mother" country. George Washington once said they represent the following: White represents liberty, red represents valor, and blue represents justice, loyalty, and perseverance.

Communicating with Diagrams

seven

What Do You Say?

English

sỗ bẩy

Vietnamese

paqallqo

Aymara

satt

Punjabi

tallimat malǵuk

eéjè

Inuit

Yoruba

Focus
Make statements about information presented in tables, graphs, and concept maps.

What do you think these people are saying? They are saying the number "7" using different languages. More than 4000 languages and dialects are spoken in the world today. Because there are so many languages, people often need diagrams instead of language to communicate.

American Sign Language

Talk it Over

1. You see these signs along a road. What do you think they mean?

   a.   b.   c.   d.

2. You see these signs in an airport. What do you think they mean?

   a.   b.   c.   d.

1-1 Communicating with Diagrams   3

PLANNING

Objectives and Strands
See pages 1A and 1B.

Spiral Learning
See page 1B.

Recommended Pacing
Section 1-1 is a one-day lesson.

Toolbox References
➤ **Toolbox Skill 17:** Finding Perimeter

Extra Practice
See pages 620–621.

Warm-Up Exercises
Warm-Up Transparency 1-1

Support Materials
➤ Practice 1
➤ Enrichment 1 in the Activity Bank
➤ Study Guide 1-1
➤ Problem Set 1
➤ Quiz 1-1
➤ Alternative Assessment 1

Answers to Talk it Over

1–2. Answers may vary. Examples are given.

1. a. slippery when wet
   b. steep hill
   c. merge left
   d. railroad crossing

2. a. food/restaurant
   b. boarding area
   c. waiting area
   d. emergency phone/security

Talk it Over

Questions 1 and 2 make the point that signs can communicate information very effectively because the graphics are universally understood by people from different cultures.

Communication: Discussion

Have students discuss which method of displaying data they prefer, a table or a graph.

Using Technology

In connection with the Sample, computer graphs can be used to display the information in the table in alternate ways. Many graphing packages display spreadsheet data as circle graphs and broken line graphs as well as bar graphs.

Additional Sample

The table and the graph show the total area of a continent as a percent of Earth and the population of the continent as a percent of the total population of the world.

Continent	% of Earth	% of Population
North America	16.2	5.1
South America	11.9	8.8
Europe	6.6	14.7
Asia	30.1	56.1
Africa	20.2	14.9
Oceania	5.7	0.4
Antarctica	9.3	0

Tables and Graphs

Tables and graphs are two ways you can show mathematical information. A table shows data numerically. A graph shows data visually. They can tell you different things about the same information.

Sample

The table and the graph show the total number of people in the world who speak each language, and the number for whom the language is their native language.

a. What are two things you can learn from the table?

b. What are two things you can learn from the graph?

c. Invent a question that is more easily answered from the table than from the graph. Explain how to find the answer.

d. Invent a question that is more easily answered from the graph than from the table. Explain how to find the answer.

Speakers (millions)

LANGUAGE	Native	Total
Mandarin	817	907
Hindi	321	383
Spanish	320	362
English	316	456
Bengali	180	189
Arabic	178	208
Russian	173	293
Portuguese	165	177
Japanese	125	126
German	98	119

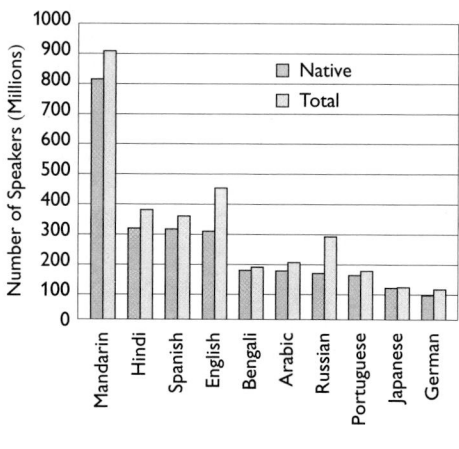

Sample Response

There are many possible answers. Examples are given.

a. Mandarin is spoken by more people than any other language. ← Compare all the totals.

Spanish is the native language of more people than English. ← Compare the numbers from two countries.

Unit 1 Exploring and Communicating Mathematics

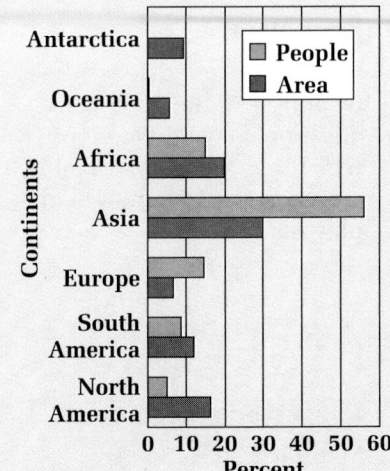

b. English has the most non-native speakers. ← Which "total" bar extends the farthest beyond the "native" bar?

There are relatively few non-native speakers of Bengali, Portuguese, and Japanese.

c. "How many non-native speakers of Arabic are there?" ← Think of a question about the numbers in the table.

To answer, subtract the data values in the row for Arabic. There are 30 million non-native speakers of Arabic.

d. "Which three languages have the greatest numbers of non-native speakers?" ← Think of a question about the differences in bar heights in the graph.

To answer, look at how far the gold bars extend beyond the blue bars. This represents the numbers of non-native speakers. English, Russian, and Mandarin have the greatest number of non-native speakers.

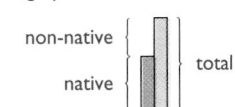

The Sample asks *open-ended questions*. There are many ways to answer an open-ended question correctly.

Concept Maps

A **concept map** is a visual summary that helps you remember the connections between ideas. You can organize ideas in a concept map to help you think about what you know.

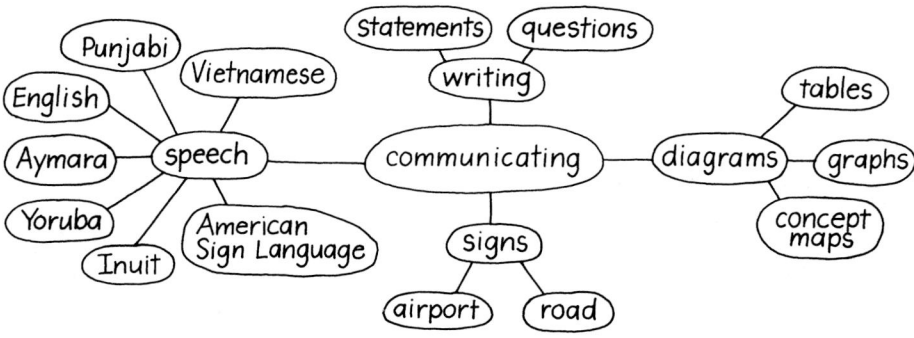

Talk it Over

3. What ideas do you think the person who made this concept map has learned?

4. List some ideas you might put in a concept map about road signs.

Answers to Talk it Over

3. Answers may vary. An example is given. The person has learned about issues involved with the subject of communication.

4. Answers may vary. An example is given. One might include ideas about shapes of signs, colors of signs, messages, proper placement, and so on.

Answers will vary. Examples are given.

a. What are two things you can learn from the table?
Of the total Earth area, Europe covers 6.6% of the land. Antarctica has no population.

b. What are two things you can learn from the graph?
Asia has the most people in the world. There are relatively few people in Oceania.

c. Invent a question that is more easily answered from the table than from the graph. Explain how to find an answer.
Which two continents have about the same percent of Earth's population? Find the two closest numbers in the % of population column. Europe and Africa have about the same percent of Earth's population.

d. Invent a question that is more easily answered from the graph. Explain how to find an answer.
Which continent's population percent is approximately double its area percent? Look at the bars for people and area for each continent. Europe's people bar length is approximately double the length of the area bar.

....

Reasoning

Concept maps can be used throughout the book by students to make connections between ideas. The ability to link together ideas demonstrates reasoning and critical thinking skills.

5

Look Back ◄

Why do you think information is sometimes presented using tables, graphs, and concept maps?

1-1 Exercises and Problems

Reading Match each number with the sign for it.

1. 6   **2.** 9   **3.** 33
4. 100   **5.** 1000

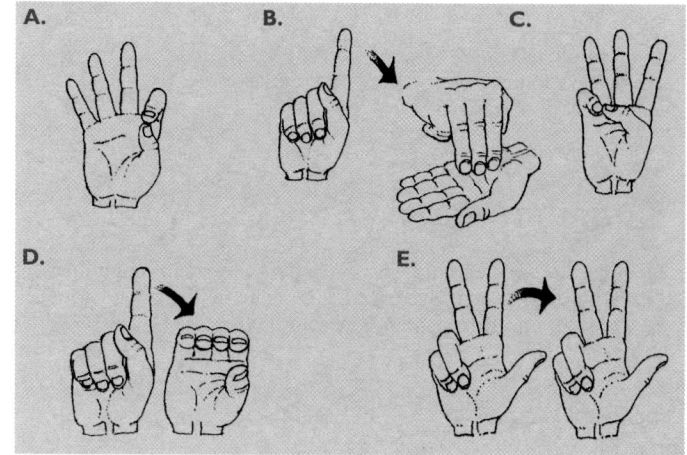

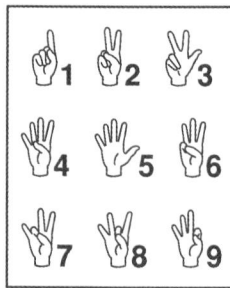

BY THE WAY...

The signs for 100 and 1000 are based on the Roman numerals C and M.

How would you show these numbers in sign language? Explain your reasoning.

6. 8   **7.** 108   **8.** 1008

You see these signs in a building. What do you think they mean?

9.  **10.**  **11.**  **12.**

You see these signs along a road. What do you think they mean?

13.  **14.**  **15.**  **16.**

Unit 1 Exploring and Communicating Mathematics

Answers to Look Back

Answers may vary. An example is given. Many people find it easier to understand things when they can see it. Tables are a good way to present information when you need to compare exact values. Graphs give you an overall picture of the information and allow you to make general conclusions. Concept maps show the connections between different parts of the information.

Answers to Exercises and Problems

1. C   **2.** A   **3.** E
4. D   **5.** B

6–8. Answers may vary. Examples are given.

6.

8

7.
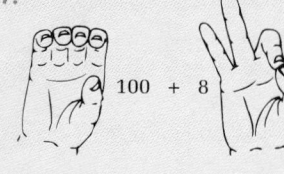
100 + 8

8.
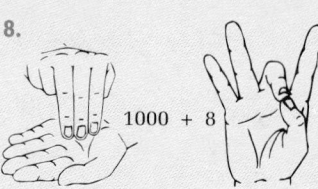
1000 + 8

9–16. Answers may vary. Examples are given.

9. restrooms
10. escalators
11. elevators
12. coatroom
13. two-way traffic
14. crosswalk
15. intersection
16. curve in road

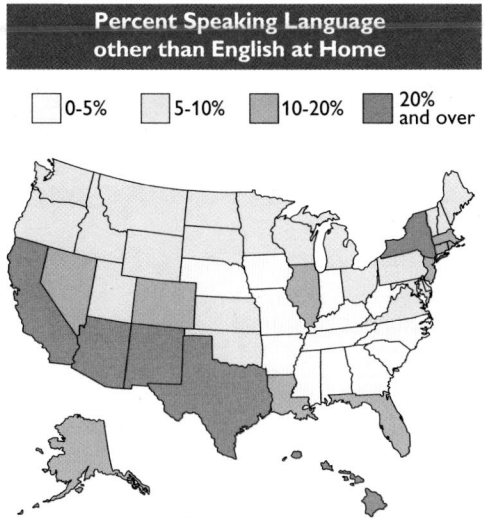

Percent Speaking Language other than English at Home

☐ 0-5% ☐ 5-10% ▨ 10-20% ■ 20% and over

Number of people five and older who speak a language other than English at home, from the 1990 census, and percent change since 1980:					
State	Non-English	Pct. Change	State	Non-English	Pct. Change
Ala.	107,866	61.6	Mont.	37,020	–1.0
Alaska	60,165	32.4	Neb.	69,872	0.7
Ariz.	700,287	38.9	Nev.	146,152	102.6
Ark.	60,781	55.7	N.H.	88,796	–0.6
Calif.	8,619,334	73.6	N.J.	1,406,148	28.3
Colo.	320,631	13.3	N.M.	493,999	10.0
Conn.	466,175	11.3	N.Y.	3,908,720	18.5
Del.	42,327	39.3	N.C.	240,866	86.5
D.C.	71,348	49.9	N.D.	46,897	–30.6
Fla.	2,098,315	73.5	Ohio	546,148	5.3
Ga.	284,546	112.9	Okla.	145,798	29.2
Hawaii	254,724	11.3	Ore.	191,710	47.1
Idaho	58,995	24.1	Pa.	806,876	5.8
Ill.	1,499,112	22.7	R.I.	159,492	8.6
Ind.	245,826	17.1	S.C.	113,163	58.5
Iowa	100,391	9.1	S.D.	41,994	–15.0
Kan.	131,604	27.9	Tenn.	131,550	59.9
Ky.	86,482	43.2	Texas	3,970,304	39.7
La.	391,994	2.1	Utah	120,404	27.9
Maine	105,441	–6.6	Vt.	30,409	–3.1
Md.	395,051	62.4	Va.	418,521	89.5
Mass.	852,228	21.0	Wash.	403,173	52.1
Mich.	569,807	1.5	W.Va.	44,203	14.2
Minn.	227,161	7.9	Wis.	263,638	4.9
Miss.	66,516	50.6	Wyo.	23,809	–11.5
Mo.	178,210	27.4			

REPORT SHOWS U.S. RESIDENTS SPEAKING MANY LANGUAGES AT HOME

WASHINGTON – Americans increasingly are speaking Spanish, Chinese, Korean or Vietnamese, but not English. The trend reflects the rise of immigration into the United States.

A new Census Bureau report shows that one in seven U.S. residents, 14 percent, spoke a language other than English at home in 1990, up from 11 percent in 1980.

Nationally, Spanish was spoken by more than half of the 31.8 million U.S. residents who didn't routinely converse in English. The next most common languages were French, German, Italian, Chinese and Tagalog, a language of the Philippines. Languages with the biggest increases were Spanish, Chinese, Tagalog, Korean and Vietnamese.

For Exercises 17–22, answer each question and tell whether you used the article, the map, or the table.

17. What is the main point of the article?

18. Which state has the largest number of people who speak a language other than English in the home?

19. Which state has the smallest number of people who speak a language other than English in the home?

20. In how many states do more than 20% of the people speak a language other than English in the home?

21. Which state had the greatest percent change from 1980 to 1990 in the number of people speaking a language other than English in the home?

22. About how many people routinely speak Spanish in the home?

23. **Writing** What are two things you can learn from the table?

24. **Writing** What are two things that are shown more clearly in the map than in the table?

25. **Open-ended** What is another appropriate title for the article?

Reading Use the table and the graph on page 4.

26. Which language has more non-native speakers, Arabic or Portuguese? Did you use the table or the graph?

27. What is an appropriate title for the table and the graph?

1-1 Communicating with Diagrams   **7**

connection to **LANGUAGE ARTS**

28. **a.** Find out how many students in your math class speak a language other than English at home. What languages do they speak?

   b. Find out how many students in your math class study a language other than English at school. What languages do they study?

   c. **Open-ended** Show the information from parts (a) and (b) in a way that others will understand. Use tables, graphs, or other diagrams.

29. **Research** Make a chart that shows you, your parents, your grandparents, your great-grandparents, and so on. How far back can you go? Why do you think this chart is sometimes called a *tree diagram?*

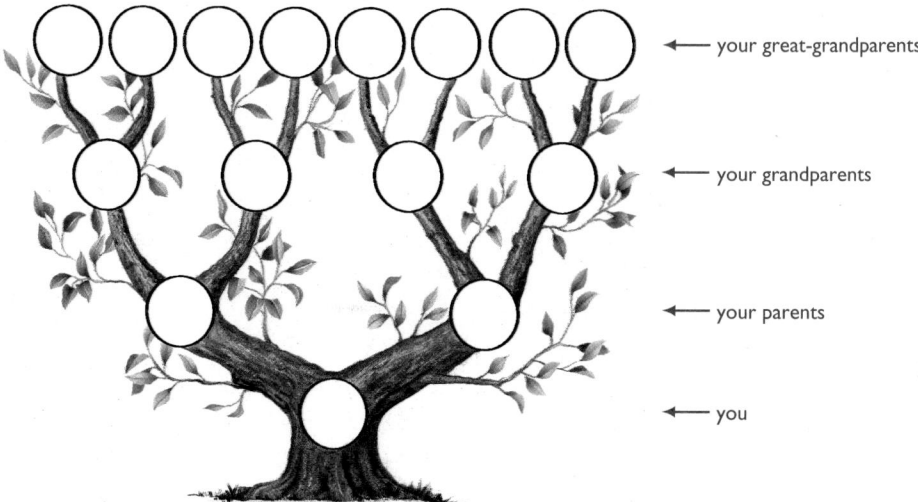

← your great-grandparents

← your grandparents

← your parents

← you

30. **Open-ended** Make a concept map about the musicians and music you like to listen to.

Ongoing **ASSESSMENT**

31. **Open-ended** Give an example of a situation that is better described with a graph than with a table or words.

Review **PREVIEW**

Student Resources Toolbox
p. 658 *Perimeter*

Find the perimeter of each rectangle. *(Toolbox Skill 17)*

32.

```
      5
  ┌───────┐
3 │       │ 3
  └───────┘
      5
```

33.

```
         6
  ┌────────────┐
2 │            │ 2
  └────────────┘
         6
```

34.

```
      4
  ┌───────┐
4 │       │ 4
  │       │
  └───────┘
      4
```

Unit 1 Exploring and Communicating Mathematics

Answers to Exercises and Problems

28–30. Answers may vary.

31. Answers may vary. An example is given. change in temperature over time

32. 16

33. 16

34. 16

Working on the Unit Project

For Ex. 37, have students share their symbols to build a larger set for the entire class.

Use the Toolbox at the back of this book. *(Toolbox, pp. 640–665)*

35. List three topics in the Toolbox that you think you do not need to study.

36. List three topics in the Toolbox that you think you may need to study.

 Working on the Unit Project

37. Find some symbols without words that are understood internationally. Sketch three examples.

Australia

United States

Spain

Practice 1 For use with Section 1-1

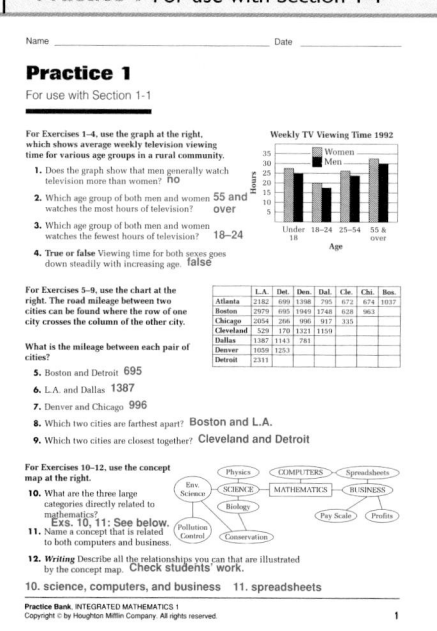

Name _____ Date _____

Practice 1

For use with Section 1-1

For Exercises 1–4, use the graph at the right, which shows average weekly television viewing time for various age groups in a rural community.

1. Does the graph show that men generally watch television more than women? no

2. Which age group of both men and women watches the most hours of television? 55 and over

3. Which age group of both men and women watches the fewest hours of television? 18–24

4. True or false Viewing time for both sexes goes down steadily with increasing age. false

Weekly TV Viewing Time 1992

For Exercises 5–9, use the chart at the right. The road mileage between two cities can be found where the row of one city crosses the column of the other city.

	L.A.	Det.	Den.	Dal.	Cle.	Chi.	Bos.
Atlanta	2182	699	1398	795	672	674	1037
Boston	2979	695	1949	1748	628	963	
Chicago	2054	266	996	917	335		
Cleveland	529	170	1321	1159			
Dallas	1387	1143	781				
Denver	1059	1253					
Detroit	2311						

What is the mileage between each pair of cities?

5. Boston and Detroit 695

6. L.A. and Dallas 1387

7. Denver and Chicago 996

8. Which two cities are farthest apart? Boston and L.A.

9. Which two cities are closest together? Cleveland and Detroit

For Exercises 10–12, use the concept map at the right.

10. What are the three large categories directly related to mathematics? Exs. 10, 11: See below.

11. Name a concept that is related to both computers and business.

12. Writing Describe all the relationships you can that are illustrated by the concept map. Check students' work.

10. science, computers, and business   11. spreadsheets

Practice Bank, INTEGRATED MATHEMATICS 1
Copyright © by Houghton Mifflin Company. All rights reserved.   1

1-1 Communicating with Diagrams

9

Answers to Exercises and Problems

35, 36. Answers may vary.   **37.** Answers may vary.

PLANNING

Objectives and Strands
See pages 1A and 1B.

Spiral Learning
See page 1B.

Materials List
➤ Algebra tiles

Recommended Pacing
Section 1-2 is a two-day lesson.
Day 1
Pages 10–12: Opening paragaraph through Talk it Over, *Exercises 1–25*
Day 2
Page 13: Exploration through Look Back, *Exercises 26–47*

Extra Practice
See pages 620–621.

Warm-Up Exercises
Warm-Up Transparency 1-2

Support Materials
➤ Practice 2
➤ Enrichment 2 in the Activity Bank
➤ Study Guide 1-2
➤ Problem Set 1
➤ Diagram Masters 6, 8 in the Explorations Lab Manual
Overhead Visual 1
➤ Quiz 1-2

Section 1-2

Investigating Patterns

Focus
Describe patterns using tables and variable expressions, and evaluate variable expressions.

Express yourself

BY THE WAY...

Kudzu was brought to the United States from Japan in 1876. By 1943 there were 20,000 people in The Kudzu Club. They held kudzu rallies to support the introduction of the plant.

"If you want to plant kudzu, drop it and run," say many farmers in the southeastern United States. Kudzu grows so fast that people call it the "mile-a-minute" vine. Kudzu does not grow quite that fast, but it can grow up to 12 in. per day.

You can use a table to show how fast kudzu grows.

variable

Days	1	2	3	4	...	d
Length of kudzu vine (inches)	12	24	36	48	...	$12d$

This means $12 \times d$

variable expression

Variable Expressions

A **variable** is a letter used to represent one or more numbers. In the table, d represents the number of days the kudzu grows. The **variable expression** $12d$ represents the length of the kudzu vine after d days.

To avoid confusing the variable x and the times sign $\times$, you write multiplication in algebra in other ways.

raised dot	*parentheses*	*no symbol*
$12 \cdot 2$	$(12)2$ or $12(2)$ or $(12)(2)$	$12d$

Unit 1 Exploring and Communicating Mathematics

Sample 1

Peter earns $12 an hour. Write a variable expression for the amount he earns in *h* hours.

Sample Response

Look for a pattern.

12(1) ← If Peter works 1 hour, he earns $12.

12(2) ← If Peter works 2 hours, he earns $24.

12*h* ← If Peter works *h* hours, he earns 12*h* dollars.

Talk it Over

1. How are the variable expressions for the length of kudzu and the amount of money Peter earns alike? How are they different?

2. Describe a situation that could be represented by the expression 4*x*. Be sure to tell what *x* represents.

Sample 2

A row of triangles is built with toothpicks. Write a variable expression for the perimeter of Shape *n*.

Shape 1 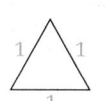 Shape 2 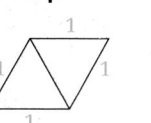 Shape 3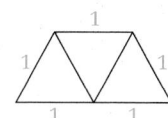

Sample Response

Problem Solving Strategy: Identify a pattern

Sketch a few more shapes.

Shape 4 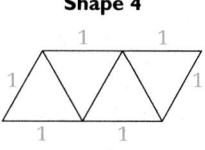 Shape 5 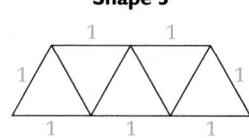 Shape 6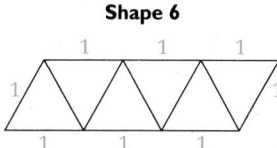

Sample response continued on next page.

1-2 Investigating Patterns

11

Answers to Talk it Over

1. The variable expressions are alike because they both multiply a variable by 12. They are different because the first expression has a *d* for days and the second one has an *h* for hours.

2. Answers may vary. An example is given. How many quarters are in *x* dollars?

Additional Sample

S1 Hitesh walks 3 miles in 1 hour. Write a variable expression for the number of miles he walks in *h* hours. **Look for a pattern.**
3(1): the number of miles in 1 hour
3(2): the number of miles in 2 hours
3*h*: the number of miles in *h* hours

Talk it Over

Questions 1 and 2 help students to understand the idea of a variable and the algebraic way of writing multiplication by not using a symbol.

Reasoning

Patterns are used frequently in mathematics to develop a generalization from a set of specific instances, as shown in the response to Sample 2.

Additional Sample

S2 A row of squares is built with toothpicks. Write a variable expression for the perimeter of Shape *n*.

Shape 1 Shape 2 Shape 3
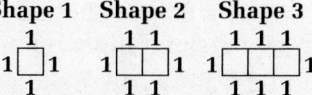

Problem Solving Strategy: Identify a Pattern. Sketch a few more shapes.

Shape 4 Shape 5
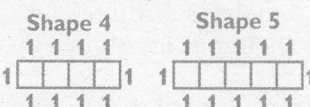

Make a table of perimeters.

Shape	1	2	3	4	5	*n*
Perimeter	4	6	8	10	12	2*n* + 2

An expression for the perimeter of Shape *n* is 2*n* + 2.

11

Make a table of the perimeters.

Shape number	1	2	3	4	5	6	n
Perimeter	3	4	5	6	7	8	?

Look for a pattern. The perimeter is 2 more than the shape number.

An expression for the perimeter of Shape n is $n + 2$.

Evaluating Variable Expressions

When you evaluate a variable expression, you substitute numbers for the variables and find the result.

Sample 3

Suppose a kudzu vine grows 12 in. a day. How long is the vine after each number of days?

  a. 7   **b.** 30   **c.** 365

Sample Response

Write the variable expression first.

a. $12d$

  $12(7)$ ◄— Substitute 7 for d.

  84

  After a week, the kudzu is 84 in. long.

b. $12d$

  $12(30)$ ◄— Substitute 30 for d.

  360

  After a month, the kudzu is 360 in. long.

c. $12d$

  $12(365)$ ◄— Substitute 365 for d.

  4380

  After a year, the kudzu is 4380 in. long.

Talk it Over

3. Sample 2 shows triangles made from toothpicks. How many toothpicks would you need to build a row of a million triangles?

4. Evaluate each expression when $a = 3$ and $b = 7$.

  a. $5a$ **b.** $\dfrac{b}{14}$

  c. ab **d.** $6a + b$

⋯► Now you are ready for:
Exs. 1–25 on pp. 14–16

12   Unit 1 Exploring and Communicating Mathematics

Answers to Talk it Over

3. 2,000,001 toothpicks

4. **a.** 15

  b. $\dfrac{1}{2}$

  c. 21

  d. 25

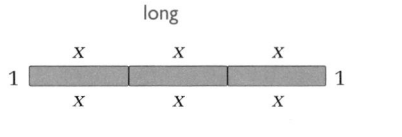

1 x 1
x

EXPLORATION

Whose rectangle has a greater perimeter?

- **Materials: ten algebra tiles, called "x-tiles"**
- **Work with another student.**

1 Make rectangles by placing the x-tiles next to each other. One of you should make long rectangles. One of you should make tall rectangles.

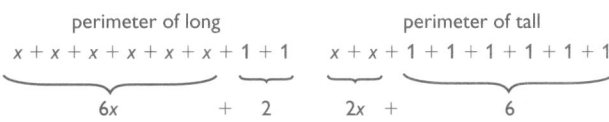

long

1 x x x 1
x x x

tall

1 x 1
1 1
1 1
x

2 Find a variable expression for the perimeter of each rectangle.

perimeter of long

$x + x + x + x + x + x + 1 + 1$

$6x + 2$

perimeter of tall

$x + x + 1 + 1 + 1 + 1 + 1 + 1$

$2x + 6$

3 Put your results in a table.

Perimeter		
Number of x-tiles	Long rectangles	Tall rectangles
1	?	?
2	?	?
3	6x + 2	2x + 6
4	?	?
5	?	?

4 Which type of rectangle do you think has a larger perimeter, a long rectangle or a tall rectangle? Why?

Look Back ←

What are the advantages of using a variable expression instead of a table to show how something changes?

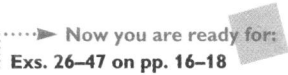 **Now you are ready for:**
Exs. 26–47 on pp. 16–18

Exploration

The goal of the Exploration is for students to begin to work with algebra tiles to create different variable expressions. Algebra tiles will be used throughout the course to model expressions and equations. This Exploration will help students to begin to gain some familiarity with the tiles.

Visual Thinking

Check students' understanding of the Exploration by asking them to visualize the sides of the two rectangles shown in step 1 as a straight line. Have them make a sketch of the two lines and discuss which is longer and why. This activity involves the visual skills of *observation* and *interpretation*.

Look Back

Ask students to write their responses to Look Back as journal entries. After they have done so, ask for volunteers to read their entries to the class.

Answers to Exploration

1, 2. Answers may vary depending on rectangles made.

3.

Perimeter		
Number of x-tiles	Long rectangles	Tall rectangles
1	2x + 2	2x + 2
2	4x + 2	2x + 4
3	6x + 2	2x + 6
4	8x + 2	2x + 8
5	10x + 2	2x + 10

4. Answers may vary. An example is given. For the number of x-tiles greater than 1, a long rectangle has a larger perimeter. In a long rectangle, the x-side or longer side is always part of the perimeter. Each perimeter increases by 2x. In a tall rectangle, the perimeter only increases by 2.

Answers to Look Back

Answers may vary. An example is given. A variable expression gives a general rule to show change for all values, not just specific ones listed in a table.

APPLYING

Suggested Assignment

Standard 1–15, 19–28, 30–32, 41–47

Extended 1–32, 34, 36–47

Integrating the Strands

Number Exs. 30–33, 43–46

Algebra Exs. 1–33, 36–41

Geometry Exs. 5, 6, 26–33, 41

Discrete Mathematics Ex. 42

Logic and Language Exs. 1, 12–18, 29, 34–36, 41, 47

Using Manipulatives

Exs. 5 and 6 may be done by having students use manipulatives to create the shapes. Have them create as many shapes as necessary in order to see the pattern.

Mathematical Procedures

The concept of a variable is the first major mathematical generalization that students encounter after studying arithmetic. It is an abstract concept that often takes time for students to fully understand. Mathematics grows as a science by making generalizations that subsume more specific concepts. Specific number concepts and operational rules (algorithms) can be generalized by the notion of variable, which lies at the center of the development of the subject of algebra.

1-2 Exercises and Problems

1. **Reading** What do variables represent?

2. **a.** The Jacksons rent an apartment for $500 a month. How much rent do they pay in 6 months? in 12 months?

   b. Write a variable expression for the amount they pay in m months.

3. **a.** There are 24 hours in one day. How many hours are there in 7 days? in 365 days?

   b. Write a variable expression for the number of hours in d days.

4. **a.** You pay for some groceries with a $20 bill. How much change do you get if your groceries cost $18? $16.95?

   b. Write a variable expression for the change you get when you buy g dollars worth of groceries.

Draw Shapes 4 and 5 in each pattern. Make a table of the perimeters of the shapes. Then write a variable expression for the perimeter of Shape n.

5.

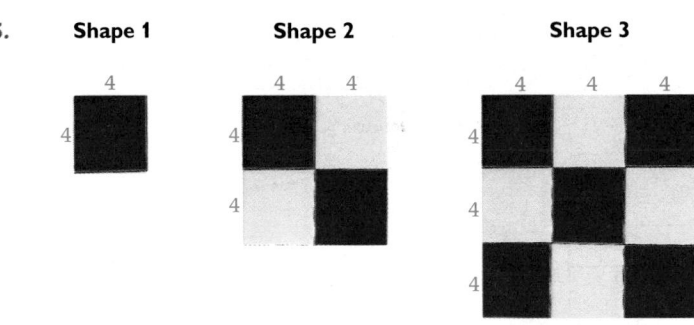

6.

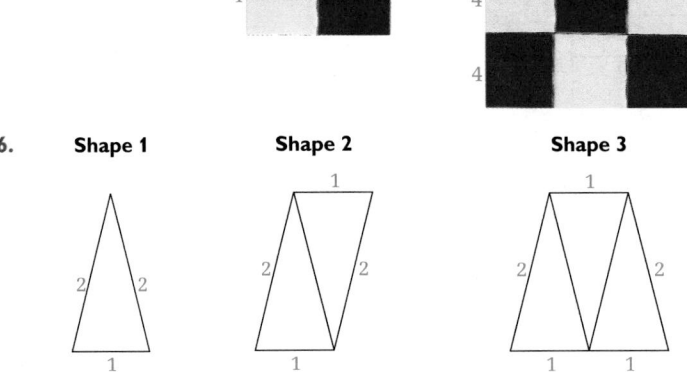

Write a variable expression for each phrase.

7. the price p plus the sales tax t

8. multiply the rate r by the time t

9. the number of parents p plus the number of teachers t

10. 55 times the distance d

11. How are the variable expressions you wrote in Exercises 7 and 9 alike? How are they different?

Answers to Exercises and Problems

1. A variable is a letter used to represent one or more numbers.

2. **a.** $3000; $6000

   b. $500m$

3. **a.** 168; 8760

   b. $24d$

4. **a.** $2; $3.05

   b. $20 - g$

5. Shape 4   Shape 5

shape	perimeter
1	16
2	32
3	48
4	64
5	80

perimeter = $16n$

6. Shape 4   Shape 5

shape	perimeter
1	5
2	6
3	7
4	8
5	9

perimeter = $n + 4$

7. $p + t$

8. rt

9. $p + t$

10. $55d$

11. Answers may vary. An example is given. They use the same variables. The variables represent different things.

Match each word phrase with the correct variable expression.

12. el producto de 5 y un número *k*

   A. $\frac{k}{5}$

13. la suma de un número *k* y 5

   B. $k - 5$

14. un número *k* menos 5

   C. $5k$

15. un número *k* dividido por 5

   D. $k + 5$

For Exercises 16–18, use the article.

TREES BY THE JAR
Scientists at the United States Department of Agriculture are developing a new method for growing apple trees.

The scientists place apple plant shoots in 16-ounce jars with a growing medium. They cover the jars with 4-in. Petri dishes. Forty shoots grow in each jar. The shoots produce roots in about four weeks. After about two more months, the plantlets can survive outside the jars.

The scientists hope this new method will help orchard owners save money. The current method used to grow new trees costs orchard owners about $5 per tree. The new method could cost orchard owners less than $2 per tree.

16. Make a list of five number facts from the article.

17. Use information from the article to write three variable expressions. Explain what your variable expressions represent.

18. **Writing** Write a letter to an orchard owner about this new method. Use variable expressions to explain how much money the orchard owner can save using the new method.

19. Suppose kudzu really grew a mile a minute. It would cover 5280 ft every minute.

   a. Write a variable expression for the length (in feet) of a kudzu vine after *m* minutes.

   b. How many feet long would the kudzu vine be after one day?

Answers to Exercises and Problems

12. C

13. D

14. B

15. A

16. Answers may vary. Examples are given. 16 oz jars; cover jars with 4 in. Petri dishes; 40 shoots grow in each jar; 4 weeks to produce roots; planted 2 months after roots grow;

current method costs $5 per tree; new method could cost less than $2 per tree.

17. Answers may vary. Examples are given. 16*j*, *j* = number of jars, expression represents number of ounces in *j* jars; 40*j*, *j* = number of jars, expression represents total number of trees grown in *j* jars;

5*n*, *n* = number of trees, expression represents cost of *n* trees at $5 per tree.

18. Letters may vary.

19. **a.** 5280*m*

   b. 7,603,200 ft

Interdisciplinary Problems
Mathematics is a subject that is used as a written scientific language to understand and solve problems in many other fields of study. Suggest that students start a journal now to keep a list of the different subjects and areas of life in which mathematics is used in this book. They can go back to page 7 and start with language arts. For Exs. 16–18, they can list agriculture.

20. According to Tom Parker's book *In One Day*, dogs in the United States bite 20 mail carriers per day. The Postal Service pays $3500 per day in medical costs for dog-bite victims.

   a. Write two variable expressions based on this information.

   b. About how many mail carriers are bitten by dogs in one year?

   c. About how much money does the Postal Service spend each year on dog-bite victims?

Evaluate each variable expression when $p = 4$ and $t = 40$.

21. $p + t$ **22.** pt **23.** $\frac{t}{p}$ **24.** $0.5t + p$

25. Which of the variable expressions in Exercises 21–24 is greatest when $p = 2$ and $t = 10$? Which is greatest when $p = 0.2$ and $t = 10$?

Use the expressions from the table you made in the Exploration on page 13.

26. Make a table showing the perimeters of the rectangles when $x = 2$.

27. Make a table showing the perimeters of the rectangles when $x = \frac{1}{2}$.

28. For what value of x are the perimeters of the long rectangles equal to the perimeters of the tall rectangles?

29. **Writing** Explain how knowing the value of x helps you decide whether the long or the tall rectangles have greater perimeters.

Write a variable expression for the perimeter of each shape.

30. 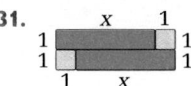   **31.** 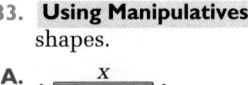   **32.**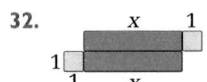

33. **Using Manipulatives** Use x-tiles to make these shapes.

   A. 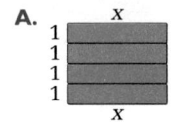   **B.**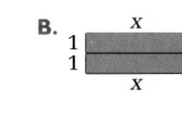

   a. Write variable expressions for the perimeters of Shapes A and B.

   b. Find a value for x so that Shape A has a larger perimeter than Shape B.

   c. Find a value for x so that Shape B has a larger perimeter than Shape A.

   d. Find a value for x so that the perimeters of the two shapes are equal.

Answers to Exercises and Problems

20. a. $20d$ = number of dog-bitten mail carriers over a period of d days, $3500d$ = number of dollars spent by the U.S. Postal Service on dog bites in d days.

  b. about 7300

  c. about $1,277,500

21. 44   **22.** 160   **23.** 10

24. 24   **25.** pt; $\frac{t}{p}$

26. $x = 2$

Perimeter

Number of x-tiles	Long rectangles	Tall rectangles
1	6	6
2	10	8
3	14	10
4	18	12
5	22	14

27. $x = \frac{1}{2}$

Perimeter

Number of x-tiles	Long rectangles	Tall rectangles
1	3	3
2	4	5
3	5	7
4	6	9
5	7	11

Letters are sometimes used to show patterns in a poem. Lines that rhyme are indicated with the same letter.

34. Writing Eve Merriam has written several "Animalimericks" like this one. The rhyme scheme of a limerick is *aabba*. Write a limerick that uses mathematical terms.

35. Research Shakespeare wrote over 150 sonnets with the rhyme scheme *abab cdcd efef gg*. Find one and write out its pairs of rhyming words.

36. How is the use of variables to describe rhyme schemes different from the use of variables in mathematical expressions?

There once was a finicky **ocelot**	*a*
Who all the year round was **cross a lot**	*a*
Except at **Thanksgiving**	*b*
When he enjoyed **living**	*b*
For he liked to eat cranberry **sauce a lot**.	*a*

—Eve Merriam

Variables are sometimes used to write general rules. Look for a pattern in the examples. Write the pattern using the given number of variables.

37. Use one variable.

$$1 \cdot 16 = 16$$
$$1 \cdot 23.5 = 23.5$$
$$1 \cdot \frac{2}{3} = \frac{2}{3}$$
$$1 \cdot 2 = 2$$

38. Use two variables.

$$5 \cdot 3 = 3 \cdot 5$$
$$8.24 \cdot 6.81 = 6.81 \cdot 8.24$$
$$\frac{3}{8} \cdot \frac{4}{7} = \frac{4}{7} \cdot \frac{3}{8}$$
$$1 \cdot 2 = 2 \cdot 1$$

39. Use three variables.

$$\frac{3}{7} + \frac{2}{7} = \frac{3 + 2}{7}$$
$$\frac{2.4}{10} + \frac{5.6}{10} = \frac{2.4 + 5.6}{10}$$
$$\frac{1}{3} + \frac{2}{3} = \frac{1 + 2}{3}$$

40. Use four variables.

$$\frac{3}{5} \cdot \frac{4}{7} = \frac{3 \cdot 4}{5 \cdot 7}$$
$$\frac{3.6}{9.1} \cdot \frac{10.8}{13.9} = \frac{3.6 \cdot 10.8}{9.1 \cdot 13.9}$$
$$\frac{1}{2} \cdot \frac{3}{4} = \frac{1 \cdot 3}{2 \cdot 4}$$

Ongoing ASSESSMENT

41. Open-ended Using x-tiles and 1-tiles, make two different shapes whose perimeters are equal no matter what value you choose for x.

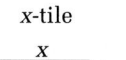

x-tile   1-tile

Working on the Unit Project

By generating a list of words in Ex. 47, students are creating a source of ideas for a possible logo. The words in the list may lead students to think of other, related words.

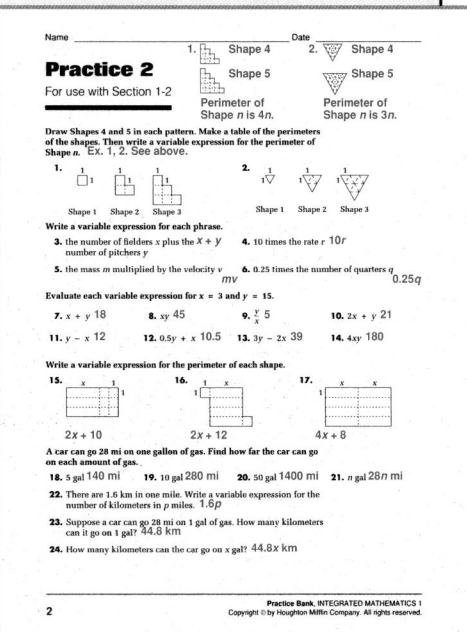
Use the excerpt about the 1990 census. *(Section 1-1)*

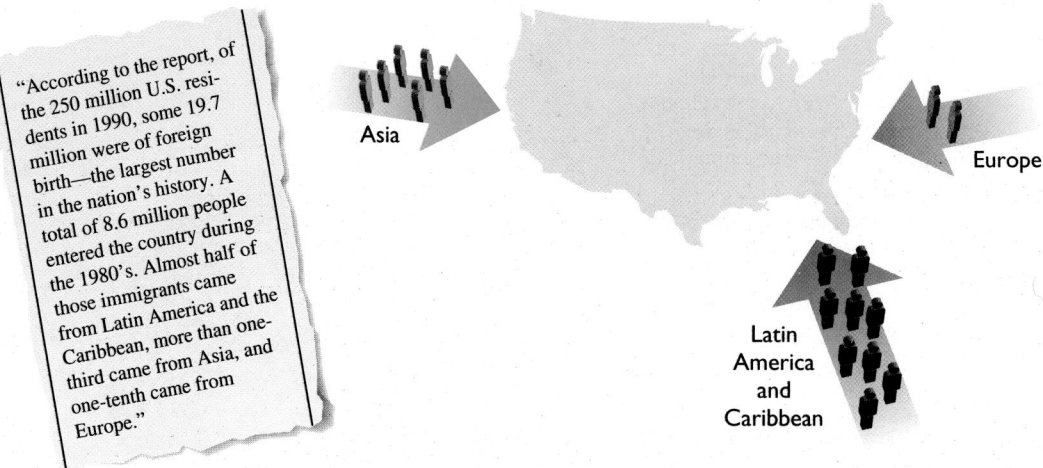

"According to the report, of the 250 million U.S. residents in 1990, some 19.7 million were of foreign birth—the largest number in the nation's history. A total of 8.6 million people entered the country during the 1980's. Almost half of those immigrants came from Latin America and the Caribbean, more than one-third came from Asia, and one-tenth came from Europe."

Asia

Europe

Latin America and Caribbean

42. Which bar graph shows the information from the excerpt?

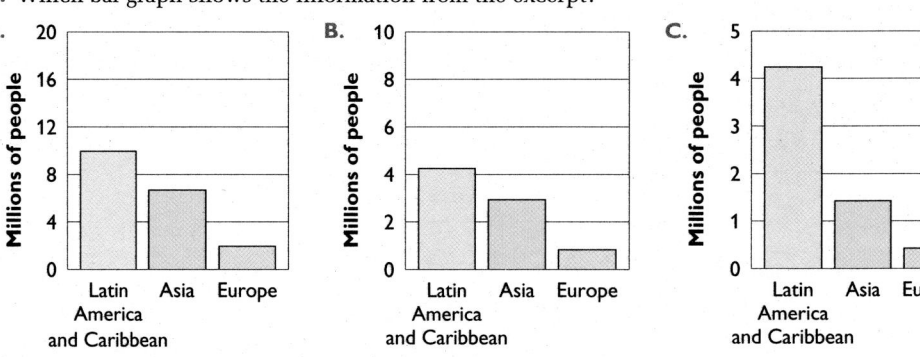

A.

B.

C.

Use the Toolbox at the back of this book. *(Toolbox, pp. 640–665)*

43. Study a topic from the Toolbox that you need to review.

Calculate each repeated multiplication.

44. $7 \times 7 \times 7 \times 7$ **45.** $9 \cdot 9 \cdot 9 \cdot 9 \cdot 9$ **46.** $(0.5)(0.5)(0.5)$

Working on the Unit Project

47. To begin to think about a logo for your math class, list some words that you think of when someone says "mathematics."

18   **Unit 1** Exploring and Communicating Mathematics

Answers to Exercises and Problems

42. B   45. 59,049
43. Choices may vary.   46. 0.125
44. 2401   47. Answers may vary.

Patterns with Powers

Flour Power

---Focus
Use exponents to express repeated multiplication, evaluate powers of variables, and make conjectures.

To make dragon's beard noodles by hand, a chef stretches dough, folds it in half, and stretches it again. This is repeated over and over. Each time the chef stretches, the noodles get thinner. Each time the chef folds, the number of noodles doubles.

A good noodle maker can fold and stretch dough eight times. An expert can fold and stretch it twelve times. How many noodles is that? To find out, you can make a table like the one below.

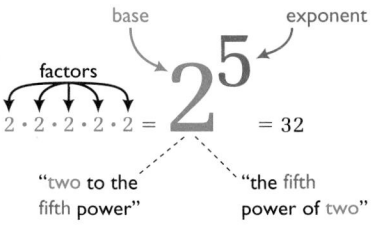

$$2 \cdot 2 \cdot 2 \cdot 2 \cdot 2 = 2^5 = 32$$

"two to the fifth power"   "the fifth power of two"

Number of folds	Repeated multiplication	Written as a power of two	Number of noodles
1	2	2^1	2
2	$2 \cdot 2$	2^2	4
3	$2 \cdot 2 \cdot 2$	2^3	8
4	$2 \cdot 2 \cdot 2 \cdot 2$	2^4	16
5	$2 \cdot 2 \cdot 2 \cdot 2 \cdot 2$	2^5	32

Numbers multiplied together are called **factors.** When the same number is repeated as a factor, you can rewrite the product as a **power** of that number. The repeated factor is the **base,** and the number of times it appears as a factor is the **exponent.**

1-3 Patterns with Powers

19

PLANNING

Objectives and Strands
See pages 1A and 1B.

Spiral Learning
See page 1B.

Materials List
➤ Calculator
➤ Algebra tiles

Recommended Pacing
Section 1-3 is a two-day lesson.
Day 1
Pages 19–21: Opening paragraph through Sample 2, *Exercises 1–27*
Day 2
Pages 21–23: Conjectures about Powers of Ten through Look Back, *Exercises 28–45*

Toolbox References
Toolbox Skill 18: Finding Area

Extra Practice
See pages 620–621.

Warm-Up Exercises
🕯 Warm-Up Transparency 1-3

Support Materials
➤ Practice 3
➤ Enrichment 3 in the Activity Bank
➤ Study Guide 1-3
➤ Problem Set 1
➤ Diagram Masters 6, 7 in the Explorations Lab Manual
🕯 Overhead Visual 1
➤ Quiz 1-3
➤ Alternative Assessments 2, 3

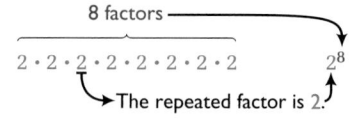
Using Manipulatives

Students can fold an 8.5" × 11" piece of paper in half repeatedly to illustrate the powers of 2. The first fold is 2^1, or 2 parts; the second fold is 2^2, or 4 parts; the third fold is 2^3, or 8 parts, and so on.

Communication: Discussion

Have students discuss the advantage of using exponents to express repeated multiplication.

Additional Sample

S1 Write the product as a power. Then write how to say it.

a. $3 \cdot 3 \cdot 3 \cdot 3 \cdot 3$
   3^5; three to the fifth power

b. $3 \cdot 3 \cdot 3 \cdot 3 \cdot 3 \cdot 3 \cdot 3 \cdot 3 \cdot 3 \cdot 3 \cdot 3$
   3^{11}; three to the eleventh power

Using Technology

Use Talk it Over question 2 to explore the exponent key on a calculator. On the TI-81 and TI-82, there is an x^2 key that can be used to square any number. There is also a key (with the symbol ^) that can be used to raise any number to any power. Scientific calculators that have an exponent key, as described in the technology note, usually have a 10^x key. Have students find 10^4 using this key. Stress that this key is used only for powers of ten.

You may wish to refer students to pages 611 and 612 of the Technology Handbook at the back of their books to explore some of the basic features of their calculators.

Write the product as a power. Then write how to say it.

a. $2 \cdot 2 \cdot 2 \cdot 2 \cdot 2 \cdot 2 \cdot 2 \cdot 2$

b. $2 \cdot 2 \cdot 2 \cdot 2 \cdot 2 \cdot 2 \cdot 2 \cdot 2 \cdot 2 \cdot 2 \cdot 2 \cdot 2$

Sample Response

a. The repeated factor is 2, so the base is 2.
   The factor appears 8 times, so the exponent is 8.
   The product is 2^8. ◄——— "Two to the eighth power."

b. The repeated factor is 2, so the base is 2.
   The factor appears 12 times, so the exponent is 12.
   The product is 2^{12}. ◄——— "Two to the twelfth power."

$$\overbrace{2 \cdot 2 \cdot 2 \cdot 2 \cdot 2 \cdot 2 \cdot 2 \cdot 2}^{\text{8 factors}} \qquad 2^8$$
The repeated factor is 2.

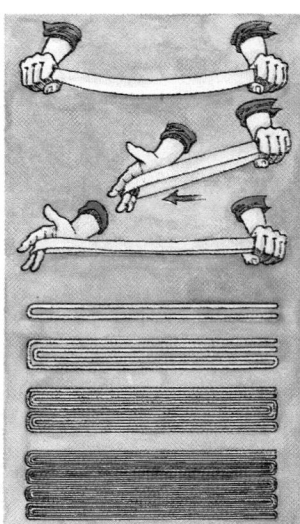

Talk it Over

1. A good noodle maker can fold and stretch dough 8 times. Use repeated multiplication to find the number of noodles.

2. An expert noodle maker can fold and stretch dough 12 times. Use the exponent key on a calculator to find the number of noodles.

3. Which method do you prefer to use, repeated multiplication or the exponent key? Why?

4. If the expert noodle maker's noodles are each about 5 ft long, how many miles would the noodles cover if they were laid end to end? (*Hint:* There are 5280 ft in 1 mi.)

Evaluating Powers of Variables

You can use a variable as the base of a power.

$x \cdot x = x^2$	$y \cdot y \cdot y = y^3$
"x squared"	"y cubed"
can represent the area of a square with sides of length x	can represent the volume of a cube with sides of length y

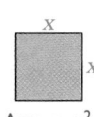

Area = x^2

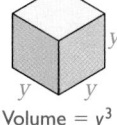

Volume = y^3

To evaluate powers of variables, you substitute a value for the variable and then simplify.

TECHNOLOGY NOTE

The exponent key on a calculator may be labeled [x^y] or [y^x] or [^]. You enter the base, press the key, and then enter the exponent. ◄

20

Unit 1 Exploring and Communicating Mathematics

Answers to Talk it Over

1. $2^8 = 256$

2. $2^{12} = 4096$

3. Answers may vary. An example is given. I prefer the exponent key because it is faster.

4. about 3.9 mi

Sample 2

Write an expression for the area covered by the tiles.

Evaluate your expression for each value of x.

a. $x = 5$ b. $x = 10$

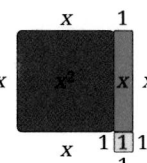

Sample Response

The tiles have areas x^2, x, and 1.

The total area they cover is $x^2 + x + 1$ square units.

a. $x^2 + x + 1$ ◄———— Write the expression.

$5^2 + 5 + 1$ ◄———— Substitute 5 for x.

$25 + 5 + 1$

31

The tiles cover 31 square units when $x = 5$.

b. $x^2 + x + 1$ ◄———— Write the expression.

$10^2 + 10 + 1$ ◄———— Substitute 10 for x.

$100 + 10 + 1$

111

The tiles cover 111 square units when $x = 10$.

►► **Now you are ready for:**
Exs. 1–27 on pp. 23–24

Conjectures about Powers of Ten

Powers of ten show an interesting pattern.

$10^3 = 1000$ ◄———— 1 with three zeros

$10^4 = 10,000$ ◄———— 1 with four zeros

$10^5 = 100,000$ ◄———— 1 with five zeros

$10^6 = 1,000,000$ ◄———— 1 with six zeros

Talk it Over

5. a. A **conjecture** is a guess based on your past experiences. Make a conjecture about the number of zeros you need to write out 10^9.

   b. Test your conjecture by using a calculator to find the decimal notation for 10^9.

6. a. Make a conjecture about the number of zeros you need to write out 10^n.

   b. Use your conjecture to predict the number of zeros you need to write out 10^{100}.

BY THE WAY...

The nine-year-old nephew of American mathematician Edward Kasner invented the name "googol" for 10^{100}.

Additional Sample

S2 Write an expression for the area covered by the tiles.

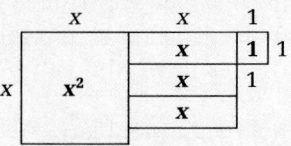

$x^2 + 3x + 1$

Evaluate your expression for each value of x.

a. $x = 4$ **29**

b. $x = 8$ **89**

Talk it Over

Questions 5–9 introduce students to the idea of making a conjecture based upon the use of a pattern. The conjectures made in questions 9 and 11 should be stated in words.

Mathematical Procedures

New mathematics is often created by making conjectures and then trying to prove them. In this sense, a conjecture is an unsolved problem. When proved, a conjecture becomes a theorem.

Answers to Talk it Over

5. a. 9 zeros

   b. $10^9 = 1,000,000,000$

6. a. n zeros

   b. 100 zeros

7. Look at the products of powers of ten. What patterns do you see in the exponents?

$$10^4 \cdot 10^1 = (10 \cdot 10 \cdot 10 \cdot 10) \cdot (10) = 10^5$$
$$10^3 \cdot 10^2 = (10 \cdot 10 \cdot 10) \cdot (10 \cdot 10) = 10^5$$
$$10^2 \cdot 10^3 = (10 \cdot 10) \cdot (10 \cdot 10 \cdot 10) = 10^5$$
$$10^1 \cdot 10^4 = (10) \cdot (10 \cdot 10 \cdot 10 \cdot 10) = 10^5$$
$$10^0 \cdot 10^5 = (10 \cdot 10 \cdot 10 \cdot 10 \cdot 10) = 10^5$$

8. Write each product as a power of ten.

   a. $10^5 \cdot 10^3 = (10 \cdot 10 \cdot 10 \cdot 10 \cdot 10) \cdot (10 \cdot 10 \cdot 10) = 10^?$

   b. $10^6 \cdot 10^2 = 10^?$ c. $10^8 \cdot 10 = 10^?$

9. Make a conjecture about multiplying powers of ten.

10. Write each quotient as a power of ten.

   a. $\dfrac{10^5}{10^3} = \dfrac{10 \cdot 10 \cdot 10 \cdot 10 \cdot 10}{10 \cdot 10 \cdot 10} = \dfrac{10 \cdot 10}{1} = 10^?$

   b. $\dfrac{10^6}{10^2} = \dfrac{?}{?} = 10^?$ c. $\dfrac{10^8}{10} = 10^?$

11. Make a conjecture about dividing powers of ten.

Counterexamples

Not all conjectures are true. A **counterexample** is an example that shows that a statement is false.

Sample 3

Larry makes a conjecture that x^2 is greater than x for all values of x. Find a counterexample.

Sample Response

The number 0 is a counterexample, because 0^2 is *not* greater than 0.

The number 1 is a counterexample, because 1^2 is *not* greater than 1.

Numbers between 0 and 1 are also counterexamples:

$$\text{If } x = \frac{1}{2}, \text{ then } x^2 = \frac{1}{2} \cdot \frac{1}{2} = \frac{1}{4}.$$

For numbers like these, x^2 is less than x.

Larry's conjecture is false.

22   **Unit 1** Exploring and Communicating Mathematics

Answers to Talk it Over

7. Answers may vary. An example is given. The sum of the exponents of the two powers of 10 on the left is equal to the power of ten on the right.

8. a. 10^8
   b. 10^8
   c. 10^9

9. Answers may vary. An example is given. To multiply powers of ten, add the exponents.

10. a. 10^2
    b. 10^4
    c. 10^7

11. Answers may vary. An example is given. To divide powers of ten, subtract the exponents.

Answers to Look Back

$4n$; n^4; Repeated addition can be written using multiplication. Multiply the quantity being added by the number of times it is being added. Repeated multiplication can be written using exponents. The base is the quantity being multiplied; the exponent is the number of times it is being multiplied.

Now you are ready for:
Exs. 28–45 on pp. 24–25

Look Back

How can you rewrite these expressions?

$$n + n + n + n$$

$$n \cdot n \cdot n \cdot n$$

How can you rewrite repeated addition? How can you rewrite repeated multiplication?

Exercises and Problems

1. **Reading** What are three other ways of writing 2^5?

Write the product as a power. Then write how to say it.

2. $2 \cdot 2 \cdot 2 \cdot 2 \cdot 2 \cdot 2$ 3. $4 \cdot 4 \cdot 4$ 4. $5 \cdot 5 \cdot 5 \cdot 5 \cdot 5$

5. $x \cdot x \cdot x$ 6. $y \cdot y$ 7. $z \cdot z \cdot z \cdot z \cdot z \cdot z$

Write using exponents.

8. two to the fifth   9. five squared   10. three to the seventh

11. seven cubed   12. y to the fourth   13. x to the sixth

14. To show 1,000,000 using sign language, you strike the sign for 1000 (fingertips in an "M" shape) into the palm twice. Explain how this sign is similar to power notation.

Arrange from smallest to largest. Explain your reasoning.

15. $4^5, 4^4, 4^6, 4^3$

16. $2^{10}, 6^{10}, 4^{10}, 3^{10}$

17. $\left(\frac{1}{2}\right)^3, \left(\frac{1}{2}\right)^2, \left(\frac{1}{2}\right)^4, \left(\frac{1}{2}\right)^5$

18. $0^5, 0^7, 0^6, 0^8$

19. $(0.1)^2, (0.3)^2, (0.4)^2, (0.8)^2$

20. A chain letter tells you to send copies to five friends.

   a. Complete the table showing the number of letters sent during each round.

   b. Write an expression for the number of letters sent during round n.

Round	1	2	3	4	5	6	7	8	9	10
Number of letters	5	25	125	625	?	?	?	?	?	?

   c. Guess which round will send out more than a billion letters. Check your guess using a calculator.

Answers to Exercises and Problems

1. Answers may vary. An example is given. $2 \cdot 2 \cdot 2 \cdot 2$; $2^3 \cdot 2^2$; $2^4 \cdot 2^1$

2. 2^6; two to the sixth power

3. 4^3; four cubed

4. 5^5; five to the fifth power

5. x^3; x cubed

6. y^2; y squared

7. z^6; z to the sixth power

8. 2^5 9. 5^2 10. 3^7

11. 7^3 12. y^4 13. x^6

14. Explanations may vary. An example is given. $1000^2 = 1,000,000$; The two strikings of the hand is the same as saying raise the number to the second power.

15–19. Explanations may vary. Examples are given.

15. $4^3, 4^4, 4^5, 4^6$; Four is used as a factor an increasing number of times. Since four is a positive number greater than one, the larger the exponent, the larger the number.

16. $2^{10}, 3^{10}, 4^{10}, 6^{10}$; Each base is multiplied by itself the same number of times (10). Since the bases are all

---

Look Back

Have students discuss their responses to the Look Back question. Then have them think of other examples of expressions that can be used to compare repeated addition and repeated multiplication.

APPLYING

Suggested Assignment

Standard 1–13, 15–19, 21–26, 28–45

Extended 1–45

Integrating the Strands

Number Exs. 1–20, 28–39

Algebra Exs. 1–42

Geometry Exs. 21–27, 43

Statistics and Probability Ex. 44

Discrete Mathematics Ex. 20

Logic and Language Exs. 1, 14, 41, 43, 45

Communication: Writing

In Exs. 2 and 3, check students' ability to write powers in symbols and in words. Students should be able to translate from one to the other.

whole numbers larger than 1, the larger the base, the larger the number.

17. $\left(\frac{1}{2}\right)^5, \left(\frac{1}{2}\right)^4, \left(\frac{1}{2}\right)^3, \left(\frac{1}{2}\right)^2$; Since $\frac{1}{2} < 1$, multiplying it by itself makes it smaller (the numerator stays the same, but the denominator gets larger). The larger the exponent, the smaller the number.

18. It does not matter what order they are in because they are all equal to zero.

19. $(0.1)^2, (0.3)^2, (0.4)^2, (0.8)^2$; Each base is multiplied by itself the same number of times (2). Since the bases are all positive, the larger the base, the larger the number.

20. See answer in back of book.

Integrating the Strands

Many of the exercises and problems in this section use numerical examples to arrive at general statements that can be expressed by using algebraic variables.

Using Manipulatives

Students can use algebra tiles to model Exs. 21–27. You may wish to have students create their own shapes as well.

Reasoning

The patterns involving powers of 10 in Exs. 28–39 can be used to make conjectures about number bases other than 10. You may wish to introduce this idea briefly now (it is discussed in Unit 10) by using a few examples, such as $5^a \cdot 5^b = 5^{a+b}$, to lead students to the conjecture that if x is any whole number, then $x^a \cdot x^b = x^{a+b}$. Similar conjectures can be based on the other exercises involving powers of 10.

Error Analysis

In Exs. 30 and 34, students may answer by writing 10^{17}. The reason is that when the exponent of a number is an unwritten 1, students may ignore it. To correct this error, remind students that any number without an exponent is raised to the first power and suggest that students write the exponent as 1 before beginning to work the exercises.

Write an expression for the area covered by each group of tiles. Evaluate each expression when $x = 7$.

21.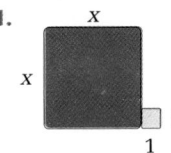

22.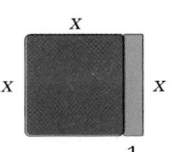

23.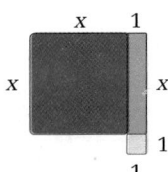

Write an expression for the area covered by each group of tiles.

24.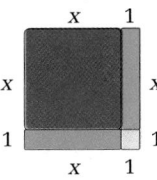

25.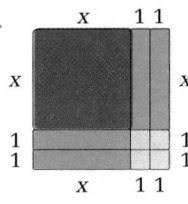

26.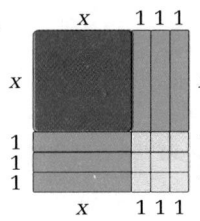

27. Sketch the next square that follows the pattern in Exercises 24–26. Write an expression for the area.

Write each expression as a power of ten.

28. $10^8 \cdot 10^7$

29. $10^6 \cdot 10^6$

30. $10^{17} \cdot 10$

31. $10^{473} \cdot 10^{27}$

32. $\dfrac{10^8}{10^7}$

33. $\dfrac{10^6}{10^6}$

34. $\dfrac{10^{17}}{10}$

35. $\dfrac{10^{473}}{10^{27}}$

36. Complete this conjecture about multiplying powers of ten.
$$10^a \cdot 10^b = 10^?$$

37. Complete this conjecture about dividing powers of ten.
$$\frac{10^a}{10^b} = 10^? \text{ when } a \text{ is greater than } b$$

38. According to the pattern on page 22, $10^0 \cdot 10^5 = 10^5$. Make a conjecture about the value of 10^0. Check your conjecture using a calculator.

39. Write as a power of ten.
   a. $10^5 \cdot 10^5 \cdot 10^5$
   b. $10^3 \cdot 10^5 \cdot 10^7$
   c. $10^2 \cdot 10^5 \cdot 10^8$
   d. Complete this conjecture about multiplying three powers of ten:
   $$10^a \cdot 10^b \cdot 10^c = 10^?$$

40. Samuel writes this statement: "$x^2 + x^5 = x^7$ for all numbers."
   a. Give a counterexample to Samuel's statement.
   b. Is there any value of x that makes the statement true?
   c. Rewrite Samuel's statement so it is true.

Answers to Exercises and Problems

21. $x^2 + 1$; 50

22. $x^2 + x$; 56

23. $x^2 + x + 1$; 57

24. $x^2 + 2x + 1$

25. $x^2 + 4x + 4$

26. $x^2 + 6x + 9$

27. $x^2 + 8x + 16$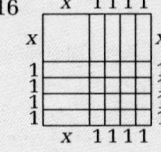

28. 10^{15}

29. 10^{12}

30. 10^{18}

31. 10^{500}

32. 10^1

33. 10^0

34. 10^{16}

35. 10^{446}

36. 10^{a+b}

37. 10^{a-b}

38. $10^0 = 1$

39. a. 10^{15}
    b. 10^{15}
    c. 10^{15}
    d. 10^{a+b+c}

40. a. Counterexamples may vary. $2^2 + 2^5 \neq 2^7$, since $4 + 32 \neq 128$
    b. Yes; $0^2 + 0^5 = 0^7$.
    c. $x^2 \cdot x^5 = x^7$

41. **Writing** Imagine that you are explaining to a friend over the telephone how to work with powers of ten.

   a. Write what you would say about multiplying powers together.

   b. Write what you would say about dividing powers.

   c. Read what you have written to a family member or a friend. If the person does not understand, rewrite your explanation.

Review **PREVIEW**

42. Latex paint comes in two sizes of cans, small and large. *(Section 1-2)*

   a. Write a variable expression for the cost of *s* small cans.

   b. Write a variable expression for the cost of *l* large cans.

   c. Evaluate the expressions in parts (a) and (b) for $s = 5$ and $l = 3$.

43. You will review polygons later in this chapter. Write three sentences about polygons based on this concept map. *(Section 1-1)*

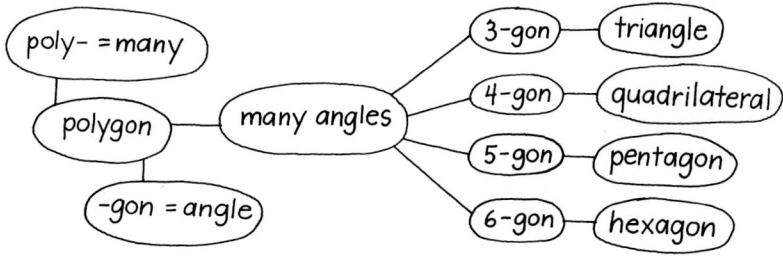

44. Find the average of these test scores: 93, 85, 85.

 Working on the Unit Project

45. a. Your knowledge of mathematics will be growing this year. Look through this book for ideas about what you will be learning.

   b. Think about including in your logo a pattern of numbers or geometric shapes that suggests growth.

   c. Sketch one of your ideas.

1-3 Patterns with Powers

Assessment: Portfolio

The written responses to Ex. 41 will make an excellent portfolio entry. You may wish to call upon a few students to read their explanations to the class. Other students can be asked to judge if the explanations are clear and easy to understand.

Working on the Unit Project

Many commercial logos combine two or more shapes into a simple and visually interesting pattern. You might wish to introduce the idea of simplicity as students work on this activity.

Practice 3 For use with Section 1-3

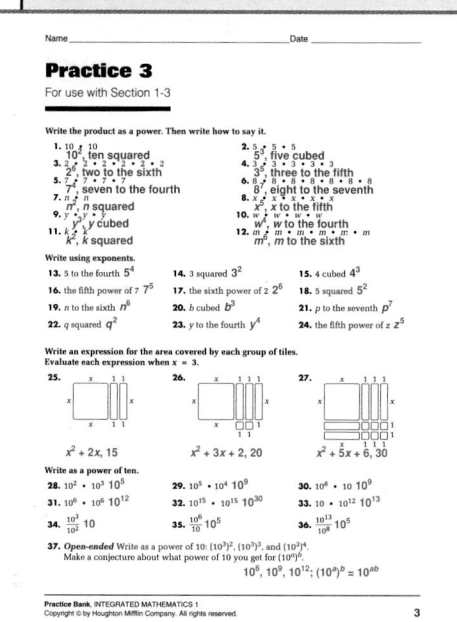

Answers to Exercises and Problems

41. a, b. Answers may vary. Examples are given.

   a. Add the exponents.

   b. Subtract the exponents.

   c. Answers may vary.

42. a. 6.59*s*

   b. 16.99*l*

   c. $32.95; $50.97

43. Answers may vary. Examples are given. The word polygon means many angles. A triangle is a 3-gon. A quadrilateral is a 4-gon.

44. about 88

45. a–c. Answers may vary.

PLANNING

Objectives and Strands
See pages 1A and 1B.

Spiral Learning
See page 1B.

Materials List
➤ Graphics calculator
➤ Algebra tiles

Recommended Pacing
Section 1-4 is a one-day lesson.

Extra Practice
See pages 620–621.

Warm-Up Exercises
💡 Warm-Up Transparency 1-4

Support Materials
➤ Practice 4
➤ Enrichment 4 in the Activity Bank
➤ Study Guide 1-4
➤ Problem Set 1
➤ Diagram Masters 6, 7 in the Explorations Lab Manual
💡 Overhead Visual 1
➤ Quiz 1-4
➤ Test 1
➤ Alternative Assessment 4

Section 1-4

Writing and Evaluating Expressions

Focus
Use the order of operations to evaluate expressions.

Talk it Over

1. Marlene wants to find the average of three test scores: 84, 75, and 75. What is a reasonable guess for her average?

2. Marlene enters 84 $+$ 2 $\times$ 75 $\div$ 3 $=$ on a calculator and gets 2150. What did the calculator do?

3. Marlene knows that 2150 is not her test average. She borrows a graphics calculator from Ron. She enters 84 $+$ 2 $\times$ 75 $\div$ 3 ENTER on the graphics calculator and gets 134. What did the graphics calculator do?

4. Marlene knows that 134 is not her test average either. Explain how she can use the parenthesis keys on her graphics calculator to find her test average. What is it?

5. Some calculators do not have parenthesis keys. Find a keystroke sequence that gives the correct average but does not use parentheses.

Ron's graphics calculator follows the **order of operations,** a set of rules people agree to use so an expression has only one answer. In algebra, the order of operations is not always from left to right.

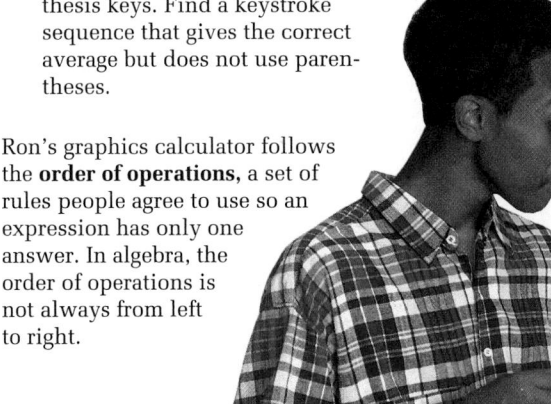

26

Answers to Talk it Over

1. Answers may vary, but should be near 78.

2. The calculator added 84 and 2, multiplied this result by 75, and then divided by 3.

3. The graphics calculator did the multiplication first, then the division, and did the addition last.

4. Enter (84 $+$ 2 $\times$ 75) $\div$ 3 ENTER; 78

5. Enter 2 $\times$ 75 $+$ 84 $=$ $\div$ 3 $=$

ORDER OF OPERATIONS

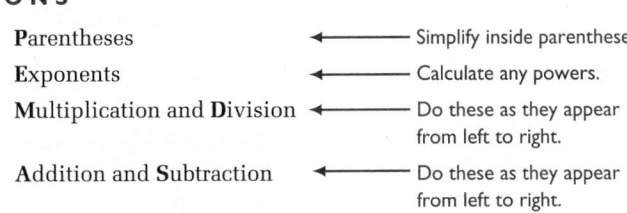

Parentheses ← Simplify inside parentheses.

Exponents ← Calculate any powers.

Multiplication and Division ← Do these as they appear from left to right.

Addition and Subtraction ← Do these as they appear from left to right.

Watch Out!
Multiplications are not done before divisions. Multiplications and divisions are done as they appear, from left to right.

Sample 1

Calculate according to the order of operations.

$48 + (12 - 8)^2 \div 8$

Sample Response

$48 + (12 - 8)^2 \div 8$

$48 + 4^2 \div 8$ ← Do operations in parentheses first.

$48 + 16 \div 8$ ← Do exponents next.

$48 + 2$ ← Do division.

50 ← Do addition.

Sample 2

Insert parentheses to make each statement true.

a. $4 + 16 \div 2 + 3 \times 5 = 20$

b. $4 + 16 \div 2 + 3 \times 5 = 59$

Sample Response

Problem Solving Strategy: Guess and check

Add parentheses in different ways and calculate.

a. $(4 + 16) \div 2 + 3 \times 5 = 25$ ← Try parentheses around first pair.

$4 + (16 \div 2) + 3 \times 5 = 27$ ← Try parentheses around second pair.

$4 + 16 \div (2 + 3) \times 5 = 20$ ← This answers part (a).

b. $(4 + 16 \div 2) + 3 \times 5 = 27$ ← Try parentheses around first three.

$4 + (16 \div 2 + 3) \times 5 = 59$ ← This answers part (b).

1-4 Writing and Evaluating Expressions

27

TEACHING

Using Technology

Scientific and graphics calculators follow the order of operations. Most nonscientific calculators do not. Students can use Talk it Over questions 1–5 to test their calculators as to whether they follow the order of operations. Questions 4 and 5 also demonstrate the importance of using parenthesis keys and what to do if a calculator does not have these keys.

Additional Samples

S1 Calculate according to the order of operations.
$72 \div (18 - 12)^2 + 9$ **11**

S2 Insert parentheses to make each statement true.

a. $2 + 8 \div 4 + 6 \times 3 = 22$
$2 + (8 \div 4) + (6 \times 3) = 22$

b. $2 + 8 \div 4 + 6 \times 3 = 3$
$(2 + 8) \div (4 + 6) \times 3 = 3$

Reasoning

Students at this level have already learned and accepted as fact the idea that the result of a numerical operation is always one correct answer. This is why they understand the need for an order of operations. However, they may have also learned that when solving some mathematical problems, it is possible to have more than one correct answer or no correct answer. Ask students if they can suggest an example of such problems and how they can reconcile these seemingly contradictory facts. Certain real-world problems may have multiple solutions or the mathematical solutions may not make sense when interpreted in a realistic context. However, the result of a computation or series of computations is always a single number.

27

Additional Sample

S3 Write an expression for the area covered by the tiles. Evaluate the expression when $x = 4$.

x	x	1	1	1
x^2	x^2	x	x	x

$2x^2 + 5x + 3$; 55 square units

Look Back

The Look Back can be used for class discussion. Have students summarize the order of operations and relate it to the given phrase.

APPLYING

Suggested Assignment

Standard 3–20, 23–29, 35–42

Extended 1–20, 23–31, 35–42

Integrating the Strands

Algebra Exs. 1, 3–31, 36–41

Geometry Exs. 32–34

Logic and Language Exs. 2, 35, 42

Communication: Drawing

Students may wish to exchange and compare drawings for Ex. 2. If two or more students have chosen the same non-mathematical task, but created different drawings, this will further demonstrate the need for an order of operations to avoid multiple answers to the same exercise.

Sample 3

Write an expression for the area covered by the tiles. Evaluate the expression when $x = 5$.

Sample Response

There are 4 x^2-tiles, 8 x-tiles, and 4 1-tiles.

An expression for the area is $4x^2 + 8x + 4$.

$4x^2 + 8x + 4$

$4 \cdot 5^2 + 8 \cdot 5 + 4$ ◄——— Substitute 5 for x.

$4 \cdot 25 + 8 \cdot 5 + 4$ ◄——— Do exponent first.

$100 + 40 + 4$ ◄——— Do all multiplications from left to right.

144 ◄——— Do all additions from left to right.

When $x = 5$, the tiles cover an area of 144 square units. This answer makes sense, because the tiles form a square. The square is 12 units on each side when $x = 5$. The area is $12 \cdot 12$, or 144 square units.

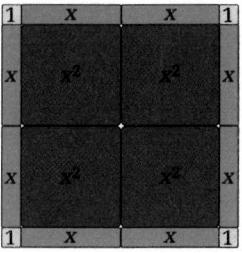

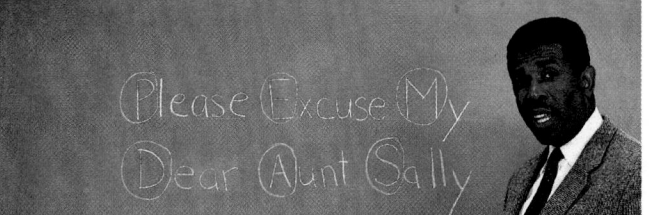

Look Back ◄

This teacher uses the phrase shown in the photograph to help students remember the order of operations. Explain how this phrase might help you.

1-4 Exercises and Problems

1. **Reading** Why do people agree on an order of operations?

2. **Open-ended** Draw a silly picture showing what happens when someone does not follow the correct "order of operations" for a nonmathematical task.

Calculate according to the order of operations.

3. $3 + 4 \cdot 5$

4. $4 \div 2 + 5$

5. $2^4 - 3$

6. $4 + 6^3$

7. $12 \div 2^2$

8. $7^5 \cdot 10$

9. $10 - 4 + 9 - 3$

10. $2 + 4^2 \cdot 3$

11. $10^3 + 36 \div 4$

12. $135 \div 3^3 - 4$

13. $17 \cdot 3 - 2^3$

14. $15 - 125 \div 5^2$

15. $15 - 3(4 - 2)$

16. $15 - (4 - 2)^3$

17. $15 + (4^3 - 2)$

18. $[(12 - 4) \cdot 2 + 11] \div 3$

19. $35 + 5[(16 + 12) \div 4]$

20. $(35 + 5) [16 + (12 \div 4)]$

Answers to Look Back

Answers may vary. An example is given. The first letter in each word tells you in which order the operations must be done: P(parentheses), E(exponents), M(multiplication), D(division), A(addition), and S(subtraction).

Answers to Exercises and Problems

1. Answers may vary. An example is given. People agree on an order of operations so that there is one agreed-upon answer for a given expression.

2. Answers may vary. An example would be to draw someone putting on shoes before socks.

3. 23

4. 7

5. 13

6. 220

7. 3

8. 168,070

9. 12

10. 50

11. 1009

12. 1

13. 43

14. 10

15. 9

16. 7

17. 77

18. 9

19. 70

20. 760

21. Enter ((15 − 3) ÷ (4 + 2)) = ; 2

22. Enter ((15 − 3 × ((4 − 2))) ÷ ((2 x^y 2 − 1)) = ; 3

23. $48 \div 4 + 8 - (4 + 2) = 14$

24. $48 \div (4 + 8 - 4) + 2 = 8$

25. $48 \div (4 + 8) - 4 + 2 = 2$

26–29. Answers may vary. Examples are given.

T E C H N O L O G Y **Write out the keystrokes you could use on a calculator with parentheses to calculate each expression.**

21. $\dfrac{15 - 3}{4 + 2}$

22. $\dfrac{15 - 3(4 - 2)}{2^2 - 1}$

Insert parentheses to make each statement true.

23. $48 \div 4 + 8 - 4 + 2 = 14$ **24.** $48 \div 4 + 8 - 4 + 2 = 8$ **25.** $48 \div 4 + 8 - 4 + 2 = 2$

Insert parentheses in the expression $6 \cdot 5^2 - 7 + 9$ to get each answer.

26. a number less than 125

27. a number between 125 and 150

28. a number between 150 and 175

29. a number greater than 175

30. [Open-ended] Write an expression using at least three different operations whose correct answer you can find by doing operations from left to right.

31. [Open-ended] Write an expression using at least three different operations whose correct answer you cannot find by doing operations from left to right.

[Using Manipulatives] **For each group of tiles, do these things.**
a. Write a variable expression for the perimeter.
b. Write a variable expression for the area.
c. Evaluate the expression when $x = 7$.

32. 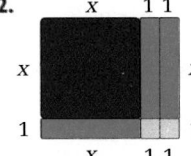   **33.** 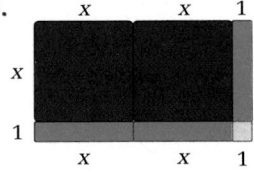   **34.**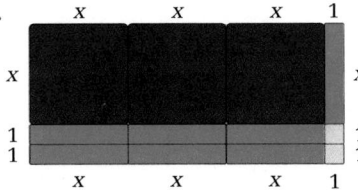

Ongoing **ASSESSMENT**

35. [Writing] Juanita wants to buy a calculator. Her teacher recommends that she buy one that follows the order of operations. Write a plan for how she might test a calculator to see if it follows the order of operations.

Review **PREVIEW**

Show each conjecture is false by finding a counterexample. *(Section 1-3)*

36. $x^2 + x = x^3$

37. $x^2 \cdot x^4 = x^8$

Tell whether each word phrase can be represented by the variable expression $d + 7$. Write *Yes* or *No*. *(Section 1-2)*

38. 7 more than d dinosaurs

39. 7 times a number of dogs

40. d days plus the number of days in a week

41. a. Evaluate the expressions $2x + 3$ and $2(x + 3)$ for three values of x.
b. For each expression, what operations did you use? In what order?
c. Which expression is larger? Why?

1-4 Writing and Evaluating Expressions 29

Answers to Problems and Exercises

26. $6 \cdot (5^2 - 7) + 9 = 117$

27. $6 \cdot 5^2 - (7 + 9) = 134$

28. $(6 \cdot 5^2 - 7) + 9 = 152$

29. $(6 \cdot 5)^2 - 7 + 9 = 902$

30. Answers may vary. An example is given.
$105^2 \div 5 + 6$

31. Answers may vary. An example is given.
$4 + 6 \div 3 - 10 \div 5$

32. a. $4x + 6$ **b.** $x^2 + 3x + 2$
c. 34; 72

33. a. $6x + 4$ **b.** $2x^2 + 3x + 1$
c. 46; 120

34. a. $8x + 6$ **b.** $3x^2 + 7x + 2$
c. 62; 198

35. Answers may vary. An example is given. Juanita could enter an expression that involves both addition

(or subtraction) and multiplication (or division) where the addition appears before the multiplication. For example, she could enter $3 + 2 \cdot 6$. If the result is 15, the calculator follows the order of operations. If the result is 30, it does not.

Error Analysis

In Exs. 21 and 22, point out that the fraction bar acts like a grouping symbol. Students should treat the numerators and denominators of these fractions as if they have parentheses around them.

Communication: Writing

In connection with Exs. 30 and 31, you might suggest that students write a journal entry to explain how they think expressions can be evaluated using the order of operations.

Using Manipulatives

Students can use algebra tiles to create the shapes shown in Exs. 32–34. Students may wish to use the tiles to create their own shapes.

Assessment: Standard

Ex. 35 assesses students' understanding of the order of operations.

36, 37. Counterexamples may vary. Examples are given.

36. When $x = 2$,
$2^2 + 2 \neq 2^3$ $(4 + 2 \neq 8)$.

37. When $x = 2$,
$2^2 \cdot 2^4 \neq 2^8$ $(4 \cdot 16 \neq 256)$.

38. Yes.

39. No.

40. Yes.

41. a. Answers may vary. Examples are given: $x = 4$: $2x + 3 = 11$, $2(x + 3) = 14$; $x = 3$: $2x + 3 = 9$, $2(x + 3) = 12$; $x = 2$: $2x + 3 = 7$, $2(x + 3) = 10$

b. addition and multiplication; For $2x + 3$, multiply first, then add. For $2(x + 3)$, add first, then multiply.

c. $2(x + 3)$; The expression $2(x + 3)$ is equivalent to $2x + 6$, which is 3 more than $2x + 3$.

Practice 4 For use with Section 1-4

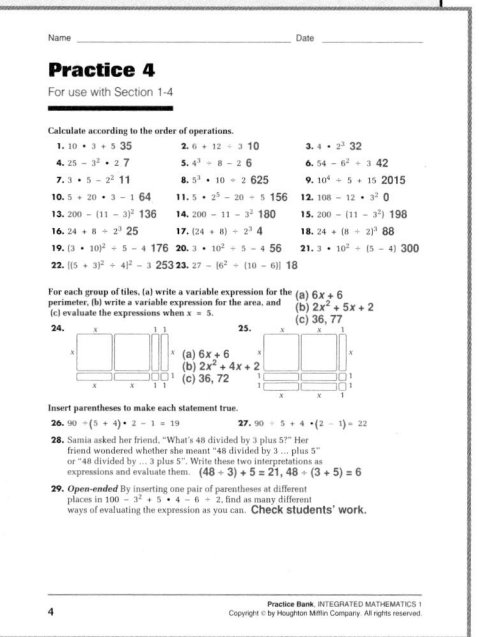

Working on the Unit Project

42. Magazines, newspapers, and telephone yellow pages are good places to find logos.

   a. Trace or cut out five different logos. Tape each logo to a separate piece of paper.

   b. Draw arrows to point out features of the logos that you think convey meaning about the product or service.

   c. Do the same with one of your sketches for your math logo.

ADAPTIVE SCUBA ASSOCIATION

Unit 1 · CHECKPOINT

Language used at home	1980	1990
Mon-Khmer	16,000	127,000
French Creole	25,000	188,000
Hindi	130,000	331,000
Vietnamese	203,000	507,000
Thai	89,000	206,000

The table shows languages that had great increases in speakers in the United States between 1980 and 1990. **1-1**

1. **Writing** Write three statements based on the information in the table.

2. **Writing** Explain why Mon-Khmer is at the top of the table and Thai is at the bottom.

Evaluate each variable expression when $x = 4$ and $y = 7$. **1-2**

3. $3x$ 4. $y + 17$ 5. $5y - x$

Write as a power of ten. **1-3**

6. $10^3 \cdot 10^5$ 7. $10 \cdot 10^{17}$ 8. $\dfrac{10^9}{10^2}$

Evaluate when $x = 9$.

9. $x^2 + 1$ 10. $x^2 + x$ 11. $x^4 + x + 1$

Calculate according to the order of operations. **1-4**

12. $36 \div 3^2 + 7 \cdot 4 + 2^4 - 11 + 6 \div 2$

BY THE WAY...

Languages in the Mon-Khmer group are spoken in Cambodia, Vietnam, and Myanmar. French Creole is spoken in Haiti, the Lesser Antilles, and Seychelles.

30   **Unit 1** Exploring and Communicating Mathematics

Answers to Exercises and Problems

42. a–c. Choices may vary.

Answers to Checkpoint

1. Answers may vary. Examples are given. The biggest increase from 1980 to 1990 is in the number of people speaking Vietnamese at home. The number of people speaking Thai at home has more than doubled from 1980 to 1990. Among languages listed, the fewest number of people still speak Mon-Khmer.

2. The languages are listed in order of highest percent change to lowest percent change from 1980 to 1990.

3. 12   4. 24

5. 31   6. 10^8

7. 10^{18} 8. 10^7

9. 82   10. 90

11. 6571   12. 40

Section 1-5

Modeling the Distributive Property

---Focus

Use the distributive property to simplify calculations, rewrite expressions, and combine like terms.

Combination Plate

Alicia is buying a sandwich and a drink for herself and the same meal for three friends. How can she figure out how much it will cost?

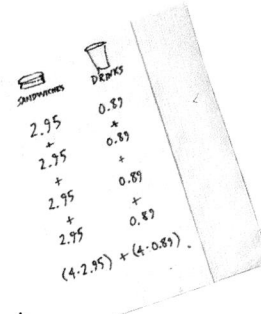

Ben's

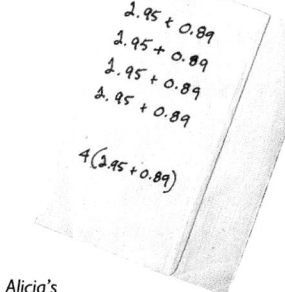

Alicia's

Talk it Over

1. What does Alicia do to figure out the cost? What does Ben do?

2. Find the answers Alicia and Ben get. What do you notice?

3. If the sandwich costs $2.25 and the drink costs $.75, whose method would you use? Why?

4. Ben uses mental math to find the cost of four sandwiches. Complete his work.

   He thinks of $2.95 as $3 − $.05.

   $4 \cdot 2.95 = 4(3 − 0.05) = 4(3) − 4(0.05) = \underline{\ ?\ } − \underline{\ ?\ } = \underline{\ ?\ }$

5. Describe a way to use mental math to find the cost of four drinks.

   $4 \cdot 0.89 = \underline{\ ?\ }$

Answers to Talk it Over ·····································

1. Alicia adds the cost of the sandwich and the drink first and then multiplies by 4. Ben multiplies the cost of the sandwich by 4 and the cost of the drinks by 4, and then adds these totals together.

2. $15.36; They are the same.

3. Answers may vary. An example is given. Alicia's; $2.25 + $.75 = $3.00, and 4(3) is easier to solve than (4 · 2.25) + (4 · 0.75).

4. 12 − 0.2 = 11.80

5. Answers may vary. An example is given. Think of $.89 as $.90 − $.01:
   4(0.9 − 0.01) =
   4(0.9) − 4(0.01) = 3.56.

PLANNING

Objectives and Strands
See pages 1A and 1B.

Spiral Learning
See page 1B.

Materials List
➤ Algebra tiles

Recommended Pacing
Section 1-5 is a two-day lesson.
Day 1
Pages 31–32: Talk it Over through Distributive Property, *Exercises 1–21*
Day 2
Page 33: Combining Like Terms through Look Back, *Exercises 22–41*

Extra Practice
See pages 620–621.

Warm-Up Exercises
 Warm-Up Transparency 1-5

Support Materials
➤ Practice 5
➤ Enrichment 5 in the Activity Bank
➤ Study Guide 1-5
➤ Problem Set 2
➤ Additional Exploration 1
➤ Diagram Masters 6, 7 in the Explorations Lab Manual
 Overhead Visual 1
➤ Function Investigator with Matrix Analyzer Activity Book: Function Investigator Activity 1
➤ Using Plotter Plus: Using Tables to Check Algebra
➤ Quiz 1-5
➤ Alternative Assessments 5, 6

31

TEACHING

Talk it Over

Questions 1–3 help students to discover and understand the distributive property.

Questions 4 and 5 help students to see how the distributive property can be used to do mental math.

Additional Samples

S1 Find each product using mental math.

  a. 9(999) **8991**

  b. 12(1003) **12,036**

S2 Illustrate the expression $4(x + 1)$ using algebra tiles. Rewrite the expression without parentheses.

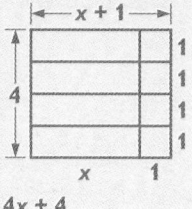

4x + 4

Using Manipulatives

Have students use algebra tiles to create more models that illustrate the distributive property. Students can discuss their models with other class members if they work together in small groups.

Sample 1

Find each product using mental math.

  a. 7(108)   **b.** 15(98)

Sample Response

a. Think of 108 as $100 + 8$.

$$7(108) = 7(100 + 8) = 7 \cdot 100 + 7 \cdot 8 = 700 + 56 = 756$$

b. Think of 98 as $100 - 2$.

$$15(98) = 15(100 - 2) = 15 \cdot 100 - 15 \cdot 2 = 1500 - 30 = 1470$$

Sample 2

Illustrate the expression $3(x + 2)$ using algebra tiles. Rewrite the expression without parentheses.

Sample Response

A rectangle of width 3 and length $(x + 2)$ has area $3(x + 2)$.

Imagine pulling the rectangle apart. What are the areas of the two smaller rectangles?

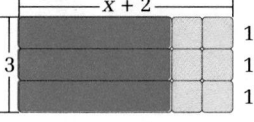 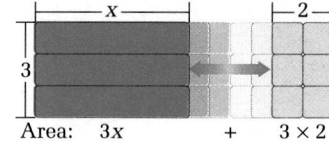 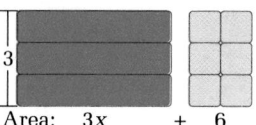

The area of the left group is $3x$. The area of the right group is 6.

The algebra tiles show that $3(x + 2) = 3 \cdot x + 3 \cdot 2 = 3x + 6$.

When you rewrite a product as a sum or a difference, you are using a mathematical rule called the *distributive property*.

DISTRIBUTIVE PROPERTY

For all numbers a, b, and c:

$$a(b + c) = ab + ac$$

$$a(b - c) = ab - ac$$

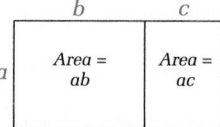

····▶ Now you are ready for:
Exs. 1–21 on pp. 34–35

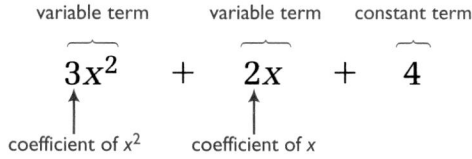

Combining Like Terms

When you add expressions, each expression is a **term** in the sum.

$$\underbrace{3x^2}_{\text{variable term}} + \underbrace{2x}_{\text{variable term}} + \underbrace{4}_{\text{constant term}}$$

coefficient of x^2 — coefficient of x

The numerical part of a variable term is called a **coefficient.**

Terms with the same variable part are called **like terms.** You use the distributive property in reverse to **combine like terms.**

variable parts are the same: $5x^2 + 3x^2 \longrightarrow (5 + 3)x^2 \longrightarrow 8x^2$

variable parts are the same: $5x + 3x \longrightarrow (5 + 3)x \longrightarrow 8x$

Talk it Over

6. How is the distributive property used in combining like terms?
7. What do you think terms with different variable parts are called?

Sample 3

Simplify $5(x + 4) - 3x$.

Sample Response

$5(x + 4) - 3x$

$5 \cdot x + 5 \cdot 4 - 3x$ ← Use the distributive property.

$5x + 20 - 3x$

$5x - 3x + 20$ ← Group like terms together.

$2x + 20$ ← Combine like terms.

Look Back

How can you use the distributive property to rewrite a product as a sum? to rewrite a sum as a product?

Show examples that use numbers and examples that use variable expressions.

Now you are ready for: Exs. 22–41 on pp. 35–36

Error Analysis

Students may add terms together that have the same variable even though the powers of the variable are not the same. Emphasize that terms with different powers are not like terms and therefore cannot be added. Also, stress that it is the distributive property that allows like terms to be added.

Talk it Over

Question 7 allows students to define unlike terms. Students should realize that terms such as $7a$ and $7b$ or $5x^2$ and $3x$ are unlike terms.

Additional Sample

S3 Simplify $4(x + 3) - 2x$.
$2x + 12$

Look Back

Ask students to write their responses to the Look Back question as journal entries. Make sure they support their statements with examples.

Answers to Talk it Over

6. Answers may vary. An example is given. When combining like terms, use the distributive property in reverse to group the coefficients of the like terms. Add the coefficients to get the new coefficient of the variable.
7. unlike terms

Answers to Look Back

Explanations may vary. An example is given. To rewrite a product as a sum, multiply each term inside the parentheses by the factor outside the parentheses and remove the parentheses. To rewrite a sum as a product, you factor out the greatest common factor of the terms and show remaining parts of the terms as a sum or a difference in parentheses.

Product as a sum:
$2(3 + 4) = 2 \cdot 3 + 2 \cdot 4 = 6 + 8$;
$9(6 - x) = 54 - 9x$

Sum as a product:
$9 + 12 = 3(3 + 4)$;
$4x + 2 = 2(2x + 1)$

Integrating the Strands

Number Exs. 1–13, 31, 36–38

Algebra Exs. 1–38

Geometry Exs. 14, 15, 39, 40

Logic and Language Exs. 20, 21, 30–32, 41

Using Manipulatives

As with Sample 2, students may wish to use algebra tiles to create their own models in conjunction with Exs. 14 and 15. Students can exchange models and write expressions for the areas.

Reasoning

Exs. 20 and 24–30 require students to not only use the distributive property and to combine like terms, but also to use these procedures to explain the reasoning behind their answers.

Using Technology

Function Investigator Activity 1 in the *Function Investigator with Matrix Analyzer Activity Book* shows students how to use a spreadsheet to check whether they have simplified a variable expression correctly. (For information about using the McDougal Littell Mathpack software, see the *Mathpack User's Guide*.)

1-5 Exercises and Problems

Rewrite each product as a sum or difference. Do not calculate.

1. $2(300 + 50)$ **2.** $5(100 - 2)$ **3.** $5(40 + 8)$ **4.** $4(1000 - 150)$

Find each product using mental math.

5. $5(205)$ **6.** $9(397)$ **7.** $7\left(4\frac{6}{7}\right)$ **8.** $3(1006)$

Use the distributive property to find each sum or difference mentally.

9. $479 \cdot 7 + 479 \cdot 3$ **10.** $280 \cdot 40 - 80 \cdot 40$

11. $250 \cdot 17 - 250 \cdot 7$ **12.** $64 \cdot 32 + 64 \cdot 68$

13. The four walls of a room are to be painted.

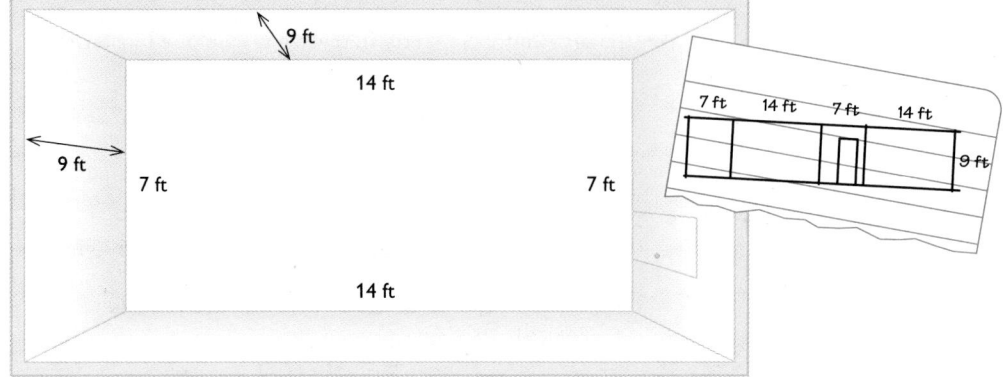

   a. Explain how you could find the total area of the four walls.

   b. Complete this extension of the distributive property:

$$a(b + c + d + e) = \underline{} + \underline{} + \underline{} + \underline{}$$

Write two expressions for the area covered by the tiles, one with parentheses and one without parentheses.

14.

15.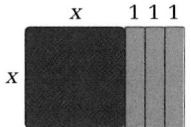

Use the distributive property to rewrite each expression without parentheses.

16. $2(l + w)$ **17.** $3(2a - b)$ **18.** $\frac{1}{3}(3x^2 - 12x)$ **19.** $\frac{1}{5}(20t^2 + 5)$

20. **Writing** When Hector did Exercises 16 and 17, he wrote

$$2(l + w) = 2l + w$$
$$3(2a - b) = 6a - b$$

Explain his mistakes and how to correct them.

Answers to Exercises and Problems

1. $600 + 100$

2. $500 - 10$

3. $200 + 40$

4. $4000 - 600$

5. 1025 **6.** 3573

7. 34 **8.** 3018

9. 4790 **10.** 8000

11. 2500 **12.** 6400

13. a. Answers may vary. Examples are given. Add together the area of each wall, or find the perimeter of the floor and multiply by the height.

   b. $ab + ac + ad + ae$

14. $2(2x + 1); 4x + 2$

15. $x(x + 3); x^2 + 3x$

16. $2l + 2w$ **17.** $6a - 3b$

18. $x^2 - 4x$ **19.** $4t^2 + 1$

20. Hector did not multiply the second term in the parentheses by the factor outside the parentheses. In the first expression, the coefficient of w should be 2. In the second expression, the coefficient of b should be 3.

21. A tile design with area $x^2 + 2x + 1$ is used four times. The new area is $4(x^2 + 2x + 1)$. Use the distributive property to rewrite this expression without parentheses. Show how your answer matches the design.

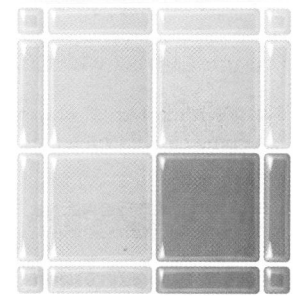

Insert parentheses to make the statement true.

22. $3x + 2 = 3x + 6$

23. $6x + 6x + 1 = 12x + 6$

Combine like terms. If you cannot, explain why.

24. $3y + 5y$

25. $3z^2 + 5z + 3$

26. $4x + 2x^3 + 3x$

27. $7 + 3(y + 4) + 6y$

28. $6y + 3(y + 4) - 7$

29. $6(y + 7) + 3(y + 4)$

30. **Writing** Alexandra writes these statements:

$$5x^2 + 3x^2 = 8x^4$$
$$2x^2 + 4x = 6x^3$$

Explain her mistakes and how to correct them.

31. **Group Activity** Have each member of your group start with a different number and follow the steps.

1) Starting number: __?__

2) Add 8.

3) Multiply by 3.

4) Subtract 12.

5) Multiply by $\frac{1}{3}$.

6) Ending number: __?__

a. Make a table showing starting and ending numbers.

b. What pattern do you notice?

c. Suppose the starting number is x. Write a variable expression for every step along the way. Simplify each expression as much as possible.

d. If the starting number is x, what is an expression for the ending number?

e. If the ending number is y, what is an expression for the starting number?

1-5 Modeling the Distributive Property

35

Answers to Exercises and Problems

21. $4x^2 + 8x + 4$; In the design, there are 4 x^2-tiles, 8 x-tiles, and 4 unit tiles.

22. $3(x + 2) = 3x + 6$

23. $6x + 6(x + 1) = 12x + 6$

24. $8y$

25. They cannot be combined because there are no like terms.

26. $2x^3 + 7x$

27. $9y + 19$

28. $9y + 5$

29. $9y + 54$

30. Explanations may vary. An example is given.
Alexandra is adding the exponents. When combining like terms you do not add exponents. Therefore, $5x^2 + 3x^2 = 8x^2$ and $2x^2 + 4x$ stays the same because there are no like terms.

31. a. Tables may vary. An example is given.

Starting number	Ending number
4	8
6	10
8	12
0	4

b. The ending number is always 4 more than the starting number.

c. x; $x + 8$; $3x + 24$; $3x + 12$; $x + 4$

d. $x + 4$ e. $y - 4$

35

Practice 5 For use with Section 1-5

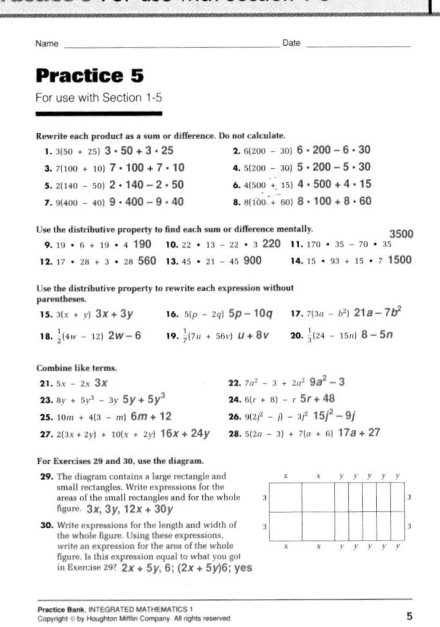

32. **Writing** Explain how the distributive property can be used to perform some calculations mentally.

Evaluate each expression when $x = 9$. *(Section 1-4)*

33. $x^2 + 2x + 1$ **34.** $4x^2 + 4x + 1$ **35.** $9x^2 + 12x + 4$

Write as a power of ten. *(Section 1-3)*

36. $10 \cdot 10$ **37.** $10^7 \cdot 10^3$ **38.** $\dfrac{10^7}{10^3}$

Write two expressions for the perimeter of each figure, one with parentheses and one without parentheses. *(Section 1-2)*

39. 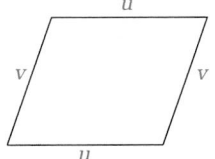   **40.**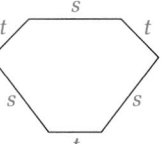

 ● ● ● ● ● ● **Working on the Unit Project**

41. Find three different logos for the same kind of product, company, or organization. How are the logos alike? How are they different?

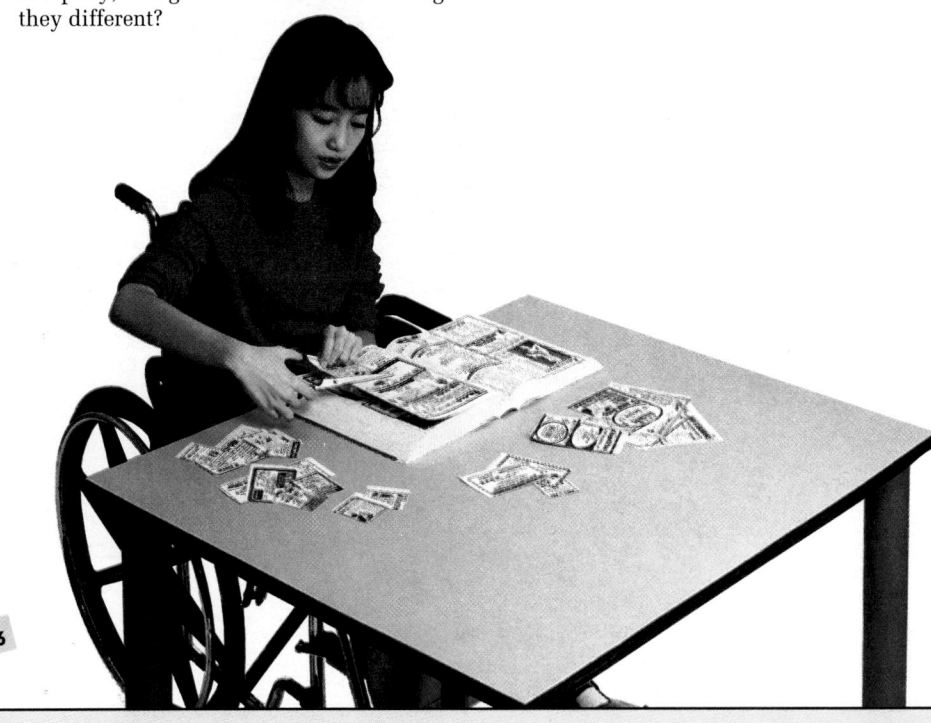

36

Answers to Exercises and Problems

32. Explanations may vary. An example is given. In a product, the distributive property can be used to break one of the numbers down into a sum or difference of two numbers which are easier to multiply, mentally, by the other number in the product. For example, think of 3(95) as

3(100 − 5), which is 300 − 15, or 285.

33. 100

34. 361

35. 841

36. 10^2

37. 10^{10}

38. 10^4

39. $2(v + u); 2v + 2u$

40. $3(s + t); 3s + 3t$

41. Answers may vary.

Section

1-6

Focus
Create and identify congruent polygons and name corresponding parts of congruent polygons.

Working Together on Congruent Polygons

MATCHED SETS

People often work together on projects like painting a mural. You may be surprised at how much you can do when you work with other people.

EXPLORATION

How many ways can you divide a square into four identical pieces?

* **Materials: dot paper or graph paper**
* **Work with another student.**

① Use a square made with 25 dots.

② Divide the square into four pieces that have the same size and shape. Use only straight lines that connect dots.

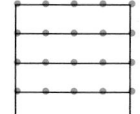 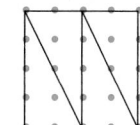 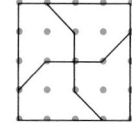

③ Draw as many different designs as you can, each in its own square.

④ Stop after five minutes.

1-6 Working Together on Congruent Polygons **37**

Answers to Exploration

1–4. Designs may vary. Examples are given.

PLANNING

Objectives and Strands
See pages 1A and 1B.

Spiral Learning
See page 1B.

Materials List
➤ Dot paper

Recommended Pacing
Section 1-6 is a two-day lesson.
Day 1
Pages 37–39: Opening paragaraph through Talk it Over, *Exercises 1–12*
Day 2
Pages 39–40: Describing Polygons through Look Back, *Exercises 13–32*

Extra Practice
See pages 620–621.

Warm-Up Exercises
Warm-Up Transparency 1-6

Support Materials
➤ Practice 6
➤ Enrichment 6 in the Activity Bank
➤ Study Guide 1-6
➤ Problem Set 2
➤ Diagram Masters 1, 9 in the Explorations Lab Manual
➤ Quiz 1-6

TEACHING

Talk it Over

1. How many designs did you and your partner find in five minutes?

2. Were you surprised at the number you found?

3. Compare these two designs. Do you think they should be considered the same or different? Explain.

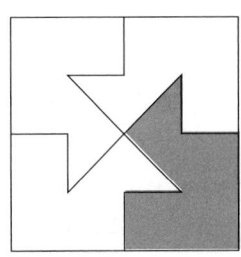 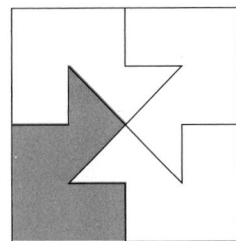

Congruent Figures

Two figures that have the same size and shape are called **congruent.** If you can slide, turn, or flip a figure so it covers another figure exactly, then the two figures are congruent.

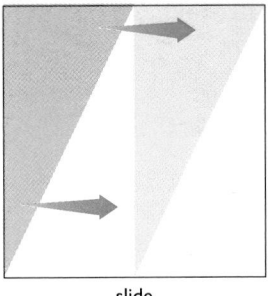
slide

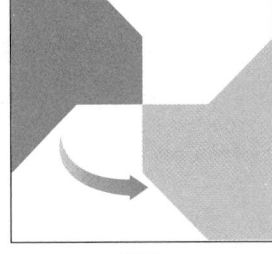
turn

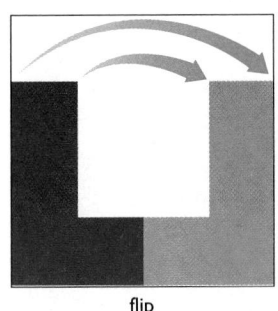
flip

Talk it Over

4. What movement will show that the blue shape is congruent to the red shape?

   a.   b.   c.

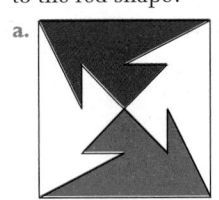 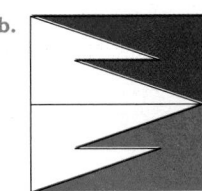 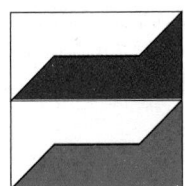

5. Are the shaded shapes in question 3 congruent?

Unit 1 Exploring and Communicating Mathematics

Answers to Talk it Over

1–2. Answers may vary.

3. Answers may vary. An example is given. I think they should be considered the same because if either one is flipped over, it will match the other one exactly. Also, the pieces are the same shape and size.

4. a. turn
   b. flip
   c. slide

5. Yes.

Can you work with others to find new ways to divide the square?

• **Materials: dot paper or graph paper**
• **Work in a group of four students.**

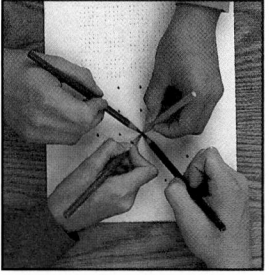 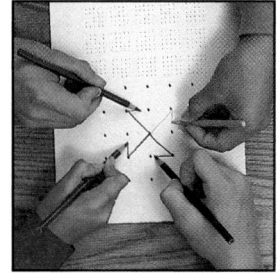 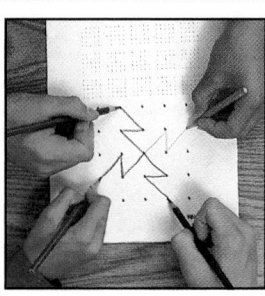

(1) Use a square made with 25 dots that measures 4 in. on each side.

(2) Start with all your pencils in the center of the square.

(3) Choose a leader to follow. Copy the leader's movements in your part of the square.

(4) Choose a new leader and make another design. Repeat.

(5) Stop after ten minutes.

(6) Shade one piece in each design you made.

(7) Tape your designs to a wall next to the other groups' designs.

Talk it Over

6. In the Exploration, how many different shaded pieces did the class find?

7. Do you think the class has found all the possible pieces?

8. How would you describe the shapes of the shaded pieces?

······► Now you are ready for:
Exs. 1–12 on p. 41

Describing Polygons

The straight-sided shapes you shaded in the Explorations are **polygons.**

3-gon	4-gon	5-gon	6-gon	7-gon	8-gon
	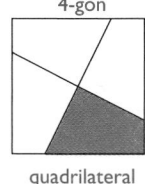	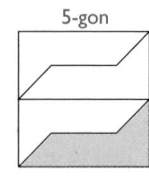	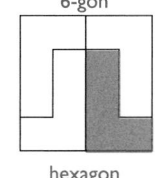		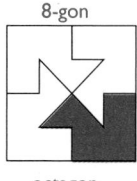
triangle	quadrilateral	pentagon	hexagon	heptagon	octagon

1-6 Working Together on Congruent Polygons

39

Exploration

The goal of the second Exploration is to demonstrate that congruent figures can be drawn by connecting dots in an identical way. This is a good activity because it encourages cooperation among students. Students should be able to tell easily if their figures are not congruent. Encourage them to start over if they are not drawing congruent figures.

Communication: Reading

The text at the bottom of this page and the top of page 40 contains terms that may be new to students. These terms are necessary to work successfully with polygons. Make certain students understand these terms. Also note the use of the congruence symbol, ≅, in the labels on the diagram.

Answers to Exploration

1–7. Designs and shadings may vary. Examples are given.

Answers to Talk it Over

6–8. Answers may vary.

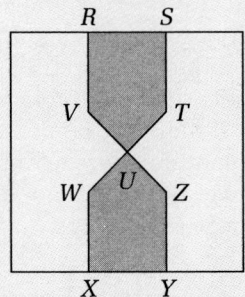
To name polygons, write a different letter at each **vertex** (corner) and list consecutive vertices in order. (*Vertices* is the plural of *vertex*.) Each **side** of the polygon is named with two consecutive vertices. Two sides that have the same length are called **congruent sides.**

The ≅ symbol means "is congruent to."

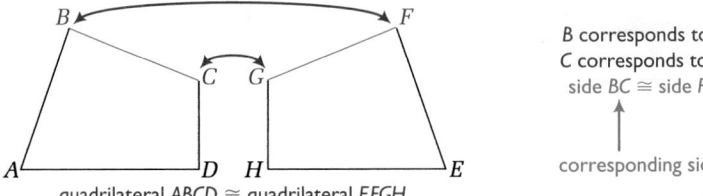

B corresponds to F

C corresponds to G

side BC ≅ side FG

corresponding sides

quadrilateral *ABCD* ≅ quadrilateral *EFGH*

In congruent polygons, matching vertices are called **corresponding vertices** and matching sides are called **corresponding sides.**

When you name congruent polygons, list the corresponding vertices in the same order.

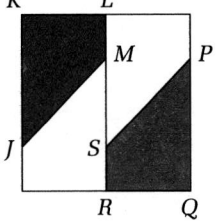

Sample

a. **Name the congruent polygons.**

b. **Name three pairs of congruent sides.**

Sample Response

a. quadrilateral *JKLM* ≅ quadrilateral *PQRS*

b. side *JK* ≅ side *PQ*; side *KL* ≅ side *QR*; side *LM* ≅ side *RS*

You can use tick marks or letters on diagrams, or written statements near the diagrams, to indicate congruent sides.

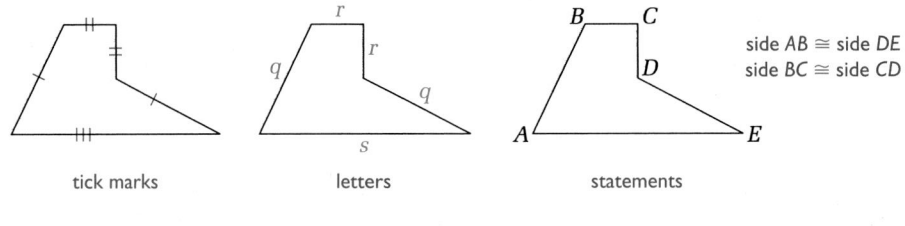

side AB ≅ side DE

side BC ≅ side CD

tick marks   letters   statements

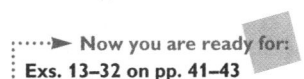 **Now you are ready for:**
Exs. 13–32 on pp. 41–43

Look Back ◄

Why is it helpful to name congruent polygons by listing corresponding vertices in order?

 40

Unit 1 Exploring and Communicating Mathematics

Exercises and Problems

What kind of movement (*slide, turn,* or *flip*) shows that the blue polygon is congruent to the red polygon?

1. 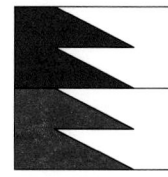   **2.** 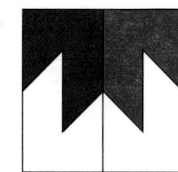   **3.**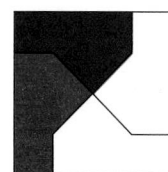

Are the letters congruent? If so, what kind of movement (*slide, turn,* or *flip*) shows this?

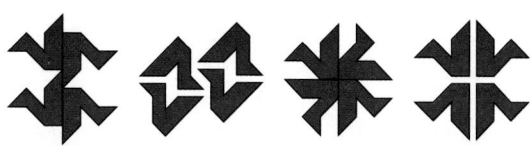

4. u n   **5.** b d   **6.** p q

For Exercises 7–11, use the designs with four congruent pieces that you made in the Explorations.

7. List some ways in which working in a group helped you in the Explorations.

8. Draw a triangle, square, and rectangle that you found in the Explorations. What other names of polygons do you know?

9. **Open-ended** Choose one of your designs. Cut out the four congruent pieces and rearrange them into a shape that is not a square. Tape your design to a separate piece of paper.

10. **Group Activity** Choose one of your designs and make an 8 in. by 8 in. version. Color your design. Join it to the squares made by your classmates to create a large mural.

11. Describe the congruent shapes you see in the quilt at the right. What slides, turns, or flips would you use to show they are congruent?

12. Choose one of your Exploration designs and repeat it to create a quilt pattern. You may want to flip or turn some of the squares for variety. Color your design.

13. Cut out two congruent polygons from the Exploration designs. Tape them to a piece of paper and label the vertices. Name the corresponding vertices. List three pairs of congruent sides.

APPLYING

Suggested Assignment
Standard 1–9, 11–24, 26–32
Extended 1–9, 11–24, 26–32

Integrating the Strands
Algebra Exs. 26–31
Geometry Exs. 1–6, 8–25, 30–32
Logic and Language Exs. 7, 14

Using Manipulatives
For Exs. 1–3, you may wish to have students use cutouts of polygons to demonstrate slides, turns, and flips. Ask students to trace around their cutouts at the original location and then trace around them again to show the final location.

Communication: Writing
In connection with Ex. 7, you might suggest that students write a journal entry to explain the advantages of working in a group. They might also record if working in a group helped them to understand the idea of congruence.

Research
Use Ex. 11 to encourage those students interested in quilt designs to research quilt making. Have students describe the congruent figures they see as slides, turns, or flips.

Answers to Exercises and Problems

1. slide

2. flip

3. turn

4. Yes; turn.

5. No.

6. Yes; flip.

7. Answers may vary. An example is given. Someone in a group may suggest an idea, a method, or a solution that you would not have suggested. Someone in a group might correct a mistaken idea of yours and set you back on a correct solution path. Working in a group allows you to get more answers to a problem situation.

8. Answers may vary. Examples are given.

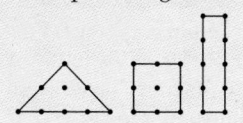

quadrilateral, pentagon, hexagon, octagon

9–13. Answers may vary.

14. **Graphic Design** Arrange four congruent polygons in a way that expresses one of these words: *order, disorder, danger, fun.* Explain why your design expresses the word.

Here are some examples.

order

disorder

danger

fun

15. **Reading** What are some ways to show that sides of a polygon have the same length?

Mark congruent lengths by using tick marks.

16.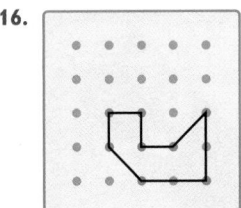

17.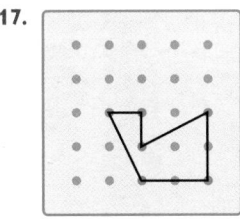

18.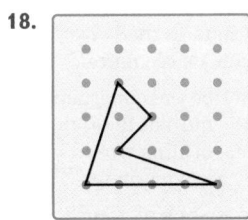

Mark congruent lengths by using letters. Then write a variable expression for the perimeter of the polygon.

19.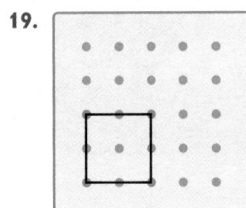

20.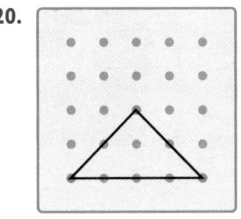

21.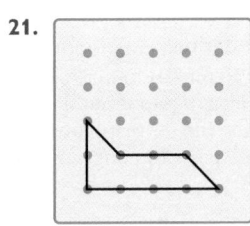

Answers to Exercises and Problems

14. Answers may vary.

15. Answers may vary. An example is given. You can use tick marks, letters on the diagram, or written statements to show that sides have the same length.

16.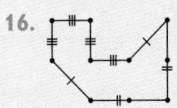

17.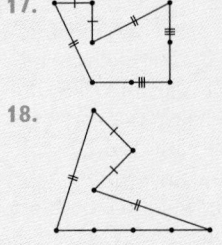

18.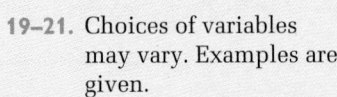

19–21. Choices of variables may vary. Examples are given.

19. $; 4s$

20. $; 2t + r$

21. 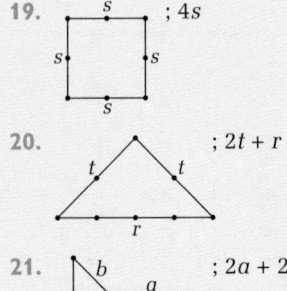 $; 2a + 2b + c$

Name the congruent polygons. Then list three pairs of corresponding sides.

22.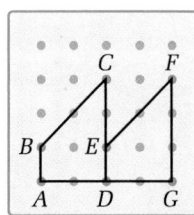

23.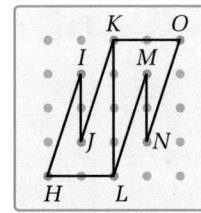

24.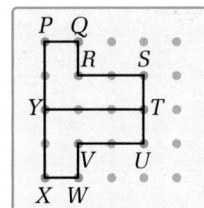

 Ongoing **ASSESSMENT**

25. **Group Activity** Organize the polygons that you found in the Explorations based on the number of sides of each polygon.

Review **PREVIEW**

Rewrite without parentheses. *(Section 1-5)*

26. $5(2x + 3)$

27. $6(y^2 - 3y + 4)$

Evaluate according to the order of operations. *(Section 1-4)*

28. $6 - 3 \cdot 2$

29. $8 \div 4 - 2$

Write a variable expression for the perimeter of each polygon. *(Section 1-2)*

30.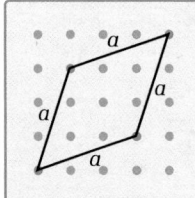

31.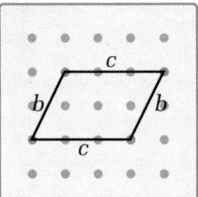

Working on the Unit Project

32. Find a logo that includes congruent pieces. Describe the slides, turns, or flips that show the pieces are congruent.

Why do you think congruent pieces are used in the logo?

The recycle logo uses three congruent arrow shapes.

A turn moves one arrow onto another.

The cycle of arrows represents an endless reuse of materials.

Find out what a Möbius Strip is. Is the recycle logo an example?

1-6 Working Together on Congruent Polygons

43

Assessment: Standard

Ex. 25 assesses students' knowledge of types of polygons. Groups should use the names of polygons. You may wish to have each group name any "missing" polygons and describe them by giving the number of sides.

Working on the Unit Project

For Ex. 32, it may be helpful if students first generate a list of commonly known corporations, television advertisers, automobile companies, and so on, that have easily identifiable logos. They can then choose a logo from the list.

Practice 6 For use with Section 1-6

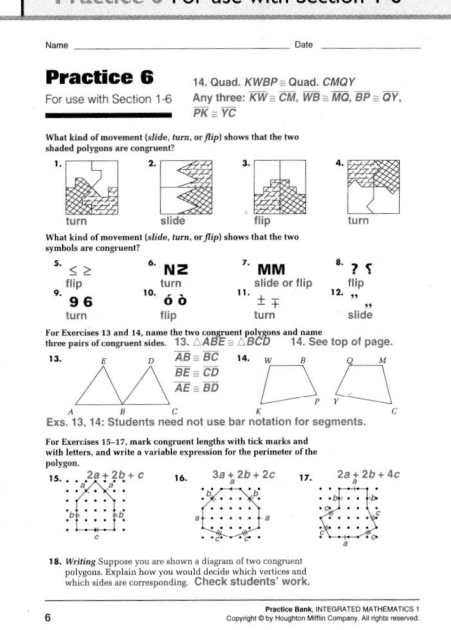

Answers to Exercises and Problems

22–24. Answers may vary. Examples are given.

22. quadrilateral *ABCD* ≅ quadrilateral *DEFG*; side *AB* corresponds to side *DE*, side *BC* corresponds to side *EF*, side *CD* corresponds to side *FG*

23. pentagon *HIJKL* ≅ pentagon *ONMLK*; side *HI* corresponds to side *ON*, side *IJ* corresponds to side *NM*, side *JK* corresponds to side *ML*

24. hexagon *PQRSTY* ≅ hexagon *XWVUTY*; side *PQ* corresponds to side *XW*, side *QR* corresponds to side *WV*, side *RS* corresponds to side *VU*

25. Answers may vary.

26. $10x + 15$

27. $6y^2 - 18y + 24$

28. 0

29. 0

30. $4a$

31. $2b + 2c$

32. Answers may vary.

PLANNING

Objectives and Strands
See pages 1A and 1B.

Spiral Learning
See page 1B.

Materials List
➤ Paper
➤ Scissors
➤ Ruler
➤ Toothpicks

Recommended Pacing
Section 1-7 is a two-day lesson.
Day 1
Pages 44–45: Exploration through the Sample, *Exercises 1–20*
Day 2
Page 46: Personal Writing through Look Back, *Exercises 21–36*

Extra Practice
See pages 620–621.

Warm-Up Exercises
Warm-Up Transparency 1-7

Support Materials
➤ Practice 7
➤ Enrichment 7 in the Activity Bank
➤ Study Guide 1-7
➤ Problem Set 2
➤ McDougal Littell Mathpack software: *Geometry Inventor*
➤ Geometry Inventor Activity Book: Activity 29
➤ Quiz 1-7
➤ Test 2
➤ Alternative Assessments 7–9

Section 1-7

Exploring Quadrilaterals and Symmetry

PUTTING IT TOGETHER

Focus
Build quadrilaterals from congruent triangles and write expression for their perimeters, find lines of symmetry, and communicate ideas through writing.

EXPLORATION

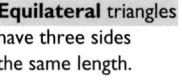 *kinds of quadrilaterals can you make with two congruent triangles?*

- **Materials: paper, scissors, ruler**
- **Work in a group of four students.**

① Use a ruler to draw two congruent triangles.
   Each of you should create a different kind of triangle: *equilateral, isosceles, scalene,* and *right*.

② Cut out your triangles and write the lengths of the sides on the front and back.

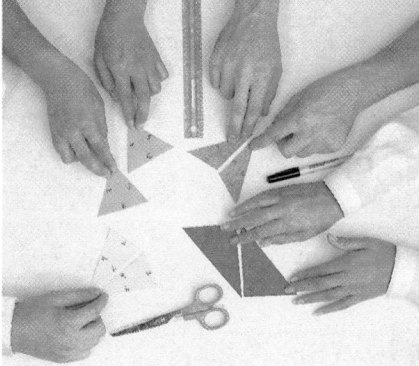

Equilateral triangles have three sides the same length.

Scalene triangles have no sides the same length.

Isosceles triangles have two sides the same length.

Right triangles have one right angle.

small squares show perpendicular sides **(right angles)**

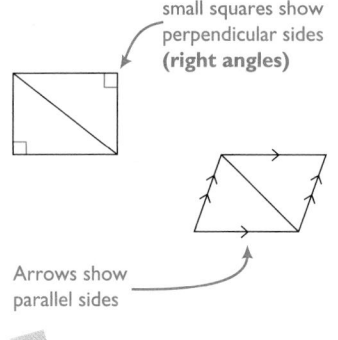

Arrows show parallel sides

③ Make as many different polygons as you can by joining your two triangles so that congruent edges match.

④ Draw each polygon you make. Use right angle marks to show **perpendicular** sides. Use arrows to show **parallel** sides.

⑤ Record the perimeter of each polygon you make.

⑥ Write a summary of the findings of your group. Include answers to these questions:
   How does the type of triangle you use affect the number of different polygons you can make? How does it affect the number of different perimeters you get?

Unit 1 Exploring and Communicating Mathematics

44

Answers to Exploration

1–5. Polygons may vary.

6. Summaries may vary. Examples of the different polygons that can be made by different triangles are given. With equilateral triangles, you can form rhombuses. With isosceles triangles, you can form parallelograms, kites, rhombuses, and squares. With scalene triangles, you can form parallelograms and kites. With right triangles, you can form isosceles triangles, parallelograms, kites, rectangles, and squares. Conclusions about perimeters may vary.

In the Exploration, you may have made these special quadrilaterals:

parallelogram **kite** **rhombus** **rectangle** **square**

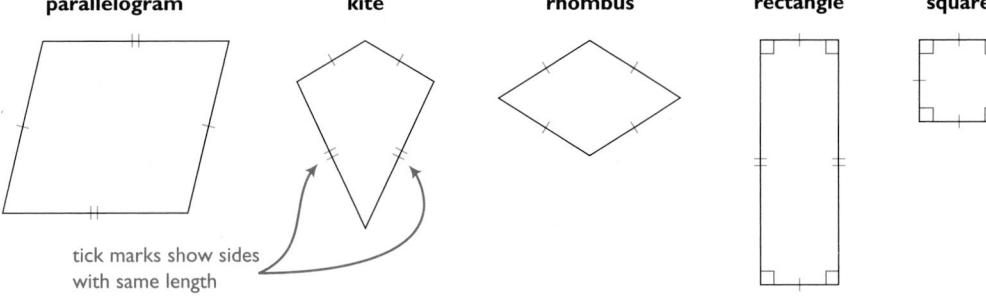

tick marks show sides with same length

If you can fold a polygon so that one half matches the other, the polygon has **symmetry**. The fold line is called a **line of symmetry**.

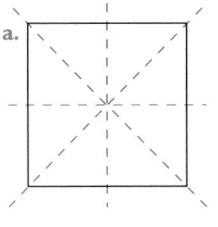

Talk it Over

1. What are some features of a parallelogram?
2. What are some features of a kite?
3. Is a rhombus a parallelogram? Is a rhombus a kite?
4. Which of your polygons have no lines of symmetry?
5. Which of your polygons have one line of symmetry?
6. Which of the polygons that you made have two lines of symmetry?

Sample

Draw the lines of symmetry for each polygon.

a. 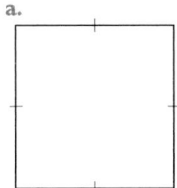   b.    c.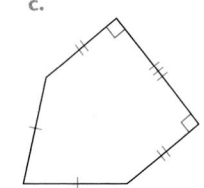

Sample Response

a.   b. 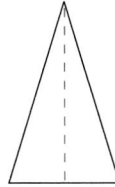   c.

····► Now you are ready for:
Exs. 1–20 on pp. 47–48

1-7 Exploring Quadrilaterals and Symmetry   **45**

Answers to Talk it Over

1. Answers may vary. An example is given. Parallelograms have opposite sides that are parallel and congruent.

2. Answers may vary. An example is given. Kites have two pairs of adjacent sides that are congruent.

3. Yes. Yes.

4. parallelograms that are not rhombuses

5. kite, isosceles triangle

6. rhombus, rectangle; The square has four lines of symmetry.

TEACHING

Exploration

After completing the Exploration, students should have an understanding of the various types of triangles defined here and should also have discovered a number of the quadrilaterals shown at the top of this page. Students should also understand the concepts of parallel sides and perpendicular sides, as well as how to indicate them on a diagram.

Talk it Over

Questions 1–6 familiarize students with the features of a parallelogram, kite, and rhombus, and introduce the concept of symmetry. Students should be aware of the features of a rectangle and a square already, but you might wish to review them at this time.

Problem Solving

Have students use the strategy of making a table to organize the properties of quadrilaterals.

Additional Sample

Draw the lines of symmetry for each polygon.

a.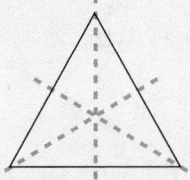

b.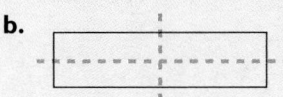

c.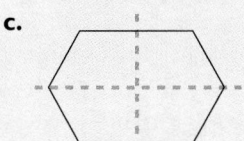

45

Communication: Writing

The communication skills of writing, reading, discussion, listening, and drawing are important in studying and learning mathematics. They play a major role in this course through the use of projects, Explorations, Talk it Over questions, and exercises and problems that encourage students to use and develop these skills. The text on this page introduces students to the ideas of personal writing and communicating ideas to others.

Limited English Proficiency

Offer students acquiring English the option of either writing down their ideas in their first language, or dictating them to a student who is fluent in English.

Look Back

The Look Back question allows students an opportunity to use the communication ideas presented on this page to write about their work in the Exploration. ....•

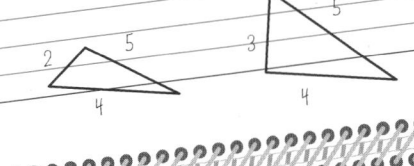

NOTES ON EXPLORATION
My triangels were 2 inchs, 4 inchs, and 5 inchs and I got shapes with different perimters like 12, 14 and 18 inchs. That's all I got. When Sondra used triangles like 3, 4, 5 she got perimters like 14, 16, 18 which are all in a row maybe 'cause 3, 4, 5 are all in a row?

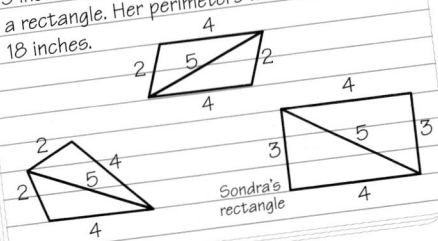

REPORT ON EXPLORATION

I made quadrilaterals by joining two congruent triangles together. My triangles had edges of length 2, 4, and 5 inches. I got parallelograms and kites. Only three perimeters were possible: 12 inches, 14 inches, and 18 inches.
When Sondra used triangles with 3, 4, and 5 inch edges, she got other shapes, including a rectangle. Her perimeters were 14, 16, and 18 inches.

Sondra's rectangle

Personal Writing

A good way to explore a problem is to write down your ideas. At first, do not worry about making it look good.

Jot down notes and ideas.

Describe the steps you try.

Write about patterns you notice.

List unanswered questions.

You can fix spelling and grammar later.

Communicating With Others

Writing or speaking to others requires more careful work. Others may expect to see words, symbols, and diagrams used in conventional ways.

Describe the problem or activity.

Use standard mathematical words that others will understand.

Organize your ideas and results clearly.

Include labeled drawings with appropriate symbols.

Talk it Over

7. In the first writing sample, the writer uses nonstandard spelling and grammar. Give some examples.

8. In the second writing sample, the writer uses mathematical words correctly. Give some examples.

9. There is an idea in the first writing sample that is not developed in the second sample. What is it? Discuss how it could be developed.

10. Compare the use of diagrams in the two writing samples.

Look Back ◄

In the Exploration, was everyone in your group able to make shapes with symmetry? Explain why or why not.

► Now you are ready for:
Exs. 21–36 on pp. 48–49

46　　**Unit 1** Exploring and Communicating Mathematics

Answers to Talk it Over

7–10. Answers may vary Examples are given.

7. Several words ("triangel," "inchs," and "perimters") are misspelled, and "maybe 'cause" is not standard grammar.

8. The writer uses the terms "quadrilateral," "congruent," "parallelogram," "kites," "perimeters," "triangles," and "rectangles" correctly.

9. The first report mentions the fact that when working with the triangle that had side lengths which

were consecutive numbers (3, 4, 5), the perimeters of the figures made were consecutive even numbers (14, 16, 18). This idea could be developed inductively by testing other triangles with consecutive numbers as side lengths. It could be developed deductively by testing a general triangle with consecutive sides a, $a + 1$, and $a + 2$.

10. The first report simply shows the two different triangles with no pictures of the shapes formed from these triangles. In the second report, the writer shows the triangles forming quadrilaterals, and labels the sides to show which congruent sides were joined. This gives the reader a visual summary of the shapes formed in the triangle exercise.

Answers to Look Back

Yes. Explanations may vary. An example is given. The only shape with no symmetry is a parallelogram. All the triangles which formed parallelograms also formed at least one other shape that does have symmetry.

46

Exercises and Problems

Exercises 1–6 are based on the Exploration. The table shows the results from one group's work.

1. Name the type of triangle that each student used.

2. Who was able to make the greatest number of different perimeters from two congruent triangles?

3. What perimeters can you make if you build quadrilaterals from two triangles with sides of length 5, 6, and 7?

	Sides of triangles	Perimeters made
Jeanine	4 4 4	16
Alain	4 5 5	18, 20
Helen	3 5 6	16, 18, 22
Duane	3 4 5	14, 16, 18

4. Make a quadrilateral using two equilateral triangles with sides of length a, a, and a. Draw a diagram. Write a variable expression for its perimeter.

5. Make two different quadrilaterals using two isosceles triangles with sides of length a, b, and b. Draw diagrams. Write variable expressions for their perimeters.

6. Make three different quadrilaterals using two scalene triangles with sides of length a, b, and c. Draw diagrams. Write variable expressions for their perimeters.

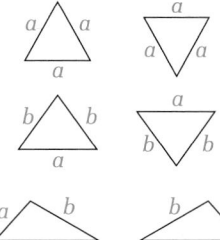

7. A quadrilateral with one pair of parallel sides is called a *trapezoid*.

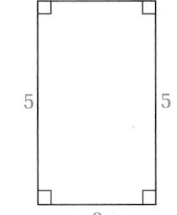

You can make a trapezoid by joining three congruent triangles.

a. Draw two other trapezoids that you can make with the triangles shown.

b. Find the perimeter of each of your trapezoids.

What name best describes each quadrilateral?

8. 9. 10. 11.

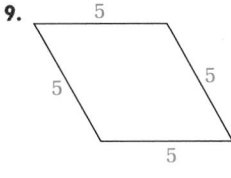 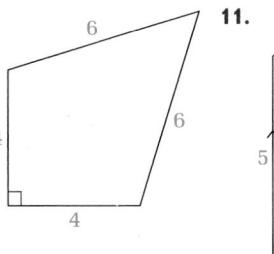

Answers to Exercises and Problems

1. Jeanine: equilateral (also isosceles); Alain: isosceles; Helen: scalene; Duane: right, scalene

2. Helen and Duane

3. 22, 24, 26

4. perimeter = $4a$

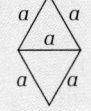

5, 6. See answers in back of book.

7. Trapezoids and perimeters may vary. Examples are given.

a.

b. 20; 18

8. rectangle

9. rhombus

10. kite

11. parallelogram

Suggested Assignment

Standard 1–16, 18–20, 24–36

Extended 1–36

Integrating the Strands

Number Ex. 23

Algebra Exs. 4–6, 21, 30–35

Geometry Exs. 1–22, 25, 28, 29, 36

Statistics and Probability Ex. 24

Logic and Language Exs. 21, 22, 25–27

Cooperative Learning

Students can work in groups on Exs. 1–6. They should discuss their work and record the answers to all exercises for discussion with other groups.

Mathematical Procedures

Ex. 7 introduces the concept, but not the definition, of trapezoid. You may wish to have students offer their own definitions of this figure.

Reasoning

Students can understand quadrilaterals better by considering questions such as the following:

1. Which quadrilaterals are parallelograms?
2. Which quadrilaterals are not parallelograms?
3. Is every square also a rhombus?
4. Is every rhombus a square?
5. Is every square a rectangle?
6. Is every rectangle a square?
7. Do the nonparallel sides of a trapezoid have to be equal in length?

Students should give reasons to support their answers.

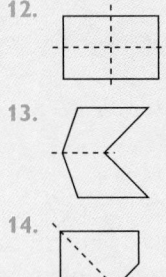
Draw all lines of symmetry for each figure or write *no symmetry*.

12. 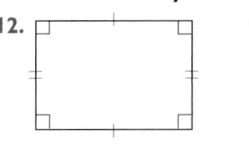   **13.** 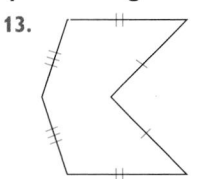   **14.** 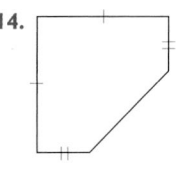   **15.**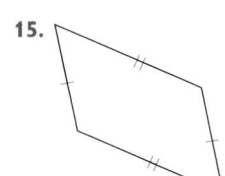

16. How many lines of symmetry does a circle have?

17. **Open-ended** List some things you see every day that have symmetry.

Write what you know about each figure, based on the marks and labels in the diagram.

18. 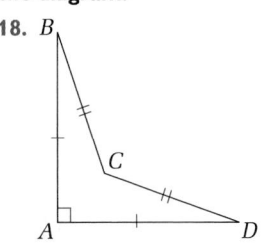   **19.** 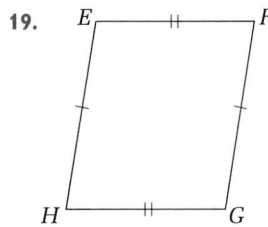   **20.**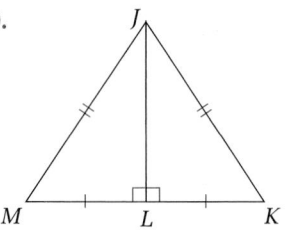

Writing Do some personal writing about each activity. Include drawings if they are helpful.

21. Make three congruent scalene triangles with sides of length a, b, and c. Build three different trapezoids with them. Draw and label the diagrams. Then write a variable expression for the perimeter of each trapezoid.

22. You can create different rectangles with the same perimeter. Choose at least three different perimeters and use toothpicks to make rectangles with these perimeters. How does the number of toothpicks affect the number of rectangles you can make?

48

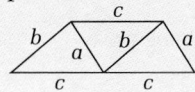
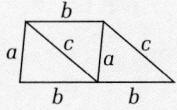
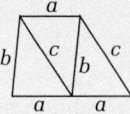

23. List some numbers that are the sum of two or more consecutive numbers. (For example, **7** = 3 + 4 and **26** = 5 + 6 + 7 + 8.) What counting numbers cannot be expressed as the sum of two or more consecutive numbers?

Writing For Exercises 24 and 25, write an answer that someone younger than you could understand.

24. How do you find the average of a set of numbers?

25. Can you draw a triangle with no line of symmetry? exactly one line of symmetry? exactly two lines of symmetry? exactly three lines of symmetry? Use diagrams to explain your answers.

26. Choose a problem from an earlier section of this unit. Write a letter to a friend explaining how you worked on the problem.

27. Explain how to predict the perimeters of the polygons you can make from two congruent triangles if you know the lengths of the sides of the triangles. You may want to use the variable expressions you wrote in Exercise 6.

Ongoing **ASSESSMENT**

28. **Open-ended** Draw a quadrilateral with the given number of lines of symmetry, if possible. If the quadrilateral has a special name, write it.

   a. exactly one   **b.** exactly two

   c. exactly three   **d.** exactly four

Review **PREVIEW**

29. Draw one of the polygons you built in the Exploration and label the vertices. Show all the congruent sides in some way. *(Section 1-6)*

Combine like terms. *(Section 1-5)*

30. $a + c + b + c + c$ **31.** $a + b + a + b$

Write each power of ten as a number and in words. *(Section 1-3)*

32. 10^2 **33.** 10^3 **34.** 10^6 **35.** 10^9

 Working on the Unit Project

36. Find some logos that have symmetry and others that do not. Why do some companies use symmetry in their logos?

Assessment: Task

Ex. 28 assesses students' knowledge of symmetry and quadrilaterals. Students can make generalizations about types of quadrilaterals and the number of lines of symmetry that they have.

Working on the Unit Project

For Ex. 36, students can look in magazines for pictures of logos. Have students show their logos to the other members of the class.

Quick Quiz (1-5 through 1-7)

See page 51.

Practice 7 **For use with Section 1-7**

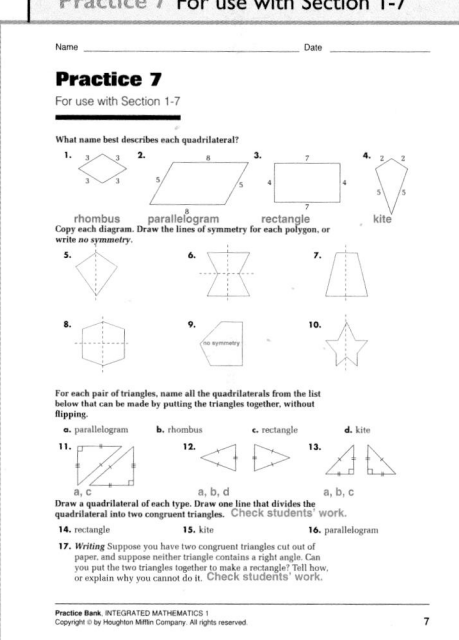

Answers to Exercises and Problems

25. Yes, a scalene triangle has no line of symmetry. Yes, an isosceles triangle that is not equilateral has exactly one line of symmetry. No, no triangle has exactly two lines of symmetry. Yes, an equilateral triangle has exactly three lines of symmetry.

26. Letters may vary.

27. Explanations may vary. An example is given. From Exercise 6, we know that for congruent scalene triangles with sides a, b, and c, the perimeters of all possible quadrilaterals are $2a + 2b$, $2a + 2c$, and $2b + 2c$. These perimeters will hold true for any type of triangle, for if the triangles are not scalene then we know that if the triangles are isosceles then $a = b$ or $a = c$ or $b = c$ or if the triangles are equilateral we know $a = b = c$.

28. a. kite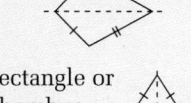

    b. rectangle or rhombus (not a square)

    c. not possible

    d. square

29. Answers may vary.

30. $a + b + 3c$

31. $2a + 2b$

32. 100; ten squared or one hundred

33. 1000; ten cubed or one thousand

34–36. See answers in back of book.

Completing the Unit Project

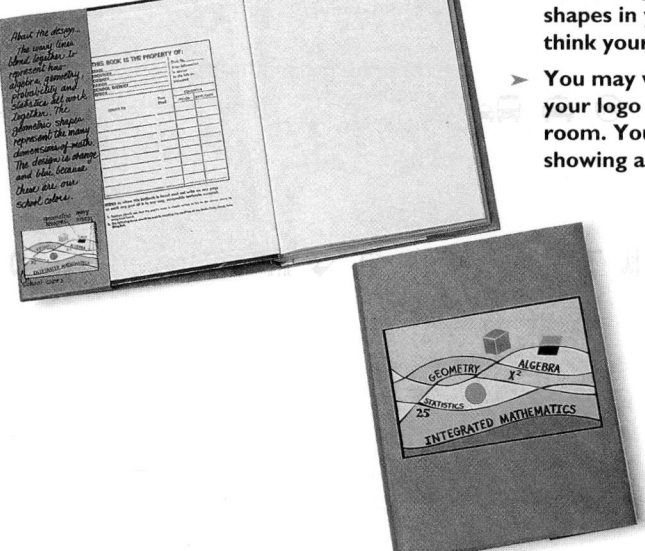

➤ Now you are ready to make your book cover.
Your group can work together to make one
cover to display or three individual versions
of your group's logo design.

➤ On the inside flap of your book cover, explain
what the patterns, symbols, numbers, and
shapes in your logo represent. Tell why you
think your design identifies your math class.

➤ You may want to make an enlargement of
your logo to display on a poster in your classroom. You may decide to make a bookmark
showing a smaller version of your logo.

Look Back

Think about how your
group worked together on
the project. What would you
recommend to a group just
starting to work on this project to improve their team-
work or their results?

Alternative Projects

Project 1: The Fibonacci Sequence

The list of numbers 1, 1, 2, 3, 5, 8, 13, 21, . . . is called the *Fibonacci sequence.*
Describe this number pattern. Research the Fibonacci sequence and write a
report about it. Describe how this pattern is displayed in nature.

Project 2: Create a Polygon Pattern

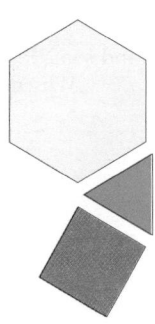

Regular polygons are shapes that have equal sides and equal angles. The
triangle, square, and hexagon shown are regular polygons. Create a few
patterns using these shapes. What kinds of patterns can you build with just
one type of polygon? with two types? with all three types? Sketch the patterns
you make. Describe the symmetry of each design. Shade the polygons with
different colors. How can the use of color affect the symmetry of your design?

Unit 1 Completing the Unit Project

The table shows the numbers of United States residents 5 years old
and over who speak a language other than English at home.

1-1

Language	1980	1990
Spanish	11,549,000	17,339,000
French	1,572,000	1,703,000
German	1,607,000	1,547,000
Italian	1,633,000	1,309,000
Chinese	632,000	1,249,000
Tagalog	452,000	843,000
Polish	826,000	723,000
Korean	276,000	626,000
Vietnamese	203,000	507,000
Portuguese	361,000	430,000

1. Which languages were spoken by fewer U.S. residents in 1990 than in 1980?

2. Explain why a bar graph of this data might be hard to read.

3. **Open-ended** What are three things you can learn from the table?

Write a variable expression for each amount. 1-2

4. 4 more than a number

5. your share of the rent in a 4-person apartment whose monthly rent is x dollars

6. the distance traveled by a car in four hours at an average speed of x miles per hour

7. Yoshi's car gets 21 miles per gallon of gasoline.

   a. How many miles can Yoshi drive on a tank of 12 gal of gasoline?

   b. Write a variable expression for the number of miles Yoshi can drive on a tank of x gallons of gasoline.

8. Complete the table of perimeters for the rectangles.

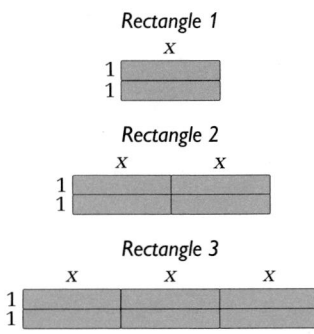

Rectangle 1
Rectangle 2
Rectangle 3

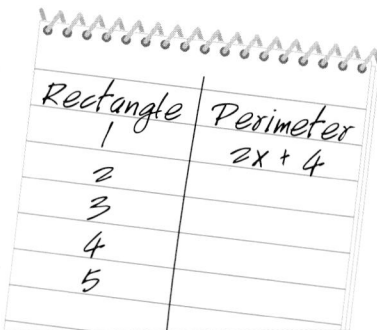

Rectangle	Perimeter
1	$2x + 4$
2	
3	
4	
5	

Write as a power of ten. 1-3

9. $10^7 \cdot 10^4$

10. $\dfrac{10^7}{10^4}$

11. $10^7 \cdot 10^4 \cdot 10$

12. a. Give a counterexample to the statement $10^a + 10^b = 10^{a+b}$.

    b. Are there any values of a and b that make the original statement true? If so, give an example.

    c. Rewrite the original statement so it is true.

Unit Support Materials
➤ Unit 1 Cumulative Practice 8
➤ Unit 1 Study Guide Review
➤ Unifying Problem 1
➤ Unit Tests 3 and 4

Quick Quiz (1-5 through 1-7)

1. Use the distributive property to find $28 \cdot 34 + 28 \cdot 66$. [1-5] 2800

2. Write an expression for the area covered by the tiles. [1-5]

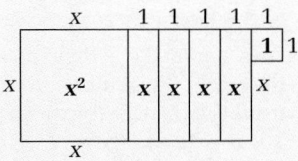

$x^2 + 4x + 1$

3. Use the distributive property to rewrite $4(3a - b)$ without parentheses. [1-5] $12a - 4b$

4. What kind of movement (a slide, a turn, or a flip) shows that polygon A is congruent to polygon B? [1-6] flip

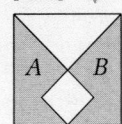

5. Name the congruent polygons. Then list four pairs of corresponding sides. [1-6]

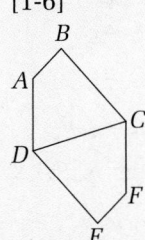

$ABCD \cong FEDC$
Side AB corresponds to side FE; side BC corresponds to side ED; side CD corresponds to side DC; side AD corresponds to side FC.

6. Draw all the lines of symmetry for the following figure. [1-7]

Answers to Unit 1 Review and Assessment

1. German, Italian, Polish

2. Explanations may vary. An example is given. A bar graph might be hard to read because the numbers are close.

3. Answers may vary. An example is given. Spanish is the language spoken most at home by United States residents 5 years old and over who speak a language other than English; Portuguese is spoken the least; speaking Polish has declined from 1980 to 1990.

4. $n + 4$

5. $x \div 4$

6. $4x$

7. a. 252 mi

   b. $21x$

8.

Rectangle	Perimeter
1	$2x + 4$
2	$4x + 4$
3	$6x + 4$
4	$8x + 4$
5	$10x + 4$

9. 10^{11}

10. 10^3

11. 10^{12}

Answers continued on next page.

12. **a.** Answers may vary. An example is given.
$10^2 + 10^3 \neq 10^{2+3}$, because $100 + 1000 \neq 100{,}000$.

   b. No.

   c. $10^a \cdot 10^b = 10^{a+b}$

13. 1041

14. 6

15. 3740

16. 1386

17. $12x^2 + 3x$

18. There are no like terms.

19. $5x^2 + 10x$

20. Amy's mistake is that she added the 4 and 6 together and then used the distributive property: $10(3x + 12) = 30x + 120$. She should have used the distributive property first, $6(3x + 12) = 18x + 72$, and then combined like terms: $4 + 18x + 72 = 18x + 76$.

21. Answers may vary. Examples are given. $ABGK \cong JEDH$; $ABGK \cong SNMQ$; $JEDH \cong SNMQ$

22. A *flip* across the line that passes through D and H will show that $ABGK \cong SNMQ$. A *slide* up two units and to the right one unit will show that $SNMQ \cong JEDH$. A *slide* up two units and to the right one unit and a *flip* will show that $ABGK \cong JEDH$.

23. Answers may vary. An example is given. Using $ABGK$ and $JEDH$, side $AB \cong$ side JE, side $BG \cong$ side ED, side $GK \cong$ side DH.

24. hexagon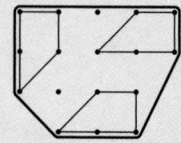

25. Choices of variables may vary. An example is given.
$a + b + c + 2d + e$

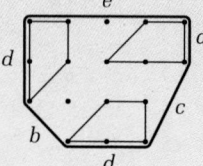

26. **a.** no perpendicular sides

   b. sides AC and MO; sides AM and CO

   c. parallelogram

27. **a.** sides LR and RJ; sides RJ and JD; sides JD and DL; sides DL and LR

   b. sides LR and DJ; sides LD and RJ

   c. rectangle

Calculate according to the order of operations. 1-4

13. $10^3 + 48 \div 3 + 25$

14. $[(10 - 6) \cdot 3 + 12] \div 4$

Use the distributive property to find each sum or product mentally. 1-5

15. $374 \cdot 6 + 374 \cdot 4$

16. $7 \cdot 198$

Combine like terms, if possible.

17. $5x^2 + 3x + 7x^2$

18. $5x^2 + 3x + 7$

19. $5x^2 + 3x + 7x$

20. **Writing** Amy made a mistake when she rewrote this expression:
$4 + 6(3x + 12) = 30x + 120$. Explain her mistake and how to correct it.

Use the diagram at the right. 1-6

21. Name two congruent polygons.

22. What kind of movement (or combination of movements) shows the trapezoids are congruent: *slide, turn,* or *flip*?

23. List three pairs of congruent sides.

24. Draw the shape a rubber band would make if placed around all three trapezoids together. Name the shape.

25. **Open-ended** Use variables to express the lengths of the sides of the shape in Question 24. Write an expression for the perimeter.

For Questions 26–29, do these things. 1-7

a. List all pairs of sides that are perpendicular.

b. List all pairs of sides that are parallel.

c. Tell what special name is given to the shape.

26. 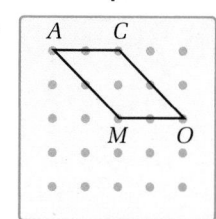 27. 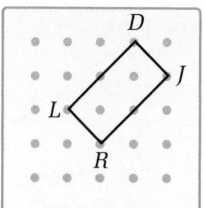 28. 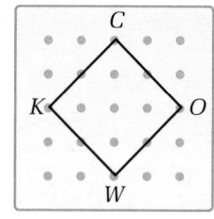 29.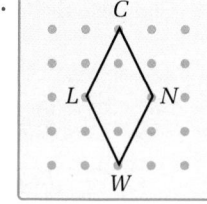

30. **Self-evaluation** Make a list of four important ideas in this unit that you think will be the most useful to you. Explain why they are important.

31. **Group Activity** Work with another student. Use x-tiles and 1-tiles.

   a. Make several different rectangles with a perimeter of $4x + 8$.

   b. Sketch the rectangles you make. Label them to show why the perimeter is $4x + 8$.

   c. Give your partner a different variable expression and challenge your partner to find more than one kind of rectangle with that perimeter. (Be sure it can be done before you give the challenge.)

---

28. **a.** sides KW and WO; sides WO and OC; sides OC and CK; sides CK and KW

   b. sides KC and WO; sides KW and CO

   c. square

29. **a.** no perpendicular sides

   b. sides LC and WN; sides LW and CN

   c. rhombus

30. Ideas may vary. Examples are given. Some useful ideas in this unit are evaluating variable expressions, using order of operations, simplifying powers of 10, using the distributive property, understanding lines of symmetry, and identifying quadrilaterals. Explanations may vary.

31. **a, b.** Answers may vary. An example is given.

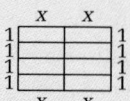

   c. Answers may vary.

IDEAS AND (FORMULAS)=x^2

ALGEBRA

➤ You can use variables to describe some patterns. *(p. 10)*

➤ You evaluate a variable expression by substituting numbers for the variables and following the order of operations. *(pp. 12, 27)*

➤ You can use exponents to write expressions involving repeated multiplication. *(p. 19)*

➤ When you multiply powers of ten, add the exponents. *(p. 22)*

➤ When you divide powers of ten, subtract the exponents. *(p. 22)*

➤ The distributive property allows you to rewrite a product as a sum or difference. *(p. 32)*

$$a(b + c) = ab + ac$$
$$a(b - c) = ab - ac$$

➤ Like terms have the same variable part. Like terms can be combined. *(p. 33)*

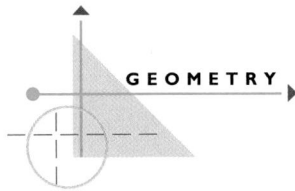

GEOMETRY

➤ Congruent figures have the same size and shape. If you can slide, turn, or flip one shape onto another so that the shapes match, they are congruent. *(p. 38)*

➤ Some special names of polygons are triangle, quadrilateral, pentagon, hexagon, heptagon, and octagon. *(p. 39)*

➤ A shape has symmetry when it can be folded so that one half matches the other. The fold line is the line of symmetry. *(p. 45)*

➤ **Reasoning** Conjectures are statements based on experience. You can show that a conjecture is false by finding a counterexample. *(pp. 21, 22)*

Key Terms

- **concept map** (p. 5)
- **evaluate** (p. 12)
- **power** (p. 19)
- **conjecture** (p. 21)
- **distributive property** (p. 32)
- **like terms** (p. 33)
- **quadrilateral** (p. 39)
- **congruent sides** (p. 40)
- **equilateral** (p. 44)
- **right triangle** (p. 44)
- **parallel** (p. 44)
- **rhombus** (p. 45)

- **variable** (p. 10)
- **substitute** (p. 12)
- **base** (p. 19)
- **counterexample** (p. 22)
- **term** (p. 33)
- **congruent** (p. 38)
- **vertex** (p. 40)
- **corresponding vertices** (p. 40)
- **isosceles** (p. 44)
- **right angle** (p. 44)
- **parallelogram** (p. 45)
- **symmetry** (p. 45)

- **variable expression** (p. 10)
- **factor** (p. 19)
- **exponent** (p. 19)
- **order of operations** (p. 26)
- **coefficient** (p. 33)
- **polygon** (p. 39)
- **side** (p. 40)
- **corresponding sides** (p. 40)
- **scalene** (p. 44)
- **perpendicular** (p. 44)
- **kite** (p. 45)
- **line of symmetry** (p. 45)

Unit 1 Review and Assessment

53

1. The table shows the most populated urban areas in the world. Numbers are given in millions. Write a statement based on the information in the table. [1-1]

Urban Area	1990	2000 (projected)
Toyko	26.95	29.97
Mexico City	20.21	27.87
São Paulo	18.05	25.35
Seoul	16.27	21.98
New York	14.62	14.65

Answers will vary. Examples are given. The urban areas of both Mexico City and São Paulo have a projected increase in population of over 7 million. The population of New York will remain about the same.

2. Evaluate $7y - x$ for $x = 3$ and $y = 8$. [1-2] **53**

3. Write $10^4 \cdot 10^7$ as a power of 10. [1-3] **10^{11}**

4. Evaluate $x^2 + 2x + 1$ when $x = 6$. [1-3] **49**

5. Calculate $72 \div 2^3 + 6^2 \div 3^2 + 8 \cdot 2$ according to the order of operations. [1-4] **29**

Using Measures and Equations

OVERVIEW

➤ **Unit 2** continues the exploration of mathematical concepts begun in Unit 1. Concepts related to negative numbers, scientific notation, estimation, and linear and angular measures are reviewed and extended. The **Student Resources Toolbox** on page 640 provides additional work with integers and integer operations for students who need further instruction. Review of measurement concepts is on pages 657–659 of the **Toolbox**.

➤ In Unit 2, students extend their study of expressions to the study of equations. They learn to solve equations by balancing and by using inverse operations. Students can review informal methods of solving one-step equations on pages 645–647 of the **Student Resources Toolbox**.

➤ Uses of numbers is the theme of the Unit Project. Students create a display of facts containing numbers used to count, measure, and order.

➤ Connections to geography, business education, literature, science, social studies, and physics are integrated into the teaching materials and exercises.

➤ Problem-solving strategies used in Unit 2 include *Break the Problem into Parts* (Section 2-4), *Use a Formula* (Sections 2-6 and 2-9), and *Use an Equation* (Sections 2-7 and 2-8).

Unit Objectives

Section	Objectives	NCTM Standards
2-1	• See different ways that numbers are used.	1, 2, 3, 4, 10, 11, 14
	• Tell whether an estimate of a population or distance is reasonable or likely.	
2-2	• Review how to do operations with negative numbers.	1, 2, 4, 5, 14
2-3	• Read and write numbers in scientific notation.	1, 2, 3, 4, 14
	• Recognize whether a result in scientific notation on a calculator is reasonable.	
2-4	• Estimate lengths, distances on a map, and areas.	1, 2, 4, 7
2-5	• Investigate and use the relationships among angles in circles, in triangles, and formed by intersecting lines.	1, 2, 3, 4, 7
2-6	• Simplify expressions for measures of geometric figures by multiplying and combining like terms.	1, 2, 3, 4, 5, 7
2-7	• Begin to solve equations about geometric figures.	1, 2, 4, 5, 7
2-8	• Use equations to solve a variety of real-life problems.	1, 2, 3, 4, 5
2-9	• Find square roots and cube roots.	1, 2, 3, 4, 5, 14
	• Distinguish a rational number from an irrational number.	

Topic Spiraling

Section	Connections to Prior and Future Concepts
2-1	**Section 2-1** reviews the ways numbers are used: for identifying, ordering, measuring, and counting. Use of estimation for counting and measures, including estimating probabilities, is investigated. Estimation is also applied in Section 2-4, in Units 3, 4, and 6 of Book 1; in Units 1 and 9 of Book 2; and in Units 1, 2, and 5–7 of Book 3.
2-2	**Section 2-2** reviews operations with positive and negative numbers. All the work in the three books of this series assumes this ability. Pages 642–644 in the Toolbox provide additional review of this skill.
2-3	**Section 2-3** extends the work with exponents from Unit 1 to include negative and zero exponents and scientific notation. The study of exponents and scientific notation is continued in Units 9 and 10 of Book 1; in Units 2, 4, 6, and 9 of Book 2; and in Units 2, 4, 5, and 8 of Book 3.
2-4	**Section 2-4** continues the review of estimation skills by developing techniques for estimating lengths and areas. Estimation is applied in a variety of mathematical and real-world settings throughout this book and in Books 2 and 3.
2-5	**Section 2-5** reviews angle measure and investigates relationships among angles in circles, in intersecting lines, and in triangles. The topic of angles and their measures recurs in Section 2-7, Units 6, 7, and 9 of Book 1; in Units 5, 7, and 8 of Book 2; and in Units 3 and 8 of Book 3. The use of a protractor for measuring angles can be reviewed in the Student Resources Toolbox on page 657.
2-6	**Section 2-6** combines writing and simplifying expressions introduced in Unit 1 with the concepts of negative numbers and area presented in Unit 2 to develop the skill of writing and simplifying expressions for measures of geometric figures.
2-7	**Section 2-7** extends the work with expressions to writing equations and solving them by balancing both sides. Students solve more complicated equations and systems of equations in the rest of Unit 2 and in Units 5 and 8–10. Solving higher-degree polynomial equations and systems of quadratic equations are introduced in Book 2, and work with logarithmic and exponential equations appears in Book 3. (A review of one-step equations appears on pages 645–647 of the Toolbox.)
2-8	**Section 2-8** uses inverse operations to solve equations. Methods of solving equations is a recurring topic in Section 2-9, Units 5 and 7–10 of Book 1; Units 2–4 and 9 of Book 2; and Units 1, 2, and 5 of Book 3.
2-9	**Section 2-9** introduces roots. Factors and roots are investigated further in Units 5, 9, and 10 of Book 1. Students will extend their understanding of roots in Units 2, 4, and 9 of Book 2 and in Units 2 and 5 of Book 3.

Integrating the Strands

Strands	Sections
Number	2-1, 2-2, 2-3, 2-5, 2-6, 2-7, 2-8, 2-9
Algebra	2-2, 2-3, 2-5, 2-6, 2-7, 2-8, 2-9
Measurement	2-1, 2-2, 2-3, 2-4, 2-5, 2-6, 2-7, 2-9
Geometry	2-4, 2-5, 2-6
Statistics and Probability	2-1, 2-3, 2-5, 2-9
Discrete Mathematics	2-1, 2-2, 2-4, 2-6, 2-7
Logic and Language	2-1, 2-6

Section Planning Guide

➤ Essential exercises and problems are indicated in boldface.
➤ Ongoing work on the Unit Project is indicated in color.
➤ Exercises and problems that require student research, group work, manipulatives, or graphing technology are indicated in the column headed "Other."

Section	Materials	Pacing	Standard Assignment	Extended Assignment	Other
2-1		Day 1 Day 2	**1–17** **19–29**, 35–41, 42	**1–17** **19–29**, 30–32, 34–41, 42	18 33, 42
2-2	calculator	Day 1 Day 2	**1–27, 31, 32** **33–44**, 45–48, **49–52**, 55–62, 63	**1–27**, 28–30, **31, 32** **33–44**, 45–48, **49–52**, 53, 55–62, 63	54, 63
2-3	scientific calculator	Day 1 Day 2	**1–14, 16–25** **27–37, 43–46**, 47–54, 55	**1–14**, 15, **16–25**, 26 **27–37**, 38–42, **43–46**, 47–54, 55	55
2-4	scissors, dot paper	Day 1	**2–4**, 5, **6–9**, 11–13, **15–20**, 23–28, 29	1, **2–4**, 5, **6–9**, 10–14, **15–20**, 21, 23–28, 29	22
2-5	protractor, scissors, geometric drawing software	Day 1 Day 2	**1–18** **19–23, 25–28**, 31–39, 40	**1–18** **19–23, 25–28**, 31–39, 40	24, 29, 30
2-6	algebra tiles	Day 1	**2–25**, 34–38, 39	1, **2–25**, 26–30, 33–38, 39	31, 32, 39
2-7	algebra tiles, geometric drawing software, protractor	Day 1	**4–16, 19–25**, 29–36, 37	2, 3, **4–16**, 17, 18, **19–25**, 26, 27, 29–36, 37	1, 28
2-8	calculator	Day 1	1, **2–13, 15–19**, 24, 25, 27–31, 32	1, **2–13**, 14, **15–19**, 20–22, 24, 25, 27–31, 32	23, 26
2-9	scientific calculator	Day 1 Day 2	1, 2, **3–15** **18–30, 32–41**, 48–54, 55–57	1, 2, **3–15**, 16, 17 **18–30**, 31, **32–41**, 42–46, 48–54, 55–57	47
Review Test		**Day 1** **Day 2**	**Unit Review** **Unit Test**	**Unit Review** **Unit Test**	

Yearly Pacing	Unit 2 Total	Units 1–2 Total	Remaining	Total
	18 days (2 for Unit Project)	34 days	130 days	164 days

Support Materials

➤ See **Project Book** for notes on Unit 2 Project: Theme Poster Contest.
➤ "UPP" and "disk" refer to **Using Plotter Plus** booklet and **Plotter Plus** disk.
➤ Warm-up exercises for each section are available on **Warm-Up Transparencies**.

Section	Study Guide	Practice Bank	Problem Bank	Activity Bank	Explorations Lab Manual	Assessment Book	Visuals	Technology
2-1	2-1	Practice 9	Set 3	Enrich 8		Quiz 2-1		
2-2	2-2	Practice 10	Set 3	Enrich 9	Add. Expl. 2	Quiz 2-2		
2-3	2-3	Practice 11	Set 3	Enrich 10	Master 10	Quiz 2-3, Test 5		
2-4	2-4	Practice 12	Set 4	Enrich 11	Masters 1, 2	Quiz 2-4		
2-5	2-5	Practice 13	Set 4	Enrich 12		Quiz 2-5	Folder 2	
2-6	2-6	Practice 14	Set 4	Enrich 13	Master 6	Quiz 2-6, Test 6		
2-7	2-7	Practice 15	Set 5	Enrich 14	Add. Expl. 3, 4 Master 6	Quiz 2-7	Folder 3	
2-8	2-8	Practice 16	Set 5	Enrich 15	Add. Expl. 4	Quiz 2-8		
2-9	2-9	Practice 17	Set 5	Enrich 16		Quiz 2-9, Test 7		UPP, page 26
Unit 2	Unit Review	Practice 18	Unifying Problem 2	Family Involve 2		Tests 8, 9		

Form A

Spanish versions of these tests are on pages 126–129 of the **Assessment Book.**

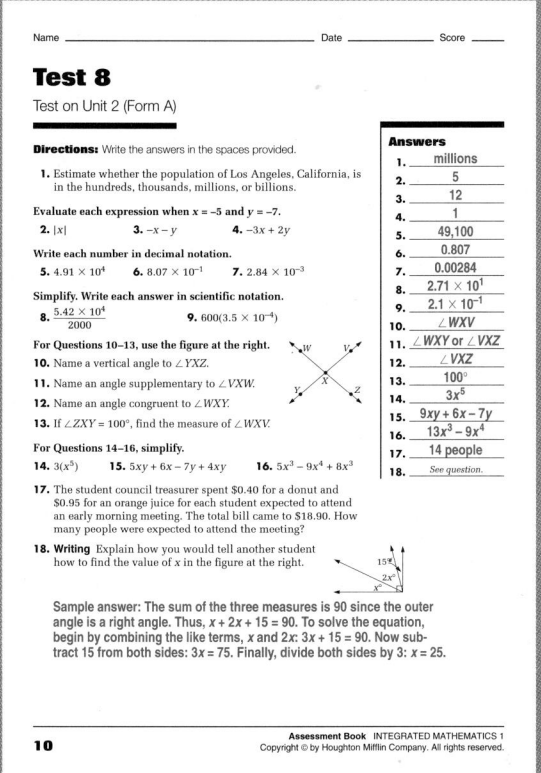

Name ______________ Date ________ Score ______

Test 8

Test on Unit 2 (Form A)

Directions: Write the answers in the spaces provided.

1. Estimate whether the population of Los Angeles, California, is in the hundreds, thousands, millions, or billions.

Evaluate each expression when $x = -5$ and $y = -7$.

2. $|x|$ 3. $-x - y$ 4. $-3x + 2y$

Write each number in decimal notation.

5. 4.91×10^4 6. 8.07×10^{-1} 7. 2.84×10^{-3}

Simplify. Write each answer in scientific notation.

8. $\frac{5.42 \times 10^4}{2000}$ 9. $600(3.5 \times 10^{-4})$

For Questions 10–13, use the figure at the right.

10. Name a vertical angle to $\angle YXZ$.
11. Name an angle supplementary to $\angle VXW$.
12. Name an angle congruent to $\angle WXY$.
13. If $\angle ZXY = 100°$, find the measure of $\angle WXV$.

For Questions 14–16, simplify.

14. $3(x^5)$ 15. $5xy + 6x - 7y + 4xy$ 16. $5x^3 - 9x^4 + 8x^3$

17. The student council treasurer spent $0.40 for a donut and $0.95 for an orange juice for each student expected to attend an early morning meeting. The total bill came to $18.90. How many people were expected to attend the meeting?

18. **Writing** Explain how you would tell another student how to find the value of x in the figure at the right.

Sample answer: The sum of the three measures is 90 since the outer angle is a right angle. Thus, $x + 2x + 15 = 90$. To solve the equation, begin by combining the like terms, x and 2x: $3x + 15 = 90$. Now subtract 15 from both sides: $3x = 75$. Finally, divide both sides by 3: $x = 25$.

Answers

1.	millions
2.	5
3.	12
4.	1
5.	49,100
6.	0.807
7.	0.00284
8.	2.71×10^1
9.	2.1×10^{-1}
10.	$\angle WXV$
11.	$\angle WXY$ or $\angle VXZ$
12.	$\angle VXZ$
13.	100°
14.	$3x^5$
15.	$9xy + 6x - 7y$
16.	$13x^3 - 9x^4$
17.	14 people
18.	*See question.*

10 Assessment Book INTEGRATED MATHEMATICS 1

Name ______________ Date ________ Score ______

Test 8 *(continued)*

Directions: Write the answers in the spaces provided.

Solve.

19. $x + 6 = -3$ 20. $\frac{x}{4} - 7 = -3$ 21. $24 = 5x + 3x$

Write and solve an equation to find each unknown value.

22.

23.

For Questions 24–28, choose the letter of the point on the number line which matches each number.

24. $\sqrt{64}$ 25. $\sqrt{24}$ 26. $\sqrt[3]{-27}$ 27. $\sqrt[3]{-1}$ 28. $-\sqrt{12}$

29. Find the area of the rectangle at the right.

30. The volume of a cube is 125 cm^3. Find the length of an edge of the cube.

31. **Open-ended** Use the elements below to make a concept map showing the relationship among rational, irrational, and real numbers.
- real numbers
- rational numbers
- irrational numbers
- fractions
- integers
- repeating decimals
- nonrepeating, non-terminating decimals
- terminating decimals
- whole numbers
- square roots of nonperfect squares

Answers

19.	–9
20.	16
21.	3
22.	45°, 45°
23.	132°
24.	E
25.	D
26.	B
27.	C
28.	A
29.	$45x^2$
30.	5 cm
31.	*See question.*

Assessment Book INTEGRATED MATHEMATICS 1

 11

Form B

Name ______________ Date ________ Score ______

Test 9

Test on Unit 2 (Form B)

Directions: Write the answers in the spaces provided.

1. Estimate whether the air distance in miles from Miami, Florida to San Diego, California, is in the hundreds, thousands, millions, or billions.

Evaluate each expression when $x = -8$ and $y = -3$.

2. $|x|$ 3. $-x - y$ 4. $-3x + 2y$

Write each number in decimal notation.

5. 2.77×10^3 6. 3.05×10^{-5} 7. 8.13×10^1

Simplify. Write each answer in scientific notation.

8. $\frac{4.68 \times 10^5}{20{,}000}$ 9. $700(2.5 \times 10^{-5})$

For Questions 10–13, use the figure at the right.

10. Name a vertical angle to $\angle WXY$.
11. Name an angle supplementary to $\angle WXY$.
12. Name an angle congruent to $\angle VXZ$.
13. If $\angle ZXY = 115°$, find the measure of $\angle WXV$.

For Questions 14–16, simplify.

14. $5(x^3)$ 15. $7xy + 3x - 9y + 5xy$ 16. $4x^2 - 6x^4 + 8x^2$

17. The student council treasurer spent $0.89 for a party hat and $0.75 for a party favor for each student expected to attend the Homecoming dance. The total bill came to $82.00. How many people were expected to attend the dance?

18. **Writing** Explain how you would tell another student how to find the value of x in the figure at the right.

Sample answer: The sum of the three measures is 180 since the outer angle is a straight angle. Thus $x + 4x + 20 = 180$. To solve the equation, begin by combining the like terms, x and 4x: $5x + 20 = 180$. Now subtract 20 from both sides: $5x = 160$. Finally, divide both sides by 5: $x = 32$.

Answers

1.	thousands
2.	8
3.	11
4.	18
5.	2770
6.	0.0000305
7.	81.3
8.	2.34×10^1
9.	1.75×10^{-2}
10.	$\angle VXZ$
11.	$\angle YXZ$ or $\angle WXV$
12.	$\angle WXY$
13.	115°
14.	$5x^3$
15.	$12xy + 3x - 9y$
16.	$12x^2 - 6x^4$
17.	50 people
18.	*See question.*

12 Assessment Book INTEGRATED MATHEMATICS 1

Name ______________ Date ________ Score ______

Test 9 *(continued)*

Directions: Write the answers in the spaces provided.

Solve.

19. $x + 8 = -3$ 20. $\frac{x}{4} - 2 = -3$ 21. $16 = 7x + x$

Write and solve an equation to find each unknown value.

22.

23.

For Questions 24–28, choose the letter of the point on the number line which matches each number.

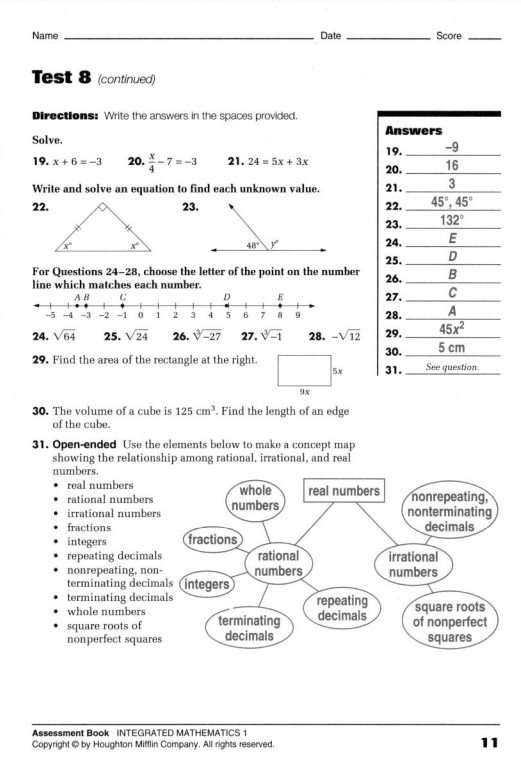

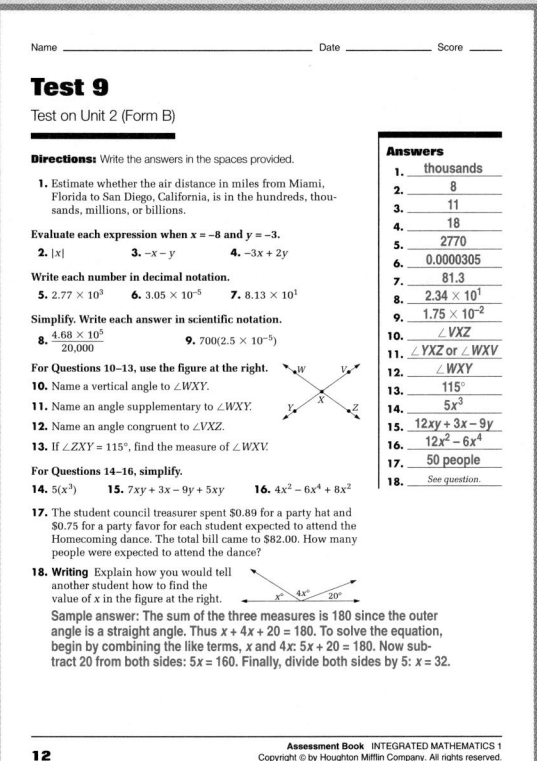

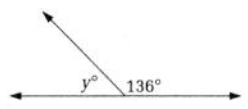

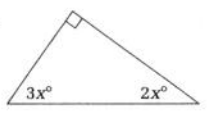

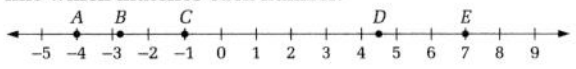

24. $\sqrt{49}$ 25. $\sqrt{20}$ 26. $\sqrt[3]{-1}$ 27. $\sqrt[3]{-64}$ 28. $-\sqrt{8}$

29. Find the area of the rectangle at the right.

30. The volume of a cube is 216 cm^3. Find the length of an edge of the cube.

31. **Open-ended** Use the elements below to make a concept map showing the relationship among rational, irrational, and real numbers.
- real numbers
- rational numbers
- irrational numbers
- fractions
- integers
- repeating decimals
- nonrepeating, non-terminating decimals
- terminating decimals
- whole numbers
- square roots of nonperfect squares

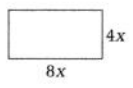

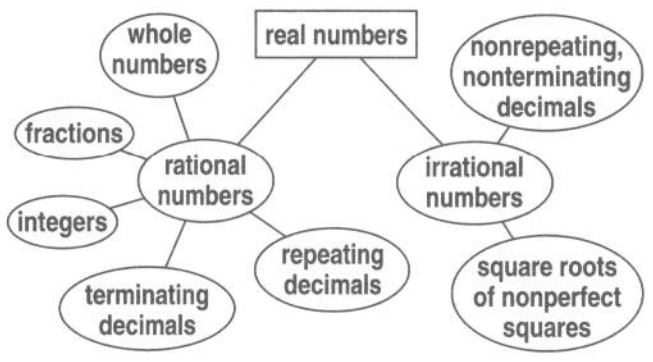

Answers

19.	–11
20.	–4
21.	2
22.	36°, 54°
23.	44°
24.	E
25.	D
26.	C
27.	A
28.	B
29.	$32x^2$
30.	6 cm
31.	*See question.*

Assessment Book INTEGRATED MATHEMATICS 1

 13

Software Support

McDougal Littell Mathpack

Geometry Inventor

Outside Resources

Books/Periodicals

Schwartz, Judah, Michal Yerushalmy, and Beth Wilson, eds. *The Geometric Supposer: What Is It a Case of?* Hillsdale, NJ: Laurence Earlbaum Associates, 1993.

Schoen, Harold L., ed. *Estimation and Mental Computation.* 1986 Yearbook. Reston, VA: NCTM, 1986.

Manipulatives

Stallings-Roberts, Virginia. "An ABSOLUTE-ly VALUE-able Manipulative." *Mathematics Teacher* (April 1991).

Shoemaker, Robert. *All You Need to Know about the Metric System for Everyday Use.* Full color metric wall chart. South Beloit, IL: Blackhawk Metric Supply, 1992.

Software

Schwartz, Judah and Michal Yerushalmy. *The Geometric Supposer: Triangles.* Scotts Valley, CA: Sunburst.

Geometer's Sketchpad. Macintosh and MS-DOS. Berkeley, CA: Key Curriculum Press.

IBM Mathematics Exploration Toolkit. IBM, Armonk, NY.

Videos

Stella Octangula: Workbook and Video. Berkeley, CA: Key Curriculum Press.

Apostol, Tom. *The Theorem of Pythagoras.* Reston, VA: NCTM.

➤ Students create a theme poster of numerical facts.

➤ When making the poster, students present four numerical facts with four questions that can be answered using these facts and an equation that can be solved to find another interesting fact.

➤ Students work in a cooperative group and appreciate each other's contribution.

PROJECT PLANNING

Materials List

➤ Markers

➤ Poster board

Project Teams

Since this project is a poster contest, have students determine as a class how the posters will be evaluated.

Have students work on the project in groups of three. One way for the individuals in the group to distribute the work is as follows:

1. Coodinator: coordinates research, collects all possible ideas from the group, and checks numerical facts for accuracy.

2. Writer: writes and checks on the mathematical operations.

3. Illustrator: draws and colors the poster.

Individual students will need a few days to find very large and very small numbers to discuss with the other team members.

unit 2

Using Measures and Equations

Did you know that Americans eat about 100 acres of pizza each day? That is enough pizza to carpet 55 floors of a 110-story tower of New York City's World Trade Center! In order to understand the magnitude of very large or very small counts and measures, you can relate them to something whose size you know.

Magnitude is size:
How **far**? How **deep**?
How **strong**? How **loud**?
How **fast**?
1.86×10^5 mi/s

Magnitude

Writers of songs and stories often make comparisons to help you understand measures . . . the corn is as HIGH as an elephant's eye . . . as *light* as a feather . . . so **HOT** you can fry an egg on it . . . so **noisy** I can't hear myself think!

compare

Iowa - Corn Capital of the World

54

Suggested Rubric for Unit Project

4 The theme poster includes a counting fact, a measurement fact, a percent fact, and either a very large or a very small number. The questions are well written and use the information on the poster. Complete solutions are provided and the poster is visually appealing and clearly relates to a theme.

3 The theme poster is adequate. Not all the facts are presented in an interesting way. The equations written may not lead to another interesting fact. The poster's theme does not relate clearly enough to the understanding of magnitude.

2 The theme poster is incomplete. The facts or details are incomplete. The questions are only minimally acceptable. This project should be returned with suggestions for improvements and a new deadline.

Theme Poster Contest

Your group project is to create a poster that presents interesting numerical facts in ways that make them easier to understand.

Your poster should have a theme. You may choose a topic from another class or from any field that you like.

When it is finished, your poster should display four numerical facts presented in interesting ways, four questions that can be answered using the numerical facts, and an equation that can be solved to find another interesting fact.

To decide how well you explained magnitudes, hold a class contest. Display all the posters in your classroom. The class will judge all the posters on standards you have chosen in advance.

← If all Earth's water fit inside a one gallon jug, the available fresh water would equal just over a tablespoon, which is less than 0.5% of the total.

↙ Sun   ↙ Earth

How **much** of the world's oxygen do the rain forests in Brazil produce?

How SMALL is the amoeba you read about in biology class?

Scientists put magnitude in perspective:

. . . if the **SUN** were the size of a basketball, EARTH would be the size of a *grain of sand* . . .

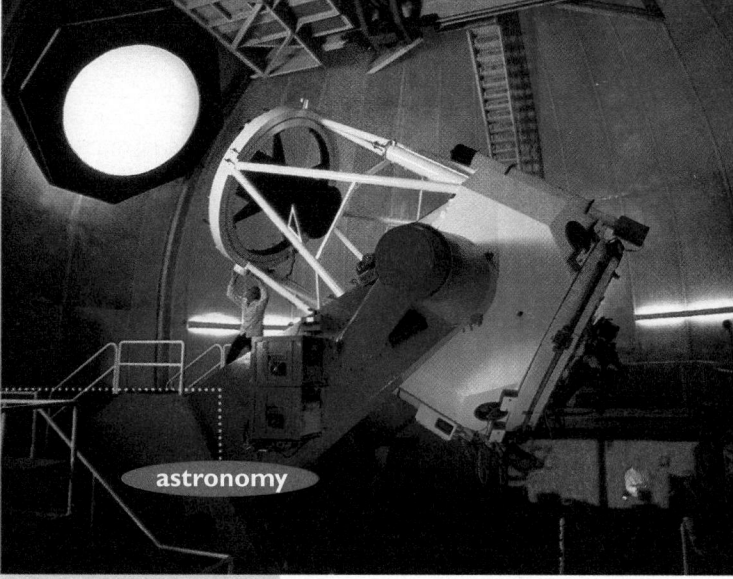

astronomy

55

Support Materials

The *Project Book* contains information about the following topics for use with this Unit Project.

➤ Project Description
➤ Teaching Commentary
➤ Working on the Unit Project Exercises
➤ Completing the Unit Project
➤ Assessing the Unit Project
➤ Alternative Projects
➤ Outside Resources

Using Technology

A scientific calculator can be used to calculate some numbers expressed exponentially with exponent values ranging between 10^{99} and 10^{-99}.

The power key is used to evaluate an expression involving exponents.

Limited English Proficiency

Making a theme poster is a valuable project for students. Have a student fluent in English work closely with a student acquiring English so that this student understands the content of the poster.

ADDITIONAL BACKGROUND

Multicultural Note

The Pan American Highway provides a number of interesting numerical facts. It is the longest drivable road in the world, extending from Alaska through Canada, the United States, and Central and South America to southern Chile. Including its east/west branches, the highway covers more that 29,525 miles.

In the United States, the highway follows the length of the Rocky Mountains southward. In Mexico, it passes through Mexico City, the largest city in the world with a population of about 20 million. At Chepo, Panama, a 250-mile stretch of

Suggested Rubric for Unit Project

1 This theme poster cannot be evaluated. It is illegible, incomplete, or not understandable. The project should be returned with a new deadline for completion. The group should speak with the teacher as soon as possible to discuss the purpose and the format of the assignment.

jungle called the Darien Gap interrupts the highway so motorists must ship their cars to Colombia or Venezuela.

The highway then follows the western coastline of South America through Ecuador and Peru all the way to Santiago, Chile, where a major branch cuts eastward to Buenos Aires, Argentina. From there, the highway follows the east coast of South America north to Rio de Janeiro, Brazil, and then turns inland to Brasilia, the capital of Brazil.

Googol

A googol is one of the largest numbers ever named. It is defined as 1 followed by 100 zeros, or 10^{100}. The number of grains of sand in the world is less than a googol. Another number is a googolplex, which is a 1 with a googol number of zeros after it.

ALTERNATIVE PROJECTS

Project 1, page 119

Redesigning a Room

Make plans to redesign a classroom in your school to make it more efficient or more attractive. Write a paragraph giving reasons for the new design and explain why it is an improvement.

Project 2, page 119

Conducting an Interview

Research some of the ways numerals, fractions, and decimals are written differently in other countries. Interview people about differences they may have noticed in other countries. Report the findings to the class.

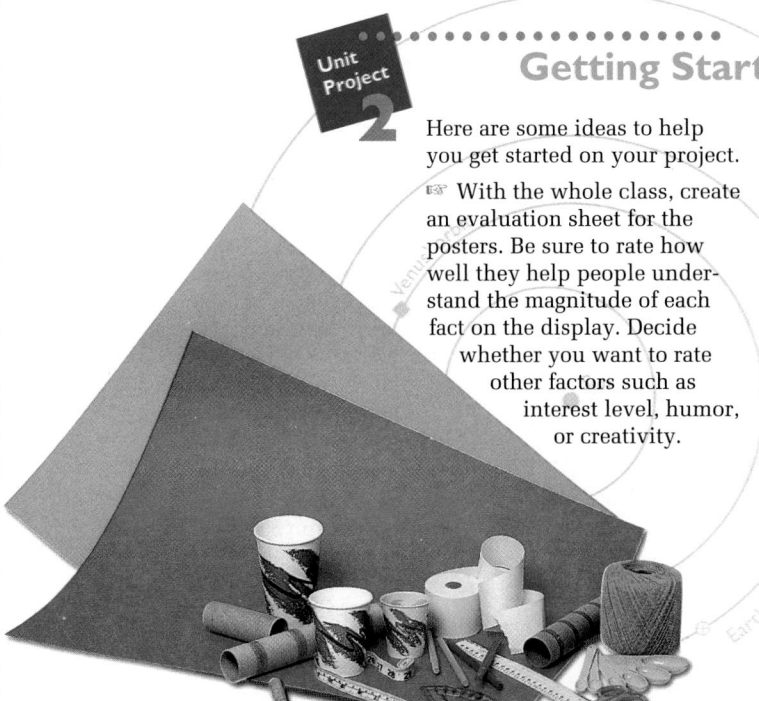

Unit Project 2

Getting Started

Here are some ideas to help you get started on your project.

☞ With the whole class, create an evaluation sheet for the posters. Be sure to rate how well they help people understand the magnitude of each fact on the display. Decide whether you want to rate other factors such as interest level, humor, or creativity.

☞ Work in a group of three students to create your theme poster.

☞ With your group, choose a theme for your poster. Brainstorm topics that use very small and very large numbers. You could look at your school from the microscopic level and from miles away.

☞ Discuss how you might add interest to your poster by using three-dimensional elements, such as a roll-out number line or different-sized measuring cups.

Working on the Unit Project

Your work in Unit 2 will help you create your theme poster.

Related Exercises:

Section 2-1, Exercise 42
Section 2-2, Exercise 63
Section 2-3, Exercise 55
Section 2-4, Exercise 29
Section 2-5, Exercise 40
Section 2-6, Exercise 39
Section 2-7, Exercise 37
Section 2-8, Exercise 32
Section 2-9, Exercises 55–57

Alternative Projects p. 119

Can We Talk

Magnitude

➤ Favorite pizza toppings include curry in Pakistan, eel and squid in Japan, shrimp and pineapple in Australia, and coconut in Costa Rica. What three toppings do you think are the most popular in the United States?

➤ In 1992, the amount of pepperoni eaten by Americans was about 318,075,000 lb. This is the same weight as 26,500 elephants! Make another comparison to describe this amount.

➤ In 1990, 52.4 million tons of paper was thrown away in the United States. What are some comparisons you can make to help you understand how big 52.4 million tons is?

➤ Suppose the distance from your home to your school were the length of a one-foot ruler. What would model the distance from your home to the nearest shopping mall?

➤ About how long would it take you to walk from one end of your town or city to the other?

Unit 2 Using Measures and Equations

Answers to Can We Talk?

➤ Besides cheese, the favorite toppings in the U.S. are pepperoni, sausage, and mushrooms.

➤ Answers may vary.

➤ Students could weigh a ream of paper and use it as a guide to understand the volume of 52.4 million tons.

➤ Answers may vary.

➤ Answers may vary.

Numbers and Estimates

figures in the CROWD

Focus
See different ways that numbers are used and tell whether an estimate of a population or distance is reasonable or likely.

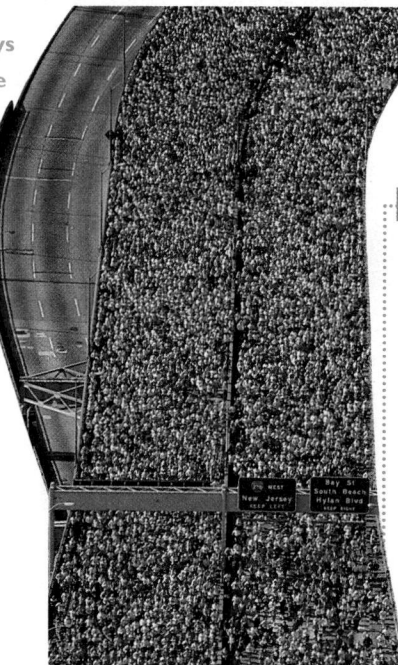

Talk it Over

1. Without counting, guess whether the number of people in the photograph is in the *tens, hundreds, thousands,* or *millions.*

2. Describe a method for estimating the number of people.

3. Use your method to estimate the number of people.

4. How does your estimate compare with your guess in question 1?

5. Can you use your method to estimate the number of hairs on your head? the size of an audience in a theater with you? a city population? How?

You see many types of numbers used in everyday life. Negative numbers are used for cold temperatures. Decimals are used to identify library books. Numbers can be used to identify, order, count, and measure. For example:

A wheelchair racer wearing the number 806 finishes first out of about 8 wheelchair racers in a 800 m race.

Numbers used for identifying and ordering are exact. ➤

IDENTIFY
The number 806 identifies the racer.

ORDER
The number *first* tells you in what position the racer finished.

COUNT
The number 8 tells you about how many racers there were.

MEASURE
The number 800 in the phrase *800 m* tells you the length of the race.

Numbers used as counts and measures can be estimates. ➤

2-1 Numbers and Estimates

57

Objectives and Strands
See pages 54A and 54B.

Spiral Learning
See page 54B.

Recommended Pacing
Section 2-1 is a two-day lesson.
Day 1
Pages 57–59: Talk it Over 1 through Talk it Over 13, *Exercises 1–18*
Day 2
Pages 59–60: Estimating Probabilities through Look Back, *Exercises 19–42*

Extra Practice
See pages 622–623.

Warm-Up Exercises
Warm-Up Transparency 2-1

Support Materials
➤ Practice 9
➤ Enrichment 8 in the Activity Bank
➤ Study Guide 2-1
➤ Problem Set 3
➤ Quiz 2-1

Answers to Talk it Over

1. Answers may vary. An example is given. thousands

2. Answers may vary. An example is given. Split the crowd into equal-size smaller sections. Count or estimate the number of people in one of these smaller sections, then multiply by the number of sections.

3. Answers may vary.

4. Answers may vary.

5. Answers may vary. Examples are given. Yes, hair could be separated into equal parts and a smaller group counted. Yes, you could choose a section of the movie theater and count the number of people seated in that section. No, you cannot see the entire population of a city in order to divide it into equal groups. You would have to conduct a survey.

Use questions 1–4 to bring out the distinction between a *wild* guess and a *reasonable* guess.

Question 5 points out the fact that selecting and gauging the size of a sample may be quite difficult, especially for large groups. It may not be possible to see all members of the group. Also, the group may fluctuate over time. All these difficulties and more may arise when we try to estimate, say, the population of a large city.

Additional Sample

S1 Estimate whether each value is in the *hundreds*, *thousands*, *millions*, or *billions*.

In 1990, the number of automobiles registered in Texas was about 8.7 million.

a. the number of automobiles registered in Rhode Island in 1990 thousands

b. the number of automobiles registered in all 50 states in the U.S. millions

---

Talk it Over

6. Give some examples of how a fraction, a decimal, and a negative number are used in your everyday life.

For questions 7–9, tell whether each number could be *estimated*, or whether it must be *exact*. Give a reason for your choice.

7. a student's school identification number

8. the population of a state capital

9. the length of time it takes to read a book

10. Look at the size of the group in the photograph on page 57. About how many groups of this size would make a total of one billion people?

11. Guess whether the number of math books that could be stacked in your classroom is in the *hundreds*, *thousands*, *millions*, or *billions*. Then explain how you could estimate this number.

The world's population in 1990 was about 5.3 billion.

The air distance from New York City to Seattle, Washington, is 2409 mi.

The maximum legal capacity of the Seattle Kingdome is 75,000.

The population of the United States in 1990 was about 250 million.

The population of New York City in 1990 was about 7.3 million. It was the largest city in the United States at that time.

Estimating by Making Comparisons

Sometimes you cannot see what you want to count or measure. You need to estimate by making comparisons using facts like those at the left.

Sample 1

Estimate whether each value is in the *hundreds*, *thousands*, *millions*, or *billions*.

a. the population of Los Angeles, California

b. attendance at the Super Bowl

Sample Response

Use facts from your experience or the list above to help you estimate.

a. Los Angeles is a very large city, like New York City. Its population is probably in the millions.

b. The Super Bowl is held in large stadiums, such as the Seattle Kingdome, and is generally well attended. Attendance at the Super Bowl is probably in the thousands.

Answers to Talk it Over

6. Answers may vary. Examples are given. A fraction might be used to indicate yardage of fabric; a decimal might be used to indicate a price; a negative number might be used to indicate a business loss.

7–9. Answers may vary. Examples are given.

7. exact; Each student has only one school identification number, and each school identification number identifies only one student.

8. estimated; The population of a state capital is constantly changing. People move in and out of the city and there are births and deaths. Generally, it is not possible to determine an exact population count and an estimate is used instead.

9. estimated; People often read books over several days and experience many interruptions.

10. Answers may vary. An example is given. about 1 million groups

Talk it Over

12. Suppose you want to know what percent of the world's population in 1990 was the United States' population. What two facts can help you find this number? Are they included in the list of facts on page 58?

13. Estimate whether the air distance in miles from Portland, Maine, to Portland, Oregon, is in the *hundreds, thousands, millions,* or *billions.*

▶ Now you are ready for:
Exs. 1–18 on pp. 60–61

Estimating Probabilities

You can use your experience to help you estimate the probability that an event will happen. The probability of an event happening is a number from 0 to 1.

Sample 2

Writing Use a number anywhere on the scale below to estimate the probability that it will rain during your school's next outdoor sports event. Give a reason for your answer.

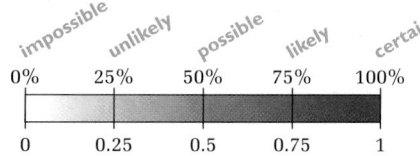

impossible   unlikely   possible   likely   certain
0%   25%   50%   75%   100%

0   0.25   0.5   0.75   1

Sample Response ·····

Use facts from your experience to help you estimate.

Meilin's answer:

"It doesn't rain much here in Los Angeles. I estimate a 25% chance of rain."

Jason's answer:

"Our school has a track meet tomorrow. The weather report said that the probability of rain tomorrow is 60%."

Talk it Over

14. Use the scale in Sample 2. Estimate the probability that it will rain during your school's next outdoor sports event. Give a reason for your answer.

15. Use the weather map. Is the probability of rain in the Northwest more likely to be 0.1 or 0.8?

2-1 Numbers and Estimates

59

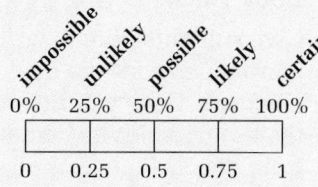

Use questions 16–19 to make the point that discrete quantities require whole numbers, whereas continuous quantities do not.

Give students time to write a journal entry for the Look Back situation. Have volunteers read their entries. Invite comments from the class and discuss the sampling and estimation techniques used.

APPLYING

Suggested Assignment

Standard 1–17, 19–29, 35–42

Extended 1–17, 19–32, 34–42

Integrating the Strands

Number Exs. 1–18, 30–35, 37–42

Measurement Exs. 5, 12, 17, 35

Statistics and Probability Exs. 19–24

Discrete Mathematics Exs. 25–29, 35, 42

Logic and Language Exs. 35, 36

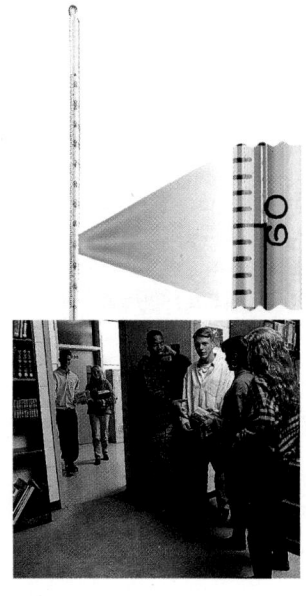

Discrete and Continuous Quantities

Continuous → Quantities that are measured are **continuous**, such as the temperature of water as it is heated from 60°F to boiling. As water is heated, it does not jump in temperature from 60°F to 61°F. It takes on every temperature between 60°F and 61°F.

Discrete → Quantities that are counted are **discrete**, such as the number of people entering a library. If you were counting, there would not be a person between the 60th person and the 61st person.

Talk it Over

Classify each quantity as *continuous* or *discrete*.

16. the water level in a sink as it is being filled

17. the speed of a car driven from your home to a shopping mall

18. the goals scored during each soccer game of the season

19. Classify each quantity in questions 16–18 as a *count* or *measure*.

Look Back ◄

Describe how you could estimate the number of people in a crowded room if you were in the room. Then describe how you could do the estimating if you were nowhere near the room.

····▶ **Now you are ready for:**
Exs. 19–42 on pp. 62–63

2-1 Exercises and Problems

1. **Career** Medical technicians use *petri dishes* to grow bacteria.

   a. Without counting any bacteria, guess whether the number of bacteria in the petri dish is in the *tens, hundreds, thousands,* or *millions*.

   b. Describe a method for estimating the number of bacteria.

   c. Use your method to estimate the number of bacteria.

   d. How does your estimate compare with your guess in part (a)?

60   Unit 2 Using Measures and Equations

Answers to Talk it Over

16. continuous

17. continuous

18. discrete

19. (16) measure;
    (17) measure; (18) count

Answers to Look Back

Answers may vary. Examples are given.

If I were in the crowded room, I would divide the room into equal areas, count the number of people in my area, and multiply by the number of equal areas to estimate the total number of people in the room.

If I were not in the room, I would try to imagine the size of a room. Could it hold tens of people? hundreds of people? thousands of people? I might estimate that a room could probably hold about 100 people. If I needed a more precise estimate, I might see how

many people could stand in a marked-off area; I could estimate how many of these areas there would be in a room and multiply to get an estimate of the total number of people in the room.

For Exercises 2–4, guess whether each number is in the *hundreds, thousands, millions,* or *billions*. Then estimate each number. Describe your method.

2. the number of printed letters, numbers, and symbols on this page

3. the number of photographs in this textbook

4. the number of tennis balls that would fill a bathtub

5. **Writing** Write one or two sentences about your school using numbers to identify, order, count, and measure. Tell the category of each number.

Estimate whether each value is in the *hundreds, thousands, millions,* or *billions*.

6. the population of Chicago, Illinois

7. the air distance in miles from Washington, D.C., to Seattle, Washington

8. the population of Mexico City, Mexico

9. the number of people in the crowd for Martin Luther King's "I have a dream" speech

10. the attendance at a sold-out rock concert

---

connection to **GEOGRAPHY**

Reading For Exercises 11–16, use the facts on page 58.

11. In 1990, was the population of North America and South America *more than* or *less than* 300 million?

12. Is the air distance from San Francisco, California, to Salt Lake City, Utah, *more than* or *less than* 1500 mi?

13. Is it *reasonable* or *unreasonable* to say that the population of Los Angeles, California, in 1990 was about 10 million? Explain.

14. Is it *reasonable* or *unreasonable* to say that the population of Europe in 1990 was about 2.5 billion? Explain.

15. In 1990, the population of China was about 20% of the world's population. What was the approximate population of China in 1990?

16. About what percent of the world's population in 1990 was the United States' population?

17. If 10 adults are standing in an elevator and a sign says "maximum capacity 4000 lb," should they be concerned?

18. **Research** Estimate the number of freshmen in your school. Do some research to check your estimate.

BY THE WAY...

On August 28, 1963, Martin Luther King, Jr., gave his "I have a dream" speech in Washington, D.C. He later received the Nobel Prize for Peace for his role in the civil-rights movement.

Multicultural Note

The 1963 March on Washington for Jobs and Freedom was an extraordinary moment in history. For the first time, more than one hundred thousand people gathered together to demonstrate peacefully in support of equal rights and justice for all Americans. Lawmakers were impressed with the size of the gathering and the determination of the participants. The next day, the *Washington Post* ran a banner headline describing the demonstration as a "solemn, orderly plea for equality." The peaceful march was a tribute to Dr. Martin Luther King's philosophy of non-violence.

A. Philip Randolph, the director of the march, drew great applause when he requested a pledge by the demonstrators to return to their homes and continue the fight for freedom. The crowd then began to disperse. Those who stayed, however, heard one of the most moving and eloquent orations of all time: Martin Luther King's "I have a dream" speech.

Reasoning

Exs. 11–17 can give insight into how students analyze problems and identify information needed for finding solutions. Go through the exercises one at a time. Each time ask which facts on page 58 are relevant and how students reasoned their way to an answer.

4. thousands; about 2016 tennis balls; Tennis balls are about $2\frac{1}{2}$ in. in diameter, and the dimensions of the inside of a bathtub are about 5 ft by $2\frac{1}{2}$ ft by $1\frac{1}{2}$ ft. About 7 layers with $12 \cdot 24 = 288$ tennis balls in each layer will fit in a bathtub; $7 \cdot 288 = 2016$ tennis balls.

5. Answers may vary. Examples are given. identify: school ID number; order: periods of the day; count: number of students in each class; measure: length of playing field

Answers continued on next page.

Answers to Exercises and Problems

1. a–d. Answers may vary. Examples are given.
   a. hundreds
   b. Split the dish into 4 parts, count the bacteria in one part and multiply by 4.
   c. about 220
   d. My guess was right.

2–4. Guesses, estimates, and methods may vary. Examples are given.

2. thousands; about 1350 characters; Count the number of characters in a typical line on the page (about 45); estimate the total number of full lines on the page, combining short lines together to make up full lines (about 30); multiply to get $45 \cdot 30 = 1350$ characters.

3. hundreds; about 130 photographs; Suppose there are 13 photographs in Unit 2 of your textbook and there are 10 units in the textbook; $13 \cdot 10 = 130$ photographs.

Interdisciplinary Problems

In connection with the questions, exercises, and problems in this section on populations, some students may be interested in researching the growth in the population of the United States during the past two hundred years. They can record the populations at important dates in American history: July 4, 1776, the beginning of Lincoln's term as President, the time of our involvement in World Wars I and II, and the census in 1990. The data will reveal the tremendous growth in the population of the U.S. during the twentieth century.

Cooperative Learning

As a follow-up to the discussion of the Uniform Product Code and Exs. 30–32, divide the class into groups of three or four students to do Exs. 33 and 34. The groups will probably discover some bar codes that differ from those in the text. They should find out what they can about these codes, especially how to verify the check digit. Each group should summarize its findings in a report that can be presented to the class.

Writing Use a number anywhere on the scale below to estimate the probability of each event. Give a reason for your answer.

19. You will have pizza for breakfast on some day this year.
20. It will rain one day next week.
21. You will find a new friend this month.
22. A woman will win the next presidential election in the United States.
23. The sun will rise tomorrow.
24. Cats will learn to fly in the year 2020.

impossible	unlikely	possible	likely	certain
0%	25%	50%	75%	100%
0	0.25	0.5	0.75	1

For Exercises 25–28, classify each quantity as _continuous_ or _discrete_.

25. the population of your school
26. the height above the ground of a wheelchair being rolled up a ramp
27. the amount of fuel in a taxi's gasoline tank
28. the number of pieces of fruit sold at a market each day
29. **Open-ended** Give your own examples of one quantity that is continuous and one quantity that is discrete.

---

connection to **BUSINESS EDUCATION**

Many store products are identified by a number called a _Uniform Product Code_ (UPC). A UPC is printed on a package in two ways: as a number and as a set of vertical lines called a _bar code_.

7 26284 12501 6

broadly classifies the product · identifies the manufacturer · gives information about the product · the check digit

Stores use scanners to read bar codes. A computer connected to the scanner calculates the check digit to be sure that the other digits scanned accurately. The computer also searches its files for the price of the product and subtracts the item from the inventory.

Here is how the computer finds the check digit for the UPC shown above, 7 26284 12501 6.

1. Add the digits in the odd positions. $\longrightarrow$ $7 + 6 + 8 + 1 + 5 + 1 = 28$

2. Multiply the answer in step 1 by 3. $\longrightarrow$ $28 \cdot 3 = 84$

3. Add the digits in the first five even positions. $\longrightarrow$ $2 + 2 + 4 + 2 + 0 = 10$

4. Add the results of steps 2 and 3. $\longrightarrow$ $84 + 10 = 94$

5. Subtract the units digit of the answer in step 4 from the number 10. $\longrightarrow$ $10 - 4 = 6$ $\longleftarrow$ The result is the check digit.

Calculate the value of the check digit. Is the check digit shown correct? Write *Yes* or *No*.

30.
9 780395 421697 9

31.
0 24182 18145 5

32.
0 70735 95500 5

33. a. **Research** Find UPCs on three different grocery-store items from one manufacturer.

b. What is the code that identifies this manufacturer?

c. Using a check digit of 2, make up a valid UPC for a product made by this manufacturer.

34. **Writing** Describe how bar codes can help a store manager be sure there are enough items in the store to meet customers' needs.

Ongoing ASSESSMENT

35. **Open-ended** Every half hour for six hours record a number you see being used. Tell whether the number is used for *identifying, ordering, counting,* or *measuring.* Also tell whether the quantity is *discrete* or *continuous.* What types of numbers appear most often?

Review PREVIEW

36. Make a concept map that summarizes the ideas you have learned in this section. *(Section 1-1)*

Write the prime factorization of each number. *(Toolbox Skill 7)*

37. 42   **38.** 36   **39.** 98   **40.** 125

41. Graph the numbers 4, −7, 3, 0, −5, and 7 on a number line. What do you notice about 7 and −7? *(Toolbox Skill 2)*

 Working on the Unit Project

42. **Research** Look in books, magazines, and newspapers for interesting numerical facts about the topic you choose for your theme poster. Write down at least 10 facts. Be sure to include counts, measurements, and percents.

63

Practice 9 For use with Section 2-1

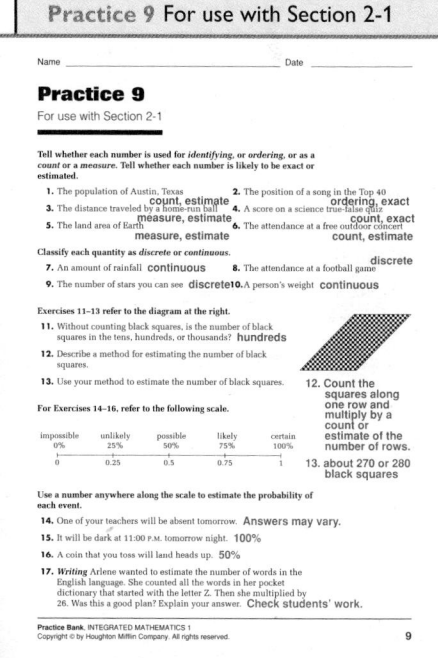

Name _____ Date _____

Practice 9
For use with Section 2-1

Tell whether each number is used for *identifying,* or *ordering,* or as a *count* or a *measure.* Tell whether each number is likely to be exact or estimated.

1. The population of Austin, Texas — count, estimate
2. The position of a song in the Top 40 — ordering, exact
3. The distance traveled by a home-run ball — measure, estimate
4. A score on a science true-false quiz — count, exact
5. The land area of Earth — measure, estimate
6. The attendance at a free outdoor concert — count, estimate

Classify each quantity as *discrete* or *continuous.*

7. An amount of rainfall — continuous
8. The attendance at a football game — discrete
9. The number of stars you can see — discrete
10. A person's weight — continuous

Exercises 11–13 refer to the diagram at the right.

11. Without counting black squares, is the number of black squares in the tens, hundreds, or thousands? — hundreds

12. Describe a method for estimating the number of black squares.

13. Use your method to estimate the number of black squares.

12. Count the squares along one row and multiply by a count or estimate of the number of rows.

13. about 270 or 280 black squares

For Exercises 14–16, refer to the following scale.

| impossible 0% | unlikely 25% | possible 50% | likely 75% | certain 100% |
| 0 | 0.25 | 0.5 | 0.75 | 1 |

Use a number anywhere along the scale to estimate the probability of each event.

14. One of your teachers will be absent tomorrow. — Answers may vary.
15. It will be dark at 11:00 P.M. tomorrow night. — 100%
16. A coin that you toss will land heads up. — 50%

17. *Writing* Arlene wanted to estimate the number of words in the English language. She counted all the words in her pocket dictionary that started with the letter Z. Then she multiplied by 26. Was this a good plan? Explain your answer. — Check students' work.

Practice Bank, INTEGRATED MATHEMATICS 1
Copyright © by Houghton Mifflin Company. All rights reserved.

9

Answers to Exercises and Problems

can easily determine if the inventory level of a popular item is lower than required. If so, the manager can restock the shelves from items in storage at the grocery store, or order the items from the distributor, if necessary.

35. Answers may vary.

36.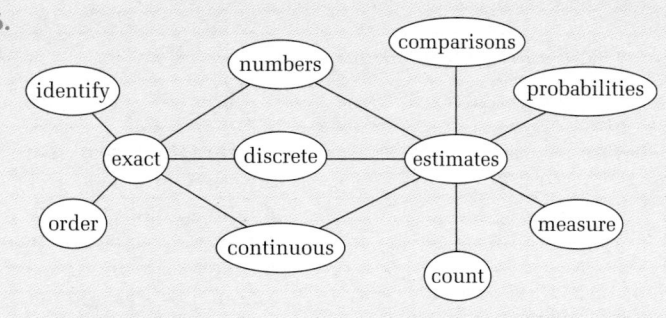

37. 2 · 3 · 7
38. 2 · 2 · 3 · 3
39. 2 · 7 · 7
40. 5 · 5 · 5
41.
-7 -5   0   3 4   7

7 and −7 are the same distance from 0.

42. Answers may vary.

Using Negative Numbers

PLUSES and minuses

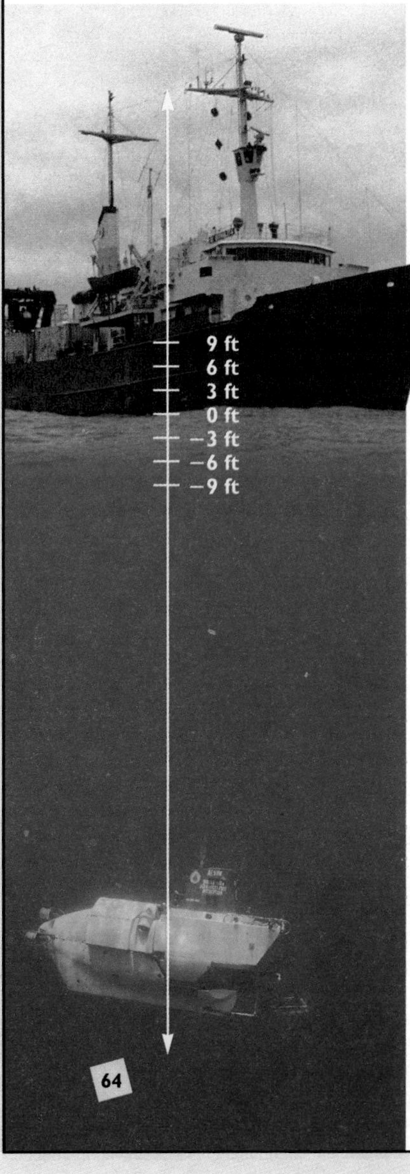

Talk it Over

1. Describe how negative numbers are used in the photo of Atlantis II and DSV Alvin. What are two other uses of negative numbers?

2. Where on the vertical number line at the left would you place −8?

3. The **absolute value** of a number is its distance from zero on a number line. You write *the absolute value of −3* as $|-3|$. An absolute value is a positive number or zero.

$$|-3| = 3 \qquad |0| = 0 \qquad |3| = 3$$

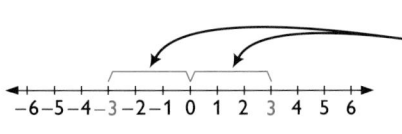

3 and −3 are located three units from 0. The absolute value of 3 is 3 and the absolute value of −3 is also 3.

   a. Read aloud and simplify $|12|$.
   b. Read aloud and simplify $|-4|$.

4. What two numbers have an absolute value of 10?

5. Two numbers are **opposites** if they are the same distance from zero on a number line, but on opposite sides of zero.

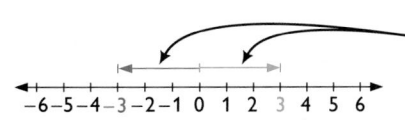

3 and −3 are the same distance from 0, but on opposite sides. They are opposites.

   What is the opposite of 7? of −2? of 0?

6. The number −5 is read *negative 5* or *the opposite of 5*. How can the number −12 be read?

7. The **+/−** key on a calculator gives the opposite of the number on the display. What is the result of pressing this key if the display shows 15? if the display shows −27?

Unit 2 Using Measures and Equations

(vertical number line at left of photo)
9 ft
6 ft
3 ft
0 ft
−3 ft
−6 ft
−9 ft

Answers to Talk it Over

1. They are used to show measurement below sea level. Answers may vary. Examples are given: temperature below zero, overdrawn bank balance

2. between −6 and −9, closer to −9

3. a. the absolute value of twelve; 12
   b. the absolute value of negative four; 4

4. 10 and −10

5. −7; 2; 0

6. *negative 12* or *the opposite of 12*

7. −15; 27

8. 0

9. 0

10. 0

11. 0

Answers to Exploration

1. Set A: a. 20   b. 13
   Set B: a. −20   b. −13
   Set C: a. −4   b. 1
   Set D: a. 4   b. −1

2. To find the sum of two negative numbers, add the numbers without their signs, then make the result negative. To find the sum of one positive and one negative number, decide which number is larger without its sign. Then find the difference of the numbers without their signs and use the sign of the number with the larger absolute value for the result.

3. Set E: Both expressions involve 42, an operation, and another number, but the operations are different and the second numbers in the expressions are oppo-

EXPLORATION

Do you remember how to operate with
negative numbers?

- **Materials: calculators**
- **Work with another student.**

Set A	Set B
a) 12 + 8	a) −12 + (−8)
b) 6 + 7	b) −6 + (−7)

Set C	Set D
a) −12 + 8	a) 12 + (−8)
b) −6 + 7	b) 6 + (−7)

Set E	Set F
a) 42 − 25	a) −14.9 − 6.4
b) 42 + (−25)	b) −14.9 + (−6.4)

Set G	Set H
a) 12.3 − (−34.9)	a) −6 − (−4)
b) 12.3 + 34.9	b) −6 + 4

① Use a calculator to simplify each expression in Sets A−D.

② Look for patterns in step 1 to figure out the rules your calculator follows to find the sum of two numbers when one or both are negative.

③ Compare the expressions in Set E. Discuss how they are alike and how they are different. Use a calculator to simplify each expression in Set E. How do the results compare?

④ Repeat step 3 for Sets F–H.

⑤ Use the results of steps 3 and 4. Replace the _?_ with the correct word.

Subtracting a number gives the same result as adding the _?_ of that number.

⑥ Use a calculator to simplify each expression in Sets J–M.

Set J	Set K
a) 6 · 3	a) (−6)(−3)
b) 50 · 5	b) (−50)(−5)

Set L	Set M
a) (−6)(3)	a) 6 · (−3)
b) (−50)(5)	b) 50 · (−5)

⑦ Look for patterns in step 6 to figure out the rules that the calculator follows to find the product of two numbers when one or both are negative.

⑧ Replace multiplication with division in each expression in Sets J–M. Use a calculator to simplify each expression.

⑨ Look for patterns in step 8 to figure out the rules that the calculator follows to find the quotient of two numbers when one or both are negative.

Talk it Over

Simplify.

8. −15 + 15   9. 8 + (−8)   10. −200 + 200

11. What do you think is the sum of any pair of opposites?

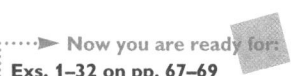
Now you are ready for:
Exs. 1–32 on pp. 67–69

2-2 Using Negative Numbers

65

Talk it Over

Use questions 3 and 4 to present the concept of *absolute value.* You may wish to give two or three more examples of reading and simplifying absolute value expressions. Also, ask students to write and simplify expressions that you give orally.

Use questions 5 and 6 to present the concept of *opposite.* Give more examples and include some nonintegers.

Using Technology

When students use calculators to compute with integers, they frequently use the [−] key rather than the [+/−] key. They may enter [− 8] or [8 −] rather than [8 +/−]. To correct these errors, go through steps 1, 3, 4, and 6 of the Exploration and have students tell which minus signs should be read *the opposite of* and which should be read *subtract.* It also may be helpful to have students write the correct keystroke sequences for several expressions before entering the expressions on their calculators.

Answers to Exploration

sites. 17; 17; The results are the same.

4. Set F: Both expressions involve −14.9, an operation, and another number, but the operations are different and the second numbers in the expressions are opposites. −21.3; −21.3; The results are the same. Set G: Both expressions involve 12.3, an operation, and

another number, but the operations are different and the second numbers in the expressions are opposites. 47.2; 47.2; The results are the same. Set H: Both expressions involve −6, an operation, and another number, but the operations are different and the second numbers in the expressions are oppo-

sites. −2; −2; The results are the same.

5. opposite

6. Set J: a. 18   b. 250
   Set K: a. 18   b. 250
   Set L: a. −18   b. −250
   Set M: a. −18   b. −250

7. To find the product of two positive or two negative numbers, multiply the numbers without their

signs, then make the result positive. To find the product of one positive and one negative number, multiply the numbers without their signs, then make the result negative.

8. Set J: a. 2   b. 10
   Set K: a. 2   b. 10
   Set L: a. −2   b. −10
   Set M: a. −2   b. −10

9. To find the quotient of two positive or two negative numbers, divide the numbers without their signs, then make the result positive. To find the quotient of one positive and one negative number, divide the numbers without their signs, then make the result negative.

65

Additional Samples

S1 Evaluate $-x$ for each value of x.

  a. $x = -7.5$ $-x = 7.5$

  b. $x = 6$ $-x = -6$

  c. $x = -\frac{2}{9}$ $-x = \frac{2}{9}$

S2 Evaluate each expression for the given values of the variables.

  a. $-7 + x$ when $x = -5$ -12

  b. $r - 3s^2$ when $r = 20$ and $s = -4$ -28

Communication: Reading

In Sample 1, the recommended way to read $-x$ is *the opposite of x*. It is best to avoid saying *negative x*, since some students may erroneously think that all values of $-x$ are negative.

Talk it Over

Question 12 leads to the important generalization that the product of -1 and a number is the opposite of the number.

Evaluating Algebraic Expressions Involving Negative Numbers

You need to be careful with negative numbers when you evaluate expressions. You may want to use parentheses to do this.

Sample 1

The expression $-x$ is read *the opposite of x*. Evaluate $-x$ for each value of x.

  a. $x = 1.3$ **b.** $x = -5$ **c.** $x = 0$

Sample Response

Substitute.

  a. $-x = -(1.3) = -1.3$ **b.** $-x = -(-5) = 5$ **c.** $-x = -(0) = 0$

Ask yourself, what is the opposite of 1.3?

Sample 2

Evaluate each expression for the given values of the variables.

  a. $-x + 4$ when $x = -9$

  b. $m^2 + 2n$ when $m = -8$ and $n = -12$

Sample Response

  a. $-x + 4 = -(-9) + 4$ ← Substitute -9 for x.

          $= 9 + 4$ ← Simplify.

          $= 13$

  b. $m^2 + 2n = (-8)^2 + 2(-12)$ ← Substitute -8 for m and -12 for n.

          $= 64 + (-24)$ ← $(-8)^2 = (-8)(-8) = 64$

          $= 40$

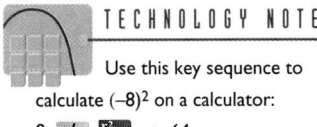

TECHNOLOGY NOTE

Use this key sequence to calculate $(-8)^2$ on a calculator:

8 [+/-] [x^2] → 64

Talk it Over

12. Evaluate $-1 \cdot x$ when $x = 1.3$, $x = -5$, and $x = 0$. Compare the results with those in Sample 1. What do you observe? Generalize your observations.

13. Suppose the expression in part (b) of Sample 2 were $m^2 - 2n$. What would the answer be?

14. Which of the samples so far can you do using mental math? Which would you prefer to do with a calculator? Why?

Unit 2 Using Measures and Equations

Answers to Talk it Over

12. $-1.3, 5, 0$; The results are the same as those in Sample 1. The opposite of a number is equal to the product of -1 and the number: $-x = -1 \cdot x$.

13. 88

14. Answers may vary. An example is given. I can do Sample 1 by mental math. I prefer to use a calculator for Sample 2 because it involves more than one calculation at a time.

Answers to Look Back

Answers may vary. An example is given. To add two numbers with the same sign, add the numbers and keep the sign. To add two numbers with different signs, decide which number is larger without its sign. Then subtract the numbers without their signs and use the sign of the number with the larger absolute value for the result. To subtract two

The type of wax you put on cross-country skis depends on the temperature of the snow. Suppose the snow is fine-grained and its temperature is −4°F. Is this the correct wax to use?

Sample Response

Find the equivalent Celsius temperature for a temperature of −4°F. To convert a Fahrenheit temperature (F) to the equivalent Celsius temperature (C), use the expression $\frac{5}{9}(F - 32)$.

$$\frac{5}{9}(F - 32) = \frac{5}{9}(-4 - 32) \longleftarrow \text{Substitute } -4 \text{ for } F.$$

$$= \frac{5}{9}(-36) \longleftarrow \begin{array}{l}\text{First work inside the}\\\text{parentheses. Then multiply.}\end{array}$$

$$= -20 \longleftarrow \text{Recall: } \frac{5}{9} \cdot \frac{-36}{1} = -20$$

A temperature of −4°F is equivalent to a temperature of −20°C. The wax shown is for a temperature between −3°C and −10°C. This is not the correct wax to use.

Look Back ◄

Summarize how to do operations with negative numbers.

⋯▶ **Now you are ready for:**
Exs. 33–63 on pp. 69–70

2-2 Exercises and Problems

Find the absolute value and the opposite of each number.

1. −17   2. 2.34   3. $-\frac{1}{2}$ 4. $6\frac{2}{3}$ 5. −3,785,200

Reading For Exercises 6–11, tell whether each statement is *True* or *False*. If the statement is false, give a counterexample.

6. The opposite of every positive number is negative.

7. The opposite of every negative number is negative.

8. The absolute value of a positive number is always negative.

9. The absolute value of a negative number is always positive.

10. The absolute value of zero is zero.

11. $|x| = x$ for every value of x.

12. Predict what you will get if you enter 8 +/− +/− on a calculator. Then try it. What does this tell you about the opposite of the opposite of a number?

2-2 Using Negative Numbers   **67**

Using Technology

Most scientific and graphing calculators have parentheses to help ensure that operations are performed in the correct order. The way expressions are written does not necessarily coincide with the way they are entered on the calculator.

Reasoning

Ask students to check whether $|a + b| = |a| + |b|$ and $|ab| = |a| \cdot |b|$. This can be done mentally or by using $\boxed{\text{abs}}$ on a calculator.

Use the result of step 5 of the Exploration on page 65. Replace each $\underline{\,?\,}$ with the correct number.

13. $86 - 53 = 86 + \underline{\,?\,}$ **14.** $-19 - (-25) = -19 + \underline{\,?\,}$

Student Resources Toolbox
p. 640 *Integers*

Simplify.

15. $|-18|$ **16.** $|0|$ **17.** $|24|$

18. $-17.4 + (-2)$ **19.** $-48 + 29$ **20.** $16 - (-4.5)$

21. $-120 - 60$ **22.** $5.78 + (-2.4)$ **23.** $12 \cdot (-2.5)$

24. $-7.2 \div (-8)$ **25.** $-9 \div 2$ **26.** $(-100)(-21)$

27. **Writing** You can do many calculations faster in your head than on a calculator. Which of the expressions in Exercises 18–26 do you think you can do as easily in your head as on a calculator? Pick two of them and explain how you could simplify them in your head.

connection to **G E O G R A P H Y**

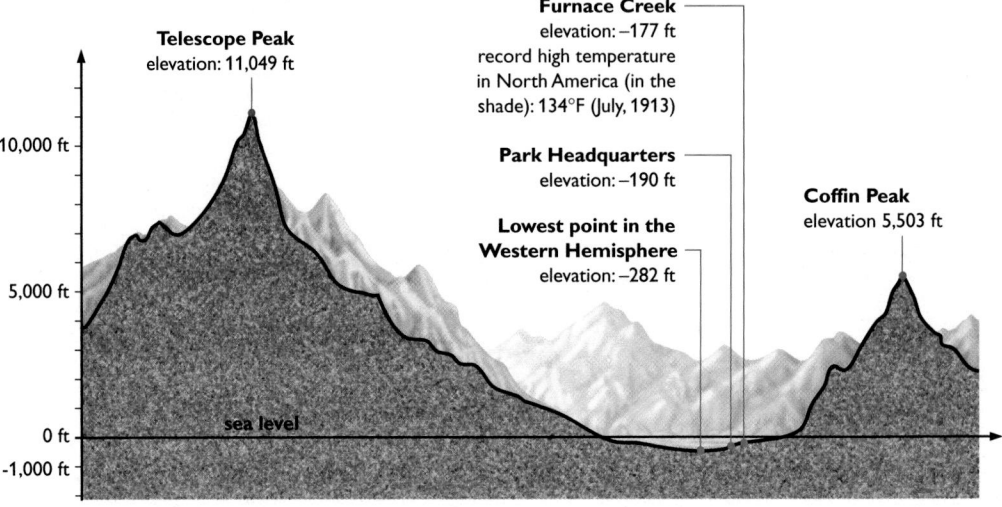

Telescope Peak elevation: 11,049 ft

Furnace Creek elevation: −177 ft record high temperature in North America (in the shade): 134°F (July, 1913)

Park Headquarters elevation: −190 ft

Lowest point in the Western Hemisphere elevation: −282 ft

Coffin Peak elevation 5,503 ft

10,000 ft

5,000 ft

sea level

0 ft

-1,000 ft

DEATH VALLEY

For Exercises 28–30, use the diagram. The elevations in the diagram are given in relation to sea level.

28. Find the difference between the elevation of Telescope Peak and the elevation of the lowest point in Death Valley.

29. Official temperatures are no longer recorded at Furnace Creek. Instead, they are measured at the Death Valley park headquarters nearby. What is the difference in elevation between these two sites?

 68   **Unit 2** Using Measures and Equations

Answers to Exercises and Problems

13. −53   **14.** 25

15. 18   **16.** 0

17. 24   **18.** −19.4

19. −19   **20.** 20.5

21. −180   **22.** 3.38

23. −30   **24.** 0.9

25. −4.5   **26.** 2100

27. Answers may vary. An example is given. Exercises 18, 20, 21, and 26 can be easily done by mental math. In Exercise 21, just add 120 and 60 together in your head and keep the negative sign; you get −180. In Exercise 26, multiplying 21 by 100 is easy and you know that the product of two negatives is a positive; the product is 2100.

28. 11,331 ft

29. 13 ft

30. a. 79°F to 101°F

    b. −3°F to 19°F

31. $300

32. 515 points

33. $3 + 18 \div 2 \cdot (-3) = 3 + 9 \cdot (-3) = 3 + (-27) = -24$

30. It is estimated that temperatures in Death Valley are 3°F to 5°F cooler for every 1000 ft you go up.

   a. Estimate what the temperature on Telescope Peak might have been when the record high temperature was recorded at Furnace Creek. Give your answer as a range of temperatures.

   b. The average temperature in Death Valley in the month of January is 52.0°F. Estimate the average temperature on Telescope Peak in the month of January. Give your answer as a range of temperatures.

31. A contestant on a television game show has a score of −$200 and then correctly answers a question worth $500. What is the contestant's new score?

32. Timmy is playing a card game with his grandfather. Timmy has a score of −15 and his grandfather has a score of 45. A score of 500 wins. How many points does Timmy need in order to win?

Simplify. Show every step.

33. $3 + 18 \div 2 \cdot (-3)$ **34.** $14 - 5(-7 + 9)$ **35.** $-9 \div 2 \cdot 10 \div 3$

36. $2.4 + (0.6)(-0.2) - 6$ **37.** $\dfrac{-20 - 5}{2}$ **38.** $\dfrac{-9 + 15}{3 \cdot (-2)}$

Evaluate each expression for the given values of the variables.

39. $-n$ when $n = 12.5$ **40.** $-p + 1.8$ when $p = -3.2$

41. $k + k$ when $k = -8$ **42.** $xy + 3$ when $x = -5$ and $y = 10$

43. $\dfrac{a + b}{2}$ when $a = -9$ and $b = 3$ **44.** $-4.9t + at$ when $t = 10$ and $a = 50$

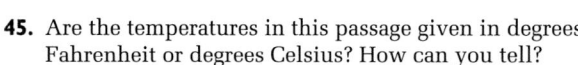

In Jack London's *To Build a Fire*, a logger's life is endangered by the extreme temperature in the Yukon.

To Build a Fire

At the man's heels trotted a dog, a big native husky, the proper wolf dog....The animal was depressed by the tremendous cold....Its instinct told it a truer tale than was told to the man by the man's judgment. In reality, it was not merely colder than fifty below zero....It was seventy-five below zero. Since the freezing point is thirty-two above zero, it meant that one hundred and seven degrees of frost obtained. The dog did not know anything about thermometers....But the brute had its instinct....The dog had learned fire, and it wanted fire...

45. Are the temperatures in this passage given in degrees Fahrenheit or degrees Celsius? How can you tell?

46. Write each temperature in this passage using symbols instead of words.

47. How many degrees below freezing is the temperature?

48. Convert the temperatures in this passage to the other temperature scale as in Sample 3.

Error Analysis

The step-by-step approach in Exs. 33–38 should help students who make errors discover whether their mistakes are computational or due to an incorrect order of operations. It is always preferable for students to discover for themselves where and how mistakes occur.

Mathematical Procedures

In Exs. 39–44, some students may find it helpful initially to use parentheses around all numbers substituted for variables, even positive numbers.

Limited English Proficiency

The task in Ex. 46, writing mathematical content of a narrative passage in symbolic form, is in general a very useful strategy for many math students acquiring English. To ensure that these students understand the content of the paragraph before converting its math to symbols, you might pair them with fluent speakers of English and have the partners orally summarize the situation described.

Answers to Exercises and Problems

34. $14 - 5(-7 + 9) = 14 - 5(2) =$
$14 - 10 = 4$

35. $-9 \div 2 \cdot 10 \div 3 =$
$(-4.5) \cdot 10 \div 3 =$
$-45 \div 3 = -15$

36. $2.4 + (0.6)(-0.2) - 6 =$
$2.4 + (-0.12) - 6 =$
$2.4 - 0.12 - 6 =$
$2.28 - 6 = -3.72$

37. $\dfrac{-20 - 5}{2} = \dfrac{-25}{2} = -12.5$

38. $\dfrac{-9 + 15}{3 \cdot (-2)} = \dfrac{6}{-6} = -1$

39. -12.5

40. 5

41. -16

42. -47

43. -3

44. 451

45. The temperatures are in degrees Fahrenheit. Freezing occurs at 32°F.

46. −50°F; −75°F; 32°F

47. 107°F

48. −50°F = −45.6°C;
−75°F = −59.4°C;
32°F = 0°C

Practice 10 For use with Section 2-2

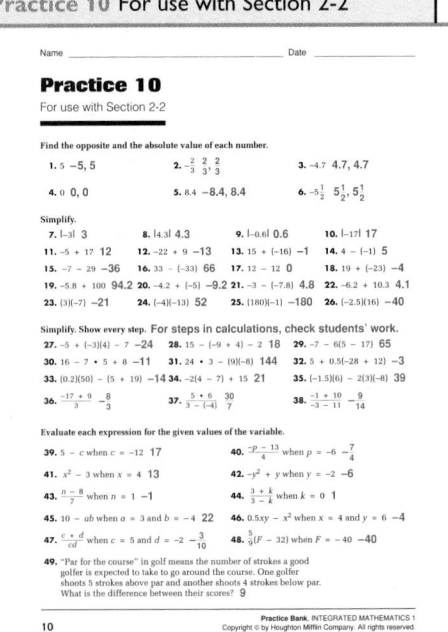

10

---

connection to SCIENCE

Here are two formulas to convert temperatures between Celsius (C) and Fahrenheit (F).

$$C = \frac{5}{9}(F - 32) \qquad F = \frac{9}{5}C + 32$$

Copy and complete the table of equivalent temperatures.

	°F	°C
49. Normal body temperature	98.6	?
50. Temperature at which water freezes	32	?
51. Average night temperature on the moon's surface	?	−153
52. Record low temperature for Antarctica (recorded on July 21, 1983)	?	−89.2

53. **Open-ended** Choose two temperatures that you come across in your daily life to add to your table in Exercises 49–52. Write the temperatures in degrees Fahrenheit and degrees Celsius.

Ongoing ASSESSMENT

54. **Group Activity** There is one very cold temperature where the temperature in degrees Fahrenheit equals the temperature in degrees Celsius. Work with another student using trial and error to find this temperature.

Review PREVIEW

For Exercises 55–57, estimate whether each value is in the *hundreds, thousands, millions,* or *billions.* (Section 2-1)

55. the population of California

56. the number of people at a World Series game

57. the distance in miles from New York City to Phoenix, Arizona

58. Replace each __?__ with the correct term. (Section 1-3)

In the power 10^{31}, 10 is the __?__ and 31 is the __?__. The number 31 in 10^{31} tells you that 10 is used as a __?__ 31 times.

Write each expression as a power of ten. (Section 1-3)

59. $10^2 \cdot 10^{12}$ 60. $10^{15} \cdot 10^{25}$ 61. $\dfrac{10^{20}}{10^5}$ 62. $\dfrac{10^{36}}{10^9}$

 Working on the Unit Project

63. **Group Activity** Share the facts you have collected with your group. Do your facts include a negative number? If not, discuss whether one can reasonably be found in a fact about your theme. Plan to include a negative number on your theme poster if you can.

Answers to Exercises and Problems

49. 37°C

50. 0°C

51. −243.4°F

52. −128.56°F

53. Answers may vary. An example is given at the right.

54. −40°F = −40°C

	Fahrenheit	Celsius
a cold winter temperature	0°F	−17.8°C
temperature at which water freezes	32°F	0°C
a spring or fall temperature	50°F	10°C
temperature on a very hot day	95°F	35°C
temperature for baking food in an oven	350°F	176.7°C

55–57. Estimates may vary. Examples are given.

55. millions

56. thousands

57. thousands

58. base, exponent or power; factor

59. 10^{14} 60. 10^{40}

61. 10^{15} 62. 10^{27}

63. Answers may vary.

Exploring Scientific Notation

SHORThand 🅕🅞🅡 Numbers

EXPLORATION

What do negative powers of 10 mean?

- **Materials: scientific calculators**
- **Work in a group of two or three students.**

1. Copy the table onto your paper.

2. Use the power key, y^x or x^y, on your calculator to complete the decimal column.

3. Complete the last lines of the fraction column.

4. Look for patterns in the data in the table. Write down the observations made by people in your group.

5. Make conjectures about the relationship between the exponent of a power of 10 and the equivalent decimal or fraction.

Talk it Over

1. a. How many zeros are there in the decimal notation for the power 10^6?

   b. How many zeros are there to the right of the decimal point in the decimal notation for 10^{-6}?

   c. How many decimal places are there in the decimal notation for 10^{-6}?

2. The power 10^{-6} can be written as the fraction $\frac{1}{1,000,000}$ or $\frac{1}{10^6}$. Write 10^{-12} as a fraction in two different ways.

2-3 Exploring Scientific Notation

71

PLANNING

Objectives and Strands
See pages 54A and 54B.

Spiral Learning
See page 54B.

Materials List
➤ Scientific calculator

Recommended Pacing
Section 2-3 is a two-day lesson.

Day 1
Pages 71–73: Exploration through Talk it Over 8, *Exercises 1–26*

Day 2
Pages 73–75: Scientific Notation on a Calculator through Look Back, *Exercises 27–55*

Extra Practice
See pages 622–623.

Warm-Up Exercises
Warm-Up Transparency 2-3

Support Materials
➤ Practice 11
➤ Enrichment 10 in the Activity Bank
➤ Study Guide 2-3
➤ Problem Set 3
➤ Diagram Master 10 in the Explorations Lab Manual
➤ Quiz 2-3
➤ Test 5
➤ Alternative Assessments 2, 3

Answers to Exploration

1–3.

Power	Decimal	Fraction
10^4	10,000	—
10^3	1000	—
10^2	100	—
10^1	10	—
10^0	1	—
10^{-1}	0.1	$\frac{1}{10}$
10^{-2}	0.01	$\frac{1}{100}$
10^{-3}	0.001	$\frac{1}{1000}$

4. Observations may vary. Examples are given. If the exponent is a positive integer, then the exponent tells how many zeros follow the 1 in the decimal form of the number. If the exponent is a negative integer, then the opposite of the exponent tells how many decimal places are in the decimal form of the number. As you go from one line in the table down to the next, the exponent decreases by 1 and the number is divided by 10.

5. Conjectures may vary. An example is given. If the exponent of a power of 10 is a positive integer, then the exponent tells the number of zeros following a 1 in the decimal form of the number. If the exponent of a power of 10 is a negative integer, then the opposite of the exponent tells the number of decimal places in the decimal form of the number and also the number of zeros following the 1 in the denominator of the fraction form. If the exponent is zero, the number is equal to 1.

Talk it Over answers are on next page.

Exploration

Use the Exploration to lead students to discover and understand the relationship between the exponent of a power of 10 and the equivalent decimal or fraction. The reasons for using the calculator are twofold: to familiarize students with the power key and to facilitate their discovery of patterns in powers of 10.

Talk it Over

Questions 1 and 2 permit students to check the conjectures they made in step 5 of the Exploration. You may wish to examine another example or two by using other positive or negative exponents.

Additional Samples

S1 Write each number in scientific notation.

a. the area of Earth's surface in square miles: 196,950,000 1.9695×10^8

b. the weight of a fairy fly beetle in grams: 0.000005 5×10^{-6}

S2 Write each number in decimal notation.

a. the speed of sound (in air) in centimeters per second: 3.43×10^4 34,300

b. the increase in length in meters of a one-meter steel bar for each degree Celsius rise in temperature: 1.2×10^{-5} 0.000012

Light travels about five trillion, nine hundred billion miles in a year! Scientists work with very small and very large numbers that have many digits. To save space and make it easier to compare numbers, scientists write them in a shorter form called **scientific notation.**

$$5{,}900{,}000{,}000{,}000 = 5.9 \times 10^{12}$$

a number that is at least one but less than 10 ——— a power of 10

multiplied by

Write each number in scientific notation.

a. the number of seconds in a week: 604,800

b. the thickness of a piece of paper in inches: 0.0039

Sample Response

a. The leading digit is 6. It is in the hundred thousands' place.

$604{,}800 = 6.048$ hundred thousands

$= 6.048 \times 100{,}000$

$= 6.048 \times 10^5$

b. The first digit that is not zero is 3. It is in the thousandths' place.

$0.0039 = 3.9$ thousandths

$= 3.9 \times 0.001$

$= 3.9 \times 10^{-3}$

Write each number in decimal notation.

a. the average radius of Earth in meters: 6.38×10^6

b. the thickness of a wire on a computer chip in meters: 3×10^{-6}

Sample Response

$6.38 \times 10^6 = 6.38 \times 1{,}000{,}000$

a. $6.38 \times 10^6 = 6{,}380{,}000$

Multiplying by 1,000,000 moves the decimal point 6 places to the right.

$3 \times 10^{-6} = 3 \times 0.000001$

b. $3 \times 10^{-6} = 0.000003$

Multiplying by 0.000001 moves the decimal point 6 places to the left.

Unit 2 Using Measures and Equations

Answers to Talk it Over

1. **a.** 6
   b. 5
   c. 6

2. $\dfrac{1}{10^{12}}, \dfrac{1}{1{,}000{,}000{,}000{,}000}$

Talk it Over

Write each number in scientific notation. Explain your steps.

3. 34,000
4. 0.00000219

Write each number in decimal notation and read it in words.

5. 9.3×10^6
6. 4.8×10^{-4}

Is each expression written in scientific notation? If not, write it in scientific notation.

7. 0.4×10^{15}
8. 12.7×10^{-6}

Now you are ready for: Exs. 1–26 on pp. 75–76

Scientific Notation on a Calculator

Results found by calculators and computers are often in scientific notation. Calculator displays have a limited number of digits. When a result is too big, some calculators display the result in scientific notation without showing the base 10.

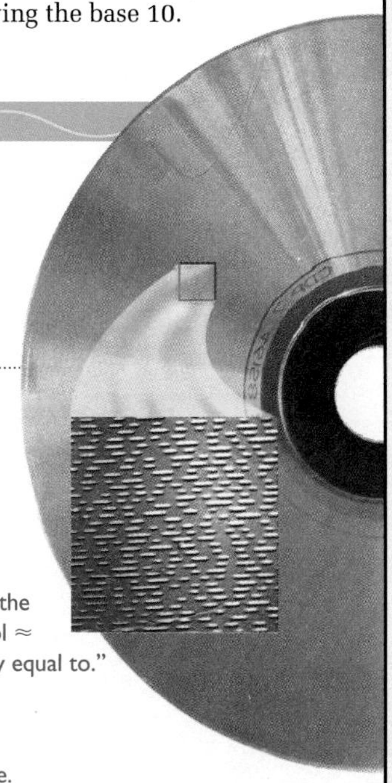

Sample 3

An audio CD contains about 3.46 million bits of information for each second of recording time. About how many bits are needed for a six-minute song?

Sample Response

Multiply by the number of seconds in six minutes.

360 [×] 3,460,000 [=] → [1.2456 09] ← This means $1.2456 \times 10^9 = 1{,}245{,}600{,}000$.

Check Estimate to see if the calculator result is reasonable.

$360 \cdot 3.46 \text{ million} \approx 400 \cdot 3{,}000{,}000$ ← Round each number to the leading digit. The symbol ≈ means "is approximately equal to."

$\approx (4 \times 10^2)(3 \times 10^6)$ ← Write each number in scientific notation.

$\approx 12 \times 10^8$ ← Use the product of powers rule. $10^2 \cdot 10^6 = 10^{2+6} = 10^8$

$\approx 1.2 \times 10^9$

Watch Out! Do more than add the powers of ten to see if a result is reasonable. Look at the product of the leading digits too.

The power of ten is the same for 1,200,000,000 and 1,245,600,000. ✓ The calculator result is reasonable.

About 1.25 billion bits on a CD are needed for a six-minute song.

Reasoning

The introduction to scientific notation mentions that this notation saves space and makes it easier to compare numbers. You may wish to show how easily comparisons can be made. Write these numbers on the chalkboard:

5.8×10^6 6.7×10^5 6.63×10^5

Ask students how the numbers can be arranged from least to greatest. Use discussion to help students see that the first number is in the millions, while the other two are in the hundred thousands. Of the two that are in the hundred thousands, 6.63×10^5 is the smaller, since 6.63 is less than 6.7. Now, write the numbers in increasing order on the chalkboard:

6.63×10^5 6.7×10^5 5.8×10^6

Additional Sample

S3 Use a calculator to help solve the problem. Then use estimation to check the reasonableness of your answer.

In the early 1990s, American passenger vehicles used about 73,526,953,000 gallons of fuel per year. At this rate, how many gallons of fuel would be used in 15 years?

about 1.1029×10^{12} or 1.1 trillion gallons

Answers to Talk it Over

3–4. Descriptions may vary. Examples are given.

3. 3.4×10^4; You can think of 34,000 as 3.4 ten thousands. Since ten thousand is the fourth power of ten, 34,000 is the same as 3.4×10^4.

4. 2.19×10^{-6}; The first digit not equal to zero, 2, is in the millionths' place. Since millionths is the sixth decimal place, 0.00000219 is the same as 2.19×10^{-6}.

5. 9,300,000; nine million, three hundred thousand

6. 0.00048; forty-eight hundred-thousandths

7. No, it should be 4×10^{14}.

8. No, it should be 1.27×10^{-5}.

S4 Use a calculator to help solve the problem. Then use estimation to check the reasonableness of your answer.

In 1991, O'Hare airport in Chicago had about 5.98×10^7 passengers. On the average, about how many passengers per day did the airport have?

about 1.6384×10^5 or 164,000 passengers

Using Technology

When you discuss Samples 3 and 4, encourage students to experiment with their calculators to see if there are different ways to enter numbers using scientific notation. They should also investigate how answers are displayed. A good approach is to use numbers that are easy to work with mentally or with paper and pencil. Students can predict what the calculator will display and check the prediction by observing the display that actually results.

distance from the sun to Pluto = 4 billion miles

Light travels four times this distance in one day.

TECHNOLOGY NOTE

Some calculators and computers use a letter E between the decimal and the exponent.

`1.6164 E10`

E stands for exponent.

74

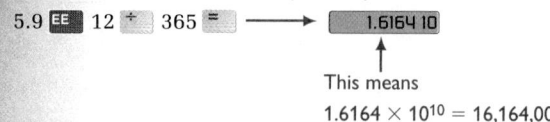

Sample 4

Light travels about 5.9×10^{12} mi in a year. About how far does light travel in a day?

Sample Response

Divide by the number of days in a year.

5.9 **EE** 12 **÷** 365 **=** ⟶ `1.6164 10`

This means
$1.6164 \times 10^{10} = 16,164,000,000.$

Check Estimate to see if the calculator result is reasonable.

$$\frac{5.9 \times 10^{12}}{365} \approx \frac{6 \times 10^{12}}{4 \times 10^2}$$

Write numbers in scientific notation and round to the leading digit.

$$\approx \frac{6}{4} \times \frac{10^{12}}{10^2}$$

$$\approx 1.5 \times 10^{10} \qquad \frac{10^{12}}{10^2} = 10^{12-2} = 10^{10}$$

The power of ten is the same for 15,000,000,000 and 16,164,000,000. ✔

The calculator result is reasonable.

Light travels about 16 billion miles in a day.

Talk it Over

9. In Sample 3, suppose you want to know how many bits of information are on a particular CD. What other information do you need?

Replace each __?__ with the correct power of ten.

10. $0.5(1.25 \times 10^6) \approx (5 \times 10^{-1})(1 \times 10^6)$

$$\approx 5 \times 10^{-1+6}$$

$$\approx 5 \times \underline{?}$$

11. $\dfrac{2.3 \times 10^3}{76,000} \approx \dfrac{2 \times 10^3}{8 \times 10^4}$

$$\approx 0.25 \times 10^{3-4}$$

$$\approx 0.25 \times \underline{?}$$

$$\approx 2.5 \times \underline{?}$$

12. What do you get when you enter 10 **yˣ** 4 **=** on your calculator? What do you get when you enter 10 **EE** 4? Why are the results different?

Unit 2 Using Measures and Equations

Answers to Talk it Over

9. You need to know how long the recording time is.

10. 10^5

11. 10^{-1}; 10^{-2}

12. 10,000; 100,000; 10 EXP 4 means $10 \cdot 10^4$ or 10^5.

Answers to Look Back

Answers may vary. An example is given. $10^3 \cdot 10^2 = 10^5$; $\dfrac{10^4}{10^3} = 10^1$; $10^{-3} = \dfrac{1}{10^3}$

RULES OF EXPONENTS

In Unit 1 you worked with these rules of exponents.

Product of Powers Rule: $10^a \cdot 10^b = 10^{a+b}$

Quotient of Powers Rule: $\dfrac{10^a}{10^b} = 10^{a-b}$

In the Exploration you discovered these rules of exponents.

Zero Exponent Rule: $10^0 = 1$

Negative Exponent Rule: $10^{-n} = \dfrac{1}{10^n}$

Look Back ◄

You have now worked with powers of ten with exponents that are positive integers, negative integers, or zero.

Choose values for a, b, and n that are positive integers. Give an example of the product of powers rule, the quotient of powers rule, and the negative exponent rule.

▶ **Now you are ready for:**
Exs. 27–55 on pp. 76–78

Look Back

Use the Look Back to review the rules of exponents. Discuss how the rules are useful in computing with numbers in scientific notation. ⋯⋯⋯⋯●

APPLYING

Suggested Assignment

Standard 1–14, 16–25, 27–37, 43–55

Extended 1–55

Integrating the Strands

Number Exs. 1–42, 47, 55

Algebra Exs. 3–9, 33, 48–51

Measurement Exs. 16–25, 52–54

Statistics and Probability Exs. 43–46

Mathematical Procedures

If students are having difficulty going from decimal notation to scientific notation, they may find it helpful to draw a box around the digits they need for the first factor in scientific notation. Then have them place the decimal point after the left-most digit and decide on the power of ten. For example:

0.0000 $\boxed{3708}$
↓
3.708

Multiplying 3.708 by 10^{-5} will move the decimal point 5 places to the left of 3. The scientific notation is 3.708×10^{-5}.

2-3 Exercises and Problems

1. **Reading** A number written in scientific notation has a specific form. What is that form?

2. **a.** Describe two methods for finding the decimal notation for 10^7.

   b. Write 10^7 in decimal notation.

3. **a.** Describe a method for finding the decimal notation for 10^{-7}.

   b. Write 10^{-7} in decimal notation.

Tell whether each power of 10 is *less than, greater than,* or *equal to* 1.

4. 10^3 _?_ 1

5. 10^0 _?_ 1

6. 10^{-8} _?_ 1

7. 10^1 _?_ 1

8. 10^{-1} _?_ 1

9. 10^{-10} _?_ 1

10. Use the results of Exercises 4–9. Describe how you can use the exponent of a power of 10 to tell whether the number is greater than or less than 1.

Write each number in exponential notation using a base of 10.

11. 10,000,000,000

12. 0.0001

13. 1

14. 0.0000001

15. Write each answer to Exercises 12 and 14 as a fraction with an exponent.

Write each number in scientific notation.

16. average radius of the sun: 695,000,000 m

17. an erg (a unit of energy): 0.0000000239 Cal

18. sound intensity of a normal conversation: 0.000001 W/m²

19. mass of propellant in a space shuttle solid rocket booster: 503,600 kg

2-3 Exploring Scientific Notation $\quad$ **75**

Answers to Exercises and Problems ⋯⋯

1. Scientific notation writes a number as the product of a number greater than or equal to one and less than 10, and a power of 10.

2. **a.** Descriptions may vary. Examples are given. Write out a product of 7 tens and multiply, use a calculator, or use the rule that 10^7 can be written as a 1, followed by 7 zeros.

   b. 10,000,000

3. **a.** Descriptions may vary. Examples are given. Use a calculator, rewrite 10^{-7} as $\frac{1}{10^7}$ and find its decimal form, or use the rule that 10^{-7} has seven decimal places (six zeros to the right of the decimal point, followed by a one).

   b. 0.0000001

4. >   5. =   6. <

7. >   8. <   9. <

10. Answers may vary. An example is given. If the exponent of a power of ten is positive, the number is greater than one. If the exponent is zero, then the number is equal to one. If the exponent is negative, then the number is less than one but greater than zero.

11. 10^{10} 12. 10^{-4}

13. 10^0 14. 10^{-7}

15. $\frac{1}{10^4}$; $\frac{1}{10^7}$ 16. 6.95×10^8 m

17. 2.39×10^{-8} Cal

Answers continued on next page.

For Exs. 31–36, ask students to separate the expressions into three categories: (1) easy to do mentally, (2) best done with paper and pencil, and (3) best done with a calculator. Ask them to explain briefly why they classified expressions as they did. Ask also for a brief description of how they would check their answers.

For Exercises 20–26, use the diagram below. Write each number in decimal notation.

20. the diameter of a cell
21. the length of a pygmy shrew
22. the height of Mount Everest
23. the length of a common dolphin
24. the length of an ant
25. the length of a blue whale

cell
1×10^{-5} m

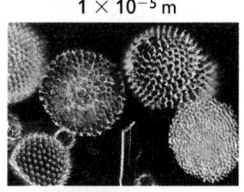

blue whale
(longest mammal)
2.6×10^1 m

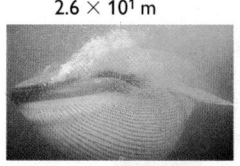

Mount Everest
8.848×10^3 m

Eastern
gray squirrel
2.5×10^{-1} m

ant
4.2×10^{-3} m

| 10^{-5} | 10^{-4} | 10^{-3} | 10^{-2} | 10^{-1} | 10^0 | 10^1 | 10^2 | 10^3 |

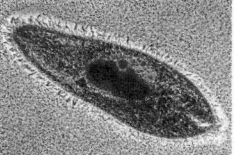

paramecium
2.10×10^{-4} m

pygmy shrew
(smallest mammal)
4×10^{-2} m

common dolphin
2.2×10^0 m

soccer field
1×10^2 m

26. **Writing** About how many times larger than a common dolphin is a blue whale? Describe how the exponents of the powers of 10 help you make this type of comparison.

Match each key sequence with the correct calculator display.

27. 10 $\boxed{y^x}$ 5 $\boxed{=}$

A. $\boxed{\qquad 50}$ E. $\boxed{5 \ {-}10}$

28. 5 $\boxed{EE}$ 10

B. $\boxed{10 \ {-}5}$ F. $\boxed{\qquad {-}50}$

29. 10 $\boxed{y^x}$ 5 $\boxed{+/-}$ $\boxed{=}$

C. $\boxed{\quad 100000}$ G. $\boxed{\quad 0.00001}$

30. 5 $\boxed{EE}$ 10 $\boxed{+/-}$

D. $\boxed{5 \ 10}$

Simplify. Write each answer in scientific notation.

31. $120(1.2 \times 10^8)$
32. $0.3(2.8 \times 10^6)$
33. $70(5.4 \times 10^{-9})$

34. $\dfrac{8.86 \times 10^7}{300}$
35. $\dfrac{7.2 \times 10^4}{36}$
36. $\dfrac{5.7 \times 10^2}{2000}$

37. **Writing** Explain how you could find the product $(3.4 \times 10^{15})(1.2 \times 10^{23})$ without using a calculator. Then write a key sequence you could use to find the product using a scientific calculator.

For Exercises 38–40, make a reasonable estimate of each answer. Then calculate a more precise answer. Write each answer in scientific notation, in decimal notation, and in words.

38. In the early 1990s, the United States Postal Service printed 118 million stamps each day. How many stamps were printed in one year? in five years?

39. Only about one fifth of the 35 million paper clips that were bought each day in the early 1980s were actually used to clip paper. How many paper clips were used in ten years?

40. The United States Bureau of Engraving and Printing used to shred 12 million pieces of worn-out money each day. At that rate, how many bills have been destroyed during your lifetime?

41. Light travels approximately 9.5×10^{15} m in one year. About how far does light travel in one day? in one minute? in one second?

42. **Chemistry** A scientist uses 1.2×10^{-5} L of an enzyme for one chemical reaction. The scientist wants to mix enough for 140 reactions. How much of the enzyme does the scientist need? Write your answer in scientific notation.

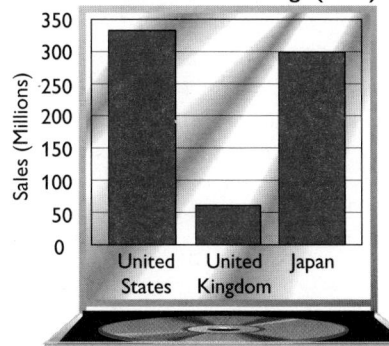

Sales of CD Recordings (1991)

Sales (Millions) — United States, United Kingdom, Japan

For Exercises 43–46, use the graph. Write each answer in scientific notation and in decimal notation.

43. About how many compact discs were sold in each country in 1991?

44. Suppose the average cost of a compact disc in the United States was $15 in 1991. About how much money was paid for compact discs in the United States in 1991?

45. About how many more compact discs were sold in the United States than in Japan in 1991?

46. About how many times more compact discs were sold in the United States than in the United Kingdom in 1991?

Ongoing **ASSESSMENT**

47. **Open-ended** Look through some newspapers and magazines. Find some examples of numbers that are written in scientific notation or that can be written more compactly using scientific notation.

Review **PREVIEW**

Simplify. *(Section 2-2)*

48. $|-1.8|$
49. $-8 - 40$
50. $-9.4 + 6.2$
51. $-2 + 3(-4 + 6)$

Replace each ? with the correct number. *(Table of Measures, page 666)*

52. 2 ft = ? in.
53. 8 mm = ? cm
54. 6 m = ? cm

2-3 Exploring Scientific Notation

77

Answers to Exercises and Problems

38. about 4.307×10^{10} = 43,070,000,000 = 43.07 billion in 1 yr; about 2.1535×10^{11} = 215,350,000,000 = 215.35 billion in 5 yr

39. about 2.56×10^{10} = 25,600,000,000 = 25.6 billion

40. Answers may vary. An example is given. A 14-year-old student would

answer: about 6.132×10^{10} = 61,320,000,000 = 61.32 billion.

41. about 2.603×10^{13} = 26,030,000,000,000 = 26.03 trillion m/day; about 1.807×10^{10} = 18,070,000,000 = 18.07 billion m/min; about 3.012×10^{8} = 301,200,000 = 301.2 million m/s

42. 1.68×10^{-3} L

43. Estimates may vary. Examples are given. United States: about 3.4×10^{8} = 340,000,000; United Kingdom: about 6×10^{7} = 60,000,000; Japan: about 3×10^{8} = 300,000,000

44. about 5.1×10^{9} = $5,100,000,000

45. about 4×10^{7} = 40,000,000

46. about 5.67×10^{0} = 5.67 times more

47. Answers may vary.

48. 1.8
49. −48
50. −3.2
51. 4
52. 24
53. 0.8
54. 600

55. **a, b.** Answers may vary.

Quick Quiz (2-1 through 2-3)

See page 122.

Working on the Unit Project

55. **a.** **Group Activity** Check your group's facts. Is any number so large that you should write it in scientific notation? Is any number so small that it has a negative exponent when written in scientific notation? If not, brainstorm some facts about your theme that would include very large or very small numbers.

b. **Research** Have each person in the group research one of the facts. Write the number in decimal notation and in scientific notation.

Practice 11 For use with Section 2-3

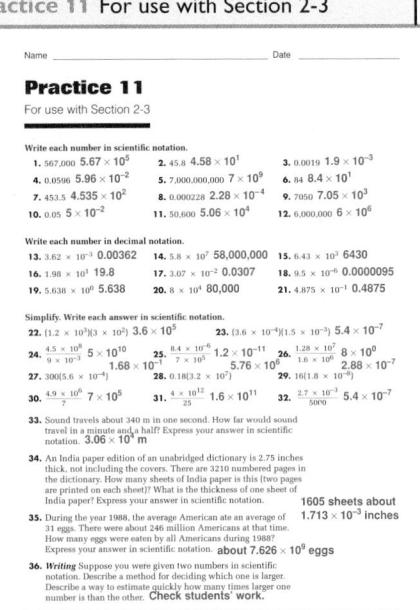

Name _____ Date _____

Practice 11

For use with Section 2-3

Write each number in scientific notation.

1. 567,000 5.67×10^5 2. 45.8 4.58×10^1 3. 0.0019 1.9×10^{-3}
4. 0.0596 5.96×10^{-2} 5. 7,000,000,000 7×10^9 6. 84 8.4×10^1
7. 453.5 4.535×10^2 8. 0.000228 2.28×10^{-4} 9. 7050 7.05×10^3
10. 0.05 5×10^{-2} 11. 50,600 5.06×10^4 12. 6,000,000 6×10^6

Write each number in decimal notation.

13. 3.62×10^{-3} 0.00362   14. 5.8×10^7 58,000,000   15. 6.43×10^3 6430
16. 1.98×10^1 19.8   17. 3.07×10^{-2} 0.0307   18. 9.5×10^{-6} 0.0000095
19. 5.638×10^0 5.638   20. 8×10^4 80,000   21. 4.875×10^{-1} 0.4875

Simplify. Write each answer in scientific notation.

22. $(1.2 \times 10^3)(3 \times 10^2)$ 3.6×10^5 23. $(3.6 \times 10^{-4})(1.5 \times 10^{-3})$ 5.4×10^{-7}
24. $\frac{4.5 \times 10^8}{9 \times 10^{-3}}$ 5×10^{10} 25. $\frac{8.4 \times 10^{-6}}{7 \times 10^5}$ 1.2×10^{-11} 26. $\frac{1.28 \times 10^7}{1.6 \times 10^6}$ 8×10^0
27. 300(5.6 × 10⁻⁴) $\frac{1.68 \times 10^{-1}}{}$ 28. 0.18(3.2 × 10⁷) $\frac{5.76 \times 10^6}{}$ 29. 16(1.8 × 10⁻⁸) $\frac{2.88 \times 10^{-7}}{}$
30. $\frac{4.9 \times 10^6}{7}$ 7×10^5 31. $\frac{4 \times 10^{12}}{25}$ 1.6×10^{11} 32. $\frac{2.7 \times 10^{-3}}{5000}$ 5.4×10^{-7}

33. Sound travels about 340 m in one second. How far would sound travel in a minute and a half? Express your answer in scientific notation. 3.06×10^4 m

34. An India paper edition of an unabridged dictionary is 2.75 inches thick, not including the covers. There are 3210 numbered pages in the dictionary. How many sheets of India paper is this (two pages are printed on each sheet)? What is the thickness of one sheet of India paper? Express your answer in scientific notation. **1605 sheets about 1.713×10^{-3} inches**

35. During the year 1988, the average American ate an average of 31 eggs. There were about 246 million Americans at that time. How many eggs were eaten by all Americans during 1988? Express your answer in scientific notation. **about 7.626×10^9 eggs**

36. **Writing** Suppose you were given two numbers in scientific notation. Describe a method for deciding which one is larger. Describe a way to estimate quickly how many times larger one number is than the other. **Check students' work.**

Unit 2

CHECKPOINT 1

Tokyo, Japan

1. **Writing** Write about how you could estimate the number of kernels of popped popcorn that will fill a shoe box.

2. Estimate the number of people in the photograph.   2-1

3. Tell whether the population of Tokyo, Japan, is in the *hundreds, thousands, millions,* or *billions.*

4. Is the population of Tokyo a continuous or discrete quantity?

5. **Writing** Use a number anywhere on the scale below to estimate the probability that it will snow in your city or town in February. Give a reason for your answer.

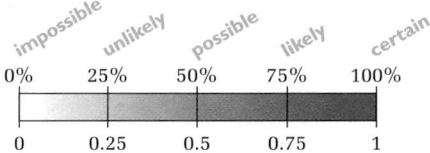

impossible	unlikely	possible	likely	certain
0%	25%	50%	75%	100%
0	0.25	0.5	0.75	1

Evaluate each expression for the given values of the variables.   2-2

6. $-a + b^2$ when $a = 9$ and $b = -4$

7. $xy + x$ when $x = -7$ and $y = -3$

Write each number in scientific notation.   2-3

8. 0.000345   9. 26,710,000

10. Simplify $2500(3.2 \times 10^8)$. Write your answer in decimal notation.

Answers to Checkpoint

1. Answers may vary. An example is given. You could estimate how many kernels fit in a handful or a measuring cup and then estimate how many of these handfuls or cupfuls would fit in a shoebox.

2. Estimates may vary. An example is given. hundreds

3. millions

4. discrete

5. Answer may vary with location of city.

6. 7

7. 14

8. 3.45×10^{-4}

9. 2.671×10^7

10. 800,000,000,000

2-4

Estimating Measures: Length and Area

MEASURING UP

GUIDE TO U.S. CUSTOMARY UNITS OF LENGTH

1 in. The length of the upper section of a thumb

1 ft The length of a shoe

1 yd The length of a step

1 mi The distance a person could walk in 20 minutes

GUIDE TO METRIC UNITS OF LENGTH

1 mm The thickness of the tip of a sharpened pencil

1 cm The width of a person's little fingertip

1 m The length of a big step

1 km The distance a person can walk in 12 minutes

Marisa Fuentes has students in her mathematics class from many different cultures. Some of the students are used to United States customary units of length. Others are used to metric units of length.

She had her students work in groups to make guides to units in these two systems of measurement.

Talk it Over

For questions 1–4, tell which length in each pair is greater.

1. 1 in., 1 mm
2. 1 in., 1 cm
3. 1 ft, 1 m
4. 1 mi, 1 km
5. Which length is greater, 1 yd or 1 m? How do you know?
6. The prefixes used in the metric system stand for powers of ten. For example, the prefix *milli-* means one thousandth, which is 10^{-3}.
   a. What does *centi-* mean? Which power of ten does it stand for?
   b. What does *kilo-* mean? Which power of ten does it stand for?
7. Estimate the length of 100 copies of this textbook laid end to end.
8. Estimate the height of the door to your classroom.
9. What units of length might a hairdresser, a carpenter, and a runner be good at estimating?

2-4 Estimating Measures: Length and Area

79

Answers to Talk it Over

1. 1 in.
2. 1 in.
3. 1 m
4. 1 mi
5. 1 m; Answers may vary. An example is given. Look in a dictionary or encyclopedia under *measurement* or *metric system*.

6. a. one hundredth; 10^{-2}
   b. one thousand; 10^{3}
7–8. Estimates may vary. Examples are given.
7. about 90 ft or about 30 m
8. about 7 ft or about 2 m

9. Answers may vary. An example is given. A hairdresser is likely to be good at estimating inches or centimeters; a carpenter, inches, feet, and yards, or millimeters, centimeters, and meters; and a runner, yards and miles, or meters and kilometers.

Talk it Over

Use questions 1–5 to help students begin to develop an intuitive understanding of metric units of length. Since the emphasis is on developing a sense for the approximate sizes of units, do not use rulers, but rely instead on information given in the guides.

Use question 6 to review the meaning of commonly used metric prefixes. Questions 7–9 help students understand that estimating a length or height requires that they first select an appropriate unit.

Additional Sample

S1 Use the map and paper ruler in Sample 1 to estimate the distance from Little Rock, Arkansas to Boise, Idaho to the nearest hundred miles.
about (7)(200) = 1400 mi

Talk it Over

In Question 10, there are many east-west and north-south lines. Part of the problem is finding one that appears to represent the greatest distance.

Visual Thinking

Most maps are prepared on a scale of miles to inches. Ask students to bring maps to class and to determine how many miles are equal to one inch. Have students then check and compare estimated distances between locations. This activity involves the visual skills of *interpretation* and *generalization*.

Map scales and grid lines can help you estimate long distances and large areas.

Sample 1

Use the map to estimate the distance from Los Angeles to New York City to the nearest hundred miles.

Sample Response

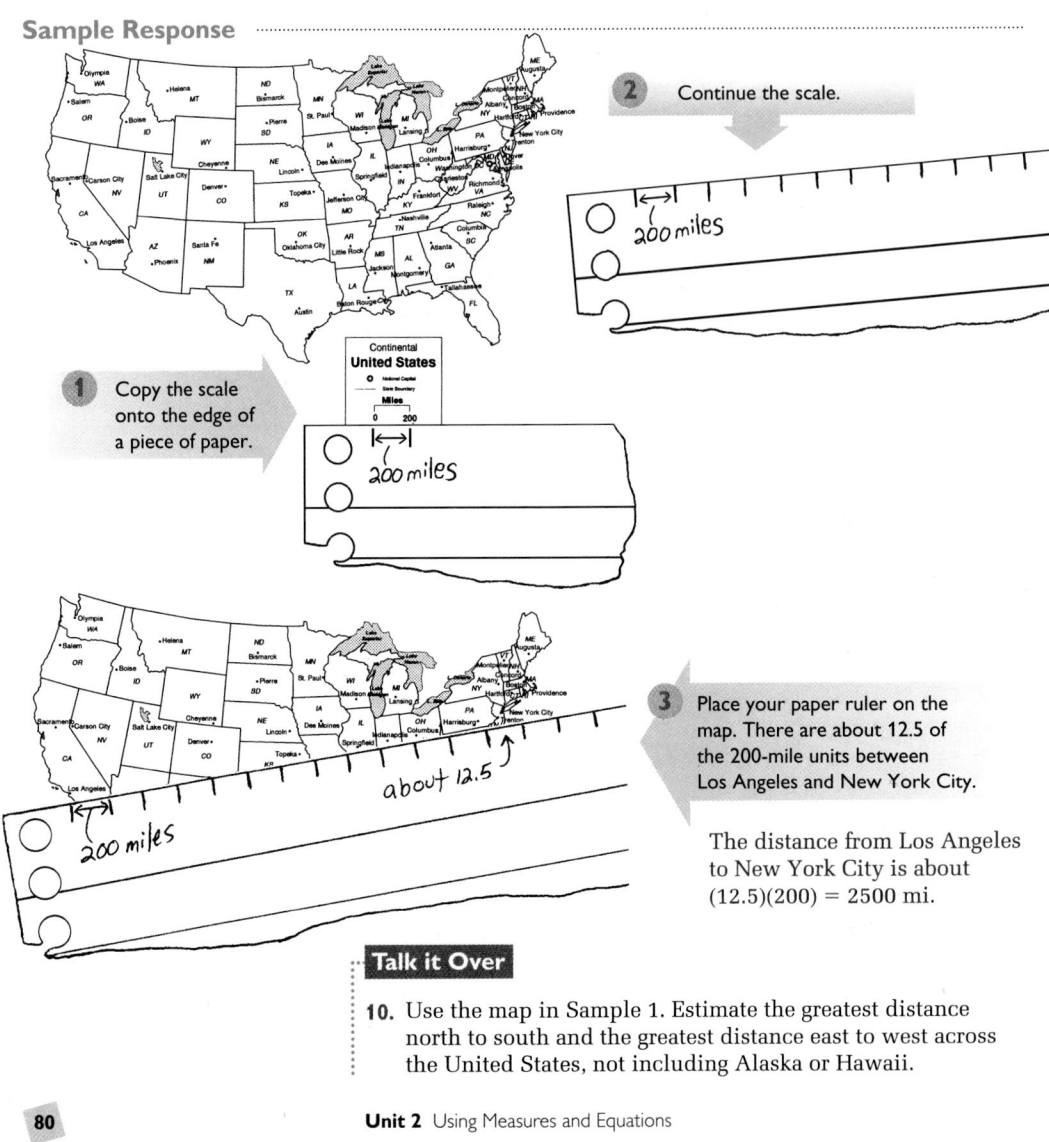

The distance from Los Angeles to New York City is about (12.5)(200) = 2500 mi.

Talk it Over

10. Use the map in Sample 1. Estimate the greatest distance north to south and the greatest distance east to west across the United States, not including Alaska or Hawaii.

Unit 2 Using Measures and Equations

Answers to Talk it Over

10. Estimates may vary. An example is given: about 1700 mi north to south; about 2800 mi east to west

Additional Sample

S2 Use the map in Sample 1 to estimate the area of the state of New Mexico.
about 120,000 mi²

Talk it Over

For question 11, let students brainstorm methods of estimating areas. Discuss the relative merits of all the approaches students propose.

For question 12, students should describe how they estimated the number of grid squares covered by the Great Salt Lake. Some may have estimated the fraction of each grid square covered by the lake. Others may have mentally rearranged parts of the lake to see about how many squares they could fill with the parts.

Sample 2

Use the map in Sample 1. Estimate the area of the state of Utah.

Sample Response

Problem Solving Strategy: Break the problem into parts

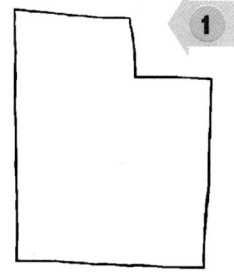

1 Make a sketch of Utah.

2 Draw a dashed line to split Utah into two rectangular areas.

3 Use the scale at the bottom of the map to estimate the dimensions of Utah. Then label your sketch.

4 Find each rectangular area.

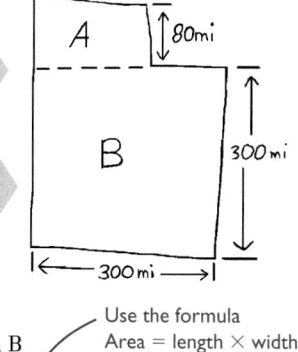

Area of Utah = Area A + Area B $\longleftarrow$ Use the formula Area = length × width.

$\approx (180)(80) + (300)(300)$

$\approx 14{,}400 + 90{,}000$

$\approx 104{,}400$

Student Resources Toolbox
p. 658 *Perimeter, Area, Volume*

5 You can think about the units this way:

mi × mi = mi².

The area of Utah is approximately 104,400 mi².

Talk it Over

11. Describe another method you could use to estimate the area of Utah in Sample 2.

12. **a.** About how many grid squares does the Great Salt Lake cover?

b. Each grid square is 20 mi on a side. Use your answer to part (a) to estimate the area of the Great Salt Lake.

This photograph of the Great Salt Lake was taken by a satellite from a distance of 475 mi.

2-4 Estimating Measures: Length and Area

Answers to Talk it Over

11. Answers may vary. An example is given. Find the total area of the large rectangle that contains both A and B and subtract the area of the small rectangle in the upper right-hand corner.

12. **a.** about 7 squares

b. 2800 mi²

Look Back

The main point of the Look Back activity is to help students develop good judgment about the need for a precise measurement or when an estimate would suffice. Have a class discussion to compare and contrast ideas.

APPLYING

Suggested Assignment

Standard 2–9, 11–13, 15–20, 23–29

Extended 1–21, 23–29

Integrating the Strands

Measurement Exs. 1–14, 22–24, 28

Geometry Exs. 15–21, 25–27

Discrete Mathematics Ex. 29

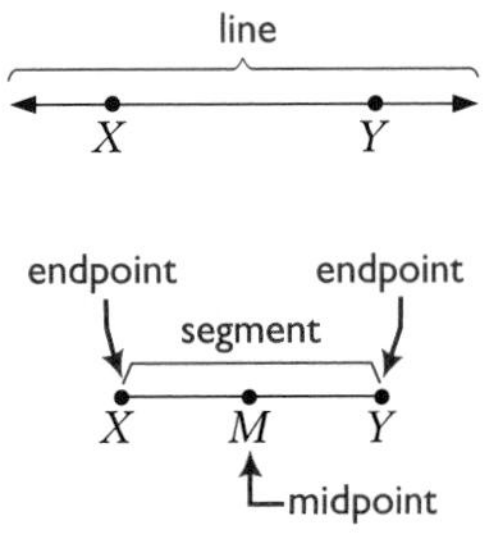

The **line** shown is identified as $\overleftrightarrow{XY}$ or $\overleftrightarrow{YX}$. It goes on forever in two directions.

The part of the line with **endpoints** X and Y is the **segment** identified as $\overline{XY}$ or $\overline{YX}$. The symbol XY means **the length of $\overline{XY}$**.

Point M on $\overline{XY}$ is called the **midpoint** of $\overline{XY}$ because it divides the segment in half. Every segment has a midpoint, but a line does not.

Segments with equal measures are congruent segments. In the diagram, $\overline{XM} \cong \overline{MY}$.

Talk it Over

13. Explain what each of the symbols $\overleftrightarrow{AB}$, $\overline{AB}$, and AB represents in this diagram.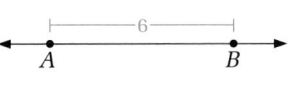

14. Estimate QP in inches.

Look Back

Give an everyday example of needing an exact length and an example of needing only an estimated length. Explain why the needs are different.

2-4 Exercises and Problems

1. **Reading** How does the information about lines and segments above lead you to the conclusion that you can measure a segment but not a line?

For Exercises 2–4, estimate each measure in both customary and metric units.

2. the width of this textbook

3. the length of each side of quadrilateral $QRST$

4. the length of your bedroom

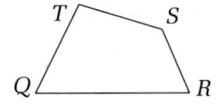

5. a. **Research** Find out the lengths of the upper sections of the thumbs of at least five people, including one adult. (Measure to the nearest $\frac{1}{8}$ in.)

   b. Is there a length that appears most often?

   c. Whose length is the closest to one inch?

   d. Suppose the length of the upper section of your thumb is not exactly one inch. How could you use your measurement to estimate lengths in inches?

Answers to Talk it Over

13. $\overleftrightarrow{AB}$ represents the line shown. $\overline{AB}$ is the segment with endpoints A and B. AB is the distance from A to B, which is 6 units.

14. Estimates may vary. An example is given. about 1.5 in.

Answers to Look Back

Answers may vary. An example is given. A precise measure is necessary in building. An estimate is good enough when trying to determine a driving distance.

Answers to Exercises and Problems

1. A line goes on forever in both directions; it is infinite. A segment has endpoints; it is finite.

2–4. Estimates may vary. Examples are given.

2. about 8 in. or about 20 cm

3. QR is about 1 in. or 2.5 cm; QT is about 0.5 in. or 1.5 cm; TS is about 0.5 in. or 1.5 cm; SR is about $\frac{3}{8}$ in. or 1 cm

4. about 12 ft or about 4 m

5. See answers in back of book.

6. about 3450 km

7. No. The distance is approximately 1425 km.

8. about 2,500,000 km^2

9. about 7,600,000 km^2

The map shows Australia.

6. Estimate the distance from Perth to Sydney.

7. Is 3500 km a reasonable estimate of the distance from Melbourne to Brisbane? Explain.

8. Estimate the area of Western Australia.

9. Each grid square is about 500 km on a side. Use the grid on the map to estimate the area of Australia.

10. **Writing** Roberto was reading about Australia. The magazine article said that the Great Sandy Desert and the Great Victoria Desert together covered 300,000 km². Roberto decided that the reported area was not reasonable, and that there must be a misprint in the article. Do you *agree* or *disagree*? Why?

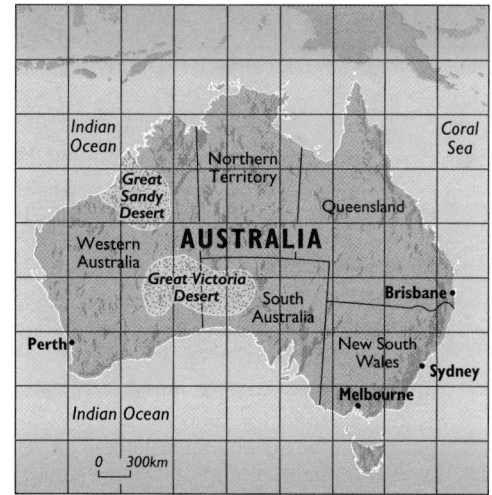

Use the double bar graph. Name each country described. Use the age scale to estimate the measure of each bar for that country.

11. the country where the average age of brides and grooms is the same

12. the country where the average age of grooms is less than the average age of brides

13. the country with the greatest difference between the average age of grooms and the average age of brides

14. Midori wants to cover one of her bedroom walls with posters. The wall is 10 ft wide by 8 ft high. She plans to use only posters that are 2 ft by 3 ft.

   a. What is the area of the wall? of a poster? Can Midori completely cover the wall without overlapping any posters? Why or why not?

   b. How many posters can Midori fit on her wall without any overlap if she buys only posters that are oriented vertically? horizontally?

   c. What combination of vertical and horizontal posters will fit on the wall and cover the most space without any overlap?

2-4 Estimating Measures: Length and Area

83

10. agree; Explanations may vary. An example is given. Both deserts together cover about $3\frac{1}{2}$ grid squares, so their area is about 875,000 km².

11. United States; Both bars measure about 27 years.

12. Dominican Republic; Bride bar measures about 26 years, groom bar measures about 24 years.

13. Egypt; Bride bar measures about 17 years, groom bar measures about 25 years.

14. a. 80 ft²; 6 ft²; No, the area of the wall is not divisible by the area of a poster.

   b. 10 posters; 12 posters

   c. 10 vertical posters and 3 horizontal posters

Land surveyors use special instruments to measure angles, distances, and elevations. Their findings are important for construction work and in establishing land and water boundaries, air space, property lines, and so on.

Mapping scientists also measure and map, but they are usually concerned with much larger areas of Earth's surface. These scientists routinely use aerial photographs, satellite data, and computers.

Research

Have students research and compare flat maps and globes. They should investigate some of the special map projections that are used to represent the three-dimensional surface of Earth on two-dimensional, flat maps. They should also investigate some of the difficulties that arise in estimating distances using a flat map that shows a very large portion of Earth's surface. Ask students to prepare a poster to display their findings in an interesting way to the class.

Problem Solving

For Ex. 14, urge students to try a variety of approaches. Discuss the approaches used to see which one works best and why. Students should be prepared to present their ideas to the class in an organized way.

Practice 12 For use with Section 2-4

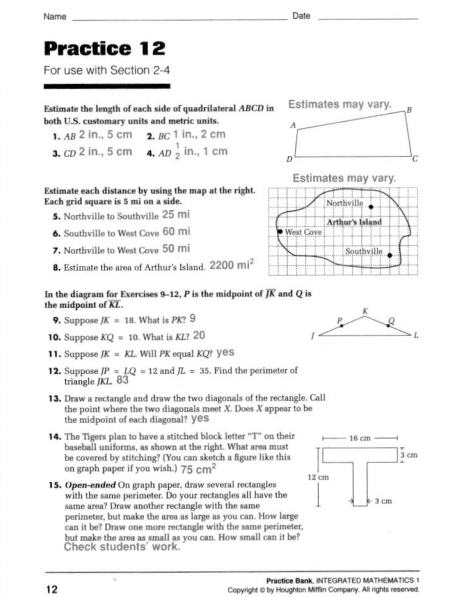

Answers to
Exercises and Problems

15. True.

16. True.

17. False.

18. False.

19. True.

20. 36

21. Answers may vary. An example is given. $LM = 6$; $XM = \frac{1}{2}KL$; $\overleftrightarrow{KN}$ is perpendicular to $\overline{NM}$; $\overleftrightarrow{KL}$ is parallel to $\overrightarrow{NM}$.

22. a. parallelogram

   b. The area of the 4 triangles is about the same as the area of the parallelogram. The area of the parallelogram is about one-half the area of the quadrilateral.

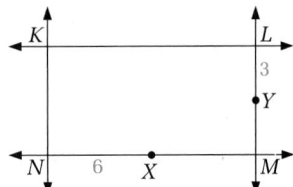

In the diagram, *KLMN* is a rectangle. *X* is the midpoint of $\overline{NM}$ and *Y* is the midpoint of $\overline{LM}$. For Exercises 15–19, tell whether each statement is *True* or *False*.

15. $XM = 6$ **16.** $KN = 2 \cdot YM$ **17.** $KL = \frac{1}{2} \cdot XM$

18. $\overline{KL}$ is perpendicular to $\overline{NM}$ **19.** $\overleftrightarrow{KN}$ is parallel to $\overleftrightarrow{LM}$

20. Find the perimeter of rectangle *KLMN*.

21. Open-ended Write four statements, like the ones in Exercises 15–19, that are true about the diagram.

22. Using Manipulatives You will need a ruler and scissors.

   a. Use a ruler to draw any quadrilateral. Connect the midpoints of the sides to form quadrilateral *ABCD*. What type of quadrilateral does *ABCD* appear to be?

   b. Cut out the quadrilateral you drew in part (a) and then cut off the four triangles. Compare the total area of the four triangles to the area of quadrilateral *ABCD* by arranging the four triangles on top of it. Then compare the area of quadrilateral *ABCD* to the area of the original quadrilateral.

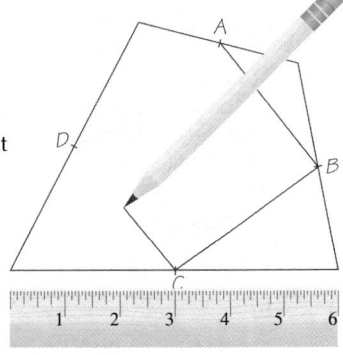

Ongoing **ASSESSMENT**

23. Open-ended Draw an eight-sided figure that has a perimeter of 42 cm. Use dot paper or graph paper if necessary. What is the area of the figure? Is your answer an estimate or is it exact? Explain.

Review **PREVIEW**

24. One angstrom is equal to 1×10^{-10} m. What is the length in meters of 2500 angstroms? Write your answer in scientific notation and in decimal notation. *(Section 2-3)*

Name each quadrilateral. *(Section 1-7)*

25. 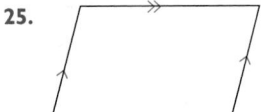   **26.**    **27.**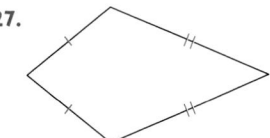

28. Define the terms *acute angle*, *right angle*, and *obtuse angle*. (*Hint:* Use the glossary.)

29. Decide on an interesting way to help people understand the magnitude of each of your numerical facts except for the percent. You may want to compare a length or an area with something in your school.

---

23. Figures and answers may vary. An example is given.

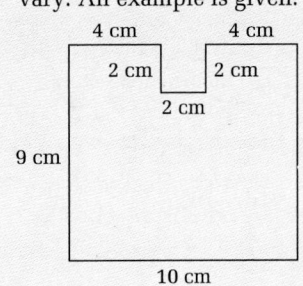

The area is 90 − 4, or 86 cm². This is the exact area because I used the formula for the area of a rectangle to calculate the area rather than estimating the number of square units inside the figure.

24. 2.5×10^{-7}; 0.00000025

25. parallelogram

26. rectangle

27. kite

28. Definitions may vary. Examples are given. An acute angle is an angle that measures between 0° and 90°. A right angle is an angle that measures 90°. An obtuse angle is an angle that measures between 90° and 180°.

29. Answers may vary.

2-5 Exploring Angle Relationships

Focus
Investigate and use the relationships among angles in circles, angles in triangles, and angles formed by intersecting lines.

You can get a moped license at age 14 in Iowa, but in California you have to wait until age $15\frac{1}{2}$. The circle graph shows the percentage of states having each age requirement. To make a circle graph, you use geometry.

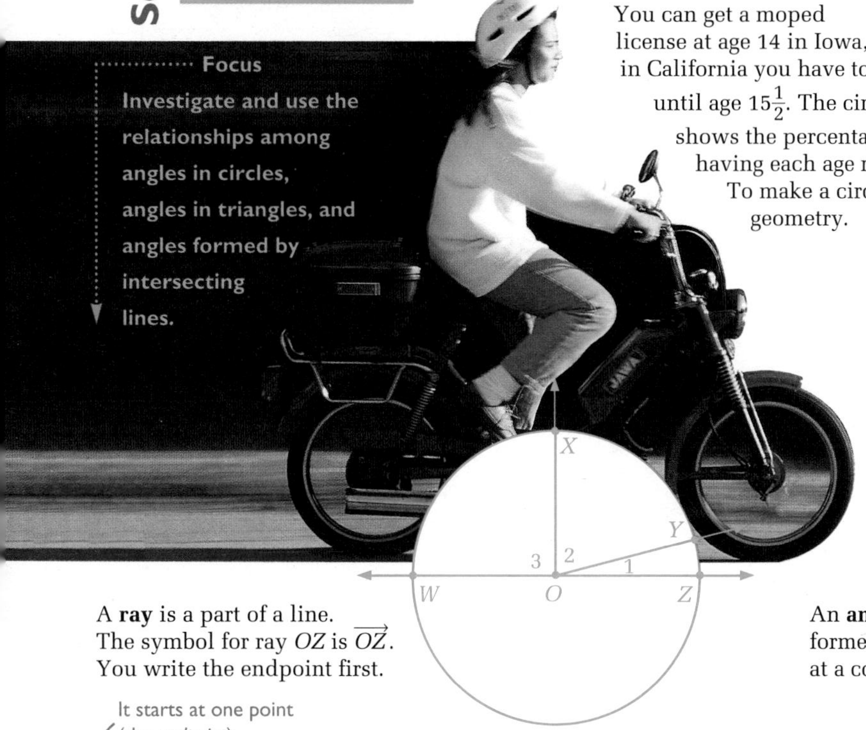

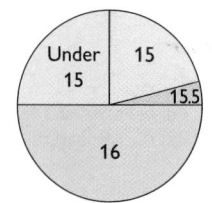

Required Age for Moped Licenses in the U.S. in 1991

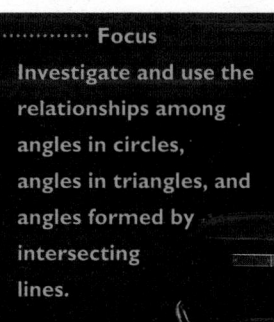

A **ray** is a part of a line. The symbol for ray OZ is $\overrightarrow{OZ}$. You write the endpoint first.

It starts at one point (the *endpoint*).

It extends forever in one direction.

O Z

An **angle** is the figure formed when two rays meet at a common endpoint.

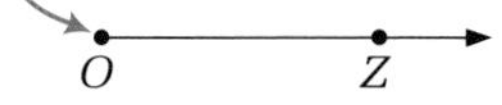

The common endpoint is called the **vertex.**

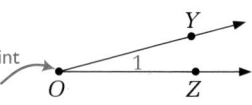

O 1 Z

The angle YOZ can be named in these ways: $\angle O$, $\angle YOZ$, $\angle ZOY$, or $\angle 1$. You can identify an angle by its vertex alone when only one angle has that vertex. Otherwise, use three letters or a number. Be sure that the middle letter names the vertex.

Talk it Over

Use the circle graph above.

1. Name three rays.

2. Name three angles.

3. Name $\angle 3$ in two other ways.

4. For each angle, point O is called the $\underline{\ ?\ }$.

5. Can any of the angles be named $\angle O$? Why or why not?

6. A **central angle** is an angle, such as $\angle YOZ$, with its vertex at the center of a circle. Name two other central angles.

Student Resources Toolbox
p. 657 *Angles*

2-5 Exploring Angle Relationships   **85**

PLANNING

Objectives and Strands
See pages 54A and 54B.

Spiral Learning
See page 54B.

Materials List
➤ Ruler
➤ Protractor
➤ Scissors
➤ Geometric drawing software

Recommended Pacing
Section 2-5 is a two-day lesson.
Day 1
Pages 85–87: Opening paragraph through Talk it Over 11, *Exercises 1–18*
Day 2
Pages 87–89: Special Angle Relationships through Look Back, *Exercises 19–40*

Toolbox References
➤ **Toolbox Skill 12:** The Percent One Number is of Another
➤ **Toolbox Skill 16:** Using a Protractor
➤ **Toolbox Skill 25:** Making a Circle Graph

Extra Practice
See pages 622–623.

Warm-Up Exercises
Warm-Up Transparency 2-5

Support Materials
➤ Practice 13
➤ Enrichment 12 in the Activity Bank
➤ Study Guide 2-5
➤ Problem Set 4
Overhead Visual 2
➤ McDougal Littell Mathpack software: *Geometry Inventor*
➤ Quiz 2-5

Answers to Talk it Over

1. Answers may vary but should include three of the following: $\overrightarrow{OW}$, $\overrightarrow{OX}$, $\overrightarrow{OY}$, $\overrightarrow{OZ}$.

2. Answers may vary but should include three different angles from among the following: $\angle 1$, $\angle 2$, $\angle 3$, $\angle WOX$, $\angle WOY$, $\angle WOZ$, $\angle XOW$, $\angle XOY$, $\angle XOZ$, $\angle YOW$, $\angle YOX$, $\angle YOZ$, $\angle ZOW$, $\angle ZOX$, $\angle ZOY$.

3. $\angle WOX$, $\angle XOW$

4. vertex

5. No. Explanations may vary. An example is given. There are many angles with vertex O, so it isn't clear which angle "$\angle O$" would identify.

6. Answers may vary but should include two different angles from among the following: $\angle 1$, $\angle 2$, $\angle 3$, $\angle WOX$, $\angle WOY$, $\angle WOZ$, $\angle XOW$, $\angle XOY$, $\angle XOZ$, $\angle YOW$, $\angle YOX$, $\angle ZOW$, $\angle ZOX$, $\angle ZOY$.

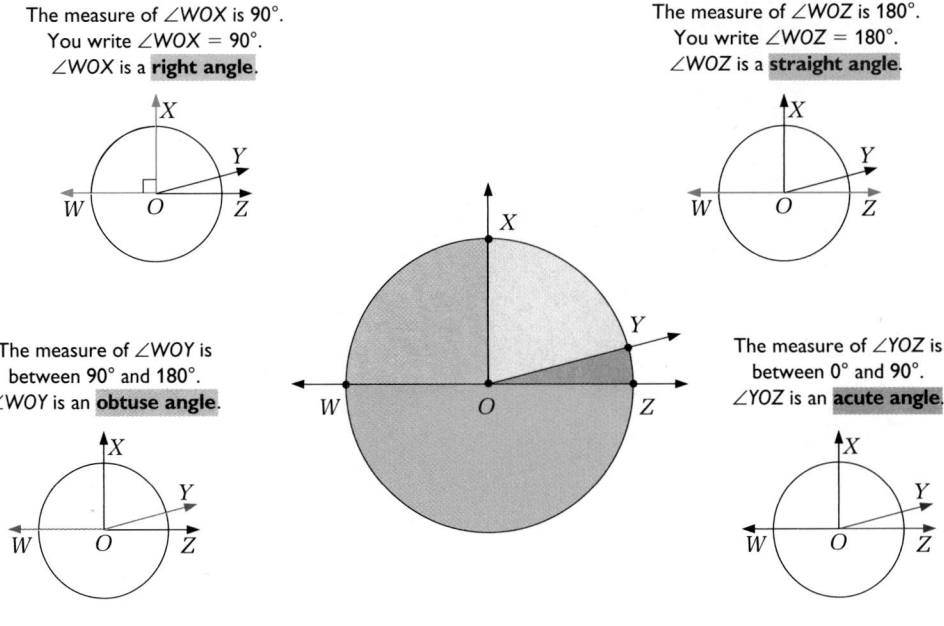

TEACHING

Talk it Over

Use questions 1–5, 7, and 8 to check students' understanding of the basic terminology, notation, and concepts for rays and angles. Also, point out the definition of central angle given in question 5.

Mathematical Procedures

This page provides a handy summary of angle relationships. Students' work with right, acute, and obtuse angles in Unit 1 is extended to include straight, complementary, and supplementary angles. Students will be using complementary and supplementary angles to solve problems again in Units 5 and 8. If your students are familiar with angle terminology from earlier courses, you can move quickly through this material. If it is new to them, you will need to take more time, and you may want to assign the Toolbox section on measuring angles.

Visual Thinking

Encourage students to make sketches of each angle. Ask them to identify which of their angles are right, straight, acute, obtuse, complementary, and supplementary and why. This activity involves the visual skills of *recognition*, *identification*, and *recall*.

The measure of an angle is the number of degrees from one ray to the other. Many angles have special names and special relationships based on their angle measures.

The measure of ∠WOX is 90°.
You write ∠WOX = 90°.
∠WOX is a **right angle**.

The measure of ∠WOZ is 180°.
You write ∠WOZ = 180°.
∠WOZ is a **straight angle**.

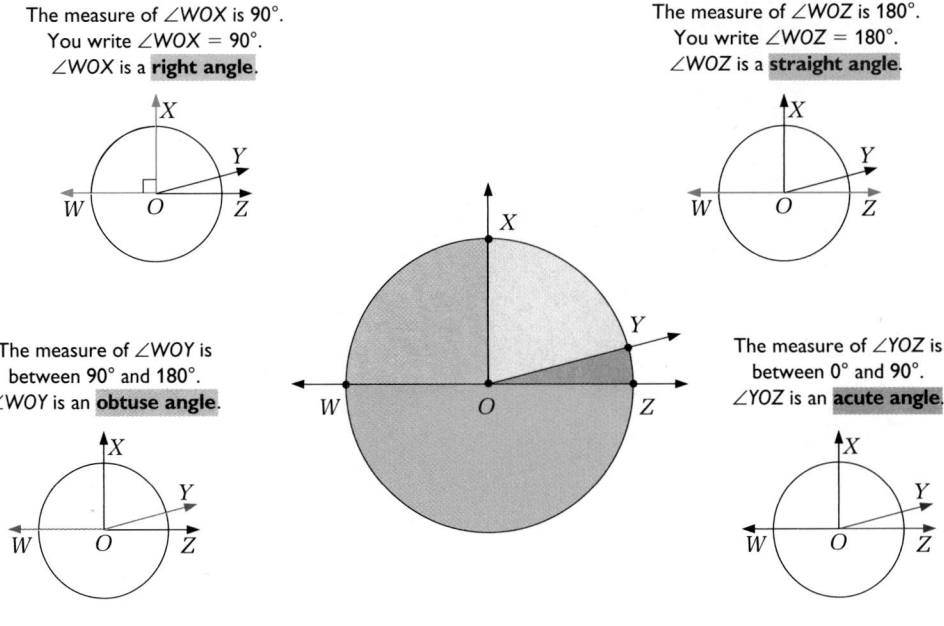

The measure of ∠WOY is between 90° and 180°.
∠WOY is an **obtuse angle**.

The measure of ∠YOZ is between 0° and 90°.
∠YOZ is an **acute angle**.

Two angles whose measures add up to 90° are called **complementary angles**.
∠XOY and ∠YOZ are complementary angles.

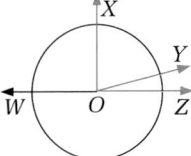

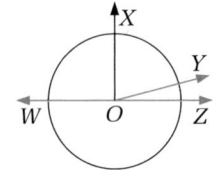

Two angles whose measures add up to 180° are called **supplementary angles**.
∠WOY and ∠YOZ are supplementary angles.

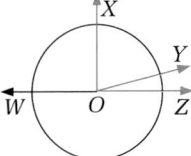

Talk it Over

7. Which combination of age groups in the circle graph has an obtuse central angle?

8. **a.** Which two parts of the circle graph together total one fourth of the circle?

   b. An entire circle has a measure of 360°. What is the angle measure of one fourth of a circle?

   c. What is the relationship between the central angles of the sectors in part (a)?

Unit 2 Using Measures and Equations

Answers to Talk it Over

7. "under 15" and "15"

8. **a.** "15" and "15.5"

   b. 90°

   c. They are complementary.

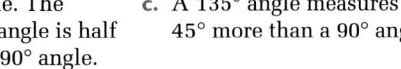
Estimate to sketch each angle.

a. $\angle A = 90°$ b. $\angle B = 45°$ c. $\angle C = 135°$

Sample Response

a. A 90° angle can be formed by sketching intersecting horizontal and vertical lines.

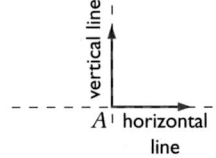

b. Sketch a 90° angle. The measure of a 45° angle is half the measure of a 90° angle.

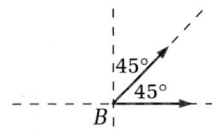

c. A 135° angle measures 45° more than a 90° angle.

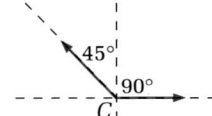

> ►► Now you are ready for:
> Exs. 1–18 on pp. 89–90

Talk it Over

Explain how you could estimate to sketch each angle.

9. $\angle Q = 60°$ 10. $\angle R = 75°$ 11. $\angle P$ is an obtuse angle.

Special Angle Relationships

Angles with equal measures are **congruent angles.**

You can mark the angles the same way to show that they are congruent.

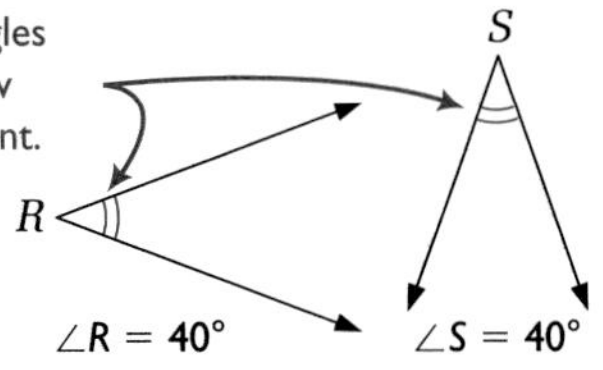

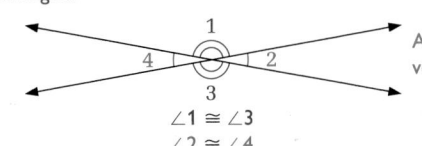

$\angle R = 40°$ $\angle S = 40°$

$\angle R \cong \angle S$

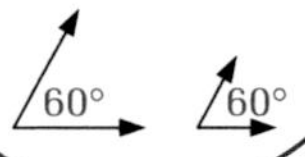

Two angles formed by intersecting lines and facing in opposite directions are called **vertical angles.**

Angles 1 and 3 are vertical angles.

Angles 2 and 4 are vertical angles.

$\angle 1 \cong \angle 3$
$\angle 2 \cong \angle 4$

The results of Exercise 13 on page 90 lead you to a special angle relationship. Vertical angles have equal measures.

The angles of a triangle also have a special relationship.

2-5 Exploring Angle Relationships

87

Watch Out!

How long you draw the rays does not affect the measure of an angle.

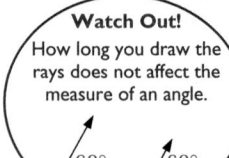

S1 Estimate to sketch each angle.

a. $\angle A = 30°$

Sketch intersecting horizontal and vertical lines to form a 90° angle. Then sketch two rays that divide the 90° angle into three equal parts.

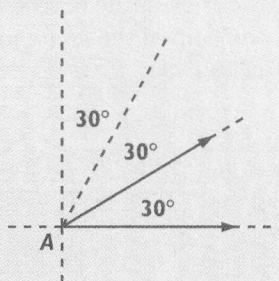

b. $\angle B = 120°$

A 120° angle measures 30° more than a right angle.

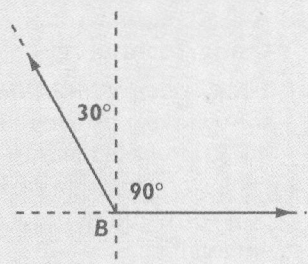

c. $\angle C = 150°$

A 150° angle measures 60° more than a 90° angle.

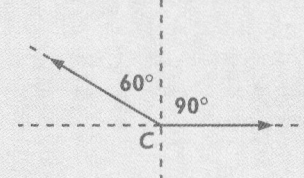

Answers to Talk it Over

9–11. Answers may vary. Examples are given.

9. Method 1: Draw a straight angle. Sketch two rays to divide the angle into three equal parts. Each part is close to a 60° angle. Method 2: Divide a 90° angle into thirds. Two thirds that are side by side are close to a 60° angle.

10. Sketch a 45° angle as described in Sample 1, part (b). Sketch a 30° angle adjacent to the 45° angle using Method 2 of question 9. The result is close to a 75° angle.

11. Draw a straight angle. From its vertex, draw a ray not perpendicular to the sides of the straight angle. The larger angle will be an obtuse angle.

Exploration

The goal of the Exploration is to have students discover that the sum of the measures of the angles of a triangle is 180°. An alternative approach is to have students tear a large triangle into three pieces, taking care to preserve the three angles. They can then fit the pieces together to arrive at the conjecture that the sum of the angle measures is 180°.

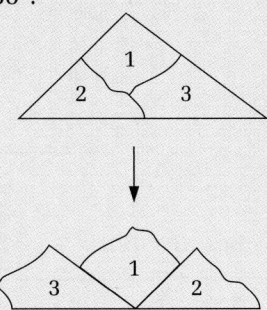

Using Technology

The *Geometry Inventor* software can be used to find the sum of the measures of the angles of a triangle and other polygons, and to draw congruent angles.

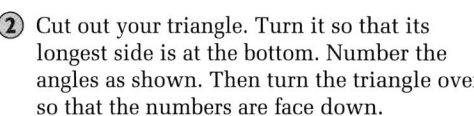

- **Materials:** rulers, protractors, scissors
- **Work in a group of three students.**

An **acute triangle** is a triangle with three acute angles.

An **obtuse triangle** is a triangle with one obtuse angle.

A **right triangle** is a triangle with one right angle.

(1) Student 1: Use a ruler to draw a large acute triangle.

Student 2: Use a ruler to draw a large obtuse triangle.

Student 3: Use a ruler to draw a large right triangle.

(2) Cut out your triangle. Turn it so that its longest side is at the bottom. Number the angles as shown. Then turn the triangle over so that the numbers are face down.

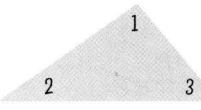
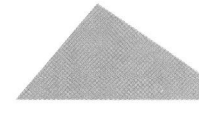

(3) Make a fold perpendicular to the longest side of the triangle and passing through the opposite vertex.

(4) Unfold the triangle. Find the point where the fold line intersects the longest side. Label it *F*.

(5) Fold each vertex of the triangle to meet point F. Leave the triangle folded.

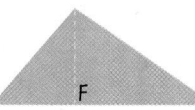

(6) With your group, make a conjecture about the sum of the measures of angles 1, 2, and 3. How can you test your conjecture?

(7) Share your conjecture with the class. Did any groups find counterexamples to your conjecture?

88   **Unit 2** Using Measures and Equations

Answers to Exploration

1–5. manual activity

6. Answers may vary. An example is given. The sum of the measures of angles 1, 2, and 3 is 180°. Fold the paper in step 5 across at *F* to see if the bottom side of ∠3 fits exactly over the bottom side of ∠2.

7. Answers may vary.

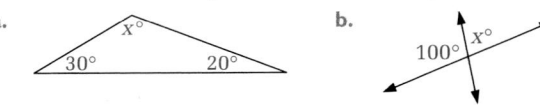

Find the unknown angle measure in each figure.

a.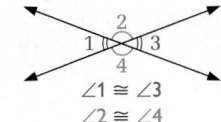

b.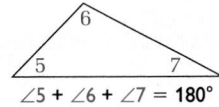

Sample Response

a. The Exploration shows that the sum of the measures of the angles of a triangle is 180°.

$$x = 180 - (30 + 20)$$ ← Subtract the sum of the known angle measures from 180.

$$= 180 - 50$$

$$= 130$$

The measure of the angle is 130°.

b. A straight angle measures 180°.

$$x = 180 - 100$$ ← Subtract the known angle measure from 180.

$$= 80$$

The measure of the angle is 80°.

SPECIAL ANGLE RELATIONSHIPS

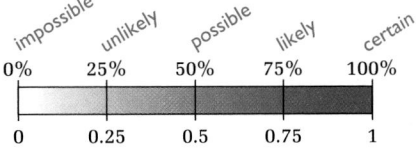

$\angle 1 \cong \angle 3$
$\angle 2 \cong \angle 4$

Vertical angles have equal measure.

$\angle 5 + \angle 6 + \angle 7 = 180°$

The sum of the measures of the angles of a triangle is 180°.

Look Back

In this section, you learned the two special angle relationships shown above. Summarize other facts about angles that you learned or reviewed in this section.

····▶ Now you are ready for:
Exs. 19–40 on pp. 91–92

2-5 Exercises and Problems

1. **Reading** Describe three different ways to name an angle.

2. Suppose you are a carpenter building a house. What is the probability that you will use a 90° angle in your work? a 30° angle? Use a number on the scale to estimate.

impossible   unlikely   possible   likely   certain
0%   25%   50%   75%   100%

0   0.25   0.5   0.75   1

2-5 Exploring Angle Relationships

89

Additional Sample

S2 Find the unknown angle measure in each figure.

a.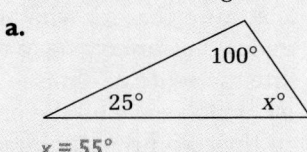

$x = 55°$

b.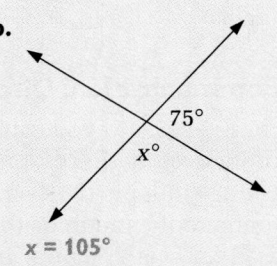

$x = 105°$

Look Back

Involve students in a class discussion to go over the main ideas of the section: names for angles, classification of angles, and special pairs of angles. Conclude with the special angle relationships for vertical angles and the sum of the measures of the angles of a triangle.

APPLYING

Suggested Assignment

Standard 1–23, 25–28, 31–40

Extended 1–23, 25–28, 31–40

Integrating the Strands

Number Exs. 36, 37, 40

Algebra Exs. 25–27, 38, 39

Measurement Exs. 8, 13–18, 21–24, 29, 30

Geometry Exs. 1–35, 40

Statistics and Probability Ex. 2

Answers to Look Back

Answers may vary. An example is given.

- Complementary angles are angles whose measures add up to 90°.
- Supplementary angles are angles whose measures add up to 180°.
- The acute angles of a right triangle are complementary.

- A central angle has its vertex at the center of a circle.
- A right angle measures 90°.
- A straight angle measures 180°.
- An obtuse angle measures between 90° and 180°.
- An acute angle measures between 0° and 90°.

Answers to Exercises and Problems

1. Answers may vary. An example is given. An angle can be named one way by a number or two ways by three letters. The middle letter must always be the vertex, but the other two letters, which represent a point on each side of the angle, can be either first or last.

2. Answers may vary. An example is given. 100%; 25%

89

Error Analysis

Students may use the wrong scale on a protractor. To correct this error, ask whether the angles they are trying to measure are acute or obtuse. Then ask which scale on the protractor they would use to draw the angles.

Communication: Discussion

For Exs. 15–17, you can have an interesting discussion by asking students to describe and compare the methods they used for sketching the angles.

For Exercises 3–9, use the circle graph.

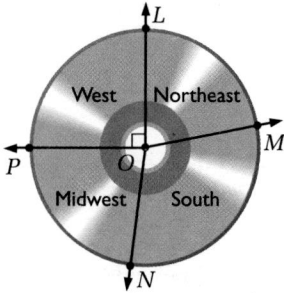

1990 Sales of Music Recordings in Regions of the United States

3. Name each ray.
4. Name a right angle.
5. Name two acute angles.
6. Name three obtuse angles.
7. Name $\angle NOL$ another way.
8. a. Estimate the measure of the angle for each region.
   b. Estimate the percent of sales for the South region.
   c. The total dollar value of recordings sold in 1990 was about \$7,541,100,000. Estimate the dollar value of sales for the Southern region.

Student Resources Toolbox pp. 653–656 *Percent*

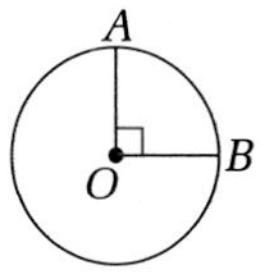

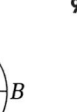

9. In the figure, $\angle AOB$ is a central angle of circle O.
   a. Are $\angle AOB$ and $\angle LOP$ in the diagram above complementary or supplementary?
   b. Does the measure of $\angle AOB$ equal the measure of $\angle LOP$? Why or why not?

10. $\angle A$ and $\angle B$ are complementary angles. Find the measure of $\angle B$.

11. $\angle C$ and $\angle D$ are supplementary angles. Find the measure of $\angle D$.

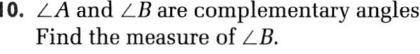

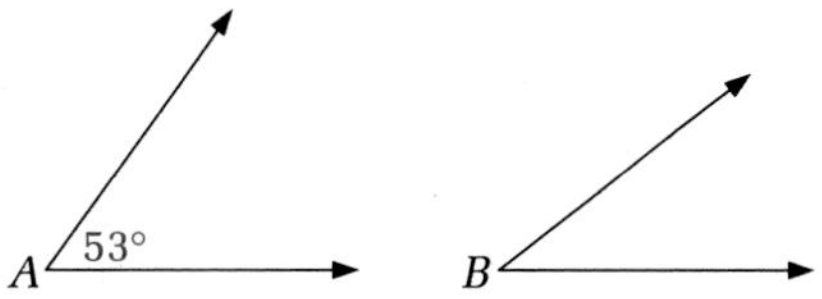

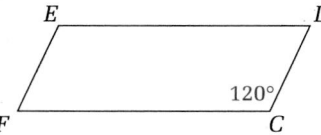

12. Use the figure in Exercise 11. Name $\angle E$ in two other ways.
13. a. Use a ruler to draw two intersecting lines. Number the angles as shown. Your angle measures do not have to match the ones in the diagram.
    b. Look at the angles and estimate their measures. Make conjectures about the relationships between the measures of the angles.
    c. Measure the angles with a protractor or trace and fold your diagram to test your conjectures.

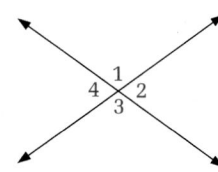

Without using a protractor, estimate to sketch each angle. Then describe how you estimated each angle.

14. $\angle T$ is an acute angle.
15. $\angle Q = 22°$
16. $\angle V = 70°$
17. $\angle G = 120°$
18. Use a protractor to measure each angle you sketched in Exercises 15–17. For which angle did you make the most accurate sketch? the least accurate sketch? How could you improve your ability to make a freehand sketch of an angle?

Answers to Exercises and Problems

3. $\overrightarrow{OP}$, $\overrightarrow{OL}$, $\overrightarrow{OM}$, $\overrightarrow{ON}$
4. $\angle POL$ ($\angle LOP$)
5. $\angle LOM$ ($\angle MOL$), $\angle PON$ ($\angle NOP$)
6. $\angle POM$ ($\angle MOP$), $\angle NOM$ ($\angle MON$), $\angle LON$ ($\angle NOL$)
7. $\angle LON$
8. a. $\angle LOP = 90°$; $\angle LOM \approx 80°$; $\angle MON \approx 105°$; $\angle PON \approx 85°$
   b. about 30%
   c. \$2,300,000,000
9. a. supplementary
   b. Yes. $\angle AOB = 90°$ and $\angle LOP = 90°$
10. 37°
11. 60°
12. $\angle FED$, $\angle DEF$
13. a. Answers may vary.
    b. Answers may vary. An example is given. $\angle 4 = \angle 2 \approx 70°$; $\angle 1 = \angle 3 \approx 110°$
    c. Answers may vary.

14–17. Sketches, methods, and descriptions may vary. Examples are given.

14. Imagine a corner of a page to estimate a 90° angle. Draw an angle with a measure less than that.

15. Sketch a 90° angle and divide it in half to estimate a 45° angle. Then divide in half one of the 45° angles to estimate a 22° angle.

16. Sketch a 90° angle and divide it in half. Then divide in half one of the 45° angles. Combine the angles, as shown, to estimate a 70° angle.

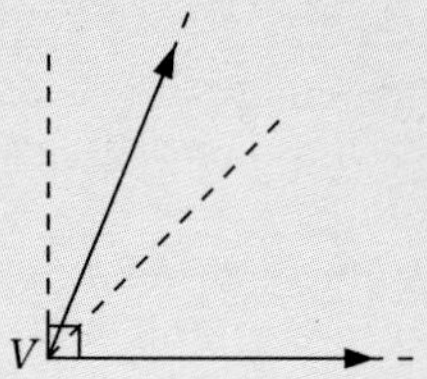

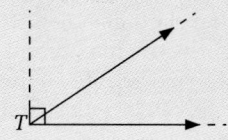

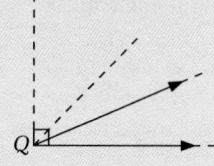

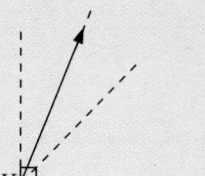

In the diagram, $\overleftrightarrow{AD}$, $\overleftrightarrow{GB}$, and $\overleftrightarrow{EH}$ are straight lines. Replace each __?__ with the name of another angle.

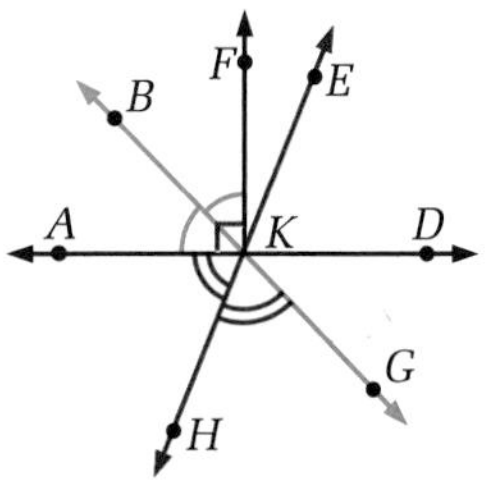

19. $\angle BKF \cong$ __?__ $\cong$ __?__

20. $\angle GKH \cong$ __?__ $\cong$ __?__

Use the diagram for Exercises 19 and 20. Find the measure of each angle without estimating or using a protractor.

21. $\angle BKF$ **22.** $\angle GKH$ **23.** $\angle EKF$

24. **Using Manipulatives** In the Exploration, you learned a relationship among the three angles of any triangle. Here you will explore a relationship between the acute angles of a right triangle.

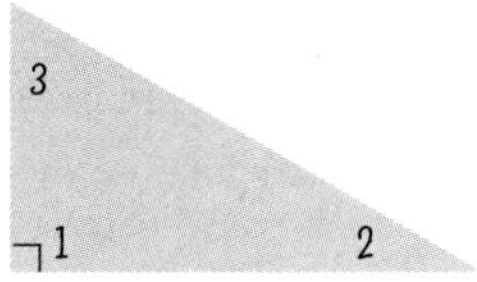

a. Use a ruler to draw three large right triangles. Cut them out. Then position the triangles and number their angles as shown.

b. Fold the vertex of $\angle 2$ and the vertex of $\angle 3$ to meet the vertex of $\angle 1$.

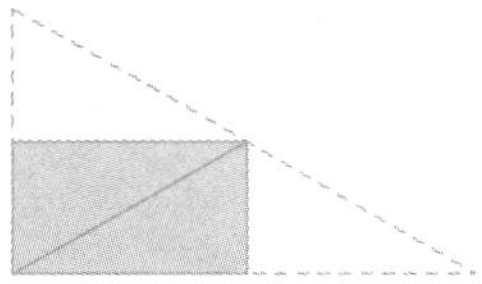

c. Make a conjecture about the relationship between the acute angles of a right triangle. Then measure to test your conjecture.

For Exercises 25–27, write and solve an equation to find each unknown angle measure.

25.

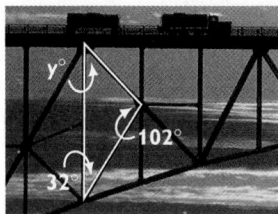

26.

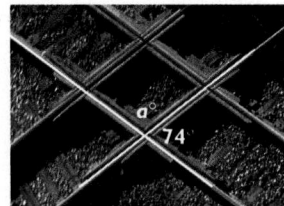

27.

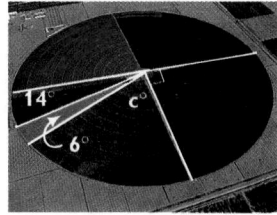

28. Tell whether the triangle in Exercise 25 is *acute, right,* or *obtuse.*

29 TECHNOLOGY Use geometric drawing software.

Alternative Approach Work in a group of three students and use rulers and protractors. Share responsibility for part (a). Have each student do parts (b) and (c). Compare results with each other in part (d).

a. Draw a large quadrilateral, a large pentagon, and a large hexagon. The figures do not have to have equal sides.

b. Measure the angles of each figure. Find the sum of the measures of the angles for each figure.

c. Compare the sums of the measures of the angles for each figure in part (b). What pattern do you see?

d. Use the pattern suggested by part (c) to predict the sum of the measures of the angles of a polygon with seven sides (a *heptagon*). How can you test your prediction?

Reasoning

After students do Ex. 24, ask if anyone can conclude *without folding* that the acute angles are complementary. Ask what must be true of the sum of *all* the angle measures of the triangle. (The sum is 180°.) How much of this sum comes from the right angle? (90°) What does this tell us about the sum of the two acute angles? (The sum must be 90°.)

Problem Solving

After students have answered all parts of Ex. 29, ask for ideas about other ways to arrive at the sum of the angle measures of a convex heptagon.

Someone may solve the problem by dividing the heptagon into five triangles. This is done by drawing all the diagonals from one vertex. The sum of the angle measures of the triangles equals the sum of the angle measures of the heptagon. Since there are 5 triangles, the total is 5 • 180° or 900°.

Answers to Exercises and Problems

17. Sketch a straight angle. Sketch two rays to divide the angle into three angles of equal measure. Combine two of the angles, as shown, to estimate a 120° angle.

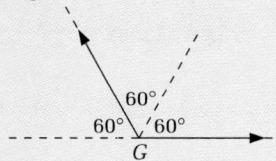

18. Answers may vary. An example is given. I drew the angle in Exercise 17 the most accurately and the angle in Exercise 16 the least accurately. I could improve my ability to make a freehand sketch of an angle by drawing many angles using a protractor to fix angle measures in my mind; then I could practice sketching angles freehand and check my results for accuracy.

19. $\angle AKB$, $\angle GKD$

20. $\angle AKH$, $\angle DKE$ or $\angle BKE$

21. 45°

22. 67.5°

23. 22.5°

24. a–b. Answers may vary.

c. Conjecture: The acute angles of a right triangle are complementary.

25. $y + 102 + 32 = 180$; 46°

26. $a + 74 = 180$; 106°

27. $c + 6 + 14 = 90$; 70°

28. obtuse

29. a. Drawings may vary.

b. The sum of the measures of the angles of the quadrilateral is about 360°. The sum of the measures of the angles of the pentagon is about 540°. The sum of the measures of the angles of the hexagon is about 720°.

c. Answers may vary. An example is given. The angle sum for the pentagon is 180° more than the angle sum for the quadrilateral, and the angle sum for the hexagon is 180° more than the angle sum for the pentagon. It seems each additional side on a polygon adds 180° to the angle sum.

Answers continued on next page.

Assessment: Standard

Ex. 31 presents students with another opportunity to explore angle measure relationships. In writing their procedures, students should draw on the knowledge gained from the Exploration and Exs. 13, 24, 29, and 30. This allows for not only assessment of this new concept, but also of these other concepts as well.

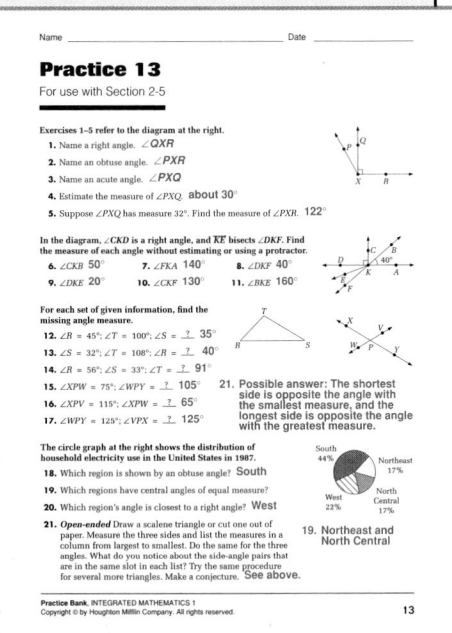

Answers to
Exercises and Problems

29. d. The sum of the measures of the angles of a heptagon should be 720 + 180, or 900°. This conjecture could be tested by drawing several heptagons, measuring the angles of each one, and finding the sum of the measures of the angles for each one.

30. a. Drawings may vary.
   b. Conjecture: Angles opposite sides of equal length are equal.
   c. Answers may vary.
   d. Answers may vary.
   e. All angles of an equilateral triangle are equal in measure.

---

30. **Group Activity** Work in a group of three students.
   a. Have each person in the group use a ruler to draw an isosceles triangle. One triangle should be acute, one triangle should be obtuse, and one triangle should be right.
   b. Write a conjecture about the angles opposite the sides of equal length in your triangle. Mark the sides and angles on your sketch to illustrate your conjecture.
   c. Measure the angles of your triangle to test your conjecture.
   d. Do you think that your conjecture is true for all isosceles triangles? Compare your conjecture with others in your group.
   e. Suppose your conjecture in part (b) is true. What does this lead you to believe about the angles of an equilateral triangle? Sketch and mark an equilateral triangle to indicate the relationships between the sides and between the angles.

Ongoing ASSESSMENT

An *exterior angle* of a triangle is an angle formed outside the triangle by extending one side of the triangle. An exterior angle is associated with a pair of angles inside the triangle called *remote interior angles.*

31. **Writing** Write a procedure for exploring the relationship between an exterior angle of a triangle and its remote interior angles. You may want to use the Exploration and Exercises 13, 24, 29, and 30 as samples.

Review PREVIEW

Describe in words what each symbol means. *(Section 2-4)*

32. $\overline{AB}$ 33. JK 34. $\overleftrightarrow{RS}$ 35. PQ

Name the property shown. *(Toolbox Skill 1)*

36. $12 \cdot 9 = 9 \cdot 12$ 37. $23 + (15 + 9) = (23 + 15) + 9$

Simplify. *(Section 1-5)*

38. $4(x + 2)$ 39. $6a + 5 + 2a + 1$

Working on the Unit Project

40. Some of your numerical facts are percents. Choose one to show in a circle graph. Tell what the whole circle represents, and tell the measure of the central angle you are using for your percent.

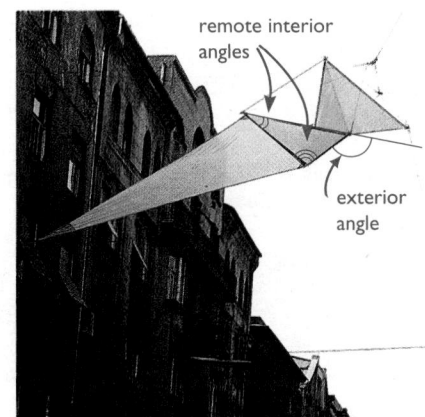
remote interior angles
exterior angle
Kite sculpture in Budapest, Hungary

Student Resources Toolbox
pp. 661–663 *Data Displays*

---

31. Procedures may vary. An example is given.
(1) Use a ruler to draw a large triangle with an exterior angle at one vertex and label the angles as shown.
(2) Tear angles 1 and 2 off the triangle and place them inside angle 3 so that the vertices of angles 1 and 2 meet the vertex of angle 3.
(3) What is the relationship between angles 1, 2, and 3?
(4) Use the result of part (3) to write a conjecture about the relationship between an exterior angle of a triangle and its remote interior angles.
(5) Test your conjecture using another triangle.

32. segment AB

33. length of segment JK

34. line RS

35. length of segment PQ

36. commutative property of multiplication

37. associative property of addition

38. $4x + 8$

39. $8a + 6$

40. Answers may vary.

2-6 Expressions for Measures

Focus
Simplify expressions for measures of geometric figures by multiplying and combining like terms.

Talk it Over

1. In baseball, home plate is a pentagon like the one shown. Write and simplify an expression for the sum of the angles of home plate.

2. Write and simplify an expression for the perimeter of home plate.

Variable expressions are sometimes used to represent measures of geometric figures. To work with these measures, you need to know how to multiply and add variable expressions.

2-6 Expressions for Measures

93

PLANNING

Objectives and Strands
See pages 54A and 54B.

Spiral Learning
See page 54B.

Materials List
➤ Algebra tiles

Recommended Pacing
Section 2-6 is a one-day lesson.

Toolbox References
➤ **Toolbox Skill 1:** Using Properties of Addition and Multiplication
➤ **Toolbox Skill 19:** Finding Volume

Extra Practice
See pages 622–623.

Warm-Up Exercises
Warm-Up Transparency 2-6

Support Materials
➤ Practice 14
➤ Enrichment 13 in the Activity Bank
➤ Study Guide 2-6
➤ Problem Set 4
➤ Diagram Master 6 in the Explorations Lab Manual
➤ Quiz 2-6
➤ Test 6

Answers to Talk it Over

1. $90 + 90 + 3x + 2x + 3x = 180 + 8x$

2. $y + 12 + 12 + y + 2y = 24 + 4y$

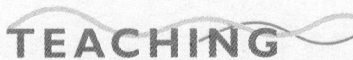
TEACHING

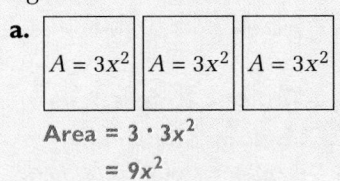

Talk it Over

Questions 1 and 2 illustrate how to write and simplify expressions that involve a single variable.

Additional Samples

S1 Write and simplify an expression for the area of each figure or group of figures.

a.

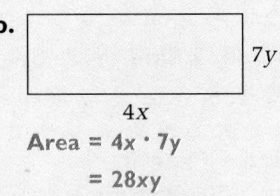

| $A = 3x^2$ | $A = 3x^2$ | $A = 3x^2$ |

$$\text{Area} = 3 \cdot 3x^2$$
$$= 9x^2$$

b.

$7y$

$4x$

$$\text{Area} = 4x \cdot 7y$$
$$= 28xy$$

S2 Write and simplify an expression for the volume of the box.

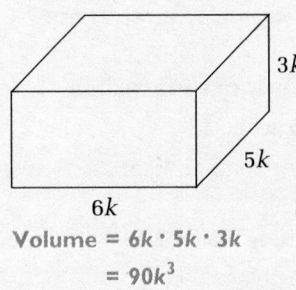

$3k$

$5k$

$6k$

$$\text{Volume} = 6k \cdot 5k \cdot 3k$$
$$= 90k^3$$

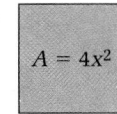

Sample 1

Write and simplify an expression for the area of each figure or pair of figures.

a.

| $A = 4x^2$ | $A = 4x^2$ |

b.

$5x$

$2y$

Sample Response

Problem Solving Strategy: Use a formula

a. Area $= 2 \cdot 4x^2$ ◄——— Total area $= 2 \times$ area of each square

$= (2 \cdot 4)x^2$ ◄——— Group numerical factors.

$= 8x^2$

b. Area $= 2y \cdot 5x$ ◄——— Area of a rectangle $=$ length $\times$ width

$= (2 \cdot 5)(y \cdot x)$ ◄——— Rearrange factors. Group coefficients and group variable factors.

$= 10xy$ ◄——— This could be written $10yx$, but variable factors are usually written in alphabetical order.

Student Resources Toolbox
p. 640 *Properties*

Sample 2

Write and simplify an expression for the volume of the box.

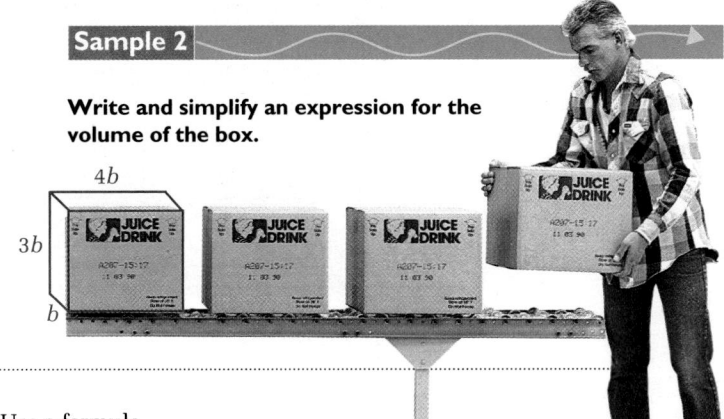

$4b$

$3b$

b

Sample Response

Problem Solving Strategy: Use a formula

Volume of a box $=$ length $\times$ width $\times$ height

$= 4b \cdot 3b \cdot b$ ◄——— The b on the end has a coefficient of 1.

$= (4 \cdot 3 \cdot 1)(b \cdot b \cdot b)$

$= 12b^3$

Student Resources Toolbox
p. 659 *Volume*

Unit 2 Using Measures and Equations

Talk it Over

Question 4 illustrates how to simplify addition expressions in which all the terms involve the same power of a variable.

Talk it Over

3. Suppose the area of each square in part (a) of Sample 1 were $9y^2$ instead of $4x^2$. What would be the simplified expression for the area?

4. Write and simplify an expression for the combined volume of the four boxes.

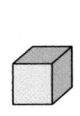 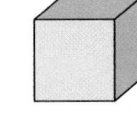 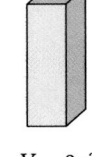

$V = x^3$ $V = 8x^3$ $V = x^3$ $V = 3x^3$

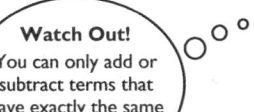
Like Terms with Two Variables

Watch Out!
You can only add or subtract terms that have exactly the same variable part.

You add like terms involving two variables, such as $-5xy$ and $9xy$, the same way you add other like terms.

Sample 3

Simplify.

a. $-5xy + 9xy$ b. $8x^3 + 7xy - 5x^3 + 3xy + 2x$

Sample Response

a. $-5xy + 9xy = (-5 + 9)xy$ ⟵ Use the distributive property.

$= 4xy$

b. $8x^3 + 7xy - 5x^3 + 3xy + 2x = (8x^3 - 5x^3) + (7xy + 3xy) + 2x$ ⟵ Group like terms.

$= 3x^3 + 10xy + 2x$

Talk it Over

Tell whether the terms in each pair are like terms.

5. $5x^2$ and $5x^3$ 6. $-8a$ and $17a$ 7. $7wz$ and $7w$

8. Tell whether $3x^2 + 7x - 9y$ can be simplified. If so, simplify it. If not, tell why not.

Look Back ⟵

You need like terms in order to do some operations. Which operations require like terms? Which operations do not?

2-6 Expressions for Measures

95

Additional Sample

S3 Simplify.

a. $-2pq - 6pq + 7pq$ $-pq$

b. $-3x^2 + 10xy + y^2 + 8x^2 - 8xy$ $5x^2 + 2xy + y^2$

Talk it Over

Use questions 5–7 to establish criteria for deciding whether two terms are like terms or not.

Look Back

The Look Back question provides an excellent opportunity for students to discuss not only which operations require like terms and which do not, but also the differences between addition/subtraction and multiplication/division. ⋯⋯⋯●

Answers to Talk it Over

3. $18y^2$

4. $x^3 + 8x^3 + x^3 + 3x^3 = 13x^3$

5. No.

6. Yes.

7. No.

8. No. It cannot be simplified, because there are no like terms.

Answers to Look Back

Addition and subtraction require like terms. Multiplication and division do not require like terms.

Suggested Assignment

Standard 2–25, 34–39

Extended 1–30, 33–39

Integrating the Strands

Number Exs. 32, 33

Algebra Exs. 1–21, 26–29, 32, 33, 36–38

Measurement Exs. 30–35

Geometry Exs. 22–25, 32, 33

Discrete Mathematics Ex. 39

Logic and Language Exs. 32, 33

Error Analysis

In Exs. 11–18, students may try to combine terms that are not like terms. If you observe this, review the meaning of *like terms*. Ask students to write out their work to show how they grouped like terms together and how they used the distributive property.

2-6 Exercises and Problems

1. **Reading** How is the distributive property used to simplify expressions?

Write and simplify an expression for the perimeter of each rectangle.

2.

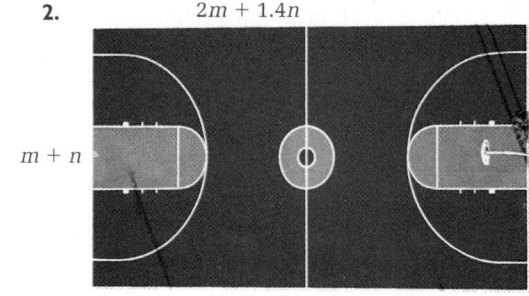

$2m + 1.4n$

$m + n$

3.

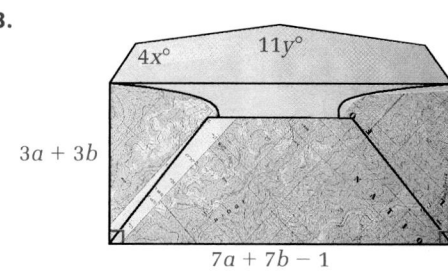

$4x°$ $11y°$

$3a + 3b$

$7a + 7b − 1$

4. Write and simplify an expression for the sum of the measures of the labeled angles in the figure in Exercise 3.

5. This is a pattern that you can use to make a box of any size. The size and proportions of the box depend on the values of x, y, and z. Write and simplify an expression for the area of the pattern.

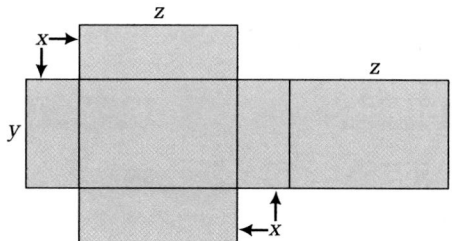

z z x y x

Simplify if possible. If it is not possible to simplify, explain why not.

6. $(13y)(13z)$ 7. $5(6m^2)$ 8. $(−8a)(6a)$ 9. $(8a)(5c)$ 10. $6(7n^3)$

11. $5h + 11k + 3hk + 9h + 4hk$ 12. $5a^2 + 4ab + 3b^2$

13. $m^3 + 4n + 10m^3$ 14. $4x^3 + 7x^2 + 3x + x^3 + 10x^2$

15. $−7x + 9xy + 7x − 7xy$ 16. $−3w^3 + 11w + 9w^2 − w + 4w^3$

17. $8cf + 6c^2 + 3 − 6f − 6c^2$ 18. $7xy + 9x^3 + 2x − 5y$

Write and simplify an expression for each measure.

19. the area of a rectangle with dimensions $9y$ and $5z$

20. the area of a rectangle with dimensions 4 and $a + 2$

21. the volume of a box with dimensions $10c$, $5c$, and $3c$

Match each expression with the figure whose area or volume it represents.

22. $2y^2$
23. $(2y)^2$
24. $2y^3$
25. $(2y)^3$

A.  y y y y

B. 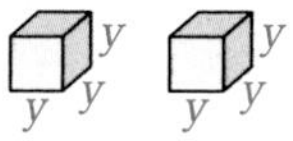 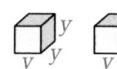 y y y y

C. 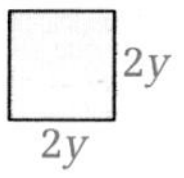 $2y$ $2y$

D. $2y$ $2y$ $2y$

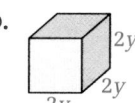

96   **Unit 2** Using Measures and Equations

Answers to Exercises and Problems

1. The distributive property lets you combine two or more like terms into a single term.

2. $2(2m + 1.4n) + 2(m + n) =$
   $6m + 4.8n$

3. $2(3a + 3b) + 2(7a + 7b − 1) =$
   $20a + 20b − 2$

4. $11y + 4x + 2(90) = 11y + 4x + 180$

5. $2(x \cdot z) + 2(y \cdot z) + 2(x \cdot y) =$
   $2xz + 2yz + 2xy$

6. $169yz$

7. $30m^2$

8. $−48a^2$

9. $40ac$

10. $42n^3$

11. $14h + 7hk + 11k$

12. It cannot be simplified because there are no like terms.

13. $11m^3 + 4n$

14. $5x^3 + 17x^2 + 3x$

15. $2xy$

16. $w^3 + 9w^2 + 10w$

17. $8cf − 6f + 3$

18. It cannot be simplified because there are no like terms.

19. $9y \cdot 5z = 45yz$

20. $4(a + 2) = 4a + 8$

21. $10c \cdot 5c \cdot 3c = 150c^3$

22. A

23. C

24. B

25. D

26. $2w + w + 2w + 2w + w + 2w + w + 2w + w = 14w$

27. $2w + 2w + w + 2w + w + 2w + 2w + w + 2w + w = 16w$

28. $3w \cdot 4w = 12w^2$

The diagram at the right is a floor plan of a Japanese house. In traditional Japanese architecture, the shape of each room is based on a rectangular mat called a *tatami*.

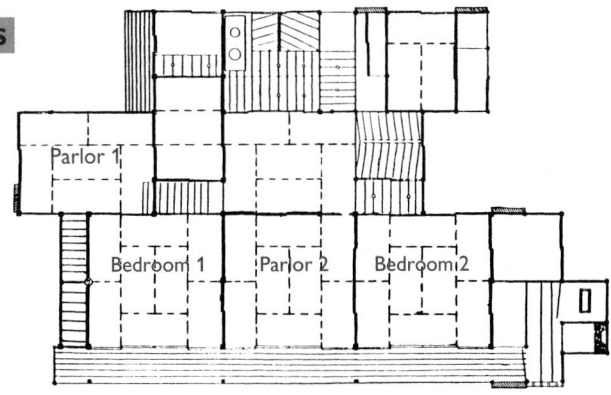

The tatami may be used as a floor covering, as a bed, and as a cushion for sitting. The length of a tatami is usually twice its width.

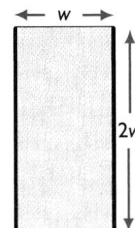

For Exercises 26–29, let w = the width of a tatami. Write and simplify an expression in terms of w for each measure.

26. the perimeter of Parlor 1
27. the perimeter of Parlor 2
28. the area of Parlor 1
29. the area of Parlor 2

30. The actual size of a tatami has varied slightly over the years. Suppose the width of a tatami is 96 cm. What is the total area of the house shown above in square meters?

31. **Research** The size and shape of a tatami are important in Japanese architecture. What measures and shapes are important in the architecture of other cultures?

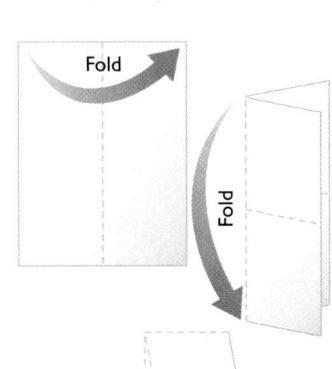

32. **Using Manipulatives** Take any rectangle of paper.
    a. Fold the paper as shown.
    b. Without measuring, predict: Is the perimeter of the small rectangle *one fourth* or *one half* of the perimeter of the big rectangle?
    c. Let *l* = the length of the big rectangle. Let *w* = the width of the big rectangle. Write an algebraic expression for the perimeter of the big rectangle and for the small rectangle.
    d. **Writing** Explain how the results of part (c) help you answer the question in part (b).

33. Use the diagram and the variables in Exercise 32(c). Use algebra to prove that the area of the small rectangle is one fourth of the area of the big rectangle.

Ongoing **ASSESSMENT**

34. **Open-ended** Draw and label a rectangle with perimeter 12*x*, a rectangle with area $12x^2$, and a box with volume $12x^3$. The variable *x* does not have to represent the same value in each figure.

2-6 Expressions for Measures **97**

Answers to Exercises and Problems

29. $4w \cdot 4w = 16w^2$

30. about 103 m^2

31. Answers may vary. An example is given. Domes and columns are found in French baroque architecture.

32. a. manual activity
    b. one half

c. big rectangle: $2l + 2w = 2(l + w)$; small rectangle: $2\left(\frac{1}{2}l\right) + 2\left(\frac{1}{2}w\right) = l + w$

d. The perimeter of the small rectangle is $\frac{1}{2}$ the perimeter of the big rectangle.

33. area of big rectangle = $l \cdot w$; area of small rectangle = $\frac{1}{2}l \cdot \frac{1}{2}w = \frac{1}{4}lw$

34. Answers may vary. An example is given.

Perimeter: $P = 12x$
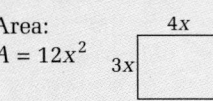

Area: $A = 12x^2$
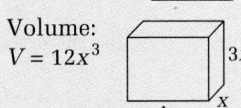

Volume: $V = 12x^3$

Quick Quiz (2-4 through 2-6)

See page 123.

Multicultural Note

Students may be interested to know that Tanzania has the highest mountain in Africa, Mount Kilimanjaro, which is 19,340 feet high. On Tanzania's borders are Lake Victoria, the largest lake in Africa, and Lake Tanganyika, the longest fresh-water lake in the world.

Some scientists believe that Tanzania may have been the original home of humankind. Dr. Louis Leakey and his wife Dr. Mary Leakey have studied Stone Age cave drawings in Tanzania, and have also dis-covered ancient footprints which they believe are 3.6 mil-lion years old.

Practice 14 For use with Section 2-6

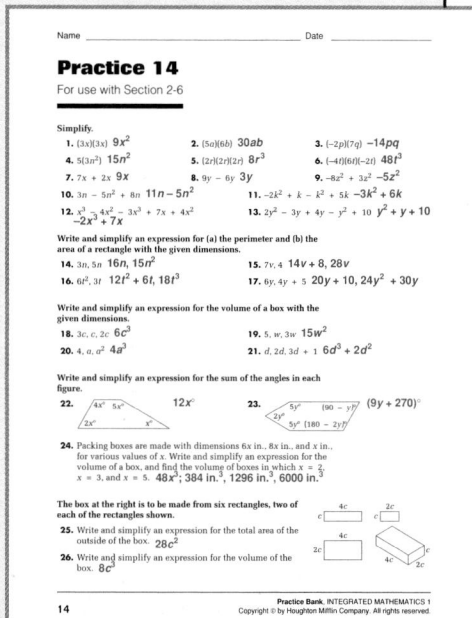

Review **PREVIEW**

35. Estimate to sketch a 40° angle. Then describe how you estimated. *(Section 2-5)*

Model each expression using algebra tiles. *(Section 1-3)*

36. $2x$

37. $x + 3$

38. $2x + 4$

Working on the Unit Project

39. **Group Activity** Show your group the circle graph you made for Exercise 40 in Section 2-5. Discuss your ideas for expressing your other numerical facts. Agree on at least four facts to illustrate on your theme poster. Include a very large number and a very small number in scientific notation, a measurement, and a percent.

Unit 2   CHECKPOINT 2

1. **Writing** Describe how you estimate distances on a map, how you estimate areas on a map, and how you estimate angle measures.

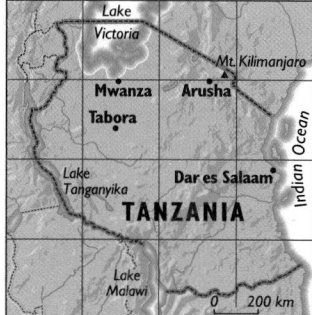

For Exercises 2–4, use the map of Tanzania, a country in Africa.   2-4

2. Estimate the distance from Tabora to Arusha.

3. Is 450 km a reasonable estimate of the distance from Mwanza to Dar es Salaam? Explain.

4. Use the grid on the map to estimate the area of Tanzania. Each grid square is 300 km on a side.

For Exercises 5–7, use triangle ABC.

5. Estimate the length of each side of triangle ABC in both customary and metric units.

6. Suppose X is the midpoint of $\overline{AC}$. Is it *True* or *False* that $AX = 2 \cdot XC$? Explain.

7. Write and solve an equation to find the unknown angle measure in the triangle.   2-5

8. Estimate to sketch a 15° angle. Then describe how you estimated.

Simplify.   2-6

9. $(2y)(7y)(y)$

10. $2c^3 - 9c^2 - 7c + 7c^3$

11. $-7mn + 5m + 4mn - 4n$

12. $11a + 7b - 4a + b$

Unit 2 Using Measures and Equations

Answers to
Exercises and Problems

35. 40° is a little less than half of 90°.

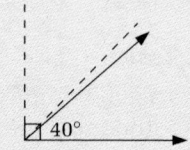

36. ⬜⬜
37. ⬜⬜⬜⬜
38. ⬜⬜⬜⬜⬜⬜
39. Answers may vary.

Answers to Checkpoint

1. Explanations may vary. An example is given. To esti-mate distances on a map, you use the given scale as a ruler to determine lengths by finding out how many of these scale units are in a given distance and then multiply the value of the scale by the number of units. To estimate areas on a map, you determine the dimensions of a given area using the method described above and then use the appropriate area formula. To estimate angle mea-sures, you compare a given angle to an angle of known measure, usually a 90° or 180° angle, and estimate depending on whether the given angle is smaller or larger than the angle it is being compared to.

2. about 500 km

3. No, it measures approxi-mately 800 km.

4. about 720,000 km²

5. $AB \approx \frac{3}{4}$ in. ≈ 2 cm;

    $AC \approx 1\frac{3}{4}$ in. ≈ 4.5 cm;

    $BC \approx 1$ in. ≈ 2.5 cm

6. False. $AX = XC$ if X is the midpoint of $\overline{AC}$.

2-7

Solving Equations: Balancing

Focus
Begin to solve equations about geometric figures.

GO FLY A KITE

BY THE WAY...

The record for the largest kite ever flown was set by a Dutch team on August 8, 1981. The area of the kite was 5952 ft², which is about three fourths the area of a baseball diamond.

$x + 2 = 5$

A mathematical statement in which one expression equals another is an **equation.**

Sometimes streamers or tassels are added to kites. To keep both sides of the kite balanced in flight, the amount added to the right side must equal the amount added to the left side.

In mathematics the equals sign works like the balance line of the stunt kite, or the pivot point of a balance scale. The equals sign tells you that the expression on the right side of it has the same value as the expression on the left side.

Your knowledge of arithmetic facts tells you that x must represent 3 in the equation.

A value of a variable that makes an equation true is a **solution** of the equation. The process of finding solutions is called **solving an equation.**

2-7 Solving Equations: Balancing

99

PLANNING

Objectives and Strands
See pages 54A and 54B.

Spiral Learning
See page 54B.

Materials List
➤ Algebra tiles
➤ Geometric drawing software or ruler and protractor

Recommended Pacing
Section 2-7 is a one-day lesson.

Toolbox References
Toolbox Skills 5, 6: One-Step Equations

Extra Practice
See pages 622–623.

Warm-Up Exercises
💡 Warm-Up Transparency 2-7

Support Materials
➤ Practice 15
➤ Enrichment 14 in the Activity Bank
➤ Study Guide 2-7
➤ Problem Set 5
➤ Additional Explorations 3, 4
➤ Diagram Master 6 in the Explorations Lab Manual
💡 Overhead Visual 3
➤ McDougal Littell Mathpack software: *Geometry Inventor*
➤ Quiz 2-7
➤ Alternative Assessments 4–6

Answers to Checkpoint

7. $22 + 19 + x = 180$,
   $x = 139°$

8. $15°$ is $\frac{1}{6}$ of a $90°$ angle.

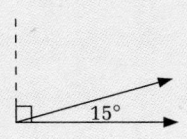

9. $14y^3$

10. $9c^3 - 9c^2 - 7c$

11. $-3mn + 5m - 4n$

12. $7a + 8b$

Mathematical Procedures

The content of this section and the next on solving equations is an informal introduction to this topic. In Unit 5, further work on solving equations is presented that builds upon the concepts developed in this unit. If you prefer, you can postpone these sections until after Section 5-1.

Reasoning

Some students may be uncomfortable with the idea that the result of solving $x + 2 = 5$ is $x = 3$, while the result of solving $2x + 3 = 11$ is $x = 4$. The question in these students' minds is, "What is x *really* equal to?" Any student who is asking this question may not fully understand the concept of a variable.

You can help students think about the nature of a variable by using an analogy with pronouns. The sentence *She weighs more than 30 pounds* is neither true nor false. If you replace *she* with the name of a student in the class, you will get a true sentence, for example, *Joanne weighs more than 30 pounds*. If you replace *she* with *Betty* and if Betty happens to be a newborn baby, you get a false sentence.

Emphasize that variables are mathematical pronouns. The equation $x + 2 = 5$ is neither true nor false. It becomes true if you replace x with 3, but it is false if you replace x with 7. The number 3 is a solution to $x + 2 = 5$ because $3 + 2 = 5$ is a true statement.

$$x + 2 = 5$$

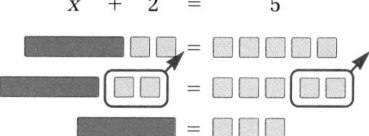

One way to solve an equation is to make changes to both sides until the variable is alone on one side and the solution is alone on the other side. The changes you make must keep the equation in balance.

You can use algebra tiles to help you solve equations like $x + 2 = 5$.

Sample 1

Solve the equation $2x + 3 = 11$.

Sample Response

Use algebra tiles to model the equation.

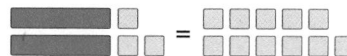

$$2x + 3 = 11$$

To get the x-tiles alone on one side, first take away three 1-tiles from both sides.

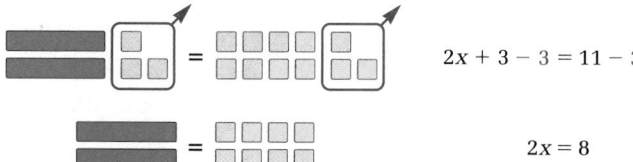

$$2x + 3 - 3 = 11 - 3$$

$$2x = 8$$

To see what a single tile alone equals, divide each side into two identical groups.

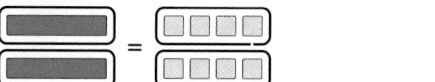

$$\frac{2x}{2} = \frac{8}{2}$$

There is one x-tile by itself in each group. It equals four 1-tiles.

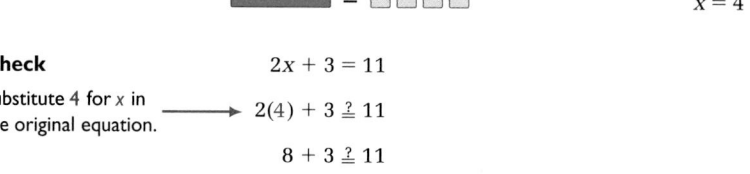

$$x = 4$$

Check

Substitute 4 for x in the original equation.

$$2x + 3 = 11$$
$$2(4) + 3 \stackrel{?}{=} 11$$
$$8 + 3 \stackrel{?}{=} 11$$
$$11 = 11 ✔$$

The solution is 4.

When you make the same change to both sides of an equation, you change the form of the equation, but not the solution. Equations that have the same solution are called **equivalent equations**. The original equation $2x + 3 = 11$ and the new equation $x = 4$ are equivalent.

100

Unit 2 Using Measures and Equations

Solve the equation $3x - 4 = 11$.

Sample Response

Use algebra tiles to model the equation.
The red tiles show numbers subtracted.

 $3x - 4 = 11$

To get the x-tiles alone on one side, add four 1-tiles
to both sides. Then you can remove four pairs of
tiles from the left side, because they add up to zero.

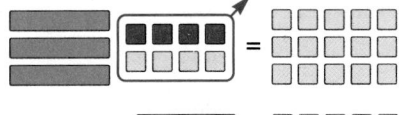 $3x - 4 + 4 = 11 + 4$

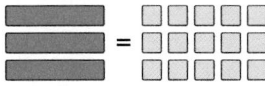 $3x = 15$

To see what a single tile alone equals, divide
each side into three identical groups.

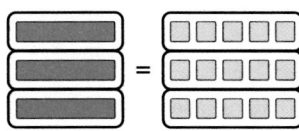 $\dfrac{3x}{3} = \dfrac{15}{3}$

There is one x-tile by itself in
each group. It equals five 1-tiles.

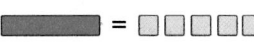 $x = 5$

Check $3x - 4 = 11$

Substitute 5 for x in
the original equation. →

$3(5) - 4 \stackrel{?}{=} 11$

$15 - 4 \stackrel{?}{=} 11$

$11 = 11$ ✔   The solution is 5.

Student Resources Toolbox
pp. 645–647 *Solving One-Step
Equations*

Talk it Over

1. Is 23 a solution of the equation $2w + 18 = 28$? Explain.

2. Are $15n - 8 = 22$ and $5n = 10$ equivalent equations? Explain.

**Tell how to get the variable alone on one side of each equation.
Then solve.**

3. $t + 9 = 54$ 4. $39 = 6b$ 5. $6r - 10 = 212$

2-7 Solving Equations: Balancing   **101**

Additional Samples

S1 Solve the equation $3x + 5 = 20$. Use algebra tiles.

Model the equation.

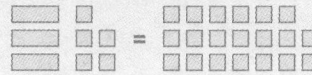

Take away five 1-tiles from
each side.

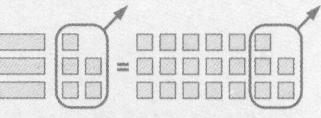

Divide each side into three
identical groups.

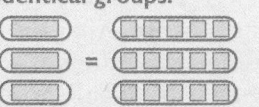

One x-tile equals five 1-tiles.

$x = 5$

S2 Solve the equation
$2x - 3 = 5$. Use algebra
tiles.

Model the equation.

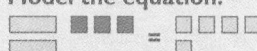

Add three 1-tiles to both
sides. Remove zero pairs from
the left side.

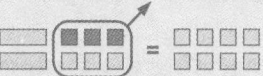

Divide each side into two
identical groups.

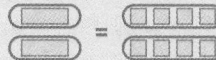

One x-tile equals four 1-tiles.

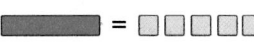

$x = 4$

Talk it Over

Question 2 can be used to
assess students' understanding
of the concept of equivalent
equations.

Answers to Talk it Over

1. No. Reasons may vary.
   An example is given.
   $46 + 18 = 64$, not 28, so 23
   is not a solution.

2. Yes. Reasons may vary. An
   example is given. Both
   equations have the solution 2. Equations with the
   same solution are
   equivalent.

3. Subtract 9 from both sides.
   45

4. Divide each side by 6.
   6.5

5. Add 10 to both sides, then
   divide both sides by 6.
   37

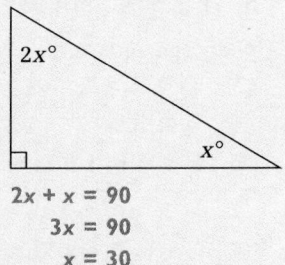

Problem Solving Strategy: Use an Equation

You can write and solve equations to find unknown measures. Some of the equations will use special angle relationships.

Sample 3

Find the value of x on the kite.

Sample Response

Problem Solving Strategy: Use an equation

$x + x + 90 = 360$ ← The angles form a complete circle. An entire circle has a measure of 360°.

Simplify the left side of the equation. → $2x + 90 = 360$

$2x + 90 - 90 = 360 - 90$ ← To get $2x$ alone on one side and keep the equation balanced, subtract 90 from both sides.

$2x = 270$

$\dfrac{2x}{2} = \dfrac{270}{2}$ ← To get x alone and keep the equation balanced, divide both sides by 2.

$x = 135$

The value of x is 135.

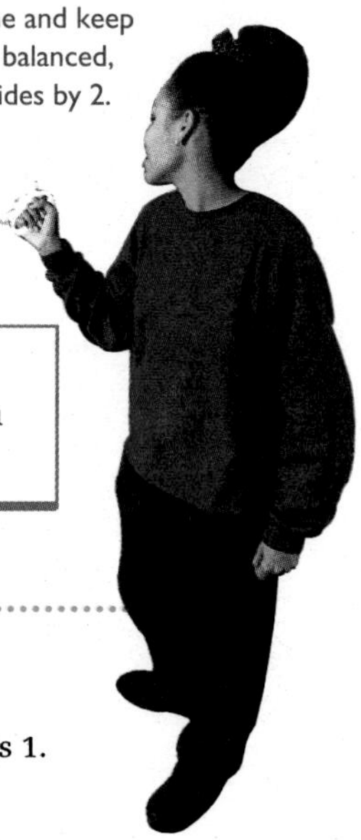

Look Back ←

What do you think are the most important ideas about equation solving for you to remember? Give at least three ideas.

2-7 Exercises and Problems

1. **Using Manipulatives** Suppose ▬ represents x and ▯ represents 1.

   a. What equation does the diagram at the right illustrate?

   b. Show how to solve the equation for ▬ using the model in part (a).

   c. Show how to solve for x using the equation you wrote in part (a).

   d. Show how to solve $2x - 1 = 3$ using algebra tiles.

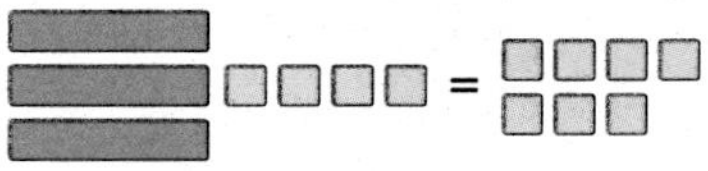

Unit 2 Using Measures and Equations

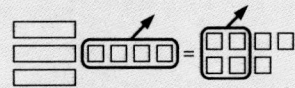
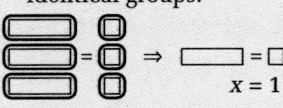

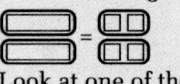
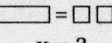

2. **Reading** It takes one more step to solve the equation in Sample 3 than it does to solve the equations in Samples 1 and 2. What is the extra step?

3. **Reading** What would happen to the balance scale shown on page 99 if you removed the two smaller weights from the left-hand pan? Describe two ways to restore the balance.

Solve.

4. $h - 7 = 48$

5. $8p = 56$

6. $0 = 16n$

7. $4a + 13 = 25$

8. $18g - 30 = 6$

9. $8 + 5x = 33$

10. $26 = 4b - 2$

11. $m + m + 7 = 15$

12. $6 + 5 + a = 18$

Two students gave the two different solutions shown for each equation. Decide which is the correct solution. Explain the mistake that may have led to the other solution.

13. $w + 6 = 17$ 23, 11

14. $x + x - 40 = 100$ 70, 140

15. $5h - 10 = 20$ 14, 6

16. $9 + 3x = 30$ 3, 7

Write and solve an equation to find each unknown measure.

17. One dimension of a rectangular lot is 330 ft. The area of the lot is 217,800 ft². What is the other dimension of the lot?

18. There is 1008 ft² of floor space in the Reboulets' house. They are planning to build a rectangular addition. The length of the addition will be 18 ft. There will be 1170 ft² of floor space in the Reboulets' house when they are done. How wide will the addition be?

Match each figure with an equation that describes a relationship among the angle measures. (*Note:* There may be more than one answer, or repeated answers.)

19.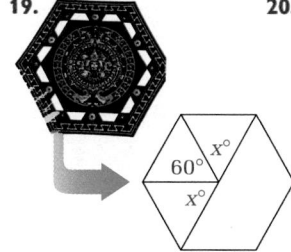

20.

21.

A. $2x + 240 = 360$

B. $2x + 60 = 180$

C. $2x + 60 = 360$

D. $2x + 90 = 180$

Write and solve an equation to find each unknown measure in each figure. (*Note:* The sum of the measures of the angles of a quadrilateral is 360°.)

22.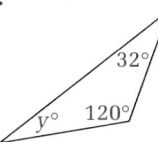

23.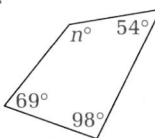

24.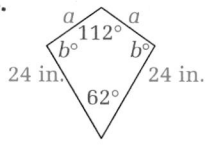

perimeter = 78 in.

25.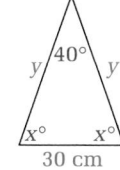

perimeter = 116 cm

2-7 Solving Equations: Balancing

103

Answers to Exercises and Problems

2. You combine like terms on either side of the equation. In this example, you add $x + x$.

3. Answers may vary. An example is given. The left-hand pan of the balance scale would be lighter than the right-hand pan, causing the left-hand pan to rise and the right-hand pan to fall. To restore the bal-

ance, replace the two smaller weights on the left side or remove two identical weights from the right side.

4. 55   5. 7   6. 0
7. 3   8. 2   9. 5
10. 7   11. 4   12. 7

13. The correct solution is 11. The other solver added 6 to both sides instead of

subtracting 6 from both sides.

14. The correct solution is 70. The other solver did not divide both sides by 2.

15. The correct solution is 6. The other solver divided by 5 first instead of adding 10 first to both sides of the equation. In doing so, the solver forgot to divide 10 by 5.

16. The correct solution is 7. The other solver chose the wrong factor of $3x = 21$ as a solution.

17. $330x = 217,800$; $x = 660$; The other dimension of the lot is 660 ft.

18. $1008 + 18x = 1170$; $x = 9$; The width will be 9 ft.

19. B   20. B   21. D

22. $y + 32 + 120 = 180$; 28°

23. $n + 69 + 98 + 54 = 360$; 139°

24. $b + b + 112 + 62 = 360$; 93°
    $a + a + 24 + 24 = 78$; 15 in.

25. $x + x + 40 = 180$; 70°
    $y + y + 30 = 116$; 43 cm

103

When polygons form a pattern that covers a plane without gaps or overlaps, the pattern is called a *tessellation*.

The sum of the measures of the angles at any vertex not on the boundary is 360°.

$$\angle 1 + \angle 2 + \angle 3 + \angle 4 = 360°$$

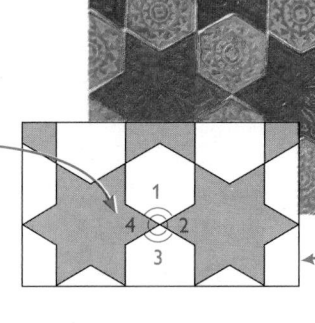

Islamic Star Pattern

boundary

For Exercises 26 and 27, write and solve an equation that describes a relationship among the angle measures in each tessellation.

26.

Midwest Amish

27.

Navajo

28 TECHNOLOGY Use geometric drawing software.

Alternative Approach Use a ruler and a protractor.

Draw a tessellation using a triangle whose angles are all 60° and a parallelogram whose angles are 60°, 120°, 60°, and 120°.

Ongoing ASSESSMENT

29. **Writing** Suppose you are helping a friend learn algebra. Explain the difference between simplifying an expression and solving an equation. Give an example of each.

Review PREVIEW

Simplify if possible. If not, explain why not. *(Section 2-6)*

30. $8(-4x^3)$ 31. $10n^3 + 5n^2 + n$ 32. $-8a^3 + 12ab + 3a^3 + 4a - 9ab$

Simplify. *(Section 2-2)*

33. $-7 + 6$ 34. $-12 - 4$ 35. $(-3)(-1.6)$ 36. $-20 \div 0.4$

2 Working on the Unit Project

37. **Writing** Write four mathematical questions that can be answered using the facts included on your display. Write one question for each fact.

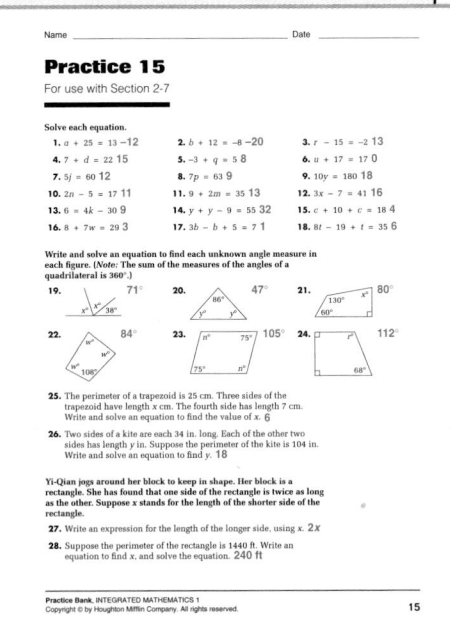

Answers to
Exercises and Problems

26. $x + x + 90 = 360$; 135°
27. $x + x + 150 + 150 = 360$; 30°
28.
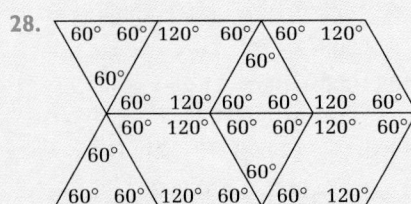

29. Explanations may vary. Examples are given. An expression has no equals sign and therefore no solution. When simplifying an expression, you combine all like terms by adding, subtracting, multiplying, or dividing. For example, the expression $2(3x) - 2 + 4x - 3(-1)$ would be simplified to $10x + 1$. An equation has an equals sign and you want to end up with the variable in the equation equal to a value. This value is called the solution to the equation. You put all the variable terms on one side of the equals sign, and all constants on the other side. Combine like terms as mentioned above. If necessary, multiply or divide both sides of the equation to get the variable alone. For example, to solve $8t - 4(5t) = -2 - t$ for t, you would do the following steps: $8t - 20t = -2 - t$
$$8t - 20t + t = -2$$
$$-11t = -2$$
$$t = \frac{2}{11}$$

30. $-32x^3$
31. The expression cannot be simplified because there are no like terms.
32. $-5a^3 + 3ab + 4a$
33. -1
34. -16
35. 4.8
36. -50
37. Answers may vary.

2-8 Solving Equations: Undoing

Focus
Use equations to solve a variety of real-life problems.

PLANNING

Objectives and Strands
See pages 54A and 54B.

Spiral Learning
See page 54B.

Materials List
➤ Calculator

Recommended Pacing
Section 2-8 is a one-day lesson.

Extra Practice
See pages 622–623.

Warm-Up Exercises
Warm-Up Transparency 2-8

Support Materials
➤ Practice 16
➤ Enrichment 15 in the Activity Bank
➤ Study Guide 2-8
➤ Problem Set 5
➤ Additional Exploration 4
➤ Quiz 2-8
➤ Alternative Assessment 7

Can you solve this number puzzle?

If you divide a number by 3 and subtract 64, you get 76. What is the number?

You could solve this puzzle using mental math, by working backward. The steps can be shown in a flow chart.

Read the puzzle again and break it into steps.

Start with the unknown number.

If you divide a number by 3…

Divide by 3.

…and subtract 64…

Subtract 64.

…you get 76.
What is the number?

76

420   That gives me 420.

Multiply by 3.   Undo dividing by 3 by multiplying by 3.

Add 64.   Undo subtracting 64 by adding 64. That makes 140.

Start with 76.

In order to solve it, you could work backward, undoing each operation.

The same kind of thinking can help you solve an equation. The method is called *solving by undoing* or *solving by inverse operations*. In Sample 1, the number puzzle is written as an equation and solved.

2-8 Solving Equations: Undoing

105

Mathematical Procedures

The procedure of solving by undoing or solving by inverse operations can be employed to solve many different kinds of mathematical problems. The flowchart on page 105 shows a list of discrete steps which can be reversed to solve the puzzle. Make sure students see how the flowchart relates to the equation in Sample 1.

Additional Sample

S1 Solve $\dfrac{5x}{7} - 4 = 6$.

$x = 14$

Using Technology

When using a calculator to undo operations to solve an equation, students should be careful to enter the inverse operations of the ones shown in the equation. Also, students should estimate to check the reasonableness of their calculator answers.

Sample 1

Solve $\dfrac{x}{3} - 64 = 76$.

Sample Response

Method ❶ Undo the operations. Remember that your goal is to get x alone on one side of the equation.

$$\dfrac{x}{3} - 64 = 76$$

$$\dfrac{x}{3} - 64 + 64 = 76 + 64 \quad \longleftarrow \quad \text{Undo subtraction of 64.}$$
$$\text{Add 64 to both sides.}$$

$$\dfrac{x}{3} = 140$$

$$3 \cdot \dfrac{x}{3} = 3 \cdot 140 \quad \longleftarrow \quad \text{Undo division by 3.}$$
$$\text{Multiply both sides by 3.}$$

$$x = 420$$

Check $\dfrac{x}{3} - 64 = 76$

Substitute 420 for x in the original equation. $\longrightarrow$ $\dfrac{420}{3} - 64 \overset{?}{=} 76$

$$140 - 64 \overset{?}{=} 76$$

$$76 = 76 ✔$$

The solution is 420.

Method ❷ Once you know how to solve equations by undoing, you can use a calculator to undo the operations. Start with 76 and work backward to x.

$$\dfrac{x}{3} - 64 = 76$$

76 [+] 64 [ENTER] [×] 3 [ENTER] → 420

Check $\dfrac{x}{3} - 64 = 76$

Substitute 420 for x and work forward. $\longrightarrow$ 420 [÷] 3 [−] 64 [ENTER] → 76

The solution is 420.

Talk it Over

1. **a.** Explain how to solve $-49 = 4n + 7$ using Method 1 from Sample 1. Then solve.

   b. How is solving $-49 = 4n - 7$ different from solving $-49 = 4n + 7$?

2. Describe a calculator key sequence you could use to solve the equation in question 1(a). Then solve using a calculator.

106   **Unit 2** Using Measures and Equations

Answers to Talk it Over

1. **a.** $-49 = 4n + 7$
   Undo addition of 7.
   Subtract 7 from both sides.
   $-49 - 7 = 4n + 7 - 7$
   $-56 = 4n$
   Undo multiplication by 4. Divide by 4.
   $-\dfrac{56}{4} = \dfrac{4n}{4}$
   $-14 = n$

   b. You would add 7 to both sides instead of subtracting 7.

2. [4][9][+/-][−][7][=][4][4][=];
   -14

Additional Samples

S2 Rosalia spent $5.12 to buy and mail eight postcards. The total cost of the stamps was $1.52. How much did she pay for each postcard? **$.45**

S3 Volker took 3.5 hours to mow the Taniyamas' lawn and 1.5 hours to mow the Leffler's lawn. He charged the same amount per hour for each job. In all, he earned $18.75. What did he charge per hour for his work? **$3.75**

Sample 2

Marcus left his bicycle at a repair shop to have seven spokes replaced. When he got home he found the phone message shown. How much did each spoke cost?

*Marcus!
The bike shop called—your bike is ready. Total bill is $9.94. The labor charge is $7.49 (they said it took about 2 hours).*

Sample Response

Problem Solving Strategy: Use an equation

Choose a variable to represent a number you do not know.

Let c = the cost of each spoke.

Cost of spokes + cost of labor = total bill

$7c$ = the cost of seven spokes. ⟶ $7c + 7.49 = 9.94$

$7c + 7.49 - 7.49 = 9.94 - 7.49$ ⟵ Undo addition of 7.49.
Subtract 7.49 from both sides.

$7c = 2.45$

$\dfrac{7c}{7} = \dfrac{2.45}{7}$ ⟵ Undo multiplication by 7.
Divide both sides by 7.

$c = 0.35$

The cost of each spoke is $0.35.

Simplifying before Solving

Sometimes you have to simplify one or both sides of an equation before you begin to solve the equation by undoing.

Sample 3

For her theme poster project, Michelle bought 3.25 yd of fluorescent blue nylon, and 2.5 yd of hot pink nylon. The price per yard was the same for both colors. The total cost for the fabric was $40.02. What was the cost per yard of the nylon?

Sample Response

Problem Solving Strategy: Use an equation

Let n = the cost per yard of nylon.

$3.25n$ = the cost of the blue nylon

$2.5n$ = the cost of the hot pink nylon

Total cost = cost of blue + cost of pink

$40.02 = 3.25n + 2.5n$ ⟵ Combine like terms to simplify the right side of the equation.

$40.02 = 5.75n$

$\dfrac{40.02}{5.75} = \dfrac{5.75n}{5.75}$

$6.96 = n$ The cost per yard of the nylon is $6.96.

2-8 Solving Equations: Undoing

Look Back

Use the Look Back to highlight the differences between evaluating expressions and solving equations. You may want to ask students to tell what keystroke sequences they would use to solve and check the equation $2x + 5 = 9$.

APPLYING

Suggested Assignment

Standard 1–13, 15–19, 24, 25, 27–32

Extended 1–22, 24, 25, 27–32

Integrating the Strands

Number Exs. 28–31

Algebra Exs. 1–27, 32

Problem Solving

When discussing Exs. 15–19, bring out the fact that the purpose of learning to solve equations is to be able to use them to solve problems.

When students are first learning to write an equation that models a real-world situation, you may wish to use a chart approach to help them write the equation. For example, in Ex. 15, students can set up a chart such as the following:

Hours	Labor	Labor + Parts
1	$35 · 1	$35 · 1 + $150.50
2	$35 · 2	$35 · 2 + $150.50
⋮		
n	$35 · n	$35 · n + $150.50

$255.50 = $35 · n + 150.50

Look Back

In the expression $2x + 5$, substitute 7 for x. What do you do first to evaluate? What do you do second?

What do you do first to solve the equation $2x + 5 = 9$? What do you do second?

Compare the order of the operations you perform to evaluate an expression and to solve an equation.

2-8 Exercises and Problems

1. **Reading** When you solve equations by undoing, you try to get the equation in a particular form. What is that form?

Solve.

2. $2x + 8 = 26$

3. $6h - 117 = -21$

4. $9h - 3h = 15$

5. $-7a - 1 = 13$

6. $\frac{m}{2} + 7 = 9$

7. $\frac{a}{3} - 2 = 1.5$

8. $-50 = 18c + 7c$

9. $17 = 0.05x - 60$

10. $-3.6y + 70 = -2$

11. $-37 = \frac{k}{6} - 19$

12. $7.49 = 0.75x + 1.39x$

13. $12 = 15 + 36d$

14. **Open-ended** Write and solve three number puzzles similar to the one at the beginning of this section. At least one puzzle should have a negative solution and at least one should have a decimal solution.

For Exercises 15–19, write and solve an equation to find each unknown value.

15. When Hanako Sakata had her car repaired, the total bill was $255.50. The bill included a cost of $150.50 for parts and a separate charge for labor. The charge for labor was $35 per hour. How many hours of labor were included in the bill?

16. Rudy bought a can of three tennis balls and a new grip for his racquet. The grip cost $7.49. The items cost $9.38 in all. How much did each tennis ball cost?

17. Michelle bought a case of motor oil and a quart of transmission fluid for $11.97. The transmission fluid sells for $1.29 per quart. There are twelve quarts of oil in a case. How much did one quart of oil cost?

18. Jason Mills bought pencils and erasers to lend to his students taking the SAT (Scholastic Assessment Test). He bought the same number of pencils as erasers. The pencils cost $.29 each and the erasers cost $.14 each. He spent $10.32 in all. How many of each did he buy?

19. Angelica bought equal amounts of two different fabrics to make costumes. The price per yard for each fabric is shown in the photograph. The total cost was $33.88. How many yards of each fabric did Angelica buy?

$4.48/yd

$3.99/yd

Unit 2 Using Measures and Equations

Answers to Look Back

First multiply 2 times 7; then add 5. First subtract 5 from both sides; then divide both sides by 2. The order is reversed.

Answers to Exercises and Problems

1. You try to get the variable alone on one side of the equals sign.

2. 9

3. 16

4. 2.5

5. −2

6. 4

7. 10.5

8. −2

9. 1540

10. 20

11. −108

12. 3.5

13. $-0.08\overline{3}$

14. Answers may vary.

15. $255.50 = 150.50 + 35x$; 3 h

16. $9.38 = 7.49 + 3x$; $.63

17. $11.97 = 1.29 + 12x$; $.89

18. $10.32 = 0.29p + 0.14p$; 24

19. $33.88 = 3.99x + 4.48x$; 4 yd

Women have made many contributions in mathematics and science. In their roles as writers and lecturers, they have combined communication skills with their knowledge of mathematics and science.

For Exercises 20–23, use the time line. Write and solve an equation to find each unknown year.

20. Emilie du Châtelet wrote a physics textbook called *Institutions de Physique*. Let y = the year this book was published. By doubling that year and then subtracting 1721, you get the year that her translation of the *Principia* was published.

21. Mary Fairfax Somerville wrote many books, including a book about chemistry and physics called *Molecular and Microscopic Science*.

    a. Let y = the year this book was published. By dividing that year by 3 and then adding 1208, you get the year her book *The Mechanism of the Heavens* was published.

    b. Use your answer to part (a) to find out how old Mary Somerville was when *Molecular and Microscopic Science* was published.

22. Women were not allowed to get university degrees in Russia in 1863. Sonya Kovalevsky went to Germany to get her Ph.D. in mathematics. Let y = the year she got her Ph.D. By tripling that year and then adding the year Winifred Merrill got her Ph.D., you get 7508.

23. a. **Research** Find the year of another important event in the life of a female mathematician. You may choose a mathematician from the time line or another female mathematician.

    b. **Writing** Using a variable to represent the year you found, write a puzzle relating the year you found to a year in the time line.

    c. Write an equation to represent the puzzle you wrote in part (b). Then solve the equation to check that the solution is the correct year.

    d. **Group Activity** Share your puzzle with a classmate. (Do not share the equation.) Solve your classmate's puzzle and compare results.

Emilie du Châtelet
(1706 – 1749) Wrote the only French translation of Newton's *Principia*, published in 1759.

Mary Fairfax Somerville
(1780 – 1872) Wrote *The Mechanism of the Heavens*, a translation of Laplace's *Mécanique céleste* with added commentary, published in 1831.

Ada Byron Lovelace
(1815 – 1852) Considered to be the first person to write about how computers are programmed, in a paper published in 1843.

Sonya Kovalevsky
(1850 – 1891) Became a mathematics lecturer at the University of Stockholm, and thus the first female lecturer in Sweden, in 1883.

Winifred Edgerton Merrill
(1862 – 1951) Was the first woman to receive a doctorate in Mathematics from an American university (Columbia) in 1886.

Emmy Noether
(1882 – 1935) Barred from academic work by the Nazis, she came to the United States to be a professor of mathematics at Bryn Mawr in 1933.

1700
1750
1800
1850
1900
1950

2-8 Solving Equations: Undoing 109

Answers to Exercises and Problems

20. $2y - 1721 = 1759$; 1740

21. a. $\frac{y}{3} + 1208 = 1831$; 1869

    b. 89 years old

22. $3y + 1886 = 7508$; 1874

23. a–d. Answers may vary.

	Cal/min
Basketball	12.5
Cross-country skiing	11.7
Dancing	10.5
Walking (5mi/h)	9.3
Tennis (singles)	7.1

24. a. Write a formula for the number of Calories Jason uses while dancing for any number of minutes, x.

b. How long should he dance to use 190 Calories?

25. Suppose Jason plans to spend equal amounts of time playing basketball and playing tennis. How much time should he allow for each if he wants to use 400 Cal?

Ongoing **ASSESSMENT**

26. Group Activity Work in a group of three students.

a. Write a five-question quiz on Sections 2-7 and 2-8. At least one question should involve a problem solving situation. At least one question should involve geometry.

b. Make enough copies of your group's quiz for students in another group. Have them take the quiz.

c. Collect and correct the quizzes. Write a summary of any errors that were made by more than one student.

Review **PREVIEW**

27. a. What equation does the diagram below illustrate? *(Section 2-7)*

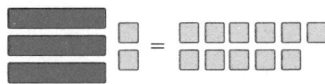

b. Show how to solve for using the model in part (a).

Write each number in decimal notation. *(Section 2-3)*

28. 6.3×10^7

29. 1.8×10^{-5}

Replace each __?__ with the correct number. *(Section 1-3)*

30. $11^2 = \underline{?} \times \underline{?} = \underline{?}$

31. $7^3 = \underline{?} \times \underline{?} \times \underline{?} = \underline{?}$

Working on the Unit Project

32. a. You should still have some numerical facts that are not yet included on your theme poster. Write an equation that someone else could solve to find one of these numerical facts. For example, to find the number of legs on a butterfly, solve the equation $5x - 11 = 19$.

b. Solve the equation you wrote in part (a) to be sure the solution matches the fact.

Answers to Exercises and Problems

24. a. $C = 10.5x$, where C is the number of Calories and x is the number of minutes.

b. about 18.1 min

25. about 20.4 min each

26. a–c. Answers may vary.

27. a. $3x + 2 = 11$

b.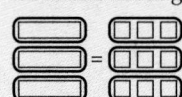

Subtract two unit tiles from each side.

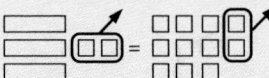

Divide each side into three identical groups.

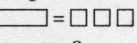

Look at one of the groups.

$\underline{\quad\quad} = \square\square\square$

$x = 3$

28. 63,000,000

29. 0.000018

30. 11, 11, 121

31. 7, 7, 7, 343

32. a, b. Answers may vary.

Square Roots and Cube Roots

Find square roots and
cube roots and distinguish
a rational number from
an irrational number.

GETTING TO THE
ROOT
OF THE PROBLEM

Talk it Over

1. In the diagram, some squares are drawn on dot paper. The
first square has an area of one square unit. Use the diagram
to complete the table below.

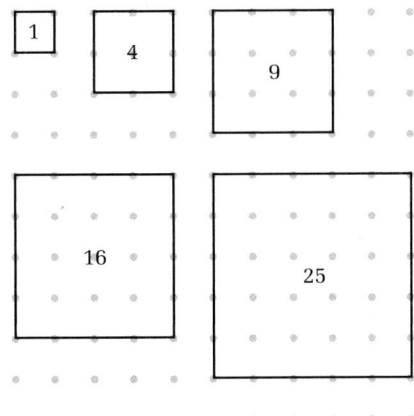

Area of square	1	4	9	16	25
Length of side	1	2	?	?	?

2. What is the length of each side of a square with an area of
36 square units? 49? 64? 100? 400?

When you use the area of a square to find the length of a side
of the square, you undo the squaring of a number. This process
is called *finding a square root*.

Answers to Talk it Over

1. 3; 4; 5

2. 6 units; 7 units; 8 units;
10 units; 20 units

PLANNING

Objectives and Strands
See pages 54A and 54B.

Spiral Learning
See page 54B.

Materials List
➤ Scientific calculator

Recommended Pacing
Section 2-9 is a two-day lesson.

Day 1
Pages 111–114: Talk it Over
through Sample 1, *Exercises 1–17*

Day 2
Pages 114–115: Cube Roots
through Look Back, *Exercises
18–57*

Extra Practice
See pages 622–623.

Warm-Up Exercises
Warm-Up Transparency 2-9

Support Materials
➤ Practice 17
➤ Enrichment 16 in the Activity
Bank
➤ Study Guide 2-9
➤ Problem Set 5
➤ **Using Plotter Plus:** Using
Tables to Solve Equations and to
Estimate Square Roots
➤ Quiz 2-9
➤ Test 7
➤ Alternative Assessments 8–10

Talk it Over

Questions 1 and 2 review the relationship between the area of a square and the length of a side of the square, thus giving students a concrete representation of what it means to find a square root.

Questions 3–7 can be used to see how students estimate the square root of a whole number that is not a perfect square. For example, in question 5, most students will be able to see that *BD* is more than 3 units but less than 4 units. Discuss the ideas students have for explaining this fact. Some students may suggest measuring with a ruler. Others may suggest the use of tracing paper to compare $\overline{BD}$ to horizontal or vertical segments that are 3 units long or 4 units long.

Some students may discover that a square having $\overline{BD}$ as a side has an area of 10 square units. They can then reason that since a 3-by-3 square has an area of 9 and a 4-by-4 square has an area of 16, a square of area 10 must have sides that are between 3 and 4 units in length.

The **square root** of a number is one of two equal factors of the number. Every positive number has both a positive and a negative square root.

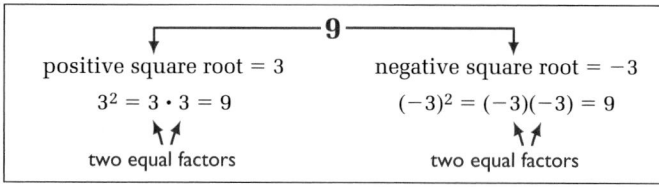

The symbol $\sqrt{}$ stands for the *nonnegative* square root of a number.

$$\sqrt{9} = 3$$

The symbol $-\sqrt{}$ stands for the negative square root.

$$-\sqrt{9} = -3$$

The symbol $\pm\sqrt{}$ stands for both square roots.

$$\pm\sqrt{9} = 3 \text{ or } -3$$

↘ plus-or-minus

When you use the [INV] [x²] on a calculator you will only get the positive square root or zero.

> **TECHNOLOGY NOTE**
>
> See page 611 of the Technology Handbook for help with finding square roots on a calculator. ←

Talk it Over

For questions 3–6, use the diagram. Suppose that *AB* is 1 unit.

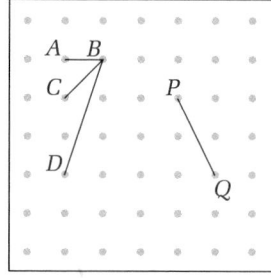

3. Is *BC greater than, less than, or equal to* 1? How do you know?

4. Is *BC greater than, less than, or equal to* 2? How do you know?

5. What can you say about *BD*?

6. What can you say about *PQ*?

7. In the diagram, some squares are drawn on dot paper and labeled with their areas. Estimate the length of a side of each square and complete the table below.

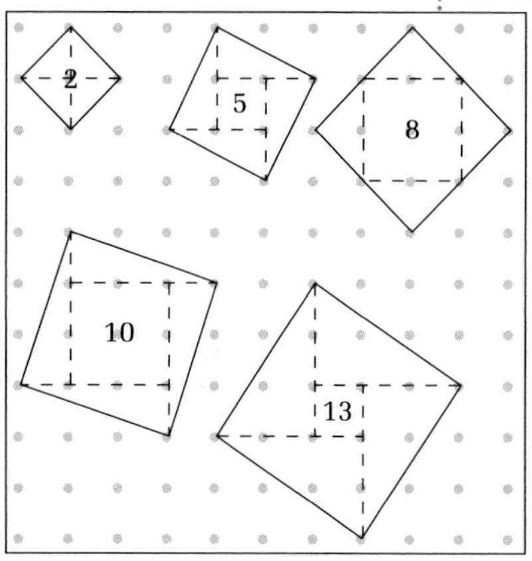

Area of square	2	5	8	10	13
Estimated length of side	?	?	?	?	?

8. Use your calculator to find the square root of each area in question 7 to the nearest hundredth.

9. Compare the results of questions 7 and 8. Should they be about the same? Are they about the same? Explain.

Unit 2 Using Measures and Equations

Answers to Talk it Over ···

3. greater than; The diagonal of a square with side 1 is greater than 1.

4. less than; One way is to measure two units on the grid and compare this to the length of the diagonal.

5. *BD* is greater than *AB* and *BC*; *BD* is greater than 3; *BD* is less than 4.

6. *PQ* is greater than *AB* and *BC*; *PQ* is less than *BD*; *PQ* is greater than 2; *PQ* is less than 3.

7. Answers may vary. Examples are given: 1.5; 2.25; 2.75; 3.25; 3.5

8. $\sqrt{2} \approx 1.41$; $\sqrt{5} \approx 2.24$; $\sqrt{8} \approx 2.83$; $\sqrt{10} \approx 3.16$; $\sqrt{13} \approx 3.61$

9. Yes, the results should be about the same. Yes, the square roots and the estimated lengths of the sides are about the same. Explanations may vary. An example is given. In the formula for the area of a square, $A = s^2$, the side, *s*, is the positive square root of the area, *A*.

Types of Numbers

When you use a calculator to find a square root, you may see any of these kinds of displays.

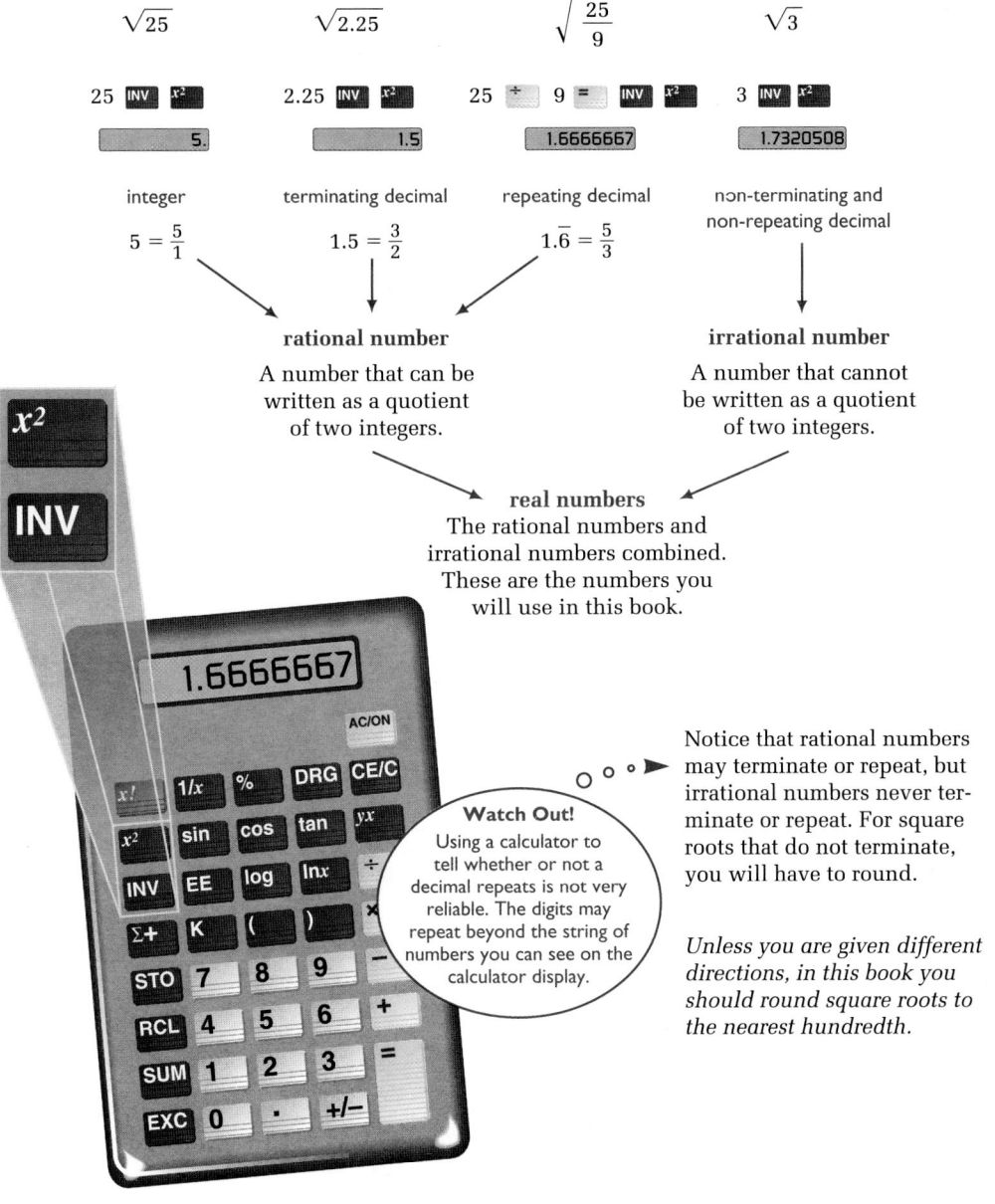

$$\sqrt{25} \qquad \sqrt{2.25} \qquad \sqrt{\dfrac{25}{9}} \qquad \sqrt{3}$$

25 INV x²	2.25 INV x²	25 ÷ 9 = INV x²	3 INV x²
5.	1.5	1.6666667	1.7320508

integer   terminating decimal   repeating decimal   non-terminating and non-repeating decimal

$$5 = \frac{5}{1} \qquad\qquad 1.5 = \frac{3}{2} \qquad\qquad 1.\overline{6} = \frac{5}{3}$$

rational number
A number that can be written as a quotient of two integers.

irrational number
A number that cannot be written as a quotient of two integers.

real numbers
The rational numbers and irrational numbers combined. These are the numbers you will use in this book.

Watch Out!
Using a calculator to tell whether or not a decimal repeats is not very reliable. The digits may repeat beyond the string of numbers you can see on the calculator display.

Notice that rational numbers may terminate or repeat, but irrational numbers never terminate or repeat. For square roots that do not terminate, you will have to round.

Unless you are given different directions, in this book you should round square roots to the nearest hundredth.

2-9 Square Roots and Cube Roots   113

Error Analysis
Caution students against relying solely on a calculator to decide whether a number is rational or irrational, or whether a fraction represents a terminating or repeating decimal.

You can illustrate the problem with an example. Ask students to find a decimal for $\frac{1}{23}$ by using a calculator. The calculator will display 0.04347826. A common conjecture is that the decimal is non-terminating and non-repeating. In fact, $\frac{1}{23}$ does give a repeating decimal. The calculator display cannot show this because the string of digits that repeats contains 22 digits.

Estimating Square Roots

Estimating Square Roots

Numbers that have square roots that are integers are **perfect squares.** The areas of the squares shown in the diagram on page 112 are perfect squares. Perfect squares may help you estimate square roots.

Sample 1

Estimate $\sqrt{51}$ within a range of two integers.

Sample Response

Find the two perfect squares closest to 51 on either side of it. They are 49 and 64. Write an inequality.

$$49 < \quad 51 \quad < 64 \quad \longleftarrow \text{Read as } 51 \text{ is between 49 and 64.}$$

$$\sqrt{49} < \sqrt{51} < \sqrt{64}$$

$$7 < \sqrt{51} < 8$$

$\sqrt{51}$ is a number between 7 and 8.

⋯▶ **Now you are ready for:**
Exs. 1–17 on p. 116

Cube Roots

$2^3 = 2 \cdot 2 \cdot 2 = 8$

three equal factors

$\sqrt[3]{8} = 2$

The **cube root** of a number is one of three equal factors of the number. For example, the cube root of 8 is 2.

A **perfect cube** is a number whose cube root is an integer. The number 8 is a perfect cube.

You can think of finding the cube root of a number as undoing the cubing of the number.

The symbol $\sqrt[3]{}$ stands for the cube root of a number. *The cube root of 8 is written $\sqrt[3]{8}$.*

Talk it Over

10. Find the volume of a cube with edge length 3 cm.

11. Find the edge length of a cube with volume 8 cm³.

Answers to Talk it Over ⋯⋯

10. 27 cm³

11. 2 cm

S1 Estimate $\sqrt{19}$ within a range of two consecutive integers. $4 < \sqrt{19} < 5$

Talk it Over

Questions 10 and 11 develop an understanding of cube roots. Remind students that every positive number has two square roots. Ask if the same is true for cube roots. (No.) Students should be asked to explain their answers.

Solving Square Root and Cube Root Equations

Sample 2

The area of a square is 15 m². Find the length of a side of the square.

Sample Response

Problem Solving Strategy: Use a formula

$$A = s^2 \longleftarrow \text{Area} = (\text{length of side})^2$$

Substitute 15 for A. $\longrightarrow$ $15 = s^2$

Undo the squaring. $\longrightarrow$ $\pm\sqrt{15} = s$

$3.87 \approx s$ $\longleftarrow$ Find the positive square root:

15 INV x²

Watch Out!
A length must be positive, so only the positive square root of 15 makes sense here.

The length of a side of the square is about 3.87 m.

Sample 3

3 in.

The volume of a cube is 27 in.³. Find the length of an edge of the cube.

Sample Response

Problem Solving Strategy: Use a formula

$$V = s^3 \longleftarrow \text{Volume} = (\text{length of side})^3$$

Substitute 27 for V. $\longrightarrow$ $27 = s^3$

Undo the cubing. $\longrightarrow$ $\sqrt[3]{27} = s$

$3 = s$ $\longleftarrow$ 27 INV yˣ 3 =

The length of an edge of the cube is 3 in.

▶ Now you are ready for:
Exs. 18–57 on pp. 117–118

Look Back

Describe the connection between a square and a square root, and between a cube and a cube root. Include an example about area and volume that shows the relationship between these measurements: 3 ft, 9 ft², 27 ft³.

Answers to Look Back

Descriptions may vary. (1) If you interpret "square" and "cube" algebraically, you might answer as follows: A square is a number that is the product of two equal factors; a square root is one of the equal factors. A cube is a number that is the product of three equal factors; a cube root is one of the equal factors. Area

involves squares and square roots; volume involves cubes and cube roots. For example, the length of each side of a square with area 9 ft² is $\sqrt{9}$ or 3 ft; the length of each edge of a cube with volume 27 ft³ is $\sqrt[3]{27}$ or 3 ft. (2) If you interpret "square" and "cube" geometrically, you might answer as follows: If the area of a square is

A, then the length of each side of the square is $\sqrt{A}$. For example, if the area of a square is 9 ft², then the length of each side is $\sqrt{9}$ or 3 ft. If the volume of a cube is V, then the length of each edge of the cube is $\sqrt[3]{V}$. For example, if the volume of a cube is 27 ft³, then the length of each edge is $\sqrt[3]{27}$ or 3 ft.

APPLYING

Suggested Assignment

Standard 1–15, 18–30, 32–41, 48–57

Extended 1–46, 48–57

Integrating the Strands

Number Exs. 1, 3–6, 12–15, 18–31, 47, 51–53

Algebra Exs. 7–11, 16, 17, 30, 31, 42–45, 48–50

Measurement Exs. 2, 32–41, 46, 55–57

Statistics and Probability Exs. 46, 54

Application

In Ex. 16, be sure students understand that h is the vertical height of the incline, not the length of the incline.

For Ex. 17, check to see if students understand the difference between velocity (speed) and acceleration by using a familiar situation. If the speedometer needle of a car is steady at 40 mi/h, then the car has a *velocity* of 40 mi/h. If the driver presses down on the gas pedal (the *accelerator*), the needle will move continuously to a higher number. While this is happening, the car is accelerating over time. Acceleration is a measure of how fast velocity is changing.

Interdisciplinary Problems

Exs. 42–45 on page 117 use mathematics in a medical situation. Students probably do not associate math and medicine, yet medical scientists are using mathematics with increasing frequency in their experiments. This is particularly true in the testing of new medicines before they are approved for public use. Mathematical statistics and probability play a very important role in these kinds of experiments.

2-9 Exercises and Problems

1. **Reading** How is $\sqrt{16}$ different from $\pm\sqrt{16}$?

2. **Writing** Suppose you were not given the area of each square outlined in the diagram at the bottom of page 112. Describe how you could use the dotted lines in the squares to find the area of each square.

Find the square roots of each number.

3. $\frac{1}{81}$　　4. 0.25　　5. $\frac{100}{9}$　　6. 0.0001

Tell whether each number is rational or irrational.

7. $8.\overline{3}$　　8. $\sqrt{10}$　　9. $\frac{22}{7}$　　10. 3.625　　11. 50

Estimate each square root within a range of two integers. Then use a calculator to find each square root.

12. $\sqrt{7}$　　13. $\sqrt{21}$　　14. $\sqrt{111}$　　15. $\sqrt{38}$

connection to **PHYSICS**

16. The photograph at the right describes the velocity of an object at the end of a slope.

　a. Find the velocity of a speed skier when the skier reaches the bottom of a ski slope that is 115 m high.

　b. **Open-ended** Look at the formula. What does the speed of the skier depend on? Explain.

Suppose that the skier starts at rest.

Suppose there is no friction as he goes down the incline.

Then at the end of the slope, the skier's velocity in meters per second will be $v = \sqrt{19.6h}$.

height h in meters

17. The photograph below describes the motion of an object traveling along the ground. Suppose a car starts at rest and then accelerates at 20 ft/s². How many seconds will it take the car to travel a distance of 100 ft?

Suppose that the car travels some distance at a constant rate of acceleration a.

Then the time in seconds that it will take to travel the distance is $t = \sqrt{\frac{2s}{a}}$.

distance s

Answers to Exercises and Problems

1. $\sqrt{16} = 4$; $\pm\sqrt{16}$ stands for two numbers, 4 and −4.

2. Descriptions may vary. An example is given. The dotted lines divide the outlined squares into smaller squares and/or triangles. To find the area of an outlined square, you can pair the triangles to form rectangles. Then it's possible to count the total number of unit squares in each outlined square to find its area. For example, in the second outlined square in the diagram, the triangles can be paired to form two rectangles that each have an area of two square units. Adding the area of the square in the middle gives a total area of five square units.

3. $\frac{1}{9}$ and $-\frac{1}{9}$ (or $0.\overline{1}$ and $-0.\overline{1}$)

4. 0.5 and −0.5

5. $\frac{10}{3}$ and $-\frac{10}{3}$ (or $3.\overline{3}$ and $-3.\overline{3}$)

6. 0.01 and −0.01

7. rational

8. irrational

9. rational

10. rational

Estimate each cube root within a range of two integers. Then use a calculator to find each cube root.

18. $\sqrt[3]{38}$ 19. $\sqrt[3]{90}$ 20. $\sqrt[3]{110}$ 21. $\sqrt[3]{1300}$

For Exercises 22–27, choose the letter of the point on the number line that matches each number.

```
                 A       B    C       D E     F   G      H
         ◄───┼───┼───┼───┼───┼───┼───┼───┼───┼───┼───►
             0   1   2   3   4   5   6   7   8   9   10
```

22. $\sqrt{40}$ 23. $\sqrt[3]{500}$ 24. $\sqrt{98}$ 25. $\sqrt[3]{98}$ 26. $\sqrt{75}$ 27. $\sqrt[3]{300}$

Rewrite the numbers in order from least to greatest.

28. $7, \sqrt{43}, 6.1, \sqrt{71}, \sqrt{86}, 8.8, \sqrt{14}, \sqrt{35}$

29. $3, \sqrt{48}, \sqrt[3]{48}, 18, \sqrt{700}, \sqrt[3]{1100}, \sqrt{169}, \sqrt[3]{14}$

30. **a.** Find $\sqrt[3]{-8}$, $\sqrt[3]{-27}$, and $\sqrt[3]{-1}$. What do you notice about your answers?

    b. Can you find $\sqrt{-4}$? Explain.

31. Both the square root and the cube root of 1 are integers. The next such number is 64. What is the next number whose square root and cube root are both integers? How many numbers like this exist from 1 to 1000?

Each measurement is the area of a square. Find the length of a side of each square.

32. 49 m²   33. 225 cm²   34. 169 in.²   35. 118 ft²

36. The length of a side of a square is an integer. The area of the square is between 50 square units and 150 square units. What are the possible lengths for a side of the square?

Each measurement is the volume of a cube. Find the length of an edge of each cube.

37. 64 cm³   38. 216 m³   39. 350 in.³   40. 0.008 ft³

41. The length of an edge of a cube is an integer. The volume of the cube is between 500 cubic units and 900 cubic units. What are possible lengths of an edge of the cube?

Career For Exercises 42–45, use the information below.

Doctors sometimes need to know a patient's body-surface area in order to decide what dose of medicine to prescribe. The formula below gives body-surface area (in square meters) for a given height (in inches) and a given weight (in pounds).

$$\text{Body-surface area} = \sqrt{\frac{\text{height} \times \text{weight}}{3131}}$$

42. Calculate the body-surface area of a person who is 72 in. tall and weighs 170 lb.

43. Calculate your own body-surface area.

44. Use your answer to Exercise 43. Imagine a square with the same area as your body-surface area. What would be the length of a side of the square?

45. Why do you suppose a doctor would want this formula to give answers in square meters instead of square inches or square feet?

2-9 Square Roots and Cube Roots   **117**

22. *D*   23. *F*   24. *H*

25. *C*   26. *G*   27. *E*

28. $\sqrt{14}, \sqrt{35}, 6.1, \sqrt{43}, 7, \sqrt{71}, 8.8, \sqrt{86}$

29. $\sqrt[3]{14}, 3, \sqrt[3]{48}, \sqrt{48}, \sqrt[3]{1100}, \sqrt{169}, 18, \sqrt{700}$

30. a. −2, −3, −1; Each cube root is a negative number.

    b. No. Explanations may vary. An example is given. There is no number you can multiply by itself to get −4. When you multiply any number by itself, you always get a positive number or zero; you cannot get a negative number.

31. The next number is 729 ($\sqrt{729} = 27$ and $\sqrt[3]{729} = 9$); there are 3 numbers like this from 1 to 1000: 1, 64, and 729.

32. 7 m   33. 15 cm

34. 13 in.   35. about 10.86 ft

36. 8 units, 9 units, 10 units, 11 units, 12 units

37. 4 cm   38. 6 m

39. about 7.05 in.

40. 0.2 ft

41. 8 units, 9 units

42. about 1.98 m²

43. Answers may vary.

44. Answers may vary.

45. This unit can be used more directly in prescribing the correct dosage of medicine since medicine is often measured in corresponding metric units.

Answers to Exercises and Problems

11. rational

12. $\sqrt{7}$ is a number between 2 and 3. 2.65

13. $\sqrt{21}$ is a number between 4 and 5. 4.58

14. $\sqrt{111}$ is a number between 10 and 11. 10.54

15. $\sqrt{38}$ is a number between 6 and 7. 6.16

16. a. about 47.48 m/s

    b. The velocity depends only on the height of the incline. Explanations may vary. An example is given.
An object sliding down either of two inclines of the same height will have the same velocity when it reaches the bottom of the incline whether the incline is long and shallow or short and steep.

17. about 3.16 s

18. $\sqrt[3]{38}$ is a number between 3 and 4. 3.36

19. $\sqrt[3]{90}$ is a number between 4 and 5. 4.48

20. $4 < \sqrt[3]{110} < 5$; $\sqrt[3]{110} \approx 4.79$

21. $10 < \sqrt[3]{1300} < 11$; $\sqrt[3]{1300} \approx 10.91$

Practice 17 For use with Section 2-9

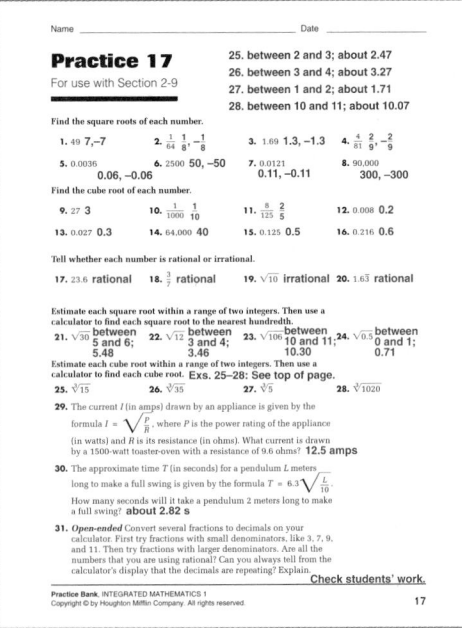

46. Use the bar graph. Suppose a cube has the same volume as each dam. What is the length of a side of each cube?

Ongoing ASSESSMENT

47. **Group Activity** Work with another student. Do parts (a) and (b) by yourself. Then work together for part (c).

   a. Make a list of 6 numbers between 1 and 1000.

   b. Graph the square root and the cube root of each number on the same number line.

   c. Exchange number lines. (Do not share your list.) Figure out the numbers on each other's list.

Review PREVIEW

Solve. *(Section 2-8)*

48. $3x + 2.7 = 5.4$ **49.** $\frac{n}{4} - 9 = 7$ **50.** $11 = -2.4a + 6.2$

51. 36 is what percent of 144? *(Toolbox Skill 12)*

52. Find 24% of 150. *(Toolbox Skill 14)*

53. 231 is 55% of what number? *(Toolbox Skill 15)*

54. Draw a bar graph. *(Toolbox Skill 22)*

Powerful Dams of the World

Dam	Itaipu (Brazil-Paraguay)	Grand Coulee (United States)	Guri (Venezuela)	Tucurui (Brazil)
Planned power capacity in megawatts	12,600	10,830	10,300	8000

Large Dams of the World

2 Working on the Unit Project

55. Choose one of the facts going on your theme poster that could be expressed as an area. Suppose you express the fact as the area of a square. What would be the length of a side of the square?

56. Choose one of the four facts for your theme poster that could be expressed as a volume. Suppose you express the fact as the volume of a cube. What would be the length of an edge of the cube?

57. Compare the area and volume you chose in Exercises 55 and 56 with some areas and volumes that you know, such as the area of the United States or the volume of your school. Plan to include these comparisons on your theme poster.

Answers to Exercises and Problems

46. Syncrude: side ≈ 814 m; Chapeton: side ≈ 667 m; Pati: side ≈ 620 m; New Cornelia Tailings: side ≈ 594 m; Tarbela side ≈ 473 m

47. a. Numbers and graphs may vary. Examples are given: 230, 343, 500, 625, 729, 850.

   b.

   c. numbers reasonably close to 230, 343, 500, 625, 729, and 850

48. 0.9   **49.** 64

50. −2   **51.** 25%

52. 36   **53.** 420

54.

Powerful Dams of the World

	Itaipu (Brazil-Paraguay)	Grand Coulee (United States)	Guri (Venezuela)	Tucurui (Brazil)

55–57. Answers may vary.

Unit Project 2

Completing the Unit Project

Quick Quiz (2-7 through 2-9)
See pages 120 and 121.

Now you are ready to make your theme poster. Keep in mind how the poster will be judged as you make your display.

➤ Show at least four facts including a count, a measurement, and a percent. One number should be very large and one very small.

➤ For each fact write one question that can be answered using the information on your poster.

➤ Write an equation that can be solved to find another fact.

Be sure to include these details.

➤ Use a circle graph to illustrate the percent.

➤ Express your very large and very small numbers in decimal notation and in scientific notation.

➤ Describe at least one fact in terms of area or volume.

Look Back

What did you learn from this project that will help you make sense of numbers that you read and hear about in everyday life?

Alternative Projects

Project 1: Redesigning a Room

Make plans to redesign a classroom in your school to make it more efficient or more attractive. Plan an arrangement of desks and workstations. Estimate how many desks will fit in the redesigned room. Write a paragraph giving your reasons for redesigning the room. Explain why your design is an improvement.

Project 2: Conducting an Interview

Numerals, fractions, and decimals are sometimes written differently in other countries. Research some of the differences. Interview people who have been to other countries to find out if they noticed any other ways numbers are written. Report your findings to your class.

Quick Quiz (2-7 through 2-9)

Write an equation for each figure that describes a relationship among the angle measures. [2-7]

1.

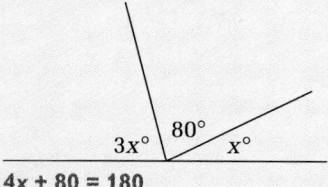

$4x + 80 = 180$

2.

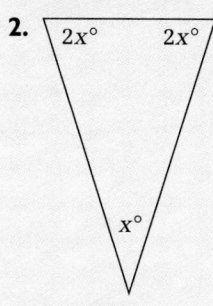

$5x = 180$

Write and solve an equation for each figure to find each unknown measure. [2-7, 2-8]

3.

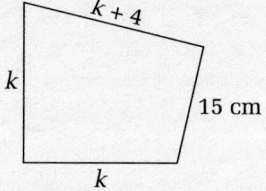

The perimeter of the figure is 79 cm.
$k + k + (k + 4) + 15 = 79$
$k = 20$ cm
$k + 4 = 24$ cm

4.

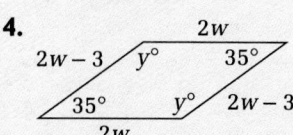

The perimeter of the figure is 106 in.
$2w + 2w + (2w - 3) + (2w - 3)$
$= 106$
$2w = 28$ in.
$2w - 3 = 25$ in.
$y + y + 35 + 35 = 360$
$y = 145°$

1. Guess whether the number of spaces between words on this page is in the *hundreds, thousands, millions,* or *billions.* Then estimate the number. Describe your method.   **2-1**

2. Estimate whether the population of Moscow, Russia, is in the *hundreds, thousands, millions,* or *billions.*

3. **Writing** Use a number anywhere along the scale at the right to estimate the probability that you will see a movie this week. Give a reason for your answer.

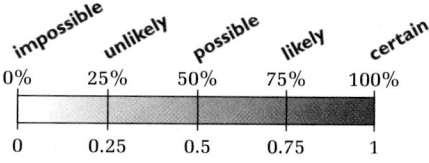

4. Is your distance from home as you travel to school *continuous* or *discrete?*

Evaluate each expression when $a = -4$ and $b = 3$.   **2-2**

5. $|a|$

6. $-a + b$

7. $\dfrac{a + 7}{b}$

8. $a^2 - ab$

Write each expression in decimal notation.   **2-3**

9. 10^{-8}

10. 10^0

Write each number in scientific notation.

11. 715,100,000

12. 0.00906

13. Simplify $\dfrac{6.13 \times 10^3}{2,000,000}$. Write the answer in scientific notation, in decimal notation, and in words.

14. The distance from Mercury to the sun is 5.79×10^{10} m. The distance from Neptune to the sun is about 78 times this distance.

   a. Make a reasonable estimate of the distance from Neptune to the sun. Write the answer in scientific notation.

   b. Calculate a more precise estimate of the distance from Neptune to the sun. Write the answer in decimal notation.

The map shows Canada.   **2-4**

15. Estimate the distance from Winnipeg to Fort Smith.

16. Is 1000 km a reasonable estimate of the distance from Yellowknife to Whitehorse? Explain.

17. Estimate the area of the Fort Smith region.

18. Use the grid on the map to estimate the area of the mainland of Canada. Each grid square is 1500 km on a side.

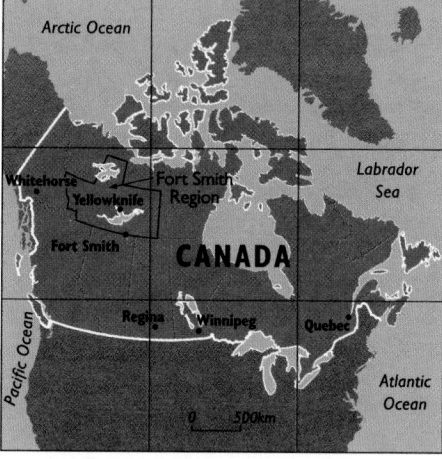

Unit 2 *Using Measures and Equations*

Answers to Unit 2 Review and Assessment

1. Answers may vary. hundreds; 340; Count the number of spaces between the words in the first line (10) and multiply by the number of lines (34).

2. millions

3. Answers may vary.

4. continuous

5. 4

6. 7

7. 1

8. 28

9. 0.00000001

10. 1

11. 7.151×10^8

12. 9.06×10^{-3}

13. 3.065×10^{-3}; 0.003065; 3065 millionths

14. a. 4.8×10^{12} m (rounding 5.79 to 6 and 78 to 80)

   b. 4,516,200,000,000 m

15. about 1250 km

16. Yes. It's between 750 km and 1000 km.

17. about 562,500 km²

18. about 5,625,000 km²

19. True.

20. True.

21. False.

In the diagram, *M* is the midpoint of $\overline{RT}$. Tell whether each statement is
True or *False*.

19. *RT* is about 4 cm.

20. $RT = 2 \cdot MT$

21. *M* is the midpoint of $\overleftrightarrow{RT}$.

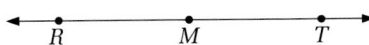

**For Questions 22–25, use this diagram of a camera lens that shows
how rays of light pass through the center of the lens.**

2-5

22. Name a pair of supplementary angles.

23. Name an angle congruent to $\angle DCE$.

24. Name a pair of vertical angles.

25. Suppose $\angle ACD = 128°$.

   a. Find the measure of $\angle ACB$.

   b. Find the measure of $\angle BCE$.

26. Estimate to sketch a 100° angle.
Then describe how you estimated.

27. Open-ended Describe what you have learned
about measurement in this unit. Be sure to include
the following points.

2-1, 2-4, 2-5

➤ why you need to estimate measurements

➤ methods for estimating measurements

➤ the need for accurate measurements

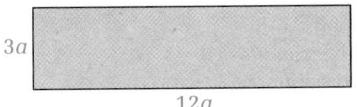

Simplify.

2-6

28. $8mn - 4m + 2mn + 9m$ **29.** $-8k^3 - 14 + 6k^3 - k^2$ **30.** $8(12a^2)$

31. Write and simplify an expression
for the area of the rectangle.

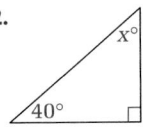

$3a$

$12a$

Write and solve an equation to find each unknown value.

2-5, 2-7, 2-8

32.

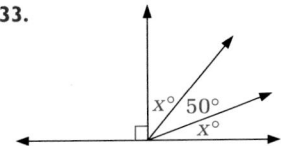

33.

Solve.

34. $t - 7 = 98$ **35.** $\frac{n}{5} + 35 = 25$ **36.** $56 = 6x + 8x$

37. Ilana Meyers plans to buy a pad of paper and a pen for each
person scheduled to attend a peer counseling workshop. Each
pad of paper costs $1.39 and each pen costs $.49. Ilana has
$45.12. How many people can she buy supplies for?

5. $7 = 0.25c + 3.75$ c = 13

6. $-8r + 3 = 19$ r = –2

**Find the value of each
expression.** [2-9]

7. $\sqrt{1600}$ 40

8. $\sqrt[3]{-64}$ –4

**Tell whether each number is
rational or *irrational*.** [2-9]

9. $\sqrt{300}$ irrational

10. $\dfrac{2\sqrt{8}}{\sqrt{8}}$ rational

Also, you can estimate an
unknown quantity by relating it
to a known quantity, such as the
population of a large city. You
can estimate distances by using a
map scale or by using a familiar
object, like a foot, whose estimated length is known. When the
measurement that you need is
limited in some way, such as the
dimensions of a door through
which you must carry a piece of
furniture or the width of a shade
that you have to order for a window, then an exact measurement
is needed. Otherwise, an estimate
is fine.

28. $10mn + 5m$

29. $-2k^3 - k^2 - 14$

30. $96a^2$

31. $(3a)(12a) = 36a^2$

32. $40 + 90 + x = 180$ or $40 + x = 90$;
$x = 50$

33. $x + 50 + x = 90$; $2x + 50 = 90$;
$x = 20$

34. 105

35. –50

36. 4

37. $45.12 = 1.39n + 0.49n$; 24 people

Answers to Unit 2 Review and Assessment

22. Answers may vary.
Examples are given. $\angle DCE$
and $\angle DCA$, $\angle ECB$ and
$\angle BCA$

23. $\angle ACB$

24. Answers may vary.
Examples are given. $\angle DCE$
and $\angle ACB$, $\angle DCA$ and
$\angle ECB$

25. a. 52°

   b. 128°

26. 100° is slightly larger than
a right angle.

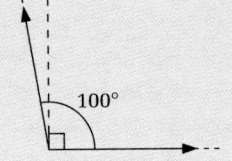

$100°$

27. Paragraphs may vary. An
example is given. Measurements can be estimated or

exact. You need to esti-
mate when the number of
objects to be counted is
very large or if you do not
have precise measuring
instruments. You can estimate a number of objects
by estimating a small portion of the whole and then
multiplying this estimate
by the number of portions
of that size in the whole.

121

1. Classify each quantity as *discrete* or *continuous*. [2-1]

   a. the time it takes a spider to spin a web **continuous**

   b. the number of telephone listings in a large telephone book **discrete**

   c. the number of red blood cells in a blood sample **discrete**

Simplify. [2-2]

2. $-14.3 + (-5)$ **–19.3**

3. $20 + (-11)$ **9**

4. $7 - (-8)$ **15**

5. $(-6)(-9)$ **54**

6. $(-100) \div 50$ **–2**

Write each number in decimal notation. [2-3]

7. 5.06×10^7 **50,600,000**

8. 4.88×10^{-4} **0.000488**

Write each number in scientific notation. [2-3]

9. 0.0000062 **6.2×10^{-6}**

10. $937,600,000$ **9.376×10^8**

---

38. Lorenzo Walker paid $8.51 for eight containers of yogurt and a loaf of bread. The bread cost $2.19. How much did each container of yogurt cost?

39. Is $\sqrt{16}$ *rational* or *irrational*?

40. Find the square roots of 121.

2-9

41. Estimate $\sqrt[3]{72}$ within a range of two integers.

42. The area of a square is 150 ft². Find the length of a side of the square.

43. The volume of a cube is 81 mm³. Find the length of an edge of the cube.

44. **Self-evaluation** Describe as much as you can about each figure. Use as many terms and symbols that you have learned in this unit as you can. Which figure do you know the most about? Which figure do you need to study more?

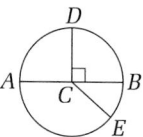 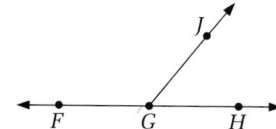 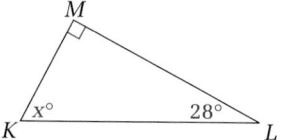

45. **Group Activity** Work with another student. You will need rulers and protractors.

   a. Have each person draw a large triangle. Measure the angles of the triangle and measure the sides in inches. Write the measurements on a separate piece of paper.

   b. Label one angle of the triangle $x°$. Use the angle measurements to write an expression in terms of $x°$ for each of the other angles. Label the triangle with your expressions.

   c. Exchange triangles. Write and solve an equation to find the measure of each angle of the triangle you receive.

   d. Compare your results with the actual measurements.

Answers to Unit 2 Review and Assessment

38. $8.51 = 8y + 2.19$; $.79

39. rational

40. 11 and –11

41. $4 < \sqrt[3]{72} < 5$

42. about 12.25 ft

43. about 4.33 mm

44. Descriptions may vary. Examples of things that may be mentioned are given: rays, central angles, supplementary angles, the sum of the measures of the angles of a triangle is 180°, and so on.

45. a–d. Answers may vary.

IDEAS AND (FORMULAS) = X²

ALGEBRA

➤ Subtracting a number gives the same result as adding the opposite of that number. *(p. 65)*

➤ $10^{-n} = \dfrac{1}{10^n}$, and $10^0 = 1$. *(p. 75)*

➤ When you simplify algebraic expressions involving addition and subtraction, you can only combine like terms. *(p. 95)*

➤ To solve an equation, you get the variable alone on one side. You must keep the equation balanced as you do this. *(p. 100)*

➤ To solve a two-step equation, undo the operations that were performed on the variable, but in the opposite order. *(p. 105)*

➤ By knowing some perfect squares, you can estimate square roots. *(p. 114)*

➤ You can solve square root equations and cube root equations by undoing the squaring and cubing. *(p. 115)*

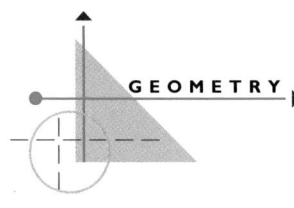
GEOMETRY

➤ You can estimate some numbers by doing an informal count or by making comparisons to facts you know. *(p. 58)*

➤ You can estimate lengths, distances, and areas using common guides for length and scales on maps. *(pp. 79–81)*

➤ An entire circle has a measure of 360°. *(p. 86)*

➤ Vertical angles have equal measures. *(p. 87)*

➤ The sum of the measures of the angles of a triangle is 180°. *(p. 88)*

STATISTICS & PROBABILITY

➤ The probability of an event is a number from 0 to 1, where 0 means impossible and 1 means certain. *(p. 59)*

Key Terms

- **continuous** (p. 60)
- **opposites** (p. 64)
- **line, $\overleftrightarrow{XY}$** (p. 82)
- **length of a segment** (p. 82)
- **ray, $\overrightarrow{OZ}$** (p. 85)
- **central angle of a circle** (p. 85)
- **complementary angles** (p. 86)
- **vertical angles** (p. 87)
- **right triangle** (p. 88)
- **solving an equation** (p. 99)
- **rational number** (p. 113)
- **perfect square** (p. 114)

- **discrete** (p. 60)
- **scientific notation** (p. 72)
- **endpoint** (p. 82, 85)
- **midpoint** (p. 82)
- **angle, ∠YOZ** (p. 85)
- **right angle** (p. 86)
- **supplementary angles** (p. 86)
- **acute triangle** (p. 88)
- **equation** (p. 99)
- **equivalent equations** (p. 100)
- **irrational number** (p. 113)
- **perfect cube** (p. 114)

- **absolute value, $|x|$** (p. 64)
- **≈** (p. 73)
- **segment, $\overline{XY}$** (p. 82)
- **congruent segments** (p. 82)
- **vertex of an angle** (p. 85)
- **straight angle** (p. 86)
- **congruent angles** (p. 87)
- **obtuse triangle** (p. 88)
- **solution** (p. 99)
- **square root, $\pm\sqrt{\ }$** (p. 112)
- **real number** (p. 113)
- **cube root, $\sqrt[3]{\ }$** (p. 114)

Unit 2 Review and Assessment

Quick Quiz (2-4 through 2-6)

Use the map for Exs. 1 and 2. [2-4]

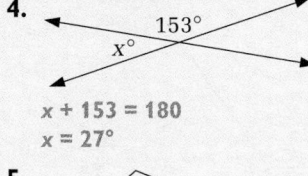

1. Estimate the distance from Sevilla to Barcelona.
   about 500 mi

2. Use the grid to estimate the area of Spain.
   180,000 mi²

3. Estimate the length of your desk in customary and in metric units. [2-4] **Answers may vary.**

Write and solve an equation to find the unknown angle measure in each figure. [2-5]

4.
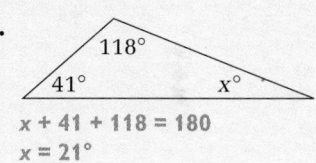
   $153°$, $x°$

   $x + 153 = 180$
   $x = 27°$

5.
   $118°$, $41°$, $x°$

   $x + 41 + 118 = 180$
   $x = 21°$

6. Write and simplify an expression for the perimeter of the figure below. [2-6]

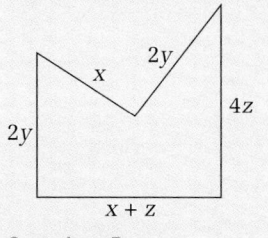

   $2y$, x, $2y$, $2y$, $4z$, $x + z$

   $2x + 4y + 5z$

Simplify each expression. [2-6]

7. $(5x)(x)(3x)$ $\quad 15x^3$

8. $-4xy + y^3 + x^3 + 11xy - 3y^3$
   $7xy - 2y^3 + x^3$

3 Representing Data

OVERVIEW

➤ **Unit 3** presents methods for systematically organizing, displaying, and analyzing data. Some prior experience with constructing and reading data displays is assumed. Students can refresh their knowledge of various kinds of statistical graphs by using the **Student Resources Toolbox** on pages 661–663.

➤ In Unit 3, students construct, read, and interpret histograms, stem-and-leaf plots, and box-and-whisker plots. They also use measures of central tendency—mean, median, and mode—to interpret data, and they analyze misleading graphs.

➤ The theme of the Unit Project is planning a music store. Students use statistical methods to make decisions about inventory and pricing.

➤ Connections to science, business, sports, and social studies are integrated into the teaching materials and exercises.

➤ Problem-solving strategies used in Unit 3 include *Guess and Check* (Section 3-2), *Work Backward* (Section 3-2), and *Use an Equation* (Section 3-2).

Unit Objectives

Section	Objectives	NCTM Standards
3-1	• Use matrices and graphs to display and interpret data.	1, 2, 3, 4, 10, 12
3-2	• Find values that are typical of a data set.	1, 2, 3, 4, 10, 12
	• Identify data values that are not typical of a data set.	
	• Use equations to find missing data values.	
3-3	• Use inequalities to describe number line graphs.	1, 2, 4, 5, 8, 10
3-4	• Use histograms, frequency tables, and stem-and-leaf plots to display data.	1, 2, 4, 10, 12
3-5	• Draw a box-and-whisker plot and use it to compare data sets.	1, 2, 4, 8, 10, 12
3-6	• Tell why a data display is a good choice for a data set.	1, 2, 3, 4, 8, 10, 12
	• Choose a good data display for a data set.	
3-7	• Recognize when graphs do not give an accurate picture of a data set.	1, 2, 3, 4, 10, 12

Section	Connections to Prior and Future Concepts
3-1	**Section 3-1** introduces matrices, spreadsheets, and graphs as ways to display and interpret data. Students have worked with data presented in tabular and graphical form in previous courses. Matrices and graphs are used to display data throughout this unit and in Units 4, 5, 7, and 8 of Book 1, as well as in Units 2, 3, 6, and 10 of Book 2 and Units 1, 2, 4, 6, 9, and 10 of Book 3. (A review of basic data displays is presented on pages 661–663 of the Toolbox.)
3-2	**Section 3-2** covers mean, median, and mode as typical values of sets of data. Students have probably studied mean, median, and mode in previous courses and will continue to use them in future studies of statistics, as in Units 1 and 6 of Book 2 and in Units 1, 6, and 9 of Book 3.
3-3	**Section 3-3** introduces the use of inequalities to describe linear intervals. Students have probably had previous experience with graphing on the number line. Inequalities are further investigated in Sections 5-4, 8-6, and 8-7 of Book 1 and are utilized throughout Books 2 and 3. Page 641 in the Toolbox provides additional number-line background.
3-4	**Section 3-4** presents histograms, frequency tables, and stem-and-leaf plots as ways to display data. These methods for presenting data will be used throughout Book 1 and in students' future work in statistics.
3-5	**Section 3-5** uses the median and the range, introduced in Section 3-2, to create box-and-whisker plots for comparing sets of data. Methods of analyzing data are the focus of this unit and will be studied further in Units 4, 6, 7, and 8 of Book 1; in Units 1, 2, and 6 of Book 2; and in Units 1, 2, 4, and 6 of Book 3.
3-6	**Section 3-6** combines the skills and techniques investigated so far in this unit, focusing on selecting the best format for presenting data. Presenting and interpreting data is frequently emphasized in Book 1, and students will use skills developed here in Books 2 and 3 as well as in future courses related to statistics.
3-7	**Section 3-7** presents ways in which graphs can present a misleading or inaccurate impression of a set of data. Reliable and accurate representation of data is stressed in Book 1 and throughout Books 2 and 3.

Integrating the Strands

Strands	Sections
Number	3-3
Algebra	3-1, 3-2, 3-3, 3-4, 3-5, 3-7
Geometry	3-1, 3-4, 3-6, 3-7
Statistics and Probability	3-1, 3-2, 3-3, 3-4, 3-5, 3-6, 3-7
Discrete Mathematics	3-1, 3-2, 3-3
Logic and Language	3-1, 3-2, 3-3, 3-4, 3-5, 3-6, 3-7

Section Planning Guide

➤ Essential exercises and problems are indicated in boldface.
➤ Ongoing work on the Unit Project is indicated in color.
➤ Exercises and problems that require student research, group work, manipulatives, or graphing technology are indicated in the column headed "Other."

Section	Materials	Pacing	Standard Assignment	Extended Assignment	Other
3-1		Day 1	**1–7**, 9, **11–13**	**1–7**, 9, 10, **11–13**, 14	8
		Day 2	**15–22**, 24–30, 31	**15–22**, 23–30, 31	31
3-2	spreadsheet software, uncooked spaghetti	Day 1	**1–9**, 10a, **11–16**	**1–9**, 10a, **11–16**	10b, 17
		Day 2	**18–23**, 27, 29–34, 35, 36	**18–23**, 24–26, **27**, 29–34, 35, 36	28, 35
3-3		Day 1	**3–13**, 14–27, 28	1, 2, **3–13**, 14–27, 28	28
3-4	index cards or self-stick, removable notes, graphics calculator or spreadsheet software	Day 1	1, **2–8**	1, **2–8**	9
		Day 2	**10, 11**, 12–14, 16, 18–24, 25	**10, 11**, 12–14, 16, 18–24, 25	15, 17, 25
3-5		Day 1	**2–15**, 16–26, 27	1, **2–15**, 16–26, 27	27
3-6	statistical graphing or spreadsheet software	Day 1	**1–17**, 19–23, 24	**1–17**, 19–23, 24	18, 24
3-7	statistical graphing or spreadsheet software	Day 1	1, 2, **5–8**, 9a, **10–15**, 17–20, 21–33, 34	1–4, **5–8**, 9a, **10–15**, 16, **17–20**, 21–33, 34	9b, 22
Review		**Day 1**	**Unit Review**	**Unit Review**	
Test		**Day 2**	**Unit Test**	**Unit Test**	

Yearly Pacing	Unit 3 Total	Units 1–3 Total	Remaining	Total
	14 days (2 for Unit Project)	48 days	116 days	164 days

Support Materials

➤ See **Project Book** for notes on Unit 3 Project: Plan a Music Store.
➤ "UPP" and "disk" refer to **Using Plotter Plus** booklet and **Plotter Plus** disk.
➤ Warm-up exercises for each section are available on **Warm-Up Transparencies**.
➤ "FI," "PC," "GI," "MA," and "Stats!" refer, respectively, to the McDougal Littell Mathpack software Activity Books for **Function Investigator, Probability Constructor, Geometry Inventor, Matrix Analyzer,** and **Stats!**.

Section	Study Guide	Practice Bank	Problem Bank	Activity Bank	Explorations Lab Manual	Assessment Book	Visuals	Technology
3-1	3-1	Practice 19	Set 6	Enrich 17	Master 11	Quiz 3-1		Stats! Acts. 3 and 6 Statistics Spreadsheets (disk)
3-2	3-2	Practice 20	Set 6	Enrich 18		Quiz 3-2		Stats! Acts. 4, 10, 12, and 18 UPP, page 27
3-3	3-3	Practice 21	Set 6	Enrich 19	Master 2	Quiz 3-3 Test 10		
3-4	3-4	Practice 22	Set 7	Enrich 20	Master 2	Quiz 3-4	Folder 4	Stats! Acts. 7 and 20
3-5	3-5	Practice 23	Set 7	Enrich 21		Quic 3-5	Folder 4	
3-6	3-6	Practice 24	Set 7	Enrich 22	Master 2	Quiz 3-6		
3-7	3-7	Practice 25	Set 7	Enrich 23	Master 2	Quiz 3-7 Test 11		
Unit 3	Unit Review	Practice 26	Unifying Problem 3	Family Involve 3		Tests 12, 13		

UNIT TESTS

Spanish versions of these tests are on pages 130–133 of the **Assessment Book**.

Form A

Name _____ Date _____ Score _____

Test 12

Test on Unit 3 (Form A)

Directions: Write the answers in the spaces provided.

Thirteen golfers were asked what their score was on their last game. The scores are shown below.

73, 79, 88, 76, 75, 77, 80, 87, 77, 74, 72, 98, 84

Use this data for Questions 1–4.

1. Find the mean of the scores.
2. Find the median of the scores.
3. Find the mode of the scores.
4. Find the range and the outlier(s), if any, of the scores.

For Questions 5 and 6, write and graph an inequality to describe each statement.

5. The cost of a box of stationery ranges from $1.65 to $2.35.

$1.65 \le c \le 2.35$;
1.65   2.35
1.60 1.80 2.00 2.20 2.40

6. Tina can type at least 40 words per minute.

$w \ge 40$;
30 40 50 60 70

7. **Writing** Explain the difference between discrete and continuous graphs on a number line. Include drawings in your explanation.

Sample answer: Graphing discrete data produces separate points on a number line while graphing continuous data produces a continuous interval.

Discrete: 10 20 30 40 50

Continuous: 10 20 30 40 50

8. The table below shows the number of runs driven in by each of twenty major league baseball players. Make a stem-and-leaf plot for the data.

Runs Batted In
105 104 103 98 98
96 95 94 94 93
91 91 91 90 90
89 87 84 83 81

8 | 1 3 4 7 9
9 | 0 0 1 1 1 3 4 4 5 6 8 8
10 | 3 4 5

Answers
1. 80
2. 77
3. 77
4. r: 26; o: 98
5. See question.
6. See question.
7. See question.
8. See question.

Name _____ Date _____ Score _____

Test 12 (continued)

Directions: Write the answers in the spaces provided.

Use the spreadsheet below to answer Questions 9–11.

National Banking Corporation: Financial Data for 1991 (Millions of Dollars)

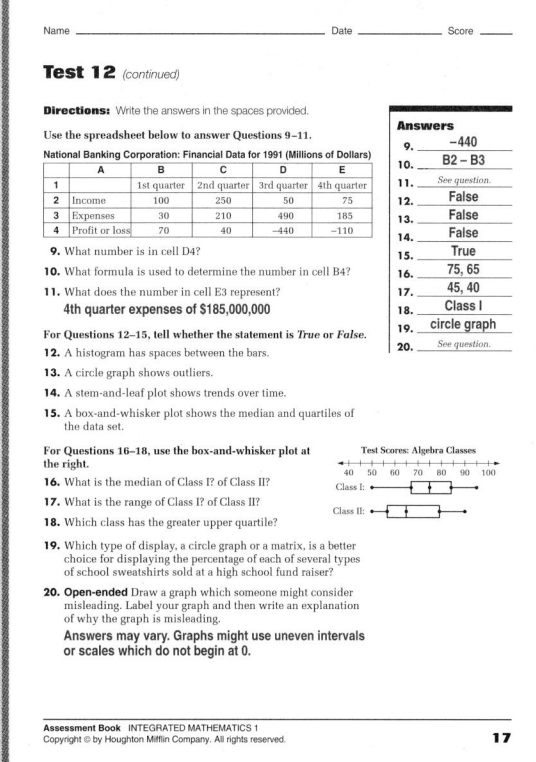

	A	B	C	D	E
1		1st quarter	2nd quarter	3rd quarter	4th quarter
2	Income	100	250	50	75
3	Expenses	30	210	490	185
4	Profit or loss	70	40	−440	−110

9. What number is in cell D4?
10. What formula is used to determine the number in cell B4?
11. What does the number in cell E3 represent?

    4th quarter expenses of $185,000,000

For Questions 12–15, tell whether the statement is *True* or *False*.

12. A histogram has spaces between the bars.
13. A circle graph shows outliers.
14. A stem-and-leaf plot shows trends over time.
15. A box-and-whisker plot shows the median and quartiles of the data set.

For Questions 16–18, use the box-and-whisker plot at the right.

Test Scores: Algebra Classes
40 50 60 70 80 90 100
Class I:
Class II:

16. What is the median of Class I? of Class II?
17. What is the range of Class I? of Class II?
18. Which class has the greater upper quartile?
19. Which type of display, a circle graph or a matrix, is a better choice for displaying the percentage of each of several types of school sweatshirts sold at a high school fund raiser?
20. **Open-ended** Draw a graph which someone might consider misleading. Label your graph and then write an explanation of why the graph is misleading.

    Answers may vary. Graphs might use uneven intervals or scales which do not begin at 0.

Answers
9. −440
10. B2 − B3
11. See question.
12. False
13. False
14. False
15. True
16. 75, 65
17. 45, 40
18. Class I
19. circle graph
20. See question.

Form B

Name _____ Date _____ Score _____

Test 13

Test on Unit 3 (Form B)

Directions: Write the answers in the spaces provided.

Thirteen bowlers were asked what their score was on their last game. The scores are shown below.

173, 179, 188, 176, 175, 177, 180, 187, 177, 174, 172, 198, 184

Use this data for Questions 1–4.

1. Find the mean of the scores.
2. Find the median of the scores.
3. Find the mode of the scores.
4. Find the range and the outlier(s), if any, of the scores.

For Questions 5 and 6, write and graph an inequality to describe each statement.

5. The cost of a 5-lb bag of dog food ranges from $4.65 to $5.35.

$4.65 \le c \le 5.35$;
4.65   5.35
4.60 4.80 5.00 5.20 5.40

6. Jack can run a mile in less than 4 minutes.

$t < 4$;
3.7 3.9 4.1

7. **Writing** Explain when it is most appropriate to use a histogram when presenting data.

Sample answer: A histogram is the most appropriate display to use when the data values are the frequencies for a given set of intervals.

8. The table below shows the number of home runs hit by each of twenty major league baseball players. Make a stem-and-leaf plot for the data.

Home Runs
42 41 40 40 39
38 37 36 35 33
32 32 31 30 29
29 29 29 27 26

2 | 6 7 9 9 9 9
3 | 0 1 2 2 3 5 6 7 8 9
4 | 0 0 1 2

Answers
1. 180
2. 177
3. 177
4. r: 26; o: 198
5. See question.
6. See question.
7. See question.
8. See question.

Name _____ Date _____ Score _____

Test 13 (continued)

Directions: Write the answers in the spaces provided.

Use the spreadsheet below to answer Questions 9–11.

National Banking Corporation: Financial Data for 1991 (Millions of Dollars)

	A	B	C	D	E
1		1st quarter	2nd quarter	3rd quarter	4th quarter
2	Income	100	250	50	75
3	Expenses	30	210	490	185
4	Profit or loss	70	40	−440	−110

9. What number is in cell B2?
10. What formula is used to determine the number in cell D4?
11. What does the number in cell C4 represent?

    2nd quarter profits of $40,000,000

For Questions 12–15, tell whether the statement is *True* or *False*.

12. A histogram is a graph of the numbers in a stem-and-leaf plot.
13. A circle graph uses a circle to represent the whole and sectors to represent the parts.
14. A stem-and-leaf plot displays every data value.
15. A box-and-whisker plot shows the mean of the data set.

For Questions 16–18, use the box-and-whisker plot at the right.

Test Scores: Algebra Classes
40 50 60 70 80 90 100
Class I:
Class II:

16. What is the median of Class I? of Class II?
17. What is the range of Class I? of Class II?
18. Which class has the greater upper extreme?
19. Which type of display, a circle graph or a matrix, is a better choice for displaying the number of miles that each of 10 different rental cars is driven on each day of the week?
20. **Open-ended** Draw a graph which someone might consider misleading. Label your graph and then write an explanation of why the graph is misleading.

    Answers may vary. Graphs might use uneven intervals or scales which do not begin at 0.

Answers
9. 100
10. D2 − D3
11. See question.
12. True
13. True
14. True
15. False
16. 75, 70
17. 45, 50
18. Class II
19. matrix
20. See question.

Software Support

McDougal Littell Mathpack

Stats!

Outside Resources

Books/Periodicals

Cibes, Margaret. "A Tale of Three Taxes." *Mathematics Teacher* (December 1992): pp. 278–281.

Data Analysis and Statistics Across the Curriculum. Grades 9–12 Addenda. Reston, VA: NCTM, 1993.

Whitmer, John C. "Spreadsheets in Mathematics and Science Teaching." *School Science and Mathematics Association.* Bowling Green, OH: Bowling Green State University, 1992.

de Lange Jzn, Jan and Helen Verhaget. *Data Visualization.* Scotts Valley, CA: Wings for Learning, 1992.

Manipulatives

Smith, Lyle R. "Multiple Solutions Involving Geoboard Problems." *Mathematics Teacher* (January 1993): pp. 278–281.

Software

Better Working Spreadsheet. Includes two $5\frac{1}{4}$" Apple IIe diskettes. Reston, VA: NCTM, 1993.

IBM Mathematics Exploration Toolkit. IBM, Armonk, NY.

Videos

Moore, David S. *Statistics: Decisions Through Data.* 5 one-hour VHS videotapes. Lexington, MA: COMAP, 1992.

124D

➤ Students develop a business brochure for a new music store.

➤ Before creating the brochure, students research the musical tastes and the buying habits of potential customers.

➤ Students work together in a cooperative group and all contribute to the project's success.

PROJECT PLANNING

Materials List

➤ Folder

➤ Paper

➤ Markers

Project Teams

Before students begin working as part of a team, help them describe the age, population, and cultural background of the community.

Have students work on the project in groups of three. One way for the individuals in the group to distribute the work is as follows:

1. Market Researcher: writes the questions for the survey.

2. Writer: describes the summary of the survey.

3. Illustrator: draws and colors the graphs and other data displays of the brochure.

Support Materials

The *Project Book* contains information about the following topics for use with this Unit Project.

➤ Project Description

➤ Teaching Commentary

➤ Working on the Unit Project Exercises

➤ Completing the Unit Project

➤ Assessing the Unit Project

➤ Alternative Projects

➤ Outside Resources

unit 3

One day you may wake up to find that your favorite station has changed its tune overnight! The style of music played by a radio station changes as new sounds develop and tastes change. In the past, the station you listen to now may have broadcast Swing music to your grandparents or Motown sounds to your parents.

Representing Data

Market researchers survey people's tastes. They may ask you about your favorite TV shows or your spending habits. Their analysis of this information helps retailers decide where to advertise, what products to sell, and where to locate stores.

Tastes and fashions have changed over the years. Car manufacturers, hair stylists, and clothes designers are among the people who try to predict and shape future trends. What do you think will be fashionable in the next decade?

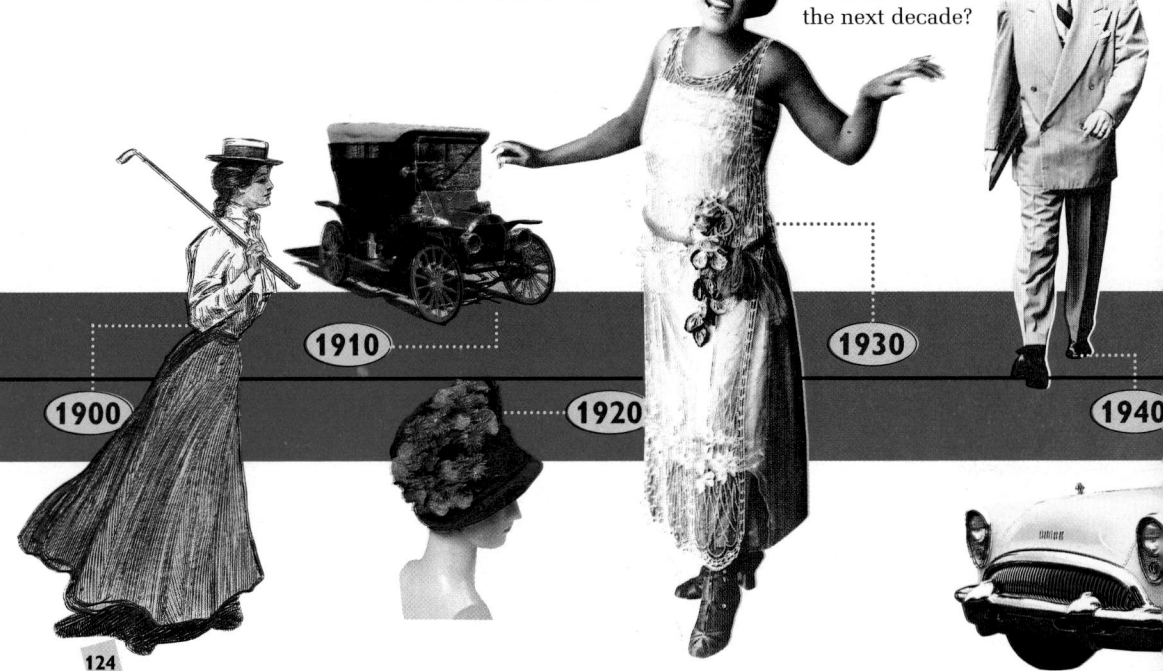

124

Suggested Rubric for Unit Project

4 The data collected are meaningful and support the plan for the business brochure. The graphs are appropriate for the presentation of the data and are drawn correctly. The brochure includes a summary of the market research and the percentages calculated are accurate. The visual design is attractive and original.

3 The research is not complete and the resulting plan is somewhat incomplete. The graphs do not fully enhance the presentation of the results of the research. The visual design is acceptable but not particularly creative.

2 The contents of the brochure are generally incomplete. The research is lacking substance and the data do not adequately support the plan. The graphs are not well drawn. The graphic design shows a lack of thought and effort.

Plan a Music Store

Your project is to develop a business brochure that presents your plan for a music store in your area. A business brochure, also known as a *prospectus*, is designed to interest investors in financially supporting your business.

Your group will need to research the musical tastes of your customers before deciding on the formats and music that you will stock. The data you collect will help you decide what is typical of your customers.

Plan to include graphs and other data displays in your brochure to show the results of your market research and support your decisions.

Malls around the country are changing their marketing focus to meet the needs of the consumers in their community. When research showed that the Hispanic community was more than five times as large as any other population near an Arizona mall, stores in the mall updated their merchandising focus. Special events, such as a festival for the Mexican holiday of Cinco de Mayo, are also held at the mall to celebrate Hispanic heritage.

1950   1960   1970   1980   1990

125

Using Technology

Have students use a computer spreadsheet and graphics software to display the data in the brochure.

Limited English Proficiency

Developing a business brochure may be a difficult project for students acquiring English. It requires good communication skills to write the survey questions and summarize the responses. Have a student fluent in English and a student acquiring English work together as partners.

ADDITIONAL BACKGROUND

Multicultural Note

Businesses use market research to help identify the needs of the consumers in their community. In some areas of the United States, businesses have adopted a successful strategy of grouping themselves into entire malls that appeal to a specific ethnic group. One such mall is the South DeKalb mall in Atlanta, Georgia, which is geared toward the neighborhood's large African-American community.

Another mall, the Tucson Mall in Arizona, advertises on three Spanish-language radio stations and uses a mariachi music group to celebrate Cinco de Mayo. Half of the mall's staff is bilingual.

The Kubuki Mall in San Francisco is operated by and for Japanese Americans. The mall advertises in Japanese-language newspapers and announces upcoming events in both Japanese and English.

Suggested Rubric for Unit Project

1   The brochure cannot be evaluated. The research is totally inadequate and the resulting plan does not support the decisions made. The group should be encouraged to speak with the teacher as soon as possible so that each student understands the purpose and the format of the project.

Compact Disc

A CD (or compact disc) reproduces sound better than any system yet available. Digital recording techniques place the sounds on the disc. A laser beam inside the CD player reads the disc. The quality of sound is excellent and is expected to remain that way, unlike records or tapes which wear down with each use.

ALTERNATIVE PROJECTS

Project 1, page 179

Supporting an Opinion

Gather data to decide whether or not the titles of popular movies in the 1970s were very short compared to titles of older movies. State the conclusion the data supports.

Project 2, page 179

Analyzing a Process

Find out how data are collected for making a Top Ten music, movie, or book list. Write a summary of how the list is determined and suggest how changes in the process could produce different results.

Answers to Can We Talk?

➤ Answers may vary.

➤ Answers may vary on the styles of music. Some include: rock, rap, jazz, soul, classical, folk, easy listening, country western, new age, heavy metal, oldies, gospel, international, and so on. Answers may vary on names of solo artist or group.

➤ Answers may vary.

➤ Answers may vary.

➤ Answers may vary. Shown in the table are the percentages of all music sales in the form of cassette tapes and CDs between 1988 and 1992 according to the Recording Industry Association of America.

126

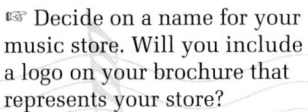

Getting Started

For this project you should work in a group of three students. Here are some ideas to help you get started.

☞ Decide on a name for your music store. Will you include a logo on your brochure that represents your store?

☞ You will need to find out which songs and music artists are the most popular. How can you find this information?

☞ Musical tastes depend on many factors, including age and cultural background. How can you find out which groups purchase the most music recordings in your area?

☞ Throughout this unit you will be gathering data and completing research. How will you organize this information? Can you use a computer to create a database?

Working on the Unit Project

Your work in Unit 3 will help you collect and organize data for your brochure.

Related Exercises:

Section 3-1, Exercise 31
Section 3-2, Exercises 35, 36
Section 3-3, Exercise 28
Section 3-4, Exercise 25
Section 3-5, Exercise 27
Section 3-6, Exercise 24
Section 3-7, Exercise 34

Alternative Projects p. 176

Can We Talk

➤ What is your favorite type of music? Do your friends like the same music? What type of music do your parents like? your grandparents?

➤ How many different styles of music can you think of? For each style that you list, can you name a popular solo artist or group?

➤ What type of products do you think would be advertised on a radio station that plays jazz? country-western music? heavy metal music? Explain your reasoning.

➤ Which music groups do you think will be popular next year? Why? Can you make predictions based on what sold well in the past?

➤ Do you buy cassettes or compact discs most often? Do you buy music videos? Do you think your buying habits are typical of teenagers?

➤ What type of music recording is the most expensive? Does the price of a recording prevent people from buying it?

➤ How has the packaging for compact discs changed since they were first introduced? Why was it changed?

126   **Unit 3** Representing Data

	1988	1989	1990	1991	1992
Cassette tapes	55.1	50.4	48.4	43.3	37.3
Compact disks	28.5	35.9	42.5	49.6	56.1
LPs	13.8	8.3	4.3	1.4	1.2
Music videos	NA	NA	NA	NA	1.6

➤ Music videos are the most expensive music format, followed by compact disks.

➤ CDs were at one time packaged in a box that was twice the size of the compact disk. This was changed in 1993 so that the disk was simply sold in a cellophane-wrapped plastic case. The packaging wasted paper and plastic and contributed to environmental pollution.

3-1

Using Matrices and Graphs

TABLE THAT...

THOUGHT

EXPLORATION

How can you use tables and graphs to understand data?

• **Work in a group of at least five students.**

The student government in your school wants to set up a new organization for volunteering. Your group's task is to gather information on student interest in various types of volunteer work.

Which types of volunteer work most interest you? Write a "1" next to your first choice and a "2" next to your second choice.

Hospital _____
Work with elderly _____
Big Brother/Big Sister _____
Homeless shelter _____
Tutoring _____
Recycling _____
Animal shelter _____

① Use a survey like the one shown. Collect responses from everyone in your class.

② Organize the data from all the surveys in a table.

③ Based on the data you collected, which type of volunteer work do you think would be the most popular? With your group, come to an agreement on the most popular volunteer activity. Support your choice with one or more graphs.

④ Compare your results with other groups.

⑤ Save your data to use in Exercise 6 on page 131.

Student Resources Toolbox
pp. 661–663 *Data Displays*

3-1 Using Matrices and Graphs

127

Answers to Exploration

1–5. Data and answers may vary.

PLANNING

Objectives and Strands
See pages 124A and 124B.

Spiral Learning
See page 124B.

Recommended Pacing
Section 3-1 is a two-day lesson.
Day 1
Pages 127–129: Exploration through Talk it Over 5, *Exercises 1–14*
Day 2
Pages 129–130: Spreadsheets through Look Back, *Exercises 15–31*

Toolbox References
➤ **Toolbox Skill 22:** Making a Bar Graph
➤ **Toolbox Skill 23:** Making a Line Graph
➤ **Toolbox Skill 24:** Making a Pictograph
➤ **Toolbox Skill 25:** Making a Circle Graph

Extra Practice
See pages 624–625.

Warm-Up Exercises
💡 Warm-Up Transparency 3-1

Support Materials
➤ Practice 19
➤ Enrichment 17 in the Activity Bank
➤ Study Guide 3-1
➤ Problem Set 6
➤ Diagram Master 11 in the Explorations Lab Manual
➤ McDougal Littell Mathpack software: *Stats!*
➤ Stats! Activity Book: Activities 3 and 6
➤ Using IBM Plotter Plus Disk: Statistics Spreadsheets
➤ Quiz 3-1
➤ Alternative Assessment 1

Exploration

The goal of the Exploration is to have students collect, organize, and display data. The systematic collection and interpretation of data form a branch of mathematics called statistics.

Talk it Over

Question 1 provides practice in identifying the correct location in a matrix. You might want to go over several examples of what number is in a certain row and column position.

Additional Sample

S1 Suppose you are doing a survey of the number of students taking a psychology course. What would you say about the line graphs of data from the school matrix?

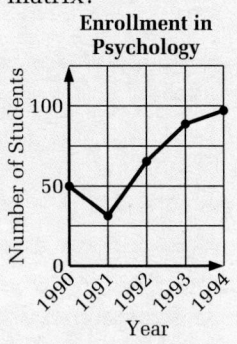

Enrollment in Psychology

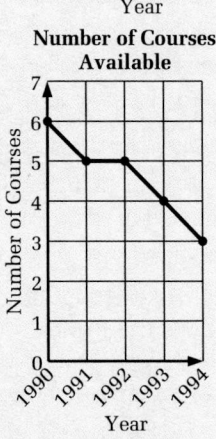

Number of Courses Available

The first graph shows that there is an increase in the number of students enrolling in a psychology course. The second graph shows that the number of courses available is decreasing.

Displaying Data in a Matrix

A table that displays data that belong to more than one category is an example of a *matrix*. The matrix below shows data from City Year, a community service organization of 17–23 year olds.

In a **matrix,** numbers are arranged in *rows* and *columns*. The number of rows and columns, in that order, are the **dimensions** of the matrix. (The plural of matrix is *matrices*.)

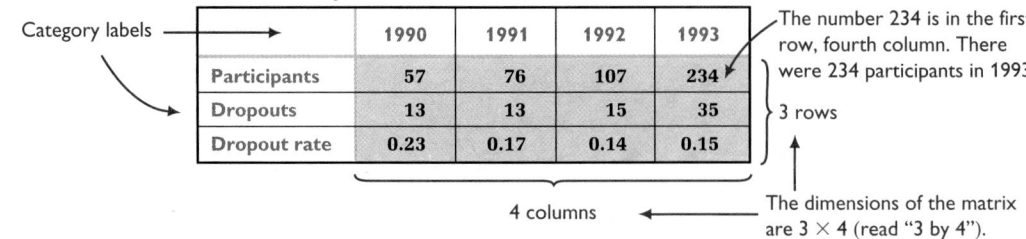

City Year Enrollment, 1990–1993

Category labels →

	1990	1991	1992	1993
Participants	57	76	107	234
Dropouts	13	13	15	35
Dropout rate	0.23	0.17	0.14	0.15

The number 234 is in the first row, fourth column. There were 234 participants in 1993.

3 rows

4 columns

The dimensions of the matrix are 3 × 4 (read "3 by 4").

BY THE WAY...

Alan Khazei and Michael Brown launched the Boston-based City Year in 1988, with the goal of providing a diverse group of young people a chance to work together in service to others.

Talk it Over

1. What number is in the third row, second column of the matrix? What does this number represent?

2. When you give the dimensions of a matrix, do you count the row and the column of category labels?

3. Is the table you made in step 2 of the Exploration a matrix? If so, what are its dimensions?

Graphs can help you compare sets of data, observe trends, draw conclusions, and support a point of view.

Sample 1

These line graphs display data from the City Year Enrollment matrix above. Use each line graph to describe a trend and to make a statement that might encourage someone to support the City Year program.

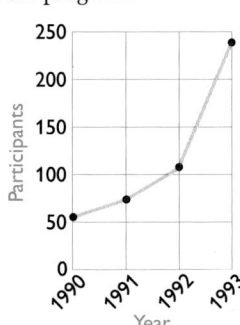

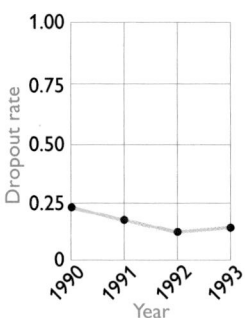

Answers to Talk it Over

1. 0.17; dropout rate in 1991
2. No.
3. Yes. Answers may vary depending on survey used.

The participants graph shows that the number of participants has increased each year. The program is popular and expanding.

The drop-out rate graph shows that the dropout rate is low and fairly stable. The program is successful in achieving its goals.

Talk it Over

4. Suppose you made a line graph of the data in the matrix row labeled *Dropouts*. How would it be different from the dropout-rate line graph?

5. Do you think it is easier to see the trends described in Sample 1 by looking at the graphs or by looking at the data displayed in the matrix? Explain your choice.

······▶ **Now you are ready for:**
 Exs. 1–14 on pp. 131–132

Spreadsheets

In a spreadsheet, columns are lettered and rows are numbered.

Computer **spreadsheet** software stores information in matrices. The spreadsheet matrix below shows the financial support received by a non-profit organization during each of three years.

Save the Forests Funding (dollars), 1990–1992			
	A	**B**	**C**	**D**
1		**1990**	**1991**	**1992**
2	**Donations from corporations**	310,512	437,015	501,329
3	**Donations from individuals**	75,216	126,005	250,100
4	**Donations of goods and services**	195,600	187,250	150,725
5	**Interest income**	21,817	29,036	42,752
6	**Total funds**	603,139	779,306	944,906

Each position in a spreadsheet is called a **cell**. The number 501,329 is in cell D2.

You can enter the formula D2 + D3 + D4 + D5 in cell D6. The software finds the sum of the numbers in cells D2 to D5.

Talk it Over

6. Which row and which column of the spreadsheet contain the category labels for the matrix?

7. What number is in cell C5? Which cell contains the number 195,600? What does each number represent?

8. In which other cells could you enter formulas like the formula in cell D6? What are some advantages of entering a formula, rather than a number, in a spreadsheet cell?

3-1 Using Matrices and Graphs

129

Answers to Talk it Over

4. Answers may vary. An example is given. The dropout rate shows that the percentage of dropouts per year declines, so the line graph is decreasing. The line graph for dropouts increases because the number of dropouts increases by year, but this is probably due to the increase in participants.

5. Answers may vary. An example is given. The trends are easier to see by looking at the graphs, because they show at a glance whether there has been an increase or decrease.

6. row 1 and column A

7. 29,036; B4; C5 represents interest income in 1991 and B4 represents donations of goods and services in 1990.

8. B6 and C6; Formulas can use information from other cells of the spreadsheet directly to compute values and save time.

Limited English Proficiency

Some students acquiring English may not make the connection between the word *matrices* in the unit title and the word *matrix* in the text. Explain that *matrices* is the plural of *matrix*: if you create one *matrix* and then create a second one, you have created two *matrices*.

Mathematical Procedures

Different types of graphs are best used to present certain kinds of data. A line graph is often used to show how data vary over time. A bar graph is generally used to show an increase or decrease in data. Circle graphs show relationships among data, usually represented in terms of percents.

Using Technology

Either you or some of your students may want to use the *Stats!* software to demonstrate how to create a spreadsheet. Point out that the way formulas are entered into a spreadsheet varies with different software packages.

Visual Thinking

Ask students to make a sketch of the spreadsheet on a separate piece of paper. Have them highlight or circle two cells in the same row and explain the relationship between the two numbers. Encourage them to discuss ways in which spreadsheets are useful. This activity involves the visual skills of *recognition* and *interpretation*.

Talk it Over

Question 7 emphasizes that only one number is put in a cell and that each cell has an address corresponding to its column and row location.

129

S2 Use the graphs below. Which graph(s) better support the statement? Explain your choice.

The percentage of students enrolled in Psychology and World History is about the same for two out of the three years.

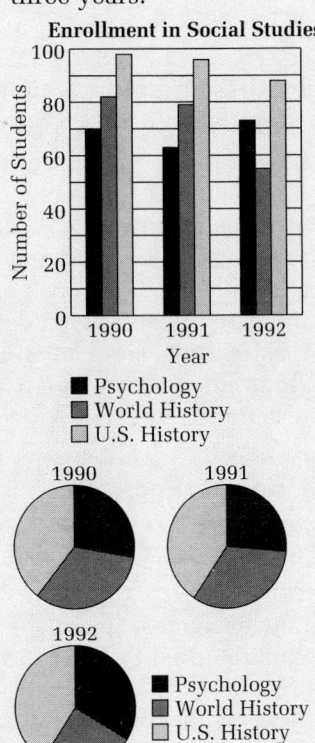

Enrollment in Social Studies

■ Psychology
■ World History
□ U.S. History

the circle graphs; **The circle graphs show best the varying percentages of the whole for each course. The bar graph does not show percentages, and it is more difficult to see the changes from year to year from the bars.**

Look Back

Use the Look Back for a class discussion about how to represent data in the most meaningful way.

Computer software can create graphs from data in a spreadsheet. These graphs display the Save the Forests Funding data shown on page 129.

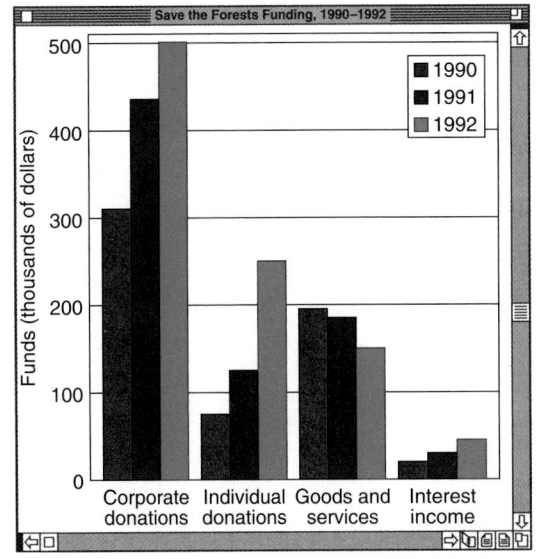

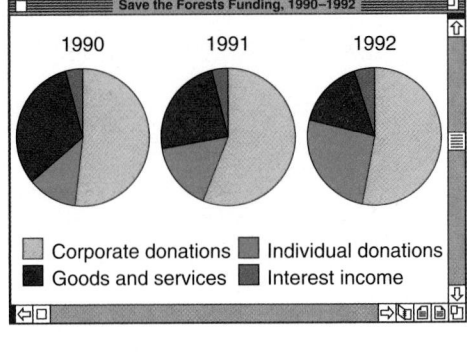

Talk it Over

9. Which rows of the spreadsheet does the bar graph represent?

10. Does each circle graph represent a *row* or a *column* of the spreadsheet?

Sample 2

Writing Use the Save the Forests Funding graphs. Which graph better supports the statement below? Explain your choice.

The amount of support from goods and services has decreased.

Sample Response

State your choice. ⟶ The bar graph. The bars represent the dollar amounts of each type of funding. The bars for goods and services get shorter, so the amount of this type of support has decreased.
◄ Explain how the graph supports the statement.

Since the circle graph does not show the dollar amounts, it does not show that the *amount* of support in goods and services has decreased.
◄ Explain why you did not use the other graph to support the statement.

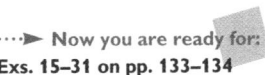
Now you are ready for:
Exs. 15–31 on pp. 133–134

Look Back ◄

What are some advantages of presenting data in a matrix? in a graph?

Answers to Talk it Over

9. rows 2, 3, 4, and 5
10. a column

Answers to Look Back

Answers may vary. Examples are given. You could use a graph because it summarizes the data in a visual form that allows you to compare the data easily, to identify trends, or to show the relationships be- tween parts of the data and the total. You could use a matrix if you need exact data, especially if you want to identify individual data items quickly and accurately.

3-1 Exercises and Problems

1. **Reading** Use the City Year Enrollment matrix on page 128. Suppose that a column is added for the year 1994. What will the dimensions of the new matrix be?

For Exercises 2–5, use the matrix and the *stacked bar graph*.

Gasoline Prices, 1992		
	Base price ($/gallon)	Tax ($/gallon)
United States	0.75	0.38
Turkey	1.02	1.80
Japan	1.99	1.71
France	0.86	2.90
Norway	1.39	3.47

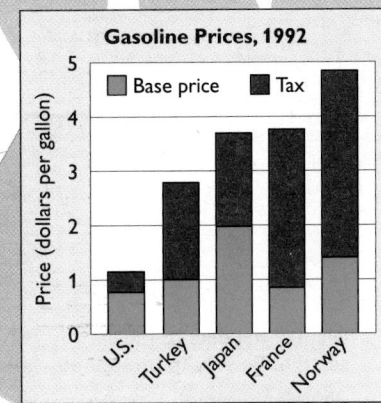

2. What are the dimensions of the matrix?

3. What does the number in the third row, second column represent?

4. Explain how the stacked bar graph shows the total price of a gallon of gasoline in each country.

5. The data in the matrix could also be displayed in a double bar graph. Do you think a *double bar graph* or a *stacked bar graph* is better for this type of data? Give a reason for your choice.

6. **a.** Display the data you collected in the Exploration in a stacked bar graph.

   b. Suppose the new student organization plans to sponsor only three volunteer activities in its first year. Which three activities should they offer? Use your graph in part (a) to support your answer.

7. Use the tide table.

   a. Describe two patterns in the tide data that are shown in the tide table.

   b. There are actually two high tides and two low tides each day. Predict the times of the other high and low tides on Friday.

8. **Research** A tide table is an example of an everyday matrix. Find at least three more examples of everyday matrices. For each example that you find, tell what kind of graph (if any) you could use to display the information in the matrix.

TIDE TABLE—AUG.

	High tide	Low tide
Friday	4:46 A.M.	11:00 A.M.
Saturday	5:35 A.M.	11:55 A.M.
Sunday	6:25 A.M.	12:51 P.M.
Monday	7:16 A.M.	1:42 P.M.
Tuesday		

3-1 Using Matrices and Graphs **131**

APPLYING

Suggested Assignment

Standard 1–7, 9, 11–13, 15–22, 24–31

Extended 1–7, 9–31

Integrating the Strands

Algebra Exs. 28–30

Geometry Ex. 25

Statistics and Probability Exs. 1–24, 26, 27, 31

Discrete Mathematics Exs. 1–24, 26, 27, 31

Logic and Language Exs. 10, 18–21, 23, 24, 31

Research

In connection with Ex. 7, if your school is in a coastal town, you may want students to gather actual tide data.

Answers to Exercises and Problems

1. 3×5

2. 5×2

3. tax, in dollars, per gallon of gas in Japan

4. The lower portion represents the base price, and the upper portion represents the tax, so that together they represent base price plus tax, which equals total price.

5. Answers may vary. An example is given. A stacked bar graph is better because it enables the reader to compare how the total cost of gas varies by country.

6. **a, b.** Answers may vary.

7. **a.** Answers may vary. An example is given. The tides occur later every day. High and low tides stay roughly the same distance apart.

   b. Answers may vary. An example is given: high tide—around 5 P.M., low tide—around 11 P.M.

8. Answers may vary. Examples are given. election result totals by precinct, bar graph; baseball scores by inning, line graph; percent of population registered to vote by age group, circle graph

131

connection to **HISTORY**

Florence Nightingale (1820–1910) used data displays to convince the British government to reform military health care. This display is based on data collected during and after the Crimean War (1854–56).

CAUSE OF DEATH	ANNUAL DEATHS PER 1000 MEN		
Contagious diseases	0.2		Civilians
	18.7		Soldiers
Other diseases	0.7		Civilians
	1.2		Soldiers
Wounds	0.1		Civilians
	3.0		Soldiers

Death Rates of English Civilians and English Soldiers, ages 15–45

9. Display the data in the second column in a matrix like the ones in this section. Be sure to include row and column category labels, and a title.

10. **Writing** A graph like the one above appeared in a report of the Royal Commission on the Health of the Army, formed in 1857. Write a part of the report, explaining the conclusions that can be drawn from the graph.

Use the matrix and the multiple line graph.

11. **a.** Which type of music recording appears to have the most steadily increasing values of shipments?

 b. Did you use the *matrix* or the *graph* to answer part (a)?

12. Which type of music recording seems most likely to disappear over the next few years? Use the graph to explain your answer.

13. **a.** Use the graph to compare the values of the shipments of different types of music recordings in 1988.

 b. Use the graph to describe how the value of cassette shipments changed from 1986 to 1989.

 c. For parts (a) and (b), you had to read the graph in different ways. How were they different?

14. Use the graph to write three statements comparing the values of shipments of albums and compact discs from 1986 to 1989.

Value of Music Shipments (millions of dollars)

		1986	1987	1988	1989
	Singles	228.1	203.3	180.4	116.4
	Albums (LPs and EPs)	983.0	793.1	532.2	220.3
	Compact discs	930.1	1593.6	2089.9	2587.7
	Cassettes	2499.5	2959.7	3385.1	3345.8

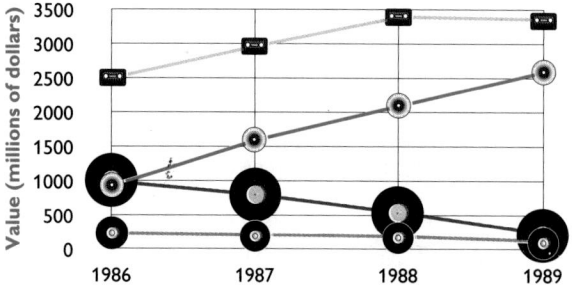

Business Use the spreadsheet and the bar graph.

	A	B	C	D	E
		1987	1988	1989	1990
1		1987	1988	1989	1990
2	Investment income	93.1	100.9	715.4	618.7
3	Expenses	106.4	127.4	413.6	412.5
4	Net income or loss	− 13.3	− 26.5	301.8	206.2

Mutual Money Fund: Financial Data (millions of dollars)

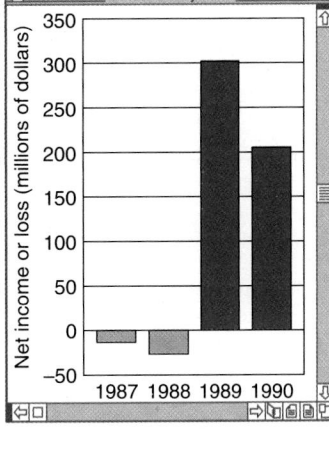

Mutual Money Fund

15. What number is in cell D4? In which cell is the number 127.4? What does each number represent?

16. Write a formula that will calculate the value in cell E4 from the values in cells E2 and E3.

17. Which row(s) of the spreadsheet does the graph display?

18. How does the graph show profits and losses?

The five breeds shown in the pictograph and the circle graph represent about one-third of all dogs registered in the United States in 1991.

Writing Which graph better supports each statement? Explain your choice.

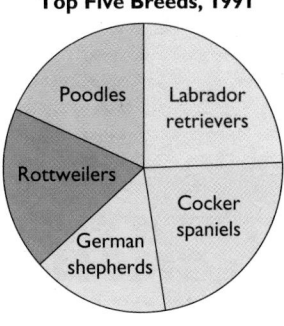

U.S. Dog Registrations Top Five Breeds, 1991

Poodles / Labrador retrievers / Rottweilers / German shepherds / Cocker spaniels

U.S. Dog Registrations: Top Five Breeds, 1991

Labrador retriever	
Cocker spaniel	
Rottweiler	
Poodle	
German shepherd	

Each represents 10,000 dogs registered.

19. Of the top five breeds, the German shepherd has the smallest number of dogs registered.

20. Of the top five breeds, the two breeds with about the same number of registered dogs are the rottweiler and the poodle.

21. The number of registered cocker spaniels and Labrador retrievers is about half the number of dogs registered from the top five breeds.

3-1 Using Matrices and Graphs

133

Interdisciplinary Problems

The development of spreadsheet software was a major advance in the application of computer technology to the needs of the business world. Vast amounts of financial data are processed every day in the United States by banks, pension funds, and brokerage firms involved in stock market transactions. During the early 1990s, the mutual fund industry also grew enormously, mostly from investments by small investors. The processing of financial data would be impossible today without the use of spreadsheet software.

Graphs also play a vital role in the presentation and interpretation of business data. Reports from financial businesses very often use bar graphs, line graphs, circle graphs, or pictographs to help their customers see and understand the data involved. These types of graphs also appear frequently in the financial pages of newspapers and in books and magazines about managing money.

Answers to Exercises and Problems

c. In part (a) you read the values vertically on the line representing 1988, while in part (b) you read the values horizontally, following the line representing cassette shipments.

14. Answers may vary. Examples are given. In 1986, the value of album shipments and compact disc shipments were about the same. From 1986 to 1989, CD shipments increased, while album shipments decreased. By 1989, the value of CD shipments was at least 2 billion dollars more than that of album shipments.

15. 301.8; C3; D4 represents net income or loss in 1989 and C3 represents expenses in 1988.

16. E2 − E3 = E4

17. row 4

18. Losses are drawn downward from "0" line, and profits are drawn upward.

19–21. Answers may vary. Examples are given.

19. pictograph; shows the slight difference in numbers better

20. pictograph; shows the slight difference in numbers better

21. circle graph; shows half the total much better

Assessment: Portfolio

Ex. 24 assesses students' understanding of how and when to use graphs. Students may wish to include their demonstrations in their portfolios.

Working on the Unit Project

For Ex. 31, students should discuss the kind of data that would be useful to follow the popularity of a song or group.

Practice 19 For use with Section 3-1

Name ____________ Date ____________

Practice 19

For use with Section 3-1

Federal Aid to Education (in billions of dollars)

	1984	1988	1990	1992
Elem. & Sec.	4.3	5.7	7.2	8.8
Handicapped	2.4	3.1	3.5	5.3
Voc. & Adult	1.0	1.0	1.1	1.9
College Aid	7.5	8.8	11.1	15.3

For Exercises 1–7 use the matrix and the stacked bar graph. The graph was drawn using the data in the matrix.

How much in federal aid was spent in each category?

1. Elementary and secondary aid in 1988 **$5.7 billion**
2. Aid for handicapped in 1992 **$5.3 billion**
3. Aid for vocational and adult education in 1984 **$1.0 billion**
4. Aid for college education in 1990 **$11.1 billion**

Tell what each number represents.

5. The second number in the fourth row **College Aid in 1988**
6. Which category of federal aid grew the least (in dollars) between 1984 and 1992? **Voc. & Adult**
7. Which category grew the most (in dollars)? **College**

For Exercises 8–12 use the matrix and graph on population of U.S. cities.

Population of U.S. cities (in millions)

	1900	1950	1970	1990
New York	3.4	7.6	7.9	7.3
L. A.	0.1	2.0	2.8	3.5
Chicago	1.7	3.6	3.4	2.8
Houston	0.04	0.6	1.2	1.6

8. Which two cities lost population between 1970 and 1990? **New York and Chicago**
9. Did the same two cities lose population between 1950 and 1970? **New York, no Chicago, yes**
10. Which city gained most in population between 1950 and 1990? **L.A.**
11. What was the ranking of the cities from largest to smallest population in 1970? **New York, Chicago, L.A., Houston**
12. Which two cities traded rankings between 1970 and 1990? **L.A. and Chicago**

Practice Bank, INTEGRATED MATHEMATICS 1

19

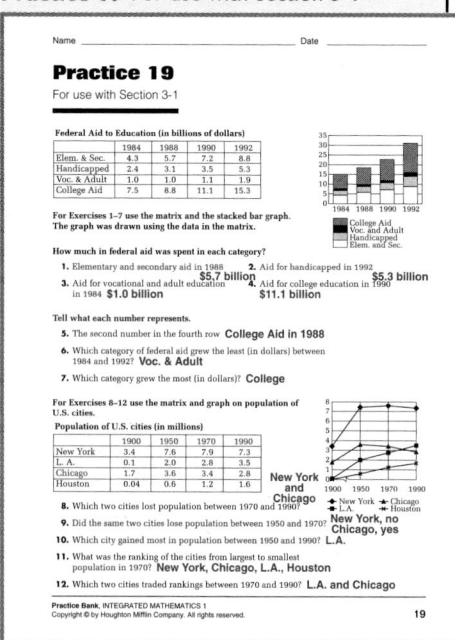

22. Suppose you enter the City Year Enrollment matrix on page 128 in a spreadsheet so that the number 57 is in cell B2. Write a formula that will calculate the value in cell B4 from the values in cells B2 and B3. Use a calculator to check your formula.

23. **Writing** Use the Save the Forests Funding graphs on page 130. Which graph better supports the statement below? Explain your choice.

> The portion of funds that comes from donations from individuals has increased each year.

Ongoing **ASSESSMENT**

24. **Open-ended** Suppose you have to show a group of fifth-grade students how to display information in a graph. You want the students to take part in the demonstration. Your materials are self-stick removable notes, felt-tipped markers, and graph paper or poster board.

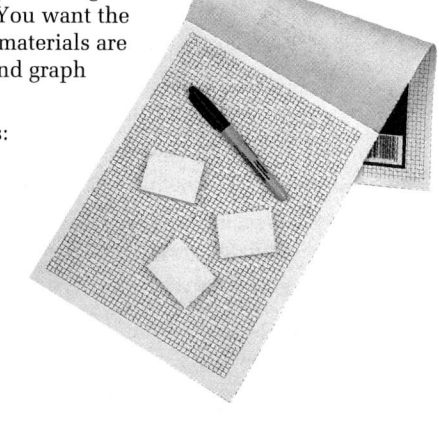

Describe your demonstration. Include these parts:

- an introduction: Why do we use graphs?
- a survey question that the students can easily answer
- a method for displaying the results of the question with your materials
- a conclusion

Review **PREVIEW**

25. The area of a square is 196 m^2. Find the length of a side of the square. *(Section 2-9)*

Use a number anywhere along the scale to estimate the probability of each event. *(Section 2-1)*

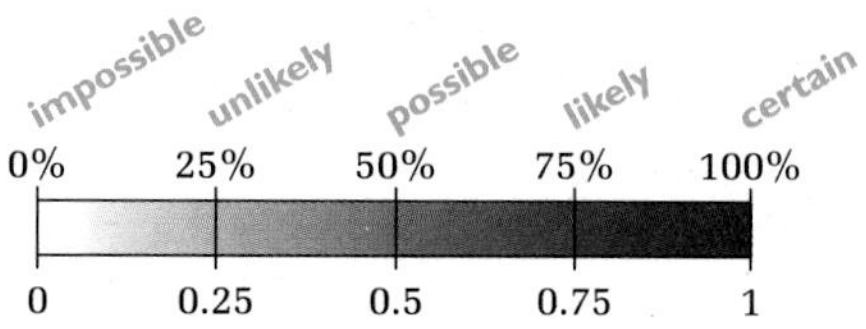

26. You will do some volunteer work this year.

27. School will be closed because of a snowstorm at least once this year.

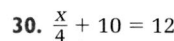

Solve each equation. *(Sections 2-7, 2-8)*

28. $3x - 8 = 13$

29. $\frac{x}{6} = -5$

30. $\frac{x}{4} + 10 = 12$

Working on the Unit Project

31. a. **Research** Follow the popularity of a song or a group using data from the last several months. You may want to use magazines such as *Billboard, Entertainment Weekly,* and *Variety.*

b. Choose a format to display the data you collected.

c. **Writing** Describe any trends you notice in the data you collected.

Answers to Exercises and Problems

22. B3 ÷ B2 = B4
23. Answers may vary. An example is given. The circle graph shows an increase in sector size as the years progress.
24. Answers may vary.
25. 14 m
26. Answers may vary. Example: 0.5
27. Answers may vary. Example: 0.75
28. 7
29. –30
30. 8
31. a–c. Answers may vary.

3-2 Mean, Median, and Mode

*The **MEDIAN** is the message.*
*I'm in the **MODE** for math.*
*But, what does it all **MEAN**?*

Focus
Find values that are typical of a data set, identify data values that are not typical of a data set, and use equations to find missing data values.

Nutritionists recommend a diet that is low in fat. You can read package labels to find the approximate fat content of many foods.

Breads and crackers	Fat content per serving* (g)
cracked wheat bread	0.9
whole wheat bread	1.1
pita bread	0.6
matzo	0.3
graham cracker	0.5
corn muffin	4.0
rice cake	0.3
tortilla	1.1
bran muffin	5.1
rye bread	0.9
pumpernickel bread	1.1

*Fat content per serving may vary by manufacturer

Talk it Over

1. What do you notice about the numbers in the table?
2. Does one value appear more often than any other?
3. Are any of the values very different from the others?
4. How many grams of fat per serving do you think is most typical of the types of breads and crackers listed?

Three values that may be typical of a data set are the *mean,* the *median,* and the *mode.* Which value is "most typical" may depend upon the type of data, and it may be a matter of opinion.

3-2 Mean, Median, and Mode

135

Answers to Talk it Over

1. Answers may vary. An example is given. The numbers are decimals that range from 0.3 to 5.1.

2. Yes, 1.1 appears three times.

3. Yes, the two types of muffins have much higher fat contents than the other breads and crackers.

4. Estimates may vary.

PLANNING

Objectives and Strands
See pages 124A and 124B.

Spiral Learning
See page 124B.

Materials List
➤ Spreadsheet software (optional)
➤ Uncooked spaghetti

Recommended Pacing
Section 3-2 is a two-day lesson.
Day 1
Pages 135–137: Opening paragraph through Talk it Over, *Exercises 1–17*
Day 2
Page 138: Sample 3 through Look Back, *Exercises 18–36*

Extra Practice
See pages 624–625.

Warm-Up Exercises
Warm-Up Transparency 3-2

Support Materials
➤ Practice 20
➤ Enrichment 18 in the Activity Bank
➤ Study Guide 3-2
➤ Problem Set 6
➤ McDougal Littell Mathpack software: *Stats!*
➤ Stats! Activity Book: Activities 4, 10, 12, and 18
➤ Using Plotter Plus: Exploring Statistics
➤ Quiz 3-2

TEACHING

Talk it Over

Questions 1–4 begin the process of students looking at data to determine measures of central tendency, as well as noticing outliers.

Communication: Discussion

Ask students if they are familiar with the terms *mean*, *median*, and *mode*. Some students may know the mean by the more commonly used name *average*.

Additional Sample

S1 Find the mean, the median, and the mode of the calories-per-serving data set shown below.

Bread and crackers	Calories per serving
cracked wheat bread	98
whole wheat bread	95
pita bread	65
matzo	34
graham cracker	55
corn muffin	150
rice cake	93
tortilla	32
bran muffin	140
rye bread	55
pumpernickel bread	110

mean: 84.3; median: 93; mode: 55

Talk it Over

Question 7 has students think about what "most typical" means in terms of data. Since it is likely that the mean, the median, and the mode will all be chosen by some students, a full-class discussion may prove useful.

Sample 1

Find the mean, the median, and the mode of the fat-per-serving data set on page 135.

Sample Response

Mean: Find the sum of all the data. Then divide by the number of data items.

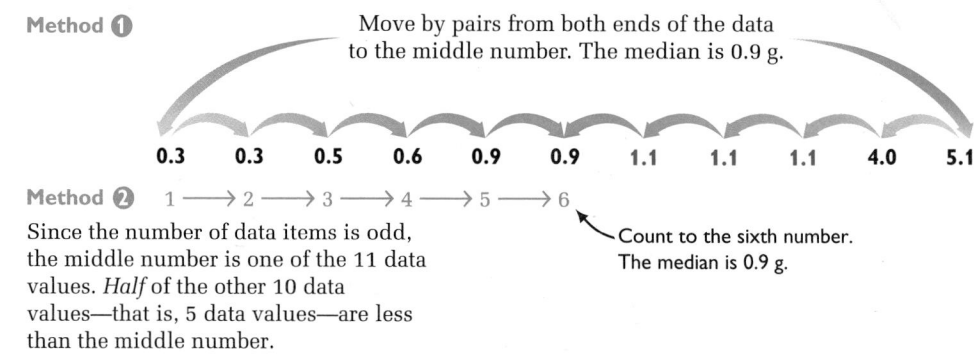

$0.9 + 1.1 + 0.6 + 0.3 + 0.5 + 4.0 + 0.3 + 1.1 + 5.1 + 0.9 + 1.1$

There are 11 data items. $\frac{15.9}{11} = 1.4\overline{45}$ To the nearest tenth, the mean is 1.4 g.

Median: List the data in order from least to greatest. Then find the middle number. *Note:* When the number of data items is even, the median is the *mean* of the *two* middle numbers.

Method ❶ Move by pairs from both ends of the data to the middle number. The median is 0.9 g.

| 0.3 | 0.3 | 0.5 | 0.6 | 0.9 | 0.9 | 1.1 | 1.1 | 1.1 | 4.0 | 5.1 |

Method ❷ 1 → 2 → 3 → 4 → 5 → 6

Since the number of data items is odd, the middle number is one of the 11 data values. *Half* of the other 10 data values—that is, 5 data values—are less than the middle number.

Count to the sixth number. The median is 0.9 g.

Mode: Find the number(s) that appear most often. *Note:* A data set may have more than one mode, or it may have no mode.

The value 1.1 appears three times. All other values appear one or two times. The mode is 1.1 g.

TECHNOLOGY NOTE

Many calculators and computer software packages can find the mean of a data set, and can sort data from smallest to largest.

Talk it Over

5. Explain why the mean of the data in Sample 1 is greater than the median and the mode.

6. Will the mean be greater than the median and the mode for any data set? Give examples to support your answer.

7. Do you think the *mean*, the *median*, or the *mode* is the most typical of the data in Sample 1? Give a reason for your choice.

136 **Unit 3** Representing Data

Answers to Talk it Over

5. The amounts 5.1 and 4.0 are very different from the other amounts. These amounts increase the total sum, which is used to find the mean, but their sizes do not affect the median or the mode.

6. No. The mean is not necessarily greater than the median or the mode(s) in a data set. For example, in the data set 2, 5, 9, 12, 12, the median, 9, and the mode, 12, are both greater than the mean, 8; in the data set 5, 5, 5, 5, the mean, the median, and the mode all equal 5.

7. Answers may vary. An example is given. The median appears most typical because it is close in value to 9 of the 11 values.

Outliers and Range

BY THE WAY...

Nutritionists recommend that less than 30% of your daily calories come from fat (1 g of fat = 9 calories).

Data values that are much larger or much smaller than the other values in a data set are called **outliers,** because they lie outside most of the other values. Outliers are *not* typical of the data set.

The difference between the smallest and largest numbers in a data set is called the **range** of the data set.

Sample 2

a. **Which values in the data set in Sample 1 are outliers? Explain.**

b. **What is the range of the data set in Sample 1?**

Sample Response

A *line plot* helps you picture the outliers and the range. In a **line plot,** you show each data value as an x above its position on a number line.

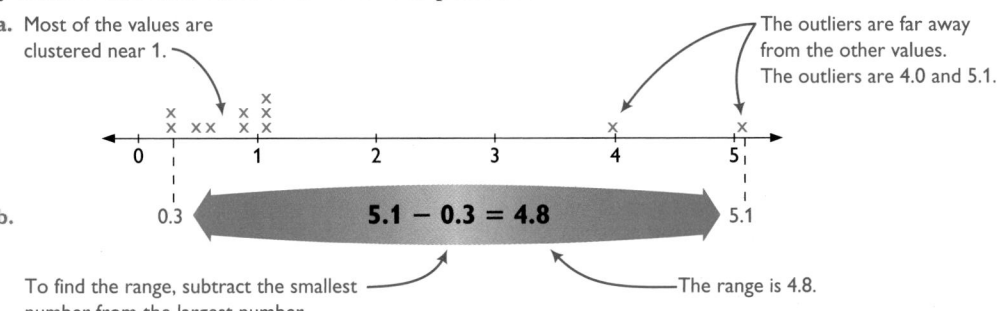

a. Most of the values are clustered near 1.

The outliers are far away from the other values.
The outliers are 4.0 and 5.1.

b.

0.3   **5.1 − 0.3 = 4.8**   5.1

To find the range, subtract the smallest number from the largest number.

The range is 4.8.

Talk it Over

8. Which breads are associated with the outliers in the fat-per-serving data set? What advice would you give to someone trying to eat a low-fat diet?

9. Replace the data for bran and corn muffins with data for breads that are lower in fat:

English muffin: 1.1 g of fat per serving

Bagel: 1.4 g of fat per serving

a. Does the new data set have any outliers? What is the range of the new data set?

b. Find the mean, the median, and the mode of the new data set. Which one is most affected by the change in the data?

> Now you are ready for:
> Exs. 1–17 on pp. 139–141

3-2 Mean, Median, and Mode   **137**

Communication: Discussion

Since not all data sets have outliers, you may wish to point out to students that the range is merely the difference between the largest and smallest data items, whether they are outliers or not.

Additional Sample

S2 a. Which values in the data set in Additional Sample S1 are outliers? Explain. The values 140 and 150 are much larger than the other values.

b. What is the range of the data set? 118

Mathematical Procedures

The Sample Response introduces the idea of a line plot. Line plots allow for a visual presentation of numerical data that usually makes analysis of the data easier.

Reasoning

Discuss with students how the various descriptions of a data set can change as new data are added to the set. Ask them to suggest some examples that show how the outliers of a data set can affect the mean of the set.

Using Technology

Students can use the *Stats!* software to sort the data and compute the mean, the median, and the mode for Talk it Over question 9.

Answers to Talk it Over

8. corn and bran muffins; avoid eating these muffins

9. a. No. 1.1

b. mean: about 0.85; median: 0.9; mode: 1.1; the mean

Visual Thinking

Encourage students to make process diagrams comparing the three methods shown in Sample 3. Ask them to use the diagrams to explain how the methods differ, to select which method they think is best, and to explain why. This activity involves the visual skills of *correlation* and *communication*.

Additional Sample

S3 Jae bought gifts that cost $24, $26, $20, and $18. She has one more gift to buy and wants her mean cost to be $24. What should she spend for the last gift?

Method 1: Guess and check.
Try $30.
$$\frac{24 + 26 + 20 + 18 + 30}{5} = 23.6$$
Try a greater price, such as $32.
$$\frac{24 + 26 + 20 + 18 + 32}{5} = 24$$
The answer is $32.

Method 2: Work backward. In order to have a mean of $24 on 5 gifts, the sum of all five gifts must be $24 · 5 or $120. The sum of the first four gifts is $88. So the last gift should cost $120 – 88 = $32.

Method 3: Use an equation. Let x = Jae's cost for the last gift.
$$\frac{88 + x}{5} = 24, \text{ or } x = 32$$
The last gift should cost $32.

Look Back

The Look Back questions can be used for class discussion. After the discussion, students can write a summary of their understanding of these ideas as journal entries.

Lin's test scores are 87, 86, 89, and 88. There is one more test in the grading period. Lin wants her mean test score to be 90. What score must she get on the last test? Explain your method.

Sample Response

Method ❶

Problem Solving Strategy: Guess and check

Lin's first four scores are less than 90, so her last test score must be greater than 90.

Try a number larger than 90, such as 95.

Find the mean of the scores:
$$\frac{87 + 86 + 89 + 88 + 95}{5} = \frac{445}{5} = 89$$
The mean is less than 90.

Try a number larger than 95, such as 100.
$$\frac{87 + 86 + 89 + 88 + 100}{5} = \frac{450}{5} = 90$$
Lin's mean test score will be 90 if her last test score is 100.

Method ❷

Problem Solving Strategy: Work backward

In order to have a mean of 90 on 5 tests, the sum of the five test scores must be
$$90 \cdot 5 = 450.$$

The sum of the first four scores is
$$87 + 86 + 89 + 88 = 350.$$

So the last test score must be
$$450 - 350 = 100.$$

Lin's mean test score will be 90 if her last test score is 100.

Method ❸

Problem Solving Strategy: Use an equation

Let x = Lin's last test score.

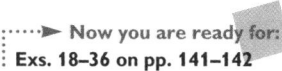

1 Write the mean of her five test scores in terms of x.
$$\frac{\text{sum of five scores}}{5} = \frac{87 + 86 + 89 + 88 + x}{5} = \frac{350 + x}{5}$$

2 Write an equation.
$$\frac{350 + x}{5} = 90 \quad \longleftarrow \text{ The mean is 90.}$$

3 Solve the equation.
$$5 \cdot \frac{350 + x}{5} = 5 \cdot 90 \quad \longleftarrow \text{ Multiply both sides by 5.}$$
$$350 + x = 450$$
$$350 + x - 350 = 450 - 350 \quad \longleftarrow \text{ Subtract 350 from both sides.}$$
$$x = 100 \quad \longleftarrow \text{ Lin's mean test score will be 90 if her last test score is 100.}$$

Now you are ready for:
Exs. 18–36 on pp. 141–142

Look Back

How are the ideas of mean, median, and mode alike? How are they different? Can you think of everyday uses of these words that can help you remember which is which?

Answers to Look Back

Answers may vary. An example is given. They are alike because each number describes what is typical for a data set. They are different because each is determined by a different method and each may yield a different value. The *mean temperature* of a city is the average of the temperatures over a period of years. The *median strip* is a line down the middle of a road, dividing it into two equal parts. A *fashion mode* is a popular clothing style worn by a large number of people.

3-2 Exercises and Problems

1. **Writing** Explain how to find the median of 23, 15, 12, 17, 43, and 56.

2. **Reading** Change the data set in Sample 1 so that it has more than one mode.

3. Is it possible for a data set to have no mode? Explain.

Find the mean, the median, and the mode(s) of each data set. What number(s) do you think is (are) most typical of the set? Give a reason for your choice.

4. Times of swimming trials (in seconds): 66, 58.9, 61.1, 69, 53.5, 64.3, 53.5, 51.1, 55.2, 54.8

5. Attendance at school assemblies: 143, 125, 122, 144, 146, 121, 123, 122, 126, 146, 144, 122, 153, 144

6.

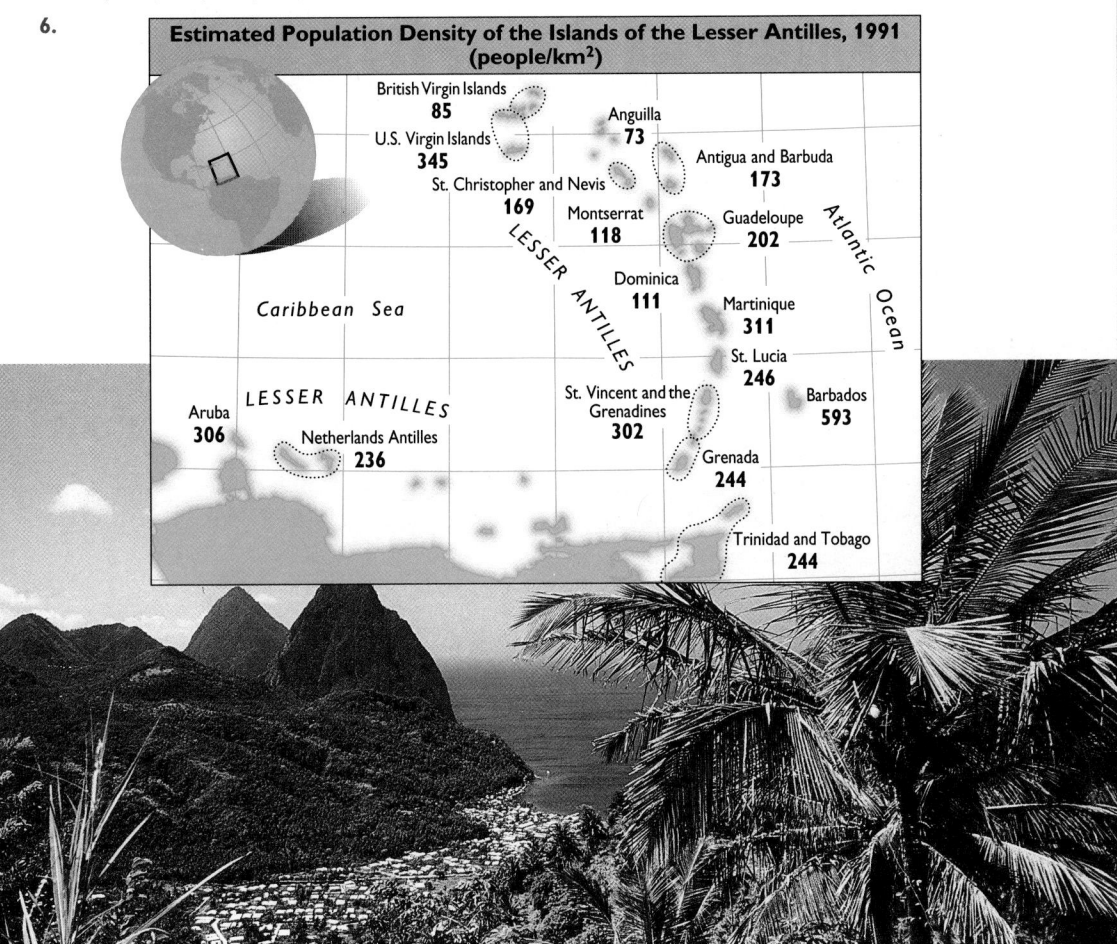

Estimated Population Density of the Islands of the Lesser Antilles, 1991 (people/km²)

British Virgin Islands **85**
Anguilla **73**
U.S. Virgin Islands **345**
Antigua and Barbuda **173**
St. Christopher and Nevis **169**
Montserrat **118**
Guadeloupe **202**
Dominica **111**
Martinique **311**
St. Lucia **246**
St. Vincent and the Grenadines **302**
Barbados **593**
Aruba **306**
Netherlands Antilles **236**
Grenada **244**
Trinidad and Tobago **244**

Caribbean Sea
LESSER ANTILLES
Atlantic Ocean

7. Find the outlier(s), if any, for each of the data sets in Exercises 4–6.

8. Find the range of each of the data sets in Exercises 4–6.

APPLYING

Suggested Assignment

Standard 1–10a, 11–16, 18–23, 27, 29–36

Extended 1–10a, 11–16, 18–27, 29–36

Integrating the Strands

Algebra Exs. 17–23, 31–34

Statistics and Probability Exs. 1–17, 21–30, 35, 36

Discrete Mathematics Exs. 1–17, 21–30, 35, 36

Logic and Language Exs. 9, 14, 16, 17, 27, 29

Multicultural Note

Students may be interested in the data shown for Ex. 6. You might explain that most of the islands of the Caribbean were formed in one of two ways: through the buildup of molten lava and ash as a result of many volcanic eruptions occurring over a long period of time, or through the buildup of limestone deposits from coral reefs under the ocean. Volcanic deposits make soil very rich and excellent for farming, and raw materials can be mined for building materials or precious metal. Agriculture is the most important way of life in the Caribbean. Sugar cane, bananas, and citrus fruits are major crops that grow well in the tropical climate. The mountainous geography of the islands makes them well-suited for coffee beans, which grow best on slopes, and tobacco, which grows in the valleys.

Answers to Exercises and Problems

1. Methods may vary. An example is given. Arrange the values in order from smallest to largest: 12, 15, 17, 23, 43, 56. There are 6 numbers in the list, so the median is the average of the third and fourth numbers, 17 and 23. The median is 20.

2. Answers may vary. An example is given. Add a bread or cracker that has 0.9 g of fat per serving. The data set will then have two modes, 1.1 and 0.9.

3. Yes, if each entry appears the same number of times.

4–6. Choices for most typical numbers and reasons may vary. Examples are given.

4. mean: about 58.7; median: 57.05; mode: 53.5; The mean and the median are most typical since they are in the middle of the data.

5. mean: about 134.4; median: 134.5; modes: 122, 144; The modes are most typical since they represent 6 of the 14 data values. The mean and median are 8 units away from the closest data value.

6. mean: about 234.9; median: 240; mode: 244; The mean, the median, and the mode are typical since they are all close to data values.

7. Answers may vary. Examples are given. There are no outliers in Exercises 4 and 5; in Exercise 6 the outlier is 593.

8. In Exercise 4, the range is 17.9; in Exercise 5, the range is 32; in Exercise 6, the range is 520.

9. The sizes of 14 shirts sold at a souvenir stand are XS, S, M, S, XL, M, XL, XL, L, XL, S, L, L, and M. Is it possible to find the *mean*, the *median*, or the *mode* of this data? Explain.

10. **Astronomy** An *astronomical unit* is about 93,000,000 mi.

    a. Write this number in scientific notation.

    b. **Research** Find the definition of astronomical unit in a dictionary. Does the definition use a *mean*, a *median*, or a *mode*?

Use these data: Maria's test scores are 78, 25, 82, 96, 91, and 82.

11. Find the mean of Maria's test scores.

12. Use the grading scale. What grade will Maria receive if her grade is the mean of her scores? if it is the median? the mode?

13. a. Give a reason to use the mean for Maria's grade.

    b. Give a reason to use the median for Maria's grade.

    c. Do you think the *mean* or the *median* is more typical of Maria's original scores? Why?

14. a. Which one of Maria's scores is an outlier?

    b. Suppose Maria's teacher decides to drop each student's lowest grade. What effect will this have on the mean of Maria's scores? on the median? on the mode?

Salaries For Exercises 15–17, use the salary data.

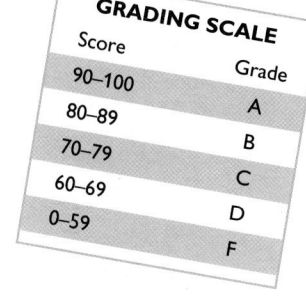

GRADING SCALE

Score	Grade
90–100	A
80–89	B
70–79	C
60–69	D
0–59	F

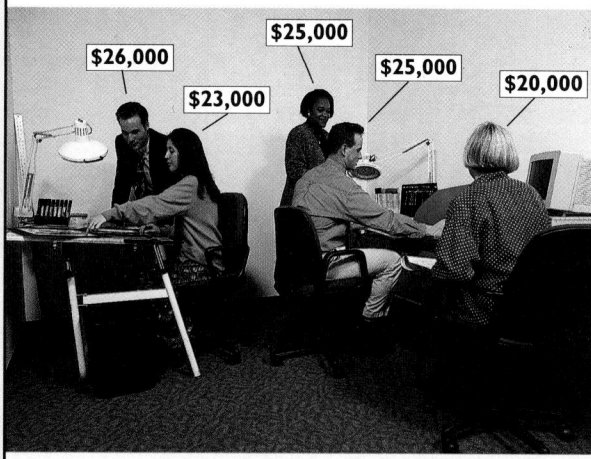

AD PROS

LUNCH FARE

15. a. Find the mean, the median, and the mode(s) of the salaries at Ad Pros and of the salaries at Lunch Fare.

    b. Which company would you think has higher salaries if you knew only the means of the salaries? if you knew only the medians? the modes?

16. **Open-ended** If you were applying for a job at Lunch Fare, would it be most helpful to know the *mean*, the *median*, or the *mode* of the salaries? Explain your choice.

140   **Unit 3** Representing Data

140

17. TECHNOLOGY Use spreadsheet software to create the matrix shown below. Use the *copy* or *fill down* command.

Alternative Approach Create the matrix with pencil and paper.

a. Create a matrix that displays the Ad Pros salaries and the results of adding $1000 and $5000 to each Ad Pros salary. Show the mean, the median, and the range of each salary data set.

Enter the salaries from smallest to largest.

In column C enter formulas that add $1000 to the salaries in column B. In column D enter formulas that add $5000 to the salaries in column B.

	A	B	C	D
1		Salary	Salary + 1000	Salary + 5000
2		20,000	= B2 + 1000	
3		23,000		
4		25,000		
5		25,000		
6		26,000		
7	mean			
8	median			
9	range			

Ad Pros Salaries

In each column enter formulas for the mean, the median, and the range of the salaries in that column.

b. What effect does adding $1000 to each salary have on the mean? on the median? on the range?

c. What effect does adding $5000 to each salary have on the mean? on the median? on the range?

d. What effect do you think adding *n* dollars to each salary will have on the mean? on the median? on the range? Test your guesses for at least three more values of *n*.

e. Revise the matrix in part (a) to display the results of *multiplying* each Ad Pros salary by some factor *n* (choose at least five values for *n*). Describe the effect this has on the mean, the median, and the range.

Solve each equation.

18. $\frac{8 + x}{7} = 12$

19. $\frac{a + 15}{3} = -6$

20. $\frac{-9 + 2n}{5} = 11$

Use a problem solving strategy such as *Guess and check*, *Work backward*, or *Use an equation*.

21. Helen bowled six games. Her mean score was 153. What was the total score for the games? How did you find the answer?

22. Dan bowls in a league. His average score for last season was 159. In the first two games of the new season, he scored 138 and 143. What must he score on a third game in order to have an average of 159 for the first three games of the new season?

23. Maggie has a mean of 88 on the first nine tests of the grading period. In order to receive an A, her mean score must be 90 or higher. There is one test remaining in this grading period. Is it possible for Maggie to receive an A? Why or why not?

> **BY THE WAY...**
>
> Ancient Egyptians played a game, similar to bowling, that used pins and balls made of stone. Two other early versions of bowling are *ula maike* (Polynesian) and *bocci* (Italian).

3-2 Mean, Median, and Mode 141

Using Technology

Ex. 17 provides another opportunity for students to use the *Stats!* spreadsheet software. Students need to know how to enter formulas into a spreadsheet.

Integrating the Strands

Ex. 17 integrates the strands of algebra, statistics and probability, discrete mathematics, and logic and language.

Using Technology

Activities 4, 10, 12, and 18 in the *Stats! Activity Book* apply the ideas taught in this section.

For Ex. 29, have three or four students read the same article and then answer the question. This may develop into a lively discussion of the meaning of the word "average."

Working on the Unit Project

For Ex. 35, allow students time to discuss the data they have found for this project. Have them share ideas about how the data can be organized and described.

Practice 20 For use with Section 3-2

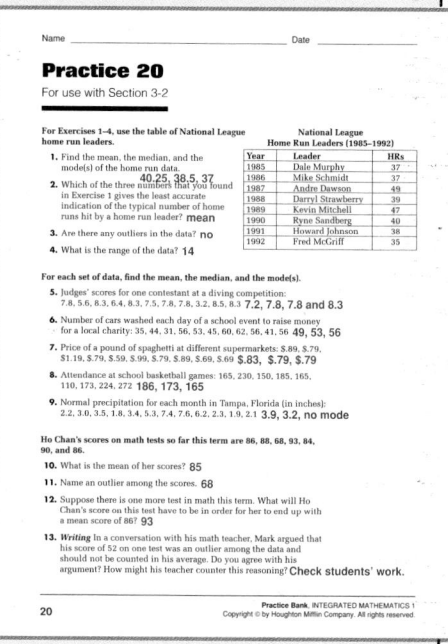

Answers to

Exercises and Problems

24–26. Data sets may vary. Examples are given.

24. 30, 34, 36, 39, 42 and 26, 29, 36, 42, 45

25. 24, 30, 35, 37, 41, 49 and 35, 46, 22, 40, 37

26. 18, 18, 10, 5, 6, 15 and 18, 12, 10, 18, 2

27. No. The median age has nothing to do with the current year, but rather with population changes and life expectancy.

28. a, b. Answers may vary. Examples are given.

28. a. The median is the length in the middle of the group. The mode is the most common length.

Open-ended Find two different data sets that fit each description. Include at least four or five data values.

24. The median is 36. 25. The mean is 36. 26. The mode is 18 and the mean is 12.

27. In 1985, the median age (in years) of the population of the United States was 31.5. Does this imply that in 2005, the median age will be 51.5? Why or why not?

28. **Using Manipulatives** Break five pieces of uncooked spaghetti into three pieces each. Do not try to make all the pieces the same length. Line up the pieces in order from shortest to longest.

   a. Without measuring the pieces, find the median and the mode. Explain how you found them.

   b. Is it possible to find the mean without measuring the pieces? Explain.

Ongoing **ASSESSMENT**

29. **Open-ended** In a newspaper or magazine, find an article that uses the word "average." Do you think that the word means the *mean*, the *median*, the *mode*, or none of these? Give a reason for your answer.

Review **PREVIEW**

30. Would you use a *matrix* or a *graph* to display the fuel-economy ratings for city and highway driving for six different cars? Explain. *(Section 3-1)*

Write each product as a power. *(Section 1-3)*

31. $3 \cdot 3 \cdot 3 \cdot 3 \cdot 3 \cdot 3 \cdot 3$ 32. $x \cdot x \cdot x$

Show each pair of integers on a number line. Then compare the integers using > or <. *(Toolbox Skill 2)*

33. 3 and -5 34. -2 and -4

Working on the Unit Project

35. a. **Research** Find the prices of at least 25 different recordings. Keep track of the type of music (for example, rock, classical, country) and the format of each recording (for example, cassette, CD, LP).

   b. Find the mean, the median, and the mode of all the price data.

   c. Find the mean, the median, and the mode of the prices of each of the different types of music. Which type has the largest mean?

   d. Find the mean, the median, and the mode of the prices of each of the different musical formats. Which format has the largest mean?

36. Suppose you ask 25 people to name their favorite type of music. Could you find the *mean*, the *median*, or the *mode* of their answers? Explain.

b. No; you need numerical values to find a mean, and the only way to get numerical values is to measure the pieces.

29. Answers may vary.

30. Answers may vary. An example is given. A matrix would be a more useful tool for displaying this data, because I could easily use the data to estimate monthly fuel costs.

31. 3^7

32. x^3

33.
$$\overset{\bullet}{\underset{-5\ -4\ -3\ -2\ -1\ \ 0\ \ 1\ \ 2\ \ 3}{\longleftarrow\!\!+\!\!+\!\!+\!\!+\!\!+\!\!+\!\!+\!\!+\!\!\longrightarrow}}$$
$-5 < 3$

34.
$$\overset{\bullet}{\underset{-5\ -4\ -3\ -2\ -1\ \ 0\ \ 1\ \ 2}{\longleftarrow\!\!+\!\!+\!\!+\!\!+\!\!+\!\!+\!\!+\!\!\longrightarrow}}$$
$-4 < -2$

35. a–d. Answers may vary.

36. Answers may vary. An example is given. You can find the mode of the answers—it is the most common answer, but you cannot find the median and the mean because the answers are not numerical and there is no accepted order in which the answers can be listed.

142

3-3

Inequalities and Intervals

Focus
Use inequalities to describe
number line graphs.

The number line graph shows data on world records set since 1900 in the men's long jump.

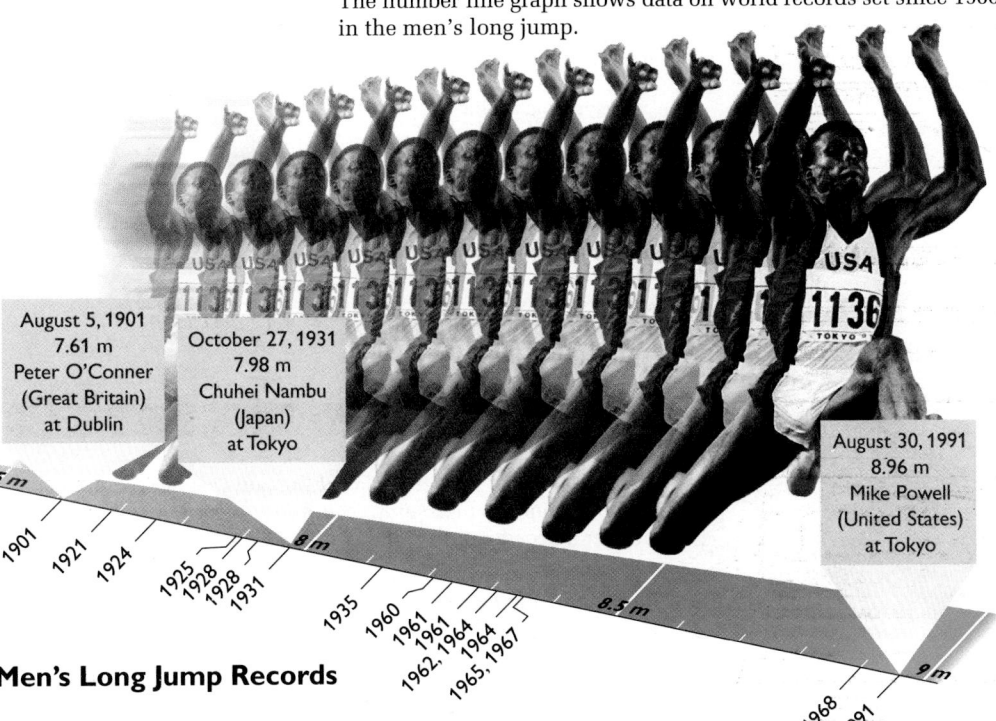

August 5, 1901
7.61 m
Peter O'Conner
(Great Britain)
at Dublin

October 27, 1931
7.98 m
Chuhei Nambu
(Japan)
at Tokyo

August 30, 1991
8.96 m
Mike Powell
(United States)
at Tokyo

7.5 m

8 m

8.5 m

9 m

1901 1921 1924 1925 1928 1928 1931 1935 1960 1961 1961 1962, 1964 1964 1965, 1967 1968 1991

Men's Long Jump Records

BY THE WAY...

The first known long jump record was 7.05 m, set by Chionis of Sparta in 656 B.C.

Talk it Over

1. Does the scale on the number line measure distance or years?

2. Estimate the record distance for 1935.

3. In which years was the record broken more than once? In which years did a record jump match the existing record?

4. Describe the lengths of record long jumps set by men after 1991.

5. Describe the lengths of all long jumps made in competition by men before October 27, 1931.

6. In which year was the *increase* in the record the greatest? the smallest? How can you use the graph to answer these questions?

3-3 Inequalities and Intervals

143

Answers to Talk it Over

1. distance

2. about 8.13 m

3. 1928, 1961, 1964; 1962 and 1964, 1965 and 1967

4. The record jumps would have to be greater than or equal to 8.96 m.

5. Long jumps before this date were less than 7.98 m.

6. greatest increase: 1968; smallest increase: 1965; The biggest interval represents the greatest increase and the smallest interval represents the smallest increase.

PLANNING

Objectives and Strands
See pages 124A and 124B.

Spiral Learning
See page 124B.

Materials List
➤ Ruler
➤ Graph paper

Recommended Pacing
Section 3-3 is a one-day lesson.

Toolbox References
➤ **Toolbox Skill 2:** Comparing Integers

Extra Practice
See pages 624–625.

Warm-Up Exercises
Warm-Up Transparency 3-3

Support Materials
➤ Practice 21
➤ Enrichment 19 in the Activity Bank
➤ Study Guide 3-3
➤ Problem Set 6
➤ Diagram Master 2 in the Explorations Lab Manual
➤ Quiz 3-3
➤ Test 10
➤ Alternative Assessment 2

Talk it Over

Questions 1–6 introduce inequality concepts using distance on a number line to show record jumps.

Additional Samples

S1 Use the number line graph on page 143.

a. Write an inequality to describe men's record long jumps after October 27, 1931.
Let *r* = length of record long jump. *r* is greater than 7.98 m, or *r* > 7.98.

b. Graph the inequality on a number line.
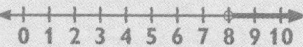

S2 Use the number line graph on page 143.

a. Write an inequality to describe the lengths of the long jumps made from October 27, 1931 through August 30, 1991. Let *j* = the length of a long jump. $7.98 \leq j \leq 8.96$

b. Graph the inequality on a number line.
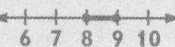

Error Analysis

Students often miss the connection between the open circle on a number line graph and the *greater than* or *less than* symbols, > and <. Stress that an open circle represents a number which is *not* a solution to the inequality.

Student Resources Toolbox
p. 641 *Comparing Integers*

In the 1988 Olympics, Louise Ritter won a gold medal for women's high jump with a jump of 2.03 m.

 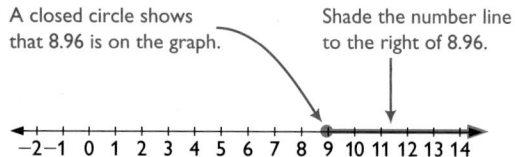
a. **Write an inequality to describe any record long jump set by a man after August 30, 1991.**

b. **Graph the inequality on a number line.**

Sample Response

a. Let *r* = the length of a record long jump set by a man after August 30, 1991.

The graph on page 143 shows that on August 30, 1991, Mike Powell set a long jump record of 8.96 m, so

r is *greater than or equal to* 8.96 m.

In symbols, write $r \geq 8.96$.

b. Graph all values equal to or greater than 8.96.

A closed circle shows that 8.96 is on the graph.

Shade the number line to the right of 8.96.

Sample 2

On July 16, 1961, Iolanda Balas of Romania set her twelfth consecutive women's high jump record. Her record of 1.91 m stood for over a decade, until October 4, 1971.

a. Write an inequality to describe the height of all high jumps made in competition by women before October 4, 1971.

b. Graph the inequality on a number line.

Sample Response

a. Let *j* = the height of a high jump made in competition by a woman before October 4, 1971.

Then *j* is *greater than* 0 m and *j* is *less than or equal to* 1.91 m.

Write $j > 0$ and $j \leq 1.91$ ← You can write these two inequalities

or $0 < j \leq 1.91$. ← as a **combined inequality**.

One way to read the combined inequality is,
"*j* is greater than 0 and less than or equal to 1.91."
You can also read it as,
"0 is less than *j* and *j* is less than or equal to 1.91."

144 **Unit 3** Representing Data

b. The combined inequality $0 < j \le 1.91$ helps you see that j is between 0 and 1.91 on the number line.

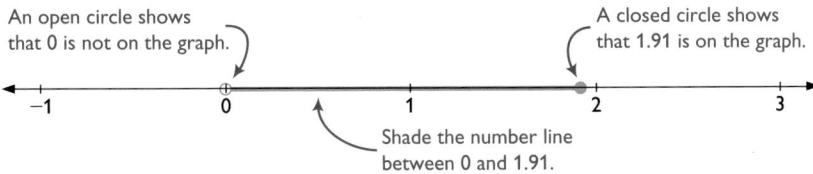

An open circle shows that 0 is not on the graph.

A closed circle shows that 1.91 is on the graph.

Shade the number line between 0 and 1.91.

Talk it Over

7. In the inequality $r \ge 8.96$ in Sample 1, why is the symbol $\ge$ used instead of the symbol $>$? Compare the meanings of the symbols $>$ and $\ge$, and the symbols $<$ and $\le$.

8. In Sample 2, the simple inequality $j > 0$ is "turned around" when it appears in the combined inequality $0 < j \le 1.91$. Explain why the inequalities $j > 0$ and $0 < j$ have the same meaning.

9. Does every point on the graph in Sample 1 represent a possible long jump record? Does every point on the graph in Sample 2 represent a past high jump record? Explain.

Intervals

The graphs of the inequalities in Samples 1 and 2 are *intervals* of the number line. Each statement below refers to an interval, as shown in the graph. Each interval is also described by an inequality.

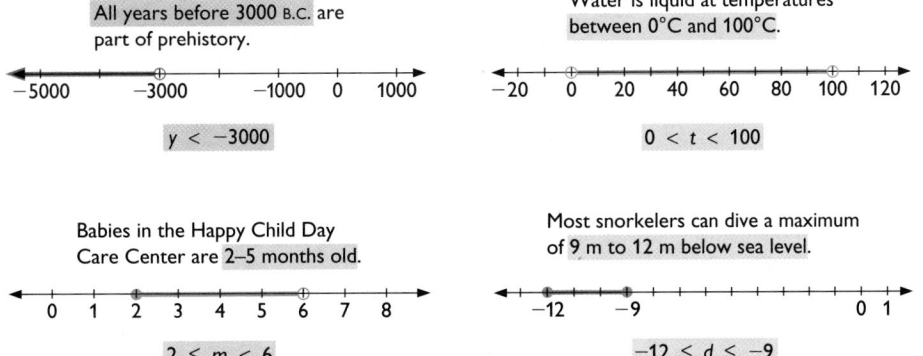

All years before 3000 B.C. are part of prehistory.

$y < -3000$

Water is liquid at temperatures between 0°C and 100°C.

$0 < t < 100$

Babies in the Happy Child Day Care Center are 2–5 months old.

$2 \le m < 6$

Most snorkelers can dive a maximum of 9 m to 12 m below sea level.

$-12 \le d \le -9$

Intervals of a number line are continuous. Data such as the record long jumps are discrete. When discrete data are displayed in a line plot, they lie inside a continuous interval of the number line.

3-3 Inequalities and Intervals   **145**

Answers to Talk it Over

7. to include 8.96 m and show the record could be tied at a future date; The symbol $>$ means "greater than," while the symbol $\ge$ means "greater than or equal to." The symbol $<$ means "less than," while the symbol $\le$ means "less than or equal to."

8. If j is greater than 0, then 0 must be less than j.

9. Answers may vary. An example is given. Yes. No. In Sample 1, any and all numbers greater than or equal to 8.96 do represent a possible long jump record, and this is what the graph shows. However, realistically speaking, the length of a long jump must be limited and cannot go into infinity. In Sample 2,

there are an infinite number of points between 0 and 1.91, yet there are not an infinite number of past high jump records.

Multicultural Note

Over the past 58 years, the men's long jump record has been held by only five individuals. The first, Jesse Owens, set a new standard in 1935 with a leap of 26 ft $8\frac{1}{4}$ in., a record that stood for 25 years. From 1960 through 1968, Ralph Boston (U.S.) and Igor Ter-Ovanesyan (U.S.S.R.) advanced the mark several times, ultimately tying for the record with a jump of 27 ft $4\frac{3}{4}$ in. Then, on October 18, 1968, at the Olympics in Mexico City, Bob Beamon made the most amazing jump in history.

On Beamon's mind was avoiding a fouled jump; he had fouled his first two tries in the qualifying round the day before. Jesse Owens and Ralph Boston, who were both present, remember knowing immediately as Beamon hit the takeoff board that it was an extraordinary jump. Beamon himself said that the jump "felt like a regular jump" until he hit the dirt. The record jump measured 29 ft $2\frac{3}{4}$ in., $21\frac{1}{2}$ in. farther than anyone had ever jumped. He had broken both the 28-ft and the 29-ft barriers in a single jump. His record stood for 23 years. U.S. long jumper Mike Powell broke Beamon's record with a leap of 29 ft $4\frac{1}{4}$ in. on August 30, 1991 in Tokyo.

Limited English Proficiency

Some students acquiring English may have difficulty visualizing the meanings of the terms *continuous* and *discrete* in the context of number lines. It may be helpful for these students if you provide the analogy with how gasoline is sold (it is pumped in continuous quantities) versus how motor oil is sold (in discrete quantities—in one-quart cans, generally).

145

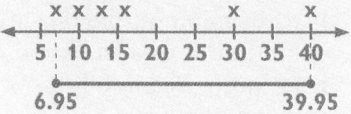

Sample 3

The photograph shows the finishing time (in seconds) of each runner of the 100 m dash in the 1992 Summer Olympic Games held in Barcelona, Spain.

a. Make a line plot of the finishing times.

b. Write an inequality to describe the shortest interval of the number line that contains all the finishing times.

Sample Response

a.

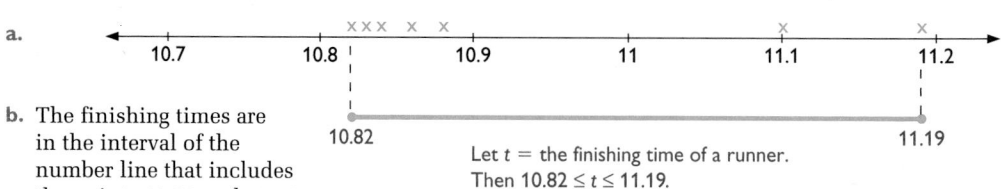

b. The finishing times are in the interval of the number line that includes the points 10.82 and 11.19 and all the points between them.

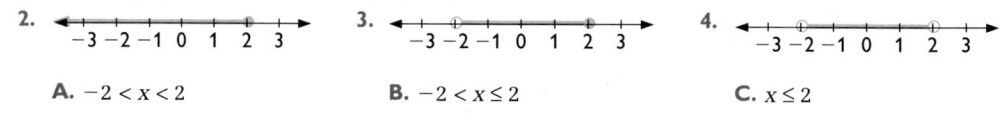

Let $t =$ the finishing time of a runner.
Then $10.82 \leq t \leq 11.19$.

Look Back

Think of and describe a real-world situation that involves data that lie in a interval. Write an inequality to describe the shortest interval that contains the data. Graph the inequality on a number line.

3-3 Exercises and Problems

1. Use the graph on men's long jump records on page 143.

   a. Estimate the men's long jump records set in 1960 and 1967.

   b. Write an inequality to describe the shortest interval that contains all the men's long jump records set from 1960 to 1967.

Match each graph with the correct inequality.

2. _(number line from −3 to 3)_   3. _(number line from −3 to 3)_   4. _(number line from −3 to 3)_

   A. $-2 < x < 2$ **B.** $-2 < x \leq 2$ **C.** $x \leq 2$

Graph each inequality on a number line. Then write each inequality in words.

5. $0 \leq x < 8$ 6. $x > 3\frac{2}{3}$ 7. $-1.6 \leq x \leq 1.1$

146   **Unit 3** Representing Data

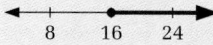
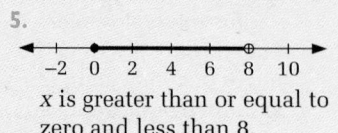
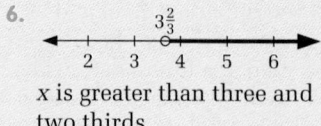
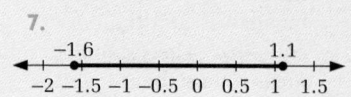

8. Reading Explain why each open or closed circle was used on the four graphs at the bottom of page 145.

Write and graph an inequality to describe the interval referred to in each statement.

9. The surface temperature of sea water may range from −2°C to 30°C.

10. Johanna has at least $80 in her savings account.

11. The measure of an obtuse angle is between 90° and 180°.

Driving Write and graph an inequality to describe the allowable driving speeds as shown by each sign.

12.

SCHOOL
SPEED LIMIT
20
WHEN CHILDREN
ARE PRESENT

13.

MAXIMUM
55
MINIMUM
40

For Exercises 14–16, use the matrix of data on women's javelin records set during the 1980s.

14. a. Write an inequality to describe the length of any record javelin throw made by a woman since 1988.

b. Graph the inequality on a number line.

15. a. Complete the increase column.

b. Make a line plot of the data in the increase column.

c. Write an inequality to describe the shortest interval that contains all the increase data.

d. What is the range of the data in the increase column? How is the range related to the shortest interval of the number line that contains all the data?

16. Career A sports statistician uses sports data to observe trends and make predictions.

Do you think a sports statistician would be more interested in knowing the mean of the data in the *distance* column or of the data in the *increase* column? Explain your answer.

17. Look at the inequalities you wrote in Exercises 9–13. Tell if the quantity each inequality describes is *discrete* or *continuous*.

increase = new record − old record
0.84 = 70.80 − 69.96

Women's Javelin Records		
Year	**Record (meters)**	**Increase (meters)**
1980	69.96	
1980	70.80	0.84
1981	71.88	1.08
1982	72.40	0.52
1982	74.20	1.80
1983	74.76	0.56
1985	75.26	
1985	75.40	
1986	77.44	
1987	78.90	
1988	80.00	

3-3 Inequalities and Intervals

147

15. a. 1985: 0.5; 1985: 0.14; 1986: 2.04; 1987: 1.46; 1988: 1.1

b.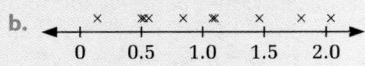

c. 0.14 ≤ x ≤ 2.04

d. 1.9; The length of the interval is equal to the range.

16. Answers may vary. An example is given. A statistician may be more interested in the mean of the data in the increase column to know what the "average" increase is.

17. (9) continuous; (10) discrete; (11) continuous; (12) continuous; (13) continuous

Answers to Exercises and Problems

8. The prehistory graph refers to time prior to 3000 B.C., not during 3000 B.C. Water is liquid between 0°C and 100°C, not at 0°C and 100°C. Babies may be 2 months old, but not 6 months old. Snorkelers can dive at 12 m and 9 m below sea level.

9. −2 ≤ t ≤ 30, t = surface temperature

10. s ≥ 80, s = Johanna's savings

11. 90 < d < 180, d = degree measure of an obtuse angle

12. 0 < l ≤ 20, l = legal school zone speed limit

13. 40 ≤ s ≤ 55, s = speed limit

14. a. j ≥ 80, j = length of record javelin throw in meters

b.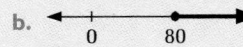

147

For Exercises 18 and 19, do these things.

a. **Open-ended** Describe a real life situation that fits each graph.

b. Write an inequality to describe each graph.

18.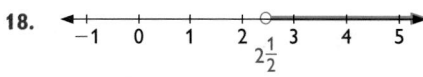

19.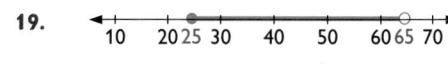

20. a. Write a description of this graph in words and in symbols.

b. Which is easiest for you to understand, the *description in words*, the *description in symbols*, or the *graph*?

connection to SCIENCE

For Exercises 21–23, do these things.

a. **Draw a number line graph to show the temperature interval(s) for each snow crystal shape.**

b. **Write one or more inequalities to describe your graph.**

21. hollow columns

22. sector plates

23. dendrites

Below −8°F
Hollow columns

−8°F to 3°F
Sector plates

21°F to 25°F
Needles

25°F to 32°F
Thin plates

3°F to 10°F
Dendrites

10°F to 14°F
Sector plates

14°F to 21°F
Hollow columns

Ongoing ASSESSMENT

24. **Writing** Write a summary of the main ideas of this section. Which ideas are new to you? How has your understanding of inequalities changed since you first learned about them?

Review PREVIEW

25. Find the mean, median, and mode of this data set. *(Section 3-2)*

Number of nations participating in Winter Olympic Games held from 1924 to 1988:
16, 25, 17, 28, 28, 30, 32, 30, 36, 37, 35, 37, 37, 49, 57

26. Find the square roots and the cube root of 225. *(Section 2-9)*

27. Make a bar graph of the data on home runs. *(Toolbox Skill 22)*

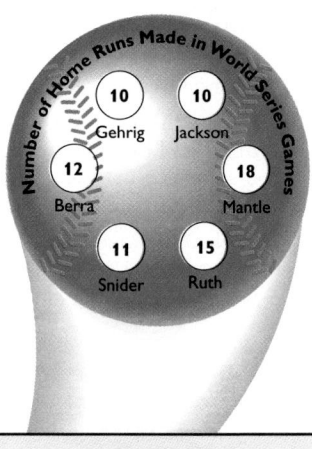

148   **Unit 3** Representing Data

 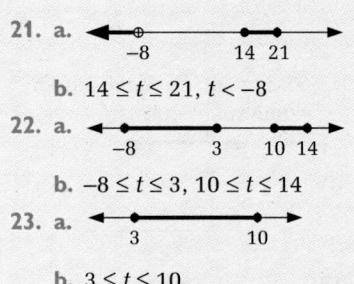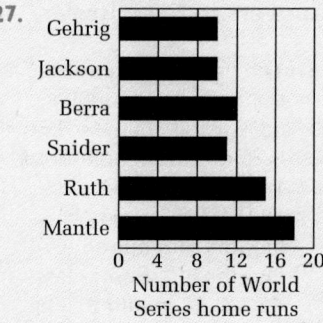
148

28. **Research** Find the prices of seven different recordings. Make a line plot of the prices. Write an inequality to describe the shortest interval of the number line that contains all the prices.

Unit 3 CHECKPOINT

Quiz	Grade
1	74
2	73
3	68
4	75
5	81
6	84
7	84
8	92

1. **Writing** Suppose the table shows your grades. In a letter to your parents, describe how your grades have improved. Include at least one graph or display, and the mean, the median, and the mode of the grades.

2. In 1968–1969, men earned 56% of the bachelor's degrees, and women earned 44%. In 1978–1979, men earned 51% and women earned 49%. In 1988–1989, men earned 47% and women earned 53%.

3-1

   a. Organize the bachelor's degree data in a matrix.

   b. Use your matrix to describe a trend in the data.

   c. Give a reason for using three circle graphs instead of a matrix to present the data.

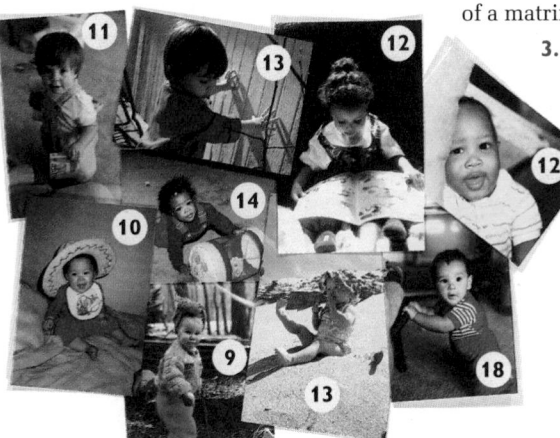

3. Each photograph shows the age (in months) when the baby first walked.

3-2

   a. Find the mean, the median, and the mode of the data set. Which number(s) would you call most typical of the set? Explain your choice.

   b. Find the range and the outlier(s), if any, of the data set.

   c. Suppose there is an error in the data set. The baby labeled "18" walked at 11 months. Is the *mean*, the *median*, or the *mode* most affected by correcting the error?

4. Lisa's mean score in chemistry on the first four tests is 86. There are two tests remaining in the grading period. What scores does she need on the last two tests in order to have a mean of at least 90?

Write and graph an inequality to describe the interval referred to in each statement.

3-3

5. The diameters of Native American tipis range from 8 ft to 40 ft.

6. Water freezes in temperatures at or below 32°F.

3-3 Inequalities and Intervals

149

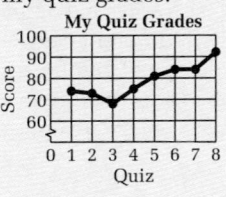

Working on the Unit Project

For Ex. 28, you may want students to share any price catalogs or advertisements they have with one another.

Quick Quiz (3-1 through 3-3)

See page 179.

Practice 21 For use with Section 3-3

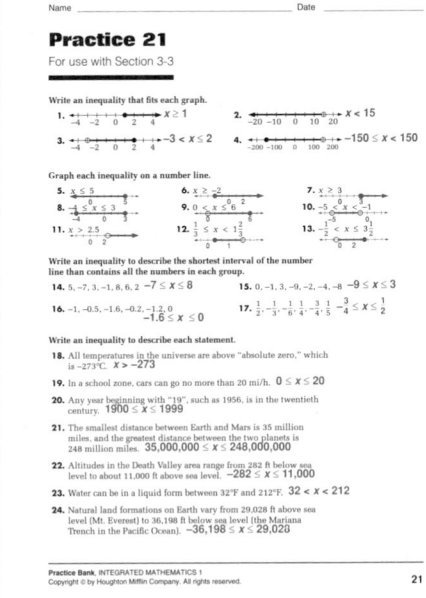

the whole for each gender, over time. One could see the trend quicker than by reading a table.

3. a. mean: about 12.4; median: 12; modes: 12, 13; Choices and explanations may vary. An example is given. Since the data are closely grouped except for one value, the mean, median, and modes are almost equivalent. I think all of them are typical for the data set.

   b. range: 9 months, outlier: 18 months

   c. Answers may vary. Examples are given. (1) The mean is most affected, decreasing from about 12.4 to about 11.7. (2) The mode is most affected because there are now three modes, 11, 12, 13, instead of two modes.

4. Answers may vary. An example is given. She needs a combined score of 196 or more on the two tests. For instance, scores of 98 and 98, 99 and 97, or 96 and 100 will give her a mean of at least 90.

5–6. See answers in back of book.

Answers to Checkpoint

1. Answers may vary. An example is given. Hi! I decided to write to tell you how my grades have improved. Although the median (middle) score is theoretically 78, it was not one of my actual scores. My first four scores were 75 or lower, my last four scores were all higher than 80. Although my mean (average) score is about 79, my mode (most common) score is 84! The line graph shows the upward trend of my quiz grades.

My Quiz Grades

2. a.

	1968–1969	1978–1979	1988–1989
Men	56%	51%	47%
Women	44%	49%	53%

   b. Women's share of awarded bachelor's degrees is increasing, while men's share is decreasing.

   c. A series of circle graphs would best show the varying percentage of

149

PLANNING

Objectives and Strands
See pages 124A and 124B.

Spiral Learning
See page 124B.

Materials List
➤ Graph paper
➤ Index cards or self-stick, removable notes
➤ Graphics calculator or spreadsheet software

Recommended Pacing
Section 3-4 is a two-day lesson.

Day 1
Pages 150–151: Opening paragraph through Talk it Over 8, *Exercises 1–9*

Day 2
Pages 152–153: Stem-and-Leaf Plots through Look Back, *Exercises 10–25*

Extra Practice
See pages 624–625.

Warm-Up Exercises
Warm-Up Transparency 3-4

Support Materials
➤ Practice 22
➤ Enrichment 20 in the Activity Bank
➤ Study Guide 3-4
➤ Problem Set 7
➤ Diagram Master 2 in the Explorations Lab Manual
Overhead Visual 4
➤ McDougal Littell Mathpack software: *Stats!*
➤ *Stats!* Activity Book: Activities 7 and 20
➤ Quiz 3-4

Section

3-4 Histograms and Stem-and-Leaf Plots

Focus
Use histograms, frequency tables, and stem-and-leaf plots to display data.

TV stations, magazines, and newspapers often report the number of viewers or readers in different age groups:

"*Galaxy* magazine is most popular with readers age 20 to 24."

Talk it Over

1. Which interval of the number line would you use to represent "readers age 20 to 24"? Explain your choice.

<!-- number lines: 20–25, 20–24, 19–24 -->

2. Write an inequality to describe the interval you chose.

Histograms

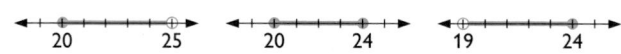

Readers of galaxy Magazine

[Histogram: Number of readers vs. Age (years) with intervals 10–14, 15–19, 20–24, 25–29, 30–34, 35–39]

There are no spaces between the bars of a histogram.

The height of a bar shows the frequency in an interval.

Each age group corresponds to an interval of the number line.

This graph, called a *histogram,* shows how many people in various age groups read *Galaxy* magazine.

A **histogram** is a bar graph that shows how many data items occur in each of one or more intervals. The number of data items in an interval is a **frequency.**

150 **Unit 3** Representing Data

Answers to Talk it Over

1. the graph on the left; It includes persons of age 20, and all persons up to but not including 25.

2. $20 \le a < 25$, $a =$ age

3. about 4500

4. Answers may vary. An example is given. Histograms and bar graphs are both ways to compare data in different categories. Each type of graph uses bars to show amounts. Each category in a his-

togram is an interval, the height of the bar shows the frequency in the interval, and there are no spaces between bars. A bar graph shows a specific value for a specific category, and has a space between bars to separate the categories.

Sample 1

Use the histogram shown on page 150 to make two statements about the ages of *Galaxy* magazine's readers.

Sample Response

Here are two possible statements based on the histogram:

➤ Most of *Galaxy* magazine's readers are in their teens and twenties.

➤ About 500 readers are in their thirties.

Talk it Over

3. Use the histogram to estimate the total number of *Galaxy* readers of ages 10–19 years.

4. Compare the histogram to the bar graphs in Section 3-1. How are they alike? How are they different?

5. Why are there no spaces between the bars of a histogram?

Frequency Tables

Sometimes you want to display the exact number of data items in an interval. This **frequency table** shows the exact number of *Galaxy* magazine readers in each age group used in the histogram on page 150.

Readers of
galaxy
Magazine

Age group	Frequency
10–14	1110
15–19	3398
20–24	4344
25–29	3215
30–34	332
35–39	112

Number of
galaxy
copies sold each week

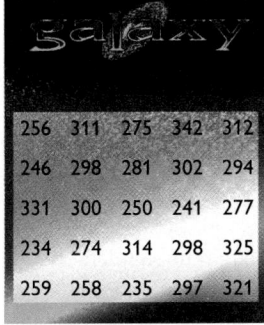

256	311	275	342	312
246	298	281	302	294
331	300	250	241	277
234	274	314	298	325
259	258	235	297	321

⋯➤ Now you are ready for:
Exs. 1–9 on pp. 154–155

Talk it Over

6. Does either the histogram or the frequency table show how many readers of *Galaxy* magazine are 24 years old? Explain.

7. How close was your estimate in question 3?

8. A newsstand owner keeps track of how many copies of *Galaxy* magazine are sold each week. The table at the left shows the data from the last 25 weeks.

   a. What is the range of the data?

   b. What intervals would you use to display the data in a histogram with five bars?

   c. Make a frequency table for the data using the intervals you chose in part (b).

3-4 Histograms and Stem-and-Leaf Plots   **151**

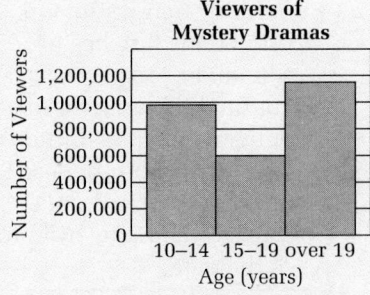
151

Stem-and-Leaf Plots

To display every number in a data set and still organize the data in intervals, you can use a *stem-and-leaf plot*. Separate each piece of data into two parts, a *stem* and a *leaf*.

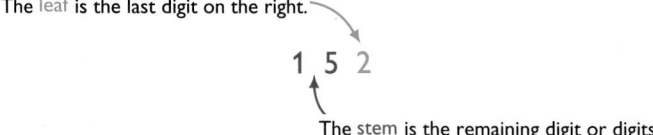

The leaf is the last digit on the right.

1 5 2

The stem is the remaining digit or digits.

Sample 2

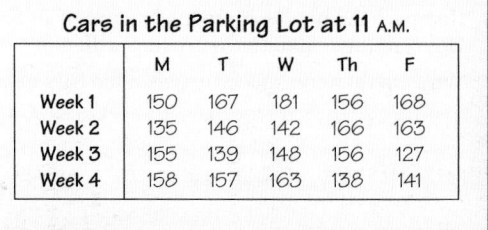

Cars in the Parking Lot at 11 A.M.

	M	T	W	Th	F
Week 1	150	167	181	156	168
Week 2	135	146	142	166	163
Week 3	155	139	148	156	127
Week 4	158	157	163	138	141

In order to decide if the high school needs a bigger parking lot, a town manager's assistant gathered the data shown in the matrix.

Make a stem-and-leaf plot of the data.

Sample Response

1 Use a vertical line to separate the stems from the leaves.

2 To the left of the line, list the stems in order from smallest to largest.

Cars in the Parking Lot at 11 A.M.

```
12 | 7
13 | 5 8 9
14 | 1 2 6 8
15 | 0 5 6 6 7 8
16 | 3 3 6 7 8
17 |
18 | 1
```

3 Write the leaf of each data value to the right of its stem. List the leaves for each stem in order from smallest to largest.

This row contains the numbers 163, 163, 166, 167, and 168.

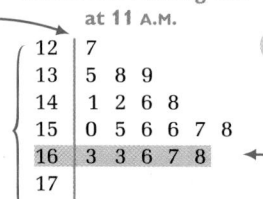

BY THE WAY...

In 1990 there were about 143 million cars registered in the United States.

Talk it Over

9. The number 156 appears twice in the parking-lot data set. Does it appear twice in the stem-and-leaf plot? Explain.

10. Describe how to use a stem-and-leaf plot to find the median of the data. What is the median of the parking-lot data?

11. Suppose the parking lot in Sample 2 contains 180 parking spaces. Should the lot be made bigger? Why or why not?

Answers to Talk it Over

9. Yes. The stem, "15," appears only once, but the leaf, "6," appears twice.

10. Answers may vary. An example is given. Count from the first leaf and the last leaf at the same time until you get to the middle number (median) or numbers (median is average of two numbers). 155.5

11. No. 181 is an outlier, since on most days there are more than enough spaces.

12. Answers may vary. An example is given. They are alike in that they each divide the data into intervals, show the frequency in each interval, and the number of "leaves" in any entry corresponds to the height of the histogram bar. They are different in that the histogram shows only the frequency of values, while the stem-and-leaf plot also shows the actual values.

Making a Histogram from a Stem-and-Leaf Plot

A stem-and-leaf plot contains the information you need to make a histogram.

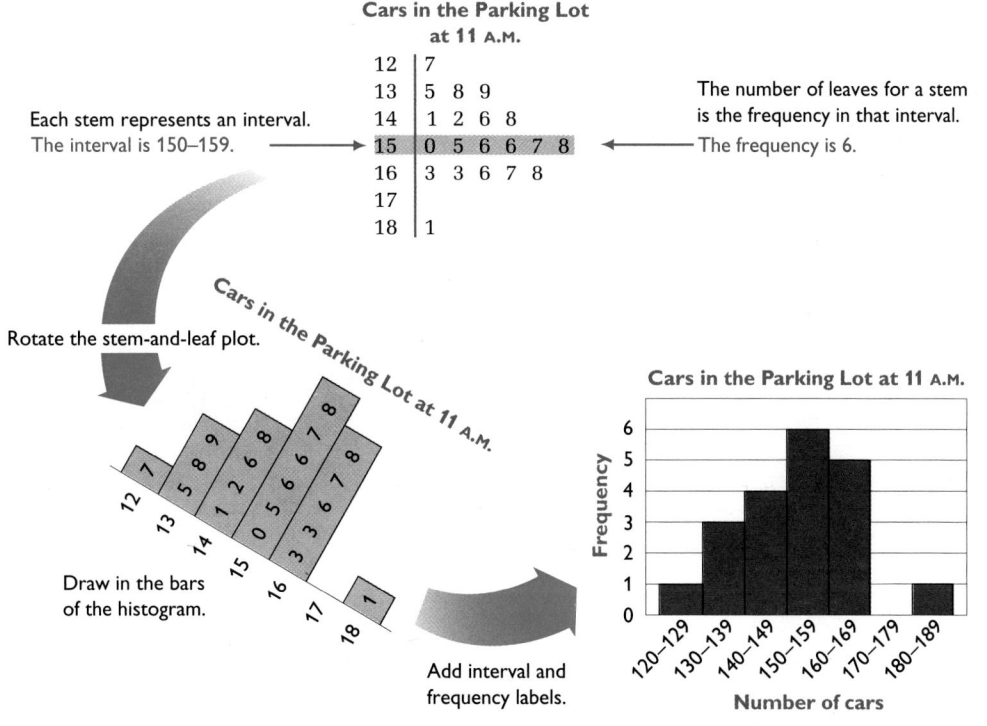

Cars in the Parking Lot at 11 A.M.

12	7
13	5 8 9
14	1 2 6 8
15	0 5 6 6 7 8
16	3 3 6 7 8
17	
18	1

Each stem represents an interval. The interval is 150–159.

The number of leaves for a stem is the frequency in that interval. The frequency is 6.

Rotate the stem-and-leaf plot.

Draw in the bars of the histogram.

Add interval and frequency labels.

Cars in the Parking Lot at 11 A.M.

······► Now you are ready for: Exs. 10–25 on pp. 155–157

Talk it Over

12. Compare the stem-and-leaf plot and the histogram made from the same data. How are they alike? How are they different?

13. Can you make a stem-and-leaf plot from either a frequency table or a histogram? Can you make a histogram from a frequency table? Explain your answers.

Look Back

Does a *histogram*, a *frequency table*, or a *stem-and-leaf plot* give you the most detailed information about a data set? Give a reason why you might *not* want to use a stem-and-leaf plot to display a data set.

3-4 Histograms and Stem-and-Leaf Plots

153

Answers to Talk it Over

13. No. Since histograms and frequency tables do not show actual data values in their intervals, it is not possible to make a stem-and-leaf plot from either. It is possible to make a histogram from a frequency table because it has all the necessary data about frequency and intervals.

Answers to Look Back

Answers may vary. Examples are given. A stem-and-leaf plot gives more detailed information about a data set than a histogram or a frequency table. A stem-and-leaf plot displays the exact data values; histograms and frequency tables show only the frequency of data values in each interval. Also, a frequency table gives the exact frequency for each interval, but you might have to estimate the frequencies from a histogram. You might not want to use a stem-and-leaf plot to display a data set if you have many values in your data set or if you are interested in certain groups within the entire data set.

153

Suggested Assignment

Standard 1–8, 10–14, 16, 18–25

Extended 1–8, 10–14, 16, 18–25

Integrating the Strands

Algebra Exs. 19, 20

Geometry Exs. 8, 21–23

Statistics and Probability Exs. 1–18, 24, 25

Logic and Language Exs. 1, 15, 16, 18, 25

Interdisciplinary Problems

Problem situations in social subjects, such as geography, can be studied and analyzed by using graphs. Exs. 2–6 ask questions about the populations in the four main regions of the United States. Graphs also could be used to illustrate the median income of the various age groups in each region, how the populations of the regions changed over time, or what kinds of industries are located in each region. You might wish to involve students in a brief discussion of some other uses of graphs in connection with geography.

3-4 **Exercises and Problems**

1. **Reading** Look at the *Galaxy* magazine readers histogram on page 150.

   a. How would the histogram change if the intervals 10–19, 20–29, and 30–39 were used instead of the intervals shown?

   b. Give a reason why the publisher of *Galaxy* magazine might want to display the data with the intervals in part (a).

connection to **GEOGRAPHY**

Use the United States population histograms.

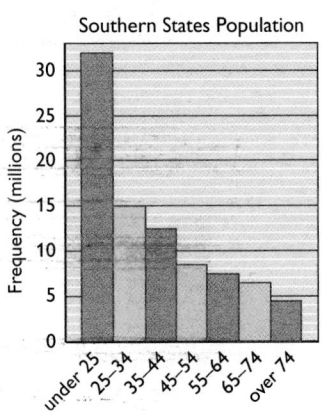

Note: Alaska is not drawn to scale.

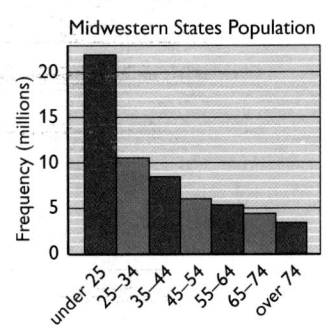

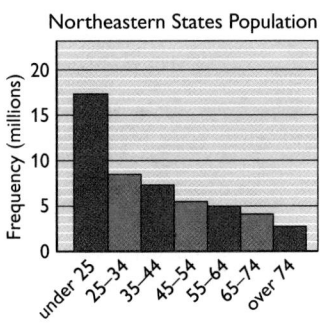

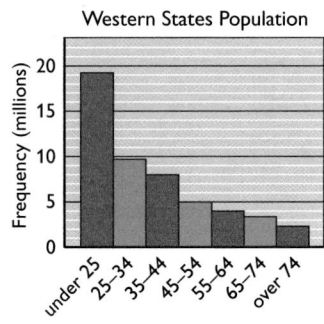

2. Make two statements about the populations in all four regions.

3. Why do you think that in each region, "under 25" is the age group with the greatest frequency?

4. Make a statement about the population of your state's region as compared to the populations of the other regions.

5. Estimate the total number of people living in your region.

6. About what percent of the population of your region is 35–64 years old?

Answers to Exercises and Problems

1. a. It would have 3 bars instead of 6. The 10–19 bar would have 4508 readers. The 20–29 bar would have 7559 readers. The 30–39 bar would have 444 readers.

   b. Answers may vary. An example is given. He may want a larger age spread for recording his data.

2. Answers may vary. An example is given. In each region, the "under 25" population group is the largest, and the "over 74" group is the smallest. The Southern States region appears to have the greatest population in each age group shown and to have the greatest total population.

3. Answers may vary. An example is given. This group covers the largest interval, 25 years, except for the "over 74" interval. It is expected that there are more people living that are under 25 than there are over 74.

4. Answers may vary by region. An example is given. The Northeast region has the smallest area and population.

5. Estimates may vary by region. Examples are given. Northeast: about 50 million; West: about 50 million; Midwest: about 60 million; South: about 85 million

6. Answers may vary by region. Examples are given. Northeast: about 35%; West: about 33%; Midwest: about 33%; South: about 33%

7. Sales The manager of a school store collected data on the amount of money spent by customers during one day. Make a histogram of the data using the intervals shown in the frequency table.

Amount (dollars)	Number of transactions
0.01 – 5.00	102
5.01 – 10.00	76
10.01 – 15.00	30
15.01 – 20.00	6
20.01 – 25.00	15
over 25	11

8. a. On graph paper, draw all the rectangles with perimeter 60 and with sides whose lengths are integers.

b. Find the area of each rectangle.

c. Make a histogram of the areas. Use intervals 0–50, 51–100, and so on.

d. Use the histogram you made in part (c) to make two statements about the rectangles.

9. Using Manipulatives Use index cards or self-stick, removable notes.

a. Research Ask at least 20 people how many magazines each one has read this month. Write each response on a separate card or note.

b. Group the cards or notes into intervals.

c. Make a histogram of the data you collected.

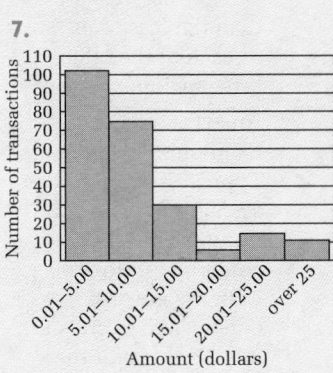

Draw a pair of axes on a large piece of paper. On the horizontal axis, write the intervals you chose in part (b).

Then put each card or note in place above the interval in which it belongs. Do not overlap the notes.

d. What is the frequency of the data in each of your intervals?

e. Writing Does your histogram give more detailed information than the Readers of *Galaxy* Magazine histogram on page 150? Explain.

10. The parking lot described in Sample 2 is also used at night. For four weeks, cars are counted at 8 P.M.

a. Make a stem-and-leaf plot of the data.

b. Writing Describe two differences between the data for 8 P.M. and the data for 11 A.M. based on the stem-and-leaf plots.

Cars in the Parking Lot at 8 P.M.

	M	T	W	Th	F
Week 1	12	14	36	35	14
Week 2	45	15	17	13	49
Week 3	51	53	146	149	35
Week 4	28	135	146	24	26

3-4 Histograms and Stem-and-Leaf Plots

155

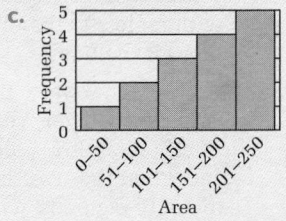

Mathematical Procedures

Exs. 11–17 allow students to practice many of the procedures involved with data analysis. Students read data, make stem-and-leaf plots and histograms, make predictions based on the available data, and then back up their predictions with reasons based on the known facts. After creating some stem-and-leaf plots and histograms by using paper and pencil in Exs. 11–16, Ex. 17 points out how technology can be used to aid in data analysis.

11. a.

Average Number of Days of Precipitation per Year

```
 3 | 5 6
 4 |
 5 | 1
 6 | 1
 7 |
 8 | 2 6 9
 9 | 0 0 6 7 8 9
10 | 1 4 4 4 7 9
11 | 2 3 4 5 5 6 7 9
12 | 2 4 4 5 5 5 6 6 7 8 9
13 | 4 4 5
14 |
15 | 2 4 4 6 6
16 | 9
17 |
18 |
19 |
20 | 9
```

b. 115.5 days

12. Los Angeles, CA; Phoenix, AZ; Reno, NV; Albuquerque, NM These are all western cities.

13. Buffalo, NY; Mount Washington, NH. These are both northeast cities.

14. Predictions may vary. An example is given. The western states appear to have the least precipitation.

15. a.
```
  West           Northeast
 3 | 5 6         11 | 2 7
 4 |             12 | 4 5 6 7 8
 5 | 1           13 | 4
 6 | 1           14 |
 7 |             15 | 4 4
 8 | 9           16 | 9
 9 | 0 0 9       17 |
10 | 1           18 |
11 | 3           19 |
12 |             20 | 9
13 |
14 |
15 | 2 6
```

connection to **S C I E N C E**

For Exercises 11–17, use the table on precipitation (rain, snow, sleet, and hail). The regional color-code matches the one used in the map on page 154.

AVERAGE NUMBER OF DAYS OF PRECIPITATION PER YEAR

West   Midwest   South   Northeast

City	Days	City	Days	City	Days
Albany, NY	134	Detroit, MI	135	Norfolk, VA	115
Albuquerque, NM	61	Duluth, MN	134	Oklahoma, OK	82
Atlanta, GA	115	Great Falls, MT	101	Omaha, NE	98
Atlantic City, NJ	112	Hartford, CT	127	Philadelphia, PA	117
Bismarck, ND	96	Houston, TX	104	Phoenix, AZ	36
Boise, ID	90	Indianapolis, IN	125	Pittsburgh, PA	154
Boston, MA	126	Kansas City, MO	104	Portland, ME	128
Buffalo, NY	169	Little Rock, AR	104	Portland, OR	152
Burlington, VT	154	Los Angeles, CA	35	Providence, RI	124
Cheyenne, WY	99	Louisville, KY	124	Reno, NV	51
Chicago, IL	126	Miami, FL	129	Salt Lake City, UT	90
Cleveland, OH	156	Milwaukee, WI	125	Seattle-Tacoma, WA	156
Columbia, SC	109	Mobile, AL	122	Sioux Falls, SD	97
Concord, NH	125	Mt. Washington, NH	209	Spokane, WA	113
Denver, CO	89	Nashville, TN	119	Wichita, KS	86
Des Moines, IA	107	New Orleans, LA	114	Wilmington, DE	116

11. a. Make one stem-and-leaf plot of all the data.

   b. What is the median number of days of precipitation in one year?

12. Which four cities have the fewest days of precipitation? What do these cities have in common?

13. Which cities have more than 160 days of precipitation per year? What do these cities have in common?

14. Predict which of the four regions has the least number of days of precipitation per year.

15. **Group Activity** Work in a group of four students.

   a. Make four stem-and-leaf plots, one for each region.

   b. Describe three ways in which the stem-and-leaf plot for the southern cities differs from the stem-and-leaf plot for the western cities.

   c. **Writing** Do the stem-and-leaf plots your group made in part (a) support the prediction you made in Exercise 14? Explain your answer.

   d. Do you think the precipitation in Norfolk is typical or unusual for its region? How is this shown on the stem-and-leaf plot?

BY THE WAY...

In 1990 the world record for the greatest rainfall in one year was still held by Cherrapunji, India, where almost 87 ft of rain fell from August 1, 1860 to July 31, 1861.

```
Midwest     South
 8 | 6        8 | 2
 9 | 6 7 8    9 |
10 | 4 7     10 | 4 4 9
11 |         11 | 4 5 5 6 9
12 | 5 5 6   12 | 2 4 9
13 | 4 5
14 |
15 | 6
```

b. Answers may vary. Examples are given. The stem-and-leaf plot for the West has a greater range (from 35 to 156) than the plot for the South (from 82 to 129);

most of the data values for the West are under 100, but most of the data values for the South are over 100; the plot for the West has more stems with no leaves than the plot for the South.

c. Answers may vary. Examples are given. Yes. The stem-and-leaf plots show that the West is the only region in which most of the data values are less than 100 and

that four of the data values for the West are less than the least data value in each of the other three regions.

d. Answers may vary. Examples are given. Typical. Norfolk's average annual precipitation, 115 days, is contained in the row with the greatest frequency and is one of the mode values for the South region data.

16. **a.** Make a histogram from the stem-and-leaf plot for the midwestern cities.

    b. Find the median number of days of precipitation for the midwestern cities.

    c. **Writing** Is it possible to find the median from the histogram? Explain.

17. TECHNOLOGY Use a graphics calculator or spreadsheet software to make a histogram for the data for the northeastern cities. What intervals are shown on the histogram?

Ongoing **ASSESSMENT**

18. **Writing** Compare the stem-and-leaf plots you made in Exercise 15 with the four United States population histograms on page 154. Write about the possible relationship, if any, between the amount of precipitation in a region and its population. Use the displays to support your position.

Review **PREVIEW**

Write and graph an inequality to describe the interval referred to in each statement. *(Section 3-3)*

19. United States citizens must be at least 18 years old in order to vote.

20. The average global temperature may rise 2°F to 9°F by 2050.

Draw an angle with each measure. *(Toolbox Skill 16)*

21. 130°   22. 40°   23. 90°

24. The scoreboard shows the total points earned in one season by each member of the North High wrestling team. Find the median, the range, and the outliers of the data. *(Section 3-2)*

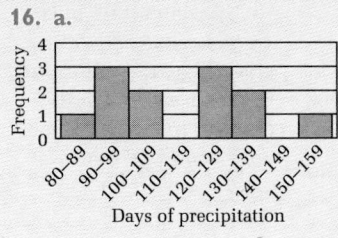

NORTH HIGH WRESTLING TEAM

202 97 103 109 156 82 148 194 163

187 279 224 181 194 199 187 170 187

 Working on the Unit Project

25. **a.** **Research** Make a list of all the radio stations that broadcast in your area. Record each station's call letters, its number, whether it is AM or FM, and its category (such as country, top 40, classical, oldies, gospel, news/talk, and so on).

    b. Use several types of data displays to look at different aspects of the information you collected in part (a). How can you use your displays to help you plan what to sell in your music store?

3-4 Histograms and Stem-and-Leaf Plots   **157**

Practice 22 For use with Section 3-4

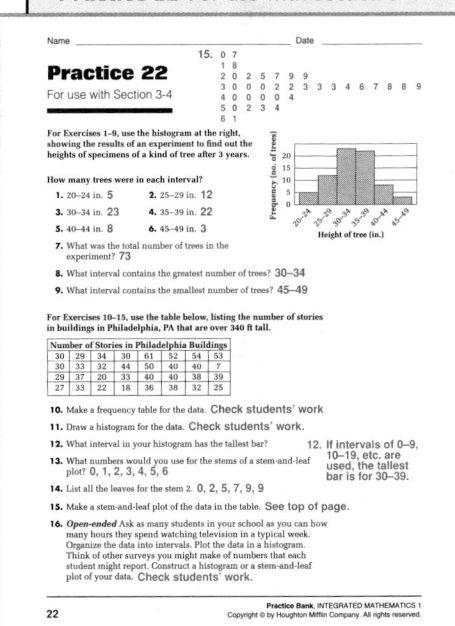

Answers to Exercises and Problems

16. **a.**

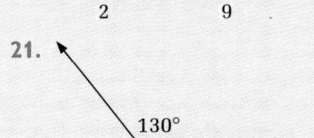

Days of precipitation

    b. 116 days

c. No. The histogram shows how many values are in a given interval but not the exact values of the data.

17. Answers may vary.

18. Answers may vary. An example is given. The least populous region has the most precipitation.

19. $v \geq 18$, v is voting age.

20. $2 \leq a \leq 9$, a is the average increase in global temperature from now until 2050.

21.

22.

40°

23.

24. median: 184; range: 197; outlier: 279

25. a, b. Answers may vary.

Objectives and Strands

See pages 124A and 124B.

Spiral Learning

See page 124B.

Materials List

➤ Ruler

Recommended Pacing

Section 3-5 is a one-day lesson.

Extra Practice

See pages 624–625.

Warm-Up Exercises

💡 Warm-Up Transparency 3-5

Support Materials

➤ Practice 23

➤ Enrichment 21 in the Activity Bank

➤ Study Guide 3-5

➤ Problem Set 7

💡 Overhead Visual 4

➤ McDougal Littell Mathpack software: *Stats!*

➤ Quiz 3-5

➤ Alternative Assessment 3

Section **3-5**

┈┈┈**Focus**

Draw a box-and-whisker plot and use it to compare data sets.

Box-and-Whisker Plots

GOING TO

EX·TRE·MES

Test Scores Integrated Math Class			
68	78	93	79
72	88	44	79
75	92	71	89
75	88	73	80
66	100	75	82

After giving a test to his Integrated Math students, LaSalle Johnson used a *box-and-whisker plot* to see how their scores are grouped.

A **box-and-whisker plot** shows the median and the range of a data set. You can use box-and-whisker plots to compare how data are grouped in two sets.

Sample 1

Make a box-and-whisker plot of the test-score data shown above.

Sample Response

Step 1 Write the scores in order from lowest to highest.

Step 2 Find the median, the *extremes,* and the *quartiles* of the ordered data set.

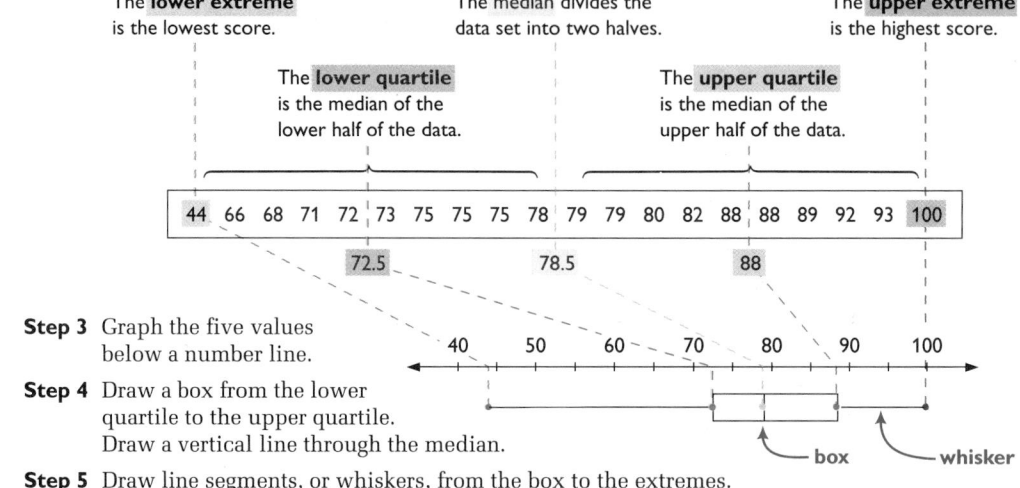

The **lower extreme** is the lowest score.

The median divides the data set into two halves.

The **upper extreme** is the highest score.

The **lower quartile** is the median of the lower half of the data.

The **upper quartile** is the median of the upper half of the data.

44 66 68 71 72 | 73 75 75 75 78 | 79 79 80 82 88 | 88 89 92 93 100

72.5   78.5   88

Step 3 Graph the five values below a number line.

Step 4 Draw a box from the lower quartile to the upper quartile. Draw a vertical line through the median.

box   whisker

Step 5 Draw line segments, or whiskers, from the box to the extremes.

158   **Unit 3** Representing Data

Answers to Talk it Over ┈┈┈┈┈┈┈┈┈┈┈┈┈┈┈┈┈┈┈┈┈┈┈┈┈

1. a. 28.5; 12

   b. These ranges are shown by the lengths of the whiskers in the plot.

2. a. 25%; 25%

   b. No. Explanations may vary. An example is given. The percents will remain about the same because the lower and upper quartiles are the medians of the bottom 50% and top 50%, so it does not matter how many scores are in the data set.

3. a. 50%

   b. These scores are where the box is drawn in the plot.

   c. No. Explanations may vary. An example is given. The quartiles and the median divide the data set into 4 parts with each part representing about 25% of the data, so the middle two parts represent about 50% for any number of scores in the data set.

Talk it Over

1. **a.** What is the difference between the lower quartile and the lower extreme? the upper extreme and the upper quartile?

    b. How are these ranges shown on the box-and-whisker plot?

2. **a.** About what percent of the scores are greater than the upper quartile? less than the lower quartile?

    b. Will the percents you found in part (a) change if the number of scores in the data set changes? Why or why not?

3. **a.** About what percent of the scores are greater than the lower quartile *and* less than the upper quartile?

    b. How are these scores shown on the box-and-whisker plot?

    c. Will the percent you found in part (a) change if the number of scores in the data set changes? Why or why not?

4. Why do you think the term *quartile* is used to describe the numbers that determine one end of each whisker?

Sample 2

Under one number line, LaSalle Johnson made box-and-whisker plots for the test scores in his two algebra classes. Write at least three statements comparing the scores of the two classes.

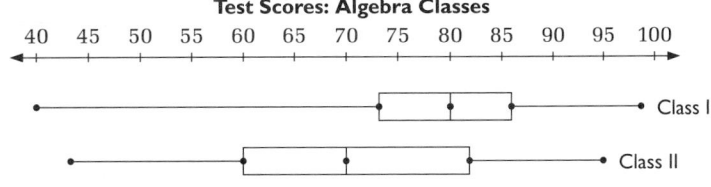

Sample Response

Here are three possible statements:

Class I has a greater median. This means that the top half of Class I generally scored higher than the top half of Class II.

Class I has a greater range, but the middle 50% of Class I's scores are closer together.

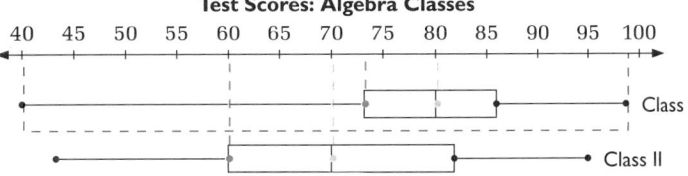

Class I's lower quartile is greater than Class II's median. This means that about 75% of the students in Class I scored higher than half the students in Class II.

3-5 Box-and-Whisker Plots

159

Additional Sample

S1 Make a box-and-whisker plot of the weight data below.

Wrestling Team Weights

98	102	143	154
98	100	132	127
110	123	141	160
128	129	141	145

See answer at bottom of the page.

Talk it Over

Questions 1–4 will help students to further develop understanding of the key ideas of quartile, lower and upper extremes, range, whisker, and box.

Additional Sample

S2 Coach Hahn made a box-and-whisker plot for the previous year's wrestling team and compared it to this year's. Write at least three statements comparing the weights of the two teams. See the box-and-whisker plots at the bottom of the page.

This year's team has a greater median. Both teams have about the same range of weights. All of this year's median, quartile, and extreme points are higher than last year's. This means that the entire team weighs more than last year's team.

Answers to Talk it Over

4. Answers may vary. An example is given. *Quartile* sounds like "quarter" or "quart," which refer to $\frac{1}{4}$ or 25% of a dollar and $\frac{1}{4}$ or 25% of a gallon, respectively. The quartiles help to divide the data into four parts, each part representing 25% of the data.

S1

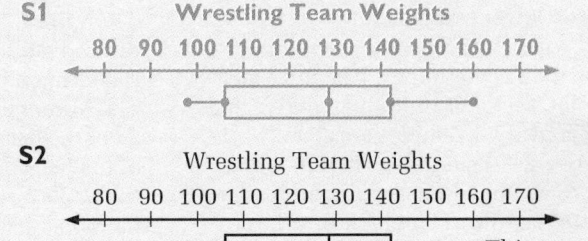

S2

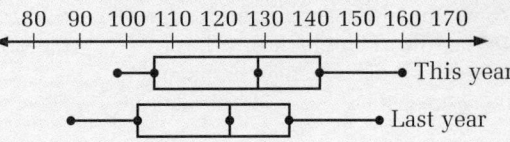

Talk it Over

5. Do you need to know the exact values of the medians, the ranges, and the lower quartiles to make the statements in Sample 2? Explain.

6. Explain how the box-and-whisker plots show that the middle 50% of Class I's scores are closer together than the middle 50% of Class II's scores are.

7. Estimate the lower quartile for each class. Write a statement comparing the classes' scores based on the lower quartiles.

Look Back ◄

What are the five numbers that determine a box-and-whisker plot? How do you find them?

3-5 Exercises and Problems

1. **Reading** The second paragraph on page 158 describes two pieces of information that a box-and-whisker plot shows. Name at least four *other* pieces of information shown by a box-and-whisker plot. In which part of the section are these ideas presented?

Kim's class collected data by asking students the ages of their brothers and sisters.

Use the data from Kim's class.

2. Tell how to find the median of the data.

3. Tell how to find the lower quartile.

4. What are the extremes of the data?

5. **a.** Make a box-and-whisker plot of the data.

 b. About what percent of the ages are greater than or equal to 11?

 c. About what percent of the ages does each whisker represent?

6. Suppose you chose an age at random from Kim's data. Is it more likely that the age is *greater than* the upper quartile or *less than* the upper quartile?

Ages of Brothers and Sisters
(Kim's Class)

17	13	12	12	5
18	11	7	18	13
12	18	14	12	4
19	15	12	11	11
12	11	8	16	9
6	10	13	11	17
11	17			

Jared's class also collected data by asking students the ages of their brothers and sisters.

Use the data from Kim's and Jared's classes.

7. a. Use the plot you made for part (a) of Exercise 5. Add a box-and-whisker plot for the data from Jared's class.

   b. Use your plots to make two statements comparing the data sets.

   c. Give an example of a statement about Jared's data set that is shown more clearly in the plot for his class than in the chart for his class.

   d. **Writing** Claudia says that the plots show that Jared's classmates have brothers and sisters who are generally older than Kim's classmates' brothers and sisters. Do you agree? Why or why not?

8. a. Jared collected four more pieces of data: 5, 24, 18, and 21. Include these in his data set. Find the new extremes, quartiles, and median. Compare them to the original extremes, quartiles, and median.

   b. Suppose Jared drew a box-and-whisker plot for the new data set. How would it differ from the original plot for his class?

9. a. Find the difference between the lower quartile and the lower extreme in Kim's data set.

   b. Find the difference between the lower quartile and the lower extreme in Jared's data set.

   c. Compare the numbers you found in parts (a) and (b). How are they shown on the box-and-whisker plot?

10. a. How many pieces of data are there in Kim's data set?

   b. How many pieces of data are there in Jared's original data set?

   c. Is there a relationship between the size of a data set and the size of a box-and-whisker plot of the data? Explain.

Ages of Brothers and Sisters (Jared's Class)

10	17	21	22
22	24	24	14
22	2	17	18
18	19	5	25
14	23	24	17
21	21	20	20

3-5 Box-and-Whisker Plots 161

Answers to Exercises and Problems

5. a.

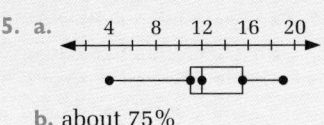

   b. about 75%

   c. about 25%

6. less than the upper quartile

7. a.

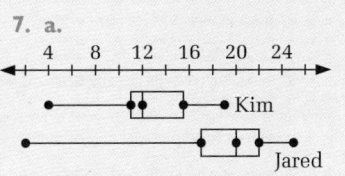

   b. Statements may vary. Examples are given. The plot for Jared's class extends farther in both directions than the plot for Kim's class, so

Jared's class has a wider range of ages than Kim's class. The median of Jared's data is greater than the upper extreme of Kim's data, so about 50% of the brothers and sisters of the students in Jared's class are older than all of the brothers and sisters of the students in Kim's class.

Communication: Writing

Ex. 7 has students use a double box-and-whisker plot to assess a statement made regarding the data presented in the plot. Students should support their reasoning with statements based on the data.

   c. Answers may vary. An example is given. More of the brothers and sisters of students in Jared's class are at least 17 years old than are less than 17 years old.

   d. Answers may vary. An example is given. Yes, agree. The box portion of the plot for Jared's data lies completely to the right of the box portion of the plot for Kim's data. This shows that 75% of the brothers and sisters of Jared's classmates are older than 75% of the brothers and sisters of Kim's classmates. Also, if you choose an age at random from Jared's data, it is more than 50% likely that the age is greater than the upper extreme of Kim's data.

8. a. new extremes: 2 and 25; new quartiles: 17 and 22; new median: 20; the extremes, quartiles, and median are the same for the new data set and the original data set.

   b. Since the extremes, quartiles, and median are the same for the two data sets, the new box-and-whisker plot would not differ from the original plot.

9. a. 7   **b.** 15

   c. Answers may vary. Examples are given. 15 > 7 (or 15 is about twice 7); the left whisker in Jared's plot is longer than (about twice as long as) the left whisker in Kim's plot.

10. a. 32 items

   b. 24 items

   c. No. Explanations may vary. An example is given. Jared has fewer items than Kim, but his box-and-whisker plot is bigger than Kim's plot. The size of a box-and-whisker plot depends more on how closely grouped or how spread out the numbers are that are in the data set than on how many numbers are in the data set.

Car Repairs For Exercises 11–15, use the box-and-whisker plots below.

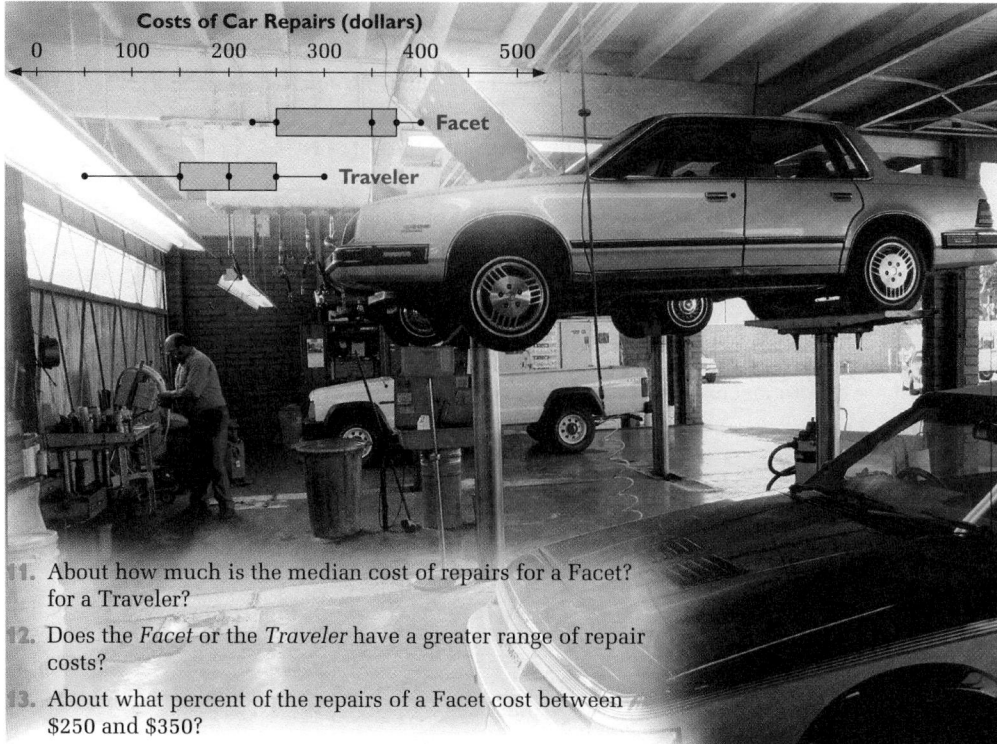

Costs of Car Repairs (dollars)

11. About how much is the median cost of repairs for a Facet? for a Traveler?

12. Does the *Facet* or the *Traveler* have a greater range of repair costs?

13. About what percent of the repairs of a Facet cost between $250 and $350?

14. A report says that more than half the repairs of a Facet cost more than any repair done on a Traveler. Do the box-and-whisker plots support the report? Why or why not?

15. Peggy Quinn's car is a Facet. The cost of repairing her car is about $300. She thinks this is higher than most repair costs for a Facet. Is she right? Explain.

16. **Writing** Explain how the word *quartile* is used in this newspaper report on standardized test scores.

17. **Writing** The lower extreme, lower quartile, median, upper quartile, and upper extreme of a data set are called the *five-number summary* of the data set. Do you think a *five-number summary* or a *box-and-whisker plot* gives more information about a data set? Explain your choice.

18. **a.** Where would the outliers of a data set be located on a box-and-whisker plot?

    b. By looking at a box-and-whisker plot, can you tell whether or not the data set has any outliers? Explain.

19. **Open-ended** Make up a data set for the box-and-whisker plot shown on this graphics calculator screen. Include at least 15 items of data. Tell why your data set fits the plot.

Scores Up 20 Points at Martin L. King, Jr., H. S.

Principal Yolanda Johnson reports that the SAT scores of this year's class are 20 points higher than last year's and 14 points above the national average. In fact, 150 students are in the upper quartile for the nation.

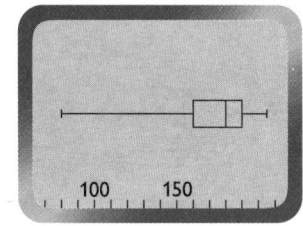

Answers to Exercises and Problems ···

11. about $350; about $200

12. Traveler

13. about 25%

14. Yes. The upper extreme value for the Traveler is less than the median value for the Facet.

15. No. The median cost of repairs for the Facet is $350, so $300 is lower.

16. The word *quartile* is used to explain that 150 students scored in the top 25% of SAT test scores nationwide.

17. Choices and explanations may vary. Examples are given. If exact data are needed, the five-number summary gives more information than the box-and-whisker plot, which shows

approximate positions for the five numbers. The box-and-whisker plot gives a visual description of the data. When two or more box-and-whisker plots are graphed together, they can be used to compare the sets of data.

18. **a.** at the lower and upper extremes

20. Use scores from your own mathematics quizzes and tests as data.

a. Find and record the median and the quartiles of your scores.

b. The next time you get a test or quiz result, add the score to your data set and repeat part (a). Continue until you have added at least three more scores to your data set. Each time, compare your results with your earlier results. Keep notes about any changes you observe.

c. **Writing** Report on your results. Your report should include these things.

➤ a box-and-whisker plot for each data set under the same number line

➤ any changes in the median and the quartiles of your data when you added a score

➤ any differences in finding the median and the quartiles with an odd or even number of scores

Use the Volunteer Network histogram that shows how many members volunteer for each range of hours during one week. *(Section 3-4)*

21. About how many Network members volunteer for 4 or more hours per week?

22. Which interval has the greatest frequency? What does this mean in terms of the data?

23. What else do you need to know in order to make a stem-and-leaf plot for this data set?

Simplify. *(Section 1-5)*

24. $4a + 5a - 3b - 9b$ **25.** $-2x^2 + 5x^2$

26. An oceanographer records the height of the tide over 24 h. Should the oceanographer display the data with a *bar graph* or a *line graph*? *(Section 3-1)*

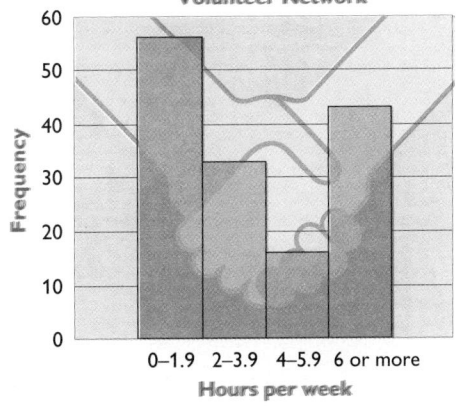

Volunteer Network

Frequency vs *Hours per week*: 0–1.9, 2–3.9, 4–5.9, 6 or more

 Working on the Unit Project

27. **Group Activity** Work in a group of four students. Have each member of the group ask at least eight people how many music recordings they bought in the last year. Keep track of the age of each person asked.

a. Combine all the data and make a box-and-whisker plot.

b. Divide the data into two groups: the answers from people under 20 years old and the answers from people 20 years old and over. Make a box-and-whisker plot for each group.

c. Compare the plots you made in part (b). Which group usually buys more recordings? Which group has a greater range?

3-5 Box-and-Whisker Plots   **163**

Assessment: Task

Ex. 20 provides students with an opportunity to analyze personal data. The exercise allows students to see how data can change over time and allows you to assess how well students have understood box-and-whisker plots and the numbers associated with them.

Practice 23 For use with Section 3-5

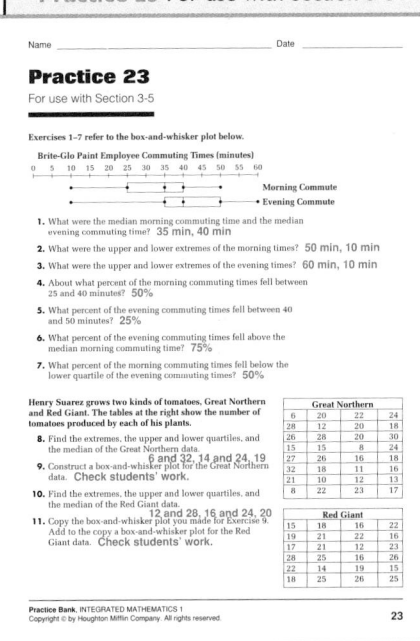

Answers to Exercises and Problems

18. b. No. The lower and upper extremes are not necessarily outliers. Every data set has a highest and lowest value, but not every data set has outliers.

19. Data sets may vary. An example is given. 80, 90, 150, 160, 165, 170, 173, 180, 180, 182, 185, 190, 193, 196, 205; This data set fits

the plot because it has the same range (125), median (180), lower extreme (80), upper extreme (205), lower quartile (160), and upper quartile (190).

20. a–c. Answers may vary.

21. about 59

22. 0–1.9. This means that most volunteers work 0–1.9 hours per week.

23. You need to know the exact amount of hours that each volunteer has worked each week.

24. $9a - 12b$ **25.** $3x^2$

26. Answers may vary. An example is given. A line graph would show specifically the change in tide over a 24-hour period.

27. a–c. Answers may vary.

Section 3-6

Choosing a Data Display

EVERY PICTURE *Tells a* STORY

After helping with hazardous-waste disposal at a recycling center, Rafael did some research on trash. He used four different graphs to display data that he found in an almanac.

Focus
Tell why a data display is a good choice for a data set and choose a good data display for a data set.

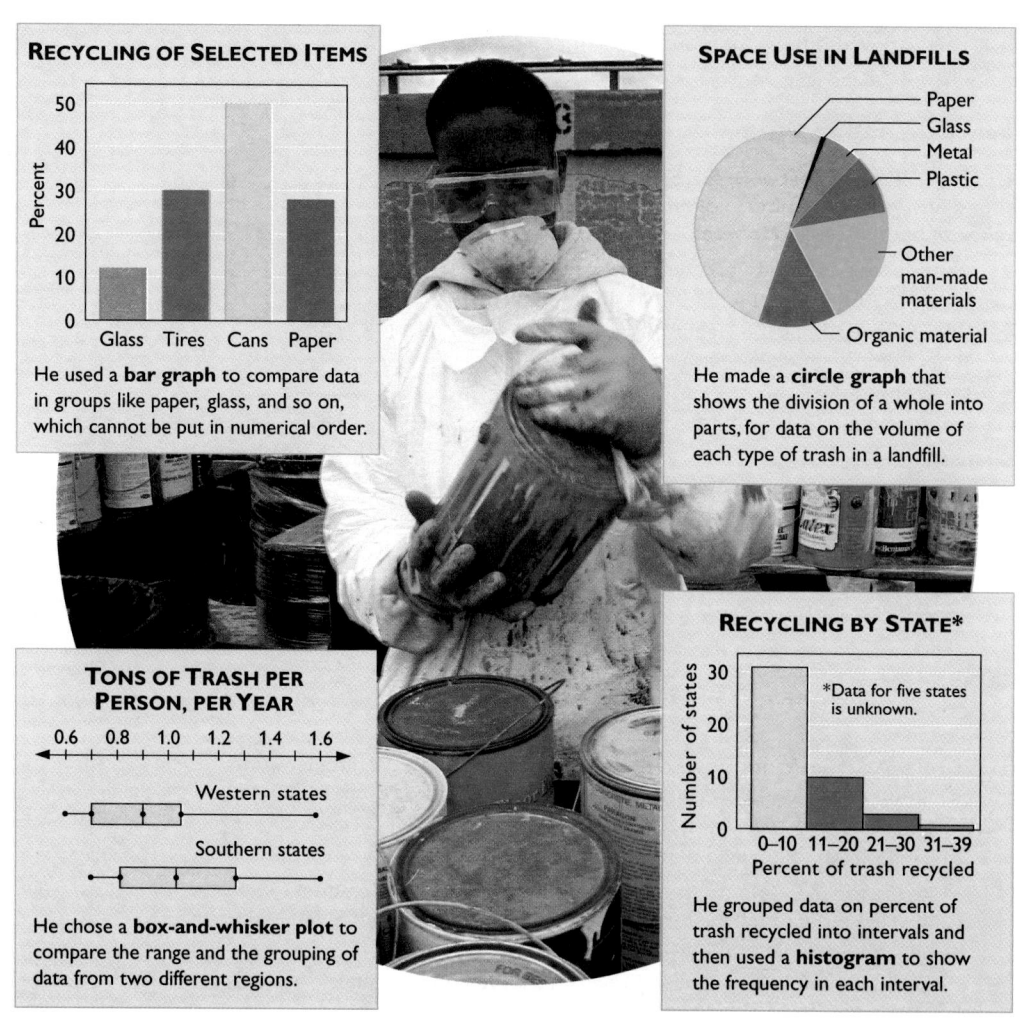

RECYCLING OF SELECTED ITEMS

He used a **bar graph** to compare data in groups like paper, glass, and so on, which cannot be put in numerical order.

SPACE USE IN LANDFILLS

Paper
Glass
Metal
Plastic
Other man-made materials
Organic material

He made a **circle graph** that shows the division of a whole into parts, for data on the volume of each type of trash in a landfill.

TONS OF TRASH PER PERSON, PER YEAR

0.6 0.8 1.0 1.2 1.4 1.6

Western states

Southern states

He chose a **box-and-whisker plot** to compare the range and the grouping of data from two different regions.

RECYCLING BY STATE*

*Data for five states is unknown.

0–10 11–20 21–30 31–39
Percent of trash recycled

He grouped data on percent of trash recycled into intervals and then used a **histogram** to show the frequency in each interval.

Talk it Over

1. Suppose you were given the exact percent of trash recycled by each state. Choose another type of display for the data and describe what it could show.

2. Could the landfill data be displayed in another type of graph? If so, describe the graph.

3. Name other types of graphs that could be used to compare the amount of trash that each person produces in the southern and western regions of the United States.

Sample 1

a. Make three statements about the data shown in the line graph on the use of agricultural chemicals in Japan.

b. **Writing** Explain why a line graph is a good choice for the data.

Sample Response

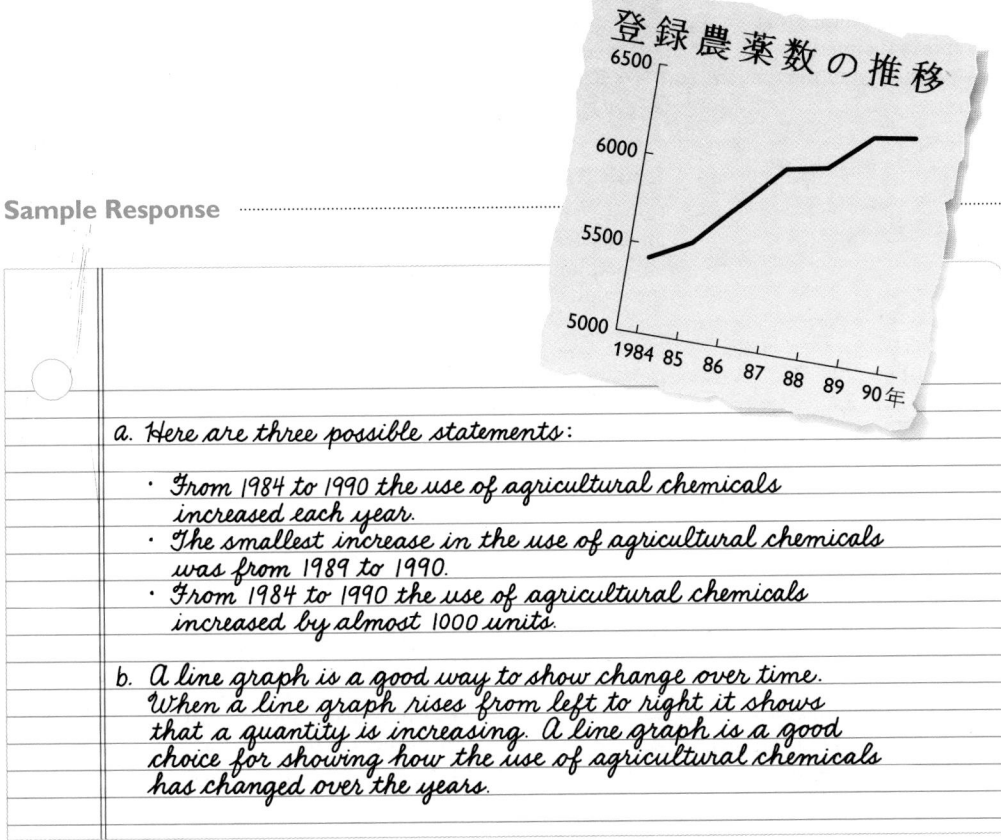

登録農薬数の推移

a. Here are three possible statements:

· From 1984 to 1990 the use of agricultural chemicals increased each year.
· The smallest increase in the use of agricultural chemicals was from 1989 to 1990.
· From 1984 to 1990 the use of agricultural chemicals increased by almost 1000 units.

b. A line graph is a good way to show change over time. When a line graph rises from left to right it shows that a quantity is increasing. A line graph is a good choice for showing how the use of agricultural chemicals has changed over the years.

3-6 Choosing a Data Display

TEACHING

Additional Sample

S1

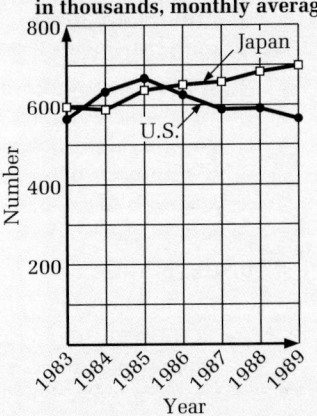

Passenger Car Production
in thousands, monthly average

a. Make three statements about the data shown in the line graph.

Japan showed an increase in its monthly average production since 1984. The U.S. had a decrease in monthly average production from 1985 to 1989. There was little change in U.S. production between 1987 and 1988.

b. Explain why a line graph is a good choice for the data. A line graph is a good way to show change over time.

Answers to Talk it Over

1. Answers may vary. An example is given. You could use a table to display each state and percent of trash recycled by that state.

2. Answers may vary. An example is given. A bar graph could be used to display these data.

3. Answers may vary. An example is given. You could use a histogram or a stem-and-leaf plot.

Additional Sample

S2 What type of graph is a good choice for displaying each data set? Explain your choice.

a. how many hours teenagers watch TV each week *a histogram or stem-and-leaf plot, since you would probably group the number of hours into intervals like 1–5, 6–10, and so on*

b. the middle 50% of your school's college entrance exam results *a box-and-whisker plot because it shows the middle 50% of a data set*

c. the number of games won by the teams in the National Basketball Association in one season *a stem-and-leaf plot since the data set is small; You can make exact comparisons and easily find the mean, median, mode, or outliers.*

Look Back

It may be helpful to have students answer the Look Back question by first describing the data. They should tell what the data show and then try to describe the data in another way. Students can use this Look Back for a journal entry.

APPLYING

Suggested Assignment

Standard 1–17, 19–24

Extended 1–17, 19–24

Integrating the Strands

Geometry Ex. 21

Statistics and Probability Exs. 1–20, 22–24

Logic and Language Exs. 8–11, 13, 14, 19

Sample 2

What type of graph is a good choice for displaying each data set? Explain your choice.

a. the middle 50% of the scores of 200 college entrance exams

b. the number of books 15 students borrowed from a library one year

c. how much the members of the junior class spend on lunch on a typical day

Sample Response

Some possible choices are:

a. a box-and-whisker plot, because the box shows the middle 50% of a data set. Also, a box-and-whisker plot helps you make sense of a large data set by turning it into a simple picture that shows the median, the quartiles, and the range.

b. a stem-and-leaf plot, since the data set is small. Since a stem-and-leaf plot shows each number in the data set, you can make exact comparisons, and you can find the mean, median, mode, range, or outliers of the data set.

c. a histogram or a stem-and-leaf plot, since you can group the lunch costs in intervals like $0–1.49 and $1.50–2.99. If the junior class is very large, use a histogram. If the class is small (under 50 students), you could use a stem-and-leaf plot.

Look Back

Choose a graph from another section in this unit. Is there another way to display the same information? If so, describe it and explain the strengths of each display. If not, explain why the graph should not be changed.

3-6 Exercises and Problems

1. **Reading** Make a statement about the data shown in each graph on page 164.

For Exercises 2–6, choose the letters of all the statements that describe each type of graph.

2. circle graph

   a. shows each number in the data set

   b. shows parts of a whole

   c. shows outliers

   d. uses the whole circle to represent the total and sectors to represent the parts

3. bar graph

   a. can represent positive and negative data

   b. shows the mean of the data set

   c. can be used for data that cannot be put in numerical order

   d. shows the range of the data set

Answers to Look Back

Answers may vary. Examples are given. I chose the histogram from Section 3-4 that shows the ages of *Galaxy* magazine subscribers. A circle graph could show the percent of the subscribers that are in each age group. The strength of the histogram is that it allows the viewer to estimate and compare the number of subscribers in each age group. The strength of the circle graph is that it allows the viewer to compare the number of subscribers in each age group to the number in other groups and to the total number of subscribers. I chose the pictograph from Section 3-1 that shows dog registrations in 1990. The data could be displayed in a bar graph with categories of dogs on the horizontal axis and numbers of registered dogs in ten thousands on the vertical axis. The strength of a bar graph is that the data can be more accurately estimated than they can with a pictograph. The strengths of a pictograph are the liveliness of the dog image and the simplicity of the graph.

4. box-and-whisker plot

  a. shows the median and the quartiles of the data set

  b. shows the range of the data set

  c. shows each number in the data set

  d. shows the mean of the data set

6. histogram

  a. has spaces between the bars

  b. is a picture of the numbers in a frequency table

  c. can be used for data that can be put into intervals

  d. shows each number in the data set

7. Write three statements similar to the ones in Exercises 2–6 to describe a line graph.

5. stem-and-leaf plot

  a. shows each number in the data set

  b. shows trends over time

  c. shows the mean of the data set

  d. groups the data in intervals

For Exercises 8–11, do these things:

a. Make a statement about the data in each display.

b. Writing Explain why this type of display was chosen for the data.

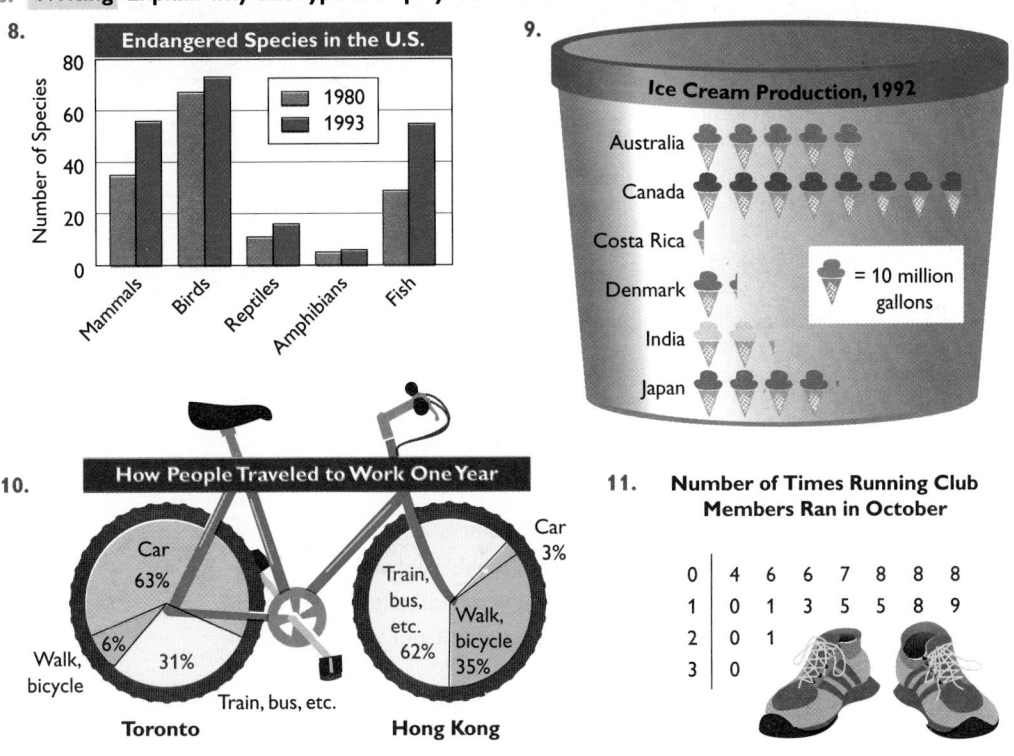

8. Endangered Species in the U.S.

9. Ice Cream Production, 1992

10. How People Traveled to Work One Year — Toronto; Hong Kong

11. Number of Times Running Club Members Ran in October

0	4 6 6 7 8 8 8
1	0 1 3 5 5 8 9
2	0 1
3	0

12. Find the mean, the median, and the mode of the data in Exercise 11. Which, if any, of these numbers are displayed in the stem-and-leaf plot?

Communication: Writing

Exs. 8–11 give students an opportunity to consider why certain types of data displays are more useful than others for displaying certain types of data. In their explanations, students may also wish to consider if another data display would be appropriate for each exercise.

Using Technology

Students can use the *Stats!* software to find the mean, the median, and the mode for Ex. 12.

8–11. Answers may vary. Examples are given.

8. a. The number of endangered species has gone up in every category from 1980–1993.

  b. The double bar graph easily compares the difference between the two years in each category.

9. a. Canada produces almost twice as much ice cream as Japan.

  b. This graph gives a quick comparison of different nations' ice cream production. The ice cream cone makes the subject matter clear.

10. a. The percentage of people who travel by car to work in Toronto is almost equal to the percentage of people who travel by public transportation to work in Hong Kong.

  b. The circle graphs are perfect to show percentages and are cleverly disguised as wheels in one of the mentioned means of transportation.

11. a. Only 3 out of 17 Running Club members ran 20 or more times in October.

  b. The plot quickly displays this small data set, and allows each data value to be seen.

12. mean: about 12.9; median: 11; mode: 8; the median and the mode

Answers to Exercises and Problems

1. Answers may vary. Examples are given. The bar graph shows that the percentage of tires recycled is about the same as the percentage of paper recycled. The circle graph shows that almost half the landfill space is taken up by paper. The box-and-whisker plot shows that about 50% of Southerners contribute more tons of trash per person, per year, than 75% of Westerners. The histogram shows that about 30 states recycle 10% or less of their trash.

2. b, d

3. a, c, d

4. a, b

5. a, d

6. b, c

7. Answers may vary. Examples are given. A line graph can be used to show trends over time. A line graph can be used to compare two sets of data, especially over time. A line graph can be used to estimate an unknown data value between two known data values.

For Exercises 13 and 14, tell what type of graph you think is a good choice for displaying each data set. Explain your choice.

13. The data below describe land use in Mexico.

Growing food 13%
Raising animals 39%
Forest 22%
Other 26%

14. In the table, the states are grouped by the dates they became states of the United States.

15. A preschool director wishes to compare the ages of the children in two classes. The director wants to display the ages of the oldest and the youngest child in each class, and the median of the ages. What type of display could the director use?

16. A coach wants to show that the members of the swim team practice from 45 to 80 minutes per day. What type of display could the coach use? Give two possible answers.

17. **a.** Which type of data display do you think shows the outliers of a data set most clearly? Give a reason for your choice.

    b. Which type of data display do you think hides the outliers of a data set? Give a reason for your choice.

18. TECHNOLOGY Use spreadsheet or statistical graphing software.

    Alternative Approach Use graph paper and work with another student.

    a. Choose one of the data sets given in Exercises 13 and 14. Make the type of graph you chose for the data set.

    b. Graph the same data, using another type of graph. If you are using software, display both graphs next to each other on the screen.

    c. Describe the differences between the graphs you made in parts (a) and (b). Which graph do you think displays the data better? Explain your choice.

Years	Number of states added
1775–1799	16
1800–1824	8
1825–1849	6
1850–1874	7
1875–1899	8
1900–1924	3
1925–1949	0
1950–1974	2

Answers to Exercises and Problems

13–17. Choices may vary. Examples are given.

13. circle graph; shows parts of a whole

14. histogram; shows frequency in each interval

15. box-and-whisker plot

16. If the amount of data is small, use a stem-and-leaf plot. Otherwise, use a histogram to show frequency.

17. a. A line plot shows outliers best. It shows exactly where the data fall and how they cluster so outliers can easily be seen.

b. A histogram conceals outliers because it shows only an interval of data. How exact data values are grouped in the interval cannot be seen.

18. a–c. Choices of data sets, graphs, descriptions, and answers may vary. Examples are given based on the data set in Exercise 13.

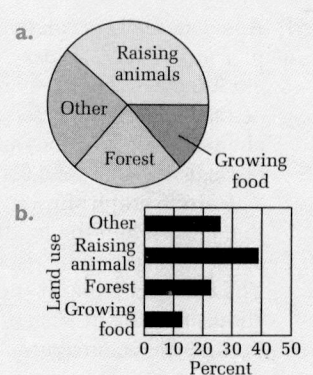

19. **Writing** Find examples of at least two types of graphs in newspapers or magazines. For each graph, tell whether or not it is a good choice for the data that it displays. Explain why or why not.

Review **PREVIEW**

20. Make a box-and-whisker plot of the data on the map of South America. *(Section 3-5)*

21. Each grid square on the map of Lake Titicaca is 20 km on a side. Estimate the area of the lake. *(Section 2-4)*

Forest Land in South America, 1991 estimates
(thousands of square miles)

115 VENEZUELA
193 COLOMBIA
28 FRENCH GUIANA
41 ECUADOR
63 GUYANA
57 SURINAME
263 PERU
214 BOLIVIA
1894 BRAZIL
51 PARAGUAY
34 CHILE
3 URUGUAY
228 ARGENTINA
0 FALKLAND ISLANDS

BOLIVIA
Lake Titicaca
PERU

Use the bar graph at the right. *(Section 3-1)*

22. Which store sold more cassettes? How many more?

23. Tell whether the statement "Record Place sold about twice as many cassettes as Music City" is *True* or *False*.

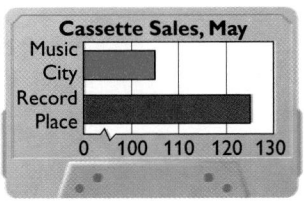

Cassette Sales, May
Music City
Record Place
0 100 110 120 130

Working on the Unit Project

24. **a. Research** Now you should have data on prices of recordings, types of radio stations, and the number of recordings people buy. What else do you want to know before completing a plan for a music store? Conduct a survey, collect data from magazines and newspapers, or contact local music organizations for information.

    b. Open-ended Choose a type of graph or display for each data set you have collected. Experiment with different types of displays.

Working on the Unit Project

For Ex. 24, it may be helpful if students make a list of all the types of information they have gathered. Encourage them to display the data in as many ways as possible and to include an explanation of why each display was chosen.

Practice 24 For use with Section 3-6

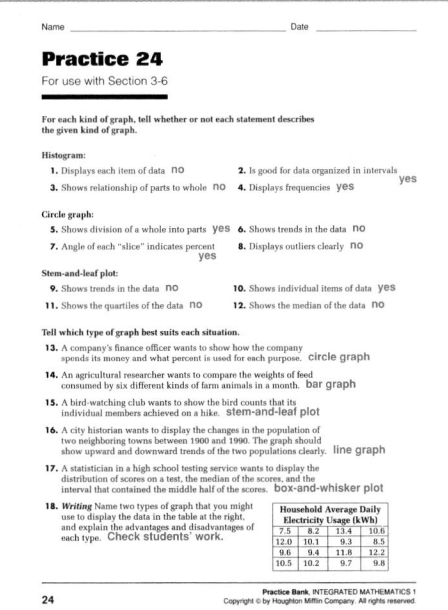

Name _____ Date _____

Practice 24
For use with Section 3-6

For each kind of graph, tell whether or not each statement describes the given kind of graph.

Histogram:
1. Displays each item of data no
2. Is good for data organized in intervals yes
3. Shows relationship of parts to whole no
4. Displays frequencies yes

Circle graph:
5. Shows division of a whole into parts yes
6. Shows trends in the data no
7. Angle of each "slice" indicates percent yes
8. Displays outliers clearly no

Stem-and-leaf plot:
9. Shows trends in the data no
10. Shows individual items of data yes
11. Shows the quartiles of the data no
12. Shows the median of the data no

Tell which type of graph best suits each situation.

13. A company's finance officer wants to show how the company spends its money and what percent is used for each purpose. circle graph
14. An agricultural researcher wants to compare the weights of feed consumed by six different kinds of farm animals in a month. bar graph
15. A bird-watching club wants to show the bird counts that its individual members achieved on a hike. stem-and-leaf plot
16. A city historian wants to display the changes in the population of two neighboring towns between 1900 and 1990. The graph should show upward and downward trends of the two populations clearly. line graph
17. A statistician in a high school testing service wants to display the distribution of scores on a test, the median of the scores, and the interval that contained the middle half of the scores. box-and-whisker plot
18. *Writing* Name two types of graph that you might use to display the data in the table at the right, and explain the advantages and disadvantages of each type. Check students' work.

Household Average Daily Electricity Usage (kWh)			
7.5	8.2	13.4	10.6
12.0	10.1	9.3	8.5
9.6	9.4	11.8	12.2
10.5	10.2	9.7	9.8

24   **Practice Bank,** INTEGRATED MATHEMATICS 1
Copyright © by Houghton Mifflin Company. All rights reserved.

Answers to Exercises and Problems

18. c. The circle graph displays the data better than the bar graph. From the circle graph you can see that the parts make up a whole. This is hard to see on the bar graph.

19. Answers may vary.

20.
0 100   1000   1900

21. about 8400 km²

22. Record Place; about 20 more cassettes

23. False.

24. a, b. Answers may vary.

Section 3-7

Analyzing Misleading Graphs

Focus
Recognize when graphs do not give an accurate picture of a data set.

LOOKS AREN'T EVERYTHING

All three line graphs below display the same data on median family income, which is shown in the table.

Median Family Income in the United States (dollars)

Year	1984	1985	1986	1987	1988	1989	1990
Income	26,433	27,735	29,458	30,970	32,191	34,213	35,353

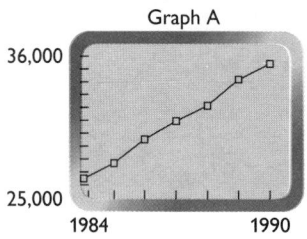

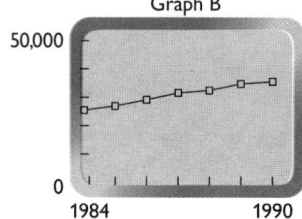

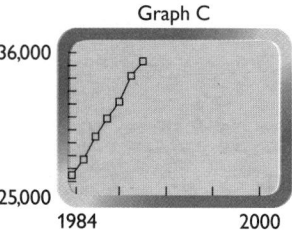

Talk it Over

1. Do you think the graphs give different pictures of how median family income changed from 1984 to 1990? If so, how would you describe the change shown by each graph?

2. a. How far apart (in dollars) are the scale marks on the vertical axis in Graph A? in Graph B? in Graph C?

   b. How far apart (in years) are the scale marks on the horizontal axis in Graph A? in Graph B? in Graph C?

   c. How do the scales affect the way the data look?

3. By what percent did median family income increase from 1984 to 1990? Which graph do you think shows this most clearly? Explain.

As the graphs above show, a *change in the scale* of a graph changes how the graph looks and it may make the graph misleading. A *misleading graph* may lead the viewer to draw a false conclusion about the data.

The graph on the next page shows how a *gap in the scale* can also change how a graph looks.

TECHNOLOGY NOTE

If you use a graphics calculator or graphing software to draw a graph, you can change the way the graph looks by changing the values you enter for the window variables. See Technology Handbook, p. 613.

170   **Unit 3** Representing Data

Answers to Talk it Over

1. Answers may vary. An example is given. Yes. The steepness of the various graphs makes it appear that income increases either slowly or quite dramatically.

2. a. $1000; $10,000; $1000

   b. 1 year; 1 year; 4 years

c. The bigger the interval between scale marks on the horizontal axis, the more vertical the graph; the bigger the interval between scale marks on the vertical axis, the more horizontal the graph.

3. about 33.7%; Answers may vary. An example is given. Graph A; Graph B suggests no change, and Graph C suggests too dramatic a change for 33.7%.

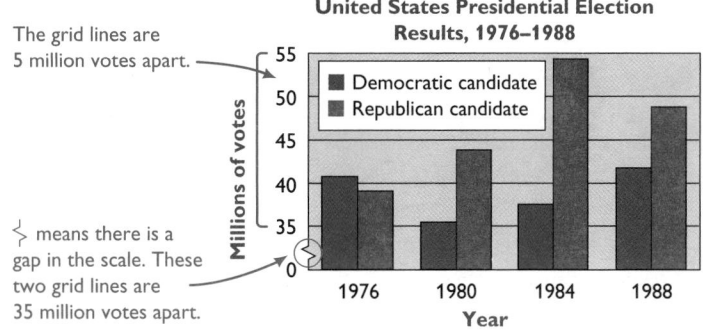

United States Presidential Election Results, 1976–1988

The grid lines are 5 million votes apart.

Millions of votes

- ■ Democratic candidate
- ■ Republican candidate

⌇ means there is a gap in the scale. These two grid lines are 35 million votes apart.

Year

Sample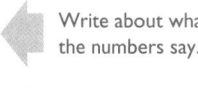

Writing Do you think the bar graph above gives an accurate picture of the 1980 election results? Explain.

Sample Response

Answer the question.

No.

The Democratic candidate received a little more than 35 million votes, and the Republican candidate received a little less than 44 million votes. The Republican received only about 9 million more votes than the Democrat. ◄ Write about what the numbers say.

The graph is misleading because the bar for the Republican is more than twice as long as the bar for the Democrat. It looks as if the Republican received over twice as many votes as the Democrat. ◄ Write about how the graph looks.

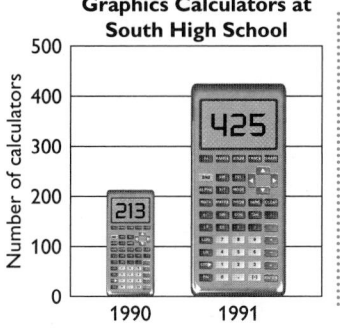

Graphics Calculators at South High School

Number of calculators

Talk it Over

4. Use the graph at the left. Compare the number of graphics calculators at South High School in 1991 to the number in 1990.

5. Compare the length and width of the "bar" for the 1991 data with the length and width of the "bar" for 1990. Then compare the areas of the "bars." What do you find?

6. Do you think the graph gives an accurate picture of the change in the number of graphics calculators from 1990 to 1991? If not, how could you make the graph less misleading?

Look Back ◄

Suppose you want to decide whether or not a graph is misleading. Describe three things to look for in the graph.

Talk it Over

Questions 1–3 discuss the importance of the vertical scale used for a graph. Changing the scale affects the steepness of the graph and therefore the perception of the viewer. Questions 4–6 discuss how changing the dimensions of a bar on a bar graph can affect a viewer's perception of the data.

Additional Sample

Do you think the line graphs below give an accurate picture of the difference in production of passenger cars in Japan and the U.S.? Explain.

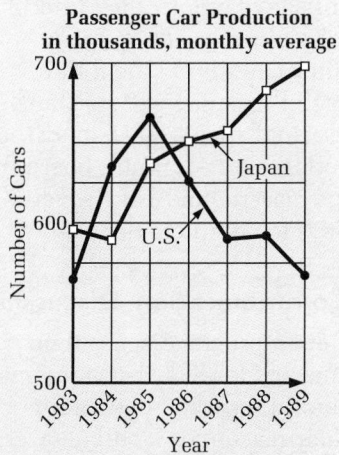

Passenger Car Production in thousands, monthly average

Number of Cars

Year

No. Compared to the graph in the side colum of page 165, this graph makes the difference in production between Japan and the U.S. seem greater than it is.

Look Back

This Look Back would make a good journal entry. Encourage students to draw diagrams to support their descriptions.

Answers to Talk it Over

4. There were about twice as many calculators in 1991 as there were in 1990.

5. The area of the 1991 calculator is about four times the area of the 1990 calculator.

6. No. Answers may vary. An example is given. The 1991 calculator should have twice the area (same width, twice the height) of the 1990 calculator.

Answers to Look Back

Answers may vary. Examples are given. Graphs may give a misleading visual impression if the scale is increased or decreased. A gap in the scale tricks viewers into thinking that bars or lines begin at a value different from the actual value. In graphs using symbols, the apparent sizes of objects may appear different from actual data values.

APPLYING

Answers to
Exercises and Problems

1. Graph C; It suggests a sharp increase in income.

2. 1988 and 1989; Graph A; The segment of the graph appears to show the steepest slope between these two years.

3. agree; The graph shows a difference of two million votes in 1976, and a difference of nine million votes for 1980, so the 1976 election was closer.

4. The bar for the Democratic vote would be about $\frac{9}{11}$ as large as the bar for the Republican vote.

5. False.

6. False.

172

3-7 Exercises and Problems

Reading Use the table and graphs on median family income on page 170.

1. Which graph would you use to convince someone that incomes rose steeply in the period 1984–1990? Explain your choice.

2. Between which two years did median family income rise the most? Which graph makes it easiest to find this information? Why?

Reading Use the graph on presidential election results on page 171.

3. A newspaper article says the graph shows that the 1976 election was much closer than the 1980 election. Do you agree or disagree? Why?

4. Suppose you redrew the graph to make it less misleading. How would the heights of the bars for 1980 compare on your new graph?

For Exercises 5–9, use the bar graph.

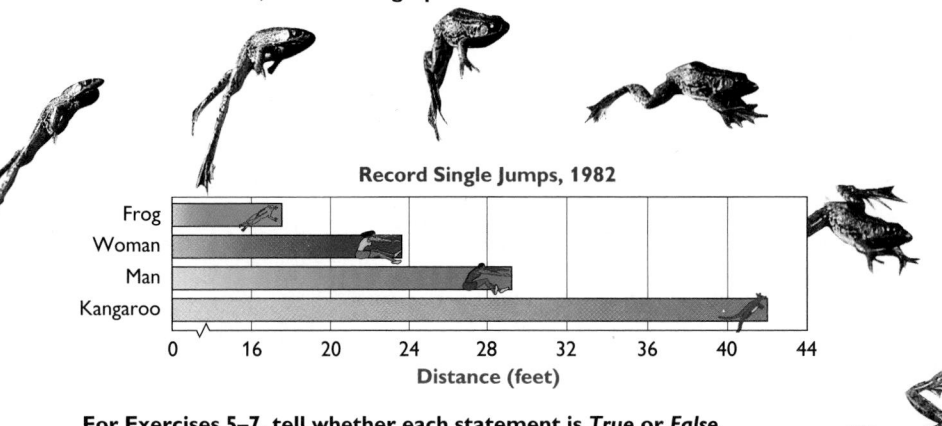

Record Single Jumps, 1982

Frog / Woman / Man / Kangaroo

Distance (feet)

For Exercises 5–7, tell whether each statement is *True* or *False*.

5. The men's record is more than twice the frog's record.

6. The kangaroo's record is about two and a half times the women's record.

7. The women's record differs from the frog's record by about 6 ft.

8. **Writing** Do you think the graph gives an accurate picture of how far the frog jumped compared to the kangaroo? Explain.

9. **a.** Use the graph to estimate the length of each record jump. Can you estimate the length to the nearest *inch* or to the nearest *foot*?

   b TECHNOLOGY Use statistical graphing or spreadsheet software.
   Alternative Approach Use paper and pencil.

   Draw a graph that gives an accurate picture of the record jumps. How is your graph different from the original graph?

BY THE WAY...

At the 1986 Calaveras Jumping Jubilee in Angels Camp, California, Rosie the Ribeter set a record of 21 ft $5\frac{3}{4}$ in. for the longest triple jump by a frog at least 4 in. long.

172 Unit 3 Representing Data

---

7. True.

8. Answers may vary. An example is given. No. There is a gap of twelve feet on the scale that makes the differences between jumpers appear greater than they are.

9. a. frog: about 17.5 ft; woman: about 23.5 ft; man: about 29 ft; kangaroo: about 42 ft; nearest foot

b.
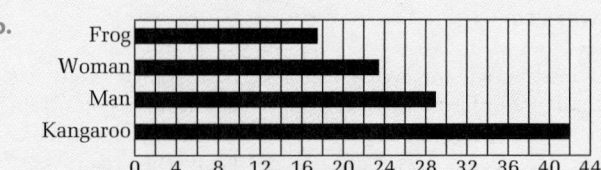

Frog / Woman / Man / Kangaroo

The original graph made the kangaroo's record appear to be about five times as long as the bullfrog's record. This graph, with no gap in the scale, makes the kangaroo's record appear to be a little more than twice as long as the bullfrog's record, which is more accurate. The new graph is in scale with the actual events.

Use Graph A, Graph B, and the table.

Graph A

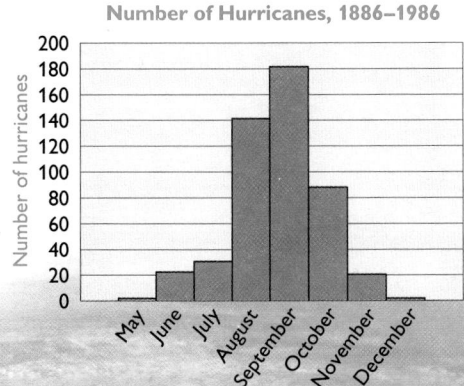

Number of Hurricanes, 1886–1986

Number of hurricanes (y-axis: 0, 20, 40, 60, 80, 100, 120, 140, 160, 180, 200)

x-axis: May, June, July, August, September, October, November, December

Graph B

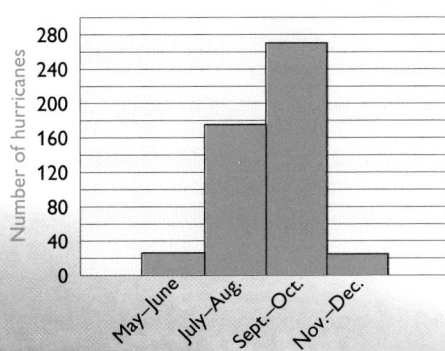

Number of Hurricanes, 1886–1986

Number of hurricanes (y-axis: 0, 40, 80, 120, 160, 200, 240, 280)

x-axis: May–June, July–Aug., Sept.–Oct., Nov.–Dec.

Number of Hurricanes, 1886–1986	
January	0
February	0
March	1
April	0
May	3
June	23
July	33
August	142
September	182
October	88
November	21
December	3

10. Both histograms display the data in the table. Do you think they give different pictures of the data? If so, how are they different?

11. Describe two pieces of information shown in the table but not in the histograms.

12. How many hurricanes occurred from 1886 through 1986? Do you think it is easier to use *Graph A* or *Graph B* to estimate the total? Explain.

13. Could the hurricane data be graphed using smaller-sized intervals than the ones used in Graph A? Why or why not?

14. Why can the hurricane data be shown in histogram form?

15. Choose the letter of the graph that better supports each statement.

   a. There were more hurricanes in fall than in summer.

   b. There were more hurricanes in September than in October.

   c. May and December had the same number of hurricanes.

16. **Open-ended** Write a question that can be answered by reading Graph A but not by reading Graph B.

BY THE WAY...

Is your name Alicia, Allen, Bob, David, Elena, Frederick, Gilbert, Gloria, Hugo, or Joan? If it is, there will never be another hurricane with your name. Storms with these names caused great damage, so the names have been "retired."

3-7 Analyzing Misleading Graphs

173

Answers to Exercises and Problems

10. Answers may vary. An example is given. Yes. Graph A displays the data from the table by single month and shows that August and September had the most hurricanes. Graph B groups two months together and shows that September and October had the most hurricanes.

11. Answers may vary. An example is given. The graph does not show the one hurricane in March. The table tells us exactly how many hurricanes occurred, but you would have to estimate this on the graph.

12. 496; Graph B; There would be fewer data values to be estimated and added since it has fewer bars.

13. No. Reasons may vary. For example, the data in the table are given per month, so one month is the smallest interval of time that could be used for a histogram.

14. The data in the table show the total number of hurricanes by month over a period of 100 years. A month is an interval of time and we are told frequency of hurricanes per month. Histograms are graphs that show frequencies during intervals.

15. a. Graph B
    b. Graph A
    c. Graph A

16. Answers may vary. Examples are given. About how many more hurricanes occurred during October than November from 1886 through 1986? In what months did the number of hurricanes exceed 100?

173

Reasoning

Exs. 17–21 present students with a different type of mis-leading graph, one involving apparent volume. Ex. 21 allows students to pull together and think about their responses to Exs. 17–20 and use those responses to give reasons as to whether the graph provides an accurate picture of the data.

Integrating the Strands

The problem situation in Exs. 17–21 integrates the strands of algebra, geometry, and statistics and probability.

Assessment: Performance

Ex. 22 gives students an opportunity to redraw and create graphs based on the same data which would lead to different conclusions. Each group can present their graphs and report to the class.

Answers to
Exercises and Problems

17. The actual number of cable systems in 1970 and 1990, 2.49 million and 9.58 million, are indicated along the horizontal axis; 9.58 > 2.49, so there was an increase from 1970 to 1990. Also, the television symbol for 1990 is much larger than the symbol for 1970, indicating a big increase from 1970 to 1990.

18. almost 4 times greater

19. b

20. a. 1970: about 15.4 cubic units; 1980: about 75.7 cubic units; 1990: about 879.2 cubic units

    b. almost 5 times greater

    c. almost 57 times greater

21. No. Reasons may vary. An example is given. The actual growth from 1970 to 1990 was 4 times, not 57 times as the graph leads us to believe.

Use the television graph.

17. Describe two ways in which the graph shows that the number of cable television systems in the United States rose from 1970 to 1990.

18. About how many times greater is the number of cable television systems in 1990 than in 1970?

19. Let x be the number of cable systems in the United States in 1980. Choose the letter of the expression that is the best estimate of the number of cable systems in the United States in 1990.

    a. x^2 **b.** $x + 5,000,000$

    c. $5x$ **d.** x^3

20. Suppose the pictures of televisions represent cubes.

    a. Find the volume of each cube.

    b. About how many times greater is the volume of the cube for 1980 than the volume of the cube for 1970?

    c. About how many times greater is the volume of the cube for 1990 than the volume of the cube for 1970?

21. **Writing** Do you think the graph gives an accurate picture of the growth in the number of cable systems from 1970 to 1990? Explain.

Ongoing ASSESSMENT

22. **Group Activity** Work in a group of four students. Use the table and the graph shown below for parts (a)–(e).

Cable Systems in the United States, 1970–1990

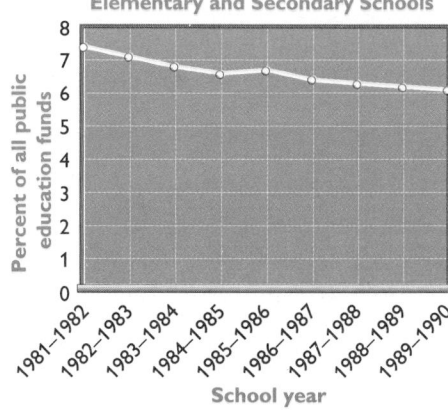

Number of systems (millions)

Federal Funding of Public Education Elementary and Secondary Schools

School year	Amount (thousands of dollars)	Percent of all public education funds
1981–1982	8,186,466	7.4
1982–1983	8,339,990	7.1
1983–1984	8,576,547	6.8
1984–1985	9,105,569	6.6
1985–1986	9,975,622	6.7
1986–1987	10,146,013	6.4
1987–1988	10,716,687	6.3
1988–1989	11,902,001	6.2
1989–1990	12,750,530	6.1

Federal Funding of Public Education Elementary and Secondary Schools

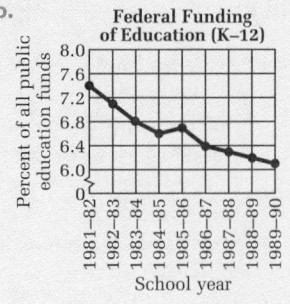

174   Unit 3 Representing Data

---

22. a–e. Answers may vary. Examples are given.

    a. The graph gives the visual impression that over the period 1981 to 1990, federal funding of elementary and secondary education in the United States went down slightly.

    The visual impression of the new graph is that federal funding of edu-

b.

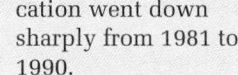

Federal Funding of Education (K–12)

c.

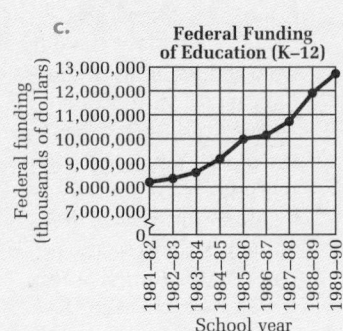

Federal Funding of Education (K–12)

cation went down sharply from 1981 to 1990.

a. Describe the change in federal funding of public education that the graph shows.

b. Choose a different scale for the vertical axis. Redraw the graph so that it gives a different picture of the data. Describe the change in federal funding of public education as shown by your graph.

c. Make a graph showing the *amount* of federal funding for the school years 1981–1982 through 1989–1990. How is the new graph different from the graph shown on page 174?

d. In the period shown in the table, the *amount* of federal funding for education went up. In the same period, the *percent* of public education funds that came from the federal government went down. Explain how both of these statements can be true.

e. **Writing** Write a report that presents the results of parts (a)–(d).

TECHNOLOGY NOTE

If you use a graphics calculator or graphing software, use the first year of each school year for the x-value. For 1981–1982, use 1981, and so on. ◀

Review **PREVIEW**

What type of graph is a good choice for displaying each data set? Explain your choice. *(Section 3-6)*

23. lengths of long jumps made in a varsity track and field meet

24. number of visitors to a museum each day of one month

25. ages of students in a graduating high school class

Write each number in scientific notation. *(Section 2-3)*

26. 700,000

27. 23,400,000

28. 0.006

29. 0.0000461

Graph each point on a coordinate plane. *(Toolbox Skill 21)*

30. $A(5, 3)$

31. $B(-2, 4)$

32. $C(0, 4)$

33. $D(-3, -5)$

Working on the Unit Project

34. Use the data you have collected about the prices of recordings of different types of music on cassette, CD, and so on.

a. Draw two graphs of the same data. Use one graph to show that cassettes, CDs, and so on have very different prices, and the other to show that cassettes, CDs, and so on have very similar prices.

b. Which graph would you use to convince customers to buy one form of recording instead of another? Explain your choice.

3-7 Analyzing Misleading Graphs

175

Working on the Unit Project

For Ex. 34, you may want to have students use their graphs to practice convincing each other to buy one form of recording instead of another. Then have them change their graphs according to the responses they get.

Quick Quiz (3-4 through 3-7)

See page 178.

Practice 25 For use with Section 3-7

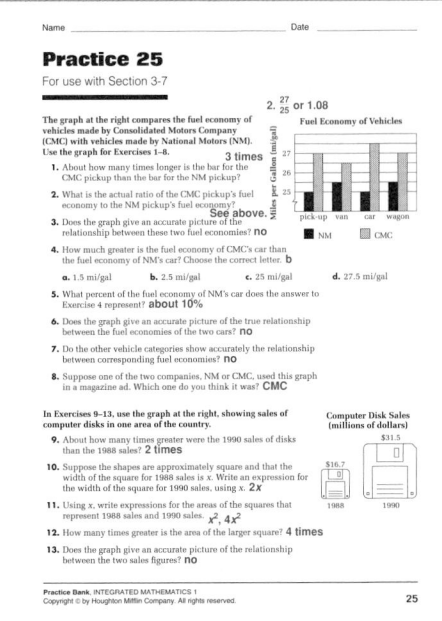

23. bar graph; Each jumper's bar would show the length of the jump and could be compared easily.

24. histogram; It shows the frequency (number of visitors) in an interval (each day of a month).

25. circle graph; It shows the percentage of total graduates for each age.

26. 7×10^5

27. 2.34×10^7

28. 6×10^{-3}

29. 4.61×10^{-5}

30–33.

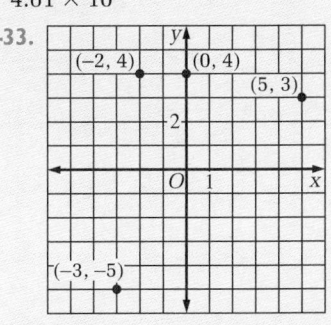

34. a, b. Answers may vary.

Answers to Exercises and Problems

22. c. The new graph shows the amount of federal funding in each of the specified school years, whereas the graph on page 174 shows the percent of elementary and secondary school funding covered by federal funding. The new graph suggests that federal funding has increased from 1981 to 1990, while the other graph suggests that federal funding has gone down.

d. Although the amount of annual federal funding for education increased in the period shown, the total amount of school funding increased even more dramatically during this period. The numerator of the fraction

$$\frac{\text{federal funding for education}}{\text{total funding for education}}$$

increased, but the denominator increased enough to make the value of the fraction decrease.

e. Answers may vary.

23–26. Choices may vary. Examples are given.

175

Completing the Unit Project

Now you are ready to make your business brochure. You may want to use color or visual designs to make your brochure more attractive.

The plan you present in your brochure should include a summary of your market research. Include the percentage of your merchandise devoted to each music style and each format. To support your figures, describe the consumers in your store's area.

Include graphs and other data displays in your brochure. You may want to use a spreadsheet or other graphics software.

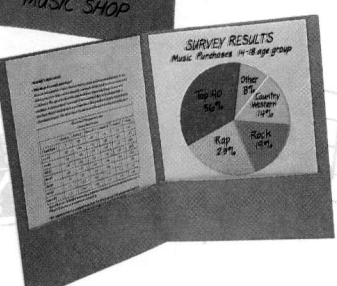

Look Back

Were the results of your research about popular styles and formats of music what you expected? Describe how your ideas about what is typical changed during the project.

Alternative Projects

Project 1: Supporting an Opinion

 One movie critic claims that the titles of popular movies in the 1970s were very short (*Jaws, Star Wars*) compared to the titles of older movies (*Casablanca, Gone with the Wind*). Gather data and decide if you agree or disagree. Include these points in your report.

➤ how you decided what "popular" means in this case

➤ how you chose your data sets

➤ a summary of the data sets, including their means, medians, or modes

➤ the process you used to make a decision

➤ the conclusion that your data supports

Project 2: Analyzing a Process

Find out how data are collected for making a Top Ten music, movie, or book list. Who supplies the data? Are any formulas used?

Write a summary of how the list is determined. Suggest how changes in the process could produce different results.

Trash Collected at State Beaches

	1989	1990
Glass beverage bottles	16,630	12,277
Plastic rope	30,439	24,537
Metal beverage cans	11,434	13,028
Cigarette filters	10,180	29,812

Trash Collected at State Beaches

Number of items (thousands)

■ 1989
■ 1990

Glass beverage bottles · Plastic rope · Metal beverage cans · Cigarette filters

Use the matrix and the double bar graph.   3-1

1. What are the dimensions of the matrix?

2. What does the number in the second row, first column of the matrix represent?

3. a. **Open-ended** Write two statements comparing the numbers of items collected on state beaches in 1989 and 1990.

   b. For each statement that you wrote for part (a), tell whether it is better supported by the *matrix* or the *graph*.

Use these scores on a magazine's "Are You Interesting?" quiz: 3-2
        18, 23, 28, 26, 22, 30, 26, 18, 27, 18, 28

4. Find the mean, the median, and the mode of the scores. Which number would you call most typical of the data set?

5. What is the range of the scores?

6. Is a score of 30 an outlier? Explain why or why not.

7. Suppose five friends take the quiz. The mean of four friends' scores is 25.5. A score of 26 is rated FASCINATING. What must the fifth score be for the mean of the five scores to be FASCINATING?

For Exercises 8 and 9, write and graph an inequality   3-3
to describe the interval referred to in each statement.

8. Laurie Chin was driving 35 mi/h when she entered a freeway on-ramp. She accelerated and merged with traffic when driving 55 mi/h.

9. In New Jersey you must be at least 17 years old to earn an unrestricted driver's license.

10. a. Use the histogram to make two statements   3-4
       about the sales of video cassette recorders.

    b. What is *not* shown in the histogram? Give two examples.

VCR Sales at Hunt's Store

Frequency

Price (dollars): 100–199, 200–299, 300–399, 400–499, 500–599

Unit 3 Review and Assessment   **177**

---

1. 4 × 2

2. plastic rope collected at state beaches in 1989

3. a, b. Answers may vary. Examples are given.

3. a. More glass beverage bottles and plastic rope were collected in 1989 than in 1990. More metal beverage cans and cigarette filters were collected in 1990 than in 1989.

   b. My statements are more easily seen by the graph.

4. mean: 24; median: 26; mode: 18; The mean and median are most typical because both are near the middle of the data set.

5. 12

6. No. It is only 2 units from the next score.

7. 28 or higher

8. $35 \le s \le 55$, s = speed of Laurie Chin;

   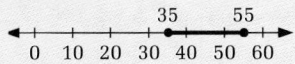
   35   55
   0 10 20 30 40 50 60

9. $a \ge 17$, a = necessary age to earn a NJ driver's license

   15 16 17 18 19 20

10. a, b. Answers may vary. Examples are given.

a. Most VCR's sold cost between $200.00 and $299.00; the number of VCR's sold in the $100.00–$199.00 price range is the same as the number of VCR's sold in the $300.00–$399.00 and the $500.00–$599.00 ranges.

b. exact cost of VCR's and how many different models there are in each price range

For Exs. 1–3, use these data.

Number of Hours per Week Watching Television

14	20	12
18	12	6
10	23	28
22	13	24
15	18	21

1. a. Make a stem-and-leaf plot of the data. [3-4]

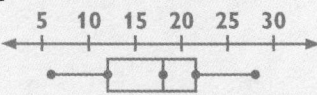

```
0 | 6
1 | 0 2 2 3 4 5 8 8
2 | 0 1 2 3 4 8
```

b. Find the median. **18**

2. Make a box-and-whisker plot for the data. [3-5]

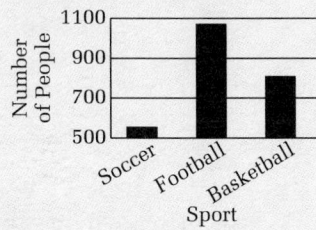

3. Which display would you choose for the data, a stem-and-leaf plot or a box-and-whisker plot? Explain. [3-6]
the box-and-whisker plot because it shows that 50% of the people surveyed watch between 12 and 22 hours of television per week

4. The table and graph show the attendance at Macon High School's first game for each sport. Does the graph fairly show the data? If not, describe how the graph is misleading. [3-7]

Attendance at First Game

Sport	No. of People
Soccer	556
Football	1073
Basketball	810

No. The display makes it appear that football attendance was twice as many as basketball and 9 times as many as soccer. The data actually show that football attendance had only 32% more people than basketball and twice as many as soccer.

---

11. a. Make a stem-and-leaf plot of the data for the first day of school.

b. Find the median of the data for the first day of school.

c. Make a general statement about the class sizes on the first day of school.

Homeroom Class Sizes

First Day of School					Last Day of School				
15	23	38	12	56	20	23	35	22	56
26	14	23	34	34	26	25	23	27	51
23	17	30	23	24	20	22	35	26	29
51	43	26	28	17	23	56	29	33	26
67	29	31	26	21	28	27	23	25	21

12. a. Make a box-and-whisker plot for each data set shown above. **3-5**
Put both plots under the same number line.

b. Use your plots to write two statements comparing the homeroom class sizes on the first day of school and the last day of school.

Explain why each type of display was chosen for the data. **3-6**

13. **Daily Attendance in August**

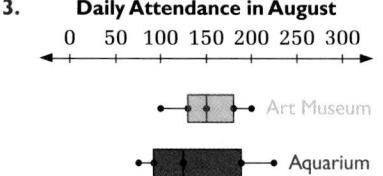

14. **Sara's Monthly Spending**

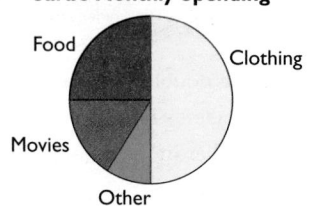

15. What type of graph would you use to display changes in the minimum wage over the last ten years? Explain your choice.

16. The graph shows the same data as the Questions 1–3. Compare the two graphs. Describe how this graph gives a misleading impression of the data. **3-7**

17. **Writing** Onika made a concept map about circle graphs. Describe three ideas that you see in her map.

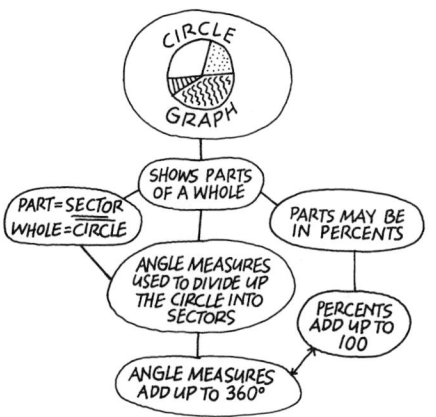

Trash Collected at State Beaches

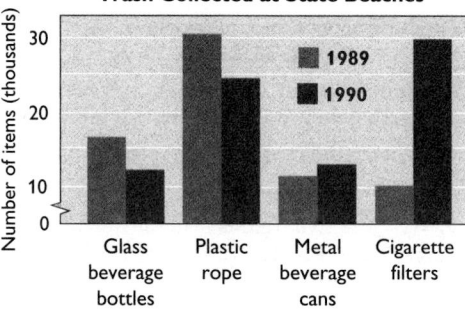

18. **Self-evaluation** You may have seen some of the data displays in this unit in other classes. Which were new to you? How have your ideas about displaying data changed?

19. **Group Activity** Work in a group of five students. Pass a sheet of paper around the group. Each student writes about a method of displaying data, saying why the method is useful. Discuss the ideas. Make a presentation about at least two types of displays.

178 Unit 3 Review and Assessment

Answers to Unit 3 Review and Assessment

11. a. Homeroom Class Sizes

```
1 | 2 4 5 7 7
2 | 1 3 3 3 3 4 6 6 6 8 9
3 | 0 1 4 4 8
4 | 3
5 | 1 6
6 | 7
```

b. 26

c. Statements may vary. An example is given. About half the class sizes were in the 20–30 student range.

12. a.

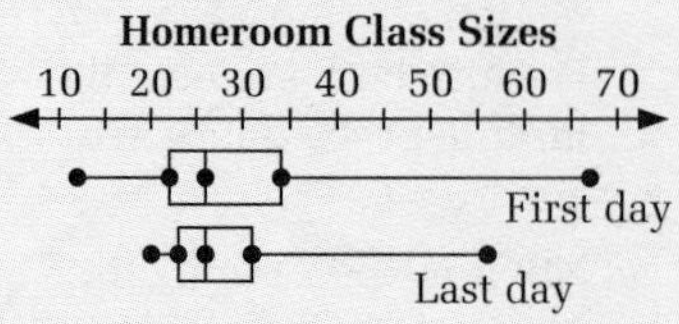

Homeroom Class Sizes

First day

Last day

b. Answers may vary. Examples are given. There was a wider range of homeroom class sizes on the first day than on the last day. The median class size on the first day was the same as the median class size on the last day.

IDEAS AND (FORMULAS)=x^2

STATISTICS & PROBABILITY

➤ You can use a matrix to organize and display data that belong to more than one category. *(p. 128)*

➤ Graphs can help you compare sets of data, observe trends, draw conclusions, and support a point of view. *(p. 128)*

➤ Computer spreadsheet software stores information in matrices. *(p. 129)*

➤ The mean, the median, and the mode are three values that may be considered typical of a data set. *(p. 135)*

➤ You can use a line plot to help you picture the outliers and the range of a data set. *(p. 137)*

➤ You can use a histogram to show the frequency of data that are grouped in intervals. *(p. 150)*

➤ Stem-and-leaf plots allow you to display every number in a data set and still organize the numbers in intervals. *(p. 152)*

➤ You can use box-and-whisker plots to compare how data are grouped in two data sets. *(p. 158)*

➤ You can use a bar graph to display data in categories that can not be put in numerical order. *(p. 164)*

➤ Circle graphs show the division of a whole into parts. *(p. 164)*

➤ A misleading graph may lead the viewer to make a false conclusion about the data. *(p. 170)*

DISCRETE MATH

➤ When discrete data are displayed on a line plot, they lie inside a continuous interval of the number line. *(p. 145)*

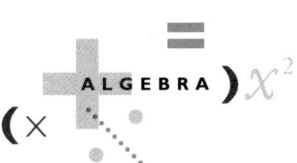
ALGEBRA $)x^2$

➤ To solve an equation with a fraction in it, begin by multiplying both sides of the equation by the denominator of the fraction. *(p. 138)*

➤ An algebraic inequality can be graphed as an interval on a number line. *(p. 145)*

Key Terms

- **matrix** (p. 128)
- **cell** (p. 129)
- **mode** (p. 136)
- **line plot** (p. 137)
- **interval** (p. 145)
- **frequency table** (p. 151)
- **lower quartile** (p. 158)

- **dimensions** (p. 128)
- **mean** (p. 136)
- **outliers** (p. 137)
- $\leq, \geq$ (p. 144)
- **histogram** (p. 150)
- **stem-and-leaf plot** (p. 152)
- **upper quartile** (p. 158)

- **spreadsheet** (p. 129)
- **median** (p. 136)
- **range** (p. 137)
- **combined inequality** (p. 144)
- **frequency** (p. 150)
- **box-and-whisker plot** (p. 158)
- **extremes** (p. 158)

For Exs. 1–3, use the following list of test scores:

1st Qtr.: 82, 84, 91, 78

2nd Qtr.: 69, 66, 90, 88

3rd Qtr.: 71, 82, 89, 91

4th Qtr.: 92, 94, 89, 92

1. Organize the data in a matrix. [3-1]

1st Qtr	2nd Qtr	3rd Qtr	4th Qtr
82	69	71	92
84	66	82	94
91	90	89	89
78	88	91	92

2. Find the mean, the median, and the mode(s) of the data for the 3rd and 4th quarters. [3-2] **mean: 87.5; median: 90; modes: 89 and 92**

3. Find the range and any outliers of the data for the 3rd and 4th quarters. [3-2] **range: 23; outliers: 71, 82**

4. The number of compact disc (CD) sales is over 30 per day at the Audio Shop.

a. Write an inequality which describes the number of CD sales per day. [3-3] $s > 30$

b. Graph the inequality. [3-3]

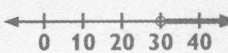

Answers to Unit 3 Review and Assessment

13–17. Answers may vary. Examples are given.

13. A box-and-whisker plot is an excellent way to compare two large sets of data. The plot allows you to compare the medians, quartiles, and ranges.

14. A circle graph shows best what portion of Sara's total monthly income was spent on various items.

15. line graph; It is best for showing change over time.

16. The new graph is misleading because of the scale gap of 5,000 units. The new graph shows that $\frac{1}{3}$ as many bottles were collected from 1989–1990, when it was actually $\frac{1}{4}$ as many.

The new graph makes it seem that 5 times as many filters were collected from 1989–1990, when it was really 3 times as many. It appears from the new graph that in 1989 twice as many bottles were collected as filters, which is false.

17. Circle graphs show parts of a whole. Circle graphs are based on percentages. Angle measures are used to divide up the circle.

18. Answers may vary.

19. Answers may vary.

OVERVIEW

➤ **Unit 4** introduces coordinates as a means of describing locations in the plane. Students learn how the coordinates of the vertices of polygons are affected by translations on the coordinate plane. Some prior experience with locating points on the coordinate plane is assumed. Students may refresh their knowledge of a coordinate system by using the **Student Resources Toolbox** on page 660.

➤ Students apply coordinate geometry to the statistical topic of scatter plots and fitted lines for making correlations and predictions.

➤ In Unit 4, students begin their study of functions. They learn to recognize functions from the shapes of their graphs, to represent functions with tables of ordered pairs, and to graph functions from equations.

➤ Animation is the theme of the Unit Project. Students employ transformations on the coordinate plane to create a flip book illustrating animated motion.

➤ Connections to social studies, art, literature, earth science, and computer science are integrated into the teaching materials and exercises.

➤ Problem-solving strategies used in Unit 4 include *Use a Diagram* (Section 4-2), *Break the Problem into Parts* (Section 4-2), and *Identify a Pattern* (Section 4-7).

Unit Objectives

Section	Objectives	NCTM Standards
4-1	• Use coordinate systems to solve problems about locations in different settings.	1, 2, 4, 5, 8
4-2	• Learn about coordinate geometry.	1, 2, 4, 5, 8
	• Identify polygons on a coordinate plane.	
	• Find the areas of polygons drawn on a coordinate plane.	
4-3	• Translate figures on a coordinate plane.	1, 2, 4, 5, 8
	• Recognize translational symmetry.	
4-4	• Recognize rotations of figures on a coordinate plane.	1, 2, 4, 5, 8
	• Rotate figures on polar graph paper.	
	• Recognize rotational symmetry.	
4-5	• Make and interpret a scatter plot of data.	1, 2, 3, 4, 5, 6, 8, 10
	• Use scatter plots to make predictions.	
4-6	• Understand what a function is.	1, 2, 4, 5, 6, 8
	• Identify control variables and dependent variables.	
	• Draw graphs of functions.	
	• Recognize functions.	
4-7	• Change from one representation of a function to another.	1, 2, 4, 5, 6, 8
	• Learn about some basic functions.	

Topic Spiraling

Section	Connections to Prior and Future Concepts
4-1	**Section 4-1** reviews the coordinate system, which many students have studied in previous courses. The coordinate system is developed and used throughout Unit 4. Coordinate graphing is also important in Units 6, 7, 8, and 10 of Book 1. Coordinate geometry is studied in Units 2, 3, 4, 5, 9, and 10 of Book 2 and in Units 1–3, 5, 9, and 10 of Book 3.
4-2	**Section 4-2** uses the coordinate system presented in Section 4-1 to introduce area concepts. Students identify polygons and find the areas of polygons drawn on the coordinate plane. Area is further studied in Sections 5-7, 7-6, 9-5, and 9-8 of Book 1 and in Unit 6 of Book 2.
4-3	**Section 4-3** introduces translations on the coordinate plane and translational symmetry. Transformation concepts are extended in Sections 4-4, 6-6, 10-1, and 10-2 of Book 1; in Units 3–5 and 8 of Book 2; and in Units 9 and 10 of Book 3.
4-4	**Section 4-4** continues the study of transformations begun in Sections 1-6 and 4-3 by introducing rotations and rotational symmetry. Transformations are studied throughout all three books of the series.
4-5	**Section 4-5** extends the study of statistical graphing begun in Unit 3. Statistical graphing and the coordinate system are combined in the exploration of scatter plots. The examination of data analysis and graphing continues throughout Book 1 and is further developed in Books 2 and 3, as already indicated.
4-6	**Section 4-6** combines the concepts presented in Sections 4-1 and 4-5 to introduce functions and graphs of functions. The study of functions and their graphs, which is a major focus of mathematics, continues throughout the remainder of Book 1 and is extended in most of the units of Book 2 and Book 3.
4-7	**Section 4-7** continues the investigation of functions begun in Section 4-6 by relating functions, graphs, and equations. Functions, graphs, and their equations are given special emphasis in Units 5, 7, 8, and 10 of Book 1; in Units 2, 4, and 9 of Book 2; and in Units 1, 2, 4, 5, 6, and 9 of Book 3.

Integrating the Strands

Strands	Sections
Number	4-2, 4-5, 4-6
Algebra	4-2, 4-3, 4-5, 4-6, 4-7
Functions	4-6, 4-7
Measurement	4-1, 4-2, 4-3, 4-5
Geometry	4-1, 4-2, 4-3, 4-4, 4-6, 4-7
Statistics and Probability	4-5, 4-6
Discrete Mathematics	4-1
Logic and Language	4-1, 4-2, 4-3, 4-4, 4-5, 4-7

Section Planning Guide

➤ Essential exercises and problems are indicated in boldface.
➤ Ongoing work on the Unit Project is indicated in color.
➤ Exercises and problems that require student research, group work, manipulatives, or graphing technology are indicated in the column headed "Other."

Section	Materials	Pacing	Standard Assignment	Extended Assignment	Other
4-1		Day 1	**1–15**, 16–23, 29–35, 36	**1–15**, 16–23, 25–35, 36	24
4-2	long pieces of string	Day 1	**1–3**, 4, 5, **6–11**	**1–3**, 4, 5, **6–11**	
		Day 2	**12–15**, **17**, **18**, 20–26, 27	**12–15**, **17**, **18**, 20–26, 27	16, 19
4-3		Day 1	1, **2–10**, 11–15, **16–22**, 24–31, 32–34	1, **2–10**, 11–15, **16–22**, 24–31, 32–34	23
4-4	polar graph paper, protractor, scissors	Day 1	1, 2, **3–7**, 8–11, **12–15**, 23, 24, 25, 26	1, 2, **3–7**, 8–11, **12–15**, 16–19, 23, 24, 25, 26	20–22
4-5	tape measure or meter stick, graphing technology, dry spaghetti	Day 1	**1**, **3–8**, 9	**1**, **3–8**, 9	2
		Day 2	**12–15**, **17–21**, 24–30, 31	10, 11, **12–15**, **17–21**, 24–30, 31	16, 22, 23, 31
4-6	geometric drawing software	Day 1	**1–11**, 13–25, 26	**1–11**, 13–25, 26	12
4-7	graphing technology	Day 1	**1–10**, 11	**1–10**, 11	
		Day 2	**12**, **16–20**	**12**, **16–20**	
		Day 3	**21–38**, 39–44, 45	**21–38**, 39–44, 45	13–15
Review		**Day 1**	**Unit Review**	**Unit Review**	
Test		**Day 2**	**Unit Test**	**Unit Test**	

Yearly Pacing	Unit 4 Total	Units 1–4 Total	Remaining	Total
	15 days (2 for Unit Project)	63 days	101 days	164 days

Support Materials

➤ See **Project Book** for notes on Unit 4 Project: Making a Flip Book.
➤ "UPP" and "disk" refer to **Using Plotter Plus** booklet and **Plotter Plus** disk.
➤ Warm-up exercises for each section are available on **Warm-Up Transparencies**.
➤ "FI," "PC," "GI," "MA," and "Stats!" refer, respectively, to the McDougal Littell Mathpack software Activity Books for **Function Investigator, Probability Constructor, Geometry Inventor, Matrix Analyzer,** and **Stats!**.

Section	Study Guide	Practice Bank	Problem Bank	Activity Bank	Explorations Lab Manual	Assessment Book	Visuals	Technology
4-1	4-1	Practice 27	Set 8	Enrich 24		Quiz 4-1		
4-2	4-2	Practice 28	Set 8	Enrich 25	Masters 2, 3	Quiz 4-2		
4-3	4-3	Practice 29	Set 8	Enrich 26	Master 2	Quiz 4-3		GI Act. 22
4-4	4-4	Practice 30	Set 8	Enrich 27	Masters 2, 4	Quiz 4-4 Test 14	Folder 5	GI Acts. 20, 21 and 30 Polar Graph Plotter (disk)
4-5	4-5	Practice 31	Set 9	Enrich 28	Masters 2, 12	Quiz 4-5		Stats! Acts. 8, 14, 16, 17, and 21
4-6	4-6	Practice 32	Set 9	Enrich 29	Master 2	Quiz 4-6		
4-7	4-7	Practice 33	Set 9	Enrich 30	Masters 2, 3, 13	Quiz 4-7 Test 15		UPP, page 28 Function Plotter (disk)
Unit 4	Unit Review	Practice 34	Unifying Problem 4	Family Involve 4		Tests 16, 17		

Form A   **Spanish versions** of these tests are on pages 134–137 of the **Assessment Book.**

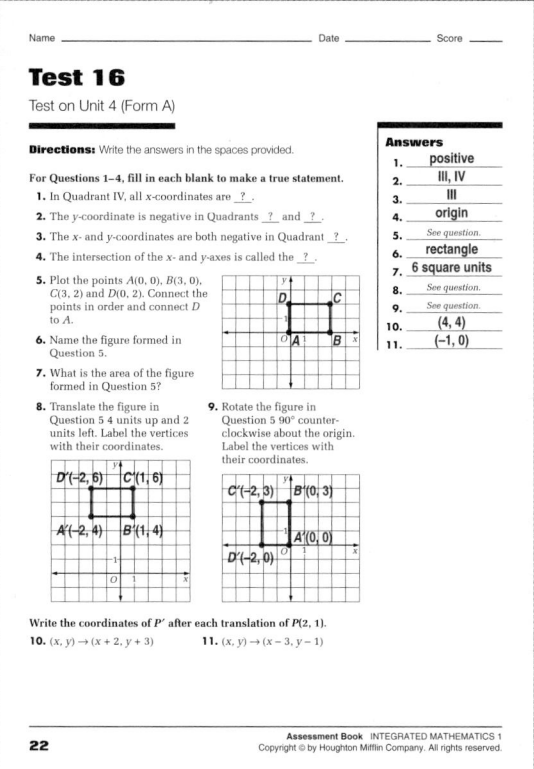

Name _____ Date _____ Score _____

Test 16
Test on Unit 4 (Form A)

Directions: Write the answers in the spaces provided.

For Questions 1–4, fill in each blank to make a true statement.

1. In Quadrant IV, all x-coordinates are ___?___ .
2. The y-coordinate is negative in Quadrants ___?___ and ___?___ .
3. The x- and y-coordinates are both negative in Quadrant ___?___ .
4. The intersection of the x- and y-axes is called the ___?___ .
5. Plot the points $A(0, 0)$, $B(3, 0)$, $C(3, 2)$ and $D(0, 2)$. Connect the points in order and connect D to A.
6. Name the figure formed in Question 5.
7. What is the area of the figure formed in Question 5?
8. Translate the figure in Question 5 4 units up and 2 units left. Label the vertices with their coordinates.
9. Rotate the figure in Question 5 90° counterclockwise about the origin. Label the vertices with their coordinates.

Write the coordinates of P' after each translation of $P(2, 1)$.

10. $(x, y) \rightarrow (x + 2, y + 3)$
11. $(x, y) \rightarrow (x - 3, y - 1)$

Answers	
1.	positive
2.	III, IV
3.	III
4.	origin
5.	See question.
6.	rectangle
7.	6 square units
8.	See question.
9.	See question.
10.	(4, 4)
11.	(−1, 0)

22 Assessment Book INTEGRATED MATHEMATICS 1
Copyright © by Houghton Mifflin Company. All rights reserved.

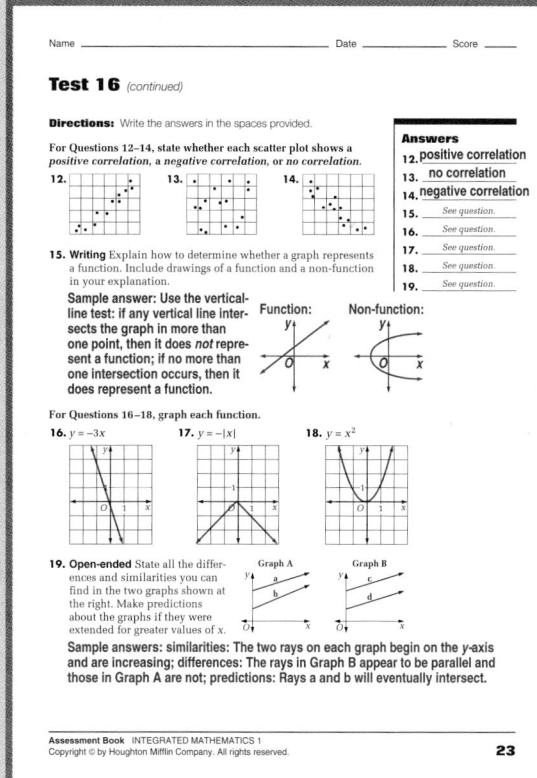

Name _____ Date _____ Score _____

Test 16 (continued)

Directions: Write the answers in the spaces provided.

For Questions 12–14, state whether each scatter plot shows a *positive correlation*, a *negative correlation*, or *no correlation*.

12.   13.   14.

15. **Writing** Explain how to determine whether a graph represents a function. Include drawings of a function and a non-function in your explanation.

Sample answer: Use the vertical-line test: if any vertical line intersects the graph in more than one point, then it does *not* represent a function; if no more than one intersection occurs, then it does represent a function.

Function: Non-function:

For Questions 16–18, graph each function.

16. $y = -3x$
17. $y = -|x|$
18. $y = x^2$

19. **Open-ended** State all the differences and similarities you can find in the two graphs shown at the right. Make predictions about the graphs if they were extended for greater values of x.

Graph A   Graph B

Sample answers: similarities: The two rays on each graph begin on the y-axis and are increasing; differences: The rays in Graph B appear to be parallel and those in Graph A are not; predictions: Rays a and b will eventually intersect.

Answers	
12.	positive correlation
13.	no correlation
14.	negative correlation
15.	See question.
16.	See question.
17.	See question.
18.	See question.
19.	See question.

Assessment Book INTEGRATED MATHEMATICS 1
Copyright © by Houghton Mifflin Company. All rights reserved.   23

Form B

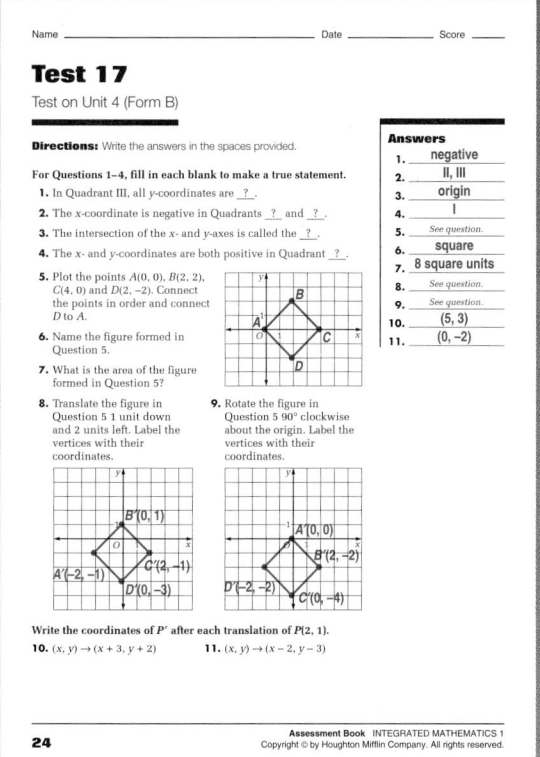

Name _____ Date _____ Score _____

Test 17
Test on Unit 4 (Form B)

Directions: Write the answers in the spaces provided.

For Questions 1–4, fill in each blank to make a true statement.

1. In Quadrant III, all y-coordinates are ___?___ .
2. The x-coordinate is negative in Quadrants ___?___ and ___?___ .
3. The intersection of the x- and y-axes is called the ___?___ .
4. The x- and y-coordinates are both positive in Quadrant ___?___ .
5. Plot the points $A(0, 0)$, $B(2, 2)$, $C(4, 0)$ and $D(2, -2)$. Connect the points in order and connect D to A.
6. Name the figure formed in Question 5.
7. What is the area of the figure formed in Question 5?
8. Translate the figure in Question 5 1 unit down and 2 units left. Label the vertices with their coordinates.
9. Rotate the figure in Question 5 90° clockwise about the origin. Label the vertices with their coordinates.

Write the coordinates of P' after each translation of $P(2, 1)$.

10. $(x, y) \rightarrow (x + 3, y + 2)$
11. $(x, y) \rightarrow (x - 2, y - 3)$

Answers	
1.	negative
2.	II, III
3.	origin
4.	I
5.	See question.
6.	square
7.	8 square units
8.	See question.
9.	See question.
10.	(5, 3)
11.	(0, −2)

24 Assessment Book INTEGRATED MATHEMATICS 1
Copyright © by Houghton Mifflin Company. All rights reserved.

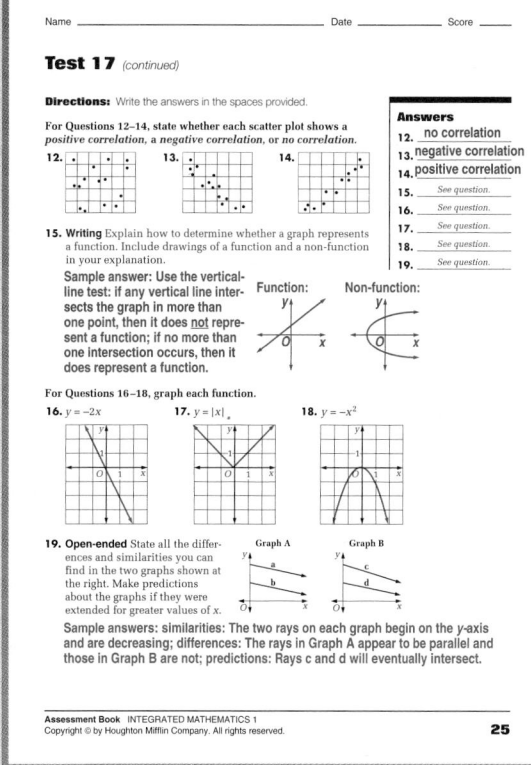

Name _____ Date _____ Score _____

Test 17 (continued)

Directions: Write the answers in the spaces provided.

For Questions 12–14, state whether each scatter plot shows a *positive correlation*, a *negative correlation*, or *no correlation*.

12.   13.   14.

15. **Writing** Explain how to determine whether a graph represents a function. Include drawings of a function and a non-function in your explanation.

Sample answer: Use the vertical-line test: if any vertical line intersects the graph in more than one point, then it does not represent a function; if no more than one intersection occurs, then it does represent a function.

Function: Non-function:

For Questions 16–18, graph each function.

16. $y = -2x$
17. $y = |x|$
18. $y = -x^2$

19. **Open-ended** State all the differences and similarities you can find in the two graphs shown at the right. Make predictions about the graphs if they were extended for greater values of x.

Graph A   Graph B

Sample answers: similarities: The two rays on each graph begin on the y-axis and are decreasing; differences: The rays in Graph A appear to be parallel and those in Graph B are not; predictions: Rays c and d will eventually intersect.

Answers	
12.	no correlation
13.	negative correlation
14.	positive correlation
15.	See question.
16.	See question.
17.	See question.
18.	See question.
19.	See question.

Assessment Book INTEGRATED MATHEMATICS 1
Copyright © by Houghton Mifflin Company. All rights reserved.   25

Software Support

McDougal Littell Mathpack

Function Investigator
Geometry Inventor
Stats!

Outside Resources

Books/Periodicals

Borlaug, Victoria. "From Algebra to Calculus." *Mathematics Teacher* (December 1992): pp. 282–287.

Geometry from Multiple Perspectives. Grades 9–12 Addenda. Reston, VA: NCTM, 1993.

Manipulatives

Algeblocks. Cincinnati, OH: Southwestern Publications.

Software

Schwartz, Judah and Michal Yerushalmy. *Geometric Supposer.* Scotts Valley, CA: Sunburst.

Hofer, Alan. *Graph Whiz.* Macintosh. Acton, MA: William K. Bradford Publishing.

Harvey, Wayne, Judah Schwartz, and Michal Yerushalmy. *Visualizing Algebra: The Function Analyzer.* Scotts Valley, CA: Sunburst.

Videos

Visualizing Algebra: A New Vision for Learning and Teaching Algebra. Reston, VA: NCTM, 1993.

PROJECT GOALS

➤ Students create an animated flip book.

➤ When making a flip book, students keep track of the movement and location of the figures on a coordinate grid.

➤ Students work together in a cooperative group and appreciate each other's strong points.

............●

PROJECT PLANNING

Materials List

➤ Graph paper

Project Teams

Have students work on the project individually or in groups of three. One way for the individuals in the group to distribute the work is as follows:

1. Editor: coordinates brainstorming of ideas and checks all aspects of the project.

2. Writer: writes the script for the flip book.

3. Designer: draws the pictures in sequence and colors the pictures.

Support Materials

The *Project Book* contains information about the following topics for use with this Unit Project.

➤ Project Description

➤ Teaching Commentary

➤ Working on the Unit Project Exercises

➤ Completing the Unit Project

➤ Assessing the Unit Project

➤ Alternative Projects

➤ Outside Resources

unit 4

Coordinates and Functions

ANIMATION

How can a single drawing of Bart Simpson turn into a moving picture? The smooth movements of cartoon characters are actually a series of drawings that appear to move when shown one after another. *Animation* is the process of making the drawings.

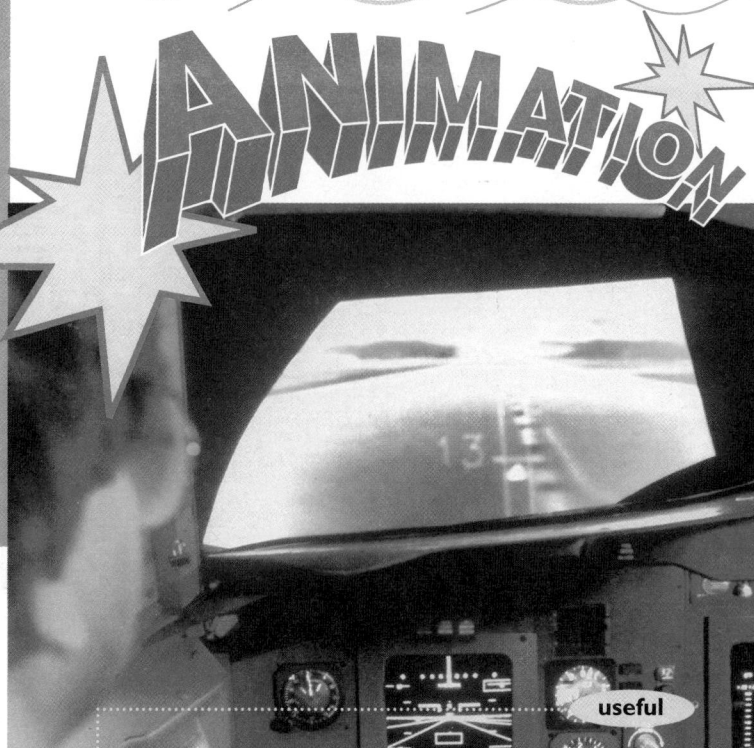

Pilots use this system to practice takeoffs.

useful

entertaining

Animation is useful as well as entertaining. Engineers use animation to simulate the flight of a new airplane design before building the airplane. Some animated programs help people learn to drive. Others help doctors practice surgery before they perform it.

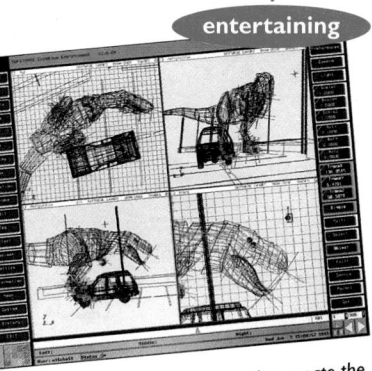

Coordinates were used to create the animated dinosaurs in Jurassic Park.

180

Suggested Rubric for Unit Project

4 Students have used their understanding of coordinates and functions successfully to create a flip book that shows a smooth continuous motion as the pages are flipped. Each picture is slightly different from the one before it. The drawings are accurate and the translations and rotations are used consistently to show the motion. The script is well written and narrates the action accurately.

3 Students understand the use of coordinates and functions to make the flip book. However, the motion is not always smooth and the positions of the drawings are not entirely consistent. The script is clear, but there is room for improvement.

Making a Flip Book

Your project is to create an animated flip book. The pages of a flip book contain pictures. Each picture is slightly different from the one before it. When you flip through the pages quickly, you see a short story showing motion.

To create a flip book, you will need to solve some of the same problems an artist or a computer programmer solves when creating animated drawings.

You will need to keep track of the location of the figures in your drawing. The movements may include slides and turns. They may follow the paths of graphs of functions. You may be able to write equations for the movements in your flip book.

$$y = -2x^2 + 2$$

Motions that repeat can be used to animate a figure. When a person or an animal is shown walking, the same leg and arm motions are shown again and again, while the background changes.

Computer programmers use equations of lines and curves that look like the motion they want to show.

An equation can describe the path of a bounced object. Using equations to define the paths of objects that are bounced or thrown makes many computer games possible.

181

Suggested Rubric for Unit Project

2 The flip book is essentially incomplete. The action is not continuous, and the use of the translations and rotations shows a lack of understanding of these concepts. The written narrative is at a minimal level. This project should be returned with suggestions for improvements and a new deadline given.

1 This project cannot be evaluated. It is illegible or incomplete. The flip book should be returned with a new deadline for completion. The group should be encouraged to speak with the teacher as soon as possible so that each student understands the purpose and the format of the project.

ALTERNATIVE PROJECTS

Project 1, page 234

Making a Map

Create a map of the neighborhood around your school or of the neighborhood in which a student lives. Use a coordinate system, make a map key, and include a scale on the map.

Project 2, page 234

Recognizing Functions in Science

Write a report on the latest findings on global warming. Discuss the evidence for global warming and what might be causing it. Include data tables and graphs with the report.

Project 3, page 234

Collecting and Graphing Data

Collect data that can be displayed in a scatter plot. Write a report that includes the method(s) used, a summary, a scatter plot, the type of correlation the scatter plot shows, and several questions that can be answered by using the plot.

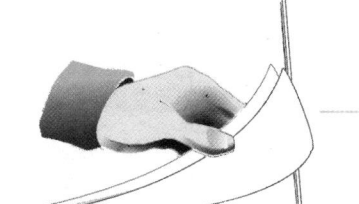

Getting Started

Here are some ideas to use when you make your flip book.

☞ Use a pad of paper or cut several sheets of paper to a convenient size.

☞ Use graph paper and a coordinate grid to keep track of the positions of your drawings.

☞ Make the first drawing on the last piece of paper. The next drawing should be slightly different.

☞ If you can see through the paper that you use, you can trace the previous drawing, then make changes to the new one.

☞ You will need to use a series of small changes to show movement. Experiment to see how many drawings are needed to show a movement.

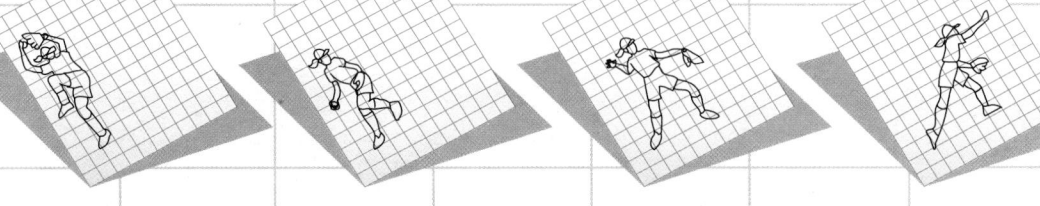

Working on the Unit Project

Your work in Unit 4 will help you create your flip book.

Related Exercises:

Section 4-1, Exercise 36
Section 4-2, Exercise 27
Section 4-3, Exercises 32–34
Section 4-4, Exercises 25, 26
Section 4-5, Exercise 31
Section 4-6, Exercise 26
Section 4-7, Exercise 45

Alternative Projects p. 234

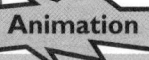 **Can We Talk**  Animation

➤ Have you seen any animated movies? If so, which ones?

➤ Do you have a favorite animated television show? If so, which one?

➤ Have you noticed any differences between animation in movies and animation in television shows? Describe the differences.

➤ Do you think it would be easier to make an animated cartoon of a car moving or of a person walking? Why?

➤ How could you show the path of an arrow being shot into the air toward a target?

➤ Suppose you want to show the motion of someone sledding down a hill. How could the background move instead of the sled?

➤ What would the path of a basketball be when it is thrown toward the hoop? Is there more than one possible path?

➤ How do objects slow down and speed up in animated movies? How do you think you could show this in a flip book?

Answers to Can We Talk?

➤ Animation examples: *Aladdin, Beauty and the Beast, Cinderella, Sleeping Beauty, The Little Mermaid.*

➤ Answers may vary.

➤ Answers may vary.

➤ a car moving; A car moving would not change but a person walking would show change in leg and arm movements.

➤ On the first page, draw an arrow at the very left and a target at the very right. On each successive page, draw the arrow a little more to the right, getting closer to the target. The path of the arrow should be a curve. On the last page, draw the arrow hitting the target.

➤ The background could show different landscape scenes for different parts of the

hill and the sled could stay the same.

➤ The basketball path would be an arc. It may be a narrow arc if the player shoots close to the basket, or a wide arc if the player shoots farther away from the basket, but it will always be an arc.

➤ The motion is picked up or slowed down by how quickly the pages are flipped.

Coordinates for Locations

Focus
Use coordinate systems to solve problems about locations in different settings.

WHERE IN THE WORLD?

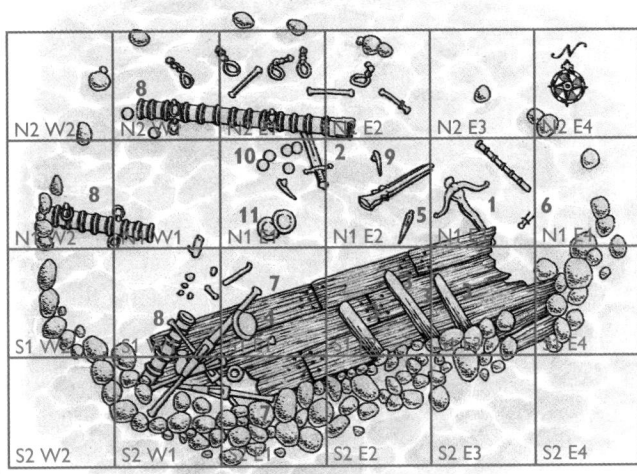

While diving in the Bahamas, you find a shipwreck. It might be a very early Spanish ship. You want to bring a sword to an expert who can tell how old it is. You also want to record where you found the sword in the wreck. What can you do?

Where objects are found in a wreck is just as important as what the objects are. Because of this, underwater explorers use a *coordinate system* to record the locations of objects.

KEY

1 Crossbow	7 Swivel
2 Swords	cannons
3 Ribs	8 Cannons
4 Helmet	9 Hand axes
5 Spearpoint	10 Cannon balls
6 Brass dividers	11 Plates

Talk it Over

1. What patterns do you see in the labels in the squares?

2. What do you think the letters in the labels represent?

3. What object is in the square labeled N1 E3?

4. Which square contains the cannon balls?

4-1 Coordinates for Locations

183

Objectives and Strands
See pages 180A and 180B.

Spiral Learning
See page 180B.

Recommended Pacing
Section 4-1 is a one-day lesson.

Toolbox Reference
➤ **Toolbox Skill 20:** Locating Points on a Coordinate Plane

Extra Practice
See pages 625–627.

Warm-Up Exercises
Warm-Up Transparency 4-1

Support Materials
➤ Practice 27
➤ Enrichment 24 in the Activity Bank
➤ Study Guide 4-1
➤ Problem Set 8
➤ Quiz 4-1
➤ Alternative Assessments 1–3

Answers to Talk it Over

1. Answers may vary but should include: Each square is labeled (N or S), (E or W), (1, 2, 3, or 4). The numbers increase as you move away from the common vertex of N1 W1, N1 E1, S1 E1, S1 W1.

2. N = North, S = South, E = East, W = West

3. crossbow or brass dividers

4. N1 E1

TEACHING

Talk it Over

Use questions 1 and 2 to help students discover the system used for labeling the grid squares. They should not have difficulty guessing that the letters indicate direction.

Questions 5–9 review the use of ordered pairs of numbers for locating points on a coordinate plane. Be sure to check that students can go in both directions, from point to ordered pair and from ordered pair to point.

Use question 10 to relate the grid system used in the map of the shipwreck to the system used for locating points in the coordinate plane.

Error Analysis

Students often give the numbers in an ordered pair in the incorrect order. They also may forget the minus sign when one or both coordinates are negative.

Correct these errors by locating the points for several ordered pairs that have the positions of the numbers reversed.

Mathematical Procedures

The idea that the order in which something is done in mathematics affects the outcome of a procedure is fundamentally important. In the context of this section, the ordered pair (2, 5) locates a specific point in the plane because the order is always to locate the horizontal first, then to locate the vertical.

Multicultural Note

Point out that the African-American mathematician and astronomer Benjamin Banneker assisted in laying out the design for the District of Columbia in 1791.

Coordinate Systems

Student Resources Toolbox
p. 660 *A Coordinate System*

In mathematics, you use an *ordered pair* of numbers to describe where a point is on a **coordinate plane.** A more general way to locate a point is to name the *quadrant* that the point is in.

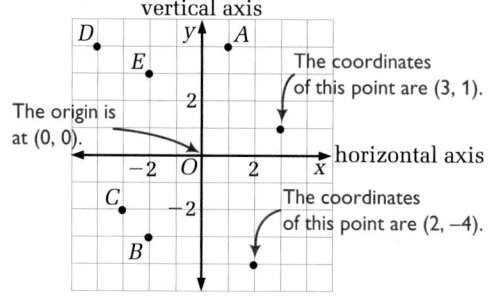

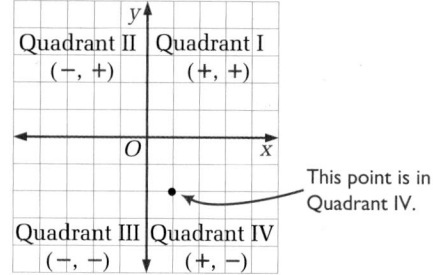

> ### Talk it Over
>
> 5. What are the coordinates of point *A*?
> 6. What point has coordinates (−2, 3)?
> 7. Name a point in quadrant III.
> 8. Which quadrant contains point *D*?
> 9. Describe how to graph the point that has coordinates (4, −3).
> 10. The origin on a shipwreck is called the *datum point.*
>   a. Where do you think the datum point is on the diagram of the wreck on page 183?
>   b. What object is near the datum point you chose in part (a)?

Maps often have coordinate systems. Their coordinates describe a square region, not a point. Usually map coordinates are a letter and a number.

WASHINGTON, D.C., LANDMARKS

1. Botanic Garden
2. Department of State
3. FBI Building
4. Lincoln Memorial
5. Martin Luther King, Jr., Memorial Library
6. National Aeronautics and Space Administration (NASA)
7. National Air and Space Museum
8. National Gallery of Art
9. Simón Bolívar Statue and Plaza
10. United States Capitol
11. Union Station
12. Vietnam Veterans Memorial
13. Washington Monument
14. White House

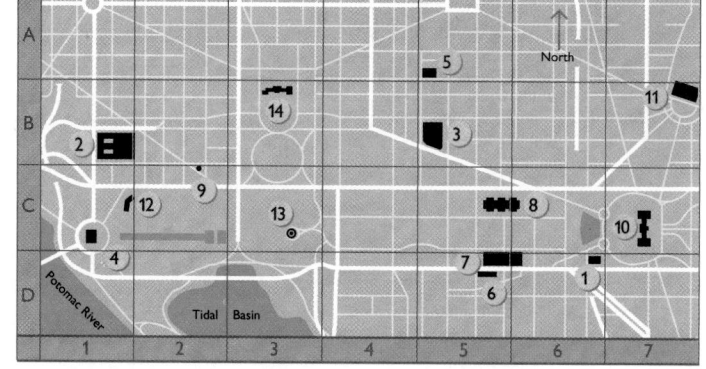

184

Unit 4 Coordinates and Functions

Answers to Talk it Over

5. (1, 4)
6. point E
7. point C or B
8. quadrant II
9. Start at the origin. Move 4 units to the right, then 3 units down and place a dot at that location.

10. a. the point two units to the right and two units below the upper left-hand corner
    b. plates

Use the map of Washington, D.C., on page 184.

a. What landmark is in square B3?

b. What are the coordinates of the Washington Monument?

Sample Response

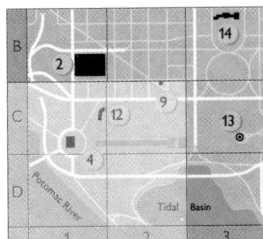

a. 1. Move across row B to column 3.

2. The building in this square is marked 14.

3. Read the list of landmarks to find what is identified by the number 14.

4. The White House is in square B3.

b. 1. Read the list of landmarks. The Monument is identified by the number 13. Find the square that contains the building marked 13.

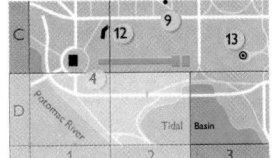

2. Move across the row to find the letter C.

3. Move down the column to find the number 3.

4. The coordinates are C3.

Look Back

Describe how using a map to locate places and using a coordinate plane to locate points are alike. How are they different?

4-1 Exercises and Problems

1. **Reading** You could call the square that contains the Washington Monument either 3C or C3. Do the coordinates (1, 3) and (3, 1) refer to the same point? Why or why not?

Use the map of Washington, D.C., on page 184 for Exercises 2–9.

What landmark is located in each square or each pair of squares?

2. B7 3. C7 4. D5, D6 5. B1

Find the coordinates of each landmark. If there is more than one pair of coordinates, list all of them.

6. Martin Luther King, Jr., Memorial Library

7. Vietnam Veterans Memorial

8. Simón Bolívar Statue and Plaza

9. National Gallery of Art

4-1 Coordinates for Locations

185

Using Technology

In connection with Talk it Over questions 5–9, you can practice and reinforce skills concerning the coordinate plane by using a graphics calculator. For the TI-81, use these settings for the RANGE screen:

Xmin = −48   Ymin = −32
Xmax = 47   Ymax = 31
Xscl = 10   Yscl = 10

For the TI-82, use these settings instead of the ones given above. The other settings remain the same.

Xmin = −47   Ymin = −31
Xmax = 47   Ymax = 31

Make sure all drawings and graphs have been cleared from the graphics screen. Students can move the screen cursor around to all the quadrants and observe the coordinates of the cursor location. With the range settings suggested above, the coordinates will always be integers.

Reasoning

Have students think about the location of points that do not belong to any of the four quadrants. Ask: If the first coordinate of a point is zero, where is the point? (on the vertical axis) If the second coordinate of a point is zero, where is the point? (on the horizontal axis)

Additional Sample

Use the street map of Washington, D.C. on page 184.

a. What landmark is in square C3? Washington Monument

b. What are the coordinates of the Lincoln Memorial? D1

Look Back

Ask students to write their responses to the Look Back as journal entries. Ask for volunteers to read their entries to the class. Use class discussion to clarify similarities and differences between a map and a coordinate plane.

Answers to Look Back

Answers may vary. An example is given. Both maps and coordinate planes specify locations by using pairs of coordinates read horizontally and vertically. But maps usually use letters and numbers to locate a particular region, whereas a coordinate plane uses a pair of numerical coordinates to locate a particular point.

Answers to Exercises and Problems

1. No; reasons may vary. An example is given. (1, 3) is the point one unit to the right and three units above the origin. This is different from (3, 1) which is three units to the right and one unit above the origin.

2. Union Station

3. the U.S. Capitol

4. National Air and Space Museum

5. Dept. of State

6. A5

7. C1, C2

8. C2

9. C5, C6

Suggested Assignment

Standard 1–23, 29–31, 32–36

Extended 1–23, 25–36

Integrating the Strands

Measurement Exs. 14–16, 20–24

Geometry Exs. 26, 32, 33, 35

Discrete Mathematics Exs. 29–31, 34

Logic and Language Exs. 1–13, 17–19, 25–28, 33, 36

Communication: Writing

In connection with Exs. 1–13, you may ask students to write their thoughts about questions such as the following: Which of the location systems studied make it possible to pinpoint a location most accurately? Why might the systems used for the shipwreck and for Washington, D.C. be better for locating objects and buildings than the ordered pair system used for coordinate planes? Which system would be best for conveying some idea of distances between locations? Which of the systems could most easily be extended to include a larger area? Ask students to explain their thinking with logical reasoning and examples.

Research

In connection with Exs. 14–25, students may find it interesting to research the International Date Line. They should relate their findings to the latitude and longitude map on this page.

They also can look for information about how boundaries for standard time zones are drawn in the U.S. and what happens when parts of the country go onto daylight saving time. Suggest that students list some common situations in which time zone differences could affect their lives.

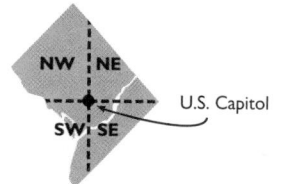

Use the diagram of Washington, D.C., at the left.

10. What is the *origin* in the coordinate system of Washington, D.C.?

11. In which direction from the Capitol is 3rd St. NE?

12. How do the four quadrants of Washington, D.C., compare with the four quadrants on a coordinate plane?

13. Which quadrant on a coordinate plane would contain G St. SW?

connection to **SOCIAL STUDIES**

Coordinates locate places on Earth. The grid lines on Earth are lines of latitude and longitude. Places are identified by latitude first, then longitude.

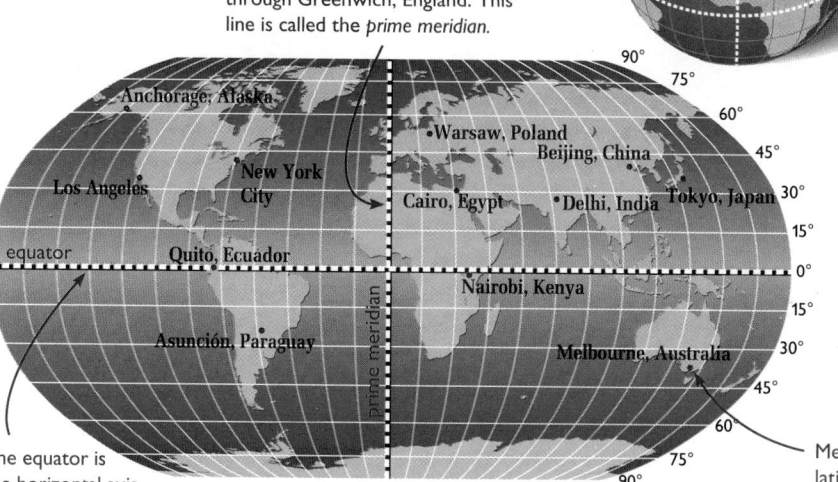

The prime meridian is a line of longitude.

The equator is a line of latitude.

The vertical axis is a line from the North Pole to the South Pole passing through Greenwich, England. This line is called the *prime meridian.*

The equator is the horizontal axis.

Melbourne has latitude 38° south and longitude 145° east. You write 38 S, 145 E.

Use the map above for Exercises 14–25.

Estimate the latitude and longitude of each city.

14. Cairo, Egypt

15. Quito, Ecuador

16. Tokyo, Japan

Answers to Exercises and Problems

10. the U.S. Capitol

11. northeast

12. Answers may vary. An example is given. The northeast quadrant is in the same direction as quadrant I in a coordinate plane, the northwest is in the same direction as quadrant II, the southwest is in the same direction as quadrant III, and the southeast is in the same direction as quadrant IV.

13. quadrant III

14. about 30 N, 31 E

15. about 0°, 78 W

16. 35 N, 140 E

What city is near each location?

17. 1 S, 37 E   **18.** 52 N, 21 E   **19.** 34 N, 118 W

20. About how many degrees west of Delhi is Asunción?

21. About how many degrees south of Los Angeles is Melbourne?

22. What line of longitude do you think is opposite the prime meridian on the other side of Earth?

23. Name two cities on the map that have about the same latitude. What is their latitude?

24. **Research** Find the latitude and longitude of your city or town.

25. **Writing** Describe how using latitude and longitude to locate places on Earth is similar to using a coordinate plane to locate points.

connection to **ART**

Chuck Close is an artist in New York who makes large paintings from small photographs. He divides the photograph and his canvas into tiny squares and uses a grid system to locate each square. Some of the grids are diagonal like the one on this painting.

26. a. Describe the grid system that Chuck Close used for this painting.

     b. How is his grid system similar to a coordinate plane? How is it different?

27. Open-ended Do you think this method could be used to make very small paintings? Explain why or why not.

28. An illness has reduced Chuck Close's mobility. For this painting, he used a chair that can move sideways and up and down to work on different areas of the painting.

Describe how the movements of the chair are related to locating a point on a coordinate plane.

Each axis is labeled with numbers.

Both the photo and the painting have a horizontal axis.

4-1 Coordinates for Locations   **187**

26. a. He has the same horizontal axis for both the photo and the painting. The scale is bigger for the painting. He looks at a grid square in the photo and paints it on a larger scale on the corresponding grid square in the painting.

     b. They both have a horizontal axis. A coordinate plane also has a vertical axis. Painting each grid square is like plotting points on the coordinate plane.

27. Answers may vary. An example is given. Yes, but it would be more difficult, since the smaller the grid square you have to paint, the harder to paint the details accurately.

28. When you plot or locate a point, you go horizontally right or left on the x-axis and then vertically up or down on the y-axis. These are the same movements the artist's chair makes.

Answers to Exercises and Problems

17. Nairobi, Kenya

18. Warsaw, Poland

19. Los Angeles

20. about 135°

21. about 70°

22. 180° longitude or the International Date Line

23. Estimates may vary. Examples: Beijing, China, and New York, about 40 N; Quito, Ecuador, and Nairobi, Kenya, about 0°; Tokyo, Japan, and Los Angeles, about 35 N.

24. Answers may vary.

25. Answers may vary. An example is given. Both systems use a pair of numbers in a specified order to locate points. Some differences are that the numbers used to locate latitude and longitude lie within a limited range, whereas numbers in a coordinate plane have no limits. On Earth, N, S, E, and W are used to indicate directions, whereas the coordinate plane uses positive and negative signs. On Earth, latitude and longitude are measured in degrees, whereas the coordinate plane uses lengths.

For Exs. 29–31, ask students to sketch and label a portion of the coordinate system that Captain Ramius is using. Ask them to shade lightly the nine squares that Ramius was thinking of in the passage that was quoted.

Assessment: Portfolio

Use Ex. 32 for a class discussion. The written response to Ex. 33 can be evaluated for inclusion in a student's portfolio.

connection to **LITERATURE**

In *The Hunt for Red October,* Soviet submarine captain Ramius reads his orders and learns that he is to participate in a tracking drill with another submarine.

---

17   **THE HUNT FOR RED OCTOBER**

Ramius chuckled. "The boys in the attack submarine directorate still have not figured out how to track our new drive system. Well, neither will the Americans. We are to confine our operations to grid square 54-90 and the immediately surrounding squares. That ought to make Viktor's task a bit easier." … "He will have a fair chance of locating us, I think. The exercise is confined to nine squares, forty thousand square kilometers. We shall see what he has learned since he has served with us."

---

29. How does Ramius identify a grid square?

30. How does Ramius know that there are only nine squares that might contain the submarine?

31. What do you think the numbers of the squares surrounding grid square 54-90 could be? Explain your reasoning.

Ongoing **ASSESSMENT**

There were about 10,000 Anasazi people living in Chaco Canyon, New Mexico, about 1000 years ago. They suddenly disappeared. Archaeologists are trying to understand why.

Archaeologists use coordinate systems to record locations of objects. As they dig deeper, the relationships of the locations can be clues to the daily life in lost civilizations.

32. Open-ended Describe how you could put a coordinate system on the sketch of the room. Where would you put the origin? the axes? Use your system to give the coordinates of the firepit.

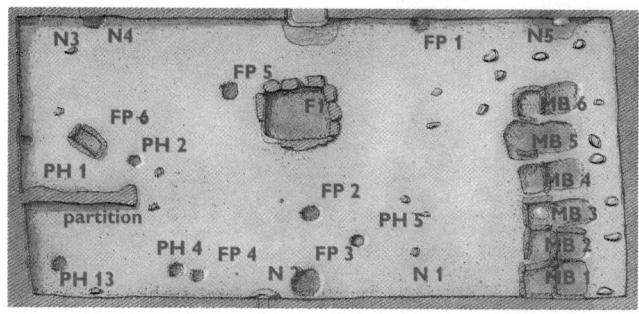

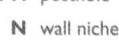

KEY

F   firepit
FP   misc. floor pit
MB   mealing bin
PH   posthole
N   wall niche

33. Writing Archaeologists also record an object's depth beneath surface level. Describe how coordinates could identify the depth of an object.

Answers to Exercises and Problems

29. with the numbers "54–90"

30. The square 54–90 and surrounding squares total nine.

53–91	54–91	55–91
53–90	54–90	55–90
53–89	54–89	55–89

31. Answers and reasons may vary. An example is given. 53–91, 54–91, 55–91, 53–90, 55–90, 53–89, 54–89, 55–89. I thought of 54–90 as the point (54, 90) on the coordinate plane. I then figured out the coordinates of the eight surrounding points.

32. Answers may vary, but should include a sketch with an origin, axes, and coordinates using letters and numbers or numbers only.

33. A three-dimensional system, using an additional coordinate for depth, should be used.

The display shows the hourly salaries for temporary jobs in high technology fields. *(Section 3-6)*

34. a. Interpret the information shown in the display.

    b. Do you think this is the best way to display the information? Explain why or why not.

35. Quadrilateral *ABCD* has two pairs of parallel sides and at least one right angle. *(Section 1-7)*

    a. Make an accurate sketch of quadrilateral *ABCD*.

    b. What is the special name for quadrilateral *ABCD*?

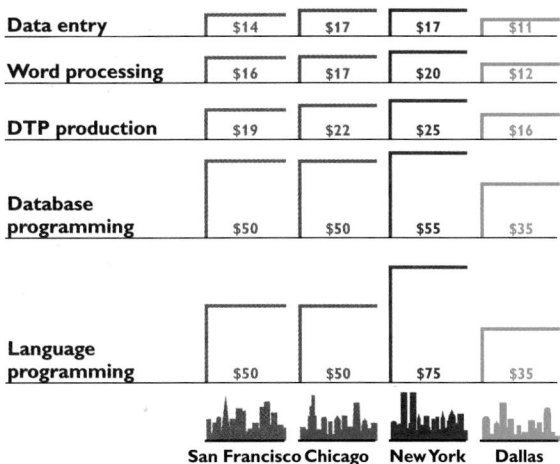

HOURLY FEES FOR TEMPS

	San Francisco	Chicago	New York	Dallas
Data entry	$14	$17	$17	$11
Word processing	$16	$17	$20	$12
DTP production	$19	$22	$25	$16
Database programming	$50	$50	$55	$35
Language programming	$50	$50	$75	$35

 Working on the Unit Project

A computer screen is made up of tiny points of light called *pixels.* Programmers use coordinates to identify which pixels to light when creating artwork. Different computers have different numbers of pixels.

pixel (0, 0)   pixel (4, 2)

pixel (1, 8)

36. Writing If you turn off the pixel at (4, 2) and turn on the one directly below it, it will look as if the pixel moved down.

    a. Describe how you could make the pixel "fall" to the bottom of the screen.

    b. How do the coordinates of the pixel change as it "falls"?

    c. Suppose you want to put a falling image in your flip book. Use your answers to parts (a) and (b) to describe how to do this.

BY THE WAY...

Pixels are called pixels because they are *pix* (pictures) *elements.*

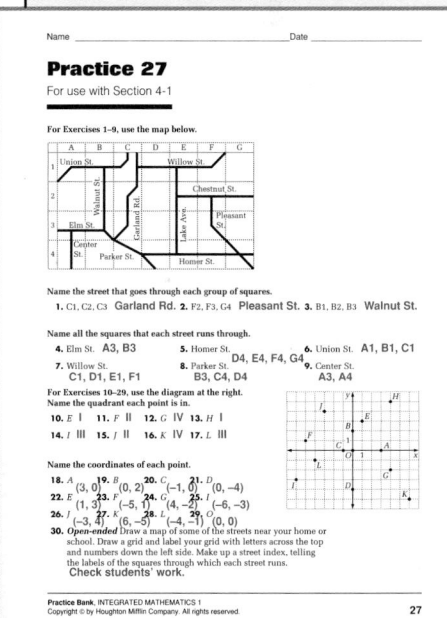

189

Working on the Unit Project

Students should understand that a computer screen illustrates another type of coordinate grid. It is similar to working in quadrant IV of a coordinate plane, with the exception that the vertical axis has positive values, not negative ones. The origin is always located at the upper left corner of the computer screen.

Practice 27 For use with Section 4-1

Name _____ Date _____

Practice 27
For use with Section 4-1

For Exercises 1–9, use the map below.

Name the street that goes through each group of squares.
1. C1, C2, C3   **Garland Rd.**   2. F2, F3, G4   **Pleasant St.**   3. B1, B2, B3   **Walnut St.**

Name all the squares that each street runs through.
4. Elm St. **A3, B3**   5. Homer St. **D4, E4, F4, G4**   6. Union St. **A1, B1, C1**
7. Willow St. **C1, D1, E1, F1**   8. Parker St. **B3, C4, D4**   9. Center St. **A3, A4**

For Exercises 10–29, use the diagram at the right.
Name the quadrant each point is in.
10. E **I**   11. F **II**   12. G **IV**   13. H **I**
14. I **III**   15. J **II**   16. K **IV**   17. L **III**

Name the coordinates of each point.
18. A **(3, 0)**   19. B **(0, 2)**   20. C **(−1, 0)**   21. D **(0, −4)**
22. E **(1, 3)**   23. F **(−5, 1)**   24. G **(4, −2)**   25. I **(−6, −3)**
26. I **(−3, 4)**   27. K **(6, −5)**   28. L **(−4, −1)**   29. **(0, 0)**

30. *Open-ended* Draw a map of some of the streets near your home or school. Draw a grid and label your grid with letters across the top and numbers down the left side. Make up a street index, telling the labels of the squares through which each street runs.
Check students' work.

27

Answers to Exercises and Problems

34. a. Answers may vary. For example, the highest salaries are earned for language programming in New York, the lowest for data entry in Dallas.

    b. Answers may vary. For example, a combined histogram using a single set of axes and color coding for the cities would make many comparisons easier.

35. a.

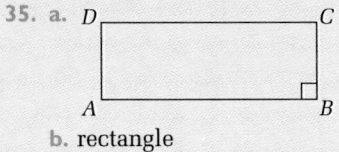

    b. rectangle

36. a. Repeatedly turn the pixel off and turn on the one below.

    b. The second coordinate increases by one each time; the first remains the same.

    c. Answers may vary. An example is given. At each successive page, draw the image a little lower than on the page before.

Objectives and Strands
See pages 180A and 180B.

Spiral Learning
See page 180B.

Materials List
➤ Graph paper
➤ Ruler
➤ Long pieces of string

Recommended Pacing
Section 4-2 is a two-day lesson.
Day 1
Pages 190–192: Exploration through Talk it Over, *Exercises 1–11*
Day 2
Pages 192–194: Finding Areas of Other Polygons through Look Back, *Exercises 12–27*

Toolbox Reference
➤ **Toolbox Skill 21:** Graphing Points on a Coordinate Plane

Extra Practice
See pages 625–627.

Warm-Up Exercises
Warm-Up Transparency 4-2

Support Materials
➤ Practice 28
➤ Enrichment 25 in the Activity Bank
➤ Study Guide 4-2
➤ Problem Set 8
➤ Diagram Masters 2, 3 in the Explorations Lab Manual
➤ McDougal Littell Mathpack software: *Geometry Inventor*
➤ Quiz 4-2

Section 4-2

Introduction to Coordinate Geometry

Focus
Learn about coordinate geometry, identify polygons on a coordinate plane, and find the areas of polygons drawn on a coordinate plane.

SHAPING UP

EXPLORATION
(How) can you describe a shape?

* **Materials: graph paper, ruler**
* **Work with a partner.**

The center of this Islamic stucco tile contains the emblem of the craftsmen who made it.

① Describe the shape of the star tile in words.

② List the coordinates of the corners of the tile in the order that they are connected by line segments.

③ What are some advantages of describing a shape in words (as you did in step 1)? What are some advantages of describing a shape with coordinates (as you did in step 2)?

④ **a.** Draw any polygon on graph paper. Don't let your partner see it.

b. Use words or coordinates to describe your polygon to your partner.

c. Have your partner draw the polygon using only your description.

d. Compare your partner's drawing with your original polygon. How could you improve your description?

Drawing figures on a coordinate plane is one part of *coordinate geometry*. When you analyze a shape on a coordinate plane, find its area, or compare it to another shape, you use coordinate geometry.

Answers to Exploration

1. an eight-pointed star

2. Moving clockwise around the star tile, the coordinates of the 16 corners, to the nearest integer, are:
(0, 4); (1, 3); (3, 3); (3, 1); (4, 0); (3, −1); (3, −3); (1, −3); (0, −4); (−1, −3); (−3, −3); (−3, −1); (−4, 0); (−3, 1); (−3, 3); (−1, 3).

3. Answers may vary. Words can give a mental picture of the star tile. Coordinates allow someone else to reproduce accurately the shape of the star tile.

4. Answers may vary.

Sample 1

For each set of points, follow these steps.

➤ **Plot the points on a coordinate plane. Connect the points in order and connect the last point to the first.**

➤ **Write the specific name of the polygon you formed.**

➤ **Explain how you know what it is.**

a. $A(5, 1)$, $B(2, 1)$, $C(2, -2)$

b. $D(3, 1)$, $E(1, 3)$, $F(-2, 3)$, $G(-3, -1)$, $H(-1, -3)$, $J(2, -3)$

Sample Response

a. The polygon is an isosceles right triangle because. . .

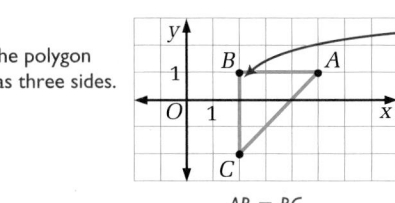

The polygon has three sides.

$\angle B$ is a right angle formed by the horizontal and vertical segments.

$AB = BC$

b. The polygon is a hexagon because. . .

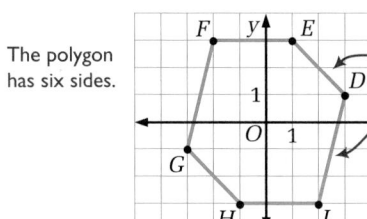

The polygon has six sides.

The sides are not the same length. The polygon is not a regular hexagon.

Talk it Over

$\triangle ABC$ means "triangle ABC."

1. How do you know that $\triangle ABC$ in Sample 1 is isosceles and not equilateral?

2. Are any sides of hexagon $DEFGHJ$ congruent? If so, which?

Finding Areas of Right Triangles

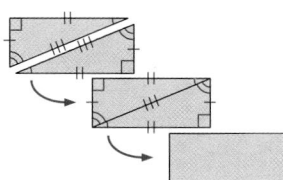

Remember that a rectangle is made up of two congruent right triangles.

You can use that fact to find the area of a right triangle on a coordinate plane.

4-2 Introduction to Coordinate Geometry

191

Answers to Talk it Over

1. Answers may vary. Examples are given. An equilateral triangle has all 60° angles; $\triangle ABC$ has one 90° angle. Also, $\overline{AC}$ is longer than $\overline{AB}$ or $\overline{BC}$, so $\triangle ABC$ is not equilateral.

2. Yes; $\overline{FE}$ and $\overline{HJ}$; $\overline{GH}$ and $\overline{ED}$; $\overline{GF}$ and $\overline{JD}$.

Exploration

The main goal of the Exploration is for students to discover how to describe polygons on a coordinate plane in terms of the coordinates of their vertices. A secondary goal is to convince students that this way of describing the polygons is generally more efficient and more accurate than a word description.

Additional Sample

S1 For each set of points, follow these steps:

• Plot the points on a coordinate plane. Connect the points in order and connect the last point to the first.

• Write the specific name of the polygon you formed.

• Explain how you know what it is.

a. $P(-1, -1)$, $Q(2, -1)$, $R(2, -3)$, $S(-1, -3)$ For the diagram, check students' drawings. The polygon is a rectangle. $\overline{PQ}$ and $\overline{SR}$ are horizontal segments 3 units long, and they are perpendicular to the vertical sides $\overline{PS}$ and $\overline{QR}$ which are each 2 units long.

b. $K(1, 2)$, $L(4, 2)$, $M(4, -2)$, $N(1, -2)$, $O(0, 0)$ For the diagram, check students' drawings. The polygon is a pentagon, but not a regular pentagon. It has five sides. $KL = MN = 3$, $LM = 4$, and OK and ON are between 2 and 3. Two angles are right angles, while the other three angles are obtuse.

Talk it Over

Use questions 1 and 2 to review the meanings of the terms *equilateral, isosceles,* and *congruent.*

191

Additional Sample

S2 Find the area of right triangle *KLM*.

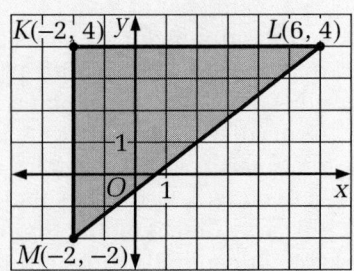

Area = 24 square units

Talk it Over

Use question 3 to check students' recall of the definition of rectangle and of the fact that horizontal and vertical lines form right angles.

Questions 4 and 5 should help students see how lengths of horizontal segments can be found by subtracting the first coordinates, and lengths of vertical segments by subtracting the second coordinates.

Question 6 reinforces the idea that the procedure in Sample 1 works for every right triangle that has a horizontal and a vertical leg.

Sample 2

Find the area of right triangle *XYZ*.

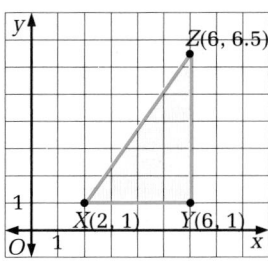

Sample Response

Problem Solving Strategy: Use a diagram

Insert helping lines to build a rectangle.

Label point W.

length 5.5

width: 4

Area of $\triangle$ = $\frac{1}{2}$ × Area of $\square$

= $\frac{1}{2} lw$

= $\frac{1}{2}(5.5)(4)$

= 11

The area of $\triangle XYZ$ is 11 square units.

Talk it Over

3. How do you know that figure *WXYZ* in Sample 2 is a rectangle?

4. How is the width 4 related to the coordinates of *X* and *Y*?

5. How is the length 5.5 related to the coordinates of *Y* and *Z*?

6. a. Plot the points $R(-1, -4)$, $S(-8, -4)$, and $T(-8, -1)$ on a coordinate plane. Connect the points in order. Connect the last point to the first.

   b. Find the area of the right triangle you formed.

▶ Now you are ready for:
Exs. 1–11 on p. 194

Finding Areas of Other Polygons

You can use what you know about the areas of rectangles and right triangles to find the areas of other polygons drawn on a coordinate plane.

Answers to Talk it Over

3. It has 4 right angles and the opposite sides are equal in length.

4. The width 4 is the same as the difference between the first coordinates of *X* and *Y*.

5. It is the difference of their second coordinates.

6. a.

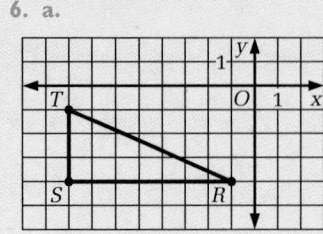

   b. area of $\triangle RST$ = 10.5 square units

Sample 3

Find the area of trapezoid LMNP.

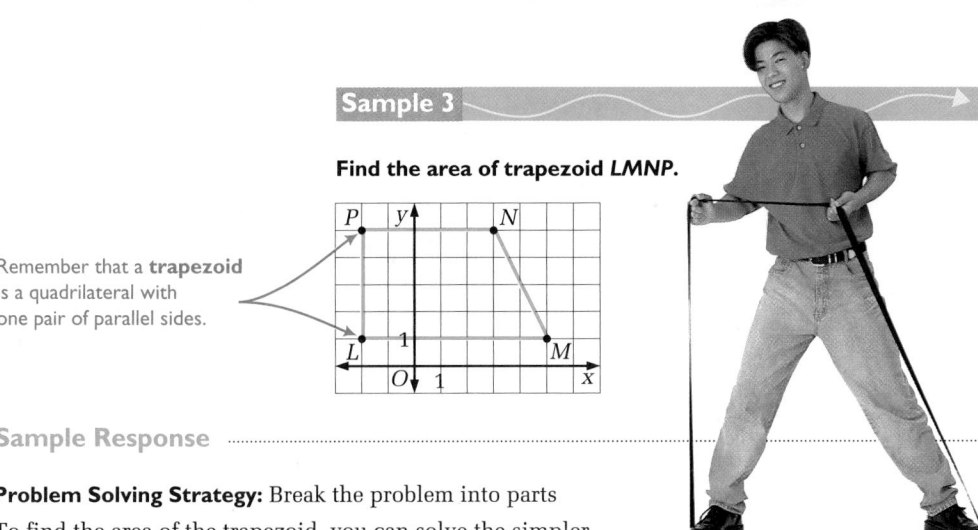

Remember that a **trapezoid** is a quadrilateral with one pair of parallel sides.

Sample Response

Problem Solving Strategy: Break the problem into parts

To find the area of the trapezoid, you can solve the simpler problems of finding the areas of a rectangle and a right triangle.

Area of ⬜ = area of ⬜ − area of ◺

First find the area of rectangle *LMRP*.

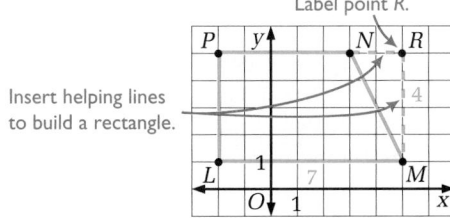

Label point *R*.

Insert helping lines to build a rectangle.

Use $A = lw$.

Area of ⬜ = 7(4) = 28

Next find the area of △*NMR*.

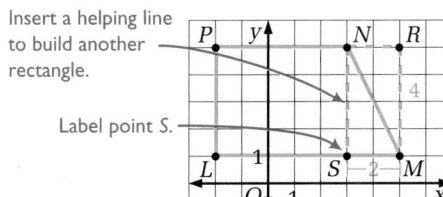

Insert a helping line to build another rectangle.

Label point *S*.

Area of rectangle *NSMR* = 4(2) = 8

Area of ◺ = $\frac{1}{2}$ (8) = 4

Then subtract.

Area of ⬜ = area of ⬜ − area of ◺

= 28 − 4

= 24

The area of trapezoid *LMNP* is 24 square units.

4-2 Introduction to Coordinate Geometry

193

Additional Sample

S3 Find the area of trapezoid *GHIJ*.

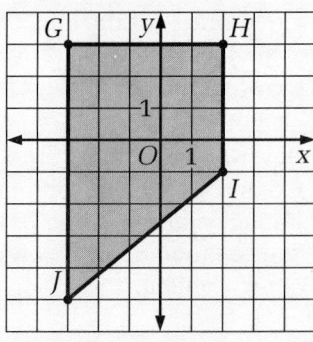

Area = 30 square units

Reasoning

After discussing Talk it Over question 8, you may wish to return to Additional Sample S3 to examine another way of finding the area of trapezoid *GHIJ*.

Problem Solving

Note that Samples 2 and 3 are handled by appealing only to the formula for the area of a rectangle. This approach helps students to understand the problem-solving strategies of using a diagram and breaking a problem into parts. It avoids using the formula for finding the area of a right triangle ($A = \frac{1}{2}bh$), which students may not yet understand, but it does provide an intuitive basis for understanding this formula.

193

Talk it Over

Use question 7 to remind students that if they cannot recall the formula for the area of a rectangle, they can use reasoning or simple counting to find the area.

Question 8 will help students see that a problem-solving strategy can sometimes be used in more than one way on the same problem.

Look Back

Students can make their solution of the Look Back problem into a good journal entry.

APPLYING

Suggested Assignment

Standard 1–15, 17, 18, 20–27

Extended 1–15, 17, 18, 20–27

Integrating the Strands

Number Exs. 22, 23

Algebra Exs. 24, 25

Measurement Exs. 10–19, 21

Geometry Exs. 1–21, 26, 27

Logic and Language Exs. 4, 5, 27

Error Analysis

In Exs. 6–9, watch for students who incorrectly plot points that have one coordinate equal to zero. Such points will lie on the *x*-axis or *y*-axis, but students often inadvertently put them on the wrong axis. Correct the error by reminding students that the first coordinate tells how far to the *left or right* of the origin the point is, while the second coordinate tells how far *above or below* the origin the point is.

Using Technology

Students can use the *Geometry Inventor* software to draw geometric figures on a coordinate plane for Exs. 6–9, 18, and 19.

Talk it Over

7. Sample 3 uses the area formula to find the area of rectangle *NSMR*. Describe another method for finding this area.
8. Try Sample 3 another way. Divide the trapezoid into rectangle *LSNP* and right triangle *NSM*. Explain how to use these polygons to find the area of trapezoid *LMNP*.

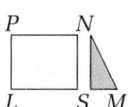

Look Back

Plot the points $K(-6, -3)$, $L(-2, -3)$, and $M(-4, 3)$ on a coordinate plane. Connect the points in order. Connect the last point to the first. Describe the polygon. Explain how you know what it is. Explain how to find the area of the polygon.

Now you are ready for: Exs. 12–27 on pp. 194–196

4-2 Exercises and Problems

Suppose you had to draw each shape on graph paper. For each shape, tell whether you would prefer to use coordinates or a description in words. Explain your choice.

1. a square
2. a circle
3. a parallelogram
4. **Writing** Draw a scalene right triangle on a coordinate plane. Then write directions that someone else could follow to draw the same triangle.
5. **Reading** Suppose all the sides of the hexagon in Sample 1 were congruent. Does this mean that all the angles must be congruent? Explain why or why not.

Follow these steps for Exercises 6–9.

- **Plot the points on a coordinate plane. Connect the points in order and connect the last point to the first.**
- **Write the specific name of the polygon you formed.**
- **Explain how you know what it is.**

6. $A(-1, -1)$, $B(0, 2)$, $C(4, -1)$
7. $M(0, 4)$, $N(3, 0)$, $P(6, 0)$, $Q(7, 5)$, $R(2, 6)$
8. $H(1, 3)$, $J(5, 3)$, $K(7, -2)$, $L(-2, -2)$
9. $S(0, 3)$, $T(-8, 3)$, $U(-8, -5)$, $V(0, -5)$

Find the area of each polygon. Describe the method you used.

10.

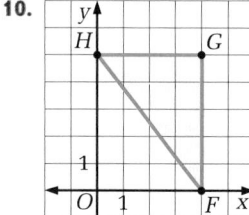

11.

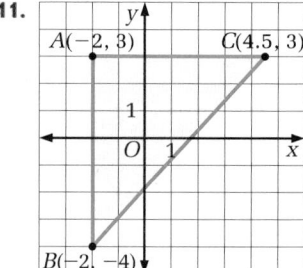

12.

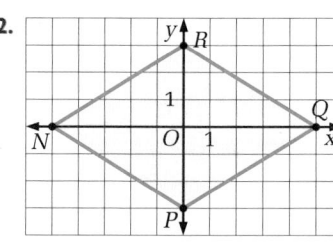

Answers to Talk it Over

7. Answers may vary. For example, another method would be to count the number of squares in rectangle *NSMR*.
8. Add the areas of rectangle *LSNP* and $\triangle NSM$ to get the area of trapezoid *LMNP*.

Answers to Look Back

The three segments form an isosceles triangle. Use a ruler to check that two sides are equal in length. To find the area, divide this triangle in half to make two right triangles. Then draw lines to make rectangles. The area of each rectangle is 12 square units. The area of the original triangle is half the total area of the rectangles. Since the total area of the rectangles is 24 square units, the area of the isosceles triangle is 12 square units.

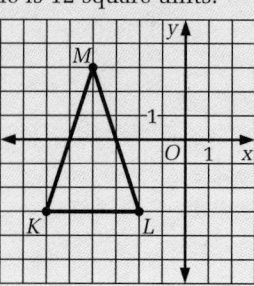

13.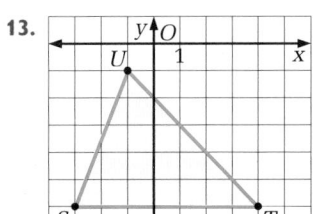

14.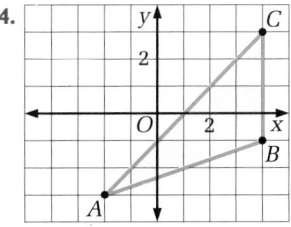

15.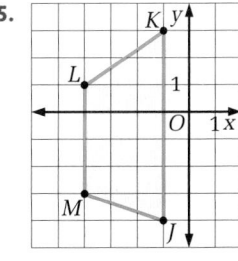

16. **Using Manipulatives** Work with another student.

   a. Choose one of the polygons in Exercises 10–15. Create a human-sized version of the polygon with a long piece of string. (Use the photographs on pages 192 and 193 as a guide.)

   b. Have your partner guess which polygon you chose.

   c. Switch places and repeat the process.

17. Four students are making flags out of checkered cloth for a school play. Which student has the most cloth left over?

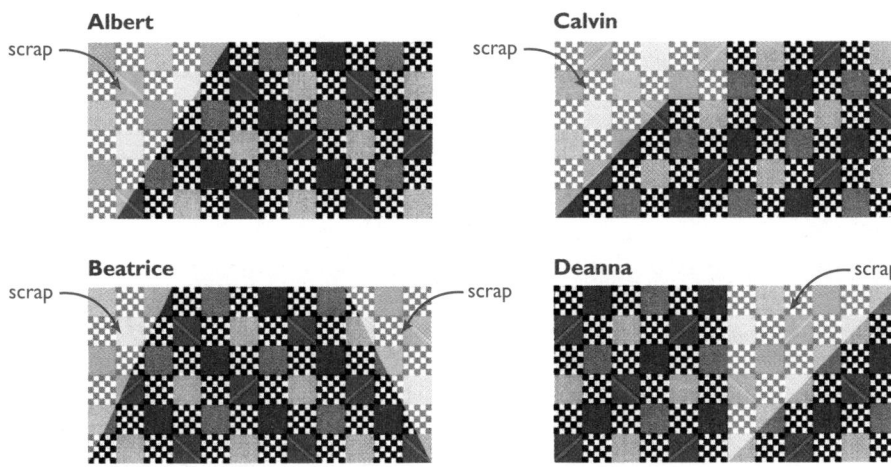

18. **Open-ended** On a coordinate plane, create two different triangles with area 20 square units.

19. **Group Activity** Each person in the group should make a different rectangle.

   a. Draw a rectangle on a coordinate plane. List the coordinates of the vertices of the rectangle and find the area of the rectangle.

   b. Double each coordinate. Graph the new rectangle. Find its area.

   c. Compare the areas of the original rectangle and the new rectangle. How are they related?

   d. Did everyone in the group get the same result in part (c)? Make a generalization about the results.

4-2 Introduction to Coordinate Geometry   **195**

Problem Solving

The problems in Exs. 10–15 can all be solved by using the problem-solving strategy of breaking the problem into parts. Nevertheless, some students are thrown off at first by the fact that the polygons for Exs. 12–15 do not have right angles. If students do not see what to do, you might give them a hint by pointing out that breaking a problem into parts may involve breaking it into *more than two* parts.

6. acute isosceles triangle; △*ABC* has 3 sides, 2 of equal length, and 3 angles each less than 90°.

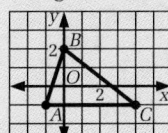

7. pentagon; The figure has 5 sides.

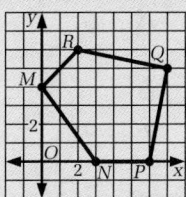

8. trapezoid; The figure has 4 sides, 2 parallel sides.

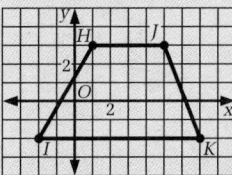

9. square; The figure has 4 equal sides and 4 right angles.

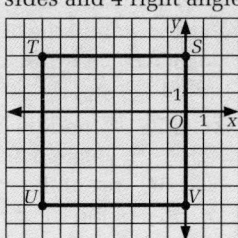

10–15. Methods may vary. Examples are given.

10. 10 square units; I found the area of a rectangle formed by the points *H, G, F* and the origin. Half this area is the area of the triangle.

11. 22.75 square units; I found the area of a rectangle with length 7 and width 6.5. Half this area is the area of the triangle.

Answers continued on next page.

Answers to Exercises and Problems

1–3. Answers may vary. Examples are given.

1. description; Describing a square is simple: a polygon with 4 right angles and 4 equal sides.

2. description; It would be difficult to give coordinates that would not result in straight lines.

3. coordinates; There are different types of parallelograms, so coordinates would probably be simpler than a description.

4. Answers may vary, but should include three coordinates and instructions to connect them in order. The figure must have a right angle and three unequal sides.

5. No. The hexagon with the following vertices has all sides congruent, but not all angles are congruent: *A*(0, 3), *B*(4, 6), *C*(9, 6), *D*(13, 3), *E*(9, 0), *F*(4, 0).

195

Practice 28 For use with Section 4-2

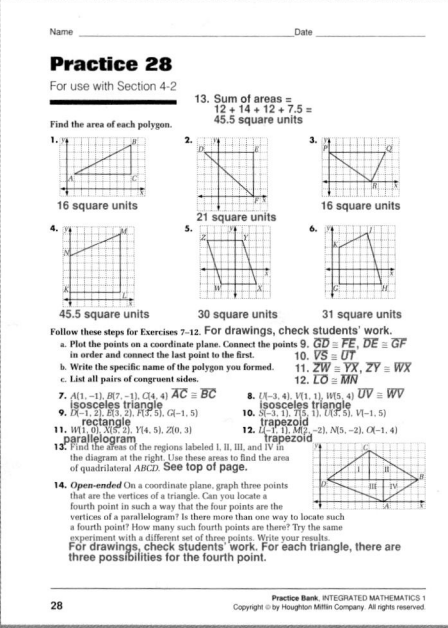

Answers to
Exercises and Problems ··············

12. 30 square units; I found the area of one triangle formed by the axes and multiplied it by 4.

13. 17.5 square units; I found the area of a rectangle with length 5 and width 7. Half the area of the rectangle is the area of the triangle.

14. 12 square units; I found the area by counting the squares inside the triangle.

15. 16.5 square units; I found the area of a rectangle with length 7 and width 3. Then I subtracted the areas of the two triangles that are not part of the trapezoid.

16. Choices of polygons may vary.

17. Calvin

18. Answers may vary, but the product of the base and related altitude of each triangle must be 40.

196

---

20. **a.** The vertices of a rectangle are $A(-6, 2)$, $B(-6, -3)$, $C(1, 2)$ and D. What are all the possible coordinates of D?

b. The vertices of a parallelogram are $W(-3, -10)$, $X(-3, 6)$, $Y(9, 6)$, and Z. What are all the possible coordinates of Z?

Ongoing **ASSESSMENT**

21. **Open-ended** Design an interesting polygon with at least six sides on a coordinate plane. Find the area of the polygon using rectangles, squares, and right triangles.

Review **PREVIEW**

Simplify each expression. *(Sections 1-4, 1-5)*

22. $3(6) + 10 - (-3)(7)$ 23. $(17 + (-3)) \div 7 - (-11)$

24. $4(x - 3) + 5x$ 25. $3a(5 + b) - 6a$

26. **a.** The coordinates of point A are $(4, 3)$. Graph point A. *(Section 4-1)*

b. What are the coordinates of a point that is 5 units below point A? Label this point B on your graph.

c. What are the coordinates of a point that is 6 units to the left of B? Label this point C on your graph.

d. Draw $\triangle ABC$. What type of triangle is it?

4 **Working on the Unit Project**

27. The diagrams below show a flip book sequence of parallelograms.

a. Describe the motion the flip book shows. Which coordinates of the parallelograms change in the sequence? Which do not change?

b. Find the areas of the parallelograms. Is one area larger than the others? If so, which one?

c. *True* or *False?* All parallelograms with sides of the same length have the same area.

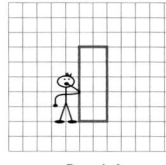
Ready?

here it goes...

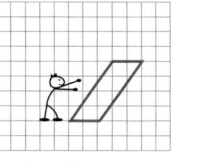
going...

ta da!

---

19. **a–d.** Answers may vary depending on choice of original rectangle. Answers are based on rectangle chosen in part (a).

a. $A(-6, 2)$; $B(-6, -2)$; $C(6, -2)$; $D(6, 2)$; 48 square units

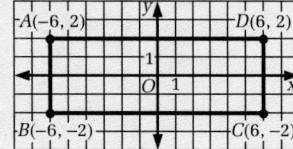

b. $A(-12, 4)$; $B(-12, -4)$; $C(12, -4)$; $D(12, 4)$; The area is 192 square units.

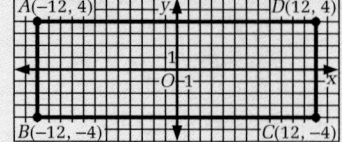

c. The area of the rectangle in part (b) is 192 square units, 4 times the area of the original rectangle.

d. When you double the coordinates, you double the length of each side of the rectangle. The area of the new rectangle is 4 times the area of the original.

20. **a.** D can only be $(1, -3)$.

b. Z can be $(9, 22)$, $(9, -10)$, or $(-15, -10)$.

21–27. See answers in back of book.

4-3 Translations

Focus
Translate figures on a coordinate plane and recognize translational symmetry.

FROM HERE TO THERE

When you ride on an escalator, only your location or position changes. You do not turn around, grow, shrink, or change shape. Escalators just slide you from place to place. Another name for a slide is a *translation*. A **translation** moves an object without changing its size or shape and without turning it or flipping it.

Here is a translation of △*ABC* on a coordinate plane.

Each vertex slides the same distance in the same direction.

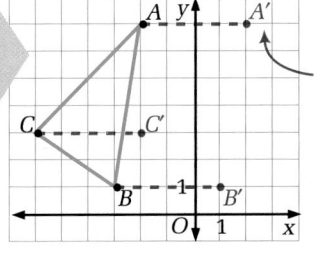

The new vertices are labeled with *primes*.

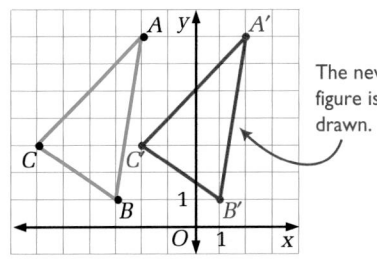

The new figure is drawn.

Talk it Over

1. Look at each vertex of △*ABC*. How far did it move? In what direction did it move?

2. Are the coordinates of every point of △*ABC* changed?

3. What are the coordinates of *B* and *B′*? How are the coordinates alike?

4-3 Translations

 197

Answers to Talk it Over

1. 4 units; right

2. Yes.

3. *B*(−3, 1) and *B′*(1, 1); They have the same *y*-coordinate.

PLANNING

Objectives and Strands
See pages 180A and 180B.

Spiral Learning
See page 180B.

Materials List
➤ Graph paper
➤ Ruler

Recommended Pacing
Section 4-3 is a one-day lesson.

Extra Practice
See pages 625–627.

Warm-Up Exercises
Warm-Up Transparency 4-3

Support Materials
➤ Practice 29
➤ Enrichment 26 in the Activity Bank
➤ Study Guide 4-3
➤ Problem Set 8
➤ Diagram Master 2 in the Explorations Lab Manual
➤ Geometry Inventor Activity Book: Activity 22
➤ Quiz 4-3
➤ Alternative Assessment 4

TEACHING

Limited English Proficiency

Kinesthetically demonstrate the concept of translation by sliding a clear plastic triangle from one position to another on a sheet of graph paper. Encourage students to carry out this action themselves while they discuss what things change and what things remain the same.

Additional Sample

Translate △HJK 4 units left and 5 units up. What are the coordinates of each vertex after the translation?

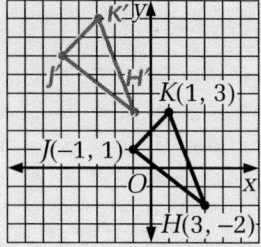

The new coordinates are H'(−1, 3), J'(−5, 6), and K'(−3, 8).

Reasoning

Ask students if a translation of 4 units left and 5 units up is the same as 5 units up and 4 units left.

Using Manipulatives

You may wish to have students use construction paper or cardboard cutouts of triangles, parallelograms, or other polygons to demonstrate translations. Ask students to trace around the shape to show the original location and then trace around it once more to show the final location.

Sample

Find the coordinates of each vertex of △DEF after the translation:

6 units right and 2 units down

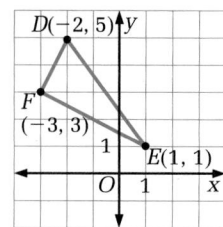

Sample Response

Translate each vertex. Then draw the triangle.

1 First, F moves 6 units right.

 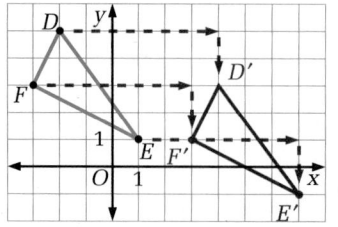 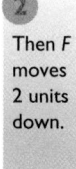

2 Then F moves 2 units down.

An arrow (→) that means "goes to" can be used to represent a translation in symbols.

The translation is written this way: $\triangle DEF \rightarrow \triangle D'E'F'$

$D(-2, 5) \rightarrow D'(4, 3)$ $E(1, 1) \rightarrow E'(7, -1)$ $F(-3, 3) \rightarrow F'(3, 1)$

The coordinates of each vertex after the translation are $D'(4, 3)$, $E'(7, -1)$, and $F'(3, 1)$.

Talk it Over

4. Suppose you change only the first coordinates of the points of a shape. What is the direction of the translation?

5. Suppose you change only the second coordinates of the points of a shape. What is the direction of the translation?

Translational Symmetry

Some computer software lets you choose the background pattern for the screen or a fill pattern for a drawing.

1. You select a simple pattern like the squares shown.

2. The computer translates this pattern again and again until it covers the screen or drawing.

198

Unit 4 Coordinates and Functions

Answers to Talk it Over

4. left or right

5. up or down

A design created by translating a pattern has **translational symmetry.** You see repeating patterns with translational symmetry on clothing, wrapping paper, and wallpaper.

Look Back

How do the coordinates of the vertices of a polygon change for each of these translations?

a. a number of units right or left

b. a number of units up or down

c. a number of units right or left *and* a number of units up or down

Exercises and Problems

1. **Reading** Is a translation most like a *flip*, a *slide*, or a *turn*? Explain your choice.

Find the coordinates of each vertex of △TDE after each translation.

2. 3 units down

3. 2 units left

4. 1 unit right and 2 units down

5. 5 units left and 4 units up

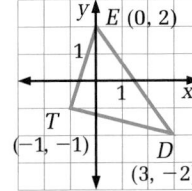

E (0, 2)
T (−1, −1)
D (3, −2)

Tell whether each picture shows a translation. If it does, describe how the vertices moved.

6.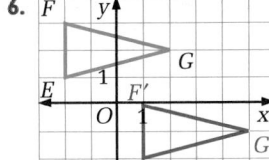

7.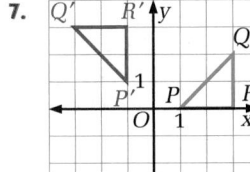

8.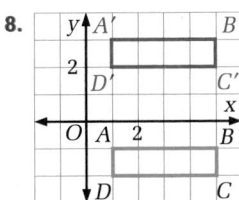

4-3 Translations

199

Look Back

The Look Back questions can be used for class discussion. Have students summarize how coordinates of a shape change when the shape is moved from one location to another by a translation. Ask students how to find the new coordinates of a point for each kind of translation mentioned in the Look Back.

APPLYING

Suggested Assignment

Standard 1–22, 24–34

Extended 1–22, 24–34

Integrating the Strands

Algebra Exs. 16–22, 26–28

Measurement Exs. 29–31

Geometry Exs. 1–25

Logic and Language Exs. 1, 24, 25, 32–34

Communication: Discussion

For Exs. 6–8, have students explain how they decided whether or not the pictures show translations.

Using Technology

Activity 22 in the *Geometry Inventor Activity Book* can be used by students to apply their skills in translating polygons.

Multicultural Note

Many examples of symmetry can be found in the traditional art of Africa, Asia, and North America. For example, Navajo weavers in the American Southwest weave blankets using the same symmetrical designs used by their ancestors. Garments and rugs woven by the peoples of northwest Africa employ symmetry using empty space: the shapes of "undecorated" areas mirror the filled-in spaces, so that the shadow images help create the repetition. The Kuba people of Zaire have created more than 100 original symmetrical geometric patterns. The design on the cup shown on this page is an example of the intricate patterns for which the Kuba are known.

Batik designs from Java, the most populous island in Indonesia, use symmetry to harmonize intricate patterns. In Ex. 13, a swatch of a cotton sarong from Pekalongan, central Java, shows delicate, waving lines that repeat to create a pattern of uniform shapes.

Reasoning

If you assign Exs. 16–19, ask students if they can tell by looking at the algebraic description of the translation just how a figure would be moved by the translation. For example, in Ex. 17, the translation $(x, y) \rightarrow (x - 3, y + 2)$ will move points 3 units to the left and up 2 units.

9. Translate △JKL 4 units left and 3 units up. What are the coordinates of each vertex after the translation?

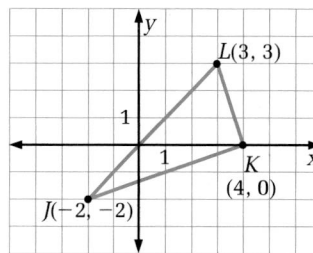

10. Translate parallelogram PQRS 5 units right. What are the coordinates of each vertex after the translation?

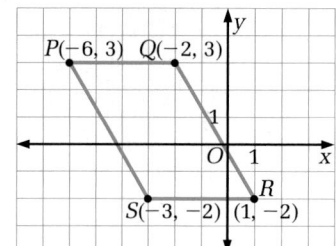

The traditional and modern designs of many cultures display translational symmetry.

Tell whether each design has translational symmetry.

11.

Hopi

12.
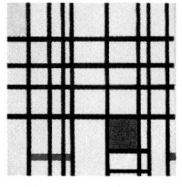
The Netherlands

13.
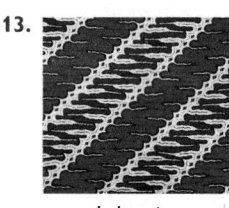
Indonesia

14.

Zaire

15. **Open-ended** Create a design. Use your design to make a pattern with translational symmetry.

Use this information for Exercises 16–22.

Another way to describe a translation is to show the change in the coordinates (x, y) of any point. For example, a translation of 2 units right and 1 unit down is written $(x, y) \rightarrow (x + 2, y - 1)$.

Write the coordinates of P′ after each translation of P(4, 5).

16. $(x, y) \rightarrow (x + 1, y + 3)$
17. $(x, y) \rightarrow (x - 3, y + 2)$
18. $(x, y) \rightarrow (x, y + 1)$
19. $(x, y) \rightarrow (x - 2, y)$

Cup from the Kuba culture of Zaire.

200   **Unit 4** Coordinates and Functions

Answers to Exercises and Problems

9. $J'(-6, 1)$, $K'(0, 3)$, $L'(-1, 6)$
10. $P'(-1, 3)$, $Q'(3, 3)$, $R'(6, -2)$, $S'(2, -2)$
11. No.   12. No.
13. Yes.   14. Yes.
15. Designs may vary. An example is given.

16. $P'(5, 8)$ 17. $P'(1, 7)$
18. $P'(4, 6)$ 19. $P'(2, 5)$
20. $(x, y) \rightarrow (x - 1, y)$
21. $(x, y) \rightarrow (x + 4, y - 6)$
22. $(x, y) \rightarrow (x - 1, y + 2)$

23. king: 1 unit right, left, up, down, or diagonally on empty squares; queen: any number of units right, left, up, down, or diagonally on empty squares; rook (castle): any number of units right, left, up, or down on empty squares; bishop: any number of units diagonally on empty squares (i.e., any number of units right or left and the same number of units up or down); knight: 2 units right or left and 1 unit up or down or 2 units up or down and 1 unit right or left; pawn: 1 or 2 units up on first move and 1 unit up thereafter; 1 unit diagonally to capture other chess pieces.

200

Describe each translation by showing the change in the coordinates (*x*, *y*) of any point.

20. 1 unit left
21. 4 units right, 6 units down
22. 1 unit left, 2 units up
23. **Research** In the game of chess, a king may move in any direction but only one square at a time. Find out what rules control the movement of the other pieces used in chess. For each chess piece, describe a move as a translation.

Ongoing ASSESSMENT

24. a. Copy the boat on graph paper. Find its area.
    b. **Writing** Suppose you want to tell a friend how to show the boat sailing across the grid. Describe to your friend how the coordinates of the boat change as it moves.
    c. Does the area of the boat change as it moves? Explain your answer.

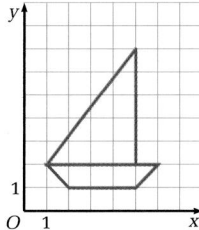

Review PREVIEW

25. a. Plot the points $W(0, 0)$, $X(3, 3)$, $Y(6, 0)$, and $Z(3, -3)$. Connect them in order and connect Z to W. *(Section 4-2)*
    b. Write the specific name of the polygon you formed. Explain how you know what it is.

Solve each equation. *(Sections 2-7, 2-8)*

26. $x + 7 = 18$
27. $3a + 15 = -32$
28. $2g - 9 = -26$

Find the measure of each angle in the diagram at the right. *(Section 2-5)*

29. $\angle ACD$
30. $\angle ACB$
31. $\angle 1$

4 Working on the Unit Project

32. The pictures show a person dropping to the ground with a parachute. How do the coordinates of the parachute change as it approaches the ground?
33. **Open-ended** Draw a cartoon figure on a coordinate plane. Sketch two movements of the figure. Describe how the coordinates of the figure change as the figure moves.

34. Decide what your flip book will show. What types of motions will it use? Write a short description of your plan for the flip book.

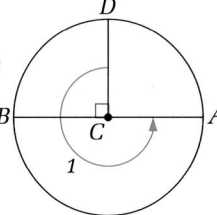

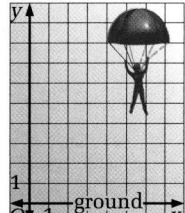

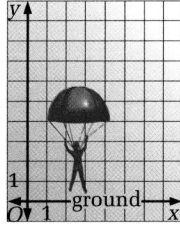

Research

A project of interest to some students would be to research the use of translational symmetry in the works of the artist M. C. Escher.

Assessment: Task

Ask two or three students to write their answers to Ex. 24(b) on the chalkboard for class discussion.

Working on the Unit Project

This might be a good time to allow a few minutes of class time for the project team members to meet and share their ideas with each other. Each team could write their response to Ex. 34 and submit it to you or read it to the class.

Practice 29 For use with Section 4-3

Name ____________ Date ____________

Practice 29
For use with Section 4-3

2. $A'(-3, -1)$, $B'(3, -5)$, $C'(2, 2)$
4. $A'(-8, 1)$, $B'(-2, -3)$, $C'(-3, 4)$

Find the coordinates of each vertex of $\triangle ABC$ after each translation.

1. 1 unit right $A'(-2, 1)$, $B'(4, -3)$, $C'(3, 4)$
2. 2 units down See above.
3. 3 units up $A'(-3, 4)$, $B'(3, 0)$, $C'(2, 7)$
4. 5 units left See above.
5. 2 units right and 3 units down $A'(-1, -2)$, $B'(5, -6)$, $C'(4, 1)$
6. 3 units left and 4 units up $A'(-6, 5)$, $B'(0, 1)$, $C'(-1, 8)$
7. 5 units right and 2 units up $A'(2, 3)$, $B'(8, -1)$, $C'(7, 6)$

Write the coordinates of P' after each translation of the given point P.

8. $P(4, 1)$; $(x, y) \to (x - 1, y)$ $P'(3, 1)$
9. $P(3, 5)$; $(x, y) \to (x, y + 2)$ $P'(3, 7)$
10. $P(2, 3)$; $(x, y) \to (x + 3, y - 5)$ $P'(5, -2)$
11. $P(1, 4)$; $(x, y) \to (x - 2, y + 4)$ $P'(-1, 8)$
12. $P(-1, 5)$; $(x, y) \to (x + 2, y - 3)$ $P'(1, 2)$
13. $P(0, -2)$; $(x, y) \to (x, y + 6)$ $P'(0, 4)$
14. $P(-3, 0)$; $(x, y) \to (x - 7, y - 2)$ $P'(-10, -2)$
15. $P(1, 6)$; $(x + 4, y - 3)$ $P'(5, 3)$
16. $P(-2, -4)$; $(x, y) \to (x + 5, y - 4)$ $P'(3, -8)$
17. $P(-5, 1)$; $(x + 4, y - 1)$ $P'(-1, 0)$

Describe each translation by showing the change in the coordinates (*x*, *y*) of any point.

18. 2 units right $(x, y) \to (x + 2, y)$
19. 3 units down $(x, y) \to (x, y - 3)$
20. 5 units up $(x, y) \to (x, y + 5)$
21. 1 unit left and 4 units up $(x, y) \to (x - 1, y + 4)$
22. 2 units right and 6 units up $(x, y) \to (x + 2, y + 6)$
23. 3 units right and 2 units down $(x, y) \to (x + 3, y - 2)$
24. 7 units left and 8 units down $(x, y) \to (x - 7, y - 8)$
25. $(3, 7) \to (4, 5)$ $(x, y) \to (x + 1, y - 2)$
26. $(2, -9) \to (5, -7)$ $(x, y) \to (x + 3, y + 2)$

Tell whether the pattern you find in each of the following has translational symmetry.

27. The rings of an archery target no
28. The checkered flag at a car race yes
29. A straight brick wall yes
30. A basketball court no
31. A maple leaf no
32. A bicycle wheel no
33. The lines on a page of ruled notebook paper yes
34. The wire fencing of a baseball backstop yes

Practice Bank, INTEGRATED MATHEMATICS 1

29

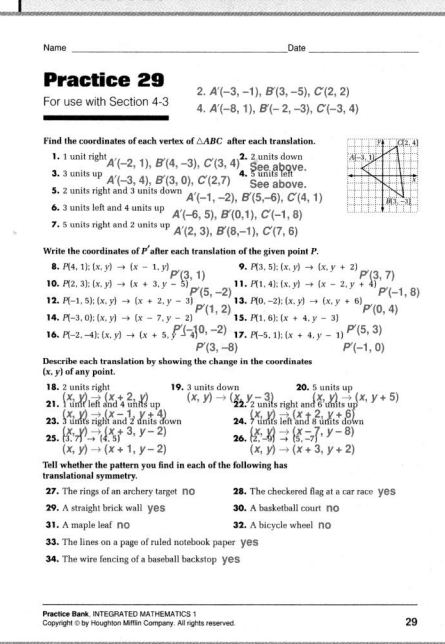

Answers to Exercises and Problems

24. a. 14 square units
    b. Descriptions may vary. An example is given. Assume the boat is sailing smoothly across the water. As it moves to the left, the first coordinates decrease. The second coordinates stay the same.
    c. The area of the boat does not change as it moves. Even though the coordinates change, the boat does not change shape or area.

25. a.

    b. square; All angles are right angles and all 4 sides are congruent.

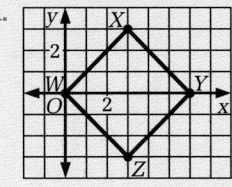

26. 11
27. $-15\frac{2}{3}$
28. $-8\frac{1}{2}$
29. 90°
30. 180°
31. 270°
32. Answers may vary. An example is given. The lower location of the parachute is 4 units down and 3 units left of the higher location. The first coordinates of the two locations differ by 3 and the second coordinates differ by 4.
33. See answer in back of book.
34. Answers may vary.

202

PLANNING

Section 4-4

Rotations

Focus
Recognize rotations of figures on a coordinate plane, rotate figures on polar graph paper, and recognize rotational symmetry.

center of rotation

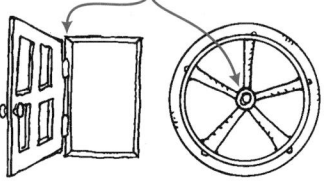

Transformations

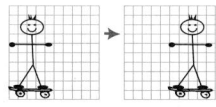

translation

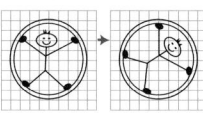

rotation

The gymnast is turning or *rotating* around the horizontal bar. A turn around a point is called a **rotation.** The point is called the **center of rotation.** Wheels, doors, and clothes dryers also rotate. A center of rotation may be at the center of an object, as on a wheel. It may also be at some other point *on* or *off* the object.

Rotating and translating are two ways to *transform* an object. A **transformation** describes a change made to an object or its position.

Unit 4 Coordinates and Functions

Talk it Over

Place a piece of paper on your desk. Press one finger down on the paper to create a center of rotation. Rotate the paper.

1. When you rotate the paper, does it change size or shape?

2. What information does someone need in order to copy a rotation? Give directions for copying your rotation.

The *direction* of a rotation may be *clockwise* (↻) or *counterclockwise* (↺). The *amount of turn* of a rotation may be stated in degrees.

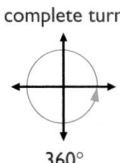

complete turn
360°

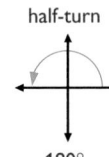

half-turn
180°

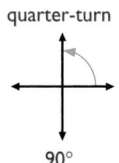

quarter-turn
90°

Sample 1

The graphs show rotations of △ABC around the origin. Describe the direction and the amount of the rotation of each graph.

a.
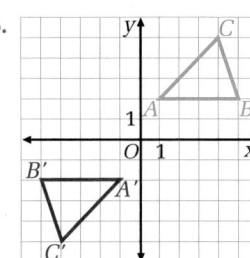

b.

c.
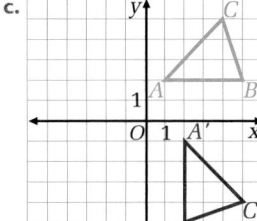

Sample Response

2 Measure ∠AOA′.

a.
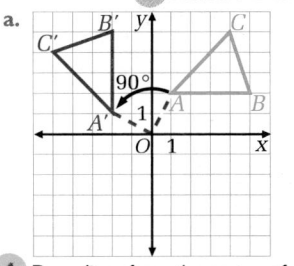

90° counterclockwise

1 Draw lines from the center of the rotation to corresponding points on each figure (OA and OA′).

b.
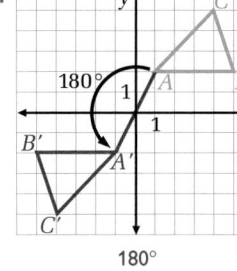

180°

c.
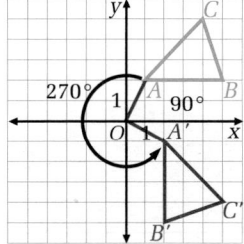

270° counterclockwise

4-4 Rotations

203

Answers to Talk it Over

1. No.

2. Answers may vary. An example is given. In order for someone to copy your rotation, you need to specify the center of the rotation and the amount and direction of the turn. Directions: Put your finger on a corner of the paper and turn it 30° clockwise.

Talk it Over

Use question 1 to help students understand that size and shape are preserved when a figure is rotated. Students may find question 1 unusual and feel that it is obvious that these features are preserved.

It is obvious if you think of the figure as a rigid object. It is less obvious if you imagine rotating a few individual points of the figure to determine whether shape and size have changed.

Additional Sample

S1 The graphs show rotations of △RST around the origin. Describe the direction and amount of the rotation of each graph.

a.
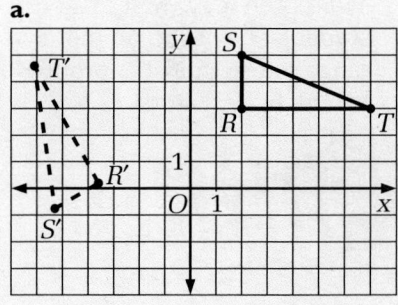

120° counterclockwise

b.
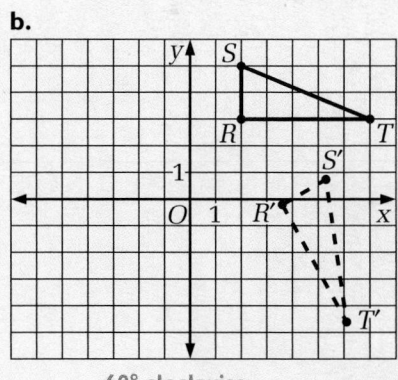

60° clockwise

c.
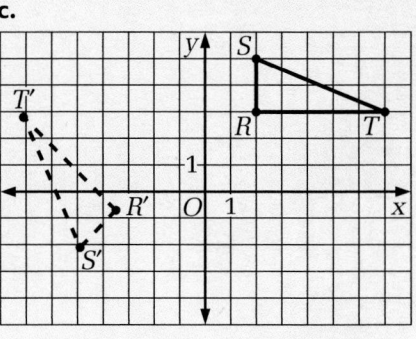

135° counterclockwise

203

Talk it Over

3. Why is the direction of turn not included in the description of a 180° rotation?

4. a. Name a clockwise rotation that gives the same results as a rotation of 90° counterclockwise.

   b. Name a clockwise rotation that gives the same results as a rotation of 270° counterclockwise.

ROTATIONS

To describe a rotation, you need to tell:

➤ the amount of turn, as a fraction or in degrees

➤ the direction—clockwise or counterclockwise

➤ the location of the center of rotation

Drawing Rotations

It is easier to draw rotations on a special type of graph paper called *polar graph paper.*

BY THE WAY...

Although an owl cannot move its eyes, it can see behind its head. An owl looks around by rotating its head up to 180° clockwise and 180° counterclockwise.

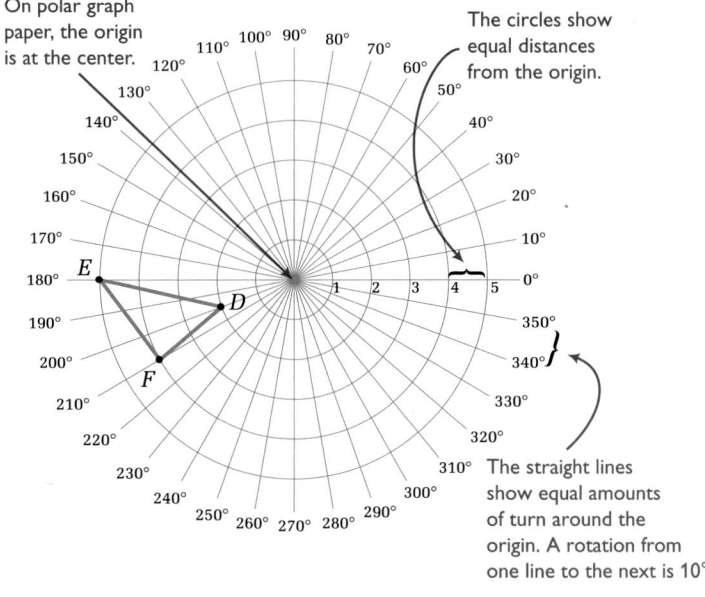

On polar graph paper, the origin is at the center.

The circles show equal distances from the origin.

The straight lines show equal amounts of turn around the origin. A rotation from one line to the next is 10°.

204   **Unit 4** Coordinates and Functions

Rotate △DEF 60° counterclockwise around the origin.

Sample Response

Move each vertex along its circle:

direction—counterclockwise

amount of turn—60° is 6 · 10°, so move six lines

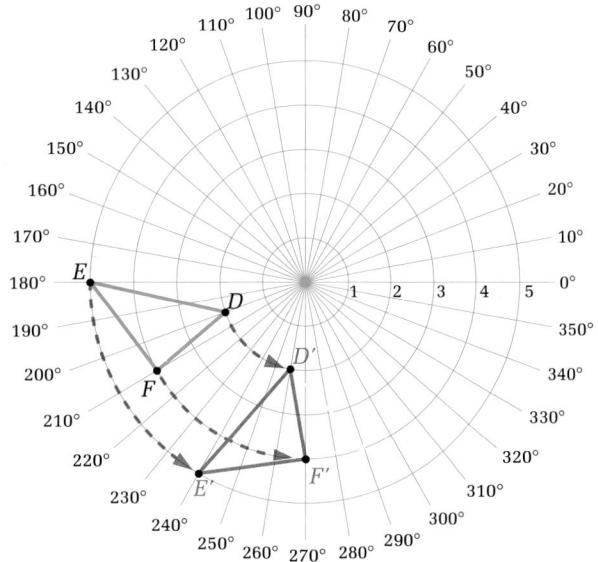

Then draw the rotated triangle.

Rotational Symmetry

Every object looks the same after a 360° rotation around its center. Some objects look the same after a rotation of less than 360°. If so, they have **rotational symmetry.** To describe rotational symmetry, tell how many degrees the object must rotate in order to look the same way it did before the rotation.

A square has 90°, 180°, and 270° rotational symmetry.

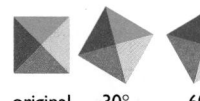 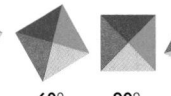  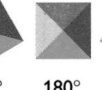  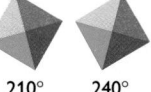 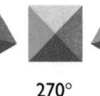 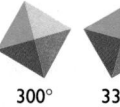

| original | 30° | 60° | 90° | 120° | 150° | 180° | 210° | 240° | 270° | 300° | 330° | original |

Tell whether or not each figure has rotational symmetry. If it does, describe the symmetry.

"Tomo" is Japanese for "friend."

a.

b.

c.

4-4 Rotations

205

S2 Rotate △*DEF* 130° clockwise around the origin.

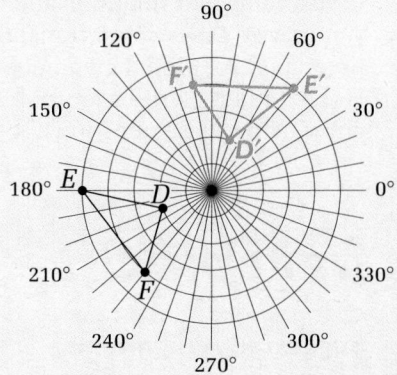

S3 Tell whether or not each figure has rotational symmetry. If it does, describe the symmetry.

a.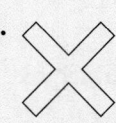

Yes; 180° symmetry.

b.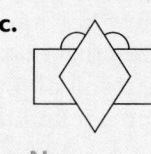

Yes; 90°, 180°, and 270° symmetry.

c.

No.

Multicultural Note

The Japanese word shown in Sample 3(a) is an example of *kanji*, a character that represents an idea or a sound. Such ideographs were brought to Japan from China long ago and have evolved over the centuries. Invite students to find other examples of Japanese ideographs and determine whether or not each has rotational symmetry.

Look Back

These questions review how figures are transformed by translations and rotations and summarize ideas about translational and rotational symmetry. Have students write a journal entry using these questions.

APPLYING

Suggested Assignment

Standard 1–15, 23–26

Extended 1–19, 23–26

Integrating the Strands

Geometry Exs. 3–8, 12–23, 25, 26

Logic and Language Exs. 1, 2, 8–11, 21, 22, 24–26

Using Manipulatives

In connection with Exs. 3–5, you may find it helpful to provide students with tracing paper and show how it can be used to confirm that one figure is indeed the result of rotating the other figure around the origin.

For example, for Ex. 4, students can place tracing paper over the diagram and trace rectangle *JKLM* and the positive portion of the horizontal axis. Without moving or sliding the tracing paper, students can put a pencil point at the origin, hold the pencil firmly in a vertical position, and turn the tracing paper until *JKLM* is observed to coincide with *J'K'L'M'*.

In this way, one can see the rotation taking place. If students hold the tracing paper carefully in place, it is easy to measure the angle between the two positions of the horizontal axis to find the amount of the rotation.

Sample Response

a. No. The character must rotate 360° in order to look the way it did before the rotation.

b. Yes. This design has 180° rotational symmetry.

c. Yes. The starfish has 72°, 144°, 216°, and 288° rotational symmetry.

Look Back ◄

How are translational symmetry and rotational symmetry alike? How are they different?

4-4 Exercises and Problems

1. **a.** Describe the rotation that occurs when a tree falls.

   b. Describe the rotation that occurs when a pair of scissors is used.

   c. Describe another rotation that you have observed in your life.

2. **Reading** What are two types of transformations? How are they alike?

The graphs show rotations around the origin. Describe the direction and the amount of the rotation in each graph.

3.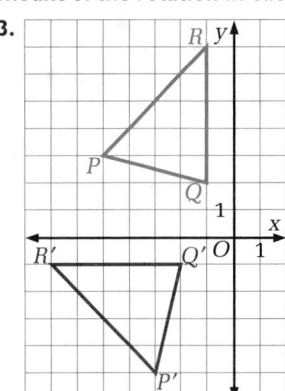

4.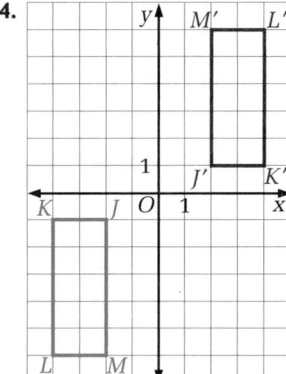

5.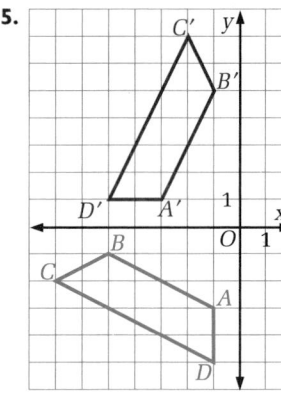

Unit 4 Coordinates and Functions

Answers to Look Back

Answers may vary. An example is given. In both translational and rotational symmetry, a figure is repeated without changing size or shape. In translational symmetry, the figure is repeated by sliding it along the plane. In rotational symmetry, the figure is turned about a center of rotation.

Answers to Exercises and Problems

1. a–c. Answers may vary. Examples are given.

   a. about a 90° rotation from a vertical to a horizontal position, with the center of rotation at the base of the trunk

   b. about a 40° rotation clockwise and counterclockwise, with the center of rotation at the hinge of the scissors

   c. The sun appears to rotate about 180° from east to west each day, with the center of rotation at my location.

2. translations and rotations; They both move a figure in a plane without changing its size or shape.

3. 90° counterclockwise or 270° clockwise

Copy each figure on polar graph paper. Draw each indicated rotation of the figure around the origin.

6. a. 50° counterclockwise

   b. 180° clockwise

7. a. 60° clockwise

   b. 120° counterclockwise

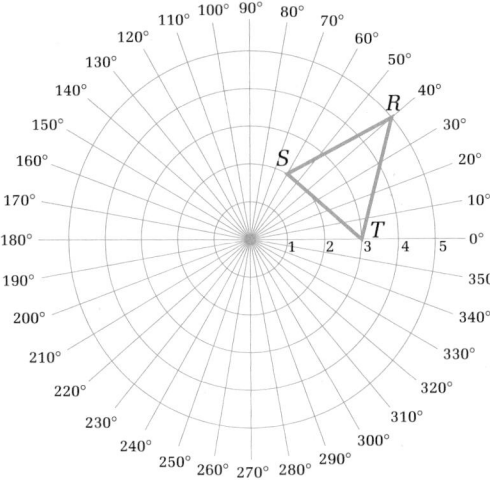

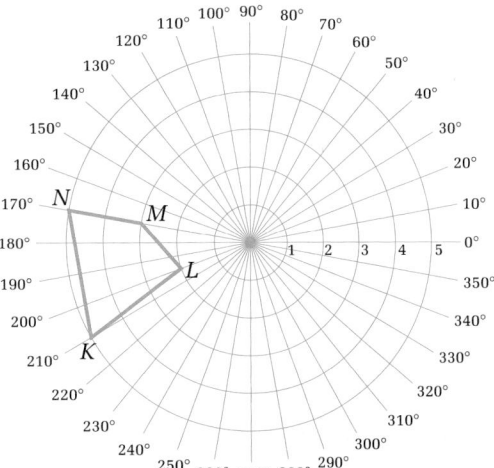

8. Career A pilot can plot the location of an airplane using its distance from a radio tower and its clockwise rotation from north on a magnetic compass.

   a. What are the coordinates of airplane *B*?

   b. What are the coordinates of airplane *C*?

   c. Writing Describe how this system of locating airplanes and the method of locating points on polar graph paper are alike and how they are different.

Airport runways are numbered by their clockwise rotation from north. To determine a runway number, follow these steps:

(1) Find the clockwise rotation from north.

(2) Divide by 10°.

(3) Round the quotient to the nearest whole number.

For example, a runway with a clockwise rotation of 32° from north is named "runway 3."

What is the number of the runway at each location?

9. 10° north of east

10. 13° south of west

11. 48° east of south

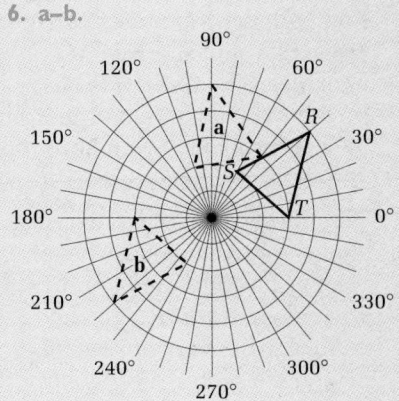

Airplane *A* has coordinates (50, 40°).

radio tower

Answers to Exercises and Problems

4. 180°

5. 90° clockwise or 270° counterclockwise

6. a–b. See side column.

7. a–b. See side column.

8. a. (70, 170°)

   b. (95, 300°)

   c. Both systems (polar coordinates and airplane navigation) locate points by distance from

a center point and degrees of rotation from a fixed ray. On polar graph paper, degrees are measured counterclockwise from the ray horizontal to the right of the pole. In airplane navigation, degrees are measured clockwise from the ray pointing north from the radio tower.

9. runway 8

10. runway 26

11. runway 13

Limited English Proficiency

Students acquiring English may benefit from being able to use real objects to demonstrate the events specified in Ex. 1 as they orally describe the rotation that occurs in each.

Career Note

When discussing Exs. 8–11, point out that air travel today would be impossible without the guidance of aircraft by air traffic controllers, whose radar screens use polar graphs like the ones in this section to locate and track airplanes.

Using Technology

Activities 20 and 21 in the *Geometry Inventor Activity Book* involve the rotation of points and polygons. Activity 30 is an exploration of rotational symmetry.

6. a–b.

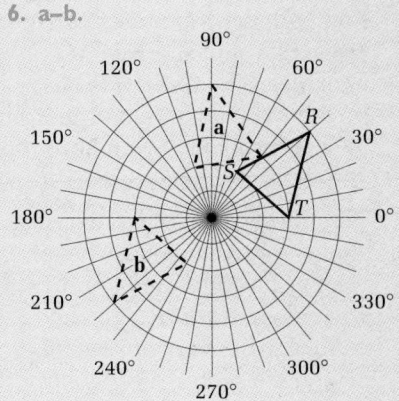

7. a–b.

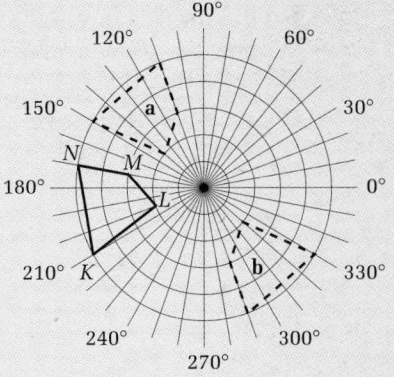

207

Multicultural Note

The design of Ex. 12 shows a square cloth from a garment woven by the Highland Maya people in Chiapas, a state in southern Mexico. This diamond pattern is said to help bring the wearer in harmony with the good forces of the universe.

The photograph below Ex. 12 shows women painting a wall of their home in the village of Nimkahera in the state of Madhya Pradesh, India to signal the upcoming marriage of their brother. The rings (one is shown in Ex. 13) represent forces in the universe which they hope will bring good fortune to the marriage.

The bird shown in Ex. 14 is a traditional design from Panama.

The photograph of the man working on an appliqué, a design created by a special cloth-layering technique, was taken in a town outside of Puri, India. The finished appliqué tapestries shown behind him and in Ex. 15 were produced for a Hindi Temple in Puri.

The piece shown in Ex. 16 is a detail of a silver Navajo medallion called a *Naja*—the end-piece of a squash blossom necklace. The pattern is believed to be an ancient Moorish design, probably brought to the Americas by the Spanish.

The photograph of Ex. 17 shows a Navajo weaver. Each weaver has total freedom to weave whatever design she wishes, from very abstract designs to pictures of pickup trucks. The rugs express the idea of harmony or balance. Because life itself is rarely in true balance, the weaver will often weave a tiny defect into each rug.

connection to **ART**

Tell whether each design has rotational symmetry. If it does, describe the symmetry.

12.
13.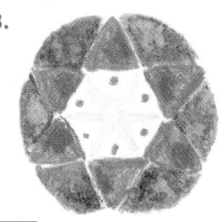

14.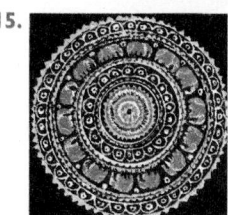
15.

In Exercises 16–18, what type(s) of symmetry does each design have? Copy the pattern that repeats and describe the transformation.

Designs with many types of symmetry are created by people all over the world. These are from India, Panama, Mexico, and the United States.

16.
17.
18.

19. **Open-ended** Create a design. Use your design to make a pattern with rotational or translational symmetry, or both.

Unit 4 Coordinates and Functions

Answers to Exercises and Problems ....

12. No.

13. Yes. The design has symmetry at 60°, 120°, 180°, 240°, and 300°.

14. No.

15. Yes. The design has symmetry at about 26°, 52°, 78°, 104°, 130°, 156°, 182°, 208°, 234°, 260°, 286°, 312°, and 338°.

16. rotational symmetry; The figure has rotational symmetry at 24°, 48°, 72°, 96°, 120°, 144°, 168°, 192°, 216°, 240°, 264°, 288°, 312°, and 336°.

17. rotation and translational symmetry; translates left to right and has 180° rotational symmetry

18. translational symmetry; translates left to right

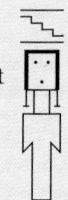

19. Answers may vary. All patterns should have translational or rotational symmetry.

20. **Using Manipulatives** Create at least three paper snowflakes. Cut out at least one snowflake each from a circle (as shown), a square, and an equilateral triangle. Describe the rotational symmetries in your favorite snowflake.

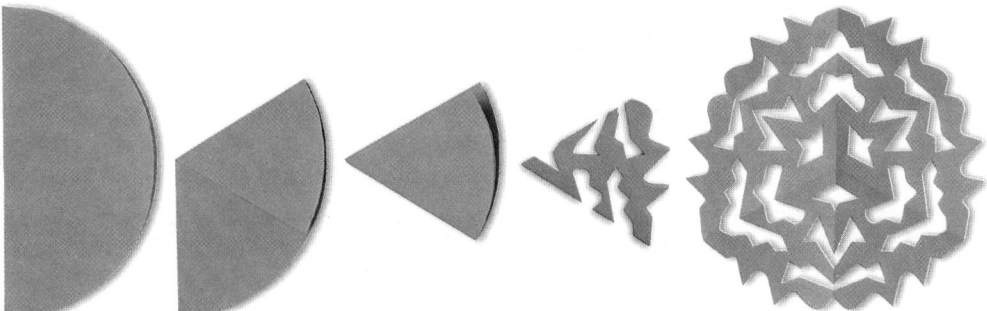

1. Fold the paper in half,   2. then fold into thirds.   3. Cut some designs out of the edges,   4. then unfold.

21. **Research** What shape does the nut on a fire hydrant have? What rotational symmetry does this nut have? Why do you think this shape was chosen?

Ongoing **ASSESSMENT**

22. **Group Activity** Work in a group of four students.
    a. Describe the rotational symmetry of a regular pentagon, hexagon, octagon, and 9-gon.
    b. Look for a pattern in your results. Then create a method for describing the rotational symmetry of any regular polygon.

Review **PREVIEW**

23. Translate $\overline{AB}$ 2 units to the right and 3 units down. Draw $\overline{A'B'}$. What are the coordinates of each point after the translation? *(Section 4-3)*

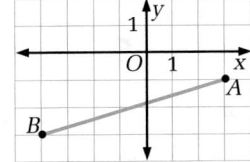

24. Make a graph to show the data. *(Section 3-6)*

RESULTS OF A POLL THAT ASKED:

"Which day of the week do you think is the worst?" Answer %	SUN	MON	TUE	WED	THU	FRI	SAT	Don't Know
	3.8	62.2	5.1	6.0	2.0	3.9	2.4	14.6

Communication: Drawing

You may wish to review with students the concept of a tessellation and then challenge them to search for and sketch tessellations that can be transformed to coincide with themselves using *different* centers of rotation. An example of such a tessellation is the familiar honeycomb pattern. This tessellation can be made to coincide with itself by using any point where three hexagons intersect as the center of rotation and by using a 120° clockwise or counterclockwise rotation. Students may find it interesting to note that this tessellation exhibits *both* rotational and translational symmetry.

Using Manipulatives

With Ex. 20, three sheets of plain paper can be used for the snowflakes or one sheet of plain paper with a circle, a square, and an equilateral triangle already drawn on it.

Assessment: Standard

For Ex. 22, each group can submit one paper for grading, with each student receiving the grade of the group.

Answers to Exercises and Problems

20. Answers may vary. An example is given. The snowflake design shown on page 209 has 120° and 240° rotational symmetry.

21. The nut on a fire hydrant is shaped like a regular pentagon and has 72°, 144°, 216°, and 288° rotational symmetry. Explanations may vary. An example is given. This shape was chosen because it is hard to grasp with fingers, a wrench, or pliers since it has no parallel sides. This shape makes it difficult for an unauthorized person to open the hydrant.

22. a. regular pentagon: 72°, 144°, 216°, 288°; regular hexagon: 60°, 120°, 180°, 240°, 300°; regular octagon: 45°, 90°, 135°, 180°, 225°, 270°, 315°; regular 9-gon: 40°, 80°, 120°, 160°, 200°, 240°, 280°, 320°

   b. Answers may vary. An example is given. Divide 360° by the number of sides to get the smallest rotational symmetry. Then add that number repeatedly to get the other rotational symmetries. Stop at 360° and do not list 360° as a rotational symmetry.

23. $A'(5, -4)$, $B'(-2, -6)$

24.

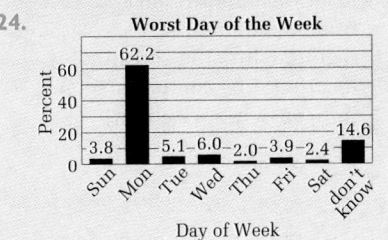

209

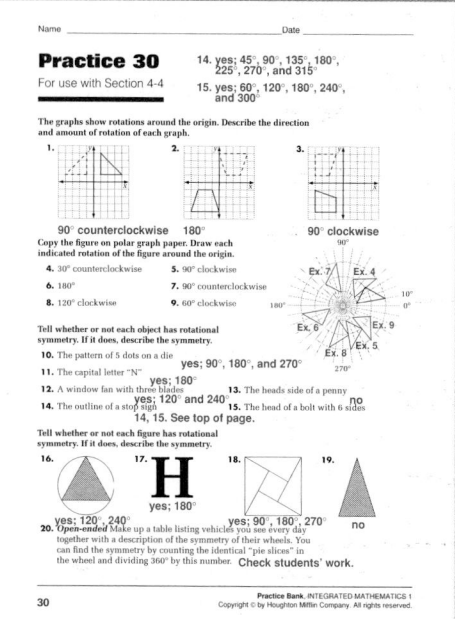
---

25. Imagine making a flip book for each situation. Would you use a *translation* or a *rotation* to show each movement?

a. a snail creeping along the ground

b. an elevator arrow switching from "up" to "down"

26. **Open-ended** Describe a movement that could be shown by each transformation.

a. translation   b. rotation

Unit 4   CHECKPOINT

1. **Writing** Choose two uses of coordinate systems. Write a paragraph describing how they are alike and how they are different.

Jo's flip book shows the Little Dipper and the Big Dipper.

2. a. Graph the points $A(-17, -6)$, $B(-14, -7)$, $C(-13, -8)$, $D(-10.5, -10)$, $E(-6, -11)$, $F(-7, -13)$, and $G(-10, -12)$ on a coordinate plane. Connect the points to form the Big Dipper. Connect G to D to complete the bowl.   **4-1**

b. What quadrant contains the Big Dipper?

c. Describe the shape of the bowl of the Big Dipper.

3. What name is used for the location of the North Star on the coordinate plane?

4. Find the area of the bowl of the Little Dipper.   **4-2**

The Little Dipper

North Star

5. Translate the Little Dipper 2 units to the left and 4 units down. What are the coordinates of each point after the translation?   **4-3**

6. Jo's flip book shows a rotation of the Little Dipper around the North Star. Describe the direction and the amount of the rotation.   **4-4**

Tell whether each pattern has *translational symmetry*, *rotational symmetry*, or *no symmetry*.

7.   8.   9.

Guatemala   *Denmark*   *United States*

210   **Unit 4** Coordinates and Functions

 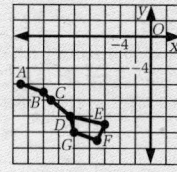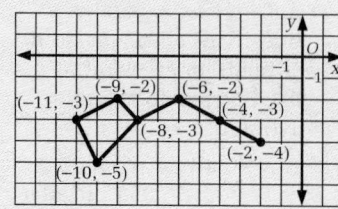
210

Section 4-5

Scatter Plots

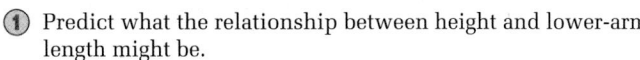

GET TO THE POINT•

Focus
Make and interpret a scatter plot of data, and use scatter plots to make predictions.

EXPLORATION

(Do) *taller people have longer arms?*

• **Materials:** tape measure or meter stick, graph paper, graphics calculator or software (optional)

• **Work in a group of at least five students.**

(1) Predict what the relationship between height and lower-arm length might be.

(2) Measure and record the height of each person in your group to the nearest centimeter.

Measure and record each person's lower-arm length.

Name	Height (cm)	Lower-arm length (cm)
Michelle	157	42
?	?	?

(3) Use graph paper or a graphics calculator to plot your data.

(4) Describe any trends you see in your graph. Does your graph support the prediction you made in step 1?

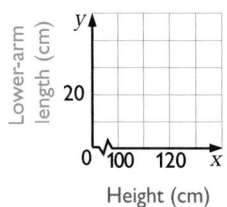

(5) Display your table and graph along with those of the other groups. How are the data and the graphs alike or different?

(6) Plot the data from all the groups on one graph.

(7) Compare this graph with the first graph your group made. Does the new graph support the prediction you made in step 1?

(8) Suppose a student is 171 cm tall. Use the graph to predict the student's lower-arm length.

(9) Measure the wrist circumference and back length of each person in your group. Save these data for use in Exercise 3.

4-5 Scatter Plots

211

Height: 157 cm
Length: 42 cm
Length: 48 cm

Answers to Exploration

1. Answers may vary. For example, the taller the person, the longer the lower-arm length.

2. Answers may vary.

3. Graphs may vary.

4. Answers may vary. An example is given. The data points tend to lie on or are close to an imaginary line that rises to the right. As height increases, lower-arm length increases, supporting the prediction in step 1.

5–6. Answers and graphs may vary.

7. Answers may vary.

8. Answers may vary.

9. Answers may vary.

PLANNING

Objectives and Strands
See pages 180A and 180B.

Spiral Learning
See page 180B.

Materials List
➤ Tape measure or meter stick (one for each group of 5)
➤ Graph paper
➤ Graphics calculator or graphing software (optional)
➤ Pieces of dry spaghetti

Recommended Pacing
Section 4-5 is a two-day lesson.

Day 1
Pages 211–212: Exploration through Talk it Over, *Exercises 1–9*

Day 2
Pages 212–214: Drawing a Fitted Line through Look Back, *Exercises 10–31*

Extra Practice
See pages 625–627.

Warm-Up Exercises
Warm-Up Transparency 4-5

Support Materials
➤ Practice 31
➤ Enrichment 28 in the Activity Bank
➤ Study Guide 4-5
➤ Problem Set 9
➤ Diagram Masters 2, 12 in the Explorations Lab Manual
➤ McDougal Littell Mathpack software: *Stats!*
➤ *Stats!* Activity Book: Activities 14, 16, 17, and 21
➤ Quiz 4-5

TEACHING

Exploration

The goal of the Exploration is to give students experience in gathering data, organizing it into a table, and graphing it as a scatter plot.

Communication: Listening

Before students get under way on the Exploration activity, explain exactly what is meant in step 1 by the term *lower-arm length*. As used here, the term means the number of centimeters from the elbow to the tip of the middle finger.

Error Analysis

It is common for students to make misleading graphs when there are repeated data points. What would you do in step 3 of the Exploration if both Jeanne and Sue have a height of 155 cm and a lower-arm length of 37 cm? How would you plot the point (155, 37) *twice* and *show* that you have done so?

Discuss this difficulty with the class. Point out the desirability of being able to count data points in the graph to be sure there are as many data points as there are students in the class (step 6). Then agree on a scheme for showing repeated points. One possibility is illustrated below.

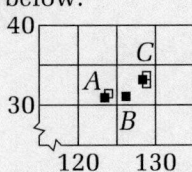

In the illustration, the data point at *A* occurs twice, at *B* once, and at *C* three times.

These considerations can be important in drawing scatter plots of real-world data and in drawing fitted lines.

Talk it Over

Use questions 1 and 2 to take a closer look at the idea of correlation. The notion of correlation has to do with how strongly one set of data is related to another.

The graph that you made in the Exploration is a *scatter plot*. A **scatter plot** is a graph that shows the relationship between two data sets.

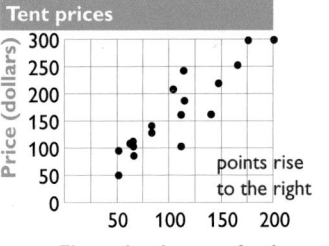

When two data sets increase together, they have a **positive correlation.**

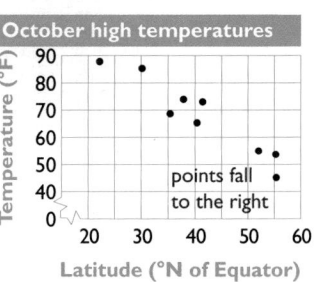

When one data set decreases as the other increases, the two data sets have a **negative correlation.**

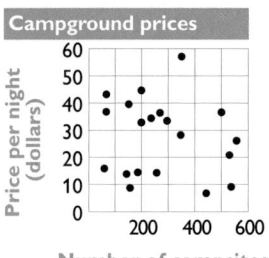

Sometimes data sets show **no correlation.**

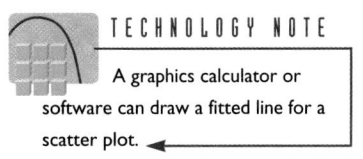

▶ Now you are ready for:
Exs. 1–9 on pp. 214–215

Talk it Over

1. Do the graphs you made for the Exploration show a positive correlation, a negative correlation, or no correlation?

2. Suppose you collect data on the length of people's longest fingernails. Do you think there is a correlation between these data and the people's heights? Why or why not?

Drawing a Fitted Line

You can use a scatter plot to make predictions. To help you make predictions from a scatter plot, draw a straight line that passes close to most of the data points. This line is called a **fitted line.**

TECHNOLOGY NOTE

A graphics calculator or software can draw a fitted line for a scatter plot.

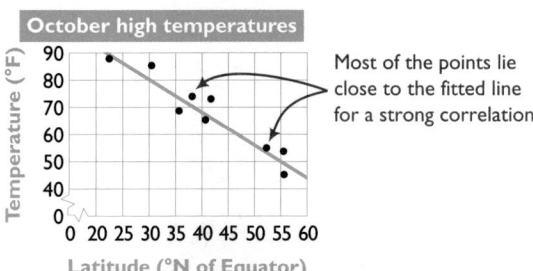

Most of the points lie close to the fitted line for a strong correlation.

A fitted line shows if the correlation between two data sets is strong or weak. If the data points come very close to the fitted line, the correlation is *strong*. If not, the correlation is *weak*. The stronger the correlation, the better your predictions will be.

Unit 4 Coordinates and Functions

Answers to Talk it Over

1. Answers may vary. A positive correlation is expected.

2. Answers may vary. One might reason there is no correlation or that there is a negative correlation— girls, who tend to be shorter, tend to have longer fingernails than boys.

Sample

The table shows the piston displacement and horsepower (hp) of nine similarly powered cars. What horsepower do you predict for a 3200 cc engine?

Car	A	B	C	D	E	F	G	H	I
Displacement (cc)	2389	1836	3405	3535	1840	2977	2986	1998	2164
Horsepower	155	134	300	310	140	270	220	119	130

Sample Response

Make a scatter plot of the data. Draw a fitted line. Then predict.

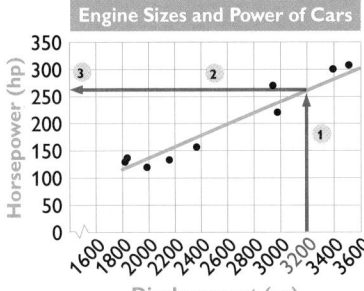

Engine Sizes and Power of Cars

3 The point where the line meets the vertical axis is the prediction.

1 From the 3200 mark on the horizontal axis, draw a vertical line to the fitted line.

2 Then draw a horizontal line to the vertical axis.

A 3200 cc engine in a similarly powered car should have about 270 hp.

Talk it Over

3. In the Sample, can you be certain that a 3200 cc engine will have 270 hp? Why or why not?

4. Use the scatter plot in the Sample to predict the displacement of an engine in a similarly powered car that has 200 hp.

5. A fitted line must lie close to the points on a scatter plot, but there is not one correct placement. Decide whether each line is appropriate as a fitted line for the scatter plot. Explain your decisions.

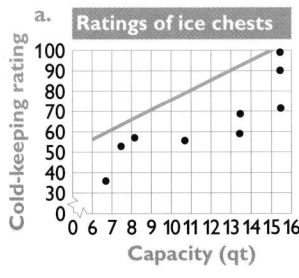

a. **Ratings of ice chests**

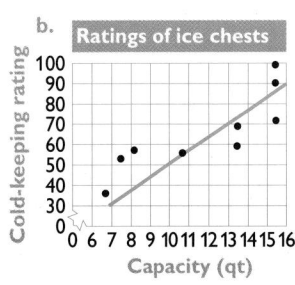

b. **Ratings of ice chests**

4-5 Scatter Plots

213

Answers to Talk it Over

3. No. 270 horsepower is a good prediction based on the data, but the actual car may have a different horsepower.

4. about 2600 cc

5. Answers may vary. Examples are given.

   a. No. All of the points lie below the line, so the line does not lie as close as it could to many of the points.

   b. Yes. About as many points lie above the line as lie below it and the line appears to be close to as many points as possible.

Communication: Discussion

Before discussing the Sample, ask students to give some examples from everyday life of quantities they think will have a strong or weak correlation. Ask them to explain their thinking.

Ask what a fitted line would look like for a scatter plot that indicates no correlation between the sets of data.

Additional Sample

The table shows the list price and quality rating (from 0 to 100) for seven brands of miniature TV sets. Predict the quality rating for a TV set that lists for $500.

TV Brand	List Price ($)	Rating
A	600	86
B	550	75
C	300	50
D	650	70
E	450	51
F	369	48
G	300	40

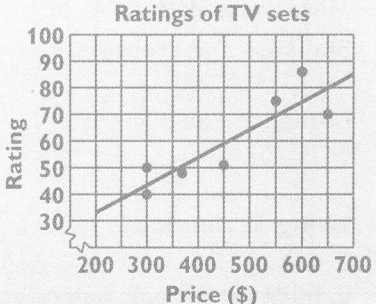

Using the scatter plot and fitted line shown above, a rating of 64 seems a reasonable prediction.

Talk it Over

Use questions 3–5 to be sure students understand what a fitted line is and how it can be used to make reasonable predictions.

213

Watch Out!
A correlation between two sets of data does *not* necessarily mean there is a cause-and-effect relationship.

► Suppose there is a positive correlation between the number of students who go to their school's basketball games and the number of games won. Here are some conclusions you might make.

➤ Stronger student support causes the team to win more often.

➤ The team's success causes more students to attend games.

➤ Something else, such as a new gymnasium causes both attendance to rise and the team to win.

Look Back ◄

Do you think that the scatter plots you made in the Exploration show a strong or a weak correlation between height and lower-arm length? How did you decide?

······► **Now you are ready for:**
Exs. 10–31 on pp. 215–217

4-5 Exercises and Problems

1. Use the graph you made in the Exploration.

   a. Suppose two people have the same height and different lower-arm lengths. How would this look on the graph?

   b. Suppose two people have different heights and the same lower-arm length. How would this look on the graph?

2 **TECHNOLOGY** Use a graphics calculator or graphing software.

   a. Plot the data that you collected in the Exploration using the statistics menu of the calculator. Does it look like the graph that you drew in Step 3 of the Exploration? If not, change the range until it does. What is the range that you need to use?

   b. Use the *linear regression* feature to find a fitted line for the data. *Trace* along the line until you find the point with x-coordinate 171. What is the y-coordinate of the point?

   c. Explain what the y-coordinate that you found in part (b) represents. How is it related to the prediction that you made in Step 8 of the Exploration?

3. Use one of the data sets that you collected in Step 9 of the Exploration.

   a. What kind of correlation, if any, do you think that there is between the data set and height? Why?

   b. Test your prediction by making a scatter plot. (Put height on the horizontal axis.)

   c. Describe the correlation shown by the scatter plot. Does your scatter plot support your prediction? Explain.

BY THE WAY...

In his first year as a professional basketball player, Shawn Bradley wore number 76 for the Philadelphia 76ers. He is 7 ft 6 in. tall. Can you predict his lower-arm length?

214   **Unit 4** Coordinates and Functions

---

4. What types of manufacturers might need to know about the relationships between heights and arm lengths? Why might they need this information?

State whether each scatter plot shows a *positive correlation*, a *negative correlation*, or *no correlation*.

5.

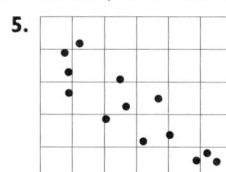

6.

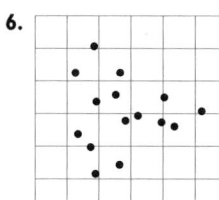

7.

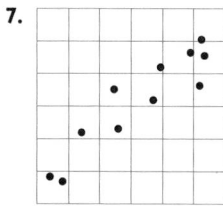

8. Choose one of these titles for each of the scatter plots in Exercises 5–7. Explain your choice.

  a. Number of Cars vs. Number of Traffic Lights in a City

  b. Amount of Time Spent Working vs. Amount of Leisure Time

  c. Temperature vs. Winter Coat Sales

  d. Shoe Size vs. Amount of Savings

9. **Open-ended** Make up your own title for each scatter plot in Exercises 5–7. Be sure your titles reflect the relationships shown in the scatter plots.

10. **Reading** Use diagrams to describe how to use a fitted line to make predictions from a scatter plot.

11. **Writing** A newspaper claims there is a strong correlation between the weather on the day of an election and voter turnout. What do you think the newspaper means? Could the newspaper use a scatter plot to support its claim? If so, what would be on each axis?

12. This graph shows a strong negative correlation.

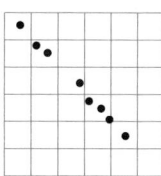

This graph shows a weak negative correlation.

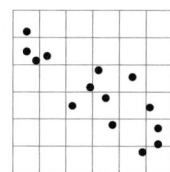

  a. Draw a graph that shows a strong positive correlation. Draw another graph that shows a weak positive correlation.

  b. When you look at a scatterplot, you can decide two things: *a positive or a negative* correlation, and *a strong or a weak* correlation. Which type of decision do you think is easier to make?

TECHNOLOGY NOTE

If you use a graphics calculator to find a fitted line for a data set, you can get a measure of how strong the correlation is.

The calculator will produce a number called "*r*". The closer *r* is to 1 or −1, the stronger the correlation.

Communication: Discussion
You may wish to ask for volunteers to read their answers for Ex. 11. Discuss the answers with the class. Ask students whether they think the newspaper claim referred to is reasonable and why.

Reasoning
In connection with Ex. 12, ask students to invent data that will lead to two special graphs, one showing five points that lie exactly on a vertical line, the other showing five points that lie exactly on a horizontal line. Ask what kind of correlation, if any, these graphs suggest.

Ask whether either of the two graphs leads to a fitted line that permits predictions.

If students know how to use a graphics calculator or computer software to produce scatter plots, you might suggest that they try using it to find fitted lines for their invented data. What happens?

10. Answers should be similar to Sample Response.

11. Answers may vary. An example is given. The newspaper means that fewer people will vote if the weather is bad; more will vote if the weather is good. In order to use a scatter plot, the newspaper would need two sets of numerical data. One axis could show temperature or amount of rain. The other could show the number of people voting or percent of voter turnout.

12. a. Answers may vary. Both graphs should rise from left to right. The strong correlation should be more collinear than the weak correlation.

  b. Positive and negative correlation is easier to decide than strong or weak correlation.

Answers to Exercises and Problems

car. Clothing manufacturers might need to know because they have to make clothing in different sizes.

5. negative correlation

6. no correlation

7. positive correlation

8. Answers may vary. Examples are given.

  a. 7; As the number of cars in a city increases, the number of traffic lights will also increase in order to control the increased traffic. There is a positive correlation.

  b. 5; The more hours a person works, the fewer leisure hours he or she has. There is a strong negative correlation.

  c. 5; As the temperature goes up, fewer people will want to buy winter coats. There is a negative correlation.

  d. 6; There is no reason why shoe size should have any correlation with a person's savings. There is no correlation.

9. Answers may vary.

connection to **EARTH SCIENCE**

Make a scatter plot for each relationship. State whether each scatter plot shows a *positive correlation*, a *negative correlation*, or *no correlation*.

City	Latitude (°N)	Average precipitation (in.)	Annual range of average temperatures (°F)	Average temperature in January (°F)	Average number of days of rain per year
Anchorage, AK	61	15.2	45.1	13	115
Austin, TX	30	31.5	35.6	49.1	83
Bismarck, ND	47	15.4	63.7	6.7	96
Charleston, WV	38	42.4	41.6	32.9	151
Grand Rapids, MI	43	34.4	49.4	22	143
Honolulu, HI	21	23.5	7.5	72.6	100
Los Angeles, CA	34	12.0	13	56	36
Madison, WI	43	30.8	55	15.6	118
Minneapolis, MN	45	26.4	61.9	11.2	115
Olympia, WA	47	51.0	25.8	37.2	164
San Francisco, CA	38	19.7	13.7	48.5	63
Tampa, FL	28	46.7	22.3	59.8	107

13. the latitude and the average precipitation

14. the annual range of average temperatures and the average temperature in January

15. the average precipitation and the days of rain per year

For Exercises 16–21, use the scatter plots you made for Exercises 13–15.

16. **Using Manipulatives** Use a piece of uncooked spaghetti to represent a fitted line in each of the scatterplots. Which scatter plot shows the strongest correlation? the weakest correlation?

17. Predict the average temperature in January in a city with an annual range of average temperatures of 30°F.

18. Which do you think is greater: the probability that a city with an average precipitation of 13 in. has an average of 80 days of rain per year or the probability that it has an average of 170 days of rain per year?

19. Predict the average precipitation of a city with 40 days of rain per year.

20. Does your graph prove that there is a cause-and-effect relationship between the annual range of average temperatures and the average temperature in January? Explain why or why not.

21. Use the scatter plot that you made for Exercise 13. Which of these conclusions, if any, can you make?

  a. The graph shows that cities that are further north have more rain.

  b. The graph shows that cities that have more rain are further north.

  c. The graph shows that latitude and rain are not related.

  d. The graph shows that cities that are further south have more rain.

216   **Unit 4** Coordinates and Functions

Answers to Exercises and Problems

13. no correlation

Precipitation and Latitude

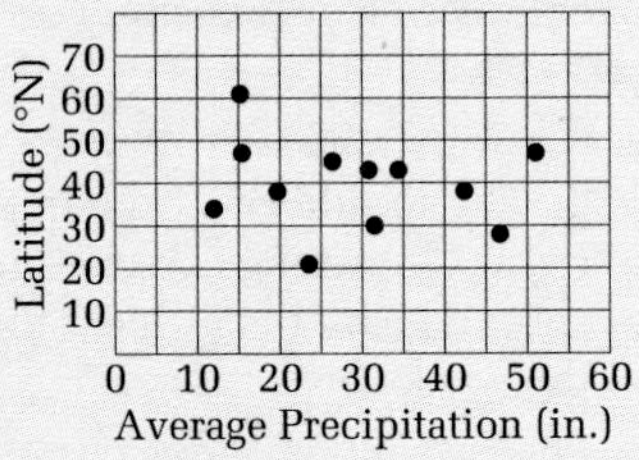

14. negative correlation

Temperatures for January and the Year

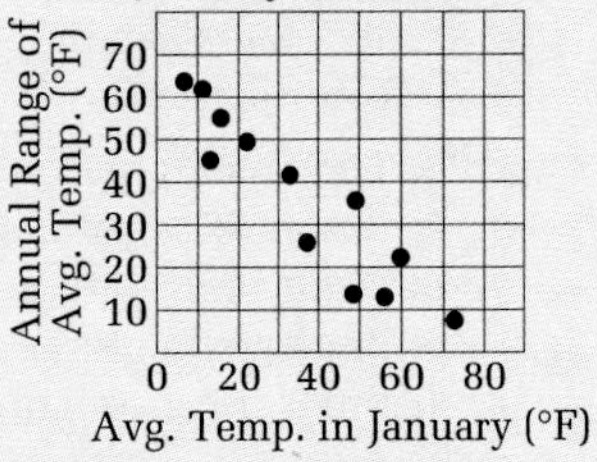

15. positive correlation

Days of Rain and Average Precipitation

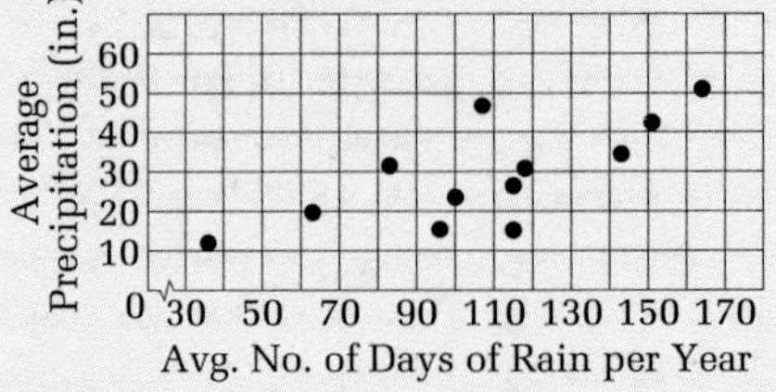

22. Research In a reference book such as an almanac, find two data sets that relate to each other like the ones in this section. The data could be about population, entertainment, sports, geography, social issues, government, sales, and so on.

a. Make a scatter plot of the data you collect.

b. What kind of correlation does the scatter plot show?

c. Write a question that can be answered by using the scatter plot.

Ongoing ASSESSMENT

23. **Group Activity** Collect data from 10 students in your class about the number of sweet snacks they eat in a week and the number of salty snacks they eat in a week.

a. What kind of correlation do you predict your data will show? Why?

b. Make a scatter plot of the data. Draw a fitted line, if possible. Does your scatter plot show the correlation you expected? Explain.

c. Compare your scatter plot with the plots of several other students. How are they alike and how are they different? Do all the scatter plots show the same correlation?

Review PREVIEW

Give the number of degrees in each type of rotation. (Section 4-4)

24. half-turn   **25.** quarter-turn   **26.** whole turn

Graph each inequality on a number line. (Section 3-3)

27. $a \geq -3$ **28.** $x \leq 5$ **29.** $-2 < c \leq 7$

30. Savannah earns \$5.75/h as a receptionist in a beauty shop. Tell how much she earns for each number of hours she works. (Section 1-2)

a. 3 hours   b. 7 hours   c. x hours

Working on the Unit Project

31. **Research** Read about how animation is created and used. As you read, you may want to take notes about animation ideas to help you as you make your flip book.

Suggested Bibliography

Laybourne, Kit. *The Animation Book.* New York: Crown Publishers, 1979.

Andersen, Yvonne. *Make Your Own Animated Movies.* Boston: Little, Brown and Company, 1970.

Platt, Richard. *Film.* (Eyewitness Guides Series). London: Dorling Kindersley Limited, 1992, pp. 50–53.

4-5 Scatter Plots

Practice 31 For use with Section 4-5

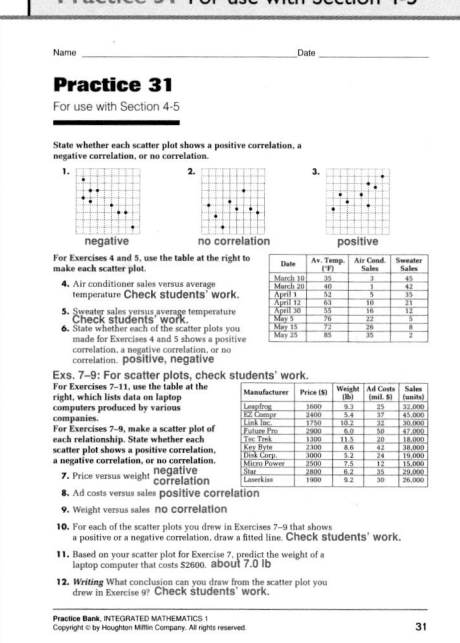

Answers to Exercises and Problems

16. Exercise 14 is strongest; Exercise 13 is weakest.

17–21. Answers may vary. Examples are given.

17. about 45°F

18. 80 days

19. about 15 in.

20. No. Correlation alone never proves causation.

21. c

22. a–c. Answers may vary, based on data collected.

23. a–c. Answers may vary, based on data collected.

24. 180°

25. 90°

26. 360°

27.

-4 -3 -2 -1 0 1 2 3

28.

-3 -2 -1 0 1 2 3 4 5 6

29.

-3 -2 -1 0 1 2 3 4 5 6 7 8

30. a. \$17.25

b. \$40.25

c. 5.75x dollars

31. Research may vary.

PLANNING

Objectives and Strands
See pages 180A and 180B.

Spiral Learning
See page 180B.

Materials List
➤ Geometric drawing software
➤ Graph paper
➤ Ruler

Recommended Pacing
Section 4-6 is a one-day lesson.

Extra Practice
See pages 625–627.

Warm-Up Exercises
Warm-Up Transparency 4-6

Support Materials
➤ Practice 32
➤ Enrichment 29 in the Activity Bank
➤ Study Guide 4-6
➤ Problem Set 9
➤ Diagram Master 2 in the Explorations Lab Manual
➤ McDougal Littell Mathpack software: *Geometry Inventor*
➤ Quiz 4-6
➤ Alternative Assessment 5

Section

4-6 Graphs and Functions

Focus
Understand what a function is, identify control variables and dependent variables, draw graphs of functions, and recognize functions.

THIS DEPENDS ON THAT

Talk it Over

1. What does this graph show?

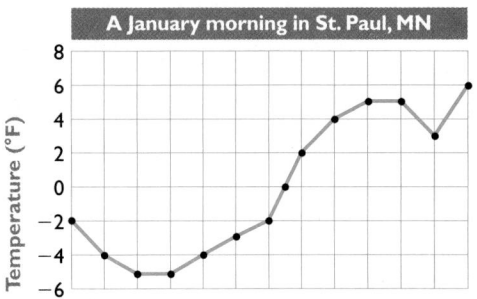

A January morning in St. Paul, MN

2. Estimate the highest and the lowest temperatures. At what time did each happen?

3. During what interval(s) of time did the temperature increase? When did it decrease?

4. Does the temperature depend on the time of day?

5. Does the time of day depend on the temperature?

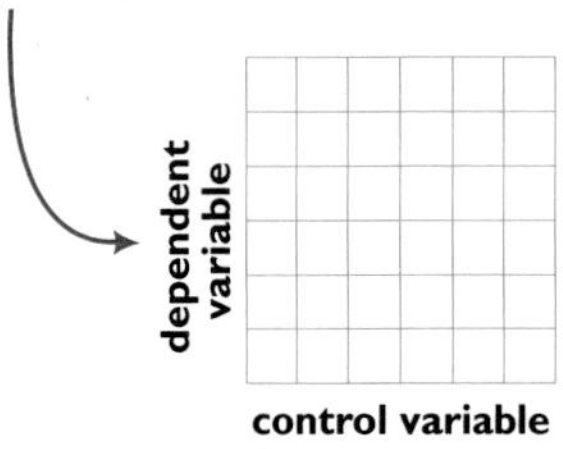

Dependent and Control Variables

The dependent variable is usually put on the vertical axis.

dependent variable

control variable

The graph shows information about two variables: time of day and temperature. The temperature *depends* on what time of day it is. The temperature is called the **dependent variable.** The time of day is called the **control variable.** (The control variable is sometimes called the *independent variable.*)

218   Unit 4 Coordinates and Functions

Answers to Talk it Over

1. Answers may vary; for example, the change in temperature on a January morning in St. Paul, MN.

2. highest: 6°F; 12:00 noon; lowest: −5°F; 2:00 or 3:00 A.M.

3. Temperature increased from 3:00 A.M. to 9:00 A.M. and from 11:00 A.M. to 12:00 noon. Temperature decreased from 12:00 midnight to 2:00 A.M. and from 10:00 A.M. to 11:00 A.M.

4. Yes.

5. No.

Sample 1

Identify the dependent variable and the control variable in each situation.

a. Cal volunteers to wrap gifts for a charity. He uses different amounts of wrapping paper to wrap boxes of different sizes.

b. Mai earns $4.50 per hour in a hardware store. When she works different numbers of hours, she earns different amounts of money.

Sample Response

a. The amount of paper *depends* on the box size.

The box size is the control variable.

The amount of paper is the dependent variable.

b. The amount of money earned *depends* on the number of hours worked.

The number of hours is the control variable.

The amount of money earned is the dependent variable.

Talk it Over

6. Suppose you graph Cal's box sizes and amounts of paper. Which variable would you plot on the horizontal axis? Why?

7. Identify the dependent variable and the control variable: When riding a Ferris wheel, you are at different heights from the ground at different times during the ride.

4-6 Graphs and Functions

219

Talk it Over

Questions 1–3 review reading and interpreting graphs. Use questions 4 and 5 to help students think of one quantity (time) as influencing the other (temperature).

Additional Sample

S1 Identify the dependent variable and the control variable in each situation.

a. Jerry is driving his car at 45 mi/h. The longer he drives, the farther away from home he is. **control variable: amount of time driving; dependent variable: distance from home**

b. Kim is reading a biography of Marie Curie. The more chapters she reads, the fewer pages she has to go to finish the book. **control variable: number of chapters read; dependent variable: number of pages to go**

Talk it Over

Use question 6 to be sure that students are aware of the convention that the values of the control variable usually go on the horizontal axis and values of the dependent variable on the vertical axis. Putting the control variable on the horizontal axis is purely a convention but one that is universal in mathematics.

Answers to Talk it Over

6. box size; The control variable is usually put on the horizontal axis.

7. dependent variable: height from the ground; control variable: time during the ride

219

S2 Tracy is going to save some of the earnings from her summer job. She plans to save $1 the first week, $2 the second week, and for the next five weeks match her savings to her total savings from the preceding weeks. Draw a graph to show her weekly savings for the seven-week period. Tell whether the graph represents a function.

Week	Savings ($)
1	1
2	2
3	3
4	6
5	12
6	24
7	48

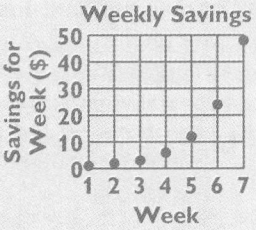

Weekly Savings

The graph represents a function.

Using Technology

You may wish to refer students to page 619 in the Technology Handbook, which discusses using a spreadsheet.

Functions

A **function** is a relationship in which there is *only one* value of the dependent variable for each value of the control variable. For example, in the graph on page 218, for each time of day, there is only one temperature. You can say "the temperature is a function of the time of day."

A function can be represented many ways: in words, in symbols, with a table of values, or with a graph.

Sample 2

At a week-long charity rummage sale, the price of each item is half the previous day's price. Draw a graph of the price of an $8 item during the sale. Does the graph represent a function?

Sample Response

First make a table of values.

To find each day's price, multiply the previous day's price by $\frac{1}{2}$.

TECHNOLOGY NOTE

You can also use spreadsheet software to make a table.

Use the "fill down" command for the cells in column B.

	A	B
1	Day	Price
2	1	8
3	2	= 0.5 * B2

The original price is $8.

Day of sale	Price (dollars)
1	8
2	$\frac{1}{2}(8) = 4$
3	$\frac{1}{2}(4) = 2$
4	$\frac{1}{2}(2) = 1$
5	$\frac{1}{2}(1) = 0.50$
6	$\frac{1}{2}(0.50) = 0.25$
7	$\frac{1}{2}(0.25) = 0.13$

Then draw a graph.

The price depends on the day of the sale. So the day of the sale is the control variable.

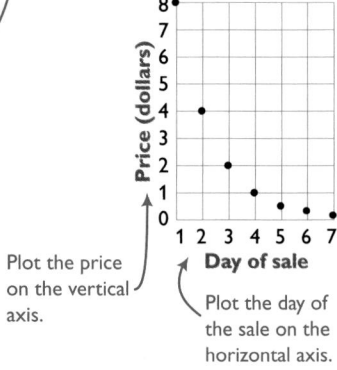

Plot the price on the vertical axis.

Plot the day of the sale on the horizontal axis.

For each day of the sale, the item has only one price. So the graph represents a function.

Vertical-Line Test for Functions

You can tell if a graph represents a function by looking at it. For each value of the control variable, there should be only one value of the dependent variable. This means that no two points on a graph of a function lie directly above one another.

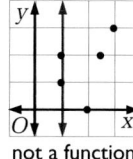

not a function

To test whether a graph represents a function, imagine drawing vertical lines through the graph. If two or more points of the graph lie in the same vertical line, then the graph does not represent a function. This test is called the *vertical-line test*.

Sample 3

Tell whether each graph represents a function.

a.
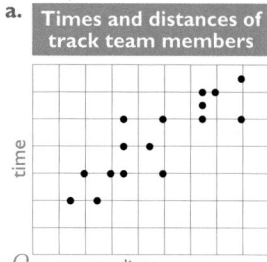
Times and distances of track team members

b.
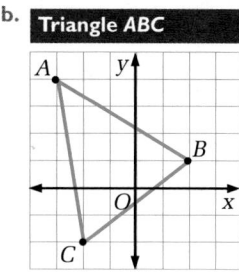
Triangle *ABC*

c.
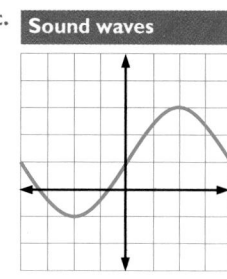
Sound waves

Sample Response

Use the side of a pencil as a vertical line. Slide the pencil across the page to check if two or more points lie in the same vertical line. You can also imagine drawing vertical lines through each graph.

a.
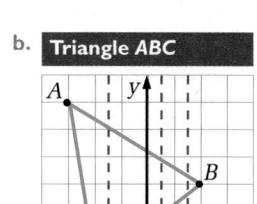
Times and distances of track team members

b.
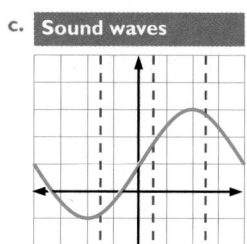
Triangle *ABC*

c.
Sound waves

Many points of the graph lie on the same vertical line.

The graph does not represent a function.

Many points of the graph lie on the same vertical line.

The graph does not represent a function.

No two points of the graph lie on the same vertical line.

The graph represents a function.

Look Back

In Sample 2 the relationship is a function. How can you tell from the table? How can you tell from the graph?

4-6 Graphs and Functions

221

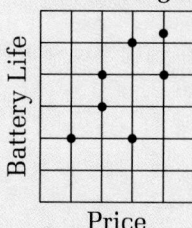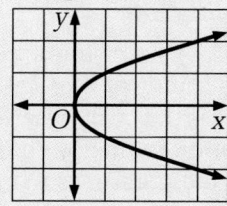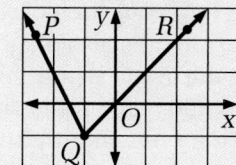
221

Suggested Assignment

Standard 1–11, 13–26

Extended 1–11, 13–26

Integrating the Strands

Number Exs. 18–21

Algebra Exs. 22–25

Functions Exs. 1–16

Geometry Ex. 26

Statistics and Probability
Ex. 17

Application

Exs. 2–4 are examples of functions from real life. Ask students if they can describe a situation in their own lives that uses the phrase "is a function of."

Reasoning

For Ex. 5, you can have an interesting discussion about what the graph would look like if you had data on the shark's length for one-month intervals, starting at birth. Ask students whether they think the graph is really a smooth curve or a series of isolated points. Discuss their reasoning.

For Ex. 7, ask students what the graph of earnings would look like if a person has a *salary* and works 40 hours a week.

4-6 Exercises and Problems

1. **Reading** Complete the statement. The control variable is usually put on the ? axis and the dependent variable is usually put on the ? axis.

For Exercises 2–4, describe each situation using the phrase "is a function of." What is the dependent variable? the control variable?

2. Felicia Ramiro owns a carpet cleaning business. She charges different prices for cleaning carpets of different sizes.

3. **Sports** Michael practices running the 100 m dash every day. The faster he runs, the shorter his time.

4. Students in a driver's education class study the night before a written test. Students who study longer earn better scores.

5. **Marine Biology** The table shows the length of a lemon shark at various ages.

   a. Identify the dependent variable and the control variable.

   b. Graph the data.

   c. Does the graph represent a function? Explain.

   d. Estimate the shark's length at age 6.

   e. Estimate the shark's age when its length was 149 cm.

Length of a lemon shark	
Age (years)	Length (cm)
1	50
2	62
3	86
4	93
5	107
7	126
8	140
10	156
11	161
12	169

6. Use the situation described in Sample 2. Suppose you want to buy an item originally priced at $20. Graph the function. How is the graph similar to the one in Sample 2? How is it different?

Choose the most reasonable graph for each situation.

7. Bob's earnings at a sub shop are a function of how many hours he works.

a.
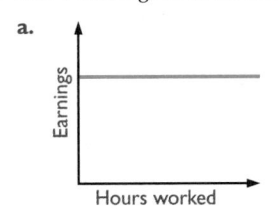

b.

c.

5. b.
Length of Lemon Shark

6.
Charity Rummage Sale: Item Originally Priced at $20

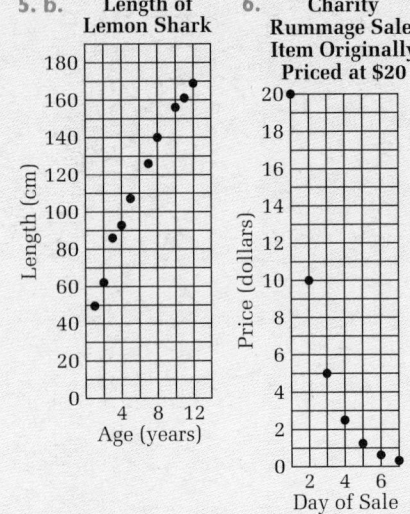

Answers to Exercises and Problems

1. horizontal; vertical

2. The price Felicia Ramiro charges is a function of the size of the carpet. dependent: price charged; control: carpet size

3. The time it takes Michael to run the 100 m dash is a function of his speed. dependent: Michael's time; control: Michael's speed

4. The score a student receives on the test is a function of the amount of time spent studying. dependent: test score; control: time spent studying

5. a. dependent: length; control: age

   b. See graph at left.

   c. Yes; passes the vertical-line test

d. about 115 cm

e. about 9 years

6. See graph at left. Answers may vary. Both graphs show a similar curve. Some differences are that the graph for the $20 item begins higher and remains higher than the graph in Sample 2.

8. An oven is turned on and set for a certain temperature. The temperature of an oven is a function of the length of time the oven has been on. (A thermostat maintains the temperature.)

a.
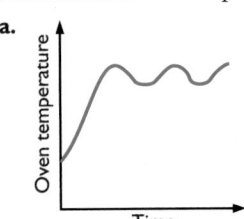

b.
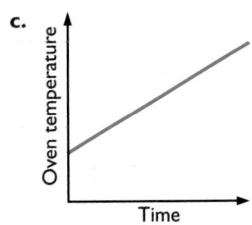

c.
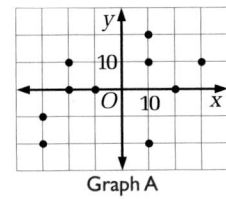

Draw a graph of each situation. Tell whether the graph represents a function. Explain.

9. The Ecology Club has a bake sale every year. This year they raised $240. They decide to try to raise 1.5 times the amount they raised the previous year for the next three years.

10. Linda Lee teaches five art classes that have 23, 26, 21, 29, and 24 students. She buys a drawing pad for each student, plus 4 extras for each class.

11. Use these two graphs.

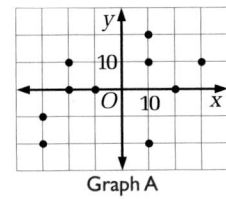
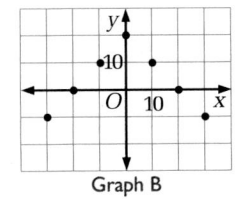

Graph A   Graph B

a. Use Graph A. If someone asks you what *y* is when *x* = 10, how many answers can you give?

b. Use Graph B. If someone asks you what *y* is when *x* = 10, how many answers can you give?

c. Writing Which graph represents a function? Explain your choice.

12 T E C H N O L O G Y Use geometric drawing software.

Alternative Approach Work in a group of three students using graph paper and rulers.

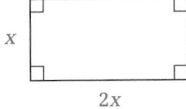
x
2*x*

a. Draw six different rectangles that are twice as long as they are wide.

b. Find the perimeter and area of each rectangle. Make a table of your results.

Width	Length	Perimeter	Area
?	?	?	?
?	?	?	?
?	?	?	?
?	?	?	?
?	?	?	?
?	?	?	?

c. Use your table to make three graphs:
➤ length as a function of width
➤ perimeter as a function of width
➤ area as a function of width

d. How are the graphs alike? How are they different?

---

12. c.
Length as a Function of Width
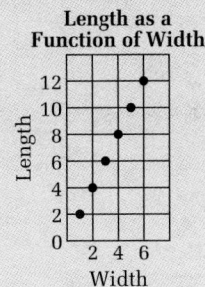

Perimeter as a Function of Width
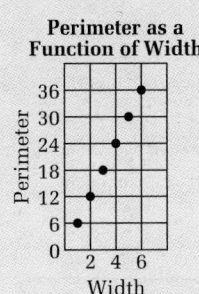

Area as a Function of Width
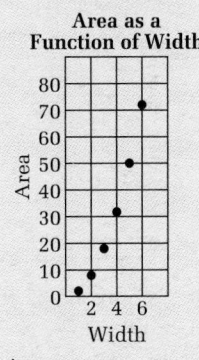

d. Answers may vary. An example is given. The graphs all represent functions. On the graphs showing length and perimeter as functions of width, the points form straight lines. The graph showing area as a function of width forms a curve. All the graphs show positive correlations.

Answers to Exercises and Problems

7. b   **8.** a

9. function; passes the vertical-line test

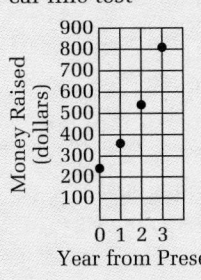

10. function; passes the vertical-line test

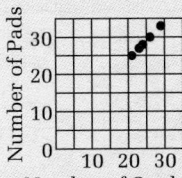

11. a. three   **b.** one

c. Graph B; There is only one *y* for each *x*.

12. a. Rectangles may vary.

b. Answers may vary. A sample table of values is given.

width	length	perimeter	area
1	2	6	2
2	4	12	8
3	6	18	18
4	8	24	32
5	10	30	50
6	12	36	72

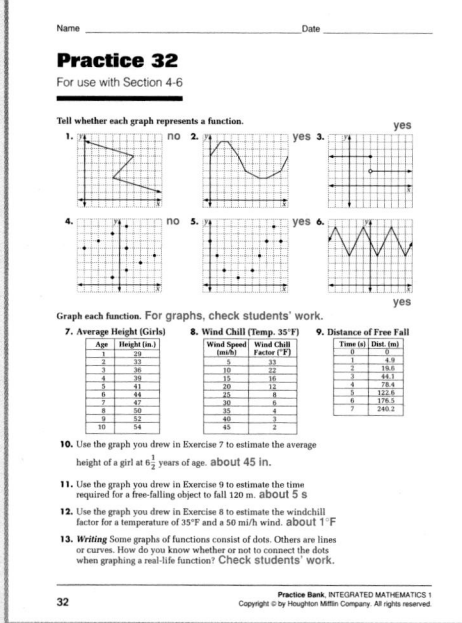
Tell whether each graph represents a function. Explain your answer.

13.
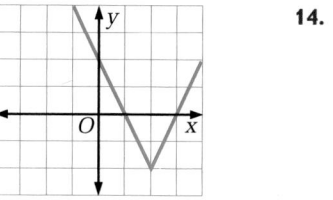

14.
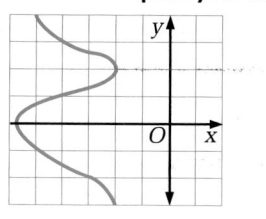

15.
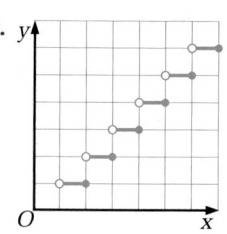

Ongoing ASSESSMENT

16. **Open-ended** Think of a situation that involves two quantities, one of which depends on the other. Describe the situation as a function. Identify the dependent variable and the control variable. Draw a graph of the function.

Review PREVIEW

17. Draw a scatter plot that shows a strong positive correlation. *(Section 4-5)*

Find the reciprocal of each number. *(Toolbox Skill 9)*

18. 4   19. -4 20. $\frac{1}{2}$ 21. $-\frac{3}{8}$

Evaluate each expression when $x = -2$, $x = 0$, and $x = 3$. *(Sections 1-4, 2-2)*

22. $-5x^2$ 23. $|x| + 8$ 24. $-x$ 25. $\frac{1}{x}$

4 Working on the Unit Project

26. Start making the drawings for your flip book. Draw on graph paper to make sure the positions of your drawings are consistent. Keep track of the types of translations and rotations you use.

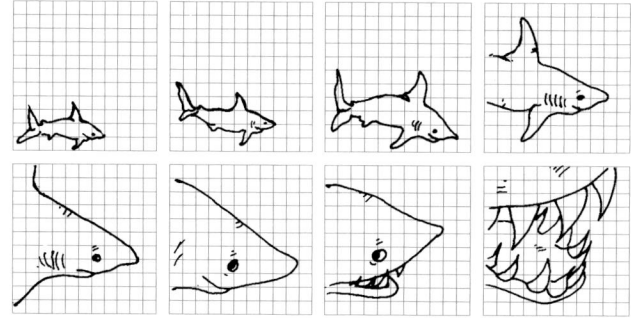

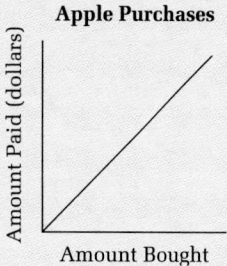
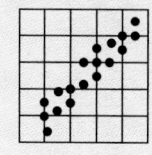

BY THE WAY...

The movie *Jaws* starred a white shark. Because it is difficult and dangerous to get near a white shark, not much is known about them. They are about 120 cm long at birth, and the maximum confirmed length of a white shark is 6 m.

224   **Unit 4** Coordinates and Functions

Answers to Exercises and Problems

13. Yes. No two points lie on the same vertical line.

14. No. Many points lie on the same vertical line.

15. Yes. No two points lie on the same vertical line.

16. Answers may vary. An example is given. The price paid for apples is a function of how many pounds are purchased. The dependent variable is the price paid; the control variable is the number of pounds.

Apple Purchases

17. Answers may vary. The data points should be close to an upward-slanting line.

18. $\frac{1}{4}$

19. $-\frac{1}{4}$

20. 2

21. $-\frac{8}{3}$

22. -20; 0; -45

23. 10; 8; 11

24. 2; 0; -3

25. $-\frac{1}{2}$; undefined; $\frac{1}{3}$

26. Answers may vary.

4-7 Functions and Equations

Focus
Change from one
representation of a function
to another, learn about
some basic functions.

ANOTHER WAY

TO LOOK AT IT

EXPLORATION

Is the relationship between a number and its square a function?

• Materials: graph paper, graphics calculator

Number	Square of the Number
X	X^2
-3	9
-2.5	6.25
-2	4
-1.7	
-1.3	
-1	
-0.5	
0	
0.5	
1	
1.3	
1.7	
2	
2.5	
3	

① Complete the table in the photo.

② Plot the ordered pairs on a coordinate plane.

(-2, 4)

③ Describe the shape the points seem to make.

④ Does the relationship between x and x^2 appear to be a function? Why or why not?

⑤ Choose four values between -3 and 3 that are not already in the table. Find their squares and plot the points.

⑥ You cannot square *all* the numbers between 3 and -3 and plot the ordered pairs. However, your graphics calculator can do many more than you can, in a much shorter time. Enter the equation $y = x^2$ on the "$y =$" screen of the calculator. Compare the result with your graph.

4-7 Functions and Equations

225

Answers to Exploration

1. values for blanks in table: 2.89, 1.69, 1, 0.25, 0, 0.25, 1, 1.69, 2.89, 4, 6.25, 9

2. See art at right.

3. a U-shaped curve (a parabola)

4. Yes. Explanations may vary. An example is given. The graph satisfies the vertical-line test because for each number there is only one number that is the square of the given number.

5. Answers may vary, but should be ordered pairs in which the second coordinate is the square of the first coordinate.

6. Graphs should be similar.

2.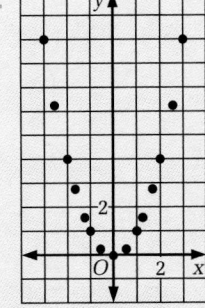

PLANNING

Objectives and Strands
See pages 180A and 180B.

Spiral Learning
See page 180B.

Materials List
➤ Graph paper
➤ Ruler
➤ Graphics calculator or graphing software

Recommended Pacing
Section 4-7 is a three-day lesson.

Day 1
Pages 225–227: Exploration through Talk it Over, *Exercises 1–11*

Day 2
Pages 227–229: Graphing Equations through Talk it Over, *Exercises 12–20*

Day 3
Pages 229–230: Shapes of Graphs through Look Back, *Exercises 21–45*

Extra Practice
See pages 625–627.

Warm-Up Exercises
Warm-Up Transparency 4-7

Support Materials
➤ Practice 33
➤ Enrichment 30 in the Activity Bank
➤ Study Guide 4-7
➤ Problem Set 9
➤ Diagram Masters 2, 3, 13 in the Explorations Lab Manual
➤ McDougal Littell Mathpack software: *Function Investigator*
➤ Using Plotter Plus: Exploring Graphs
➤ Using Mac or IBM Plotter Plus Disk: Function Plotter
➤ Quiz 4-7
➤ Test 15
➤ Alternative Assessments 6, 7

TEACHING

Exploration

The goal of the Exploration is to have students understand that the relationship between a number and its square is a function, and that we can show this by making a table and drawing a graph. The last part of the Exploration has students use a graphics calculator to graph the function. This leads naturally to representing functions by equations.

Communication: Discussion

For the last part of the Exploration, have a brief discussion with the class so that students understand the effect that range settings can have on the shape of the graph displayed on the screen.

Additional Samples

S1 Write an equation to represent the function given in the table.

x	y
−5	−10
−3	−6
−2	−4
−1	−2
0	0
1	2
4	8
6	12

$y = 2x$

S2 Write an equation to represent the function.

To mail a letter to Mexico, Felipe pays $.35 for the first half ounce and $.10 for each additional half ounce.

Let c = cost, w = number of half ounces over first half ounce. The equation is $c = 0.35 + 0.10w$.

Writing Equations

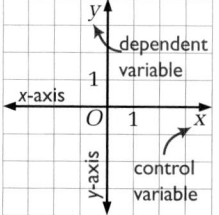

For a function, x often represents the control variable and y represents the dependent variable. In the Exploration:

The *number* is x

y is the *square of the number*

so $y = x^2$.

The equation, the table of values that you made, the graph, and the description in words all represent the same relationship. You can change one representation to another.

Sample 1

Write an equation to represent the function given in the table.

x	y
−3	−5
−2	−4
−1	−3
0	−2
1	−1
2	0
3	1

Sample Response

$$-3 - 2 = -5$$
$$-2 - 2 = -4$$
$$-1 - 2 = -3$$
$$0 - 2 = -2$$
$$1 - 2 = -1$$
$$2 - 2 = 0$$
$$3 - 2 = 1$$

Problem Solving Strategy: Identify a pattern

Think about how the y-values relate to the x-values in the table. Make a conjecture about the pattern. Write a rule for the pattern: Each y-value is two less than its x-value.

Test your conjecture using the values in the table.

Write the rule as an equation: $y = x - 2$.

Sample 2

Write an equation to represent the function.

The cost of a telephone call depends on the number of minutes that it lasts. The first minute costs $.26 and each additional minute costs $.12.

Unit 4 Coordinates and Functions

Talk it Over

Use questions 1 and 2 to give additional experience in writing equations for functions that are described in words.

Sample Response

The number of additional minutes after the first is the control variable. Represent it by m.

→ The number of additional minutes is m.

The cost is the dependent variable. Use a different letter to represent it.

→ The total cost in dollars of the telephone call is c.

Because each additional minute costs $.12, the cost of m minutes is $.12 times m.

→ The cost of the additional minutes is $0.12m$.

The total cost is the sum of the first minute's cost and the additional minutes' costs. Write an equation to represent the function.

→ Then $c = 0.26 + 0.12m$.

Talk it Over

Identify the control variable and the dependent variable for each function. Then write an equation to represent the function.

1. A videotape movie is 108 min long. The amount of time you have left to watch depends on how many minutes you already watched.

2. The senior class play costs $500 to produce. Each ticket to the play sells for $5. The profit the senior class earns depends on the number of tickets that are sold.

······► Now you are ready for:
Exs. 1–11 on p. 231

Graphing Equations

You can graph a function represented by an equation by using *graph paper*, a *graphics calculator*, or *graphing software*.

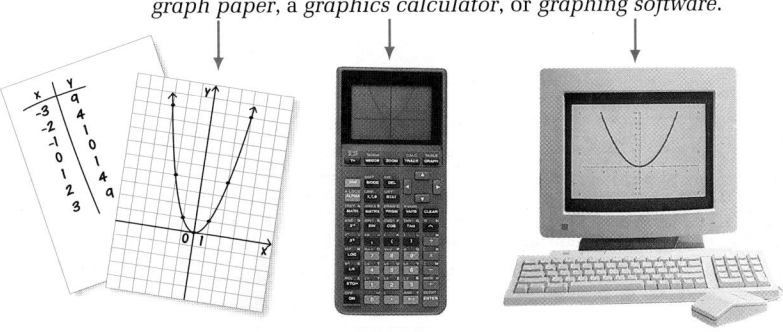

To graph an equation on graph paper, first make a table of values. You can write the values as ordered pairs (x, y). Then plot the ordered pairs. If all values between the ordered pairs also make the equation true, then the points can be connected.

4-7 Functions and Equations

227

Answers to Talk it Over

1, 2. Choices of variables may vary.

1. control: the number of minutes you have already watched, w; dependent: the number of minutes left on the tape, t; $t = 108 - w$

2. control: number of tickets sold, t; dependent: profit earned by senior class, p; $p = 5t - 500$

S3 Graph each equation.

a. $y = |x - 1|$

x	y	(x, y)
–3	|–3 –1| = 4	(–3, 4)
–2	|–2 –1| = 3	(–2, 3)
–1	|–1 –1| = 2	(–1, 2)
0	|0 –1| = 1	(0, 1)
1	|1 –1| = 0	(1, 0)
2	|2 –1| = 1	(2, 1)
4	|4 –1| = 3	(4, 3)

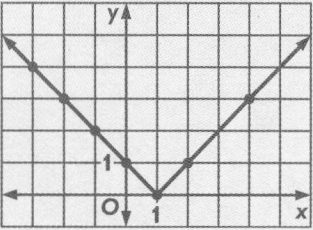

b. $y = -2x$

x	y	(x, y)
–2	–2(–2) = 4	(–2, 4)
–1	–2(–1) = 2	(–1, 2)
0	–2(0) = 0	(0, 0)
1	–2(1) = –2	(1, –2)
2	–2(2) = –4	(2, –4)

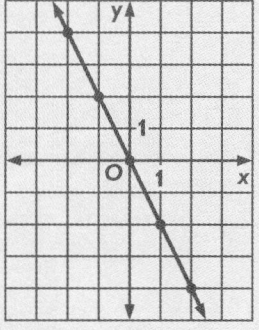

Using Technology

Students can use the *Function Investigator* software to graph functions in this section. For Sample 3, urge students who have graphics calculators or graphing software to check the graphs in the text by graphing them with the calculator. Do the same for Additional Sample S3.

Sample 3

Graph each equation.

Carmella burns 5 Calories per minute when she rides her bicycle.

a. $y = |x|$

b. The equation $y = 5x$ gives the number of Calories (y) that Carmella burns in x minutes.

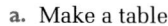

Sample Response

➤ Make a table to find the coordinates of several points.

➤ Choose both positive and negative values for x. Some easy values to evaluate are −1, 0, and 1. Use other values to show the shape of the graph better.

➤ Graph the ordered pairs and connect them.

a. Make a table.   Plot the points and connect them.

y = |x|		
x	y	(x, y)
–3	|–3| = 3	(–3, 3)
–2	|–2| = 2	(–2, 2)
–1	|–1| = 1	(–1, 1)
0	|0| = 0	(0, 0)
1	|1| = 1	(1, 1)
2	|2| = 2	(2, 2)
3	|3| = 3	(3, 3)

If x is negative, then its absolute value is the opposite of x.

If x is positive or zero, its absolute value is itself.

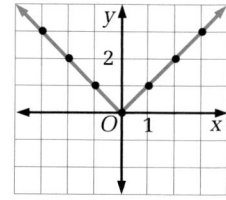

The shape of the graph is a V.

b. Make a table.   Plot the points and connect them.

TECHNOLOGY NOTE

If you use a graphics calculator or software to graph the equation, be sure that the RANGE includes all the x- and y-values from your table.

y = 5x		
x	y	(x, y)
–1	5(–1) = –5	(–1, –5)
0	5(0) = 0	(0, 0)
1	5(1) = 5	(1, 5)
2	5(2) = 5	(2, 10)

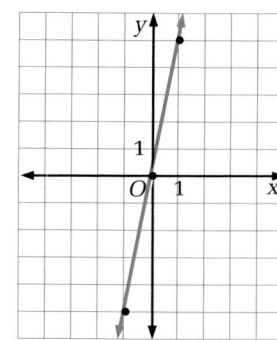

The shape of the graph is a line.

Unit 4 Coordinates and Functions

3. All the points on the line in Sample 3(b) make the equation true, but only the points in the first quadrant model the situations. Why is this?

4. Choose a point on the graph in Sample 3(b) that is between two of the plotted points. Find its coordinates. Does this ordered pair make the equation true?

5. Elena made this chart to help her remember which axis to use. Complete the chart.

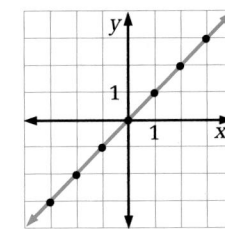

Now you are ready for:
Exs. 12–20 on pp. 231–232

horizontal axis	<u>?</u> axis
x	y
<u>?</u> variable	dependent variable

Shapes of Graphs

The U-shaped graph of the squaring function in the Exploration is called a **parabola**. In Sample 3 you saw a function whose graph is a line and another function whose graph is V-shaped. Other functions have special shapes when they are graphed.

Sample 4

Graph each equation.

a. $y = x$

b. $y = \dfrac{1}{x}$

Sample Response

Make a table to find the coordinates of several points. Graph the ordered pairs and connect them.

a. Make a table.

$y = x$		
x	**y**	**(x, y)**
−3	−3	(−3, −3)
−2	−2	(−2, −2)
−1	−1	(−1, −1)
0	0	(0, 0)
1	1	(1, 1)
2	2	(2, 2)
3	3	(3, 3)

Plot the points and connect them.

The shape of the graph is a line.

When discussing question 5, stress the importance of keeping in mind which is the control variable and which is the dependent variable.

Additional Sample

S4 Graph each equation.

a. $y = x^2 - 3$

x	**y**	**(x, y)**
−3	6	(−3, 6)
−2	1	(−2, 1)
0	−3	(0, −3)
2	1	(2, 1)
3	6	(3, 6)

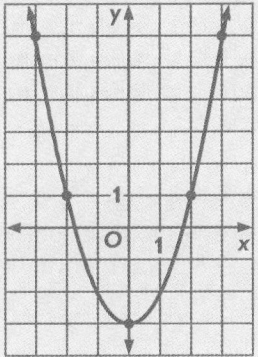

b. $y = -x - 1$

x	**y**	**(x, y)**
−2	1	(−2, 1)
−1	0	(−1, 0)
0	−1	(0, −1)
1	−2	(1, −2)
2	−3	(2, −3)

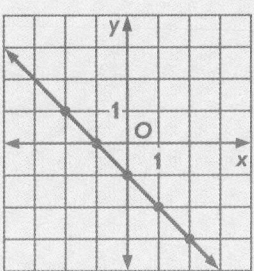

Answers to Talk it Over

3. Carmella cannot burn negative calories in negative minutes. Both calories and minutes have to be positive quantities and this is true only in quadrant I.

4. Answers may vary, but the point should be of the form $(x, 5x)$.

5. vertical; control

b. Make a table.

Plot the points and connect them.

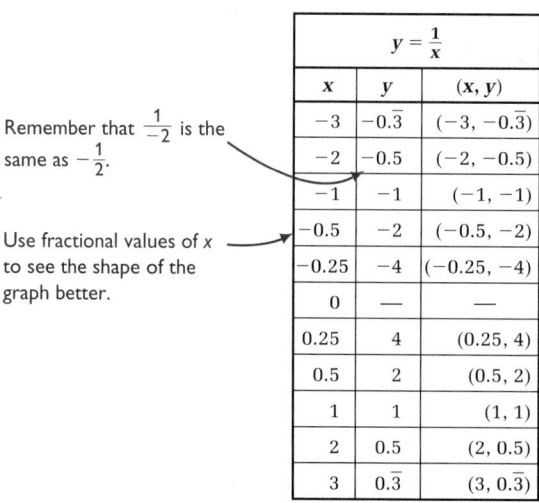

Remember that $\frac{1}{-2}$ is the same as $-\frac{1}{2}$.

Use fractional values of x to see the shape of the graph better.

$y = \frac{1}{x}$		
x	y	(x, y)
-3	$-0.\overline{3}$	$(-3, -0.\overline{3})$
-2	-0.5	$(-2, -0.5)$
-1	-1	$(-1, -1)$
-0.5	-2	$(-0.5, -2)$
-0.25	-4	$(-0.25, -4)$
0	—	—
0.25	4	$(0.25, 4)$
0.5	2	$(0.5, 2)$
1	1	$(1, 1)$
2	0.5	$(2, 0.5)$
3	$0.\overline{3}$	$(3, 0.\overline{3})$

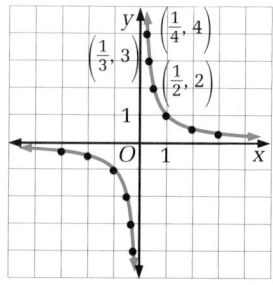

The shape of the graph is two curves. This graph is called a **hyperbola**.

Talk it Over

6. The function $y = x$ is called the *identity function*. Why do you think this function has this name?

7. The function $y = \frac{1}{x}$ is called the *reciprocal function*. Why do you think this function has this name?

SOME BASIC FUNCTIONS

Equation	$y = x^2$	$y = |x|$	$y = x$	$y = \frac{1}{x}$
Words	squaring function	absolute value function	identity function	reciprocal function
Graph	shape: parabola	shape: V	shape: line	shape: hyperbola

Look Back

Suppose you are given a function described by an equation. How can you represent it with a table? with a graph?

▶ Now you are ready for:
Exs. 21–45 on pp. 232–233

230

Unit 4 Coordinates and Functions

Answers to Talk it Over

Answers to Look Back

6. Each y-value is the same as its x-value.

7. y is the reciprocal of x.

Answers may vary. An example is given. Find values for the table by substituting values for the control variable and determining the values of the dependent variable. List these values in the table. Graph the ordered pairs from the table.

4-7 Exercises and Problems

1. **Reading** What are three different ways to represent a function?

Write an equation to represent each function.

2.

x	y
−3	−1
−2	0
−1	1
0	2
1	3
2	4
3	5

3.

x	y
−3	−1.5
−2	−1
−1	−0.5
0	0
1	0.5
2	1
3	1.5

4.

x	y
−3	−14
−2	−13
−1	−12
0	−11
1	−10
2	−9
3	−8

5. **Commissions** Rehema Strong works at a shoe store. Her weekly base pay is $350. She also receives a 4.5% commission on her total sales.

   a. Identify the control variable and the dependent variable.

   b. Write an equation to represent the situation.

   c. One week Rehema Strong sold $5348 worth of shoes. Find her total pay for the week.

Write an equation to represent each function. Tell what your variables represent.

6. **Health** An estimate of normal systolic blood pressure depends on a person's age. A general guideline estimates that normal blood pressure is 110 more than one half a person's age.

7. Use the photo. Sound travels at an average rate of 1000 ft/s.

8. On a summer day, the temperature at 3 A.M. is about 16°F less than the temperature at 3 P.M.

9. A group of people plan to rent a vacation cottage for $800. The amount of money each person pays depends on the number of people who go on vacation at the cottage.

10. The number of times a cricket chirps per minute depends on the air temperature. The number of chirps is thirty less than seven times the temperature in degrees Celsius.

11. **Writing** Think of a real life situation that the equation $y = 1.25x$ could represent. Describe the situation in words and tell what x and y represent.

12. The amount of change received from a $20 bill depends on the total cost of the items bought.

   a. Identify the control variable and the dependent variable.

   b. Write an equation to represent the situation.

   c. What limits are there on the values of the dependent variable? the control variable?

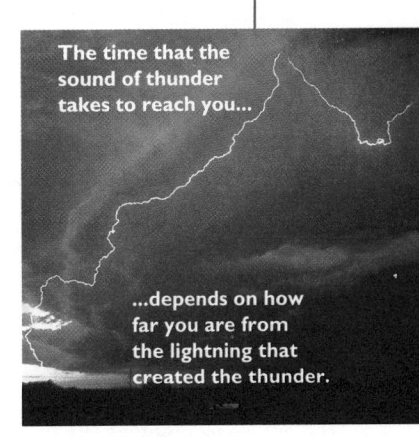

The time that the sound of thunder takes to reach you...

...depends on how far you are from the lightning that created the thunder.

4-7 Functions and Equations

231

Answers to Exercises and Problems

1. table of values, equation, graph, verbal description

2. $y = x + 2$

3. $y = \frac{x}{2}$ or $y = 0.5x$

4. $y = x - 11$

5–10. Choices of variables may vary.

5. a. control variable: total sales, s; dependent variable: total pay, p

   b. $p = 0.045s + 350$

   c. $590.66

6. Let a = a person's age; let p = normal systolic blood pressure; $p = \frac{1}{2}a + 110$

7. Let t = time (seconds); let d = distance (feet); $t = \frac{d}{1000}$

8. Let y = A.M. temperature; let x = P.M. temperature; $y = x - 16$

9. Let n = number of people; let c = amount each person pays; $c = \frac{800}{n}$

10. Let t = air temperature in °C; let c = number of chirps per minute; $c = 7t - 30$

APPLYING

Suggested Assignment

Standard 1–12, 16–45

Extended 1–12, 16–45

Integrating the Strands

Algebra Exs. 1–37, 39–44

Functions Exs. 1–12, 16–29, 34–41

Geometry Exs. 38, 45

Logic and Language Exs. 1, 11, 39, 40

Communication: Writing

In connection with Exs. 6–10, have students write brief descriptions of situations in their own experience that involve functions. Ask them to identify the control variable and dependent variable for each situation and to write an equation for the situation if they have enough information to do so.

11. Answers may vary. An example is given. A person working in Payville gets a raise each year of 25% of their previous year's salary. If José Alvarez earned $20,000 last year, how much will he earn this year? x = salary last year; y = salary this year

12. a. control variable: total cost of items bought, t; dependent variable: amount of change, c

   b. $c = 20 - t$

   c. Assuming that the lowest price you can pay is $.01, the most change you can receive is $19.99. The upper limit for the dependent variable is $19.99. The total cost of the items could be $20, in which case you would get no change. The upper and lower limits for the dependent variable, then, are $19.99 and $0. The upper and lower limits for the control variable are $20 and $.01.

231

Using Technology

For Exs. 13–15, students may need a reminder that the range settings may change the appearance of their graphs.

To be sure that the graphs students get on their calculators look like the choices for these exercises, students can request an automatic range setting to equalize the scales on the axes. For example, with the TI-81 or TI-82, this can be done by entering the equation, pressing zoom, and selecting Square (TI-81) or ZSquare (TI-82).

Answers to
Exercises and Problems

13. A   **14.** C   **15.** B

16. $y = x + 25$ **17.** $y = 5.3x$

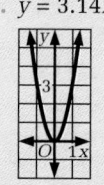   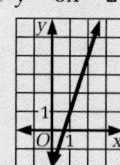

18. $y = 3.14x^2$ **19.** $y = 30x$

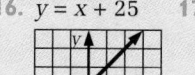

20. $y = 3x - 2$

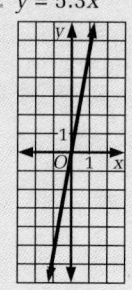

21. a. $5.24

b. $x + 0.05x = 1.05x$

c. Let x = total price before tax; let y = total cost; $y = x + 0.05x$ or $y = 1.05x$

d.

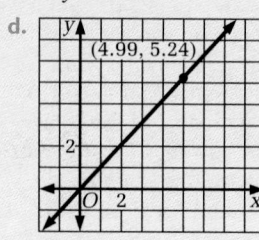

(4.99, 5.24)

e. Yes.

22. squaring function or $y = x^2$

TECHNOLOGY Use a graphics calculator or graphing software.

Match each equation with its graph.

13. $y = -4x - 1$ **14.** $y = \dfrac{3}{x}$ **15.** $y = -3x^2$

A.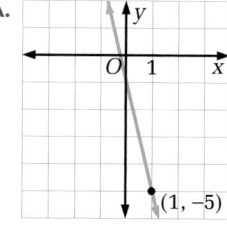
(1, −5)

B.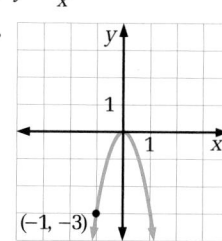
(−1, −3)

C.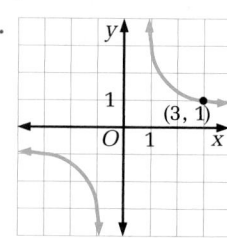
(3, 1)

Write each function as an equation. Then graph the function.

16. Your age 25 years from now depends on your current age.

17. A worker earns $5.30 per hour. The worker's gross pay depends on the number of hours worked.

18. The area of a circle depends on the length of its radius. The area is about 3.14 times the square of the radius.

19. A car travels at an average rate of 30 mi/h. The number of miles traveled depends on the number of hours the car traveled.

20. The length of a rectangle is two less than three times the width.

21. Consumerism A 5% sales tax is charged on some items. The total cost of taxable items depends on the total price of the items before the tax.

a. What is the total cost of a taxable bracelet that costs $4.99 before tax?

b. What is the total cost of an item that costs x dollars before tax?

c. Write an equation to represent the function.

d. Graph the equation in part (c).

e. Plot the values from part (a) on the same coordinate plane you used in part (d). Does this point appear to be on your graph? (Extend your graph if necessary.)

For Exercises 22–37, use the summary of basic functions on page 230.

Name the basic function that has each shape.

22. parabola   **23.** V-shaped   **24.** line   **25.** hyperbola

State whether the graph of each function includes the origin.

26. $y = |x|$ **27.** $y = \dfrac{1}{x}$ **28.** $y = x^2$ **29.** $y = x$

In which quadrant(s) is the graph of each equation?

30. $y = x$ **31.** $y = x^2$ **32.** $y = \dfrac{1}{x}$ **33.** $y = |x|$

For each function, make a conjecture about the shape of its graph. Test your conjecture by graphing the function.

34. $y = -2x^2$ **35.** $y = 6x - 3$ **36.** $y = \dfrac{1}{x} + 5$ **37.** $y = 4|x|$

Unit 4 Coordinates and Functions

23. absolute value function or $y = |x|$

24. identity function or $y = x$

25. reciprocal function or $y = \dfrac{1}{x}$

26. Yes.   **27.** No.

28. Yes.   **29.** Yes.

30. quadrants I, III

31. quadrants I, II

32. quadrants I, III

33. quadrants I, II

34–37. Answers may vary.

34. conjecture: a parabola opening down

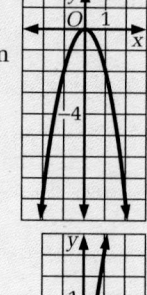

35. conjecture: line

36. conjecture: hyperbola

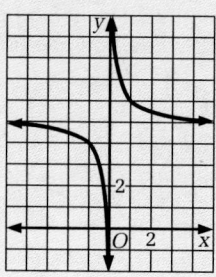

38. a. Make a table of widths x and lengths y for rectangles with area 60 m². Find at least six rectangles.

b. Use your table to make a graph.

c. What shape does the graph seem to have?

d. Use the formula for the area of a rectangle to write an equation that shows the length as a function of the width.

x	y
60	1
30	2
?	?

Ongoing ASSESSMENT

39. a. Graph the equation $y = -x$.

b. How do you know that it is a function?

c. **Writing** What name would you give to the function you graphed? Why?

40. Writing Investigate the function $y = x^3$. Include:

➤ a graph of the function

➤ the name you would give to this function

➤ a comparison of the graph to the other basic graphs in the section

Review PREVIEW

41. Draw a graph that does not represent a function. *(Section 4-6)*

Solve each equation. Check each solution. *(Section 2-7)*

42. $2t + 6 = 48$

43. $3n + 5 = 17$

44. $7x \div 3 = -9$

Working on the Unit Project

45. Suppose you want your flip book to show a spaceship landing, then taking off again.

a. **Open-ended** Which graph would you use to show the path of the spaceship?

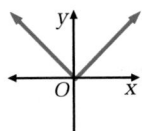
absolute value

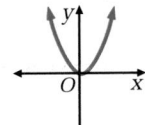
parabola

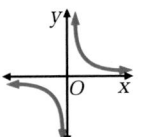
hyperbola

b. Find the coordinates of five points on the graph that you chose.

c. Make five separate diagrams of the spaceship, using the points that you found in part (b).

d. Cut out the diagrams and make a 5-page flip book with them.

4-7 Functions and Equations

233

Assessment: Standard

Ex. 39 can be graded as a standard problem. Ex. 40 can be used for a class presentation by a student.

Quick Quiz (4-5 through 4-7)

See page 235.

Practice 33 For use with Section 4-7

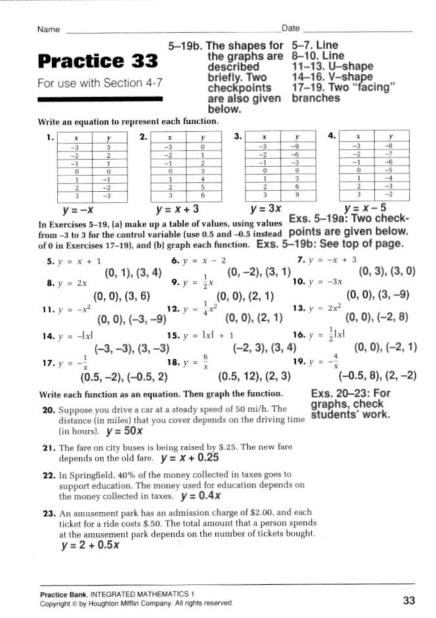

38. d. $y = \dfrac{60}{x}$

39. a.

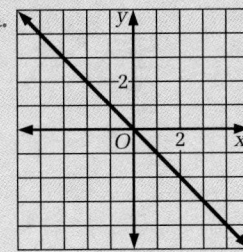

b. It passes the vertical-line test.

c. The graph of the function $y = -x$ is a line. The graph looks like the graph of the identity function, $y = x$, but is flipped over the y-axis so that it lies in quadrant II and quadrant IV. The function $y = -x$ could be named the negative identity function.

40. See answer in back of book.

41. Answers may vary. There should be an x-value with two y-values.

42. 21

43. 4

44. $\dfrac{-27}{7}$

45. Answers may vary.

Answers to Exercises and Problems

37. conjecture: V-shaped

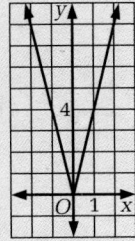

38. a. Answers may vary. Answers should include at least six pairs of values. A sample table is given.

x	y
60	1
30	2
20	3
15	4
12	5
10	6
6	10
5	12
4	15
3	20
2	30
1	60

b.

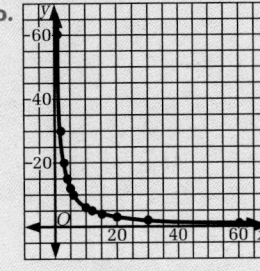

c. Answers may vary. Example: half of a hyperbola

233

Completing the Unit Project

➤ **Now you are ready to finish your flip book.**

➤ **Complete your drawings. Put the pages in order and make sure that they are all the same size. Clip or staple the pages together.**

➤ **Write a script to narrate while the pages are being flipped.**

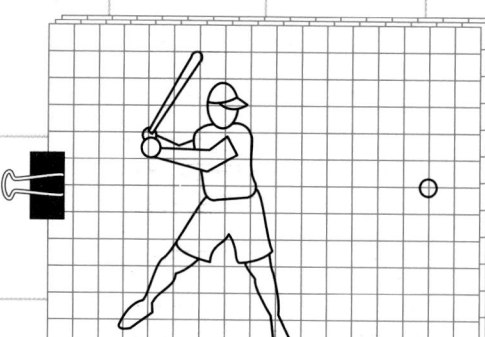

Look Back

Describe how you used a coordinate system when making the drawings for your flip book. Did you have any problems making the flip book? If so, describe how you solved them.

Alternative Projects

 Project 1: Making a Map

Create a map of the neighborhood around your school or of the neighborhood in which you live. Use a coordinate system and make a map key by listing the coordinates for the locations of the places on the map. Include a scale on your map.

Project 2: Recognizing Functions in Science

Read about the scientific research on global warming. Write a report on the latest findings. Discuss the evidence for global warming and what might be causing it. Include data tables and graphs with your report. Use the phrases "is a function of," "control variable," and "dependent variable" to describe the data.

Project 3: Collecting and Graphing Data

Collect data that can be displayed in a scatter plot. Write a report and include:

➤ the method(s) you used to gather the data

➤ a summary of the data

➤ a scatter plot of the data

➤ the type of correlation the scatter plot shows and the reasons for it

➤ several questions that can be answered by using the scatter plot

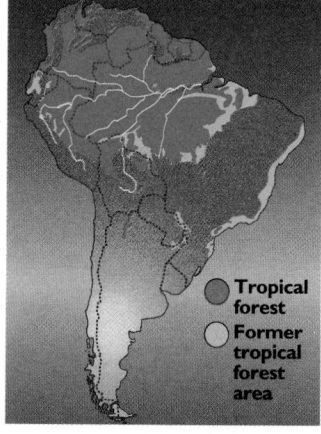

● Tropical forest
○ Former tropical forest area

1. **Open-ended** Use graph paper to make a map of your classroom. Put yourself at the origin. Mark the locations of desks, windows, and classroom "landmarks" on the plan. Give coordinates of the locations.

4-1

2. The Yoruba of Nigeria stencil *adire* cloth with geometric designs like this one. A coordinate system has been placed on the design.
   a. Find the coordinates of points *A*, *B*, and *C*.
   b. Which quadrant contains point *B*?

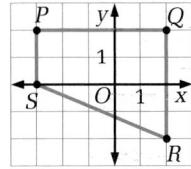

3. a. Plot the points $W(-2, 2)$, $X(4, 2)$, $Y(2, -3)$, and $Z(-4, -3)$ on a coordinate plane. Connect the points in order and connect *Z* to *W*.

4-2

   b. Write the specific name of the polygon you formed.
   c. Find the area of *WXYZ*.

4. Translate quadrilateral *PQRS* 3 units left and 4 units up. What are the coordinates of each vertex after the translation?

4-3

The graphs show rotations of △*ABC* around the origin. Describe the direction and the amount of the rotation in each graph.

4-4

5.
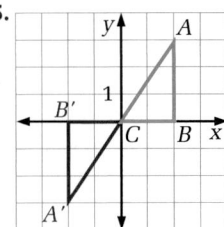

6.
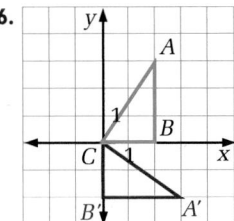

7.
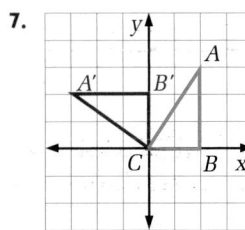

Tell whether each pattern has *translational symmetry*, *rotational symmetry*, or *no symmetry*.

8.

Japan

9.

Panama

10.

United States

Answers to Unit 4 Review and Assessment ·······················

1. Answers may vary.
2. a. $A(-1, 1)$; $B(-1, -1)$; $C(1, -1)$
   b. quadrant III
3. a.
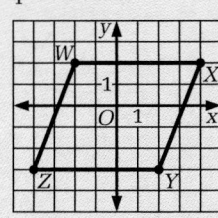

   b. parallelogram
   c. 30 square units
4. $P'(-6, 6)$, $Q'(-1, 6)$, $R'(-1, 2)$, $S'(-6, 4)$
5. 180 °
6. 90° clockwise
7. 90° counterclockwise
8. no symmetry
9. translational
10. rotational

Unit Support Materials
➤ Unit 4 Cumulative Practice 34
➤ Unit 4 Study Guide Review
➤ Unifying Problem 4
➤ Unit Tests 16 and 17

·······················

Quick Quiz (4-5 through 4-7)

1. Draw a scatter plot that shows a weak negative correlation. [4-5] **Answers will vary. An example is given.**

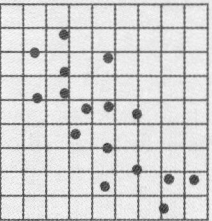

2. Leona has been hired for a new job at $450 per week. She has been promised a raise of $25 per week each year on the anniversary of the date she was hired. Tell what the control variable and the dependent variable are for this situation. Then tell whether the situation represents a function. [4-6] **control variable: years on the job; dependent variable: weekly salary; The situation represents a function.**

3. Write an equation for this situation. Be sure you tell what your variables mean. The price of gasoline is $1.25 per gallon. The cost of filling the tank depends on the number of gallons purchased. [4-7] **$c = 1.25g$, where c is the cost of filling the tank and g is the number of gallons of gasoline purchased.**

4. Name the shape of the graph of the basic function $y = x^2$. [4-7] **parabola**

5. Write an equation for the basic function whose graph looks like a right angle with vertex at (0, 0). [4-7] **$y = |x|$**

Answers to
Unit 4 Review and Assessment

11.

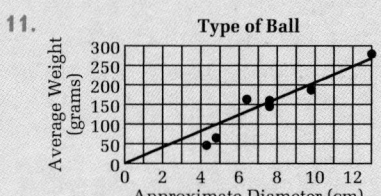

Type of Ball

12. positive correlation

13–14. Estimates may vary. Examples are given.

13. about 110 g

14. about 8.5 cm

15. No.   **16.** Yes.   **17.** No.

18. a. The amount of money left is the dependent variable. The number of tapes bought is the control variable.

b. The amount of money left is a function of the number of tapes bought.

c. $a = 30 - 3.5t$, where a = amount of money left (in dollars) and t = number of tapes bought.

d.

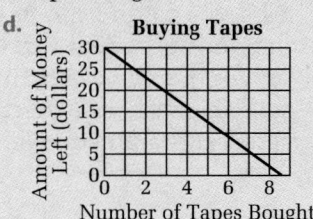

Buying Tapes

19. $y = x + 2$

20.

x	y
−2	7
−1	5
0	3
1	1
2	−1
3	−3

21. a parabola

x	y
−2	1
−1.5	−0.75
−1	−2
−0.5	−2.75
0	−3
0.5	−2.75
1	−2
1.5	−0.75
2	1

22. Answers may vary. An example is given. Coordinates can be used to show movement. By changing the coordinates you can move shapes up or down, to the left or to the right, and diagonally.

Use the data in the table.

TYPE OF BALL	Approximate diameter (cm)	Average weight (g)
baseball	7.6	145
billiard	6.4	163
field hockey	7.6	160
golf	4.3	46
handball	4.8	65
softball, large	13	279
softball, regular	9.8	187

4-5

11. Make a scatter plot of the data. Draw a fitted line.

12. Is there a *positive correlation*, a *negative correlation*, or *no correlation* between the diameter and weight of the balls?

13. Predict the weight of a sports ball with diameter 5.5 cm.

14. Predict the diameter of a sports ball with weight 175 g.

Tell whether each graph represents a function.

4-6

15.

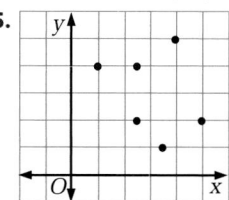

16.

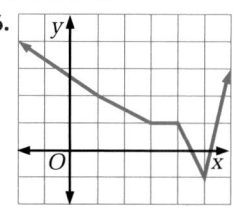

17.

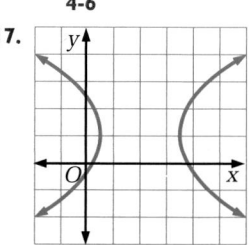

18. You have $30. You buy some tapes on sale for $3.50 each. The money you have left depends on how many tapes you buy.

a. Identify the dependent variable and the control variable.

b. Describe the situation using the phrase "is a function of."

c. Write an equation to represent the function.

d. Draw a graph of the function.

19. Write an equation to represent the function in the table.

20. Graph $y = -2x + 3$ on a coordinate plane.

21. Graph $y = x^2 - 3$ on a coordinate plane. Describe its shape.   **4-7**

22. **Writing** Identify at least three different uses of coordinates and explain how they are helpful in each situation.

23. **Self-evaluation** Think back to your work on this unit. Which section did you like working on best? Explain what you liked about it.

24. **Group Activity** Your school is planning an International Festival. You are helping to plan it. Here are some variables that you might consider:

➤ the amount of space available   ➤ the size of the booths   ➤ the costs of materials

Use two variables from the list (or use other variables) that are related.

a. Identify the control variable and the dependent variable.

b. Make a table of possible values for these variables. Explain why you think your values are reasonable.

c. Use your table to make a graph of the function.

d. Use your graph to make a recommendation for the festival plans.

x	y
−3	−1
−2	0
−1	1
0	2
1	3
2	4
3	5

236   **Unit 4** Coordinates and Functions

Coordinates can be used to show locations on maps by identifying squares where buildings and other places can be found. Coordinates can be used to make graphs that show relationships between two quantities.

23. Answers may vary.

24. a–d. Answers may vary.

IDEAS AND (FORMULAS) $= x^2$

ALGEBRA

➤ The control variable is put on the horizontal axis of a graph, and the dependent variable is put on the vertical axis. *(p. 218)*

➤ You can represent a function in words, as an equation, with a table of values, or with a graph. *(pp. 220, 226)*

➤ A vertical-line test shows if a graph represents a function. *(p. 221)*

➤ In a function, x often represents the control variable, and y often represents the dependent variable. *(p. 226)*

➤ Some basic functions have distinct shapes of graphs and equations. *(p. 230)*

GEOMETRY

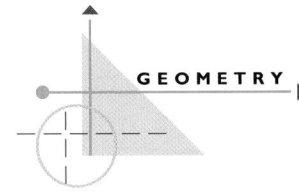

➤ Polygons can be described using coordinates. *(p. 191)*

➤ You can find the areas of some geometric figures on a coordinate plane using the areas of rectangles and right triangles. *(p. 193)*

➤ A translation moves an object without rotating or flipping it. *(p. 197)*

➤ When you translate or rotate an object, its size and shape do not change. *(pp. 197, 203)*

➤ A transformation is a change, such as a translation or a rotation, made to an object. *(p. 202)*

➤ To describe a rotation, you need to give the center of the turn, and the amount and direction of the turn. *(p. 204)*

STATISTICS & PROBABILITY

➤ A fitted line on a scatter plot suggests the correlation between the sets of data. *(p. 212)*

➤ A correlation does not mean that there is a cause-and-effect relationship between two sets of data. *(p. 214)*

Key Terms

- **coordinate plane** (p. 184)
- **quadrant** (p. 184)
- **vertical axis** (p. 184)
- **translation** (p. 197)
- **center of rotation** (p. 202)
- **scatter plot** (p. 212)
- **dependent variable** (p. 218)
- **x-axis** (p. 226)
- **hyperbola** (p. 230)
- **absolute value function** (p. 230)

- **ordered pair** (p. 184)
- **coordinates** (p. 184)
- **coordinate geometry** (p. 190)
- **translational symmetry** (p. 199)
- **transformation** (p. 202)
- **correlation** (p. 212)
- **control variable** (p. 218)
- **y-axis** (p. 226)
- **identity function** (p. 230)
- **reciprocal function** (p. 230)

- **origin** (p. 184)
- **horizontal axis** (p. 184)
- **trapezoid** (p. 193)
- **rotation** (p. 202)
- **rotational symmetry** (p. 205)
- **fitted line** (p. 212)
- **function** (p. 220)
- **parabola** (p. 229)
- **squaring function** (p. 230)

Quick Quiz (4-1 through 4-4)

1. Draw axes on a piece of graph paper to set up a coordinate plane. Label the quadrants. Then mark one point in each quadrant and tell the coordinates of each point you marked. [4-1] Answers may vary.

Refer to the following figure for Exs. 2 and 3.

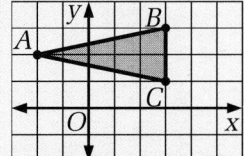

2. Classify polygon ABC. [4-2] isosceles triangle

3. Find the area of polygon ABC. [4-2] 5 square units

4. Draw a triangle with vertices $X(-1, 4)$, $Y(1, -2)$, and $Z(3, 1)$ on graph paper. Draw the triangle that results from translating $\triangle XYZ$ to the right 4 units and down 1 unit. Label the vertices of the translated triangle with their coordinates. [4-3] For diagrams, see students' work. The new coordinates are $X'(3, 3)$, $Y'(5, -3)$, and $Z'(7, 0)$.

5. Sketch a figure that has 180° rotational symmetry. [4-4] Answers may vary.

Equations for Problem Solving

OVERVIEW

➤ **Unit 5** extends the work with equations and solving equations introduced in Unit 2. Additional techniques for solving equations are presented and using equations to model problem situations is emphasized.

➤ Students can review finding reciprocals on page 650 of the **Student Resources Toolbox** in preparation for using reciprocals to solve equations.

➤ In Unit 5, students learn to solve formulas for specific variables. Inequalities with one variable are introduced, and students learn how to modify equation-solving techniques in order to solve inequalities. The idea of solving a system of equations is introduced.

➤ The theme of the Unit Project is setting up a car wash to raise money. Students use equations to represent situations and solve mathematical problems to determine income, expenses, schedule, location, break-even point, and possible profit.

➤ Connections to literature, science, and music are integrated into the teaching materials and exercises.

➤ Problem-solving strategies used in Unit 5 include *Use a Formula* (Sections 5-3 and 5-7), *Solve a Simpler Problem* (Section 5-5), and *Use a System of Equations* (Section 5-8).

Unit Objectives

Section	Objectives	NCTM Standards
5-1	• Use a table, graph, or equation to model situations and solve problems.	1, 2, 4, 5, 6, 8
5-2	• Simplify expressions and solve equations involving opposites, the distributive property, and combining like terms.	1, 2, 4, 5, 14
5-3	• Model situations and solve equations that have a variable on both sides of the equals sign.	1, 2, 4, 5
5-4	• Understand how adding, subtracting, multiplying, and dividing affect an inequality. • Model situations with inequalities.	1, 2, 4, 5
5-5	• Learn to rewrite formulas to make them easier to use.	1, 2, 4, 5
5-6	• Use reciprocals to solve equations.	1, 2, 4, 5, 14
5-7	• Use the formulas for the area of a parallelogram, a triangle, and a trapezoid to solve problem situations.	1, 2, 4, 5, 7
5-8	• Model situations and solve equations with two variables using substitution.	1, 2, 4, 5, 7

Section	Connections to Prior and Future Concepts
5-1	**Section 5-1** applies skills learned in earlier units to model and solve problem situations by using a table, a graph, an equation, or technology. Selecting a method for solving a problem is a basic mathematical skill. Students are presented with techniques and tools for problem solving throughout Books 1, 2, and 3. Also, some problem-solving review appears on pages 664 and 665 of the Toolbox.
5-2	**Section 5-2** builds on the models for solving equations presented in Sections 2-7, 2-8, and 4-7 and expands them to include using the distributive property and opposites, as well as combining like terms. Equations are further studied in this unit and in Units 7, 8, and 10 of Book 1. Equations are also the focus of Units 2, 3, 4, and 9 of Book 2 and of Units 1, 2, and 5 of Book 3.
5-3	**Section 5-3** extends the concept of using equations to model situations, introduced in Sections 2-7, 2-8, and 5-1, to encompass modeling situations in which there is a variable on both sides of the equation. The skill of writing and solving equations to solve problems is refined throughout Book 1 and in Books 2 and 3.
5-4	**Section 5-4** combines the technique of modeling situations and the concept of inequalities to illustrate how to model and solve situations with inequalities. Work with inequalities is continued in Unit 8 of Book 1 and throughout Books 2 and 3.
5-5	**Section 5-5** continues the study of equations to include specialized equations called formulas. Students have encountered formulas in previous courses and in exercises for earlier sections. Formulas are introduced and applied in Units 5, 7, 9, and 10 of Book 1 and throughout Books 2 and 3.
5-6	**Section 5-6** extends equation-solving techniques to include reciprocals. Reciprocals will be used to solve equations in the remaining units of Book 1 and throughout Books 2 and 3. Reciprocals are reviewed on page 650 of the Toolbox.
5-7	**Section 5-7** continues the study of area to include formulas for the area of a parallelogram, a triangle, and a trapezoid. Many problems can be solved using area formulas; examples can be found in Units 6, 7, and 9 of Book 1 as well as throughout Books 2 and 3.
5-8	**Section 5-8** extends the skill of modeling by using a system of equations in two variables to solve problems. Solving such systems is studied further in Unit 8 of Book 1, as well as in Units 3, 4, and 9 of Book 2 and in Unit 1 of Book 3.

Integrating the Strands

Strands	Sections
Number	5-1, 5-2, 5-3, 5-5, 5-6, 5-8
Algebra	5-1, 5-2, 5-3, 5-4, 5-5, 5-6, 5-7, 5-8
Measurement	5-5, 5-6, 5-7, 5-8
Geometry	5-1, 5-2, 5-5, 5-7, 5-8
Statistics and Probability	5-2, 5-5
Discrete Mathematics	5-1, 5-5, 5-6
Logic and Language	5-1, 5-2, 5-3, 5-4, 5-5, 5-6, 5-7, 5-8

Equations for Problem Solving

Section Planning Guide

➤ Essential exercises and problems are indicated in boldface.
➤ Ongoing work on the Unit Project is indicated in color.
➤ Exercises and problems that require student research, group work, manipulatives, or graphing technology are indicated in the column headed "Other."

Section	Materials	Pacing	Standard Assignment	Extended Assignment	Other
5-1	graphics calculator, spreadsheet software	Day 1	**2–7**, 8–10, **11–17**, 18–35, 36	1, **2–7**, 8–10, **11–17**, 18–35, 36	
5-2		Day 1	1, **2–7**	1, **2–7**	
		Day 2	**8–24, 26**, 32–36, 37, 38	**8–24, 26**, 27–30, 32–36, 37, 38	25, 31, 37, 38
5-3	algebra tiles	Day 1	**3–22**, 25–33, 34–36	**3–22**, 23, 25–33, 34–36	1, 2, 23c, 24
5-4	scientific calculator	Day 1	**2–15**	1, **2–15**	
		Day 2	**16–22, 25–31**, 32–41, 42–45	**16–22**, 23, 24, **25–31**, 32–41, 42–45	24a
5-5		Day 1	**1–18**, 23–31, 32, 33	**1–18**, 19, 20, 22–31, 32, 33	19c, 21, 33
5-6	graphing technology	Day 1	**1–15**, 16–18, 21–29, 30, 31	**1–15**, 16–20, 21–29, 30, 31	31
5-7	rectangular sheets of paper, scissors	Day 1	**1–8**	**1–8**	
		Day 2	9, 10, 14, **15–19**, 21–27, 28–30	9, 10, 12–14, **15–19**, 21–27, 28–30	11, 20
5-8		Day 1	**2–12**, 13, **14, 16–18**, 19–27, 28	1, **2–12**, 13, **14, 16–18**, 19–27, 28	15
Review		**Day 1**	**Unit Review**	**Unit Review**	
Test		**Day 2**	**Unit Test**	**Unit Test**	

Yearly Pacing	Unit 5 Total	Units 1–5 Total	Remaining	Total
	15 days (2 for Unit Project)	78 days	86 days	164 days

Support Materials

➤ See **Project Book** for notes on Unit 5: Planning a Car Wash.
➤ "UPP" and "disk" refer to **Using Plotter Plus** booklet and **Plotter Plus** disk.
➤ Warm-up exercises for each section are available on **Warm-Up Transparencies**.
➤ "FI," "PC," "GI," "MA," and "Stats!" refer, respectively, to the McDougal Littell Mathpack software Activity Books for **Function Investigator, Probability Constructor, Geometry Inventor, Matrix Analyzer,** and **Stats!**.

Section	Study Guide	Practice Bank	Problem Bank	Activity Bank	Explorations Lab Manual	Assessment Book	Visuals	Technology
5-1	5-1	Practice 35	Set 10	Enrich 31	Add. Expl. 5 Master 2	Quiz 5-1		FI Act. 2
5-2	5-2	Practice 36	Set 10	Enrich 32		Quiz 5-2		
5-3	5-3	Practice 37	Set 10	Enrich 33	Add. Expl. 5, 6 Master 6	Quiz 5-3	Folder 3	
5-4	5-4	Practice 38	Set 10	Enrich 34		Quiz 5-4 Test 18		FI Act. 9 UPP, page 29 Inequality Plotter (disk)
5-5	5-5	Practice 39	Set 11	Enrich 35		Quiz 5-5		
5-6	5-6	Practice 40	Set 11	Enrich 36		Quiz 5-6		
5-7	5-7	Practice 41	Set 11	Enrich 37	Master 2	Quiz 5-7	Folder 6	
5-8	5-8	Practice 42	Set 11	Enrich 38		Quiz 5-8 Test 19		
Unit 5	Unit Rev.	Practice 43	Unif. Prob. 5	Fam. Inv. 5		Tests 20–22		

UNIT TESTS

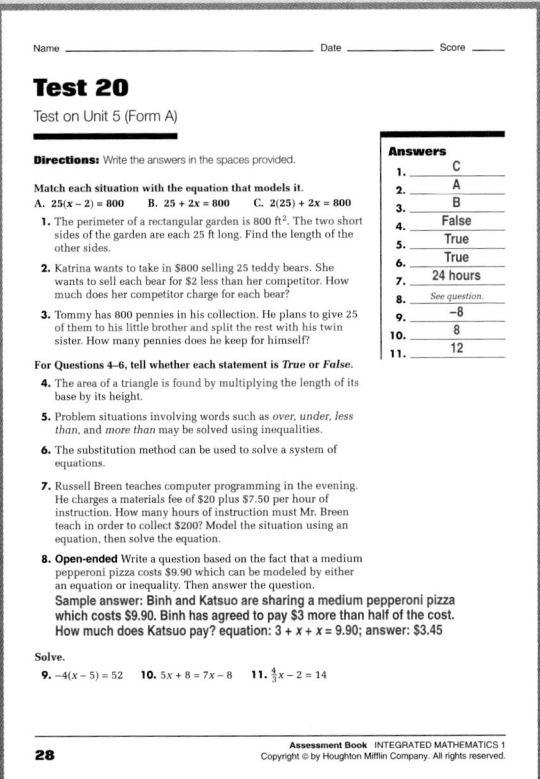

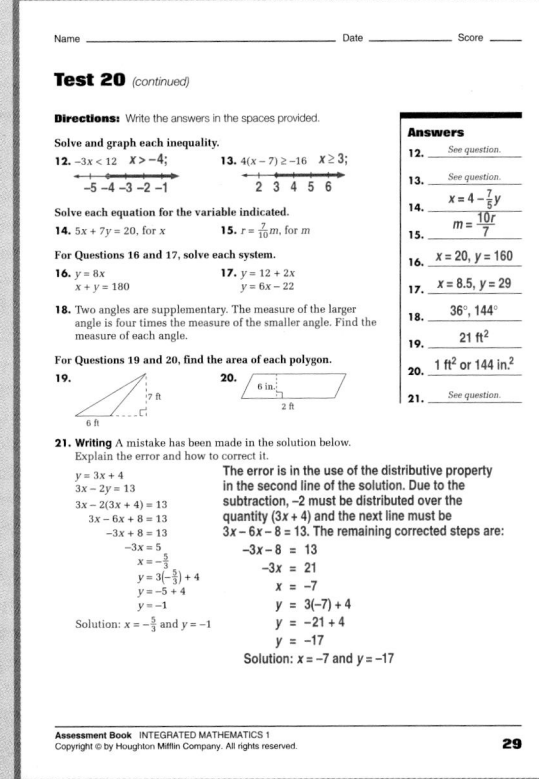

Test 20

Test on Unit 5 (Form A)

Directions: Write the answers in the spaces provided.

Match each situation with the equation that models it.
A. $25(x - 2) = 800$ B. $25 + 2x = 800$ C. $2(25) + 2x = 800$

1. The perimeter of a rectangular garden is 800 ft². The two short sides of the garden are each 25 ft long. Find the length of the other sides.

2. Katrina wants to take in $800 selling 25 teddy bears. She wants to sell each bear for $2 less than her competitor. How much does her competitor charge for each bear?

3. Tommy has 800 pennies in his collection. He plans to give 25 of them to his little brother and split the rest with his twin sister. How many pennies does he keep for himself?

For Questions 4–6, tell whether each statement is *True* or *False*.

4. The area of a triangle is found by multiplying the length of its base by its height.

5. Problem situations involving words such as *over, under, less than,* and *more than* may be solved using inequalities.

6. The substitution method can be used to solve a system of equations.

7. Russell Breen teaches computer programming in the evening. He charges a materials fee of $20 plus $7.50 per hour of instruction. How many hours of instruction must Mr. Breen teach in order to collect $200? Model the situation using an equation, then solve the equation.

8. **Open-ended** Write a question based on the fact that a medium pepperoni pizza costs $9.90 which can be modeled by either an equation or inequality. Then answer the question.
Sample answer: Binh and Katsuo are sharing a medium pepperoni pizza which costs $9.90. Binh has agreed to pay $3 more than half of the cost. How much does Katsuo pay? equation: $3 + x + x = 9.90$; answer: $3.45

Solve.
9. $-4(x - 5) = 52$ 10. $5x + 8 = 7x - 8$ 11. $\frac{4}{3}x - 2 = 14$

Answers
1. C
2. A
3. B
4. False
5. True
6. True
7. 24 hours
8. See question.
9. −8
10. 8
11. 12

Test 20 (continued)

Directions: Write the answers in the spaces provided.

Solve and graph each inequality.

12. $-3x < 12$ $x > -4$;

13. $4(x - 7) \geq -16$ $x \geq 3$;

Solve each equation for the variable indicated.
14. $5x + 7y = 20$, for x   15. $r = \frac{7}{10}m$, for m

For Questions 16 and 17, solve each system.
16. $y = 8x$
$x + y = 180$

17. $y = 12 + 2x$
$y = 6x - 22$

18. Two angles are supplementary. The measure of the larger angle is four times the measure of the smaller angle. Find the measure of each angle.

For Questions 19 and 20, find the area of each polygon.
19.

20.

21. **Writing** A mistake has been made in the solution below. Explain the error and how to correct it.

$y = 3x + 4$
$3x - 2y = 13$
$3x - 2(3x + 4) = 13$
$3x - 6x + 8 = 13$
$-3x + 8 = 13$
$-3x = 5$
$x = -\frac{5}{3}$
$y = 3\left(-\frac{5}{3}\right) + 4$
$y = -5 + 4$
$y = -1$
Solution: $x = -\frac{5}{3}$ and $y = -1$

The error is in the use of the distributive property in the second line of the solution. Due to the subtraction, −2 must be distributed over the quantity $(3x + 4)$ and the next line must be $3x - 6x - 8 = 13$. The remaining corrected steps are:
$-3x - 8 = 13$
$-3x = 21$
$x = -7$
$y = 3(-7) + 4$
$y = -21 + 4$
$y = -17$
Solution: $x = -7$ and $y = -17$

Answers
12. See question.
13. See question.
14. $x = 4 - \frac{7}{5}y$
15. $m = \frac{10r}{7}$
16. $x = 20, y = 160$
17. $x = 8.5, y = 29$
18. $36°, 144°$
19. 21 ft^2
20. $1 \text{ ft}^2 \text{ or } 144 \text{ in.}^2$
21. See question.

Form B

Test 21

Test on Unit 5 (Form B)

Directions: Write the answers in the spaces provided.

Match each situation with the equation that models it.
A. $2(20) + 2x = 600$ B. $20(x - 2) = 600$ C. $20 + 2x = 600$

1. The perimeter of a rectangular garden is 600 ft². The two short sides of the garden are each 20 ft long. Find the length of the other sides.

2. Maria wants to take in $600 selling 20 teddy bears. She wants to sell each bear for $2 less than her competitor. How much does her competitor charge for each bear?

3. Miguel has 600 pennies in his collection. He plans to give 20 of them to his little brother and split the rest with his twin sister. How many pennies does he keep for himself?

For Questions 4–6, tell whether each statement is *True* or *False*.

4. Inequalities are solved the same way as equations except when multiplying or dividing both sides by a negative number, which reverses the inequality.

5. The area of a trapezoid is found by multiplying the height by the sum of the lengths of the two bases.

6. The product of a nonzero number and its reciprocal is always 1.

7. Mireya Jimenez teaches SAT preparation classes on weekends. She charges a materials fee of $30 plus $20.00 per hour of instruction. How many hours of instruction must Ms. Jimenez teach in order to collect $450? Model the situation using an equation, then solve the equation.

8. **Open-ended** Write a question based on the fact that a typical school day is seven hours long which can be modeled by either an equation or inequality. Then answer the question.
Sample answer: During her 7-hour school day, Kaitlin gets one-half hour to eat lunch. This lunch period divides her school day in half. How long does she attend classes before lunch?   equation: $x + x + 0.5 = 7$; answer: 3.25 hours

Solve.
9. $-8(x - 5) = 96$ 10. $3x + 9 = 7x - 7$ 11. $\frac{4}{5}x - 2 = 18$

Answers
1. A
2. B
3. C
4. True
5. False
6. True
7. 21 hours
8. See question.
9. −7
10. 4
11. 25

Test 21 (continued)

Directions: Write the answers in the spaces provided.

Solve and graph each inequality.

12. $-2x \leq 14$ $x \geq -7$;

13. $3(x - 5) < -21$ $x < -2$;

Solve each equation for the variable indicated.
14. $6x + 2y = 18$, for x   15. $m = \frac{3}{10}z$, for z

For Questions 16 and 17, solve each system.
16. $y = 4x$
$x + y = 180$

17. $y = 16 + 2x$
$y = 4x - 22$

18. Two angles are complementary. The measure of the larger angle is five times the measure of the smaller angle. Find the measure of each angle.

For Questions 19 and 20, find the area of each polygon.
19.

20.

21. **Writing** A mistake has been made in the solution below. Explain the error and how to correct it.

$y = 3x - 4$
$3x - 2y = 13$
$3x - 2(3x - 4) = 13$
$3x - 6x - 8 = 13$
$-3x - 8 = 13$
$-3x = 21$
$x = -7$
$y = 3(-7) - 4$
$y = -21 - 4$
$y = -25$
Solution: $x = -7$ and $y = -25$

The error is in the use of the distributive property in the second line of the solution. Due to the subtraction, −2 must be distributed over the quantity $(3x - 4)$ and the next line must be $3x - 6x + 8 = 13$. The remaining corrected steps are:
$-3x + 8 = 13$
$-3x = 5$
$x = -\frac{5}{3}$
$y = 3\left(-\frac{5}{3}\right) - 4$
$y = -5 - 4 = -9$
Solution: $x = -\frac{5}{3}$ and $y = -9$

Answers
12. See question.
13. See question.
14. $x = 3 - \frac{1}{3}y$
15. $z = \frac{10m}{3}$
16. $x = 36, y = 144$
17. $x = 19, y = 54$
18. $15°, 75°$
19. 30 cm^2
20. 56 in.^2
21. See question.

Software Support

McDougal Littell Mathpack
Function Investigator
Stats!

Outside Resources

Books/Periodicals

Parish, Charles R. "Inequalities, Absolute Value, and Logical Connectives." *Mathematics Teacher* (December 1992).

Philip, Randolph A. "The Many Uses of Algebraic Variables." *Mathematics Teacher* (October 1992): pp. 557–561.

Hawkings, Vincent. "Ring-Around-A-Trapezoid." *Mathematics Teacher* (September 1984): pp. 450–451.

Witkowski, Joseph C. "Mathematical Modeling and the Presidential Elections." *Mathematics Teacher* (October 1992): pp. 520–525.

Software

Schwartz, Judah and Michal Yerushalmy. *Geometric Supposer: Quadrilaterals or Geometric Super-Supposer.* Scotts Valley, CA: Sunburst.

Geometric Probabilty. Materials and Software. Department of Mathematics and Computer Science, North Carolina School of Science and Mathematics. Reston, VA: NCTM, 1988.

Geometer's Sketchpad. Macintosh and MS-DOS. Berkeley, CA: Key Curriculum Press.

PROJECT GOALS

➤ Students plan and organize a car wash as a fund-raising event.

➤ When planning the car wash, students decide which group or organization will receive the money raised, a location for it, supplies needed, and the fee to be charged.

➤ Students work together in a cooperative group and implement the goals of the project successfully.

PROJECT PLANNING

Materials List

➤ Poster board
➤ Markers

Project Teams

Before students begin working as part of a team, have individual students make a list of possible sites for the car wash, supplies needed, and possible recipients of the funds raised.

Have students work on the project in groups of four to six. This project gives students valuable experience in the type of group work that they will encounter as adults. The project provides the student groups with an opportunity to analyze a problem situation to determine what tasks are involved, and then to decide how to distribute the work.

unit 5

One person really can make a difference! In 1983, 11-year-old Trevor Ferrell watched a news report on homeless people and decided to try to help. He asked his parents to drive him into Philadelphia, where he gave his favorite pillow and a blanket to a homeless man. Since then, Trevor's Campaign for the Homeless has become a $1.4 million non-profit organization.

Equations for Problem Solving

COMM**UNITY** SERVICE

Have you ever asked people to pledge a certain amount of money for every mile you walk for a **cause**? Special events raise millions of dollars each year to fund organizations that fight homelessness, hunger, AIDS, and cancer, or promote community development.

One of the best ways to support a cause is to donate your **time**. People who are familiar with computer spreadsheets and databases can help groups keep accurate records of donors, donations, and expenses.

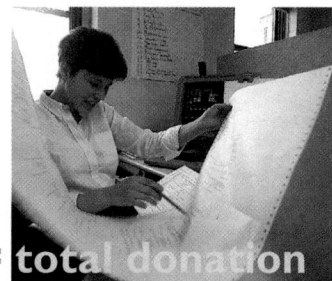

20 mi x pledge = total donation

About 100 students from the Chicago-area Niles West High School belong to the volunteer club West Helping Others. The students visit schools, hospitals, and shelters to help in any way they can. The group frequently organizes bake sales and car washes to finance its activities.

238

Suggested Rubric for Unit Project

4 The presentation is well organized and uses the mathematics developed in the unit appropriately. The plan of action includes the location and a diagram of the car wash, the charges for each car, supplies needed and where to purchase them, the number of cars necessary to break even and reach the fund-raising goal, sample advertisements, and the jobs and number of people needed to run the car wash. Reasons for these recommendations are included in the presentation.

3 The presentation is adequate but is not presented in a very well-organized manner. The details of the fund-raising car wash are not as thorough as they should be.

Planning a Car Wash

Your group project is to plan and organize a car wash. The money you raise will be used to help you provide a service to your school or community.

When disasters such as hurricanes, earthquakes, and floods strike, teenagers are among the first to reach out and help.

Students are often in the news, collecting food and clothing, filling sandbags, or even helping to rescue stranded people and pets.

You will need to consider the conditions that will affect the success of your car wash and recommend a location, a time, and a price.

When your plan is complete, your group will present it to the class. Your plan should include the reasons for your recommendations and a prediction of the amount of money you will raise.

The class will decide on the best plan of action. The groups may want to work together to put the plan into action.

Support Materials

The *Project Book* contains information about the following topics for use with this Unit Project.

➤ Project Description
➤ Teaching Commentary
➤ Working on the Unit Project Exercises
➤ Completing the Unit Project
➤ Assessing the Unit Project
➤ Alternative Projects
➤ Outside Resources

Using Technology

Have students create a data file to organize the car wash.

ADDITIONAL BACKGROUND

Community Involvement

One program that tries to teach teenagers the value of giving time and money to their communities is Youth Engaged in Service (YES!) in Washington, DC. The participants try to raise money so that a corporation or individual can match the amount up to $1000. The fund raising is just part of the program. The main emphasis is on educating teens regarding community responsibility and the importance of philanthropy.

Suggested Rubric for Unit Project

2 The presentation is generally incomplete. The description of the location and the details of the necessary charges, supplies needed, or the number of cars needed to break even may be lacking. This project should be returned with suggestions for improvements and a new deadline.

1 This presentation cannot be evaluated. It is unorganized, illegible, or incomplete. The project should be returned with a new deadline for completion. The group should be encouraged to speak with the teacher as soon as possible so that students understand the purpose of the project.

ALTERNATIVE PROJECTS

Project 1, page 294

Working on Commission

Interview someone in a business to find out the salaries and commission rates for new salespeople and experienced salespeople. Collect typical weekly or monthly sales figures, choose an income goal, and figure out how many sales a new salesperson and an experienced salesperson need to earn that income.

Project 2, page 294

Researching Telephone Costs

Choose two locations to call long distance. Decide on the time of day, the day of the week, and the length of each call. Find out the cost for each call by contacting two or three long-distance telephone services. Use the results to choose a long-distance service for making calls.

Unit Project 5

Getting Started

For this project, work in a group of four to six students. Here are some ideas to help you get started.

☞ As a class, consider which groups or organizations you may want to receive the funds you raise in your car wash.

☞ In your group discuss possible car wash locations.

Where will your car wash attract the most customers?

Where will your water supply come from? Will you pay for water?

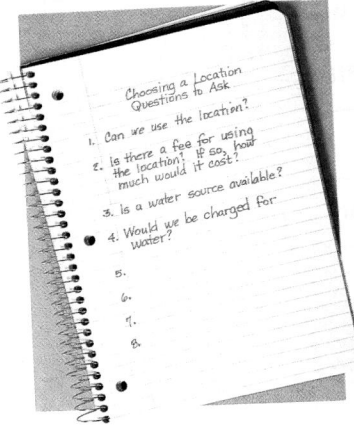

Will you have room for several cars? If a line forms, how can you avoid blocking traffic?

Will there be a fee for the location?

☞ List some questions you might ask a property owner or business manager about holding a car wash at their place of business.

☞ Plan to contact the managers of several possible car wash locations and ask them your questions.

☞ In your group, decide on three locations that are most likely to meet your needs.

Working on the Unit Project

Your work in Unit 5 will help you plan your car wash.

Related Exercises:
Section 5-1, Exercise 36
Section 5-2, Exercises 37, 38
Section 5-3, Exercises 34–36
Section 5-4, Exercises 42–45
Section 5-5, Exercises 32, 33
Section 5-6, Exercises 30, 31
Section 5-7, Exercises 28–30
Section 5-8, Exercise 28

Alternative Projects p. 294

Can We Talk COMMUNITY SERVICE

➤ In his 1961 Inaugural Address, President John Kennedy inspired many people when he said,

"... ask not what your country can do for you— ask what you can do for your country."

What are some things that you can do to serve your community?

➤ How often do you think people your age volunteer?

➤ What ads or promotions catch your attention? Are any of them for volunteer or service groups?

➤ How often do you think people have their cars washed? Do you think car owners would rather have their cars washed at a professional car wash or by students? Why?

➤ How can you estimate the number of customers you will have?

Answers to Can We Talk?

➤ Examples of how students can serve their community: work in a soup kitchen, visit the elderly, help with special needs children, gather materials to be recycled, collect and dispose of litter.

➤ Teenagers who volunteer donate about 5 hours of their time per week. About 13.4% of teens (1,902,000 teens) volunteer. The largest number of teens volunteer for religious organizations, followed by educational institutions and health organizations.

➤ Answers may vary.

➤ Answers may vary. In a recent survey, it was found that 78% of car owners wash their cars at least once a month. Many people prefer hand-washing to mechanized commercial car washes. In fact, the commercial service used most often is self-service, coin operated (41%), followed by 23% conveyor (exterior only), 23% full-service conveyor, and 13% stationary cleaning (roll-over or high pressure spray). Car washes are often successful because people know their car will be hand-washed, and they will be contributing to a worthy cause.

➤ Students could estimate the number of customers by counting the number of cars at a prospective car wash area.

Section 5-1

Modeling Problem Situations

Focus
Use a table, graph, or equation to model situations and solve problems.

WHAT ARE THE OPTIONS? OPTIONS? OPTIONS?

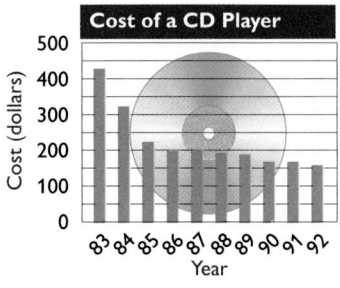

Cost of a CD Player

Cost (dollars) vs. Year (83–92)

Would you pay $428 for a CD player? That was the cost in 1983, when only 35,000 players were sold!

The graph shows the cost of CD players from 1983–1992. Graphs, tables, functions, equations, or inequalities that describe a situation are called **mathematical models.** Using mathematical models is called **modeling.**

Sample 1

Model and solve this situation. A CD player costs $195, including tax. You already have $37 and can save $9 per week. After how many weeks can you buy the player?

Sample Response

Method ❶ Model the situation using a table.

Number of weeks	Total saved ($) $37 + 9 \times$ (no. of weeks)
0	37
1	$37 + 9(1) = 46$
2	$37 + 9(2) = 55$
3	$37 + 9(3) = 64$
•	•
•	•
16	$37 + 9(16) = 181$
17	$37 + 9(17) = 190$
18	$37 + 9(18) = 199$

After 18 weeks, more than $195.

Method ❷ Model the situation using a spreadsheet.

Add $9 for each week.

	Savings Table	
	A	**B**
1	Number of Weeks	Total saved <dollars>
2	0	37
3	1	9 + B2
4	2	9 + B3
5	3	9 + B4
•	•	•
•	•	•
18	16	9 + B17
19	17	9 + B18
20	18	9 + B19

You can buy the player the week that the total saved is at least $195.
You can buy the player after 18 weeks.

5-1 Modeling Problem Situations

241

PLANNING

Objectives and Strands
See pages 238A and 238B.

Spiral Learning
See page 238B.

Materials List
➤ Graph paper
➤ Graphics calculator
➤ Spreadsheet software

Recommended Pacing
Section 5-1 is a one-day lesson.

Extra Practice
See pages 628–629.

Warm-Up Exercises
Warm-Up Transparency 5-1

Support Materials
➤ Practice 35
➤ Enrichment 31 in the Activity Bank
➤ Study Guide 5-1
➤ Problem Set 10
➤ Additional Exploration 6
➤ Diagram Master 2 in the Explorations Lab Manual
➤ Function Investigator with Matrix Analyzer Activity Book: Function Investigator Activity 2
➤ Quiz 5-1

Problem Solving

This section focuses on the use of mathematical models to solve real-world problems. Having students examine alternate models helps sharpen their analytical thinking skills and their ability to judge the reasonableness of solutions.

Communication: Discussion

Have a brief class discussion about the term *mathematical model*. Students are familiar with physical models, such as model cars and model airplanes. Point out that mathematical models, like physical models, are a substitute for reality. Good mathematical models preserve the key features of the situations on which they are based.

Using Technology

In Method 1 of Sample 1, students can use a graphics calculator to help build the table. Discuss how to use special features of the calculator to do this as efficiently as possible.

In Method 5, students can use the TRACE and ZOOM features to come very close to the point with *y*-coordinate 195. These features are discussed in the Technology Handbook on pages 615–617.

Additional Sample

S1 Suppose you are reading a novel that has 378 pages. You are now on page 62. Starting tomorrow, you plan to read 20 pages a day. How many days will you need to finish? **16 days**

Method ❸

Model the situation using a graph.

On graph paper, first label the number of weeks on the horizontal axis. Label the total amount saved on the vertical axis. Then follow steps 1 through 6.

3 Add $9 to each new total. Plot the points (2, 55), (3, 64), and so on. Draw a line through the points.

2 To find the next point, add $9 to $37. After one week the total saved is $46. Plot the point (1, 46).

1 You start with $37. Mark this point on the vertical axis as (0, 37).

4 Estimate 195 on the vertical axis. Read across to the line you drew.

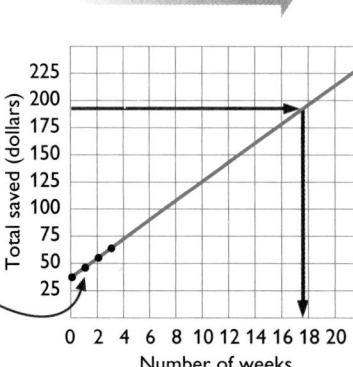

5 Read down to the horizontal axis to find the number of weeks.

6 Read the estimate of the solution, 18 weeks, on the horizontal axis.

Method ❹

Model the situation using an equation.

The problem asks for the number of weeks until you will have $195.

Let w = the number of weeks.

Then $9w$ = the amount in dollars you can save in w weeks, and $37 + 9w$ = the total amount you will have saved in w weeks.

You will have enough money to buy the compact disc player when

the total amount saved = the amount you need.

$$37 + 9w = 195$$

Solve the equation using algebra:

$$37 + 9w = 195$$
$$37 + 9w - 37 = 195 - 37 \quad \longleftarrow \text{Subtract 37 from both sides.}$$
$$9w = 158$$
$$\frac{9w}{9} = \frac{158}{9} \quad \longleftarrow \text{Divide both sides by 9.}$$
$$w \approx 17.6$$

You can buy the CD player after 18 weeks.

Method ⑤

Model the situation using a graphics calculator.

First model the situation with an equation as in Method 4. Use *y* for the amount of money you need and *x* for the number of weeks.

Set an appropriate viewing window on a graphics calculator and graph the equation $y = 37 + 9x$.

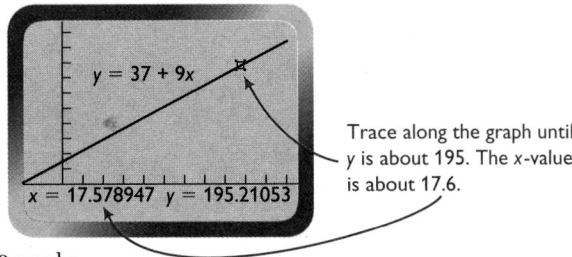

Trace along the graph until *y* is about 195. The *x*-value is about 17.6.

You can buy the CD player after 18 weeks.

> ### Talk it Over
>
> 1. Which method do you like best? Why?
> 2. Suppose you could save $10 per week. What would the equation in Methods 4 and 5 be?

One Model for Different Situations

The same equation can be used to model different situations.

Sample 2

Writing Describe a situation that could be modeled by the equation $5 = 3 + 0.25x$.

Sample Response

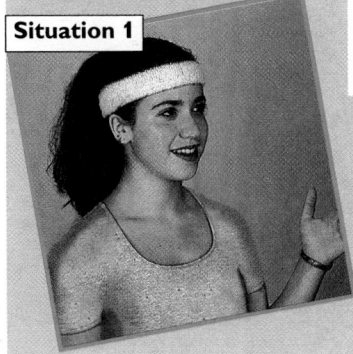

Situation 1

There are many possible answers. For the expression $0.25x$, think of a situation that involves 25% or a quarter of something.

"I ran 3 miles. How many times (x) around a quarter-mile track must I run to complete a 5 mile run?"

5-1 Modeling Problem Situations

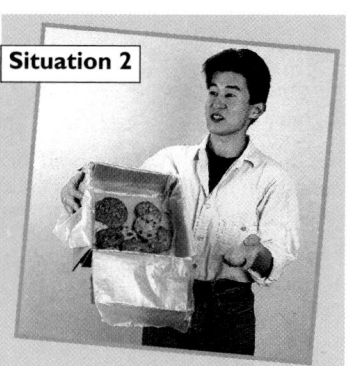

Situation 2

"I have sold three dollars worth of cookies at the neighborhood bake sale. How many 25-cent cookies (x) do I need to sell to bring my sales total up to five dollars?"

Talk it Over

Describe a situation that could be modeled by each equation.

3. $15 = 4p$ **4.** $100 = 50 + 2a$

Look Back ◄

Model and solve this situation using two different methods. In chemistry class, Inez needs to estimate how long it will take for her mystery solution's temperature to reach the freezing point of water. She measures the starting temperature at 24°C and notices it drops about 4°C per hour. About how long will it take the temperature to reach 0°C?

5-1 Exercises and Problems

1. **Reading** In Method 4 of Sample 1, the expression $37 + 9w$ is used for the total amount saved. What expression is used for the total amount saved in Method 5?

For Exercises 2–7, model each situation using one of the five methods from Sample 1. Do not use the same method for all the exercises.

2. Denby High School has 278 students signed up to take a bus to the regional basketball finals. Each bus can hold 48 people. How many buses are needed to take the students to the game?

3. The expenses for a dance are $131. Tickets cost $6.75 each. How many tickets must be sold to cover the expenses?

4. Between 1991 and 2000, the population of Santiago, Chile is expected to increase by about 86,048 people each year. In what year will the population be more than 6,000,000?

5. Between 1980 and 1991, the population of Argentina increased by about 419,783 people each year. What is the most recent year that the population was less than 30,000,000?

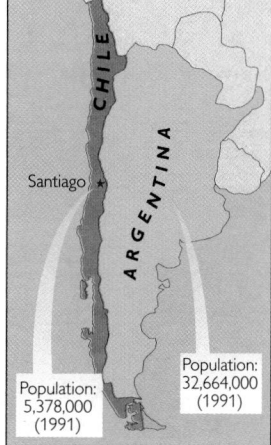

Santiago ★

CHILE

ARGENTINA

Population: 5,378,000 (1991)

Population: 32,664,000 (1991)

Answers to Talk it Over   **Answers to** Look Back

3, 4. Situations may vary. Examples are given.

3. Four copies of Shakespeare's *The Tempest* cost $15. How much does one copy cost?

4. A 100-point quiz consists of a 50-point essay and two short-answer questions. If the short-answer questions are worth an equal number of points, how many points is each short-answer question worth?

Answers may vary. An example is given.

Method: Table

Number of Hours	Temperature 24°C – 4°C/h
0	24° C
1	20° C
2	16° C
3	12° C
4	8° C
5	4° C
6	0° C

It will take 6 hours to reach 0° C.

Method: Equation

$$24 - 4h = 0$$
$$24 - 4h + 4h = 0 + 4h$$
$$\frac{24}{4} = \frac{4h}{4}$$
$$6 = h$$

It will take 6 hours to reach 0° C.

6. The perimeter of an isosceles triangle is 47 in. The length of the longest side is 17 in. Find the length of each of the congruent sides.

7. A rectangular park has a perimeter of 1950 ft. One side is 800 ft long. What are the lengths of the other three sides?

Writing **Describe a situation that could be modeled by each equation.**

8. $x + 15 = 35$

9. $250 = 10d$

10. $100 = 20 + 5x$

17 in.

The book *Rules of Thumb 2* by Tom Parker contains rules from common wisdom, including the ones in Exercises 11–15. Model each rule using an equation. Tell what each variable represents.

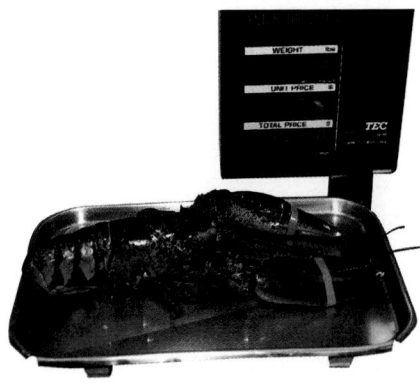

11. When microwaving solid vegetables, allow seven minutes per pound.

12. To estimate the age of a large lobster, multiply its weight in pounds by seven.

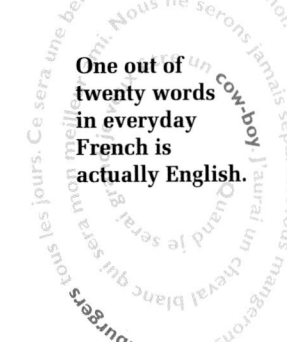

13. **One out of twenty words in everyday French is actually English.**

14. On any day the high temperature in degrees Fahrenheit is 18 degrees higher than the temperature at 6 A.M.

15. The number of minutes you will wait in a line at the bank is equal to 2.75 times the number of people ahead of you, divided by the number of tellers.

5-1 Modeling Problem Situations

245

Research

In connection with Exs. 4 and 5, suggest that students research the populations of the world's ten largest cities. Ask them to find out how the populations of the cities have changed since the early 1900s, and to write some problems suggested by the data they discover.

Using Technology

For Exs. 11–15, ask students if they can graph the equations using a graphics calculator or the *Function Investigator* software. If so, have them graph the equations. If any of the equations cannot be graphed with the calculator or software, ask students to explain why this is so.

5. The population was less than 30,000,000 in 1984.

	A	B
1	Year	Population
2	1991	32,664,000
3	1990	B2 − 419,783
4	1989	B3 − 419,783
5	1988	B4 − 419,783
6	1987	B5 − 419,783
7	1986	B6 − 419,783
8	1985	B7 − 419,783
9	1984	B8 − 419,783
⋮	⋮	⋮

6. Solve $2x + 17 = 47$. The length of each congruent side is 15 in.

7. Solve $2(800) + 2x = 1950$. The lengths of the other sides are 800 ft, 175 ft, and 175 ft.

8–10. Answers may vary. Examples are given.

8. If the original cost of a shirt is increased by $15, then the new price will be $35. What is the original cost?

9. Ayesha has a collection of dimes worth $2.50. How many dimes are in her collection?

10. Carlos now has $20. He receives an allowance of $5 per week. If he saves all his money, in how many weeks will he be able to buy a class ring that costs $100?

Answers continued on next page.

Answers to Exercises and Problems

1. $9x + 37$

2–7. Methods may vary. Examples are given.

2. Six buses are needed.

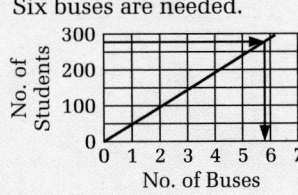

No. of Students / No. of Buses

3. Twenty tickets must be sold.

Number of Tickets Sold	Value of Tickets
16	$108.00
17	$114.75
18	$121.50
19	$128.25 ← < $131
20	$135.00 ← > $131

4. In 1999 the population will be more than 6,000,000.

Year	Population
1991	5,378,000
1992	5,464,048
1993	5,550,096
1994	5,636,144
1995	5,722,192
1996	5,808,240
1997	5,894,288
1998	5,980,336
1999	6,066,384

Practice 35 For use with Section 5-1

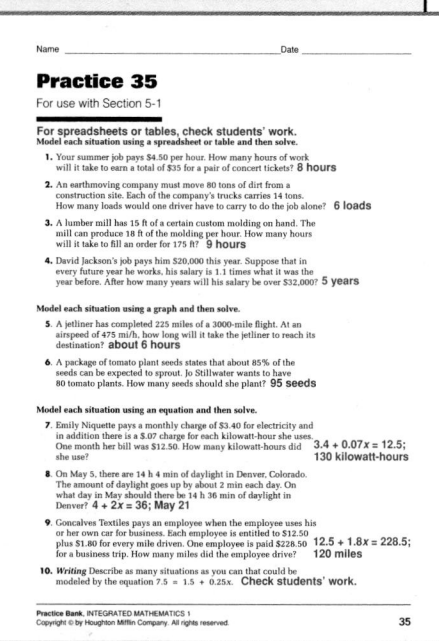

Tell which rule or rules in Exercises 11–15 are modeled by each graph.

16.

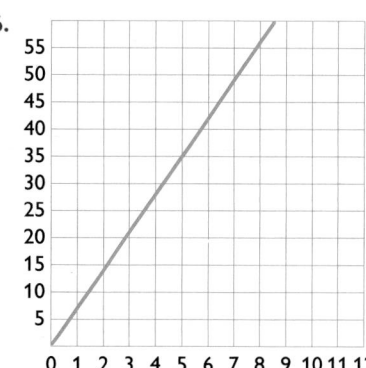

17.

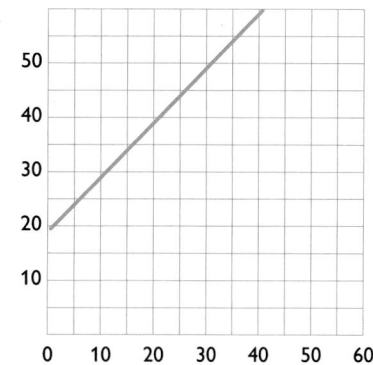

 Ongoing **ASSESSMENT**

18. **Open-ended** Write a "rule of thumb" like the ones in Exercises 11–15 based on some experience in your life. Model the rule using an equation.

Review **PREVIEW**

Graph each equation. *(Section 4-7)*

19. $y = |x|$

20. $y = x^2 - 3$

21. $y = \frac{1}{x}$

Write the opposite of each number. *(Section 2-2)*

22. 7

23. -7

24. $\frac{3}{5}$

25. 0

26. -0.75

27. -6.3

Simplify. *(Sections 1-5, 2-2)*

28. $5 - 8$

29. $5 + (-8)$

30. $-4 - (-2)$

31. $-4 + 2$

32. $2(x + 4)$

33. $3(2k - 7) - 10k$

34. $\frac{2}{3}(t + 12) + 2$

35. $-6 + 5(3x - 8)$

Working on the Unit Project

36. **a.** In your group, set a reasonable goal of how much money your class wants to raise at the car wash.

b. Make a table or spreadsheet to model your income if you charge $3 for each car you wash. Show the number of cars you need to wash to reach your goal.

c. What is the effect on the number of cars you need to wash to reach your goal if you raise the price by $2? lower the price by $2?

d. What do you think is the highest price you can charge and still attract customers?

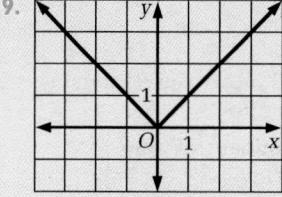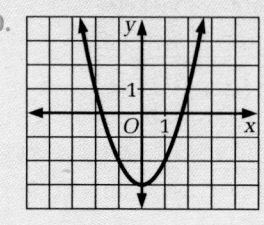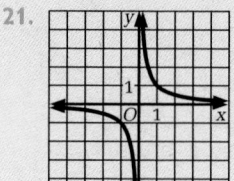
246

Opposites and the Distributive Property

---Focus

Simplify expressions and solve equations involving opposites, the distributive property, and combining like terms.

CHANGING + − SIGNS

Talk it Over

(−)   +/−

1. Read each expression aloud using the words "the opposite of" for the negative symbol.

   a. $-(-18)$ b. $-n$ c. $-(-n)$ d. $-(7 + 13)$

2. Keith thinks that it makes sense to use (−) to label the calculator key that gives the opposite of a number. Alison thinks +/− makes more sense. What do you think? Why?

3. Evaluate each pair of expressions. Are they equal?

   a. $-(7 + 13) \stackrel{?}{=} -7 - 13$ b. $-(11 - 6) \stackrel{?}{=} -11 + 6$

   c. $-(-2 - 5) \stackrel{?}{=} 2 + 5$ d. $-(-8 + 9) \stackrel{?}{=} 8 - 9$

4. Look at the expressions in question 3. How do you find the opposite of an expression in parentheses?

5-2 Opposites and the Distributive Property   **247**

PLANNING

Objectives and Strands
See pages 238A and 238B.

Spiral Learning
See page 238B.

Recommended Pacing
Section 5-2 is a two-day lesson.
Day 1
Pages 247–248: Talk it Over through Ideas about Opposites, *Exercises 1–7*
Day 2
Pages 249–250: Multi-step Equations through Look Back, *Exercises 8–38*

Extra Practice
See pages 628–629.

Warm-Up Exercises
Warm-Up Transparency 5-2

Support Materials
➤ Practice 36
➤ Enrichment 32 in the Activity Bank
➤ Study Guide 5-2
➤ Problem Set 10
➤ Quiz 5-2
➤ Alternative Assessments 5–7

Talk it Over

For question 3, ask students to say what each minus sign means: the *opposite of* or *subtract*.

Additional Samples

S1 Simplify $-(c + 6)$.

Method 1
$$-(c + 6) = -1(c + 6)$$
$$= (-1)c + (-1)6$$
$$= -c + (-6)$$
$$= -c - 6$$

Method 2
$$-(c + 6) = -c + (-6)$$
$$= -c - 6$$

S2 Simplify each expression.

a. $7d - (5d - y)$
$$7d - (5d - y) = 7d + (-5d + y)$$
$$= 7d - 5d + y$$
$$= 2d + y$$

b. $26 - 3(h + 5)$
$$26 - 3(h + 5) = 26 - (3h + 15)$$
$$= 26 - 3h - 15$$
$$= -3h + 11$$

Sample 1

Simplify $-(a - 5)$.

Sample Response

Method ❶ Use the distributive property.

$$-(a - 5) = -1(a - 5)$$ ⟵ Use the fact that $-x = -1x$.
$$= (-1)(a) - (-1)(5)$$ ⟵ Use the distributive property.
$$= -a - (-5)$$
$$= -a + 5$$ ⟵ Subtracting a number is the same as adding its opposite.

Method ❷ Find the opposite of each term.

$$-(a - 5) = -a - (-5)$$ ⟵ To find the opposite of a difference, find the opposite of each term.
$$= -a + 5$$ ⟵ Subtracting a number is the same as adding its opposite.

Sample 2

Simplify each expression.

a. $5m - (3m - n)$ **b.** $19 - 2(x + 4)$

Sample Response

a. $5m - (3m - n) = 5m + (-3m + n)$ **b.** $19 - 2(x + 4) = 19 - (2x + 8)$
$$= 5m - 3m + n$$ $$= 19 - 2x - 8$$
$$= 2m + n$$ $$= -2x + 11$$

IDEAS ABOUT OPPOSITES

➤ Subtracting a number is the same as adding its opposite.

➤ $-(-n) = n$

➤ To find the opposite of a sum or difference, find the opposite of each term.

➤ Patterns: $-(a + b) = -a - b$

➤ $-(a - b) = -a + b$

·····➤ **Now you are ready for:**
Exs. 1–7 on p. 250

Multi-step Equations

Solving some equations may require many steps. The equation may involve finding the opposite of a number or an expression, combining like terms, or using the distributive property.

Sample 3

The temperature at noon today was 6°C. At 9 P.M., the temperature was −2°C. How many degrees did the temperature drop today?

Sample Response

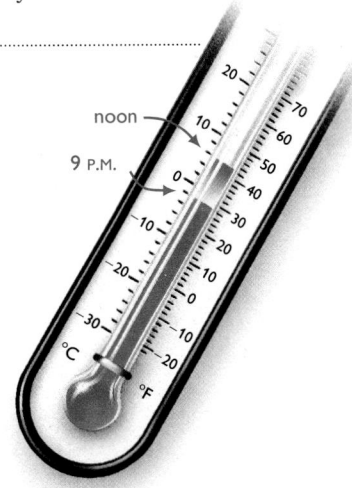

Let x = the number of degrees the temperature dropped today.

$$6 - x = -2$$

$$6 + (-x) = -2 \quad \longleftarrow \quad \text{Subtracting } x \text{ is the same as adding the opposite of } x.$$

$$6 + (-x) - 6 = -2 - 6 \quad \longleftarrow \quad \text{Subtract 6 from both sides.}$$

$$-x = -8$$

$$-1x = -8 \quad \longleftarrow \quad \text{Use the fact that } -x = -1x.$$

$$\frac{-1x}{-1} = \frac{-8}{-1} \quad \longleftarrow \quad \text{Divide both sides by } -1.$$

$$x = 8$$

The temperature dropped 8 degrees today.

Sample 4

Solve the equation $12 + 6x - 8x = 20$.

Sample Response

$$12 + 6x - 8x = 20$$

$$12 - 2x = 20 \quad \longleftarrow \quad \text{Combine like terms.}$$

$$12 - 2x - 12 = 20 - 12 \quad \longleftarrow \quad \text{Subtract 12 from both sides.}$$

$$-2x = 8$$

$$\frac{-2x}{-2} = \frac{8}{-2} \quad \longleftarrow \quad \text{Divide both sides by } -2.$$

$$x = -4$$

The solution is −4.

S3 The sea-level elevation for the entrance to a cave was −50 m. An exploration team reported 3 hours after entering the cave that they had descended to a level of −130 m. They did this in two equal stages. How many meters did they descend in each stage?

Let d = distance descended in each stage.

$$-50 - 2d = -130$$
$$-50 + (-2d) = -130$$
$$-50 + (-2d) + 50 = -130 + 50$$
$$-2d = -80$$
$$\frac{-2d}{-2} = \frac{-80}{-2}$$
$$d = 40$$

They descended 40 m during each stage.

S4 Solve the equation $-7x + 18 + 2x = 33$.

$$-7x + 18 + 2x = 33$$
$$-5x + 18 = 33$$
$$-5x + 18 - 18 = 33 - 18$$
$$-5x = 15$$
$$\frac{-5x}{-5} = \frac{15}{-5}$$
$$x = -3$$

Additional Sample

S5 Two integers are *consecutive integers* if one comes immediately before or after the other. Can the sum of three consecutive integers be equal to 41?

Suppose the largest of three consecutive integers is *n* and that the sum of *n*, *n* – 1, and *n* – 2 is 41. Write an equation.

$$n + (n - 1) + (n - 2) = 41$$
$$n + n - 1 + n - 2 = 41$$
$$3n - 3 = 41$$
$$3n - 3 + 3 = 41 + 3$$
$$3n = 44$$
$$\frac{3n}{3} = \frac{44}{3}$$
$$n = 14\frac{2}{3}$$

An integer *n* cannot be a mixed number. The sum of three consecutive integers cannot be 41.

Reasoning

Sample 5 uses a form of reasoning called *indirect proof,* in which you begin by assuming temporarily that what you want to prove is *not true.* Then you reason logically until you reach a contradiction of a known fact, which shows your initial assumption to be false. Lawyers often use such reasoning to discredit a statement made by a witness in court. They assume that what the witness says is true, then use logic to reach a conclusion that contradicts a known fact. This shows the statement made by the witness is false.

Problem Solving

Sample 5 uses an equation to solve the problem. Ask how other models can be used to arrive at a solution.

Look Back

Ask students to work in pairs. Each student can write his or her description of how to solve the equation. Then they can compare what they wrote and decide whether both approaches give the same answer.

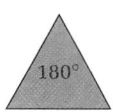

$n = 3$ $n = 4$

Now you are ready for: Exs. 8–38 on pp. 250–254

Sample 5

The sum of the measures of the angles of a convex polygon with *n* sides is $(n - 2)180$. Can the sum of the measures of the angles of a convex polygon be 450°?

Sample Response

Suppose there is such a polygon. Write an equation.

$(n - 2)180 = 450$ ◄—— The sum of the measures of the angles is 450°.

$180n - 360 = 450$ ◄—— Use the distributive property.

$180n - 360 + 360 = 450 + 360$ ◄— Add 360 to both sides.

$180n = 810$

$\dfrac{180n}{180} = \dfrac{810}{180}$ ◄—— Divide both sides by 180.

$n = 4.5$

Although 4.5 is the solution of the equation, a polygon cannot have 4.5 sides. The sum of the measures of the angles of a convex polygon *cannot* be 450°.

Talk it Over

5. Another method for solving the equation $6 - x = -2$ in Sample 3 is to add the *x* term to both sides first. Show the steps of this method.

6. Another method for solving Sample 5 is to divide both sides of the equation $(n - 2)180 = 450$ by 180 first. Show the steps that follow when you divide by 180 first.

Look Back ◄——

Describe the steps you can use to solve the equation $3(x + 2) - 5x = 10$.

5-2 Exercises and Problems

1. **Reading** Read "Ideas About Opposites" on page 248. Give an example of each idea.

Simplify.

2. $-(b - 8)$

3. $-(x + 6)$

4. $-(-3y + 4)$

5. $9 - (5k + 12)$

6. $5 - (-x + 2)$

7. $4x - (6x + 5)$

8. The temperature at 7 A.M. on a cold day was -4°C. At 3 P.M. the temperature was 1°C. Write and solve an equation to find how many degrees the temperature rose between 7 A.M. and 3 P.M.

Answers to Talk it Over

5. $6 - x = -2$
$6 - x + x = -2 + x$
$6 = -2 + x$
$6 + 2 = -2 + x + 2$
$8 = x$

6. $(n - 2)180 = 450$
$\dfrac{(n - 2)180}{180} = \dfrac{450}{180}$
$n - 2 = 2.5$
$n - 2 + 2 = 2.5 + 2$
$n = 4.5$

Answers to Look Back

$3(x + 2) - 5x = 10$

$3x + 6 - 5x = 10$ ← Distributive property

$-2x + 6 = 10$ ← Combine like terms.

$-2x + 6 - 6 = 10 - 6$ ← Subtract 6.

$-2x = 4$ ← Simplify.

$\dfrac{-2x}{-2} = \dfrac{4}{-2}$ ← Divide by –2.

$x = -2$

Solve.

9. $12 - x = -5$

10. $55 - 3n = 70$

11. $-23 = -13 - g$

12. $4x + 2 + 3x = 58$

13. $7 + 0.6y + 1.2y = 19.6$

14. $8n + 7 - 10n = 11$

15. $6(x - 3) = 48$

16. $-(3 + m) = 2$

17. $3(2 - x) = 12$

18. $4(x + 12) + 7x = 26$

19. $2(k + 5) - 3k = 4.3$

20. $9y - 4(y - 2) = 3$

For Exercises 21–23, use the fact that the sum of the measures of the angles of a convex polygon with *n* sides is $(n - 2)180$.

21. **Writing** Can the value of *n* be 3? less than 3? 3.5? 1000? Explain what kinds of numbers are reasonable in this situation.

22. Can the sum of the measures of the angles of a polygon be 800°?

23. Can the sum of the measures of the angles of a polygon be 1080°?

24. a. Write an equation that describes the relationship between the measures of $\angle ABD$ and $\angle DBC$.

b. Use your equation to find the measure of each angle.

25. **Group Activity** Work in a group of three students.

a. **Open-ended** Each student should create an equation that will take several steps to solve, as in Sample 4 or Sample 5.

b. Write your equation and single steps of the solution on separate strips of paper. Write "original equation" on the first strip.

c. Distribute one person's strips of paper among your group. Have students work together to put them in a correct order for solving the original equation. Repeat the process with the other equations.

26. To raise money for school activities, Central High School is collecting newspapers for recycling. A local business has promised to pledge enough money per pound to help the school reach its fundraising goal of $2000.

a. Students and parents have pledged a total of $32 per pound of newspapers collected. Let $x =$ the business's pledge per pound. Write an expression for the total amount pledged per pound.

b. The school collects 50 lb of newspapers. Write an expression for the amount of money collected.

c. Write and solve an equation to find the amount per pound that the business must pay to help the school reach its fundraising goal.

Goal:....................$2,000

Student and Parent pledges $32.00/lb

Business pledges.........$X/lb

5-2 Opposites and the Distributive Property

251

251

Research

Formulas are used in many applications of mathematics to sports. You might wish to ask the sports fans in your class to research formulas for their favorite sport.

Sports For Exercises 27–30, use this information.

Shina, Aiyana, Terry, and Hanna are playing in a 9-hole golf tournament. They have completed 6 holes and their strokes for each hole so far are listed on the score card below.

Your score for the entire golf course is the sum of your strokes on all of the holes you played. Your score can be over par, under par, or equal to par.

Each hole on a golf course has a *par*, which is the number of strokes a very good golfer should take to get the ball from the tee to the hole.

These numbers tell how many yards from the red tee to the hole.

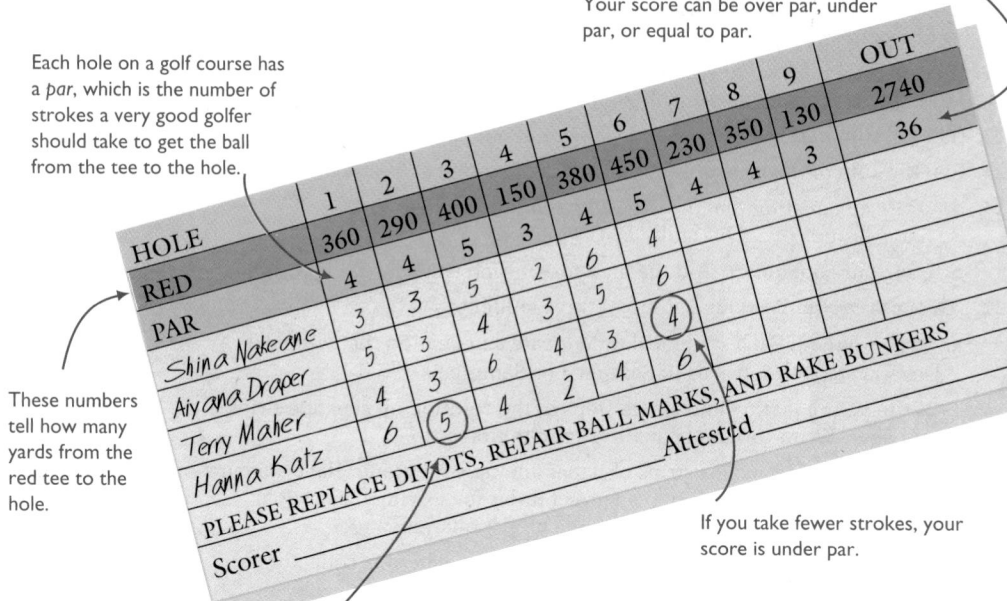

If you take fewer strokes, your score is under par.

If you take more strokes, your score for that hole is over par.

2 under par

Standings after 6 holes	
Shina Nakeane	−2
Aiyana Draper	+1
Terry Maher	?
Hanna Katz	?

27. **a.** What is par for the first six holes?

  b. What is each golfer's score for the first six holes?

  c. Complete the chart.

  d. In golf, the player who takes the fewest strokes wins. Who is the leader of the four players on the chart?

28. Shina Nakeane's foursome is heading toward the seventh hole. Some teams have finished the course. They must wait until all the players finish nine holes to see who wins.

  a. The current leader of the tournament has finished all nine holes and is 6 under par. What is her total score?

  b. Suppose Shina scores par on the last three holes. What will her total score be?

  c. How many strokes under par must Shina shoot on the next three holes to tie with the current leader?

252   **Unit 5** Equations for Problem Solving

Answers to Exercises and Problems

27. **a.** 25

  b. Shina: 23; Aiyana: 26; Terry: 24; Hanna: 27

  c. Terry: −1; Hanna: + 2

  d. Shina

28. **a.** 30

  b. 34

  c. 4

29. Two students are using equations to answer Exercise 28, part (c).

a. Solve Lindsay's equation.

b. Solve Wei-Hwa's equations.

Lindsay Hackstock

Let p = the number of strokes under par Shina must shoot on the next three holes.

$$-2 - p = -6$$

Wei-Hwa Chan

Let p = the number of strokes under par Shina must shoot on the next three holes.

Let t = the total score Shina must get to tie the current leader.

$$t = 30 - 23$$
$$p = 11 - t$$

c. **Writing** Choose either Lindsay's or Wei-Hwa's work and explain the meaning of his or her equation(s). Compare the student's work to your own method.

30. To even up the game when golfers of different abilities play against each other, each player may use a *handicap* to adjust his or her score. A handicap differential is used to find the player's handicap. Here is the formula for the handicap differential:

$$\frac{113 (\text{player's score} - \text{course rating})}{\text{slope rating}} = \text{handicap differential}$$

a. Suppose that you submitted a score of 97 on a course with a 70.1 course rating and a slope rating of 116. What is the handicap differential for that score?

b. The handicap differential for the first score on Terry Maher's handicap card is 18.1. If the slope rating for the course is 123, find the course rating to the nearest tenth.

NAME TERRY MAHER
GOLF HANDICAP INFORMATION
CLUB Los Incas Golf Club   477
SCORES POSTED 20   USGA   HOME
                          HCP INDEX
SCORE HISTORY–MOST RECENT FIRST • IF USED   14.1   16

1	92				
6	90*	89	85*A	89*	
11	89 A	88*	86*A	90*	85 A
16	99	87*	92 A	100	85*
		88*A	91	96 A	88 A
					95 A

BY THE WAY...

Nancy Lopez is the only golfer, male or female, to be selected Rookie of the Year and Player of the Year in the same year.

Multicultural Note

Students who are golfers would be interested to know that Nancy Lopez, who is of Mexican-American descent, was born in Torrance, California in 1957. She began playing golf as a child, and turned pro before she completed college. Lopez broke several records during her first year on the professional circuit, and throughout her career she has ranked at the very top of her sport. In 1987, she became the youngest woman named to the LPGA Hall of Fame. To date, Lopez has won more than 40 tournaments.

Answers to Exercises and Problems

29. a.
$$-2 - p = -6$$
$$-2 - p + 2 = -6 + 2$$
$$-p = -4$$
$$p = 4$$

b. $t = 7$
$$p = 11 - 7$$
$$p = 4$$

c. Answers may vary. An example is given. Lindsay's equation shows that Shina is currently 2 under par (−2) and needs to be a total of 6 under par (−6). Solve to find the value of p, which will give her the necessary score.

30. a. 26.2

b. 72.3

Name _____   Date _____

Practice 36

For use with Section 5-2

Simplify.

1. $-(-4r)$ $4r$
2. $-(5 - y)$ $-5 + y$
3. $-(10 - 2z)$ $-10 + 2z$
4. $6 - (-3p + 1)$ $5 + 3p$
5. $x - (0.4 + 0.7x)$ $0.3x - 0.4$
6. $12 - (-m - 8)$ $m + 20$
7. $a^2 - (5 - 9a^2)$ $10a^2 - 5$
8. $3 + c - (4 + c)$ -1
9. $3 - b - (b - 3)$ $6 - 2b$
10. $n - 2(-n + 4)$ $3n - 8$
11. $6t - t^2 - 5(t - t^2)$ $t + 4t^2$
12. $10v - v(6 - 4v)$ $4v + 4v^2$

Solve.

13. $9 - x = 16$ -7
14. $-32 - n = -7$ -25
15. $2z - 5 + z = 43$ 16
16. $y + 17 - 4y = -4$ 7
17. $-2d + 6 - 5d = 62$ -8
18. $-(4 - t) = 15$ 19
19. $-(7 + 6w) = -19$ 2
20. $-(5 - 2g) = -11$ -3
21. $4(8 - m) = 56$ -6
22. $-7(5 - y) = 49$ 12
23. $6 - (3 + 2n) = 25$ -11
24. $5 - 2(a + 7) = 32$ -20.5
25. $8 + 3(5 - b) = -10$ 11
26. $k - 3(6 - k) = -14$ 1
27. $6(3 - z) - 5z = -4$ 2
28. $(5 + t) - (6 - t) = 13$ 7
29. $-2(r - 5) - (n + 1) = -18$ 9

Find the measure of each angle.

30. $78°$ $102°$ $2x°$ $3(x - 5)°$
31. $132°$ $48°$ $11x°$ $6(x - 4)°$
32. $30°$ $110°$ $40°$ $2x°$ $5(x - 7)°$

33. The sum of the measures of the angles of a convex polygon is $1980°$. How many sides does the polygon have? 13 sides

Lin Hsia plans to invest $1400 in two stocks: Consolidated Industries (CI) and Amalgamated Manufacturing (AM).

34. Let $x =$ the amount she invests in CI. Write an expression for the amount Lin has left to invest in AM. $1400 - x$

35. Suppose that after 5 years Lin's CI stock does not change in value, but her AM stock triples, making her stocks worth $3300. Write an equation that expresses this fact. $x + 3(1400 - x) = 3300$

36. Find out how much she invested in each company by solving the equation you wrote in Exercise 35. $450 in CI, $950 in AM$

36

Practice Bank, INTEGRATED MATHEMATICS 1
Copyright © by Houghton Mifflin Company. All rights reserved.

---

31. a. **Group Activity** Work in a group of three students. Write a five-question quiz on Section 5-2. At least one question should involve geometry.

   b. Make enough copies of your group's quiz for students in another group. Have them take the quiz.

   c. Correct the quizzes. Did any answers surprise you? What was the average of the scores?

32. Model and solve this situation. *(Section 5-1)*

   Sharon Hall is an electrician. She charges a fee of $45 to come to a person's house and then $38 for each hour she works. How much would she charge for a job that took her 4 h to complete?

33. Two players have the same final golf score. The graph shows their strokes on each hole. Make two statements about the skill of each golfer. *(Section 1-1)*

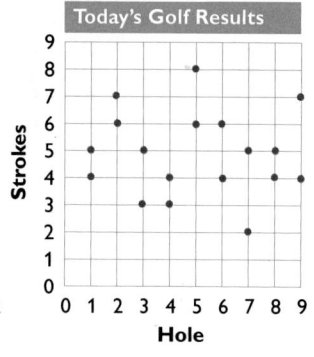

Today's Golf Results

• player 1
• player 2

Solve. *(Section 2-8)*

34. $x + (-4x) = 30$

35. $6t + 3t = 117$

36. $8x - 5x = -7 + 4$

Working on the Unit Project

37. a. In your group discuss ways you can advertise your car wash. Where are potential customers most likely to find out about the car wash?

   b. **Research** Find out the costs of your advertising options.

   c. Based on the costs and the number of potential customers you could reach, decide on two or three ways to advertise your car wash.

38. a. Make a list of supplies you will need for your car wash. Include advertising materials as well as cleaning supplies.

   b. **Research** Call or visit several stores to find the least expensive places to buy your supplies. Make a table to record the information.

254   **Unit 5** Equations for Problem Solving

Answers to Exercises and Problems

31. a–c. Answers may vary.

32. Choice of variables may vary. $T = 45 + 38h$, where $T =$ total earnings, $h =$ number of hours. If $h = 4$, then $T = \$197$.

33. Answers may vary. Player 1 is more consistent in his or her scores. Player 2 received better scores than Player 1 on 5 of the 9 holes.

34. -10

35. 13

36. -1

37. a. Answers may vary. Examples are given: ads in local newspapers, parking lot flyers, posters in stores, word of mouth, and so on

   b, c. Answers may vary.

38. a. Answers may vary. Examples are given: buckets, hoses, sponges, soap, towels, window cleaner, paper towels, poster boards, markers, and so on

   b. Answers may vary.

5-3 Variables on Both Sides

Have you ever heard the expression "Slow but steady wins the race"? That is the moral of a classic fable about a race between a tortoise and a hare. The hare knows it is the faster runner and stops during the race to nap. The slower but steady tortoise wins the race.

Focus

Model situations and solve equations that have a variable on both sides of the equals sign.

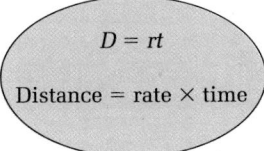

$D = rt$

Distance = rate × time

Talk it Over

Suppose the tortoise averaged 0.2 mi/h and the hare averaged 35 mi/h. Use these rates and the formula $D = rt$ to answer questions 1–6.

1. How far would each animal go in 10 h?

2. If the hare stops for 9 h 30 min to take a nap, he will be running for only 30 min or $\frac{1}{2}$ h. How far will the hare go in 10 h if he takes a nap?

3. Suppose the race lasts for t hours and the tortoise runs continuously but the hare rests for 12 h during the t hours. Write an algebraic expression to represent each situation.

   a. the number of hours the hare actually runs

   b. the number of hours the tortoise runs

   c. the distance the hare runs in the race

   d. the distance the tortoise runs in the race

4. Suppose the race is a tie. What can you say about the distance each animal has traveled when they complete the race?

5-3 Variables on Both Sides　　　　　　　　**255**

PLANNING

Objectives and Strands
See pages 238A and 238B.

Spiral Learning
See page 238B.

Materials List
➤ Algebra tiles

Recommended Pacing
Section 5-3 is a one-day lesson.

Extra Practice
See pages 628–629.

Warm-Up Exercises
💡 Warm-Up Transparency 5-3

Support Materials
➤ Practice 37
➤ Enrichment 33 in the Activity Bank
➤ Study Guide 5-3
➤ Problem Set 10
➤ Additional Explorations 5, 6
➤ Diagram Master 6 in the Explorations Lab Manual
💡 Overhead Visual 3
➤ Quiz 5-3

Answers to Talk it Over

1. tortoise: 2 mi; hare: 350 mi

2. 17.5 mi

3. a. $(t - 12)$ h

   b. t h

   c. $35(t - 12)$ mi

   d. $0.2t$ mi

4. The distances are equal.

Use questions 1 and 2 to be sure students understand the distance formula and that they can apply it.

Questions 3–6 are structured to show that even though the rates of the tortoise and hare are very different, the variable *t,* actual running time, affects the outcome of the race because the distance from start to finish is the same for both.

Using Manipulatives

In working through Sample 1, you may wish to have students use the actual tiles to model the situation.

Mathematical Procedures

Learning to solve equations that have a variable on both sides of the equals sign is simply an extension of procedures students already know.

Additional Sample

S1 Solve $3x = x + 16$.

$$3x = x + 16$$
$$3x - x = x + 16 - x$$
$$2x = 16$$
$$x = 8$$

Check: $3x = x + 16$
$$3(8) \stackrel{?}{=} 8 + 16$$
$$24 = 24$$

5. Using the expressions you wrote in questions 3(c) and 3(d), write an equation to show the relationship between the two animals' distances.

6. What information about the race can you find if you solve the equation in question 5?

Solving Equations with Variables on Both Sides

Many situations, like the tortoise and hare fable, are modeled by equations with variables on both sides of the equation. To solve these equations you first move the **variable terms** to one side of the equals sign and then solve.

Sample 1

Solve $4x = 2x + 6$.

Sample Response

You can model the equation with tiles.

Let ⬜ represent x and ◻ represent 1.

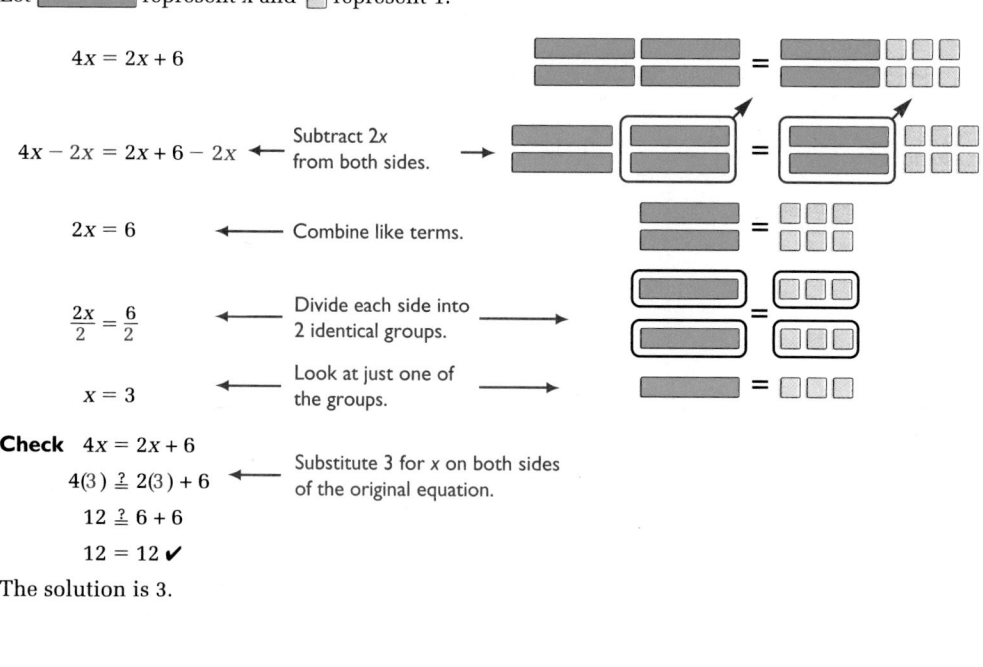

$$4x = 2x + 6$$

$$4x - 2x = 2x + 6 - 2x \quad \longleftarrow \text{Subtract } 2x \text{ from both sides.}$$

$$2x = 6 \quad \longleftarrow \text{Combine like terms.}$$

$$\frac{2x}{2} = \frac{6}{2} \quad \longleftarrow \text{Divide each side into 2 identical groups.}$$

$$x = 3 \quad \longleftarrow \text{Look at just one of the groups.}$$

Check $4x = 2x + 6$
$$4(3) \stackrel{?}{=} 2(3) + 6 \quad \longleftarrow \text{Substitute 3 for } x \text{ on both sides of the original equation.}$$
$$12 \stackrel{?}{=} 6 + 6$$
$$12 = 12 ✔$$

The solution is 3.

Unit 5 Equations for Problem Solving

Answers to Talk it Over

5. $35(t - 12) = 0.2t$

6. You can find out how many hours the race lasted.

Sample 2

Use the tortoise and hare situation and *Talk it Over* questions 3–6 on pages 255 and 256. About how long did the race last?

Sample Response

Problem solving strategy: Use a formula

Use the formula $D = rt$ to write an expression for each animal's distance.
Let $t = $ the number of hours the race lasts.

Distance in miles run by the tortoise $= 0.2t$
Distance in miles run by the hare $= 35(t - 12)$

Write an equation to find out how long the race lasts.

The animal's distances are equal.	→	$0.2t = 35(t - 12)$
Use the distributive property.	→	$0.2t = 35t - 420$
Subtract $35t$ from both sides.	→	$0.2t - 35t = 35t - 420 - 35t$
Combine like terms.	→	$-34.8t = -420$
Divide both sides by -34.8.	→	$\dfrac{-34.8t}{-34.8} = \dfrac{-420}{-34.8}$
		$t \approx 12.1$

The race took about 12.1 h.

Talk it Over

7. How can you use the answer to Sample 2 to find the length of the race course?

8. In Sample 2 the race took 12.1 h. About how many minutes is 0.1 h?

Look Back

Compare the equations in Samples 1 and 2. How are they alike? How are they different? What do the solutions have in common?

5-3 Exercises and Problems

1. **Using Manipulatives** Suppose ▭ represents x and ▢ represents 1.
   a. Use algebra tiles to make this model.
   b. What equation do your tiles model?
   c. Solve for ▭ using your model. Explain your steps.
   d. Write an equation for your result in part (c).

5-3 Variables on Both Sides

257

Additional Sample

S2 Refer to the hare and tortoise situation. Suppose the tortoise had an accident, landed on its back, and had to wait for 10 hours for someone to put it on its feet. During this time, the hare rested again. If the race were still a tie, about how long did it last?

$$0.2(t - 10) = 35 (t - 12 - 10)$$
$$0.2t - 2 = 35t - 770$$
$$0.2t - 2 - 35t = 35t - 770 - 35t$$
$$-34.8t - 2 = -770$$
$$-34.8t - 2 + 2 = -770 + 2$$
$$-34.8t = -768$$
$$t = \frac{-768}{-34.8} \approx 22.07$$

The race lasted about 22 hours.

Look Back

Discuss the questions in class. Students should see that the methods used are the same even though the numbers in Sample 2 are more complicated. You may wish to ask whether the equation in Sample 2 can be solved with algebra tiles. Students should explain their answers.

APPLYING

Suggested Assignment
Standard 3–22, 25–36
Extended 3–23, 25–36

Integrating the Strands
Number Exs. 29–31
Algebra Exs. 1–21, 25–36
Logic and Language Exs. 3–7, 20–25

Answers to Talk it Over

7. Use $t \approx 12.1$ and $r = 0.2$ in $D = rt$ to get $D = r \cdot t \approx 2.42$ mi.

8. 6 min

Answers to Look Back

Answers may vary. An example is given. The equations are alike because both equations have variables on both sides. The equation in Sample 2 requires using the distributive property, while the equation in Sample 1 does not. Both solutions require combining like terms and dividing both sides by a number.

257

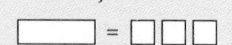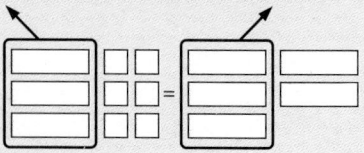
2. **Using Manipulatives** Model and solve the equation $3x + 6 = 5x$ using algebra tiles.

Tell whether the information shown is a *rate*, a *time*, or a *distance*.

3. 12 mi/h   4. 16 km   5. 64 h   6. $(x + 2)$ ft/s

7. **Reading** Which numbers in Sample 2 on page 257 are the rates at which the animals ran?

Solve.

8. $12x = 9x + 7$

9. $7x - 24 = 15x$

10. $x = 2x - 10$

11. $-c = 9(c + 40)$

12. $-4a + 6 = 2a - 36$

13. $5(t - 4) = 3t$

14. $-4(x - 4) = 6x + 36$

15. $0.5(4 - k) = 2.5k$

16. $3(6 + 3.2m) + 4 = 60.4$

17. Suppose the tortoise and the hare had another race. Again the tortoise averaged 0.2 mi/h and the hare averaged 35 mi/h. This time the hare rested 6 h and the race was still a tie. About how long did the race last?

18. The Quesadas and the Kemps leave a campsite at 10:00 A.M. and meet later at a rest stop that is 150 mi away. The Quesadas travel at 50 mi/h. Because of their trailer, the Kemps travel at only 30 mi/h.

    a. Copy the number lines below. Show each car's position at hourly intervals until the time the families meet.

    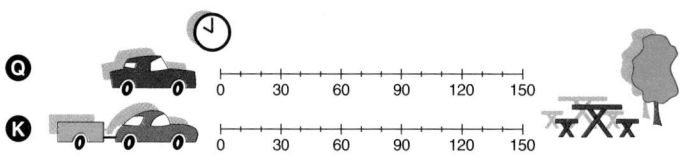

    b. How long will the Quesadas need to wait at the rest stop for the Kemps?

    c. Use the formula $D = rt$ to find how long it will take each family to drive 150 mi. Use your solutions to tell if your answer to part (b) is correct.

19.

    Salaries

    Candace
    Salary: $25,000/yr.
    Annual raises of $1000.

    Craig
    Salary: $20,000/yr.
    Annual raises of $1500.

    a. Write an expression to model Candace's salary in y years.

    b. Write an expression to model Craig's salary in y years.

    c. How long will it be before Candace and Craig earn the same salary?

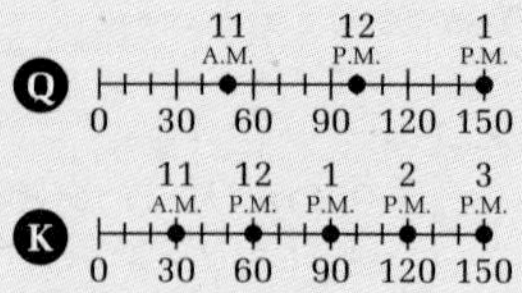

258   **Unit 5** Equations for Problem Solving

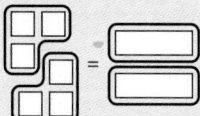
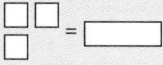

Jack London's story "To Build a Fire" describes a man walking near the Yukon trail on a bitterly cold day. This is part of that story from the book *The Best Short Stories of Jack London*.

To Build a Fire

He was a newcomer in the land, a *chechaquo*, and this was his first winter. Undoubtedly it was colder than fifty below—how much colder he did not know. But the temperature did not matter. He was bound for the old claim on the left fork of Henderson Creek, where the boys were already. He held on through the level stretch of woods for several miles. . . and dropped down a bank to the frozen bed of a small stream. This was Henderson Creek, and he knew he was ten miles from the forks. He looked at his watch. It was ten o'clock. He was making four miles an hour, and he calculated that he would arrive at the forks at half-past twelve. He decided to celebrate that event by eating his lunch there.

20. **a.** Suppose the man used the formula $D = rt$ to find out how long it would take him to get to the forks. What information did he know?

  b. Write and solve an equation to find the time it would take him to get to the forks.

  c. According to the man's calculation, how long would it take him to reach the forks? Do you agree with his answer?

21. Suppose the man's rate slowed down when he reached a hilly, icy path. He did not arrive at the forks until three o'clock.

  a. How many miles per hour did the man average for the ten miles?

  b. Suppose the man spent 3.75 hours on the hilly, icy path. How slowly did he walk on that part of the path?

22. **Reading** Compare this story with the fable of the tortoise and the hare. What ideas do they share?

23. **a.** **Open-ended** Write your own story about a race or a walk. Include information like rates, times, and distances.

  b. **Writing** Write a question that someone can answer using the information in your story.

  c. **Group Activity** Give your story and question to another student to answer. Make changes to your story or question if necessary.

24. **Research** What do you think the word *chechaquo* means? Find a definition for the word. Were you right? What language does the word come from?

5-3 Variables on Both Sides | 259

Interdisciplinary Problems

Mathematical ideas often find their way into the works of published novelists. Jack London's story "To Build a Fire" is one example. Detective stories, such as those written by Arthur Conan Doyle (about Sherlock Holmes), Agatha Christie, and many others, make frequent use of situations that can be unraveled using logical thinking skills. Books on science fiction often involve physical concepts about the universe that are described in mathematical terms, such as universes that have different dimensions. Two very famous books that involve mathematical ideas are *Gulliver's Travels* by Jonathan Swift and *Alice in Wonderland* by Lewis Carroll.

Communication: Drawing

For Exs. 20 and 21, have students sketch a map to illustrate the situation described in the passage from "To Build a Fire."

Answers to Exercises and Problems

20. **a.** He knew he was 10 mi from the forks, he was making 4 mi/h, the time was now 10 o'clock.

  b. $10 = 4t$; $t = 2.5$ h

  c. 2.5 h; Yes.

21. **a.** 2 mi/h

  b. $10 = 3.75r + 1.25(4)$; $r = 1\frac{1}{3}$ mi/h

22. Answers may vary. An example is given. Both stories involve the relationship between rate, time, and distance. In both cases the distance is known.

23. **a–c.** Answers may vary.

24. *Chechaquo* means newcomer and comes from the Chinook language (Native American, Pacific Northwest).

Assessment: Task

Have students read their paragraphs for Ex. 25(a) to the class. Discuss each paragraph in relation to Carmen's work.

Have a few volunteers read their answers to Ex. 25(b). If no volunteers can come up with a satisfactory answer, decide whether to leave the problem open or to give a hint that would lead to an answer.

Working on the Unit Project

You may wish to remind students that in considering break-even points, they still have the option of adjusting the price. However, a change in price may affect the number of customers they are likely to get.

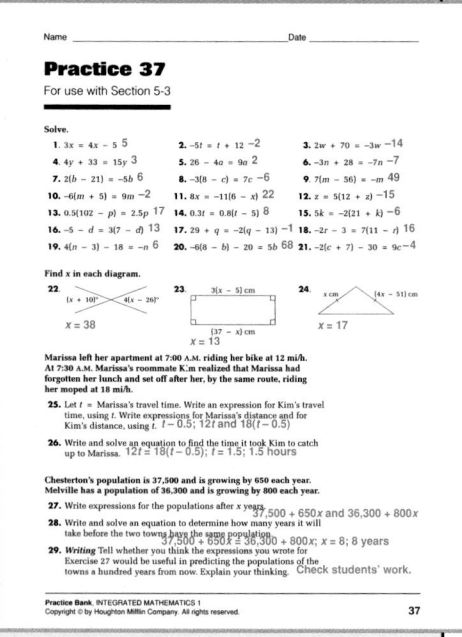
25. a. **Writing** Carmen modeled a situation with the equation $3x - 9 = 3(x - 3)$ and solved for x. She says that all numbers are solutions of this equation. Do you agree? Write a short paragraph explaining why you agree or disagree with Carmen.

b. **Open-ended** Make up an equation that has many solutions.

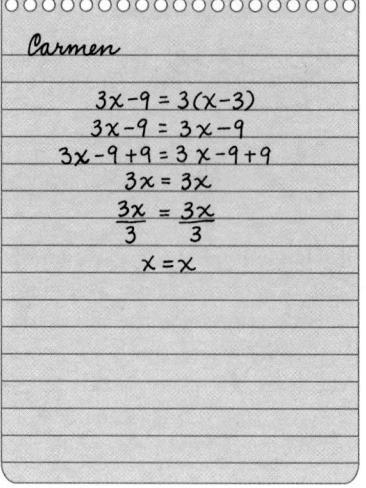

Solve. *(Section 5-2)*

26. $18 - 4n = 6$

27. $-3(2x + 5) = 9$

28. $8 - (3t + 7) = 19$

Write using exponents. *(Section 1-3)*

29. eight cubed

30. five to the second power

31. $3 \cdot 3 \cdot 3 \cdot 3$

32. Choose the statement represented by the graph. *(Section 3-3)*

a. $x \geq 3$ b. $x = 3$

c. $x < 3$ d. $x \leq 3$

$$\xleftarrow{\quad} -5\ -4\ -3\ -2\ -1\ \ 0\ \ 1\ \ 2\ \ 3\ \ 4\ \ 5 \xrightarrow{\quad}$$

33. The average life of a one-dollar bill is at least 13 months but no more than 18 months. Write an inequality to represent the situation. *(Section 3-3)*

Working on the Unit Project

34. Write and solve an equation to find how many cars you have to wash to reach your goal if you charge:

a. $3 per car

b. $5 per car

c. another price you are considering

35. a. Find the total cost for advertising and supplies based on your research.

b. Write an equation to model your total income.

36. a. When your income equals your total expenses, you *break even*. Write and solve an equation to find out how many cars you need to wash at the price you are considering in order to break even.

b. Is the number of cars you have to wash to break even a reasonable number? If you have to wash too many or too few cars, adjust your price and repeat part (a) until the number of cars you have to wash is reasonable.

Answers to Exercises and Problems

25. a. Paragraphs may vary. An example is given. I agree with Carmen. After using the distributive property you can write the equation as $3x - 9 = 3x - 9$. Both sides of the equation are exactly alike. I can substitute any value for x and the sides will be equal, so 0.1, $-\frac{1}{3}$, 1, and any other numbers are solutions of the equation.

b. Equations may vary. An example is given.
$3x + x = 4(x - 1) + 4$

26. 3   27. −4   28. −6

29. 8^3 30. 5^2 31. 3^4

32. c

33. Let a = age in months of a dollar bill; $13 \leq a \leq 18$

34. a–c. Answers may vary. Examples are given.

a. Goal: $300; $300 = 3c$: 100 cars

b. Goal: $300; $300 = 5c$: 60 cars

c. Goal: $300; try $6/car, $300 = 6c$: 50 cars

35. a, b. Answers may vary.

36. a, b. Answers may vary.

5-4 Inequalities with One Variable

Focus
Understand how adding, subtracting, multiplying, and dividing affect an inequality and model situations with inequalities.

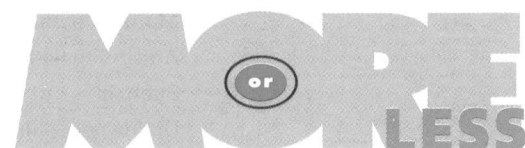

EXPLORATION

(How) do adding, subtracting, multiplying, and dividing change an inequality?

• **Materials: scientific calculators**
• **Work with a partner.**

① Write down any two different numbers. Leave a space between them.

② Write the symbol < or > in the space between your numbers to make a correct statement about which is greater.

③ Apply each rule below to your original two numbers and simplify. Write the symbol < or > between the two new numbers.

   a. Add 2.   **b.** Subtract 2.

   c. Multiply by 2.   **d.** Multiply by −2.

   e. Divide by 2.   **f.** Divide by −2.

④ Make a note about any changes in the direction of the inequality symbol in your results in parts (a)–(f) of step 3.

⑤ Repeat steps 1–4 at least two more times. Each time choose a different pair of numbers in step 1.

⑥ Repeat steps 1–4, but this time change the quantity 2 in parts (a)–(f) of step 3 to any number you wish.

⑦ Look for a pattern in the inequalities you wrote. Discuss how different operations affect inequalities. From your discussion, write a rule for inequalities that you and your partner agree on.

5-4 Inequalities with One Variable

261

PLANNING

Objectives and Strands
See pages 238A and 238B.

Spiral Learning
See page 238B.

Materials List
➤ Scientific calculator

Recommended Pacing
Section 5-4 is a two-day lesson.
Day 1
Pages 261–264: Exploration through Solving Inequalities, *Exercises 1–15*
Day 2
Pages 264–265: Modeling Situations with Inequalities through Look Back, *Exercises 16–45*

Extra Practice
See pages 628–629.

Warm-Up Exercises
Warm-Up Transparency 5-4

Support Materials
➤ Practice 38
➤ Enrichment 34 in the Activity Bank
➤ Study Guide 5-4
➤ Problem Set 10
➤ Function Investigator with Matrix Analyzer Activity Book: Function Investigator Activity 9
➤ Using Plotter Plus: Graphing Inequalities with One Variable
➤ Using Mac or IBM Plotter Plus Disk: Inequality Plotter
➤ Quiz 5-4
➤ Test 18

Answers to Exploration

1–7. Answers may vary. Examples are given.

1. 7   16

2. 7 < 16

3. a. 9 < 18

   b. 5 < 14

   c. 14 < 32

   d. −14 > −32

   e. 3.5 < 8

   f. −3.5 > −8

4. The direction of the inequality symbol changes for parts (d) and (f).

5. The result in Step 4 will be the same for different numbers.

6. Numbers chosen may vary.

7. Multiplying or dividing by negative numbers will change the direction of the inequality symbol. Adding or subtracting any numbers will not change the direction of the inequality symbol. Multiplying or dividing by positive numbers will not change the direction of the inequality symbol.

261

Exploration

The goal of the Exploration is to discover general patterns about how operations affect the direction of an inequality. Therefore, it is important for students to try combinations of different kinds of numbers.

In step 1, for example, it is fine for students to start with two positive numbers. When they get to step 5, however, they should try a combination of positive and negative, negative and negative, and so on.

Move around the room and look at the work of each pair of partners. Any time it appears that they are not trying a variety of combinations, give a suggestion in the form of a question. ("Have you tried a positive number and a negative number?" or "Have you tried two negatives?")

After everyone has written a rule for step 7, discuss the rules to compare and summarize.

Talk it Over

Use question 1 to ensure that students can apply the rules discovered in the Exploration to an inequality containing a variable.

Additional Samples

S1 Solve and graph the inequality $x - 5 > -1$.

$$x - 5 > -1$$
$$x - 5 + 5 > -1 + 5$$
$$x > 4$$

$$\begin{array}{ccc} & & \circ \\ \hline -4 & 0 & 4 \end{array}$$

S2 Solve and graph the inequality $-3x \geq -6$.

$$-3x \geq -6$$
$$\frac{-3x}{-3} \leq \frac{-6}{-3}$$
$$x \leq 2$$

$$\begin{array}{ccc} & & \bullet \\ \hline -2 & 0 & 2 \end{array}$$

Talk it Over

1. From your experience in the Exploration, which operation(s) change the direction of the inequality symbol in the sentence $25x > 50$?

   a. Multiply both sides by -4.

   b. Add -5 to both sides.

   c. Subtract -2 from both sides.

   d. Divide both sides by -5.

2. Why do you think you were not asked to add -2 or subtract -2 in step 3 of the Exploration?

Sample 1

Solve and graph the inequality $x + 2 < -1$.

Sample Response

$$x + 2 < -1$$
$$x + 2 - 2 < -1 - 2 \quad \longleftarrow \quad \text{Subtract 2 from both sides.}$$
$$x < -3$$

Check You cannot check every point on the graph, but you should check at least one. Choose a value for x that is less than -3. Try -7.

$$x + 2 < -1$$
$$-7 + 2 \stackrel{?}{<} -1 \quad \longleftarrow \quad \text{Substitute } -7 \text{ for } x \text{ in the original inequality.}$$
$$-5 < -1 \; \checkmark$$

The solution is all numbers less than -3.

Use an open circle. The graph does not include -3.

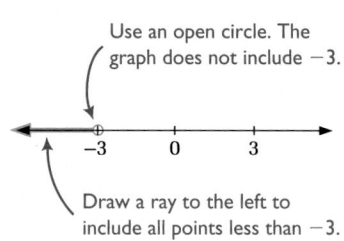

Draw a ray to the left to include all points less than -3.

Sample 2

Solve and graph the inequality $-8x \leq 40$.

Unit 5 Equations for Problem Solving

Answers to Talk it Over

1. a and d

2. Answers may vary. An example is given. Adding -2 is the same as subtracting 2, subtracting -2 is the same as adding 2.

Sample Response

$$-8x \leq 40$$
$$\frac{-8x}{-8} \geq \frac{40}{-8}$$
$$x \geq -5$$

◄── When you divide both sides by –8, reverse the inequality symbol.

Check Choose a value for x that is greater than or equal to -5. Try 0.

$$-8x \leq 40$$
$$-8(0) \overset{?}{\leq} 40$$
$$0 \leq 40 \; ✔$$

◄── Substitute 0 for x in the original inequality.

The solution is all numbers greater than or equal to -5.

Use a closed circle. The graph includes -5.

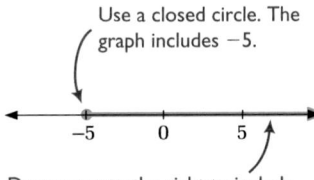

Draw a ray to the right to include all points greater than -5.

Talk it Over

3. Is -3 a solution to Sample 1? Why or why not? Is -5 a solution to Sample 2? Why or why not?

4. Why is it impossible to check *every* point on the graph of an inequality?

5. The inequalities $7x > 35$ and $x > 5$ have the same solutions. They are **equivalent inequalities.** Write an inequality that is equivalent to $8x < 88$.

6. Write an inequality that is equivalent to $5 < -x$.

Sample 3

Solve and graph the inequality $4(x + 1.5) \leq 70$.

Sample Response

$$4(x + 1.5) \leq 70$$
$$4x + 4(1.5) \leq 70$$ ◄────── Use the distributive property.
$$4x + 6 \leq 70$$
$$4x + 6 - 6 \leq 70 - 6$$ ◄────── Subtract 6 from both sides.
$$4x \leq 64$$
$$\frac{4x}{4} \leq \frac{64}{4}$$ ◄────── Divide both sides by 4.
$$x \leq 16$$

The solution is all numbers less than or equal to 16.

Use a closed circle. The graph includes 16.

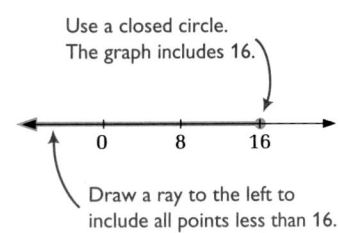

Draw a ray to the left to include all points less than 16.

5-4 Inequalities with One Variable

263

Talk it Over

When discussing the definition in question 5, point out that to solve an inequality, it is necessary to perform only those operations to both sides that yield equivalent inequalities.

Additional Sample

S3 Solve and graph the inequality $36 > -3(x - 5)$.

$$36 > -3(x - 5)$$
$$36 > (-3)x - (-3)5$$
$$36 > -3x + 15$$
$$36 - 15 > -3x + 15 - 15$$
$$21 > -3x$$
$$\frac{21}{-3} < \frac{-3x}{-3}$$
$$-7 < x$$

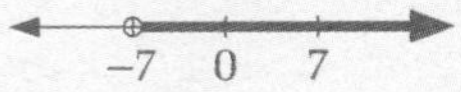

Answers to Talk it Over

3. No. -1 is not less than -1.
   Yes. $-8(-5) \leq 40$ is true.

4. There are an infinite number of points on the graph.

5. Answers may vary. An example is given. $x < 11$

6. Answers may vary. An example is given. $25 < -5x$

S4 Katerina has $123 in her checking account. With the extra earnings from her job, she can deposit $50 a week. When she reaches $500 in the checking account, she plans to open a savings account. Model this situation.

Let w = the number of weeks she deposits $50.
The inequality $123 + 50w \geq 500$ models the situation.

····► Now you are ready for:
Exs. 1–15 on p. 266

Talk it Over

7. Explain how to check the solution in Sample 3.

8. Name the operation used to rewrite the first inequality to get the second one.

    a. $5x < 25$ b. $10 \geq a - 4$ c. $-3r > -3$
        $x < 5$ $14 \geq a$ $r < 1$

SOLVING INEQUALITIES

An inequality *stays the same* when you:

➤ add, subtract, multiply by, or divide by the same positive number on both sides of an inequality

➤ add or subtract the same negative number on both sides of an inequality

An inequality is *reversed* when you:

➤ multiply or divide by the same negative number on both sides of an inequality

Modeling Situations with Inequalities

Sample 4

Jamal can afford to spend $50 to buy some concert tickets and pay for parking. Model this situation.

Sample Response

Let n = the number of tickets he can buy.
Then $18n$ is the total cost of the tickets.

The amount he can spend	is less than or equal to	$50
parking + ticket cost	$\leq$	50
$5 + 18n$	$\leq$	50

The inequality $5 + 18n \leq 50$ models the situation.

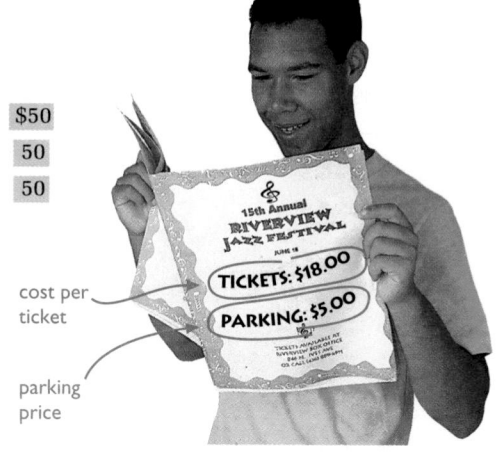

cost per ticket

parking price

TICKETS: $18.00
PARKING: $5.00

Answers to Talk it Over

7. Answers may vary. An example is given. Choose any x less than or equal to 16, for example, $x = 0$. Substitute this value into the original inequality. Simplify to see if the result is true.

8. a. Divide by 5.

   b. Add 4.

   c. Divide by -3.

Sample 5

Jamal borrowed $50 from his mom for the concert. He repays the loan at the rate of $6 per week. When will his debt be under $20?

Sample Response

Model the situation using an inequality and solve.

Let w = the number of weeks Jamal has paid his mother.

Then $6w$ = the amount he has paid in w weeks and

$50 - 6w$ = the amount that Jamal still owes after w weeks.

Write an inequality to show when the amount Jamal owes is under $20.

$$50 - 6w < 20$$

$$50 - 6w - 50 < 20 - 50 \quad \longleftarrow \quad \text{Subtract 50 from both sides.}$$

$$-6w < -30$$

$$\frac{-6w}{-6} > \frac{-30}{-6} \quad \longleftarrow \quad \text{When you divide both sides by } -6, \text{ reverse the inequality symbol.}$$

$$w > 5$$

Jamal's debt will be under $20 after the fifth week, or in 6 weeks.

Talk it Over

9. In Sample 5, why are the inequality symbols different in the inequality $50 - 6w < 20$ and the equivalent inequality $w > 5$?

10. Model each sentence with an inequality.

  a. The maximum number of tickets that can be sold for the concert is 12,000.

  b. The report is due in less than 2 weeks.

  c. Claire Rooney's hourly wage is at least $8 over the minimum wage.

  d. Kwai's test score is more than 10 points above the average.

11. The words *over*, *under*, *less than*, *more than*, *maximum*, and *minimum* can describe an inequality.

  a. Tell which one of the symbols $<$, $>$, $\leq$, or $\geq$ is described by each word.

  b. Write three sentences that use words describing inequalities.

Look Back

How is solving an inequality like solving an equation? How is it different?

▶ **Now you are ready for:**
Exs. 16–45 on pp. 266–269

5-4 Inequalities with One Variable   **265**

Additional Sample

S5 Yuri has 17 minutes of music on a 90-minute cassette. How many 5-minute songs can he still get onto the cassette?

Let s = the number of 5-minute songs.

$$17 + 5s \leq 90$$

$$17 + 5s - 17 \leq 90 - 17$$

$$5s \leq 73$$

$$\frac{5s}{5} \leq \frac{73}{5}$$

$$s \leq 14\frac{3}{5}$$

Yuri can still get 14 songs onto the cassette.

Talk it Over

For question 11, ask students to give examples of other words that describe inequalities. (Some possibilities are: at least, at most, no more than, no less than.)

Look Back

Well-stated answers to the Look Back questions, with examples, would make a good journal entry.

Answers to Talk it Over

9. In the third step, dividing both sides by –6 reversed the inequality symbol.

10. **a.** $t \leq 12,000$; t = no. of tickets

  b. $w < 2$; w = weeks to complete report

  c. $8 + w \leq s$; w = minimum wage, s = Claire's hourly wage

  d. $s > 10 + a$; a = average test score, s = Kwai's score

11. **a.** over: $>$; under: $<$; less than: $<$; more than: $>$; maximum: $\leq$; minimum: $\geq$

  b. Answers may vary. An example is given. The maximum amount that she can spend is $100.

Answers to Look Back

Answers may vary. An example is given. To solve an equation or an inequality, undo the operations to get the variable alone on one side. When you multiply or divide both sides of an equation by a negative number, you get an equivalent equation, but when you multiply or divide both sides of an inequality by a negative number, you reverse the inequality.

5-4 Exercises and Problems

1. **Reading** Read the Solving Inequalities section on page 264. Give an example to support one of the ideas in the summary. Use $x > 0$ as the original inequality.

Solve each inequality and write the letter of its graph.

2. $3x \le -15$

3. $-6x < 42$

4. $x - 4 \ge 1$

A.
B.
C.

Solve and graph each inequality.

5. $2q < -3$

6. $-5x < 105$

7. $8 \ge 5x + 10$

8. $7 - a < 25$

9. $6(x - 4) \le 3$

10. $-2(x + 5) > 6$

11. $18 \le 4(x + 2)$

12. $12 - x \le 7x$

13. $3t \ge 8t - 75$

14. Which inequality is equivalent to $-5x > 10$?
   a. $x > -2$ b. $x < 2$ c. $x < -2$ d. $x = -2$

15. Which inequality is equivalent to $-3 > -p$?
   a. $p < -3$ b. $p < 3$ c. $p > 3$ d. $-3 = p$

For Exercises 16–19, choose the letter of the inequality that models each situation.

16. In 7 more years my car will be over 10 years old. Let y = my car's age in years.
   a. $y + 7 < 10$
   b. $y + 7 > 10$
   c. $y + 7 \ge 10$

17. The 237 tickets sold at tonight's game more than tripled the number sold at the last game. Let t = the number of tickets sold at the last game.
   a. $3t > 237$
   b. $3t < 237$
   c. $3t \ge 237$

18. If I buy one CD for $12, I won't have enough money left to buy another. Let m = the amount of money I have.
   a. $m - 12 \le 12$
   b. $m - 12 \ge 12$
   c. $m - 12 < 12$

19. Five balcony seats for the concert cost over $20 more than three balcony seats. Let b = the cost of one balcony seat.
   a. $5b > 3b + 20$
   b. $5b + 20 > 3b$
   c. $5b \ge 3b + 20$

20. **Open-ended** Find a maximum or minimum number related to your school life and write an inequality to describe the number. (For example, is there a minimum number of credits for graduation? a maximum number of people allowed in a school bus?)

266   **Unit 5** Equations for Problem Solving

Answers to Exercises and Problems

1. Answers may vary. An example is given. $x > 0$; multiply by -3: $-3x < 0$

2. B   3. C   4. A

5. $q < -\frac{3}{2}$

6. $x > -21$

7. $x \le -\frac{2}{5}$

8. $a > -18$

9. $x \le 4.5$

10. $x < -8$

11. $x \ge 2.5$

12. $x \ge 1.5$

13. $t \le 15$

14. c

15. c

16. b

17. b

18. c

19. a

20. Answers may vary. An example is given. I must take at least three years of mathematics in order to graduate; let y = the number of years of mathematics; $y \ge 3$.

21. The temperature of lightning is about 50,000°F. This is more than 4.5 times the temperature of the sun's surface. Write and solve an inequality to find a maximum possible temperature for the sun's surface.

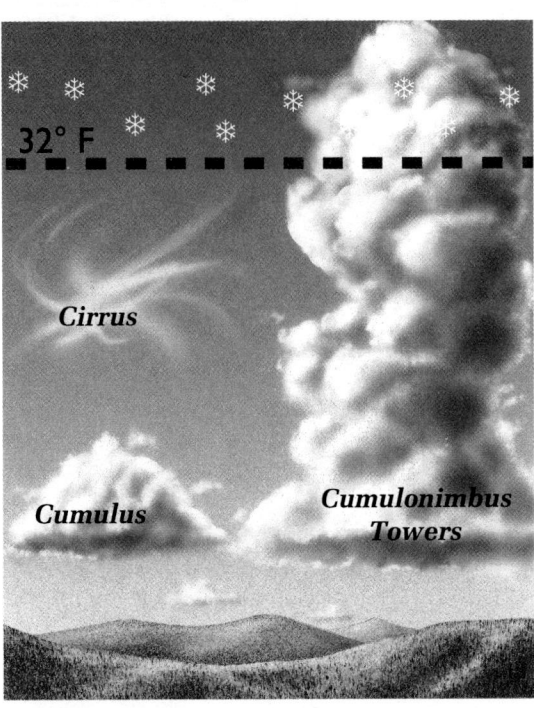

32° F

Cirrus

Cumulus

Cumulonimbus Towers

22. Cirrus clouds are found above 7 km. This is 3.5 times as high as cumulus clouds. Write and solve an inequality to determine the heights of cumulus clouds.

Weather Clouds

U.S. WEATHER SERVICE—Cumulus clouds that grow higher than 50,000 ft are called Cumulonimbus Towers. When these clouds reach a temperature between 32°F and 39°F, ice crystals start to form and fall to Earth as snow or rain. Cumulonimbus Towers are responsible for up to three-fourths of the world's rain and some of its most dangerous weather.

23. Reading Write two inequalities about the information in this article. Be sure to tell what your variables represent.

24. Career To study clouds, Joanne Simpson, a meteorologist and chief scientist at NASA, once flew directly into them in a single engine plane. Today she uses satellites, airplanes, and computers to study weather patterns. Simpson was the first person to write a paper about using math models to study cloud growth.

Joanne Simpson, Meteorologist and NASA scientist.

  a. Research Find the names and height ranges for two types of clouds other than cumulus, cirrus, or cumulonimbus towers. Write an inequality to show the relationship between their heights. Draw a picture of each type of cloud.

  b. Writing Joanne Simpson says that "science isn't just cold, hard facts. The crux is fitting unfitted things together and making them hang together. It's similar to composing music or art or poetry." Do you think this statement is true for mathematics? Explain why or why not.

5-4 Inequalities with One Variable   **267**

Interdisciplinary Problems

Mathematical models are used to study physical phenomena in order to understand them and also to make predictions. The basic reason why it is difficult to predict weather patterns for more than a few days in advance is that the models developed so far are not adequate to the task. Meteorologists and mathematicians know that weather is a *dynamic system,* and that such systems may behave in an extremely unpredictable way. A very new branch of mathematics called *chaos theory* has been developing recently to study dynamic systems in order to learn more about them.

Research

In connection with Exs. 21–24, have students research different classes of storms to see what facts about storms can be described by using inequalities. Students can present their findings by writing a report, making a poster, or presenting an oral report to the class.

Answers to Exercises and Problems

21. $50{,}000 > 4.5t$; $t < 11{,}111\frac{1}{9}$; maximum temperature is about 11,111°F.

22. $7 < 3.5c$; $c > 2$; cumulus clouds are at heights greater than 2 km.

23. Answers may vary. An example is given.
c = Cumulonimbus Towers' height, $c > 50{,}000$;
r = rain caused by Cumulonimbus Towers,
R = all rain, $r \le \frac{3}{4}R$

24. a. Answers may vary.

    b. Answers may vary.

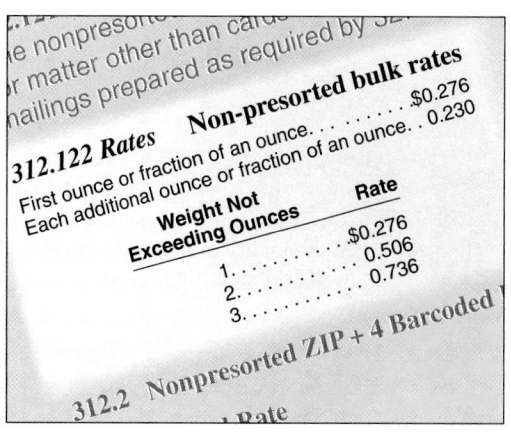

Use the postal chart for Exercises 25 and 26. Model each situation with an inequality and solve.

25. The Drum Corps club wants to send a newsletter to its members. If each newsletter weights 1.7 oz, how many newsletters can they afford to send if they can spend up to $150?

26. The admissions office at Arlington University wants to mail information packets to students. The budget allows them to spend a maximum of $1.75 to mail each packet. How heavy a packet can be sent to the students?

27. A town requires that a one-story house built on a standard lot be 2000 ft² or less in area. A builder plans to build a rectangular house on a standard lot. The width of the house is 20 ft. How long can the house be?

28. Fiona Wilson earns $8.50 an hour plus time-and-a-half for overtime as a pastry chef. Her normal work week is 37 h. How many full hours of overtime must she work to earn over $400 in a week?

29. Annette has $65 in her bank account. She can earn $9 for each lawn she mows. How many lawns must she mow in order to save at least $200?

Solve and graph each inequality.

30. $5x + 13 > 4(x + 3) + x$
31. $6x + 15 < 6(x + 1)$

Ongoing ASSESSMENT

32. **Writing** Write an inequality like the ones in Samples 1–3. Write a paragraph to describe a situation that it models. Solve the inequality. Explain what the solution means.

Review PREVIEW

Solve. *(Sections 2-7, 2-8, 5-3)*

33. $4(5a + 1) = 104$
34. $7c + 5 = 68$
35. $-2y = 38$
36. $4x - 10 + 5x = -55$
37. $x = 3x + 16$
38. $3a + 4 = a - 8$

For each formula, tell what each variable represents. *(Sections 2-6, 5-3)*

39. $D = rt$
40. $P = 2l + 2w$
41. $V = lwh$

Answers to Exercises and Problems

25. up to 296 newsletters
26. at most 7 ounces
27. 100 ft or less
28. at least 7 h
29. at least 15 lawns
30. all real numbers

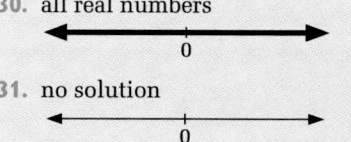

31. no solution

32. Answers may vary. An example is given. $40 < 1.75r$. A bus ride in Jamesburg costs $1.75 and a monthly bus pass costs $40. How many rides per month would you need in order for the cost of a bus pass to be less than the cost of individual rides? Solution: $22.86 < r$; You would need at least 23 rides per month.

33. 5   34. 9   35. −19
36. −5   37. −8   38. −6
39. *D:* distance; *r:* rate; *t:* time
40. *P:* perimeter; *l:* length; *w:* width
41. *V:* volume; *l:* length; *w:* width; *h:* height
42. $5c ≥ 300$; at least 60 cars
43. Answers may vary.
44. Answers may vary.
45. Answers may vary.

Working on the Unit Project

42. Write and solve an inequality to show how many cars you would have to wash at $5 per car to raise at least $300.

43. Write and solve an inequality to show how many cars you would have to wash at your price per car to at least reach your goal.

44. Write and solve an inequality to show how many cars you would have to wash at your price per car to raise more than you spend on supplies and advertising.

45. Decide on the price you will charge for each car you wash.

Unit 5   CHECKPOINT

Quick Quiz (5-1 through 5-4)

See page 297.

1. Writing How would you teach someone to solve the equation $(x - 5)5 + 2x = 17$? Be sure to include the distributive property and combining like terms in your explanation.

PREMIUM ETR/CASSETTE
W/6 SPKERS INC: EQUALIZER/
DIVERSITY ANT
6-DISC CHANGER. . . 1,200

2. Alma wants to save $1200 so that she can get a deluxe stereo system for her new car. She has already saved $500 and plans to save $22 each week. After how many weeks will she have saved at least $1200? Model the situation using a table or a graph and then solve.   5-1

3. Describe a situation that can be modeled by the equation $29 + 3x = 47$.

Write and solve an equation to find the measure of each angle shown.   5-2

4.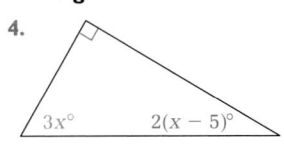

5.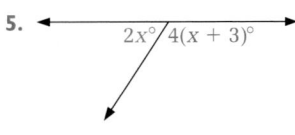

Solve each equation.   5-3

6. $-(t + 5) = 8$

7. $6y - 4(y - 1) = 10$

8. $14x - 18 = 2x$

9. $2x = -4(x - 3)$

Solve and graph each inequality.   5-4

10. $9y \leq 36$ **11.** $-15m > 45$ **12.** $3c + 10 < 4$

13. Chris earns $7 an hour plus $10 for each hour of overtime. If he normally works 40 h each week, how many full hours of overtime must he work to earn over $380?

Practice 38 For use with Section 5-4

Name _____ Date _____

Practice 38
For use with Section 5-4

Solve and graph each inequality. For graphs, check students' work.
1. $p + 3 < 5$ $p < 2$ 2. $c - 4 \geq 2$ $c \geq 6$ 3. $x - 1 > -6$ $x > -5$
4. $7 - y \leq 2$ $y \geq 5$ 5. $-9 - q \geq -5$ $q \leq -4$ 6. $2 - v > -3$ $v < 5$
7. $5w \geq -35$ $w \geq -7$ 8. $-3n > 9$ $n < -3$ 9. $-4k \leq 24$ $k \geq -6$
10. $-12b < 0$ $b > 0$ 11. $0.27 < -0.06h$ $h < -4.5$ 12. $10d \leq -35$ $d \leq -3.5$
13. $2u - 5 < 7$ $u < 6$ 14. $8 - 2r \geq 10$ $r \leq -1$ 15. $6t + 5 > -7$ $t > -2$
16. $-3m + 22 > 4$ $m < 6$ 17. $9 \leq 24 - 5z$ $z \leq 3$ 18. $18 \leq -8 - 13a$ $a \leq -2$
19. $2(x - 5) \leq -12$ $x \leq -1$ 20. $-4(7 - t) > -28$ $t > 0$ 21. $3 > -3(y + 7)$ $y > -8$
22. $-15 \leq -3(8 + v)$ $v \leq -3$ 23. $6(w - 9) < -21$ $w < 5.5$ 24. $-10(5 - c) \leq 25$ $c \leq 7.5$
25. Fashion Statement, Inc. makes men's shirts for department stores. A store needs at least 525 shirts. The company has 84 shirts in its stock. Write and solve an inequality to find the number of shirts the company will have to make to fill the store's order. $x + 84 \geq 525$; $x \geq 441$; at least 441 shirts
26. An elevator has an inspection certificate stating that the maximum weight the elevator can carry is 2100 lb. Suppose each person who takes the elevator weighs 140 lb. Write and solve an inequality for the number of persons the elevator can carry. $140n \leq 2100$; $n \leq 15$; at most 15 people
27. Miguel Santos saves $65 a week out of his salary toward a vacation trip. Suppose the trip will cost at least $1430. Write and solve an inequality to find the number of weeks it will take him to pay for his vacation trip. $65w \geq 1430$; $w \geq 22$; at least 22 weeks
28. Dugungi's hardware store has 17 pitchforks in its inventory. The store manager estimates that the store will need over 65 pitchforks for the coming growing season. Pitchforks are packed 4 to a box. Write and solve an inequality to find the number of boxes of pitchforks Dugungi's should order. $17 + 4b > 65$; $b > 12$; more than 12 boxes
29. Mei Ling Won wants to keep her local telephone bill under $20. The phone company charges a base rate of $5.60 each month and $.24 for each message unit used. Write and solve an inequality to find how many message units she can use. $5.6 + 0.24m < 20$; $m < 60$; fewer than 60 message units
30. A supermarket manager wants to price a box of cereal. During a sale in which each box is marked $1.50 off, the manager wants 5 boxes to sell for less than 3 boxes did before the sale. Write and solve an inequality to find the price the supermarket should charge for a box of cereal. $5(p - 1.5) < 3p$; $p < 3.75$; less than $3.75

38   **Practice Bank,** INTEGRATED MATHEMATICS 1
Copyright © by Houghton Mifflin Company. All rights reserved.

Answers to Checkpoint

1. Answers may vary. An example is given. First use the distributive property to get $5x - 25 + 2x = 17$. Then combine the like terms $5x$ and $2x$ to get $7x - 25 = 17$. Next, add 25 to both sides to undo the subtraction of 25; this gives $7x - 25 + 25 = 17 + 25$ or $7x = 42$. To undo the multiplication by 7, divide both sides by 7; this gives $\frac{7x}{7} = \frac{42}{7}$; $x = 6$.

2.

Number of weeks	Total saved ($)
0	500
1	$500 + 22(1) = 522$
2	$500 + 22(2) = 544$
⋮	⋮
31	$500 + 22(31) = 1182$
32	$500 + 22(32) = 1204$

Alma needs 32 weeks.

3. Answers may vary. An example is given. On Saturday I bought a $29 jacket and three small plants and spent $47 in all. How much did each plant cost?

4. $3x + 2(x - 5) = 90$; $x = 20$; 60° and 30°

5. $2x + 4(x + 3) = 180$; $x = 28$; 56° and 124°

6. −13   **7.** 3

8. 1.5   **9.** 2

10. $y \leq 4$

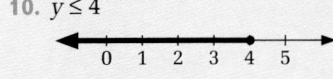

11. $m < -3$

12. $c < -2$

13. $7(40) + 10x > 380$; $x > 10$; 11 h

269

Section 5-5

Rewriting Equations and Formulas

Focus
Learn to rewrite formulas
to make them easier to use.

How many formulas can you think of in two minutes?

• **Work in a group of four students.**

① Write at least four formulas you know. Try to relate at least one formula to geometry, one to science, one to consumerism or business, and one to sports or hobbies.

② Choose one of your formulas and explain what the variables represent.

③ Give an example of how each formula is used in everyday life.

Sometimes you need to rewrite a formula or an equation to use it more easily. You used the formula $D = rt$ to find the distances run by the tortoise and the hare. You can write this formula in other ways.

Sample 1

Solve $D = rt$ for t.

Sample Response

Problem Solving Strategy: Solve a simpler problem

Solve the formula $D = rt$ for t just as you would solve the equation $100 = 50t$. Undo any operations to get t alone on one side of the equals sign.

270   **Unit 5** Equations for Problem Solving

Answers to Exploration

1–3. Answers may vary. Examples are given.

1. geometry: $A = bh$ (area of a rectangle); science: $C = \frac{5}{9}(F - 32)$ (conversion from Fahrenheit temperature to Celsius temperature); business: $I = prt$ (interest formula); sports: $A = \frac{h}{b}$ (baseball batting averages)

2. For $A = bh$, A, b, and h represent area, base, and height, respectively.

3. Use $C = \frac{5}{9}(F - 32)$ to compute the temperature in degrees Celsius if you are in a country that uses the Fahrenheit scale to measure temperatures.

$$100 = 50t \quad \longleftarrow \text{Divide both sides by the coefficient of } t. \longrightarrow \quad D = rt$$

$$\frac{100}{50} = \frac{50t}{50} \quad \longleftarrow \text{Divide by 50.} \qquad \text{Divide by } r. \longrightarrow \quad \frac{D}{r} = \frac{rt}{r}$$

$$2 = t \quad \longleftarrow t \text{ is a number.} \qquad \begin{array}{c} t \text{ is an algebraic} \\ \text{expression.} \end{array} \longrightarrow \quad \frac{D}{r} = t$$

Sample 2

Solve $2x + y = 180$ for y.

Sample Response

$$2x + y = 180 \quad \longleftarrow \text{To solve for } y, \text{ get } y \text{ alone on one side.}$$
$$2x + y - 2x = 180 - 2x \quad \longleftarrow \text{Undo addition of } 2x. \text{ Subtract } 2x \text{ from both sides.}$$
$$y = 180 - 2x$$

Sample 3

Solve $P = 2l + 2w$ for w.

Sample Response

$$P = 2l + 2w \quad \longleftarrow \text{To solve for } w, \text{ get } w \text{ alone on one side.}$$
$$P - 2l = 2l + 2w - 2l \quad \longleftarrow \text{Undo addition of } 2l. \text{ Subtract } 2l \text{ from both sides.}$$
$$P - 2l = 2w$$
$$\frac{P - 2l}{2} = \frac{2w}{2} \quad \longleftarrow \text{Undo multiplication by 2. Divide both sides by 2.}$$
$$\frac{P - 2l}{2} = w$$

Talk it Over

1. The formula $\frac{D}{r} = t$ is read, "distance divided by rate equals time." Solve the formula $D = rt$ for r. How would you read the rewritten formula?

2. Choose one of the formulas that your group wrote for the Exploration. Tell how you can solve the formula for a different variable.

Look Back

Compare solving the equation $158 = 18(x - 4)$ with solving the equation $y = 18(x - 4)$ for x.

5-5 Rewriting Equations and Formulas

271

Answers to Talk it Over

1. $\frac{D}{t} = r$; Distance divided by time equals rate.

2. Answers may vary. An example is given. To solve $I = prt$ for r, divide each side by pt: $\frac{I}{pt} = r$.

Answers to Look Back

Answers may vary. An example is given. The method for solving the equations is very similar. To solve either one you use the distributive property, combine like terms, add 72 to both sides, and divide both sides by 18. The difference is that the first equation has a single solution, whereas the solution of the second equation is an equivalent equation involving two variables.

TEACHING

Exploration

The purpose of the Exploration is to have students realize how many formulas they already know and how often formulas are used. Ask the groups to share their formulas with one another.

Limited English Proficiency

Consider allowing more than two minutes for groups to think and respond, in order to allow LEP students to understand the task and to respond without feeling rushed and frustrated. Peer tutors in each group can work with LEP students to express everyday situations in the form of formulas with variables.

Additional Samples

S1 Solve $A = \frac{1}{2}bh$ for b.
$$A = \frac{1}{2}bh$$
$$2A = 2 \cdot \frac{1}{2}bh$$
$$2A = bh$$
$$\frac{2A}{h} = \frac{bh}{h}$$
$$\frac{2A}{h} = b$$

S2 Solve $5f - 9c = 160$ for f.
$$5f - 9c = 160$$
$$5f - 9c + 9c = 160 + 9c$$
$$5f = 160 + 9c$$
$$\frac{5f}{5} = \frac{160 + 9c}{5}$$
$$f = \frac{160 + 9c}{5}$$

S3 Solve $A = \pi rs + \pi r^2$ for s.
$$A = \pi rs + \pi r^2$$
$$A - \pi r^2 = \pi rs + \pi r^2 - \pi r^2$$
$$A - \pi r^2 = \pi rs$$
$$\frac{A - \pi r^2}{\pi r} = \frac{\pi rs}{\pi r}$$
$$\frac{A - \pi r^2}{\pi r} = s$$

Look Back

You may wish to ask a few students to solve each equation at the chalkboard. Then compare and discuss the solutions.

271

APPLYING

Suggested Assignment

Standard 1–18, 23–33

Extended 1–20, 22–33

Integrating the Strands

Number Exs. 23, 28–31

Algebra Exs. 1–26, 32

Measurement Ex. 23

Geometry Ex. 23

Statistics and Probability Ex. 27

Discrete Mathematics Ex. 23

Logic and Language Exs. 1, 22, 23, 32, 33

Reasoning

For Ex. 3, ask whether it is possible to solve the equation for π. Have students discuss whether π is a variable or a constant.

For Ex. 5, ask if it is possible to solve the equation for c.

5-5 Exercises and Problems

1. **Reading** Read Sample 1 on pages 270 and 271. Follow the format of the Sample Response and solve the equations $128 = r \cdot 32$ and $D = rt$ for r. Label the steps of each solution as in Sample 1.

Solve each equation for the variable shown in red.

2. The formula to determine postage cost: $c = 29 + 23w$

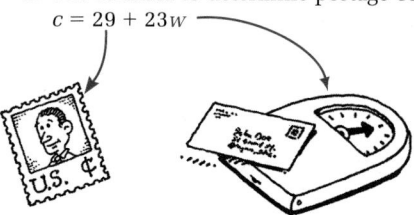

3. The formula for the circumference of a circle: $C = \pi d$

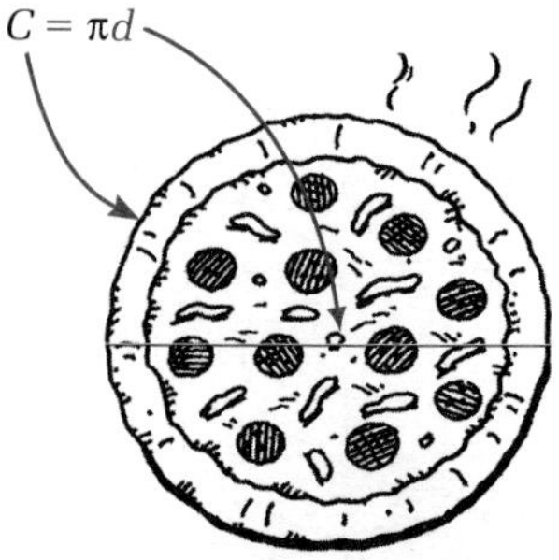

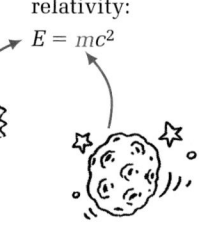

4. Lumber is sold by the board foot. The formula to find the board-foot measure for a piece of wood: $B = \frac{t \times w \times l}{12}$

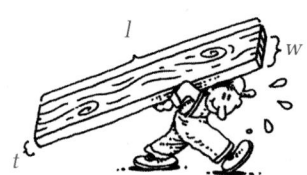

5. Einstein's formula for the theory of relativity: $E = mc^2$

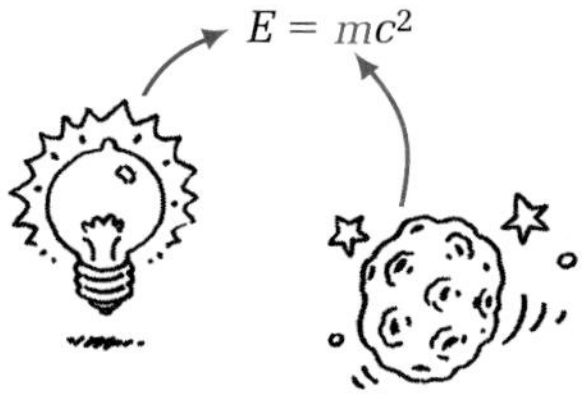

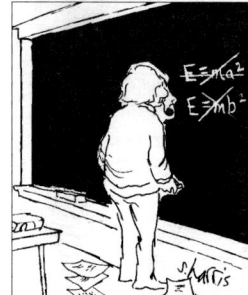

6. The formula for finding the octane rating of a gasoline: $O = \frac{R + M}{2}$

Research Method
Motor Method
Octane

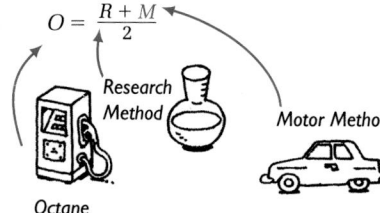

7. The formula for optimum heart rate while exercising: $r = p(220 - a)$

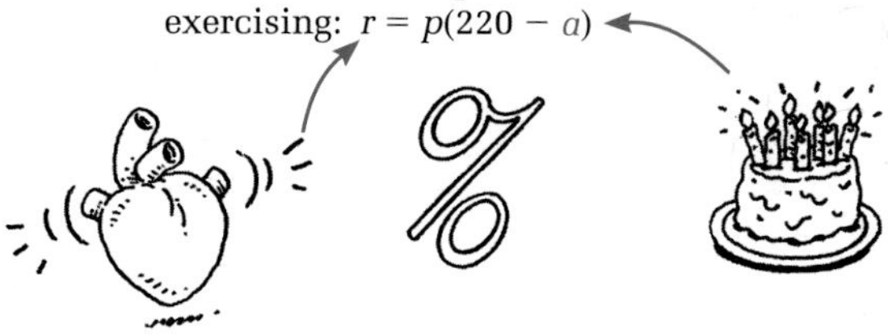

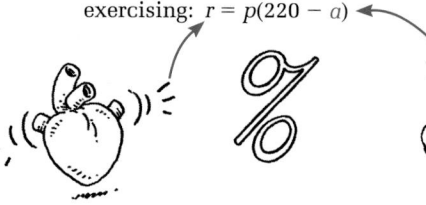

Solve each equation for the variable shown in red.

8. $E = IR$
9. $b = s - c$
10. $I = prt$
11. $x + y = 100$
12. $360 = 2y + x$
13. $5x + y = 75$
14. $c = ax - b$
15. $ax + by = c$
16. $s = 2\pi rh + 2\pi r^2$

Answers to Exercises and Problems

1. $128 = r \cdot 32$ $\qquad D = rt$
   Divide both sides by the coefficient of r.
   $\frac{128}{32} = \frac{r \cdot 32}{32}$ $\qquad \frac{D}{t} = \frac{rt}{t}$
   $4 = r$ $\qquad \frac{D}{t} = r$
   r is a number. $\qquad r$ is an algebraic expression.

2. $\frac{C - 29}{23} = w$

3. $d = \frac{C}{\pi}$

4. $\frac{12B}{w \times l} = t$

5. $\frac{E}{c^2} = m$

6. $2O - R = M$

7. $\frac{-r + 220p}{p} = a$

8. $I = \frac{E}{R}$

9. $s = b + c$

10. $r = \frac{I}{pt}$

11. $y = 100 - x$

12. $x = 360 - 2y$

13. $y = 75 - 5x$

14. $x = \frac{c + b}{a}$

15. $y = \frac{c - ax}{b}$

16. $h = \frac{s - 2\pi r^2}{2\pi r}$

17. Every Adventure Travel agent earns a base pay of $200 a week, plus 12% commission for each airline ticket the agent sells.

   a. Write a formula for an agent's weekly salary, s, when the agent sells d dollars worth of airline tickets.

   b. Rewrite your formula to show the amount, in dollars, of airline tickets an agent needs to sell in order to make a weekly salary of s dollars.

   c. How many dollars worth of airline tickets will Frances Medved have to sell each week in order to make a weekly salary of $600?

   d. The Tappan School band orders $8650 worth of airline tickets from Frances for their upcoming trip. What will her salary be if she does not sell any other tickets that week?

18. Saburo is traveling with the school band. He flies between two cities in Hawaii. A 10% tax is added to the base cost of a ticket for flights within Hawaii.

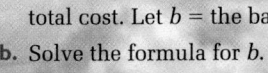

   a. Write a formula for finding the total cost of a plane ticket between cities within Hawaii. Let t = the total cost. Let b = the base cost.

   b. Solve the formula for b.

   c. Find the base cost for a ticket that costs $49.

19. Gus Uchida, the band's equipment manager, parked his car at the airport. Parking costs $12 the first day and $9 each additional day.

   a. Write a formula for the cost, c, to park d days at the airport.

   b. Gus paid $120 for parking. How many days was Gus' car parked at the airport?

   c. TECHNOLOGY Make a spreadsheet that shows the daily parking costs for 1 to 30 days.

20. Winona Birch, the director of the school band, took a taxicab from the airport to her home. She had to pay a flat fee of $1.50 plus $1.60 per mile.

   a. Write a formula for the cost, c, for a trip m miles long.

   b. The total cost of Winona's trip was $8.50. How many miles was her trip?

21. a. **Research** Find out what the minimum wage is.

   b. Write a formula for the amount earned for h hours worked at minimum wage.

   c. Solve the formula you wrote in part (b) for h.

22. **Writing** Josie says that $s = \dfrac{n}{n+1}$ and $\dfrac{s}{1-s} = n$ are two ways to write the same formula. Decide whether or not you agree with Josie. Explain how you made your decision.

Answers to Exercises and Problems

17. a. $s = 200 + 0.12d$

   b. $d = \dfrac{s - 200}{0.12}$

   c. $3333.33

   d. $1238.00

18. a. $t = b + 0.1b$ or $t = 1.1b$

   b. $\dfrac{t}{1.1} = b$

   c. $44.55

19. a. $c = 12 + 9(d - 1)$

   b. He parked for 13 days.

   c. A sample spreadsheet showing cell formulas and a sample table of values are given.

	A	B
1	1	=12+9*(A1−1)
2	=A1+1	=12+9*(A2−1)
3	=A2+1	=12+9*(A3−1)
4	=A3+1	=12+9*(A4−1)
⋮	⋮	⋮
28	=A27+1	=12+9*(A28−1)
29	=A28+1	=12+9*(A29−1)
30	=A29+1	=12+9*(A30−1)

Using Technology

In connection with Ex. 19(c), students can refer to page 619 in the Technology Handbook which discusses using a spreadsheet.

Students can use the spreadsheet feature of the *Stats!* or the *Function Investigator* software to complete Ex. 19(c).

Integrating the Strands

In Ex. 23 on page 274, students use concepts from the strands of number, algebra, measurement, geometry, discrete mathematics, and logic and language to answer the questions.

Parking at the Air Terminal

Number of Days	Total Parking Costs
1	$12
2	$21
3	$30
4	$39
⋮	⋮
28	$255
29	$264
30	$273

20. a. $c = 1.5 + 1.6m$

   b. 4.375 mi

21. a–c. Answers may vary. An example is given.

   a. $4.25/h

   b. $s = 4.25h$

   c. $h = \dfrac{s}{4.25}$

22. Answers may vary. An example is given. I agree. If I solve $s = \dfrac{n}{n+1}$ for n, I get $s(n + 1) = n$; $sn + s = n$; $s = n - sn$; $s = n(1 - s)$; $\dfrac{s}{1-s} = n$; that is, I get the second formula when I solve for n. The first formula gives s in terms of n, and the second formula gives n in terms of s.

273

Assessment: Portfolio

In evaluating answers for Ex. 23, give special attention to the written report for part (e), which may qualify for inclusion in a student's portfolio.

Working on the Unit Project

The research in Ex. 33 may present difficulties if students want to study traffic patterns at times when they are in class. In that event, have students brainstorm ways to get information other than gathering it themselves. Have them gather information from commercial car washes about peak business times.

Practice 39 For use with Section 5-5

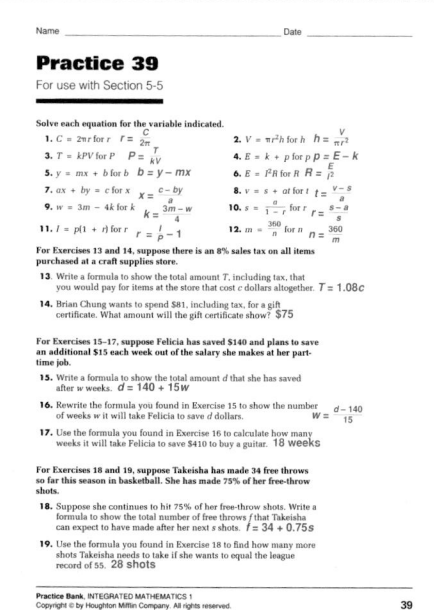

Answers to
Exercises and Problems

23. a. $l = \dfrac{P - 2w}{2}$

b. 17 rectangles meet the conditions.

width	length	width	length
1 in.	33 in.	10 in.	24 in.
2 in.	32 in.	11 in.	23 in.
3 in.	31 in.	12 in.	22 in.
4 in.	30 in.	13 in.	21 in.
5 in.	29 in.	14 in.	20 in.
6 in.	28 in.	15 in.	19 in.
7 in.	27 in.	16 in.	18 in.
8 in.	26 in.	17 in.	17 in.
9 in.	25 in.		

---

23. Leo Hernandes told his mathematics class, "I am thinking of a rectangle. Its perimeter is 68 in. The measure of each side is a whole number of inches. I want you to write about such a rectangle."

a. Solve the formula $P = 2l + 2w$ for l.

b. Use the formula you wrote for part (a) to complete the table in the photograph. How many different rectangles meet Leo Hernandes' conditions?

c. Sketch the rectangle with the largest area and the rectangle with the smallest area.

d. How could you use the problem solving strategy *Identify a pattern* to do part (b)?

e. **Writing** How could you describe the rectangles to Leo Hernandes?

Review **PREVIEW**

Solve and graph each inequality. *(Section 5-4)*

24. $-6x > 12$

25. $15y \le -60$

26. $6(x - 1.5) > -21$

27. Find the mean of 87%, 96%, 75%, 81%, and 64%. *(Section 3-2)*

Find the reciprocal of each number. *(Toolbox Skill 9)*

28. 3

29. $\dfrac{2}{5}$

30. $-\dfrac{11}{6}$

31. $\dfrac{9}{7}$

---

5 Working on the Unit Project

32. Your *profit* from the car wash is your income minus your expenses. Your profit will depend on the number of customers you get.

Write a formula for your profit after expenses. Let $n =$ the number of cars you wash.

33. One way to estimate which location will have the most customers is to study traffic patterns. The number of cars on a certain street may depend on the day of the week or on the time of day.

a. **Research** Design and carry out an experiment at your three location choices to estimate the number of customers you will have. Include different locations, days, and times.

b. Summarize the results of your experiment. You may want to use a table or spreadsheet.

c. Based only on the number of potential customers, which location is best? What are some reasons to choose another location?

d. Choose one location to be the place for your car wash.

---

c. Maximum area = 289 in.²

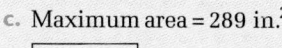

17 in.

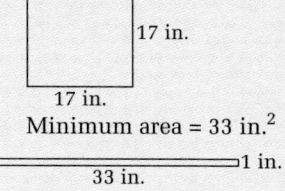

17 in.

Minimum area = 33 in.²

33 in.

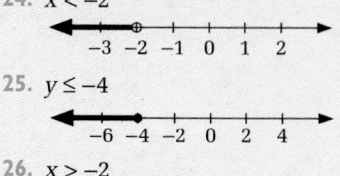

1 in.

d. Answers may vary. Examples are given. (1) Notice that the length decreases by 1 in. and the width increases

by 1 in. from one row to the next. (2) Notice that the sum of the length and the width is always 34 in.

e. Answers may vary.

24. $x < -2$

25. $y \le -4$

26. $x > -2$

27. 80.6%

28. $\dfrac{1}{3}$

29. $\dfrac{5}{2}$

30. $-\dfrac{6}{11}$

31. $\dfrac{7}{9}$

32. Answers may vary. An example is given. $P = cn - E$, where $P =$ profit, $c =$ charge per car, $n =$ no. of cars washed, and $E =$ expenses.

33. a–d. Answers may vary.

Using Reciprocals

PLANNING

FLIP THE SIDE

People who make quilts like to include interesting fabrics in their projects. To get patterns that they may not be able to find in their towns, quilters often exchange quarter-yard pieces of fabric with people around the world.

Some of the most desired fabrics are the uniquely-designed blue and white Japanese fabrics used to make *yukatas*, the traditional Japanese summer kimonos.

BY THE WAY...

When you cut 44 in. wide fabric into half-yard strips and then cut those strips in half, you get a "fat quarter." Quilt stores often sell fat quarters because the pieces are more practical to use than the 9 in. × 44 in. regular quarter-yard strips.

Talk it Over

1. Kimberly Quandt has three yards of fabric to share with the members of her quilt exchange group. How many quarter-yard pieces can she cut from the 3-yd piece of fabric?
$3 \div \frac{1}{4} = \underline{?}$.

2. You have learned that when the product of two numbers is 1, the numbers are called *reciprocals* of each other. Give an example of a number and its reciprocal.

3. What is the reciprocal of 1?

4. Give a reason why *zero* has no reciprocal.

5. In arithmetic, $\frac{5}{6} \div 2 = \frac{5}{6} \cdot \frac{1}{2}$ and $7 \div \frac{2}{3} = 7 \cdot \frac{3}{2}$. Complete the general statement: "Dividing by a number gives the same result as $\underline{?}$ by the $\underline{?}$ of the number."

6. Explain how you can rewrite the product $\frac{1}{6} \cdot 6h$ as h.

A goal in solving equations is to get the variable alone on one side of the equals sign. Sometimes the best way to do this is to use the fact that the product of a number and its reciprocal is 1.

5-6 Using Reciprocals   **275**

Answers to Talk it Over

1. 12

2. Answers may vary. Examples are given. $\frac{2}{5}$ and $\frac{5}{2}$, -3 and $-\frac{1}{3}$, -1 and -1

3. 1

4. Division by zero is undefined.

5. multiplying, reciprocal

6. Explanations may vary. An example is given. Since the factors in a product can be grouped in any way, $\frac{1}{6} \cdot 6h = \left(\frac{1}{6} \cdot 6\right)h$. Since $\frac{1}{6} \cdot 6 = 1$ and multiplying by 1 does not change a number, $\left(\frac{1}{6} \cdot 6\right)h = 1 \cdot h = h$.

PLANNING

Objectives and Strands
See pages 238A and 238B.

Spiral Learning
See page 238B.

Materials List
➤ Scientific or graphics calculator

Recommended Pacing
Section 5-6 is a one-day lesson.

Toolbox References
➤ **Toolbox Skill 9:** Finding Reciprocals

Extra Practice
See pages 628–629.

Warm-Up Exercises
💡 Warm-Up Transparency 5-6

Support Materials
➤ Practice 40
➤ Enrichment 36 in the Activity Bank
➤ Study Guide 5-6
➤ Problem Set 11
➤ Quiz 5-6
➤ Alternative Assessments 1–4

Talk it Over

Use questions 1–4 to review the meaning of reciprocals.

After discussing question 5, have students think about how division, multiplication, and reciprocals are similar to subtraction, addition, and opposites.

After question 6, ask the class how they can write and solve an equation to find the reciprocal of a given number.

Additional Sample

S1 Solve.

a. $14w = 49$

Method 1: Divide both sides by 14.

$14w = 49$

$\dfrac{14w}{14} = \dfrac{49}{14}$

$w = 3.5$

The solution is 3.5.

Method 2: Multiply both sides by the reciprocal of 14.

$14w = 49$

$\dfrac{1}{14} \cdot 14w = \dfrac{1}{14} \cdot 49$

$w = 3.5$

The solution is 3.5.

b. $\dfrac{11}{9}x = 165$

$\dfrac{11}{9}x = 165$

$\dfrac{9}{11}\left(\dfrac{11}{9}x\right) = \dfrac{9}{11}(165)$

$1 \cdot x = 135$

$x = 135$

The solution is 135.

c. $-10 = -\dfrac{4}{5}y + 6$

$-10 = -\dfrac{4}{5}y + 6$

$-10 - 6 = -\dfrac{4}{5}y + 6 - 6$

$-16 = -\dfrac{4}{5}y$

$-\dfrac{5}{4}(-16) = -\dfrac{5}{4}\left(-\dfrac{4}{5}y\right)$

$20 = 1 \cdot y$

$20 = y$

The solution is 20.

Sample 1

Solve.

a. $15 = 6h$ **b.** $-\dfrac{2}{3}x = 97$ **c.** $12 = \dfrac{4}{5}x + 4$

Sample Response

a. Method ❶ Use division.

Divide both sides by 6.

$15 = 6h$

$\dfrac{15}{6} = \dfrac{6h}{6}$

$2.5 = h$

The solution is 2.5.

Method ❷ Use multiplication.

Multiply both sides by the reciprocal of 6.

$15 = 6h$

$\dfrac{1}{6} \cdot 15 = \dfrac{1}{6} \cdot 6h$

$2.5 = h$

The solution is 2.5.

b. $-\dfrac{2}{3}x = 97$

$-\dfrac{3}{2}\left(-\dfrac{2}{3}x\right) = -\dfrac{3}{2}(97)$ ⟵ Multiply both sides by the reciprocal of $-\dfrac{2}{3}$.

$1 \cdot x = -145.5$ ⟵ The product of the reciprocals is 1.

$x = -145.5$ ⟵ $1 \cdot x = x$.

The solution is –145.5.

c. $12 = \dfrac{4}{5}x + 4$

$12 - 4 = \dfrac{4}{5}x + 4 - 4$ ⟵ Subtract 4 from both sides.

$8 = \dfrac{4}{5}x$

$\dfrac{5}{4} \cdot 8 = \dfrac{5}{4} \cdot \dfrac{4}{5}x$ ⟵ Multiply both sides by the reciprocal of $\dfrac{4}{5}$.

$10 = 1 \cdot x$ ⟵ The product of the reciprocals is 1.

$10 = x$ ⟵ $1 \cdot x = x$.

The solution is 10.

Student Resources Toolbox
p. 650 *Fractions*

Talk it Over

7. Are the numbers 4 and 0.25 reciprocals? Explain why or why not.

8. Are the numbers $\dfrac{1}{10}$ and –10 reciprocals? Explain why or why not.

9. Explain how you would solve the equation $-\dfrac{3}{4}x - 19 = -40$.

10. Describe two ways to solve the equation $0.4x + 12 = -34$. Which method do you like the best? Why?

Answers to Talk it Over

7. Yes. $4(0.25) = 1$

8. No. The product is not 1, it is –1.

9. Add 19 to both sides of the equation. Then multiply both sides by $-\dfrac{4}{3}$, the reciprocal of $-\dfrac{3}{4}$.

10. Methods may vary. Examples are given. (1) Subtract 12 from both sides and divide both sides by 0.4, or (2) subtract 12 from both sides, rewrite 0.4 as $\dfrac{4}{10}$ or $\dfrac{2}{5}$, and multiply both sides by the reciprocal $\dfrac{5}{2}$. I prefer (2), using the reciprocal, because dividing by decimals takes too long.

 **Sample 2**

Leela wants to use her new graphics calculator to see the graph of the equation $2x + 3y = 6$. She pressed the ▩ key to enter the equation. Here is what she saw:

:Y1 =
:Y2 =
:Y3 =
:Y4 =

Rewrite the equation so that Leela can enter it on her calculator.

Sample Response

Solve $2x + 3y = 6$ for y.

$$2x + 3y = 6$$
$$2x + 3y - 2x = 6 - 2x \quad \longleftarrow \text{ Subtract } 2x \text{ from both sides.}$$
$$3y = 6 - 2x$$
$$\tfrac{1}{3}(3y) = \tfrac{1}{3}(6 - 2x) \quad \longleftarrow \begin{array}{l}\text{Multiply both sides by the}\\ \text{reciprocal of 3. Then use the}\\ \text{distributive property.}\end{array}$$
$$y = 2 - \tfrac{2}{3}x$$

Leela entered the rewritten equation into her calculator and hit the **GRAPH** key. Here is what she saw:

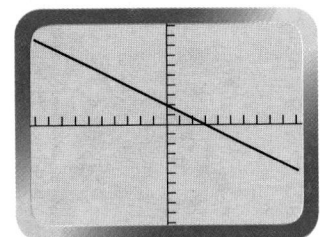

Look Back ◄

How are the equations $\frac{x}{3} = 7$ and $\frac{1}{3}x = 7$ alike? How are they different? Explain how to solve each equation.

5-6 Exercises and Problems

1. **Reading** In part (b) of Sample 1, what is the goal of multiplying both sides by the reciprocal of $-\frac{2}{3}$?

Solve.

2. $-80 = -32x$

3. $\frac{5}{8}y = 20$

4. $-\frac{3}{5}t = -9$

5. $\frac{1}{2}x + 7 = 15$

6. $10 = \frac{1}{5}m + 14$

7. $30 - \frac{2}{3}y = 10$

5-6 Using Reciprocals   **277**

Answers to Look Back

Answers may vary. An example is given. The equations are alike because they have the same solution, 21; each involves the variable x, one operation, and has 7 on the right side. They are different in that $\frac{x}{3} = 7$ involves division by 3, whereas $\frac{1}{3}x = 7$ involves multiplication by $\frac{1}{3}$. Each equation can be solved by multiplying both sides by 3.

Answers to Exercises and Problems

1. The goal is to get the variable alone.
2. 2.5
3. 32
4. 15
5. 16
6. −20
7. 30

Additional Sample

S2 Xiong wants to graph the equation $-3x + 4y = 12$ on his graphics calculator. How can he rewrite the equation so that he can enter it? Xiong can solve the equation for y.

$$-3x + 4y = 12$$
$$-3x + 4y + 3x = 12 + 3x$$
$$4y = 12 + 3x$$
$$\tfrac{1}{4}(4y) = \tfrac{1}{4}(12 + 3x)$$
$$y = 3 + \tfrac{3}{4}x$$

Look Back

Use the Look Back questions to achieve closure on the idea that dividing by a number gives the same result as multiplying by its reciprocal. Ask students if the given equations are equivalent and why.

APPLYING

Suggested Assignment
Standard 1–18, 21–31
Extended 1–31

Integrating the Strands
Number Exs. 1, 14, 15, 21
Algebra Exs. 1–31
Measurement Exs. 17, 18
Discrete Mathematics Exs. 30, 31
Logic and Language Exs. 1, 15, 16, 21

Solve each equation for the variable shown in red.

8. $5x + 6y = 30$

9. $-4y + 9x = 18$

10. $x - \frac{2}{3}y = 10$

11. $7x + 2y = 15$

12. $y = -\frac{1}{3}x - 8$

13. $y = \frac{5}{2}x + \frac{5}{2}$

14. Find a positive number that is equal to its reciprocal. Is there a negative number that is equal to its reciprocal? If so, which one?

15. Explain how you could use the division key on your calculator instead of the 1/x or x⁻¹ key to find the reciprocal of 2.5.

16. **Writing** Kim tried to solve the equation $-3w = 12$ by adding 3 to both sides. She got $w = 15$. Explain what is wrong with her method. Show how you would solve the equation.

17. The formula $y = 12x$ converts x feet to y inches. Find a formula that converts y inches to x feet.

18. The formula $y = \frac{3}{5}x$ converts x kilometers to approximately y miles. Find a formula that converts y miles to approximately x kilometers.

connection to **MUSIC**

19. Sound waves vibrate at a certain frequency when instruments produce musical notes. Two notes are a "perfect fifth" apart when the frequency of the lower note (l) is $\frac{2}{3}$ the frequency of the higher note (h).

    a. Write a formula to describe this situation.

    b. Solve the formula for h.

    c. Find the frequency of the lower note when the higher note has a frequency of 440 vibrations per second.

    d. Find the frequency of the higher note when the lower note has a frequency of 440 vibrations per second.

20. The sound that you hear when a violin string is played is determined by the length of the string that is allowed to vibrate.

    a. Write a formula to describe this situation.

    b. Rewrite the formula to show the part of the E string that vibrates in relation to the length of the string that vibrates for an F note.

Length for an F note = $\frac{9}{10}$ of the length for an E note

G D A E

278   **Unit 5** Equations for Problem Solving

Answers to Exercises and Problems

8. $x = 6 - \frac{6}{5}y$

9. $y = \frac{18 - 9x}{-4}$

10. $y = -15 + \frac{3}{2}x$

11. $y = \frac{15 - 7x}{2}$

12. $x = -3y - 24$

13. $x = \frac{2}{5}y - 1$

14. 1; Yes; −1

15. Enter "1" and then press the ÷ key. Then enter "2.5" and press the = key. This gives 0.4, the reciprocal of 2.5.

16. Explanations may vary. An example is given. Kim probably used addition to undo multiplication, so her method does not produce the correct result. I would solve the equation

$-3w = 12$ by dividing each side by −3 or by multiplying each side by $-\frac{1}{3}$. Either method gives $w = -4$, so the solution of the equation is −4.

17. $x = \frac{y}{12}$ or $x = \frac{1}{12}y$

18. $x = \frac{5}{3}y$

19. a. $l = \frac{2}{3}h$

    b. $h = \frac{3}{2}l$

c. $293.\overline{3}$

d. 660

20. a. $F = \frac{9}{10}E$

    b. $E = \frac{10}{9}F$

21. Answers may vary. An example is given. Reciprocals are two numbers whose product is 1; for example, 2 and $\frac{1}{2}$, $\frac{2}{3}$ and $\frac{3}{2}$, and −8 and $-\frac{1}{8}$. The reciprocal of a positive number

Ongoing ASSESSMENT

21. Writing Write all you know about using reciprocals. Include two examples of reciprocals and an example showing how to use reciprocals to solve equations and formulas.

Review PREVIEW

Solve each equation for the variable shown in red. *(Section 5-5)*

22. $P = 2s + b$

23. $-6x + 5y = 46$

24. $4x - 2y = 28$

25. Describe the pattern in the table. *(Section 1-2)*

x	−5	−2	0	3	7	10
y	−10	−4	0	6	14	20

Evaluate each expression when $a = 5$, $b = 10$, and $c = \frac{1}{3}$. *(Section 2-2)*

26. $a^2 b$

27. $\frac{1}{2}ab$

28. $\frac{1}{5}(a + b)$

29. $c(b + 32)$

Working on the Unit Project

30. Suppose one student can wash a car in 20 min. In one minute the student can wash $\frac{1}{20}$ of the car.

rate of washing $= \frac{1}{20}$ car per minute

a. What part of a car can 10 students working together wash in one minute?

b. Do you think 20 students working together can wash a whole car in one minute? Why or why not?

31. a. Research Suppose a group of students is washing a car together.

group's rate of washing $= \dfrac{1}{\text{group's time to wash a car}}$

Experiment with different numbers of students. Find the different group rates of car washing. Decide how many people are needed to wash a car quickly and efficiently.

b. Using a group the size you choose in part (a), how long will it take you to wash enough cars to reach your goal?

c. Will there be enough time in one day for your group to reach your fund-raising goal if only one group of students washes cars? Why or why not?

d. If you do not have enough time, what are some things you could do to reach your goal?

5-6 Using Reciprocals

279

Practice 40 For use with Section 5-6

Name _____ Date _____

Practice 40
For use with Section 5-6

Solve.

1. $-6x = 15$ −2.5
2. $\frac{7}{8}w = -24$ −32
3. $-\frac{2}{3}k = -18$ 27
4. $14 = \frac{3}{5}y - 1$ 25
5. $-\frac{5}{6}p + 2 = -23$ 30
6. $8 = -\frac{3}{7}c - \frac{1}{7}$ −19

Solve each equation for the indicated variable.

7. $A = \frac{1}{2}bh$ for b $b = \frac{2A}{h}$
8. $V = \frac{1}{3}b^2h$ for h $h = \frac{3V}{b^2}$
9. $p = \frac{1}{2}mv^2$ for m $m = \frac{2p}{v^2}$
10. $V = \frac{1}{6}lwh$ for w $W = \frac{6V}{lh}$
11. $A = \frac{2}{5}k$ for k $k = \frac{5}{2}A$
12. $m = -\frac{3}{4}v$ for v $v = -\frac{4}{3}m$
13. $y = -\frac{2}{3}x + b$ for x $x = -\frac{3}{2}(y - b)$
14. $A = \frac{1}{2}(a + b)h$ for a $a = \frac{2A}{h} - b$
15. $g = \frac{v^2}{2h}$ for h $h = \frac{v^2}{2g}$

Roberto and two roommates ordered take-out shrimp, and the three agreed to split the cost (c) equally. Let $s =$ Roberto's share of the cost.
16. Write an equation that describes this situation. $s = \frac{c}{3}$
17. Rewrite the formula to solve for c. $c = 3s$
18. Suppose Roberto's share was $2.63. Find the total cost of the shrimp. $7.89

Carlotta Mendez overhauled her tractor and mowed two of her five equal-sized fields. Overhauling the tractor took her 1.5 h. Let $t =$ the time it takes her to mow all five fields and let $s =$ the time for the work she has already done.
19. Write an equation that describes this situation. $s = 1.5 + \frac{2}{5}t$
20. Rewrite the equation to solve for t. $t = \frac{5}{2}(s - 1.5)$
21. Suppose the work she has already done took Carlotta 6.5 h. How long does it take her to mow all five fields? 12.5 h
22. Open-ended On a calculator, enter any number other than 0 or 1. Press the reciprocal key. Press the reciprocal key again. What do you notice? Try this with other numbers. What can you say in general about pressing the reciprocal key twice? Try starting with the number $\frac{\sqrt{5}+1}{2}$. What is unusual about the reciprocal of this number and the number itself?

22. If you enter a nonzero number and then press the reciprocal key twice, you return to the original number.
$\frac{\sqrt{5}+1}{2}$ and its reciprocal differ by exactly 1.

Answers to Exercises and Problems

is a positive number, and the reciprocal of a negative number is a negative number. Reciprocals can be used to solve equations and formulas. Dividing by a number is the same as multiplying by its reciprocal. For example, to solve $\frac{5}{3}x - 4 = 11$, add 4 to both sides, then multiply both sides by $\frac{3}{5}$, the reciprocal of

$\frac{5}{3}$. To solve $C = \frac{22}{7}d$ for d, multiply both sides by $\frac{7}{22}$.

22. $\dfrac{P - b}{2} = s$

23. $\dfrac{46 - 5y}{-6} = x$

24. $2x - 14 = y$

25. $y = 2x$ **26.** 250

27. 25   **28.** 3   **29.** 14

30. a. $\frac{1}{2}$ of a car

b. Answers may vary. An example is given. Yes, if they work at the same rate. In any event, they would get in each others' way!

31. a–d. Answers may vary.

PLANNING

Objectives and Strands
See pages 238A and 238B.

Spiral Learning
See page 238B.

Materials List
- Rectangular sheets of paper
- Scissors
- Graph paper

Recommended Pacing
Section 5-7 is a two-day lesson.

Day 1

Pages 280–283: Exploration through Talk it Over 3, *Exercises 1–8*

Day 2

Pages 283–284: Area of a Trapezoid through Look Back, *Exercises 9–30*

Extra Practice
See pages 628–629.

Warm-Up Exercises
- Warm-Up Transparency 5-7

Support Materials
- Practice 41
- Enrichment 37 in the Activity Bank
- Study Guide 5-7
- Problem Set 11
- Diagram Master 2 in the Explorations Lab Manual
- Overhead Visual 6
- Quiz 5-7
- Alternative Assessments 5, 6

Section 5-7

Area Formulas

THE INSIDE SCOOP

Focus

Use the formulas for the area of a parallelogram, a triangle, and a trapezoid to solve problem situations.

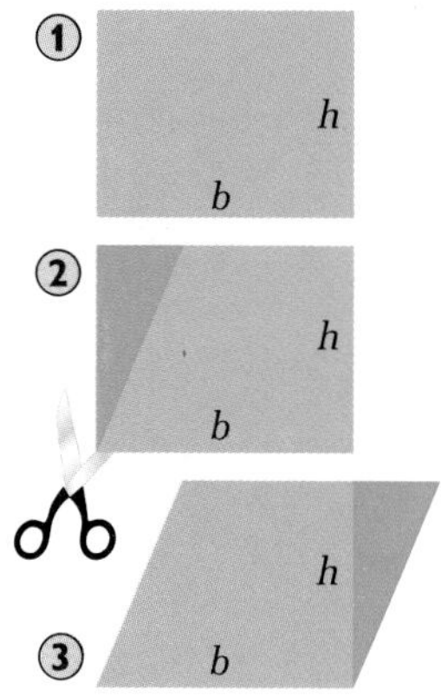

EXPLORATION

How can you cut a parallelogram to find its area?

- **Materials: rectangular sheets of paper, scissors**
- **Work with another student.**

① The length and width of a rectangle are also called the base (b) and height (h). Label your paper with b and h as shown. What is a formula for the area of your rectangle?

② Make a fold in your paper to form a triangle. Cut along the fold line.

③ Slide or translate the triangle across your paper. What kind of quadrilateral is formed?

④ What is a formula for the area of the quadrilateral you created in step 3? Explain how you arrived at your answer.

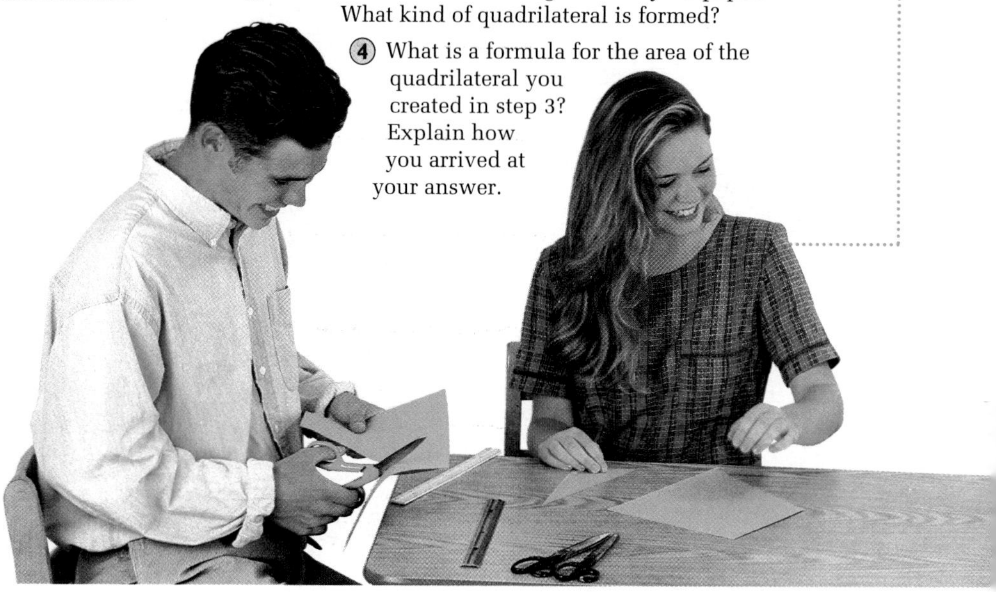

Answers to Exploration

1. $A = bh$
2. manual activity
3. parallelogram
4. $A = bh$; explanations may vary. An example is given. The parallelogram and rectangle have different shapes. But their areas must be the same since the figures are made from the same pieces of paper.

Any side of a parallelogram may be called a **base.** The **height** of a parallelogram is the distance to the base from a point on the opposite side.

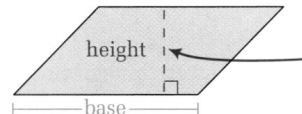

The *distance from a point to a line* is the length of the perpendicular segment from the point to the line.

Sample 1

Find the area of the parallelogram.

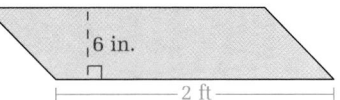

6 in.

2 ft

Sample Response

Problem Solving Strategy: Use a formula

All the measures used must be in the same units, so write 2 ft as 24 in.

$A = bh$ ⟵ Area of a parallelogram = base × height.

$= 24(6)$ ⟵ Substitute 24 for b and 6 for h.

$= 144$

The area is 144 in.²

Sample 2

Find the height of a parallelogram with base 10 cm and area 25 cm².

Sample Response

Problem Solving Strategy: Use a formula

$A = bh$ ⟵ Area of a parallelogram = base × height.

$25 = 10h$ ⟵ Substitute 25 for A and 10 for b.

$\dfrac{25}{10} = \dfrac{10h}{10}$ ⟵ Divide both sides by 10.

$2.5 = h$

The height is 2.5 cm.

Talk it Over

1. **a.** Solve the formula $A = bh$ for h.

   b. Explain how to use the formula you wrote in part (a) to solve Sample 2.

2. After doing the Exploration, Brandon decides that the width of a rectangle is the same as the height of a parallelogram. Do you agree? Why or why not?

5-7 Area Formulas

281

TEACHING

Exploration

The purpose of the Exploration is to have students find a formula for the area of a parallelogram. This is done by cutting a rectangle apart and reassembling it to form a parallelogram.

After doing this activity, ask students whether they could cut a parallelogram apart and reassemble it to get a rectangle.

Additional Samples

S1 Find the area of the parallelogram.

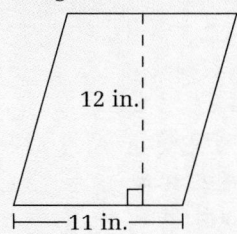

12 in.

11 in.

$A = bh$
$A = 11 \times 12$
$A = 132$
The area is 132 in².

S2 Find the base of a parallelogram with height 7 m and area 112 m².

$A = bh$
$112 = b \cdot 7$
$\dfrac{112}{7} = \dfrac{b \cdot 7}{7}$
$16 = b$
The base is 16 m.

Answers to Talk it Over

1. **a.** $\dfrac{A}{b} = h$

   b. Replace A with 25 and b with 10, then divide.

2. Answers may vary. An example is given. I agree because the width of a rectangle is the perpendicular distance between two opposite sides. This is the same as the height of the parallelogram (the distance to the base from a point on the opposite side).

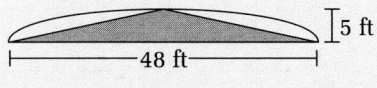
Area of a Triangle

You can think of a triangle as half of a parallelogram. The parallelogram and the yellow triangle have the same base and height.

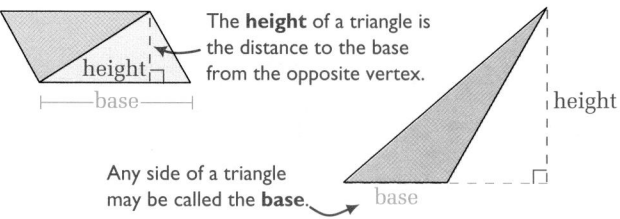

The **height** of a triangle is the distance to the base from the opposite vertex.

height

base

Any side of a triangle may be called the **base**.

base

height

Sometimes you need to extend the base to draw the height.

AREA OF A PARALLELOGRAM

Area of a parallelogram = base × height

$A = bh$

AREA OF A TRIANGLE

Area of a triangle = $\frac{1}{2}$ × base × height

$A = \frac{1}{2}bh$

Sample 3

A neighborhood committee wants to grow wildflowers on the traffic island shown. How many cans of wildflower seeds do they need to cover the traffic island?

SUMMER STREET
AUTUMN DRIVE
WINTER AVENUE
10'

Sample Response

Problem Solving Strategy: Use a formula

$A = \frac{1}{2}bh$ ← Area of a triangle = $\frac{1}{2}$ × base × height.

$= \frac{1}{2}(20)(30)$ ← Substitute 20 for b and 30 for h.

$= 300$

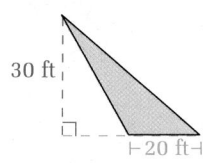

30 ft

20 ft

The area of the traffic island is 300 ft². Each can of wildflower seeds covers 80 ft².

$300 \div 80 = 3.75$

The committee needs 4 cans of wildflower seeds.

3. a. In Sample 3, why is the height drawn outside the triangle?

   b. The triangle in Sample 3 is obtuse. Can the height of an obtuse triangle lie inside the triangle?

   c. Can the height of a right triangle or an acute triangle lie outside the triangle?

······► Now you are ready for:
Exs.1–8 on p. 285

Area of a Trapezoid

Architect I. M. Pei was asked to design the new East Wing of the National Gallery of Art. The plot for the new building was in the shape of a trapezoid. This presented a design problem for the architect.

Pei came up with a solution when sketching a trapezoid on the back of an envelope. He drew a diagonal of the trapezoid. When he saw the two pieces, he had the idea to use the smaller triangle for the study center and the larger triangle for exhibits.

In Exercise 11, you will divide a trapezoid into two pieces to get the formula for the area.

AREA OF A TRAPEZOID

Area of a trapezoid $= \frac{1}{2} \times$ the sum of the bases $\times$ height

$$A = \frac{1}{2}(b_1 + b_2)h$$

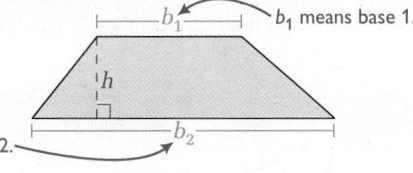

b_1 means base 1.

b_2 means base 2.

Answers to Talk it Over ······:

3. a. A height drawn from a vertex must be perpendicular to the base opposite the vertex. Sometimes the base must be extended in order to meet the height at a right angle.

   b. Yes.

   c. No.

Question 3 points out that heights of triangles can lie outside or on the triangle. It also shows that different sides may be the base, which leads to different heights.

Multicultural Note

Ieoh Ming Pei was born in Canton, China in 1917, came to the U.S. in 1935, and became a U.S. citizen in 1954. Drawing on memories of Chinese gardens of his childhood, he used simple, often sculptural, geometric forms in his designs. Besides the East Wing of the National Gallery of Art in Washington, D.C., he also designed the John F. Kennedy Library and the John Hancock Tower, both located in Boston, and the famous glass pyramid entrance to the Louvre Museum in Paris.

S4 Find the area of the trapezoid.

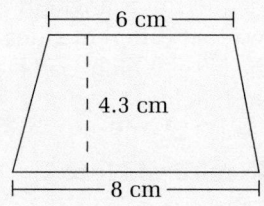

$A = \frac{1}{2}(b_1 + b_2)h$

$A = \frac{1}{2}(8 + 6)(4.3)$

$A = \frac{1}{2}(14)(4.3)$

$A = 30.1$

The area is 30.1 cm².

S5 The side faces of the control tower at an airport are trapezoidal. Each side face has an area of 425 ft². The edges along the floor each measure 18 ft. The trapezoids each have a height of 17 ft. What is the measure of each edge along the ceiling?

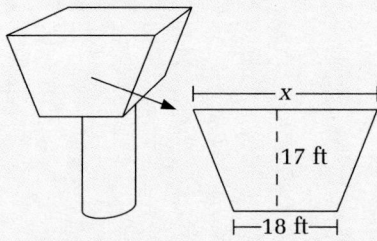

Let x = length in feet of edge along the ceiling.

$A = \frac{1}{2}(b_1 + b_2)h$

$425 = \frac{1}{2}(18 + x)17$

$850 = (18 + x)17$

$850 = 306 + 17x$

$544 = 17x$

$32 = x$

The edges along the ceiling each measure 32 ft.

Look Back

Students can use this activity as a journal entry. Look for correct labeling and for figures that are not special cases; for example, a rectangle should not be represented by a square.

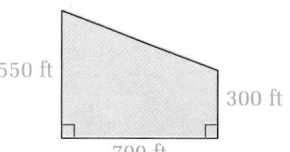

Find the approximate area of the plot of land where the East Wing of the National Gallery was built.

Sample Response

Problem Solving Strategy: Use a formula

$A = \frac{1}{2}(b_1 + b_2)h$ ←— Area of a trapezoid = $\frac{1}{2}$ × sum of the bases × height.

$A = \frac{1}{2}(550 + 300)(700)$ ←— Substitute 550 for b_1, 300 for b_2, and 700 for h.

$A = \frac{1}{2}(850)(700)$

$A = 297,500$

The area is about 297,500 ft².

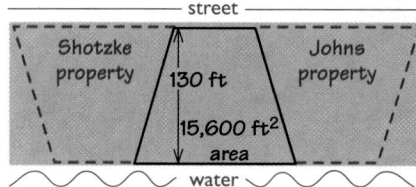

Janelle Rose wants to buy the trapezoidal plot of land shown. She knows that the border along the water is twice as long as the border along the street. How long is the border along the water?

Sample Response

Problem Solving Strategy: Use a formula

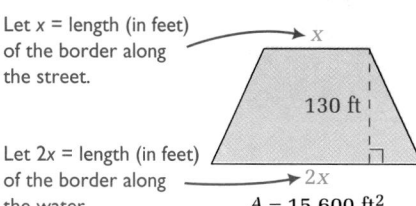

Let x = length (in feet) of the border along the street.

Let 2x = length (in feet) of the border along the water.

$A = 15,600 \text{ ft}^2$

$A = \frac{1}{2}(b_1 + b_2)h$ ←— Area of a trapezoid

$15,600 = \frac{1}{2}(x + 2x)130$ ←— Substitute 15,600 for A and 130 for h. Substitute x for b_1 and 2x for b_2.

$15,600 = \frac{1}{2}(3x)130$ ←— Combine like terms. Since x = 1x, x + 2x = 3x.

$15,600 = 195x$ ←— Simplify: $\frac{1}{2} \cdot 3 \cdot 130 = 195$

$\frac{15,600}{195} = \frac{195x}{195}$ ←— Divide both sides by 195.

$80 = x$

The border along the street is 80 ft. The border along the water is twice as long as the border along the road, or 2 × 80. The border along the water is 160 ft.

Look Back ←—

Write all the area formulas you know. Draw diagrams to help someone else understand the formulas.

⋯⋯▶ Now you are ready for:
Exs. 9–30 on pp. 285–288

Answers to Look Back ⋯⋯⋯⋯⋯⋯⋯⋯⋯⋯⋯⋯

Answers may vary. Examples are given.

Area Formulas

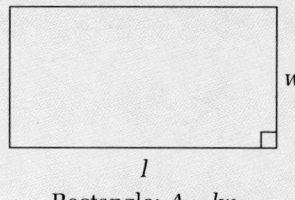

Rectangle: $A = lw$

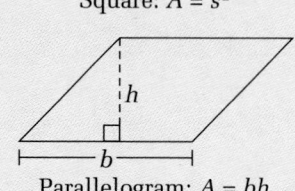

Parallelogram: $A = bh$

Square: $A = s^2$

Triangle: $A = \frac{1}{2}bh$

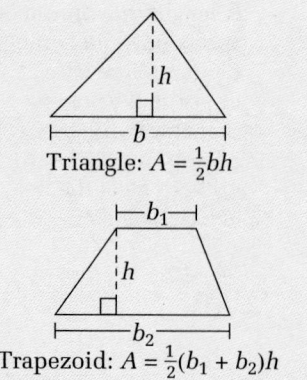

Trapezoid: $A = \frac{1}{2}(b_1 + b_2)h$

Exercises and Problems

1. **Reading** On page 281, read about the distance from a point to a line. Describe the segment you would measure to find the distance from point *P* to segment $\overline{AB}$.

a.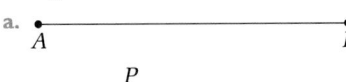

b.

Find the area of each parallelogram.

2.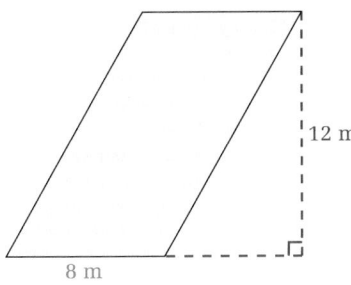

3.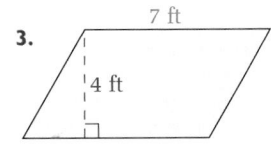

4. Find the area of a parallelogram with base 2.8 in. and height 6 in.

5. Find the base of a parallelogram with area 127.4 cm² and height 24.5 cm.

6. Find the area of a triangle with base 6.8 cm and height 3 cm.

7. Find the height of a triangle with area 28.5 cm² and base 6 cm.

8. Leela and Corey are finding the area of the same parallelogram. Leela calls the side closest to her the base. Corey calls the side closest to him the base. Both students use the formula $A = bh$.

   a. Are Leela's values for *b* and *h* the same as Corey's? Why or why not?

   b. Will they get the same area? Explain.

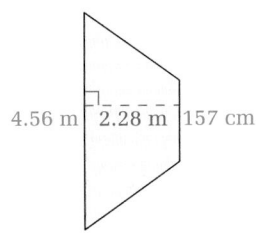
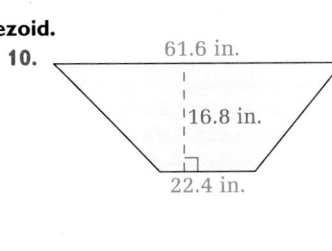

For Exercises 9 and 10, find the area of each trapezoid.

9.
4.56 m   2.28 m   157 cm

10.
61.6 in.
16.8 in.
22.4 in.

APPLYING

Suggested Assignment
Standard 1–10, 14–19, 21–30
Extended 1–10, 12–19, 21–30

Integrating the Strands
Algebra Exs. 16–19, 21–23, 25–27
Measurement Exs. 1–10, 12, 14, 15, 20
Geometry Exs. 1–18, 20, 24
Logic and Language Exs. 1, 13, 15, 20, 30

Reasoning
In Ex. 8, students learn to reason that in a parallelogram any side may be labeled as the base. Figures placed in different orientations will also help students to think in general terms and not draw conclusions based upon position.

Answers to Exercises and Problems

1. a. the segment drawn from *P* perpendicular to $\overline{AB}$
   b. same as (a)

2. 96 m²

3. 28 ft²

4. 16.8 in.²

5. 5.2 cm

6. 10.2 cm²

7. 9.5 cm

8. a. No. The students chose different sides to use as the base, and these sides are not equal in length. Leela's value for *b* is greater than Corey's, and her value for *h* is less than Corey's.

   b. Yes. Explanations may vary. An example is given. The area of a parallelogram does not depend on which side is chosen as the base.

9. 69,882 cm² or 6.9882 m²

10. 705.6 in.²

11. a. **Using Manipulatives** Draw any trapezoid ABCD on a piece of paper and label it as shown. Cut out the trapezoid.

    b. Find the midpoint of $\overline{BC}$ by folding and label it M. Then cut the trapezoid into two pieces by cutting along $\overline{DM}$. Rotate △DCM, as shown, to form a large triangle.

    c. Write an expression for the length of the base of the large triangle you made. Write an expression for the area of this triangle.

    d. **Writing** Explain how you can get the formula for the area of a trapezoid by using the formula for the area of a triangle.

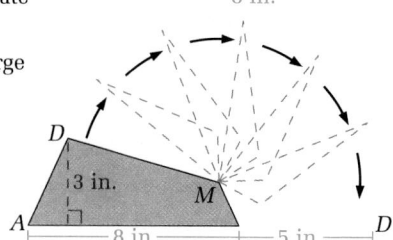

12. A sign manufacturer needs to fill an order for 200 No Passing Zone signs. The diagram shows the dimensions required by the United States Department of Transportation.

    a. The signs are made from a sheet of metal. What is the maximum number of signs the manufacturer can cut from a 45 in. by 72 in. sheet? Make a sketch on graph paper to support your answer.

    b. How many 45 in. by 72 in. sheets are needed to make 200 signs?

    c. In part (b) how many square inches of metal will be left over after making the 200 signs?

13. **Open-ended** The United States Department of Transportation chooses the shape for a sign based on the danger of the traffic situation. Think of as many different traffic signs as you can. Give the shape and message of each sign. Describe any patterns you notice.

14. Kenisha and Joel are trying to find the area of a School Crossing sign. Each student started by drawing a diagram.

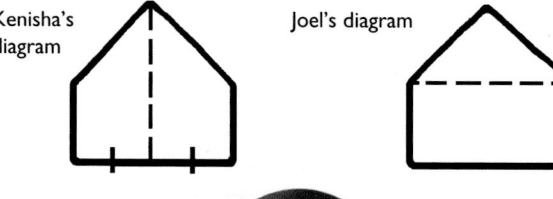

Kenisha's diagram   Joel's diagram

    a. Find the area of the School Crossing sign using Kenisha's method.

    b. Find the area of the School Crossing sign using Joel's method.

    c. **Writing** Describe how the methods are different. Which method do you think is easier? Why?

286   Unit 5 Equations for Problem Solving

Answers to
Exercises and Problems

11. a, b. manual activity

   c. $8 + 5$; $\frac{1}{2}(8 + 5)(3)$

   d. Explanations may vary. An example is given. The formula for the area of a triangle is $A = \frac{1}{2}bh$. By cutting a trapezoid in the manner given, you can create a triangle that has a base formed from the two bases of the trapezoid. The triangle and trapezoid have equal heights and equal areas. So $A = \frac{1}{2}bh = \frac{1}{2}(b_1 + b_2)h$, where b_1 and b_2 represent the bases of the trapezoid.

12. a. 3 signs (see sketch below)

   b. 67 sheets

   c. 56,880 in.^{2}

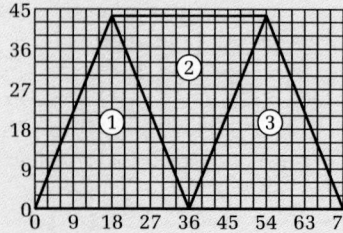

13. Answers may vary. An example is given. Yield signs have 3 sides (triangle), caution signs have 4 sides (rhombus), stop signs have 8 sides (octagon); it seems the more dangerous the situation, the more sides the sign has.

14. a. $2 \cdot \left[\frac{1}{2}(b_1 + b_2)h\right] =$
       $2 \cdot \left[\frac{1}{2}(30 + 15)15\right] =$
       675 in.^{2}

   b. $15 \cdot 30 + \frac{1}{2}(15 \cdot 30) =$
      675 in.^{2}

   c. Explanations may vary. An example is given. Kenisha used two trapezoids; Joel used a rectangle and a triangle. Joel's way is easier. It's easier to find the area of a triangle and a rectangle than to find the area of two trapezoids.

15. The town council of Massapoag wants to build a bandstand. Architects were asked to submit plans. In the plan shown, the floor has the shape of a regular octagon.

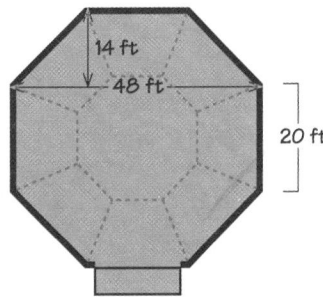

14 ft
48 ft
20 ft

a. What is the area of the floor of the bandstand in this plan?

b. Compare the perimeter of the bandstand in this plan with the perimeter of a square bandstand with the same area.

c. **Writing** Suppose the cost of the bandstand depends on the length of the railing. Which shape costs the least? Write a proposal to the town council about the shape of the bandstand. Give reasons to support your proposal.

For Exercises 16–18, find the missing dimension(s).

16.
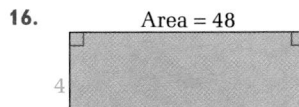
Area = 48
4
$2(x - 3)$

17.
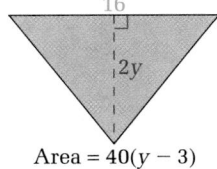
16
$2y$
Area = $40(y - 3)$

18.
$x + 2$
$A = 28$
4
$4x - 1$

19. Explain how to solve $A = \frac{1}{2}(b_1 + b_2)h$ for h.

Ongoing **ASSESSMENT**

20. **Group Activity** Work with another student. Use the diagram to answer parts (a)–(e).

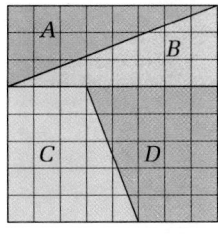
A
B
C
D

a. Find the areas of triangles A and B and trapezoids C and D.

b. What is the area of the big square?

c. **Using Manipulatives** Draw the triangles and trapezoids on graph paper. Carefully cut out each piece. Rearrange the pieces to form a rectangle 13 units long and 5 units wide. What is the area of the rectangle you formed?

d. Compare the area of the original square and the area of the rectangle. Are the areas the same or different? Did the areas of the triangle and trapezoid pieces change?

e. **Writing** Look up the word *paradox* in a dictionary. Do you think this type of problem is a paradox? Why?

5-7 Area Formulas

287

Career Note

As Ex. 15 points out, architects draw plans for structures that are used by people. They learn their skills by studying for four years or more in departments of architecture at colleges or universities.

Architects must be skilled in mathematics, mechanical drawing, many different engineering subjects, design, and financial matters that relate to the cost of materials. Architects generally specialize in designing residential homes or commercial office buildings. The major buildings in the great cities of the United States and throughout the world demonstrate the influence that architects can have on the environment and culture.

Assessment: Task

For Ex. 20, the written responses of pairs of students can be read to the class on a voluntary basis and discussed. Have students think about the source of the paradoxical results of the exercise. Ask students who have pinpointed the source of the problem to share their ideas with the class.

Answers to Exercises and Problems

15. a. 1912 ft^2

b. The perimeter of the square bandstand would be about 15 ft greater.

c. Recommendations may vary. An example is given. The perimeter of the octagonal bandstand is less than that of a square bandstand. The octagonal shape will probably cost less.

16. length: 12 units

17. height: 10 units; area: 80 square units

18. bases: 4.6 units and 9.4 units

19. Explanations may vary. An example is given. Multiply both sides by 2, then divide both sides by $(b_1 + b_2)$ to get $h = \dfrac{2A}{b_1 + b_2}$

20. a. $A = B = 12$ units^2; trap $C =$ trap $D = 20$ units^2

b. 64 units^2

c. Area = 65 units^2

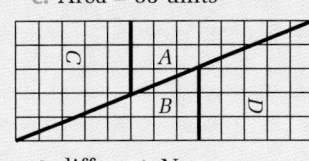
C
A
B
D

d. different; No.

e. Answers may vary. An example is given. A paradox is a seemingly self-contradicting situation, and this problem seems to be an example. It seems that the total area of the pieces depends on their arrangements. However, when I examine the rectangle more closely I see that the pieces do not quite fit together along the diagonal. This small gap must account for the extra square unit.

287

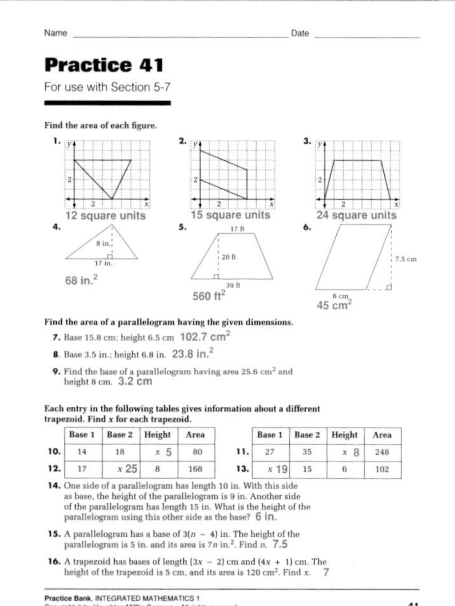

Review PREVIEW

Solve. *(Section 5-6)*

21. $\frac{2}{3}y = 50$

22. $-6 = \frac{1}{2}x + 2$

23. $14 = -\frac{3}{4}x + 2$

24. A quadrilateral has vertices at (0, 4), (−5, 1), (−1, −6), and (2, 4). If the quadrilateral is translated 3 units to the left, find the new vertices. *(Section 4-3)*

Solve each equation for y. *(Sections 5-5, 5-6)*

25. $2x + y = 20$

26. $x + 6y = -36$

27. $16 = \frac{4}{5}y - 4x$

Working on the Unit Project

28. Make a list of the number of students you need to work at the car wash. You may want to consider people for these jobs: advertisers, car washers, fee collectors, traffic controllers, and "runners" who bring dry towels and other supplies.

29. Make a diagram of the car wash area. Show where students will work. Also mark where you will have signs and the location of supplies (including the location of the water supply). Make sure each student has a big enough work area.

30. Decide on the hours of your car wash. Make a schedule listing the jobs and the time periods. Write in the number of students needed for each job during each time period.

288   Unit 5 Equations for Problem Solving

Answers to Exercises and Problems

21. 75

22. −16

23. −16

24. (−3, 4), (−8, 1), (−4, −6), (−1, 4)

25. $y = 20 - 2x$

26. $y = -6 - \frac{x}{6}$

27. $y = \frac{5}{4}(16 + 4x)$ or $y = 20 + 5x$

28. Answers may vary.

29. Answers may vary.

30. Answers may vary.

5-8

Systems of Equations in Geometry

Focus
Model situations and solve equations with two variables using substitution.

THE ANGLE CONNECTION

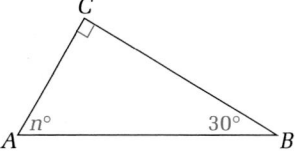

PLANNING

Objectives and Strands
See pages 238A and 238B.

Spiral Learning
See page 238B.

Recommended Pacing
Section 5-8 is a one-day lesson.

Extra Practice
See pages 628–629.

Warm-Up Exercises
Warm-Up Transparency 5-8

Support Materials
➤ Practice 42
➤ Enrichment 38 in the Activity Bank
➤ Study Guide 5-8
➤ Problem Set 11
➤ Quiz 5-8
➤ Test 19

Talk it Over

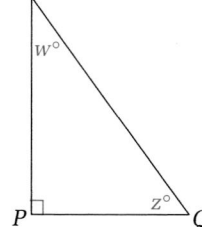

1. a. How are the measures of the acute angles of △ABC related?

   b. Write an equation that models this relationship.

   c. What is the measure of ∠A?

2. a. Write an equation that models the relationship between the measures of the acute angles of △PQR.

   b. Can you use this equation to find the unknown angle measures? Why or why not?

   c. Suppose that the measures of the acute angles of △PQR are equal. Write an equation that models this relationship. Find the measures of ∠Q and ∠R.

Some situations can be modeled by writing one equation that has one variable. To model other situations you may need more than one variable and write a *system* of two equations.

5-8 Systems of Equations in Geometry **289**

Answers to Talk it Over

1. a. Their sum is 90°.

   b. $n + 30 = 90$

   c. 60°

2. a. $w + z = 90$

   b. No; reasons may vary. The equation has two variables, and to find the unknown measures you need to write the equation in terms of only one variable.

   c. Answers may vary. An example is given.
   $w + w = 90$ or $2w = 90$; 45°, 45°

Mathematical Procedures

The procedure of using substitution to solve a system of equations can be related to the procedure of evaluating a variable expression by substituting numbers for the variables.

Additional Sample

The measure of one acute angle of a right triangle is 12° more than the measure of the other acute angle. Find the measure of each acute angle.

Let x be the measure of the smaller angle and y be the measure of the larger angle.

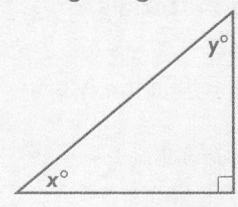

$$y = x + 12$$
$$x + y = 90$$

Substitute $x + 12$ for y in the equation $x + y = 90$.

$$x + (x + 12) = 90$$
$$2x + 12 = 90$$
$$2x + 12 - 12 = 90 - 12$$
$$2x = 78$$
$$\frac{2x}{2} = \frac{78}{2}$$
$$x = 39$$

Substitute 39 for x in $y = x + 12$.

$$y = 39 + 12$$
$$y = 51$$

The measures of the two acute angles, x and y, are 39° and 51°, respectively.

Sample

The measure of an acute angle of a right triangle is four times the measure of the other acute angle. Find the measure of each angle.

Sample Response

Problem Solving Strategy: Use a system of equations

Sketch and label a right triangle.

Let x and y be the measures of the two acute angles.

Think about what you know about the relationship between x and y.
Write two equations to show these relationships.

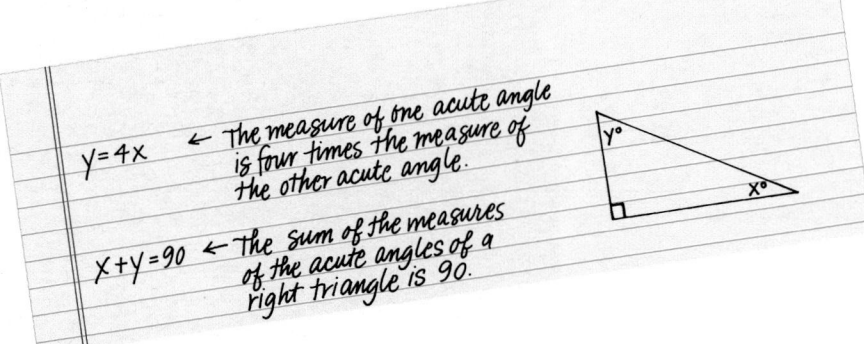

You can solve this system of equations with a method called *substitution*.
Notice that the equation $y = 4x$ tells you that $4x$ is another name for y.

$$x + 4x = 90 \qquad \longleftarrow \text{Substitute } 4x \text{ for } y \text{ in the equation } x + y = 90.$$
$$5x = 90 \qquad \longleftarrow \text{Combine like terms.}$$
$$\frac{5x}{5} = \frac{90}{5} \qquad \longleftarrow \text{Divide both sides by 5.}$$
$$x = 18$$

To find y, substitute 18 for x in either of the original equations.

$$y = 4x \qquad \longleftarrow \text{Suppose you chose the equation } y = 4x.$$
$$y = 4(18) \qquad \longleftarrow \text{Substitute 18 for } x.$$
$$y = 72$$

Check Substitute the values of x and y in the other equation.

$$x + y = 90 \qquad \longleftarrow \text{Since you used } y = 4x \text{ to substitute 18 for } x,$$
$$\text{check the solution with } x + y = 90.$$
$$18 + 72 \stackrel{?}{=} 90 \qquad \longleftarrow \text{Substitute 18 for } x \text{ and 72 for } y.$$
$$90 = 90 \; \checkmark$$

Both equations are true when $x = 18$ and $y = 72$.

The measures of the two acute angles are 18° and 72°.

Solutions of Systems

Two or more equations that state relationships that must all be true at the same time are called a **system of equations.** The values of the variables that make both equations true at the same time are the **solution of a system.**

$$y = 4x$$
$$x + y = 90$$

a system of equations

$$x = 18, y = 72$$

the solution of the system

Talk it Over (sidebar)

Question 4 shows that different methods can be used to solve the same problem. Point out that any method that works is acceptable. Some methods, however, may be quicker or easier to use than others.

Look Back (sidebar)

In groups, have students discuss the relationship between the one-variable equations they used in Section 5-3 and the systems of two-variable equations used here.

Talk it Over

3. Suppose the system of equations used to solve a problem has no solution. If the equations have been written correctly, what does this tell you about the conditions of the problem?

4. Cliff and Emma used different methods to solve the problem in the Sample.

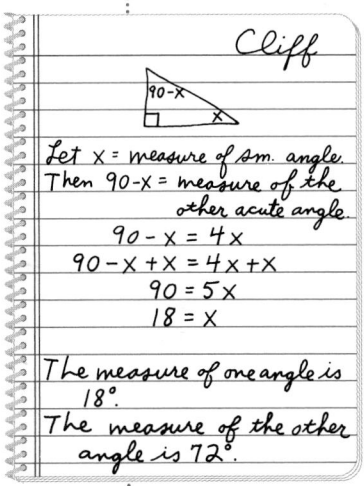

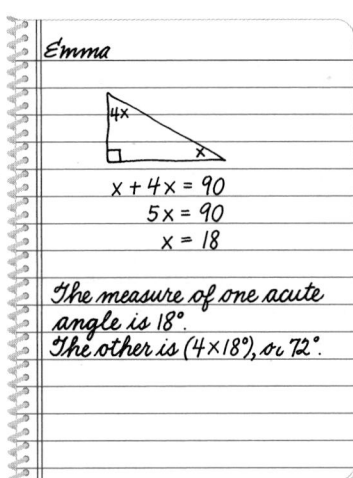

a. Describe Cliff's method.

b. Describe Emma's method.

c. Suppose you are showing a friend how to solve the Sample. Which method would you use? Why?

Look Back

Look back at the tortoise and hare problem in Sample 2 of Section 5-3. Explain how you can solve this problem using a system of two equations.

5-8 Systems of Equations in Geometry

291

Answers to Talk it Over

3. The conditions cannot all be met at the same time.

4. a. Cliff labeled the acute angles by using the definition of complementary. His equation relates the complement of the angle with four times the angle.

   b. Emma used x for one angle and the condition that the second angle is 4 times the first to label the angles. Her equation shows the sum of these two to be 90°.

   c. Answers may vary. An example is given. Emma's method; it seems more straightforward.

Answers to Look Back

Explanations may vary. An example is given. Let h represent the hare's racing time in hours, and let t represent the tortoise's racing time in hours. Since the hare takes a 12-hour rest, $h = t - 12$. Since each travels the same distance, $0.2t = 35h$. You can substitute $t - 12$ for h in this second equation and then solve.

291

APPLYING

5-8 Exercises and Problems

1. **Reading** Read the Sample on page 290. What information is used to write the system of equations?

Check Matthew's, Zena's, and Corazon's work. If they solved incorrectly, find the correct solution.

2. $a = 3b$
   $4a + 2b = 28$

   Matthew's solution:
   $a = 2, b = 10$

3. $12r - 8t = -52$
   $5t = r$

   Zena's solution:
   $t = -1, r = -5$

4. $y = 13 - 2x$
   $y = 5x + 41$

   Corazon's solution:
   $y = 5, x = -4$

Solve each system of equations.

5. $y = 8x$
   $4x + y = -108$

6. $6a + 5c = 46$
   $a = 3c$

7. $m = 2p + 3$
   $3p - 2m = -12$

For Exercises 8–14, solve using a system of equations.

8. In $\triangle ABC$, the measures of $\angle B$ and $\angle C$ are equal, and $\angle B$ is twice as large as $\angle A$. Find the measure of each angle.

9. Two angles form a straight angle. The measure of one angle is five times the measure of the other angle. What are the measures of the angles?

10. Two angles are supplementary. One angle is 40° larger than the other. Find the measures of the angles.

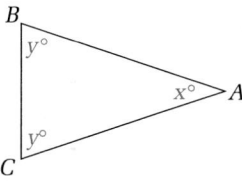

11. In a triangle, the measure of the largest angle is twice the measure of the smallest. The measure of the third angle is 20° more than the measure of the smallest.

    a. Sketch the triangle and label the angles.

    b. Find the measures of the angles of the triangle.

12. The perimeter of a rectangle is 48 in. The width is 10 in. shorter than the length. What are the dimensions of the rectangle?

13. Solve this puzzle:

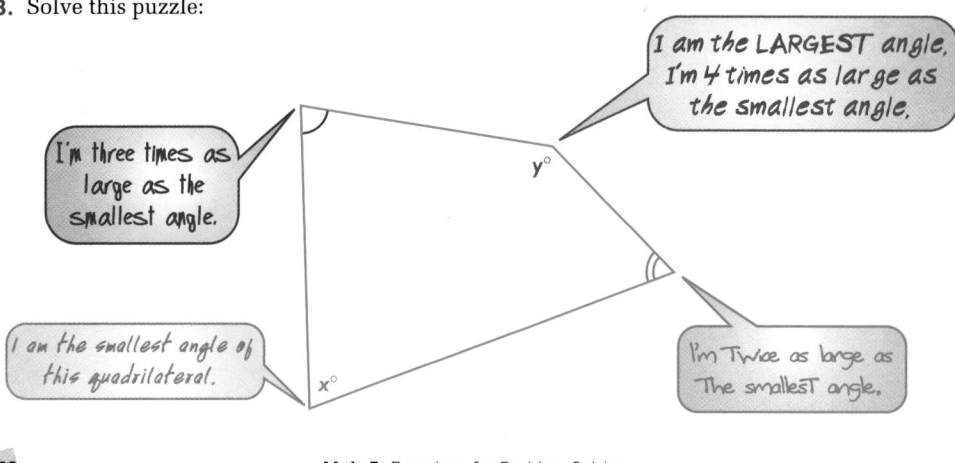

Answers to Exercises and Problems

1. The relationship between two acute angles in a right triangle and the relationship between these two particular acute angles are used.

2. $a = 6, b = 2$

3. correct

4. $y = 21, x = -4$

5. $x = -9, y = -72$

6. $a = 6, c = 2$

7. $p = 6, m = 15$

8. $\angle A = 36°; \angle B = \angle C = 72°.$

9. 30° and 150°

10. 70° and 110°

11. a.

b. 40°, 60°, and 80°

12. The length is 17 in. and the width is 7 in.

13. $x = 36; y = 144$

14. Delia and Gilberto are sharing the cost of a $99 camera. Delia paid $35 more than Gilberto. How much did each pay?

15. **Group Activity** Work in a group of three students.

   a. Each person should write a system of equations like the ones you have used in this section.

   b. Pass your paper to the person on your left. Solve the system you receive.

   c. Pass your paper to the left again and check the solution you receive. Help each other until everyone in your group can solve systems successfully.

In Exercises 16 and 17, rewrite one of the equations in each system to get one of the variables alone on one side of the equation. Then solve the system of equations.

16. $r + 2v = -5$
    $4r + v = 8$

17. $3a + b = 2$
    $2a - 5b = 7$

18. If you buy two hair clips and three head bands, you will get $3 change from your $10 bill. If you buy six hair clips and one head band, you will get $1 change. How much does each hair clip and head band cost?

Ongoing **ASSESSMENT**

19. **Writing** Write a set of directions for a student who was absent on the day of this lesson. The student should be able to use your directions to solve a system of equations by substitution and check the solution.

Review **PREVIEW**

20. Compare the areas of a rectangle, a parallelogram, and a triangle that each have $h = 3$ and $b = 8$. *(Section 5-7)*

Solve and graph each inequality. *(Section 5-4)*

21. $6q < 24$

22. $-5a \geq 25$

23. $m + 4 \geq 13$

Write each fraction or decimal as a percent. *(Toolbox Skill 11)*

24. 0.78

25. 1.56

26. $\frac{13}{30}$

27. $\frac{7}{8}$

Working on the Unit Project

28. a. The equation $P = I - E$ shows your profit after you subtract the expenses for the car wash from your income. Substitute your goal for P and your total expenses for E.

   b. Write an equation to represent your income after washing n cars at your price.

   c. Write a system of equations using your answers to parts (a) and (b).

   d. Solve the system you wrote in part (c) to find out how many cars you need to wash for your profit to equal your goal.

5-8 Systems of Equations in Geometry

293

Assessment: Performance

Call on volunteers to read the instructions they wrote for Ex. 19. Have others follow the instructions by working at the chalkboard.

Working on the Unit Project

The effectiveness of the car wash project is demonstrated through the use of systems of equations. Students need to analyze their results to determine whether the number of cars that must be washed to reach their goal is reasonable.

Quick Quiz (5-5 through 5-8)

See page 296.

Practice 42 For use with Section 5-8

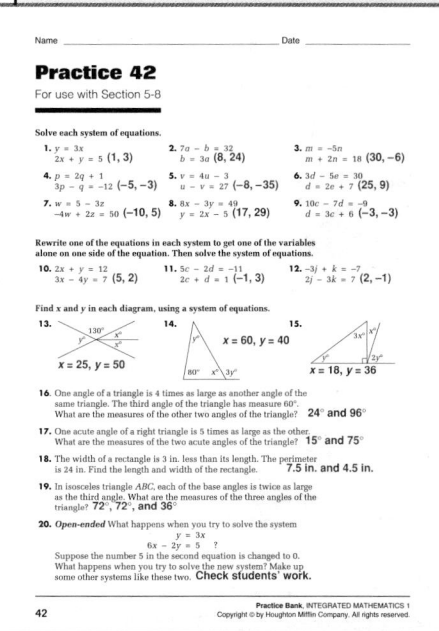

Practice 42
For use with Section 5-8

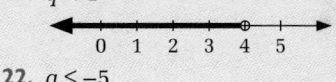

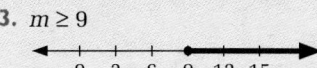

Answers to Exercises and Problems

14. Gilberto, $32; Delia, $67

15. a–c. Answers may vary.

16. Answers may vary. An example is given.
$v = 8 - 4r$; $r = 3$, $v = -4$

17. Answers may vary. An example is given.
$b = 2 - 3a$; $a = 1$, $b = -1$

18. hair clip, $1.25; head band, $1.50

19. Directions may vary. An example is given. You need to find values of the variables that makes both equations true. To do this, follow these steps: (1) Solve one equation for one of the variables. (2) Substitute the expression you found in step 1 for the same variable in the other equation. You get an equation that contains just one variable. (3) Solve the equation in step 2. (4) Substitute this value in the expression you found in step 1 to find the value of the other variable. To check, substitute the values of the variables in both of the original equations and see if you get true statements.

20. The rectangle's area equals the parallelogram's area (24 units^2). The triangle's area is 12 units^2.

21. $q < 4$

22. $a \leq -5$

23. $m \geq 9$

24. 78%

25. 156%

26. $43\frac{1}{3}$%

27. $87\frac{1}{2}$%

28. a–d. Answers may vary.

293

Completing the Unit Project

Now you are ready to present your plan of action. Your presentation should include reasons for your recommendations. Include these things in your presentation.

➤ a description of the best location for the car wash

➤ the price you have decided to charge for each car

➤ a list of the supplies you need and where you will purchase these items

➤ a discussion of how many cars you need to wash to break even

➤ a table showing your income and how many cars you need to wash to meet your fund-raising goal

➤ samples of advertisements and posters

➤ a diagram of your car wash area

➤ a list of the jobs and the number of people needed to run the car wash efficiently

CAR WASH
to support
CENTRAL HIGH SCHOOL
VOLUNTEER CLUB
$5.00

Look Back

What did you learn from this project that can help you plan for future fund-raising events?

Alternative Projects

Project 1: Working on Commission

Interview someone in a business to find out the salaries and commission rates for new salespeople and experienced salespeople.

Ask for some typical weekly or monthly sales figures. Make a table and a graph and write an equation for several different amounts of sales. Choose an income goal and figure out how many sales a new salesperson and an experienced salesperson need to earn that income.

Project 2: Researching Telephone Costs

Choose two locations you would like to call long distance. Decide on the time of day, the day of the week, and the length of each call. Contact two or three long-distance telephone services to find out the cost for each call. Make a table and graph the data. Use the results to choose a long-distance service for your calls. Explain your choice.

REVIEW AND ASSESSMENT

1. Karin types students' papers on a computer. She charges $6 for supplies and $2 per page.

   5-1

  a. Model the situation using a table.

  b. Model the situation using a graph.

  c. Model the situation with an equation.   **5-2**

  d. How long a paper does Karin need to type to earn $30?

2. The sum of the measures of the angles of a convex polygon with n sides is $(n-2)180$. Can the sum of the angles of a polygon be 1800°? Explain why or why not.

3. Evelyn and Kelsey have a bike race. Evelyn bikes at 18 ft/s and   **5-3** Kelsey bikes at 14 ft/s. Evelyn falls. She stops for 30 s before continuing the race. She and Kelsey tie. How many seconds long was the race?

4. **Open-ended** Create a problem situation that you can model with a table, a graph, and an equation. Choose one of your models to solve the problem.

Solve.   **5-2, 5-3**

5. $13 - y = 39$ **6.** $3x + 12 - x = 2$ **7.** $(n-2)180 = 1080$

8. $x - 3(x - 5) = -9$ **9.** $2t = 5t + 6$ **10.** $10x - 2 = 20 - 34x$

11. **Writing** John and Dwayne each made a mistake in their work. Check their solutions. Explain each mistake and tell how to correct it.

  a. John's work: **b.** Dwayne's work:

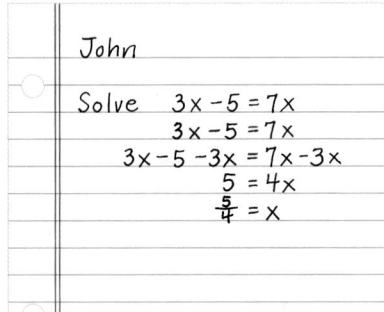

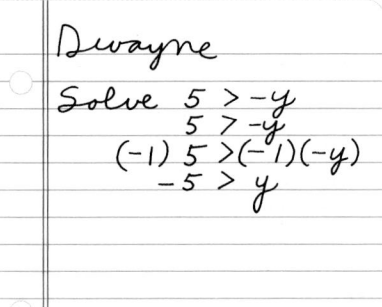

Solve and graph each inequality.   **5-4**

12. $-5x > 15$ **13.** $8 - a \le 12$ **14.** $4(x-2) > 8$

15. Raoul has $82 in his bank account. He can earn $9 an hour working at the library. How many full hours must he work in order to have at least $250?

Unit 5 Review and Assessment   **295**

Answers to Unit 5 Review and Assessment

1. a.

Number of Pages	Total Charges
1	6 + 2(1) = 8
2	6 + 2(2) = 10
3	6 + 2(3) = 12
⋮	⋮
n	6 + 2(n)

b.

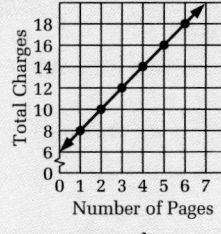

Number of Pages

c. $6 + 2n = t$, where n = number of pages, t = total charges

d. 12 pages

2. Yes. The polygon would have 12 sides.

3. 135 s

4. Answers may vary.

5. −26   **6.** −5   **7.** 8

8. 12   **9.** −2   **10.** $\frac{1}{2}$

11. a. Explanations may vary. An example is given. John forgot to put a negative sign in front of the 5. He should have had $-5 = 4x$ and then $-\frac{5}{4} = x$.

b. Dwayne forgot to reverse the inequality sign when he multiplied by −1. The result should be $-5 < y$.

12. $x < -3$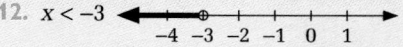

13. $a \ge -4$

14. $x > 4$

15. 19 h

Quick Quiz (5-5 through 5-8)

1. Solve the equation
   $a = 180n - 360$ for n. [5-5]
   $$n = \frac{a + 360}{180}$$

2. For the numbers 0.8 and
   1.5 to be reciprocals, what
   would their product have
   to be equal to? Are these
   numbers reciprocals? [5-6]
   1; No.

Find the area of each figure.
[5-7]

3.
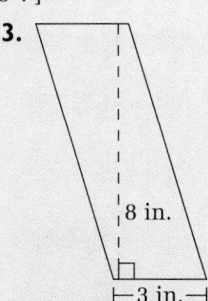

8 in.

3 in.

24 in²

4.
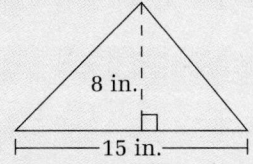

8 in.

15 in.

60 in²

5.
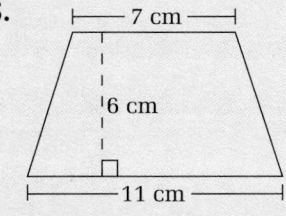

7 cm

6 cm

11 cm

54 cm²

6. Solve using a system of
   equations. Two angles are
   supplementary. One of the
   angles is 35° smaller then
   the other. What are the
   measures of the angles?
   [5-8] **72.5° and 107.5°**

Solve for the variable in red.   **5-5, 5-6**

16. $I = prt$ 17. $4x + y = 180$ 18. $P = (n - 2)T$

19. $5 - \frac{4}{7}x = -3y$ 20. $\frac{3}{8}x + 6 = y$ 21. $\frac{5}{4}y + 5x = 1$

22. The formula $y = 4x$ converts x half-gallons to y pints. Write a
    formula that converts y pints to x half-gallons.

Identify the area formula represented by each diagram.   **5-7**

23.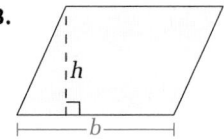

h

b

24.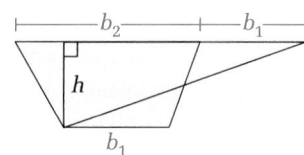

b_2 b_1

h

b_1

0   Miles   200

· Reno

NEVADA

25. The shape of Nevada is nearly a trapezoid. Estimate
    the area of Nevada using the formula for the area of a
    trapezoid.

26. Solve the system. $3x + 2y = 10$
                             $x = y + 5$

27. In an isosceles triangle, the measure of one angle is   **5-8**
    three times the measure of one of the congruent
    angles. Write and solve a system of equations to find
    the measure of each angle.

28. **Self-evaluation** Make a diagram that shows the
    types of equations that you can now solve. (See the
    concept map at the right for an idea.) Use a colored
    pencil or different shapes to show the types that
    you feel confident about and the types with which
    you need more experience.

29. a. **Group Activity** Work in a group of four students.
       Write four different questions using the informa-
       tion below. At least one question should involve
       an equation and one should involve an inequality.

    ➤ The length of the football field is 100 yd.

    ➤ The fastest runner on the football team runs 6 yd/s.

    ➤ The time it takes to pass a football ranges from 0.5 s to 8 s.

    ➤ The slowest runner on the football team runs 3 yd/s.

    b. Answer your questions on a separate sheet of paper.

    c. Give another group your questions to answer. Compare answers
       and discuss any differences.

One Step
$3x = 12$
$\frac{x}{3} = 4$

Two
Step

Distributive
Property

Combining
Like Terms

Answers to Unit 5 Review and Assessment ···

16. $\frac{I}{pr} = t$

17. $y = 180 - 4x$

18. $\frac{P}{T} + 2 = n$ or $\frac{P + 2T}{T} = n$

19. $x = \frac{21}{4}y + \frac{35}{4}$

20. $x = \frac{8}{3}(y - 6)$ or $x = \frac{8}{3}y - 16$

21. $y = \frac{4}{5}(1 - 5x)$ or $y = \frac{4}{5} - 4x$

22. $\frac{y}{4} = x$

23. Area of a parallelogram =
    bh

24. Area of a trapezoid =
    $\frac{1}{2}(b_1 + b_2)h$

25. about 105,000 mi²

26. $x = 4$, $y = -1$

27. $y = 3x$, $y + 2x = 180$; two
    angles of 36°, 1 angle of
    108°

28. Diagrams may vary.

29. Answers may vary. An
    example is given. What is
    the greatest distance that
    the fastest runner on a
    football team can go in
    order to receive a pass if
    his speed is 6 yd/s and if
    the pass takes from 0.5 to
    8 s? How much time
    would the slowest runner
    need to go this distance if
    his rate of speed is 3 yd/s?

IDEAS AND (FORMULAS) $= x^2$

ALGEBRA

➤ **Problem Solving** Situations can be modeled with tables, graphs, and equations. The pattern in a table can help you write an equation for the situation. *(p. 241)*

➤ Different situations may have the same model. *(p. 243)*

➤ To solve multi-step equations, you may need to combine like terms, use the distributive property, or divide both sides by −1. *(pp. 256, 257)*

➤ **Problem Solving** Problem situations involving words such as *over*, *under*, *less than*, and *more than* may be solved using inequalities. *(pp. 264, 265)*

➤ You solve inequalities the same way as equations *except* that when you multiply or divide by a negative number, the inequality symbol is reversed. *(p. 264)*

➤ **Problem Solving** You can solve some problem situations by first writing a formula. *(p. 270)*

➤ To solve an equation or a formula for a variable, you rewrite the equation or formula so the variable is alone on one side. You use the same steps as in solving equations. *(p. 271)*

➤ You can use the fact that the product of reciprocals is 1 to get a variable alone on one side of an equals sign. *(p. 276)*

➤ **Problem Solving** You can solve problems by using more than one variable, then solving a system of equations. *(p. 289)*

➤ A system of equations may be solved using substitution. *(p. 290)*

GEOMETRY ➤

➤ **Reasoning** The area formulas of several polygons can be derived from one another. *(pp. 280–283)*

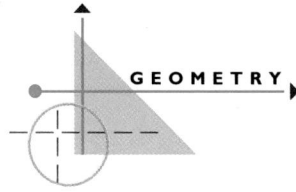

$A = bh$

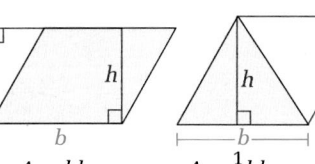

$A = \frac{1}{2}bh$

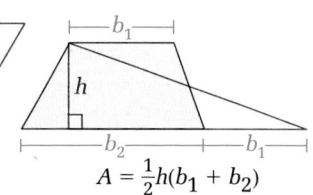

$A = \frac{1}{2}h(b_1 + b_2)$

Key Terms

- **mathematical model** (p. 241)
- **equivalent inequalities** (p. 263)
- **height** (pp. 280–283)

- **modeling** (p. 241)
- **reciprocal** (p. 275)
- **system of equations** (p. 291)

- **variable terms** (p. 256)
- **base** (pp. 280–283)
- **solution of a system** (p. 291)

Unit 5 Review and Assessment

297

Quick Quiz (5-1 through 5-4)

Use this situation for Exs. 1 and 2.

A one-day rental car costs \$30, with an additional \$.15 charged for each mile driven. Rolando Perez plans to rent a car and drive as far as \$60 will allow. How many miles can he drive the car?

1. Model and solve the situation with an equation. [5-1]
Let m = the number of miles he drives.
$30 + 0.15m = 60$
He can drive 200 miles.

2. Suppose you modeled the situation with a graph. Name two points that would be on the graph. [5-1]
Answers will vary. Possible answer: (100, 45) and (200, 60).

3. Solve the equation $6p - 3(p - 1) = -6$. [5-2]
$p = -3$

4. Solve the equation $-7(n - 3) = -3n + 1$. [5-3]
$n = 5$

5. Selma Lewis and her brother Sam drove their cars over the same road from Clayton to Rayburn. Selma drove 55 mi/h. Sam drove 60 mi/h, with half an hour spent having a minor car repair. They left Clayton at the same time and arrived at Rayburn at the same time. Model the situation with an equation. [5-3] Let t = the number of hours Selma drove.
$55t = 60(t - 0.5)$

6. Solve and graph the inequality $3x + 8 > 10 + 5x$. [5-4]
$x < -1$

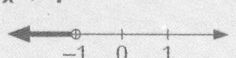
$$-1 \quad 0 \quad 1$$

7. Choose the letter of the inequality that best models the situation. In 5 more years, Lee Chan will still be under 20 years old. Let d = Lee's present age. [5-4]

(a) $d + 5 > 20$

(b) $d + 5 < 20$

(c) $d + 5 \le 20$ (b)

297

OVERVIEW

➤ **Unit 6** introduces ratios and solving proportions as skills needed for investigating probability, for making predictions, and for analyzing similar figures. Students can review factoring and finding the GCF and the LCM on page 649 and finding percents on pages 653–656 of the **Student Resources Toolbox** to prepare for these topics.

➤ In Unit 6, students use proportions to predict events based on a sample and to solve problems involving similar polygons, scale drawings, and dilations. They also extend the concept of ratio to include sine and cosine, two special right triangle ratios.

➤ Planning an advertising campaign is the theme of the Unit Project. Students select a product or service, determine where and how to advertise, and estimate costs.

➤ Connections to geography, health, social studies, biology, and literature are integrated into the teaching materials and exercises.

➤ Problem-solving strategies used in Unit 6 include *Use a Formula* (Section 6-2) and *Use a Proportion* (Sections 6-3, 6-4, and 6-5).

Unit Objectives

Section	Objectives	NCTM Standards
6-1	• Write ratios in simplest form.	1, 2, 4, 5
	• Solve problems involving ratios.	
	• Compare unit rates.	
6-2	• Describe how likely it is that something will happen.	1, 2, 3, 4, 10, 11, 12
6-3	• Solve proportions.	1, 2, 4, 5
6-4	• Use sampling to make predictions.	1, 2, 4, 5, 10, 11
6-5	• Apply proportions to problems using similar figures and scale drawings.	1, 2, 3, 4, 5, 7
6-6	• Change the size of a figure without changing its shape.	1, 2, 4, 8
6-7	• Use two special right triangle ratios to solve problems.	1, 2, 4, 6, 7, 9

Section	Connections to Prior and Future Concepts
6-1	**Section 6-1** introduces ratios and rates. Students will have studied these topics in previous courses, and familiarity with the concepts is assumed. Ratios and rates are used throughout Unit 6 and in Units 7 and 9 of Book 1; in Units 1, 2, 3, 6, and 8 of Book 2; and in Units 2, 4, 5, 7, and 8 of Book 3.
6-2	**Section 6-2** builds on the concept of ratio introduced in Section 6-1 to describe experimental and theoretical probabilities. Probability is used in many real-world and mathematical situations. The study of probability is continued in Sections 6-4 and 9-4 of Book 1, in Units 1 and 6 of Book 2, and in Unit 7 of Book 3.
6-3	**Section 6-3** extends the concept of ratio introduced in Section 6-1 to include proportions. Proportions are used to solve problems throughout Unit 6 and in Units 7 and 9 of Book 1. They are used whenever appropriate to solve problems throughout Books 2 and 3.
6-4	**Section 6-4** introduces sampling as a method of making predictions. Proportions, which were presented in Section 6-3, and estimation, which was studied in Unit 2, are used to predict events based on sampling. The idea of linear intervals, presented in Section 3-3, is discussed in the context of sampling as a margin of error. Related topics are taught in Section 9-4 of Book 1, in Unit 1 of Book 2, and in Unit 7 of Book 3.
6-5	**Section 6-5** applies proportions, introduced in Section 6-3, to problems involving similar figures and to scale drawings. Similar figures are further studied in the remainder of this unit and in Units 7 and 9 of Book 1, in Units 5 and 8 of Book 2, and in Units 9 and 10 of Book 3.
6-6	**Section 6-6** combines the concepts of similar figures and scale from Section 6-5 with the concept of transformations from Unit 4 to introduce dilations. Other transformation concepts are presented in Sections 4-3, 4-4, 10-1, and 10-2 of Book 1; in Units 3–5 and 8 of Book 2; and in Units 9 and 10 of Book 3.
6-7	**Section 6-7** uses the concept of ratio, introduced in Section 6-1, to present two special right triangle ratios, sine and cosine. These and other right triangle ratios are further studied in Unit 7 of Book 1, in Unit 8 of Book 2, and in Units 8 and 10 of Book 3.

Integrating the Strands

Strands	Sections
Number	6-1, 6-2, 6-3, 6-6
Algebra	6-1, 6-2, 6-3, 6-4, 6-5, 6-6
Functions	6-3
Measurement	6-5
Geometry	6-1, 6-4, 6-5, 6-6, 6-7
Trigonometry	6-7
Statistics and Probability	6-1, 6-2, 6-3, 6-4, 6-5, 6-7
Discrete Mathematics	6-1, 6-2, 6-4
Logic and Language	6-1, 6-3, 6-4, 6-5

Section Planning Guide

➤ Essential exercises and problems are indicated in boldface.
➤ Ongoing work on the Unit Project is indicated in color.
➤ Exercises and problems that require student research, group work, manipulatives, or graphing technology are indicated in the column headed "Other."

Section	Materials	Pacing	Standard Assignment	Extended Assignment	Other
6-1	geometric drawing software	Day 1	**1**, 2, **3–6, 8,** 10, 13–16, 17	**1**, 2, **3–6,** 7, **8,** 10, 13–16, 17	7d, 9, 11, 12, 17
6-2	paper clips, red and blue markers, protracter, compass, graphics calculator, spreadsheet software, coin	Day 1 / Day 2 / Day 3	**1–4, 6** / **7, 9–13** / **14–23,** 25–33, 34–36	**1–4, 6** / **7,** 8, **9–13** / **14–23,** 25–33, 34–36	5 / / 24
6-3		Day 1	**1–20,** 21, 22, 28–37, 38–40	**1–20,** 21–25, 27–37, 38–40	26, 39
6-4	bags of dried beans, opaque containers (bags or bowls), nontoxic markers	Day 1 / Day 2	**1, 2, 4** / **5–7,** 9, 11–16, 17	**1, 2,** 3, **4** / **5–7,** 9, 11–16, 17	/ 8, 10, 17
6-5	protractor, geoboard, rubber bands, geometric drawing software, penny	Day 1 / Day 2	**1, 3–7** / **10–13, 15, 16, 19, 20,** 25–31, 32, 33	**1,** 2, **3–7** / **10–13, 15, 16,** 17, **19, 20,** 21–23, 25–31, 32, 33	8, 9 / 14, 18, 24
6-6	protractor	Day 1	**2–12,** 13, 16–23, 24	1, **2–12,** 13, 16–23, 24	14, 15
6-7	protractor, scientific calculator, graphing technology	Day 1 / Day 2	**3–13** / **16–23,** 25–28, 29	1, **3–13,** 14 / **16–23,** 24–28, 29	2, 15 / 29
Review / Test		**Day 1** / **Day 2**	Unit Review / Unit Test	Unit Review / Unit Test	

Yearly Pacing	Unit 6 Total	Units 1–6 Total	Remaining	Total
	16 days (2 for Unit Project)	94 days	70 days	164 days

Support Materials

➤ See **Project Book** for notes on Unit 6: Plan an Advertising Campaign.
➤ "UPP" and "disk" refer to **Using Plotter Plus** booklet and **Plotter Plus** disk.
➤ Warm-up exercises for each section are available on **Warm-Up Transparencies.**
➤ "FI," "PC," "GI," "MA," and "Stats!" refer, respectively, to the McDougal Littell Mathpack software Activity Books for **Function Investigator, Probability Constructor, Geometry Inventor, Matrix Analyzer,** and **Stats!.**

Section	Study Guide	Practice Bank	Problem Bank	Activity Bank	Explorations Lab Manual	Assessment Book	Visuals	Technology
6-1	6-1	Practice 44	Set 12	Enrich 39	Master 2	Quiz 6-1		
6-2	6-2	Practice 45	Set 12	Enrich 40	Masters 2, 14,	Quiz 6-2	Folder 7	PC Acts. 1, 2, 6, 9, 11, 12, and 16
6-3	6-3	Practice 46	Set 12	Enrich 41		Quiz 6-3 Test 23		FI Act. 23
6-4	6-4	Practice 47	Set 13	Enrich 42	Masters 2, 16	Quiz 6-4		Sampling Experiment (disk)
6-5	6-5	Practice 48	Set 13	Enrich 43	Masters 1, 2, 17	Quiz 6-5		
6-6	6-6	Practice 49	Set 13	Enrich 44	Add. Expl. 7 Master 2	Quiz 6-6		GI Acts. 22 and 24
6-7	6-7	Practice 50	Set 13	Enrich 45	Masters 2, 18	Quiz 6-7 Test 24		
Unit 6	Unit Rev.	Practice 51	Unif. Prob. 6	Fam. Inv. 6		Tests 25, 26		

Form A

Spanish versions of these tests are on pages 142–145 of the **Assessment Book**.

Software Support

McDougal Littell Mathpack

Function Investigator
Geometry Inventor
Probability Constructor
Stats!

Name _____ Date _____ Score _____

Test 25

Test on Unit 6 (Form A)

Directions: Write the answers in the spaces provided.

1. What is the ratio of the perimeter of a regular pentagon to the length of one side?

2. While attending a school carnival, you estimate the ratio of children to adults as 5:2. If there are 525 people at the carnival, about how many children are in attendance?

For Questions 3–5, a fair die is rolled. Tell whether each statement is *True* or *False*.

3. $P(\boxed{\cdot}) = \frac{1}{3}$ **4.** $P(\text{not } \boxed{\because}) = \frac{5}{6}$

5. Rolling a multiple of 3 and rolling a multiple of 2 are complementary events.

In Questions 6 and 7, solve each proportion for x.

6. $\frac{6}{7.5} = \frac{x}{12.5}$ **7.** $\frac{75}{0.6} = \frac{200}{x}$

8. It takes 1.5 c of milk to make 8 pancakes for breakfast. How many pancakes can be made using 3 qt of milk?

9. Writing Explain what theoretical probability is and how to find it. Be sure to include the formula.

Sample answer: Theoretical probability is applied when all the possible outcomes of an experiment are equally likely to occur. To find the theoretical probability of an event E under these circumstances, use the ratio $P(E) = \frac{\text{number of favorable outcomes}}{\text{number of possible outcomes}}$.

For Questions 10 and 11, use the following situation.

Two hundred students were surveyed one week prior to the election for student body president regarding their choice of three candidates. The table at the right shows the results of the survey.

Candidate	A	B	C
Votes	102	64	34

10. There are 1200 students in the school. According to the survey, about how many votes can each candidate expect to receive?

11. Assume that the survey has a 5% margin of error. Estimate an interval for the number of votes expected to go for Candidate A.

Answers

1. 5 to 1, 5:1, or $\frac{5}{1}$
2. 375
3. False
4. True
5. False
6. 10
7. 1.6
8. 64
9. *See question.*
10. A: 612; B: 384; C: 204
11. about 552 to 672

Name _____ Date _____ Score _____

Test 25 (continued)

Directions: Write the answers in the spaces provided.

For Questions 12–14, use the figure at the right. Find each of the following measures.

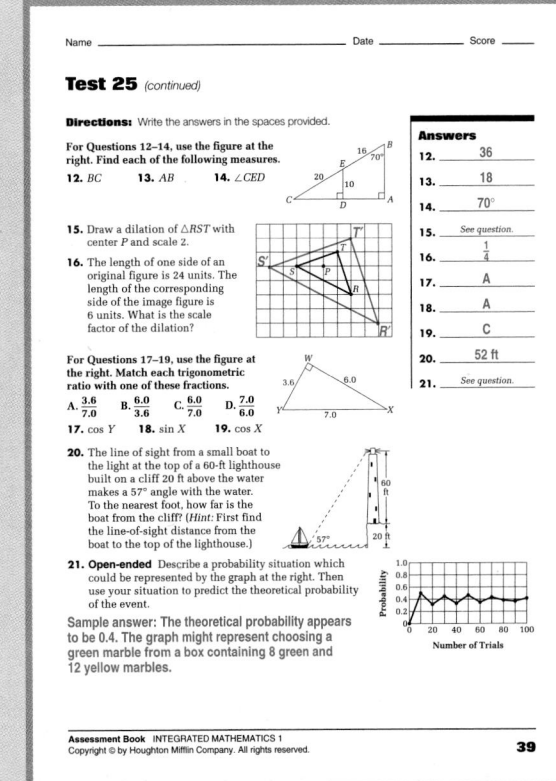

12. BC **13.** AB **14.** $\angle CED$

15. Draw a dilation of $\triangle RST$ with center P and scale 2.

16. The length of one side of an original figure is 24 units. The length of the corresponding side of the image figure is 6 units. What is the scale factor of the dilation?

For Questions 17–19, use the figure at the right. Match each trigonometric ratio with one of these fractions.

A. $\frac{3.6}{7.0}$ B. $\frac{6.0}{3.6}$ C. $\frac{6.0}{7.0}$ D. $\frac{7.0}{6.0}$

17. cos Y **18.** sin X **19.** cos X

20. The line of sight from a small boat to the light at the top of a 60-ft lighthouse built on a cliff 20 ft above the water makes a 57° angle with the water. To the nearest foot, how far is the boat from the cliff? (*Hint:* First find the line-of-sight distance from the boat to the top of the lighthouse.)

21. Open-ended Describe a probability situation which could be represented by the graph at the right. Then use your situation to predict the theoretical probability of the event.

Sample answer: The theoretical probability appears to be 0.4. The graph might represent choosing a green marble from a box containing 8 green and 12 yellow marbles.

Answers

12. 36
13. 18
14. 70°
15. *See question.*
16. $\frac{1}{4}$
17. A
18. A
19. C
20. 52 ft
21. *See question.*

Outside Resources

Books/Periodicals

Weiss, Ann E. *Lotteries: Who Wins, Who Loses?* Hillsdale, NJ: Enslow Publishers, 1991.

Software

Statistics Workshop. BBN Laboratories, Inc. Scotts Valley, CA: Sunburst.

Yerushalmy, Michal and Daniel Chazan. *Supposer Solutions; Making the Most of Your Classroom Computer.* Apple and MS-DOS. Newton, MA: Education Development Center.

Kemeny-Kurtz Probability Theory. Macintosh or MS-DOS. Developed by TrueBASIC. Acton, MA: William K. Bradford Publishing.

Videos

Apostol, Tom. *Similarity.* Reston, VA: NCTM.

Form B

Name _____ Date _____ Score _____

Test 26

Test on Unit 6 (Form B)

Directions: Write the answers in the spaces provided.

1. What is the ratio of the perimeter of a regular octagon to the length of one side?

2. While visiting a daycare center, you estimate the ratio of toddlers to infants as 3:2. If the center has an enrollment of 25 children, about how many of them are infants?

For Questions 3–5, a fair die is rolled. Tell whether each statement is *True* or *False*.

3. $P(\boxed{\vdots}) = \frac{1}{6}$ **4.** $P(\text{not } \boxed{\because}) = \frac{5}{6}$

5. Rolling a number less than 6 and rolling a multiple of 3 are complementary events.

In Questions 6 and 7, solve each proportion for x.

6. $\frac{6}{4.5} = \frac{x}{16.5}$ **7.** $\frac{70}{0.8} = \frac{210}{x}$

8. It takes 10 in. of wire to form a bubble wand. How many feet of wire will be needed to form 9 bubble wands?

9. Writing Explain what experimental probability is and how to find it. Be sure to include the formula.

Sample answer: Experimental probability is based on the observed results of an experiment. To find the experimental probability of an event E under these circumstances, use the ratio $P(E) = \frac{\text{number of times event } E \text{ happens}}{\text{number of times the experiment is done}}$.

For Questions 10 and 11, use the following situation.

Two hundred students were surveyed one week prior to the election for student body president regarding their choice of three candidates. The table at the right shows the results of the survey.

Candidate	A	B	C
Votes	98	35	67

10. There are 1200 students in the school. According to the survey, about how many votes can each candidate expect to receive?

11. Assume that the survey has a 5% margin of error. Estimate an interval for the number of votes expected to go for Candidate A.

Answers

1. 8 to 1, 8:1, or $\frac{8}{1}$
2. 10
3. True
4. True
5. False
6. 22
7. 2.4
8. 7.5 ft
9. *See question.*
10. A: 588; B: 210; C: 402
11. about 528 to 648

Name _____ Date _____ Score _____

Test 26 (continued)

Directions: Write the answers in the spaces provided.

For Questions 12–14, use the figure at the right. Find each of the following measures.

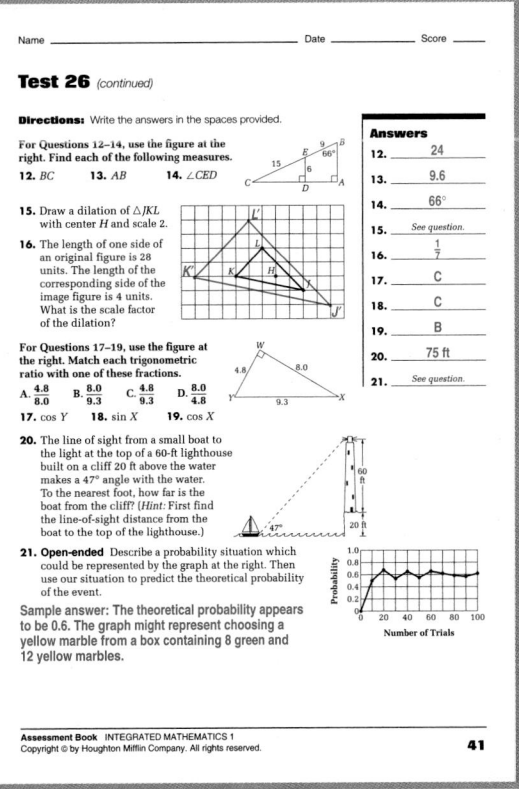

12. BC **13.** AB **14.** $\angle CED$

15. Draw a dilation of $\triangle JKL$ with center H and scale 2.

16. The length of one side of an original figure is 28 units. The length of the corresponding side of the image figure is 4 units. What is the scale factor of the dilation?

For Questions 17–19, use the figure at the right. Match each trigonometric ratio with one of these fractions.

A. $\frac{4.8}{8.0}$ B. $\frac{8.0}{9.3}$ C. $\frac{4.8}{9.3}$ D. $\frac{8.0}{4.8}$

17. cos Y **18.** sin X **19.** cos X

20. The line of sight from a small boat to the light at the top of a 60-ft lighthouse built on a cliff 20 ft above the water makes a 47° angle with the water. To the nearest foot, how far is the boat from the cliff? (*Hint:* First find the line-of-sight distance from the boat to the top of the lighthouse.)

21. Open-ended Describe a probability situation which could be represented by the graph at the right. Then use your situation to predict the theoretical probability of the event.

Sample answer: The theoretical probability appears to be 0.6. The graph might represent choosing a yellow marble from a box containing 8 green and 12 yellow marbles.

Answers

12. 24
13. 9.6
14. 66°
15. *See question.*
16. $\frac{1}{7}$
17. C
18. C
19. B
20. 75 ft
21. *See question.*

- Students plan an advertising campaign for a product or service that is appealing to students in their school.

- When planning the advertisements, students choose what to advertise, decide which media to use, compare the unit costs for different media, and then either write a script or draw a sketch of the ad.

- Students work together in a cooperative group to create a successful project. ....●

PROJECT PLANNING

Materials List

- Folder

Project Teams

Before students begin working as part of a team, have individual students make a list of advertisements they have seen on television, heard on the radio, or seen in newspapers and magazines recently.

Have students work on the project in groups of four. One way for the individuals in the group to distribute the work is as follows:

1. Coordinator: collects all possible ideas of products or services to advertise and coordinates market research.

2. Writer: writes the description of the product or service and the script for the ad.

3. Illustrator: draws and colors the ad.

4. Analyst: finds and compares the costs of advertising using the different media.

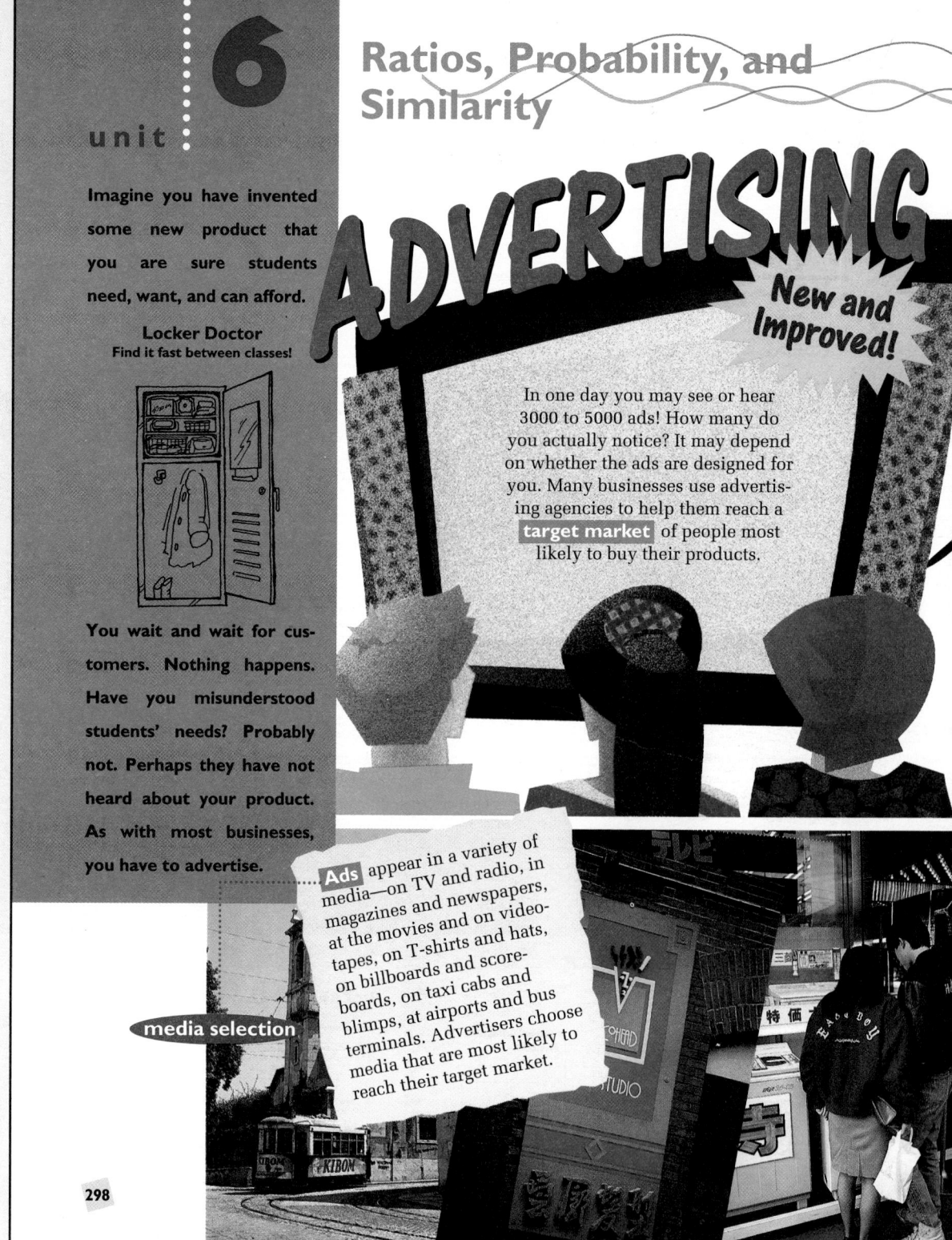

unit **6**

Imagine you have invented some new product that you are sure students need, want, and can afford.

Locker Doctor
Find it fast between classes!

You wait and wait for customers. Nothing happens. Have you misunderstood students' needs? Probably not. Perhaps they have not heard about your product. As with most businesses, you have to advertise.

Ratios, Probability, and Similarity

ADVERTISING
New and Improved!

In one day you may see or hear 3000 to 5000 ads! How many do you actually notice? It may depend on whether the ads are designed for you. Many businesses use advertising agencies to help them reach a target market of people most likely to buy their products.

media selection

Ads appear in a variety of media—on TV and radio, in magazines and newspapers, at the movies and on videotapes, on T-shirts and hats, on billboards and scoreboards, on taxi cabs and blimps, at airports and bus terminals. Advertisers choose media that are most likely to reach their target market.

298

Suggested Rubric for Unit Project

4 The advertising campaign is well organized. The written plan describes the product or service thoroughly and has supporting data. The unit costs and use of probability concepts are applied correctly. The choice of advertising media, its schedule, and the estimated cost is appropriate. The media ad sketch or script is appealing.

3 The advertising campaign is acceptable but is not fully supported by the data. Some key features of the plan are missing. The written plan is not as thorough as possible. The media ad may be appealing, but some improvements are possible.

2 The advertising campaign is incomplete. The written plan is incomplete and not supported by the data. The media ad is incomplete or unclear. This project should be returned with suggestions for improvements and a new deadline.

Plan an Advertising Campaign

Your project is to plan an advertising campaign for a product or service that you think will appeal to students in your school. You may use a product or service that already exists or invent a new one.

Your group's plan should explain how you decided whether to advertise on television or radio, or in newspapers or magazines. You will need to do research to determine the target markets and advertising costs for the different media. You will compare unit costs for different media and use probability to help you match up your target market with a media audience.

Your group's plan should also include a schedule of when and how often your ad will appear, as well as an estimate of costs.

If your group chooses to run its ad on radio or TV, you will write a script for the ad. If you choose a newspaper or magazine, you will make a sketch of the ad sized to fit the available space.

HOW MANY PEOPLE DOES IT TAKE TO PRODUCE AN AD?

Every ad is the result of the organized efforts of a creative team.

Account Executives
Plan and direct the ad campaign for a business.

Creative Directors

Photographers, Writers, and Artists

Media Planners and Buyers
Bring the ad to the consumer.
Media planners decide when and where ads run.
Media buyers purchase time or space for ads.

299

Suggested Rubric for Unit Project

1 The advertising campaign cannot be evaluated. It is illegible or incomplete. The plan should be returned with a new deadline for completion. The group should be encouraged to speak with the teacher so that they understand the purpose and the format of the project.

Support Materials

The *Project Book* contains information about the following topics for use with this Unit Project.

➤ Project Description
➤ Teaching Commentary
➤ Working on the Unit Project Exercises
➤ Completing the Unit Project
➤ Assessing the Unit Project
➤ Alternative Projects
➤ Outside Resources

Using Technology

Have students create a data file to organize the information for the ad campaign.

Limited English Proficiency

Creating an advertising campaign is a valuable project for students acquiring English. It helps reinforce the English words that students hear on TV and radio or see in print in magazines and newspapers. Pair a student fluent in English with a student acquiring English to help that student understand the language used in the ad campaign.

ADDITIONAL BACKGROUND

Multicultural Note

Some international newspapers, such as the *Wall Street Journal* and *USA Today*, publish an international edition, a European edition, and an Asian edition. The European edition of the *Wall Street Journal* is printed in the Netherlands, Switzerland, and the U.K., and circulates throughout Europe and the Middle East. The Asian edition is printed in Hong Kong, Singapore, and Tokyo, and circulates throughout Asia and Australia.

Advertising

Advertising costs money, so advertisers want to reach particular people. People can be grouped by age, sex, cultural background, income, and lifestyle. Programs on radio and TV, as well as newspapers and magazines, have a profile, or rough idea, of which people are in their audience. Advertisers can choose the media to reach those people who are possible customers.

ALTERNATIVE PROJECTS

Project 1, page 351

Changing the Rules

Select a game in which dice are thrown to determine players' moves around a game board. Investigate how changing the rules of the game affects the probabilities of several events in the game.

Project 2, page 351

Finding Ratios

Research the history of the golden ratio. Find examples of its use in art, architecture, and interior design.

Getting Started

For this project you should work in a group of four students. Here are some ideas to help you get started.

☞ In your group, talk about the kinds of products and services students in your school may like.

☞ Decide whether to use an existing product or service or invent a new one.

☞ Discuss whether controlling costs or reaching your target audience is more important in deciding where to place your ad.

☞ Your group may wish to think about whether it is better to have a long ad that appears only a few times or a shorter ad that runs many times.

☞ You will be keeping a log of your TV viewing and newspaper and magazine reading. Think about how you can keep good records.

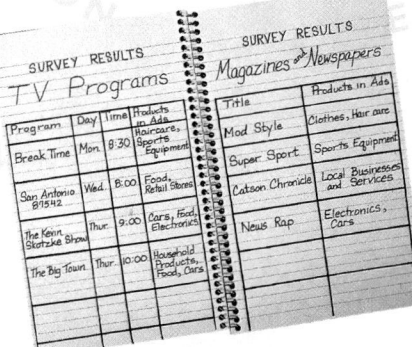

Working on the Unit Project

Your work in Unit 6 will help you plan your advertising campaign.

Related Exercises:

Section 6-1 Exercise 17
Section 6-2, Exercises 34–36
Section 6-3, Exercises 38–40
Section 6-4, Exercise 17
Section 6-5, Exercises 32, 33
Section 6-6, Exercise 24
Section 6-7, Exercise 29

Alternative Projects p. 351

Can We Talk

> Where else have you seen ads? Which places do you think are the most effective? Which do you think are the most expensive?

> Think of a product that is popular at your school. Have you seen any advertisements for the product? Where did the ads appear?

> Have you ever bought a product because it was advertised on television or radio? in a magazine or a newspaper? If so, how did the ad convince you?

> How do you think advertising affects the costs of products?

> What kinds of knowledge and skills do you think media planners and media buyers need?

> A company proposed placing a giant billboard in orbit around Earth. The ads on the billboard would be clearly visible in the night sky. The proposal was very controversial. What do you think of this idea?

Answers to Can We Talk?

> Ads are found on billboards, in airports, on TV, in newspapers, in magazines, in theater Playbills, on public transportation, and so on. TV commercials and color magazine ads are the most expensive form of advertising. TV commercials, however, have among the lowest *per person* cost due to the large viewing audience.

> Answers may vary.

> Answers may vary.

> Planning, developing, producing, and placing ads can be very expensive and those costs are considered when the price of an item is calculated.

> Media planners must know what consumers are interested in. They must be organized, determine the appropriate budget and ad distribution, and have good communication skills.

> Answers may vary.

Ratios and Rates

·····Focus
Write ratios in simplest
form, solve problems
involving ratios, and
▼ compare unit rates.

The **Big** Picture

Talk it Over

1. The size of a television screen is usually given by the length of the diagonal. What size is this TV screen?

2. A **ratio** is a quotient of two numbers or two quantities. You can write the ratio comparing quantity a to quantity b in three ways:

$$\frac{a}{b} \quad a{:}b \quad a \text{ to } b$$

What is the ratio of the length to the width of this TV screen?

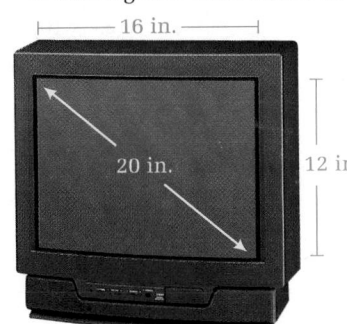

├── 16 in. ──┤

20 in.   12 in.

3. The ratio of length to width of a TV screen is called the *aspect ratio*. Standard TV screens have an aspect ratio of 4:3. Is this TV screen a standard screen? Why or why not?

Student Resources
p. 666 *Table of Measures*
p. 649 *Factors*

Sample 1

Find the aspect ratio of a movie screen that is 24 ft long and 13 ft 6 in. wide. Write the ratio in simplest form.

Sample Response

1 ft = 12 in.
24 ft = 24 · 12 in.

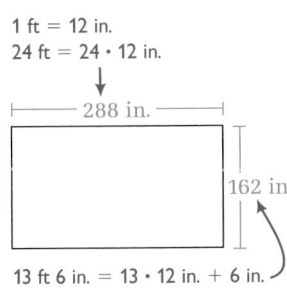

├── 288 in. ──┤

162 in.

13 ft 6 in. = 13 · 12 in. + 6 in.

First write the ratio as a fraction.

$$\frac{\text{length}}{\text{width}} = \frac{288 \text{ in.}}{162 \text{ in.}} \qquad \longleftarrow \quad \text{Write each measure with the same unit.}$$

Then simplify the ratio.

$$\frac{288}{162} = \frac{288 \div 18}{162 \div 18} = \frac{16}{9} \qquad \longleftarrow \quad \text{Divide the numerator and denominator by the greatest common factor.}$$

The ratio is $\frac{16}{9}$ *or* 16:9 *or* 16 to 9.

PLANNING

Objectives and Strands
See pages 298A and 298B.

Spiral Learning
See page 298B.

Materials List
➤ Geometric drawing software
➤ Graph paper

Recommended Pacing
Section 6-1 is a one-day lesson.

Toolbox References
➤ **Toolbox Skill 8:** Finding GCF and LCM
➤ **Toolbox Skills 11–15:** Percent

Extra Practice
See pages 629–631.

Warm-Up Exercises
Warm-Up Transparency 6-1

Support Materials
➤ Practice 44
➤ Enrichment 39 in the Activity Bank
➤ Study Guide 6-1
➤ Problem Set 12
➤ Diagram Master 2 in the Explorations Lab Manual
➤ McDougal Littell Mathpack software: *Geometry Inventor* and *Stats!*
➤ Quiz 6-1
➤ Alternative Assessments 1–3

Answers to Talk it Over

1. 20 in.

2. $\frac{16}{12}$

3. Yes. The aspect ratio, $\frac{16}{12}$, can be simplified to $\frac{4}{3}$.

TEACHING

Talk it Over

For questions 1–3, you might want to have students measure the diagonal of a TV screen owned by the school. Have them also measure the length and width of the screen and calculate its aspect ratio.

Additional Samples

S1 Find the aspect ratio of a TV screen that is 30 in. long and 24 in. wide. Write the ratio in simplest form.

$$\frac{\text{length}}{\text{width}} = \frac{30 \text{ in.}}{24 \text{ in.}} = \frac{5}{4}, \text{ or } 5:4$$
or 5 to 4

S2 The ratio of computer desks to regular desks sold at Furniture World is $8:3$. Of the next 100 desks sold, how many do you estimate will be computer desks?

$$8x + 3x = 100$$
$$11x = 100$$
$$x \approx 9.1$$
$$8x = 8(9.1) \approx 73$$

About 73 of the next 100 desks sold will be computer desks.

Error Analysis

Students sometimes make errors in comparing measurements because they do not use the same units. Illustrate this error by asking students to compare 3 yd to 1 ft. The ratio is 9:1, not 3:1.

Talk it Over

Use question 4 to stress the fact that a percent always involves the comparison of a number to 100. Remind students that the answer, 15%, means that for every 100 TVs sold, 15 are black and white sets.

Sample 2

The ratio of color TVs to black-and-white TVs sold at Electronics City is about 11:2. Of the next 100 TVs sold, how many do you estimate will be color sets?

Sample Response

The ratio 11:2 means that for every 13 sets sold, 11 are color and 2 are black and white.

Total sold	Color	BW
13	11	2
26	22	4
39	33	6
⋮	⋮	⋮
$13x$	$11x$	$2x$

Problem Solving Strategy: Use an equation

color TVs + BW TVs = total TVs

$$11x + 2x = 100$$
$$13x = 100 \qquad \longleftarrow \text{Combine like terms.}$$
$$\frac{13x}{13} = \frac{100}{13} \qquad \longleftarrow \text{Divide each side by 13.}$$
$$x \approx 7.7$$

Evaluate the expression $11x$ for $x = 7.7$.

$$11x = 11(7.7) \approx 85$$

About 85 of the next 100 TVs sold will be color sets.

Talk it Over

4. Remember that a *percent* is a special ratio in which the second number is 100. In Sample 2, about what percent of the TVs sold will be black-and-white sets?

5. List at least three situations where percents are used.

Student Resources Toolbox
pp. 653–656 *Percent*

Rates

A commercial shown during the Olympics may cost $110,000 for 30 seconds. The ratio 110,000 : 30 is an example of a *rate*. A **rate** is a ratio of two different types of measurements.

A **unit rate** is a rate for one unit of a given quantity. Some examples of unit rates are *55 miles per hour, 33 miles per gallon,* and *85 cents per pound*. A *unit price* is a unit rate that tells the price for one unit of measure.

Unit 6 Ratios, Probability, and Similarity

Answers to Talk it Over

4. about 15%

5. Answers may vary. Examples are given: discounts at stores, newspaper survey results, test grades.

Answers to Look Back

Answers may vary. An example is given. A rate is a special kind of ratio because it is the quotient of two numbers having different types of measurements. Examples of rates in everyday life are speed limits, such as 55 mi/h; wages, such as $8.25/h; and costs, such as $.79/lb.

Sample 3

During a regular prime-time TV program, 1.5 min of commercial time may cost $240,000. During the Olympics, a commercial may cost $110,000 for 30 seconds. Compare the unit prices for commercials during the Olympics and during a regular prime-time program.

Sample Response

Find the unit price for each commercial.

Olympics:

$$\frac{\text{cost}}{\text{seconds}} = \frac{\$110,000}{30 \text{ s}}$$

$$\approx \$3670 \text{ per second}$$

Regular program:

$$\frac{\text{cost}}{\text{seconds}} = \frac{\$240,000}{90 \text{ s}} \longleftarrow 1.5 \text{ min} = 90 \text{ s}$$

$$\approx \$2670 \text{ per second}$$

A commercial during the Olympics costs about $1000 more per second than a commercial during a regular prime-time program.

Look Back

In what way is a rate a special kind of ratio? Give examples of rates that you have used.

6-1 Exercises and Problems

1. **Open-ended** Make a concept map for this section. Include the terms *ratio, percent, rate, unit rate,* and *unit price,* and examples of each.

2. **Movie Screens** One of the world's largest movie screens is in the Keong Emas Theater in Jakarta, Indonesia, and is 92.75 ft long and 70.5 ft wide.

   a. What is the aspect ratio of this screen?

   b. How does this aspect ratio compare with the standard TV screen ratio of 4:3?

3. What is the ratio of the length of a side of a square to its perimeter?

4. What is the ratio of the length of a side of an equilateral triangle to its perimeter?

Answers to Exercises and Problems

1. Concept maps may vary. An example is given.

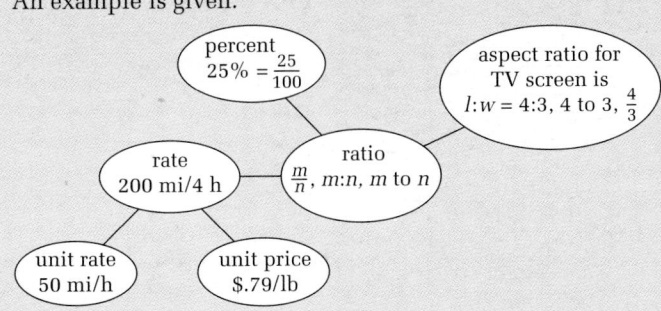

2. a. $\frac{92.75}{70.5} = \frac{3.71}{2.82}$, or about $\frac{4}{3}$

   b. This aspect ratio is about the same as the standard TV screen ratio.

3. 1:4   4. 1:3

Additional Sample

S3 The regular price for certain audiotapes is $18 for 3 tapes. During a special promotion, the same audiotapes cost $9 for 2 tapes. Compare the unit prices for audiotapes during a special promotion and at regular prices.

Regular: $\frac{\text{cost}}{\text{number}} = \frac{\$18}{3} =$ $6 per tape;

Promotion: $\frac{\text{cost}}{\text{number}} = \frac{\$9}{2} =$ $4.50 per tape

An audiotape purchased during a promotion cost $1.50 less than the regular price.

Look Back

You can use the Look Back to have a class discussion about rates that are familiar to students. Make sure students understand that in a rate both units of measure are always mentioned.

APPLYING

Suggested Assignment

Standard 1–6, 8, 10, 13–17

Extended 1–8, 10, 13–17

Integrating the Strands

Number Exs. 1–12

Algebra Exs. 13–15

Geometry Exs. 3, 4, 9

Statistics and Probability Exs. 12, 16

Discrete Mathematics Ex. 12, 17

Logic and Language Exs. 1, 7, 12, 17

Multicultural Note

For Ex. 2, point out that Jakarta, on the island of Java, is the capital and largest city in Indonesia. Indonesia consists of more than 13,600 islands, and ranks fourth in population size in the world. Nearly three-fifths of all Indonesians live on the island of Java.

303

5. What is the ratio of the length of the part of the telephone pole above the ground to the length of the part underground?

 ← $\frac{1}{5}$ of the pole is underground.

6. A package of two 120-page notebooks sells for $2.29. A package of three 100-page notebooks sells for $2.49.

  a. What is the price per page for each package of notebooks? Which is the better buy? Why?

  b. Suppose the package of two notebooks goes on sale for 20% off. Which package will be the better buy? Why?

7. **Automobiles** Use the magazine article.

  a. What is the height in millimeters of the sidewall of a tire marked like this?

    P195/75R14 90H

  b. What is the total height in inches, including the rim, of the tire in part (a)? (*Hint:* Use the Table of Measures on page 666.)

  c. **Writing** Some drivers prefer a tire with a lower sidewall. Do you think that these drivers should buy tires with a larger or smaller aspect ratio? Explain.

  d. **Research** When you buy tires, you can choose the tire width and the aspect ratio. Look at an ad for tires. Choose one type of tire. Make a scatter plot comparing the tire width to the price. What conclusions can you draw?

8. Choose the letter of the rate of pay that gives you the highest yearly earnings. Assume you work 40 h per week.

  a. $20,000 per year

  b. $1600 per month

  c. $395 per week

  d. $9.85 per hour

9 TECHNOLOGY Use geometric drawing software.

  Open-ended Draw several triangles with the same ratio for the lengths of their sides. What is the ratio of the measures of the angles?

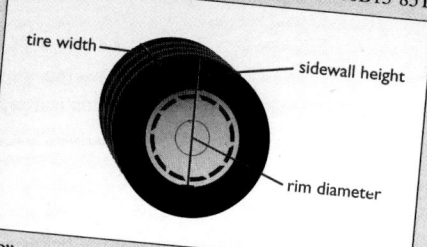

What Your Tires Can Tell You

The sidewall of a tire can make some interesting reading when your car needs new tires. A typical sidewall may contain the marking P205/60B15 85T.

tire width — sidewall height — rim diameter

"P" means the tire is for a passenger car.

"205" is the width of the tire in millimeters.

"60" is the aspect ratio, or the sidewall height as a percent of the tire width.

"B" stands for bias-belted. Other letters that indicate the type of construction include "R" for radial and "D" for diagonal bias.

"15" is the diameter of the rim in inches.

"85" is a code for the maximum weight the tire can carry.

"T" is the speed symbol. Other letters used are H, S, U, V, and Z.

Answers to Exercises and Problems

5. 4:1

6. a. 0.95¢/page; 0.83¢/page; The package of 3 notebooks is the better buy because the cost per page is lower.

  b. The package of 2 notebooks would be the better buy. The sale would result in a new price of 0.76¢/page.

7. a. 146.25 mm

  b. about 25.5 in.

  c. Answers may vary. An example is given. Drivers should buy tires with smaller aspect ratios; if the aspect ratio is larger, then the height of the tire would be greater, and the advantages of more stability and better control might be reduced.

  d. Answers may vary. In general, the price per tire increases as the tire width increases.

8. c

GEOGRAPHY

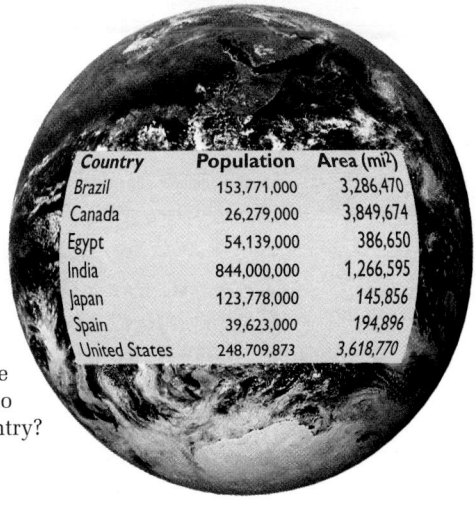

Country	Population	Area (mi²)
Brazil	153,771,000	3,286,470
Canada	26,279,000	3,849,674
Egypt	54,139,000	386,650
India	844,000,000	1,266,595
Japan	123,778,000	145,856
Spain	39,623,000	194,896
United States	248,709,873	3,618,770

10. a. Find the population per square mile for each country listed in the table.

b. Which country has the largest number of people per square mile? the smallest number of people per square mile?

c. Does the country with the highest rate have the most people? Does the country with the lowest rate have the least area?

11. Research Find the population per square mile for your city or town and for your state. How do these rates compare with the rate for your country?

Ongoing ASSESSMENT

12. Group Activity Work in a group of three students.

a. Research Find the population and the number of United States representatives for the United States and for each state.

b. Social Studies For each state, calculate the ratio *people : representatives*. Describe the lowest and highest ratios.

c. Make a scatter plot showing the relationship between population and number of representatives. What conclusions, if any, can you draw?

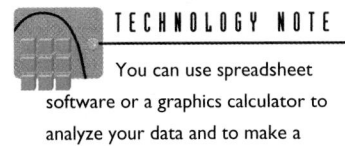

TECHNOLOGY NOTE

You can use spreadsheet software or a graphics calculator to analyze your data and to make a scatter plot.

Review PREVIEW

For Exercises 13–15, solve each system of equations. *(Section 5-8)*

13. $b = 3a$
$34 = 25a + 3b$

14. $r = -s$
$-5r + 18s = 115$

15. $d = 6e + 1$
$2d - 14e = -2$

16. Open-ended Use a number anywhere along the scale to estimate the probability that on your way home from school today you will meet someone who speaks Spanish. Give a reason for your answer. *(Section 2-1)*

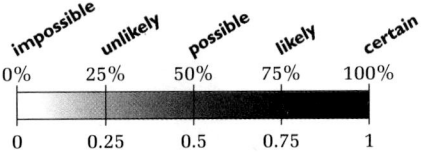

impossible	unlikely	possible	likely	certain
0%	25%	50%	75%	100%

| 0 | 0.25 | 0.5 | 0.75 | 1 |

Working on the Unit Project

17. Research Keep a daily record of the TV programs you watch and the magazines and newspapers you read.

a. List the name of each TV program you watch, the day and time you watch it, and the products advertised during commercial breaks.

b. List each magazine or newspaper you look at or read, and the products in the ads you look at or read.

Answers to Exercises and Problems

9. If triangles *ABC* and *XYZ* have their vertices matched so that *AB*:*XY* = *BC*:*YZ* = *AC*:*XZ*, then the measures of matching angles will have the ratio 1:1.

10. a. Brazil: about 47 people/mi²; Canada: about 7 people/mi²; Egypt: about 140 people/mi²; India: about 666 people/mi²; Japan: about 849 people/mi²; Spain: about 203 people/mi²; United States: about 69 people/mi²

b. Japan; Canada

c. No; No.

11. Answers may vary.

12. a. Answers may vary.

b. Paragraphs may vary.

An example is given. Montana has the highest ratio for representatives, about 799,000:1 (since Montana has only 1 United States representative). Wyoming has the lowest ratio for representatives, about 454,000:1 (since Wyoming also has only 1 United States representative).

c. Scatter plots may vary. You can conclude that the number of representatives is a function of the population. The greater the state's population, the greater the number of representatives for that state.

13. $a = 1, b = 3$ **14.** $r = -5, s = 5$

15. $d = 13, e = 2$

16. Answers may vary.

17. Answers may vary.

305

PLANNING

Objectives and Strands
See pages 298A and 298B.

Spiral Learning
See page 298B.

Materials List
➤ Paper clips
➤ Red and blue markers
➤ Protractor
➤ Compass
➤ Graphics calculator (optional)
➤ Spreadsheet software (optional)
➤ Coin
➤ Graph paper

Recommended Pacing
Section 6-2 is a three-day lesson.
Day 1
Pages 306–307: Exploration,
Exercises 1–6
Day 2
Pages 308–309: Experimental
Probability through Talk it Over 2,
Exercises 7–13
Day 3
Pages 309–311: Theoretical
Probability through Look Back,
Exercises 14–36

Extra Practice
See pages 629–631.

Warm-Up Exercises
Warm-Up Transparency 6-2

Support Materials
➤ Practice 45
➤ Enrichment 40 in the Activity
Bank
➤ Study Guide 6-2
➤ Problem Set 12
➤ Diagram Masters 2, 14, 15 in the
Explorations Lab Manual
Overhead Visual 7
➤ McDougal Littell Mathpack
software: *Probability Constructor*
➤ Probability Constructor Activity
Book: Activities 1, 2, 6, 9, 11, 12,
and 16
➤ Quiz 6-2
➤ Alternative Assessments 4, 5

306

Section

6-2 Investigating Probability

Focus
Describe how likely it is
that something will happen.

Take a SPIN

EXPLORATION

*Can you predict how often a spinner lands
on red?*

• **Materials:** paper clips, red and blue markers, protractors,
compasses
Optional: graphics calculator or spreadsheet software

• **Work in a group of three students.**

① To get ready for this experiment, assign each person in
your group one of these tasks:

Student 1:
Draw a large circle with eight equal
wedges numbered 1–8 and colored
as shown.

Student 2:
Straighten one end of a paper clip
as shown and use a pencil to hold
it at the center of
the circle.

306

Answers to Exploration

1. manual activity

2, 3. a. A sample table of values is given.

	Number of Spins	Number of Landings on Red	Number of Landings on Red / Number of Spins
Student 1	10	4	$\frac{4}{10} = 0.40$
Students 1 and 2	20	8	$\frac{8}{20} = 0.40$
Students 1, 2, and 3	30	11	$\frac{11}{30} \approx 0.37$

	Number of spins	Number of landings on red	$\dfrac{\text{Number of landings on red}}{\text{Number of spins}}$
Student 1	10	?	$\dfrac{?}{10} = ?$
Students 1 and 2	20	?	$\dfrac{?}{20} = ?$
Students 1, 2, and 3	30	?	$\dfrac{?}{30} = ?$

Student 3:
Make a table like this one to record your results.

② Have each person in your group spin the spinner 10 times. For each person, keep track of the number of times that the spinner lands on red.

  a. Record Student 1's results in the first row of the table.

  b. Add Student 2's results to Student 1's results. Then record the combined results in the second row.

  c. Add Student 3's results to the results of the other two students. Then record the combined results for your group in the third row.

③ **a.** Write the three ratios in the third column of the table as decimals. Round your answers to the nearest hundredth.

  b. Discuss how the three ratios compare.

④ Graph your group's results on a coordinate plane. Plot the points and connect them with segments.

landings on red
to ⟶
number of spins

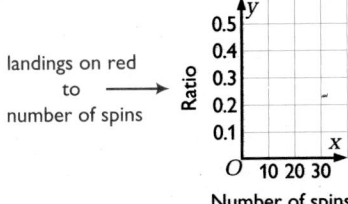

⑤ Combine your group's results for the number of spins and the number of landings on red for Students 1, 2, and 3 with the results of the other groups. Plot the combined results on your graph.

⑥ Do you see any pattern in the points that were plotted? If so, describe it.

⑦ What conclusions, if any, can you draw about how the number of spins affects the ratio?

⋯▶ **Now you are ready for:**
  Exs. 1–6 on p. 311

Answers to Exploration

3. b. The three ratios are all about the same.

4. The graph shown is based on the sample table given in steps 2–3a.

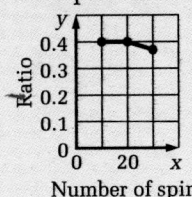

5. The results from the other groups should be about the same.

6, 7. Answers may vary. Examples are given.

6. Yes. The points are almost on the same level, with y-coordinates around 0.375.

7. As the number of spins gets larger, the ratio gets closer to 0.375.

Exploration

The goal of the Exploration is to have students generate the experimental probability of having a spinner land on a red sector of a circle. Students are then asked to graph the number of spins versus the ratio of red landings. This leads naturally to representing the experimental probability of landing on red as a function of the number of spins. This experimental approach to probability gives students an intuitive feeling for assigning probabilities to events.

During the Exploration, students should discuss the results of the experiment. A key idea is that as the number of spins increases, the ratio gets closer to 0.375. You can generalize this idea, if you wish, by pointing out that in any probability experiment, the number of times an event happens *stabilizes* as the number of times the experiment is done increases to a very large number.

Using Technology

Students can use the *Probability Constructor* software to simulate spinning the spinner in the Exploration.

Experimental Probability

Event: landing on red

some possible outcomes

Surveys used to collect data and activities like the one in the Exploration in which you observe and record data are *experiments*. There are one or more possible results or **outcomes** of an experiment that can make an **event** happen.

To describe how likely an event is, you use a number on a scale from 0 to 1. This number is called the **probability** of the event. Probability that is based on the result of an experiment is called **experimental probability**.

X₍.Δab

EXPERIMENTAL PROBABILITY

To find the experimental probability of an event E, you observe how often the event happens and calculate this ratio:

$$P(E) = \frac{\text{number of times an event } E \text{ happens}}{\text{number of times the experiment is done}}$$

You read $P(E)$ as, "the probability of event E."

Sample 1

A telemarketer types these requirements into a computer:

Age: 25 years or over

Last schooling: 4 years of high school

This means that everyone in the database has an equal chance of being selected.

The software selects a name *at random* from a database that contains the names and phone numbers of every person in the United States.

Last Schooling of U.S. Residents 25 Years or Older, 1991 (numbers in thousands)			
Last years of schooling	Total	Male	Female
Elementary: 0–8 years	16,849	8317	8531
High School: 1–3 years	17,379	7887	9491
High School: 4 years	61,272	27,189	34,083
College: 1–3 years	29,170	13,720	15,449
College: 4 or more years	34,026	18,373	15,652
TOTALS	158,694	75,487	83,207
Note: Numbers may not add correctly due to rounding.			

Use the table to estimate the probability that the person called is female. Round decimal answers to the nearest hundredth.

Unit 6 Ratios, Probability, and Similarity

Problem Solving Strategy: Use a formula

Use the formula for experimental probability.

$$P(E) = \frac{34{,}083}{61{,}272} \longleftarrow \text{female residents at least 25 years old with 4 years of H.S.} \\ \longleftarrow \text{total residents at least 25 years old with 4 years of H.S.}$$

$$\approx 0.556$$

The probability is about 0.56, or 56%.

Talk it Over

1. How likely is it that the person called in Sample 1 is a female? How did you get your answer?

2. Suppose the telemarketer in Sample 1 omits the "last schooling" requirement. Estimate the probability of selecting a male whose last schooling is 1–3 years of college.

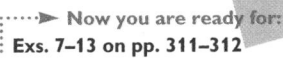

▶ Now you are ready for:
Exs. 7–13 on pp. 311–312

Theoretical Probability

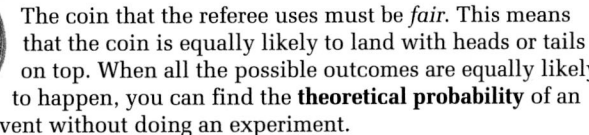

Possible outcomes

heads   tails

Before a field hockey game, the referee tosses a coin to see which team has the ball first. For the team that chose heads, heads is a *favorable outcome.*

The coin that the referee uses must be *fair.* This means that the coin is equally likely to land with heads or tails on top. When all the possible outcomes are equally likely to happen, you can find the **theoretical probability** of an event without doing an experiment.

If you toss a fair coin over and over, the experimental probability of getting heads will get closer and closer to the theoretical probability, 0.5.

THEORETICAL PROBABILITY

To find the theoretical probability of an event E when all outcomes are equally likely, you find this ratio:

$$P(E) = \frac{\text{number of favorable outcomes}}{\text{number of possible outcomes}}$$

Answers to Talk it Over

1. Since the probability is a little more than 50%, it is just slightly more likely the person is a female than not.

2. $P(E) =$

$$\frac{\text{male residents at least 25 yr old with 1–3 years of college}}{\text{total residents at least 25 yr old}}$$

$$= \frac{13{,}720}{158{,}694} \approx 0.086. \text{ The}$$

probability is about 0.09 or 9%.

There are two ideas in this section that would benefit from a class discussion, namely a random event and equally likely outcomes. These ideas provide a conceptual foundation for an understanding of both experimental and theoretical probability.

Problem Solving

Ask students how they would solve the problem of finding the probability that a thumbtack dropped onto the floor from a height of three feet will land on its head or on its side.

Limited English Proficiency

Activities in the kinesthetic domain can help students acquiring English gain an understanding of key concepts. If you observe that some students are having difficulty understanding the concepts of ratio, proportion, and probability, consider having them use real objects or manipulatives to carry out activities presented in symbolic form in the unit. Students should work in small groups and discuss what they are doing with one another.

Using Technology

To reinforce the basic concepts of probability and the relationship between experimental and theoretical probability, you may wish to assign some or all of Activities 6, 9, 11, and 12 in the *Probability Constructor Activity Book.*

Sample 2

Suppose you roll a fair die. Find each probability.

a. $P(\boxdot)$ **b.** $P(\boxdot \text{ or } \boxdot)$ **c.** $P(\text{not } \boxdot)$

Sample Response

a roll of 5 on a die

There are six possible outcomes: ⚀ ⚁ ⚂ ⚃ ⚄ ⚅

a. There is one favorable outcome: ⚂

$P(\boxdot) = \dfrac{1}{6} \approx 0.17 \approx 17\%$

b. There are two favorable outcomes: ⚁ ⚄

$P(\boxdot \text{ or } \boxdot) = \dfrac{2}{6} \approx 0.33 \approx 33\%$

c. There are five favorable outcomes: ⚀ ⚁ ⚂ ⚄ ⚅

$P(\text{not } \boxdot) = \dfrac{5}{6} \approx 0.83 \approx 83\%$

Complementary Events

Sample 2(c) shows the event of not rolling a ⚅.
The events "not rolling a ⚅" and "rolling a ⚅" are **complementary events**.

←	All Possible Events	→
E	E does not happen	

COMPLEMENTARY EVENTS

You can find the probability of the *complement* of an event by using this formula:

$$P(\text{not } E) = 1 - P(E)$$

The probability that E does **not** happen.

BY THE WAY...

The numbers of dots on dominoes represent the outcomes of rolling two dice. Dominoes first appeared in China about 800 years ago.

Talk it Over

3. For the spinner used in the Exploration, an event may be defined as landing on a color or on a number. How many "color" events are there?

4. Sample 2(b) shows the probability of getting a ⚁ or a ⚄ when a die is rolled. What do you think the probability is of getting a ⚁ *and* a ⚄ when a die is rolled? Explain.

Unit 6 Ratios, Probability, and Similarity

5. Why is there a "1" in the formula for the complement of an event?

6. How can the formula for the complement of an event be rewritten so it contains a sum rather than a difference?

Look Back ←

How are experimental probability and theoretical probability different? How are they alike?

► **Now you are ready for:**
Exs. 14–36 on pp. 312–313

6-2 Exercises and Problems

For Exercises 1–4, use the spinner in the Exploration.

Tell whether each event is impossible or certain to happen.

1. landing on green
2. landing on a number greater than 10
3. landing on a single-digit number
4. landing on red, white, or blue

Tosses	Heads	$\frac{\text{Heads}}{\text{Tosses}}$
10	?	?
20	?	?
⋮	⋮	⋮
100	?	?

5. **Using Manipulatives** Toss a coin 100 times.
   a. Keep track of your results. Complete a table like this for 10, 20, 30, 40, 50, 60, 70, 80, 90, and 100 tosses.
   b. Graph your results on a coordinate grid. Put the number of tosses on the horizontal axis and the ratio of heads to tosses on the vertical axis.
   c. **Writing** What conclusion can you draw about the chance of tossing a head? Explain your reasoning.

6. **Reading** When is the probability of an event 0? When is it 1?

7. Suppose there are 2000 students in your school and 1158 of them are female. A reporter for the school paper chooses a student at random to interview for the next issue.
   a. Is it more likely that a male or female is chosen? Why?
   b. What is the probability that the student chosen is male?

8. **Open-ended** Find a graph in a newspaper, a magazine, another textbook, or this book. Write two probability problems that can be answered by the data in the graph.

9. **Writing** Suppose you toss a coin and get 6 heads in a row. Do you think the probability is greater than 50% that you will get a tail in the next toss? Explain.

> **BY THE WAY...**
>
> The probability of winning a typical state lottery may be about 0.0000001. This is one tenth the probability of being hit by lightning.

6-2 Investigating Probability **311**

Look Back

Ask students to write their Look Back responses as journal entries. After they have done so, ask for volunteers to read their entries to the class.

APPLYING

Suggested Assignment
Standard 1–4, 6, 7, 9–23, 25–36
Extended 1–4, 6, 7–23, 25–36

Integrating the Strands
Number Exs. 25–30
Algebra Exs. 31–33
Statistics and Probability Exs. 1–24, 34–36
Discrete Mathematics Exs. 1–24, 34–36

Communication: Listening
Encourage students to discuss the answers to the exercises and problems. They will benefit from hearing their classmates' explanations, and may wish to ask their classmates questions based on what they have heard.

Cooperative Learning
For Ex. 5, you may want students to do the experiment in groups.

Using Technology
Students can use the *Probability Constructor* software to simulate tossing a coin for Ex. 5.

Answers to Exercises and Problems

1. impossible
2. impossible
3. certain   4. certain
5. a–c. Results and explanations may vary. The theoretical probability of tossing a head is 0.5. The experimental probabilities should cluster around 0.5 and get closer with the increase in tosses.

6. The probability of an event is 0 when there is no possibility that the event can happen. The probability of an event is 1 when it happens with certainty.

7. a. Answers and explanations may vary. An example is given. The probability of choosing a female is about 58%, and the probability of choosing a male is about 42%. Therefore, there is a slightly better probability of a female student being chosen.
   b. about 0.42 or 42%

8. Graphs and problems may vary.
9. Answers may vary. An example is given. No, the probability for the event will remain at 50%. Tossing a fair coin in an example of theoretical probability and the probability of a tail is equally as likely as the probability of a head regardless of any previous outcomes.

Interdisciplinary Problems

The study of probability had its origin with games of chance. Today mathematical probability is a subject whose concepts and procedures are used to solve problems from a wide range of real-world situations. The exercises and problems on pages 311–313 touch upon a number of interesting topics that illustrate the focus of the section. In doing Ex. 8, students will uncover additional situations related to probability. You may wish to point out that the mathematics of probability plays a crucial role in explaining the structure of the basic building blocks of all matter, namely atoms by means of a theory called *quantum theory*.

For Exercises 10–13, use the graph.

The manager of Pizza Place kept a record of the types of crusts used for the last 1000 pizza orders. The manager uses the results to decide how much pizza dough to make.

10. Which type of crust is a customer chosen at random most likely to order?

11. a. What is the probability that a customer chosen at random orders a thick crust?

b. What is the probability that a customer chosen at random does not order a thick crust?

12. Of the next 500 pizzas ordered from Pizza Place, about how many will have thin crusts?

13. A cheese pizza with a stuffed crust costs $2 more than one with a thin crust. Of the next 500 cheese pizzas ordered, does Pizza Place expect to earn more money on stuffed crusts or thin crusts?

Number of pizzas

Types of crusts

- Thin — 581
- Thick — 239
- Pan — 156
- Stuffed — 24

For Exercises 14–22, use the information about playing cards.

Games A standard deck of playing cards consists of 52 cards, with 13 cards in each of four *suits:* clubs, spades, diamonds, and hearts. *Face cards* are jacks, queens, and kings.

Clubs (♣):	ace, 2, 3, 4, 5, 6, 7, 8, 9, 10, jack, queen, king
Spades (♠):	ace, 2, 3, 4, 5, 6, 7, 8, 9, 10, jack, queen, king
Diamonds (♦):	ace, 2, 3, 4, 5, 6, 7, 8, 9, 10, jack, queen, king
Hearts (♥):	ace, 2, 3, 4, 5, 6, 7, 8, 9, 10, jack, queen, king

Find the probability of choosing each card at random.

14. $P(\text{diamond})$ **15.** $P(\text{face card})$ **16.** $P(\text{ace})$

17. $P(\text{not a club})$ **18.** $P(\text{not a 10 of diamonds})$ **19.** $P(\text{club or spade})$

20. $P(\text{red or black card})$ **21.** $P(\text{5 of hearts})$ **22.** $P(\text{numbered card and club})$

23. A bag contains marbles of different colors. Use the clues below to find the probability of selecting each color. Complete the table.

Color	Probability
brown	0.3
red	?
yellow	?
green	?
orange	?
tan	?

 Clue 1: Choosing brown is three times as likely as choosing tan.

 Clue 2: Choosing green and choosing orange are each as likely as choosing tan.

 Clue 3: Choosing red is as likely as choosing yellow.

312

Unit 6 Ratios, Probability, and Similarity

Answers to Exercises and Problems

10. thin crust

11. a. about 0.24 or 24%

    b. about 0.76 or 76%

12. about 291 thin crusts

13. thin crusts

14. $\frac{13}{52} = \frac{1}{4} = 0.25 = 25\%$

15. $\frac{12}{52} = \frac{3}{13} \approx 0.23$ or 23%

16. $\frac{4}{52} = \frac{1}{13} \approx 0.08$ or 8%

17. $\frac{39}{52} = \frac{3}{4} = 0.75 = 75\%$

18. $\frac{51}{52} \approx 0.98$ or 98%

19. $\frac{26}{52} = \frac{1}{2} = 0.5 = 50\%$

20. $\frac{52}{52} = 1 = 100\%$

21. $\frac{1}{52} \approx 0.02$ or 2%

22. $\frac{9}{52} \approx 0.17$ or 17%

23. red: 0.2; yellow: 0.2; green: 0.1; orange: 0.1; tan: 0.1

24. a–e. Answers may vary.

25. $\frac{8}{5}$ **26.** $\frac{37}{3}$

27. $\frac{3}{5}$

28. square roots ±5; cube root 2.92

29. square roots ±7.62; cube root 3.87

24. **Group Activity** Work in a group of four students. Design an experiment. Choose an object to toss or roll, or choose a situation about which you can collect data. Some suggestions are tossing a paper cup or counting the colors of cars at an intersection.

    a. List the possible outcomes.

    b. Have each person in the group perform the experiment at least 30 times. Record all the results in a table.

    c. Choose one possible event for your experiment. What is its experimental probability?

    d. Suppose your group performed the experiment 200 more times. How many times do you expect the event in part (c) to happen?

    e. **Writing** Describe your experiment and summarize your results.

Review **PREVIEW**

Write each ratio as a fraction in simplest form. *(Section 6-1)*

25. 40 to 25   **26.** 37:3   **27.** 9 to 15

Find the square roots and the cube root of each number. *(Section 2-9)*

28. 25   **29.** 58   **30.** 72

Solve each equation. *(Section 5-6)*

31. $\frac{x}{19.5} = 37$ **32.** $27 = \frac{m}{-6}$ **33.** $\frac{n}{7} = -0.82$

Working on the Unit Project

Use the table.

34. a. Suppose a movie is showing on television. Find the probability that a movie viewer picked at random is an adult woman and the probability that the viewer is an adult man.

    b. Which event from part (a) has the greater probability? Are these complementary events? Why or why not?

35. Based on the data in the table, what kind of show do you think is most popular with teenagers? Give a reason for your answer.

36. Give two examples of how an advertiser could use the data shown.

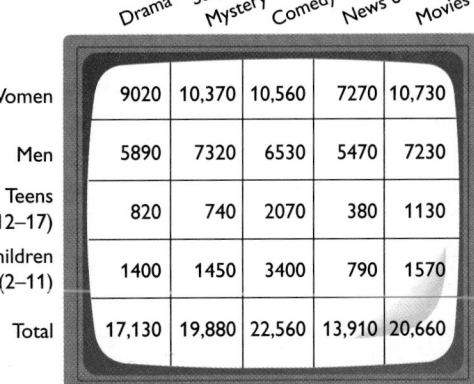

Viewers (Thousands)

	Drama	Suspense/Mystery	Comedy	News 6–7 P.M.	Movies
Women	9020	10,370	10,560	7270	10,730
Men	5890	7320	6530	5470	7230
Teens (12–17)	820	740	2070	380	1130
Children (2–11)	1400	1450	3400	790	1570
Total	17,130	19,880	22,560	13,910	20,660

6-2 Investigating Probability   **313**

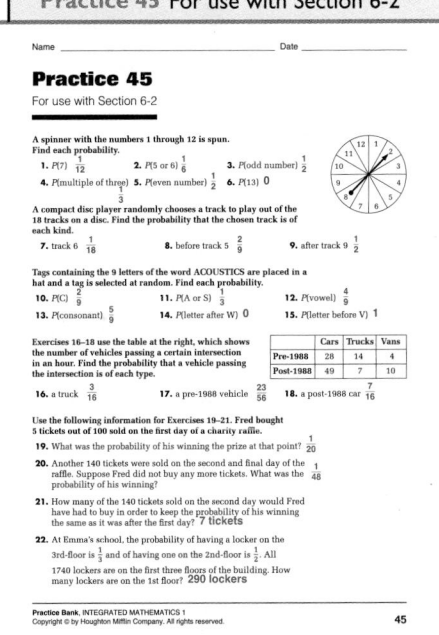

Answers to Exercises and Problems

30. square roots ±8.49; cube root 4.16

31. 721.5   32. –162

33. –5.74

34. a. about 0.52 or 52%; about 0.35 or 35%

    b. adult women; No. Explanations may vary. An example is given. The two events are not complementary because there were other viewers—teens and children.

35. comedy; Reasons may vary. An example is given. The largest number in the row for teens is 2070, the entry for teens who watched comedies.

36. Answers may vary. Examples are given. For drama programs, advertisers would want to promote products that appeal to adult viewers, particularly women, since the majority of viewers are adults. Advertisers would not want to promote products that appeal to teens or children during news programs, since they are the least likely to watch these shows.

6-3 Solving Proportions

Focus
Solve proportions.

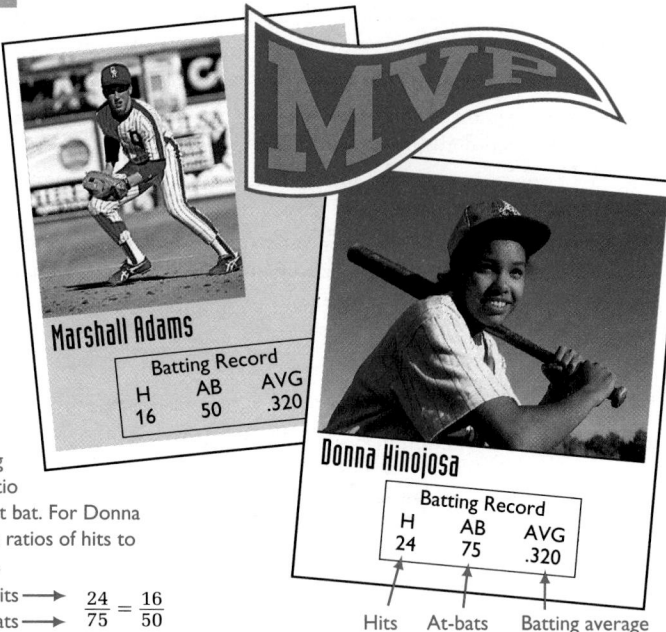

Marshall Adams

Batting Record		
H	AB	AVG
16	50	.320

Donna Hinojosa

Batting Record		
H	AB	AVG
24	75	.320

Hits At-bats Batting average

A player's batting average is the ratio of hits to times at bat. For Donna and Marshall, the ratios of hits to at-bats are equal.

number of hits ⟶
number of at-bats ⟶ $\dfrac{24}{75} = \dfrac{16}{50}$

A statement that two ratios are equal is a **proportion.** The batting averages proportion is read "24 is to 75 as 16 is to 50." The numbers 24, 75, 16, and 50 are the **terms of the proportion.**

Sometimes one of the terms of a proportion is not known. When you *solve* a proportion, you find the unknown term.

Sample 1

Solve the proportion $\dfrac{x}{15} = \dfrac{3}{4}$.

Sample Response

To get *x* alone on one side, undo the division.

$$\frac{x}{15} = \frac{3}{4}$$

$$\frac{x}{15} \cdot 15 = \frac{3}{4} \cdot 15 \quad \longleftarrow \text{Multiply both sides by 15.}$$

$$x = 11.25$$

The solution is 11.25.

Check

$$\frac{x}{15} = \frac{3}{4}$$

$$\frac{11.25}{15} \stackrel{?}{=} \frac{3}{4} \quad \longleftarrow \text{Substitute 11.25 for } x.$$

$$0.75 = 0.75 \ ✔$$

Talk it Over

1. Describe a method for solving the proportion $\frac{x}{12} = \frac{3}{4}$ using mental math.

2. a. What is the lowest common denominator of the fractions in $\frac{x}{15} = \frac{3}{4}$?

   b. Multiply both sides of the proportion by the lowest common denominator of the fractions. Simplify the result. What is the new equation?

   c. Look at the expressions on both sides of the equation you wrote in part (b). How are the factors of these expressions related to the terms of the original proportion?

CROSS PRODUCTS OF A PROPORTION

For any proportion, the two **cross products** formed by multiplying the numerator of one fraction by the denominator of the other fraction are equal.

$$\frac{6}{2} \bowtie \frac{3}{1} \qquad\qquad \frac{a}{b} \bowtie \frac{c}{d}$$

$$6 \cdot 1 = 2 \cdot 3 \quad \longleftarrow \text{cross products} \longrightarrow \quad ad = bc$$

$$6 = 6$$

Sample 2

Solve the proportion $\frac{5}{6} = \frac{12}{y}$.

Sample Response

$$\frac{5}{6} = \frac{12}{y}$$

$$5 \cdot y = 6 \cdot 12 \quad \longleftarrow \text{Use cross products.}$$

$$5y = 72$$

$$\frac{5y}{5} = \frac{72}{5} \quad \longleftarrow \text{Divide both sides by 5.}$$

$$y = 14.4$$

The solution is 14.4.

Talk it Over

3. Describe another way to solve the proportion in Sample 2.

4. Can you use cross products to solve $5 = \frac{12}{y}$? If so, how?

6-3 Solving Proportions   315

TEACHING

Communication: Discussion

When introducing the idea of a proportion, you may want to use the example *24 is to 75 as 16 is to 50* to identify the numbers 24 and 50 as the *extreme* terms and 75 and 16 as the *means*. Point out also that *to solve a proportion* has the same meaning as to solve an equation, because a proportion is simply a special type of equation.

Additional Samples

S1 Solve the proportion $\frac{x}{12} = \frac{2}{3}$.

$$\frac{x}{12} \cdot 12 = \frac{2}{3} \cdot 12$$
$$x = 8$$

S2 Solve the proportion $\frac{5}{8} = \frac{14}{s}$.

$$5 \cdot s = 8 \cdot 14$$
$$5s = 112$$
$$\frac{5s}{5} = \frac{112}{5}$$
$$s = 22.4$$

Using Technology

Using the $\boxed{\text{MATH}}$ ▶ Frac command, it should be rather easy for some students to write a program to find the LCD of two integers.

Answers to Talk it Over

1. Descriptions may vary. An example is given. The first ratio must equal the second ratio. Multiplying the denominator of the second ratio, 4, by 3 gives the denominator of the first ratio, 12. So multiply the numerator of the second ratio, 3, by 3 to get 9 as the numerator of the first ratio. Then $x = 9$.

2. a. 60

   b. $4x = 45$

   c. The factors of the left side of the equation in part (b), 4 and x, are the numerator of the first ratio and the denominator of the second ratio. The factors of the right side, 15 and 3 ($45 = 3 \cdot 15$), are the denominator of the first ratio and the numerator of the second ratio.

3. Descriptions may vary. An example is given. Multiply both sides by the LCD of the fractions, $6y$. This gives $5y = 72$. Divide both sides by 5; $y = 14.4$.

4. Yes. Explanations may vary. An example is given.
   Rewrite 5 as $\frac{5}{1}$ to get $\frac{5}{1} = \frac{12}{y}$.
   Then use cross products to solve.

Additional Sample

S3 Kareem can type 2 pages in 18 minutes. At this rate, how long will it take him to type an 11-page paper?

Let t = the time it will take Kareem to type 11 pages.

$$\frac{2}{18} = \frac{11}{t}$$

$$18 \cdot 11 = 2 \cdot t$$
$$198 = 2t$$
$$99 = t$$

It will take him 99 minutes.

Error Analysis

Students often set up proportions like the one in Sample 3 incorrectly by writing one fraction upside down. Encourage them to label what each term of a proportion represents. Then if, as in Sample 3, the number of pages is the numerator in one fraction, it must be the numerator in the other fraction.

Look Back

Call upon each of three volunteers to write one method for solving a proportion. Have them make up an example to illustrate their method. ....•

APPLYING

Suggested Assignment

Standard 1–22, 28–40

Extended 1–25, 27–40

Integrating the Strands

Number Exs. 35–37

Algebra Exs. 1–29, 33, 34, 38

Functions Exs. 33, 34, 38

Statistics and Probability Exs. 30–32

Logic and Language Exs. 14–19, 26–29, 39, 40

Sample 3

Latasha checks out a 145-page book from the library. She can read about 15 pages in 25 min. At this rate, about how long will it take Latasha to read the book?

Sample Response

Problem Solving Strategy: Use a proportion

The reading rates are the same for 15 pages and 145 pages. Therefore you can write a proportion.

Let t = the time it will take Latasha to read 145 pages.

number of pages $\longrightarrow \dfrac{15}{25} = \dfrac{145}{t}$ $\longleftarrow$ time in minutes

Solve the proportion for t.

$$15 \cdot t = 25 \cdot 145 \qquad \longleftarrow \text{ Use cross products.}$$
$$15t = 3625$$
$$t \approx 242$$

60 min = 1 h
242 min = 4 h 2 min

Write 242 min as hours and minutes. It will take Latasha about 4 h to read the book.

Look Back ◄

What is the difference between a ratio and a proportion? Describe three methods for solving a proportion.

6-3 Exercises and Problems

1. **Reading** What are the terms of the proportion $\frac{2}{3} = \frac{8}{12}$? Use cross products to rewrite the proportion.

2. **Writing** Describe how you could solve the proportion $\frac{37}{134} = \frac{23}{x}$ with a calculator.

Solve each proportion.

3. $\dfrac{x}{15} = \dfrac{60}{125}$

4. $\dfrac{2.59}{y} = \dfrac{5.92}{16}$

5. $\dfrac{18}{75} = \dfrac{300}{m}$

6. $\dfrac{1000}{18} = \dfrac{d}{0.036}$

7. $\dfrac{7}{18} = \dfrac{a}{6}$

8. $\dfrac{10}{t} = \dfrac{15}{32}$

Unit 6 Ratios, Probability, and Similarity

Answers to Look Back ··································

Answers may vary. An example is given. A ratio is a quotient of two numbers or quantities. A proportion is a statement that two ratios are equal. There are three methods for solving a proportion: (1) When the variable is in the numerator of one of the ratios, you can undo the division by multiplying both sides by the denominator of that ratio; (2) you can multiply both sides of the proportion by the LCD of the ratios in the proportion; (3) you can use cross products.

For Exercises 9–12, use proportions A–D.

A. $\frac{6}{x} = \frac{84}{7}$ B. $\frac{x}{6} = \frac{84}{7}$

C. $\frac{6}{x} = \frac{7}{84}$ D. $\frac{x}{6} = \frac{7}{84}$

9. Which proportion has the same solution as proportion A?

10. Which proportion has the same solution as proportion B?

11. Write two other proportions that have the same solution and the same terms as proportion A.

12. Write two other proportions that have the same solution and the same terms as proportion B.

13. Write as many different proportions as you can that have a solution of 16 and whose terms are *n*, 36, 9, and 4.

Writing **Explain why proportional reasoning *is* or *is not* appropriate for each situation. If it is appropriate, write a proportion.**

14. If Cecilia walks 3 mi in 60 min, then Cecilia will walk 4.5 mi in 90 min.

15. If Tai is 86 cm tall when he is 3 years old, then Tai will be 430 cm tall when he is 15 years old.

16. If an 8 in. round pizza costs $6, then a 12 in. round pizza will cost $9.

17. If a car's gas mileage is 20 mi/gal at a speed of 30 mi/h, then it will be 40 mi/gal at a speed of 60 mi/h.

18. If a turkey weighs 25 lb standing on one leg, then it will weigh 50 lb standing on two legs.

19. If it takes 12 seconds to saw a log into 3 pieces, then it will take 16 seconds to saw a similar log into 4 pieces.

Solve using the problem solving strategy *Use a proportion.*

20. At the beginning of a basketball season, the Warriors won 11 games and lost 7. At this rate, how many games will they win in a normal 82-game season?

21. Running at a pace of 1 mi in 8 min, the average person burns 120 Cal/mi. How many minutes would a person have to run to burn off a 600 Cal meal?

6-3 Solving Proportions

317

Reasoning

Exs. 14–19 provide an opportunity for students to think about the appropriate use of proportions to model real-world situations. These exercises would be interesting to discuss in class.

Problem Solving

It is important for students to be able to apply proportions to solving everyday problems, as in Exs. 20–22. In working these problems and others involving proportions, students need to remember that when two ratios are compared in a proportion, the same correspondence must exist between the first and second terms of each ratio. Emphasize the point that a ratio is a comparison of two numbers in a definite order.

15. Proportional reasoning is not appropriate because people do not grow at a constant rate.

16. Proportional reasoning is not appropriate because the cost of the pizza depends on the area of the pizza, not the diameter.

17. Proportional reasoning is not appropriate because many factors determine gasoline mileage, such as road surface, tire wear, and weather conditions.

18. Proportional reasoning is not appropriate because a turkey's weight is the same whether it is standing on one leg or two.

19. Proportional reasoning is appropriate but the wrong proportion is used in the problem. The number of seconds depends on the number of cuts, not the number of pieces. Two cuts are needed to get 3 pieces. So $\frac{12 \text{ s}}{2 \text{ cuts}} = 6$ s/cut. To get 4 pieces, 3 cuts are needed. At 6 s/cut, 3 cuts take $3(6) = 18$ s. So, it would take 18 s (not 16 s) to saw a log into 4 pieces.

20. about 50 games

21. 40 min

Answers to Exercises and Problems

1. 2, 3, 8, 12; $2 \cdot 12 = 3 \cdot 8$, 24 = 24

2. Multiply 134 and 23, then divide by 37.

3. 7.2   4. 7

5. 1250   6. 2

7. about 2.3

8. about 21.3

9. D   10. C

11. Answers may vary. Examples are given.
$\frac{x}{7} = \frac{6}{84}, \frac{7}{x} = \frac{84}{6}$

12. Answers may vary. Examples are given.
$\frac{x}{84} = \frac{6}{7}, \frac{84}{x} = \frac{7}{6}$

13. $\frac{n}{36} = \frac{4}{9}, \frac{36}{n} = \frac{9}{4}, \frac{n}{4} = \frac{36}{9}$,
$\frac{4}{n} = \frac{9}{36}, \frac{4}{9} = \frac{n}{36}, \frac{9}{4} = \frac{36}{n}$,
$\frac{36}{9} = \frac{n}{4}, \frac{9}{36} = \frac{4}{n}$

14–19. Explanations may vary. Examples are given.

14. Proportional reasoning is appropriate because the walking rates can be the same for both distances.
$\frac{3}{60} = \frac{4.5}{90}$

317

22. Marla wanted to predict how many miles her car would travel on a full tank of gasoline. Her car's tank holds 14 gal of gasoline. She kept the record at the right and then averaged the amounts. How many miles did Marla predict her car would travel on a full tank?

gallons	miles
3.7	62.5
6.8	118.7
5.1	93.5

connection to **HEALTH**

To maintain good health it is important to eat properly and exercise regularly.

23. Nutritionists recommend that no more than 30% of total Calories consumed in a day come from fat. One gram of fat provides 9 Cal.

   a. A chicken sandwich that provides 427 Cal contains 19 g of fat. What percent of the Calories in the sandwich come from fat?

   b. Suppose that a person consumes 2000 Calories in one day. During lunch that day, the person eats two of the chicken sandwiches described in part (a). What percent of the total Calories for the day come from the fat in the chicken sandwiches?

   c. A different chicken sandwich has 40 g of fat and 55% of the Calories in the sandwich come from fat. How many Calories does the sandwich provide?

   d. Repeat part (b) for the chicken sandwich in part (c).

24. A single serving of Lite Recipe Fried Chicken that provides 199 Cal contains 13 g of fat. A single serving of Special Recipe Fried Chicken that provides 250 Cal contains 16 g of fat. Compare the percent of Calories from fat in a serving of Lite Recipe and a serving of Special Recipe.

   a. Complete this statement: The Lite Recipe is _?_ % lower in Calories than the Special Recipe.

   b. Complete this statement: The Lite Recipe is _?_ % lower in fat than the Special Recipe.

   c. What percent of the Calories in the Lite Recipe come from fat? in the Special Recipe?

25. **Fitness Training** The recommended maximum target heart rate for aerobic exercise is 80% of the maximum heart rate. To calculate the maximum heart rate, subtract age in years from 220.

   a. What is the maximum target heart rate for aerobic exercise for a person who is 22 years old?

   b. At what age is 140 the maximum target heart rate for aerobic exercise?

Unit 6 Ratios, Probability, and Similarity

Answers to Exercises and Problems

22. about 246.5 mi

23. a. about 40%

    b. about 17%

    c. about 654.5 Calories

    d. 36%

24. a. 20.4%

    b. 18.75%

    c. about 59%, about 58%

25. a. 158.4

    b. 45 years old

26. Translations may vary. An example is given. *Solve using a proportion.* A photocopy machine copied 60 pages in 1.5 min. At this rate, how much time will the machine take to copy 200 pages? 5 min; cinco minutos

27. Answers may vary.

28, 29. Problems may vary. Examples are given.

28. Aircraft carriers use 1 gal of fuel to travel 6 in. The distance from San Francisco to Honolulu is about 2400 mi. About how many gallons of fuel does an aircraft carrier need to make this trip?

26. Group Activity Work in a group of four students. Translate the problem from Spanish into English. Then solve the problem. Try to answer the problem in both English and Spanish.

Resuelve usando una proporción.

Una máquina fotocopió 60 páginas en 1.5 min. A este ritmo, ¿cuánto tiempo tardará la máquina en fotocopiar 200 páginas?

27. Open-ended If someone in your group speaks a language other than English, write a proportion problem in that language. Work together to translate the problem into English. Answer the problem in both languages.

Ongoing **ASSESSMENT**

Open-ended Write a problem you could solve by using a proportion, the given information, and any additional information you want to include.

28. Aviation Aircraft carriers use 1 gal of fuel to travel 6 in.

29. Access Standards On signs indicating accessibility for the physically challenged, letters and numbers must have a width-to-height ratio between 3:5 and 1:1.

Review **PREVIEW**

A paper clip is chosen at random from a bag that contains 4 red paper clips, 3 green paper clips, 3 white paper clips, and 2 yellow paper clips. Find each probability. *(Section 6-2)*

30. P(yellow)   **31.** P(white or red)   **32.** P(not green)

Identify the control variable and the dependent variable for each function. *(Section 4-6)*

33. Sanjay is practicing for a race. He runs a mile in six minutes. The distance he runs on each practice run depends on how much time he runs.

34. Beryl charges $15 to mow a lawn. The more lawns she mows, the more money she earns.

Find each number. *(Toolbox Skill 14)*

35. 12% of 750   **36.** 25% of 4700   **37.** 5% of 1250

Answers to Exercises and Problems

29. On signs indicating accessibility for the physically handicapped, letters and numbers must have a width:height ratio between 3:5 and 1:1. A sign is planned using letters that are 1 in. high. Find the range of acceptable letter widths.

30. $\frac{2}{12} = \frac{1}{6} \approx 0.17$ or 17%

31. $\frac{7}{12} \approx 0.58$ or 58%

32. $\frac{9}{12} = \frac{3}{4} = 0.75 = 75\%$

33. control variable: time; dependent variable: distance

34. control variable: number of lawns cut; dependent variable: earnings

35. 90

36. 1175

37. 62.5

Assessment: Task

For Exs. 28 and 29, the task of writing a problem will demonstrate students' understanding of the concept of proportion and how to use a proportion to solve a problem.

Visual Thinking

Ask students to draw a sign that adheres to the standards listed in Ex. 29. Have them demonstrate their drawings and discuss how the specified proportions were considered in their design activity. This activity involves the visual skills of *perception* and *self-expression*.

Practice 46 For use with Section 6-3

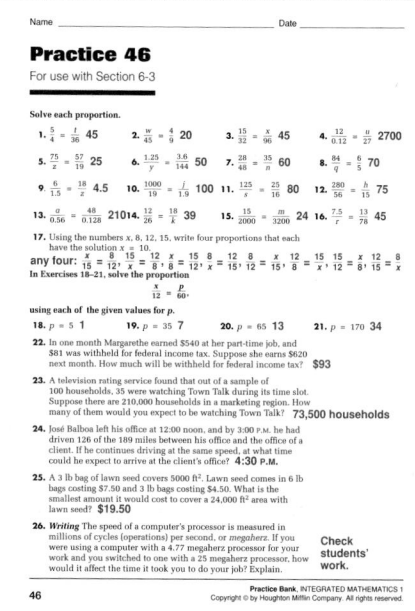

 Working on the Unit Project

38. a. A typical cost for a 30-second commercial during prime time on network television is $100,000. At this rate, what would you expect the cost of a 45-second commercial to be?

b. Suppose a 30-second commercial will reach 300,000 teenage viewers. What is the cost per viewer?

39. Research Find the cost of a full-page ad in a local newspaper and how many copies of the newspaper are sold in a day. What is the cost per reader?

40. a. What do you think are some advantages of advertising on television? in a newspaper?

b. Why is it important for the advertiser to know the reading and viewing habits of the people who will buy the product?

Unit 6　　　　**CHECKPOINT**

1. Writing Describe two types of ratios you have studied in this unit. In what ways are the types of ratios alike? How are they different?

2. The ratio of the measures of the acute angles of a right triangle is 13 : 5.　　**6-1**

a. Find the measures of the two acute angles.

b. Find the ratio of the measure of the larger acute angle to the measure of the right angle.

Use the table from Sample 1 on page 308.　　**6-2**

3. Suppose you pick a person at random from a database containing the names of all United States residents who are at least 25 years old. What is the probability that the person is male?

Find each probability.

The whole numbers from 10 through 49 are printed on separate slips of paper that are put into a box. Suppose you select one slip of paper at random from the box.

4. P(odd number)　　　**5.** P(multiple of 5)

6. P(not a multiple of 5)

Solve each proportion.　　**6-3**

7. $\frac{m}{3} = \frac{9}{30}$　　**8.** $\frac{24}{15} = \frac{r}{2.5}$　　**9.** $\frac{450}{800} = \frac{9}{z}$

10. In a 150-mile test drive, a new car model used 5.5 gal of gas. At this rate, how far can the new model travel on a full tank of 16.5 gal?

320　　**Unit 6** Ratios, Probability, and Similarity

Answers to
Exercises and Problems

38. a. $150,000

b. about $.33

39. Answers may vary.

40. Answers may vary. Examples are given.

a. Television can reach a larger audience in a shorter period of time. Also, knowing the type of viewers most likely to watch particular programs allows advertisers to target specific audiences. Newspaper advertising may permit more information to be given to the reader than television advertising. Also, readers can save the advertisement.

b. It is important to know the reading and viewing habits of the people who will buy your product so that you can advertise at the appropriate times and in the appropriate ways to reach the most people.

320

Answers to Checkpoint

1. Answers may vary. An example is given. Two types of ratios I have studied in this unit are experimental probability and theoretical probability. For each ratio, the possible values are any number on a scale from 0 to 1, including 0 and 1. The ratios are alike because they describe the probability of an event.

They are different because experimental probability describes events based on actual observations, and theoretical probability describes events based on outcomes that are likely to occur without doing an experiment.

2. a. 65°, 25°

b. 13:18

3. about 0.48 or 48%

4. $\frac{1}{2} = 0.5 = 50\%$

5. $\frac{1}{5} = 0.2 = 20\%$

6. $\frac{4}{5} = 0.8 = 80\%$

7. 0.9

8. 4

9. 16

10. 450 mi

6-4

Sampling and Making Predictions

HEAD COUNT

Focus
Use sampling to make predictions.

Talk it Over

1. Describe some problems you might face if you tried to count all the whales in an ocean or all the squirrels in a city park.

2. a. What methods do you think you could use to estimate the size of a crowd?

   b. Could you use those methods to estimate the number of deer in a forest?

One way to estimate the number of deer in a forest is the *capture-recapture method*.

4. The biologists wait until the tagged deer have mixed with the other deer. Then they find a second group of deer and count how many in this group are tagged.

1. Suppose this diagram represents all the deer in a forest.

2. Wildlife biologists first find some deer and tag them.

3. Then the biologists release the tagged deer back into the forest.

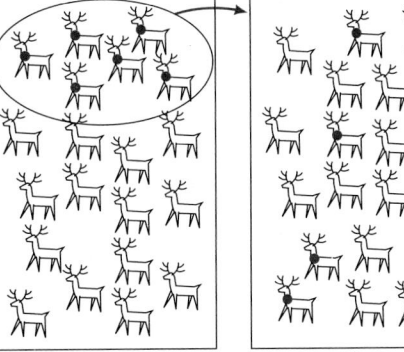

Biologists use a proportion to estimate the total number of deer in the forest:

$$\frac{\text{tagged deer in second group}}{\text{total deer in second group}} = \frac{\text{tagged deer in forest}}{\text{total deer in forest}}$$

A whole group is called a **population.** A part of a group is a **sample.** When biologists study a group of deer, they are choosing a sample. The population is all the deer in the forest.

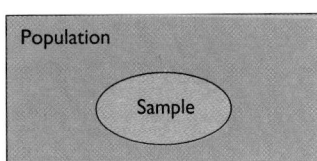

Population

Sample

6-4 Sampling and Making Predictions

321

Answers to Talk it Over

1. Answers may vary. An example is given. An ocean can be very large, which would make it difficult to be sure all the whales in every part are counted. Since whales are constantly in motion, it would be hard to tell if a whale had already been counted or not. Usually a city park has a very large squirrel population. Squirrels can move rapidly and easily climb trees, hiding above eye level. Also, the squirrel population is constantly changing as squirrels run in and out of the park.

2. Answers may vary. Examples are given.

   a. Estimate the area that the crowd takes up. Count the number of people in one part of the area. Estimate how many of these parts are in the total area and multiply this number by the number of people who can fit in this part.

   b. It would be difficult to use this method. A deer moves among the trees, whereas crowds are usually still.

PLANNING

Objectives and Strands
See pages 298A and 298B.

Spiral Learning
See page 298B.

Materials List
➤ Bags of dried beans
➤ Containers (opaque)
➤ Nontoxic markers
➤ Graph paper

Recommended Pacing
Section 6-4 is a two-day lesson.
Day 1
Pages 321–322: Talk it Over through Exploration, *Exercises 1–4*
Day 2
Pages 323–325: Sampling through Look Back, *Exercises 5–17*

Extra Practice
See pages 629–631.

Warm-Up Exercises
Warm-Up Transparency 6-4

Support Materials
➤ Practice 47
➤ Enrichment 42 in the Activity Bank
➤ Study Guide 6-4
➤ Problem Set 13
➤ Diagram Masters 2, 16 in the Explorations Lab Manual
➤ Using IBM Plotter Plus Disk: Sampling Experiment
➤ Quiz 6-4
➤ Alternative Assessment 8

Talk it Over

Questions 1 and 2 prepare students for understanding the idea of a *sample*.

Exploration

The goal of the Exploration is to give students hands-on experience in using the *capture-recapture* method. Students perform the experiment five times and then find the mean of their five estimates. This activity helps students to understand the concepts of population, sample, and making predictions.

Mathematical Procedures

The procedure discussed in this section is fundamental to the science of making statistical predictions. In most real-world situations, populations can only be studied by using samples because the populations themselves are too large. Underscore the point that predictions based upon samples are always estimates and therefore contain some degree of error. The error can be quantified, however, by stating the *margin of error* along with the error. These ideas are explored on pages 323–325.

How can you estimate what you cannot see?

- **Materials: bags of dried beans, containers (bags or bowls that are not clear), nontoxic markers**
- **Work in a group of two to four students.**

(1) Pour a bag of beans into a container.

(2) Take out a handful of beans, mark them all, and record how many you marked.

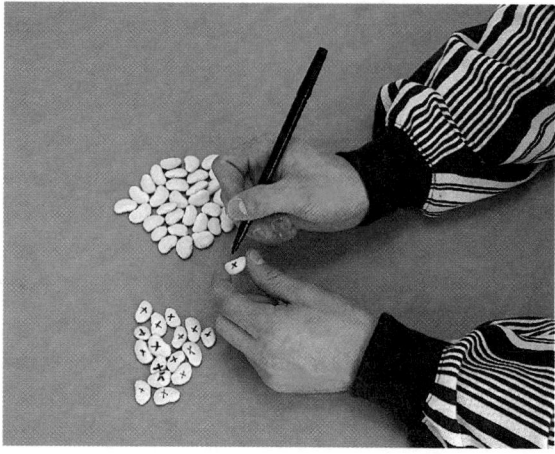

(3) Put the beans back in the container. Mix the beans well.

(4) Take out another handful of beans. Record the total number of beans in the sample and how many are marked.

(5) Use this proportion to estimate the total number of beans (*x*) in the container.

$$\frac{\text{marked beans in handful}}{\text{total beans in handful}} = \frac{\text{marked beans in container}}{x}$$

Round your answer to the nearest whole number and record it.

(6) Repeat steps 3, 4, and 5 four times, to get five estimates in all.

(7) Find the mean of your five estimates.

(8) Count all the beans in the container. Compare your count with your estimates and the mean.

(9) What do you think would happen to the mean if you repeated steps 3, 4, and 5 one hundred times?

▶ Now you are ready for:
Exs. 1–4 on p. 325

Unit 6 Ratios, Probability, and Similarity

Answers to Exploration

1–3. Answers may vary.

4. The number of marked beans found may vary.

5–9. Answers may vary. Examples are given.

5. There are 4 marked beans out of the 30 chosen at random: $\frac{4}{30} = \frac{20}{x}$; *x* is about 150.

6. (1) marked beans: 3; estimate: 200 (2) marked beans: 3; estimate: 200 (3) marked beans: 4; estimate: 150 (4) marked beans: 5; estimate: 120

7. 164

8. There are 196 beans in the container. Three estimates were too low. Two estimates were too high. The mean of the five estimates is lower than the actual number by 32 beans.

9. The mean would get closer to 196.

Sampling

Choosing a sample is called **sampling.** Data collected about a sample can be used to make predictions about a population. Samples are often used to take opinion polls, test the quality of a product, and find out what people buy.

Sample 1

A media rating service found that 860 of a sample of 4000 households with a TV watched a certain program one night. Estimate how many of the 93.1 million households in the United States with a TV watched this program that night.

Sample Response

Problem Solving Strategy: Use a proportion

Let n = millions of households with a TV that watched the program.

$$\frac{\text{households in sample that watched the program}}{\text{total households in sample}} = \frac{\text{households with a TV that watched the program (millions)}}{\text{total households with a TV (millions)}}$$

$$\frac{860}{4000} = \frac{n}{93.1}$$

$$\left(\frac{860}{4000}\right)(93.1) = \left(\frac{n}{93.1}\right)(93.1) \quad \longleftarrow \text{ Multiply both sides by 93.1.}$$

$$20.0 \approx n$$

About 20 million households watched the program.

Margin of Error

When polls are based on samples, the results are estimates. Estimates always contain some error. The **margin of error** estimates the interval that is most likely to include the exact result for the population.

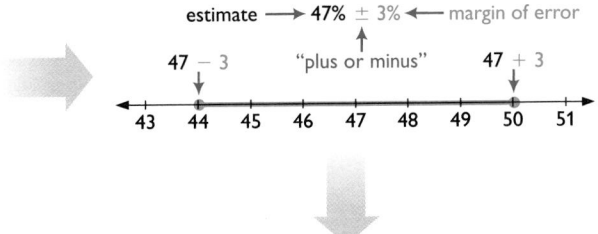

estimate ⟶ **47%** $\pm$ 3% ⟵ margin of error

47 − 3   "plus or minus"   47 + 3

The interval is "about 44% to 50%."

Additional Sample

S2 A quality control expert finds 6 new cars out of 150 need to have the tires re-aligned. Suppose the margin of error is 3%. Estimate the interval that contains the number of faulty cars in the lot of 1200.

Method 1: Let x = the number of faulty cars in the lot.

$$\frac{6}{150} = \frac{x}{1200}$$

$$\frac{6}{150} \cdot 1200 = \frac{x}{1200} \cdot 1200$$

$$48 = x$$

3% of 1200 = 36

48 − 36 = 12

48 + 36 = 84

About 12 to 84 cars are faulty.

Method 2: $\frac{6}{150} = 4\%$

1% of 1200 = 12

7% of 1200 = 84

About 12 to 84 cars are faulty.

Problem Solving

The Sample Response shows two methods for finding the interval that contains the number of faulty bulbs. While Method 1 is more direct, Method 2 does have the advantage of giving the interval in terms of percent as well as actual numbers.

Sample 2

A quality tester finds 3 faulty bulbs in a sample of 50 light bulbs. Suppose the margin of error is 2%. Estimate the interval that contains the number of faulty bulbs in the lot of 2000.

Sample Response

Method ❶

Problem Solving Strategy: Use a proportion

Let x = the number of faulty bulbs in the lot.

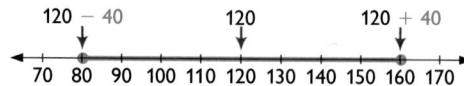

faulty bulbs in sample ⟶ $\frac{3}{50} = \frac{x}{2000}$ ⟵ faulty bulbs in lot
total bulbs in sample ⟶ ⟵ total bulbs in lot

$$\frac{3}{50} \cdot 2000 = \frac{x}{2000} \cdot 2000 \quad \longleftarrow \text{Multiply both sides by 2000.}$$

$$120 = x$$

Use the margin of error to estimate the interval.

2% of 2000 = (0.02)(2000) = 40

```
      120 − 40          120          120 + 40
         ↓               ↓               ↓
  ├───●───┼───┼───┼───┼───┼───┼───┼───●───┤
   70  80  90 100 110 120 130 140 150 160 170
```

About 80 to 160 light bulbs are faulty.

Method ❷

Use percents.

First find what percent of the sample are faulty.

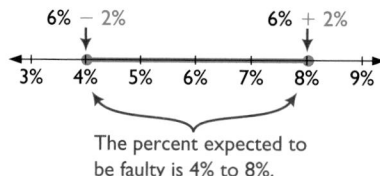

faulty bulbs in sample ⟶ $\frac{3}{50} = 0.06 = 6\%$
total bulbs in sample ⟶

Next use the 2% margin of error to estimate the interval.

```
       6% − 2%              6% + 2%
          ↓                    ↓
  ├─┼─┼───●───┼───┼───┼───●───┼─┤
    3%  4%  5%  6%  7%  8%  9%
           ⎧_____⎫
     The percent expected to
     be faulty is 4% to 8%.
```

Then estimate the number of faulty bulbs.

4% of 2000 = (0.04)(2000) = 80

8% of 2000 = (0.08)(2000) = 160

About 80 to 160 light bulbs are faulty.

Answers to Talk it Over

3. Multiply 3 · 2000 and 50 · x. Then divide by 50.

4. The estimate would be less certain due to the increased overall range of numbers of faulty bulbs. A larger margin gives more room for error.

Answers to Look Back

Answers may vary. An example is given. The representative could select a sample of all the voters and find out what percent of the sample is in favor of the bill. This percent could be used to estimate the number of voters who support the bill. A margin of error would help the representative determine an interval around the estimate. The actual number of voters who favor the bill would most likely be found in this interval.

3. What would change in Method 1 if you used cross products?

4. If the margin of error in Sample 2 were 4%, would the estimate be more or less certain? Why?

▸ **Now you are ready for:**
Exs. 5–17 on pp. 325–327

Look Back ◂

Suppose you are a member of Congress trying to decide whether to support a bill. How could you use a sample to estimate how many voters in your district favor the bill? Why would you want to know the margin of error?

6-4 Exercises and Problems

1. a. In the Exploration, what is the sample?

   b. In the Exploration, what is the population?

2. Writing Explain how the Exploration is like the method used by biologists to estimate the number of deer in a forest.

3. Reading Compare how the word *population* is used in this section with the way it is used in everyday language.

4. Biologists captured 400 penguins, tagged them, and released them back into the same region. Later, the penguin population of the region was sampled once a month for four months.

	Month 1	Month 2	Month 3	Month 4
Size of sample	200	100	150	100
Tagged animals in sample	50	30	45	24

   a. Estimate the size of the population of penguins in the region for each month.

   b. Find the mean of the four estimates you made in part (a).

   c. Suppose the biologists combine the four samples into one large sample. Estimate the size of the penguin population in the region.

   d. Compare your answers in parts (b) and (c).

5. A town's high school has 627 students. A sample of 200 people living in the town is picked at random. Of the 200, 40 are students at the high school.

   a. About how many people live in the town?

   b. Estimate an interval for the number of people living in the town. Assume a 4% margin of error.

6-4 Sampling and Making Predictions   **325**

Look Back

The Look Back can be used for class discussion. Have students give reasons for choosing their methods of estimation. ●

APPLYING

Suggested Assignment

Standard 1, 2, 4–7, 9, 11–17

Extended 1–7, 9, 11–17

Integrating the Strands

Algebra Exs. 11–13

Geometry Exs. 14–16

Statistics and Probability Exs. 1–10, 17

Discrete Mathematics Exs. 1–10, 17

Logic and Language Exs. 3, 17

Communication: Discussion

For Ex. 5, discuss the meaning of a random sample. As used in this exercise and others, a sample that is drawn from a finite population is said to be random if each member of the population has an equal probability of being selected.

Reasoning

Ask students to explain why nonrandom samples cannot be used to make predictions.

Answers to Exercises and Problems

1. a. the second handful of dried beans

   b. all the dried beans

2. Explanations may vary. An example is given. All the beans in the bag are the population the same as all the deer in a forest. A random amount of beans is marked the same as a random number of deer is tagged. The marked beans are put back into the container same as the tagged deer are put back in the forest. A sample of beans, marked and unmarked, is used to estimate the total population same as a sample of deer, tagged and untagged, is used to estimate the total deer population. In both cases, a proportion is used.

3. Answers may vary. An example is given. In this section, the word population is used to refer to an entire group. This can be a group of deer, a group of households, or a group of light bulbs. In everyday usage, the word population is used to refer to the people living in a certain area.

4. a. about 1600; about 1333; about 1333; about 1667

   b. about 1483 penguins

   c. about 1477 penguins

   d. Answers may vary. An example is given. The answers are very close. The answer in part (b) estimates there are 6 more penguins in the population than the answer in part (c).

5. a. about 3135 residents

   b. about 2610 to 3920 residents

Interdisciplinary Problems

When discussing Ex. 7, students need to understand that the sample of 400 voters out of about 483,000 people and the sample in part (c) may be very different samples, and therefore may result in different predictions. The first sample was taken at one voting place and thus reflects the views of people who actually voted. A poll taken before an election usually involves some people who do not vote and therefore its prediction would be much less reliable.

Application

Manufacturing operations in businesses usually employ quality control procedures based upon random samples of a product. The results allow the managers of the business to make informed decisions about the quality of their products. A certain level of nonstandard products can be defined so as not to hurt the sales of the product. Clearly, zero defects is the ideal goal, but this may be unattainable without incurring additional and substantial costs which would impact the price adversely. Customers might reject the product if the price is too high.

---

6. **Writing** A bag contains 200 cubes of various colors. A person picked 15 cubes at random. The colors of the cubes are shown in the table. About how many green cubes do you think are in the bag? How did you get your answer?

red	⊥⊥⊥	
green	||||
blue	⊥⊥⊥

connection to SOCIAL STUDIES

Exit polling is used to predict who will win an election. In an exit poll, people leaving a voting place are asked how they voted.

7. In one exit poll, a sample of 400 voters was polled out of about 483,000 people voting in the election. The table shows the results. The margin of error is 6%.

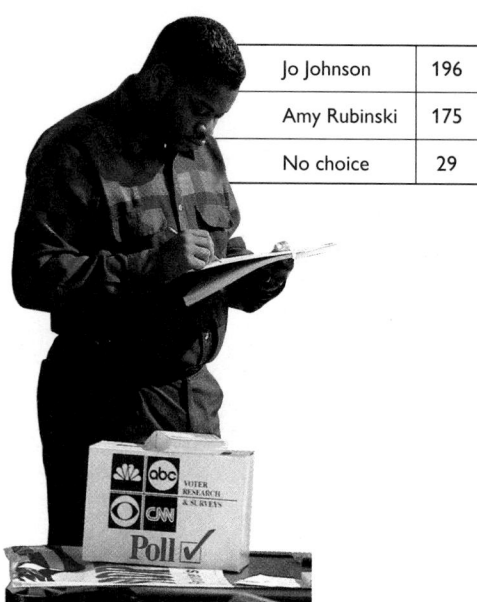

Jo Johnson	196
Amy Rubinski	175
No choice	29

a. Estimate the interval for the number of votes expected to go to Jo Johnson.

b. Whom do you expect to win the election? Explain your choice.

c. In a poll taken before the election, a sample of 400 out of about 1,295,000 registered voters were asked which candidate they supported. This poll predicted that Amy Rubinski would win by a wide margin. Give as many reasons as you can why the result does not agree with the exit poll.

8. **Research** Find out what percent of the people in your state who are eligible to vote are registered voters, and what percent of the registered voters voted in the last national election. Do the results surprise you?

9. **Career** A quality tester tests a random sample of 20 items. The graph shows the results. Item numbers are on the horizontal axis and item weights in grams are on the vertical axis. Each item should weigh 10 g, but any weight in the interval 10 ± 0.1 g is acceptable. There are 4000 items in the lot. Estimate the number of unacceptable items in the lot.

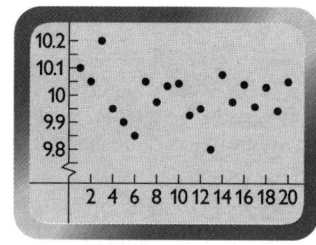

Unit 6 Ratios, Probability, and Similarity

Answers to Exercises and Problems

6. about 53; Explanations may vary. An example is given. I used the proportion $\frac{4}{15} = \frac{x}{200}$ and solved for x.

7. a. about 207,690 to 265,650 votes

   b. Answers may vary. An example is given. Johnson's and Rubinski's intervals overlap, so it is possible either could win, but the results of the exit poll show Johnson is favored.

c. Reasons may vary. Examples are given. The margin of error may be larger than originally anticipated. Only 483,000 or about 37% of registered voters voted, so perhaps the pre-election poll included people who did not end up

voting at all. The pre-election poll counts changeable opinions, whereas the exit poll counts unchangeable votes.

8. Answers may vary with area.

9. about 600 items

10. a–d. Research may vary. Examples are given.

10. a. Brand A: 1; Brand B: 10; Brand C: 5; Brand D: 11; Brand E: 2; Brand F: 1

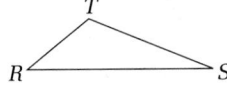
10. **Open-ended** Market researchers use samples to predict which products people will buy.

  a. **Research** Sample 30 students in your school. Ask which brand of sneaker or other product they prefer.

  b. Make a bar graph to show the results.

  c. Which brand did most students prefer?

  d. **Research** Find out how many students are in your school. Use the results shown in part (b) to predict the number of students in your school who prefer the brand in part (c).

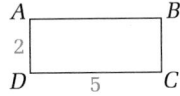 *Review* **PREVIEW**

Solve each proportion. *(Section 6-3)*

11. $\frac{4}{5} = \frac{x}{45}$ 12. $\frac{x}{34} = \frac{6}{13}$ 13. $\frac{42}{70} = \frac{18}{x}$

14. List the pairs of corresponding vertices. *(Section 1-6)*

$\triangle RST \cong \triangle XYZ$

Find the perimeter of each rectangle. *(Toolbox Skill 17)*

15. rectangle $ABCD$, side 2, side 5

16. rectangle $EFGH$, side 4, side 10

 Working on the Unit Project

17. **Group Activity** Work with the whole class.

  a. Use the daily records of television and newspaper or magazine ads kept by you and your classmates. Which TV programs, magazines, and newspapers are the most popular? What types of products are advertised on or in them?

  b. **Research** Find out the total number of students in your school. Using your class as a sample, estimate the total number of students in the school who watch the most popular programs and read the most popular newspapers and magazines.

  c. Do you think the daily records kept by you and your classmates is a good sample for predicting the viewing and reading habits of teenagers throughout the country? Why or why not?

BY THE WAY...

Every time a song is played on the radio or TV, the songwriter receives a royalty payment. The royalties are estimated by sampling.

6-4 Sampling and Making Predictions   **327**

Assessment: Task

Ex. 10 allows students to conduct their own "market research" based on sampling. Students can compare their results. If the predictions vary widely, try to find out if any of the samples were nonrandom.

Working on the Unit Project

Ex. 17 allows students some class time to meet and share their ideas with their classmates.

Practice 47 For use with Section 6-4

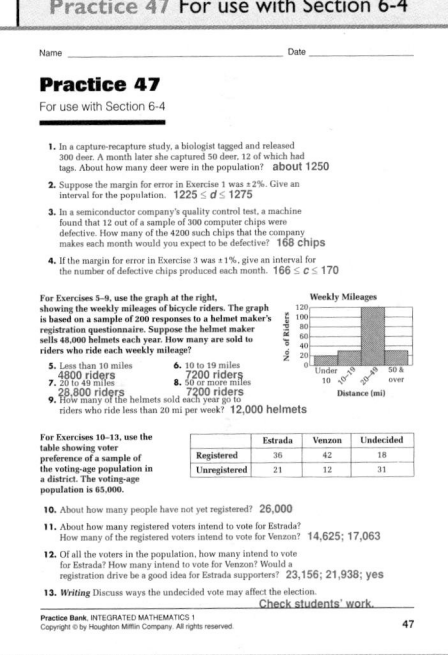

15. 14   16. 28

17. a, b. Answers may vary.

  c. Answers may vary. An example is given. No; newspapers are different from city to city. Also, many towns have their own local newspapers. Certain television programs may not be available in all parts of the country. Certain groups of teenagers, such as those who live in cities or in suburban or rural areas, may have different interests that would affect what is popular in that particular group.

Answers to Exercises and Problems

10. b.
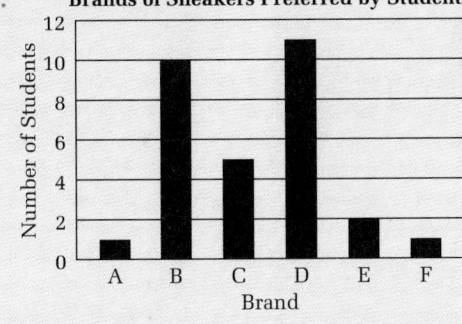

c. Brand D

d. Out of 200 students in the school, 73 would be expected to prefer Brand D sneakers.

11. 36

12. about 15.7

13. 30

14. R corresponds to X, S corresponds to Y, T corresponds to Z

327

Objectives and Strands

See pages 298A and 298B.

Spiral Learning

See page 298B.

Materials List

➤ Ruler
➤ Protractor
➤ Geoboard
➤ Rubberbands
➤ Graph paper
➤ Geometric drawing software
➤ Penny

Recommended Pacing

Section 6-5 is a two-day lesson.

Day 1

Pages 328–330: Opening paragraph through Talk it Over, *Exercises 1–9*

Day 2

Pages 330–332: Scale Drawings through Look Back, *Exercises 10–33*

Extra Practice

See pages 629–631.

Warm-Up Exercises

Warm-Up Transparency 6-5

Support Materials

➤ Practice 48
➤ Enrichment 43 in the Activity Bank
➤ Study Guide 6-5
➤ Problem Set 13
➤ Diagram Masters 1, 2, 17 in the Explorations Lab Manual
➤ McDougal Littell Mathpack software: *Geometry Inventor*
➤ Quiz 6-5
➤ Alternative Assessments 9

Section 6-5

Similar Polygons

Focus
Apply proportions to problems using similar figures and scale drawings.

SIZE IT UP

Photograph editors change the size of a photograph to fit in a column on a textbook, newspaper, or magazine page.

For a magazine article, this photograph of Ida B. Wells is enlarged to fill a $2\frac{1}{2}$ in. wide column.

To determine the new size, the editor may use a ruler to extend the diagonal of the photograph across the width of the column.

$2\frac{1}{2}$ in.

BY THE WAY...

Ida B. Wells lived from 1862 to 1931. She took part in the meeting that led to the founding of the NAACP. The Ida B. Wells apartment complex in Chicago is named for her.

EXPLORATION

What happens to a photograph when you enlarge it?

• **Work with another student.**
• **Materials: rulers, protractors**

	Height	Width
Enlarged Photo	?	?
Original Photo	?	?
Enlarged Photo / Original Photo	?	?

① Measure the widths and the heights of the original photo and the enlarged photo. Record your results in a table.

② Complete the third column. What do you notice about the ratios?

③ Measure the four angles of the original photo. Then measure the corresponding angles of the enlarged photo. What do you notice about the measures of the corresponding angles?

Answers to Exploration

1–2. The ratios are equal.

	Enlarged Photo	Original Photo	Enlarged Photo / Original Photo
Width	2.5 in.	1 in.	$\frac{2.5}{1}$
Height	3 in.	1.2 in.	$\frac{2.5}{1}$

3. The measures of the corresponding angles are equal.

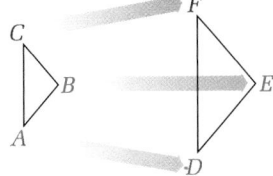

The original photo and the enlarged photo have the same shape. Two figures with the same shape are called **similar figures.** Similar figures may be the same size or different sizes.

You say, "triangle *ABC* is similar to triangle *DEF*."

You write △*ABC* ~ △*DEF*.

The symbol ~ means "is similar to."

Remember that the order of the vertices in the names of the triangles shows the corresponding parts.

PROPERTIES OF SIMILAR POLYGONS

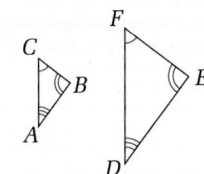

➤ Corresponding angles are congruent.

➤ Corresponding sides are *in proportion.*   Their lengths have the same ratio.

Sample 1

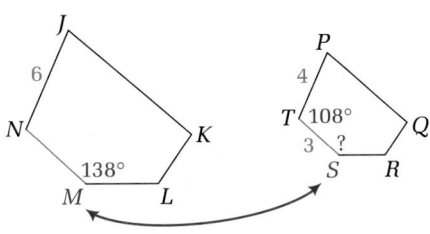

Polygon *JKLMN* ~ polygon *PQRST*. Find each measure.

a. ∠*S*   b. *NM*

Sample Response

a. ∠*S* corresponds to ∠*M*, so they have the same measure.
∠*M* = 138°, so ∠*S* = 138°.

b. **Problem Solving Strategy:** Use a proportion

Write a proportion using the lengths of $\overline{NM}$ and the corresponding sides whose lengths are given.

smaller polygon ⟶ $\dfrac{PT}{JN} = \dfrac{TS}{NM}$
larger polygon ⟶

Solve the proportion for *NM*.

$\dfrac{4}{6} = \dfrac{3}{NM}$ ⟵ Substitute the lengths you know.

$4 \cdot NM = 18$ ⟵ Use cross products.

$NM = 4.5$ ⟵ Divide both sides by 4.

6-5 Similar Polygons

329

TEACHING

Multicultural Note

When reading the By the Way note on page 328, point out that Ida B. Wells was a teacher, a writer, and a civil rights activist. She spent her life fighting inequity and exposing injustice.

In 1950, Chicago named her one of the 25 outstanding women in the city's history. In 1990, as part of its observance of Black History Month, the U.S. Postal Service issued a stamp honoring Wells. Today in Memphis, Tennessee, a historical marker stands at the site of a former newspaper office where Wells worked as a journalist.

Exploration

The goal of the Exploration is to have students verify that the corresponding sides of two similar figures are proportional and the corresponding angles have equal measure.

Additional Sample

S1 Polygon *QRST* ~ polygon *EFGH*. Find each measure.

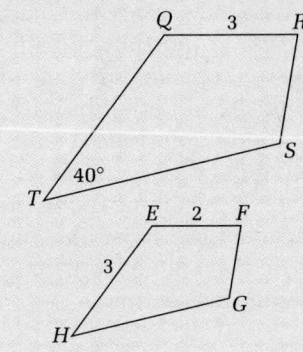

a. ∠*H* 40°

b. *QT* $\dfrac{QR}{EF} = \dfrac{QT}{EH}$

$\dfrac{3}{2} = \dfrac{QT}{3}$

$2(QT) = 9$

$QT = 4.5$

Communication: Drawing

Have students draw several examples of pairs of similar figures. Ask them to match up the corresponding sides and angles.

► **Now you are ready for:**
Exs. 1–9 on pp. 332–333

Talk it Over

1. What type of polygon is shown in Sample 1 on page 329?

2. List all the corresponding angles and corresponding sides of the similar polygons in Sample 1.

3. These triangles are congruent.

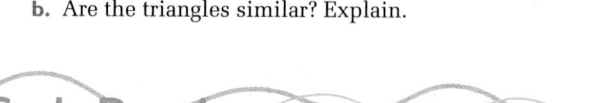

   a. What is the ratio of the lengths of the corresponding sides? What can you say about the measures of the corresponding angles?

   b. Are the triangles similar? Explain.

Scale Drawings

A **scale drawing** represents an actual object. A scale drawing of your classroom floor *is similar to* the floor of the room. The **scale** is the ratio of the size of the drawing to the actual size of the object. You might see scales written in several ways.

1 in.: 40 ft   1 in. = 40 ft

Sample 2

What is the actual length of Store 10 in the Palm Court mall to the nearest foot?

scale: 1 in. = 40 ft

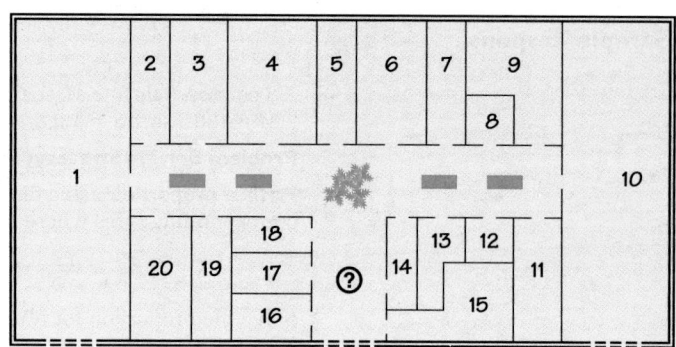

Answers to Talk it Over

1. pentagon

2. ∠J corresponds to ∠P,
   ∠K corresponds to ∠Q,
   ∠L corresponds to ∠R,
   ∠M corresponds to ∠S,
   ∠N corresponds to ∠T;

   $\overline{JK}$ corresponds to $\overline{PQ}$,
   $\overline{KL}$ corresponds to $\overline{QR}$,
   $\overline{LM}$ corresponds to $\overline{RS}$,
   $\overline{MN}$ corresponds to $\overline{ST}$,

   $\overline{NJ}$ corresponds to $\overline{TP}$

3. a. 1:1; the measures of the corresponding angles are equal.

   b. Yes. Since the corresponding angles are congruent and the corresponding sides are in proportion, the triangles are similar.

Problem Solving Strategy: Use a proportion

Write a proportion using the scale and the length of Store 10 in the drawing.

Use a ruler to measure Store 10. The scale shows that 1 in. on the drawing represents 40 ft in the actual mall.

Let a = the actual length of the store.

scale drawing measurements (in.) $\longrightarrow$ $\dfrac{1}{40} = \dfrac{1\frac{15}{16}}{a}$ $\longleftarrow$ In the scale drawing, the
actual measurements (ft) $\longrightarrow$ store is $1\frac{15}{16}$ in. long.

Solve the proportion for a.

$$1 \cdot a = \left(1\frac{15}{16}\right)(40) \qquad \longleftarrow \text{Use cross products.}$$
$$a = (1.9375)(40) \qquad \longleftarrow 1\frac{15}{16} = 1.9375$$
$$a = 77.5$$

Store 10 is about 78 ft long.

Talk it Over

4. Describe how to find the actual area of Store 18.

5. Suppose the scale of the drawing in Sample 2 were $\frac{3}{4}$ in. = 40 ft. How would the proportion be different?

Indirect Measurement

Would you ever consider measuring the height of a tree or building using a ruler or tape measure? Probably not. Similar triangles give you a way to measure an object indirectly. To use indirect measurement, you need to know this important fact:

A TEST FOR SIMILAR TRIANGLES

Two triangles are similar whenever two angles of one triangle are congruent to two angles of the other triangle.

Answers to Talk it Over

4. Descriptions may vary. An example is given. Use a ruler to find the dimensions of store 18 on the drawing. Use the scale and the drawing dimensions to write proportions that can be solved to find the actual dimensions. Multiply the actual dimensions to find the area in square feet.

5. The proportion would have 1 in. replaced with $\frac{3}{4}$ in.

$$\dfrac{\frac{3}{4}}{40} = \dfrac{1\frac{15}{16}}{a}; \quad 77.5 = \frac{3}{4}a;$$
$$a \approx 103 \text{ ft}$$

Using Technology

The *Geometry Inventor* or other geometric drawing software can be used to study the effects of scaling. Students can draw a polygon or some other simple figure and then use the scaling options of the software to change the size of the figure. Students can use the software to take measurements before and after the scaling and compare the measurements.

Visual Thinking

Ask students to make a sketch of the drawing in Sample 2 at twice the size of the drawing. Have them describe the scale of their sketch. This activity involves the visual skills of *interpretation* and *perception*.

Reasoning

When discussing a test for similar triangles, ask students what conclusions they can draw about the third angle of the triangles. Students should explain the reasoning that supports their conclusion.

Additional Sample

S3 Fire Company #9 is planning to buy a new ladder that would reach the top of the tallest building in their district. To estimate the height of the building, the company measured the length of a meter stick's shadow and the length of the building's shadow. Estimate the height of the building.

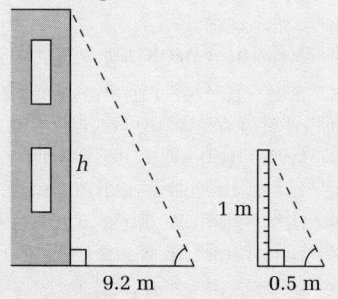

9.2 m   0.5 m

$$\frac{\text{height of building}}{\text{height of stick}} =$$

$$\frac{h}{1} = \frac{9.2}{0.5}$$

$$0.5h = 9.2$$

$$h = 18.4$$

The building is 18.4 m tall.

Look Back

Have students support their work by explaining how to find each of the corresponding sides and angles. They should also explain why the figures are similar. ....●

APPLYING

Suggested Assignment

Standard 1, 3–7, 10–13, 15, 16, 19, 20, 25–33

Extended 1–7, 10–13, 15–17, 19–23, 25–33

Integrating the Strands

Algebra Ex. 23

Measurement Ex. 14–18, 21–28

Geometry Exs. 1–28, 31–33

Statistics and Probability Exs. 29, 30

Logic and Language Exs. 21–24

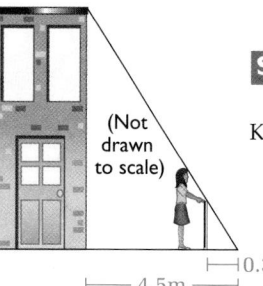

(Not drawn to scale)

├── 4.5m ──┤ ├─0.3m

Karen, Lydia, and Dion are planning a mural for the side of a building. To estimate the height of the building, the students measured the length of a meter stick's shadow and the length of the building's shadow. Estimate the height of the building.

Sample Response

Problem Solving Strategy: Make a sketch

Your sketch will probably not be in the correct scale.

Write a proportion using the heights and the shadows.

Let h = the height of the building.

height of building ⟶ $\dfrac{h}{1} = \dfrac{4.5}{0.3}$ ⟵ building's shadow
height of stick ⟶ ⟵ stick's shadow

Solve the proportion for h.

$$h = 15$$

The building is about 15 m tall.

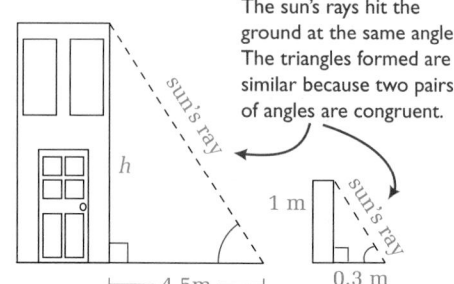

The sun's rays hit the ground at the same angle. The triangles formed are similar because two pairs of angles are congruent.

├── 4.5m ──┤   0.3 m

Look Back

Measure each side and each angle in the polygons at the right. Record the measurements and compare them. Which polygons are similar?

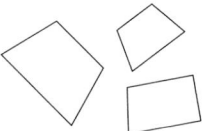

......► Now you are ready for:
Exs. 10–33 on pp. 333–336

6-5   Exercises and Problems

1. **a.** **Reading** The symbol ≅ means "is congruent to." What is the symbol for "is similar to"?

   b. **Writing** Discuss how similarity and congruence are alike and how they are different. How do the symbols show this relationship?

2. Other proportions could have been used in part (b) of Sample 1 on page 329. Which of these could *not* have been used? Why?

   A. $\dfrac{JN}{PT} = \dfrac{MN}{TS}$ **B.** $\dfrac{PT}{TS} = \dfrac{JN}{NM}$ **C.** $\dfrac{PT}{TS} = \dfrac{NM}{JN}$ **D.** $\dfrac{TS}{PT} = \dfrac{NM}{JN}$

Unit 6 Ratios, Probability, and Similarity

Answers to Look Back

Measurement estimates and units may vary. Figures I and II are similar.

Answers to Exercises and Problems

1. **a.** ~

   b. Answers may vary. An example is given. Congruent figures have corresponding sides of equal length and corresponding angles of equal measure. Similar figures have corresponding angles of equal measure and corresponding sides in proportion, not necessarily equal. Congruent figures are always similar. Therefore, the symbol for congruence contains an equals sign and the symbol for similarity. Similar figures are not always congruent, so the symbol for similarity does not contain the equals sign.

Pentagon *JKLMN* ~ pentagon *PQRST*.
Find each measure.

3. $\angle P$ 4. $\angle L$

5. PQ 6. KL

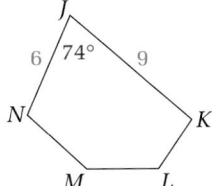

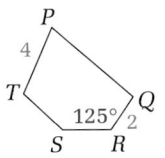

7. A basketball player "bounce passes" a ball to a player 6 m away as shown in the diagram. Can the other player easily catch the pass? Explain why or why not.

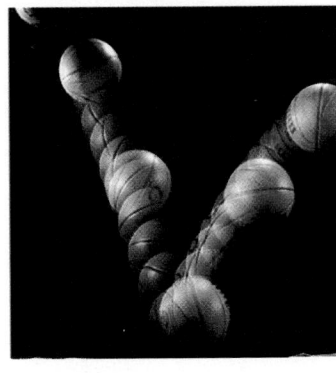

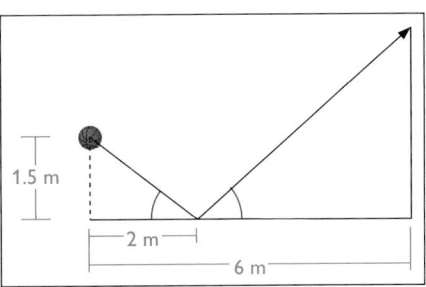

(*Hint:* From physics, the angles that the ball makes with the floor are congruent.)

1.5 m

2 m

6 m

8. **Using Manipulatives** Use a geoboard and rubber bands.

   Alternative Approach Use graph paper.

   How many rectangles similar to the one shown can be made on a 5 × 5 geoboard?

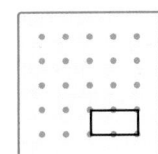

9. T E C H N O L O G Y   Use geometric drawing software.

   Alternative Approach Work in groups using rulers and graph paper.

   Find the dimensions of a rectangle that can be cut into two congruent rectangles so that each is similar to the original rectangle.

connection to **B I O L O G Y**

Use the drawing of a grasshopper.

10. Is the actual grasshopper larger or smaller than the scale drawing?

11. What does the scale 2:1 mean?

12. Describe how to find the actual length of the grasshopper.

13. About how long is one of the actual grasshopper's antennae?

14. **Research** Find a scale drawing in one of your textbooks. Explain what the scale means. Then use the scale to find an actual measurement.

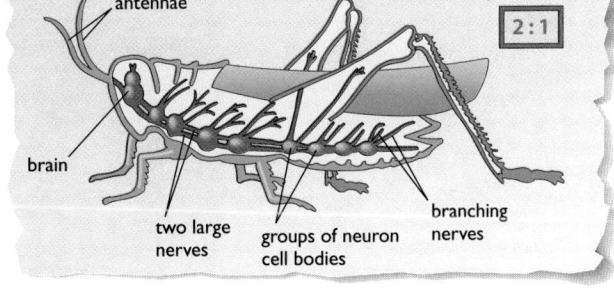

antennae

2:1

brain

two large nerves

groups of neuron cell bodies

branching nerves

Students in general find the concept of similarity more difficult to understand than the concept of congruence. Use Ex. 1 to discuss the two concepts in geometric terms only (same size and shape versus same shape). Similarity becomes more difficult for students when it is quantified using the concept of equal ratios, that is, proportions.

Application
Practical applications that involve scale drawings, as in Exs. 10–14 on the grasshopper and Exs. 15–17 on houses and furniture, help give meaning to the ideas of similarity, proportion, and scale. These exercises should be discussed thoroughly. Listen to students' answers and explanations carefully in order to correct any misunderstandings they may have.

9. Answers may vary. An example is given. A rectangle with dimensions $6\sqrt{2}$ units by 6 units can be cut into two congruent rectangles with dimensions 6 units by $3\sqrt{2}$ units, each of which is similar to the original rectangle.

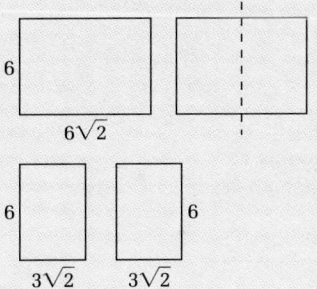

6

$6\sqrt{2}$

6   6

$3\sqrt{2}$ $3\sqrt{2}$

10. smaller

11. A distance of 2 units on the drawing represents a distance of 1 unit on an actual grasshopper. The drawing is twice as large as an actual grasshopper.

12. Descriptions may vary. An example is given. Measure the length of the grasshopper in the drawing and divide by 2.

13. Answers may vary. An example is given: about $\frac{1}{4}$ in.

14. Answers may vary.

Answers to Exercises and Problems ·············

2. C; Reasons may vary. An example is given. This proportion could not have been used since it does not involve the corresponding sides in the correct way.

3. 74°

4. 125°

5. 6

6. 3

7. No. Explanations may vary. An example is given. Solving the proportion $\frac{1.5}{2} = \frac{x}{4}$, you get $x = 3$. The ball would reach the player 3 m above the floor. The player might catch the pass, but not easily.

8. 38; 30 vertical and horizontal rectangles and 8 slanted rectangles.

Some examples are given.

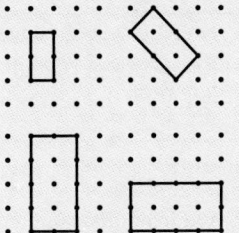

333

Application

When discussing Exs. 15–18, emphasize that scale models and drawings preserve the shapes of the original objects, but reduce or enlarge them in size by a fixed ratio, or scale.

Use the information below.

Miniature houses and furniture are usually built in one of three scales:

one-inch scale	half-inch scale	quarter-inch scale
1 in.:1 ft	$\frac{1}{2}$ in.:1 ft	$\frac{1}{4}$ in.:1 ft

These scales show the length of the miniature that represents 1 ft of the length of the full-size object.

15. Replace each ___?___ with the correct number of feet.

   a. Half-inch scale is the same as 1 in.: ___?___ ft.

   b. Quarter-inch scale is the same as 1 in.: ___?___ ft.

16. How much longer is an object when it is built in one-inch scale than when it is built in half-inch scale?

17. The miniature Chinese table is being built in quarter-inch scale. A scale drawing of the miniature tabletop is shown.

   a. What are the length and width of the miniature table?

   b. How wide would the full-size table be?

   c. How long would the full-size table be?

18. **Using Manipulatives** Measure a penny at its widest part. Then measure the penny in the photograph. Measure the length and width of the painting in the photograph. (Do not include the frame.) Estimate the length and width of the miniature painting.

$\frac{15}{32}$ in.

Miniature tabletop

$\frac{11}{32}$ in.

334   **Unit 6** Ratios, Probability, and Similarity

Answers to
Exercises and Problems

15. a. 2

  b. 4

16. two times longer

17. a. $\frac{15}{32}$ in. long, $\frac{11}{32}$ in. wide

  b. $1\frac{3}{8}$ ft

  c. $1\frac{7}{8}$ ft

18. Measurement estimates and units may vary. A penny is $\frac{3}{4} = \frac{12}{16}$ in. in diameter. The penny in the photograph is $\frac{11}{16}$ in. in diameter. The length of the painting (not counting the frame) is about

$1\frac{9}{16}$ in. The width of the painting (not counting the frame) is about $1\frac{1}{4}$ in. Since a real penny is $\frac{12}{11}$ the size of the penny in the photograph, then the real miniature painting's length is about $1\frac{31}{44}$ in. and its width is about $1\frac{4}{11}$ in.

19. Explanations may vary. An example is given. Each triangle contains a 90° angle. The vertical angles are congruent. Since two pairs of corresponding angles are congruent, the triangles are similar. 50 m

20. Explanations may vary. An example is given. Each triangle contains a 90° angle because the sign and the person are perpendicular to the ground. The angle at the other end of the shadows is the same for both triangles. Since two pairs of corresponding angles are congruent, the triangles are similar. 10 ft

Writing Explain why each pair of triangles is similar.
Then find the unknown length.

19.
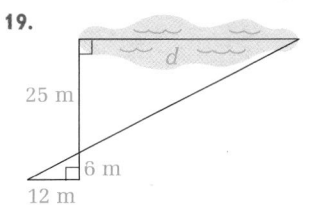
25 m
6 m
12 m
d

20.
MAIN ST.
h
6 ft 3 in.
⊢10 ft shadow⊣
⊢— 16 ft shadow —⊣

connection to **LITERATURE**

In *The Musgrave Ritual*, Sherlock Holmes uses mathematics to find some hidden objects. Here is his description of the process.

THE MUSGRAVE RITUAL

There were two guides given us to start with, an oak and an elm. Right in front of the house, ... stood a patriarch among oaks...

'Have you any old elms?' I asked.

'There used to be a very old one over yonder ... It was sixty-four feet.' [said Musgrave]

'How do you come to know it?' I asked in surprise.

'When my old tutor used to give me an exercise in trigonometry, it always took the shape of measuring heights.' [said Musgrave]

This was an unexpected piece of luck. ... I had, then, to find where the far end of the shadow would fall when the sun was just clear of the oak.

I ... whittled ... this peg, to which I tied this long string ... Then I took ... a fishing-rod, which came to just six feet, and I went back ... to where the elm had been. The sun was just grazing the top of the oak. I fastened the rod on end, marked out the direction of the shadow, and measured. It was nine feet in length.

Of course the calculation now was a simple one. If a rod of six feet threw a shadow of nine [feet], a tree of sixty-four feet would throw one of ninety-six [feet], and the line of the one [shadow] would of course be the line of the other. I measured out the distance ... and I thrust a peg into the spot.

21. Sherlock Holmes used shadows and indirect measurement to find an unknown length. Describe the objects and the method he used.

22. Make a sketch of the objects and the shadows in the story. Label all measurements. Be sure that the sketch is drawn to scale.

23. Show why Holmes's calculation of the unknown length is correct.

24. **Research** Read the entire story *The Musgrave Ritual*. What other measurements did Sherlock Holmes make? What were the objects that were hidden? What was their place in history?

Integrating the Strands

Exs. 21–24 involve concepts of measurement, geometry, and logic and language. In Ex. 23, an algebraic proportion is used to find an unknown length.

Answers to Exercises and Problems

21. Descriptions may vary. An example is given. Sherlock Holmes used a 6 ft fishing rod to represent the missing elm tree. He noticed that the rod cast a 9 ft shadow at the time the sun's rays were at the top of the oak. He used the height of the elm, 64 ft, to create similar triangles and calculate the length of the elm's shadow.

22. Sketches may vary. An example is given.

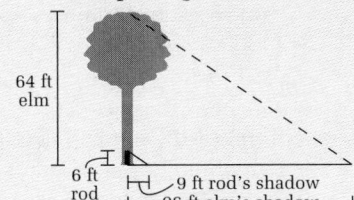
64 ft elm
6 ft rod
9 ft rod's shadow
⊢— 96 ft elm's shadow —⊣

23. The triangles are similar, so using the proportion $\frac{64}{6} = \frac{s}{9}$, you get $6s = 576$; $s = 96$. The shadow of the elm is 96 ft.

24. Answers may vary. An example is given. Following the words of the ritual, Sherlock Holmes took 10 steps north with each foot, 5 east with each foot, 2 south with each foot, 2 west with each foot, and then went into the cellar. The hidden objects were several coins and the crown of Charles the First. The crown was being held for the return of Charles the Second.

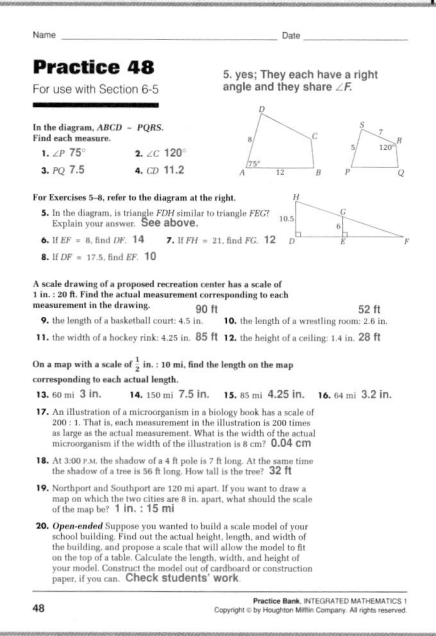
Writing Suppose you want to make a scale model of each object described. What scale would you use to make a model of reasonable size? Explain.

25. a stadium that contains a rectangular field that is 110 yd by 80 yd

26. a car that is 13 ft 4 in. long and 5 ft high

27. the planet Jupiter, whose diameter is about 142,700 km

28. a red blood cell whose "thickness" is 9×10^{-6} m

29. A marketing survey found that 525 people out of 2200 prefer clear dishwashing liquid. Estimate how many people out of a population of 2.5 million prefer clear dishwashing liquid. *(Section 6-4)*

30. Find the mean, median, and mode of these test scores. *(Section 3-2)*

76	84	65	87	84	58
92	74	76	84	94	70
68	80	84	76	82	90

31. Plot the points $(-3, 3)$, $(5, 3)$, $(0, -1)$, and $(-8, -1)$ on a coordinate plane. Connect the points in order and connect the last point to the first. Write the name of the polygon you formed. *(Section 4-2)*

 Working on the Unit Project

The cost of a newspaper ad is usually related to its size. The size is measured in *column inches*. One column inch is $2\frac{1}{16}$ in. wide (the width of one column of print in the newspaper) and 1 in. long.

32. a. An ad spreads across 2 columns and is 4 in. deep. Another ad spreads across 4 columns and is 2 in. deep. Are the ads similar? Explain.

b. Which of the ads described in part (a) has a greater area?

33. In one newspaper, a full-page ad is 216 column inches. Is it similar to a quarter-page ad? to a half-page ad? Explain why or why not.

1 in.

1 column inch

$2\frac{1}{16}$ in.

seems a reasonable size to represent the small cell.

29. about 597,000 people

30. mean: about 79; median: 81; mode: 84

31. parallelogram

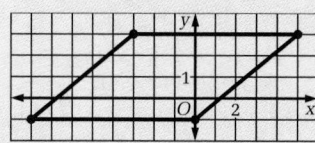

32. a. No. The first ad is almost square. The second ad is a long, narrow rectangle. The sides are not in proportion.

b. The two ads have the same area, 8 column inches, which is 16.5 in.^{2}.

33. If the quarter-page ad is half the width and half the length of the full-page ad, the ads are similar because

the corresponding sides are in proportion. If the quarter-page ad is the length of the page and one fourth the width of the page, the ads are not similar because the corresponding sides are not in proportion. A full-page ad is not similar to a half-page ad because the corresponding sides are not in proportion.

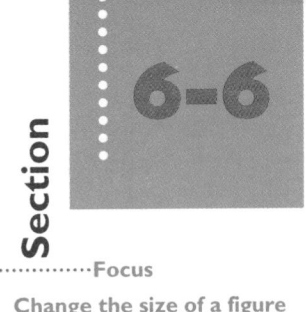

⌐⌐⌐⌐Focus
Change the size of a figure without changing its shape.

Dilations

IT'S A STRETCH

A drawing created on a computer is called a *graphic*. With most computer drawing software, you stretch or shrink a graphic by dragging a *handle* on a side or corner of its *bounding box*.

Lines drawn through corresponding points on the original figure and its image meet in a point called the **center of dilation**.

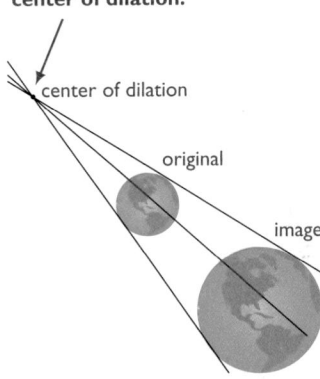

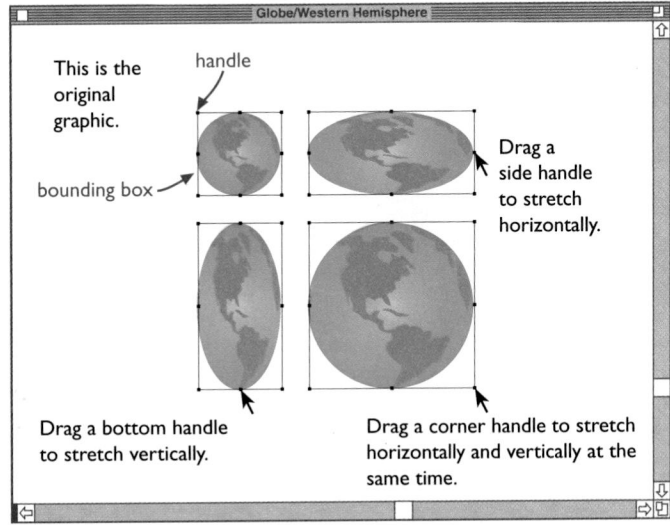

Globe/Western Hemisphere

This is the original graphic.

handle

bounding box

Drag a side handle to stretch horizontally.

Drag a bottom handle to stretch vertically.

Drag a corner handle to stretch horizontally and vertically at the same time.

Talk it Over

1. Compare the horizontal and vertical dimensions of the globe in the original graphic with its dimensions in each stretched graphic.

2. Are any of the stretched graphics congruent to the original? Are any of them similar to the original?

The result of a transformation is called an **image**. A **dilation** is a transformation in which the original figure and its image are similar.

The ratio of any length on the image to the corresponding length on the original figure is the **scale factor** of the dilation.

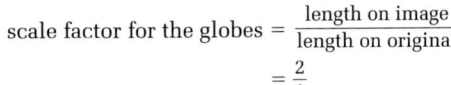

$$\text{scale factor for the globes} = \frac{\text{length on image}}{\text{length on original}}$$
$$= \frac{2}{1}$$

6-6 Dilations

337

PLANNING

Objectives and Strands
See pages 298A and 298B.

Spiral Learning
See page 298B.

Materials List
➤ Graph paper
➤ Ruler
➤ Protractor

Recommended Pacing
Section 6-6 is a one-day lesson.

Extra Practice
See pages 629–631.

Warm-Up Exercises
Warm-Up Transparency 6-6

Support Materials
➤ Practice 49
➤ Enrichment 44 in the Activity Bank
➤ Study Guide 6-6
➤ Problem Set 13
➤ Additional Exploration 7
➤ Diagram Master 2 in the Explorations Lab Manual
➤ McDougal Littell Mathpack software: *Geometry Inventor*
➤ Geometry Inventor Activity Book: Activities 23 and 24
➤ Quiz 6-6
➤ Alternative Assessments 10

Answers to Talk it Over ⋯⋯⋯⋯⋯⋯⋯⋯⋯⋯⋯

1. Answers may vary. An example is given. The top right globe has the same height but twice the width. The bottom left globe has the same width but twice the height. The bottom globe has twice the height and twice the width.

2. No. Yes, the bottom right graphic is similar to the original.

TEACHING

Talk it Over

A discussion of questions 1 and 2 can be enhanced by doing an actual demonstration on a computer of stretching or shrinking a graphic. Print out the different graphics as you change them so that they can be compared to the original.

Communication: Writing

Encourage students to use the new terms of this section in their writing: transformation, image, dilation, center of dilation, and scale factor. Relate these new concepts with those of the previous section by reviewing what it means for figures to be similar: corresponding angles are congruent and corresponding sides are in proportion.

Additional Sample

S1 Draw a dilation of △*ABC* with center *P* and scale factor 2.

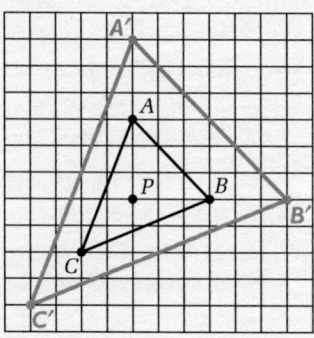

Using Technology

Students can use the *Geometry Inventor* software to draw dilations of geometric figures.

Enlargements and Reductions

When the image of a dilation is larger than the original figure, the dilation is an *enlargement*. The dilation on page 337 with scale factor $\frac{2}{1}$ is an enlargement. When the image is smaller than the original figure, the dilation is a *reduction*.

Sample 1

Draw a dilation of △ ABC with center P and scale factor 3.

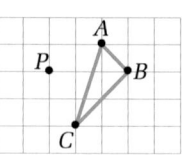

Sample Response

1 Copy the figure onto graph paper.

2 Draw rays from *P* through each vertex of △*ABC*.

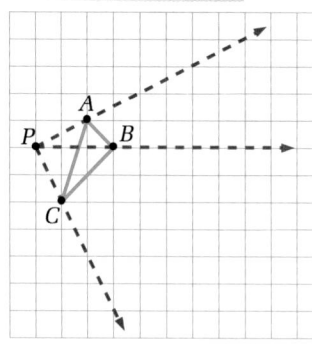

3 The scale factor is 3. Mark A' on $\overrightarrow{PA}$, B' on $\overrightarrow{PB}$, and C' on $\overrightarrow{PC}$.

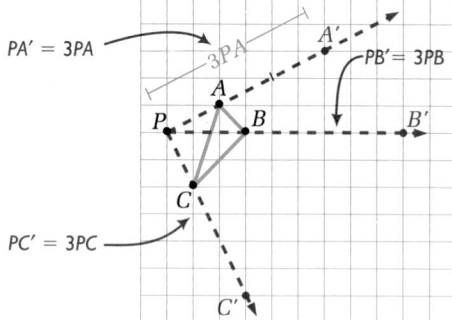

$PA' = 3PA$
$PB' = 3PB$
$PC' = 3PC$

4 Connect A', B', and C' to form the image, △A'B'C'.

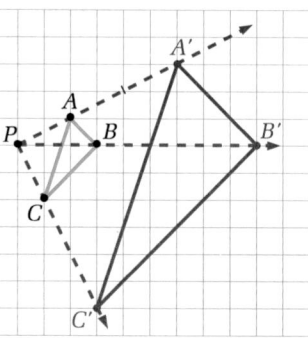

Unit 6 Ratios, Probability, and Similarity

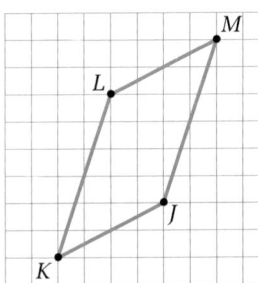

Sample 2

**Draw a dilation of JKLM
with center J and
scale factor $\frac{1}{2}$.**

Sample Response

1 Copy the figure
onto graph paper.

2 Draw rays from J
through K, L, and M.

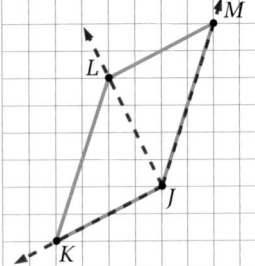

$JL' = \frac{1}{2} JL$

3 The scale factor is $\frac{1}{2}$.
Mark K' on $\overrightarrow{JK}$, L' on $\overrightarrow{JL}$,
and M' on $\overrightarrow{JM}$.

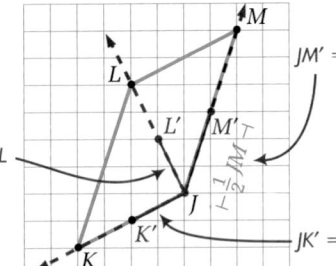

$JM' = \frac{1}{2} JM$

$JK' = \frac{1}{2} JK$

4 Connect J', K', L', and
M' to form the image,
quadrilateral J'K'L'M'.

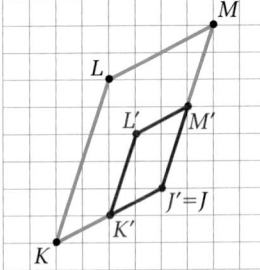

Talk it Over

Use Samples 1 and 2 to answer questions 3 and 4.

3. Check that $\triangle ABC$ and $\triangle A'B'C'$ are similar. Measure the
   sides and angles of the original figure and its image. Are the
   corresponding sides in proportion? Are the corresponding
   angles congruent? Also check JKLM and J'K'L'M'.

4. **a.** Which dilation is an enlargement?

   b. Which dilation is a reduction?

   c. Describe the scale factors that result in enlargements.

   d. Describe the scale factors that result in reductions.

5. In the dilations shown on page 337 and in Sample 1, the
   center of dilation is *outside* the original figure. In Sample 2,
   the center of dilation is *on* the original figure.

   Copy this figure onto graph paper. Draw a dilation where the
   center of dilation is *inside* the original figure.

6-6 Dilations

339

Additional Sample

S2 Draw a dilation of $\triangle ABC$
with center P and scale fac-
tor $\frac{1}{2}$.

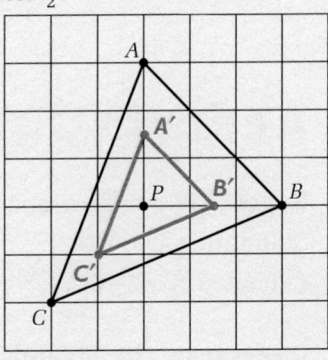

Talk it Over

Question 5 points out that
regardless of where the center
of dilation is located, the image
is always found in the same
way.

Visual Thinking

Encourage students to visualize
and discuss multiple enlargements and reductions. How big
would they be if they were
twice as big? three times as big?
How small would they be if
they were half as small? three
times as small? This activity
involves the visual skills of
exploration and *perception*.

Answers to Talk it Over

3. Answers may vary. An
   example is given.
   $\frac{A'C'}{AC} = \frac{3}{1}$, $\frac{A'B'}{AB} = \frac{3}{1}$, $\frac{B'C'}{BC} = \frac{3}{1}$;
   $\angle A \cong \angle A'$, $\angle B \cong \angle B'$,
   $\angle C \cong \angle C'$; the triangles
   are similar. $\frac{J'K'}{JK} = \frac{1}{2}$,
   $\frac{K'L'}{KL} = \frac{1}{2}$, $\frac{L'M'}{LM} = \frac{1}{2}$, $\frac{M'J'}{MJ} = \frac{1}{2}$;
   $\angle J \cong \angle J'$, $\angle K \cong \angle L'K'J'$,
   $\angle L \cong \angle L'$, $\angle M \cong \angle L'M'J'$;

   the quadrilaterals are
   similar.

4. **a.** The dilation in Sample
   1 is an enlargement.

   b. The dilation in Sample
   2 is a reduction.

   c. A scale factor that is
   greater than 1 results in
   an enlargement.

   d. A scale factor between
   0 and 1 results in a
   reduction.

5. Drawings may vary. An
   example is given. The center of dilation is at P and
   the scale factor is $\frac{1}{2}$.

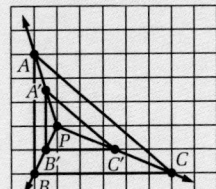

....
Look Back

The Look Back question should help to solidify the idea that dilations produce similar figures.

........................●

![wave graphic] **APPLYING**

....
Suggested Assignment

Standard 2–13, 16–24

Extended 1–13, 16–24

....
Integrating the Strands

Number Exs. 21–23

Algebra Exs. 18–20

Geometry Exs. 1–17, 24

....
Communication: Discussion

When discussing Exs. 2–4, ask students to state the essential difference between a dilation and a rotation and a translation. (Rotating and translating preserve size and shape, while a dilation preserves shape only.)

When discussing Ex. 8, ask students why proportions can be used to find the missing lengths. (The figures are similar under a dilation.)

Look Back ←

In Section 6-5 on page 328, a rectangular photograph was enlarged by drawing a diagonal through the rectangle. Where is the center of the dilation? What is the scale factor?

....

6-6 Exercises and Problems

1. **Reading** In the diagram at the top of page 337, which of the stretched graphics is an image of the original figure under a dilation?

Does each diagram show a *dilation*, a *rotation*, or a *translation*? There may be more than one answer.

2.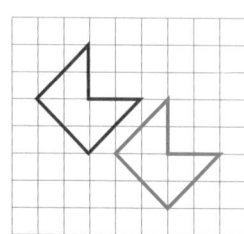

3.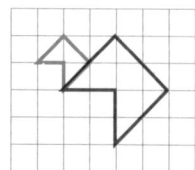

4.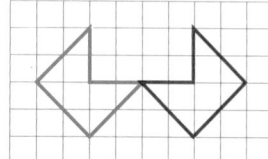

For Exercises 5–7, the blue figure is the original figure.

a. **State the coordinates of the center of the dilation.**

b. **Find the scale factor.**

5.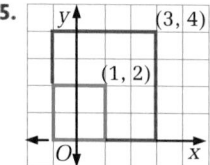

6.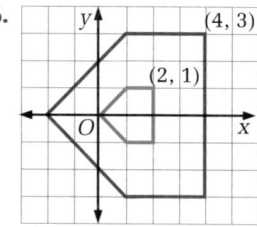

7.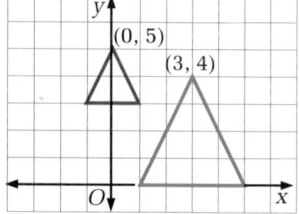

8. Use proportions to find the missing lengths in this dilation.

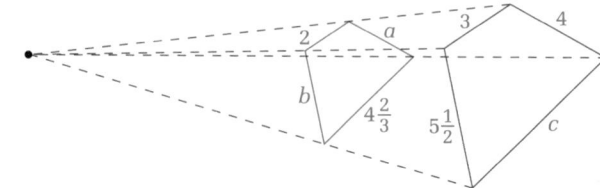

Unit 6 Ratios, Probability, and Similarity

Answers to Look Back ·········

The center of the dilation is the upper left corner of the original photograph. The scale factor is 2.5.

Answers to Exercises and Problems ························

1. The lower right-hand figure is a dilation. It has been stretched both vertically and horizontally.

2. translation

3. dilation

4. rotation

5. a. (−1, 0)
   b. 2

6. a. (1, 0)
   b. 3

7. a. (−3, 6)
   b. 2

8. $a = 2\frac{2}{3}$, $b = 3\frac{2}{3}$, $c = 7$

Draw a dilation of each polygon with the given center of dilation and scale factor.

9. center $R(2, 5)$; scale factor 2

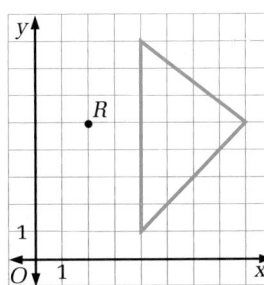

10. center $X(-4, -1)$; scale factor $\frac{1}{2}$

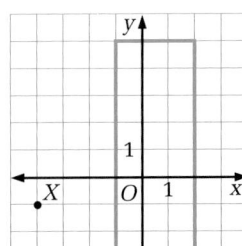

11. vertices $A(7, -1)$, $B(-1, 7)$, $C(-5, 3)$, $D(-1, -9)$; center $A(7, -1)$; scale factor $\frac{1}{4}$

12. vertices $A(3, 1)$, $B(2, 3)$, $C(0, 0)$; center $P(5, 5)$; scale factor 3

connection to **A R T**

13. One way to create the illusion of a three-dimensional object is to draw it in *one-point perspective.*

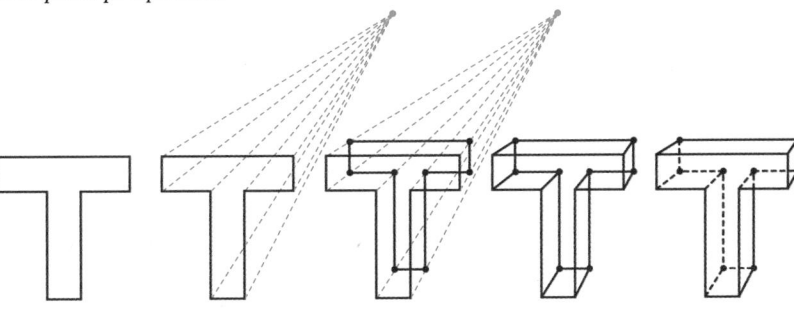

a. On graph paper, draw your initials using block letters.

b. Choose a center of dilation. This is called the *vanishing point.*

c. Draw rays from the vanishing point through the vertices of the letters.

d. Draw a reduction of the letter by locating image points on the rays. (*Hint:* Corresponding edges should be parallel.)

e. Connect the original and image letters by joining corresponding vertices.

f. Erase hidden lines.

g. Color your three-dimensional initials in a creative way.

14. **Research** Measure the dimensions of a photographic negative and a photograph that has been made from it. Is the process a dilation? How can you tell?

6-6 Dilations

341

Answers to Exercises and Problems

9.

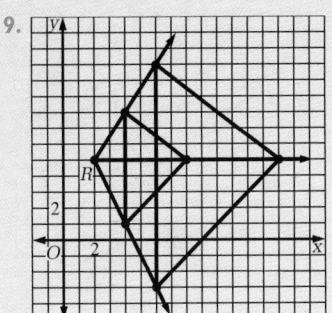

10.

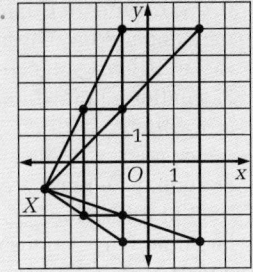

11.

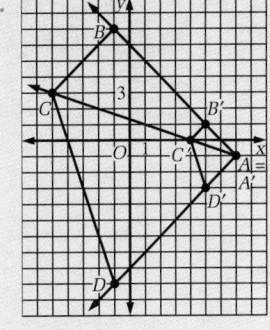

12.

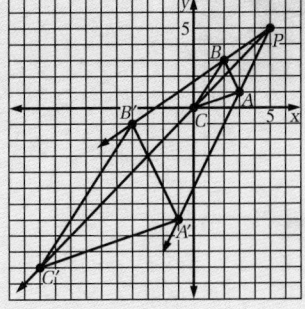

13. a–g. Drawings may vary.

14. Answers may vary.

Practice 49 For use with Section 6-6

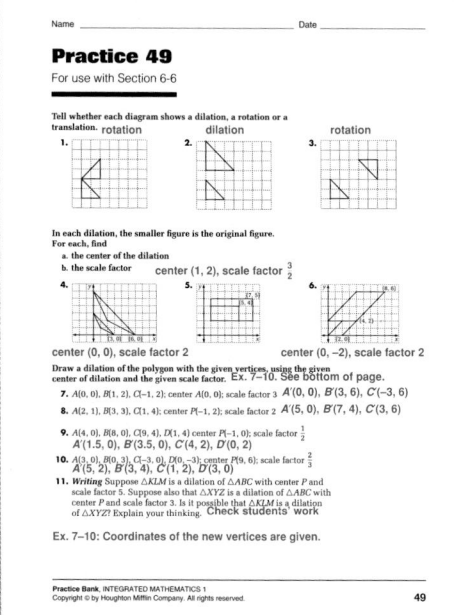

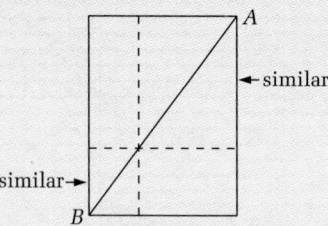

15. **Using Manipulatives** Draw a diagonal across a rectangular sheet of paper. Mark any point on the diagonal. Make horizontal and vertical folds through the point. Unfold the paper.

   a. The two folds divide your paper into four rectangles. Which of the rectangles are similar to the original sheet of paper?

   b. Suppose that each rectangle you found in part (a) is the image of a dilation of the original sheet of paper. For each dilation, find the center of dilation and the scale factor.

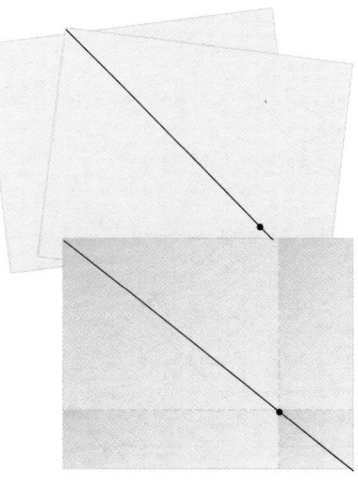

Ongoing **ASSESSMENT**

16. **Open-ended** Draw a polygon on a coordinate plane. Mark a center of dilation.

   a. Draw two different dilations of your polygon. Use the same center of dilation but use two different scale factors. Label the new polygons with primes.

   b. Name some similar polygons in your diagram.

   c. Write some proportions that compare the lengths of sides in your diagram.

Review **PREVIEW**

17. $\triangle ABC \sim \triangle DEF$. Find all the missing side and angle measures. *(Section 6-5)*

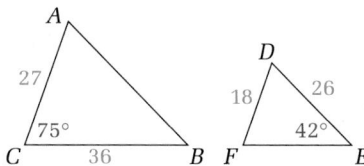

Solve. *(Section 5-3)*

18. $2x - 8 = x + 16$ **19.** $e + 8 = 3e - 14$ **20.** $3a + 7 = -2a - 38$

Write each ratio in simplest form. *(Section 6-1)*

21. $\dfrac{6}{36}$ **22.** 14 to 21   **23.** 63:84

 Working on the Unit Project

24. a. Make a sketch of an ad for your product or trace a simple ad. Dilate the sketch with a scale factor of 2 and a center of dilation at one corner of the ad. Which size is more effective, the original or the enlargement?

   b. Suppose you want to put the ad on a billboard. What scale factor might you use?

Unit 6 Ratios, Probability, and Similarity

Sine and Cosine Ratios

RIGHT ON

EXPLORATION

What relationships are there between the sides of similar right triangles?

- **Materials:** graph paper, protractors
- **Work with another student.**

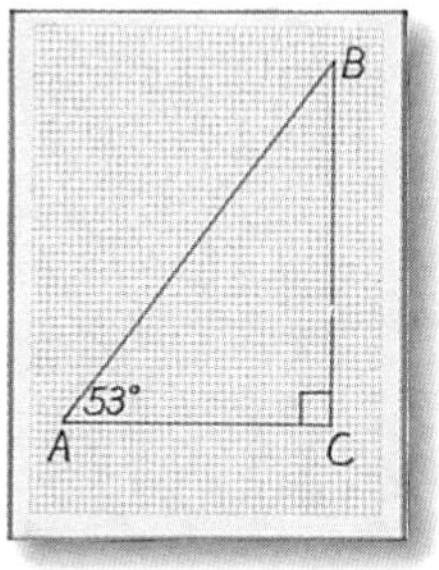

(1) Draw a right triangle *ABC* with $\angle C = 90°$, *AC* = 3, and $\angle A = 53°$. Draw straight lines. Call this triangle △1.

(2) Draw three more right triangles similar to △1 using different lengths for side *AC*. Call the triangles △2, △3, and △4.

(3) Use a strip of graph paper to measure the lengths of the sides of each triangle to the nearest half unit. Find the ratios $\frac{AC}{AB}$ and $\frac{BC}{AB}$ to the nearest hundredth. Make and complete a table like this one.

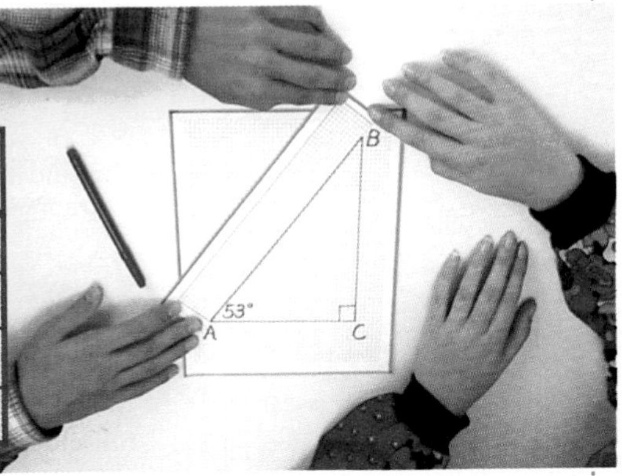

	AC	BC	AB	$\frac{AC}{AB}$	$\frac{BC}{AB}$
△1	3	4	5	?	?
△2	?	?	?	?	?
△3	?	?	?	?	?
△4	?	?	?	?	?

(4) Do you see any pattern in the ratios $\frac{AC}{AB}$? in the ratios $\frac{BC}{AB}$?

(5) Suppose you draw any size right triangle △*ABC* with $\angle A = 53°$. Predict the ratios $\frac{AC}{AB}$ and $\frac{BC}{AB}$ for the triangle.

6-7 Sine and Cosine Ratios

343

Answers to Exploration

1. Drawings may vary.

2–5. Drawings, tables, and estimates may vary. Examples are given.

2.

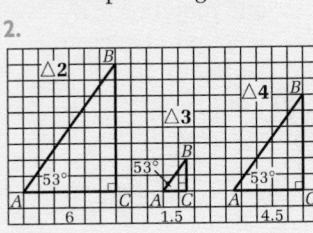

3.

	AC	*BC*	*AB*	$\frac{AC}{AB}$	$\frac{BC}{AB}$
△1	3.0	4.0	5.0	0.60	0.80
△2	6.0	8.0	10.0	0.60	0.80
△3	1.5	2.0	2.5	0.60	0.80
△4	4.5	6.0	7.5	0.60	0.80

4. Yes, 0.60. Yes, 0.80.

5. $\frac{AC}{AB} = 0.60$; $\frac{BC}{AB} = 0.80$

PLANNING

Objectives and Strands
See pages 298A and 298B.

Spiral Learning
See page 298B.

Materials List
➤ Graph paper
➤ Protractor
➤ Ruler
➤ Scientific calculator
➤ Graphics calculator or graphing software

Recommended Pacing
Section 6-7 is a two-day lesson.
Day 1
Pages 343–345: Exploration through Talk it Over, *Exercises 1–15*
Day 2
Pages 346–348: Sample 2 through Look Back, *Exercises 16–29*

Extra Practice
See pages 629–631.

Warm-Up Exercises
Warm-Up Transparency 6-7

Support Materials
➤ Practice 50
➤ Enrichment 45 in the Activity Bank
➤ Study Guide 6-7
➤ Problem Set 13
➤ Diagram Masters 2, 18 in the Explorations Lab Manual
➤ McDougal Littell Mathpack software: *Geometry Inventor* and *Function Investigator*
➤ Quiz 6-7
➤ Test 24

TEACHING

Exploration

The goal of the Exploration is to have students discover that the ratios of each leg to the hypotenuse of a right triangle (the cosine and sine ratios) depend only on the measure of the given acute angle of the right triangle and not on the particular right triangle. These constant ratios are then defined as sine A and cosine A for the given angle A.

Using Technology

Students can use the *Geometry Inventor* software in the Exploration to draw triangles and find the lengths of the sides.

Talk it Over

As they answer questions 1 and 2, students should realize that the ratios found in the Exploration are based on measurements, which, by their nature, are inexact. The displayed number on the calculator is a much more exact answer.

Using Technology

Stress to students that their calculators must be set in degree mode. Most scientific calculators also operate in radian and gradient modes. Should their calculators be set in radian or gradient mode, students will be finding and using incorrect sine and cosine values.

Error Analysis

Students may find it difficult to identify the leg adjacent to and the leg opposite an acute angle in a right triangle. You may want to have them practice on several examples of right triangles.

For each right triangle in the Exploration, the measure of $\angle A$ is 53°. In each triangle, $\overline{AB}$ is the **hypotenuse** because it is opposite the right angle. $\overline{AC}$ and $\overline{BC}$ are the **legs** of the right triangle.

The ratios $\frac{AC}{AB}$ and $\frac{BC}{AB}$ depend only on the measure of $\angle A$ and not on the lengths of the sides of the triangle. These ratios have special names.

The **sine** of an acute angle of a right triangle is the ratio of the length of the opposite leg to the length of the hypotenuse. The **cosine** of an acute angle of a right triangle is the ratio of the length of the adjacent leg to the length of the hypotenuse.

SINE AND COSINE RATIOS

The sine of $\angle A$ is written *sin A*.

$$\sin A = \frac{\text{leg opposite } \angle A}{\text{hypotenuse}} = \frac{BC}{AB}$$

The cosine of $\angle A$ is written *cos A*.

$$\cos A = \frac{\text{leg adjacent to } \angle A}{\text{hypotenuse}} = \frac{AC}{AB}$$

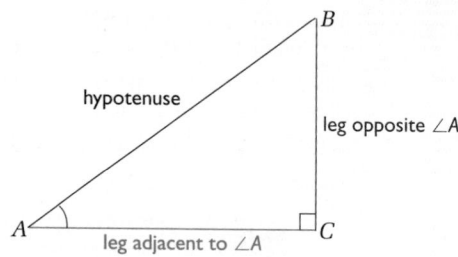

The sine and cosine ratios are called **trigonometric ratios.** They are programmed into your scientific calculator.

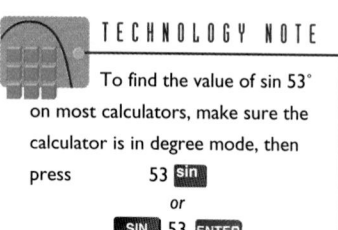

TECHNOLOGY NOTE

To find the value of sin 53° on most calculators, make sure the calculator is in degree mode, then press

53 **sin**

or

SIN 53 **ENTER**

Talk it Over

1. **a.** Use a calculator to find sin 53°. (Be sure your calculator is in degree mode.)

   b. Compare your answer in part (a) with the ratios $\frac{BC}{AB}$ in the Exploration. Why might the values be slightly different?

2. **a.** Use a calculator to find cos 53°.

   b. Compare your answer in part (a) with the ratios $\frac{AC}{AB}$ in the Exploration.

3. In $\triangle ABC$ above, which leg is opposite $\angle B$? Which leg is adjacent to $\angle B$?

344

Unit 6 Ratios, Probability, and Similarity

Answers to Talk it Over ··

1. **a.** sin 53° ≈ 0.79863551

   b. Answers may vary. An example is given. If the calculator answer were rounded to the nearest hundredth, it would be the same as the ratio I found in the Exploration. The values may be slightly different due to small measuring errors in measuring the angle

   and rounding the lengths to the nearest half unit.

2. **a.** cos 53° ≈ 0.6018150232

   b. Answers may vary. An example is given. If the calculator answer were rounded to the nearest hundredth, it would be the same as the ratio I found in the Explora-

   tion. The values may be slightly different due to small measuring errors in measuring the angle and rounding the lengths to the nearest half unit.

3. $\overline{AC}$; $\overline{BC}$

Sample 1

Write each trigonometric ratio as a fraction and as a decimal rounded to the nearest hundredth.

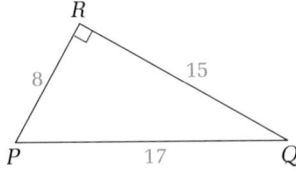

a. sin P

b. cos P

c. sin Q

d. cos Q

Sample Response

a. $\sin P = \dfrac{\text{leg opposite } \angle P}{\text{hypotenuse}} = \dfrac{15}{17} \approx 0.88$

b. $\cos P = \dfrac{\text{leg adjacent to } \angle P}{\text{hypotenuse}} = \dfrac{8}{17} \approx 0.47$

c. $\sin Q = \dfrac{\text{leg opposite} \angle Q}{\text{hypotenuse}} = \dfrac{8}{17} \approx 0.47$

d. $\cos Q = \dfrac{\text{leg adjacent to } \angle Q}{\text{hypotenuse}} = \dfrac{15}{17} \approx 0.88$

Talk it Over

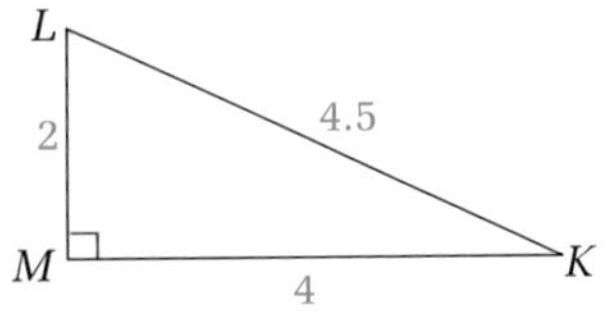

4. Find the values of cos K, sin K, cos L, and sin L for this triangle. What do you notice about the ratios?

5. Suppose you know only the sine of an acute angle of a right triangle. How can you tell what the cosine of the other acute angle is?

6. In the Exploration, $\angle B = 37°$. Predict the values of sin 37° and cos 37°. Test your predictions on your calculator.

7. Do you think the values for the sine or the cosine of an angle in a right triangle can be greater than 1? Explain your reasoning.

▶ **Now you are ready for:**
Exs. 1–15 on pp. 348–349

6-7 Sine and Cosine Ratios

345

Additional Sample

S1 Write each trigonometric ratio as a fraction and as a decimal rounded to hundredths.

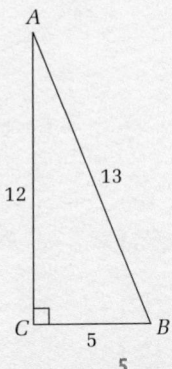

a. $\sin A \ \dfrac{5}{13} \approx 0.38$

b. $\cos A \ \dfrac{12}{13} \approx 0.92$

c. $\sin B \ \dfrac{12}{13} \approx 0.92$

d. $\cos B \ \dfrac{5}{13} \approx 0.38$

Talk it Over

Questions 4–6 lead students to the conjecture that the sine of one acute angle of a right triangle is equal to the cosine of the other acute angle.

Answers to Talk it Over

4. cos $K \approx 0.89$, sin $K \approx 0.44$, cos $L \approx 0.44$, sin $L \approx 0.89$ Statements may vary. An example is given. The cosine of $\angle K$ is the same as the sine of $\angle L$, and the sine of $\angle K$ is the same as the cosine of $\angle L$.

5. Explanations may vary. An example is given. Question 4 suggests that the sine of one acute angle of a right triangle is equal to the cosine of the other acute angle.

6. sin 37° = cos 53° ≈ 0.6018150232; cos 37° = sin 53° ≈ 0.79863551

7. No, the hypotenuse will always be the longest side and its value will always be in the denominator of the ratio.

345

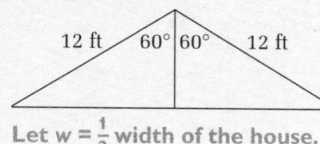
Sample 2

To be used safely, a ladder should make an angle of about 75° with the ground. What is the maximum height that a 12-foot ladder can reach safely when it leans against a building?

Sample Response

Problem Solving Strategy: Make a sketch

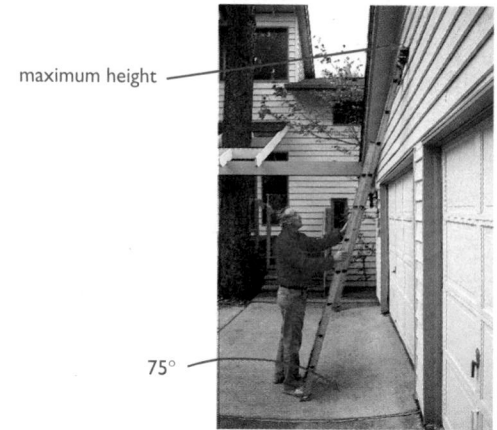

maximum height

75°

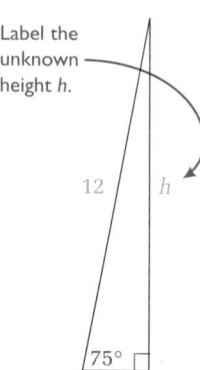

Label the unknown height h.

12   h

75°

Use the sketch to decide which trigonometric ratio to use to solve the problem. In this problem, you know the measure of an acute angle and the length of the hypotenuse (the ladder). You need to find the height, h, which is the leg *opposite* the angle. The leg opposite the angle is part of the sine ratio.

$$\sin 75° = \frac{h}{12}$$ ⟵ Use the sine ratio: $\frac{\text{opposite}}{\text{hypotenuse}}$.

$$0.9659 \approx \frac{h}{12}$$ ⟵ Use the **SIN** key on your calculator.

$$(0.9659)(12) \approx \left(\frac{h}{12}\right)(12)$$ ⟵ Multiply both sides by 12.

$$11.6 \approx h$$ ⟵ Round to the nearest tenth.

The maximum safe height is about 11.6 ft, or about 11 ft 7 in.

Talk it Over

8. Suppose the measure of the angle that the ladder in Sample 2 makes with the ground were greater than 75°. Would the foot of the ladder be farther from or closer to the base of the building?

9. How can you use the cosine ratio to find the distance between the foot of the ladder and the base of the building?

Unit 6 Ratios, Probability, and Similarity

Answers to Talk it Over

8. closer to

9. Answers may vary. An example is given. Let d be the distance between the foot of the ladder and the base of the building. You can find d by using the equation $\cos 75° = \frac{d}{12}$.

A wheelchair ramp makes a 4° angle with the ground. The beginning of the ramp is 366 cm away from the end of the ramp. What is the length of the ramp to the nearest centimeter?

Sample Response

Problem Solving Strategy: Make a sketch

Label the unknown length m.

A right triangle is formed by the ramp and the ground. Use the sketch to decide which trigonometric ratio to use to solve the problem.

You want to find the ramp length, which is the hypotenuse of the triangle and is represented by m. You know the measure of the angle and the length of the leg adjacent to the angle. The leg adjacent to the angle is part of the cosine ratio.

$\cos 4° = \dfrac{366}{m}$ ← Use the cosine ratio: $\dfrac{\text{adjacent}}{\text{hypotenuse}}$.

$0.9976 \approx \dfrac{366}{m}$ ← Use the **COS** key on your calculator.

$(0.9976)(m) \approx \left(\dfrac{366}{m}\right)(m)$ ← Multiply both sides by m to remove m from the denominator.

$0.9976\, m \approx 366$

$\dfrac{0.9976\, m}{0.9976} \approx \dfrac{366}{0.9976}$ ← Divide both sides by 0.9976.

$m \approx 367$ ← Round to the nearest centimeter.

The length of the ramp is about 367 cm.

Talk it Over

10. Suppose you know the length of a wheelchair ramp and the measure of the angle it makes with the ground. Which trigonometric ratio would you use to find the height of the ramp at its highest point? Explain your reasoning.

6-7 Sine and Cosine Ratios

347

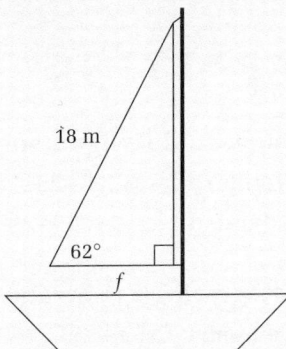
Answers to Talk it Over

10. Answers may vary. An example is given. I would use the sine ratio since it is the ratio of the leg opposite the angle over the hypotenuse. The leg opposite has a length equal to the height of the ramp, and the hypotenuse has a value equal to the length of the ramp.

Look Back

The answer to this Look Back would be a good journal entry. Encourage students to draw diagrams to support their descriptions. ....●

APPLYING

Suggested Assignment

Standard 3–13, 16–23, 25–29

Extended 1, 3–14, 16–29

Integrating the Strands

Geometry Ex. 14, 26

Trigonometry Exs. 1–25

Statistics and Probability Exs. 27–29

Communication: Drawing

Exs. 12 and 13 can be extended by having students draw other right triangles. Have them measure the sides and find the values of the sine and cosine ratios of the acute angles.

......► **Now you are ready for:**
Exs. 16–29 on pp. 349–350

> **Look Back** ◄
>
> Review the problem in Sample 2. What made it possible to solve the problem by using the sine ratio? What other measurements can you find from the given information?

6-7 Exercises and Problems

1. **Reading** Look back at the Sherlock Holmes story on page 335. How did Musgrave use trigonometry when he was a boy? What trigonometric ratios did you learn about in this section?

2. **Research** The Exploration with triangles led you to some special ratios. Use a dictionary to find the origin of the word "trigonometry." Why do you think sine and cosine are called trigonometric ratios?

Use a calculator to write each ratio as a decimal rounded to hundredths.

3. sin 32°   4. sin 85°   5. cos 10°   6. cos 63°

Write each ratio as a fraction and as a decimal rounded to hundredths.

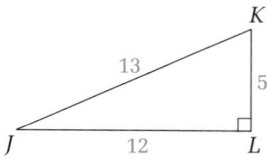

7. sin J 8. sin K 9. cos K 10. cos J

11. **Open-ended** Use a ruler and protractor.

   a. Draw any right triangle. Label the right angle $\angle C$ and the other two angles $\angle A$ and $\angle B$. Measure the sides of your triangle.

   b. Use your measurements to find sin A and cos A to the nearest hundredth.

   c. Measure $\angle A$ with a protractor. Use a calculator to find sin A and cos A to the nearest hundredth.

   d. Compare your results for parts (b) and (c). Why might the values be slightly different?

For Exercises 12 and 13, draw and label a right triangle that fits each description.

12. $\triangle ABC$ with $\sin A = \dfrac{\text{leg opposite } \angle A}{\text{hypotenuse}} = \dfrac{6}{10}$ 13. $\triangle XYZ$ with $\cos X = \dfrac{\text{leg adjacent to } \angle X}{\text{hypotenuse}} = \dfrac{16}{34}$

14. **Writing** $\triangle P'Q'R'$ is a dilation of $\triangle PQR$. Describe the relationship between the sine and the cosine values of $\angle P'$ and $\angle P$.

348   **Unit 6** Ratios, Probability, and Similarity

Answers may vary. An example is given. I was able to use the sine ratio to solve this problem because the measure of an acute angle of a right triangle was given along with the length of the hypotenuse. Also, I could find the measure of the other acute angle, $90° - 75° = 15°$, and the approximate length along the ground from the ladder to the building, by using either $\cos 75° = \frac{g}{12}$ or $\sin 15° = \frac{g}{12}$.

1. Answers may vary. An example is given. When he was a boy, Musgrave used trigonometry to measure heights. I learned to use the sine and cosine ratios to find unknown lengths of right triangles.

2. Answers may vary. An example is given. Trigonometry is a branch of mathematics dealing with the

relations between the sides and angles of triangles. The word *trigonometry* comes from the Greek words *trigonon*, meaning triangle, and *metria*, meaning measurement. The sine and cosine ratios are called trigonometric ratios because each relates the measure of an angle of a right triangle to the lengths of its sides.

3. 0.53

4. 0.996 ≈ 1.00

5. 0.98   6. 0.45

7. $\frac{5}{13}$, 0.38   8. $\frac{12}{13}$, 0.92

9. $\frac{5}{13}$, 0.38   10. $\frac{12}{13}$, 0.92

11. a–d. Answers may vary.

15 TECHNOLOGY Use a graphics calculator or graphing software. Make sure you work in degree mode. Use the RANGE key to set the x values between 0° and 90°.

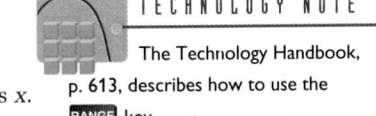

TECHNOLOGY NOTE

The Technology Handbook, p. 613, describes how to use the RANGE key.

a. Graph y = sin x. Do not clear the screen. Graph y = cos x.

b. How can you use the graph to tell when sin x = cos x?

c. When sin x = cos x, what is the value of x? What is the value of y?

Write two equations you could use to find the length x. Then find x.

16.
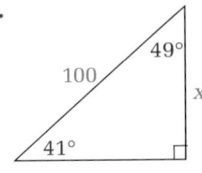

17.
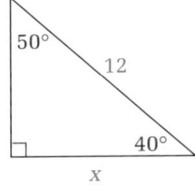

18.
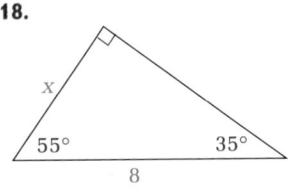

For Exercises 19–22, use the problem solving strategy *Make a sketch*.

19. What is the maximum height that a 15-foot ladder can reach safely when it leans against the side of a building? (*Note:* Use the safety information in Sample 2.)

20. Suppose an escalator makes a 30° angle with the ground floor. The part that people stand on is 15 m in length. How high is the top of the escalator from the ground floor?

21. A kite is on a 150-meter string that makes a 40° angle with the ground. Estimate how far the kite is above the ground.

22. **History** In a Chinese legend the famous general Han Xin flew a kite over his enemy's palace to calculate how far his troops were from the palace.

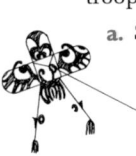

a. Suppose the general measured the length of the kite string and the angle the kite string made with the ground. Draw a sketch to model this situation. Label the information General Han Xin knew and what he was trying to find.

b. How do you think General Han Xin calculated the distance to the palace?

c. Suppose the string made a 28° angle with the ground and was 1800 ft long. About how far away were General Han Xin's forces from the palace?

BY THE WAY...

The best angle for flying a kite in mild wind is 45° to 60° from the ground. In this range, the forces on the kite are balanced.

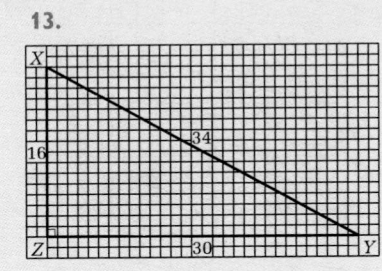

6-7 Sine and Cosine Ratios

349

Using Technology

In Ex. 15, the text refers to the RANGE key. On the TI-82, you use the WINDOW key. Note that ZOOM ZTrig shows the x-axis in multiples of 90° (if the MODE is degrees), or in multiples of π/2.

Students can use the *Function Investigator* software to graph the functions in Ex. 15.

Cooperative Learning

If students go over the answers to Exs. 19–22 in groups, have them check each other's diagrams and answers.

Multicultural Note

In reference to Ex. 22, General Han Xin was the leader who helped set up the powerful Han dynasty in China about 2200 years ago. The Han dynasty lasted from 207 B.C. to A.D. 220, during which time many mathematical and astronomical advances were made. The Han dynasty system of counting was more advanced than the Babylonian, Egyptian, Greek, or Roman methods of counting. In these cultures, certain higher numbers could only be expressed by using addition and subtraction; for example, XIX or 19 = 10 + 10 − 1. In the Chinese system, the value of a numeral depended entirely on its placement; thus, they could express any number, however large.

Answers to Exercises and Problems

12, 13. Drawings may vary. Examples are given.

12.
13.

14. The triangles are similar, so corresponding angles are equal and sin P′ = sin P and cos P′ = cos P.

15. a.

15. b. Use the TRACE and ZOOM features to find coordinates of the point where the sin x and cos x graphs intersect.

c. Answers may vary. An example is given. x = 44.975069; y = 0.70679904

16. $\sin 41° = \frac{x}{100}$, $\cos 49° = \frac{x}{100}$; about 65.6

17. $\sin 50° = \frac{x}{12}$, $\cos 40° = \frac{x}{12}$; about 9.2

18. $\sin 35° = \frac{x}{8}$, $\cos 55° = \frac{x}{8}$; about 4.6

19–22. See answers in back of book.

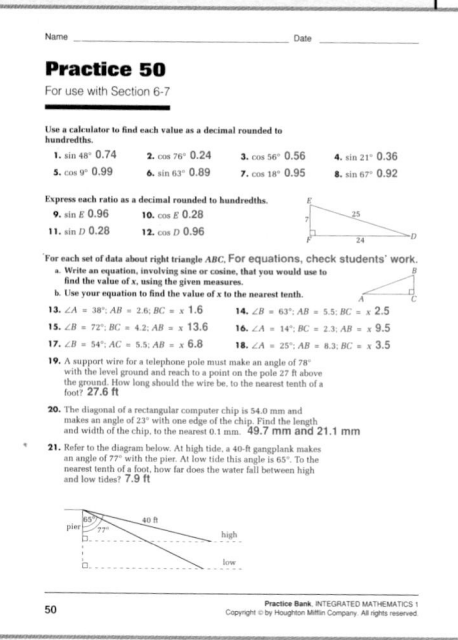
---

23. **Aviation** An airplane is at an altitude of 33,000 ft when the pilot starts the descent to the airport. The pilot wants the plane to descend at an angle of 3°. How many miles away from the airport must the descent begin?

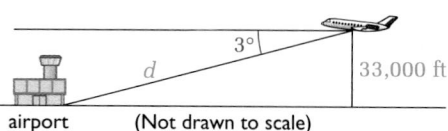

airport   (Not drawn to scale)

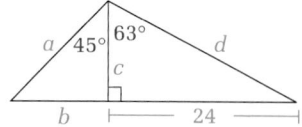

24. Find the lengths *a*, *b*, *c*, and *d*. (First decide in which order they can be found.)

Ongoing **ASSESSMENT**

25. **Writing** Describe how the sine and cosine ratios are alike and how they are different.

Review **PREVIEW**

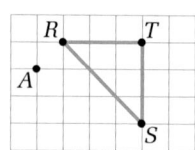

26. Draw a dilation of △RST with center A and a scale factor of 2. *(Section 6-6)*

27. Is the graph misleading? Explain. *(Section 3-7)*

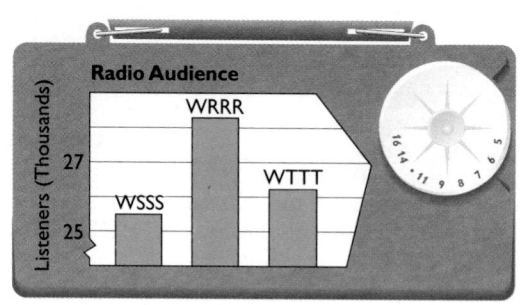

28. A grab bag contains 12 key rings. There are 6 silver-toned, 4 gold-toned, and 2 bronze-toned key rings. If you pick a key ring at random, what is the probability that it is gold-toned? *(Section 6-2)*

Working on the Unit Project

29. **Research** Find the costs of advertising in at least three of these media.

➤ your local newspaper   ➤ the magazines you read   ➤ a local TV station

➤ your school newspaper   ➤ a local radio station   ➤ some other source

---

Answers to Exercises and Problems

23. about 119.4 mi

24. $d \approx 26.9$, $c \approx 12.2$, $a \approx 17.3$, $b \approx 12.2$

25. Paragraphs may vary. An example is given. The sine and cosine ratios are alike because they are both ratios that involve the lengths of the legs of a right triangle. They are different because the sine ratio uses the side opposite the angle, and the cosine ratio uses the side adjacent to the angle. Also, as the measure of the angle increases, the sine ratio increases while the cosine ratio decreases.

26.
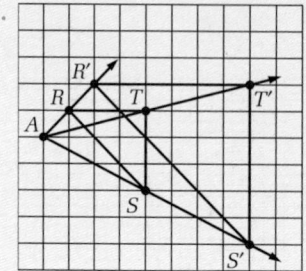

27. Yes. The broken scale gives the impression that station WRRR is much more popular than WSSS or WTTT when it actually only has about 3000 more listeners than WSSS and about 2000 more listeners than WTTT.

28. $\frac{4}{12} = \frac{1}{3} \approx 0.33$

29. Answers may vary.

Completing the Unit Project

Now you are ready to use the data you have collected to plan your advertising campaign. Your plan should include the following information.

➤ a description of the product or service you are advertising and the data that supports your choice of product

➤ your decision about where to advertise your product or service and the data that supports your choice of media

➤ a schedule of when and how often your ad will run

➤ an estimate of the cost of running your ad on the media you selected

➤ either a sketch of a print ad or a script for a TV or radio ad

Look Back

Did your classmates all make the same decisions or were there many different results? Discuss why people may develop different plans even though they all target the same market segment.

Alternative Projects

Project 1: Changing the Rules

Use a game in which dice are thrown to determine the players' moves around a game board. Find the probabilities of several events in the game.

Change the rules of the game. How does this affect the probabilities? Present your conclusions in an oral report. Display the results on a poster.

Project 2: Finding Ratios

In a golden rectangle, the ratio $\frac{length}{width}$ is about 1.618. This is called the *golden ratio*.

➤ Research the history of the golden ratio. Find its exact value.

➤ Find examples of the golden ratio in art and architecture.

➤ Measure rooms, pictures, and furniture. Do any use the golden ratio?

➤ Summarize your results. Make a visual display.

Quick Quiz (6-4 through 6-7)

1. A consumer survey group found that 3 out of 10 households use Brand X laundry detergent. Estimate how many of the 93.3 million households in the United States would use Brand X laundry detergent. [6-4] **about 28 million**

For Exs. 2–4, use the pair of similar quadrilaterals.

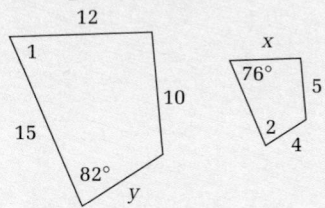

2. Find the measures of $\angle 1$ and $\angle 2$. [6-5] $\angle 1 = 76°$; $\angle 2 = 82°$

3. Find x and y. [6-5] **6, 8**

4. What is the scale factor between the two quadrilaterals? [6-6] **2:1**

5. Draw a dilation of $\triangle ABC$ with vertices $A(0, 0)$, $B(0, 3)$, $C(5, 0)$ and scale factor 2. Use A as the center of dilation. [6-6]

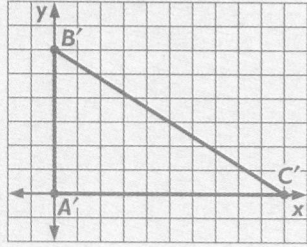

For Exs. 6–8, use the triangle below. [6-7]

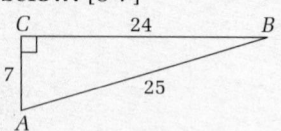

6. Find sin A. $\frac{24}{25}$

7. Find cos A. $\frac{7}{25}$

8. Find cos B. $\frac{24}{25}$

352

1. What is the ratio of the perimeter of a regular hexagon to the length of a side?

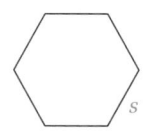

2. A 15 oz box of cereal costs $2.39. What is the unit price?

3. From the people seated around you at a baseball game, you estimate the ratio of men to women is 3:2. If the attendance at the game is 48,000, about how many of the people at the game are women?

6-1

Estimate each probability of what will happen on Joan Laredo's next at-bat. The table shows her record against right-handed and left-handed pitchers.

6-2

	Right-handed pitchers	Left-handed pitchers
Hits	35	10
Outs	65	40
Total	100	50

4. P(gets a hit against a left-handed pitcher)

5. P(gets a hit)

6. P(does *not* get a hit against a right-handed pitcher)

7. Suppose Joan gets a hit the next time she is at bat. Estimate the probability that the pitcher is right-handed.

Suppose you spin this spinner.

8. How many possible number outcomes are there?

9. Is landing on each number equally likely?

10. How many possible color outcomes are there?

11. Is landing on each color equally likely?

Find each probability for a roll of a fair die.

12. $P(\cdot)$

13. P(even number)

14. $P(\text{not } \vdots)$

15. P(a number less than 4)

16. **Writing** Write at least four ideas about probability. Be sure to include these points.

➤ the difference between experimental and theoretical probability

➤ how to use a data table or graph to find the probability of an event

Solve each proportion for y.

6-3

17. $\frac{5}{6.5} = \frac{y}{19.5}$

18. $\frac{30}{0.4} = \frac{100}{y}$

19. It takes 9 ft of twine to tie one stack of newspapers for recycling. How many stacks can be tied by a 60 yd ball of twine?

352

Unit 6 Ratios, Proportions and Similarity

Answers to Unit 6 Review and Assessment

1. 6:1

2. $.16/oz

3. There are about 19,200 women attending the game.

4. 0.2 or 20%

5. 0.3 or 30%

6. 0.65 or 65%

7. Since 35 of 45 hits were against a right-handed

pitcher, the probability of the next hit being against a right-handed pitcher is about 0.78 or 78%.

8. 8 **9.** Yes.

10. 3 **11.** No.

12. $\frac{1}{6} \approx 0.17$ or 17%

13. $\frac{1}{2} = 0.5 = 50\%$

14. $\frac{5}{6} \approx 0.83$ or 83%

15. $\frac{1}{2} = 0.5 = 50\%$

16. Answers may vary. An example is given. Probability has many important uses, from interpreting the results of polls to telling the fairness of games. There are two types of probability, experimental probability and theoretical probability. Both tell

For Exercises 20–22, use this exit poll data.

A sample of 1500 voters took part in an exit poll during a recent election. There were 3 million voters in this election. The poll has a 5% margin of error. The table shows the results.

| ★ | Candidate A 690 | ★ | ★ | Candidate B 560 | ★ | ★ | Candidate C 250 | ★ |

20. About how many votes did each candidate expect to receive?

21. In this election, a candidate needed a majority (at least 50%) of the votes to win. Did any candidate expect to receive a majority?

22. Estimate an interval for the number of votes each candidate could expect.

Tell whether each statement is *True* or *False*. If it is false, give a counterexample.

23. Two quadrilaterals that are similar must be congruent.

24. Corresponding sides of similar rectangles are in proportion.

25. **Open-ended** Write a problem about a situation in which indirect measurement can be used to find an unknown length or distance.

26. a. Explain why $\triangle ABC \sim \triangle DEC$.
   b. Find BC.
   c. Find AB.

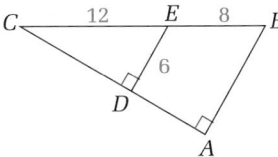

27. A scale model of a bridge was built to a scale of 1 in.:240 ft. The actual bridge is 2000 ft long. How long is the scale model?

Draw a dilation of $\triangle ABC$ with $A(2, 4)$, $B(10, 4)$, and $C(7, 8)$, using each center of dilation and scale factor.

28. center $P(0, 0)$; scale factor 3   **29.** center $A(2, 4)$; scale factor $\frac{1}{2}$

30. A length on an original figure is 15. The corresponding length on the image is 5. What is the scale factor of the dilation?

31. Solve $\frac{15}{x} = 0.25$ for x.

Express each trigonometric ratio as a fraction in simplest form.

32. $\cos Y$
33. $\sin X$
34. $\cos X$
35. $\sin Y$

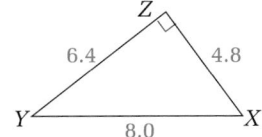

Unit 6 Review and Assessment

---

23. False; any two squares are similar, but one square may be smaller, so they would not be congruent.

24. True.

25. Answers may vary. An example is given. Christie's father wants to measure the ropes for a tree swing from a branch about halfway up a tree in the backyard. He needs to know the height of the tree. He is 6 ft tall and his shadow is 4 ft long. The tree's shadow is 12 ft long. How tall is the tree?

26. a. Explanations may vary. An example is given. Both triangles have $\angle C$ in common, and each triangle has a right angle. Since two pairs of corresponding angles are congruent, the triangles are similar.
   b. 20   **c.** 10

27. $8\frac{1}{3}$ in.

28.

Vertices of Original	Vertices of Image
$A(2, 4)$	$A'(6, 12)$
$B(10, 4)$	$B'(30, 12)$
$C(7, 8)$	$C'(21, 24)$

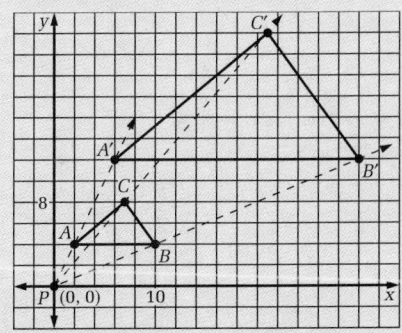

29.

Vertices of Original	Vertices of Image
$A(2, 4)$	$A'(2, 4)$
$B(10, 4)$	$B'(6, 4)$
$C(7, 8)$	$C'(4.5, 6)$

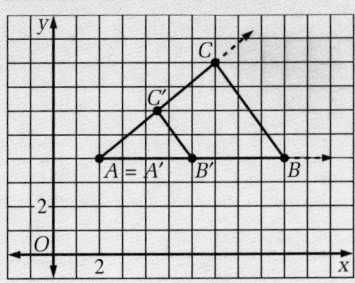

Answers to Unit 6 Review and Assessment

the chances that a certain event is likely to happen. However, experimental probability is based on observations and recorded results, while theoretical probability is based on the knowledge that the possible outcomes of an event are equally likely to happen without doing an experiment. You can use

data from a table to find the probability of an event by using the formula for experimental probability:
$$P(E) = \frac{\text{number of times } E \text{ is observed to happen}}{\text{total number of observations}}.$$

17. 15

18. $\frac{4}{3}$ or about 1.33

19. 20 stacks

20. Candidate A: about 1,380,000 votes; Candidate B: about 1,120,000 votes; Candidate C: about 500,000 votes

21. No.

22. Candidate A: 1,230,000 to 1,530,000 votes; Candidate B: 970,000 to 1,270,000 votes; Candidate C: 350,000 to 650,000 votes

30. $\frac{1}{3}$ **31.** 60

32. $\frac{4}{5}$ **33.** $\frac{4}{5}$

34. $\frac{3}{5}$ **35.** $\frac{3}{5}$

36. A TV tower is 1800 ft high and is supported by wires that make an angle of 63° with the ground.

   a. How long is each wire?

   b. How far from the base of the tower is each wire anchored in the ground?

37. **Self-evaluation** When you began this unit, which applications of ratios and proportions were you already familiar with? Now that you have completed this unit, do you know any more about these applications? Do you know any new uses for ratios and proportions?

38. **Group Activity** Work in a group of three students.

   a. Have each person in your group write about three different situations where ratios are used.

   b. For each situation, have each group member tell whether the ratio is a percent, a rate, a probability, a scale factor, or a trigonometric ratio.

   c. As a group, discuss each member's list. Make a combined group list and report back to the class.

IDEAS AND (FORMULAS)

➤ Three ways to write the ratio of a to b are: $\frac{a}{b}$, $a:b$, or a to b. *(p. 301)*

➤ When you solve a proportion, you find the value of the variable that represents the unknown quantity. *(p. 314)*

➤ For any proportion, the cross products are equal. *(p. 315)*

$$\text{If } \frac{a}{b} = \frac{c}{d}, \text{ then } ad = bc.$$

➤ For any similar polygons, the corresponding angles are congruent and the corresponding sides are in proportion. *(p. 329)*

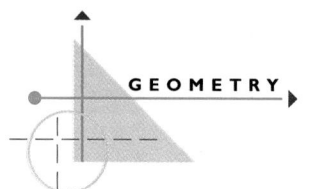 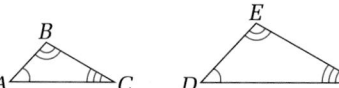

$$\frac{AB}{DE} = \frac{BC}{EF} = \frac{CA}{FD}$$

➤ A scale drawing of an object is similar to the object. *(p. 330)*

➤ Two triangles are similar whenever two angles of one triangle are congruent to two angles of the other triangle. *(p. 331)*

➤ You can use similar triangles to estimate distances or heights that cannot be measured directly. *(p. 331)*

➤ A dilation image is similar to the original figure. *(p. 337)*

    Unit 6 Ratios, Proportions and Similarity

Answers to Unit 6 Review and Assessment

36. a. The wires are about 2020 ft each.

   b. The distance from the base is about 917 ft.

37. Answers may vary.

38. Answers may vary. An example is given. (1) Ratios are used when people want to find how many miles per gallon their car gets. This ratio is a rate. (2) An ad in a newspaper may also be used on a billboard. To enlarge the ad, a scale factor ratio is used. (3) I can use a ratio to find the height of a building when I know the length of a ladder against the building and the angle the ladder makes with the ground. The ratio used is the sine ratio.

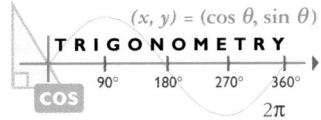

TRIGONOMETRY

$$\sin A = \frac{\text{length of the leg opposite } \angle A}{\text{length of the hypotenuse}}$$

$$\cos A = \frac{\text{length of the leg adjacent to } \angle A}{\text{length of the hypotenuse}}$$

(p. 344)

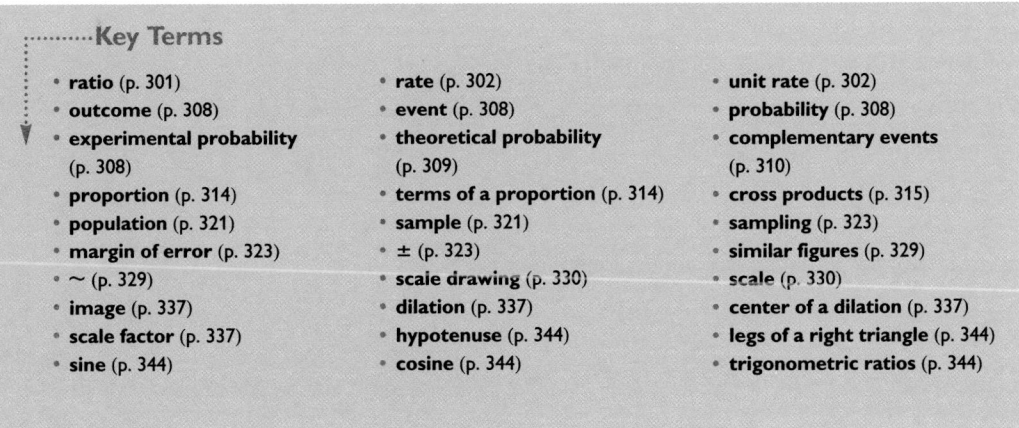

➤ You can use trigonometric ratios to solve problems involving right triangles. *(p. 346)*

STATISTICS & PROBABILITY

➤ The experimental probability that an event E will happen is

$$P(E) = \frac{\text{number of times an event } E \text{ happens}}{\text{number of times the experiment is done}}. \text{ (p. 308)}$$

➤ The theoretical probability that an event E will happen is

$$P(E) = \frac{\text{number of favorable outcomes}}{\text{number of possible outcomes}}. \text{ (p. 309)}$$

➤ The probability of the complement of an event E is

$$P(\text{not } E) = 1 - P(E). \text{ (p. 310)}$$

➤ By collecting data about a sample of a population, you can make predictions about the population. *(p. 323)*

Key Terms

- **ratio** (p. 301)
- **outcome** (p. 308)
- **experimental probability** (p. 308)
- **proportion** (p. 314)
- **population** (p. 321)
- **margin of error** (p. 323)
- **~** (p. 329)
- **image** (p. 337)
- **scale factor** (p. 337)
- **sine** (p. 344)

- **rate** (p. 302)
- **event** (p. 308)
- **theoretical probability** (p. 309)
- **terms of a proportion** (p. 314)
- **sample** (p. 321)
- **±** (p. 323)
- **scale drawing** (p. 330)
- **dilation** (p. 337)
- **hypotenuse** (p. 344)
- **cosine** (p. 344)

- **unit rate** (p. 302)
- **probability** (p. 308)
- **complementary events** (p. 310)
- **cross products** (p. 315)
- **sampling** (p. 323)
- **similar figures** (p. 329)
- **scale** (p. 330)
- **center of a dilation** (p. 337)
- **legs of a right triangle** (p. 344)
- **trigonometric ratios** (p. 344)

1. If an 8 oz carton of fruit juice costs 38 cents and a 16 oz carton of orange juice costs 72 cents, which carton of juice is the better buy? [6-1] **the 16 oz carton**

2. Find the probability of getting a five when you roll a die. [6-2] **1 out of 6, or about 17%**

3. Find the probability of getting a three or a six when you roll a die. [6-2] **1 out of 3, or about 33%**

4. The ratio of the heights of two students is $8:7$. If the taller student is 62 in. tall, what is the height of the shorter student? [6-3] **54.25 in.**

5. Solve $\frac{d}{16} = \frac{18}{48}$. [6-3] **d = 6**

OVERVIEW

➤ In **Unit 7,** students reinforce their earlier work with ratios, variables, and formulas while exploring the concepts of direct variation, slope, and tangent.

➤ Students gain experience in making predictions based on models, in writing equations to model direct variation, and in drawing and interpreting graphs. They are also introduced to dimensional analysis and investigate linear and square measures related to circles.

➤ The Unit Project is based on skateboarding, with a goal of having students demonstrate at least two uses of direct variation in the design of a skating arena.

➤ Connections to consumer topics, sports, social studies, and science are integrated into the teaching materials and exercises.

➤ Problem-solving strategies used in Unit 7 include *Make a Sketch* (Section 7-1), *Use a Formula* (Sections 7-3 and 7-6), *Use a Proportion* (Sections 7-3, 7-4, and 7-6), and *Use Dimensional Analysis* (Section 7-5).

Unit Objectives

Section	Objectives	NCTM Standards
7-1	• Explore three relationships that involve constant ratios.	1, 2, 4, 5, 6, 7, 9
7-2	• Model a direct variation situation and use the model to make predictions.	1, 2, 4, 5, 6, 8
	• Find the slope of a line in a coordinate plane.	
7-3	• Find the circumference and the length of an arc of a circle.	1, 2, 4, 5, 7, 14
7-4	• Recognize general characteristics of direct variation equations and graphs.	1, 2, 4, 5, 6, 8
	• Understand negative slope.	
7-5	• Use dimensional analysis to work with the units in rates and conversion factors.	1, 2, 4, 5
7-6	• Use the formula for the area of a circle.	1, 2, 4, 5, 7
	• Find the area of a sector.	

Section	Connections to Prior and Future Concepts
7-1	**Section 7-1** uses the concept of variable from previous units, and students will continue to use variables in their future study of mathematics. The topics of ratios and similar polygons presented in Unit 6 will be extended in this unit and in Unit 9 of Book 1. These topics will recur in Units 1, 2, 3, 6, and 8 of Book 2 and in Units 2, 4, 5, 7 and 8 of Book 3. The tangent ratio joins the sine and cosine ratios presented in Section 6-7. Trigonometric ratios will be studied again in Unit 8 of Book 2 and in Units 8 and 10 of Book 3.
7-2	**Section 7-2** builds on earlier experiences in Unit 4 of gathering data, plotting data, and using a fitted line. Direct variation models will be used throughout the remainder of Book 1 and will be emphasized in Unit 2 of Book 2 and in Unit 2 of Book 3.
7-3	**Section 7-3** builds on Sections 7-1 and 7-2 by having students confirm that the relationship between the circumference and diameter of a circle is a geometric example of direct variation. Proportions, studied in Section 6-3, are used to define arc length for a circle. Students will use their knowledge of circumference and arc length in Unit 9 of Book 1 and frequently in Books 2 and 3.
7-4	**Section 7-4** continues the study of direct variation in the context of linear equations and graphs. Linear equations and their graphs are important and recurring topics throughout Books 1, 2, and 3. There is special emphasis on these topics in Units 2 and 3 of Book 2 and Units 1, 2, and 6 of Book 3.
7-5	**Section 7-5** uses rates and conversion factors as examples of constants of variation, introduced in Section 7-1. Converting rates and measurements is a skill used throughout Books 1, 2, and 3.
7-6	**Section 7-6** continues the study of circles and direct variation through investigation of the relationship between the area of a circle and its radius. This relationship is an example of direct variation with respect to the square of the control variable. Proportions, studied in Section 6-3, are used to define the area of a sector of a circle. Students will use their knowledge of the area of a circle and a sector of a circle in Unit 9 of Book 1 and throughout Books 2 and 3.

Integrating the Strands

Strands	Sections
Number	7-4
Algebra	7-1, 7-2, 7-3, 7-4, 7-5, 7-6
Functions	7-1, 7-2, 7-3, 7-4, 7-5, 7-6
Measurement	7-1, 7-3, 7-5, 7-6
Geometry	7-1, 7-2, 7-3, 7-4, 7-5, 7-6
Trigonometry	7-1, 7-2
Statistics and Probability	7-2, 7-3
Logic and Language	7-2, 7-3, 7-4

Section Planning Guide

➤ Essential exercises and problems are indicated in boldface.
➤ Ongoing work on the Unit Project is indicated in color.
➤ Exercises and problems that require student research, group work, manipulatives, or graphing technology are indicated in the column headed "Other."

Section	Materials	Pacing	Standard Assignment	Extended Assignment	Other
7-1	scientific calculator	Day 1 Day 2	**1–8**, 12 **13–20**, 21–26, 27	**1–8**, 11, 12 **13–20**, 21–26, 27	9, 10
7-2	tape measure or meter stick, tape, 5–6 different types of balls, scientific calculator, spreadsheet software, 3 coins	Day 1 Day 2	**1–3, 5–12** **13–18**, 22, **23**, 25–31, 32	**1–3**, 4, **5–12** **13–18**, 19, 20, 22, **23**, 25–31, 32	21, 24, 32
7-3	string or tape measure, metric ruler, scissors, circular shapes (such as a coffee can, jar lid, CD), scientific calculator, globe	Day 1 Day 2	**4–6, 7a, 8–13** **14–16**, 17, 19, 23–28, 29	1–3, **4–6, 7a, 8–13** **14–16**, 17, 19–21, 23–28, 29	7b 18, 22
7-4	graphing technology	Day 1 Day 2	**1–7** **8–13, 16–20, 22–31**, 34–42, 43	**1–7** **8–13**, 14, **16–20**, 21, **22–31**, 32–42, 43	15
7-5	inch ruler, metric ruler	Day 1 Day 2	**2–8** **9–14a, 15–17, 20–23**, 27–35, 36, 37	1, **2–8** **9–14a, 15–17**, 18, 19, **20–23**, 24–35, 36, 37	14b, 37
7-6	scientific or graphics calculator	Day 1 Day 2	**2–11** **12–22**, 24–33, 34, 35	1, **2–11** **12–22**, 23–33, 34, 35	
Review Test		**Day 1** **Day 2**	**Unit Review** **Unit Test**	**Unit Review** **Unit Test**	

Yearly Pacing	Unit 7 Total	Units 1–7 Total	Remaining	Total
	16 days (2 for Unit Project)	110 days	54 days	164 days

Support Materials

➤ See **Project Book** for notes on Unit 7 Project: Design a Sports Arena.
➤ "UPP" and "disk" refer to **Using Plotter Plus** booklet and **Plotter Plus** disk.
➤ Warm-up exercises for each section are available on **Warm-Up Transparencies**.
➤ "FI," "PC," "GI," "MA," and "Stats!" refer, respectively, to the McDougal Littell Mathpack software Activity Books for **Function Investigator, Probability Constructor, Geometry Inventor, Matrix Analyzer,** and **Stats!**.

Section	Study Guide	Practice Bank	Problem Bank	Activity Bank	Explorations Lab Manual	Assessment Book	Visuals	Technology
7-1	7-1	Practice 52	Set 14	Enrich 46		Quiz 7-1	Folder 5	
7-2	7-2	Practice 53	Set 14	Enrich 47	Masters 2, 19	Quiz 7-2		
7-3	7-3	Practice 54	Set 14	Enrich 48	Masters 2, 20	Quiz 7-3 Test 27		GI Act. 12
7-4	7-4	Practice 55	Set 15	Enrich 49	Master 2	Quiz 7-4		FI Act. 3 UPP, page 30 UPP, pages 13–16
7-5	7-5	Practice 56	Set 15	Enrich 50		Quiz 7-5		
7-6	7-6	Practice 57	Set 15	Enrich 51	Master 2	Quiz 7-6 Test 28		
Unit 7	Unit Review	Practice 58	Unifying Problem 7	Family Involve 7		Tests 29, 30		

UNIT TESTS

Spanish versions of these tests are on pages 146–149 of the **Assessment Book**.

McDougal Littell Mathpack

Function Investigator
Geometry Inventor
Stats!

Outside Resources

Books/Periodicals

de Lange Jzn, Jan. *Meaningful Mathematical Matrices.* Scotts Valley, CA: Wings for Learning, 1992.

Hirsch, Christian R. and Robert E. Laing, eds. *Activities for Active Teaching and Learning: Selections from the Mathematics Teacher.* Reston, VA: NCTM, 1993.

Software

Rosenberg, Jon. *Math Connections: Algebra I.* Prometheus Software. Scotts Valley, CA: Sunburst.

Schwartz, Judah and Michal Yerushalmy. *Geometric Supposer: Circles.* Scotts Valley, CA: Sunburst.

Geometer's Sketchpad. Macintosh and MS-DOS. Berkeley, CA: Key Curriculum Press.

Masalski, William. *How to Use a Spreadsheet in the Secondary Mathematics Classroom.* Apple IIe. Reston, VA: NCTM, 1990.

Videos

Apostol, Tom. *The Story of Pi.* Reston, VA: NCTM.

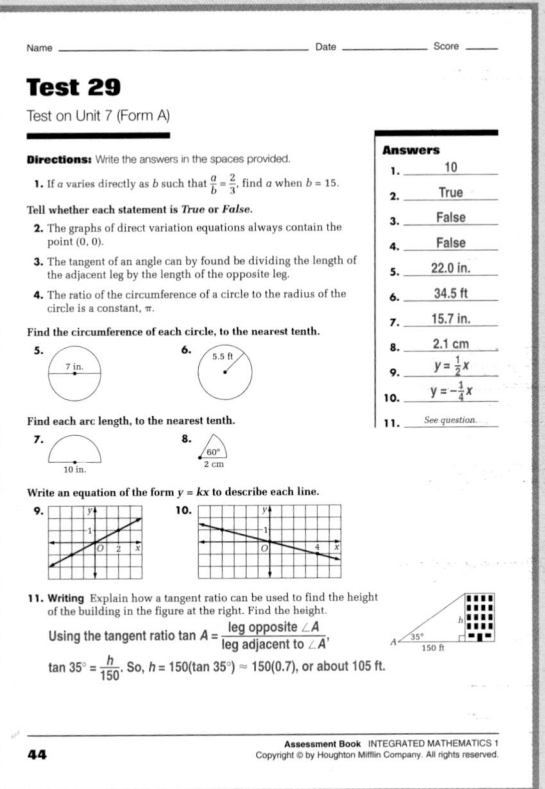

Test 29

Test on Unit 7 (Form A)

Directions: Write the answers in the spaces provided.

1. If a varies directly as b such that $\frac{a}{b} = \frac{2}{3}$, find a when $b = 15$.

Tell whether each statement is *True* or *False*.

2. The graphs of direct variation equations always contain the point $(0, 0)$.

3. The tangent of an angle can be found by dividing the length of the adjacent leg by the length of the opposite leg.

4. The ratio of the circumference of a circle to the radius of the circle is a constant, π.

Find the circumference of each circle, to the nearest tenth.

5.   6.

Find each arc length, to the nearest tenth.

7.   8.

Write an equation of the form $y = kx$ to describe each line.

9.   10.

11. **Writing** Explain how a tangent ratio can be used to find the height of the building in the figure at the right. Find the height.

Using the tangent ratio $\tan A = \frac{\text{leg opposite } \angle A}{\text{leg adjacent to } \angle A}$,

$\tan 35° = \frac{h}{150}$. So, $h = 150(\tan 35°) \approx 150(0.7)$, or about 105 ft.

Answers
1. 10
2. True
3. False
4. False
5. 22.0 in.
6. 34.5 ft
7. 15.7 in.
8. 2.1 cm
9. $y = \frac{1}{2}x$
10. $y = -\frac{1}{4}x$
11. *See question.*

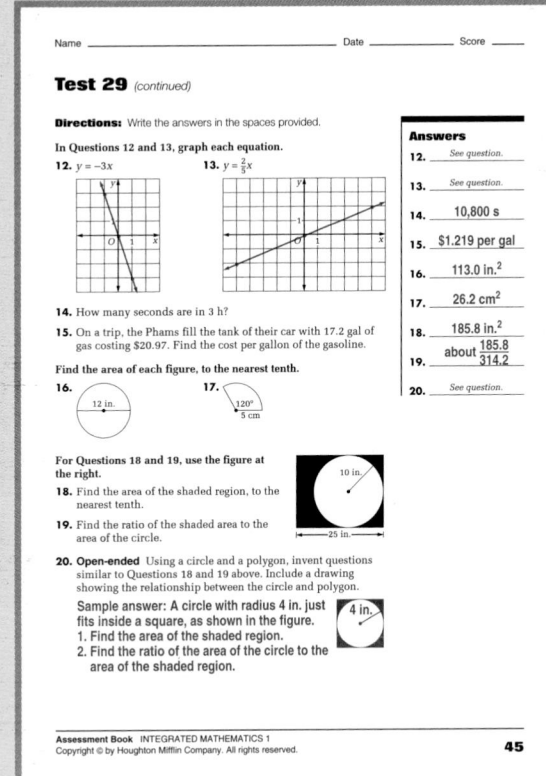

Test 29 *(continued)*

Directions: Write the answers in the spaces provided.

In Questions 12 and 13, graph each equation.

12. $y = -3x$ 13. $y = \frac{2}{5}x$

14. How many seconds are in 3 h?

15. On a trip, the Phams fill the tank of their car with 17.2 gal of gas costing $20.97. Find the cost per gallon of the gasoline.

Find the area of each figure, to the nearest tenth.

16.   17.

For Questions 18 and 19, use the figure at the right.

18. Find the area of the shaded region, to the nearest tenth.

19. Find the ratio of the shaded area to the area of the circle.

20. **Open-ended** Using a circle and a polygon, invent questions similar to Questions 18 and 19 above. Include a drawing showing the relationship between the circle and polygon.

Sample answer: A circle with radius 4 in. just fits inside a square, as shown in the figure.
1. Find the area of the shaded region.
2. Find the ratio of the area of the circle to the area of the shaded region.

Answers
12. *See question.*
13. *See question.*
14. 10,800 s
15. $1.219 per gal
16. 113.0 in.²
17. 26.2 cm²
18. 185.8 in.²
19. about $\frac{185.8}{314.2}$
20. *See question.*

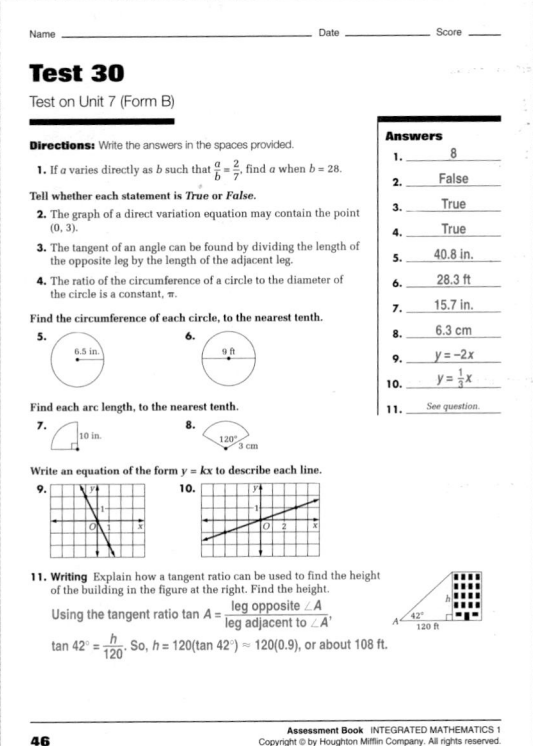

Test 30

Test on Unit 7 (Form B)

Directions: Write the answers in the spaces provided.

1. If a varies directly as b such that $\frac{a}{b} = \frac{2}{7}$, find a when $b = 28$.

Tell whether each statement is *True* or *False*.

2. The graph of a direct variation equation may contain the point $(0, 3)$.

3. The tangent of an angle can be found by dividing the length of the opposite leg by the length of the adjacent leg.

4. The ratio of the circumference of a circle to the diameter of the circle is a constant, π.

Find the circumference of each circle, to the nearest tenth.

5.   6.

Find each arc length, to the nearest tenth.

7.   8.

Write an equation of the form $y = kx$ to describe each line.

9.   10.

11. **Writing** Explain how a tangent ratio can be used to find the height of the building in the figure at the right. Find the height.

Using the tangent ratio $\tan A = \frac{\text{leg opposite } \angle A}{\text{leg adjacent to } \angle A}$,

$\tan 42° = \frac{h}{120}$. So, $h = 120(\tan 42°) \approx 120(0.9)$, or about 108 ft.

Answers
1. 8
2. False
3. True
4. True
5. 40.8 in.
6. 28.3 ft
7. 15.7 in.
8. 6.3 cm
9. $y = -2x$
10. $y = \frac{1}{3}x$
11. *See question.*

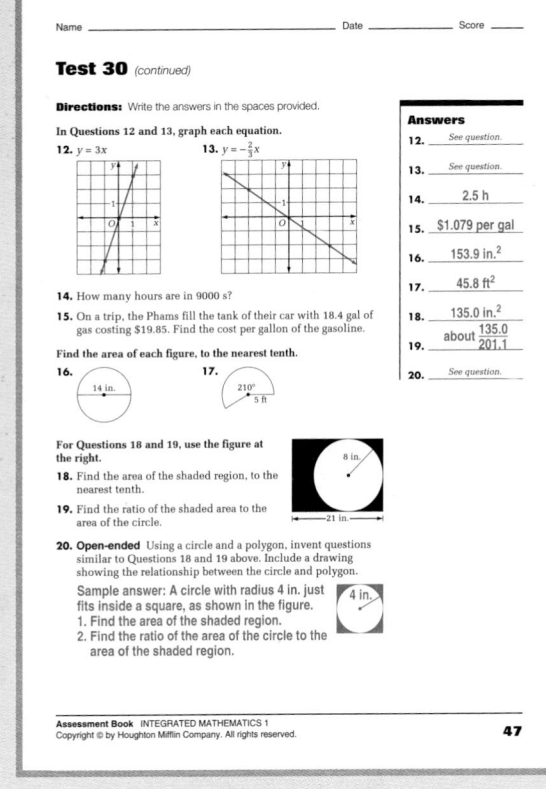

Test 30 *(continued)*

Directions: Write the answers in the spaces provided.

In Questions 12 and 13, graph each equation.

12. $y = 3x$ 13. $y = -\frac{2}{3}x$

14. How many hours are in 9000 s?

15. On a trip, the Phams fill the tank of their car with 18.4 gal of gas costing $19.85. Find the cost per gallon of the gasoline.

Find the area of each figure, to the nearest tenth.

16.   17.

For Questions 18 and 19, use the figure at the right.

18. Find the area of the shaded region, to the nearest tenth.

19. Find the ratio of the shaded area to the area of the circle.

20. **Open-ended** Using a circle and a polygon, invent questions similar to Questions 18 and 19 above. Include a drawing showing the relationship between the circle and polygon.

Sample answer: A circle with radius 4 in. just fits inside a square, as shown in the figure.
1. Find the area of the shaded region.
2. Find the ratio of the area of the circle to the area of the shaded region.

Answers
12. *See question.*
13. *See question.*
14. 2.5 h
15. $1.079 per gal
16. 153.9 in.²
17. 45.8 ft²
18. 135.0 in.²
19. about $\frac{135.0}{201.1}$
20. *See question.*

PROJECT GOALS

➤ Students make a two-dimensional scale drawing or a three-dimensional model of a sports arena for skateboard or in-line skating competition and include a cost estimate for building the arena.

➤ When designing an arena, students decide on the number and slope of the ramps, the arc lengths of any circular parts, the materials needed, and the estimated cost of the arena.

➤ Students work together in a cooperative group and appreciate each other's contribution to the successful completion of the project.

PROJECT PLANNING

Materials List

➤ Ruler
➤ Poster board
➤ Cardboard for model

Project Teams

Before students begin working as part of a team, have individual students write down ideas they think need to be taken into consideration. These suggestions may be the result of personal experiences or events they may have seen.

Have students work on the project in groups of four. One way for the individuals in the group to distribute the work is as follows:

1. Coordinator: collects and summarizes all possible ideas from the group.

2. Writer: describes the model sports arena.

3. Designer: draws and/or makes the scale model.

4. Analyst: analyzes and calculates the estimated cost of building the arena.

7
unit

Direct Variation

Test your sports savvy with this trivia quiz!

① Why is a tennis ball fuzzy?

② How did home plate get its shape?

③ Why does a race car have wings?

④ Why is the rim of a basketball hoop ten feet high?

⑤ How did an airplane change the sport of diving?

⑥ Why is the surface of a soccer ball formed by a pattern of polygons?

⑦ How many dimples are on a golf ball?

See page 358 for answers.

SPORTS DESIGN

aerodynamics

Cyclists, skiers, and speed skaters know that one of the biggest obstacles to winning a race is the **WIND.** Sports scientists use computers and wind tunnels to design equipment to decrease the wind's effect. In a 25-mi race, an aerodynamic handlebar can make a three-minute difference in a cyclist's time—the difference between finishing first or fifty-first!

The outer seams on a **baseball** may help a pitcher's curve ball, but the inside of the ball is on the batter's side. The lightweight materials found inside a baseball allow the ball to go farther when it is hit.

- cowhide
- blue-gray woolen yarn
- white woolen yarn
- blue-gray woolen yarn
- cork
- black rubber
- red rubber

layers

356

Suggested Rubric for Unit Project

4 The scale drawing or model is drawn or made correctly. The parts and dimensions are labeled correctly. The slopes of the ramps are realistic, as is the slope of the bleachers. The overall design of the arena is functional and the cost estimates are appropriate.

3 The scale drawing or model meets the criteria for the project, but certain details are lacking or are somewhat incomplete. There are minor mathematical errors involving the slopes of the ramps or in the calculations of the arc lengths. The cost estimates may be slightly too high or too low.

2 The scale drawing or model of the arena and its description contain serious errors or are incomplete. Not all of the questions asked are answered, and there is little understanding of the mathematics involved in the project.

Design a Sports Arena

Construction of the skateboard
bowl arena at The Hanger in
Charleston, N.C.

Your project is to make a scale
drawing or a model of an arena
for skateboarding or in-line
skating competitions.

Your group's design should
include at least one ramp for
launching jumps. The ramp
may be straight or curved. You
may want to create several
"stations," each with a different kind of ramp.

Along with your design, you
should include an estimate of
the cost of building your arena.

design

One arena can be the site of
a concert on Friday, a basketball game Saturday afternoon,
a hockey game Saturday night,
and a trip to the circus on
Sunday! With a good design, a
multi-event arena can change
from a floor of ice to hardwood
within a few hours. The design
of other arenas, like those for
skateboarding events, affects
the difficulty level of the
competition.

357

Suggested Rubric for Unit Project

1 This project cannot be evaluated. It is illegible, incomplete, or not understandable.
The scale drawing or model
should be returned with a new
deadline for completion. The
groups should be encouraged
to speak with the teacher as
soon as possible so that they
understand the purpose and
format of the project.

Support Materials

The *Project Book* contains
information about the following topics for use with this
Unit Project.

➤ Project Description

➤ Teaching Commentary

➤ Working on the Unit Project
Exercises

➤ Completing the Unit Project

➤ Assessing the Unit Project

➤ Alternative Projects

➤ Outside Resources

Limited English Proficiency

Designing a sports arena is a
valuable project for students
acquiring English. It provides a
visual display of mathematical
concepts and requires written
descriptions of these concepts.

ADDITIONAL
BACKGROUND

Multicultural Note

Sports design is very important
at the Olympic games. Athletes
from all over the world come to
compete in one place. That
place needs to be able to
accommodate each event with
state-of-the-art equipment and
facilities. The Winter Olympic
Games often pose a special
problem for Olympic designers
because many events are held
indoors. An entire facility must
be built to hold a number of
different events and a great
number of spectators.

For the 1972 Olympic Games,
the International Olympic
Committee selected the first
Asian site in Winter Olympic
history, Sapporo, Japan. For
the 1984 Olympic Games, the
committee chose the first East
European site, Sarajevo,
Yugoslavia.

Aerodynamics

In 1973, an American cyclist, Dr. Allan Abbott, achieved a speed of 104.5 miles per hour, the highest speed ever by a cyclist, at the Bonneville Salt Flats in Utah.

Dr. Abbott rode behind a car with a wind shield mounted on it to cut air drag. Some other things that help cyclists include the following:

- sleek helmets allow the wind to pass smoothly,
- tight, ribbed clothing prevents fabric from grabbing the wind,
- a deep tuck can cut drag and increase speed by about 5%,
- certain handlebars can allow a deeper tuck and cut drag.

ALTERNATIVE PROJECTS

Project 1, page 408

A Trip to Another Country

Write a story about a trip to another country and include how direct variation might be used during the trip.

Project 2, page 408

The History of π

Research the number π and identify some of the approximations for π that have been used by different cultures.

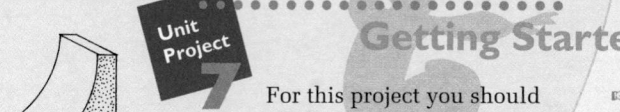

Unit Project 7

quarter pipe

curved ramp

straight ramp

Answers to trivia quiz from page 356.

Getting Started

For this project you should work in a group of four students. Here are some ideas to help you get started.

☞ Consider the overall design of the arena. Do you want a shape that will include a track for long-distance in-line skating? Where will spectators watch the skaters?

☞ How many ramps will you include in your design?

☞ Books about skateboarding often include plans for ramps. You may want to refer to one for ideas.

☞ Think about what makes one skating arena more challenging than another. What level of difficulty do you want?

☞ You can build a model or make a scale drawing of your arena. Which one is most appealing to you?

① The fuzz on a tennis ball allows the racquet to grip the ball so that it can be hit more easily.

② These legs define the foul lines, but limit the pitcher's and umpire's ability to see the strike zone clearly. This addition lets the pitcher and the umpire see when a pitch "hits the corner" of the strike zone.

③ The wind hits the wings and then pushes the car to the ground, allowing it to go faster, especially around turns.

④ When James Naismith created the game, peach baskets were hung from the gym's balcony, 10 feet from the floor.

⑤ Surplus parts from a World War II airplane led to the design of a flexible diving springboard.

⑥ The polygonal shapes control the distance that a soccer ball can be hit or kicked.

⑦ There is an average of 400 dimples on a golf ball. Without dimples, a ball that would normally go 250 yards would go only 125 yards.

Working on the Unit Project

Your work in Unit 7 will help you create your model or scale drawing.

Related Exercises:
Section 7-1, Exercise 27
Section 7-2, Exercise 32
Section 7-3, Exercise 29
Section 7-4, Exercise 43
Section 7-5, Exercises 36, 37
Section 7-6, Exercises 34, 35

Alternative Projects p. 408

Can We Talk

SPORTS DESIGN

➤ Why do you think skateboarding and in-line skating are popular?

➤ What protective equipment should skaters wear? Why?

➤ If you were building an arena in your neighborhood, where would you put it?

➤ Where else have you seen ramps? What other sports use ramps?

➤ What are some ways to measure the steepness of a straight ramp?

Answers to Can We Talk?

➤ Skateboarding is challenging and thrilling. It has a variety of skills and moves that can be learned.

➤ Skaters should wear helmets, knee pads, elbow pads, and so on, to protect themselves from injury.

➤ Answers may vary.

➤ Ramps provide access to buildings, allow vehicles to exit the highway, and are used in sports. Ramps are used in the sports of ski jumping, water skiing, snow skiing, and stunt car racing.

➤ A ramp's steepness can be measured using slope, angle measurements, or estimated by experimentation.

Direct Variation, Slope, and Tangent

Talk it Over

1. To build a roof a carpenter puts posts every 2 ft along a beam. Describe a pattern you see in the heights of the posts.

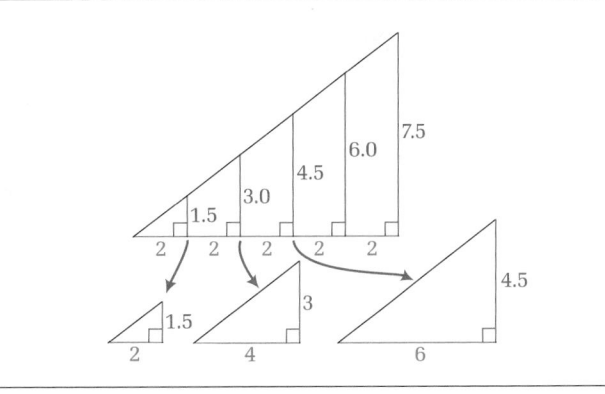

2. Copy and complete the table. For the last column, use a calculator to find the decimal ratio of height to horizontal distance. What do you notice about these ratios?

Post	Horizontal distance d of post from eaves (ft)	Height h of post (ft)	$\frac{h}{d}$
1	2	1.5	0.75
2	4	3	?
3	?	4.5	?
4	?	6	?
5	?	7.5	?

3. Use similar triangles to explain why all the ratios $\frac{h}{d}$ are the same.

7-1 Direct Variation, Slope, and Tangent

359

Answers to Talk it Over

1. The heights of the posts increase by 1.5 ft as the distance from the eaves increases by 2 ft.

2. For each triangle, the ratio $\frac{h}{d}$ equals 0.75.

Post	Horizontal distance d of posts from eaves (ft)	Height h of post (ft)	$\frac{h}{d}$
1	2	1.5	0.75
2	4	3	0.75
3	6	4.5	0.75
4	8	6	0.75
5	10	7.5	0.75

3. Explanations may vary. An example is given. Two triangles are similar when two pairs of corresponding angles are congruent. The triangles formed by the posts all have right angles and all have the angle formed with the eaves in common. Since the triangles are similar, corresponding sides are proportional, so the ratios $\frac{h}{d}$ are the same.

PLANNING

Objectives and Strands
See pages 356A and 356B.

Spiral Learning
See page 356B.

Materials List
➤ Scientific calculator

Recommended Pacing
Section 7-1 is a two-day lesson.
Day 1
Pages 359–361: Talk it Over through Sample 2, *Exercises 1–12*
Day 2
Pages 361–362: The Tangent Ratio through Look Back, *Exercises 13–27*

Extra Practice
See pages 631–633.

Warm-Up Exercises
💡 Warm-Up Transparency 7-1

Support Materials
➤ Practice 52
➤ Enrichment 46 in the Activity Bank
➤ Study Guide 7-1
➤ Problem Set 14
💡 Overhead Visual 5
➤ Quiz 7-1
➤ Alternative Assessments 1, 2

359

TEACHING

Talk it Over

If students have difficulty with question 3, refer them back to the triangles shown in question 1. You may need to remind students that when triangles are similar, their corresponding sides are proportional.

Additional Sample

S1 Suppose the carpenter of the roof shown on page 359 wants only three support posts between the eaves and the post at the peak. Assuming that the posts are evenly spaced, how long should the shortest support post be? **The distance d from the eaves to the first post will be $\frac{10}{4}$ or 2.5 ft.**

$$h = 0.75d$$
$$h = 0.75(2.5)$$
$$h = 1.875$$

The shortest post should be 1.875 or $1\frac{7}{8}$ ft long.

Talk it Over

For question 6, it may help students to imagine keeping d constant and changing only h. It is fairly easy to see that an increase in h will increase both the ratio $\frac{h}{d}$ and the steepness of the roof. This observation leads to the next idea of the lesson, that the ratio $\frac{h}{d}$ can be used to quantify our intuitive idea of steepness.

Application

Make sure students understand intuitively that in a direct variation two quantities are related in such a way that as one quantity changes, the other quantity changes also. Ask students if they can suggest some geometric applications of direct variation. For example, the perimeter of a square varies directly as the length of a side.

Direct Variation

For the roof, the height of a post depends on its horizontal distance from the eaves. Although the height increases as the distance increases, the ratio of these two quantities is constant.

When two variable quantities have a constant ratio, their relationship is called **direct variation.** The constant ratio is called the **variation constant.** You can write an equation to show the relationship.

$$\frac{h}{d} = 0.75$$

ratio of quantities = variation constant

The equation $\frac{h}{d} = 0.75$ can also be written as $h = 0.75d$.

Both equations show that the height of the post varies directly with its horizontal distance from the eaves. You can use either equation to find the height of a post needed to support the roof at another point.

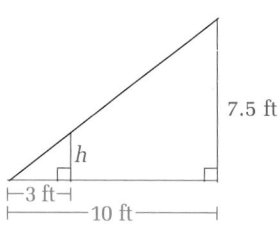

Sample 1

Suppose the carpenter of the roof shown on page 359 decides that a post should be 3 ft from the eaves. What should the height of the post be?

Sample Response

The equation $h = 0.75d$ describes the direct variation of the height h of a post and its distance d from the eaves.

Find the value of h when d is 3 ft.

$$h = 0.75d$$
$$h = 0.75(3) \quad \longleftarrow \quad \text{Substitute 3 for } d.$$
$$h = 2.25$$

The post should be 2.25 ft long.

BY THE WAY...

Carpenters sometimes use the word "pitch" to describe the slope of a roof.

Talk it Over

4. What step would you use to rewrite the equation $\frac{h}{d} = 0.75$ in the form $h = 0.75d$?

5. Suppose you know that $h = 5$ ft. Explain how can you find the value of d.

6. For a steeper roof, would the ratio $\frac{h}{d}$ be larger or smaller?

Unit 7 Direct Variation

Answers to Talk it Over

4. Multiply both sides of the equation by d.

5. Substitute 5 for h in the equation $h = 0.75d$; $5 = 0.75d$; divide both sides by 0.75; $\frac{5}{0.75} = \frac{0.75}{0.75}d$; $d = 6\frac{2}{3}$.

6. larger

Marble Temple, Bangkok, Thailand

Slope

Some roofs are steeper than others. In mathematics, a number called *slope* is a measure of the steepness of a line. The **slope of a line** is the ratio of *rise* to *run* for any two points on the line.

Roofs of Rheinland Pfalz Freudenberg, Germany

rise = vertical change between two points

run = horizontal change between two points

$$\text{slope} = \frac{\text{rise}}{\text{run}}$$

Sample 2

Find the slope of the roof shown on page 359. Use the eaves and the peak as the two points.

Sample Response

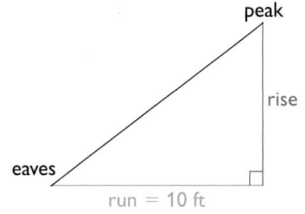

Problem Solving Strategy: Make a sketch

The roof rises 7.5 ft across a distance of 10 ft.

$$\text{slope} = \frac{\text{rise}}{\text{run}} = \frac{7.5}{10} = 0.75$$

The slope of the roof is 0.75.

▶ Now you are ready for:
Exs. 1–12 on pp. 363–364

The Tangent Ratio

Remember the sine ratio and the cosine ratio of an acute angle in a right triangle? Each of these ratios compares a leg and the hypotenuse of a right triangle.

$$\sin A = \frac{\text{leg opposite } \angle A}{\text{hypotenuse}} = \frac{BC}{AB} \qquad \cos A = \frac{\text{leg adjacent to } \angle A}{\text{hypotenuse}} = \frac{AC}{AB}$$

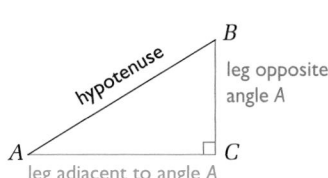

The **tangent ratio** of an acute angle compares the two legs of a right triangle.

$$\tan A = \frac{\text{leg opposite } \angle A}{\text{leg adjacent to } \angle A} = \frac{BC}{AC}$$

7-1 Direct Variation, Slope, and Tangent

361

Additional Sample

S2 Suppose the post from the horizontal beam to the peak of a roof is 9 ft high and the length of the beam from the eaves to the post is 18 ft. Draw a sketch to illustrate the situation and then find the slope of the roof.

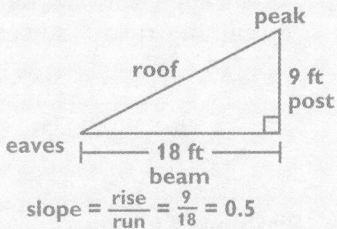

$$\text{slope} = \frac{\text{rise}}{\text{run}} = \frac{9}{18} = 0.5$$

The slope of the roof is 0.5.

Communication: Discussion

After introducing the idea of slope of a line, draw a line at random on the chalkboard. Ask students: What is the slope of the line? The point of the question is to have students understand that the line must be oriented to a second, horizontal line in order to discuss its slope.

Reasoning

Ask students what other quantities might be used to measure the steepness of a roof. (Possible answers: the size of the angle that the roof makes with a beam from the eaves to the center post. The sine and cosine of that angle also could be used.)

Consider the slope of the roof and the measure of the angle that the roof makes with a horizontal beam. Ask which quantity gives the more useful idea of the steepness of the roof. You can have students imagine that someone has to climb the roof to make repairs. Does doubling the slope make the climb twice as difficult? What about doubling the angle?

361

Additional Sample

S3 Consider the roof shown in the following diagram.

What is the tangent ratio of the angle between the roof and the horizontal beam?

$\tan A = \frac{BC}{AC} = \frac{6}{15} = \frac{2}{5}$ or 0.4

The tangent ratio for the angle is 0.4.

Problem Solving

You may wish to ask students to suggest different ways of finding the approximate measure of an angle if you know the value of the tangent ratio for the angle. Ask them to explain their answer using an angle A, whose tangent ratio is 0.875. (Possible answers: Use the inverse tangent key of a scientific calculator. Since $0.875 = \frac{7}{8}$, draw a right triangle whose legs have lengths 7 cm and 8 cm, then measure the angle opposite the 7 cm leg with a protractor.)

Look Back

Use students' responses to conduct a brief class discussion that will recap the important new concepts and terms of the section.

Sample 3

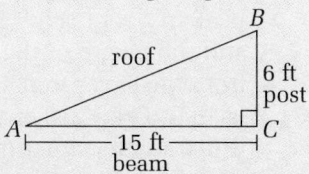

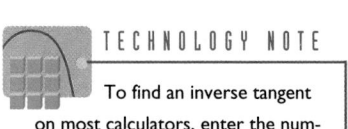

Look again at the roof shown on page 359. What is the tangent ratio of the angle between the roof and the horizontal beam?

Sample Response

Use the right triangle formed by the roof, the beam, and the 7.5 ft post. The angle between the roof and the beam is $\angle A$.

$$\tan A = \frac{\text{leg opposite } \angle A}{\text{leg adjacent to } \angle A} = \frac{7.5}{10} = 0.75$$

The tangent ratio of the angle is 0.75.

Talk it Over

7. Use the triangle in Sample 3. What is the tangent ratio of $\angle B$? How is the tangent ratio of $\angle B$ related to the tangent ratio of $\angle A$?
8. Luisa made a scale drawing of the roof in Sample 3. Using a protractor, she measured $\angle A$ to be 37°.
   a. Use a scientific calculator to find the tangent ratio of a 37° angle.
   b. Explain why your answer in part (a) is not exactly 0.75.
9. Use the *inverse tangent* key of a scientific calculator to find the angle that has a tangent ratio of exactly 0.75. To the nearest tenth of a degree, what is the measure of $\angle A$? Compare this measure of $\angle A$ to the measure given in question 8.
10. Do you think the tangent ratio of a steeper roof is larger or smaller than the tangent ratio of this roof? Explain.

TECHNOLOGY NOTE

To find an inverse tangent on most calculators, enter the number, then press INV tan.

Now you are ready for: Exs. 13–27 on pp. 365–366

Look Back

Explain how these three ideas are related: variation constant, slope, and tangent ratio.

Answers to Talk it Over

7. $\frac{4}{3}$; The tangent ratio of $\angle B$ is the reciprocal of the tangent ratio of $\angle A$.
8. a. about 0.7536
   b. Answers may vary. An example is given. It is difficult with a protractor to find the exact angle measure. While 37° is close, the actual angle measure is slightly smaller.
9. about 36.9°; This measure is smaller by one tenth of a degree.
10. larger; Explanations may vary. An example is given. For a steeper roof, the height of the roof would be greater. This would make the tangent ratio, $\frac{\text{height}}{\text{horizontal distance}}$, larger.

Answers to Look Back

Answers may vary. An example is given. The diagrams show that the variation constant, tangent ratio of $\angle A$, and slope are different ways to express the ratio of the length of a vertical leg to the length of a horizontal leg of a right triangle.

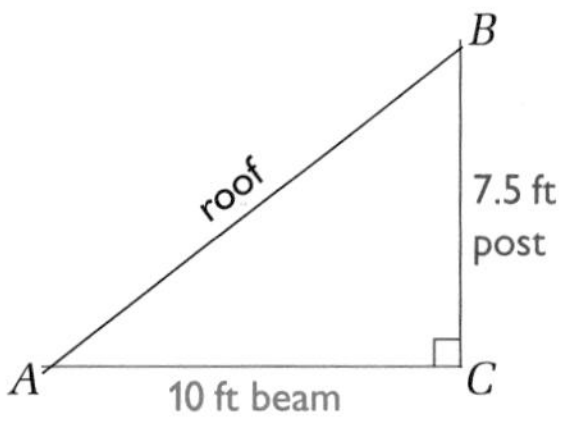

Exercises and Problems

Use the direct variation equation $\frac{a}{b} = 1.25$.

1. What is the variation constant?

2. Rewrite the equation so that it does not have a fraction in it.

3. Find the value of a when $b = 8$.

4. Find the value of b when $a = 12$.

For Exercises 5 and 6, use the photograph at the right. The parasailor is 150 ft behind the boat and 200 ft above the water.

5. **a.** Write a direct variation equation relating height above the water and distance behind the boat.

   b. What is the variation constant?

6. If the parasailor is pulled in so he is 90 ft behind the boat, how far above the water will he be?

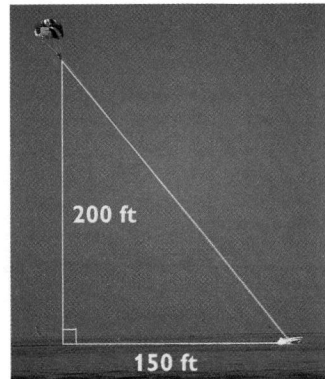

200 ft

150 ft

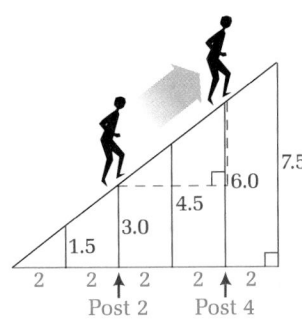

7.5
6.0
4.5
3.0
1.5

2 2 2 2 2
Post 2   Post 4

7. The definition of slope on page 361 tells you that you can use any two points on a line to calculate the slope of the line.

   a. The roof shown is the same as the roof shown on page 359. Suppose the carpenter walks along the roof from post 2 to post 4. What is the rise and the run of the roof from post 2 to post 4?

   b. Use your answers in part (a) to find the slope of the roof. Is your answer the same as the slope of the roof calculated in Sample 2?

   c. Find the slope of the roof using the rise and run from post 1 to post 2.

   d. Find the slope of the roof using the rise and run from post 3 to post 5.

8. The roof shown is the same as the roof shown on page 359. Suppose the carpenter decides to extend the roof so it reaches a height of 9 ft instead of 7.5 ft. Find the length of the beam needed.

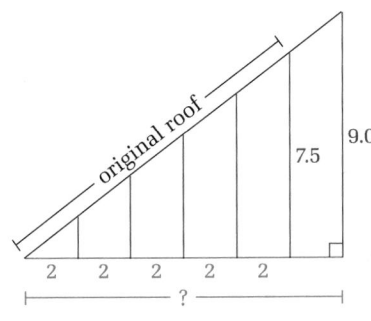

original roof

9.0
7.5

2 2 2 2 2
?

9. **Research** Estimate the slopes of roofs at your school or in the neighborhood. Discuss how slopes of roofs in areas that get a lot of snow in the winter might be different from slopes of roofs in areas that get very little snow.

7-1 Direct Variation, Slope, and Tangent

363

APPLYING

Suggested Assignment
Standard 1–8, 12–27
Extended 1–8, 11–27

Integrating the Strands
Algebra Exs. 1–6, 12, 23–26
Functions Exs. 1–5, 12, 21, 26
Measurement Exs. 10, 11
Geometry Exs. 7–11, 13, 14
Trigonometry Exs. 15–22, 27

Communication: Discussion
When discussing the exercises and problems in this section, continue to reinforce the idea that the variation constant, slope, and tangent ratio of an angle are three different ways to express the ratio of the length of a vertical leg to the length of a horizontal leg in a right triangle.

Application
Invite students to suggest ways to get good estimates in Ex. 9. You may wish to suggest some possible techniques if students have difficulty coming up with ideas. One possibility is to make a device something like a carpenter's square (see Ex. 11). Mark the scales on the *insides* of the horizontal and vertical arms. Then sight the roofs using the device. Note that the eaves should be sighted along the horizontal scale, and the peak of the roof along the vertical scale. Read the scales to get the lengths of the legs of a right triangle.

Answers to Exercises and Problems

1. 1.25

2. $a = 1.25b$

3. 10

4. 9.6

5. **a.** $h = \frac{4}{3}d$
   b. $\frac{4}{3}$

6. 120 ft

7. **a.** rise = 3 ft, run = 4 ft
   b. slope = $\frac{3}{4}$ = 0.75; Yes.
   c. $\frac{1.5}{2} = 0.75$
   d. $\frac{3}{4} = 0.75$

8. 12 ft

9. Answers may vary. The slopes of roofs in areas that get a lot of snow would probably be steeper so that the snow would not build up and put a lot of weight on the roof. Also, when melting, the snow would fall off more easily.

Application

Exs. 10 and 11 provide practical applications of the concept of slope as the measure of the steepness of a line. They reinforce the definition of slope as the ratio of rise to run.

Visual Thinking

Ask students to study some steps or a stairway at school or in their homes. Have them make a sketch of what they have observed about the relationship of the risers, the treads, and the slope. This activity involves the visual skills of *observation* and *generalization*.

Communication: Discussion

When discussing Ex. 12, use the fact that Tables A and B show a direct variation to reinforce the functional relationship between units and seconds and between numbers and price. Students should see that a change in units results in a change in seconds and that a change in numbers results in a change in price. In each case, the rate of change, $\frac{\text{seconds}}{\text{unit}}$ and $\frac{\text{price}}{\text{number}}$, is constant. In Table C, the rate of change, $\frac{\text{income}}{\text{age}}$, is not constant.

10. **Research** Measure the slopes of wheelchair access ramps at your school or in the neighborhood. What seems to be the steepest acceptable ramp? How does steepness affect the ease of using a ramp?

11. **Carpentry** A carpenter is asked to replace a staircase with one that is less steep. The steepness of a staircase depends upon the lengths of its *risers* and *treads*.

    a. A *carpenter's square* looks like two rulers joined at the 0 point to form a right angle. Use the carpenter's square shown in the diagram to read the riser height and tread length of the staircase.

    b. Find the ratio $\frac{\text{riser height}}{\text{tread length}}$. Explain why this ratio can be called the slope of the staircase.

    c. **Open-ended** Give three possible pairs of riser heights and tread lengths that the carpenter might use for a staircase that is less steep.

12. Each table shows a relationship between two quantities.

Table A			Table B			Table C		
units	seconds	$\frac{\text{seconds}}{\text{unit}}$	numbers	price	$\frac{\text{price}}{\text{number}}$	age	income	$\frac{\text{income}}{\text{age}}$
5	60	?	10	1.50	?	20	25,000	?
10	120	?	12	1.80	?	25	30,000	?
15	180	?	16	2.40	?	35	40,000	?

    a. Find the ratios.

    b. Which of the tables does not show direct variation? Explain.

Answers to Exercises and Problems

10. Answers may vary. Wheelchair ramps usually have slopes between $\frac{1}{20}$ and $\frac{1}{12}$, that is, between 0.05 and 0.08. The steeper the ramp, the more difficult it is to travel up the ramp, and the more dangerous it is to travel down the ramp.

11. a. riser height = 5 in.; tread length = 6.25 in.

    b. 0.8; The slope is the rise over the run.

    c. Answers may vary. An example is given. riser height 7.5 in. with a tread length of 10 in.

12. a. Table A: 12, 12, 12; Table B: 0.15, 0.15, 0.15; Table C: 1250, 1200, about 1143

    b. Table C; the ratio $\frac{\text{income}}{\text{age}}$ is not the same for each age in the table. Therefore, there is no variation constant.

For Exercises 13–15, use the diagram at the right.

13. What is the hypotenuse of $\triangle RST$?

14. Which leg is opposite the 63° angle? Which leg is adjacent to it?

15. **a.** Use $\triangle RST$ to find the value of tan 63°.

    b. Use your calculator to find the value of tan 63°.

    c. Do your answers in parts (a) and (b) agree exactly? Explain.

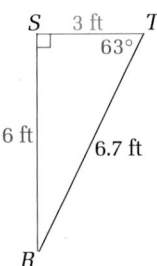

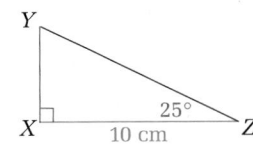

16. **a.** Express tan 25° as the ratio of the lengths of two sides of the triangle.

    b. Use a calculator to find tan 25°.

    c. Find the length of $\overline{XY}$ to the nearest 0.1 cm.

17. **a.** Use a calculator to complete a table of values for the tangent, sine, and cosine ratios of the following acute angles. Round values to four decimal places.

ANGLE	10°	20°	30°	40°	50°	60°	70°	80°
tangent	?	?	?	?	?	?	?	?
sine	?	?	?	?	?	?	?	?
cosine	?	?	?	?	?	?	?	?

    b. Describe any patterns you see in the table.

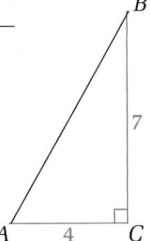

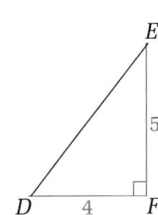

TECHNOLOGY NOTE

To find an inverse tangent on most calculators, enter the number, then press **INV** **tan**.

18. **a.** Which has the larger tangent ratio, $\angle A$ or $\angle D$? Explain.

    b. Use the inverse tangent key on a calculator to find the measures of $\angle A$ and $\angle D$.

    c. Which has the larger tangent ratio, $\angle B$ or $\angle E$? Explain.

    d. Use the inverse tangent key on a calculator to find the measures of $\angle B$ and $\angle E$.

19. A right triangle has a 20° angle and a leg of length 14 cm opposite it.

    a. Draw a picture of the triangle.

    b. **Writing** Explain how to use sine, cosine, or tangent ratios to find the lengths of the other leg and the hypotenuse.

    c. Find the lengths of the other leg and the hypotenuse to the nearest tenth of a centimeter.

7-1 Direct Variation, Slope, and Tangent   **365**

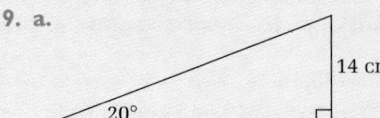
Answers to Exercises and Problems

13. $\overline{RT}$ 14. $\overline{RS}$; $\overline{ST}$ 16. **a.** $\frac{XY}{XZ}$ **b.** about 0.466   **c.** 4.7 cm

15. **a.** 2

    b. about 1.963

    c. No. Explanations may vary. An example is given. If a triangle is formed using the given lengths, the measure of $\angle T$ would be close to, but not exactly equal to, 63°.

17. **a.**

Angle	10°	20°	30°	40°	50°	60°	70°	80°
Tangent	0.1763	0.3640	0.5774	0.8391	1.1918	1.7321	2.7475	5.6713
Sine	0.1736	0.3420	0.5000	0.6428	0.7660	0.8660	0.9397	0.9848
Cosine	0.9848	0.9397	0.8660	0.7660	0.6428	0.5000	0.3420	0.1736

365

Practice 52 For use with Section 7-1

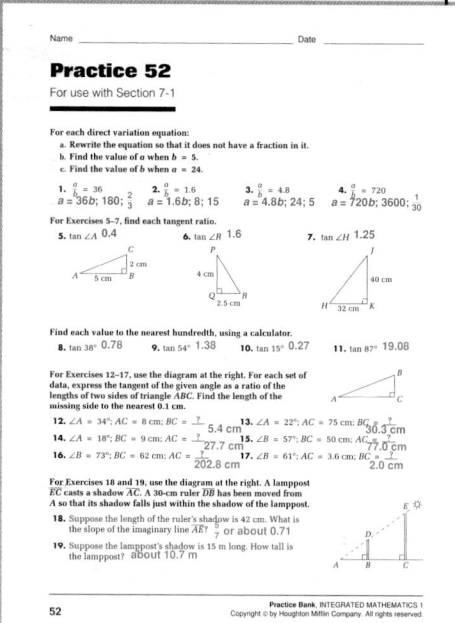

20. A tourist is 150 ft from the base of the Leaning Tower of Pisa. From the tourist's position, the *angle of elevation* to the top of the tower is 50°. Find the height of the tower to the nearest tenth of a foot.

Ongoing **ASSESSMENT**

21. The height *h* of a post that supports the peak of a roof varies directly with the horizontal distance *d* from the peak to the eaves. The roof and the beam form a 30° angle.

  a. Write a direct variation equation relating *h* and *d*. (*Hint:* Use the tangent ratio.)

  b. **Open-ended** Give three possible pairs of values for *h* and *d*.

  c. **Writing** Explain why the slope of the roof is tan 30°.

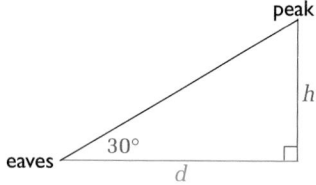

Review **PREVIEW**

22. Use △*ABC* to find sin *A* and cos *A* to four decimal places. *(Section 6-7)*

Solve. *(Sections 2-7, 2-8)*

23. $10x = 24$

24. $\dfrac{x}{24} = 10$

25. $10 = 24x$

26. Write an equation to model this statement: The length of an infant's arm is about one-third of the infant's height. *(Section 5-1)*

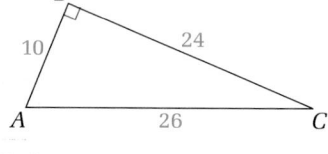

Working on the Unit Project

27. Suppose you decide to construct a straight skateboard ramp that makes an angle of 8° with the ground.

  a. How long would the base of the ramp have to be in order to reach a height of 3 ft above the ground?

  b. If you do not have enough space for a ramp that long, should you increase or decrease the angle? What is the disadvantage for the skater if you design the ramp this way?

  c. Explain why a curved ramp might be better than a straight ramp.

Answers to Exercises and Problems

20. 178.8 ft

21. a. $h = (\tan 30°)d \approx 0.577d$

  b. Answers may vary. Examples are given. $d = 10$, $h \approx 5.77$; $d = 20$, $h \approx 11.55$; $d = 15$, $h \approx 8.66$

  c. Explanations may vary. An example is given. The slope is the ratio of rise to run. For the roof, the height is the rise and the horizontal distance from the peak to the eave is the run. Thus, $\dfrac{\text{rise}}{\text{run}} = \dfrac{h}{d}$. Since the ratio $\dfrac{h}{d}$ is the ratio for the tangent of 30°, the tan 30° is equal to the slope.

22. sin $A \approx 0.9231$; cos $A \approx 0.3846$

23. 2.4

24. 240

25. $0.41\overline{6}$, or $\frac{5}{12}$

26. $l = \frac{1}{3}h$

27. a. about 21.3 ft

  b. increase the angle; The steeper the ramp, the more difficult it is to skate up the ramp.

  c. Explanations may vary. An example is given. A curved ramp allows a skater to begin along a gently sloped portion, and gradually increases the slope. A steep, straight ramp might be too abrupt.

A Direct Variation Model

That's the Way The **BALL** Bounces

Focus

Model a direct variation situation and use the model to make predictions. Find the slope of a line in a coordinate plane.

EXPLORATION

How is the bounce of a ball related to the height of the drop?

- **Materials: tape measure or meter sticks, tape, five or six types of balls (such as a tennis ball, a rubber ball, a table tennis ball, a golf ball, a baseball, a soccer ball)**

- **Work in a group of three or four students.**

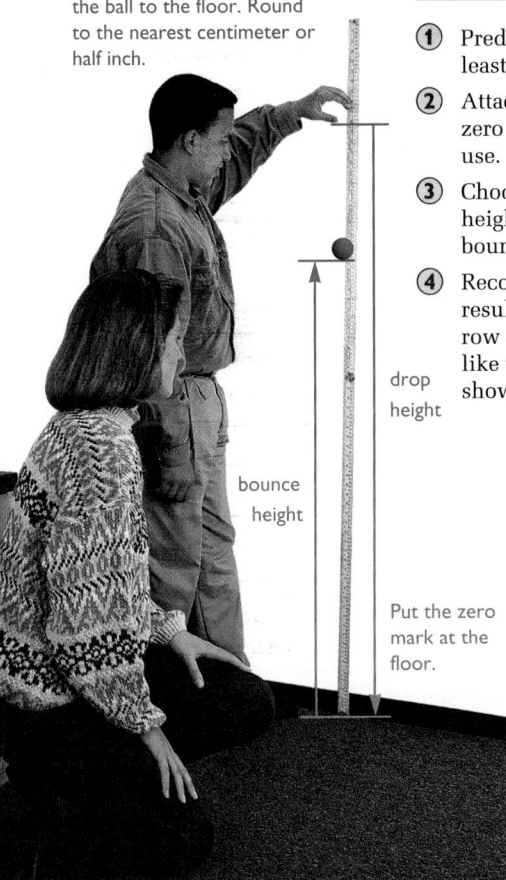

Measure from the bottom of the ball to the floor. Round to the nearest centimeter or half inch.

drop height

bounce height

Put the zero mark at the floor.

① Predict which ball has the most bounce and which has the least bounce. Discuss how you might measure this.

② Attach a tape measure or meter sticks to a wall. Put the zero mark at the floor. Choose one ball for your group to use.

③ Choose a drop height. Drop the ball four times from that height. Each time, measure the height of the ball's first bounce.

④ Record your results in one row of a table like the one shown.

Type of Ball:				
Drop height D	Bounce height			Mean bounce height B	Ratio $\frac{B}{D}$

  a. Record the drop height D and the four heights of the first bounce.

  b. Find the mean B of the bounce heights.

  c. Find the ratio of B to D.

⑤ Repeat step 4 with four different drop heights. Complete the table.

⑥ What patterns do you notice in your table?

⑦ Make a scatter plot of your data. Use D as the control variable and B as the dependent variable.

⑧ Compare your scatter plot with the scatter plots of other groups. Check your predictions in step 1.

367

Objectives and Strands
See pages 356A and 356B.

Spiral Learning
See page 356B.

Materials List
➤ Tape measure or meter sticks
➤ Tape
➤ Five or six types of balls (such as a tennis ball, a rubber ball, a table tennis ball, a golf ball, a baseball, a soccer ball)
➤ Graph paper
➤ Scientific calculator
➤ Spreadsheet software
➤ Three coins

Recommended Pacing
Section 7-2 is a two-day lesson.
Day 1
Pages 367–368: Exploration through Sample 1, *Exercises 1–12*
Day 2
Pages 368–370: Graphing Direct Variation through Look Back, *Exercises 13–32*

Extra Practice
See pages 631–633.

Warm-Up Exercises
Warm-Up Transparency 7-2

Support Materials
➤ Practice 53
➤ Enrichment 47 in the Activity Bank
➤ Study Guide 7-2
➤ Problem Set 14
➤ Diagram Masters 2, 19 in the Explorations Lab Manual
➤ McDougal Littell Mathpack software: *Stats!* or *Function Investigator*
➤ Quiz 7-2
➤ Alternative Assessment 3

Answers to Exploration

1. Answers may vary. An example is given. I predict that the tennis ball has the most bounce and the baseball has the least. You can test the bounce of a ball by measuring how far up it bounces when dropped from a specified height.

2–5. Drop heights and data may vary.

6. Answers may vary. An example is given. The greater the drop height, the greater the bounce height, and for all drop heights, the ratio $\frac{B}{D}$ is approximately the same for a given ball.

7. Scatter plots may vary.

8. Comparisons may vary. Lines should come close to the origin.

Exploration

Be sure students are clear on all aspects of the procedure. Discuss why the *mean* bounce heights are used for computing the ratios in the table on page 367. (Using the mean helps smooth out errors in estimating how high a ball bounces.) Stress the importance of measuring bounce heights to the *bottom* of each ball. The crucial nature of this point can be discussed *after* students have completed the Exploration activity.

Additional Sample

S1 Adrian hung a new spring from a hook on the edge of a workbench. He attached different masses to the spring and measured how far the spring stretched each time.

He recorded the masses in kilograms and the stretch in centimeters. Each time, he computed the ratio of stretch to mass. His results are shown in the table. (Ratios are rounded.)

Mass (kg)	Stretch (cm)	Ratio $\left(\frac{s}{m}\right)$
1.2	2.9	2.42
1.5	3.8	2.53
2.0	5.1	2.55
2.6	6.4	2.46
4.0	9.9	2.48

a. Adrian thinks the relationship between stretch (*s*) and mass (*m*) is an example of direct variation. Does this seem so? If you answer *yes,* what is the variation constant? Yes; about 2.5.

b. Model the behavior of the spring with a direct variation equation.

$s = 2.5m$

A Mathematical Model: Direct Variation

To model the behavior of a bouncing ball with an equation, look for a relationship between the two variables, the drop height *D* and the mean bounce height *B*.

Sample 1

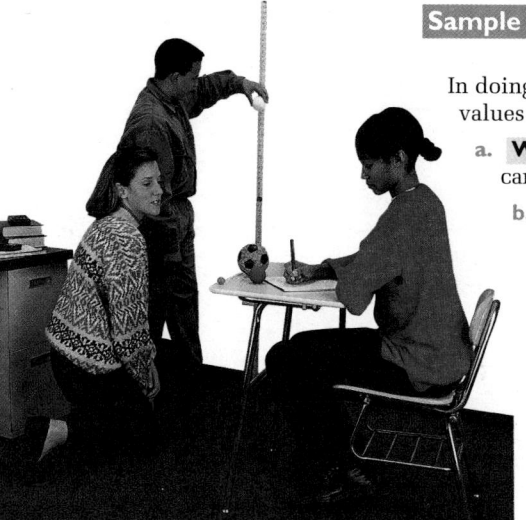

In doing the Exploration, Julio, Tara, and Linda recorded the values shown in the table.

a. **Writing** Julio thinks the relationship between *D* and *B* can be modeled by direct variation. Explain his reasoning.

b. Model the behavior of the ball with a direct variation equation relating *D* and *B*. What is the variation constant for the ball?

Drop height D (in.)	Mean bounce height B (in.)	Ratio $\frac{B}{D}$
20	15.0	0.75
30	24	0.8
40	32.5	0.8125
50	41	0.82
60	48.5	0.80$\overline{83}$

Sample Response

a. When two variables have a constant ratio, you can say that one variable varies directly with the other.

Although the ratios $\frac{B}{D}$ are not all exactly the same, to the nearest tenth each ratio is 0.8. Direct variation is a good model for the relationship between *D* and *B*.

> State the meaning of direct variation.

> Explain why the data fit a direct variation model.

b. The direct variation equation $\frac{B}{D} = 0.8$ models the behavior of the bouncing ball. The variation constant for the ball is about 0.8.

▶ Now you are ready for:
Exs. 1–12 on pp. 370–372

Graphing Direct Variation

You can also model the direct variation relationship between *D* and *B* with a graph, and then use your graph to make predictions.

Because of errors in measurement, real-world results may not match exactly the predictions of a mathematical model. However, a good model should give reasonable approximations.

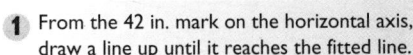 **Sample 2**

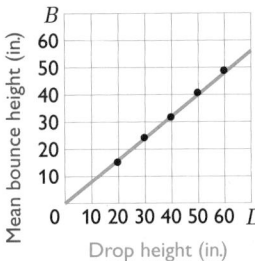

To model the data from the table in Sample 1, Tara made a scatter plot of the five data points (D, B). Then she drew a fitted line.

Use Tara's graph to predict how high the ball will bounce if she drops it from a height of 42 in.

Sample Response

1 From the 42 in. mark on the horizontal axis, draw a line up until it reaches the fitted line.

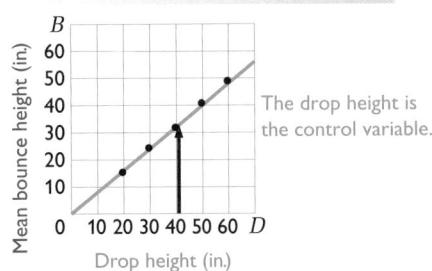

The drop height is the control variable.

2 Then draw a line to the left until it meets the vertical axis at about 34 in.

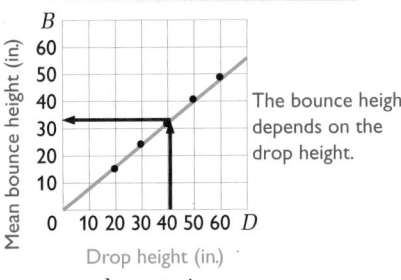

The bounce height depends on the drop height.

If Tara drops the ball from a height of 42 in., it will bounce to about 34 in.

BY THE WAY...

The fuzzy felt on tennis balls slows them down and gives them a better "grip" on different surfaces. The dimples on a golf ball lift them farther through the air.

Talk it Over

1. Use the equation in part (b) of Sample 1 to predict how high the ball will bounce if the group drops it from a height of 42 in. Compare your answer to the prediction in Sample 2.

2. Instead of using the equation $\frac{B}{D} = 0.8$ in question 1, Linda used the equation $B = 0.8D$. Do you think this equation is correct? Explain.

3. Julio wants the ball to bounce to a height of 12 in. How can Tara's graph help Julio estimate the drop height he should use?

4. What algebraic equation could Julio solve to find the drop height that will give a bounce height of 12 in.? Show how to solve it.

Finding Slope Using Coordinates

In Section 7-1, you learned about the slope of a line. You can use the coordinates of two points on a line to find the line's slope.

7-2 A Direct Variation Model

Make sure that students understand that using a graph to make a prediction will give answers that are only approximations to the actual result. Different students' results will probably vary slightly, but this is to be expected. Such variations do not imply that errors have been made.

Additional Sample

S2 To model the data in the first two columns of the table in Additional Sample S1, Adrian made a scatter plot of the five data points (m, s). Then he drew a fitted line.

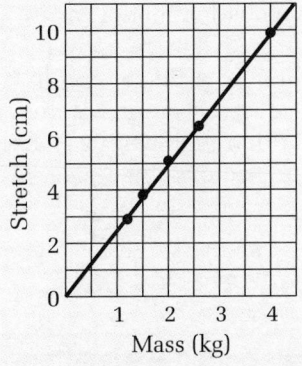

Use Adrian's graph to predict how far the spring would stretch if he attached a mass of 3.5 kg.

about 8.8 cm

Talk it Over

Point out to students that equations such as those in the Samples and Additional Samples do have their limits. For example, a ball with a bounce factor of 0.8 would not really bounce up 0.8 mile if it were dropped onto a highway from a plane 1 mile up.

Answers to Talk it Over

1. about 33.6 in.; This answer is close to the prediction in Sample 2.

2. Yes. Explanations may vary. An example is given. When you multiply each side of the equation $\frac{B}{D} = 0.8$ by D, you obtain the equivalent equation $B = 0.8D$.

3. From the 12 in. mark on the vertical axis, draw a horizontal line across until it reaches the fitted line. Then draw a vertical line down until it reaches the horizontal axis. Estimate the drop height at the crossing point.

4. Answers may vary. An example is given. Rashid could use the equation $\frac{B}{D} = 0.8$ or $B = 0.8D$; substitute 12 for B and then solve for D; $\frac{12}{D} = 0.8$; 12 = 0.8D; $D = \frac{12}{0.8} = 15$; he should use a drop height of about 15 in.

Additional Sample

S3 Use Adrian's graph in Additional Sample S2. Find the slope of the fitted line. Answers may vary. If you use the points (2.6, 6.4) and (1.5, 3.8), the result will be slope = $\frac{\text{rise}}{\text{run}} = \frac{2.6}{1.1} = 2.\overline{36}$.

Limited English Proficiency

Encourage students acquiring English to refer to drawings and charts to help them describe the two ways to model data in the Look Back that have a direct variation relationship.

Look Back

Use the Look Back to discuss also which model can be used to make a more accurate prediction. A discussion of key features of graphs of direct variation relationships is also appropriate at this time.

APPLYING

Suggested Assignment

Standard 1–3, 5–18, 22, 23, 25–32

Extended 1–20, 22, 23, 25–32

Integrating the Strands

Algebra Exs. 1–24, 26–28

Functions Exs. 1–24

Geometry Exs. 17, 18, 23, 29–31

Trigonometry Exs. 23, 25

Statistics and Probability Exs. 2, 3, 17, 18, 21, 24

Logic and Language Ex. 32

Sample 3

Use Tara's graph in Sample 2. Find the slope of the fitted line.

Sample Response

Choose any two points on the line. Compare the x-coordinates and the y-coordinates of the points.

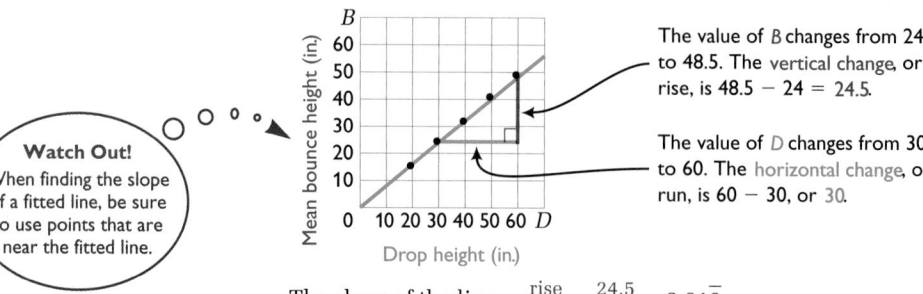

The value of B changes from 24 to 48.5. The vertical change, or rise, is $48.5 - 24 = 24.5$.

The value of D changes from 30 to 60. The horizontal change, or run, is $60 - 30$, or 30.

Watch Out! When finding the slope of a fitted line, be sure to use points that are near the fitted line.

The slope of the line $= \frac{\text{rise}}{\text{run}} = \frac{24.5}{30} = 0.81\overline{6}$.

Talk it Over

5. Compare the slope of the fitted line in Sample 3 to the variation constant found in part (b) of Sample 1.

6. Find the slope of the line that contains the points (2, 5) and (5, 9).

► Now you are ready for:
Exs. 13–32 on pp. 372–374

Look Back

Describe two ways to model data that have a direct variation relationship. Which type of model do you find easier to use to make a prediction?

7-2 Exercises and Problems

1. **Reading** What are the two variables used in the Exploration? What does each represent?

2. **a.** In the Exploration, Kathryn gathers data using a lacrosse ball. Use her data to find the four ratios $\frac{B}{D}$ to the nearest hundredth.

   b. Estimate a variation constant for Kathryn's ball to the nearest tenth.

Lacrosse ball

Drop height D (in.)	Mean bounce height B (in.)	Ratio $\frac{B}{D}$
12	10	?
24	19	?
36	29	?
48	38	?

Answers to Talk it Over

5. The variation constant, 0.8, and the slope, $0.81\overline{6}$, are close.

6. $\frac{4}{3} \approx 1.3$

Answers to Look Back

Answers may vary. An example is given. Data that have a direct variation relationship can be modeled through an equation with a variation constant or with a graph using a fitted line. I like to use the equation to make a prediction because it seems more accurate and it is simpler to find the value.

Answers to Exercises and Problems

1. B and D; B represents the bounce height, D represents the drop height.

2. **a.** $\frac{10}{12} \approx 0.83$, $\frac{19}{24} \approx 0.79$, $\frac{29}{36} \approx 0.81$, $\frac{38}{48} \approx 0.79$

   b. 0.8

3. ► No. No.

   ► Answers may vary. Most balls display direct variation.

► Answers may vary. The variation constants should be nearly the same for the same type of ball. Differences among groups' results could be due to measurement errors or rounding differences.

► Variation constants depend on the balls used. The ball with the most

3. **Writing** Summarize the results of the Exploration for someone who missed the class. Use the data gathered by all the groups. Include these points.

➤ Did all groups use the same drop heights? Is this important?

➤ Did every ball display direct variation?

➤ Did each group have the same variation constant? What might cause any differences among groups' results?

➤ Which type of ball had the most bounce? the least bounce?

4. **Writing** In the Exploration, the variation constants are all less than 1. Write a story about an impossible ball for which $\frac{B}{D}$ is greater than 1. How would this ball act? How could people use the ball? Would it be hard to control?

Decide whether direct variation is a good model for the data in each table. Explain your reasoning.

5.

width	height
2	0.14
4	0.26
6	0.43
8	0.56
10	0.69

6.

age	weight
5	100
10	141
15	173
20	201
25	224

7.

errors	cost
3	63
6	127
9	188
12	251
15	314

For Exercises 8–10:

a. **Determine whether direct variation is a good model for each situation. Explain your reasoning.**

b. **Tell whether the data are discrete or continuous.**

Control variable	Dependent variable
8. number of batches of cookie dough	number of cookies
9. time	water level in a lake
10. gallons of gasoline bought	cost of the gasoline bought

11. **Driver Education** The faster you drive a car, the longer it takes to stop in an emergency.

a. Which is the control variable and which is the dependent variable?

b. Find the ratio $\frac{\text{dependent variable}}{\text{control variable}}$ for each pair of values.

c. Is direct variation a good model for the data? Explain.

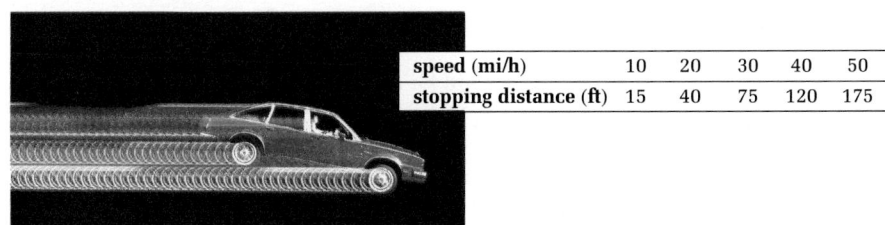

speed (mi/h)	10	20	30	40	50	60	70
stopping distance (ft)	15	40	75	120	175	240	315

7-2 A Direct Variation Model **371**

Application

For Ex. 3, have two or three students read their answers to the class. Have the class compare and discuss the answers that are read.

In this connection, you may wish to discuss why it was important to measure from the *bottom* of the ball to the floor and not from the *top* of the ball. For example, suppose a data chart had displayed a perfect direct variation relationship for a ball that had a diameter of 5 cm.

Drop Height D (cm)	Bounce Height B (cm)	Ratio $\frac{B}{D}$
200	100	0.5
140	70	0.5
100	50	0.5
50	25	0.5
20	10	0.5
5	2.5	0.5

If measurements had been made using the distance from the top of the ball to the floor, the ratios would steadily increase as the drop height decreased.

Measuring from the Top of the Ball to the Floor

Drop Height D (cm)	Bounce Height B (cm)	Ratio $\frac{B}{D}$
205	105	0.512
145	75	0.517
105	55	0.524
55	30	0.545
25	15	0.6
10	7.5	0.75

Answers to Exercises and Problems

bounce has the largest variation constant. The ball with the least bounce has the smallest variation constant.

4. a. Answers may vary. An example is given. A ball with a variation constant greater than 1 would bounce to a greater height than the drop height. If such a

ball escaped, it would bounce higher and higher, leaping over buildings, passing airplanes, and finally rebounding so high it would leave Earth's atmosphere. This ball would be very hard to control.

5. Yes. For each data point, the ratio $\frac{\text{height}}{\text{width}}$ is close to 0.07.

6. No. The ratio $\frac{\text{weight}}{\text{age}}$ decreases steadily from 20 to about 9.

7. Yes. For each data point, the ratio $\frac{\text{cost}}{\text{errors}}$ is close to 21.

8–10. Reasons may vary. Examples are given.

8. a. Yes. The number of cookies you can make depends directly on the

number of batches of cookie dough you make.

b. discrete

9. a. No. The water level does not change steadily over time.

b. continuous

10. a. Yes. The cost of one gallon of gasoline is the variation constant.

b. discrete

Answers continued on next page.

Research

In connection with Exs. 13–16, students can see what information is available from the library, sporting goods stores, or other sources about the dangers of diving too deep.

Multicultural Note

Students who are interested in scuba diving and oceanography might enjoy reading articles in *National Geographic* magazine by Dr. Eugenie Clark, a professor of zoology at the University of Maryland. Dr. Clark, who is of Japanese ancestry, also has coauthored a book called *The Desert Beneath the Sea*.

Having an avid interest in sea life as a young girl, she later earned a Ph.D. in zoology. Dr. Clark is a highly respected expert on sharks. She and her scientific teams study sharks both by scuba diving and in underwater submersible observation tanks. While studying in Israel and diving in the Red Sea, she discovered that a fish known as the Moses sole secretes a substance that is an effective shark repellant. Today, scientists are looking for a way to duplicate this chemical in the laboratory.

Using Technology

Students can use the *Stats!* software to plot the data and draw a fitted line for Ex. 17(d).

12. Using a table tennis ball, Jennifer found that the ratio of bounce height to drop height is given by the equation $\frac{B}{D} = 0.7$.

 a. Suppose $D = 4$ ft. Find B.

 b. Suppose $B = 4$ ft. Find D.

 c. Suppose $D = 7.5$ ft. Find B.

 d. Suppose $B = 7.5$ ft. Find D.

 e. Which is easier, finding the bounce height when given the drop height, or finding the drop height when given the bounce height?

Scuba Diving **For Exercises 13–16, use the graph.**

13. A sea diver's pressure gauge shows how many atmospheres (atm) of pressure are caused by the weight of the water on the diver. The pressure varies directly with the depth below the surface. Estimate the variation constant.

14. Estimate the water pressure on a diver at a depth of 70 ft.

15. The recommended maximum depth for recreational diving is 130 ft. What is the approximate water pressure at that depth?

16. Sami's water pressure gauge reads 2 atm. About how many feet below sea level is she?

Pressure from Water

17. **Veterinary Medicine** A veterinarian uses a table to determine how many milligrams of medicine to give a dog to prevent heartworm.

Weight of dog, W (lb)	15	30	45	60
Dosage, D (180 mg tablets)	0.25	0.5	0.75	1.0

 a. Find the ratio $\frac{D}{W}$ for each weight in the table.

 b. Which quantity varies directly with the other? What is the variation constant? Write an equation relating D and W.

 c. What dosage should the veterinarian recommend for a 75 lb dog? Explain your answer.

 d. Plot the data for weight and dosage on a graph. Draw a fitted line through the four points.

 e. Find the slope of the fitted line.

372   **Unit 7** Direct Variation

Answers to
Exercises and Problems

11. a. control: speed; dependent: stopping distance

 b. $\frac{15}{10} = 1.5$, $\frac{40}{20} = 2$, $\frac{75}{30} = 2.5$,

 $\frac{120}{40} = 3$, $\frac{175}{50} = 3.5$, $\frac{240}{60} = 4$,

 $\frac{315}{70} = 4.5$

 c. No. Explanations may vary. An example is given. The graph is a curve, not a straight line, and the ratio $\frac{\text{dependent variable}}{\text{control variable}}$ increases steadily.

12. a. 2.8 ft   b. about 5.7 ft

 c. 5.25 ft   d. about 10.7 ft

 e. Answers may vary. An example is given. Finding the bounce height when given the drop height is easier because you just have to multiply each side by the drop height to find the bounce height.

13. about $\frac{1}{33}$, or about 0.03

14. about 2.1 atm

15. about 3.9 atm

16. about 66 ft

17. a. Each ratio $\frac{D}{W}$ is equal to about 0.017.

 b. The dosage varies directly with the weight. about 0.017; $D = \frac{1}{60}W$ or $D \approx 0.017\ W$.

 c. 1.25 tablets; Methods may vary. For example, substitute 75 for W in the equation in part (b) and solve to find the value of D, or continue the pattern established in the table, noting that the values of W are increasing by 15 and the corresponding values of D are increasing by 0.25.

Speed of bat swing (mi/h)	Distance ball travels (ft)
50	270
60	325
70	378
75	405

18. **Baseball** The table shows how far a baseball player can hit an 85 mi/h pitch.

   a. Do these data show direct variation? If so, write an equation to model the situation.

   b. Plot these data on a graph and draw a fitted line.

   c. Find the slope of the line you have drawn. Explain how you calculated the slope.

   d. Suppose the speed of a player's swing is 64 mi/h. Predict how far the ball will go.

19. A sky diver's speed varies directly with the amount of time spent falling. A sky diver's speed 2.5 s after jumping from a plane is 24.5 m/s.

   a. Draw a graph with time on the horizontal axis and speed on the vertical axis.

   b. Use your graph to predict the sky diver's speed after 3 s.

   c. Use your graph to predict the sky diver's speed after 4 s.

20. To solve Exercise 19 using an equation, you may write $\frac{v}{t} = k$, where v represents speed, t represents elapsed time, and k represents a constant.

   a. Use the data given in Exercise 19 to determine k.

   b. Find v when $t = 3$.

   c. Find v when $t = 4$.

   d. Compare the results of parts (b) and (c) with the results in Exercise 19.

2.5 s

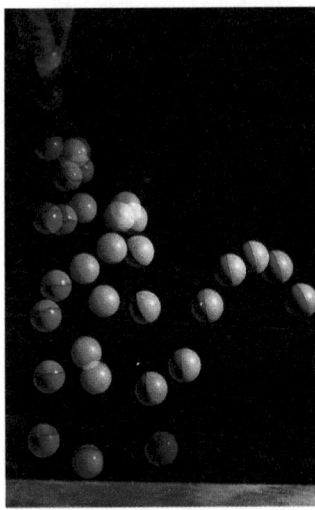

21. **Group Activity** Work in a group of four students.

   a. Select a ball. Decide on a drop height. Predict how many bounces the ball will make and the height of each bounce.

   b. Drop the ball from the given drop height. Measure the height of the first three bounces.

   c. How are the drop height, the first bounce height, the second bounce height, and the third bounce height related to each other mathematically?

   d. TECHNOLOGY Record your results in a spreadsheet. Use your spreadsheet to predict what might happen to the heights of the next five bounces, using various drop heights.

   e. Choose a drop height modeled in your spreadsheet. Drop the ball from this height. How closely does your model predict what actually happens?

7-2 A Direct Variation Model

373

Communication: Discussion

When discussing the answers to Ex. 18, observe if any students understand that the constant of variation 5.4 in the formula $D = 5.4s$ is the slope of the fitted line. If not, remind them that the constant of variation, slope, and the tangent ratio are three different ways of expressing the same idea.

Using Technology

The use of a spreadsheet is discussed in the Technology Handbook on page 619.

Students can use the spreadsheet feature of the *Stats!* or the *Function Investigator* software for Ex. 21.

Integrating the Strands

On page 374, Ex. 24 does a nice job of integrating the strands of algebra, functions, and statistics and probability.

18. c. 5.4; Explanations may vary. An example is given. I found the slope using the coordinates of two points. The value of D changes from 378 to 405, so the rise is $405 - 378$, or 27. The value of S changes from 70 to 75, so the run is $75 - 70$, or 5. The slope is $\frac{27}{5}$, which equals 5.4.

   d. about 346 ft

19. a.
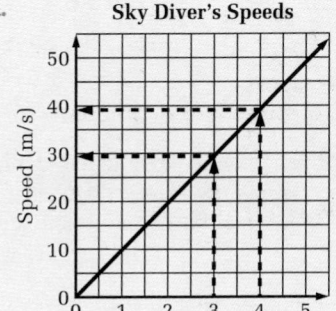
Sky Diver's Speeds

   b. about 29 m/s   c. about 39 m/s

20. a. $k = 9.8$

   b. 29.4 m/s   c. 39.2 m/s

   d. The results from using an algebraic approach in parts (b) and (c) of this exercise are more precise than those from using a graphing approach in Exercise 19, but the results are approximately the same.

21. See answer in back of book.

Answers to Exercises and Problems

17. d.
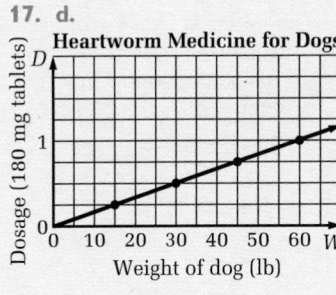
Heartworm Medicine for Dogs

   e. $\frac{1}{60}$ or about 0.017

18. a. Yes.; $\frac{D}{S} = 5.4$ or $D = 5.4S$, where D is the distance the ball travels and S is the speed of the bat swing.

   b.
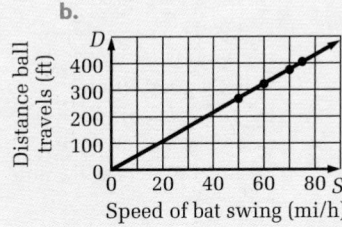

373

Assessment: Peformance

You will need to allow class time for the completion of Ex. 24. Evaluate each group's work on its own merits. Then invite class discussion so groups can compare results and conclusions.

Working on the Unit Project

Allow time for students to research Ex. 32. Perhaps a weekend or several days might be needed to visit the library, consult with science teachers, or interview local engineers or skateboard experts.

Practice 53 For use with Section 7-2

Name ______________________ Date ____________

Practice 53
For use with Section 7-2

11. yes; Equations may vary. Possible answer: $y = 0.85x$, where x = number of seeds planted, y = number of plants produced
13. Answers may vary. Possible answer: slope is about 0.85.
14. Answers may vary. Possible answer: about 162 seeds

For Exercises 1–6, decide whether direct variation is a good model for the data in each table. If it is, write an equation that includes the variation constant.

1. $y = 1.5x$

x	y
3	4.5
4	6
5	7.5
10	15

2. $y = 0.4x$

x	y
15	6
20	8
28	11.2
36	14.4

3. no

x	y
1	3
2	6
4	18
8	66

4. $y = \frac{7}{3}x$

x	y
4.5	10.5
15	35
21	49
102	238

5. no

x	y
7	17.5
20	30.5
15.5	26
19.5	30

6. $y = 3.6x$

x	y
5	18
15	54
17.5	63
21.5	77.4

For each direct variation relationship, find the missing value.

7. $\frac{a}{b} = 2.4$; $b = 35$; $a = \underline{?}$ 84
8. $\frac{p}{q} = 0.6$; $p = 33$; $q = \underline{?}$ 55
9. $\frac{c}{d} = 2.2$; $c = 18.7$; $d = \underline{?}$ 8.5
10. $\frac{v}{w} = 16.8$; $w = 3.5$; $v = \underline{?}$ 58.8

For Exercises 11–14, use the table at the right, which shows the results of an experiment in re-introducing a certain type of plant into a region.

Seeds Planted	Plants Produced
50	41
75	63
100	88
125	100
150	127
175	149

11. Do these data seem to show direct variation? If so, write an equation to model the situation. See top of page.
12. Plot these data on a graph and draw a fitted line. Check students' work.
13. Find the slope of the line you have drawn. See top of page.
14. About how many seeds would have to be planted to produce 136 plants? See top of page.

Exs. 15, 16: Answers may vary.
For Exercises 15 and 16, use the graph at the right.

15. Estimate the variation constant. about $\frac{4}{3}$
16. Estimate the weight of a pane of glass whose area is 25 ft^2. about $33\frac{1}{3}$ lb

22. What happens to the first bounce height of a ball if you double the drop height? if you triple the drop height? Explain your answers.

23. a. Find the slope of the line shown at the right.

b. Use the tangent ratio to find the angle the line makes with the x-axis.

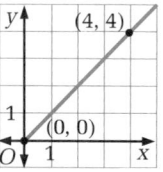

Ongoing ASSESSMENT

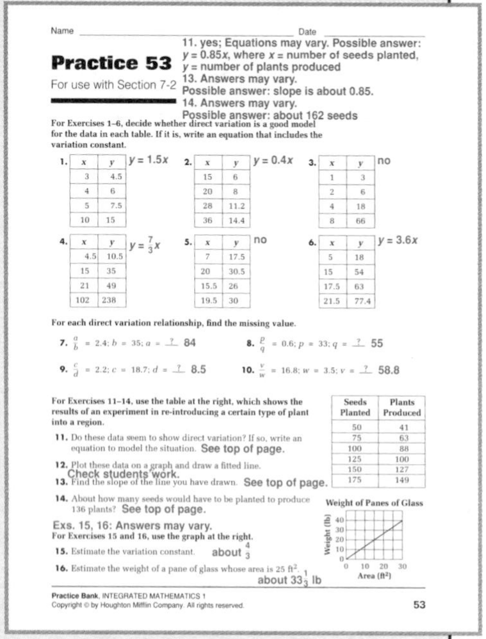

24. a. **Group Activity** Toss three coins 50 times. Count the number of times exactly two heads appear. Copy and complete the table.

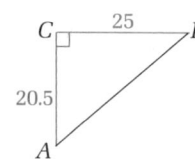

Number of tosses	Number of times exactly two heads appear	$\frac{\text{Two heads appear}}{\text{Number of tosses}}$
10	?	?
20	?	?
30	?	?
40	?	?
50	?	?

b. Graph the data from the table. Put the number of tosses on the horizontal axis and the number of times exactly two heads appear on the vertical axis.

c. Is direct variation a useful model for predicting the number of times exactly two heads will appear when three coins are tossed 10 or 20 times? 50 times? Why or why not?

d. Gather all the data from the groups in your class. What is the total number of tosses? What is the total number of times exactly two heads appeared? Is the ratio what you expect?

Review PREVIEW

25. Find the tangent ratio of $\angle A$. *(Section 7-1)*

Graph each equation. *(Section 4-7)*

26. $y = 3x$ **27.** $y = -5x$ **28.** $y = -|x|$

Give the formula for the perimeter of each figure. *(Toolbox Skill 17)*

29. rectangle **30.** square **31.** equilateral triangle

Working on the Unit Project

32. **Research** Why is urethane used in making wheels for skateboards and in-line skates? How are the properties of urethane related to a bouncing ball?

Answers to Exercises and Problems

22. The bounce height also doubles. The bounce height triples. Explanations may vary. Examples are given. Since $\frac{B}{D}$ is constant, if you double (or triple) D, then B must also double (or triple) for $\frac{B}{D}$ to remain unchanged. Algebraically, we have $\frac{B}{D} = \frac{2}{2} \cdot \frac{B}{D} = \frac{2B}{2D}$, so the bounce height doubles when the drop height doubles.

23. a. 1 b. 45°

24. a. Answers may vary. As the number of tosses increases, the ratio $\frac{\text{two heads appear}}{\text{number of tosses}}$ should be close to $\frac{3}{8}$, or 0.375.

b. Graphs may vary.

c. No. Yes. Explanations may vary. An example is given. Direct variation is not a good model when three coins are tossed just a few times, because widely different ratios are likely. However, if the coins are tossed many times, points on the graph are likely to fall closer and closer to a fitted line through the origin with slope 0.375.

d. Data may vary. The ratio $\frac{\text{two heads appear}}{\text{number of tosses}}$ should be close to 0.375.

25. about 1.22

26–28. See answers in back of book.

29. $P = 2l + 2w$

30. $P = 4s$

31. $P = 3s$

32. Urethane is a tough, flexible plastic that resists chemicals. These properties cause the rolling wheels to respond predictably over time, just like a bouncing ball does.

Section 7-3

Circumference and Arc Length

Focus
Find the circumference and the length of an arc of a circle.

EXPLORATION

(What) is the relationship between the circumference and the diameter of a circle?

- **Materials: string or tape measures, metric rulers, scissors, circular shapes of different sizes (such as a coffee can, a jar lid, a round waste basket, a CD, a small wheel, or a plate)**
- **Work with another student.**

Use a ruler to find the diameter of the circular shape. The **diameter** of a circle is the distance across the circle at its widest part.

Record the diameters and circumferences in a table like the one shown below.

(1) Select five circular shapes of different sizes.

(2) Find and record the *diameter* and the *circumference* of each circular shape:

Wrap a string around the circular shape. When you unwrap it you can measure the circle's circumference. The **circumference** of a circle is the distance around the circle.

Circle	Diameter *d* (mm)	Circumference *C* (mm)	$\frac{C}{d}$
1	?	?	?
2	?	?	?
3	?	?	?
4	?	?	?
5	?	?	?

(3) For each circle find the ratio $\frac{C}{d}$ to the nearest hundredth. Write it in the last column of the table.

(4) What do you notice about the ratios in the last column? Does the relationship between diameter and circumference seem to be an example of direct variation?

7-3 Circumference and Arc Length

375

Answers to Exploration

1–4. Answers may vary based on choice of five circular shapes. Examples are given.

2–3. A sample table of values is given at right.

4. The ratios are all approximately equal. Yes.

2–3.

Circle	Diameter (*d*) in mm	Circumference (*C*) in mm	$\frac{C}{d}$
1	88	282	3.20
2	75	239	3.19
3	112	352	3.14
4	150	470	3.13
5	55	176	3.20

PLANNING

Objectives and Strands
See pages 356A and 356B.

Spiral Learning
See page 356B.

Materials List
➤ String or tape measure
➤ Metric ruler
➤ Scissors
➤ Circular shapes (such as a coffee can, a jar lid, a round waste basket, a CD, a small wheel, or a plate)
➤ Scientific calculator
➤ Graph paper
➤ Globe

Recommended Pacing
Section 7-3 is a two-day lesson.
Day 1
Pages 375–376: Exploration through Sample 1, *Exercises 1–13*
Day 2
Pages 377–378: Arc Length through Look Back, *Exercises 14–29*

Extra Practice
See pages 631–633.

Warm-Up Exercises
Warm-Up Transparency 7-3

Support Materials
➤ Practice 54
➤ Enrichment 48
➤ Study Guide 7-3
➤ Problem Set 14
➤ Diagram Masters 2, 20
➤ Geometry Inventor Activity Book: Activity 12
➤ Quiz 7-3
➤ Test 27
➤ Alternative Assessments 4, 5

Exploration

The goal of the Exploration is to use measurements of real-world objects to demonstrate that there is a direct variation relationship between the diameters of circles and their circumferences.

Metric tape measures are ideal for this activity. If they are not available, students can make their own. They need only cut long, narrow strips from notebook paper. They can mark the strips in centimeters (using a metric ruler), then tape them together with a small piece of transparent tape.

Using Manipulatives

You may wish to discuss techniques for using tape measures or other materials for getting accurate measures of diameters. Remind students that diameters pass through the center of a circular object and therefore are the longest segments that join two points on the object.

An instructive activity is to have students measure the diameter using a tape measure. They can then measure the circumference, put that number into the circumference formula, and solve for d. In this way, students can check on how well they estimated the diameter when they measured.

Additional Sample

S1 Find the circumference of the top rim of the lampshade if the diameter is 22 cm. Give the circumference to the nearest centimeter.

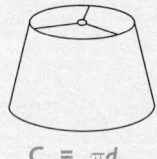

$$C = \pi d$$
$$= \pi \cdot 22$$
$$\approx 69.08$$

The circumference of the top rim of the lampshade is about 69 cm.

The Ratio Pi

For all circles, the ratio of circumference to diameter is a constant. This ratio is named by the Greek letter π (**pi**).

Common approximations for π are 3.14 and $\frac{22}{7}$. The exact value is a nonterminating and nonrepeating decimal between these two approximations.

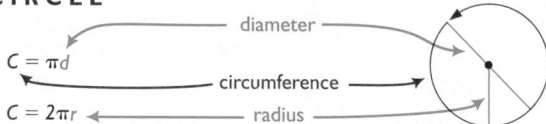

$$\frac{314}{100} \qquad \pi \qquad \frac{22}{7}$$

3.140000000... 3.141592653... 3.142857142...

a terminating decimal a nonterminating decimal a repeating decimal

To see the value your calculator uses for π, press [π]. For calculations involving π, give your answers to the nearest tenth unless told otherwise.

You can use the equation $\frac{C}{d} = \pi$, or the form $C = \pi d$, to find the circumference of any circle when you know its diameter.

CIRCUMFERENCE OF A CIRCLE

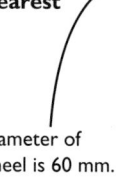

$C = \pi d$ diameter

    circumference

$C = 2\pi r$ radius

Sample 1

Find the circumference of the skateboard wheel to the nearest millimeter.

The diameter of the wheel is 60 mm.

Sample Response

Problem Solving Strategy: Use a formula

$$C = \pi d$$
$$= \pi \cdot 60 \quad \longleftarrow \text{Substitute 60 for } d.$$
$$\approx 188 \quad \longleftarrow \quad [\pi]\ [\times]\ 60\ [ENTER]$$

The circumference of the wheel is about 188 mm.

▶ Now you are ready for:
Exs. 1–13 on pp. 378–379

376

Unit 7 Direct Variation

Arc Length

When you cut a circle and straighten it to form a line segment, the length of the line segment is the circumference of the circle. An **arc** is a part of a circle. The length of an arc is a fraction of the circumference. This measurement is called **arc length.**

arc length

circumference

ARC LENGTH

The endpoints of an arc determine a central angle called the central angle of the arc.

You can use this proportion to find the length of an arc:

$$\frac{\text{arc length}}{\text{circumference}} = \frac{\text{measure of central angle of arc}}{360°}$$

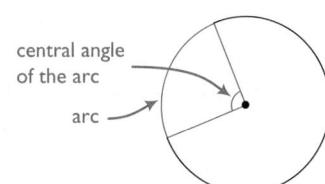

central angle of the arc

arc

Talk it Over

1. A circle graph with five central angles is "unwrapped" into a straight line of length 1. Complete the diagram to find the arc lengths.

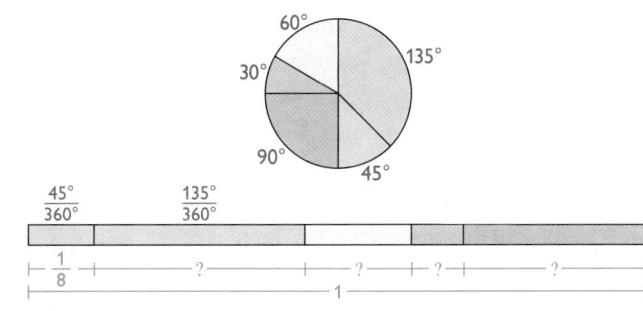

2. What is the measure of the central angle of an arc whose arc length is one fifth the circumference of a circle?

7-3 Circumference and Arc Length

377

Answers to Talk it Over

1.

$\frac{45°}{360°}$	$\frac{135°}{360°}$	$\frac{60°}{360°}$	$\frac{30°}{360°}$	$\frac{90°}{360°}$
$\frac{1}{8}$	$\frac{3}{8}$	$\frac{1}{6}$	$\frac{1}{12}$	$\frac{1}{4}$

2. 72°

Integrating the Strands

Mathematics is an integrated body of knowledge. Its concepts are often closely interrelated. This is apparent from the topics presented on this page about angle measures, proportions, percents, and circle graphs. The procedure of making connections among mathematical ideas can help students to understand them better and at a deeper level.

Talk it Over

Questions 1 and 2 reinforce the key fact that the ratio of arc length to circumference is equal to the ratio of the measure of the central angle of the arc to 360°. The equality of these ratios can be used to find arc length, circumference, or the measure of the central angle, provided the other two measures are known.

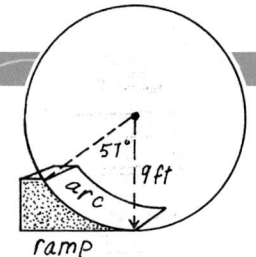

Sample 2

A launch ramp for skateboarders is formed by an arc with a central angle of 57° and a radius of 9 ft. Find the length of the arc to the nearest foot.

Sample Response

Problem Solving Strategy: Use a proportion

Let x = the unknown arc length. ⟶ $\dfrac{x}{56.5} \approx \dfrac{57}{360}$ ⟵ The measure of the central angle is 57°.

$d = 18$, so $C = \pi \cdot 18 \approx 56.5$ ⟶ ⟵ The measure of a complete circle is 360°.

$$\left(\dfrac{x}{56.5}\right)(56.5) \approx \left(\dfrac{57}{360}\right)(56.5)$$ ⟵ Multiply both sides by 56.5.

$$x \approx 8.9$$

The arc length is about 9 ft.

........► **Now you are ready for:**
Exs. 14–29 on pp. 379–382

Look Back ◄

For any circle, how are diameter, circumference, and arc length related?

7-3 Exercises and Problems

1. **Writing** Suppose Congress passed a law saying that the value of π is exactly 3. Would this law have any effect on the actual relationship between the circumference and the diameter of any circle? Use the results of the Exploration to help you explain why or why not.

2. Use the data you gathered in the Exploration.
   a. Plot your diameter and circumference data on a graph.
   b. Draw a fitted line and find its slope.
   c. How does the slope compare with π?

3. **Reading** Look at the number line on page 376 that shows π between the values 3.14 and $\frac{22}{7}$. How do you know that π is not exactly halfway between these values?

4. The five circles are congruent. Which of these segments is closest in length to the circumference of one of the circles: $\overline{AB}$, $\overline{AC}$, $\overline{AD}$, $\overline{AE}$, or $\overline{AF}$? Why?

Answers to Look Back ·······················

Answers may vary. Examples are given. For any circle, if you divide the circumference by the diameter, you will always get a value that is about 3.14; the bigger the diameter, the bigger the circumference. The circumference of a circle is always about three times its diameter. To determine the length of an arc, you also need information that indicates what part of the circle the arc represents and one of the following: the circumference of the circle, the diameter, or the radius.

Find the circumference of each circle.

5.

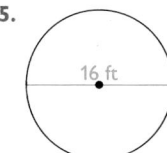

16 ft

6.

3 m

7. a. The Santa Maria del Tule tree in Oaxaca, Mexico, is a Montezuma cypress whose trunk is about 117 ft around. Suppose the trunk is circular. Estimate its diameter.

   b. **Research** Measure the circumference of at least two tree trunks or columns. Estimate their diameters.

A gardener decides to plant tulip bulbs around the edge of a circular flower bed with a diameter of 12 ft.

8. About how many tulip bulbs can the gardener plant if the bulbs are spaced about 6 in. apart?

9. About how much space should the gardener leave between the bulbs if 50 bulbs are planted?

For Exercises 10–12, find the variation constant. Sketch and label a diagram to help you.

10. The perimeter of an equilateral triangle varies directly with the length of one side.

11. The perimeter of a square varies directly with the length of one side.

12. The circumference of a circle varies directly with the radius.

13. Choose the letter(s) of the equation(s) that correctly show the relationship between the circumference and diameter of a circle.

   a. $C = \pi d$ **b.** $\dfrac{C}{d} = \pi$ **c.** $d = \dfrac{C}{\pi}$ **d.** $d = \pi C$

Find each arc length.

14.

45 in.

15.

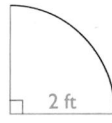

2 ft

16. Six equilateral triangles form the regular hexagon drawn in the circle. The radius of the circle is 8 cm.

   a. Find the perimeter of the hexagon.

   b. Find the circumference of the circle.

   c. How do your answers in parts (a) and (b) compare?

   d. Find the length of one of the six arcs shown.

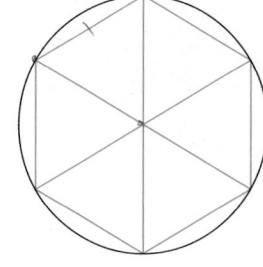

7-3 Circumference and Arc Length

379

Answers to Exercises and Problems

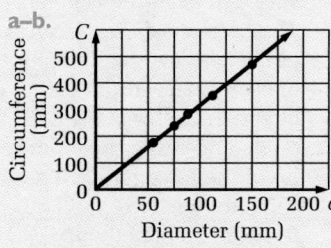
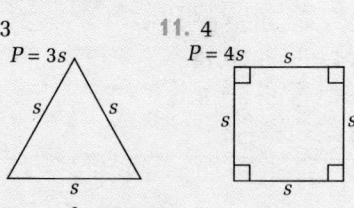
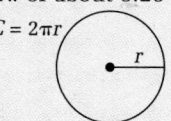

Research

In connection with Ex. 17, students can write to NASA, do library research, and look for information through other sources to see how satellite data are processed to help geographers gather measurement data about Earth's surface. Students who undertake such research should also inquire about the uses to which the data are put. This research will extend beyond the last class session for this unit. Results can be presented in the form of posters or oral reports to the class.

Visual Thinking

Ask students to prepare a sketch of their findings in Ex. 18. Have them explain why the shortest distance is not the line of latitude of the two cities.

This activity involves the visual skill of *perception*.

Interdisciplinary Problems

When discussing Ex. 19, you may wish to point out that astronomy has made many contributions to mathematics. In fact, many centuries ago, the name *mathematician* meant an astronomer.

connection to **GEOGRAPHY**

For Exercises 17 and 18, use the fact that the diameter of Earth is about 7900 mi.

17. **a.** Find the length of the equator. Round the answer to the nearest hundred miles.

   b. Quito and Nairobi are two cities located very close to the equator. Quito is 79° west of the prime meridian, and Nairobi is 37° east of the prime meridian. Estimate the distance from Quito to Nairobi along the equator. Round the answer to the nearest 100 mi.

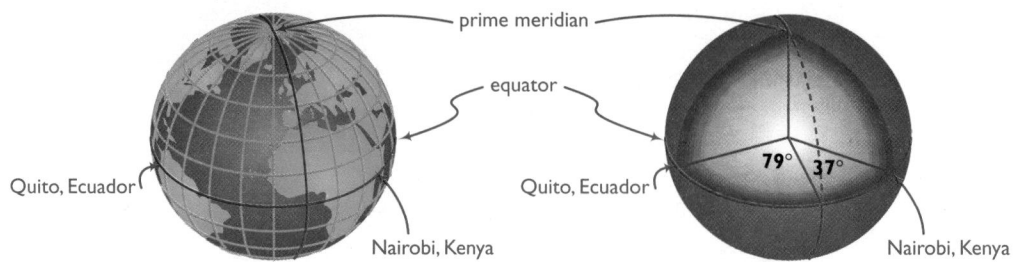

18. **Using Manipulatives** On a globe, choose any two cities that have the same latitude but are not on the equator. (*Note:* For part (b) you will need to use your answer to part (a) of Exercise 17.)

   a. Using the globe and a string, try to find the shortest path between the two cities. Is it along their line of latitude?

   b. Use a proportion to find the length of the shortest path between the two cities. (*Hint:* You will need to measure the string length that represents the shortest path between the cities. What other distance on your globe will you need to measure?)

New York and Istanbul have about the same latitude.

19. **Astronomy** It takes Earth one year to travel around the sun. The distance from Earth to the sun is about 9.3×10^7 mi.

   a. About how far does Earth travel in its orbit in one year? Assume the Earth's orbit is circular.

   b. About how far does Earth travel in its orbit in one day? (Use 365.25 days for 1 year.)

   c. Find the length of an arc of Earth's orbit with a central angle of 30°.

   d. About how long does it take Earth to travel the length of the arc in part (c)?

380   **Unit 7** Direct Variation

Answers to Exercises and Problems

13. a, b, and c

14. about 70.7 in.

15. about 3.1 ft

16. **a.** 48 cm

   b. about 50.2 cm (using $\pi \approx 3.14$)

   c. Answers may vary. An example is given. The circumference of the circle is about 2 cm greater than the perimeter of the hexagon.

   d. about 8.4 cm

17. **a.** about 24,800 mi

   b. 8000 mi

18. **a.** No.

   b. Answers may vary.

19. **a.** about 5.84×10^8 mi

   b. about 1,600,000 mi

   c. about 4.87×10^7 mi

   d. about 30 days, or 1 month

20. The curves are the same length. Explanations may vary. An example is given. The diameter of the large circle is twice the diameter of each small circle. Let x be the diameter of a small circle and $2x$ be the diameter of the large circle. You can slide the two parts of the blue S-shaped curve together to form a circle of circumference $\pi \cdot x$. The red arc is half the circumference of the circle with diameter $2x$; the length of the red arc is $\frac{1}{2}(\pi \cdot 2x) = \left(\frac{1}{2} \cdot 2\right)(\pi \cdot x) = \pi \cdot x$. Each curve has length $\pi \cdot x$.

20. A Chinese yin/yang symbol shows the relationship between two opposite forces. To make the symbol you draw circles and arcs. Is the blue curve longer than the red arc? Explain your answer.

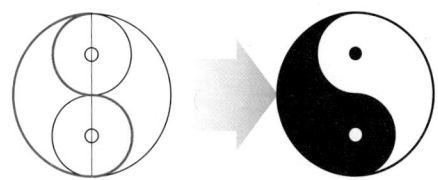

21. **Open-ended** A track has two straight portions of length *l* and two half-circle parts with a diameter *d*.

  a. Find values of *l* and *d* that result in a quarter-mile (1320 ft) track.

  b. Show that there is more than one possible answer.

  c. Which track do you think is best?

Multicultural Note

When discussing Ex. 20, you might further explain the yin/yang symbol. According to traditional Chinese religious belief, yin represents Earth (darkness, cold, or wetness) and yang represents heaven (light, heat, or dryness). The universe is one whole, huge organism in which sometimes yin, and sometimes yang takes the lead for a time. Since harmony is regarded as the ideal state, the universe can function properly only if the forces are in balance. The symbol shows this perfect balance and conveys the idea of cyclical change. At the top and bottom are points where one force starts diminishing and the other starts expanding. According to traditional belief, nothing is constant except this slow, ceaseless change.

Ongoing **ASSESSMENT**

22. a. Research Gather data about the arm span of everyone in your class.

  b. What is the greatest length your class can cover if everyone joins hands in a line?

  c. Predict the diameter of the largest circle your class can make if everyone joins hands.

  d. Group Activity Test your prediction by making a circle with all your classmates and measuring its diameter.

Assessment: Performance

For Ex. 22, time will be needed for students to gather the data from each other. Students could work in pairs, one measuring the arm span of the other, and then record their measurements on a transparency for all to see. Another possibility is to collect and list the data on the first day of this section, then pass out copies of it on the second day. Remind students that the diameter of a circle passes through the center of the circle.

Review **PREVIEW**

Draw a line through the points and find its slope. *(Section 7-2)*

23. (0, 0) and (5, 2)

24. (3, 2) and (7, 10)

25. Draw a box-and-whisker plot for the SAT scores. *(Section 3-5)*

SAT Scores (Math)				
553	602	407	346	701
778	475	530	427	307
621	602	575	525	522

Evaluate each expression for the given value of the variable. *(Section 2-2)*

26. $-3x$ when $x = -6$ **27.** $\frac{1}{2}x$ when $x = -2$ **28.** $0.75x$ when $x = 4$

7-3 Circumference and Arc Length   **381**

Answers to Exercises and Problems

21. a–c. Answers may vary. Examples are given.

  a. Using $\pi \approx \frac{22}{7}$, when $l = 440$, $d = 140$, and when $l = 110$, $d = 350$.

  b. The equation for this situation is $2l + \pi d = 1320$, which has 2 variables so there are many pairs of values that will satisfy the situation. The only limit is what is reasonable. That is, *l* and *d* cannot be negative.

  c. I think the best track would be one where the turns are not too sharp so as to hinder a runner and also where you can see the whole track from all spectators' seats.

22. a–d. Answers may vary.

23. slope $= \frac{2}{5}$

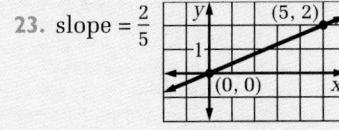

24. slope $= 2$

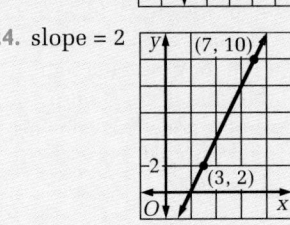

25. SAT Scores

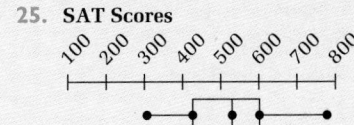

26. 18

27. −1

28. 3

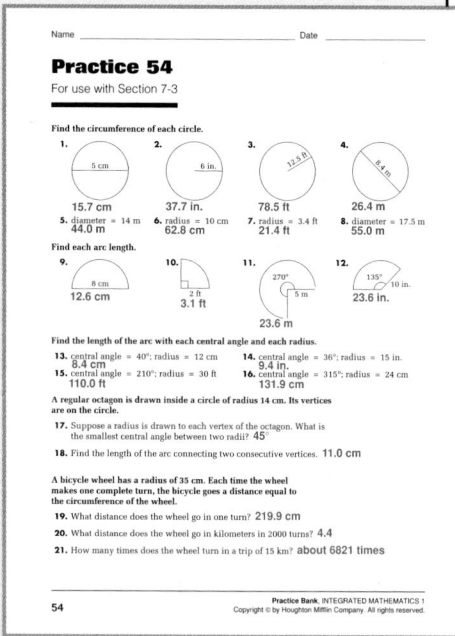

Working on the Unit Project

29. A "half-pipe" skateboard ramp is formed by two quarter-circle ramps with flat space in between. The ramps reach a height of 11 ft and they are 16 ft apart. Find the distance a skateboarder travels from the top of one ramp to the top of the other ramp.

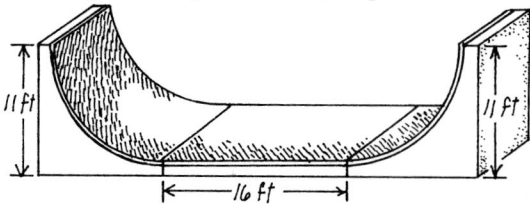

Unit 7 — CHECKPOINT

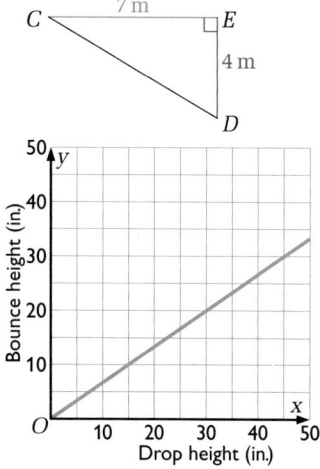

1. **Writing** Suppose you have data involving two vari- **7-1**
ables. How could you tell if one varies directly with the other?

2. In right $\triangle CDE$, find each tangent.

   a. tan D b. tan C

3. An escalator makes an angle of 30° with the horizontal. The horizontal length of the escalator is 24 ft. How high does the escalator rise?

4. The graph shows the relationship between the drop **7-2**
height and the bounce height of a rubber ball. Use the graph to estimate each height.

   a. the bounce height when the drop height is 20 in.

   b. the drop height when the bounce height is 25 in.

5. Estimate the variation constant for the graph.

6. In a circle, the relationship between the circumfer- **7-3**
ence and the diameter can be written $\frac{C}{d} = \pi$. ($\pi \approx 3.14$)

   a. Suppose $d = 400$ cm. Find C.

   b. Suppose $C = 60$ ft. Find d.

   c. Rewrite the equation in another form.

7. The Mather family orders a pizza. For toppings, they order $\frac{1}{3}$ vegetarian, $\frac{1}{2}$ ham and mushroom, and $\frac{1}{6}$ cheese.

   a. What is the circumference of the pizza?

   b. What is the length of the crust for the cheese section?

   c. What is the length of the crust for the vegetarian section?

Answers to Exercises and Problems

29. about 50.5 ft
(using $\pi \approx 3.14$)

Answers to Checkpoint

1. Answers may vary. An example is given. For one variable to vary directly with another, the two quantities must have a constant ratio. Find the ratio for each data pair. If the ratio is constant or almost constant, one variable varies directly with the other.

2. a. $\frac{7}{4}$ or 1.75

   b. $\frac{4}{7}$ or about 0.571

3. about 13.9 ft

4. Estimates may vary. Examples are given.

   a. about 13 in.

   b. about 37 in.

5. Estimates may vary; for example, about 0.7.

6. a. about 1256 cm

   b. about 19.1 ft

   c. $C = \pi d$

7. a. about 44 in. (using $\pi \approx 3.14$)

   b. about 7.3 in.

   c. about 14.7 in.

7-4 Direct Variation with $y = kx$

Talk it Over

1. The table shows three examples of direct variation you have seen in this unit. How are the equations different? How are they alike?

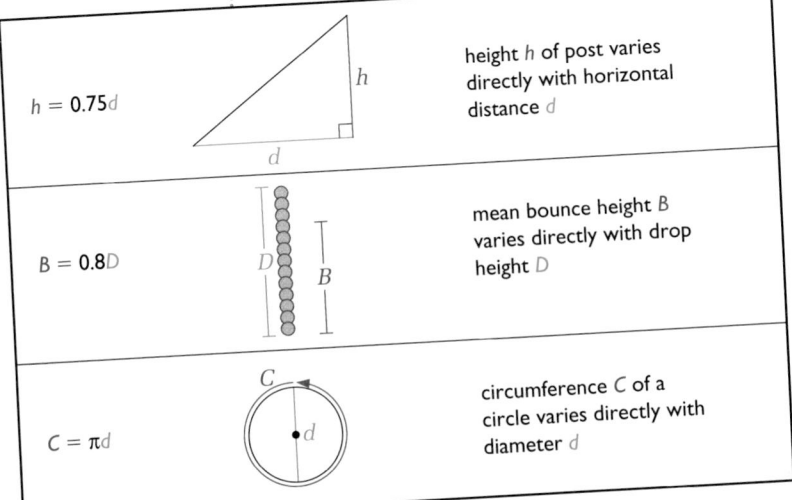

$h = 0.75d$	height h of post varies directly with horizontal distance d
$B = 0.8D$	mean bounce height B varies directly with drop height D
$C = \pi d$	circumference C of a circle varies directly with diameter d

2. For each equation, identify the control variable, the dependent variable, and the variation constant.

You can model direct variation with an equation of the form $y = kx$.

one quantity **varies directly with** another quantity

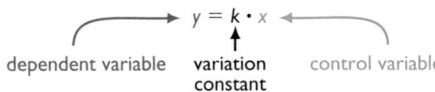

$$y = k \cdot x$$

dependent variable   variation constant   control variable

The equation $y = kx$ is called the **general form for direct variation**.

7-4 Direct Variation with $y = kx$ **383**

Answers to Talk it Over

1. Answers may vary. An example is given. The equations are different because they involve different quantities, variables, and constants, and they have different solution pairs. They all have the form variable = constant × another variable, so each is a direct variation equation.

2.
equation	$h = 0.75d$	$B = 0.8D$	$C = \pi d$
control variable	d	D	d
dependent variable	h	B	C
variation constant	0.75	0.8	π

PLANNING

Objectives and Strands
See pages 356A and 356B.

Spiral Learning
See page 356B.

Materials List
➤ Graphics calculator
➤ Graph paper
➤ Graphing software

Recommended Pacing
Section 7-4 is a two-day lesson.

Day 1
Pages 383–385: Talk it Over through Sample 2, *Exercises 1–7*

Day 2
Pages 386–387: Sample 3 through Look Back, *Exercises 8–43*

Extra Practice
See pages 631–633.

Warm-Up Exercises
Warm-Up Transparency 7-4

Support Materials
➤ Practice 55
➤ Enrichment 49 in the Activity Bank
➤ Study Guide 7-4
➤ Problem Set 15
➤ Diagram Master 2 in the Explorations Lab Manual
➤ McDougal Littell Mathpack software: *Function Investigator*
➤ Function Investigator with Matrix Analyzer Activity Book: Function Investigator Activity 3
➤ Using Plotter Plus: Direct Variation with $y = kx$
➤ Using Plotter Plus: Exploring Slopes
➤ Quiz 7-4
➤ Alternative Assessments 6–8

TEACHING

Talk it Over

Questions 1 and 2 relate the general form for direct variation, $y = kx$, to the three examples given in the table, namely $h = 0.75d$, $B = 0.8D$, and $C = \pi d$. Students need to understand that every direct variation has the form $y = kx$, regardless of the specific quantities, variables, or constants that are involved.

Additional Sample

S1 Efren set an empty fish tank on a bathroom scale. When he added 20 quarts of water, the weight increase was $41\frac{3}{4}$ lb. He plans to have 28 quarts of water in the tank when he sets the tank up for his fish. If the weight of the water varies directly with the number of quarts used, how much will the water in the tank weigh?

Method 1
Find the variation constant.
$$y = kx$$
$$41\frac{3}{4} = k \cdot 20$$
$$\frac{41.75}{20} = k$$
$$2.0875 = k$$

Use the variation constant to find the weight of 28 qt of water.
$$y = 2.0875(28)$$
$$y = 58.45$$
The water will weigh about 58 lb.

Method 2
Use a proportion.
$$\frac{41.75}{20} = \frac{w}{28}$$
$$28 \cdot \frac{41.75}{20} = 28 \cdot \frac{w}{28}$$
$$58.45 = w$$
The water will weigh about 58 lb.

Sample 1

The amount Cindy earns varies directly with the number of hours she works. She earns $336 in a 40-hour work week. How much will Cindy earn in 130 h?

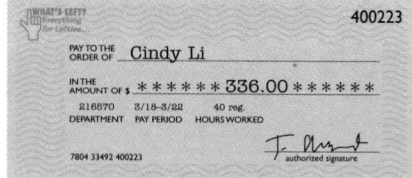

Sample Response

Method ❶

Model the situation with a direct variation equation in general form.

The amount earned varies directly with the number of hours worked.
Let $y =$ the amount earned and $x =$ the number of hours worked.

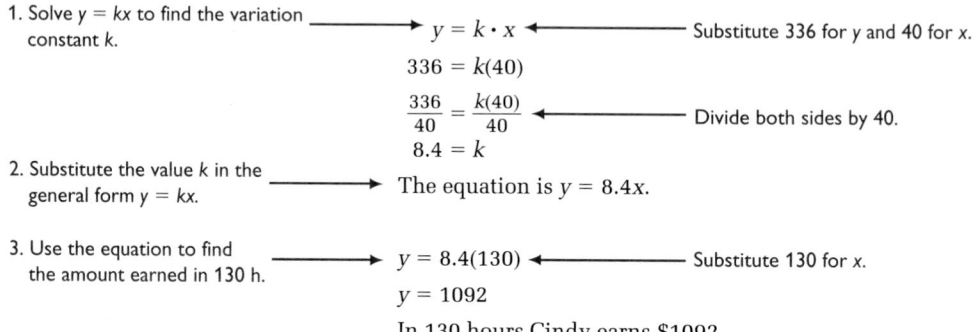

1. Solve $y = kx$ to find the variation constant k. ⟶ $y = k \cdot x$ ⟵ Substitute 336 for y and 40 for x.
$$336 = k(40)$$
$$\frac{336}{40} = \frac{k(40)}{40}$$ ⟵ Divide both sides by 40.
$$8.4 = k$$

2. Substitute the value k in the general form $y = kx$. ⟶ The equation is $y = 8.4x$.

3. Use the equation to find the amount earned in 130 h. ⟶ $y = 8.4(130)$ ⟵ Substitute 130 for x.
$$y = 1092$$
In 130 hours Cindy earns $1092.

Method ❷

Problem Solving Strategy: Use a proportion

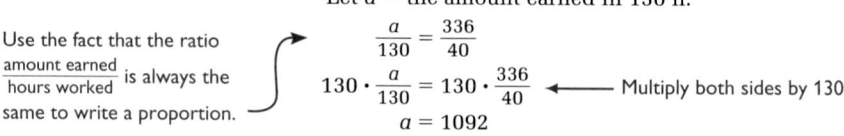

Use the fact that the ratio $\frac{\text{amount earned}}{\text{hours worked}}$ is always the same to write a proportion.

Let $a =$ the amount earned in 130 h.
$$\frac{a}{130} = \frac{336}{40}$$
$$130 \cdot \frac{a}{130} = 130 \cdot \frac{336}{40}$$ ⟵ Multiply both sides by 130.
$$a = 1092$$
In 130 h Cindy earns $1092.

Talk it Over

3. Suppose you want to find the amount Cindy earns in five different times, such as in 100 h, in 150 h, in 200 h, in 250 h and in 300 h. Which method in Sample 1 is easier to use? Explain your choice.

Answers to Talk it Over

3. Method 1; Explanations may vary. An example is given. Method 1 is easier because you just need to multiply each time by the variation constant, 8.4, to find the amount earned, rather than setting up a proportion.

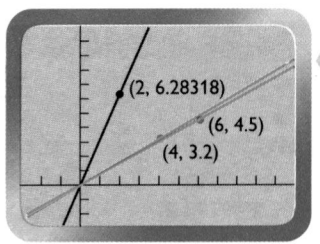
(2, 6.28318)
(6, 4.5)
(4, 3.2)

4. With a graphics calculator, Elia graphed the three equations shown in the table on page 383. She used TRACE to find coordinates of some points on the lines.

a. What is the *y*-intercept of each graph?

b. Find the slope of each line. (*Hint:* Use the *y*-intercept and the point Elia found with TRACE.)

c. Match each graph with its equation. What do you notice?

Direct variation graphs share some general characteristics.

DIRECT VARIATION GRAPHS

The graph of a direct variation equation $y = kx$ is a line that passes through the origin.

The variation constant k is the slope of the line.

| Sample 2 |

Is direct variation a good model for the data shown in each graph? Explain why or why not.

a. 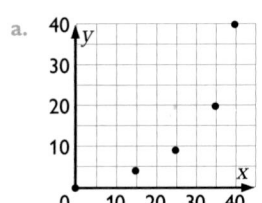   b. 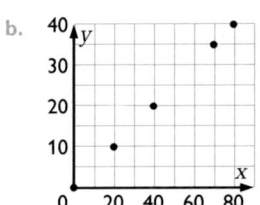   c.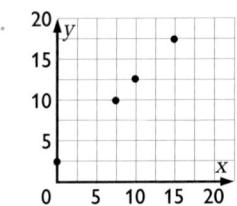

Sample Response

a. 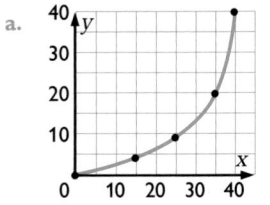   b. 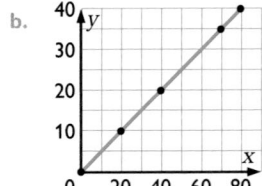   c.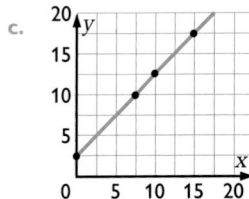

No; the data points fit on a curve instead of a straight line.

Yes; the data points fit on a line that goes through the origin.

No; the data points fit on a line, but the line does not go through the origin.

⋯⋯▶ Now you are ready for:
Exs. 1–7 on p. 388

7-4 Direct Variation with $y = kx$ **385**

Answers to Talk it Over ⋯⋯:

4. a. 0

b. 3.14159; 0.75; 0.8

c. Point (2, 6.28318) is on the graph of $C = \pi d$; point (6, 4.5) is on the graph of $h = 0.75d$; point (4, 3.2) is on the graph of $B = 0.8D$. The slope of the line is the same as the constant of variation in each case.

Using Technology

Talk it Over question 4 involves graphs done on a graphics calculator. If students try to graph the equations on a graphics calculator, the range settings must be just right or the TRACE cursor will not go exactly to the point for the *y*-intercept. You can simply have students note that the *y*-intercept must be 0. If you prefer, suggest range settings that will permit the cursor to go exactly to the point (0, 0).

Students can use Function Investigator Activity 3 in the *Function Investigator with Matrix Analyzer Activity Book* to explore how changing the value of *k* affects the graph of $y = kx$.

Additional Sample

S2 Is direct variation a good model for the data shown in each graph? Explain why or why not.

a.
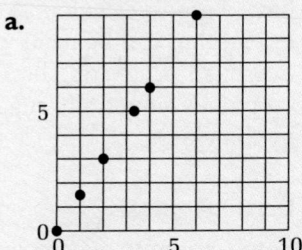

Yes. The data points fit on a line that goes through the origin.

b.
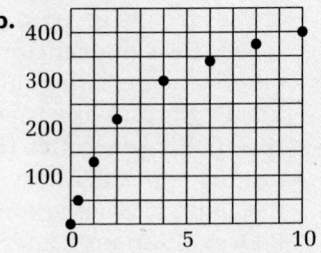

No. The data points fit on a curve instead of a straight line.

c.
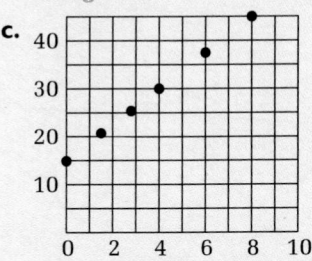

No. The data points fit on a line, but the line does not go through the origin.

385

S3 The direct variation graph shown is a line with slope 1.5. Write an equation for the graph.

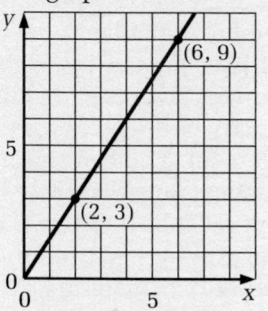

The variation constant k is the slope of the line, 1.5. Using 1.5 for k in $y = kx$ gives the equation $y = 1.5x$.

Using Technology

In conjunction with Talk it Over question 5, students who have graphics calculators or access to the *Function Investigator* software can use the calculators or software to graph other equations with negative values of k. If you have a graphics calculator model that can be used with an overhead projector, show the graphs for equations suggested by the class.

Error Analysis

Some students may erroneously think that in a direct variation $y = kx$, an increase or decrease in one of the variables always results in a corresponding increase or decrease in the other variable. Point out that this is true only if k is positive ($k > 0$). If k is negative ($k < 0$), just the opposite is true. Use the graph in Sample 4 and Talk it Over questions 8 and 9 to illustrate this point.

386

Sample 3

The direct variation graph shown is a line with slope 3. Write an equation of the line.

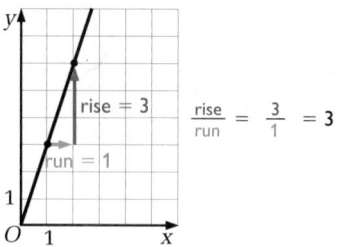

Sample Response

The equation of a direct variation graph is in the form $y = kx$.

The variation constant k is the slope of the line.

The slope of the line is 3, so the equation of the line is $y = 3x$.

Negative Slope

Every direct variation graph you have seen in this unit so far is a line with positive slope. If the value of k is negative, then the graph of $y = kx$ is a line with negative slope.

Talk it Over

5. With a graphics calculator, Elia investigated the graph of $y = kx$ for three negative values of k.

   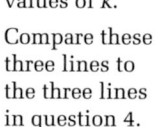
   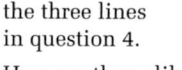

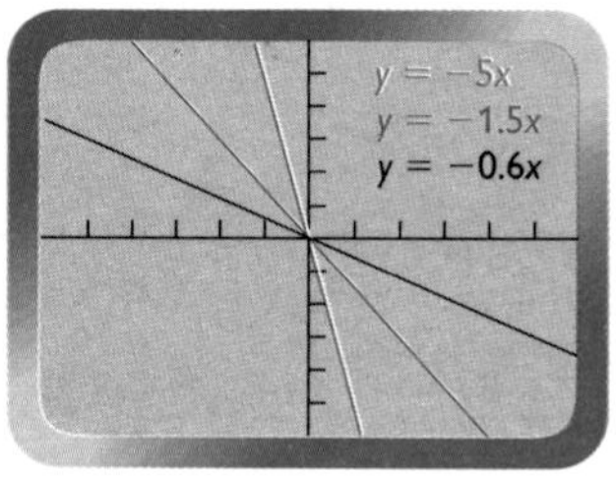

   Compare these three lines to the three lines in question 4.

   How are they alike?

   How are they different?

6. How do you think the sign of the slope affects the graph of a line?

7. Estimate the slope of the ground in the photograph of the biker.

Unit 7 Direct Variation

Answers to Talk it Over

5. Answers may vary. An example is given. The three lines are similar because they all have the general form for direct variation, $y = kx$, and all pass through the origin. They are different because the three lines in question 4 have positive slopes or variation constants, while these three have negative slopes or variation constants.

6. Lines with negative slope slant downward from left to right. Lines with positive slopes slant upward from left to right.

7. −1

Sample 4

Graph y = −2x.

Sample Response

$y = -2x$ is a direct variation equation in the form $y = kx$.

Make a table and plot points.

y = −2x	
x	**y**
−2	4
−1	2
0	0
1	−2
2	−4

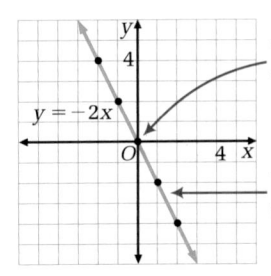

$y = -2x$

The graph is a line that goes through the origin.

The slope k is negative, so the graph falls from left to right.

Talk it Over

8. To check that the line $y = -2x$ has slope -2, two students find the ratio $\frac{\text{rise}}{\text{run}}$ between the points $(-1, 2)$ and $(2, -4)$.

   a. Explain why Regan says the run is positive and the rise is negative.

   b. Explain why Lianon says the run is negative and the rise is positive.

Regan moves from the upper point to the lower point.

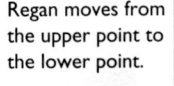
run +3
rise −6

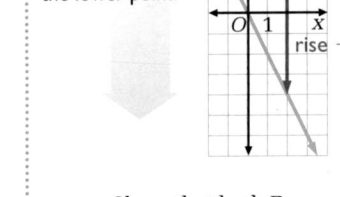
rise +6
run −3

Lianon moves from the lower point to the upper point.

   c. Show that both Regan and Lianon get -2 when they find the slope.

9. George looks at the table in Sample 4. He notices that as you go from row to row, the x-value increases by 1 and the y-value decreases by 2. He says that this also shows that the slope of the line is -2. Do you agree? Explain.

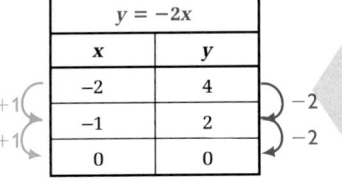

y = −2x	
x	**y**
−2	4
−1	2
0	0

+1 → −2
+1 → −2

Look Back

Does an equation of the form $y = kx$ always represent a function? What special names are given to y, k, and x? What does the value of k tell you about the graph?

> ▶ Now you are ready for:
> Exs. 8–43 on pp. 388–391

7-4 Direct Variation with $y = kx$

387

Additional Sample

S4 Graph $y = -\frac{2}{3}x$.

Make a table.

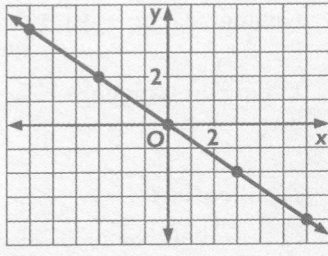

$y = -\frac{2}{3}x$	
x	**y**
−6	4
−3	2
0	0
3	−2
6	−4

Plot the points and draw the line.

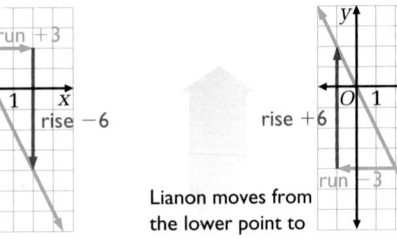

Using Technology

Students who have access to the *Function Investigator* software can use it to graph direct-variation functions for this section.

Look Back

The Look Back questions can be used for class discussion. Point out that the concept of direct variation is inseparable from the concept of a function. The general notion of dependency, as expressed by the equation $y = kx$ (y is dependent upon x), is an essential aspect of functionality. Students can summarize the results of the discussion in a journal entry. They should illustrate what they write with graphs of $y = kx$ for positive, negative, and zero values of k.

Answers to Talk it Over

8. a, b. Explanations may vary. Examples are given.

   a. To go from $(-1, 2)$ to $(2, -4)$, you move 3 units to the right (a run of 3) and 6 units down (a rise of -6).

   b. To go from $(2, -4)$ to $(-1, 2)$, you move 3 units to the left (a run of -3) and 6 units up (a rise of 6).

   c. Regan: $\frac{\text{rise}}{\text{run}} = \frac{-6}{3} = -2$;

   Lianon: $\frac{\text{rise}}{\text{run}} = \frac{6}{-3} = -2$

9. Yes. The rise has a change of -2, which is down, for every run change of 1 to the right.

Answers to Look Back

Yes. In $y = kx$, y is the dependent variable, k is the variation constant, and x is the control variable. A positive value of k means that the line slopes upward from left to right, a negative value of k that it slopes downward.

Suggested Assignment

Standard 1–13, 16–20, 22–31, 34–43

Extended 1–14, 16–43

Integrating the Strands

Number Exs. 41, 42

Algebra Exs. 1–15, 21–34, 37–40, 43

Functions Exs. 1–11, 14–34, 43

Geometry Exs. 8–10, 16–20, 22–32, 35, 36

Logic and Language Exs. 1, 34, 43

Application

For each of Exs. 5–7, ask students to say what the variation constant represents in the given situation.

Communication: Writing

After students have done Exs. 1–7, you may wish to have them work in pairs. Ask each student to write two problems involving direct variation situations. Students in each pair can trade problems, solve them, and discuss the results.

7-4 Exercises and Problems

1. **Reading** In Sample 1, what does the variation constant 8.4 represent in terms of the situation?

2. The cost of a soft drink varies directly with the number of ounces you buy. It costs 75 cents to buy a 12 oz bottle. How much does it cost to buy a 16 oz bottle?

3. The cost of gasoline varies directly with the number of gallons bought. It costs $10.29 to buy 7.4 gal of gasoline. How much does it cost to buy 9.3 gal of gasoline?

4. The mass of any substance varies directly with its volume. The mass of 5 cm³ of platinum is 107.5 g. Find the volume of 200 g of platinum.

Model each situation with a direct variation equation in general form.

5. Shelby can type 250 words in 4 minutes.
   a. How many words can Shelby type in half an hour?
   b. How long does it take Shelby to type 20,000 words?

6. A worker earns $490 in a 40-hour work week.
   a. How much does the worker earn after working 100 hours?
   b. How many hours does it take the worker to earn $5000?

7. An in-line skater covers 5 mi in 18 min.
   a. How many minutes does it take the skater to cover 16 mi?
   b. The skater skates for 10 min. Does the skater cover more than 2 mi? Explain.

Write an equation of the form y = kx to describe each line.

8.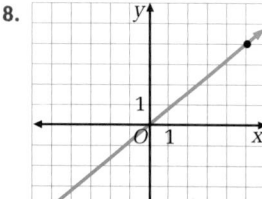

9.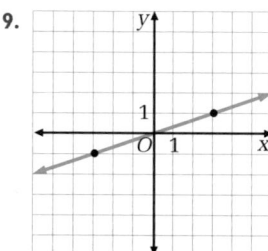

10.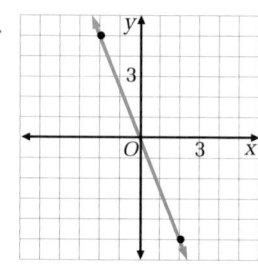

For Exercises 11–13:
  a. Does the graph show direct variation? Why or why not?
  b. If the graph shows direct variation, write an equation for the line.

11.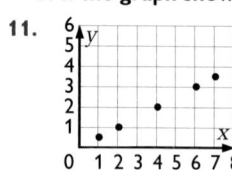

12.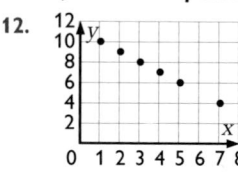

13.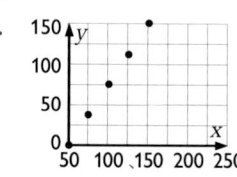

1. 8.4 represents the hourly rate of pay, $8.40.

2. $1.00

3. $12.93

4. about 9.3 cm³

5. a. 1875 words
   b. 320 min, or 5 h 20 min

6. a. $1225   b. about 408 h

7. a. 57.6 min
   b. Yes, the skater covers about 2.8 mi.

8. $y = \frac{4}{5}x$ 9. $y = \frac{1}{3}x$ 10. $y = \frac{-5}{2}x$

11. a. Yes. All the points lie close to a fitted line passing through the origin.
   b. $y = \frac{1}{2}x$

12. a. No. The line connecting the points does not pass through the origin.

13. a. No. The line connecting the points does not pass through the origin.

14. Answers may vary. Examples are given.
   (11) A ball bounces to one half of the drop height.
   (12) The amount of money Tim has left every time he spends $1 of his $11.

(13) The income the Art Club makes by buying $75 worth of pens and selling them for $1.50 each.

15. a.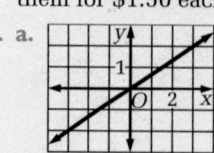
   b. about 7.5
   c. 7.5; The answer obtained by solving the equation is more accurate.

16.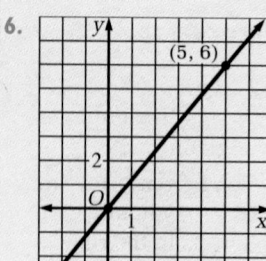

14. **Open-ended** Describe a situation that could be modeled by each of the graphs in Exercises 11–13.

15. TECHNOLOGY Use a graphics calculator or graphing software.

   a. Graph the line $y = \frac{2}{3}x$.

   b. Trace along your graph until y is about 5. What is x?

   c. Solve the equation $5 = \frac{2}{3}x$. How does your answer compare with part (b)? Which is more accurate?

The rise and run from (0, 0) to another point are given. Plot the other point using the directions below. Then draw the line.

16. rise = 6, run = 5

17. rise = −8, run = 3

18. rise = 5, run = −2

19. rise = −4, run = −8

20. a. For each rise and run in Exercises 16–19, find the slope of the line.

   b. **Writing** Look for a pattern in the signs of your results for part (a). Describe how the signs of the rise and run determine whether the slope is positive or negative.

21. **Writing** Describe the graphs of $y = x$ and $y = -x$. How are they alike? How are they different?

Match each equation with its graph.

22. $y = 0.5x$

23. $y = -0.5x$

24. $y = 2.5x$

25. $y = -2.5x$

A.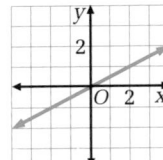

B.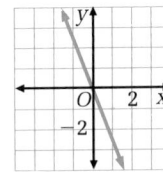

C.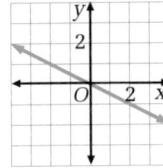

D.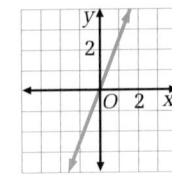

Match each graph with its equation and explain how you arrived at your answer.

26. $y = 6x$

   $y = \frac{1}{6}x$

A.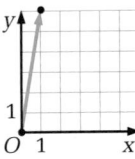

B.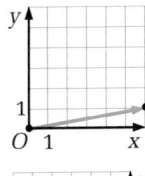

27. $y = -\frac{3}{5}x$

   $y = \frac{5}{3}x$

A.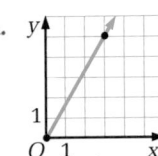

B.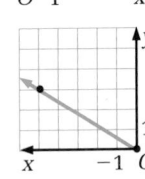

7-4 Direct Variation with $y = kx$ 389

Mathematical Procedures

Ex. 21 hints strongly at the relationship between perpendicularity and slope: Two lines (neither of which is horizontal) are perpendicular if and only if the slope of one is the opposite of the reciprocal of the other.

If you wish to explore this idea with students, have them draw graphs of at least two pairs of equations, each pair on its own set of axes. Use $y = \frac{2}{3}x$, $y = -\frac{3}{2}x$ and $y = \frac{4}{5}x$, $y = -\frac{5}{4}x$.

Using Technology

The graphs suggested in the note above can be done with a graphics calculator or with the *Function Investigator* software. Graphing the lines with a graphics calculator or software speeds up the graphing so that you can concentrate less on the graphing and more on the relationship between the graphs. *It is crucial to use the same scale on both axes.* The quickest way to do this on a graphics calculator is to go to the zoom menu and select Square (TI-81) or ZSquare (TI-82).

20. b. If the signs of the rise and the run are the same, then the slope of the line is positive. If the signs are different, then the slope is negative.

21. Answers may vary. An example is given. Both graphs are lines showing direct variation. Both pass through the origin. The graph of $y = x$ has a positive slope and the graph of $y = -x$ has a negative slope. The graphs are reflections of each other across the y-axis.

22. A   23. C   24. D   25. B

26, 27. Explanations may vary. Examples are given.

26. A: $y = 6x$; B: $y = \frac{1}{6}x$; A rise of 6 to a run of 1 is a very steep line.

27. A: $y = \frac{5}{3}x$; B: $y = -\frac{3}{5}x$; Graph A shows positive slope and Graph B shows negative slope.

389

Answers to Exercises and Problems

17.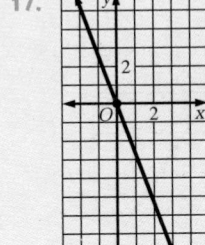

18.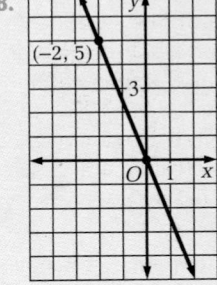

19.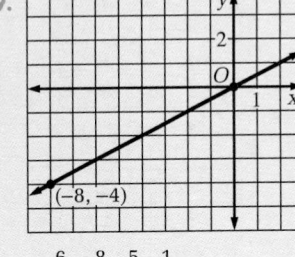

20. a. $\frac{6}{5}, \frac{-8}{3}, \frac{5}{-2}, \frac{1}{2}$

Ex. 34 gives students the opportunity to pull together what they know about the direct variation equation $y = kx$. Answers to this exercise would make an excellent entry into students' portfolios.

Answers to
Exercises and Problems

28.

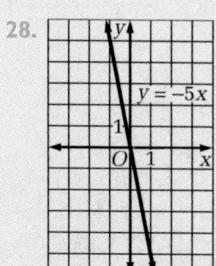

29.

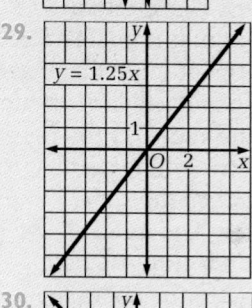

30.

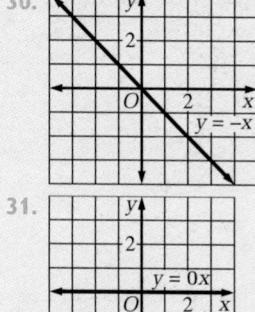

31.

32. Answers may vary based on choice of k. Answers in part (a) use $k = 4$.

a.

$y = x + 4$	
x	y
−6	−2
−4	0
−2	2
0	4
2	6

$y = x - 4$	
x	y
−4	−8
−2	−6
0	−4
2	−2
4	0

$y = \dfrac{4}{x}$	
x	y
−8	−0.5
−4	−1
−2	−2
−1	−4
1	4
2	2
4	1
8	0.5

b. See answers in back of book.

390

Graph each equation.

28. $y = -5x$ **29.** $y = 1.25x$ **30.** $y = -x$ **31.** $y = 0x$

32. In a direct variation equation of the form $y = kx$, the constant k is *multiplied* by the variable x. Now you will see what happens when k is added to x, subtracted from x, or divided by x.

  a. Choose a nonzero value of k. Create tables for the equations
  $y = x + k$, $y = x - k$, and $y = \dfrac{k}{x}$, using your value of k.
  List at least five ordered pairs in your tables.

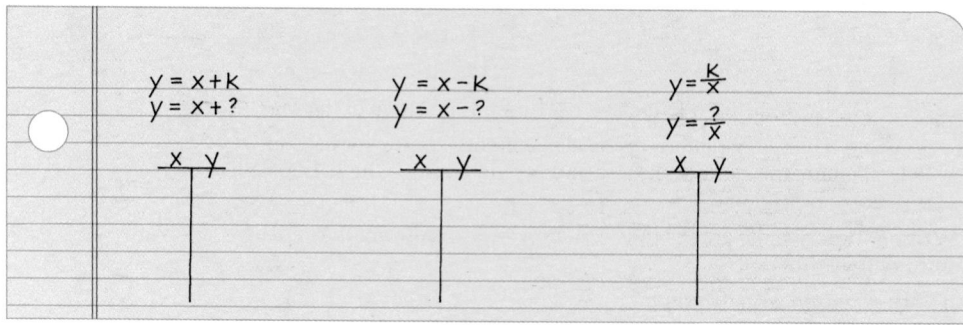

  b. Use your tables to graph the equations.

  c. Describe the graphs in words.

  d. Do any of the graphs represent direct variation? Explain.

33. On a cold day, a home furnace breaks down. For the first few hours, the drop in temperature varies directly with the time elapsed.

  a. Graph the data.

Time elapsed (h)	Temperature change (°F)
0.5	−1.5
0.75	−2.25
1	−3
1.5	−4.5

  b. Find the variation constant.

  c. Write an equation for the graph.

Ongoing ASSESSMENT

34. **Writing** Explain how changing the value of k affects the appearance of the graph of $y = kx$. Include both positive and negative values of k. Use diagrams to support your writing. Explain why the point $(0, 0)$ is found on all the graphs of equations of the form $y = kx$.

---

c. The first graph is a line with positive slope that does not go through the origin. The second graph is a line parallel to the first line but 8 units below it. The third graph is two curved lines (called a hyperbola) that do not cross either axis.

d. No. Explanations may vary. An example is given. In each table, the ratios $\dfrac{y}{x}$ vary, and none of the graphs goes through the origin.

33. a.

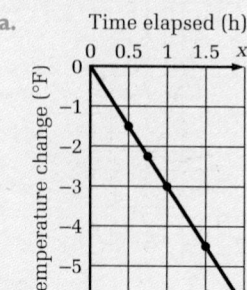

Find the circumference of each circle. *(Section 7-3)*

35. diameter = 14 in.

36. radius = 11 cm

Simplify. *(Section 5-2)*

37. $-2(3x + 2y)$

38. $-4(5a - 2b)$

39. $-7(-c + 5d)$

40. $-(-m - n)$

Multiply. *(Toolbox Skill 10)*

41. $\dfrac{2}{3} \times \dfrac{4}{5} \times \dfrac{1}{4} \times \dfrac{9}{10}$

42. $\dfrac{50}{63} \times \dfrac{7}{12} \times \dfrac{9}{11} \times \dfrac{33}{5}$

 Working on the Unit Project

43. The frictional force on an in-line skater's wheels varies directly with the skater's weight.

   a. A 150 lb skater experiences 40.5 lb of friction when skating on a dry steel surface. Find the variation constant.

   b. Write a direct variation equation for this situation.

   c. Find the frictional force on a 130 lb skater.

   d. The variation constant for skating on a wet steel surface is 0.23. Is this greater or less than the variation constant you found in part (a)? Does this make sense?

   e. What kind of surfaces should you consider in your plans for a skateboard ramp?

7-4 Direct Variation with $y = kx$

391

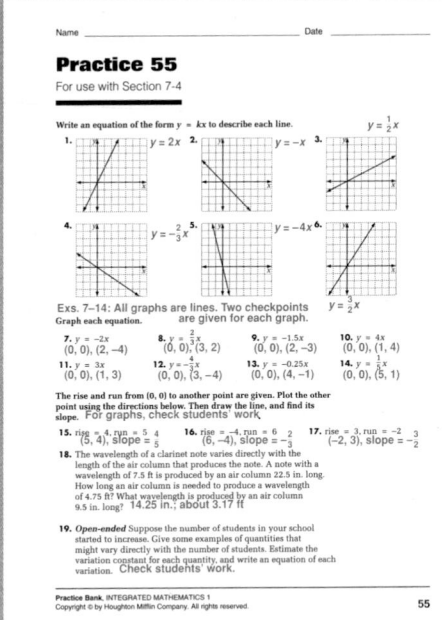

Working on the Unit Project

Ex. 43 allows students to apply their knowledge of direct variation and variation constants to skateboarding situations involving friction. This is the first time they are asked to think about the type of surface they would use for the ramp in their design project.

Practice 55 For use with Section 7-4

Name _____ Date _____

Practice 55
For use with Section 7-4

Write an equation of the form $y = kx$ to describe each line.

1. $y = 2x$ **2.** $y = -x$ **3.** $y = \frac{1}{2}x$

4. $y = -\frac{2}{3}x$ **5.** $y = -4x$ **6.** $y = \frac{3}{2}x$

Exs. 7–14: All graphs are lines. Two checkpoints are given for each graph. Graph each equation.

7. $y = -2x$ (0, 0), (2, −4)
8. $y = \frac{2}{3}x$ (0, 0), (3, 2)
9. $y = -1.5x$ (0, 0), (2, −3)
10. $y = 4x$ (0, 0), (1, 4)
11. $y = 3x$ (0, 0), (1, 3)
12. $y = -\frac{4}{3}x$ (0, 0), (3, −4)
13. $y = -0.25x$ (0, 0), (4, −1)
14. $y = \frac{1}{5}x$ (0, 0), (5, 1)

The rise and run from (0, 0) to another point are given. Plot the other point using the directions below. Then draw the line, and find its slope. For graphs, check students' work.

15. rise = 4, run = 5   (5, 4), slope = $\frac{4}{5}$
16. rise = −4, run = 6   (6, −4), slope = $-\frac{2}{3}$
17. rise = 3, run = −2   (−2, 3), slope = $-\frac{3}{2}$

18. The wavelength of a clarinet note varies directly with the length of the air column that produces the note. A note with a wavelength of 7.5 ft is produced by an air column 22.5 in. long. How long an air column is needed to produce a wavelength of 4.75 ft? What wavelength is produced by an air column 9.5 in. long? 14.25 in.; about 3.17 ft

19. *Open-ended* Suppose the number of students in your school started to increase. Give some examples of quantities that might vary directly with the number of students. Estimate the variation constant for each quantity, and write an equation of each variation. Check students' work.

Answers to Exercises and Problems

33. b. −3

  c. Choices of variables may vary. An example is given. $y = -3x$

34. Answers may vary. An example is given. The graph of $y = kx$ is a line with slope k that passes through the origin. When $k > 0$, the slope is positive, so the line slopes up from left to right. When $k < 0$, the slope is negative, so the line slopes down from left to right. When $k = 0$, the line is horizontal (the x-axis). If the value of k is close to 0, say $-1 < k < 0$ or $0 < k < 1$, then the line is not very steep. If the value of k is not close to 0, say $k \le -1$ or $k \ge 1$, then the line is steep.

The point (0, 0) is found on all graphs of equations of the form $y = kx$, because if you substitute 0 for x in the equation, you get $y = kx = k(0) = 0$.

Diagrams may vary.

35. about 44.0 in.

36. about 69.1 cm

37. $-6x - 4y$

38. $-20a + 8b$

39. $7c - 35d$

40. $m + n$

41. $\dfrac{3}{25}$

42. $2\dfrac{1}{2}$

43. a. 0.27   **b.** $f = 0.27w$

  c. 35.1 lb

  d. less than; Yes. A wet surface is more slippery than a dry surface. The wetness reduces the resistance or friction.

  e. Answers may vary.

PLANNING

Objectives and Strands
See pages 356A and 356B.

Spiral Learning
See page 356B.

Materials List
➤ Inch ruler
➤ Metric ruler

Recommended Pacing
Section 7-5 is a two-day lesson.
Day 1
Pages 392–394: Opening paragraph through Talk it Over, *Exercises 1–8*
Day 2
Pages 394–396: Converting the Units of a Rate through Look Back, *Exercises 9–37*

Extra Practice
See pages 631–633.

Warm-Up Exercises
💡 Warm-Up Transparency 7-5

Support Materials
➤ Practice 56
➤ Enrichment 50 in the Activity Bank
➤ Study Guide 7-5
➤ Problem Set 15
➤ Quiz 7-5
➤ Alternative Assessment 9

Section 7-5

Using Dimensional Analysis

Focus
Use dimensional analysis to work with the units in rates and conversion factors.

THE GOING RATE

At a gas station, self-serve unleaded gasoline sells for $1.189 per gallon. Here is what the gasoline pump might look like at various times as you fill your car's tank.

TOTAL SALE	TOTAL SALE	TOTAL SALE	TOTAL SALE
000.00	**003.57**	**005.95**	**009.93**
GALLONS	GALLONS	GALLONS	GALLONS
00.000	**03.000**	**05.000**	**08.350**
01.189	**01.189**	**01.189**	**01.189**
$Price Per Gallon	$Price Per Gallon	$Price Per Gallon	$Price Per Gallon

Here is how to calculate the amount you owe:

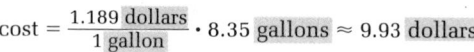

$$\text{cost} = \frac{1.189 \text{ dollars}}{1 \text{ gallon}} \cdot 8.35 \text{ gallons} \approx 9.93 \text{ dollars}$$

Notice that the gallons units cancel out. The result is measured in dollars. When you cancel units of measurement as if they are numbers, you are using a problem solving strategy called **dimensional analysis.**

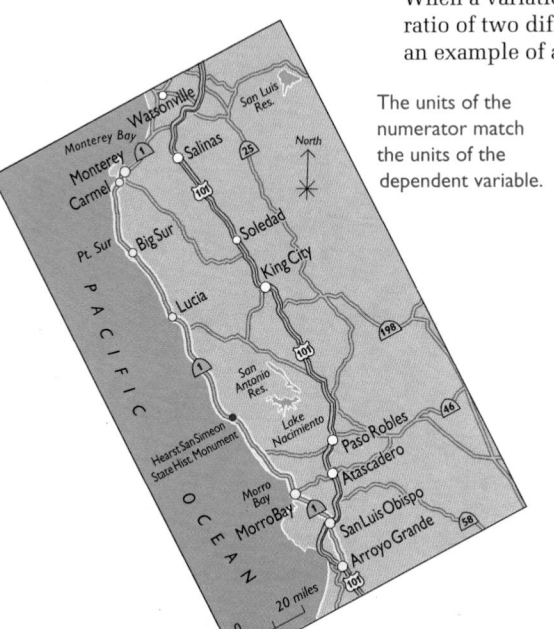

Talk it Over

1. Estimate to check that the dollar amounts shown for 3.00 gal and 5.00 gal are correct. Explain your method.

2. The amounts shown on the gas pump fit a direct variation model. What is the control variable? What is the dependent variable? What is the variation constant?

When a variation constant is a ratio of two different units it is an example of a rate.

$$\frac{1.189 \text{ dollars}}{1 \text{ gallon}}$$

$$\text{cost} = 1.189 \cdot \text{amount}$$

The units of the numerator match the units of the dependent variable.

$$(\text{dollars}) = \left(\frac{\text{dollars}}{\text{gallon}}\right) \cdot (\text{gallons})$$

The units of the denominator match the units of the control variable.

Sample 1

The distance Milo travels varies directly with the numbers of hours he drives. Milo drives 145 mi in 3 h.

a. Identify the control variable and the dependent variable.

b. Express the variation constant as a rate.

Sample Response

a. The control variable is the time traveled (h).
   The dependent variable is the distance traveled (mi).

b. The variation constant is $\frac{145 \text{ mi}}{3 \text{ h}} \approx 48.3$ mi/h.

Conversion Factors

A **conversion factor** is a ratio of two equal (or approximately equal) quantities that are measured in different units. For example, the ratio $\frac{2.54 \text{ cm}}{1 \text{ in.}}$ is a conversion factor, because 2.54 cm = 1 in. You use a conversion factor to convert a measurement from one unit to another.

7-5 Using Dimensional Analysis

Answers to Talk it Over

1. Estimates may vary. An example is given.
   3 × 1.2 = 3.6, which is about equal to 3.57;
   5 × 1.2 = 6, which is about equal to 5.95.

2. number of gallons of gasoline; the cost of the gasoline; 1.189

S2 Pedro Zepeda weighs 70.5 kg.

a. How many pounds are in 70.5 kg? (The conversion factor is $\frac{2.205 \text{ lb}}{1 \text{ kg}}$.)

Multiply 70.5 kg by the conversion factor.

$70.5 \text{ kg} \times \frac{2.205 \text{ lb}}{1 \text{ kg}}$

$= 155.4525 \text{ lb}$

70.5 kg is about 155.5 lb.

b. Write a direct variation equation for converting kilograms to pounds. What is the variation constant?

Let x = the number of kilograms. Let y = the number of pounds. The equation is $y = 2.205x$. The variation constant is the conversion factor $\frac{2.205 \text{ lb}}{1 \text{ kg}}$, or $2.205 \frac{\text{lb}}{\text{kg}}$.

Reasoning

After discussing Talk it Over questions 3–5, have students think about and explain how to go the other way with conversion factors. For example, to convert inches to centimeters, we multiply the number of inches by the conversion factor $2.54 \frac{\text{cm}}{\text{in.}}$. What conversion factor can we use to convert centimeters to inches? (Find the reciprocal of 2.54 and "invert" the unit expressed in the original conversion factor. The conversion factor is about $0.3937 \frac{\text{in.}}{\text{cm}}$.)

Sample 2

a. How many centimeters are in 3 in.?

b. Write a direct variation equation that helps you convert inches to centimeters. What is the variation constant?

Sample Response

a. Method ❶ Use two rulers, one marked in inches, the other marked in centimeters. Place them so their zero marks line up.

The 3 in. mark lines up with about 7.6 cm, so 3 in. ≈ 7.6 cm.

Method ❷ Use the conversion factor $\frac{2.54 \text{ cm}}{1 \text{ in.}}$.

Multiply the original amount by the conversion factor. After the inches cancel, only centimeters are left.

$3 \text{ in.} \times \frac{2.54 \text{ cm}}{1 \text{ in.}} = 7.62 \text{ cm}$

Using the conversion factor gives a more accurate answer than the estimation from rulers in Method 1.

b. ➤ Let x = the number of inches.

➤ Let y = the number of centimeters.

The equation $y = 2.54x$ helps you convert inches to centimeters. The variation constant is the conversion factor $\frac{2.54 \text{ cm}}{1 \text{ in.}}$, or 2.54 cm/in.

Talk it Over

➤ Now you are ready for:
Exs. 1–8 on p. 396

3. Do you change the value of an expression when you multiply by a conversion factor? Why or why not?

4. What conversion factor would you use to convert centimeters to inches? Convert 12 cm to inches.

5. Another well-known conversion factor is $\frac{365 \text{ days}}{1 \text{ year}}$. What other commonly-used conversion factors can you think of?

Converting the Units of a Rate

You can use conversion factors to convert the units of a rate.

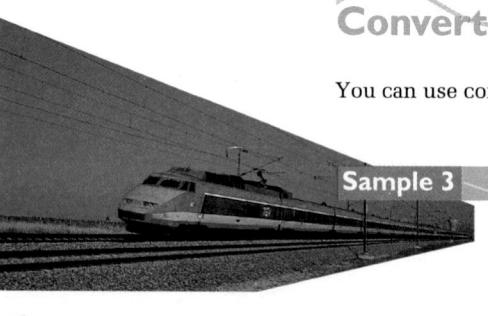

Sample 3

In 1990 the French Train à Grande Vitesse (TGV) set a speed record of 515 km/h on the Atlantique line. Find its speed in miles per hour.

Answers to Talk it Over

3. No. Reasons may vary. An example is given. Because the amounts in a conversion factor are equal, the ratio represents the number 1. So when you multiply by a conversion factor, you are changing the units of the expression but not the value.

4. $\frac{1 \text{ in.}}{2.54 \text{ cm}}$; about 4.7 in.

5. Answers may vary. Examples are given.
$\frac{60 \text{ s}}{1 \text{ h}}, \frac{7 \text{ days}}{1 \text{ week}}, \frac{3 \text{ ft}}{1 \text{ yd}},$
$\frac{100 \text{ cm}}{1 \text{ m}}, \frac{36 \text{ in.}}{1 \text{ yd}}$

Sample Response

Problem Solving Strategy: Use dimensional analysis

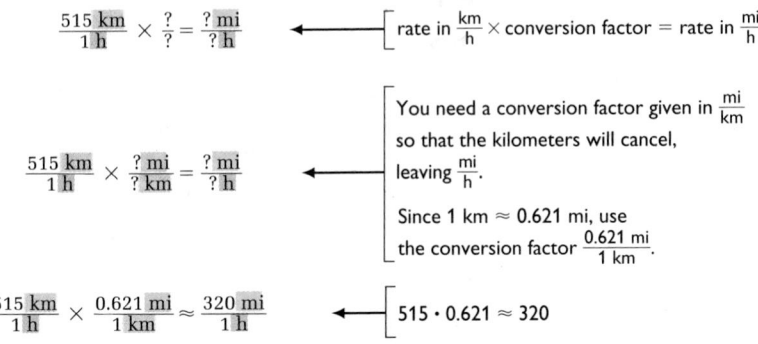

$$\frac{515 \text{ km}}{1 \text{ h}} \times \frac{?}{?} = \frac{? \text{ mi}}{? \text{ h}} \longleftarrow \left[\text{ rate in } \frac{\text{km}}{\text{h}} \times \text{conversion factor} = \text{rate in } \frac{\text{mi}}{\text{h}} \right.$$

$$\frac{515 \text{ km}}{1 \text{ h}} \times \frac{? \text{ mi}}{? \text{ km}} = \frac{? \text{ mi}}{? \text{ h}} \longleftarrow \left[\begin{array}{l} \text{You need a conversion factor given in } \frac{\text{mi}}{\text{km}} \\ \text{so that the kilometers will cancel,} \\ \text{leaving } \frac{\text{mi}}{\text{h}}. \\ \text{Since 1 km} \approx 0.621 \text{ mi, use} \\ \text{the conversion factor } \frac{0.621 \text{ mi}}{1 \text{ km}}. \end{array} \right.$$

$$\frac{515 \text{ km}}{1 \text{ h}} \times \frac{0.621 \text{ mi}}{1 \text{ km}} \approx \frac{320 \text{ mi}}{1 \text{ h}} \longleftarrow \left[515 \cdot 0.621 \approx 320 \right.$$

The TGV's speed was about 320 mi/h.

Sometimes you need many conversion factors to convert a rate to different units.

Sample 4

Convert 30 mi/h to ft/s.

Sample Response

Problem Solving Strategy: Use dimensional analysis

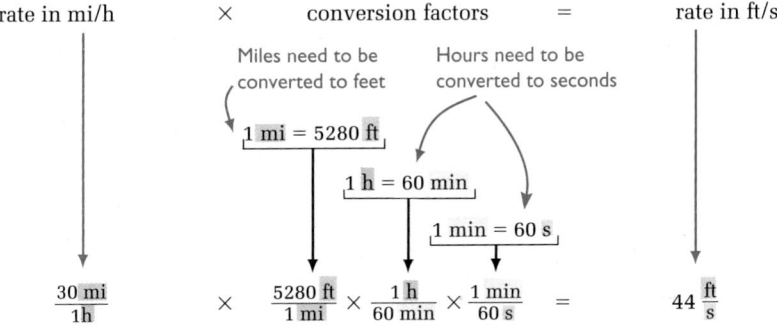

rate in mi/h　　×　　conversion factors　　=　　rate in ft/s

Miles need to be converted to feet　　Hours need to be converted to seconds

1 mi = 5280 ft

1 h = 60 min

1 min = 60 s

$$\frac{30 \text{ mi}}{1 \text{ h}} \times \frac{5280 \text{ ft}}{1 \text{ mi}} \times \frac{1 \text{ h}}{60 \text{ min}} \times \frac{1 \text{ min}}{60 \text{ s}} = 44 \frac{\text{ft}}{\text{s}}$$

The rate 30 mi/h converts to the rate 44 ft/s.

Additional Samples

S3 A large crude oil pipeline from Canada to the United States has a flow of 8.3 million gallons per day. How many liters per day does the pipeline handle?

The needed conversion factor must be $\frac{\text{L}}{\text{gal}}$. Use $\frac{3.78531 \text{ L}}{1 \text{ gal}}$.

$$\frac{(8.3 \times 10^6) \text{ gal}}{1 \text{ day}} \times \frac{3.78531 \text{ L}}{1 \text{ gal}} = \frac{(3.1418073 \times 10^7) \text{ L}}{\text{day}}$$

The pipeline handles about 3.142×10^7, or 31.42 million liters of crude oil per day.

S4 One of the fastest passenger elevators in the world is in the "Sunshine 60" building in Tokyo, Japan. It operates at 2,000 ft/min. Find its speed in miles per hour.

There are 60 min per h and 1 mi per 5280 ft.

$$\frac{2000 \text{ ft}}{1 \text{ min}} \times \frac{1 \text{ mi}}{5280 \text{ ft}} \times \frac{60 \text{ min}}{1 \text{ h}} \approx \frac{22.73 \text{ mi}}{1 \text{ h}}, \text{ or about } 22.73 \text{ mi/h.}$$

Look Back

Have a class discussion of the Look Back questions to be sure students understand how ratios, rates, and conversion factors are related to one another.

APPLYING

Suggested Assignment

Standard 2–14a, 15–17, 20–23, 27–37

Extended 1–14a, 15–37

Integrating the Strands

Algebra Exs. 1–35

Functions Exs. 6–8, 15–17, 26, 28, 29

Measurement Exs. 36, 37

Geometry Exs. 33–35

Application

When discussing the application problems in Exs. 6–8, ask students to identify the functional relationships that exist. For example, in Ex. 6, the cost paid for gasoline is a function of the price of one gallon.

Talk it Over

6. In Sample 3, suppose you do not know that 1 km ≈ 0.621 mi. Instead you know that 1 mi ≈ 1.61 km. Explain how can you use the conversion factor $\frac{1 \text{ mi}}{1.61 \text{ km}}$ instead of the conversion factor $\frac{0.621 \text{ mi}}{1 \text{ km}}$.
7. Sam drives at a speed of 30 mi/h.
   a. Write a direct variation equation that shows the relationship between the number of seconds Sam drives and the number of feet Sam drives. (*Hint:* Use the results of Sample 4.)
   b. How many feet does Sam's car travel in 30 s?

Look Back

In what way is a rate a special kind of ratio? In what way is a conversion factor a special kind of rate?

Now you are ready for: Exs. 9–37 on pp. 396–399

7-5 Exercises and Problems

1. **Reading** What rates were used in the Samples of this section? What other rates can you think of?

For Exercises 2–5, each expression shows the units of a conversion problem. Use dimensional analysis to find the unit(s) of the answer.

2. $\frac{\text{yen}}{\text{dollars}} \times \text{dollars} = \underline{?}$

3. $\frac{\text{h}}{\text{min}} \times \frac{\text{min}}{\text{s}} = \underline{?}$

4. $\frac{\text{cups}}{\text{pt}} \times \frac{\text{pt}}{\text{qt}} = \underline{?}$

5. $\frac{\text{mi}}{\text{h}} \times \frac{\text{h}}{\text{min}} \times \frac{\text{ft}}{\text{mi}} = \underline{?}$

For Exercises 6–8:

a. Identify the control variable and the dependent variable.

b. Express the variation constant as a rate.

6. The cost paid for gasoline varies directly with the amount bought. Marshall pays \$9.52 for 7 gal of gasoline.
7. The number of gallons of gasoline used varies directly with the distance traveled in miles. Mina's car uses 20 gal of gasoline when it travels 440 mi.
8. The cost of carpet varies directly with its area in square yards. Weston Carpets sells 9 yd² of a certain carpet for \$324.
9. How many kilometers are in 50 mi? (1 mi ≈ 1.6 km)
10. How many gallons of juice are in a 2 L bottle? (1 gal ≈ 3.8 L)
11. How many feet are in 7.2 mi? (1 mi = 5280 ft)

Answers to Talk it Over

6. Answers may vary. An example is given. The conversion factor $\frac{1 \text{ mi}}{1.61 \text{ km}}$ is the same as the conversion factor $\frac{0.621 \text{ mi}}{1 \text{ km}}$ because $\frac{1}{1.61} \approx \frac{0.621}{1}$.
7. a. $d = 44t$
   b. 1320 ft

Answers to Look Back

A rate is a special kind of ratio because a rate is a ratio that uses two different units of measurement. A conversion factor is a special kind of rate because a conversion factor is the ratio of two equal quantities measured in different units.

Answers to Exercises and Problems

1. 48.3 mi/h; 2.54 cm/in.; 515 km/h; 0.621 mi/km; 320 mi/h; 30 mi/h; 5280 ft/mi; $\frac{1 \text{ h}}{60 \text{ min}}$; $\frac{1 \text{ min}}{60 \text{ s}}$; 44 ft/s
2. yen
3. $\frac{\text{h}}{\text{s}}$
4. $\frac{\text{cups}}{\text{qt}}$
5. $\frac{\text{ft}}{\text{min}}$
6. a. number of gallons, cost of gasoline
   b. \$1.36/gal
7. a. distance traveled, number of gallons
   b. 22 mi/gal
8. a. area of carpet, cost of carpet
   b. \$36/yd²
9. about 80 km
10. about 0.5 gal
11. 38,016 ft
12. \$9428

12. On one day, one Japanese yen was equal to 0.009428 United States dollars. How many United States dollars equal 1,000,000 yen?

13. Before leaving for a vacation to Mexico, the Nguyen family exchanged $1250 for Mexican pesos. Each dollar was worth 3.10 pesos. How many pesos did they get?

14. Centavo coins are used in Mexico in amounts of 5, 10, 20, and 50 centavos. (100 centavos = 1 peso)

  a. How many centavos are in 53 pesos?

  b. **Research** What do you think the word *centavos* means? Look up the word in a Spanish-English dictionary. Why do you think it was chosen?

15. On their vacation, the Nguyens fill the tank of their rental car with gasoline *sin plomo*. They buy 32 L for 41.6 pesos.

  a. Find the cost per liter of the gasoline they bought.

  b. Write a direct variation equation that shows the relationship between the cost of gasoline and the amount bought.

  c. How much would it cost to buy 25 L of gasoline?

16. You can buy 9.8 gal of super unleaded gasoline for $13.23. You can buy 12 gal of regular unleaded at the same station for $12.60.

  a. What is the cost per gallon of super unleaded?

  b. Write a direct variation equation that shows the relationship between cost of super unleaded gasoline and number of gallons bought.

  c. Find the cost of 15 gal of super unleaded gasoline.

  d. How much will it cost to buy 10 gal of regular unleaded?

  e. Suppose you have $20 to spend on gasoline. How much super unleaded can you buy for this amount? How much regular unleaded?

17. In the graphs, one line represents super unleaded gasoline, and the other represents regular unleaded gasoline.

  a. Does the green line represent super unleaded or regular unleaded?

  b. Explain the comparison that the graph at the left shows.

  c. Explain the comparison that the graph at the right shows.

  d. What information does the length of the gray segment in each graph tell you?

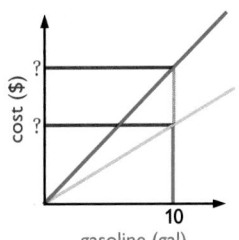
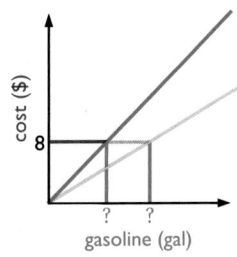

7-5 Using Dimensional Analysis

397

397

Interdisciplinary Problems

From ancient times to today, astronomers and mathematicians have wondered about the origins of Earth, the stars, galaxies, and the universe itself. On page 380, Exs. 17–19, students solved problems about Earth and its rotation around the sun. In Ex. 24 on this page, they learn how far the nearest star, Alpha Centauri, is from Earth. On page 409, Ex. 5 involves the diameter and circumference of the sun, and Ex. 14 on page 410 talks about the planet Neptune.

Some students may wish to explore the field of astronomy further by researching topics of interest to themselves. They can organized their work in a written report, with drawings, questions, and answers based upon what they have learned in this unit.

Problem Solving

Ex. 26 provides an excellent opportunity to review some of the ideas from Unit 3 about misleading graphs. Have students think about and describe how they would change the graphs in Ex. 26 to get a fair comparison of the slopes of the lines.

18. Exchange rates on one day are shown at the right. Find out how many yen equaled one mark on that day. Use conversion factors.

COUNTRY	Foreign currency in U.S. dollars	U.S. dollar in foreign currency
Germany (Mark)	0.6083	1.6440
Japan (Yen)	0.009428	106.07

19. Regular unleaded gasoline costs 115 yen per liter in Japan. Find its cost in dollars per gallon. Assume that 106 yen equal one dollar and use the fact that there are 3.8 liters in one gallon.

20. Convert 4000 s to hours.

21. Convert 32,500 in. to miles.

22. How long is one million seconds? When will you be one billion seconds old?

23. Suppose you are driving at 76 km/h. How many miles per hour is this?

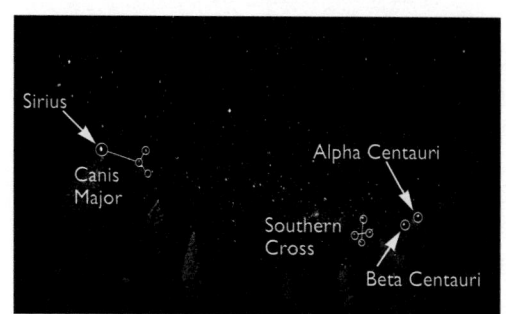

24. Aside from our own sun, the nearest star to Earth is Alpha Centauri. It takes light 4.3 years to reach Earth from that star. The speed of light is about 186,000 mi/s. Find the distance, in miles, from Earth to Alpha Centauri. Give your answer in scientific notation.

Alpha Centauri and the Southern Cross are visible in this photograph taken in Queenstown, New Zealand.

25. In a game against the Detroit Tigers, Nolan Ryan threw a pitch at a speed of 148 ft/s. Find the speed in miles per hour.

26. The scale you use for the vertical axis of a graph affects how steep a line looks. Here are three graphs of the equation $c = 1.5a$. Each has a different vertical scale.

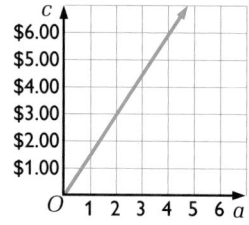 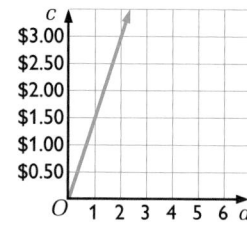 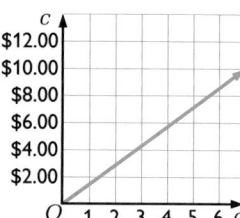

a. Which line *looks as if* it has the steepest slope?

b. Which line *looks as if* it has the shallowest slope?

c. Are the variation constants the same or different for the three graphs?

d. Are the slopes the same or different for the three lines?

e. **Writing** Describe how the scale used for the vertical axis affects the way a graph of a line looks.

Answers to Exercises and Problems

18. about 64.5 yen

19. about $4.12/gal

20. $1.\overline{1}$ h, or $1\frac{1}{9}$ h

21. about 0.513 mi

22. Answers may vary. Examples are given. One million seconds is $277\frac{7}{9}$ h, or about 11.57 days. I will be 1 billion seconds old after I am 31 years old and before I am 32 years old.

23. about 47.2 mi/h

24. about 2.5×10^{13}

25. about 100.9 mi/h

26. a. the line in the middle graph

b. the line in the graph at the far right

c. the same

d. the same

e. When the distance between the units on the vertical axis is the same as the distance between the units on the horizontal axis, then the graph shows the true steepness of the line. Using larger or smaller numbers on the vertical axis makes the graph appear less or more steep.

Ongoing ASSESSMENT

27. **Open-ended** Use the table at the right.

   a. Find the exchange rates from United States dollars to the money of three countries of your choice.

   b. List three consumer items whose price in United States dollars you know.

   c. Use conversion factors to find the price of these items in the countries you have chosen. (Assume that the prices are equivalent.)

Review PREVIEW

Graph. *(Section 7-4)*

28. $y = 5x$

29. $y = -3x$

Solve each proportion. *(Section 6-3)*

30. $\frac{x}{240} = \frac{120}{360}$

31. $\frac{x}{50} = \frac{40}{360}$

32. $\frac{20}{x} = \frac{40}{360}$

Find each value for x. *(Sections 2-5, 2-7)*

33.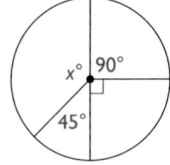

34.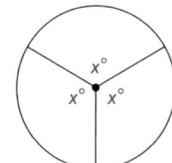

35.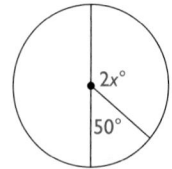

EXCHANGE RATES

THURSDAY, SEPTEMBER 23

	Fgn. currency in dollars	Dollar in fgn. currency
Australia (Dollar)	.625	1.5326
Brazil (CruzeiroR)	.0089	111.94
Britain (Pound)	1.5045	.6647
Canada (Dollar)	.7560	1.3227
China (Yuan)	.1733	5.7703
Denmark (Krone)	.1508	6.6300
France (Franc)	.1749	5.7175
Germany (Mark)	.6083	1.6440
Hong Kong (Dollar)	.1293	7.7335
India (Rupee)	.0319	31.330
Israel (Shekel)	.3598	2.7797
Italy (Lira)	.000629	1589.25
Japan (Yen)	.009428	106.07
Mexico (Peso)	.321957	3.1060
N. Zealand (Dollar)	.5536	1.8064
Peru (New Sol)	.4950	2.020
Russia (Ruble)	.000770	1299.00
Saudi Arabia (Riyal)	.2667	3.7500
Thailand (Baht)	.03975	25.16

7 Working on the Unit Project

36. One plan for building a skateboard launch ramp calls for four sheets of $\frac{5}{8}$ in. plywood, one sheet of $\frac{3}{8}$ in. plywood, and one sheet of $\frac{1}{8}$ in. plywood. All the sheets measure 4 ft by 8 ft. Find the total cost of purchasing the plywood needed for this ramp.

37. **Research** List the materials you would need to build the ramps for your skating arena. Check with a local lumber yard to find current prices. Estimate the cost of the materials you would need.

Thickness of plywood	Cost per square foot
$\frac{5}{8}$ in.	$.66
$\frac{3}{8}$ in.	$.47
$\frac{1}{8}$ in.	$.34

7-5 Using Dimensional Analysis

399

Assessment: Task

Ex. 27 provides a natural and interesting application of the ideas about dimensional analysis that have been studied in this section. Allow time for students to complete the exercises out of class.

Working on the Unit Project

The research required for Ex. 37 may take students several days to gather. Be sure to allow extra time, perhaps a weekend.

Practice 56 For use with Section 7-5

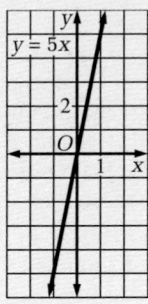

Answers to Exercises and Problems

27. **a–c.** Answers may vary. Each item's price in dollars should be multiplied by the "Dollar in fgn. currency" number.

28.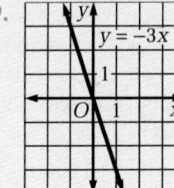

29.

30. 80

31. $5\frac{5}{9}$ or about 5.6

32. 180

33. 135

34. 120

35. 65

36. $110.40

37. Answers may vary.

Objectives and Strands
See pages 356A and 356B.

Spiral Learning
See page 356B.

Materials List
➤ Graph paper
➤ Scientific or graphics calculator

Recommended Pacing
Section 7-6 is a two-day lesson.

Day 1
Pages 400–401: Exploration through Talk it Over, *Exercises 1–11*

Day 2
Pages 402–403: Area of a Sector through Look Back, *Exercises 12–35*

Extra Practice
See pages 631–633.

Warm-Up Exercises
Warm-Up Transparency 7-6

Support Materials
➤ Practice 57
➤ Enrichment 51 in the Activity Bank
➤ Study Guide 7-6
➤ Problem Set 15
➤ Diagram Master 2 in the Explorations Lab Manual
➤ Quiz 7-6
➤ Test 28
➤ Alternative Assessments 10, 11

Section **7-6**

Areas of Circles and Sectors

Focus
Use the formula for the area of a circle and find the area of a sector.

Pieces Of Pi

EXPLORATION

What is the relationship between the area of a circle and its radius?

• **Work in a group of three students. Have each person in your group use one of the diagrams on this page.**

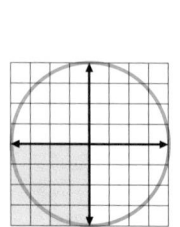
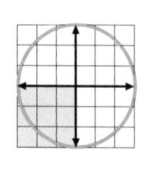
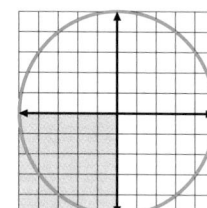

(1) Estimate the area of the circle in your diagram by counting the squares inside the circle. Include parts of squares in your estimate.

(2) Find the area of the large shaded square in your diagram.

(3) Find the ratio of the area you estimated in step 1 to the area you found in step 2:

$$\frac{\text{estimated area of circle}}{\text{area of shaded square}} = \frac{?}{?}$$

(4) Write your ratio as a decimal.
➤ Compare your ratio with the ratios found by the other students in your group.
➤ What do you notice?
➤ Let A represent your estimate of the area of the circle in your diagram.
➤ Discuss in your group why the ratio you wrote in step 3 can be represented by $\frac{A}{r^2}$.

(5) Find the mean of the three ratios your group found in step 3. Use that number to complete this formula: $\frac{A}{r^2} = \frac{?}{?}$

(6) Compare your group's formula with the known formula for the area of a circle, $A = \pi r^2$. How close was your estimate to the value of π?

Unit 7 Direct Variation

Answers to Exploration

1. Estimates may vary. Examples are given. about 28 square units, about 48 square units, and about 80 square units

2. 9 square units; 16 square units; 25 square units

3. Answers may vary, depending on the estimates in step 1. For the estimates given, the ratios are about 3.11, 3, and 3.2, respectively. Each ratio is 3 or a little more than 3.

4. The length of a side of each shaded square is the same as the radius of the circle in that diagram.

5. Answers may vary. For example, using the ratios listed in step 3, $\frac{A}{r^2} = 3.10$.

6. Comparisons may vary. The average in step 5 should be close to 3.14.

Sample 1

Find the area of a circle whose radius is 3 units. Round the answer to the nearest square unit.

Sample Response

Problem Solving Strategy: Use a formula

$$A = \pi r^2$$

$$= \pi \cdot 3^2 \quad \longleftarrow \text{Substitute 3 for } r.$$

$$\approx 28.3 \quad \longleftarrow \boxed{\pi} \; \boxed{\times} \; \boxed{3} \; \boxed{X^2} \; \boxed{\text{ENTER}}$$

The area is about 28 square units.

Sample 2

In a center-pivot irrigation system, a moving arm sprinkles water over a circular region. How long must the arm be to water an area of 586,000 m²?

Round the answer to the nearest meter.

Sample Response

Problem Solving Strategy: Use a formula

The arm forms the radius of a circle with area 586,000 m².

$$A = \pi r^2$$

$$586,000 = \pi r^2 \quad \longleftarrow \text{Substitute 586,000 for } A.$$

$$\frac{586,000}{\pi} = \frac{r^2}{\pi} \quad \longleftarrow \text{Divide both sides by } \pi.$$

$$186,529.6 \approx r^2$$

$$\sqrt{186,529.6} \approx r \quad \longleftarrow \text{Find the positive square root.}$$

$$431.9 \approx r$$

The arm must be about 432 m long.

Talk it Over

······▶ **Now you are ready for:**
Exs. 1–11 on pp. 404–405

1. In Sample 2, why is only the positive square root considered as an answer to the problem?

2. Explain why the equation $\frac{A}{\pi} = r^2$ is equivalent to $A = \pi r^2$.

7-6 Areas of Circles and Sectors   **401**

Answers to Talk it Over ·······:

1. The variable r represents the length of an irrigation arm; it cannot be a negative number.

2. If you multiply both sides of $\frac{A}{\pi} = r^2$ by π, you get the equivalent equation $A = \pi r^2$.

TEACHING

Exploration

The purpose of the Exploration is to have students discover the relationship between the area, A, of a circle and its radius, r. By using different size circles, students can see that the relationship $\frac{A}{r^2} = \pi$ is independent of the size of the circle as is the ratio $\frac{C}{d} = \pi$. In fact, you may wish to have students compare what they have learned in this Exploration to what they learned in the Exploration on page 375.

Additional Samples

S1 Find the area of a circle whose radius is 7.5 units. Round the answer to the nearest tenth of a square unit. **Use $A = \pi r^2$. Substitute 7.5 for r.**

$$A = \pi r^2$$
$$= \pi (7.5)^2$$
$$= 56.25 \, \pi$$
$$\approx 176.7$$

The area is about 176.7 square units.

S2 The amount of water a pipe can carry depends on the area of the opening at the end. An engineer wants a large sewer pipe to have an opening of 13 ft². What should the radius of the pipe be? Round the answer to the nearest tenth of a foot. **Use $A = \pi r^2$. Substitute 13 for A and solve for r.**

$$A = \pi r^2$$
$$13 = \pi r^2$$
$$\frac{13}{\pi} = \frac{\pi r^2}{\pi}$$
$$4.13803 = r^2$$
$$\sqrt{4.13803} = r$$
$$2.03 \approx r$$

The radius of the pipe should be about 2.0 ft.

401

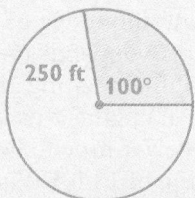
Area of a Sector

The region of a circle formed by a central angle and its arc is called a **sector**. The area of a sector is a fraction of the area of the circle.

AREA OF A SECTOR

You can use this proportion to find the area of any sector:

$$\frac{\text{area of a sector}}{\text{area of the circle}} = \frac{\text{measure of the central angle}}{360°}$$

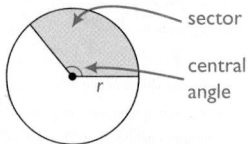

Sample 3

A circular quilt pattern has a radius of 12 in. What is the area of the blue panel? Round the answer to the nearest square inch.

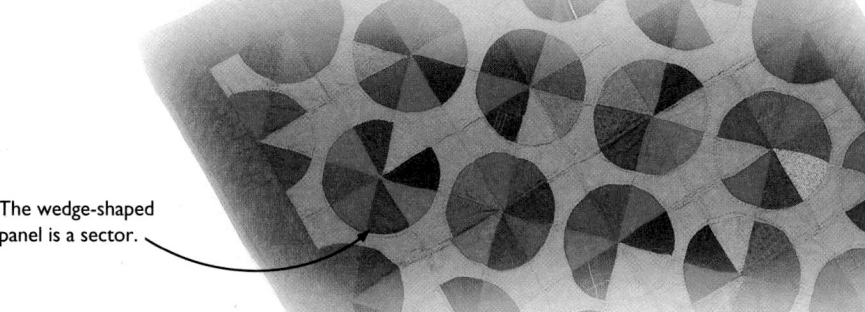

The wedge-shaped panel is a sector.

Sample Response

Problem Solving Strategy: Use a proportion

Let x = the area of the sector. ⟶   ⟵ The measure of the central angle is 45°.

$$\frac{x}{452} = \frac{45}{360}$$

The area of the circle is $\pi \cdot 12^2$. ⟶   ⟵ The measure of a complete circle is 360°.

$$\frac{x}{452} \cdot 452 = \frac{45}{360} \cdot 452 \longleftarrow \text{Multiply both sides by 452.}$$

$$x = 56.5$$

The area of the blue panel is about 57 in.².

Expressing Areas in Terms of π

Sometimes you can solve problems without substituting a value for π.

About how many times more pizza do you get when you buy the large size instead of the small size?

The thickness of both pizzas is about the same.

Sample Response

First find the area of each pizza in terms of π.

The radius of the large pizza is 8 in.

$$A = \pi r^2 = \pi \cdot 8^2 = 64\pi$$

$$A = \pi r^2 = \pi \cdot 6^2 = 36\pi$$

The radius of the small pizza is 6 in.

Then write an equation:

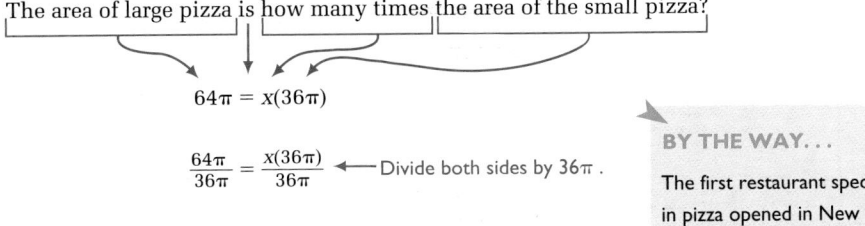

The area of large pizza is how many times the area of the small pizza?

$$64\pi = x(36\pi)$$

$$\frac{64\pi}{36\pi} = \frac{x(36\pi)}{36\pi} \quad \longleftarrow \text{Divide both sides by } 36\pi.$$

$$1.8 \approx x$$

> **BY THE WAY...**
>
> The first restaurant specializing in pizza opened in New York City in 1895.

You get about 1.8 times more pizza when you buy the large size instead of the small size.

▶ **Now you are ready for:**
Exs. 12–35 on pp. 405–407

Look Back ◀

How is finding the area of a sector like finding its arc length? How is it different?

Answers to Look Back

Answers may vary. An example is given. Finding the area of a sector is like finding its arc length because each can be computed by using a proportion and the ratio $\frac{\text{central angle}}{360}$. Finding the area of a sector is different from finding its arc length because when you find the area of a sector you use the ratio $\frac{\text{sector area}}{\text{circle area}}$, while for arc length you use the ratio $\frac{\text{arc length}}{\text{circumference}}$.

S4 The camera in Additional Sample S3 replaced an older camera that had a viewing angle of 110° but a range of only 175 ft. How many times greater is the viewing area of the newer camera than that of the older camera?

Set up proportions and solve for the areas of the sectors.

$$\text{Area (new)} = \pi(250)^2\left(\frac{100}{360}\right)$$
$$\approx 17{,}361\pi$$

$$\text{Area (old)} = \pi(175)^2\left(\frac{110}{360}\right)$$
$$\approx 9358\pi$$

$$17{,}361\pi = x(9358\pi)$$
$$\frac{17{,}361\pi}{9358\pi} = \frac{x(9358\pi)}{9358\pi}$$
$$1.86 \approx x$$

The new camera has a viewing area about 1.86 times that of the older camera.

Look Back

Use the Look Back questions for class discussion. Be sure students understand how to use proportions to solve problems about arc length and the area of a sector of a circle. Check to be sure that students correctly recall the formulas for finding the circumference and the area of a circle. ●

Integrating the Strands

Algebra Exs. 31–33

Functions Exs. 8, 9, 25, 27

Measurement Exs. 1–30, 34, 35

Geometry Exs. 1–30, 34, 35

Research

In connection with Ex. 1, students may want to research other methods for determining how the area of circle is related to the radius.

Error Analysis

In part (d) of Example 9, some students may use r instead of r^2 along the horizontal axis. Correct this error by having these students think about the area relationship $A = \pi r^2$. Does it involve r or r^2? What should be graphed on the horizontal axis, r or r^2?

Answers to
Exercises and Problems

1. Answers may vary. An example is given. The estimate for π in the Exploration could be improved by using smaller squares, such as dividing each square into 4 smaller identical squares.

2. about 78.5 mm^2

3. about 113.0 ft^2 (using $\pi \approx 3.14$)

4. about 4.2 m

5. about 28.3 square units

6. about 201.0 square units (using $\pi \approx 3.14$)

7. Answers may vary. An example is given. I agree. The circumference is 4π units and the area is 4π square units. The kinds of units are different, but the number of units is the same.

7-6 Exercises and Problems

1. **Writing** How could you improve your estimate for π in the Exploration?

For Exercises 2–4, find the unknown measurement for each circle.

2. radius = 5 mm
   area = _?_

3. diameter = 12 ft
   area = _?_

4. area = 56 m^2
   radius = _?_

5. Find the area of the circle at the right.

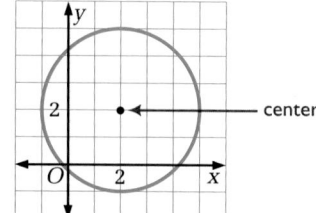

6. A circle on a coordinate plane has its center at $(1, -1)$. The circle goes through the point $(1, 7)$. What is the area of the circle?

7. Tanya says that when the radius of a circle is 2 units, the number of units in the circumference of the circle is equal to the number of square units in its area. Do you agree or disagree? Why?

Radius r (cm)	Area A (cm^2)	$\frac{A}{r}$ (cm)
1	?	?
2	?	?
3	?	?
4	?	?
5	?	?
6	?	?

8. **a.** Complete the table of values for a circle of radius r.

   b. Is the ratio $\frac{A}{r}$ constant?

   c. Graph the data from the table with r on the horizontal axis and A on the vertical axis.

   d. Is the graph a straight line through the origin?

   e. Does the area of a circle vary directly with the radius? Why or why not?

9. **a.** Complete the table of values for a circle of radius r.

   b. What do you notice about the ratio column?

   c. Find the variation constant for the relationship between the area of a circle and the square of its radius.

   d. Suppose you make a graph with r^2 on the horizontal axis and A on the vertical axis. What do you think the graph will look like? Draw the graph. Was your prediction correct?

Radius r (cm)	r^2 (cm^2)	Area A (cm^2)	$\frac{A}{r^2}$
1	?	?	?
2	?	?	?
3	?	?	?
4	?	?	?
5	?	?	?
6	?	?	?

---

8. **a.**

r (cm)	A (cm^2)	$\frac{A}{r}$ cm
1	3.14	3.14
2	12.56	6.28
3	28.26	9.42
4	50.24	12.56
5	78.50	15.70
6	113.04	18.84

b. No.

c.

Areas of Circles

(graph: Area (cm^2) on vertical axis 0–120, Radius (cm) on horizontal axis 0–6, curve rising)

d. No.

e. No. Explanations may vary. An example is given. The table shows that $\frac{A}{r}$ does not have a constant ratio, so the area of a circle does not vary directly with the radius. Also, the graph of the data is not well represented by a fitted line through the origin. Therefore, the area of a circle does not vary directly with the radius.

For Exercises 10 and 11, find the area of the circle in each figure.

10. Area of square = 64 cm²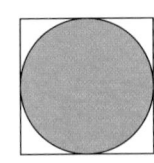

11. Perimeter of rectangle = 28 in.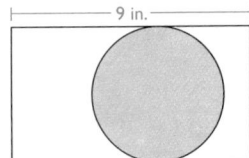

12. Reading In Sample 4, why is substituting a value for π not necessary?

For Exercises 13–15, find the area of each sector.

13.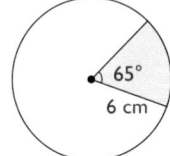

14.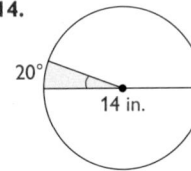

15. The area of the circle is 25π ft².

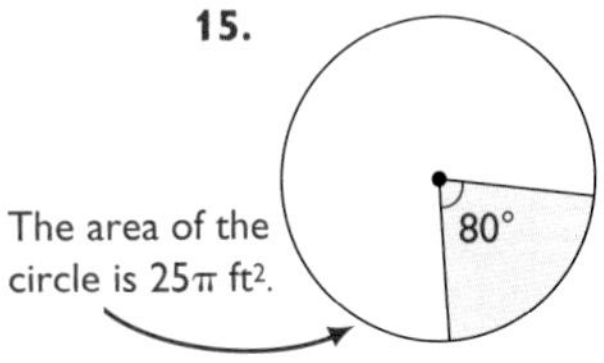

For Exercise 16, use the menu shown. Assume that large and small round pizzas have about the same thickness.

16. a. About how many times more pizza do you get when you order the large size instead of the small size?

b. Which size cheese pizza is the better buy? Why?

c. Do you think the prices of extra toppings are fair? Why or why not?

17. a. Tom cuts a 12 in. pizza into 6 slices of equal size. What is the area of one slice?

b. Vera cuts a 12 in. pizza into 8 slices of equal size. What is the area of one slice?

c. Compare the sizes of a slice of Vera's pizza and a slice of Tom's pizza.

18. a. Compare the areas of the sectors.

b. Compare the arc lengths of the sectors.

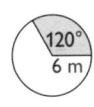 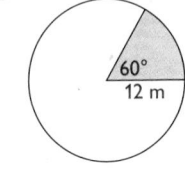

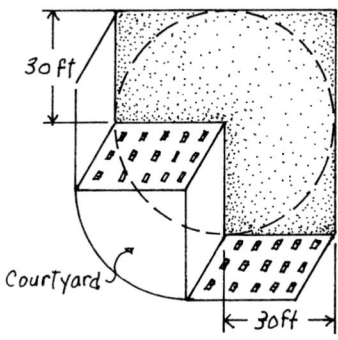

19. The courtyard of an apartment building is shaped like one fourth of a circle. Cement that covers an area of one square foot costs $2.60. How much will it cost to repave the courtyard?

SLICES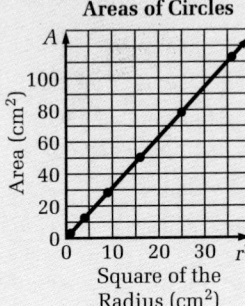
House of Pizza

Small Cheese $5.65
(11 in. diameter)
$.95 per extra topping

Large Cheese $8.75
(14 in. diameter)
$1.80 per extra topping

Problem Solving

In connection with Exs. 13–15, you may wish to have students think about how they would find the central angle for a sector if they know the area of the sector and the radius of the circle of which it is a part. Have students explain their thinking by using examples illustrated with drawings.

---

12. Answers may vary. An example is given. After dividing both sides of the equation by 36π, you can cancel π on both sides. Therefore, you do not need to substitute a value for π.

13. about 20.4 cm²

14. about 8.5 in.²

15. about 17.4 ft²

16. a. about 1.6 times

b. The large cheese pizza is the better buy. The price of the large pizza is less than 1.6 times the price of the small pizza. Also, the cost per square inch for the large pizza, about 5.7¢, is lower than the cost per square inch for the small pizza, about 5.9¢.

c. Answers may vary. An example is given. I don't think the prices are fair because the topping cost for the large pizza is about 2 times the topping cost for the small pizza, even though the large pizza is only about 1.6 times as large as the small pizza.

17. a. about 18.8 in.²

b. about 14.1 in.²

c. A slice of Tom's pizza is about $1\frac{1}{3}$ times as large as a slice of Vera's pizza.

18. a. The area of the 60° sector is twice the area of the 120° sector.

b. Each arc length is 4π m, or about 12.6 m; they are equal.

19. $1836.90 (using π ≈ 3.14)

Answers to Exercises and Problems

9. a.

r (cm)	r²(cm²)	A (cm²)	$\frac{A}{r^2}$
1	1	3.14	3.14
2	4	12.56	3.14
3	9	28.26	3.14
4	16	50.24	3.14
5	25	78.50	3.14
6	36	113.04	3.14

b. There is a constant ratio, 3.14.

c. 3.14.

d. Answers may vary. An example is given. I predict a straight line through the origin with a slope equal to 3.14. See graph at right.

10. about 50.2 cm² (using π ≈ 3.14)

11. about 19.6 in.²

Areas of Circles

Area (cm²) vs *Square of the Radius (cm²)*

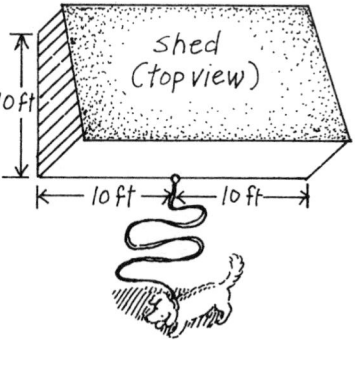

20. The end of a dog's leash is tied to a hook on the side of a shed.

   a. Over what area can the dog roam if the dog's leash is 10 ft long? 18 ft long? 25 ft long?

   b. **Open-ended** Suppose the leash is 25 ft long. Will the dog have more room to roam if the hook is moved? If so, where do you think the hook should be placed?

21. A circular herb garden is divided into three equal sections. The radius of the garden is 4 ft.

   a. Find the area of one sector.

   b. A gardener decides to plant marigolds on the outside edge of two of the sectors. The marigolds should be planted 1 ft apart. How many marigold plants does the gardener need?

22. To estimate the area of a circle, the ancient Egyptians measured the diameter of a circle and drew a square with sides that are $\frac{8}{9}$ as long as the diameter.

   a. Compare the areas of a circle and a square with these dimensions:

   circle: diameter = 9 in., *square:* side length = 8 in.

   b. Compare the areas of a circle and a square with these dimensions:

   circle: diameter = d, *square:* side length = $\frac{8}{9}d$

   c. What value for π results from the Egyptian method? How close is this estimate to the value of π that you use?

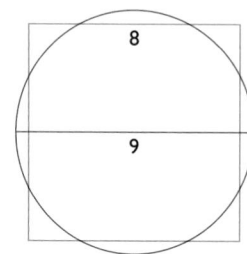

23. **Writing** The area S of a sector can be found by using the formula $S = \frac{x}{360}(\pi r^2)$, where $x°$ is the measure of the central angle for the sector. Explain why this formula works.

Ongoing **ASSESSMENT**

Writing **Tell whether each statement is *True* or *False*. Explain your choice.**

24. The area of a circle varies directly with the diameter.

25. The area of a circle varies directly with the square of the diameter.

26. The area of a circle varies directly with the circumference.

27. The area of a sector varies directly with the arc length of the central angle for the sector.

Review **PREVIEW**

For Exercises 28 and 29, use conversion factors. *(Section 7-5)*

28. Convert 1100 ft/s to miles per hour.

29. Convert $22/ft² to dollars per square yard.

406

30. The graph shows a rotation around the origin. What is the direction of the rotation and the amount of turn? *(Section 4-4)*

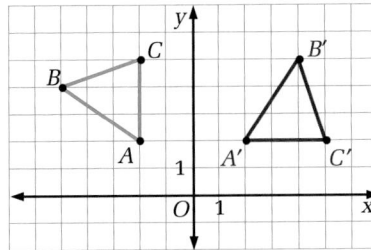

Solve. *(Sections 2-7, 2-8)*

31. $14 = 0.7x + 28$

32. $\frac{1}{3}y - 12 = 9$

33. $-4 = 12a + 8$

Working on the Unit Project

34. To make the sides of a skateboard ramp, two corner-to-corner arcs are drawn from opposite corners of a plywood square. The wood where the sectors overlap is not used.

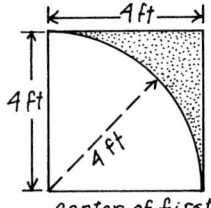
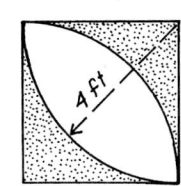
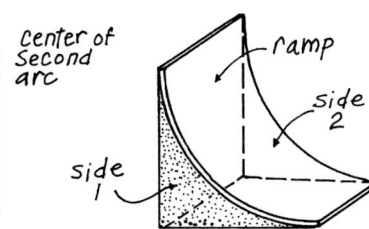

a. What is the area of each side of the ramp?

b. What is the area of the plywood that is not used?

c. What fraction of the plywood square is not used?

35. A circular arena for skaters includes a skating rink with a radius of 30 ft. There are benches in a circle 10 ft beyond the edge of the rink.

a. Find the total area of the arena, including the space between the rink and the benches.

b. Find the area of the space between the rink and the benches.

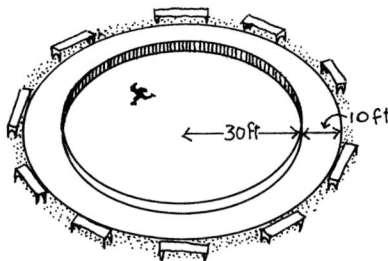

7-6 Areas of Circles and Sectors

407

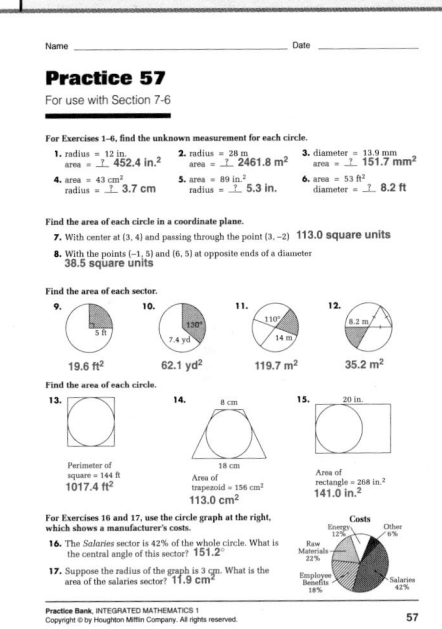
Answers to Exercises and Problems

28. 750 mi/h

29. $198/yd^2

30. 90° clockwise or 270° counterclockwise

31. −20

32. 63

33. −1

34. a. about 3.44 ft^2
(using $\pi \approx 3.14$)

b. about 9.12 ft^2
(using $\pi \approx 3.14$)

c. about 0.57 or $\frac{57}{100}$

35. a. about 5024 ft^2
(using $\pi \approx 3.14$)

b. about 2198 ft^2
(using $\pi \approx 3.14$)

Completing the Unit Project

Unit
Project
7

Now you are ready to complete a scale drawing or a model of your skating competition arena. Label its parts and the dimensions.

Include answers to these questions on your drawing or in a description of your model.

➤ **What are the slopes of the ramps used in your arena?**

➤ **What are the arc lengths of any of the circular parts of your arena?**

➤ **What is the slope of the bleachers?**

➤ **What materials are needed to make your arena?**

➤ **What is the estimated cost of building your arena?**

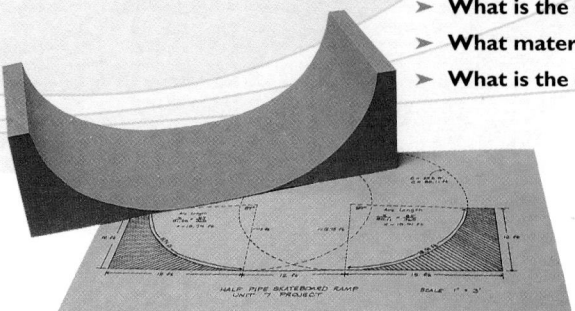

Look Back ◄

Write an evaluation of each arena designed by your class. What features do many of the designs have in common? What features are unique? Which designs do you like the best? Why?

Alternative Projects

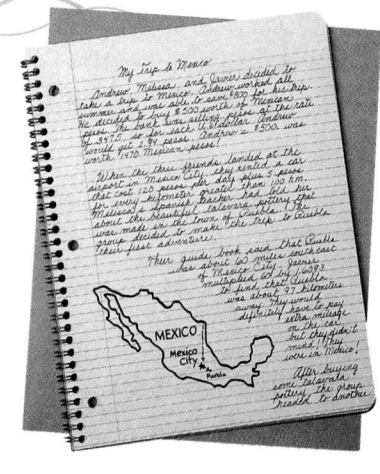

...... **Project 1:** **A Trip to Another Country**

Write a story about a trip to another country. Include how you might use direct variation during the trip. For example, show how the cost of a car rental varies with mileage, how to convert distances between cities from kilometers to miles, and how to change dollars into the currency of another country.

...... **Project 2:** **The History of π**

Research the number π. What are some of the approximations for π that have been used by different cultures? How many digits of π are known?

408   Unit 7 Completing the Unit Project

Answers to Unit 7 Review and Assessment ···

1. Answers may vary. An example is given. The slope, tangent ratio, and direct variation ratio are different ways to express the ratio of the length of a vertical leg to the length of a horizontal leg in a right triangle. The tangent can be thought of as the slope of the hypotenuse and the slope as the "rise" of the opposite leg over the "run" of the adjacent leg. A straight line passing through the origin can be expressed by an equation with a variation constant. This variation constant is the same as the slope. While the tangent ratio involves an angle, the slope is a rate and does not necessarily involve angle. You might find the slope of the roof to determine its steepness, the tangent of an angle in order to find the length of a side or an angle measure for a triangle, and the variation constant to write an equation relating two variables.

2. a. $h = s \tan 35°$

   b. 0.700

   c. about 14 ft

   d. about 60 ft

   e. tan 35° or 0.700

3. a. For each central angle,
      $\frac{M}{W} = 0.025$.

   b. $\frac{M}{W} = 0.025$, or
      $M = 0.025W$

   c. 7 m

408

1. **Writing** Compare slope, tangent ratios, and direct variation ratios. How are they alike? How are they different? Give an example of a situation where you would use each.

2. **a.** Set up an equation to relate the height of a tree to its shadow when the sun is 35° above the horizon.

   b. Use a calculator to find tan 35° to the nearest thousandth.

   c. Find the height of a tree that casts a 20 ft shadow.

   d. Find the length of a shadow cast by a tree that is 42 ft tall.

   e. What is the variation constant in this situation?

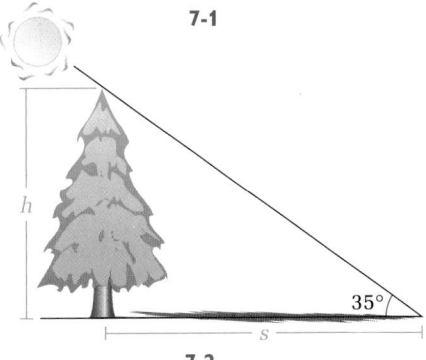

7-1

7-2

3. Arc lengths for a circle are related to the central angles as shown in the table.

Central angle W (°)	180	240	300	360
Arc length M (m)	4.5	6.0	7.5	9.0

   a. Find the ratio $\frac{M}{W}$ for each central angle in the table.

   b. Write an equation relating M and W.

   c. What is the arc length for a 280° central angle?

4. Keisha and Kevin performed the bouncing ball exploration. Here are their data.

Drop height D (cm)	40	60	80	100	120	140
Bounce height B (cm)	17	24	31	47	48	57

   a. Make a graph of the data.

   b. Keisha and Kevin decide that one of their measurements is wrong. They cross it off their graph. Which one do they cross off?

   c. Is direct variation a good model for the set of data that remains? Explain.

   d. Estimate the variation constant from these data.

   e. Use your graph to predict the bounce height when the ball is dropped from a height of 70 cm.

5. The diameter of the sun is approximately 865,000 miles. Find its circumference.

7-3

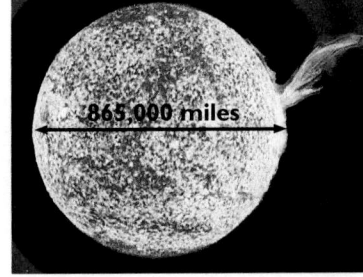

865,000 miles

Answers to Unit 7 Review and Assessment

4. **a.**

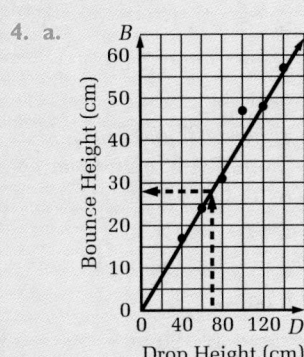

   Bounce Height (cm) vs Drop Height (cm)

   b. bounce height of 47 cm from a drop height of 100 cm

   c. Yes. Explanations may vary. An example is given. The remaining points are close to a straight line that passes through the origin. Also,

   for the remaining data each ratio, $\frac{B}{D}$ is close to 0.4.

   d. about 0.4

   e. about 28 cm

5. about 2,716,100 mi (using π ≈ 3.14)

Unit Support Materials
➤ Unit 7 Cumulative Practice 58
➤ Unit 7 Study Guide Review
➤ Unifying Problem 7
➤ Unit Tests 29 and 30

Quick Quiz (7-4 through 7-6)

1. An actor memorizes 3 pages of a new script each day, working 5 hours a day. [7-4]

   a. Model the situation with a direct variation equation. $p = \frac{3}{5}h$

   b. How many hours will it take the actor to memorize 39 pages of the script? **65 h**

2. Write an equation of the form $y = kx$ for the graph. [7-4]

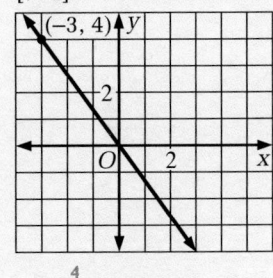

   (−3, 4)

   $y = -\frac{4}{3}x$

For Exs. 3–5, give the expression for the units in the answer. You do not have to give the number part of the answer. [7-5]

3. $\frac{12 \text{ in.}}{1 \text{ ft}} \times 9.5 \text{ ft}$ **in.**

4. $\frac{2 \text{ c}}{1 \text{ pt}} \times \frac{2 \text{ pt}}{1 \text{ qt}}$ $\frac{\text{c}}{\text{qt}}$

5. $\frac{5 \text{ mi}}{1 \text{ h}} \times \frac{24 \text{ h}}{1 \text{ da}} \times \frac{365 \text{ da}}{1 \text{ yr}}$ $\frac{\text{mi}}{\text{yr}}$

6. What is the area of a circle whose diameter is 32 cm? Round to the nearest tenth. [7-6] **804.2 cm²**

7. A sector of a circle has a central angle of 18°. The circle of which it is a sector has a radius of 10 in. What is the area of the sector? Round to the nearest tenth. [7-6] **15.7 in.²**

6. A can with radius 2.5 in. rolls down a ramp a total distance of 96 in. How many complete turns will it make?

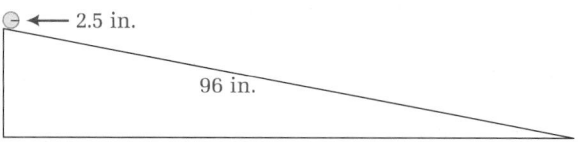
2.5 in.
96 in.

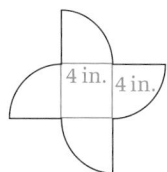
4 in. | 4 in.

7. Four quarter-circles are attached to a 4 in. square to form a pinwheel design. Find the perimeter of the design.

8. The length (L) that a spring stretches varies directly with the force (F) that is applied. When a force of 3.5 lb is applied to a spring, it stretches 2.3 in.

  a. Find the variation constant. What are its units?

  b. How far will the spring stretch when a 5 lb force is applied?

  c. How much force is needed to stretch the spring 3 in.?

7-4

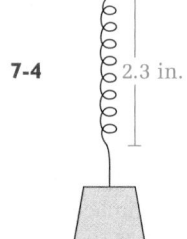
2.3 in.
3.5 lbs

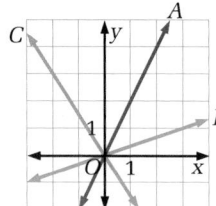

9. **a.** Which line has negative slope? Explain.

  b. Which line has the smallest positive slope? Explain.

  c. Which line has the greatest variation constant? Explain.

10. **a.** Graph the equation $y = -1.5x$.

  b. What is the slope of this line?

  c. What quadrants does the line pass through?

  d. For what value of x does $y = -9$?

11. How many days are in 16 years?

7-5

12. Write a direct variation equation that helps you convert days to years. What is the variation constant?

13. A marathon is 26.2 mi. Convert this distance to meters.

14. It takes about 15,000 s for light to travel from the sun to the planet Neptune. Convert this amount to hours.

Answers to Unit 7 Review and Assessment ···

6. 6 complete turns

7. about 41.1 in.

8. **a.** about 0.657; inches per pound

  b. about 3.3 in.

  c. about 4.6 lb

9. **a.** C; The line slants down from left to right.

  b. B; Lines A and B have positive slope because they slant up from left to right. B has the smallest slope because it has a small rise compared to the run.

  c. A; The slope of line A is positive and greater than that of line B.

10. **a.**

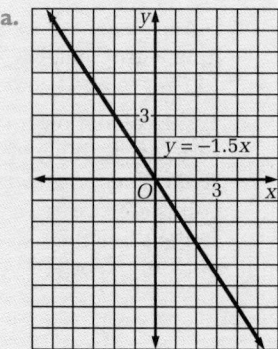
$y = -1.5x$

  b. -1.5

  c. quadrants II and IV

  d. 6

11. about 5840 days (4 to 5 more if you include an extra day for each leap year)

12. $y = \frac{1}{365}d$; $\frac{1}{365}$, or about 0.0027

13. about 42,190 m

14. about 4.17 h

410

15. The speed limit on some highways is 65 mi/h. Convert to feet per second.

16. For Thanksgiving dinner, Grandpa makes his favorite recipe for apple pie. There are 7 people who want an equal share.

7-6

a. Find the size of each slice of pie in degrees.

b. If the pie has a diameter of 12 in., find the area of each slice.

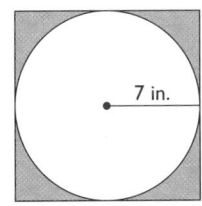

7 in.

17. A circle with a 7 in. radius just fits inside a square.

a. Find the area of the shaded region.

b. Find the ratio of the shaded area to the area of the circle.

c. Do you think the ratio you found in part (b) would be different if the circle and square were each twice as large? Explain why or why not.

18. **Self-evaluation** What do you think are the important ideas of this unit? Make a concept map of these ideas and show how they relate to one another.

19. **Group Activity** Work in a group of four students. You will need a tape measure or meter stick scaled in both inches and centimeters.

a. Measure the height of everyone in your group. For half the group, use centimeters. For the other half, use inches.

b. Write an equation that relates height in centimeters (*y*) to height in inches (*x*). What is the variation constant? What are the units of the variation constant?

c. Use the equation to find each group member's height in both inches and centimeters.

d. Verify your results by measuring each member's height using the other unit of measurement.

Unit 7 Review and Assessment

411

Answers to Unit 7 Review and Assessment

15. about 95.3 ft/s

16. a. $51\frac{3}{7}°$, or about 51.4°

b. about 16.1 in.^{2} (using $\pi \approx 3.14$)

17. a. about 42.1 in.^{2}

b. about 0.27

c. No. Explanations may vary. For example, multiplying both parts of a ratio by 2 does not change the ratio.

18. Concept maps may vary. An example is given at right.

19. a. Heights may vary.

b. $y = 2.54x$; 2.54; $\frac{\text{cm}}{\text{in.}}$

c, d. Results may vary.

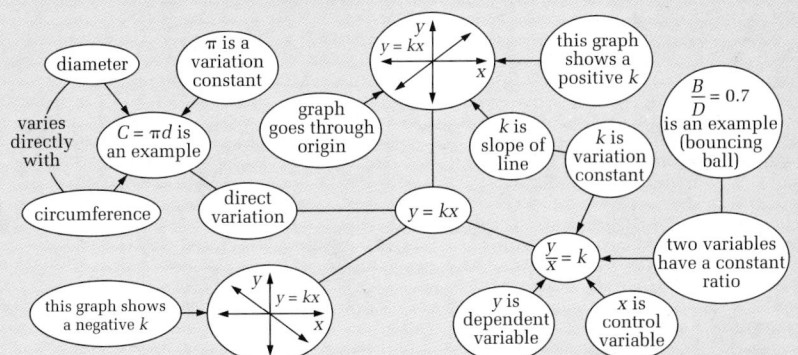

411

IDEAS AND (FORMULAS) $=x^2$

- ➤ **Measurement** You can cancel units of measurement as if they are numbers when dividing. *(p. 392)*

- ➤ **Measurement** If two variables in direct variation have different units of measurement, then the variation constant will be a rate. *(p. 393)*

- ➤ **Measurement** You can use conversion factors to convert from one measurement or rate to another with different units. *(p. 393)*

- ➤ **Measurement** Sometimes you need to use more than one conversion factor. *(p. 395)*

- ➤ When two variables have a constant ratio, they are in direct variation. *(p. 360)*

- ➤ When two variables vary directly, you can find the value of one variable when given a value of the other variable. *(p. 360)*

- ➤ The slope of a line is the ratio of the rise to the run between any two points on the line. *(p. 361)*

- ➤ Direct variation can be used to model real-life behavior, like the bouncing of a ball. *(p. 368)*

- ➤ You can use a direct variation graph to make predictions. *(p. 368)*

- ➤ The letter k is sometimes used to represent a variation constant in the general form for direct variation: $y = kx$. *(p. 383)*

- ➤ The graph of direct variation is a straight line through the origin. *(p. 385)*

- ➤ The slope of a direct variation graph is the same as the variation constant. *(p. 386)*

- ➤ If the variation constant of a direct variation equation is negative, the slope of the line is negative. Lines with negative slope go down from left to right. *(p. 386)*

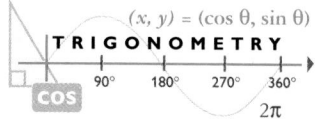

- ➤ The tangent ratio compares the legs of a right triangle.

$$\tan A = \frac{\text{length of the leg opposite } \angle A}{\text{length of the leg adjacent } \angle A} \quad (p. 361)$$

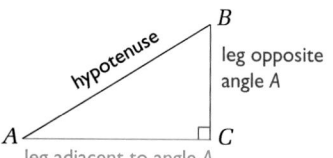

412   **Unit 7** Direct Variation

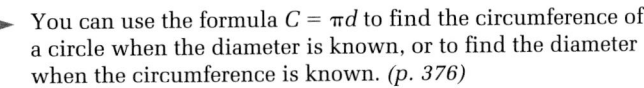

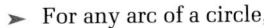
GEOMETRY

➤ The circumference, C, of a circle varies directly with the diameter, d.
$$C = \pi d \text{ (p. 376)}$$

➤ The value of π can be approximated by 3.14 or $\frac{22}{7}$. *(p. 376)*

➤ You can use the formula $C = \pi d$ to find the circumference of a circle when the diameter is known, or to find the diameter when the circumference is known. *(p. 376)*

➤ For any arc of a circle,
$$\frac{\text{arc length}}{\text{circumference}} = \frac{\text{central angle}}{360°} \text{ (p. 377)}$$

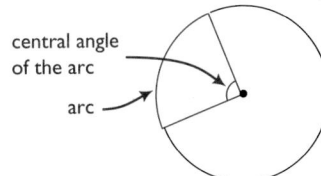

central angle of the arc

arc

➤ The area, A, of a circle varies directly with the *square* of the radius of the circle.
$$A = \pi r^2 \text{ (p. 400)}$$

➤ You can use the formula $A = \pi r^2$ to find the area of a circle when the radius is known, or to find the radius when the area is known. *(p. 401)*

➤ For any sector,
$$\frac{\text{area of sector}}{\text{area of circle}} = \frac{\text{central angle}}{360°} \text{ (p. 402)}$$

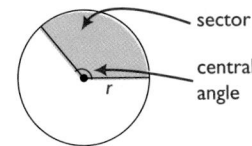

sector

central angle

r

➤ You can express areas of circles in terms of π. *(p. 403)*

········Key Terms

- **direct variation** (p. 360)
- **rise** (p. 361)
- **diameter** (p. 375)
- **arc** (p. 377)
- **dimensional analysis** (p. 392)

- **variation constant** (p. 360)
- **run** (p. 361)
- **circumference** (p. 375)
- **arc length** (p. 377)
- **conversion factor** (p. 393)

- **slope** (p. 361)
- **tangent ratio** (p. 361)
- **pi (π)** (p. 376)
- **general form for direct variation** (p. 383)
- **sector** (p. 402)

Unit 7 Review and Assessment

413

For Exs. 1 and 2, suppose that y varies directly with x and that $\frac{y}{x} = 0.65$. [7-1]

1. Find the value of y when $x = 4$. **$y = 2.6$**

2. Find the value of x when $y = 13$. **$x = 20$**

3. Find tan P. Round to four decimal places. [7-1]

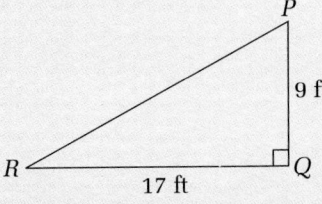

1.8889

4. Find the slope of the line passing through (6, 3) and (1, 1). [7-1] **$\frac{2}{5}$ or 0.4**

5. Decide whether direct variation is a good model for the data in the table. If you say that it is, write a direct variation equation. [7-2]

Number of Books	Weight of Shipment (lb)
10	25.0
17	42.5
25	62.5
40	100.0
48	120.0

Yes. $w = 2.5b$

6. Find the circumference of the circle to the nearest tenth of a centimeter. [7-3]

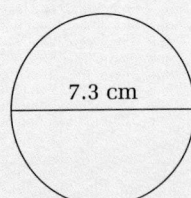

7.3 cm

22.9 cm

7. Find the length of the arc to the nearest inch. [7-3]

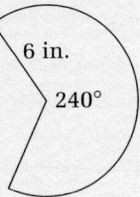

6 in.

240°

25 in.

8 Linear Equations as Models

OVERVIEW

➤ **Unit 8** introduces linear equations and inequalities and systems of linear equations and inequalities. Students learn to represent real-life situations with tables, equations, and graphs. They learn to write linear equations in slope-intercept form and in standard form and to write equations for lines from given information. In this unit, students learn to solve linear systems of equations and inequalities by graphing.

➤ Trends in Olympic swimming data is the theme of the Unit Project. Students analyze past Olympic swimming records and use mathematical models in the form of linear equations to predict future records.

➤ Connections to sports, consumer topics, social studies, business, and science are integrated into the teaching materials and exercises.

➤ Problem-solving strategies used in Unit 8 include *Use a Graph* (Sections 8-5 and 8-7).

Unit Objectives

Section	Objectives	NCTM Standards
8-1	• Recognize how the rate and the initial value in a real-life situation appear in an equation and a graph representing the situation.	1, 2, 4, 5, 6, 8
8-2	• Use the standard form of an equation for a line to model real-life situations. • Change standard form to slope-intercept form.	1, 2, 4, 5, 6, 8
8-3	• Describe slopes of horizontal and vertical lines. • Write equations for horizontal and vertical lines.	1, 2, 4, 5, 8
8-4	• Use two points or the slope and one point to write an equation for a line. • Use an equation for a line to solve problems.	1, 2, 4, 5, 6, 8
8-5	• Find the point where two lines intersect.	1, 2, 4, 5, 6, 8
8-6	• Graph linear inequalities on a coordinate plane.	1, 2, 4, 5, 8
8-7	• Solve problems by writing and graphing systems of inequalities.	1, 2, 4, 5, 8, 12

Section	Connections to Prior and Future Concepts
8-1	**Section 8-1** builds on the concepts of coordinates and graphs of equations introduced in Unit 4 and extended in Unit 7 of this book. Students will continue their study of these topics in the rest of this unit and in Unit 10 of Book 1 as well as in most of the units of Books 2 and 3.
8-2	**Section 8-2** continues the study of linear graphs from Section 8-1, focusing on models of linear combinations and the standard and slope-intercept forms of equations. These topics are extended throughout Unit 8 of Book 1; in Units 2, 3, 5, and 9 of Book 2; and in Units 1, 2, and 9 of Book 3.
8-3	**Section 8-3** extends the study of linear graphs from Sections 8-1 and 8-2 to include writing equations for horizontal and vertical lines. Knowledge of these special linear equations will be assumed throughout Books 2 and 3.
8-4	**Section 8-4** builds on the topics from Sections 8-1, 8-2, and 8-3 to introduce writing equations for lines by using two points or the slope and one point. The skill of writing equations is developed throughout this unit and in Unit 10 of Book 1 and extended throughout Books 2 and 3.
8-5	**Section 8-5** presents graphing systems of linear equations and finding the point where the two lines intersect. Section 5-8 and previous lessons in Unit 8 provide the foundation for this instruction. Systems of equations are studied again in Units 3, 4, and 9 of Book 2 and in Unit 1 of Book 3.
8-6	**Section 8-6** uses the skills for graphing linear equations to graph linear inequalities. Inequalities were explored earlier in Sections 3-3 and 5-4. They will be further examined in Section 8-7 and in Units 4 and 9 of Book 2 and Units 1 and 2 of Book 3.
8-7	**Section 8-7** extends the study of linear inequalities begun in Section 8-6 to introduce graphing systems of inequalities. This topic is extended in Book 2 and in Book 3, where linear programming is introduced in Unit 1.

Integrating the Strands

Strands	Sections
Number	8-7
Algebra	8-1, 8-2, 8-3, 8-4, 8-5, 8-6, 8-7
Functions	8-1, 8-4
Geometry	8-1, 8-2, 8-3, 8-4, 8-5, 8-6, 8-7
Statistics and Probability	8-4, 8-6, 8-7
Discrete Mathematics	8-1, 8-2, 8-3, 8-6
Logic and Language	8-2, 8-4, 8-5

Section Planning Guide

➤ Essential exercises and problems are indicated in boldface.
➤ Ongoing work on the Unit Project is indicated in color.
➤ Exercises and problems that require student research, group work, manipulatives, or graphing technology are indicated in the column headed "Other."

Section	Materials	Pacing	Standard Assignment	Extended Assignment	Other
8-1	graphing technology	Day 1	**1–9, 11–14**, 15, 16	**1–9**, 10, **11–14**, 15, 16	17, 18
		Day 2	**19–21**, 23a, **24–32**, 34–43, 44	**19–21**, 22, 23a, **24–32**, 34–43, 44	23b–c, 33
8-2	construction paper, scissors,	Day 1	**1–5**, 6	**1–5**, 6	7, 8
	graphing technology	Day 2	**9–19**, 23–26, 27	**9–19**, 22–26, 27	20, 21, 27
8-3	scissors	Day 1	**1–3, 5–21**, 25, 27–35, 36	**1–3**, 4, **5–21**, 23–25, 27–35, 36	22, 26
8-4	graphing technology	Day 1	**2–11, 13, 14**	1, **2–11, 13, 14**	12
		Day 2	**15–24**, 25–30, 31	**15–24**, 25–30, 31	31
8-5	white and colored paper,	Day 1	**1–5**	**1–5**	
	protractor, scissors, tape	Day 2	**6–16, 18–20**, 24–29, 30	**6–16**, 17, **18–20**, 22–29, 30	21
8-6		Day 1	**2–14**, 15	1, **2–14**, 15	16
		Day 2	**17–26**, 31–38, 39	**17–26**, 27–29, 31–38, 39	30
8-7		Day 1	1, **2–16**, 18, 19, **22–24**, 25–38, 39	1, **2–16**, 17–21, **22–24**, 25–38, 39	39
Review		**Day 1**	Unit Review	Unit Review	
Test		**Day 2**	Unit Test	Unit Test	

Yearly Pacing	Unit 8 Total	Units 1–8 Total	Remaining	Total
	16 days (2 for Unit Project)	126 days	38 days	164 days

Support Materials

➤ See **Project Book** for notes on Unit 8 Project: Predicting the Future.
➤ "UPP" and "disk" refer to **Using Plotter Plus** booklet and **Plotter Plus** disk.
➤ Warm-up exercises for each section are available on **Warm-Up Transparencies**.
➤ "FI," "PC," "GI," "MA," and "Stats!" refer, respectively, to the McDougal Littell Mathpack software Activity Books for **Function Investigator, Probability Constructor, Geometry Inventor, Matrix Analyzer**, and **Stats!**.

Section	Study Guide	Practice Bank	Problem Bank	Activity Bank	Explorations Lab Manual	Assessment Book	Visuals	Technology
8-1	8-1	Practice 59	Set 16	Enrich 52	Master 2	Quiz 8-1	Folder 8	FI Acts. 4 and 5 UPP, page 31 UPP, page 32 Line Plotter (disk)
8-2	8-2	Practice 60	Set 16	Enrich 53	Masters 2, 3	Quiz 8-2		
8-3	8-3	Practice 61	Set 16	Enrich 54	Masters 2, 3	Quiz 8-3 Test 31	Folder 8	
8-4	8-4	Practice 62	Set 17	Enrich 55	Masters 2, 3	Quiz 8-4	Folder 8	Line Quiz (disk)
8-5	8-5	Practice 63	Set 17	Enrich 56	Add. Expl. 8 Masters 2, 3	Quiz 8-5 Test 32	Folder 8	FI Act. 10 UPP, pages 16–17 UPP, page 33
8-6	8-6	Practice 64	Set 18	Enrich 57	Masters 2, 3	Quiz 8-6	Folder 9	UPP, pages 17–19
8-7	8-7	Practice 65	Set 18	Enrich 58	Masters 2, 3	Quiz 8-7 Test 33	Folder 9	UPP, pages 19–20 UPP, page 34
Unit 8	Unit Review	Practice 66	Unifying Problem 8	Family Involve 8		Tests 34, 35		

UNIT TESTS

Form A

Spanish versions of these tests are on pages 150–153 of the Assessment Book.

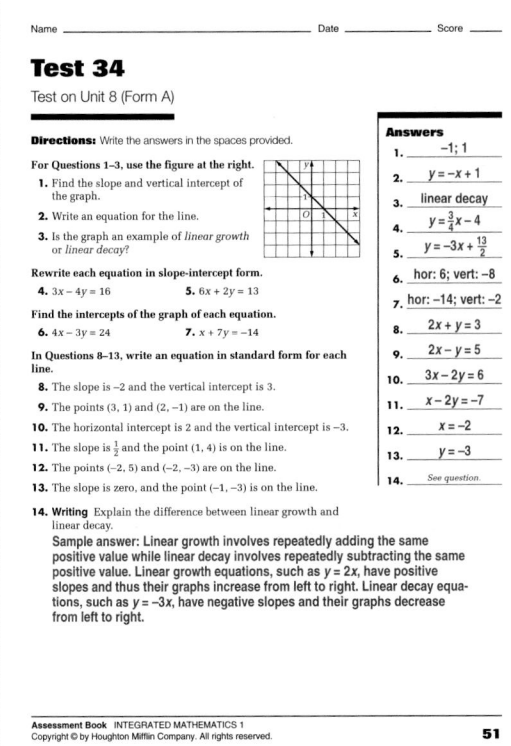

Name _____ Date _____ Score _____

Test 34

Test on Unit 8 (Form A)

Directions: Write the answers in the spaces provided.

For Questions 1–3, use the figure at the right.

1. Find the slope and vertical intercept of the graph.
2. Write an equation for the line.
3. Is the graph an example of *linear growth* or *linear decay*?

Rewrite each equation in slope-intercept form.

4. $3x - 4y = 16$ 5. $6x + 2y = 13$

Find the intercepts of the graph of each equation.

6. $4x - 3y = 24$ 7. $x + 7y = -14$

In Questions 8–13, write an equation in standard form for each line.

8. The slope is -2 and the vertical intercept is 3.
9. The points $(3, 1)$ and $(2, -1)$ are on the line.
10. The horizontal intercept is 2 and the vertical intercept is -3.
11. The slope is $\frac{1}{2}$ and the point $(1, 4)$ is on the line.
12. The points $(-2, 5)$ and $(-2, -3)$ are on the line.
13. The slope is zero, and the point $(-1, -3)$ is on the line.

14. **Writing** Explain the difference between linear growth and linear decay.
Sample answer: Linear growth involves repeatedly adding the same positive value while linear decay involves repeatedly subtracting the same positive value. Linear growth equations, such as $y = 2x$, have positive slopes and thus their graphs increase from left to right. Linear decay equations, such as $y = -3x$, have negative slopes and their graphs decrease from left to right.

Answers
1. $-1; 1$
2. $y = -x + 1$
3. linear decay
4. $y = \frac{3}{4}x - 4$
5. $y = -3x + \frac{13}{2}$
6. hor: 6; vert: -8
7. hor: -14; vert: -2
8. $2x + y = 3$
9. $2x - y = 5$
10. $3x - 2y = 6$
11. $x - 2y = -7$
12. $x = -2$
13. $y = -3$
14. See question.

51

Name _____ Date _____ Score _____

Test 34 (continued)

Directions: Write the answers in the spaces provided.

In Questions 15 and 16, without graphing tell whether each system of equations has a *solution* or *no solution*.

15. $-5x + y = 3$
    $-10x + 2y = 8$

16. $3x + 3y = 6$
    $y = -3x + 6$

17. On the coordinate plane at the right, graph the system of equations
$y = -3x$
$y = 3x + 3$.

18. Use substitution to find the exact solution of the system given in Question 17.

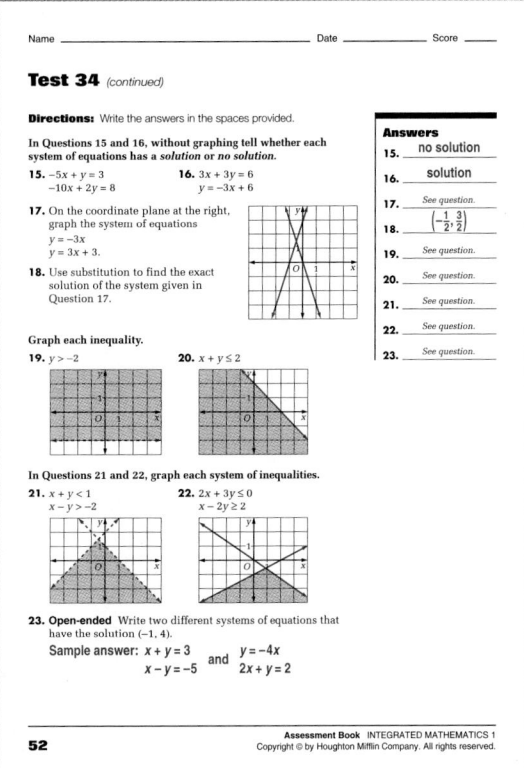

Graph each inequality.

19. $y > -2$ 20. $x + y \le 2$

In Questions 21 and 22, graph each system of inequalities.

21. $x + y < 1$
    $x - y > -2$

22. $2x + 3y \le 0$
    $x - 2y \ge 2$

23. **Open-ended** Write two different systems of equations that have the solution $(-1, 4)$.
Sample answer: $x + y = 3$ and $y = -4x$
              $x - y = -5$ $2x + y = 2$

Answers
15. no solution
16. solution
17. See question.
18. $\left(-\frac{1}{2}, \frac{3}{2}\right)$
19. See question.
20. See question.
21. See question.
22. See question.
23. See question.

52

Form B

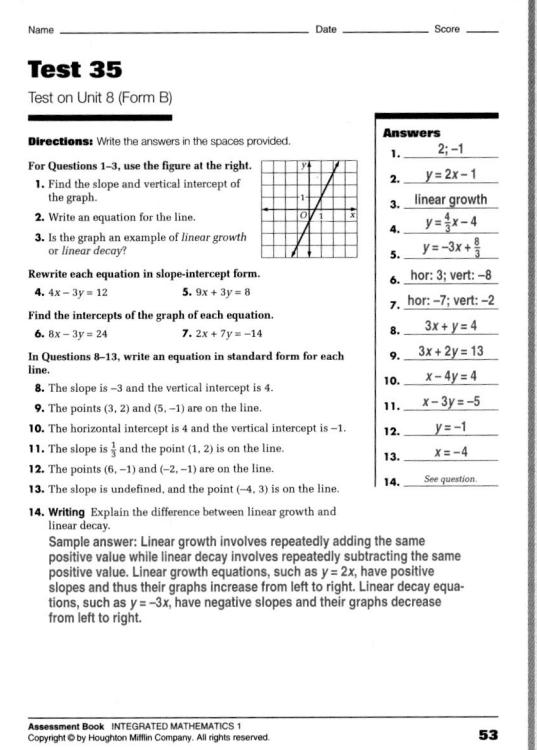

Name _____ Date _____ Score _____

Test 35

Test on Unit 8 (Form B)

Directions: Write the answers in the spaces provided.

For Questions 1–3, use the figure at the right.

1. Find the slope and vertical intercept of the graph.
2. Write an equation for the line.
3. Is the graph an example of *linear growth* or *linear decay*?

Rewrite each equation in slope-intercept form.

4. $4x - 3y = 12$ 5. $9x + 3y = 8$

Find the intercepts of the graph of each equation.

6. $8x - 3y = 24$ 7. $2x + 7y = -14$

In Questions 8–13, write an equation in standard form for each line.

8. The slope is -3 and the vertical intercept is 4.
9. The points $(3, 2)$ and $(5, -1)$ are on the line.
10. The horizontal intercept is 4 and the vertical intercept is -1.
11. The slope is $\frac{1}{3}$ and the point $(1, 2)$ is on the line.
12. The points $(6, -1)$ and $(-2, -1)$ are on the line.
13. The slope is undefined, and the point $(-4, 3)$ is on the line.

14. **Writing** Explain the difference between linear growth and linear decay.
Sample answer: Linear growth involves repeatedly adding the same positive value while linear decay involves repeatedly subtracting the same positive value. Linear growth equations, such as $y = 2x$, have positive slopes and thus their graphs increase from left to right. Linear decay equations, such as $y = -3x$, have negative slopes and their graphs decrease from left to right.

Answers
1. $2; -1$
2. $y = 2x - 1$
3. linear growth
4. $y = \frac{4}{3}x - 4$
5. $y = -3x + \frac{8}{3}$
6. hor: 3; vert: -8
7. hor: -7; vert: -2
8. $3x + y = 4$
9. $3x + 2y = 13$
10. $x - 4y = 4$
11. $x - 3y = -5$
12. $y = -1$
13. $x = -4$
14. See question.

53

Name _____ Date _____ Score _____

Test 35 (continued)

Directions: Write the answers in the spaces provided.

In Questions 15 and 16, without graphing tell whether each system of equations has a *solution* or *no solution*.

15. $2x + 2y = 6$
    $y = -2x + 6$

16. $-7x + y = 3$
    $-14x + 2y = 8$

17. On the coordinate plane at the right, graph the system of equations
$y = 3x$
$y = -x + 2$.

18. Use substitution to find the exact solution of the system given in Question 17.

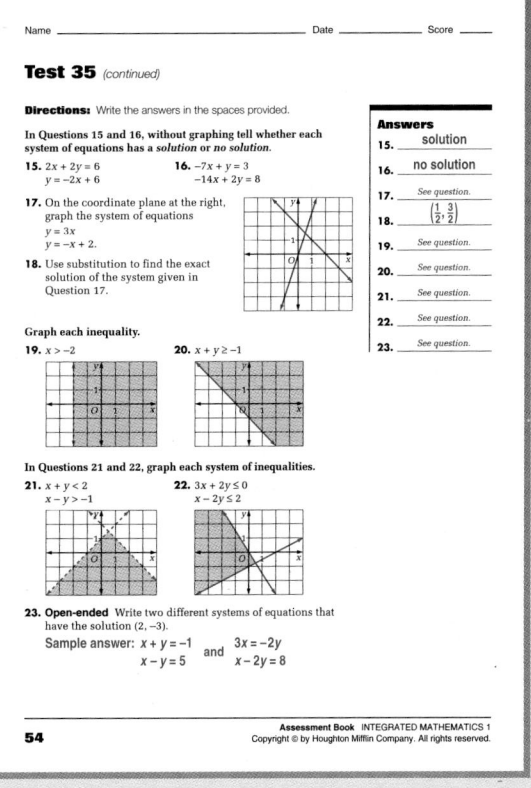

Graph each inequality.

19. $x > -2$ 20. $x + y \ge -1$

In Questions 21 and 22, graph each system of inequalities.

21. $x + y < 2$
    $x - y > -1$

22. $3x + 2y \le 0$
    $x - 2y \le 2$

23. **Open-ended** Write two different systems of equations that have the solution $(2, -3)$.
Sample answer: $x + y = -1$ and $3x = -2y$
              $x - y = 5$ $x - 2y = 8$

Answers
15. solution
16. no solution
17. See question.
18. $\left(\frac{1}{2}, \frac{3}{2}\right)$
19. See question.
20. See question.
21. See question.
22. See question.
23. See question.

54

Software Support

McDougal Littell Mathpack
Function Investigator
Stats!

Outside Resources

Books/Periodicals

Seymour, Dale. *Introduction to Line Designs.* Palo Alto, CA: Dale Seymour Publications, 1992.

McArthur, David. *Algebraic Thinking Tools: Supports for Modeling Situations and Solving Problems in Kids' Worlds.* Santa Monica, CA: Rand Corporation, 1989.

Austin, Joe Dan, Arthur Howard, and R. D. Thomas. *Mathematics of Money.* Palo Alto, CA: West Publishing, 1992.

Manipulatives

Britton, Jill and Walter Britton. *Teaching Tessellating Art: Activities and Transparency Masters.* Palo Alto, CA: Dale Seymour Publications, 1992.

Software

Yerushalmy, Michal. *Algebraic Patterns.* Macintosh and MS-DOS. Pleasantville, NY: Sunburst, 1992.

Confrey, Jere, Francis Modugus, and Adam Miller. *Function Finder.* San Antonio, TX: Blohm & Associates, 1991.

Math Tools for Algebra. Spring Branch Software. Acton, MA: William K. Bradford Publishing.

IBM Mathematics Exploration Toolkit. IBM, Armonk, NY.

➤ Students write a news article and predict a new Olympic record for a swimming event.

➤ When writing the article, students analyze past Olympic results to help predict the new record.

➤ Students work together with a partner and both contribute to the success of the project.

PROJECT PLANNING

Project Teams

Have students work on the project with a partner. Students can work together to analyze the data, make the prediction, write the article, and create any graphs or illustrations.

Support Materials

The *Project Book* contains information about the following topics for use with this Unit Project.

➤ Project Description

➤ Teaching Commentary

➤ Working on the Unit Project Exercises

➤ Completing the Unit Project

➤ Assessing the Unit Project

➤ Alternative Projects

➤ Outside Resources

Using Technology

Have students create and use a data file to support the information in the news article.

unit 8

Linear Equations as Models

The secret to success may be to flop! Dick Fosbury won a gold medal in the 1968 Olympic high jump competition by flopping over the bar head first, landing on his back. Fosbury's "backwards" approach led to a major breakthrough in sports. Most high jump competitors now use his innovative style—the Fosbury flop!

What do **MATH**, **COMPUTER SCIENCE**, and **PHYSICS** have to do with coaching **SPORTS**?

Sports trainers use scientific breakthroughs to make athletes' training more efficient and productive. Gymnasts and ice skaters have their moves captured on film and analyzed by a computer. Mike Powell set an Olympic high jump record after initiating a scientific training plan that included pulling an open parachute behind him when he ran.

training

414

Suggested Rubric for Unit Project

4 The news article predicting a new swimming record is clearly written and supported by data displays. The trends of past Olympic swimming events are analyzed and incorporated in the article. The mathematical models chosen are used correctly to make a prediction. The writing style is creative and appropriate for the readers of the newspaper or magazine.

3 The news article is adequate. The data displays have not been analyzed as thoroughly as they could have been. The data file has minor problems in supporting the information included in the article.

Predicting the Future

Your project is to write a news article predicting the new record-breaking time for the Olympic women's 400 m freestyle swimming event.

To write the article, you will use mathematical models to analyze past Olympic results and make your prediction. A completed project will include a data file you create to support the information you include in the article.

record breaker

One Massachusetts Institute of Technology group developed an innovative design for a **human-powered boat** that lifts out of the water and flies when pedaled fast enough. The boat's speed of 18.5 knots beat the former world record of 14.4 knots to become the fastest human-powered water vehicle.

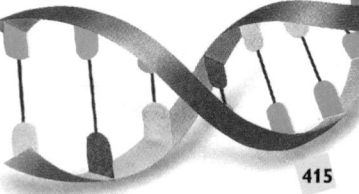

A revolutionary **scientific** breakthrough occurred when James Watson, Francis Crick, and Rosalind Franklin discovered that DNA could be modeled by a double helix. Their work revealed the structure of genetic material and paved the way for modern medical research.

415

Suggested Rubric for Unit Project

2 The news article is incomplete. It is not supported in large part by the data displays or analysis of past Olympic events. This project should be returned with suggestions for improvements and a new deadline.

1 The news article cannot be evaluated. It is illegible, incomplete, or not understandable. The project should be returned with a new deadline for completion. Students should be encouraged to speak with the teacher as soon as possible so that they understand the purpose and the format of the project.

Limited English Proficiency

Writing a news article is a valuable project for students acquiring English. Have a student fluent in English be a partner to a student acquiring English. The project gives students an opportunity to develop further their skills in written communication, as well as giving them experience in analyzing real-world data.

ADDITIONAL BACKGROUND

Multicultural Note

Dr. Charles Richard Drew was an African-American who discovered how to preserve blood in blood banks until it was needed. In 1939, while doing research at Columbia University in New York City, Dr. Drew discovered that plasma could be stored for a long time without spoiling, and that blood typing was unnecessary when using plasma. These discoveries were very important during World War II and helped the American Red Cross bring blood to the armed forces on battlefields throughout the war.

Penicillin

Another medical breakthrough was the discovery in 1928 of penicillin by Sir Alexander Fleming, a British doctor. Penicillin is able to stop the growth of and kill hundreds of different kinds of bacteria that cause such illnesses as sore throats, diphtheria, and pneumonia. It was one of the most widely used medicines for many years.

Playing the Market

Pretend to invest $1000 for one week in a stock chosen from a stock newspaper report. Write a report comparing the stock investment with the interest earned on $1000 in a bank savings account for one week.

Projected Savings

Research a long-term payment plan for a car or other item. Calculate how much money would be saved by paying the total cost all at once, how long it would take to save the money, and decide which plan is best.

Working on the Unit Project

Your work in Unit 8 will help you make your predictions.

Related Exercises:
Section 8-1, Exercise 44
Section 8-2, Exercise 27
Section 8-3, Exercise 36
Section 8-4, Exercise 31
Section 8-5, Exercise 30
Section 8-6, Exercise 39
Section 8-7, Exercise 39

Alternative Projects p. 469

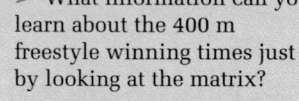

Unit Project 8

Getting Started

For this project you should work with a partner. Here are some ideas to help you get started on your project.

☞ Think about how you will organize the data you collect. Will you be able to use a computer database or spreadsheet?

☞ You should include graphs in your article. You may want to review the various types of data displays in Section 3-6.

☞ Think about a newspaper or magazine you may want to write for. Will you suit the style of your article to its readers?

Can We Talk BREAKTHROUGHS

➤ What information can you learn about the 400 m freestyle winning times just by looking at the matrix?

➤ Suppose you want to predict the winning time for the men's 400 m freestyle in the year 2012. What information do you need to know?

➤ What factors do you think have caused Olympic records to improve?

➤ Records for Olympic sports have been improving ever since modern-day Olympics were first held in 1896. Do you think the records can continue to improve forever? Why or why not?

➤ Describe some of the advantages that modern Olympic athletes have over earlier Olympic athletes.

Winning Times Olympic 400 m Freestyle Swimming		
	Men	Women
Year	Time (seconds)	Time (seconds)
1960	258.3 AUS	290.6 USA
1964	252.2 USA	283.3 USA
1968	249.0 USA	271.8 USA
1972	240.27 USA	259.44 AUS
1976	231.93 USA	249.89 E.Ger
1980	231.31 USSR	248.76 E.Ger
1984	231.23 USA	247.10 USA
1988	226.95 E.Ger	243.85 USA
1992	225.00 UT	247.18 GER

➤ Do you think events occurring now help you predict future events? Explain.

➤ What types of predictions have you made in your life? What information have you used to help you make a prediction?

Answers to Can We Talk?

➤ For the Olympic years between 1960 and 1992, the winning times and the winner's country are shown for men and women.

➤ The information necessary to predict winning times would involve how the times have decreased over the years; look for a pattern.

➤ Training, swimsuit design, and so on, have contributed to improved performance.

➤ Records can improve, however the time can never decrease forever because it will always take some amount of time to swim the 400 m freestyle.

➤ Advantages include help from technology, and better understanding of diet and training experience.

➤ Yes. Studying past and present events help us to notice trends and patterns and help us to make informed guesses about the future.

➤ Answers may vary.

8-1 Linear Growth and Decay

RISING AND FALLING

Talk it Over

At a hot air balloon festival in Albuquerque, New Mexico, a balloon takes off from a spot 6000 ft above sea level. The balloon rises 110 ft every minute.

? ft
t = 3 min

? ft
t = 2 min

6110 ft
t = 1 min

6000 ft
t = 0 min

6000 ft above sea level

Time in minutes (t)	Height of balloon in feet (h)
0	6000
1	6110
2	?
⋮	⋮

1. Model the situation with a table of values.

2. Write an equation for the height of the balloon after t minutes.

3. Which variable in the table is the control variable? Which is the dependent variable?

4. Plot the points. What pattern do you see? Connect the points.

5. Your graph is an example of *linear growth*. What does *linear* mean? What is *growing* in this situation?

6. What is the slope of the graph? How is the slope related to the situation?

7. Where does the graph meet the vertical axis? How is this point related to the height of the balloon?

8-1 Linear Growth and Decay   **417**

Answers to Talk it Over

1.

Time in minutes (t)	Height of balloon in feet (h)
0	6000
1	6110
2	6220
3	6330
4	6440
5	6550
6	6660

2. $h = 6000 + 110t$

3. control: t; dependent: h

4. Graphs may vary. An example is given at right. The points lie on a straight line.

5. Answers may vary. Examples are given. *Linear* means in a straight line. In this situation, *growing* means increasing in height above the ground.

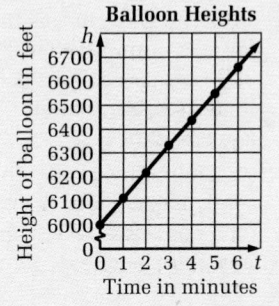

Balloon Heights

Answers continued on next page.

PLANNING

Objectives and Strands
See pages 414A and 414B.

Spiral Learning
See page 414B.

Materials List
➤ Graph paper
➤ Ruler
➤ Graphics calculator or graphing software

Recommended Pacing
Section 8-1 is a two-day lesson.
Day 1
Pages 417–419: Talk it Over 1 through Talk it Over 13, *Exercises 1–18*
Day 2
Page 420: Linear Growth and Decay through Look Back, *Exercises 19–44*

Extra Practice
See pages 633–635.

Warm-Up Exercises
Warm-Up Transparency 8-1

Support Materials
➤ Practice 59
➤ Enrichment 52 in the Activity Bank
➤ Study Guide 8-1
➤ Problem Set 16
➤ Diagram Masters 2, 3 in the Explorations Lab Manual
Overhead Visual 8
➤ McDougal Littell Mathpack software: *Function Investigator*
➤ Function Investigator with Matrix Analyzer Activity Book: Function Investigator Activities 4 and 5
➤ Using Plotter Plus: Slope-Intercept Form
➤ Using Plotter Plus: Parallel and Perpendicular Lines
➤ Using Mac or IBM Plotter Plus Disk: Line Plotter
➤ Quiz 8-1
➤ Alternative Assessments 1–3

417

Talk it Over

Use question 3 to review the meaning of *control variable* and *dependent variable*. In question 5, point out that the word *line* is the root of the word *linear*.

For question 6, relate the slope of the line to the ascension of the balloon.

Encourage students to use the terms *slope* and *vertical intercept* as they compare equations for question 8.

Communication: Discussion

You may wish to take a few minutes to discuss the equation $y = mx + b$. Ask students to consider what happens when b remains a fixed value (say 2) but m takes on different values. (A set of lines is generated that all pass through $(0, 2)$). Then look at the situation when m remains a fixed value and different values are assigned to b. (A set of parallel lines with slope m is generated, one line for each value of b.) Such considerations will deepen students' understanding of the slope-intercept equation for a line.

Research

Point out that the slope or pitch of a roof is described by the ratio of its rise to its run. Have students research slopes commonly used for roofs and the conditions that determine an appropriate slope.

Another balloon takes off from 8200 ft above sea level and rises at a rate of 60 ft/min. The graph shows the height of this balloon over time. The equation below also shows the height (h) of the balloon after t minutes.

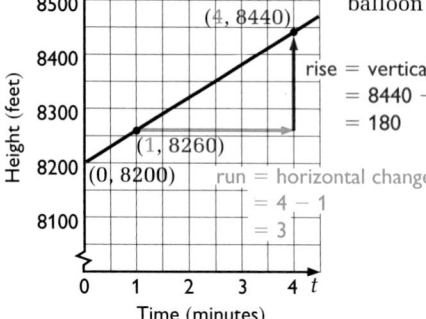

rise = vertical change
= $8440 - 8260$
= 180

run = horizontal change
= $4 - 1$
= 3

$h = 60t + 8200$

vertical intercept:
The line meets the vertical axis at $(0, 8200)$. Its vertical intercept is 8200.

This is the height of the balloon when $t = 0$.

slope of the line:
$\dfrac{\text{rise}}{\text{run}} = \dfrac{180}{3} = 60$

This is the rate at which the balloon rises in feet per minute.

Often the variable m is used for the slope, b is used for the vertical intercept, x is the control variable, and y is the dependent variable. Below is the **slope-intercept form of an equation** for a line.

$$y = mx + b$$

slope ⟵ ⟶ vertical intercept

Watch Out!
The coefficient of the x-term is the slope, even if the equation is written in the form $y = b + mx$.

BY THE WAY...
Since hot air is lighter than cold air, hot air balloons rise when the air inside the balloon is warmer than the air outside. The hottest air is at the top of a hot air balloon. The air ranges in temperature from 180°F to 225°F.

Talk it Over

8. Compare the equation you wrote in question 2 with the equation $h = 60t + 8200$. How are they alike? How are they different?

9. Suppose you redraw the graph of the balloon's height so that there is no gap in the scale. Would this change how the slope of the line looks? Would the slope still be 60? Explain.

10. Does the graph represent a function? How can you tell?

11. a. Does the graph represent direct variation? Why or why not?

    b. Explain how the equation $h = 60t + 8200$ is different from a direct variation equation.

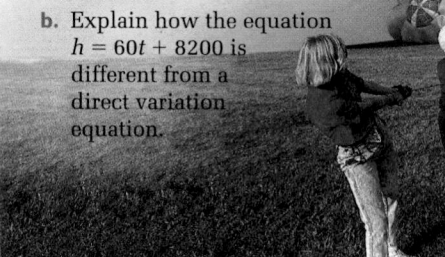

Unit 8 Linear Equations as Models

Answers to Talk it Over

6. 110; The slope means that for every minute after the balloon takes off, the balloon rises 110 ft more above sea level.

7. $(0, 6000)$; This point represents the height of the balloon above sea level before it begins rising.

8. Answers may vary. Examples are given. Both equations have the variables h for height and t for time and model rise of a balloon. The slope in the equation for question 2 is 110 and the vertical intercept is 6000, while the slope in the equation $h = 60t + 8200$ is 60 and

the vertical intercept is 8200.

9. Explanations may vary. An example is given. If you redraw the graph so there is no gap in the scale, but still let two units on the scale equal 100 ft, the graph of the line will look the same (only higher up). If you decide to let one

Sample 1

When you travel up from Earth's surface, the air temperature decreases by about 11°F for each mile you rise above the ground. Suppose the temperature of the air at ground level is 68°F.

a. Model the situation with a table of values.

b. Use your table to make a graph. Find the slope and the vertical intercept of the line that contains the points. Explain what they mean in terms of the situation.

c. Write an equation for the Fahrenheit temperature of the air (f) as a function of the height in miles above the ground (h).

Sample Response

a. The starting temperature is 68°F. Because the temperature is decreasing, subtract 11 for each mile above the ground.

Height above ground (mi)	Temperature (°F)
0	68
1	57
2	46
3	35
4	24

b. Plot the points and connect them.

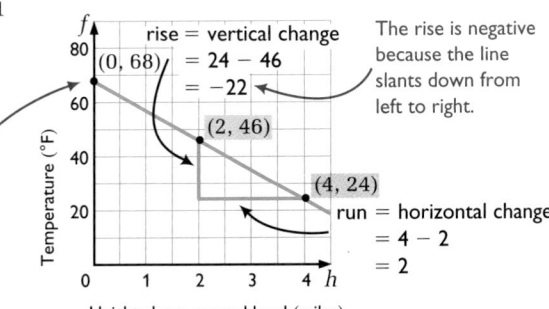

rise = vertical change
= 24 − 46
= −22

The rise is negative because the line slants down from left to right.

run = horizontal change
= 4 − 2
= 2

vertical intercept = 68
The vertical intercept is the temperature at ground level.

$$\text{slope} = \frac{\text{rise}}{\text{run}} = \frac{-22}{2} = -11$$

The slope is the rate at which the temperature changes in degrees per mile.

Watch Out!
Subtract in the same order to find the *run* as you do to find the *rise*.

second point   first point
$$\frac{\text{rise}}{\text{run}} = \frac{24 - 46}{4 - 2}$$

c. The equation is $f = -11h + 68$.

Talk it Over

12. Jena wrote the equation $f = 68 - 11h$ for the temperature in Sample 1. Compare Jena's equation and the equation $f = -11h + 68$. Do both equations have the same table of values? Do they have the same graph?

13. Suppose you want to graph $f = -11h + 68$ on a graphics calculator, using x and y as the control and dependent variables. Which variable represents f? Which represents h?

····▶ Now you are ready for:
Exs. 1–18 on pp. 421–422

8-1 Linear Growth and Decay

419

Answers to Talk it Over

unit on the scale equal 100 ft, then the graph of the line rises more steeply. Either way, the slope will not change.

10. Yes. The graph passes the vertical-line test.

11. a. No. Reasons may vary. An example is given. The graph does not pass through the origin.

b. Explanations may vary. An example is given. The vertical intercept in the equation $h = 60t + 8200$ is 8200. The vertical intercept in a direct variation equation is always zero.

12. Yes; Yes.

13. y; x

Additional Sample

S1 Frankie has $500 in her savings account at the beginning of the year. Each month she saves $150.

a. Model the situation with a table of values and a graph.

Month	Savings ($)
0	500
1	650
2	800
3	950
4	1100

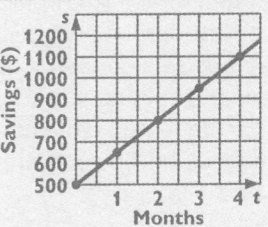

b. Find the slope and the vertical intercept of the line. Explain what they mean in terms of the situation.

slope = 150;
vertical intercept = 500;
The slope is the rate at which the savings change in dollars per month. The vertical intercept is the savings at the beginning of the year.

c. Write an equation for the savings (s) as a function of the time in months (t). $s = 150t + 500$

Using Technology

Graphing equations with technology can help students to develop a better understanding of the behavior of graphs as equations change.

Invite students to graph the equation for Sample 1 on their graphics calculators or with the *Function Investigator* software. Let them use a calculator or *Function Investigator* Activities 4 and 5 in the *Function Investigator with Matrix Analyzer Activity Book* to explore what happens to the graph as m and b change. Entering and graphing a function on a graphics calculator is discussed on page 613 of the Technology Handbook.

419

Talk it Over

Use question 13 on page 419 to relate the specific equation $f = -11h + 68$ to the general form $y = mx + b$. Stress that when specific numerical values are assigned to m and b, particular lines are determined. Also, in practical situations such as in Sample 1, letters other than x and y are used to relate to the conditions of the problem. Thus, temperature in degrees Fahrenheit is f (to replace y) and height is h (to replace x). Some students may get confused by this changing of variables in problem situations, so continue to make the connection back to the general form $y = mx + b$.

Additional Sample

S2 Classify each situation as *linear growth, linear decay,* or *neither.*

a. A dog owner buys a 50-pound bag of dog food. Each day she uses 2 pounds of dog food to feed her dogs.
linear decay

b. A potter makes 2 bowls each day. linear growth

c. Postage is $.29 for the first ounce and $.20 for each additional ounce.
linear growth

Look Back

Use the Look Back questions for class discussion. Ask students to give some examples of linear growth and linear decay. Then ask them to summarize how slope and vertical intercept are used in graphs of linear growth and linear decay.

The graphs modeling the height of a balloon and the temperature of the air are lines. They are examples of **linear growth** and **linear decay.** They are also *linear functions.*

	Linear Growth	**Linear Decay**
Characteristic	same value repeatedly added	same value repeatedly subtracted
Slope	positive	negative
Sketch		

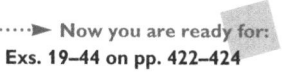
Sample 2

Writing Classify each situation as *linear growth*, *linear decay*, or *neither*. Explain your choice.

a. Eleanor Basave put $5 in a piggy bank for her granddaughter when she was born. Every day after that Eleanor put a quarter in the bank.

b. During each year, the shortest distance between Earth and the sun is 9.14×10^7 mi and the greatest distance is 9.45×10^7 mi.

c. A teacher on a remote Arctic island buys 144 cans of fruit for the year. Each week the teacher eats 4 cans of fruit.

Sample Response

a. Linear growth. The value 0.25 is repeatedly added to the starting value, 5.

b. Neither. The distance increases and decreases during one year, then follows the same pattern the following year. There is not one number that is repeatedly added or subtracted.

c. Linear decay. The value 4 is repeatedly subtracted from the starting value, 144.

▶ **Now you are ready for:**
Exs. 19–44 on pp. 422–424

Look Back ◄

What are two aspects of a linear growth or decay situation? How are they represented in the situation, on the graph, and in the equation?

Unit 8 Linear Equations as Models

Answers to Look Back

Two aspects of a linear growth or decay situation are the original value of the dependent variable and the rate of change. In the situation, the original value of the dependent variable is the value of the variable before any change occurs. The rate of change indicates how quickly the original value of the dependent variable increases or decreases. On the graph, the original value of the dependent variable is represented by the vertical intercept, and the rate of change is represented by the slope of the line. In the slope-intercept form of an equation for a line, $y = mx + b$, b represents the original value of the dependent variable, and m represents the rate of change.

Exercises and Problems

For Exercises 1–4, match each equation with its graph.

1. $y = 3x + 2$
2. $y = 3x - 2$
3. $y = 2 - 3x$
4. $y = -3x$

A.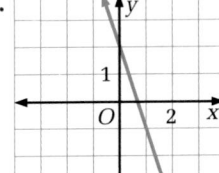

B.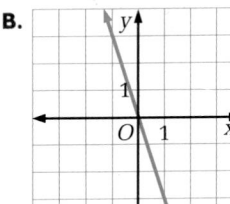

C.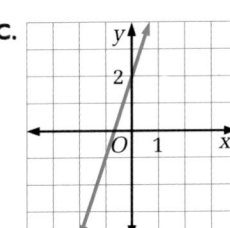

D.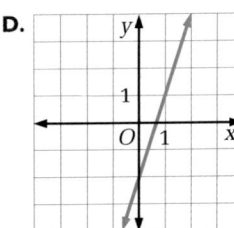

For Exercises 5–8, do these things.

a. Model each situation with a table and a graph.

b. Find the slope and the vertical intercept of each line. Explain how they are related to the original situation.

c. Model each situation with an equation.

5. Better Building Supplies sells sand for $19/ton, plus $25 for delivery.

6. For each kilometer beneath Earth's surface, the temperature increases by 25°C. Suppose the temperature at Earth's surface is 20°C.

7. The gas tank of a van contains 12 gal. The van uses $\frac{1}{15}$ gal/mi.

8. A truck rental company charges $45 per day, plus $.39 per mile.

9. Nancy noticed that in Exercises 5–8, the slope of the line is always the rate given in the situation, and the vertical intercept is always the starting value. Do you *agree* or *disagree* with her? Explain.

10. **Open-ended** Describe a situation that could be modeled by the equation $y = 0.25x + 1.75$.

Without graphing, find the slope and the vertical intercept of the line modeled by each equation.

11. $y = 3x - 5$

12. $y = 4.5 + 0.5x$

13. $y = -x - \frac{1}{2}$

14. $y = -1.2x + 6$

8-1 Linear Growth and Decay

Suggested Assignment

Standard 1–9, 11–16, 19–21, 23–32, 34–44

Extended 1–16, 19–32, 34–44

Integrating the Strands

Algebra Exs. 1–33, 35–44

Functions Exs. 23, 32, 33

Geometry Exs. 1–10, 17, 18, 24–27, 33, 34

Discrete Mathematics Ex. 23

Communication: Discussion

You may need to review the meaning of rate when discussing Ex. 9.

Reasoning

Ex. 10 challenges students to show they understand the meaning of slope and vertical intercept by using an example of linear growth from their own experience.

Communication: Reading

The direction line for Exs. 11–14 refers to the *line modeled by each equation.* As students work with equations and the lines that represent them, it tends to become a natural use of language to refer to the equation as the line. For example, it is acceptable to say *graph the line $y = 3x - 5$.*

Answers to Exercises and Problems

1. C   2. D   3. A   4. B

5. a.

Weight in tons (w)	Total cost in dollars (c)
0	25
1	44
2	63
3	82

Sand Sales at Better Building Supplies

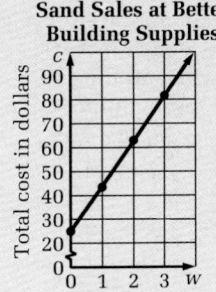

Weight in tons

b. slope: 19; vertical intercept: 25; Explanations may vary. An example is given. The slope shows that the cost increases $19 for each additional ton of sand sold. The vertical intercept is the delivery charge.

c. $c = 19w + 25$

6–8. See answers in back of book.

9. Answers and explanations may vary. An example is given. I agree with Nancy that the slope of the line is always the rate given in the situation because the rate shows how the dependent variable changes based on the control variable. This is the same as the slope or the ratio $\frac{\text{rise}}{\text{run}}$. I also agree that the vertical intercept is always the starting value. It is the point where the value of the control variable is zero.

10. Answers may vary. An example is given. The cost of a phone call is $1.75 for the first minute and $.25 for each additional minute.

11. 3, −5   12. 0.5, 4.5

13. −1, −$\frac{1}{2}$ 14. −1.2, 6

15. The table at the right shows the coordinates of points that lie on a line. Rupert says that the equation for the line is $y = 2x + 6$. Georgia says the equation is $y = 2(x + 3)$. Are they both right? Explain.

x	-6	-3	0	3	6
y	-6	0	6	12	18

16. The table at the right shows the coordinates of points that lie on a line. Write an equation for the line.

x	-4	-2	0	2	4
y	-11	-5	1	7	13

TECHNOLOGY **For Exercises 17 and 18, use a graphics calculator or software.**

Alternative Approach **Work in a group of two to four students.**

17. a. Use one set of axes to graph these four equations.

   $y = 2x + 5$ $y = 5 + 0.75x$ $y = 5 - x$ $y = -\frac{1}{4}x + 5$

   b. What property do the graphs share? Why? Write an equation for another line that has the same property and graph it with the others.

18. a. Use one set of axes to graph these four equations.

   $y = 2x + 3$ $y = 2x - 1$ $y = 5 + 2x$ $y = -4 + 2x$

   b. What property do the graphs share? Why? Write an equation for another line that has the same property and graph it with the others.

Writing **For Exercises 19–21, tell whether the amount spent on movies during a year is an example of linear growth. If it is, tell whether it is also an example of direct variation. Explain both answers.**

19.   20.   21.

Unit 8 Linear Equations as Models

22. **Reading** Linear growth and linear decay situations have a rate of change. What are three rates of change for the real-life situations on pages 417–419? How was each rate of change related to the slope of the line that modeled each situation?

23. **a.** Describe a situation that you think is a linear function. (For example, the cost of hiring a disc jockey for a school dance might be a function of the length of the dance.)

   b. **Research** Find some specific data related to the situation you described in part (a). Is it a linear function? Explain why or why not.

   c. If your situation is a linear function, is it *linear growth* or *linear decay*?

For Exercises 24–27, do these things.

a. Find the slope and the vertical intercept of each graph.

b. Write an equation for each line.

c. Tell whether each graph is an example of *linear growth* or *linear decay*.

d. Tell whether each graph is an example of direct variation. Write *Yes* or *No*. Explain why or why not.

24.   25.   26.   27.

  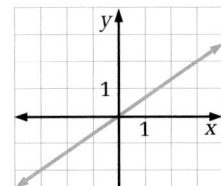 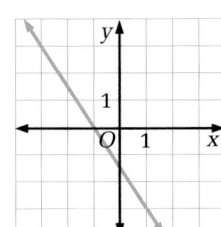 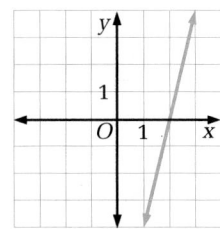

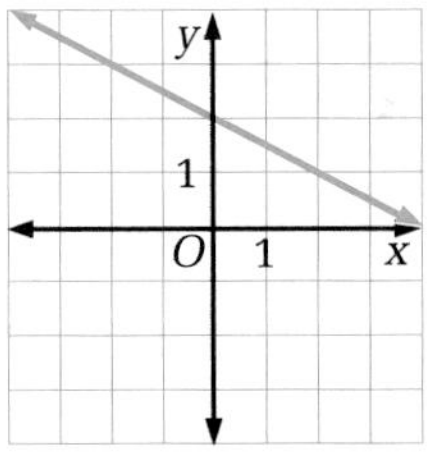

Decide whether each graph shows *linear growth*, *linear decay*, or *neither*. Explain your choice.

28.   29.   30.   31.

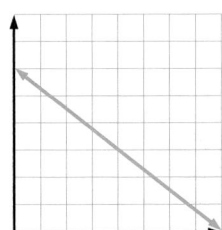 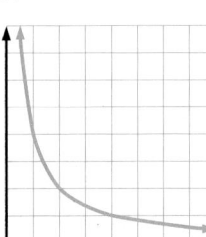 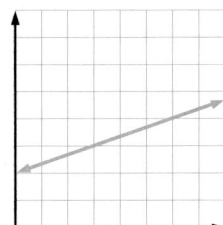 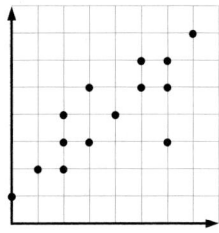

32. Which of the graphs in Exercises 28–31 are functions? Which are linear functions?

Integrating the Strands

For Ex. 23, students need to use concepts from the strands of algebra, functions, and discrete mathematics to answer the questions.

Reasoning

When discussing Exs. 24–27, in particular part (b), ask students why there is only one equation that represents each line. Why are there not two equations? Ask students how many points in a plane are needed to determine a unique line. Then try to have them relate this fact to knowing the slope and vertical intercept of a line. The vertical intercept is one point. The slope locates a second point by using the ratio of rise to run. These two points determine the line and therefore the equation of the line is unique.

25. a. $\frac{2}{3}, 0$

   b. $y = \frac{2}{3}x$

   c. linear growth

   d. Yes. The line goes through the origin.

26. a. $-\frac{3}{2}$ or -1.5, $-\frac{3}{2}$ or -1.5

   b. $y = -\frac{3}{2}x - \frac{3}{2}$ or $y = -1.5x - 1.5$

   c. linear decay

   d. No. The line does not go through the origin.

27. a. $4, -8$

   b. $y = 4x - 8$

   c. linear growth

   d. No. The line does not go through the origin.

28–31. Explanations may vary. Examples are given.

28. linear decay; The graph is a straight line that has a negative slope.

29. neither; The graph is not a straight line.

30. linear growth; The graph is a straight line that has a positive slope.

31. neither; The graph is not a line.

32. functions: 28, 29, 30; linear functions: 28, 30

Answers to Exercises and Problems

22. a balloon rising 110 ft for every minute, a balloon rising 60 ft for every minute, air temperature decreasing 11°F for every mile above ground; In every situation the rate of change is equal to the slope of the line that models it.

23. a–c. Answers may vary. Examples are given.

   a. The cost of renting a room is a function of the number of weeks you stay.

   b. The cost of renting a certain room in Somerville is $60 per week plus a $100 security deposit. The equation for this situation is $c = 60w + 100$. This equation is in slope-intercept form, so its graph is a nonvertical straight line, which makes it a linear function.

   c. linear growth

24. a. $-\frac{1}{2}, 2$ b. $y = -\frac{1}{2}x + 2$

   c. linear decay

   d. No. The line does not go through the origin.

Practice 59 For use with Section 8-1

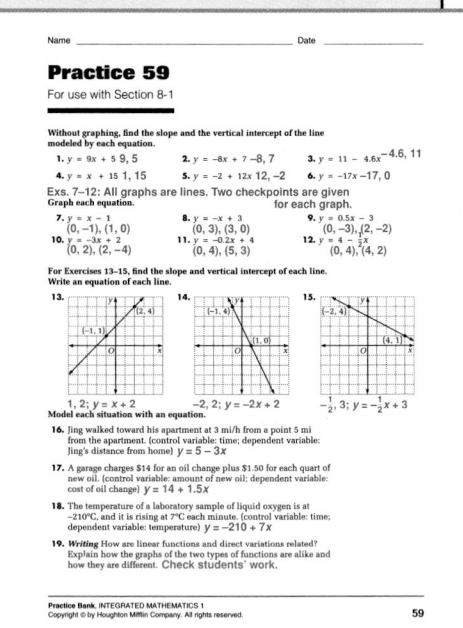

33. **Group Activity** Work with one or two other students.

a. Complete the table.

b. Graph the data in the second and fourth columns of the table.

c. Use your graph to find the sum of the degrees of the angles in a decagon (10-sided polygon).

d. The sum of the degrees of the angles of a polygon is 1800°. Find the number of sides of the polygon.

e. Are the points of your graph on a line? If so, find the slope and the vertical intercept of the line and write its equation.

f. Describe the relationship between the numbers in the middle two columns of the table. Is it linear? Is it direct variation? Explain.

You can draw *diagonals* to form triangles within a polygon. A **diagonal** is a segment other than a side that connects two vertices.

POLYGON	Number of sides	Number of triangles formed	Sum of degrees in all triangles
	?	?	?
	?	?	?
	?	?	?
	?	?	?
	?	?	?
	?	?	?

Review **PREVIEW**

34. Regina cuts a circular 10-inch pizza into three slices of equal size. What is the area of one slice? *(Section 7-6)*

35. A line contains the origin and the point $(6, -9)$. Find its slope. *(Section 7-2)*

Evaluate each expression when $x = 5$ and $y = -2$. *(Section 2-2)*

36. $x + 8y$ 37. $6x + 2y$ 38. $4x - 3y$ 39. $-x + \frac{1}{2}y$

Solve for y. *(Section 5-5)*

40. $by = c$ 41. $x + y = 60$ 42. $ax + y = b$ 43. $150 = 3y - x$

Working on the Unit Project

44. Use the data on page 416 for the Olympic 400 m freestyle swimming race.

a. Find the slope of the line between 1972 and 1976 for the women.

b. Find the slope of the line between 1972 and 1976 for the men.

c. **Writing** Did the men or the women show more improvement in their times between 1972 and 1976? Describe two different ways you can justify your choice.

d. **Writing** *True* or *False*? The rates of change in men's and women's times have been negative for the Olympic data. Describe how this is shown in a graph of data. How is it shown in the table?

424   **Unit 8** Linear Equations as Models

Answers to Exercises and Problems

33. See answers in back of book.

34. about 26.2 in.^{2}

35. $-\frac{3}{2}$ or -1.5

36. -11

37. 26

38. 26

39. -6

40. $y = \frac{c}{b}$

41. $y = 60 - x$

42. $y = b - ax$

43. $y = \frac{1}{3}x + 50$

44. a. about -2.4

b. about -2.1

c. women; Descriptions may vary. An example is given. From the data, the women decreased their time by 9.55 s in four years, while the

men only decreased their time by 8.34 s. Using the slopes, the women's rate of change is a decrease of about 2.4 s per year, while the men's rate of change is a decrease of only about 2.1 s per year. Therefore, the women showed greater improvement in the four-year period.

d. True. Descriptions may vary. An example is given. The fitted lines for both sets of data have negative slopes. For each year from 1960 to 1988, the times decrease for both men and women. This shows that the rates of change have been negative.

424

8-2 Linear Combinations

Objectives and Strands
See pages 414A and 414B.

Focus

Use the standard form of an equation for a line to model real-life situations.

Change standard form to slope-intercept form.

The Right Mix

Spiral Learning
See page 414B.

EXPLORATION

(**Is**) *there a pattern in how you can spend the same amount of money to buy different combinations of two items?*

- **Materials:** graph paper

- **Work in a group of at least four students. Each group has a different amount of money to "spend" at a music store: $72, $96, or $120. At this store tapes cost $8 and compact discs cost $12.**

1. How much does it cost to buy these combinations?
   - **a.** 1 tape and no CDs
   - **b.** no tapes and 1 CD
   - **c.** 3 tapes and 2 CDs
   - **d.** 1 tape and 3 CDs

2. **a.** Use the lattice to find the coordinates and the cost of each combination in step 1.

   b. Copy and fill out the lattice so that it contains all the combinations you can buy with the full amount of your group's spending money. (*Hint:* Look for patterns to help you fill the lattice faster.)

3. **a.** Circle all the numbers that equal the full amount of money your group can spend. What pattern do you see in the placement of these numbers?

   b. Draw a line through your group's numbers. What is the slope of your line? Write an equation for the line in slope-intercept form.

4. To compare your line with the other groups' lines, draw all the lines on one lattice. What do you notice about the lines?

5. **a.** Write a variable expression for the cost of *t* tapes and *d* compact discs.

   b. Use the variable expression to write an equation that relates *t*, *d*, and the full amount of money your group can spend.

This graph is a *lattice*. In this lattice, the cost of buying a given number of tapes and compact discs is written at points on the graph. A rule relates the cost to the number of items bought.

The cost of three tapes and two compact discs is written at (3, 2).

Number of Compact Discs

```
3  $36 —$44 —$52 —$60 —
2  $24 —$32 —$40 —$48 —
1  $12 —$20 —$28 —$36 —
0  $0 — $8 — $16 —$24 —
   0    1    2    3   t
```

Number of Tapes

8-2 Linear Combinations

Materials List
➤ Graph paper and ruler
➤ Construction paper
➤ Scissors
➤ Graphics calculator or software

Recommended Pacing
Section 8-2 is a two-day lesson.
Day 1
Pages 425–426: Exploration through Talk it Over, *Exercises 1–8*
Day 2
Pages 427–429: Standard Form of an Equation for a Line through Look Back, *Exercises 9–27*

Extra Practice
See pages 633–635.

Warm-Up Exercises
Warm-Up Transparency 8-2

Support Materials
➤ Practice 60
➤ Enrichment 53 in Activity Bank
➤ Study Guide 8-2
➤ Problem Set 16
➤ Diagram Master 2 in the Explorations Lab Manual
➤ McDougal Littell Mathpack software: *Function Investigator*
➤ Quiz 8-2
➤ Alternative Assessments 4, 5

Answers to Exploration

1. **a.** $8   **b.** $12
   c. $48   **d.** $44

2–4. Lattices and answers may vary depending on the group's amount of money to "spend." See lattice at right.

2. **a.** (1, 0): $8; (0, 1): $12; (3, 2): $48; (1, 3): $44
   b. See lattice.

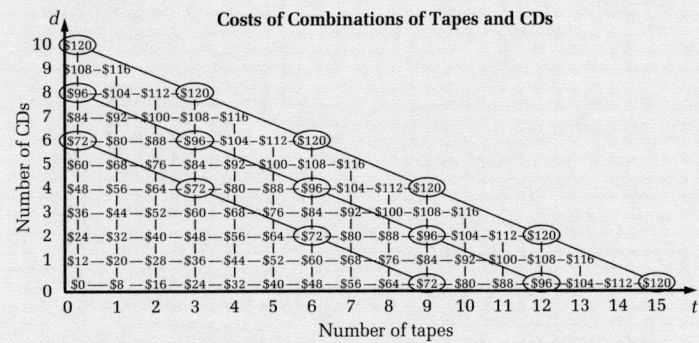

Costs of Combinations of Tapes and CDs

Number of CDs vs. *Number of tapes*

3. **a.** All the numbers lie on a straight line.

   b. for $72: $-\frac{2}{3}$; $d = -\frac{2}{3}t + 6$;

   for $96: $-\frac{2}{3}$; $d = -\frac{2}{3}t + 8$;

   for $120: $-\frac{2}{3}$; $d = -\frac{2}{3}t + 10$

4. The lines are all parallel.

5. **a.** $8t + 12d$

   b. for $72: $8t + 12d = 72$;
   for $96: $8t + 12d = 96$;
   for $120: $8t + 12d = 120$

The goal of the Exploration is to introduce students to the concept of linear combinations and their graphs through the use of a real-life situation.

Limited English Proficiency

Students acquiring English may not be familiar with the term *lattice*. Showing them a picture of a lattice in a garden book or a home improvement book will clarify its meaning.

Error Analysis

Watch for students who read or write ordered pairs incorrectly. Remind them to read the x-axis first and then to read the y-axis. They should associate the number of tapes (t) with the x-axis and the number of discs (d) with the y-axis.

Talk it Over

Questions 1–3 point out the difference between a lattice and a graph for a linear combination, thus moving students from a real-life, concrete representation to one that is purely mathematical.

Graphs of Linear Combinations

An expression like $8t + 12d$ is an example of a *linear combination*. The combination of values of t and d that result in the same cost, say $48, lie on a line in the lattice.

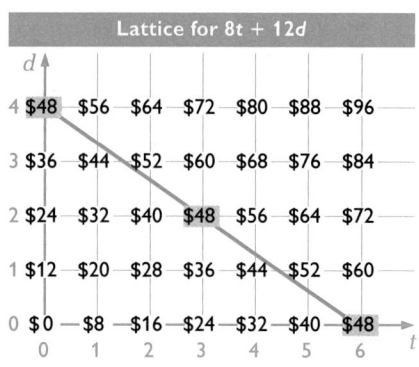

Lattice for $8t + 12d$

A **solution of an equation with two variables** is an ordered pair of numbers that makes the equation true. For example, $(3, 2)$ is a solution of $8t + 12d = 48$ because $8(3) + 12(2) = 48$. All the points whose coordinates are solutions of an equation form the **graph of the equation**.

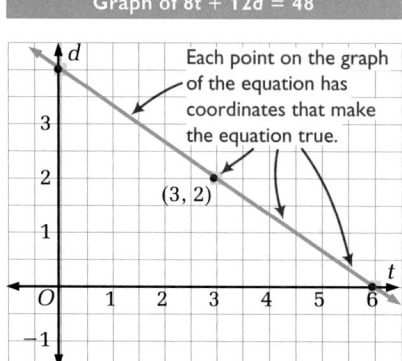

Graph of $8t + 12d = 48$

Each point on the graph of the equation has coordinates that make the equation true.

$(3, 2)$

Talk it Over

1. In the Exploration, suppose your group has $48 and you buy only compact discs. How many can you buy? How is this shown on the lattice? on the graph?

2. Is the point $(1.5, 3)$ on the graph of $8t + 12d = 48$? Can it represent numbers of tapes and compact discs you can buy? Explain why or why not.

3. Is $(2, 3)$ a solution of $8t + 12d = 48$? Explain why the order of the coordinates of a point is important when testing them in an equation.

4. a. Compare the equations that all the groups wrote for step 3(b) of the Exploration. How are the equations alike? How are they different?

   b. Compare the equations that all the groups wrote for step 5(b) of the Exploration. How are the equations alike? How are they different? How are they different from the equations in step 3?

▶ Now you are ready for:
Exs. 1–8 on pp. 429–430

Answers to Talk it Over

1. 4; by the circle around $48 at the point $(0, 4)$; by the point $(0, 4)$

2. Yes; No. Explanations may vary. An example is given. It is not possible to buy a fraction of a tape.

3. No. Explanations may vary. An example is given. The order of the coordinates is important because each coordinate represents a specific variable. If you change the order, the coordinates no longer represent the correct variable.

4. a, b. Answers may vary. Examples are given.

   a. All the equations have the same slope, $-\frac{2}{3}$. However, they all have different vertical intercepts.

   b. All the equations have the same variable expression, $8t + 12d$, but they each equal different amounts of money. The equations in step 3 are in slope-intercept form, while the equations in step 5 (b) are not.

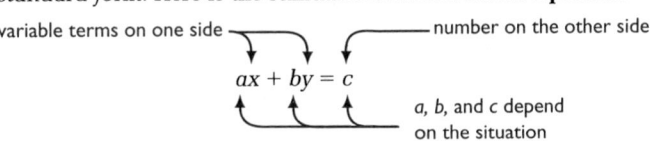

Standard Form of an Equation for a Line

The equation $8t + 12d = 48$ is called a **linear equation** because it is an equation for a line. One form of a linear equation is slope-intercept form. The equation $8t + 12d = 48$ is written in *standard form*. Here is the **standard form of a linear equation.**

variable terms on one side ⎯⎯⎯⎯⎯⎯⎯⎯⎯ number on the other side

$$ax + by = c$$

a, b, and c depend on the situation

You can change an equation from standard form to slope-intercept form using algebra.

TECHNOLOGY NOTE

You need to use slope-intercept form, not standard form, to input an equation on most graphics calculators.

Sample 1

Several times a week, Dynah Colwin runs part of the way and walks part of the way on a trail that is 4 mi long. Her running speed is 6 mi/h and her walking speed is 4 mi/h.

a. Write an equation for this situation.

b. Rewrite the equation in slope-intercept form.

Sample Response

a. Let r = the time in hours Dynah Colwin spends running.
Let w = the time in hours she spends walking.

the distance she runs at 6 mi/h for r hours
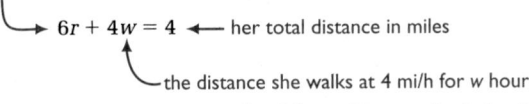
$6r + 4w = 4$ ⎯ her total distance in miles
the distance she walks at 4 mi/h for w hours

b. The equation is in standard form. To rewrite it in slope-intercept form, solve the equation for the dependent variable. In this situation, either r or w can be the dependent variable. Suppose it is w.

$$6r + 4w = 4$$
$$6r + 4w - 6r = 4 - 6r \qquad \longleftarrow \text{Subtract } 6r \text{ from both sides of the equation.}$$
$$4w = 4 - 6r$$
$$\tfrac{1}{4} \cdot 4w = \tfrac{1}{4}(4 - 6r) \qquad \longleftarrow \text{Multiply both sides by the reciprocal of 4.}$$
$$w = \tfrac{1}{4} \cdot 4 - \tfrac{1}{4} \cdot 6r \qquad \longleftarrow \text{Use the distributive property.}$$
$$w = 1 - \tfrac{3}{2}r$$
$$w = -\tfrac{3}{2}r + 1 \qquad \longleftarrow \text{Rewrite the equation in slope-intercept form.}$$

8-2 Linear Combinations

427

Mathematical Procedures

You may wish to point out that it is customary in mathematical applications for letters at the beginning of the alphabet to represent constants and letters at the end of the alphabet to represent variables.

Additional Sample

S1 Each week, Will swims the backstroke part of the way and the crawl part of the way for 30 laps of the swimming pool. His backstroke speed is 3 laps per minute and his crawl speed is 2 laps per minute.

a. Write an equation for this situation.

Let b = time it takes to swim the backstroke.
Let c = time it takes to swim the crawl.
$$3b + 2c = 30$$

b. Rewrite the equation in slope-intercept form.

$$3b + 2c = 30$$
$$3b + 2c - 2c = 30 - 2c$$
$$3b = 30 - 2c$$
$$\tfrac{1}{3} \cdot 3b = \tfrac{1}{3} \cdot (30 - 2c)$$
$$b = \tfrac{1}{3} \cdot 30 - \tfrac{1}{3} \cdot 2c$$
$$b = 10 - \tfrac{2}{3}c$$
$$b = -\tfrac{2}{3}c + 10$$

428

Talk it Over

Use question 7 to emphasize the fact that it is sometimes possible for an equation to have solutions that are not logical in real-world situations. For the equation $w = -\frac{3}{2}r + 1$, $(-\frac{4}{3}, -1)$ is a solution; however, it is not possible for time to be a negative number. Emphasize that solutions to real-world problems should always be checked against the conditions of the problem to make sure they are realistic.

Additional Sample

S2 Find the intercepts of the graph of $y = -2x + 2$. Use them to graph the equation. **Let $x = 0$. The vertical intercept is 2. Let $y = 0$. The horizontal intercept is 1.**

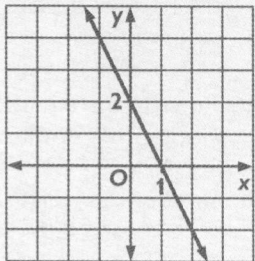

Mathematical Procedures

Remind students that a straight line is uniquely determined by two points. As Sample 2 shows, the vertical and horizontal intercepts are easy to find and provide two points through which the line passes. Thus, the intercepts can always be used to graph the equation.

For questions 5–7, use the situation in Sample 1.

5. What is the slope of the graph of $w = -\frac{3}{2}r + 1$? Does the graph show linear decay or linear growth?

6. Suppose $r = 0$. What is the value of w? How are these values related to the time Dynah Colwin spends running and walking?

7. **a.** Can r and w be negative numbers? Explain.

   b. Can r and w be fractions? Explain.

   c. Write an inequality to describe the possible values for r.

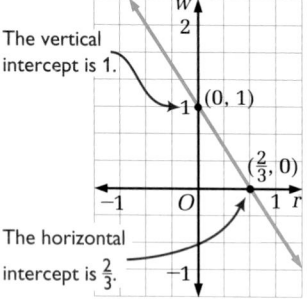

The vertical intercept is 1.

The horizontal intercept is $\frac{2}{3}$.

Intercepts of a Line

The graph of $w = -\frac{3}{2}r + 1$ crosses the vertical axis at $(0, 1)$. Its **vertical intercept** is 1. The graph crosses the horizontal axis at $(\frac{2}{3}, 0)$. Its **horizontal intercept** is $\frac{2}{3}$. One coordinate of each of these points is 0.

Sample 2

Find the intercepts of the graph of $2x - 5y = 15$. Use them to graph the equation.

Sample Response

To find the vertical intercept, substitute 0 for x.

$$2x - 5y = 15$$
$$2(0) - 5y = 15$$
$$-5y = 15$$
$$y = -3$$

The vertical intercept is -3. The line crosses the vertical axis at $(0, -3)$.

To find the horizontal intercept, substitute 0 for y.

$$2x - 5y = 15$$
$$2x - 5(0) = 15$$
$$2x = 15$$
$$x = 7.5$$

The horizontal intercept is 7.5. The line crosses the horizontal axis at $(7.5, 0)$.

Use what you found to make a graph.

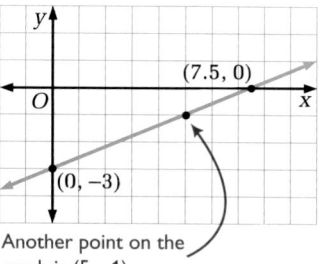

Another point on the graph is $(5, -1)$.

Unit 8 Linear Equations as Models

Answers to Talk it Over

5. $-\frac{3}{2}$; linear decay

6. 1; If $r = 0$, then Dynah does not spend any time running. She would spend 1 h walking the 4 mi.

7. **a, b.** Explanations may vary. Examples are given.

   a. No. You cannot have a negative amount of time.

   b. Yes. It is possible to run or walk for part of an hour.

   c. $0 \le r \le \frac{2}{3}$

8. **a.** Substitute 0 for x and solve to find the value for y.

   b. Substitute 0 for y and solve to find the value for x.

   c. Substitute any value for x (or y) other than those for the intercepts and solve to find the value for y (or x).

Talk it Over

Use the equation 5x + 4y = 20.

8. **a.** How can you find the vertical intercept of its graph?

   b. How can you find the horizontal intercept of its graph?

   c. How can you find another point on its graph?

> ····► **Now you are ready for:**
> Exs. 9–27 on pp. 431–432

Look Back ◄

Describe two ways to graph the equation $5x + 3y = 2$.

8-2 Exercises and Problems

1. **Reading** Tell the meaning of each part of the expression $8t + 12d$ on page 426.

   a. 8 b. 12 c. d d. t e. $8t$ f. $12d$

2. An art department raises money by selling students' pottery work. Mugs cost \$4 each and bowls cost \$3 each. Write an expression for the cost of m mugs and b bowls.

3. On one section of the SAT (Scholastic Assessment Test), you receive 1 point for every right answer, lose one fourth of a point for every wrong answer, and receive no points for blank answers. You have r right answers, w wrong answers, and b blank answers.

   a. Write a variable expression for your score.

   b. Suppose your score is 32. Use the expression from part (a) to write an equation for your score.

4. $\angle A$ and $\angle B$ are supplementary angles. Write an equation relating their measures.

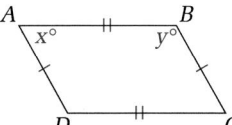

5. The perimeter of a pool is 260 m.

   a. Make a table of possible widths (w) and lengths (l) for the rectangle.

   b. Write an equation relating w, l, and the perimeter, 260 m.

   c. Graph the equation you wrote in part (b). Put the widths on the horizontal axis.

   d. Explain why your graph should or should not cross either axis.

> **BY THE WAY...**
>
> Of all the students taking the SAT from 1990 to 1993, 5453 achieved either perfect 800 Verbal scores or perfect 800 Math scores. Almost 92% of these perfect scores were in Math.

8-2 Linear Combinations

429

Answers to Look Back··········

Answers to Exercises and Problems·················

Descriptions may vary. An example is given. (1) Find the slope-intercept form and use this to make a table of values. Graph the values and connect the points with a line. (2) Graph the vertical and horizontal intercepts and connect them.

1. a. the cost of 1 tape
   b. the cost of 1 compact disc
   c. the number of compact discs bought
   d. the number of tapes bought
   e. the total cost of t tapes
   f. the total cost of d compact discs

2. $4m + 3b$

3. a. $r - \frac{1}{4}w + 0b$ or $r - \frac{1}{4}w$

   b. $r - \frac{1}{4}w = 32$

4. $x + y = 180$

Communication: Discussion

In connection with Talk it Over question 8 (c), you may wish to have students discuss why finding a third point would help to graph $5x + 4y = 20$.

Look Back

The Look Back question may be used as a class discussion. You may wish to have students graph the equation at the board. ················•

APPLYING

Suggested Assignment

Standard 1–6, 9–19, 21, 23–27
Extended 1–6, 9–19, 21–27

Integrating the Strands

Algebra Exs. 1–26
Geometry Exs. 4, 5, 7, 8, 16–21, 24–26
Discrete Mathematics Exs. 22, 27
Logic and Language Exs. 6, 22

5. a. Tables may vary. An example is given.

w	l
10	120
20	110
30	100
40	90
50	80
60	70

   b. $2l + 2w = 260$

   c. Graphs may vary. They should represent part of the line below.

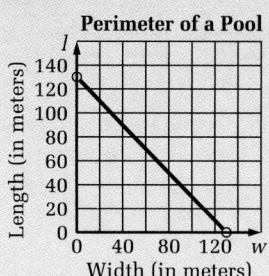

   d. Explanations may vary. An example is given. The graph should not cross either axis because if $l = 0$ or $w = 0$, then you would have a line segment instead of a rectangle and a pool cannot be a line segment.

429

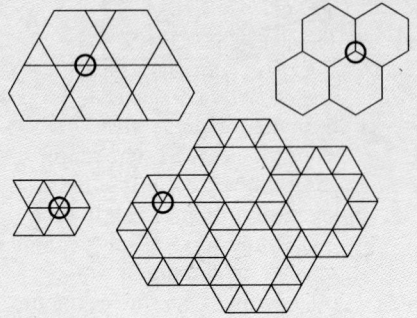
6. **Writing** After completing the Exploration on page 425, Erica decided that every time you buy two more compact discs, you can buy three fewer tapes. Erica's group had $72 to spend. Do you think her conjecture depends on the amount of spending money? Write a paragraph explaining why or why not.

Using Manipulatives For Exercises 7 and 8, use the information below about *tessellations.*

Honeycomb

A *regular tessellation* is a pattern made of one regular polygon that covers the plane without gaps or overlaps.

Dome of the Rock, Jerusalem

A *semiregular tessellation* is made of two or more regular polygons, and the arrangement of polygons at every vertex is identical.

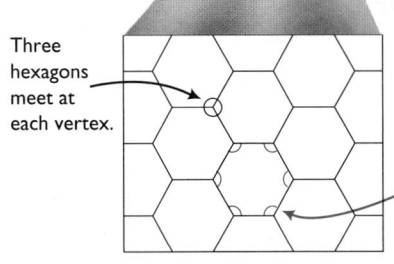

Three hexagons meet at each vertex.

The measures of the angles at any vertex not on the boundary add up to 360°.

All angles in a regular polygon have the same measure.

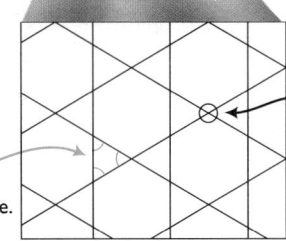

Two regular hexagons and two equilateral triangles meet at each vertex.

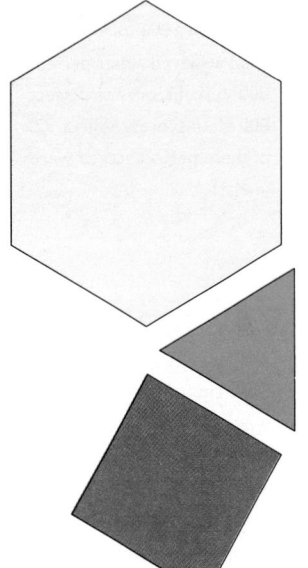

7. a. Trace and cut out several hexagons and triangles. Use them to make as many different regular and semiregular tessellations as you can. Make a sketch of each tessellation.

   b. Make a table showing the number of hexagons (*h*) and triangles (*t*) you used at one vertex in each tessellation.

   c. Write an equation for the sum of the measures of the angles of *h* hexagons and *t* triangles at a vertex.

   d. Find four solutions of the equation you wrote for part (c). Can all four solutions be included in the table for part (b)? Why or why not?

8. a. Write an equation for the sum of the measures of the angles of *t* equilateral triangles and *s* squares at each vertex of a tessellation.

   b. Find three solutions of your equation.

   c. Using cutouts, try to make the tessellations suggested by the solutions you found in part (b). Are all of them possible? Why or why not?

   d. One solution of the equation you found in part (b) leads to two different semiregular tessellations. Find the solution and sketch the tessellations.

430

Unit 8 Linear Equations as Models

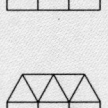
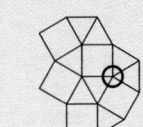

Find three solutions of each equation.

9. $x + 3y = 11$

10. $x - \frac{1}{2}y = 3\frac{1}{2}$

11. $-5x - 2y = 2$

Rewrite each equation in slope-intercept form.

12. $5x + 8y = 48$

13. $\frac{2}{3}x + y = -2$

14. $6x - 2y = 11$

15. $9x - \frac{3}{4}y = 1$

Find the intercepts of the graph of each equation.

16. $-x - 8y = 16$

17. $0.2x + 0.3y = 3.6$

18. $-3x + \frac{3}{4}y = 21$

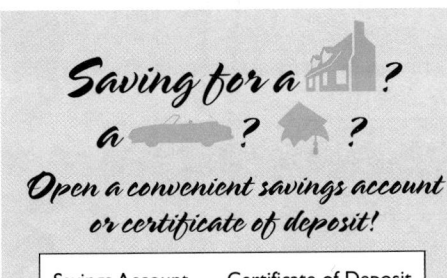

Saving for a 🏠?
a 🚗? 🎓 ?

Open a convenient savings account
or certificate of deposit!

Savings Account	Certificate of Deposit
5.4%	**7.2%**
annual interest	annual interest

19. **Banking** A bank offers a savings account and a certificate of deposit that pay the annual interest rates shown in this promotional advertisement. Suppose you have s dollars in a savings account and a certificate of deposit for c dollars.

   a. Suppose your total interest in one year is $18. Write an equation in standard form for your interest.

   b. Find the intercepts of the graph of the equation. What do they represent in terms of the situation?

▓ T E C H N O L O G Y **For Exercises 20 and 21, use a graphics calculator or software.**

Alternative Approach Work in a group of four students.

20. For parts (a)–(c), graph each group of four equations on its own set of axes.

   a. $x + y = 6$ $x + y = 2$ $x + y = -2$ $x + y = -6$

   b. $3x - y = 6$ $3x - y = -3$ $3x - y = 0$ $3x - y = -6$

   c. $2x - y = 12$ $4x - 2y = 24$ $6x - 3y = 36$ $-2x + y = -12$

   d. What do you observe about the equations in each of parts (a)–(c)? What do you observe about their graphs?

21. a. Graph these equations on the same set of axes.

   $x + 4y = 0$ $6x - y = 0$ $2x + 3y = 0$

   b. What property do the graphs share? Do they represent direct variation? Make a conjecture about the graph of an equation with the form $ax + by = 0$.

   c. Test your conjecture by writing another equation with the form $ax + by = 0$ and graphing it.

---

20. See answers in back of book.

21. a.

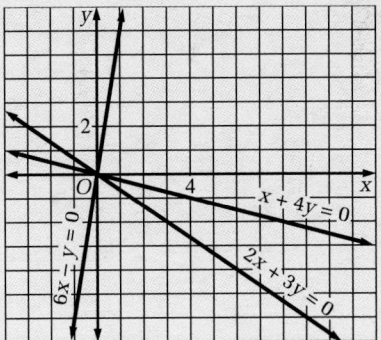

   b. Each line passes through the origin. Yes. The graph of an equation in the form $ax + by = 0$ passes through the origin. (Of course, for this to be true, a and b cannot both be zero.)

   c. Equations and graphs may vary. An example is given.
      $x - 2y = 0$

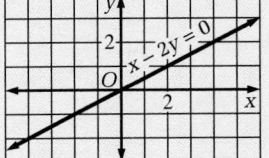

Answers to Exercises and Problems

9–11. Answers may vary. Examples are given.

9. $(11, 0)$, $\left(0, \frac{11}{3}\right)$, $(2, 3)$

10. $\left(3\frac{1}{2}, 0\right)$, $(0, -7)$, $(1, -5)$

11. $\left(-\frac{2}{5}, 0\right)$, $(0, -1)$, $(2, -6)$

12. $y = -\frac{5}{8}x + 6$

13. $y = -\frac{2}{3}x - 2$

14. $y = 3x - \frac{11}{2}$

15. $y = 12x - \frac{4}{3}$

16. horizontal intercept: −16; vertical intercept: −2

17. horizontal intercept: 18; vertical intercept: 12

18. horizontal intercept: −7; vertical intercept: 28

19. a. $0.054s + 0.072c = 18$

   b. savings intercept: about $333.33, certificate of deposit intercept: $250; If you put about $333.33 in a savings account and nothing in certificates of deposit, you will earn $18 interest in one year. If you put $250 in certificates of deposit and nothing in a savings account, you will earn $18 interest in one year.

431

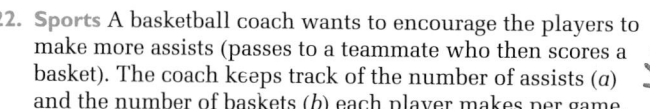

Practice 60 For use with Section 8-2

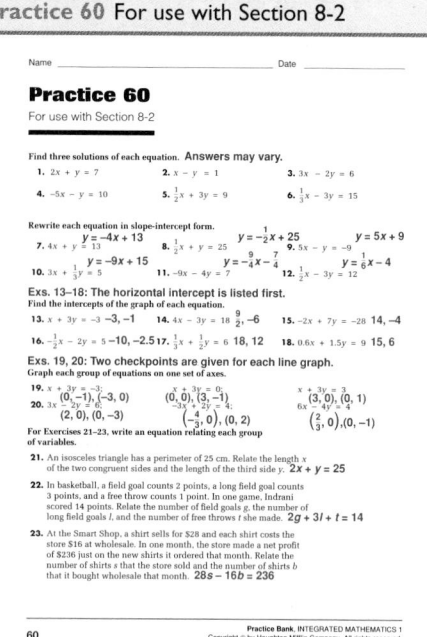

22. **Sports** A basketball coach wants to encourage the players to make more assists (passes to a teammate who then scores a basket). The coach keeps track of the number of assists (a) and the number of baskets (b) each player makes per game.

   a. The coach counts each assist as four points and each basket as two points. Write a linear combination for the total number of points the coach gives a player.

   b. The coach rewards the players who earn a total of at least 32 points per game. Suppose a player earns exactly 32 points. Write an equation that relates a, b, and 32.

   c. If Bonnie makes no baskets in a game, how many assists must she make in order to have a total of 32 points?

   d. **Open-ended** The table shows Janine's record. Do you think that she has improved? Would the coach think so? Explain why or why not.

Game	Number of assists made	Number of baskets made
1	3	10
2	5	6
3	6	4

 Ongoing **ASSESSMENT**

23. **Writing** Compare linear growth situations and linear combination situations. How are they alike? How are they different? How are they represented with equations?

 Review **PREVIEW**

24. A tow truck driver charges $32 to hook up your car and $1.25 per *(Section 8-1)* mile towed.

   a. Model the situation with a graph.

   b. Find the slope and the vertical intercept of the line. Explain how they are related to the original situation.

25. a. Graph the points $A(7, 4)$, $B(-1, 4)$, $C(0, 4)$, $D(4, 4)$, and $E(-2, 4)$. *(Section 4-6)*

   b. What pattern do you see in the graph? in the coordinates?

   c. Is the graph a function? Explain why or why not.

26. Repeat Exercise 25 using the points $P(-3, 1)$, $Q(-3, -5)$, $R(-3, -2)$, $S(-3, 0)$, and $T(-3, 4)$.

8 Working on the Unit Project

27. **Research** Find another set of winning times for the 400 m freestyle for the years shown in the table on page 416. You might look for world records, college records, or high school records. Find both men's and women's records.

Answers to Exercises and Problems

22. a. $4a + 2b$

   b. $4a + 2b = 32$

   c. 8 assists

   d. Answers may vary. An example is given. If you look only at the point total, she has not improved, because the total points for each game is 32. However, basketball is a team sport, and since Janine's number of assists has improved, it would appear that she is becoming more of a team player. Therefore, I think she has improved. I also think the coach would agree because the coach places a greater value on assists than on baskets.

23. Answers may vary. An example is given. Linear growth situations and linear combination situations are alike because they both lead to graphs that are straight lines. Linear growth is more easily represented with slope-intercept form because it shows the rate and the initial value of the situation. Linear combinations are more easily represented with standard form because it shows the value of the items being combined.

24–26. See answers in back of book.

27. Answers may vary.

Horizontal and Vertical Lines

Focus
Describe slopes of horizontal and vertical lines and write equations for horizontal and vertical lines.

What happens when one of the coefficients in an equation for a line in standard form is zero? Here are two ways that could happen.

1 The coefficient of *x* could be zero.

$$0x + 3y = 6$$
$$3y = 6$$
$$y = 2$$

All points with a *y*-coordinate of 2 are on the graph of the equation $y = 2$.

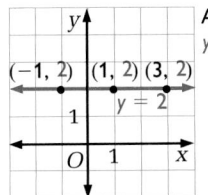

All points have *y*-coordinate 2.

When the coefficient of *x* is zero the graph is a horizontal line.

To find the slope, use any two points on the line, such as (1, 2) and (3, 2).

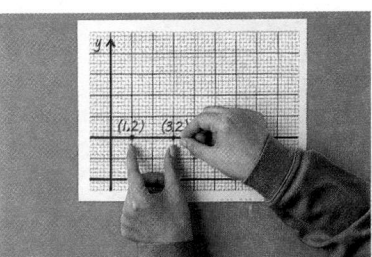

$$\frac{\text{change in } y}{\text{change in } x} = \frac{2 - 2}{3 - 1} = \frac{0}{2} = 0$$

The slope of a horizontal line is 0.

2 The coefficient of *y* could be zero.

$$2x + 0y = 6$$
$$2x = 6$$
$$x = 3$$

All points with an *x*-coordinate of 3 are on the graph of the equation $x = 3$.

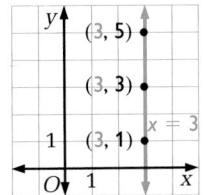

All points have *x*-coordinate 3.

When the coefficient of *y* is zero the graph is a vertical line.

To find the slope, use any two points on the line, such as (3, 1), and (3, 5).

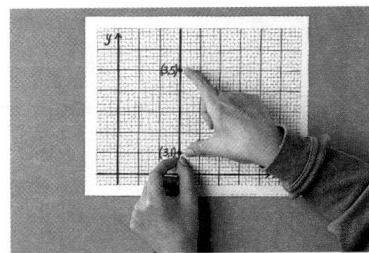

$$\frac{\text{change in } y}{\text{change in } x} = \frac{5 - 1}{3 - 3} = \frac{4}{0} \text{ undefined}$$

Because division by zero is undefined, the slope of a vertical line is undefined.

PLANNING

Objectives and Strands
See pages 414A and 414B.

Spiral Learning
See page 414B.

Materials List
➤ Graph paper
➤ Ruler
➤ Scissors

Recommended Pacing
Section 8-3 is a one-day lesson.

Extra Practice
See pages 633–635.

Warm-Up Exercises
Warm-Up Transparency 8-3

Support Materials
➤ Practice 61
➤ Enrichment 54 in the Activity Bank
➤ Study Guide 8-3
➤ Problem Set 16
➤ Diagram Masters 2, 3 in the Explorations Lab Manual
Overhead Visual 8
➤ Quiz 8-3
➤ Test 31

TEACHING

Additional Samples

S1 Find the slope of each line and write an equation for each line.

a.

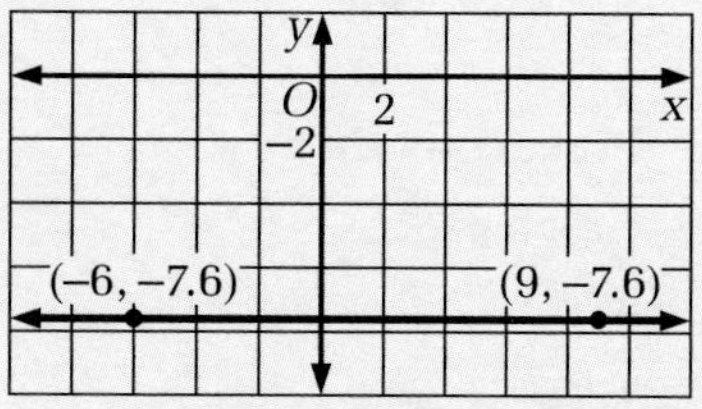

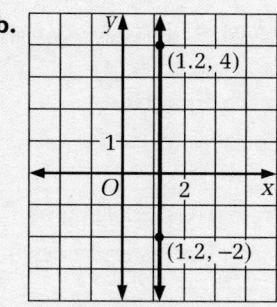

The slope is 0. The equation is $y = -7.6$.

b.

The slope is undefined. The equation is $x = 1.2$.

S2 Write an equation for each line.

a. The points (9, 3) and (−1, 3) are on the line. $y = 3$

b. The slope is undefined, and the point (−1, 4) is on the line. $x = -1$

Using Technology

The TI-82 calculator can draw vertical lines by means of a special command, although not, of course, as part of a function-graphing procedure.

Look Back

After graphing the given equations, you may want to give students some other pairs of equations and ask them to find the intersection. Students should see that the intersection of a pair of equations $x = a$ and $y = b$ is (a, b).

Sample 1

Find the slope of each line and write an equation for each line.

a.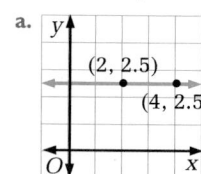

b.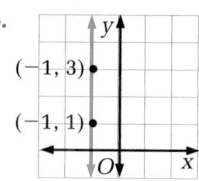

Sample Response

a. The graph is a horizontal line. The slope is 0.

Every point on the line has a y-coordinate of 2.5.

An equation for the line is $y = 2.5$.

b. The graph is a vertical line. The slope is undefined.

Every point on the line has an x-coordinate of −1.

An equation for the line is $x = -1$.

Sample 2

Write an equation for each line.

a. The points (−2, 4) and (−2, 0) are on the line.

b. The slope is zero, and the point (2, 3) is on the line.

Sample Response

a. The x-coordinates are both −2. The graph is a vertical line 2 units left of the vertical axis. The slope is undefined.

Every point on the line has the same x-coordinate, −2.

An equation for the line is $x = -2$.

b. The slope is zero. The graph is a horizontal line.

Every point on the line has the same y-coordinate, 3.

An equation for the line is $y = 3$.

Talk it Over

Graph each equation and find the slope of each line.

1. $x = -4$ 2. $y = 2$ 3. $2y = -3$

4. Describe how the equation in question 3 is different from the other two. Does the difference affect the graph? Explain.

Look Back

On the same set of axes, graph the equations $x = 3$ and $y = 2$. Where do the lines intersect?

434

Unit 8 Linear Equations as Models

Answers to Talk it Over

1. The slope is undefined.

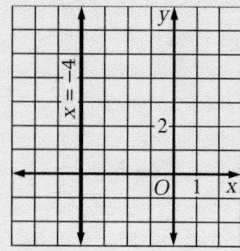

2. The slope is 0.

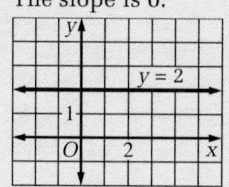

3–4. See answers in back of book.

Answers to Look Back

The lines intersect at the point (3, 2).

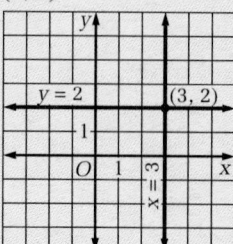

8-3 Exercises and Problems

Estimate the slope of the land in each picture.

1.

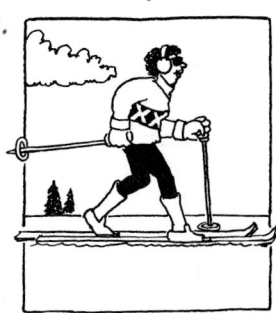

2.

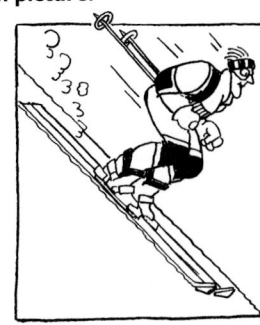

3.

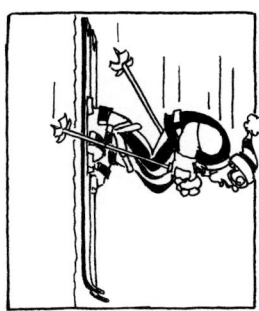

4. **Reading** Is the horizontal change equal to zero for a *horizontal* line or a *vertical* line? Why? Is the vertical change equal to zero for a *horizontal* line or a *vertical* line? Why?

Find the slope of each line and write an equation for each line.

5.

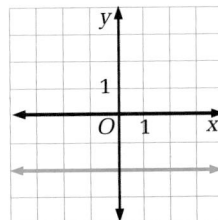

6.

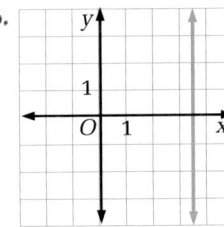

7.

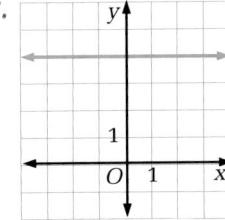

Graph each equation and find the slope of each line.

8. $y = 3$ 9. $x = -2.5$ 10. $6x = 12$ 11. $-4y = 10$

For Exercises 12–15, write an equation for each line.

12. The points (3, 7) and (−2, 7) are on the line.
13. The points (4, −8) and (4, 0) are on the line.
14. The slope is undefined, and the point (5, 3) is on the line.
15. The slope is zero, and the point (4, 6) is on the line.
16. List the coordinates of some points that lie on the x-axis. What do they have in common? Write an equation for the x-axis.
17. List the coordinates of some points that lie on the y-axis. What do they have in common? Write an equation for the y-axis.

For Exercises 18–21, tell whether each statement is *True* or *False*.

18. The graph of $x = 8$ is parallel to the x-axis.
19. The graph of $y = 12$ is parallel to the x-axis.
20. The graph of $x = -3$ is parallel to the y-axis.
21. The graph of $y = 1.1$ is parallel to the y-axis.

APPLYING

Suggested Assignment

Standard 1–3, 5–21, 25, 27–36
Extended 1–21, 23–25, 27–36

Integrating the Strands

Algebra Exs. 1–35
Geometry Exs. 1–22
Discrete Mathematics Exs. 23, 36

Reasoning

After students do Exs. 18–21, ask them to describe the geometric relationship between two lines, one parallel to the x-axis and the other parallel to the y-axis. (The lines are perpendicular.)

Answers to Exercises and Problems

1–3. Estimates may vary. Examples are given.
1. about 0
2. about −1
3. Since the line appears to be vertical, the slope is undefined.
4. Reasons may vary. Examples are given. The horizontal change is zero for a vertical line because all of its points have the same first coordinate. The vertical change is zero for a horizontal line because all of its points have the same second coordinate.
5. 0; $y = -2$
6. undefined; $x = 3\frac{1}{2}$ or $x = 3.5$
7. 0; $y = 4$
8. The slope is 0.

9. The slope is undefined.

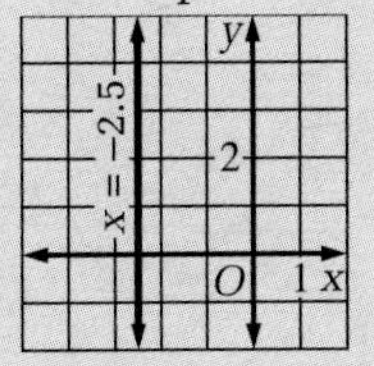

10. The slope is undefined.

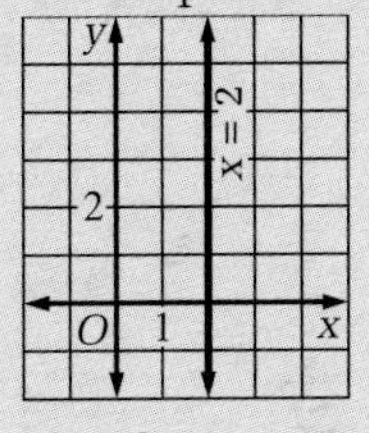

11. The slope is 0.

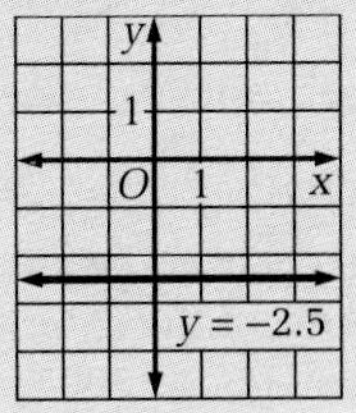

12. $y = 7$
13. $x = 4$
14. $x = 5$
15. $y = 6$
16. Choices of coordinates may vary. An example is given: (−4, 0), (−1, 0), (0, 0), (3, 0). They each have y-coordinate 0. $y = 0$
17. Choices of coordinates may vary. An example is given: (0, −4), (0, −2), (0, 1), (0, 6). They each have x-coordinate 0. $x = 0$
18. False.
19. True.
20. True.
21. False.

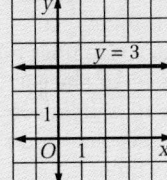

Answers to
Exercises and Problems

22. a–c. manual activity

d. Estimates may vary. The sum of the measures of the exterior angles of a polygon is 360°.

e. Graphs may vary. An example is given.

Sum of Exterior Angles of Polygons

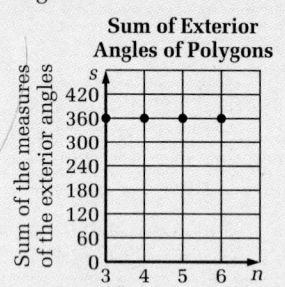

f. Choice of variable may vary. An example is given. $s = 360$, where s represents the sum of the measures of the exterior angles.

---

22. **Using Manipulatives** The *exterior angles* of a polygon are formed by extending the sides of the polygon.

a. On four pieces of paper, draw a 3-gon, 4-gon, 5-gon, and 6-gon.

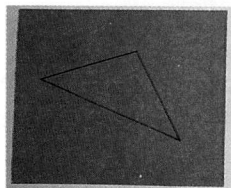 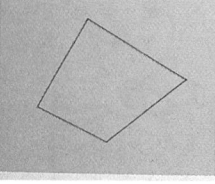 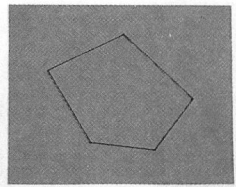 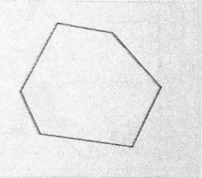

b. Extend the sides to the edge of the paper. Shade the exterior angles.

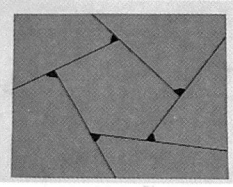

c. Cut along the lines and set aside the polygon.

d. Estimate the sum of the measures of the exterior angles. (*Hint:* arrange the exterior angles around a point to see what part of a circle they fill.)

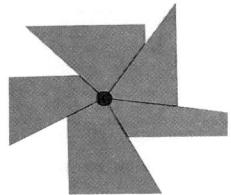

e. Make a graph of your results in part (d). Put the *number of sides of the polygon* on the horizontal axis and the *sum of the measures of the exterior angles* on the vertical axis.

f. Write an equation that describes your results from part (e).

23. **Government** The graph shows the number of senators for each state in the United States.

a. What is the slope of the fitted line?

b. Does the number of United States senators for a state vary directly as the population of the state?

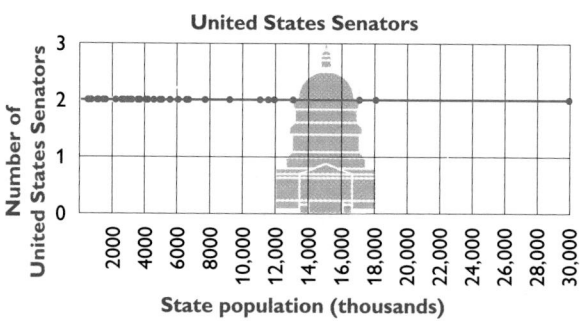

United States Senators

24. **Writing** Describe a situation that can be graphed with a horizontal line. Write an equation that describes the situation. Include the meaning of each variable that you use.

436   **Unit 8** Linear Equations as Models

---

23. a. 0

b. No.

24. Answers may vary. An example is given. An adult's shoe size does not change as he or she gets older. $s = 10$, where s represents a shoe size and t represents time.

25. Answers may vary. An example is given. All three ratios can be used to describe the slopes of hori-

zontal and vertical lines. For horizontal lines, the rise, the change in y, and the vertical change would all be 0. This would make all the ratios equal to 0. For vertical lines, the run, the change in x, and the horizontal change would all be 0. Since you cannot divide by 0, this would make all the ratios undefined. The rate of change can describe the slope of a

horizontal line in the sense that there is no change because the slope is 0.

26. b; Explanations may vary. An example is given. The equation $x = -5$ has more than one value of y when $x = -5$. Therefore, $x = -5$ is not a function and cannot be input on my graphics calculator.

25. Writing The list below gives several different meanings of the word "slope." Which meanings can be used to describe the slopes of horizontal and vertical lines? Explain how.

Meanings of Slope

$\dfrac{\text{rise}}{\text{run}}$ $\dfrac{\text{vertical change}}{\text{horizontal change}}$ $\dfrac{\text{change in } y}{\text{change in } x}$

rate of change

variation constant

value repeatedly added or subtracted (linear growth and decay)

26 TECHNOLOGY Many graphics calculators graph only equations that represent functions. Which of these equations can *not* be input on these graphics calculators? Explain.

a. $y = 2x + 5$ **b.** $x = -5$ **c.** $y = 6$

Ongoing **ASSESSMENT**

27. Open-ended Rosa started making this list to help her remember the differences between equations that have only x and equations that have only y.

	equations with only x	equations with only y
example	X = 2	y = 2
description	?	horizontal
slope	undefined	?
vertical intercept	doesn't have one	?
horizontal intercept	?	?
is parallel to	?	the x-axis
graph	?	?
other	can't be written in slope-intercept form	?

Copy and complete the list. What other ideas would you add?

Review **PREVIEW**

Rewrite each equation in slope-intercept form. If it is not possible to do so, explain why not. *(Section 8-2)*

28. $2x + 3y = 12$ **29.** $4x - 9y = 72$ **30.** $0x + 4y = 20$ **31.** $4x + 0y = 40$

Solve. *(Sections 2-7, 5-2, 5-3)*

32. $3(x - 8) = 14$ **33.** $5x + 9 = 2x - 15$ **34.** $415 = 37 + b$

35. Graph the line that contains the points $(-2, 1)$ and $(1, -5)$. Find the slope and the vertical intercept of the graph. Write an equation for the line. *(Section 8-1)*

8-3 Horizontal and Vertical Lines

437

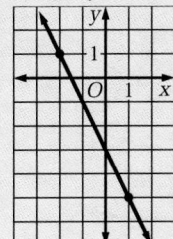
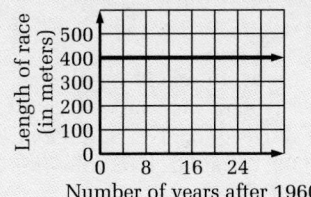
Answers to Exercises and Problems

27.

	equations with only x	equations with only y
example	$x = 1$	$y = 1$
description	vertical line	horizontal line
slope	undefined	0
vertical intercept	does not have one	equals the number
horizontal intercept	equals the number	does not have one
is parallel to	the y-axis	the x-axis
graph	a vertical line	a horizontal line
other	cannot be written in slope-intercept form	is in slope intercept form with $m = 0$

Additional ideas may vary. An example is given. Equations with only x are not functions and equations with only y are functions. Equations with only x represent lines that are perpendicular to lines represented by equations with only y.

437

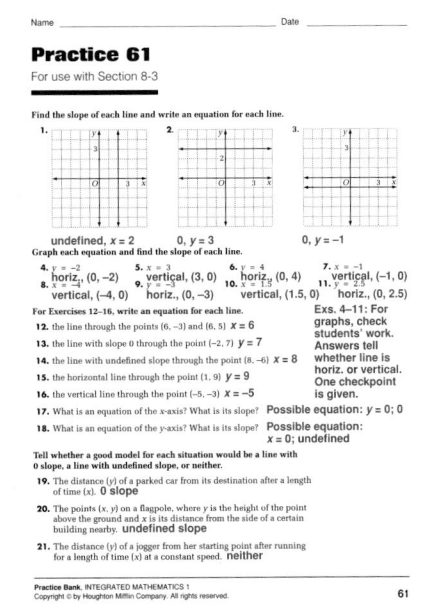
 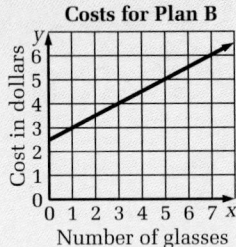
438

Working on the Unit Project

36. The winning times in the 400 m swimming event at the Olympics have changed over the years. What numerical aspect of the event has *not* changed over time? Draw a graph that shows this.

Ethelda Bleibtrey, USA   *Debbie Meyer, USA*   *Dagmar Hase, Germany*
1920   Time: 4:34.0   1968   Time: 4:31.8   1992   Time: 4:07.18

Unit 8   CHECKPOINT 1

1. **Writing** How are linear growth and direct variation situations alike? How are they different?

Students can choose from three different weekly beverage plans.   **8-1**

   Plan A. Pay $.75 per glass.
   Plan B. Pay $2.50 and $.50 per glass.
   Plan C. Pay $10 and drink all you like.

2. Which plan(s) are examples of linear growth?

3. Which plan(s) are examples of direct variation?

4. Model Plan B with a graph and an equation.

5. Find the slope of the line that represents Plan C.

Mark's favorite store is having a sale. Suppose there is no sales tax.   **8-2**

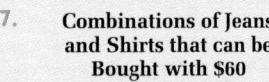

6. Write an expression for how much it would cost to buy j pairs of jeans and s shirts on sale.

7. Mark spent $60 on the items shown. Graph the possible numbers of jeans and shirts he bought.

8. Does your graph have a vertical intercept and a horizontal intercept? What do they represent?

9. **a.** Write an equation for the situation in Exercise 7.

   b. What is the slope of the line whose equation you wrote in part (a)? What does this slope mean in terms of the original situation?

Find the intercepts of the graph of each equation.

10. $2x - 3y = 12$ 11. $7x + 14y = 42$

12. Graph the equation $y = -4$. Find the slope of the line.   **8-3**

13. The slope of a line is undefined, and the point $(-3, 9)$ is on the line. Write an equation for the line.

Unit 8 Linear Equations as Models

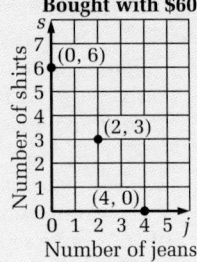
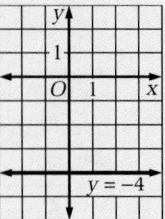

Writing Equations for Lines

THE LINE FORMS HERE

Olympic swimmers keep getting faster. In other words, the winning time for the 400 m freestyle race keeps getting shorter.

Year	Country of winner	Years after 1960	Time (seconds)
1960	AUS	0	258.3
1964	USA	4	252.2
1968	USA	8	249.0
1972	USA	12	240.27
1976	USA	16	231.93
1980	USSR	20	231.31
1984	USA	24	231.23
1988	E.GER	28	226.95
1992	UT	32	225.00

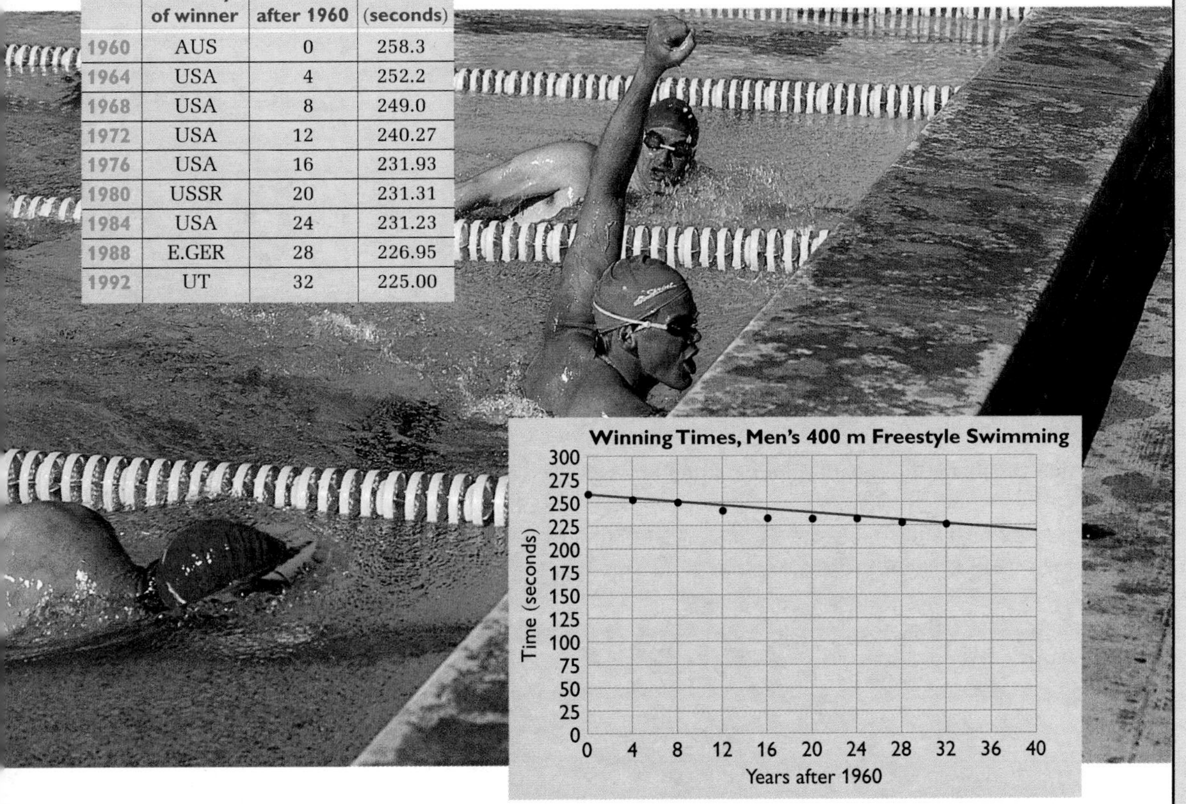

Winning Times, Men's 400 m Freestyle Swimming

Time (seconds) vs. Years after 1960

Suppose you want to predict the winning time for the year 2000. The points do not lie exactly on a line, but a line that passes close to most of the points (a fitted line) is drawn. You can estimate an equation for a fitted line and use it to make predictions.

8-4 Writing Equations for Lines

439

PLANNING

Objectives and Strands
See pages 414A and 414B.

Spiral Learning
See page 414B.

Materials List
➤ Graph paper
➤ Ruler
➤ Graphics calculator or graphing software

Recommended Pacing
Section 8-4 is a two-day lesson.
Day 1
Pages 439–441: Opening paragraph through Sample 2, *Exercises 1–14*
Day 2
Pages 442–443: Equations from Other Facts through Look Back, *Exercises 15–31*

Extra Practice
See pages 633–635.

Warm-Up Exercises
Warm-Up Transparency 8-4

Support Materials
➤ Practice 62
➤ Enrichment 55 in the Activity Bank
➤ Study Guide 8-4
➤ Problem Set 17
➤ Diagram Masters 2, 3 in the Explorations Lab Manual
Overhead Visual 8
➤ McDougal Littell Mathpack software: *Stats!*
➤ Using Mac or IBM Plotter Plus Disk: Line Quiz
➤ Quiz 8-4
➤ Alternative Assessment 6

TEACHING

Additional Sample

S1 Estimate an equation for the fitted line for the data. Interpret Week 0 to mean the beginning of the first week.

Week	Number of Sit-ups
0	15
1	19
2	22
3	27
4	33
5	35
6	38

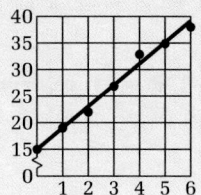

$\text{slope} = \dfrac{\text{rise}}{\text{run}} = \dfrac{19 - 15}{1 - 0} = \dfrac{4}{1} = 4$

$y = 4x + b$

$15 = 4(0) + b$

$15 = b$

An estimate for the equation is $y = 4x + 15$.

Estimate an equation for the fitted line for the Olympic data.

Sample Response

To write your equation, you can use slope-intercept form and the coordinates of two data points that are on or nearly on the fitted line. The closer the points are to the fitted line, the better your estimate will be.

1 Find the slope. Use the points (8, 249) and (32, 225).

$\text{slope} = \dfrac{\text{rise}}{\text{run}}$

$= \dfrac{\text{change in winning times}}{\text{change in years}}$

$= \dfrac{225 - 249}{32 - 8}$ ⟵ 32 and 8 represent the number of years after 1960.

$= \dfrac{-24}{24}$

$= -1$ ⟵ The slope is –1.

In the equation $y = mx + b$, you now know m: $y = -1x + b$.

2 Find the vertical intercept. Substitute the coordinates of either point for x and y in the equation and solve for b.

dependent variable = (slope · control variable) + vertical intercept

$y = -1x + b$

$249 = (-1)(8) + b$ ⟵ For the point (8, 249), $x = 8$ and $y = 249$.

$249 = -8 + b$

$257 = b$ ⟵ The vertical intercept is 257.

3 Substitute -1 for m and 257 for b in the equation $y = mx + b$.

An estimate for the equation for the fitted line is $y = -1x + 257$, or $y = -x + 257$.

Check Check the other point that you used to find the slope, (32, 225). Are its coordinates a solution of the equation?

$y = -x + 257$

$225 \overset{?}{=} -32 + 257$ ⟵ Substitute 32 for x and 225 for y.

$225 = 225$ ✔

Talk it Over

1. Does the fitted line on the graph on page 439 have a vertical intercept of about 257?

2. Cover up the scales on the graph on page 439. Does the fitted line look like it has a slope of -1? How do the scales affect the appearance of the graph?

Unit 8 Linear Equations as Models

Answers to Talk it Over

1. Yes.

2. No. Answers may vary. An example is given. Covering up the scales, it might seem as if each line on the graph represents 1 unit, making the slope appear to be $-\dfrac{1}{6}$. The unequal scales make the rate of change appear to be more gradual than it would appear with equal scales.

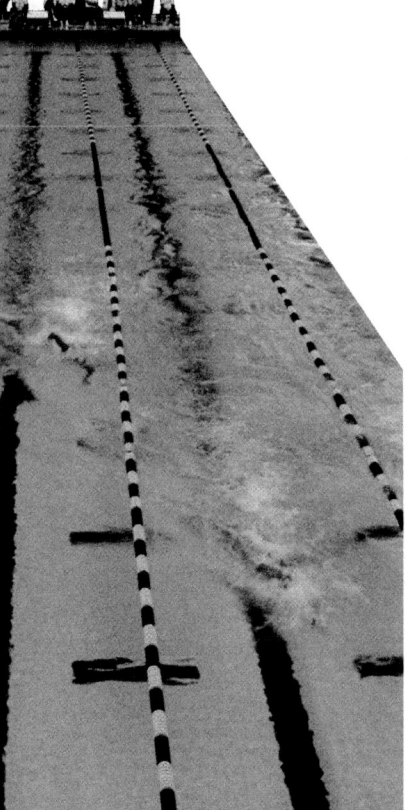

3. Use the equation $y = -x + 257$ from Sample 1.

   a. Substitute 16 in the equation to find the predicted winning time for the race in 1976. How close is the prediction to the actual time?

   b. Predict the winning time for the race in the year 2000.

4. Suppose you use two different data points in Sample 1 to estimate the equation for the fitted line. Will the result be exactly the same?

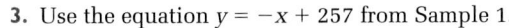

Exact Equations from Two Points

You can estimate an equation for a fitted line using two points that are *close* to the fitted line. You can find an exact equation for any line if you know the coordinates of any two points *on* the line.

Sample 2

Two points on a line are (4, 3) and (−6, −2). Write an equation for the line.

Sample Response

First find the slope.

$$m = \frac{\text{change in } y}{\text{change in } x}$$

$$= \frac{-2 - 3}{-6 - 4}$$

$$= \frac{-5}{-10}$$

$$= \frac{1}{2}$$

Next find the vertical intercept. To do this, substitute $\frac{1}{2}$ for m and the coordinates of either point for x and y in the equation $y = mx + b$.

$$y = \frac{1}{2}x + b$$

$$3 = \frac{1}{2}(4) + b \longleftarrow \text{For the point (4, 3), } x = 4 \text{ and } y = 3.$$

$$3 = 2 + b$$

$$1 = b$$

> **Now you are ready for:**
> **Exs. 1–14 on pp. 443–445**

You now know that $m = \frac{1}{2}$ and $b = 1$. The equation is $y = \frac{1}{2}x + 1$.

8-4 Writing Equations for Lines

441

Talk it Over

Questions 3 and 4 should alert students to the fact that fitted lines only allow for predictions, that is, approximations, of actual data. The closer the fitted line is to the actual points, the better the prediction will be.

Additional Sample

S2 Two points on a line are (1, −2) and (5, 4). Write an equation for the line.

$$m = \frac{(-2 - 4)}{(1 - 5)} = \frac{-6}{-4} = \frac{3}{2}$$

$$y = mx + b$$

$$-2 = \frac{3}{2}(1) + b$$

$$-2 - \frac{3}{2} = b$$

$$-\frac{7}{2} = b$$

The equation is $y = \frac{3}{2}x - \frac{7}{2}$.

Mathematical Procedures

Students should become thoroughly familar with the procedure presented in Sample 2. You may wish to put the three main steps on the board for students to copy into their journals.

1. Use the two points to find the slope of the line.

2. Use the coordinates of one point and the slope to find the vertical intercept.

3. Use the slope and vertical intercept to write the equation of the line.

Answers to Talk it Over

3. a. 241 s; The prediction is 9.07 s more than the actual time.

   b. 217 s

4. No, but the results should be close.

S3 The slope of a line is $-\frac{1}{2}$. One point on the line is (3, 4). Write an equation for the line.

Method 1: $y = mx + b$

$$4 = -\frac{1}{2}(3) + b$$
$$4 = -\frac{3}{2} + b$$
$$\frac{11}{2} = b$$
$$y = -\frac{1}{2}x + \frac{11}{2}$$

Method 2:

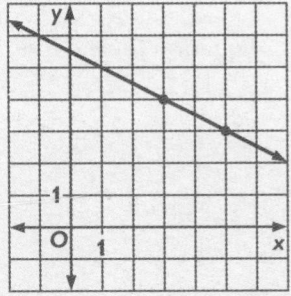

Graph the point (3, 4). Use the slope: move down 1 unit and to the right 2 units. Connect the points to graph the line. The vertical intercept is $5\frac{1}{2}$ or $\frac{11}{2}$. An equation for the line is $y = -\frac{1}{2}x + \frac{11}{2}$.

Reasoning

After discussing Method 2 of Sample 3, ask students if graphing the line to find the vertical intercept is always a reliable method. Why or why not? (No; if the vertical intercept is not an integer, then only an estimate can be read from the graph.)

Equations from Other Facts

You can use two points to find an equation for a line. You also know how to find an equation for a line from its graph. You can use other types of information to find an equation for a line.

Sample 3

The slope of a line is $\frac{3}{2}$. One point on the line is (−4, 1). Write an equation for the line.

Sample Response

Method ❶ Use substitution to find the vertical intercept.

You know the slope and the coordinates of a point. First, substitute these numbers in the slope-intercept form of an equation for a line.

$$y = mx + b$$
$$1 = (\tfrac{3}{2})(-4) + b \longleftarrow \text{Substitute } (-4, 1) \text{ for } (x, y) \text{ and } \tfrac{3}{2} \text{ for } m.$$
$$1 = -6 + b$$
$$7 = b \longleftarrow \text{The vertical intercept is 7.}$$

Then substitute $\frac{3}{2}$ for m and 7 for b in the equation $y = mx + b$.

An equation for the line is $y = \frac{3}{2}x + 7$.

Method ❷ Graph the line to find the vertical intercept.

1 Graph the point (−4, 1).

2 Use the slope: Move 3 units up and 2 units to the right.

3 Connect the points to graph the line.

4 The vertical intercept is 7.

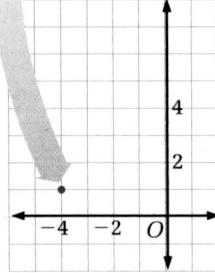
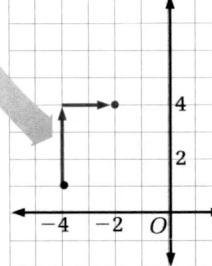
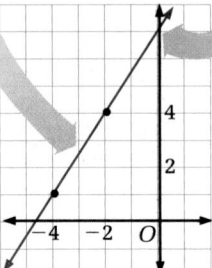

5 Substitute $\frac{3}{2}$ for m and 7 for b in the equation $y = mx + b$.

An equation for the line is $y = \frac{3}{2}x + 7$.

Unit 8 Linear Equations as Models

Answers to Talk it Over

5. substituting the values for the slope and the vertical intercept in the equation $y = mx + b$

6. $y = -\frac{3}{5}x + 8$

7. a. dependent variable

   b. (12, 41)

   c. a decrease of 7 dollars per week; negative; m; Explanations may vary. An example is given. The rate of change is the same as the slope. Since Kemitra is spending money every week, the amount she has decreases, so the slope is negative.

   d. $y = -7x + 125$; See graph at right.

   e. The vertical intercept represents the amount of money Kemitra is given for her birthday. The horizontal intercept represents the number of weeks it takes for her to spend all her money.

   f. $104

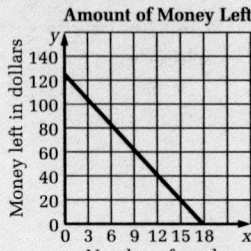

Amount of Money Left

5. Which step is the same for both methods in Sample 3?

6. The slope of a line is $-\frac{3}{5}$. One point on the line is $(15, -1)$. Find an equation for the line.

7. Kemitra is given some money for her birthday. She spends $7 every week. After 12 weeks she has $41 left.

   a. Is the amount of money she has left the *control variable* or the *dependent variable*?

   b. Write "after 12 weeks she has $41" as an ordered pair.

   c. What is the rate of change in this situation? Is it *positive* or *negative*? What does the rate of change represent in the equation $y = mx + b$? Explain your answer.

   d. Write an equation for the amount of money Kemitra has left after x weeks of spending. Graph your equation.

   e. In this situation, what is represented by the vertical intercept? by the horizontal intercept?

   f. Use the graph to find out how much money she had left three weeks after her birthday.

Look Back ◄

Describe how to find an equation for a line if you have each type of information.

   a. a graph of the line

   b. two points on the line

   c. the slope and one point on the line

> ······► **Now you are ready for:**
> **Exs. 15–31 on pp. 445–446**

8-4 Exercises and Problems

1. **Reading** How is estimating an equation of a fitted line different from finding an exact equation of a line?

For Exercises 2–7, write an equation for each line. If there is too little information, say so.

   2. The slope is -3 and the vertical intercept is 1.

   3. The points $(5, 2)$ and $(-1, 8)$ are on the line.

   4. The point $(3, -8)$ is on the line.

   5. The points $(-8, 2)$ and $(5, 2)$ are on the line.

   6. The horizontal intercept is 6 and vertical intercept is -7.

   7. The points $(3, 7)$ and $(-5, -9)$ are on the the the line.

Talk it Over

Question 7 provides an opportunity for students to relate the skills learned in this section to a real-life situation.

Look Back

Have each student write a journal entry to describe how to find an equation given each type of information. Students may want to illlustrate each description with an example.

······●

APPLYING

Suggested Assignment
Standard 2–11, 13–31
Extended 1–11, 13–31

Integrating the Strands
Algebra Exs. 1–31
Functions Exs. 8, 9, 11, 23, 24
Geometry Exs. 1–8, 13–23, 26, 27
Statistics and Probability Exs. 1, 10–12
Logic and Language Exs. 1, 25

Reasoning

For Exs. 2–7, students must use logical thinking and their knowledge of the concepts taught in this section to determine whether or not they have enough information to write an equation. If students have difficulty with any of these exercises, remind them of the basic fact that two points determine a line.

Answers to Look Back ································

Answers may vary. Examples are given.

a. Use two points on the graph to find the slope. Find the vertical intercept on the graph. Substitute the slope for m and the vertical intercept for b in the equation $y = mx + b$.

b. Use the coordinates of the two points to find the slope: $\frac{\text{change in } y}{\text{change in } x}$. Substitute the slope and the coordinates of one of the points in the equation $y = mx + b$. Solve for b to find the vertical intercept. Substitute the slope for m and the vertical intercept for b in the equation $y = mx + b$.

c. Substitute the slope and the coordinates of the point in the equation $y = mx + b$. Solve for b to find the vertical intercept. Substitute the slope for m and the vertical intercept for b in the equation $y = mx + b$.

Answers to Exercises and Problems begin on page 444.

8. The Chu family is driving home. Because they are using cruise control, their speed is constant. After 3 h, they are 350 mi from home. After 6 h, they are 200 mi from home.

   a. What is the control variable? What is the dependent variable?

   b. Represent the given information as two points on a graph.

   c. Write an equation for the line through the two points.

   d. What do the slope and the vertical intercept of the equation mean in terms of the original situation?

9. **Science** Water freezes at 0°C, or 32°F. Water boils at 100°C, or 212°F.

   a. Use these two points to write an equation relating degrees Fahrenheit to degrees Celsius.

   b. Use the equation to convert 68°F to degrees Celsius.

10. **Economics** Use the median family income data.

    a. Model the data with an equation for the median family income in terms of the year.

Year	1984	1985	1986	1987	1988	1989
Median family income in the United States	$26,433	27,735	29,458	30,970	32,191	34,213

    b. Use your model to predict the median family income in 1999.

11. **Geology** The length of time between eruptions of Old Faithful, a geyser in Yellowstone Park, can be predicted with a linear equation. The table below gives some data values (in minutes) for the geyser's eruptions.

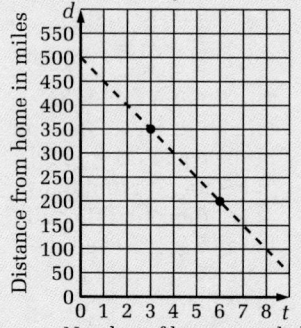

> **BY THE WAY...**
>
> Old Faithful's eruption schedule is affected by earthquakes. In 1983, the average amount of time between eruptions increased from 69 min to 76 min because of an earthquake in Idaho.

Length of eruption (l)	3	4.5	4	2.5	3.5	5	2
Time until next eruption (t)	68	89	83	62	75	92	57

    a. Graph the data and draw a fitted line.

    b. Model the data with an equation for t in terms of l.

    c. Suppose an eruption lasts 4.2 min. Predict the number of minutes until the next eruption.

Unit 8 Linear Equations as Models

Answers to Exercises and Problems

1. Answers may vary. An example is given. Estimating an equation of a fitted line means also estimating the slope and the vertical intercept, plus all the real-life data will not be solutions to the equation. Finding the exact equation of a line results in the actual slope and vertical intercept, and all the real-life data will be solutions.

2. $y = -3x + 1$

3. $y = -x + 7$

4. too little information

5. $y = 2$

6. $y = \frac{7}{6}x - 7$

7. $y = 2x + 1$

8. a. control variable: time spent driving; dependent variable: distance from home

b. **The Chu Family's Drive Home**

Distance from home in miles vs *Number of hours traveled*

c. $d = -50t + 500$

d. The slope shows that the distance from home is decreasing 50 mi for every hour of driving time. The vertical intercept means that initially they were 500 mi from home.

12. TECHNOLOGY Use the data from Exercise 10 or 11 and a graphics calculator or software.

   a. Enter the data points and make a scatter plot of the data.

   b. Use the calculator or software to find an equation for a fitted line for the data. (It may be called the line of best fit, or linear regression.) Graph the calculator's equation and the one that you found.

   c. Compare the graph of the equation that you found to the one from the calculator or software. How are they alike? How are they different?

13. In Exercise 18 on page 422, you saw that lines with the same slope are parallel. Use this fact and the diagram.

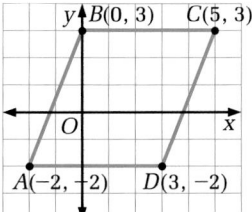

   a. Find the slopes of $\overline{AB}, \overline{BC}, \overline{CD},$ and $\overline{AD}.$

   b. Which pairs of segments are parallel?

   c. Write equations for $\overleftrightarrow{AB}, \overleftrightarrow{BC}, \overleftrightarrow{CD},$ and $\overleftrightarrow{AD}.$

14. a. How many lines contain the point $(-1, 3)$?

   b. How many lines contain the point $(6, 5)$?

   c. How many lines contain the point $(-1, 3)$ *and* the point $(6, 5)$?

For Exercises 15–22, write an equation for each line. If there is too little information, say so.

15. The slope is $\frac{4}{5}$ and the point $(-10, 3)$ is on the line.

16. The slope is $-0.5.$

17. The slope is $-\frac{3}{2}$ and the point $(6, -1)$ is on the line.

18. The slope is 4 and the point $(0, -3)$ is on the line.

19. The points $(-4, 0)$ and $(4, 6)$ are on the line.

20.
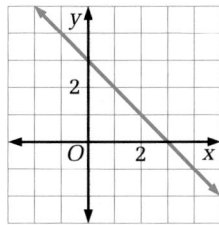

21.
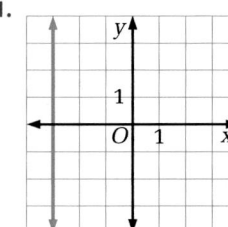

22.
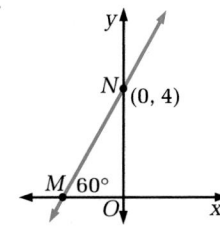

Interdisciplinary Problems

The focus of this section, using an equation for a line to solve problems, is a powerful problem-solving tool because it can be applied to a wide-ranging set of problems from many different disciplines. Exs. 8 and 23 are related to transportation. Exs. 9–11 touch upon ideas from science, economics, and geology, respectively. In any real-world situation that involves a linear functional relationship between two variables, the equation for a line can be used to model the situation and answer questions about it.

Using Technology

Scatter plots are discussed on page 618 of the Technology Handbook.

The *Stats!* software can be used to do Ex. 12.

13. c. $\overleftrightarrow{AB}: y = \frac{5}{2}x + 3; \overleftrightarrow{BC}: y = 3;$
   $\overleftrightarrow{CD}: y = \frac{5}{2}x - \frac{19}{2}; \overleftrightarrow{AD}: y = -2$

14. a. an infinite number

   b. an infinite number

   c. one

15. $y = \frac{4}{5}x + 11$

16. too little information

17. $y = -\frac{3}{2}x + 8$

18. $y = 4x - 3$

19. $y = \frac{3}{4}x + 3$

20. $y = -x + 3$

21. $x = -3$

22. $y = \sqrt{3}x + 4$

Answers to Exercises and Problems

9. a. $F = \frac{9}{5}C + 32$ or
   $C = \frac{5}{9}F - \frac{160}{9}$

   b. 20°C

10. a, b. Equations and predictions may vary. Examples are given.

   a. $i = 1617.5t + 26,117.5,$ where t is the number of years after 1984 and i is the median income.

   b. about $50,380

11. a–c. Graphs, equations, and predictions may vary. Examples are given.

   a. Activity at Old Faithful
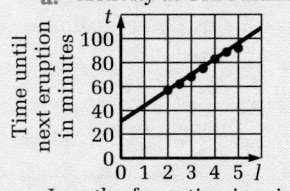
   Length of eruption in minutes

   b. $t = 13l + 31$

   c. about 85.6 min

12. a, b. Equations and graphs may vary.

   c. Answers may vary.

13. a. $\overline{AB}: \frac{5}{2}; \overline{BC}: 0; \overline{CD}: \frac{5}{2};$
   $\overline{AD}: 0$

   b. $\overline{AB}$ and $\overline{CD}; \overline{BC}$ and $\overline{AD}$

Practice 62 For use with Section 8-4

Practice 62
For use with Section 8-4

Name _____ Date _____

Write an equation for the line that has each slope and each vertical intercept.

1. slope = 5; intercept = −1 $y = 5x - 1$
2. slope = −2; intercept = 1 $y = -2x + 1$
3. slope = 0.5; intercept = 3 $y = 0.5x + 3$
4. slope = $\frac{2}{3}$; intercept = −4 $y = \frac{2}{3}x - 4$
5. slope = −6; intercept = 0 $y = -6x$
6. slope = $-\frac{1}{3}$; intercept = $\frac{4}{3}$ $y = -\frac{1}{3}x + \frac{4}{3}$

Write an equation for the line that has each slope and has each point on it.

7. slope = 2; (−1, 5) on line $y = 2x + 7$
8. slope = −3; (2, −1) on line $y = -3x + 5$
9. slope = −4; (−3, 9) on line $y = -4x - 3$
10. slope = $\frac{1}{2}$; (2, −7) on line $y = \frac{1}{2}x - 8$
11. slope = $-\frac{3}{4}$; (−6, 2) on line $y = -\frac{3}{4}x - \frac{5}{2}$
12. slope = $-\frac{3}{2}$; (5, −4) on line $y = -\frac{3}{2}x + \frac{7}{2}$

Write an equation for the line that has each pair of points on it.

13. (2, 5), (6, 13) $y = 2x + 1$
14. (−1, 4), (1, 10) $y = 3x + 7$
15. (3, 6), (−5, −2) $y = x + 3$
16. (−1, 3), (5, 0) $y = -\frac{1}{2}x + \frac{5}{2}$
17. (−7, −2), (5, −6) $y = -\frac{1}{3}x - \frac{13}{3}$
18. (1, 0.8), (5, −4) $y = -1.2x + 2$

Find an equation of the line with each pair of intercepts.

19. vertical: 3; horizontal: 2 $y = -\frac{3}{2}x + 3$
20. vertical: −5; horizontal: 10 $y = \frac{1}{2}x - 5$

21. A long-distance phone call costs $1.60 for the first minute and a fixed charge for each minute after that. A 7-minute call costs $2.92. Write an equation for the cost of a call as a function of time after the first minute. $y = 1.6 + 0.22x$

22. A computer repair shop charges $25 to test an out-of-order computer plus an hourly charge for actual repairs. The shop charged a customer $92.50 for a job that took 1.5 h. Write an equation for the cost of a repair as a function of time. $y = 25 + 45x$

23. *Open-ended* Estimate how far your school building is from your home. Then time your trip home. Assume that you travel at a constant speed, and write an equation for your distance from home as a function of time during your trip. Check students' work

---

23. **Transportation** A taxi charges a base fee plus $.75 per mile. A 10 mi trip costs $8.70.
    a. Is the length of the trip the *control variable* or the *dependent variable*?
    b. Write "a 10 mi trip costs $8.70" as an ordered pair.
    c. What is the rate of change in this situation?
    d. Write an equation for the cost of hiring the taxi in terms of the length of the trip.
    e. Graph the equation.
    f. What is the cost of a 5 mi trip? a 17 mi trip? How are they shown on your graph?
    g. What does the vertical intercept represent in this situation?

24. The height of the water in a bathtub decreases at a constant rate of 3 cm per minute. After 11 min, the height of the water left in the tub is 2 cm. Write an equation for the height of the water left in the tub as a function of time.

Ongoing ASSESSMENT

25. **Writing** Suppose you are writing a "how to do math" guide for your friends. What would you include in the chapter called "How to Write Equations for Lines"? Write an outline and an introduction for the chapter.

Review PREVIEW

Write an equation for each line. *(Section 8-3)*

26. The points (3, 5) and (−2, 5) are on the line.

27. The slope is undefined, and the point (1, 8) is on the line.

Solve each system of equations. *(Section 5-8)*

28. $y = 5x$
    $3x + y = 44$

29. $y = 2x + 7$
    $-5x + y = -5$

30. $y = 2x - 4$
    $y = x - 6$

Working on the Unit Project

31. **Group Activity** Use the winning times for the women's 400 m freestyle swimming race in the Olympics on page 416. Use *Years after 1960* as the control variable. Have each person write an equation to model the data. Compare equations. Does one equation fit the data better than the other? Decide which equation you will use to model the data.

Answers to Exercises and Problems

23. a. control variable
    b. (10, 8.7)
    c. $.75 per mile
    d. $c = 0.75m + 1.2$
    e.
    Taxi Charges

    Cost in dollars / Distance in miles

    f. $4.95; $13.95; by the points (5, 4.95) and (17, 13.95)
    g. The vertical intercept represents the initial charge before driving any distance.

24. $h = -3t + 35$

25. See answers in back of book.

26. $y = 5$

27. $x = 1$

28. $x = 5.5, y = 27.5$

29. $x = 4, y = 15$

30. $x = -2, y = -8$

31. Equations may vary. An example is given. $t = -1.81n + 290.6$, where n is the number of years after 1960 and t is a woman's winning time.

Section 8-5

Focus
Find the point where two lines intersect.

Graphing Systems of Linear Equations

WHERE SHALL WE MEET?

EXPLORATION

Can one triangle meet two conditions?

- Materials: $8\frac{1}{2}$-by-11 inch paper (**two different colors**), protractors, rulers, scissors, tape
- Divide the class into two groups.

(1) Assign Group I one color of paper and Group II the other color.

(2) Cut out a triangle that meets your group's rule. Make sure that no two triangles in your group are the same.

Group I rule:
Make angle y twice as big as angle x.

Use any value between 0 and 45 for x.

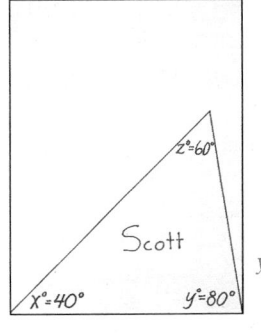

Scott

$y = 2x$

$z° = 60°$
$x° = 40°$
$y° = 80°$

Group II rule:
Make angle y and angle x complementary.

Use any value between 0 and 90 for x.

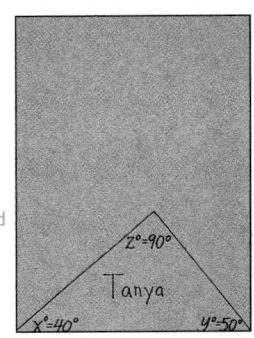

Tanya

$y = 90 - x$

$z° = 90°$
$x° = 40°$
$y° = 50°$

Both Groups:

- Use the $8\frac{1}{2}$ in. edge of your paper as the base of your triangle.

- Put angle x on the left and angle y on the right.

- Write your name and all three angle measures on your triangle.

Continued on next page.

8-5 Graphing Systems of Linear Equations

447

Answers to Exploration

1, 2. Answers may vary.

PLANNING

Objectives and Strands
See pages 414A and 414B.

Spiral Learning
See page 414B.

Materials List
- White and colored $8\frac{1}{2} \times 11$ in. paper
- Protractor
- Ruler
- Scissors
- Tape
- Graph paper

Recommended Pacing
Section 8-5 is a two-day lesson.

Day 1
Pages 447–448: Exploration, *Exercises 1–5*

Day 2
Pages 448–450: Sample 1 through Look Back, *Exs. 6–30*

Extra Practice
See pages 633–635.

Warm-Up Exercises
Warm-Up Transparency 8-5

Support Materials
- Practice 63
- Enrichment 56 in the Activity Bank
- Study Guide 8-5
- Problem Set 17
- Additional Exploration 8
- Diagram Masters 2, 3 in the Explorations Lab Manual
- Overhead Visual 8
- McDougal Littell Mathpack software: *Function Investigator*
- Function Investigator with Matrix Analyzer Activity Book: Function Investigator Activity 10
- Using Plotter Plus: Graphing Systems of Linear Equations
- Using Plotter Plus: Estimating Solutions of Linear Systems
- Quiz 8-5
- Test 32
- Alternative Assessments 7, 8

447

Exploration

The goal of the Exploration is to use manipulatives to have students discover how to solve a system of equations by graphing. The concrete approach presented here should make students more comfortable with the standard mathematical approach presented in Sample 1.

Additional Sample

S1 The service charge on a checking account at Hometown Bank is $5 per month plus $.15 for each check written. The service charge at Twentieth Century Bank is $.25 per check.

a. Write and solve a system of equations to model this situation.
  Let *c* = the cost of the service charge. Let *n* = the number of checks written. Hometown Bank (solid line): c = 0.15n + 5 Twentieth Century Bank (dashed line): c = 0.25n The lines intersect at (50, 12.50).

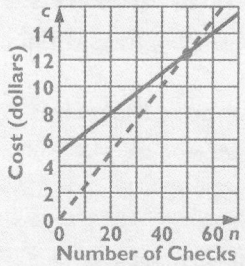

b. What advice would you give to someone trying to decide which bank to use? If you plan to write fewer than 50 checks per month, Twentieth Century Bank is less expensive. If you plan to write more than 50 checks per month, Hometown Bank is less expensive.

③ Tape all the triangles from both groups on a large graph in the front of the room.

④ Look for patterns in the graph. Describe any patterns you see.

⑤ Is there any triangle that both groups created? If so, what are the measures of its angles and where is it on the graph? If not, make a triangle that meets each group's rule. Tell its angle measures and its location on the graph.

⑥ What equation relates *x* and *y* for all the Group I triangles? What equation relates *x* and *y* for all the Group II triangles?

⑦ Solve the system of equations in step 6 by substitution.

⑧ How is the answer to step 7 related to the answer to step 5? Explain.

The graph of triangles you made in the Exploration illustrates a *system of equations.* You can estimate the solution of a system of equations by graphing the equations and looking for a point of intersection.

······► Now you are ready for:
Exs. 1–5 on p. 451

Sample 1

TECHNOLOGY NOTE

You can zoom in on the point where the lines intersect with a graphics calculator.

MoviesPlus rents videos for $2.50 each and has no membership fee. Videobusters rents videos for $2 each but has a $10 membership fee.

a. Write and solve a system of equations to model this situation.

b. What advice would you give to someone trying to decide which video store to use?

Unit 8 Linear Equations as Models

Answers to Exploration

3. Answers may vary.

4. Answers may vary. An example is given. The triangles in Group I seem to lie on a line, and the triangles in Group II seem to lie on a line.

5. Answers may vary. An example is given. Yes. $x = 30$, $y = 60$, $z = 90$; at the point (30, 60)

6. $y = 2x$; $y = 90 - x$

7. $x = 30$, $y = 60$

8. The solution of the system of equations has the same values for x and y as the triangles created by both groups.

Sample Response

Problem Solving Strategy: Use a graph

a. Write an equation to describe the cost at each video store.

➤ Let c = the total cost of renting videos.

➤ Let n = the number of videos rented.

➤ MoviesPlus: $c = 2.50n$

➤ Videobusters: $c = 2.00n + 10$

Graph the system of equations.

b. Here is some advice you can give: "If you plan to rent fewer than 20 videos, then MoviesPlus is less expensive. If you plan to rent more than 20 videos, then Videobusters is less expensive."

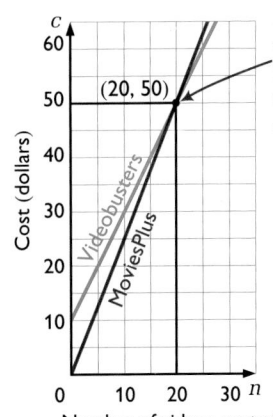

The lines intersect at (20, 50). This means that if you rent exactly 20 videos, the cost is $50 at either video store.

(20, 50)

Cost (dollars)

Videobusters

MoviesPlus

Number of videos rented

Talk it Over

1. In Sample 1, how can you tell from the graph that Videobusters is less expensive if you rent more than 20 movies?

2. Check the solution in Sample 1 by substituting 20 for n and 50 for c in both equations.

Systems without Solutions

In Sample 1, the graph of the system of equations shows that the lines intersect. The coordinates of the point of intersection are the solution of the system.

Sometimes two lines do not intersect. In this case the system of equations has no solution.

Sample 2

Sandy and Rita are practicing for a 100 m dash competition. Sandy gives Rita a 10 m head start.

Write and graph a system of equations to model this situation.

Sandy's speed is 8 m/s.

Rita's speed is 8 m/s.

Answers to Talk it Over

1. For any value of n that is greater than 20, the line representing Videobusters is below the line representing Movies Plus. This means that the cost at Videobusters is below or less than the cost at Movies Plus.

2. For $c = 2.50n$, $50 = 2.5(20)$, and $50 = 50$.
For $c = 2.00n + 10$,
$50 = 2.00(20) + 10$,
$50 = 40 + 10$, and $50 = 50$.

Reasoning

For part (b) of Sample 1 and Talk it Over question 1, students should reach the conclusion that whichever graph is below the other represents the less expensive alternative.

Additional Sample

S2 James left the campground riding his bicycle at 15 miles per hour. When he was 5 miles away from camp, Lucie left the camp riding at the same speed. Write and graph a system of equations to model this situation.

Let x = the time Lucie started. Let y = their distance from the camp.
James: $y = 15x + 5$ (solid line)
Lucie: $y = 15x$

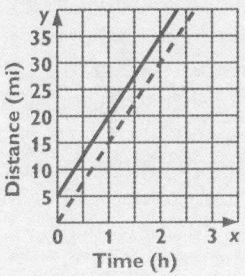

Distance (mi)

Time (h)

Using Technology

Some students may have difficulty with the concept of a system of equations not having a solution. To help overcome this, have students graph the equations from Sample 1 on their graphics calculators and use the TRACE feature to find the intersection point. Students will agree that this point is the solution, as shown in Sample 1. Now ask students to do the same for the equations in Sample 2. The cursor will trace along one equation or the other, but never touch both at the same time. Thus, there is no solution.

Talk it Over

Question 5 makes clear the idea that equations with the same slope are parallel. Students should use questions 5 and 6 to come to the realization that equal slopes means the lines are parallel and therefore the system has no solution.

Limited English Proficiency

For the Look Back question, students acquiring English might each be offered the help of a peer tutor who can help them state clearly their ideas on the usefulness of each method.

Look Back

Have students write a paragraph in their journals describing when to use each method of solving a system of linear equations. The graphical method illustrates very nicely why a solution of a system of linear equations consists of a pair of numbers rather than a single number. Its disadvantage is that it is a relatively slow process to implement, and, in general, only approximate solutions can be found, not exact ones. Substitution is a more efficient method that yields exact solutions. ·············•

Sample Response

Write an equation for each runner's distance from the starting line.

➤ Let x = the time since the runner started.

➤ Let y = the runner's distance from the starting line.

➤ Sandy's distance from the starting line: $y = 8x$

➤ Rita's distance from the starting line: $y = 8x + 10$

The graph shows each runner's distance from the starting line as a function of time.

Because the lines are parallel, there is no point of intersection. The system of equations has *no solution*. The runners are never at the same point at the same time.

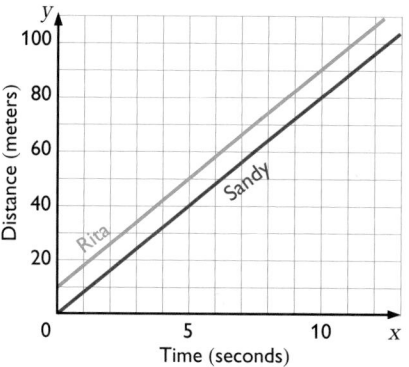

Check You can check the solution of a system of equations by using substitution.

$y = 8x$ and $y = 8x + 10$

$8x = 8x + 10$ ◄——— Substitute $8x$ for y in the second equation.

$8x - 8x = 8x + 10 - 8x$ ◄——— Subtract $8x$ from both sides.

$0 = 10$ ◄——— This statement is *false*.

Because $0 \neq 10$, there is no solution that makes the equation true. The system of equations has no solution.

Talk it Over

3. In Sample 2, the graph shows diagonal lines. Are Rita and Sandy running diagonally across the race track? Explain your answer.

4. Use the graph in Sample 2 to estimate the runners' race times. Who wins the race?

5. The graphs of two equations with the same slope are parallel. Use the equations $y = 8x$ and $y = 8x + 10$. Without graphing, how can you tell that the system of equations has no solution?

6. How can you tell from a graph that a system of equations has either a *solution* or has *no solution*?

Look Back ◄

You can find the solution of a system of linear equations by graphing the equations or by using substitution. When is each method useful?

·····➤ **Now you are ready for:**
Exs. 6–30 on pp. 451–454

Answers to Talk it Over

3. No. Explanations may vary. An example is given. The graph shows diagonal lines because the distance increases as the time increases. This comparison of distance and time does not involve the direction in which they are running.

4. Estimates may vary. An example is given. Rita; about 11 s; Sandy: about 12.5 s; Rita wins.

5. Both equations have the same slope, 8.

6. If the lines intersect, then the system has a solution. If the lines are parallel, then the system has no solution.

Answers to Look Back

Answers may vary. An example is given. Graphing is useful when you need only an estimate of the common solution or when you want to compare situations before or after the point of intersection, such as in Sample 1 with the video stores. Using substitution is useful when you need to find the exact common solution. For instance, in the Exploration, you could have used substitution to find the measures of angles that fit both rules without graphing. Also, you can use one method to check the other method.

8-5 Exercises and Problems

1. Look back at the Exploration on pages 447 and 448.

   a. Does a 25°-50°-105° triangle meet the rule for *Group I* or *Group II*?

   b. Is there more than one triangle that meets the rule for both groups?

For Exercises 2–4, do these things.

a. Estimate the solution of each system of equations.

b. Use substitution to find the exact solution of each system.

2.

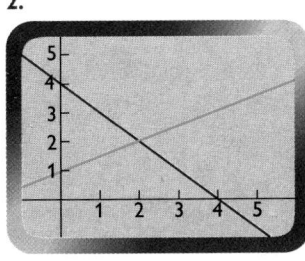

$y = \frac{1}{2}x + 1$

$y = -x + 4$

3.

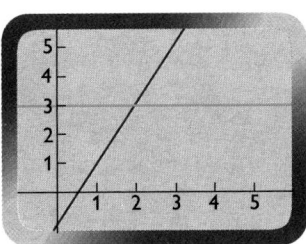

$y = 3$

$y = 2x - 1$

4.

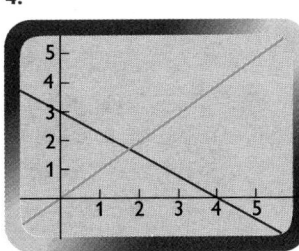

$y = -\frac{3}{4}x + 3$

$y = x$

5. **Open-ended** Write two different systems of equations that have the solution (5, 3).

6. **Reading** How can you tell just by looking at the equations whether a system of equations has a solution or no solution?

Graph each system of equations. Estimate the solution of the system, or write *no solution*.

7. $y = x - 2$

   $y = -2x + 5$

8. $y = 0.5x$

   $y = 0.5x + 3$

9. $y = x$

   $y = \frac{1}{2}x - 3$

A dog notices a squirrel 5 m away and starts chasing it. Describe what happens in each case.

10.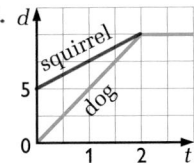

11.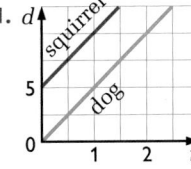

12.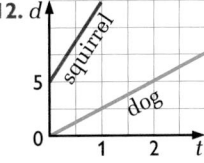

8-5 Graphing Systems of Linear Equations

451

---

8. no solution

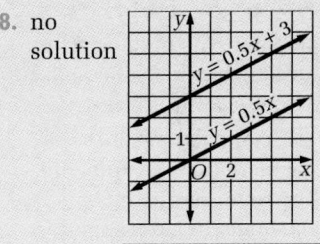

9. (−6, −6)

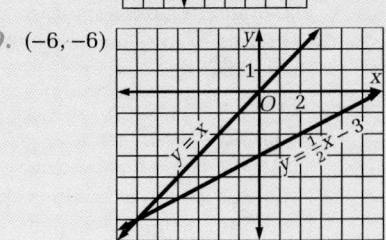

10–12. Descriptions may vary. Examples are given.

10. The dog is running faster than the squirrel and catches up with it after about 2 units of time.

11. The dog chases the squirrel but never catches up with it because they are both running at the same rate, and the squirrel had a 5 m head start.

12. The squirrel must see the dog and runs at a faster rate than the dog, so the dog does not catch up with it.

Answers to Exercises and Problems

1. **a.** Group I **b.** No.

2–4. **a.** Estimates may vary. Examples are given.

2. **a.** (2, 2) **b.** (2, 2)

3. **a.** (2, 3) **b.** (2, 3)

4. **a.** $\left(1\frac{2}{3}, 1\frac{1}{2}\right)$ **b.** $\left(1\frac{5}{7}, 1\frac{5}{7}\right)$

5. Answers may vary. An example is given.

  (1) $y = 3$, $y = x - 2$

  (2) $y = -x + 8$, $y = 2x - 7$

6. Answers may vary. An example is given. When the equations are in slope-intercept form and the values of *m* are the same but the values of *b* are different, then there is no solution. Otherwise there is at least one solution.

7. Estimates may vary. An example is given: about $\left(2\frac{1}{3}, \frac{1}{3}\right)$.

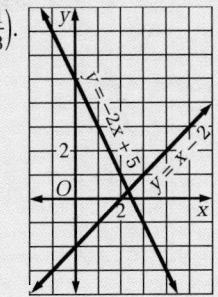

451

Interdisciplinary Problems

Exs. 16, 17, 22, and 23 illustrate how systems of linear equations can be used to analyze three quite different problem situations. In Ex. 16, the break-even point is a crucial concept for a business, and the expense and income lines show exactly the location of this point. In Exs. 17, 22, and 23, intersecting lines are also used to model the situations involved. The graphs are then used to reach conclusions.

Using Technology

Solving problems with a graphics calculator and break-even points are discussed on page 617 of the Technology Handbook.

The *Function Investigator* software can be used to do Ex. 21.

Answers to

Exercises and Problems ·············

13. solution

14. no solution

15. no solution

16. a. (30, 600); the number of units for which expenses equal income

b. The company profits after selling more than 30 units; the line representing income is above the line representing expenses, so income is greater than expenses.

c. The company loses money when it sells less than 30 units; the line representing expenses is above the line representing income, so expenses are greater than income.

17. a. $c = 31n + 40$; $c = 18n + 170$

b. Estimates may vary. An example is given: (10, 350). See graph at right.

c. (10, 350)

d. Advice may vary. An example is given. If you want a system for more than 10 months, then renting videos would be less expensive. If you want a system for less than 10 months, then cable TV would be less expensive.

452

Without graphing, tell whether each system of equations has a *solution* or *no solution*.

13. $y = 3x + 4$
    $y = 5x - 2$

14. $y = 2x - 5$
    $y = 2x + 6$

15. $y = 100 - x$
    $y = 50 - x$

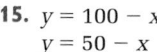

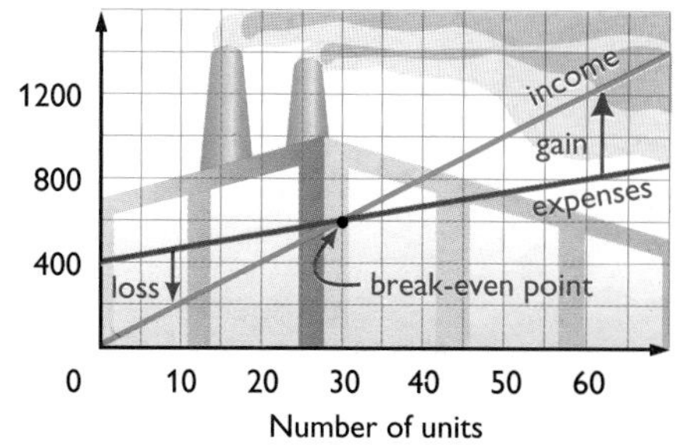

16. **Business** The graph shows the income and expenses of a company. The company makes a profit if income is greater than expenses. It loses money if expenses are greater than income.

a. What are the coordinates of the break-even point? What does this point represent?

b. When does the company make a profit? How can you tell this from the graph?

c. When does the company lose money? How can you tell this from the graph?

17. **Entertainment** A friend is trying to choose between cable TV and renting videos. Here is the information your friend gives you.

a. Write an equation for the cost of installing cable TV with a movie channel. Write an equation for the cost of buying a video cassette player and renting movies.

b. Graph the two equations on the same axes. Estimate the point of intersection of the graphs.

c. Solve the system of equations by substitution.

d. What advice would you give to your friend?

> "Cable TV costs $40 to install and then $20 each month. Adding a movie channel costs an extra $11 each month."
>
> "On the other hand, the cost of a video cassette player is $170. Renting a video costs $3, and I think I will rent 6 videos a month."

For Exercises 18–20, rewrite each equation in slope-intercept form. Then tell whether each system of equations has a *solution* or *no solution*.

18. $2x + 3y = 4$
    $4x + 6y = 4$

19. $x - 3y = 2$
    $x - 4y = 2$

20. $4x - 2y = 6$
    $6x - 2y = 3$

21. **TECHNOLOGY** Use a graphics calculator or software.

You can solve an equation like $2(x + 4) = -3x - 7$ by making it into a system of equations and graphing the two equations.

a. Set the expression on each side of the equation equal to y. The equations are $y = 2(x + 4)$ and $y = -3x - 7$. Graph both equations.

b. Use the zoom and trace features to find the coordinates of the point of intersection. The value of x is the solution of the original equation.

c. Use this method to solve the equation $-\frac{1}{2}x + 11 = 4(-3 + x)$.

d. **Open-ended** Write another equation that has x on both sides and use this method to solve it.

452   **Unit 8** Linear Equations as Models

Home Entertainment Costs

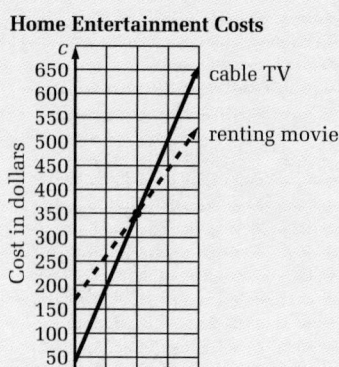

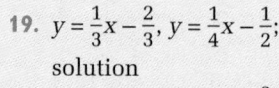

18. $y = -\frac{2}{3}x + \frac{4}{3}$, $y = -\frac{2}{3}x + \frac{2}{3}$;
    no solution

19. $y = \frac{1}{3}x - \frac{2}{3}$, $y = \frac{1}{4}x - \frac{1}{2}$;
    solution

20. $y = 2x - 3$, $y = 3x - \frac{3}{2}$;
    solution

21. a.
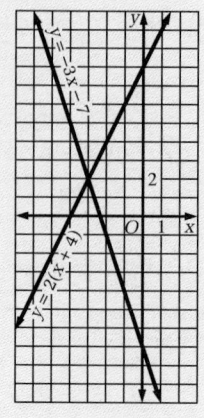

In the book *Address Unknown,* James D. Wright writes about the causes of homelessness. He found that between 1977 and 1983, the population needing low-income housing in twelve United States cities increased from 2,522,000 to 3,425,000. During that time, the number of low-income housing units in those cities decreased from 1,607,000 to 1,128,000.

22. Suppose two people can be housed in each low-income housing unit. Then the number of available spaces would be 3,214,000 in 1977 and 2,256,000 in 1983.

a. Interpret the information shown in the graph.

b. What does the point of intersection represent? What was the housing situation before this time? What was the housing situation after this time?

c. Write equations for each line. Use the coordinates of the points given on the graphs.

d. Solve the system of equations by substitution. Does your solution agree with the point of intersection on the graph?

A group of students help build a house as part of a Habitat for Humanity project in Coahoma, Mississippi.

23. Suppose three people can be housed in each available low-income housing unit. Then the number of available spaces would be 4,821,000 in 1977 and 3,384,000 in 1983.

a. Redraw the graph from Exercise 22 using the new data for the number of available spaces.

b. Estimate the point of intersection. Explain why it appears later in time than the point of intersection in Exercise 22.

c. Write a new system of equations and solve it by substitution to check your estimate.

d. What do the graphs in Exercise 22 and part (a) above suggest is one cause of homelessness?

8-5 Graphing Systems of Linear Equations   **453**

22. c. people who need housing: $y = 150.5x + 2522$; available spaces: $y = -159.7x + 3214$

d. Solutions may vary due to rounding. An example is given. (2.23, 2858); Yes.

23. a.

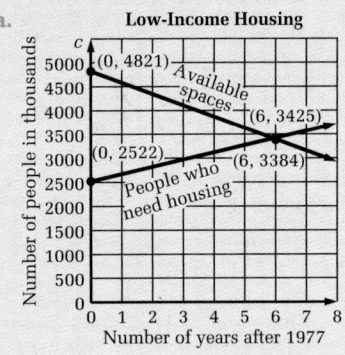

b. Estimates may vary. An example is given. (5.8, 3400); Explanations may vary. An example is given. Having more available spaces will mean that the need for housing will be met for a longer period of time. Therefore, the point of intersection occurs later in time.

c. people who need housing: $y = 150.5x + 2522$; available spaces: $y = -239.5x + 4821$; Solutions may vary due to rounding. An example is given. (5.9, 3410)

d. There is not enough low-income housing.

Answers to Exercises and Problems

21. b. (−3, 2); −3   **c.** about 5.1

  d. Equations may vary.

22. a. Interpretations may vary. An example is given. Between 1977 and 1983, the number of people who needed housing increased while the number of available spaces decreased. In 1977, there were more than enough available spaces for people needing housing. At some time during 1979, the number of available spaces equaled the number of people who needed housing. For the rest of 1979 through to 1983, the number of people who needed housing outnumbered the number of available spaces.

b. the year in which the number of available housing spaces equaled the number of people who needed housing; There were more housing spaces available than people who needed housing. There were more people who needed housing than available spaces.

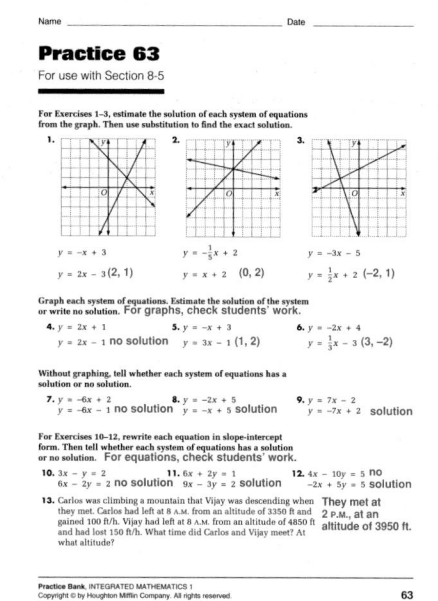

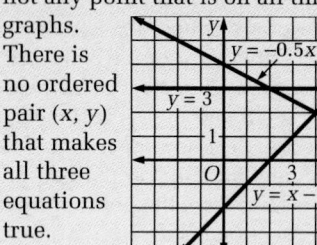
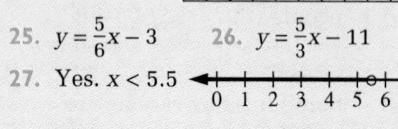
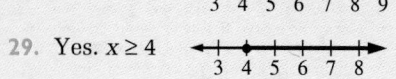
454

Ongoing ASSESSMENT

24. **Writing** Graph this sytem of three equations. Explain why you think this system of three equations does or does not have a solution.

$$y = -0.5x + 4$$
$$y = 3$$
$$y = x - 2$$

Review PREVIEW

Write an equation for each line. *(Section 8-4)*

25. The points (12, 7) and (6, 2) are on the line.

26. The slope is $\frac{5}{3}$ and the point (6, −1) is on the line.

Tell whether 5 is a solution of each inequality. Write *Yes* or *No*. Then solve and graph each inequality. *(Section 5-4)*

27. $-2x + 13 > 2$ 28. $4(x - 5) \geq 0$ 29. $-3 \leq x - 7$

8 Working on the Unit Project

30. The equations of the fitted lines for the graphs of some California high school record times are given. In the equations, x is the number of years since 1957 and y is the record time in seconds.

    a. Use a graph to estimate a solution of the system of equations.

    b. Use substitution to find a solution.

    c. What do the coordinates of the solution mean?

    Girls' 400 m Individual Medley:
    $y = -1.95x + 335.05$

    Boys' 400 m Individual Medley:
    $y = -1.68x + 299.1$

Unit 8   CHECKPOINT 2

1. **Writing** Describe two ways to tell that a system of equations has no solution.

Write an equation for each line.   8-4

2. The points (3, −4) and (4, 1) are on the line.

3. The horizontal intercept is 1 and the slope is $-\frac{1}{4}$.

Without graphing, tell whether each system has a solution or *no solution*.   8-5

4. $y = -2x + 5$ 5. $y = 4 - 3x$ 6. $y = -2x$
   $y = x + 5$ $y = 8 - 3x$ $y = x$

7. Natural Springs sells bottled water for $1.00 per gallon and charges $10 as a deposit. Water Fresh charges $1.50 per gallon, with no deposit. Write and solve a system of equations to model this situation.

454   **Unit 8** Linear Equations as Models

---

30. a. Estimates may vary. An example is given. (132, 75)

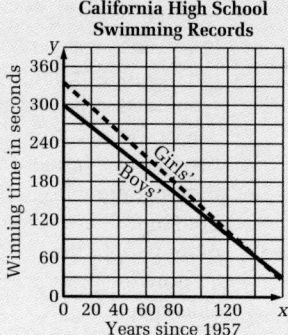

California High School Swimming Records

b. Solutions may vary due to rounding. Example: (133.15, 75.41)

c. The x-coordinate means that after about 133 years after 1957, or in the year 2090, the record times will be the same for both boys and girls. The y-coordinate means that this record time will be 75.41 s.

Answers to Checkpoint

1. Write the equations in slope-intercept form. If the values for m are equal and the values for b are not, then there is no solution. Graph the equations. If the lines are parallel, then there are no solutions.

2. $y = 5x - 19$ 3. $y = -\frac{1}{4}x + \frac{1}{4}$

4–7. See answers in back of book.

Graphing Linear Inequalities

WHICH SIDE ARE YOU ON?

Focus
Graph linear inequalities on a coordinate plane.

EXPLORATION

What does an inequality graphed on a coordinate plane look like?

• **Work with the whole class.**

Jeremy is (0, 4). He is standing because 4 is greater than or equal to 0.

(0, 4) (1, 4) (2, 4) (3, 4) (4, 4)

(0, 3) (1, 3) (2, 3) (3, 3) (4, 3)

(0, 2) (1, 2) (2, 2) (3, 2) (4, 2)

(0, 1) (1, 1) (2, 1) (3, 1) (4, 1)

(0, 0) (1, 0) (2, 0) (3, 0) (4, 0)

Origin

1. Arrange the desks in rows. Make the number of desks in each row equal the number of rows. For example, there might be 5 rows of 5 desks each. When the desks are arranged, return to your seat.

2. Choose a student at a corner desk to represent the origin so that your class forms the first quadrant of a coordinate plane. Who is (0, 0) in your class?

3. Each student represents an ordered pair. Have each student write down his or her coordinates.

Michelle is (3, 1). She is sitting because 1 is not greater than or equal to 3.

4. At a signal, stand up if your *y*-value is greater than or equal to your *x*-value. What inequality does this represent? Describe the shape formed by the people who are standing. Then sit down.

5. Repeat step 4 for a *y*-value that is less than twice your *x*-value.

6. Repeat step 4 for a *y*-value that is greater than 3.

7. Repeat step 4 for a *y*-value that is less than or equal to 2.

455

Answers to Exploration

1–3. Arrangements and origins may vary.

4–7. Descriptions may vary. Examples are given based on the diagram in the Exploration.

4. $y \geq x$; a triangle with vertices at (0, 0), (4, 4), and (0, 4)

5. $y < 2x$; a staircase going up from left to right with 2 rows of 4 people, 2 rows of 3 people, and 1 row of 2 people

6. $y > 3$; a line formed by the last row of people

7. $y \leq 2$; a rectangle formed by the first 3 rows of people

PLANNING

Objectives and Strands
See pages 414A and 414B.

Spiral Learning
See page 414B.

Materials List
➤ Graph paper
➤ Ruler

Recommended Pacing
Section 8-6 is a two-day lesson.
Day 1
Pages 455–457: Exploration through Talk it Over, *Exercises 1–16*
Day 2
Pages 457–459: Sample 2 through Look Back, *Exercises 17–39*

Extra Practice
See pages 633–635.

Warm-Up Exercises
♦ Warm-Up Transparency 8-6

Support Materials
➤ Practice 64
➤ Enrichment 57 in the Activity Bank
➤ Study Guide 8-6
➤ Problem Set 18
➤ Diagram Masters 2, 3 in the Explorations Lab Manual
♦ Overhead Visual 9
➤ McDougal Littell Mathpack software: *Stats!*
➤ Using Plotter Plus: Graphing Linear Inequalities
➤ Quiz 8-6
➤ Alternative Assessment 9

TEACHING

Exploration

The goal of the Exploration is for students to model linear inequalities physically and thus develop a visual and intuitive grasp of what an inequality graphed on a coordinate plane looks like.

Communication: Listening

Before students begin the Exploration, review the terms *greater than or equal to* and *less than or equal to* and their symbols. For steps 4–7, read the inequality to be modeled and write the expression on the chalkboard.

Additional Sample

S1 Graph each inequality.

a. $y \geq -x$

Graph the boundary line as a solid line. Shade the region above the line.

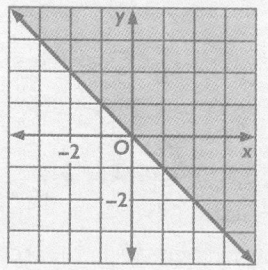

b. $y > \frac{1}{2}x - 1$

Draw the boundary line as a dashed line. Shade the region above the line.

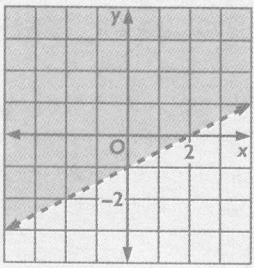

Linear Inequalities on a Coordinate Plane

You just modeled four *linear inequalities*. Another way to model a linear inequality is to draw its graph on the coordinate plane.

The graph of a **linear inequality** is a region on a coordinate plane whose edge is a line. This line is called a **boundary line.**

Sample 1

Graph each inequality.

a. $y \geq x$ **b.** $y < -\frac{1}{2}x + 4$

Sample Response

a. First graph the boundary line $y = x$. Draw it as a solid line.

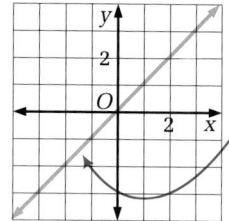

If the inequality has the symbol $\leq$ or $\geq$, use a solid line to show that the line is part of the shaded region.

Then shade the region above the line.

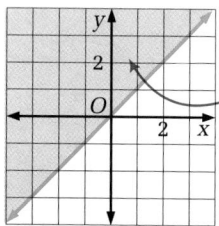

The shaded portion of the graph is the **solution region**.

All points in the shaded region have coordinates that make the inequality $y \geq x$ true.

b. First graph the boundary line $y = -\frac{1}{2}x + 4$. Draw it as a dashed line.

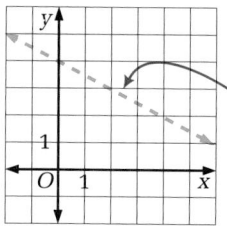

If the inequality has the symbol $>$ or $<$, use a *dashed* line to show that the line is *not* part of the shaded region.

Then shade the region below the line.

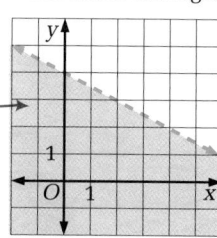

All points in the shaded region have coordinates that make the inequality $y < -\frac{1}{2}x + 4$ true.

Talk it Over

1. Which inequality in Sample 1 did you model in step 4 of the Exploration?

2. Describe how to graph $y \geq 3$.

Unit 8 Linear Equations as Models

Answers to Talk it Over ·····

1. (a) $y \geq x$

2. First, graph the boundary line $y = 3$ as a solid line. Then shade the region above the line.

3. **a.** Describe the graph of $x = 2$.

   b. Describe the graph of $x \le 2$.

Tell whether the boundary line for each inequality is *solid* or *dashed*. Then tell whether you would shade the region *above* or *below* the line.

4. $y \le 9x + 6$

5. $y > \frac{2}{3}x$

6. $y \ge -4x - 11$

7. $y < -x + \frac{1}{4}$

 Now you are ready for:
Exs. 1–16 on pp. 459–460

The inequalities in Sample 1 have the dependent variable, y, alone on one side of the inequality symbol. Inequalities written this way are in *slope-intercept form*.

Sample 2

Derek has saved $48 by baby-sitting. He plans to use the money to buy tapes and compact discs at a music store. Tapes cost $8 and compact discs cost $12. The inequality $8x + 12y \le 48$ represents the amount Derek can spend at the store.

Graph the inequality $8x + 12y \le 48$.

Sample Response

First rewrite the inequality in slope-intercept form.

$$8x + 12y \le 48$$

$$8x + 12y - 8x \le 48 - 8x \quad \longleftarrow \text{Subtract 8x from both sides.}$$

$$12y \le 48 - 8x$$

$$\frac{1}{12} \cdot 12y \le \frac{1}{12}(48 - 8x) \quad \longleftarrow \text{Multiply both sides by the reciprocal of 12.}$$

$$y \le 4 - \frac{2}{3}x$$

Next graph the boundary line $y = 4 - \frac{2}{3}x$.

Draw it as a solid line.
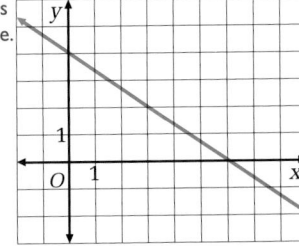

Then shade the region below the line.
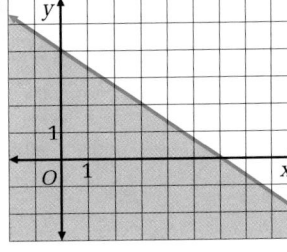

All points in the shaded region have coordinates that make the inequality $8x + 12y \le 48$ true.

8-6 Graphing Linear Inequalities

457

Answers to Talk it Over

3. **a, b.** Descriptions may vary. Examples are given.

   a. The graph of $x = 2$ is a vertical line, parallel to the y-axis, with all points having x-coordinate 2.

   b. The graph of $x \le 2$ includes the graph of $x = 2$, and the region to the left of the line is shaded.

4. solid; below

5. dashed; above

6. solid; above

7. dashed; below

Talk it Over

Use questions 4–7 to reinforce these generalizations for inequalities written in slope-intercept form.

(1) The boundary is a solid line for linear inequalities with signs $\le$ or $\ge$ and a dashed line for inequalities with signs $<$ or $>$.

(2) The region above the boundary line is shaded for inequalities with signs $\ge$ or $>$. The region below the boundary line is shaded for linear inequalities with signs $\le$ or $<$.

Limited English Proficiency

Complementary pairs of locational terms that are second-nature to native speakers of English (*above, below; near, far*) cause problems for some students acquiring English.

Error Analysis

If students are having difficulty determining which region to shade, tell them to check a point in one of the regions by substituting its coordinates into the inequality to see whether it is a solution or not. Regions that contain solution points are shaded. The most simple point to check is usually the origin.

Additional Sample

S2 Graph the inequality $5y - 3x > 15$.

$$5y - 3x + 3x > 15 + 3x$$

$$5y > 15 + 3x$$

$$\frac{1}{5} \cdot 5y > \frac{1}{5}(15 + 3x)$$

$$y > 3 + \frac{3}{5}x$$

Draw the boundary line as a dashed line. Shade the region above the line.

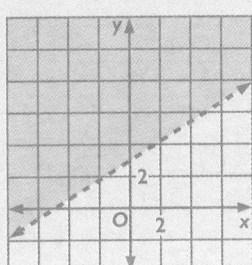

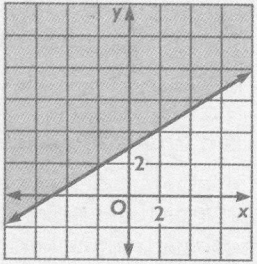
BY THE WAY...

Thomas Harriot (1560–1621) was a British mathematician and astronomer. He may have been the first person to use the inequality symbols > and <.

Graphing Inequalities in Standard Form

In Sample 2, the inequality $8x + 12y \leq 48$ has both the dependent variable and the control variable on the same side of the inequality symbol. You can graph inequalities in this *standard form* without rewriting them in slope-intercept form.

Sample 3

Graph the inequality $8x - 12y < 48$.

Sample Response

1 First use the intercepts to graph the boundary line $8x - 12y = 48$. Draw it as a dashed line.

2 Next evaluate the inequality at a point *not* on the line, such as $(0, 0)$.
$$8x - 12y < 48$$
$$8(0) - 12(0) \overset{?}{<} 48 \quad \longleftarrow \text{Substitute } x = 0 \text{ and } y = 0.$$
$$0 < 48 \; \checkmark \quad \longleftarrow \text{The inequality is true.}$$

$(0, 0)$ is part of the solution.

3 Then shade the region on the *same side* of the line as $(0, 0)$.

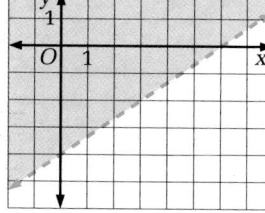

Talk it Over

10. Look back at the inequalities and graphs in Samples 2 and 3. Does the symbol < always mean you should shade below the boundary line? Explain.

11. Why do you need to pick only one point not on the line?

12. Why do you have to pick a point that is not on the line?

13. Rewrite the inequality in Sample 3 in slope-intercept form and then graph it. Is your graph the same as the one shown?

Answers to Talk it Over

8. Answers may vary. An example is given. The coordinates of all the points in the shaded region make the inequality $y \leq 4 - \frac{2}{3}x$ true. Since this inequality is the slope-intercept form of the original inequality, $8x + 12y \leq 48$, they both have the same solutions.

9. all points with whole-number coordinates in the region bounded by the x-axis, the y-axis, and the line $y = 4 - \frac{2}{3}x$

10. No. Explanations may vary. An example is given. If the inequality is in slope-intercept form, then the symbol < means to graph below the boundary line. However, if the inequality is in standard form, you need to check a point on one side of the boundary line to determine which region to shade.

11. If one point in the region makes the inequality true, then all the points in the region make the inequality true and this is the correct region to shade.

12. The boundary line is a boundary for the solution region. A point on this line may be a solution to the inequality, but it will not help you determine which

region is the solution region.

13. $y > \frac{2}{3}x - 4$; Yes.

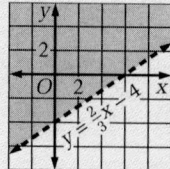

HOW TO GRAPH AN INEQUALITY

1. Graph the boundary line first. Decide whether to use a solid or a dashed line.

2. Pick a point that is not on the line. Evaluate the inequality for values of the coordinates of this point.

3. a. If the inequality is true for this point, shade the region on the same side of the line.

   b. If the inequality is false for this point, shade the region on the opposite side of the line.

······▶ **Now you are ready for:**
Exs. 17–39 on pp. 460–462

Look Back ◀——

In Samples 1 and 2 you graphed inequalities in slope-intercept form. Write a general method for doing this.

Communication: Reading

Be sure students understand the method presented in How to Graph an Inequality. You can relate the three steps given here to Exs. 3–6.

Look Back

Students may wish to write a journal entry explaining a general method for graphing inequalities in slope-intercept form.

APPLYING

Suggested Assignment

Standard 2–15, 17–26, 31–39

Extended 1–15, 17–29, 31–39

Integrating the Strands

Algebra Exs. 1–34, 37–39

Geometry Exs. 1–34, 37–39

Statistics and Probability Exs. 35, 36

Discrete Mathematics Exs. 15, 16, 35, 36

8-6 Exercises and Problems

1. Look back at the Exploration on page 455. Which ordered pairs in the diagram have y-values less than their x-values? What inequality does this represent?

2. **Reading** How can you tell when to use a solid boundary line and when to use a dashed boundary line?

Match each inequality with its graph.

3. $y \le 2.5x + 1$

4. $y < 2.5x + 1$

5. $y > 2.5x + 1$

6. $y \ge 2.5x + 1$

A.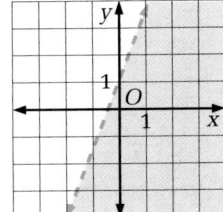

B.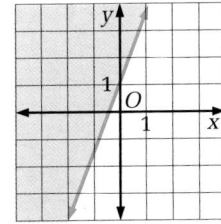

C.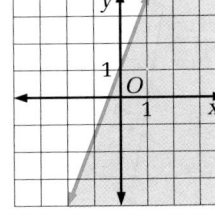

D.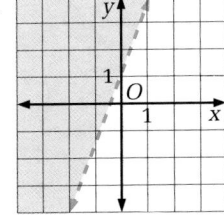

Answers to Look Back ········

Answers may vary. An example is given. If the inequality symbol is $\ge$ or $\le$, the boundary line is solid; if the inequality symbol is $>$ or $<$, the boundary line is dashed. If the inequality is in the form $y > mx + b$ or $y \ge mx + b$, shade the region above the boundary line. If the inequality is in the form $y < mx + b$ or $y \le mx + b$, shade the region below the boundary line.

Answers to Exercises and Problems ··········

1. the ordered pairs below the line $y = x$; $y < x$

2. Use a solid boundary line for $\ge$ and $\le$; use a dashed line for $>$ and $<$.

3. C

4. A

5. D

6. B

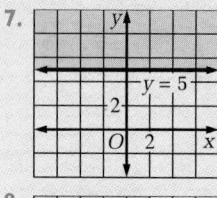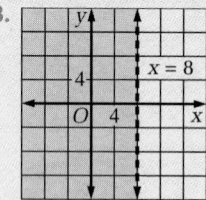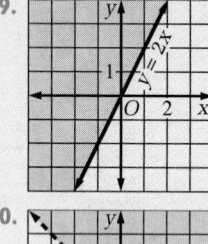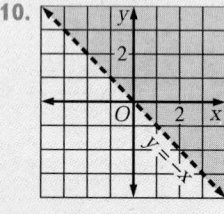
Graph each inequality.

7. $y \geq 5$ **8.** $x < 8$ **9.** $y \geq 2x$ **10.** $y > -x$

11. $y < \frac{1}{3}x + 2$ **12.** $y \leq -3x + 4$ **13.** $y > -\frac{2}{3}x - 5$ **14.** $y \leq \frac{1}{2}x - 1$

15. Four different brands of wheat flakes are sold at Super Shop.

  a. Find the unit price for each brand.

  b. Which brand is the best buy?

  c. Graph the data in the table. Put the weight on the horizontal axis and the price on the vertical axis.

  d. Draw a line from the origin through the point for Treat-O-Wheat.

Brand	Weight	Price
Flakies	300 g	$1.45
Nutri-Flakes	355 g	$1.65
Wheat Delight	380 g	$1.99
Treat-O-Wheat	420 g	$2.15

  e. Which data points lie above the line drawn? Which lie below the line?

  f. Interpret the results of part (e).

16. a. **Research** At a supermarket, find a single product that is packaged in several different sizes by the manufacturer. Make a table of size and cost data.

  b. Make a scatter plot that shows the cost and size data. Is the unit price constant?

  c. Describe any patterns you observe in your scatter plot.

  d. Why is unit price usually less for larger sizes?

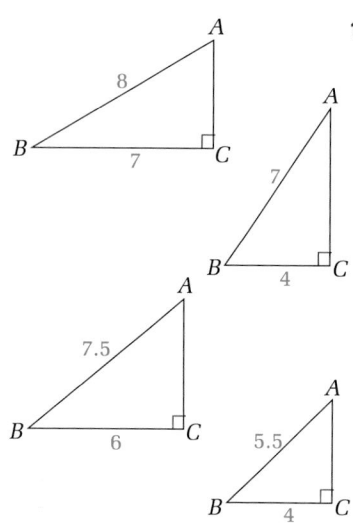

17. Use these right triangles.

  a. Write each triangle's given measurements as an ordered pair (AB, BC).

  b. Graph the ordered pairs. Put the length of the hypotenuse (AB) on the horizontal axis.

  c. Graph the line $y = x$ on the same set of axes. What is the slope of this line?

  d. Do the points you found in part (a) lie *above* or *below* the line?

  e. Complete each ? with >, <, or =.

   In all four triangles, AB ? BC.

   In all four triangles, $\frac{BC}{AB}$? 1.

  f. The ratio $\frac{BC}{AB}$ represents cos B. In part (e), you showed that cos $B < 1$. Explain how your graph supports this.

Graph each inequality.

18. $2x - 5y \leq 10$ **19.** $3x + 3y > -6$ **20.** $4y - 3x < 12$

21. $-y + 2x \geq 8$ **22.** $y - 2x < 1$ **23.** $y + 3x \geq -2$

Unit 8 Linear Equations as Models

 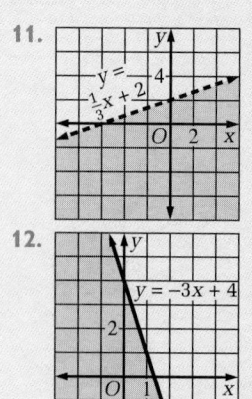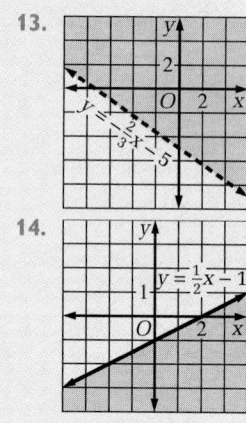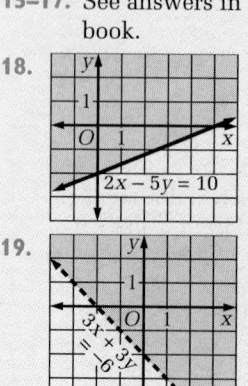

Lattice for x + 2y	

3	4	5	y 6	7	8	9
1	2	3	4	5	6	7
−1	0	1	2	3	4	5
−3	−2	−1	0	1	2	3 x
−5	−4	−3	−2	−1	0	1
−7	−6	−5	−4	−3	−2	−1
−9	−8	−7	−6	−5	−4	−3

Use the lattice to help you graph the inequalities in Exercises 24–26.

24. $x + 2y > 2$

25. $x + 2y \le 6$

26. $x + 2y \ge 0$

connection to **SCIENCE**

27. According to legend a king asked the Greek mathematician Archimedes (287?–212 B.C.) for help. The king worried that his crown might not be pure gold. Archimedes weighed the crown and measured its volume. He then found the density of the crown.

   a. The *density* $\left(\frac{\text{mass}}{\text{volume}}\right)$ of gold is 19.3 g/cm³. The volume of the king's crown was 300 cm³. If the crown were made of pure gold, how many grams would it weigh?

   b. Suppose Archimedes measured three objects that he knew were made of pure gold and plotted their data points on the graph at the right. Why do the three points lie on a line through the origin?

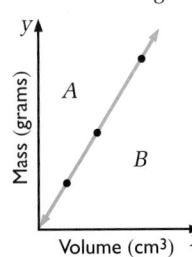

   c. Archimedes found that the king's crown had a lower density than gold. Would a data point for the king's crown lie in region *A* or region *B* of this graph? Explain.

Archimedes is pictured in this mosaic found at Pompeii, a Roman city near Naples, Italy, that was destroyed by the eruption of Mt. Vesuvius.

28. The density of silver is 10.5 g/cm³.

   a. Draw a graph that models the situation in which the ratio $\frac{\text{mass}}{\text{volume}}$ is *less* than the density of silver. Put the mass on the vertical axis.

   b. Draw a graph that models the situation in which the ratio $\frac{\text{mass}}{\text{volume}}$ is *greater* than the density of silver. Put the mass on the vertical axis.

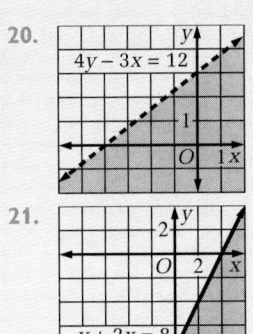

8-6 Graphing Linear Inequalities   **461**

Interdisciplinary Problems

For Ex. 27, you may want to mention that among Archimedes' other contributions to mathematics was his discovery that the volume of a sphere is two thirds the volume of the smallest cylinder that will enclose it. He applied his knowledge of mathematics and engineering to pulleys and levers to hold off a Roman navy for three years using catapults and levered claws. His knowledge of levers induced him to assert that given a lever long enough and a place on which to stand, he could move Earth.

27. c. region B; Explanations may vary. An example is given.

   Since the density ratio $\frac{\text{mass}}{\text{volume}}$ was less than 19.3 g/cm³, the data point would be located in the region $y < 19.3x$, or region B.

28. a, b.

$\frac{\text{mass}}{\text{volume}} < 10.5$	$\frac{\text{mass}}{\text{volume}} > 10.5$

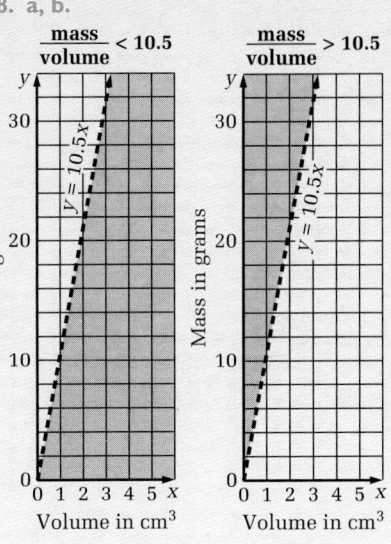

Answers to Exercises and Problems ·······

20.

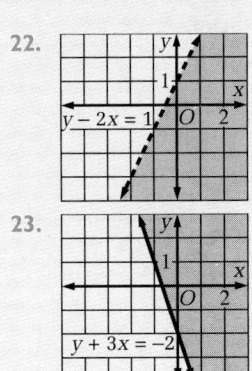

21.

22.

23.

24–26. See answers in back of book.

27. a. 5790 g

   b. Answers may vary. An example is given. The relationship between mass (y) and volume (x) is direct variation with the density as the variation constant: $y = 19.3x$.

Practice 64 For use with Section 8-6

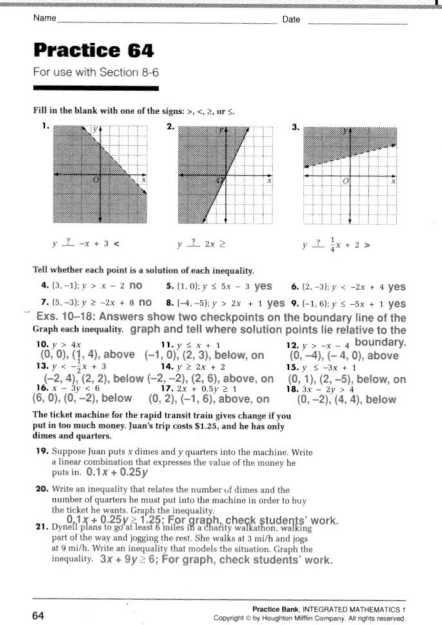

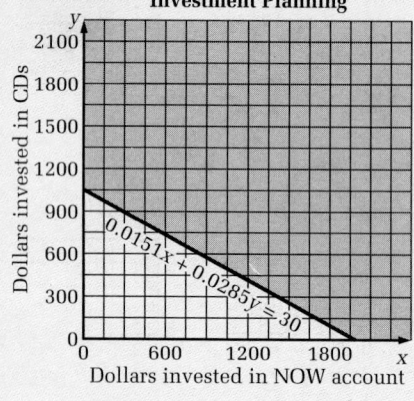

Answers to
Exercises and Problems

29, 30. Assume that interest rates are not compound.

29. a. $0.0151x + 0.0285y$

b. $0.0151x + 0.0285y \geq 30$

Investment Planning

---

29. **Banking** A bank offers a NOW account with an annual interest rate of 1.51%. The bank also offers a 12-month certificate of deposit (CD) with an annual interest rate of 2.85%.

a. Write an expression for the interest you receive in one year if you invest x dollars in the NOW account and y dollars in the CD.

b. You want to receive $30 or more interest each year. Write an inequality that expresses this and graph the inequality.

c. Choose a point in the solution region and check that the interest you would receive each year is greater than or equal to $30.

d. What is the smallest amount of money you could invest to earn $30 in interest? How would you have to invest it?

30. **Research** Find annual interest rates for a NOW account and a 12-month CD in a newspaper or at a bank. Using the rates you found, repeat Exercise 29.

Ongoing **ASSESSMENT**

31. **Writing** Write a quiz on graphing linear inequalities. Include at least six questions. Explain why you think your questions are good.

Review **PREVIEW**

Without graphing, tell whether each system has a *solution* or *no solution*. *(Section 8-5)*

32. $y = 2x - 7$
    $y = -2x + 3$

33. $y = 0.5x + 9$
    $y = 0.5x + 10$

34. $y = x + 1$
    $3x - 3y = 2$

35. An inspector checked 40 books at random from a group of 1500 books that had just been printed. The inspector found errors in 5 books. About how many of the printed books had errors? *(Section 6-2)*

36. There are 36 nuts in a bag. Six of the nuts are almonds, twelve are cashews, and the rest are peanuts. Find the probability of not getting an almond.

Graph each inequality on a number line. *(Section 3-3)*

37. $-3 \leq x < 1$

38. $5 < x < 20$

8 Working on the Unit Project

39. Look at the graph of the winning times for men's 400 freestyle swimming on page 439. Which points clearly lie below the fitted line? Which years do they represent? Write an inequality to describe the region below the fitted line.

462   **Unit 8** Linear Equations as Models

---

c. Choices for points may vary. An example is given. (2000, 400):
$0.0151(2000) + 0.0285(400) = 30.2 + 11.4 = 41.6$

d. $1052.63; in the CD

30. Answers may vary.

31. See answers in back of book.

32. solution

33. no solution

34. no solution

35. about 188 books

36. $\frac{5}{6} \approx 83\%$

37.

38.

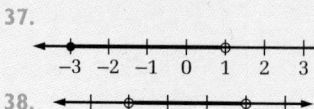

39. (12, 240.27), (16, 231.93), (20, 231.31); 1972, 1976, 1980; $y < -x + 257$

8-7 Systems of Linear Inequalities

········Focus
Solve problems by writing
and graphing systems of
inequalities.

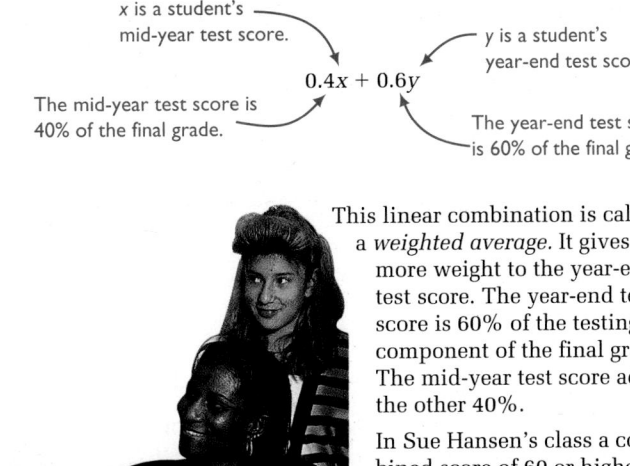

COMMON GROUND

Sue Hansen uses a mid-year test score and a year-end test score to calculate the testing component of her students' final grades. To reward improvement, she uses this formula.

x is a student's
mid-year test score.

y is a student's
year-end test score.

$$0.4x + 0.6y$$

The mid-year test score is
40% of the final grade.

The year-end test score
is 60% of the final grade.

This linear combination is called a *weighted average*. It gives more weight to the year-end test score. The year-end test score is 60% of the testing component of the final grade. The mid-year test score adds the other 40%.

In Sue Hansen's class a combined score of 60 or higher is a passing score. The linear inequality

$$0.4x + 0.6y \geq 60$$

describes the pairs of test scores that give a passing weighted average.

8-7 Systems of Linear Inequalities

463

PLANNING

Objectives and Strands
See pages 414A and 414B.

Spiral Learning
See page 414B.

Materials List
➤ Graph paper
➤ Ruler

Recommended Pacing
Section 8-7 is a one-day lesson.

Extra Practice
See pages 633–635.

Warm-Up Exercises
Warm-Up Transparency 8-7

Support Materials
➤ Practice 65
➤ Enrichment 58 in the Activity Bank
➤ Study Guide 8-7
➤ Problem Set 18
➤ Diagram Masters 2, 3 in the Explorations Lab Manual
Overhead Visual 9
➤ Using Plotter Plus: Systems of Linear Equalities
➤ Using Plotter Plus: Applications of Linear Inequalities
➤ Quiz 8-7
➤ Test 33
➤ Alternative Assessments 10, 11

Additional Sample

S1 The Trail Riders Club sponsors a competition in which 70% of a participant's score is based on riding and 30% of the score is based on jumping. Every participant with a score of 60 points or more gets a ribbon. Is it possible for a rider to get a ribbon without participating in the jumping? Graph the system of inequalities to find out. **Graph $0.7x + 0.3y \geq 60$ and $0.7x \geq 60$.**

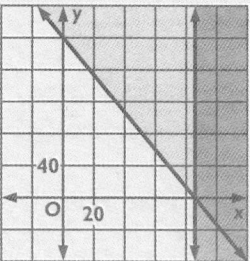

It is possible to get a ribbon without participating in the jumping.

Talk it Over

Questions 1–4 check students' understanding of Sample 1. For question 4, there are three regions outside the doubly shaded region. Have students pick a point in each of these regions and discuss its coordinates as suggested in the question.

Multicultural Note

Indian music has had a great influence on western music. Many famous rock and roll musicians have studied Indian music, including members of the Beatles and the Rolling Stones. This music has also influenced such great jazz saxophonists as John Coltrane, Sonny Rollins, and Pharoah Sanders. Indian artists have also influenced classical musicians and composers. Yehudi Menuhin and Andre Previn have recorded and performed Indian music.

464

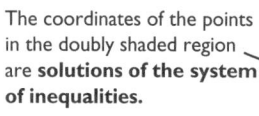

Is it possible for a student in Sue Hansen's class to fail the mid-year test and still have a passing weighted average? Only if both of these inequalities are true at the same time.

$$x < 60 \quad \longleftarrow \text{ failing mid-year test score}$$
$$0.4x + 0.6y \geq 60 \quad \longleftarrow \text{ passing weighted average}$$

Graph the **system of inequalities** to find out.

Sample Response

Problem Solving Strategy: Use a graph

Graph both inequalities on the same set of axes. The shaded regions may overlap. If they do, darken the region of overlap.

The coordinates of the points in the doubly shaded region are **solutions of the system of inequalities.**

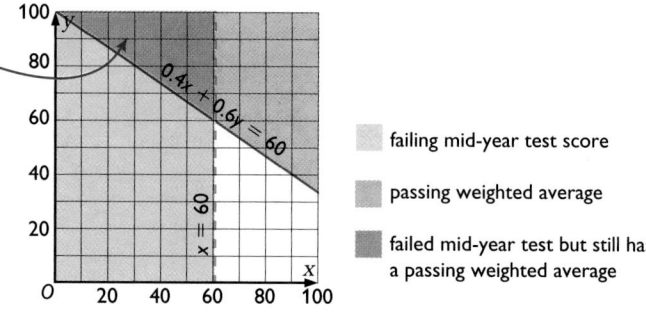

☐ failing mid-year test score

☐ passing weighted average

☐ failed mid-year test but still has a passing weighted average

There is a region on the graph where the shading overlaps. This doubly shaded region contains the possible test scores for students who fail the mid-year test but still have a passing weighted average. This shows that it is possible to fail the mid-year test and still have a passing weighted average.

Talk it Over

Use Sample 1 for questions 1–4.

1. Why is the boundary line of the graph of $x < 60$ dashed?

2. Is the point (60, 60) part of the solution of the system of inequalities?

3. Suppose one of Sue Hansen's students scores 50 on the mid-year test. What grade does he or she have to get on the year-end test in order to have a passing weighted average?

4. Pick a point outside the doubly shaded region on the graph for the weighted average. Tell whether its coordinates make *none, one,* or *both* of the inequalities true and what the coordinates mean in terms of the situation.

Answers to Talk it Over

1. The boundary line is dashed because the graph represents only failing grades, and 60 is a passing grade. Also, 60 is not a solution of $x < 60$.

2. No.

3. at least 67 (assuming scores must be whole numbers)

4. Answers may vary. An example is given. (80, 10); none; 80 is the mid-year test score, and 10 is the year-end score. A student with these grades will not have a passing weighted average.

3 rupees each

5 rupees each

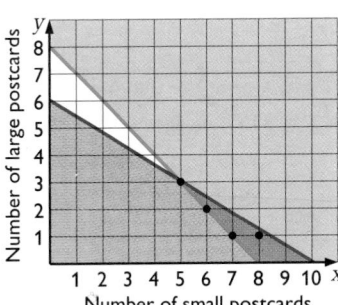

Raj is visiting relatives in India. He wants to send postcards to at least 8 friends. He has 30 rupees to spend. Large postcards cost 5 rupees each, and small postcards cost 3 rupees each.

a. Write and graph a system of inequalities to model the situation.

b. Suppose Raj wants to buy at least one large postcard. What combination of small and large postcards should he buy?

Sample Response

a. Write two inequalities.

➤ Let x = the number of small postcards Raj buys.

➤ Let y = the number of large postcards Raj buys.

➤ Total number of items: $x + y \geq 8$

➤ Total cost: $3x + 5y \leq 30$

Graph the system of inequalities.

b. Raj can only buy whole numbers of postcards. The combinations in the doubly shaded region that include at least one large postcard are (5, 3), (6, 2), (7, 1), and (8, 1). To get the most for his money, he should buy 8 small postcards and 1 large postcard.

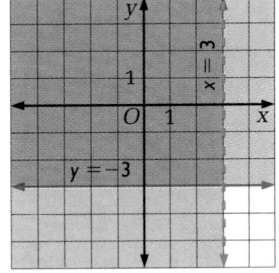

Number of large postcards

Number of small postcards

Look Back

How is graphing a system of linear inequalities different from graphing a system of linear equations? How is it similar?

8-7 Exercises and Problems

1. **Reading** In Sample 2, why isn't the combination (10, 1) a possible solution?

For Exercises 2–7, use the graph to tell whether each ordered pair is a solution of this system. Write Yes or No.

$y \geq -3$
$x < 3$

2. (0, 0) 3. (3, 1) 4. (3, −3)

5. (5, −5) 6. (2, −4) 7. (−4, 4)

$y = -3$

8-7 Systems of Linear Inequalities

465

Answers to Look Back

Answers may vary. An example is given. When you graph a system of linear inequalities, you need to graph the boundary line for each inequality, and then shade the appropriate region. When you graph a system of linear equations, you only graph a line for each equation. When a system of equations has a solution, it is only one point unless the equations represent the same line. When a system of inequalities has a solution, it is a region that contains an infinite number of points. Graphing both systems involves graphing lines.

Answers to Exercises and Problems begin on page 466.

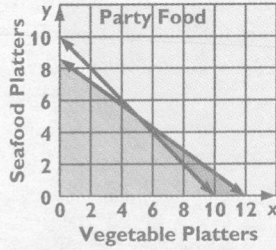

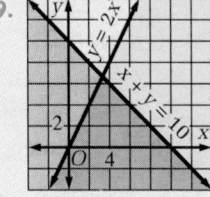

Integrating the Strands

Number Exs. 34–38

Algebra Exs. 1–29

Geometry Exs. 1–32

Statistics and Probability
Exs. 17, 39

Communication: Reading

Ex. 1 checks students' ability to read and interpret a system of linear inequalities.

Problem Solving

Ask students how they would solve this problem. The sum of two numbers is greater than 21 and twice the first number is less than the second number. Find three solutions that meet both conditions. Have students graph the inequalities to find the solutions.

Answers to
Exercises and Problems

1. The combination (10, 1) means Raj would buy 10 small postcards for 30 rupees and 1 large postcard for 5 rupees. This adds up to 35 rupees, but he can only spend 30 rupees.

2. Yes.   3. No.   4. No.

5. No.   6. No.   7. Yes.

8. a, b. Choices of points may vary. Examples are given.

 a. (2, 2); $y \geq 1 - 2x$: $2 \overset{?}{\geq} 1 - 2(2)$, $2 \overset{?}{\geq} 1 - 4$, $2 \geq -3$, true; $y \leq 2 + x$: $2 \overset{?}{\leq} 2 + 2$, $2 \overset{?}{\leq} 4$, true

 b. (−3, 2); $y \geq 1 - 2x$: $2 \overset{?}{\geq} 1 - 2(-3)$, $2 \overset{?}{\geq} 1 + 6$; $2 \overset{?}{\geq} 7$, not true; $y \leq 2 + x$: $2 \overset{?}{\leq} 2 + (-3)$, $2 \overset{?}{\leq} -1$, not true

9.
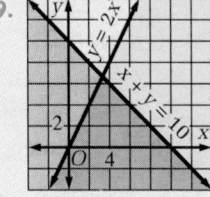

8. The graph of this system is shown.

$$y \geq 1 - 2x$$
$$y \leq 2 + x$$

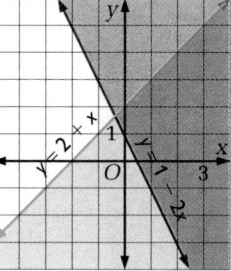

 a. Choose a point in the solution region and show that it makes both inequalities true.

 b. Choose a point that is *not* in the solution region and show that it does *not* make both inequalities true.

For Exercises 9–14, graph each system of inequalities.

9. $x + y \leq 10$
   $y \leq 2x$

10. $y < -3$
    $x > 3$

11. $2x + 3y > 18$
    $y > x$

12. $y < 2x + 6$
    $y > 5$

13. $x > -1$
    $y > 5$

14. $3x + y \geq 15$
    $3x - y < 15$

15. a. Graph the inequality $x + y < 8$.

 b. On the same set of axes, graph $y > -x + 10$.

 c. What is special about the lines?

 d. **Writing** Where do the two regions overlap? Explain.

16. a. Graph this system: $x + y > 8$
    $y < -x + 10$

 b. Describe the shape of the solution region.

 c. **Writing** Compare this system with the system in Exercise 15. How are the systems alike? How are they different?

17. Suppose the teacher mentioned on page 463, Sue Hansen, uses the mean of the mid-year and end-year test grades to determine the testing component of her students' scores. She uses the linear combination $0.5x + 0.5y$.

 a. To find the mean of x and y, she could also use the formula $\frac{x+y}{2}$. Explain why this is the same as $0.5x + 0.5y$.

 b. What percent of the testing component of the score comes from the mid-year test now? What percent comes from the year-end test?

 c. Graph the inequality $0.5x + 0.5y \geq 60$.

 d. **Writing** Describe the differences between your graph and the graph on page 464.

18. This system has one equation and one inequality.

$$y = 2x + 4$$
$$x + y \leq 10$$

 a. Draw a graph to show the points that make both sentences true.

 b. What is the geometric name for the solution?

466   **Unit 8** Linear Equations as Models

10.
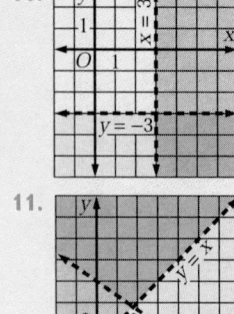

11.

12.
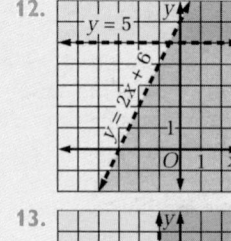

13.
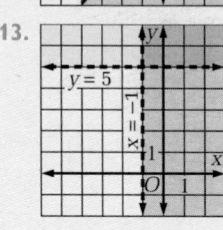

14.
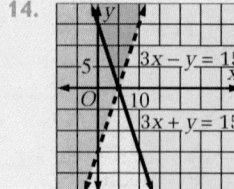

15. a, b.
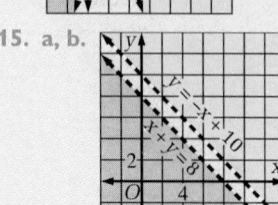

19. Budgeting Nadine has two part-time jobs. A neighbor hires her to do odd jobs for no more than 4 h per week and pays her $6 per hour. Nadine also works with a photographer in a darkroom for $5 per hour. Nadine wants to earn enough money in a week to buy a new flash for her camera. The flash will cost more than $90.

a. Write and graph a system of inequalities to model this situation.

b. How can Nadine divide her time so that she works the minimum number of hours needed to buy the flash for her camera?

20. Budgeting Cheryl and Jim Sherman got a $50 gift certificate to a department store as a wedding gift. They decided to use it to buy at least 4 bowls and some plates to add to the dishes that Cheryl already had.

a. Write and graph a system of inequalities to model this situation.

b. How many of each item should they buy to minimize the amount of change they get back?

21. Career Suppose you are a customer service representative at the bank described in Exercise 29 on page 462. A customer has $1500 to invest and wants to decide how much to invest in a CD, how much to invest in a NOW account, and how much to keep as cash.

a. Copy the graph for part (b) of Exercise 29 on page 462.

b. On the same set of axes, graph the inequality $x + y \leq 1500$. The solution region for this inequality shows the possible ways to invest $1500 or less in the two accounts.

c. Describe the region that is doubly shaded. What are its vertices?

d. List the vertices of the solution region in a table like this. Complete the table.

Vertex	Amount in NOW	Amount in CD	Amount in cash	Interest
(0, 1500)	$0	$1500	$0	$42.75
?	?	?	?	?
?	?	?	?	?

e. Customers must pay a penalty if they withdraw money from a 12-month CD before the year is over. The money is "locked up" for a 12-month period. Money kept in a NOW account is always available. How should your customer split the $1500 if he or she wants to earn at least $30 in interest, but wants to minimize the amount locked up in a CD?

Communication: Writing

After students complete Exs. 19 and 20, ask them to write a problem that can be solved by using a system of inequalities. The ability to write problems indicates that students understand the circumstances for which a particular problem-solving skill can be used.

Research

In connection with Ex. 21, students may wonder if the penalty for withdrawing money from a CD before its term is up is significant. Have them gather information from a local bank about the terms for CD and NOW accounts and discuss circumstances under which each type of account might be preferable. Tell students to suppose they have $3000 to invest. Ask how they would invest the money for the next year and to state their goals.

17. a. Explanations may vary. An example is given:
$$\frac{x+y}{2} = \frac{1}{2}(x+y) = 0.5(x+y) = 0.5x + 0.5y$$

b. 50%; 50%

c.

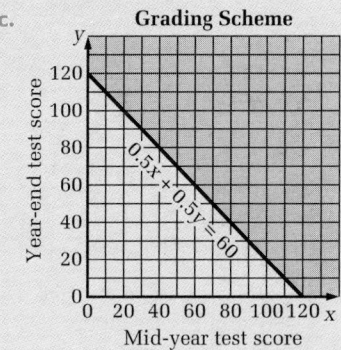

Grading Scheme

d. Descriptions may vary. An example is given. This graph shows one solution region of one inequality, where the graph on page 464 shows a solution region of a system of inequalities, which is an intersection of the solution regions for two inequalities.

18. a.

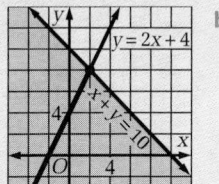

b. a ray

19–21. See answers in back of book.

Answers to Exercises and Problems

15. c. The lines are parallel.

d. The regions do not overlap. Explanations may vary. An example is given. The boundary lines do not overlap since they are parallel, so the only way the regions could overlap is if the region between the two parallel lines contained solution points and this does not happen.

16. a.

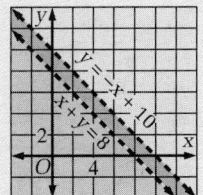

b. Descriptions may vary. An example is given. The solution region is the region between the two parallel lines.

c. Answers may vary. An example is given. The systems are alike because the boundary lines are the same. The systems are different because the inequality symbols for each inequality are reversed, so that unlike the system in Exercise 15, this system has a solution.

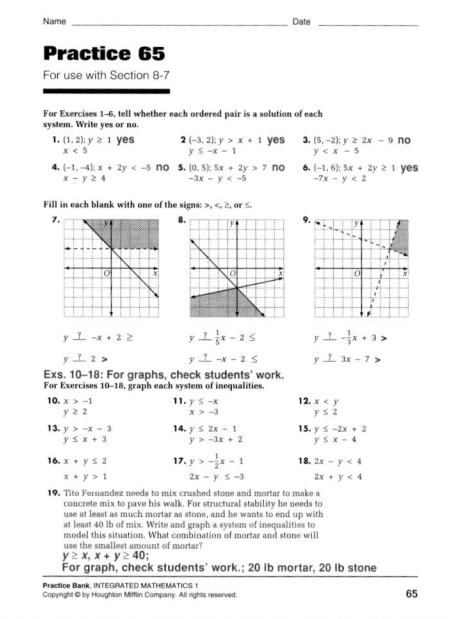

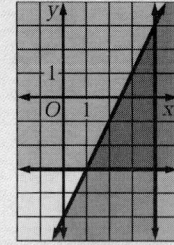

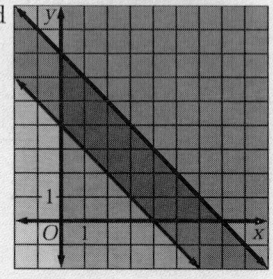

Assessment: Task

Ex. 25 assesses students' understanding of what a solution region is. Ask students to discuss the methods they used to create their system of inequalities.

Working on the Unit Project

Review the meaning of a negative correlation.

Quick Quiz (8-6 through 8-7)

See page 470.

Practice 65 For use with Section 8-7

Graph each system of three or more inequalities. Describe the shape of the region where the shading for _all_ the inequalities overlaps.

22. $y \geq -3$
$x \leq 4$
$y \leq 2x - 5$

23. $x + y \leq 7$
$x + y \geq 4$
$x \geq 0$
$y \geq 0$

24. $y > 2x - 5$
$y > -2x - 5$
$2x + 2y < 5$
$-2x + 2y < 5$

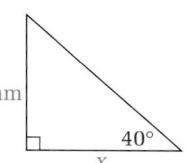

25. Open-ended Create a system of inequalities whose solution region is a polygon.

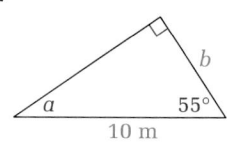

Graph each inequality. _(Section 8-6)_

26. $x < 4$ **27.** $y \leq -2$ **28.** $y \geq \frac{1}{4}x + 3$ **29.** $4x - 8y > 24$

Find each unknown measure. _(Sections 2-5, 6-7, 7-1)_

30.   **31.**   **32.**

33. Which angles in Exercise 32 are complementary? Why? _(Section 2-5)_

Estimate each square root within a range of two integers. _(Section 2-9)_

34. $\sqrt{50}$ **35.** $\sqrt{58}$ **36.** $\sqrt{193}$ **37.** $\sqrt{0.18}$ **38.** $\sqrt{0.0136}$

8 Working on the Unit Project

39. Group Activity Use the data you researched in Section 8-2.

a. Compare the data you found to the Olympic data on page 416. Which values are greater? Which are less? Which are equal?

b. Have one person graph the men's winning times and the other graph the women's. Put _Years after 1960_ on the horizontal axis.

c. For each graph, decide whether there is a negative correlation between the winning times and the number of years since 1960. If so, draw a fitted line and write an equation to model your data.

Answers to

Exercises and Problems

22–24. Descriptions may vary. Examples are given.

22. right triangle

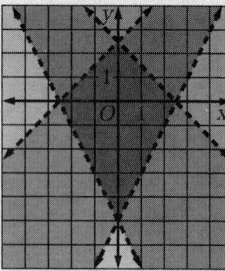

23. trapezoid

24. quadrilateral or kite

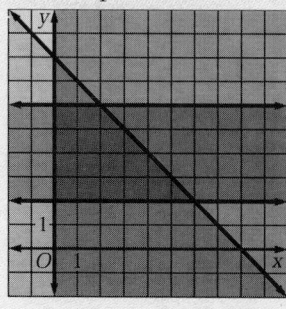

25. Answers may vary. An example is given.
$y \leq -x + 8,\ y \geq 2,\ y \leq 6,$
$x \geq 0;$ trapezoid

26.

27.

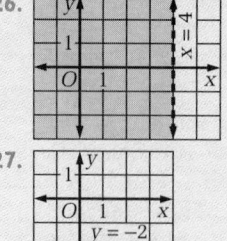

Completing the Unit Project

OLYMPIC UPDATE

New World Record Set!

Now you can analyze the trends in the Olympic swimming data. Make each of these predictions.

➤ In what year will the men's and women's winning times for the 400 m freestyle event be equal?

➤ How will the men's and women's times compare in the years after their times are equal?

➤ In what year will the women's 400 m freestyle record be broken? What do you think the new record will be?

Write your news article predicting the new record-breaking time for the women's 400 m freestyle swimming event. Use graphs to present some of the information in your article.

Look Back ◄

How accurate do you think your predictions are? List some reasons why the model you used may not be suitable for making long-term predictions. Keep your news article and data file. Watch the news to see whether your predictions come true!

Alternative Projects

Project 1: **Playing the Market**

Use a stock market report from a newspaper and "invest" $1000 in a stock of your choice. Check the paper every day and record the value of your stock. At the end of a week, graph the data and find an equation for a fitted line.

Find out what the interest rate is for a savings account at a bank and calculate how much interest $1000 will earn in one week.

Write a report to present your findings. Include answers to these questions: Do you think investing money in stock is better than leaving it in a bank? Why or why not? Is this a good way to decide how to invest?

Project 2: **Projected Savings**

Research a long-term payment plan for a car or another item. Find out the initial payment, the monthly payments, and the number of months in the plan.

Model the plan with a graph and an equation. Write a report that describes your model. Include answers to these questions: Suppose you saved the amount of money needed to pay the total cost all at once. When could you buy the item? How much money would you save by not paying interest with monthly payments? Which payment plan is the best? Why?

Answers to Exercises and Problems

28.

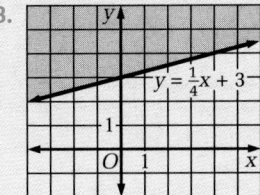

$y = \frac{1}{4}x + 3$

29.

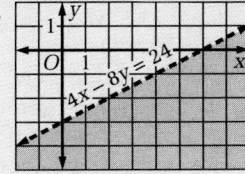

$4x - 8y = 24$

30. about 6 mm

31. about 17.5 cm

32. $a = 35°$; $b \approx 5.7$ m

33. angle a and 55°; Their sum is 90°.

34. $7 < \sqrt{50} < 8$

35. $7 < \sqrt{58} < 8$

36. $13 < \sqrt{193} < 14$

37. $0 < \sqrt{0.18} < 1$

38. $0 < \sqrt{0.0136} < 1$

39. a–c. Answers may vary.

Unit Support Materials

➤ Unit 8 Cumulative Practive 66
➤ Unit 8 Study Guide Review
➤ Unifying Problem 8
➤ Unit Tests 34 and 35

Quick Quiz (8-6 through 8-7)

Graph each inequality. [8-6]

1. $y \le 2x - 1$

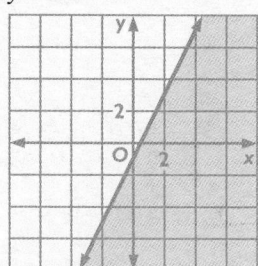

2. $x + 2y > 1$

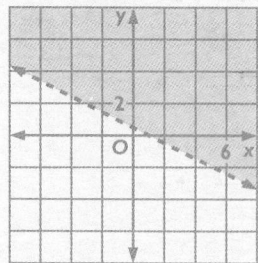

3. Graph the system of inequalities. [8-7]

$y \ge -x + 10$
$5x + 7y \le 60$

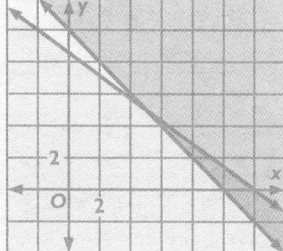

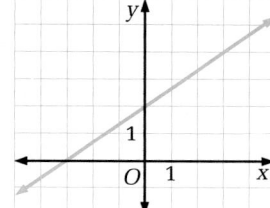

1. Use the graph at the left.   8-1

  a. Find the slope and the vertical intercept.

  b. Write an equation for the line.

  c. Tell whether the graph is an example of *linear growth* or *linear decay*.

  d. Tell whether the graph is an example of *direct variation*. Explain.

2. The Levy family keeps their cement fish pond filled to a height of 20 in. At the end of the summer, they drain the pond. The height of the water drops at a rate of 1.5 in. per minute.

  a. Model the situation with a graph.

  b. Find the slope and the vertical intercept of the line and explain how they are related to the original situation.

  c. Model the situation with an equation.

  d. How long will it take to empty the pond? Where does this information appear on your graph?

3. When triathlete Jim MacLaren prepares for an event,   8-2
his biking speed might be 20 mi/h and his running speed might be 7.5 mi/h. After swimming for a while, he might ride his bicycle part of the way and run the rest of the way on a 30 mi trail.

  a. Write an equation for this situation.

  b. Rewrite the equation in slope-intercept form.

  c. Graph the equation.

  d. Find the intercepts of the graph and explain what each means in terms of the situation.

Graph each equation and find the slope of each line.   8-3

4. $x = -3$ **5.** $y = 4$

6. A line has a slope of zero and includes the point $(-5, 7)$.

  a. Graph the line.   **b.** Write an equation for the line.

7. **Writing** Suppose you plan to graph $y = 2x - 3$ and $5x - 4y = 12$ on graph paper. For each equation, describe a method you could use and explain your choice. If you have a graphics calculator, also describe how you can use it to graph each line.

For Exercises 8 and 9, write an equation for each line.

8. The horizontal intercept is 4 and the vertical intercept is -2.

9. The slope is 2 and the point $(-3, 5)$ is on the line.   8-4

10. **Open-ended** Write an equation for a line that goes through quadrants two, three, and four on a coordinate plane.

Unit 8 Linear Equations as Models

Answers to Unit 8 Review and Assessment

1. **a.** $\frac{2}{3}$; 2

  b. $y = \frac{2}{3}x + 2$

  c. linear growth

  d. No. The line does not pass through the origin.

2. **a.**

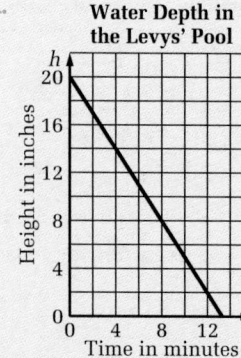

Water Depth in the Levys' Pool

  b. slope: -1.5; vertical intercept: 20; The slope means that the water level decreases 1.5 in. per minute. The vertical intercept is the height of the water before the pool is drained, 20 in.

  c. $h = -1.5t + 20$

  d. $13\frac{1}{3}$ min; horizontal intercept

3. **a–e.** Answers may vary depending on the choice of control variable. Examples are given. Let r = running distance and b = biking distance.

  a. $7.5r + 20b = 30$

  b. $r = -2\frac{2}{3}b + 4$

11. Monthly dues for a union are a fixed amount plus a fraction of a member's hourly pay. A member who earns $10 per hour pays $23.75 in union dues each month. A member who earns $12 per hour pays $25.25.

   a. Model the situation with an equation.

   b. What are the values of the slope and the vertical intercept? What does each mean in terms of the original situation?

   c. A union member earns $14.50/h. How much does this person pay in monthly dues?

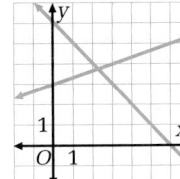

12. a. Use the graph to estimate the solution to the system of equations.   8-5

   b. Use substitution to find the exact solution.

   $$y = \frac{1}{3}x + 3$$
   $$y = -x + 6$$

Without graphing, tell whether each system has a *solution or no solution*.

13. $y = 3x - 8$
    $y = 3x + 8$

14. $y = x$
    $y = 2x$

15. $y = 5 - x$
    $y = 5 + x$

16. OverKnight Xpress charges $5.00 plus $1.00 per ounce for sending a package to another state. NextDay Express charges $3.50 plus $1.25 per ounce.

   a. Write a system of equations to model this situation.

   b. Graph both equations on the same set of axes.

   c. For what weight package is the cost the same?

   d. When is OverKnight Xpress less expensive than NextDay Express?

For Exercises 17–20, graph each system of inequalities.   8-6, 8-7

17. $y \ge -2$
    $x < 1$

18. $y > x$
    $x + y < 20$

19. $\frac{1}{2}x - y > 10$
    $y > -x$

20. $y \ge 2x - 3$
    $x + 2y < 6$

21. An interior designer plans to put plants in an office building. Ferns cost $20 each. Palms cost $25 each. The designer may need as many as 65 plants and can spend no more than $1200 for the plants.

   a. Let x = the number of ferns. Let y = the number of palms. Write an inequality for the total number of plants.

   b. Write an inequality for the cost of the project in terms of x and y.

   c. Graph the system of inequalities.

   d. What is the largest number of palms that the designer can buy? How is this shown on the graph?

22. **Self-evaluation** Without graphing, describe as much as you can about this system of inequalities.
    $y > 4$
    $y < -3x + 4$

Unit 8 Review and Assessment   471

---

6. a. 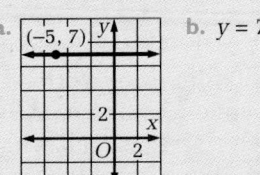   b. $y = 7$

7. Descriptions and explanations may vary. An example is given. To graph $y = 2x - 3$, I would graph the vertical intercept, -3, which is obvious in the equation because it is in slope-intercept form. I would then use the slope and move up 2 and right 1 to graph another point, $(1, -1)$. I'd check this point in the equation before drawing the line. To graph $5x - 4y = 12$, I would substitute 0 for x to find the vertical intercept, $(0, -3)$, and then substitute 0 for y to find the horizontal intercept, $\left(2\frac{2}{5}, 0\right)$. I would then draw a line through the two intercepts. I like the method for using intercepts because they are usually easy points to calculate.

8. $y = \frac{1}{2}x - 2$ 9. $y = 2x + 11$

10. Equations may vary. An example is given. $y = -x - 11$

11. a. $y = 0.75x + 16.25$

    b. slope: 0.75; vertical intercept: 16.25; The slope means that the fraction of the hourly pay (one hour per month) that goes towards union dues is 0.75. The vertical intercept is the fixed amount each member must pay toward union dues, $16.25, regardless of the rate of hourly pay.

    c. $27.13

12. a. Estimates may vary. An example is given. $\left(2\frac{1}{3}, 3\frac{2}{3}\right)$

    b. $\left(2\frac{1}{4}, 3\frac{3}{4}\right)$

13. no solution   14. solution

15. solution

16. a. Let c = total cost and w = weight of package in ounces. $c = w + 5$ and $c = 1.25w + 3.50$

    b.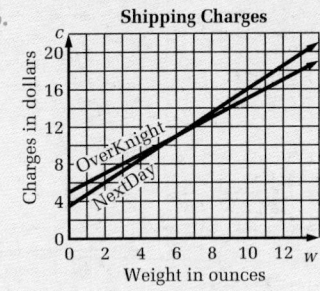
    Shipping Charges

    c. 6 oz

    d. for packages that weigh more than 6 oz

Answers continue on next page.

Answers to Unit 8 Review and Assessment

c.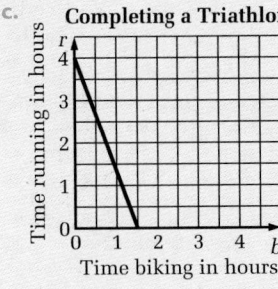
   Completing a Triathlon

d. horizontal intercept: 1.5; vertical intercept: 4; The horizontal intercept means that Jim will spend 1.5 h biking the 30 mile trail if he does not run. The vertical intercept means that Jim will spend 4 h running the 30 mile trail if he does not ride his bicycle.

4. The slope is undefined.

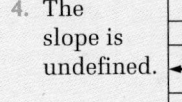

5. The slope is 0.

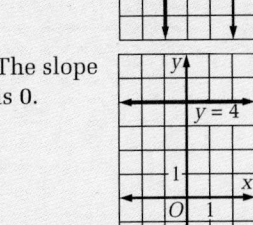

472

Quick Quiz (8-1 through 8-3)

Find the slope and vertical intercept of the line modeled by each equation. Tell whether the equation shows *linear growth* or *linear decay*. [8-1]

1. $y = 5x - 3$
   5; –3; linear growth

2. $y = -0.5x + 1$
   –0.5; 1; linear decay

Find the intercepts of the graph of each equation. [8-2]

3. $3x - 4y = 12$ $x = 4$, $y = -3$

4. $x + 2y = -14$ $x = -14$, $y = -7$

Write an equation for each situation. [8-2]

5. Rose started a savings account with $45. She added $3 to her savings account each week.
   $y = 3x + 45$

6. Richard has some nickels and some dimes. The total value of the coins is $.75.
   $0.05n + 0.10d = 0.75$

Find the slope of each line. [8-3]

7. $x = -3$ undefined

8. $y = 17$ zero

Describe the graph of the equation. [8-3]

9. $x = -3$ The graph is a vertical line. Each point on the line has an *x*-coordinate of –3.

10. $y = 17$ The graph is a horizontal line. Each point on the line has a *y*-coordinate of 17.

Answers to
Unit 8 Review and Assessment ····

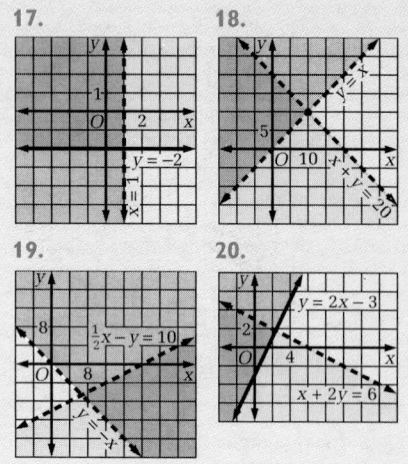

17.   18.

19.   20.

21. a. $x + y \le 65$
    b. $20x + 25y \le 1200$

---

23. **Group Activity** The table shows the winning times for the men's and women's 100 m dash in the Olympics. Work with another student. Have one student use the women's data and the other use the men's data.

   a. Make a scatter plot for the data and draw a fitted line. Put years after 1896 on the horizontal axis.

   b. Estimate an equation for your fitted line.

   c. Does your graph show *linear growth* or *linear decay*? Explain.

   d. Exchange papers with your partner and check each other's work.

   e. Have one student find the solution of the system of equations by substitution. Have the other student find the solution by graphing. Compare solutions.

   f. Tell what the solution of the system means in terms of the situation.

Year	Men's 100 m Dash (seconds)	Women's 100 m Dash (seconds)	Year	Men's 100 m Dash (seconds)	Women's 100 m Dash (seconds)
1896	12.0	—	1948	10.3	11.9
1900	11.0	—	1952	10.4	11.5
1904	11.0	—	1956	10.5	11.5
1906	11.2	—	1960	10.2	11.0
1908	10.8	—	1964	10.0	11.4
1912	10.8	—	1968	9.95	11.0
1920	10.8	—	1972	10.14	11.07
1924	10.6	—	1976	10.06	11.08
1928	10.8	12.2	1980	10.25	11.06
1932	10.3	11.9	1984	9.99	10.97
1936	10.3	11.5	1988	9.92	10.54

IDEAS AND (FORMULAS) = x^2

STATISTICS & PROBABILITY

ALGEBRA x^2

➤ The equation of a fitted line can be estimated using two points close to or on the fitted line. The equation can be used to analyze a set of data points and make predictions. *(p. 440)*

➤ The slope-intercept form of an equation is $y = mx + b$. The variable m represents the slope and the variable b represents the vertical intercept. *(p. 418)*

➤ Graphs of linear functions model linear growth if the slope is positive. If the slope is negative, the graph of a linear function models linear decay. *(p. 420)*

➤ The standard form of a linear equation is $ax + by = c$. *(p. 427)*

➤ The horizontal and vertical intercepts can be used to graph a line. To find the horizontal intercept, substitute 0 for y. To find the vertical intercept, substitute 0 for x. *(p. 428)*

---

c.

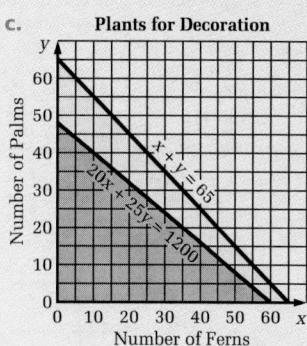

Plants for Decoration

d. 48; This number is the vertical intercept of the boundary line $20x + 25y \le 1200$.

22. Answers may vary. An example is given. The graph of $y > 4$ will be the region above the dashed boundary line $y = 4$, with slope 0 and vertical intercept 4. The graph of $y < -3x + 4$ will be the region below the dashed boundary line $y = -3x + 4$, with slope –3 and vertical intercept 4. Since both inequalities' boundary lines have the same vertical intercept 4, I know the system has a solution region which I would say starts at 4 on the *y*-axis and goes left and up.

- All points on a horizontal line have the same y-coordinate. All points on a vertical line have the same x-coordinate. *(p. 433)*

- When the control variable is x and the dependent variable is y, you can express slope this way: *(p. 440)*

$$\frac{\text{change in } y}{\text{change in } x}$$

- The equation of a line can be found in different ways: from the coordinates of any two points on the line, from the slope and the coordinates of one point on the line, and from the graph. *(p. 441)*

- You can graph a line using one point on the line and the slope. *(p. 442)*

- To estimate the solution to a system of linear equations, find the point of intersection on the graph of the lines. *(p. 448)*

- A system of linear equations has no solution when the lines do not intersect. *(p. 449)*

- When you graph a linear inequality, use a solid boundary line for $\geq$ and $\leq$, and use a dashed boundary line for $>$ and $<$. *(p. 456)*

- To graph a linear inequality in slope-intercept form, shade the region above the boundary line if the inequality symbol is $>$ or $\geq$. Shade below the boundary line if it is $<$ or $\leq$. *(p. 456)*

- To graph a linear inequality in standard form, find a point whose coordinates make the inequality true. Then shade the region on that side of the boundary line. *(p. 458)*

- To solve a system of inequalities, you graph each inequality to find out what solutions they have in common. *(p. 464)*

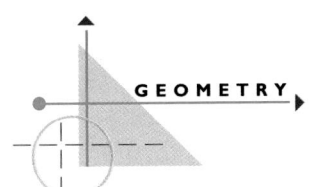
GEOMETRY

- The slope of a horizontal line is 0. The slope of a vertical line is undefined. *(p. 433)*

- Lines that have the same slope are parallel. *(p. 450)*

Key Terms

- **slope** (p. 418)
- **linear growth** (p. 420)
- **graph of an equation** (p. 426)
- **horizontal intercept** (p. 428)
- **solution region** (p. 456)

- **vertical intercept** (p. 418)
- **linear decay** (p. 420)
- **linear equation** (p. 427)
- **linear inequality** (p. 456)
- **system of inequalities** (p. 464)

- **slope-intercept form of an equation** (p. 418)
- **solution of an equation with two variables** (p. 426)
- **standard form of a linear equation** (p. 427)
- **boundary line** (p. 456)
- **solution of a system of inequalities** (p. 464)

Quick Quiz (8-4 through 8-5)

Write an equation for each line. [8-4]

1. The slope is -6 and the vertical intercept is -2.
   $y = -6x - 2$

2. The points $(5, 2)$ and $(7, -4)$ are on the line.
   $y = -3x + 17$

3. The slope of the line is 4 and $(0, 8)$ is on the line.
   $y = 4x + 8$

For Exs. 4–8, use the following situation. [8-4]

A train is traveling at a constant speed. After 1 hour, it is 460 miles from its destination. After 3 hours, it is 340 miles from its destination.

4. What is the control variable? time

5. What is the dependent variable? distance

6. Write an equation for the line between the two points. $d = -60t + 520$

7. What does the slope mean in this situation? the distance traveled per hour

8. What does the vertical intercept mean in this situation? the distance from its destination at the beginning of the trip

Tell whether each system of equations has a *solution* or *no solution*. [8-5]

9. $y = 7x - 2$
   $y = \frac{1}{3}x + 1$ solution

10. $y = -4x - 4$
    $y = 4x - 4$ solution

11. $y = -0.5x + 3.2$
    $y = -0.5x - 3.2$ no solution

Answers to Unit 8 Review and Assessment

23. a.

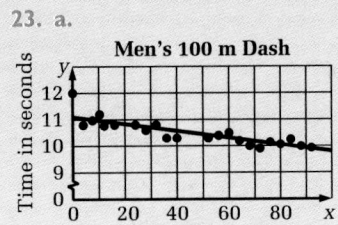
Men's 100 m Dash
Time in seconds — Years after 1896

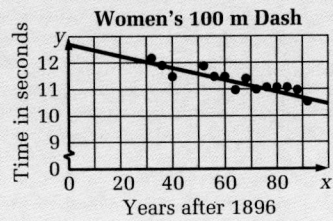
Women's 100 m Dash
Time in seconds — Years after 1896

b. Equations may vary. An example is given. men's: $y = -0.0132x + 11.116$; women's: $y = -0.0215x + 12.704$

c. Both show linear decay. Explanations may vary. An example is given. The times are decreasing as the years are in-

creasing, which gives us a negative rate of change or slope.

e. Solutions may vary. An example is given. $(191.33, 8.59)$

f. Answers may vary depending on rounding and solutions from part (e). For the Olympics held in 2090, the winning times for both men's and women's 100 m dash should be about the same, about 8.59 s.

Reasoning and Measurement

OVERVIEW

➤ **Unit 9** applies logical reasoning to numerical and geometric relationships. Students learn to recognize and employ inductive and deductive reasoning and to understand and utilize geometric concepts related to the Pythagorean theorem, surface area, volume, and similar space figures. Area and volume concepts can be reviewed by using the **Student Resources Toolbox** on pages 658 and 659.

➤ Building a boat is the theme of the Unit Project. Students use reasoning skills and learn about area, surface area, volume, and buoyancy as they construct boats from milk cartons.

➤ Connections to sports, social studies, consumer topics, history, and literature are integrated into the teaching materials and exercises.

➤ Problem-solving strategies used in Unit 9 include *Make a Sketch* (Sections 9-1 and 9-5), *Break the Problem into Parts* (Sections 9-2, 9-5, 9-6, and 9-8), *Use a Formula* (Sections 9-4 and 9-7), *Draw a Diagram* (Section 9-4), and *Use a Proportion* (Section 9-8).

Unit Objectives

Section	Objectives	NCTM Standards
9-1	• Investigate and use the relationship among the lengths of the sides of a right triangle.	1, 2, 3, 4, 5, 7
	• Recognize different kinds of reasoning.	
9-2	• Simplify and multiply square roots.	1, 2, 3, 4, 5, 7, 14
	• Solve equations like $2x^2 = 36$.	
9-3	• Write if-then statements and their converses and determine whether they are true or false.	1, 2, 3, 4, 5, 7
9-4	• Use lengths and areas to determine theoretical geometric probabilities.	1, 2, 4, 5, 7, 11
9-5	• Find the surface areas of prisms, cylinders, and pyramids.	1, 2, 4, 5, 7
9-6	• Find the volumes of prisms and cylinders.	1, 2, 4, 5, 7
9-7	• Find the volumes of pyramids and cones.	1, 2, 4, 5, 7
9-8	• Use ratios of corresponding lengths, areas, and volumes in similar figures.	1, 2, 4, 5, 7

Section	Connections to Prior and Future Concepts
9-1	**Section 9-1** investigates the relationship among the lengths of the sides of a right triangle and explores the difference between inductive and deductive reasoning. Right triangle relationships were introduced in Sections 6-7 and 7-1. These relationships will be further explored in Units 5 and 8 of Book 2 and Unit 8 of Book 3. Inductive and deductive reasoning will be studied throughout this unit, in Units 1 and 7 of Book 2, and in Unit 3 of Book 3.
9-2	**Section 9-2** examines the properties of square roots. Square roots were introduced in Section 2-9 and will be used in Unit 10 of Book 1 and throughout Books 2 and 3.
9-3	**Section 9-3** builds on the reasoning strategies introduced in Section 9-1 to present if-then statements and their converses. Reasoning and logic are used in all areas of mathematics and will be discussed and applied throughout Books 2 and 3, with emphasis in Units 1 and 7 of Book 2 and in Unit 3 of Book 3.
9-4	**Section 9-4** explores the use of segment lengths and areas of regions to determine theoretical geometric probabilities. Probability was first studied in Section 6-2 of Book 1 and will be further investigated in Units 1 and 6 of Book 2 and in Unit 7 of Book 3.
9-5	**Section 9-5** extends the concept of area studied in Sections 2-4, 5-7, and 7-6 to include surface areas of prisms, cylinders, and pyramids. Area concepts will be further studied in Section 9-8 of Book 1. Applications involving surface areas will arise frequently in Books 2 and 3.
9-6	**Section 9-6** continues the study of space figures begun in Section 9-5. In this section, students learn to find the volumes of prisms and cylinders, introduced informally in Unit 1. Volume is further explored in Sections 9-7 and 9-8 of Book 1, in Unit 10 of Book 2, and will be found in applications in Book 3.
9-7	**Section 9-7** extends the study of volume begun in Section 9-6 to include volumes of pyramids and cones. The exploration of volume continues in Section 9-8 and in Unit 10 of Book 2.
9-8	**Section 9-8** combines the concepts of area and volume studied in this unit, the concepts of ratio and proportion studied in Unit 6, and the concept of similar figures studied in Section 6-5 to introduce the idea of using proportions for finding areas and volumes of similar figures.

Integrating the Strands

Strands	Sections
Number	9-1, 9-2, 9-3
Algebra	9-1, 9-2, 9-3, 9-4, 9-5, 9-6, 9-7
Functions	9-6
Measurement	9-1, 9-6, 9-7
Geometry	9-1, 9-2, 9-3, 9-4, 9-5, 9-6, 9-7, 9-8
Trigonometry	9-3, 9-8
Statistics and Probability	9-3, 9-4, 9-5, 9-8
Discrete Mathematics	9-5, 9-8
Logic and Language	9-1, 9-2, 9-3, 9-4, 9-5, 9-6, 9-7, 9-8

Section Planning Guide

➤ Essential exercises and problems are indicated in boldface.
➤ Ongoing work on the Unit Project is indicated in color.
➤ Exercises and problems that require student research, group work, manipulatives, or graphing technology are indicated in the column headed "Other."

Section	Materials	Pacing	Standard Assignment	Extended Assignment	Other
9-1	scissors, protractor	Day 1	**1–3**	**1–3**	4
		Day 2	**5–13**, 17–21	**5–13**, 14–21	
		Day 3	**22–26**, 28–35, 36	**22–26**, 27–35, 36	36
9-2	scientific calculator, milk carton or other container, scissors, tape or stapler	Day 1	**2–21**, 26–35, 36, 37	1, **2–21**, 22–24, 26–35, 36, 37	25
9-3	small objects (such as a coin, pencil, paperclip)	Day 1	**1–18**	**1–18**, 20	19
		Day 2	**21–34**, 39–46, 47–51	**21–34**, 35–46, 47–51	
9-4	$\frac{1}{4}$ in. diameter disks of paper	Day 1	**1–5**	**1–5**	
		Day 2	**7–10**, 15–18, 19, 20	6, **7–10**, 11–18, 19, 20	19, 20
9-5		Day 1	**2–8**, **10–14**	1, **2–8**, 9, **10–14**, 15, 16	
		Day 2	**17–19**, 20, 24–30, 31, 32	**17–19**, 20–22, 24–30, 31, 32	23
9-6	tin cans	Day 1	**2–7**, 8, **9**, 10	1, **2–7**, 8, **9**, 10	
		Day 2	**11–16**, 17, **18**, 19, **20**, 21, 23–28, 29	**11–16**, 17, **18**, 19, **20**, 21, 23–28, 29	22
9-7	stiff paper, scissors, masking tape, sand or uncooked rice, protractor, metric ruler	Day 1	**2–7**, 8, **9**, **14**, **15**, 17–22, 23–25	**2–7**, 8, **9**, 10, 11, 13, **14**, **15**, 17–22, 23–25	1, 12, 16
9-8		Day 1	**1–7**, 9, **10**, **11**	**1–7**, 8, 9, **10**, **11**	
		Day 2	12, **13**, **18–27**, 29–31, 32, 33	12, **13**, 14–17, **18–27**, 29–31, 32, 33	28
Review		**Day 1**	**Unit Review**	**Unit Review**	
Test		**Day 2**	**Unit Test**	**Unit Test**	

Yearly Pacing	Unit 9 Total	Units 1–9 Total	Remaining	Total
	19 days (2 for Unit Project)	145 days	19 days	164 days

Support Materials

➤ See **Project Book** for notes on Unit 9 Project: Build Your Own Boat.
➤ "UPP" and "disk" refer to **Using Plotter Plus** booklet and **Plotter Plus** disk.
➤ Warm-up exercises for each section are available on **Warm-Up Transparencies**.
➤ "PC" refers to the McDougal Littell Mathpack **Probability Constructor** Activity Book.

Section	Study Guide	Practice Bank	Problem Bank	Activity Bank	Explorations Lab Manual	Assessment Book	Visuals	Technology
9-1	9-1	Practice 67	Set 19	Enrich 59	Masters 2, 21	Quiz 9-1		
9-2	9-2	Practice 68	Set 19	Enrich 60		Quiz 9-2		UPP, page 35
9-3	9-3	Practice 69	Set 19	Enrich 61		Quiz 9-3, Test 36		
9-4	9-4	Practice 70	Set 20	Enrich 62	Masters 2, 22	Quiz 9-4		PC Acts. 4, 7, 10, 14, and 15
9-5	9-5	Practice 71	Set 20	Enrich 63	Add. Expl. 9 Master 2	Quiz 9-5		
9-6	9-6	Practice 72	Set 20	Enrich 64	Master 2	Quiz 9-6, Test 37		
9-7	9-7	Practice 73	Set 21	Enrich 65	Add. Expl. 10 Master 2	Quiz 9-7		
9-8	9-8	Practice 74	Set 21	Enrich 66	Master 2	Quiz 9-8, Test 38		
Unit 9	Unit Rev.	Practice 75	Unif. Prob. 9	Fam. Inv. 9		Tests 39, 40		

UNIT TESTS

Form A   **Spanish versions** of these tests are on pages 154–157 of the **Assessment Book.**

Name _____ Date _____ Score _____

Test 39
Test on Unit 9 (Form A)

Directions: Write the answers in the spaces provided.

1. Find the missing length in the right triangle at the right.

Simplify.

2. $\sqrt{48}$ 3. $3\sqrt{2}\cdot 5\sqrt{12}$

For Questions 4–7, use the statement below.
A number is divisible by 2 if the number is divisible by 4.

4. What is the hypothesis of the statement?
   A number is divisible by 4.
5. What is the conclusion of the statement?
   The number is divisible by 2.
6. Write the converse of the statement. **If a number is divisible by 2, then the number is divisible by 4.**
7. Tell whether the converse is *True* or *False*. If it is false, give a counterexample.
   False; sample counterexample: 6

In Questions 8–10, solve for *x*.

8. $5x^2 = 100$ 9. $0.3x - 8 = 1$ 10. $4x(x - 3) = 0$

11. A dartboard is designed in the shape of a square 15 in. long on each side. Extra points are earned when a dart hits the bull's-eye zone which is a circle of diameter 2 in. If a player throws a dart which hits the dartboard, what is the probability that the dart hits the bull's-eye? Leave your answer in terms of π.

12. **Writing** Explain the difference between inductive reasoning and deductive reasoning.
   Sample answer: Inductive reasoning involves making conjectures based upon several observations, while deductive reasoning involves using known facts, definitions, and accepted principles and properties to prove a general statement.

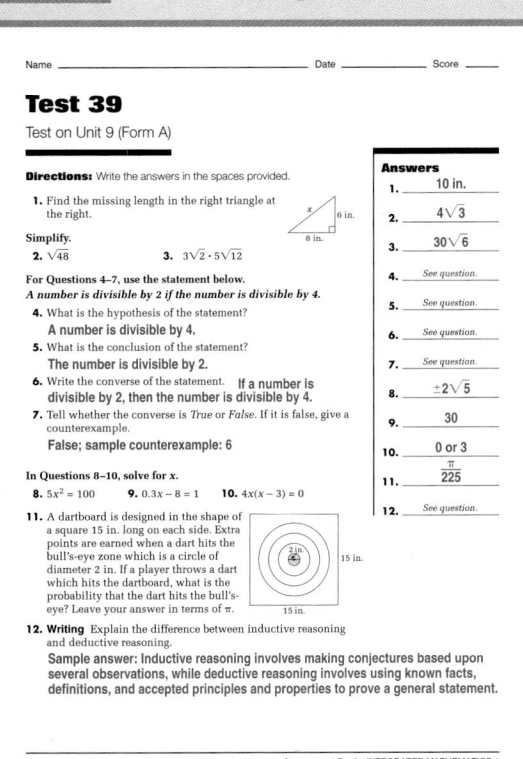

Answers
1. _____ 10 in.
2. _____ $4\sqrt{3}$
3. _____ $30\sqrt{6}$
4. _____ *See question.*
5. _____ *See question.*
6. _____ *See question.*
7. _____ *See question.*
8. _____ $\pm 2\sqrt{5}$
9. _____ 30
10. _____ 0 or 3
11. _____ $\dfrac{\pi}{225}$
12. _____ *See question.*

Name _____ Date _____ Score _____

Test 39 *(continued)*

Directions: Write the answers in the spaces provided.

13. The surface area of a cube is 150 in.². What is the length of one edge of the cube?
14. Find the surface area of a regular square pyramid with base edge 6 in. and slant height 8 in.
15. How many cubic feet of concrete are needed to build a rectangular driveway that is 20 ft wide, 40 ft long, and 6 in. thick?
16. How much canvas is needed to make the tent shown at the right? The tent will *not* have a bottom.
17. To the nearest tenth, what is the volume of a cylindrical can with diameter 6 in. and height 10 in.?

In Questions 18 and 19, find each volume.

18.   19.

20. The ratio of the lengths of the corresponding sides of two similar polygons is 3 : 4 and the area of the larger polygon is 48 ft². Find the area of the smaller polygon.
21. The ratio of the lengths of the corresponding sides of two similar space figures is 2 : 5 and the volume of the smaller figure is 64 cm³. Find the volume of the larger figure.
22. **Open-ended** Create an interesting package design for a new candy. Draw a picture of your package, including the dimensions. Then find the volume and surface area of your package.
   Sample answer: The package is a plastic "treasure chest" (a half-cylinder on top of a rectangular prism) as shown at the right.
   volume: ≈57.2 cm³; surface area: ≈89.3 cm²

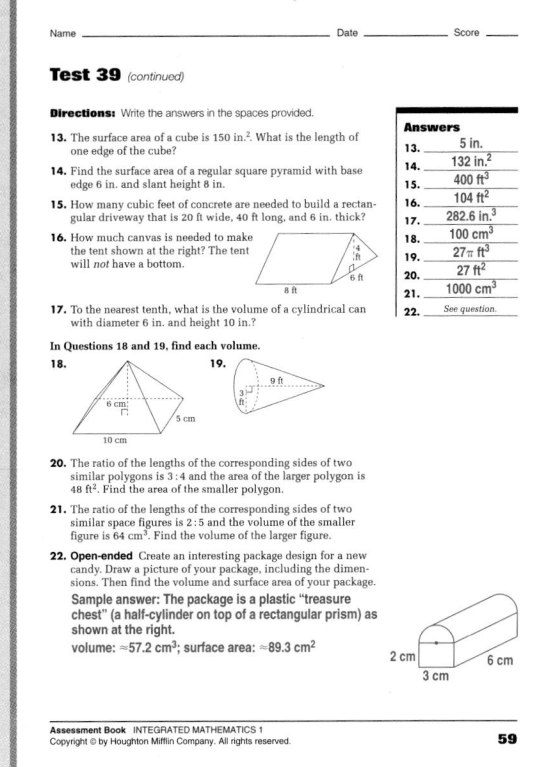

Answers
13. _____ 5 in.
14. _____ 132 in.²
15. _____ 400 ft³
16. _____ 104 ft²
17. _____ 282.6 in.³
18. _____ 100 cm³
19. _____ 27π ft³
20. _____ 27 ft²
21. _____ 1000 cm³
22. _____ *See question.*

Form B

Name _____ Date _____ Score _____

Test 40
Test on Unit 9 (Form B)

Directions: Write the answers in the spaces provided.

1. Find the missing length in the right triangle at the right.

Simplify.

2. $\sqrt{75}$ 3. $4\sqrt{3}\cdot 3\sqrt{20}$

For Questions 4–7, use the statement below.
A number is divisible by 5 if the last digit of the number is 0.

4. What is the hypothesis of the statement?
   The last digit of a number is 0.
5. What is the conclusion of the statement?
   The number is divisible by 5.
6. Write the converse of the statement. **If a number is divisible by 5, then the last digit of the number is 0.**
7. Tell whether the converse is *True* or *False*. If it is false, give a counterexample.
   False; sample counterexample: 15

In Questions 8–10, solve for *x*.

8. $7x^2 = 84$ 9. $0.5x - 9 = 3$ 10. $9x(x - 2) = 0$

11. A dartboard is designed in the shape of a square 16 in. long on each side. Extra points are earned when a dart hits the bull's-eye zone which is a circle of diameter 2 in. If a player throws a dart which hits the dartboard, what is the probability that the dart hits the bull's-eye? Leave your answer in terms of π.

12. **Writing** Explain the difference between inductive reasoning and deductive reasoning.
   Sample answer: Inductive reasoning involves making conjectures based upon several observations, while deductive reasoning involves using known facts, definitions, and accepted principles and properties to prove a general statement.

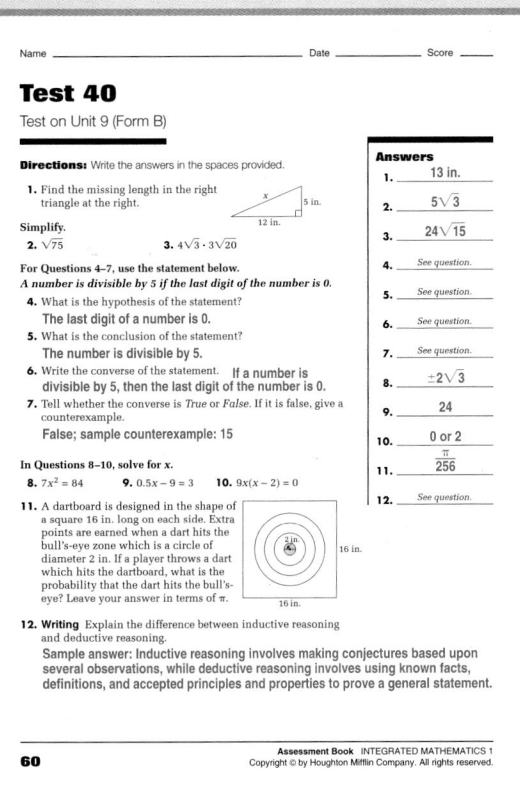

Answers
1. _____ 13 in.
2. _____ $5\sqrt{3}$
3. _____ $24\sqrt{15}$
4. _____ *See question.*
5. _____ *See question.*
6. _____ *See question.*
7. _____ *See question.*
8. _____ $\pm 2\sqrt{3}$
9. _____ 24
10. _____ 0 or 2
11. _____ $\dfrac{\pi}{256}$
12. _____ *See question.*

Name _____ Date _____ Score _____

Test 40 *(continued)*

Directions: Write the answers in the spaces provided.

13. The surface area of a cube is 216 in.². What is the length of one edge of the cube?
14. Find the surface area of a regular square pyramid with base edge 8 in. and slant height 6 in.
15. How many cubic feet of sand are needed to fill a rectangular sandbox that is 5 ft wide, 4 ft long, and 6 in. deep?
16. How much canvas is needed to make the tent shown at the right? The tent will *not* have a bottom.
17. To the nearest tenth, what is the volume of a cylindrical can with diameter 8 in. and height 9 in.?

In Questions 18 and 19, find each volume.

18.   19.

20. The ratio of the lengths of the corresponding sides of two similar polygons is 5 : 7 and the area of the larger polygon is 98 ft². Find the area of the smaller polygon.
21. The ratio of the lengths of the corresponding sides of two similar space figures is 3 : 4 and the volume of the smaller figure is 108 cm³. Find the volume of the larger figure.
22. **Open-ended** Create an interesting package design for a new candy. Draw a picture of your package, including the dimensions. Then find the volume and surface area of your package.
   Sample answer: The package is a plastic "treasure chest" (a half-cylinder on top of a rectangular prism) as shown at the right.
   volume: ≈57.2 cm³; surface area: ≈89.3 cm²

Answers
13. _____ 6 in.
14. _____ 160 in.²
15. _____ 10 ft³
16. _____ 94 ft²
17. _____ 452.2 in.³
18. _____ 80 cm³
19. _____ 48 ft³
20. _____ 50 ft²
21. _____ 256 cm³
22. _____ *See question.*

Software Support

McDougal Littell Mathpack
Probability Constructor

Outside Resources

Books/Periodicals

Miller, William and Linda Wagner. "Pythagorean Dissection Puzzles." *Mathematics Teacher* (April 1993): pp. 302–314.

Boles, Martha and Rochelle Newman. *Universal Patterns.* Bedford, MA: Pythagorean Press, 1987.

Brooks, David L. *Problems for Puzzlebusters.* Washington, DC: Enigmatics Press.

Manipulatives

Fetter, Ann E., Cynthia Schmalzried, Nancy Eckert, Doris Schattenschneider, and Euguene Klotz. *The Platonic Solids Activity Book.* Berkeley, CA: Key Curriculum Press, 1991.

Wenninger, Magnus J. *Polyhedron Models for the Classroom.* Reston, VA: NCTM, 1975.

Software

Hoffer, Alan and Richard Koch. *3-D IMAGES.* Macintosh. Acton, MA: William K. Bradford Publishing.

Schwartz, Judah and Michal Yerushalmy. *Geometric Supposer: Circles.* Scotts Valley, CA: Sunburst.

Geometer's Sketchpad. Macintosh and MS-DOS. Berkeley, CA: Key Curriculum Press.

➤ Students build a cardboard boat using a milk or juice carton.

➤ After making the boat, students predict the draft and the amount of cargo the boat will hold without sinking and then check their predictions.

➤ Students work with a partner to build a boat that meets the criteria for the project. ....●

PROJECT PLANNING

Materials List

➤ Milk or juice carton
➤ Objects for boat cargo

Project Teams

Have students work on the project with another student. Students can work together to build and test their boat and describe the results of the cargo test.

Support Materials

The *Project Book* contains information about the following topics for use with this Unit Project.

➤ Project Description
➤ Teaching Commentary
➤ Working on the Unit Project Exercises
➤ Completing the Unit Project
➤ Assessing the Unit Project
➤ Alternative Projects
➤ Outside Resources

unit 9

Reasoning and Measurement

Can a concrete boat float? University students from around the world participate in national concrete canoe competitions to answer this question. The contestants are judged on their canoe's design and racing ability, as well as on the team's oral and written reports. To qualify for the competition, the canoe must float when filled with water!

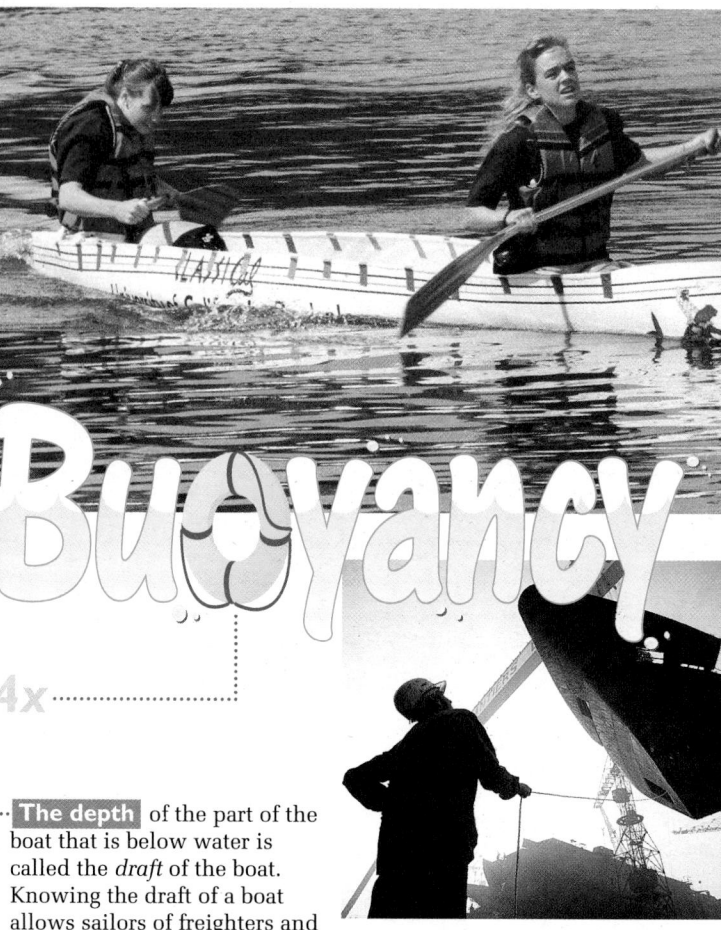

Buoyancy

$$s^2 = 1.44x$$

Queen Elizabeth 2

draft

32 ft

474

39 ft

The depth of the part of the boat that is below water is called the *draft* of the boat. Knowing the draft of a boat allows sailors of freighters and pleasure boats to pass over sandbars and reefs safely.

The speed of a boat affects the draft. In 1992, the Queen Elizabeth II hit a ridge of rock when the ship's speed caused its draft to change from 32 ft to 39 ft.

Ship designers use a boat's planned weight and volume to determine how much cargo the boat will be able to carry safely. The weight of the boat's cargo affects the ship's draft. The temperature and the amount of salt in the water also determine how low in the water a boat will float.

Suggested Rubric for Unit Project

4 The boat is well made. Student predictions of the boat's actual draft and how much cargo it can carry are supported by appropriate data. All calculations are error free. A thorough description of the observations of the cargo tests is included in the folder. The project notes are thorough and complete.

3 The boat is adequately made. The predictions of the draft and cargo capacity are not thoroughly supported by the data. Some of the calculations contain minor errors. The written report does not describe the cargo test observations as completely as it should.

2 The boat is incomplete. The predictions of the draft and cargo capacity are incomplete or inaccurate. The report of the cargo tests may need more elaboration. This project should be returned with suggestions for improvements and a new deadline.

Unit Project 9

Build Your Own Boat

Your project is to build a cardboard boat using a milk or juice carton.

Before you float your boat, you will use the math skills you learn in this unit to predict the draft of your boat. You will also predict the amount of cargo your boat will hold before it sinks.

You will launch your boat on water to test your calculations. Keep a folder of your sketches, notes, and observations for a group discussion at the end of the unit.

It may seem odd to make a boat out of concrete, but many other materials have been used to make boats—reeds, tree trunks, and even caribou skins!

Clipper ships often brought tea to America in the mid-1800s.

The Inuit peoples of the Arctic built the first *kayaks* thousands of years ago by stretching seal or caribou skins over a wooden frame.

In about A.D. 700, the Vikings made ships with overlapping planks, like siding on a house.

The Chinese developed *junks*, which are still used today.

The Carib Indians of the Caribbean Islands used the term *kanu* for these boats dug out of large tree trunks.

The ancient Egyptians made riverboats by tying bundles of reeds together.

Dhows originated in Arabia. Today, adaptations are common in ports of India.

475

ADDITIONAL BACKGROUND

Multicultural Note

In China, a wooden boat called a *junk* is used on rivers and seas to transport cargo and also as a place to live. Junks can still be seen in Hong Kong harbor today. The Chinese were skillful boat designers who first aligned sails along the length of a ship. Because the sails were not set at right angles, the Chinese junks could sail into the wind and travel very long distances. This enabled the Chinese to trade with people in the Persian Gulf, the Bay of Bengal, and the coast of East Africa.

Density

Density determines whether something will sink or float. Water has a density that varies with its temperature. The density of an object equals its weight divided by its volume. A solid that is less dense than water floats, while a solid that is more dense sinks. A hollow object, such as a boat, floats if its overall density (its total weight divided by its total volume) is less than the density of water.

Suggested Rubric for Unit Project

1   This project cannot be evaluated. It is illegible, incomplete, or not understandable. The report should be returned with a new deadline for completion. Students should be encouraged to speak with the teacher as soon as possible so that they understand the purpose and the format of the project.

ALTERNATIVE PROJECTS

Project 1, page 539

Boating in Various Seas

Write a story describing how the cooling and freezing of water affects the draft of a boat passing through the Arctic, through the North Atlantic, through the Panama Canal, and across the equator.

Project 2, page 539

Design a Container

Design a container to keep things hot or cold. Decide upon the objects the container will carry and plan a shape to fit those objects.

Getting Started

For this project you should work with another student. Here are some ideas to help you get started on your project.

☞ You will need to collect an empty milk or juice carton to make your boat. It may be a good idea to have each partner bring a carton to school.

☞ You will be testing your boat's ability to carry cargo. Think about what objects you can use as cargo and where you can get them. Here are some suggestions: marbles, table tennis balls, beads, coins, lipstick tubes, dried beans.

☞ Label a folder in which you will keep your project notes.

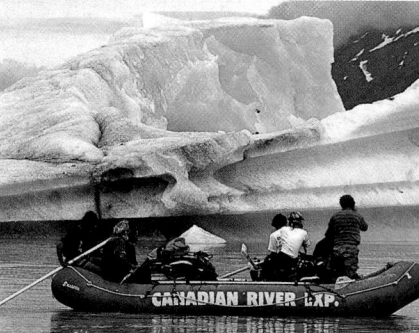

Can We Talk

Buoyancy

 Working on the Unit Project

Your work in Unit 9 will help you build and test your boat.

Related Exercises:

Section 9-1, Exercise 36
Section 9-2, Exercises 36, 37
Section 9-3, Exercises 47–51
Section 9-4, Exercises 19, 20
Section 9-5, Exercises 31, 32
Section 9-6, Exercise 29
Section 9-7, Exercises 23–25
Section 9-8, Exercises 32, 33

Alternative Projects p. 539

➤ Oil tankers, garbage barges, icebergs, and surfboards all have something in common with you—they float in water. Have you ever tried to float in fresh water? in salt water? What happened?

➤ If a fish did not have a gas-filled swim bladder, it would sink instead of swim! Sharks do not have them and need to swim constantly to stay afloat. What do you have in your chest cavity that is similar to a swim bladder?

➤ An inflatable raft weighs about the same whether or not it is inflated. Does inflating it affect its ability to float? Why do you think so?

➤ Why will a boat with a hole in the bottom start to sink when you put it into a pond?

➤ Do you think a glass marble will float? What about a sealed glass bottle with a message inside it?

➤ About 70% of Earth is covered with water. How might history have been different if everything put into water sank to the bottom?

Answers to Can We Talk?

➤ People are able to float in both fresh and salt water, though it is quite a bit easier in salt water. The density of salt water is greater than the density of fresh water, and so salt water is less easily displaced.

➤ Humans have lungs in their chest cavity which allow them to store a limited supply of air under water.

➤ Rafts float when they are inflated because the weight is more evenly distributed. The inflated raft is lighter than the amount of water it displaces; the non-inflated raft is heavier than the water it displaces, so it sinks.

➤ The boat will take on water as the pressure outside the boat pushes through the hole to create equal pressure on the inside of the boat.

➤ A glass marble will sink because it is denser than a bottle. Like the non-inflated raft, the marble is heavier than the water it displaces, so it sinks; the bottle is lighter than the amount of water it displaces, so it floats.

➤ Answers may vary. An example is given. No ships would have been able to transport people and products from continent to continent. Cultures would not have been able to share customs, inventions, or discoveries. People would not be able to swim. No one would be able to eat fish, because they would all sink to the bottom.

The Pythagorean Theorem and Reasoning

Focus

Investigate and use the relationship among the lengths of the sides of a right triangle and recognize different kinds of reasoning.

THE RIGHT REASON

EXPLORATION

How are the lengths of the sides of a right triangle related?

• **Work with another student.**

1 Measure the horizontal and vertical sides of △A by counting the number of grid units in their lengths.

2 Make a unit ruler by marking off the distance between grid lines along the edge of a piece of paper. Use this ruler to measure the remaining side of △A.

	Lengths of sides			Squares of lengths of sides		
Triangle	x	y	z	x^2	y^2	z^2
△A	3	4	5	9	16	25
△B	?	?	?	?	?	?
△C	?	?	?	?	?	?
△D	?	?	?	?	?	?

3 Do your measurements in steps 1 and 2 agree with the numbers recorded for △A in the columns labeled x, y, and z?

4 For each of the other right triangles on the grid, measure and record the lengths of the sides in a table like the one shown. For each triangle, find the square of the length of each side. Enter your results in columns x^2, y^2, and z^2.

5 What do you observe from your results? Make a conjecture about how the lengths of the sides of a right triangle are related.

9-1 The Pythagorean Theorem and Reasoning

477

PLANNING

Objectives and Strands
See pages 474A and 474B.

Spiral Learning
See page 474B.

Materials List
➤ Graph paper
➤ Ruler
➤ Scissors
➤ Protractor

Recommended Pacing
Section 9-1 is a three-day lesson.

Day 1
Pages 477–478: Exploration through The Pythagorean Theorem, *Exercises 1–4*

Day 2
Pages 478–479: Sample 1 through Sample 2, *Exercises 5–21*

Day 3
Pages 479–480: Inductive and Deductive Reasoning through Look Back, *Exercises 22–36*

Extra Practice
See pages 635–638.

Warm-Up Exercises
Warm-Up Transparency 9-1

Support Materials
➤ Practice 67
➤ Enrichment 59 in the Activity Bank
➤ Study Guide 9-1
➤ Problem Set 19
➤ Diagram Masters 2, 21 in the Explorations Lab Manual
➤ Quiz 9-1
➤ Alternative Assessment 1

Answers to Exploration

1, 2. Check work.

3. Yes.

4. Choice of variable for each side may vary.

	Length of sides			Squares of lengths of sides		
Triangle	x	y	z	x^2	y^2	z^2
△A	3	4	5	9	16	25
△B	5	12	13	25	144	169
△C	4	7.5	8.5	16	56.25	72.25
△D	6	8	10	36	64	100

5. Answers may vary. An example is given. For each triangle with shorter sides x and y and longest side z, $x^2 + y^2 = z^2$. Conjecture: The sum of the squares of the lengths of the legs of a right triangle is equal to the square of the length of the hypotenuse.

477

TEACHING

Exploration

The goal of the Exploration is to have students discover the Pythagorean theorem through the use of inductive reasoning. This helps to tie together the two major concepts of this section. After learning the Pythagorean theorem, students can use it deductively to reach conclusions.

Using Technology

The $\boxed{x^2}$ and $\boxed{\sqrt{}}$ keys are very useful when working with the Pythagorean theorem. For example, the following key sequence can be used to solve Sample 1: 21.6 $\boxed{x^2}$ $\boxed{+}$ 16.2 $\boxed{x^2}$ $\boxed{=}$ $\boxed{\sqrt{}}$.

Additional Sample

S1 Laura left her home and drove 15 miles due south and then 20 miles due east. How far is she from her home?

Make a sketch.

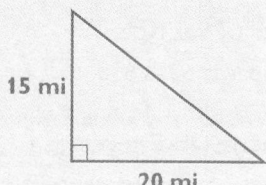

15 mi

20 mi

Use the Pythagorean Theorem.
$c^2 = a^2 + b^2$
$c^2 = (20)^2 + (15)^2$
$c^2 = 400 + 225$
$c^2 = 625$
$c = \sqrt{625}$
$c = 25$

Laura is 25 miles from her home.

Error Analysis

Some students may memorize the formula $c^2 = a^2 + b^2$ without understanding the relationship between the legs and the hypotenuse of a right triangle. Then, for example, if the letters x, y, and z are used, they may not be able to write the correct relationship. Emphasize the written statement of the Pythagorean theorem to help students avoid this problem.

478

Writings from China, Babylonia, India, and Greece show that the relationship among the lengths of the sides of a right triangle was known and used by many civilizations thousands of years ago. One of the earliest *proofs* of the relationship may have been given about 2500 years ago by the Greek mathematician Pythagoras. Today the Pythagorean theorem continues to be one of the best-known and most useful theorems in mathematics.

THE PYTHAGOREAN THEOREM

In any right triangle, the square of the length of the hypotenuse is equal to the sum of the squares of the lengths of the legs.

$$c^2 = a^2 + b^2$$

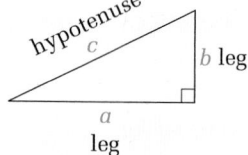

hypotenuse
c
b leg
a
leg

▸ **Now you are ready for:**
Exs. 1–4 on p. 481

Sample 1

The size of a rectangular TV screen is given by the length of its diagonal. What is the size of this screen?

16.2 in.

21.6 in.

Sample Response

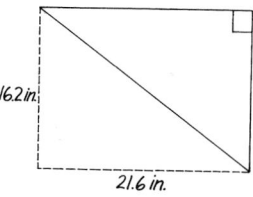

16.2 in.

21.6 in.

Problem Solving Strategy:
Make a sketch

The diagonal divides the screen into two congruent right triangles with legs of 21.6 in. and 16.2 in. The diagonal forms the hypotenuse of both triangles.

$c^2 = a^2 + b^2$ ◂——— Use the Pythagorean theorem.
$c^2 = (21.6)^2 + (16.2)^2$ ◂— Substitute 21.6 for a and 16.2 for b.
$c^2 = 466.56 + 262.44$
$c^2 = 729$
$c = \sqrt{729}$ ◂——— Find the positive square root.
$c = 27$

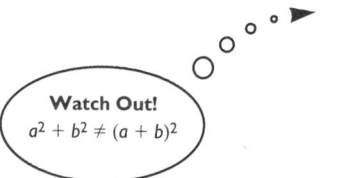

Watch Out!
$a^2 + b^2 \neq (a + b)^2$

The length of the diagonal is 27 in. The TV has a 27-inch screen.

478

Unit 9 Reasoning and Measurement

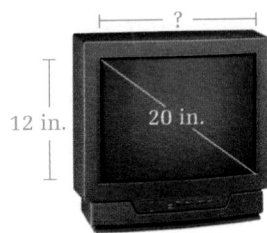

12 in. | 20 in.

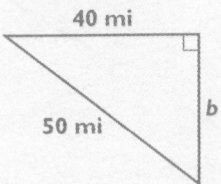
Sample 2

An appliance store advertises a sale on 20 in. TVs. The ad says that the screen is 12 in. tall but does not mention the width. Find the width.

Sample Response

Problem Solving Strategy: Make a sketch

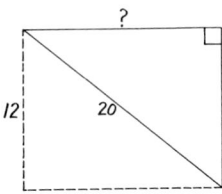

? | 12 | 20

The 20 in. diagonal is the hypotenuse of a right triangle with a 12 in. leg. The length of the other leg is the width of the screen.

$$c^2 = a^2 + b^2$$ ← Use the Pythagorean theorem.

$$20^2 = a^2 + 12^2$$ ← Substitute 20 for c and 12 for a or b.

$$400 = a^2 + 144$$

$$400 - 144 = a^2 + 144 - 144$$ ← Subtract 144 from both sides.

$$256 = a^2$$

$$\sqrt{256} = a$$ ← Find the positive square root.

$$16 = a$$

The width of the screen is 16 in.

▶ Now you are ready for:
Exs. 5–21 on pp. 481–483

Inductive and Deductive Reasoning

In the Exploration you made a conjecture that the relationship you saw for *four* right triangles would hold true for *all* right triangles. When you make a conjecture based on several observations, you are using **inductive reasoning.** It would be impossible to test all right triangles to prove your conjecture.

Next year you will prove the Pythagorean theorem. When mathematicians want to show that a statement is always true, they use *deductive reasoning.* **Deductive reasoning** uses facts, definitions, and accepted principles and properties to prove general statements.

BY THE WAY…

James A. Garfield (1831–1881) wrote a deductive proof of the Pythagorean theorem 5 years before he was elected the 20th President of the United States.

Additional Sample

S3 Do you think that vertical angles are equal in measure? Explain your reasoning.

Method 1: Use inductive reasoning. Draw and measure several pairs of vertical angles.

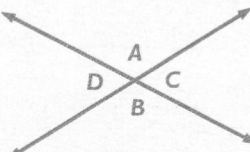

Method 2: Use deductive reasoning.

Given: ∠A and ∠B are vertical angles.

Prove: ∠A and ∠B are equal in measure.

∠A and ∠C are supplementary. ∠B and ∠C are supplementary.

∠A + ∠C = 180
∠B + ∠C = 180
∠A + ∠C = ∠B + ∠C
    ∠A = ∠B

Talk it Over

Students should understand the point of question 2, which is that in mathematics inductive reasoning does not prove a conjecture. Students should realize that a conjecture based on inductive reasoning can be proved false if one counterexample can be found. Thus, to prove a conjecture, one must use deductive reasoning, as in Method 2.

Communication: Writing

Ask students to write about an experience in which they made a conjecture based on several observations and then found out later that their conjecture was incorrect.

Look Back

The Look Back may be used for class discussion. The differences between inductive and deductive reasoning will be clarified as students share examples of each.

Writing Do you think the acute angles of a right triangle are complementary? Explain your reasoning.

Sample Response

Method ❶ Use inductive reasoning.

TECHNOLOGY NOTE

For Sample 3, Charles could have used geometric drawing software to draw triangles and measure angles.

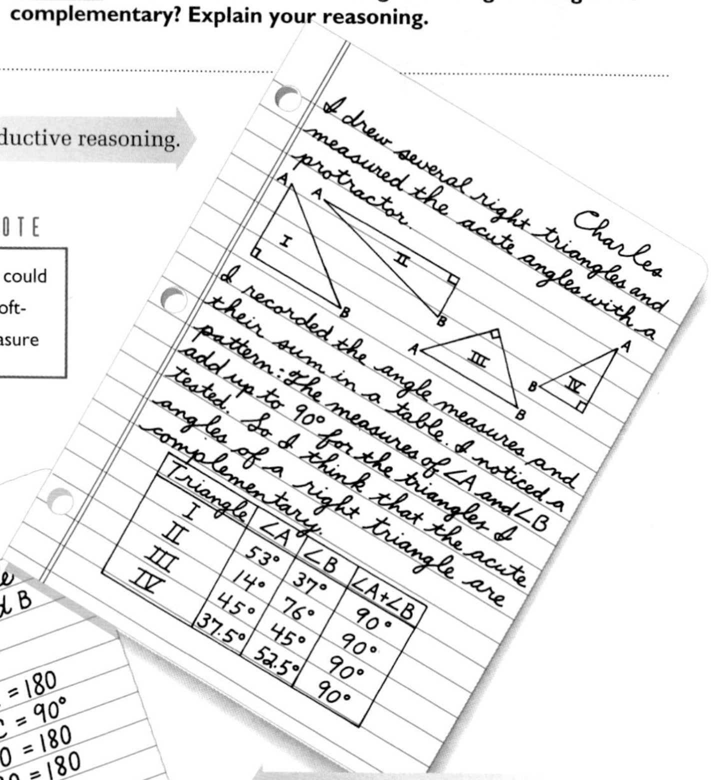

Charles

I drew several right triangles and measured the acute angles with a protractor.

I recorded the angle measures and their sum in a table. I noticed a pattern: The measures of ∠A and ∠B add up to 90° for the triangles I tested. So I think that the acute angles of a right triangle are complementary.

Triangle	∠A	∠B	∠A+∠B
I	53°	37°	90°
II	14°	76°	90°
III	45°	45°	90°
IV	37.5°	52.5°	90°

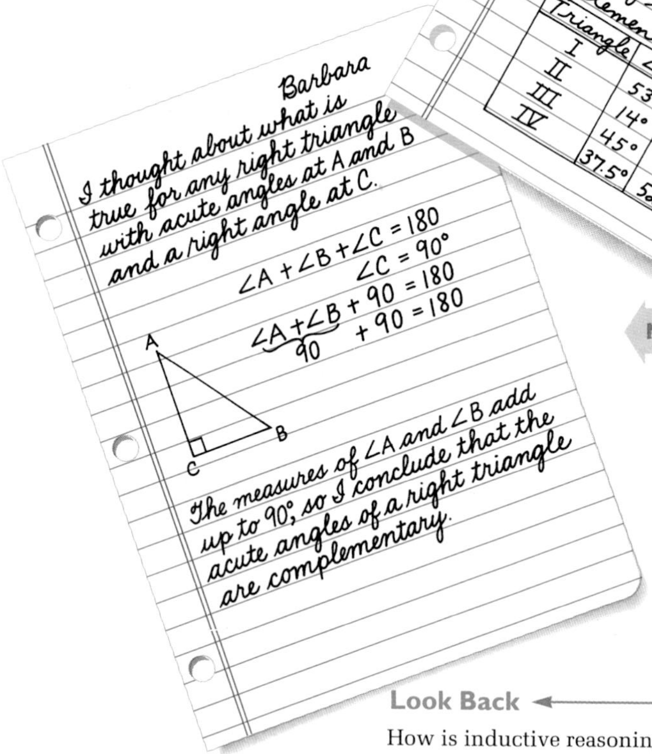

Method ❷ Use deductive reasoning.

Barbara

I thought about what is true for any right triangle with acute angles at A and B and a right angle at C.

∠A + ∠B + ∠C = 180
∠C = 90°
∠A + ∠B + 90 = 180
$\underset{90}{\underline{∠A + ∠B}}$ + 90 = 180

The measures of ∠A and ∠B add up to 90°, so I conclude that the acute angles of a right triangle are complementary.

Talk it Over

1. What definition do you need to know for Sample 3?

2. Does Method 1 prove that the acute angles of a right triangle are complementary? Does Method 2? Explain.

Look Back

How is inductive reasoning different from deductive reasoning? Describe situations in which you have used inductive or deductive reasoning.

➤ Now you are ready for:
Exs. 22–36 on pp. 483–485

Answers to *Talk it Over*

1. the definition of complementary angles: two angles whose measures add up to 90°

2. No. Yes. Explanations may vary. An example is given. The inductive method shows that for only 4 right triangles, the acute angles are complementary. Maybe if Charles had tried a fifth triangle, he would have had a different result. The deductive method uses a general right triangle, not a specific one. Barbara uses known definitions and properties to show and prove that the acute angles of a right triangle are complementary.

Answers to *Look Back*

See answers in back of book.

9-1 Exercises and Problems

1. **a.** On graph paper, draw three right triangles of different sizes.

   b. Make a table like the one you used in the Exploration on page 477. Measure the lengths of the sides of each triangle you drew in part (a). Record your results in the table.

   c. Find the square of the length of each side. Enter your results in the table. Do your results support the conjecture you made in the Exploration?

2. The Pythagorean theorem states that in a right triangle $c^2 = a^2 + b^2$. What does the Pythagorean theorem tell you about this triangle?

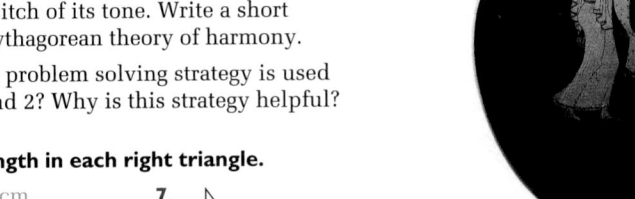

3. Which statements are true for this triangle?

   a. $y^2 = x^2 + z^2$
   b. $x^2 = y^2 + z^2$
   c. $y^2 = x^2 - z^2$
   d. $z^2 = x^2 + y^2$

4. **Research** As a young man, the Greek mathematician and philosopher Pythagoras studied in Egypt and Babylonia. Around 530 B.C. he settled in southern Italy, where he founded a group known as the Pythagoreans. Pythagoras and his followers developed a theory of harmony when they studied the relationship between the length of a musical instrument's string and the pitch of its tone. Write a short report on the Pythagorean theory of harmony.

5. **Reading** What problem solving strategy is used in Samples 1 and 2? Why is this strategy helpful?

Find the missing length in each right triangle.

6. 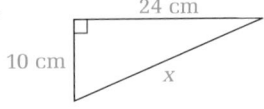 24 cm, 10 cm, x

7. 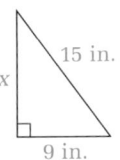 15 in., x, 9 in.

8. 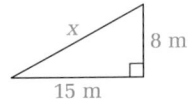 x, 8 m, 15 m

9. 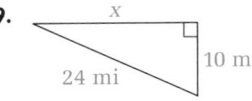 x, 10 mi, 24 mi

Find the area of each triangle by first finding the height h.

10. 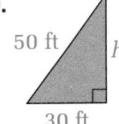 50 ft, h, 30 ft

11. 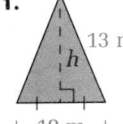 13 m, h, 10 m

12. 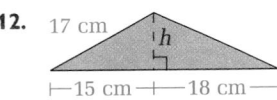 17 cm, h, 15 cm, 18 cm

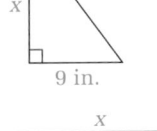

The portrait on this Greek vase from about 490 B.C. shows a man singing and playing a kithara, a stringed instrument similar to a lyre.

9-1 The Pythagorean Theorem and Reasoning   481

Answers to Exercises and Problems

1. **a–c.** Drawings and tables may vary. The results should support the conjecture made in the Exploration.

2. Answers may vary. An example is given. The Pythagorean theorem says that, for the triangle shown, $5^2 = 3^2 + 4^2$. This equation is true since $5^2 = 25$ and $3^2 + 4^2 = 9 + 16 = 25$.

3. b and c

4. Pythagoras saw many ways in which pitch and length were directly related. An example of the relationship is that if one string is twice as long as another, the sound it produces will be one octave lower. Two strings whose lengths are in a ratio of 3:2 create a musical interval called a *fifth*. Pythagoras also proved that the ratio 4:3 forms a *fourth*. In addition, he found that increasing string tension raises pitch.

Answers continued on next page.

481

Some students acquiring English may not be familiar with the rules and the terminology of baseball. You might ask a student acquiring English who is knowledgeable about baseball to "coach" the "rookies" through Exs. 13 and 14.

Application

Exs. 13–16 illustrate an application of the Pythagorean theorem to solving two problems involving baseball and two home-related practical problems. Because the Pythagorean theorem expresses an abstract relationship about *any* right triangle, it can serve as a model in *every* practical situation in which a right triangle appears. In other words, it is the abstract nature of mathematics that empowers it to model real-world situations.

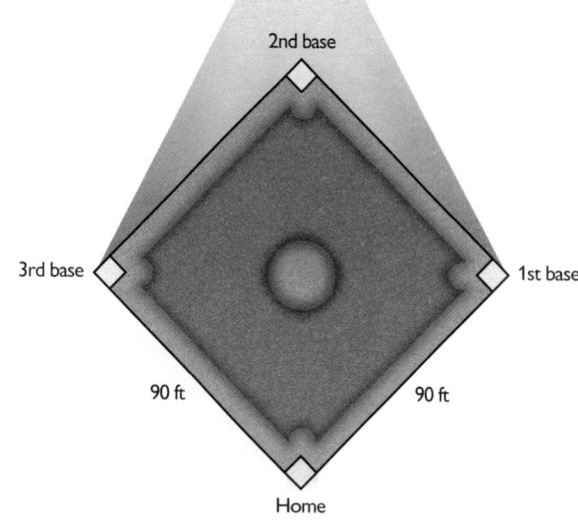

John is planning what furniture to move up to his new room in the attic of his family's home. The only entrance to the attic is a pull-down stairway in the ceiling of the second floor hallway. The opening in the ceiling is a rectangle that is 54 in. long and $25\frac{1}{2}$ in. wide.

15. John wants to move his stereo cabinet up to his new room. The cabinet is 26 in. wide, 48 in. tall, and 20 in. deep. Will the cabinet fit through the opening in the ceiling? Explain.

16. The top of John's desk can be removed from the legs of the desk. The top of the desk is 56 in. long, 36 in. deep, and 1 in. thick. Will John be able to fit the top of his desk through the opening in the ceiling? Explain.

Unit 9 Reasoning and Measurement

13. **Baseball** A major league baseball diamond is a square that is 90 ft on each side. When a runner on first base tries to steal second base, the catcher has to throw from home to second. How far is that throw?

14. **Writing** Suppose the shortstop is standing right on the line between second and third. Describe how the Pythagorean theorem could be used to find the distance from the shortstop to home plate.

Answers to Exercises and Problems

5. making a sketch; Answers may vary. An example is given. Each problem gives measurements of figures. Making a sketch of each figure helps you understand the problem and makes it easier to solve.

6. 26 cm

7. 12 in.

8. 17 m

9. about 21.8 mi

10. $h = 40$ ft; area = 600 ft^2

11. $h = 12$ m; area = 60 m^2

12. $h = 8$ cm; area = 132 cm^2

13. about 127.3 ft

14. Find the distance from the shortstop to third base. This distance and the distance from third base to home plate are the lengths of the legs of a right triangle. The hypotenuse is the distance from the shortstop to home plate. You can use the Pythagorean theorem to find this distance.

15. Yes. It will fit if he carries it through the opening with the 20 in. side of the cabinet parallel to the $25\frac{1}{2}$ in. side of the opening.

The lengths of two sides of a right triangle are given. The length of the missing side is an integer. Find the length of the missing side and tell whether the missing side is a leg or a hypotenuse.

17. 9, 15, _?_

18. 15, 36, _?_

19. 21, 72, _?_

20. 75, 85, _?_

21. Find the length of the diagonal d of this rectangular box. (*Hint:* You will need to use two different right triangles.)

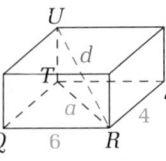

For Exercises 22–26, tell whether each person is using *inductive* or *deductive* reasoning.

22. Kim thinks that the product of two even numbers is always even. Here is her evidence:

$$2 \cdot 4 = 8 \qquad 12 \cdot 20 = 240 \qquad 32 \cdot 16 = 512 \qquad 100 \cdot 200 = 20,000$$

23. Sheila agrees that the product of two even numbers is always even. Here is her argument: Let the two even numbers be $2m$ and $2n$, where m and n are integers. Then their product is $4mn$, which equals $2(2mn)$. Since $4mn$ is the product of 2 and an integer, it is always even.

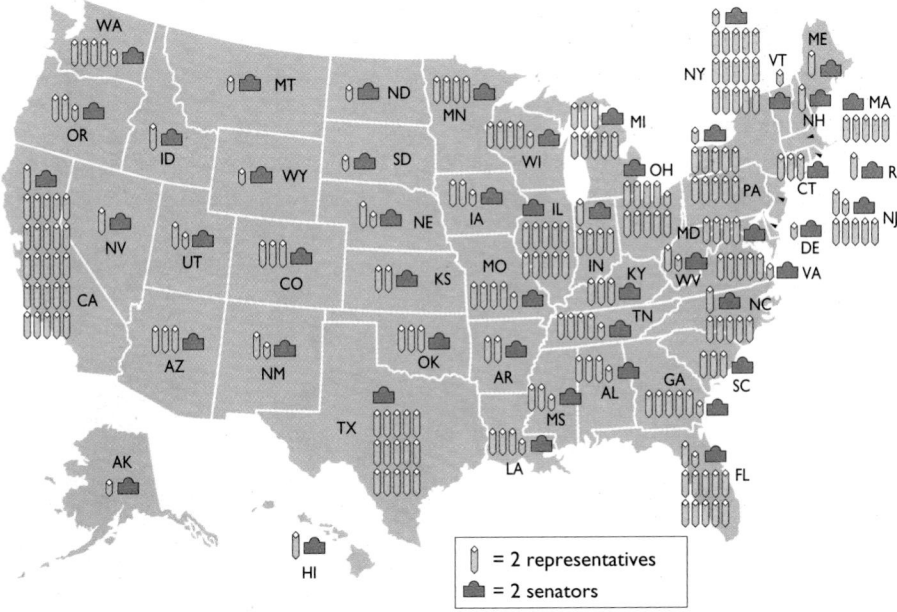

= 2 representatives

= 2 senators

24. Government Sara's history class is studying the presidential election. In the Electoral College, the number of electors from each state equals the number of representatives it has in the House plus the number of senators it has in the Senate. The Constitution requires each state to have two senators. Sara says, "If n is the number of representatives from a state, then $n + 2$ is the number of presidential electors from that state."

Answers to Exercises and Problems

16. Yes. Descriptions may vary. An example is given. The diagonal of the entrance is almost 60 in. John should hold the desk top parallel to himself (with either the 36 in. side or the 56 in. side up) and tilt it so it can be carried through the opening on the diagonal.

17. 12; leg

18. 39; hypotenuse

19. 75; hypotenuse

20. 40; leg

21. about 7.8

22. inductive

23. deductive

24. deductive

483

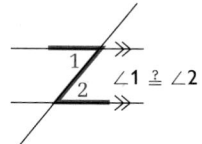

$\angle 1 \stackrel{?}{=} \angle 2$

25. Javier draws several pairs of parallel lines and draws a third line across each pair. He conjectures that the angles inside the letter Z are always equal in measure.

26. Kira claims that the sum of two odd numbers is even. Here is her argument:

Kira J.

Suppose m and n are any integers.
Then 2m is even.
2n is even.
2m + 1 is odd.
2n + 1 is odd.
Add two odd numbers:
(2m+1) + (2n+1) = 2m + 2n + 2
= 2(m+n+1)
This number is evenly divisible by 2. It's even!
So the sum of any two odd numbers is even.

27. **Writing** Nilu claims that the sum of the measures of the angles of any quadrilateral is 360°.

   a. Use inductive reasoning to support Nilu's claim.

   b. Use deductive reasoning to support Nilu's claim. (*Hint:* Draw a diagonal.)

Ongoing ASSESSMENT

28. **Writing** Use deductive reasoning to show that $(\sin A)^2 + (\cos A)^2 = 1$ for the triangle at the right. (*Hint:* Use the sine and cosine ratios.)

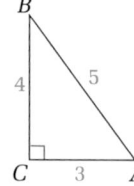

Review PREVIEW

Graph each system. (*Section 8-5, 8-7*)

29. $y = 5x$
    $y = -x + 6$

30. $2x + y \leq 1$
    $y > 2$

Find the square roots of each number. (*Section 2-9*)

31. 0.04   32. $\frac{1}{9}$ 33. 64   34. 196

35. The area of a square is 400 ft². Find the length of a side of the square. (*Section 2-9*)

Answers to Exercises and Problems ···

25. inductive   26. deductive

27. **a, b.** Answers may vary. Examples are given.

   a. When you draw any quadrilateral and measure its angles, the measures of the angles add up to 360°.

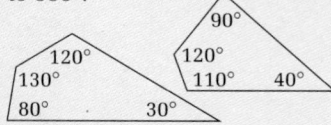

   b. Choose any quadrilateral and draw a diagonal to form two triangles. The sum of the measures of the angles of a triangle is 180°, so the sum of the measures of the angles of both triangles is 180° + 180° = 360°. Since the angles of the quadrilateral are composed of the angles

of the triangles, the sum of the measures of the angles of the quadrilateral is also 360°.

28. Answers may vary. An example is given.

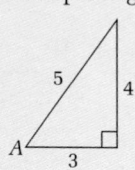

$\sin A = \frac{4}{5}$ and $\cos A = \frac{3}{5}$, so
$(\sin A)^2 + (\cos A)^2 =$
$\left(\frac{4}{5}\right)^2 + \left(\frac{3}{5}\right)^2 = \frac{16}{25} + \frac{9}{25} = \frac{25}{25} = 1.$

29.

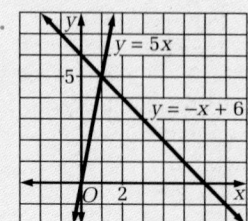

Working on the Unit Project

36. **Group Activity** Work with another student.

You will need a ruler, scissors, and a protractor. You will use your results in Exercise 37 on page 491.

a. Draw any isosceles triangle.

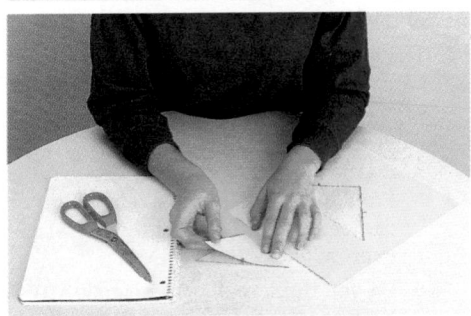

b. **Using Manipulatives** Cut out the isosceles triangle. Fold the triangle to line up the sides of equal length.

c. Open up your triangle. Measure all the angles and segments. You may want to record the measurements in a table.

d. Draw at least three other isosceles triangles. Repeat parts (a)–(c) for each new triangle.

e. **Writing** Describe what you have learned about isosceles triangles from parts (a)–(d). Did you use inductive or deductive reasoning?

f. Write a formula for finding h if you know a and b. Write any fraction in the formula in decimal form for easy use on a calculator.

9-1 The Pythagorean Theorem and Reasoning

485

Working on the Unit Project

You may want to have students draw their isosceles triangles on graph paper.

Students can test their formulas by exchanging triangles with another group and then using their formula to find the heights.

Practice 67 For use with Section 9-1

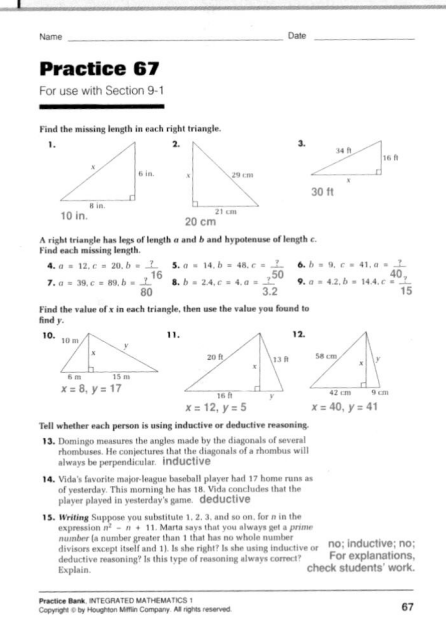

Answers to Exercises and Problems

30.

31. 0.2

32. $\frac{1}{3}$

33. 8

34. 14

35. 20 ft

36. a–d. Triangles may vary.

e. Descriptions may vary but should include the following conjecture: The segment that falls from the vertex of an isosceles triangle and is perpendicular to the base (the height to the base of the triangle) divides the base into two congruent segments; inductive reasoning.

f. using the Pythagorean theorem: $\left(\frac{1}{2}b\right)^2 + h^2 = a^2$;

$(0.5b)^2 + h^2 = a^2$;
$0.25b^2 + h^2 = a^2$;
$h^2 = a^2 - 0.25b^2$;
$h = \sqrt{a^2 - 0.25b^2}$

485

PLANNING

Objectives and Strands
See pages 474A and 474B.

Spiral Learning
See page 474B.

Materials List
➤ Scientific calculator
➤ Milk carton or other container
➤ Scissors
➤ Tape or stapler
➤ Ruler

Recommended Pacing
Section 9-2 is a one-day lesson.

Extra Practice
See pages 635–638.

Warm-Up Exercises
Warm-Up Transparency 9-2

Support Materials
➤ Practice 68
➤ Enrichment 60 in the Activity Bank
➤ Study Guide 9-2
➤ Problem Set 19
➤ Using Plotter Plus: Exploring Square Roots
➤ Quiz 9-2

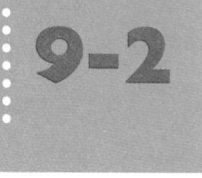

Section 9-2

Investigating Properties of Square Roots

Focus
Simplify and multiply square roots and solve equations like $2x^2 = 36$.

A √ Radical Idea

Talk it Over

1. Give a convincing argument that the value of x in the larger triangle is $\sqrt{18}$.

2. The two triangles are similar. What do you know about the sides of similar triangles? Give a convincing argument that the value of x in the larger triangle is $3\sqrt{2}$. ($3\sqrt{2}$ means $3 \cdot \sqrt{2}$.)

3. a. From questions 1 and 2, what can you conclude about the expressions $\sqrt{18}$ and $3\sqrt{2}$?

   b. Do you think that $\sqrt{9 \cdot 2}$ equals $\sqrt{9} \cdot \sqrt{2}$? Why or why not?

   c. On your calculator, compare the results of these two key sequences:

   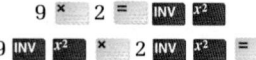

4. Choose any positive number for a. On your calculator, find the results of this key sequence:

PRODUCT PROPERTIES OF SQUARE ROOTS

For nonnegative numbers a and b:

$\sqrt{ab} = \sqrt{a} \cdot \sqrt{b}$

$\sqrt{a} \cdot \sqrt{b} = \sqrt{ab}$

$\sqrt{a} \cdot \sqrt{a} = a$

Example:

$\sqrt{39} = \sqrt{13} \cdot \sqrt{3}$

$\sqrt{13} \cdot \sqrt{3} = \sqrt{39}$

$\sqrt{3} \cdot \sqrt{3} = 3$

486   **Unit 9** Reasoning and Measurement

Answers to Talk it Over

1. Arguments may vary. An example is given. Since the triangle is a right triangle, you can use the Pythagorean theorem to find the value of x. The hypotenuse is x, so $x^2 = 3^2 + 3^2 = 9 + 9 = 18$. $x^2 = 18$, so $x = \sqrt{18}$.

2. Corresponding sides of similar triangles are in proportion; arguments may vary. An example is given.

Since the triangles are similar, the corresponding sides have the same ratio. Therefore, $\frac{1}{3} = \frac{\sqrt{2}}{x}$. Using cross products, $x = 3\sqrt{2}$.

3. a. They are equal.

   b. Yes; explanations may vary. An example is given. Since $18 = 9 \cdot 2$, $\sqrt{18} = \sqrt{9 \cdot 2}$. Also,

$\sqrt{9} = 3$, so $\sqrt{9} \cdot \sqrt{2} = 3\sqrt{2}$. From part (a), $\sqrt{18} = 3\sqrt{2}$. Therefore, $\sqrt{9 \cdot 2} = \sqrt{9} \cdot \sqrt{2}$.

   c. They both equal about 4.24.

4. The result is the number a.

When you use your calculator to evaluate a square root like $\sqrt{18}$, your calculator gives you a decimal approximation. When you write an answer using the $\sqrt{}$ symbol, you are writing the answer in **radical form.**

You can simplify an expression in radical form when the number under the $\sqrt{}$ symbol has a perfect square factor.

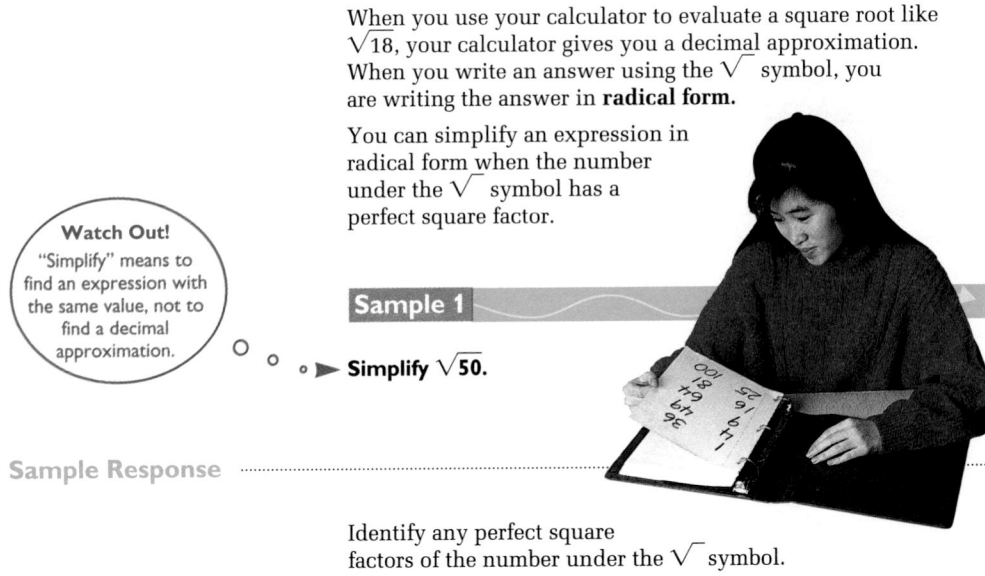

Sample 1

▶ Simplify $\sqrt{50}$.

Sample Response

Identify any perfect square factors of the number under the $\sqrt{}$ symbol.

$\sqrt{50} = \sqrt{25 \cdot 2}$ ◀──── 25 is a perfect square factor of 50. Write 50 as the product of 25 and 2.

$= \sqrt{25} \cdot \sqrt{2}$ ◀──── Use the property $\sqrt{ab} = \sqrt{a} \cdot \sqrt{b}$.

$= 5\sqrt{2}$ ◀──── Write $\sqrt{25}$ as 5.

Sample 2

Simplify $3\sqrt{2} \cdot 5\sqrt{10}$.

Sample Response

$3\sqrt{2} \cdot 5\sqrt{10} = 3 \cdot 5 \cdot \sqrt{2} \cdot \sqrt{10}$ ◀── Group radical factors.

$= 15\sqrt{20}$ ◀──────── Use the property $\sqrt{a} \cdot \sqrt{b} = \sqrt{ab}$.

$= 15\sqrt{4 \cdot 5}$ ◀──── 4 is a perfect square factor of 20. Write 20 as the product of 4 and 5.

$= 15 \cdot \sqrt{4} \cdot \sqrt{5}$ ◀──── Use the property $\sqrt{ab} = \sqrt{a} \cdot \sqrt{b}$.

$= 15 \cdot 2\sqrt{5}$ ◀──────── Write $\sqrt{4}$ as 2.

$= 30\sqrt{5}$

9-2 Investigating Properties of Square Roots

487

Talk it Over

Questions 1–4 develop an inductive rationale for the general statements of the product properties of square roots. Of course, these questions use the Pythagorean theorem to support the reasoning behind them. When discussing the property that $\sqrt{a} \cdot \sqrt{a} = a$, use the previous property to show that $\sqrt{a} \cdot \sqrt{a} = \sqrt{a \cdot a} = \sqrt{a^2} = a$. The example of $\sqrt{3} \cdot \sqrt{3} = \sqrt{3 \cdot 3} = \sqrt{9} = 3$ makes the statement of the property clear.

Communication: Discussion

The small right triangle at the top of page 486, whose legs are 1 and whose hypotenuse is $\sqrt{2}$, can provide the basis for a brief discussion about the history of numbers. It was during the time of Pythagoras that the number $\sqrt{2}$ was discovered. At that time, it was believed that all numbers were counting numbers or fractions. Greek mathematicians thought all points on a line could be matched with these rational numbers. The following diagram shows this not to be true. The point P corresponds to $\sqrt{2}$, which is now called an irrational number.

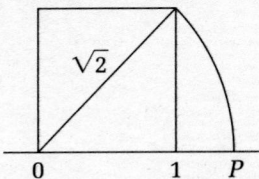

Additional Samples

S1 Simplify $\sqrt{98}$.

$\sqrt{98} = \sqrt{49 \cdot 2}$

$= \sqrt{49} \cdot \sqrt{2}$

$= 7\sqrt{2}$

S2 Simplify $2\sqrt{3} \cdot 7\sqrt{6}$.

$2\sqrt{3} \cdot 7\sqrt{6} = 2 \cdot 7 \cdot \sqrt{3} \cdot \sqrt{6}$

$= 14\sqrt{18}$

$= 14 \cdot \sqrt{9} \cdot \sqrt{2}$

$= 14 \cdot 3\sqrt{2}$

$= 42\sqrt{2}$

487

Question 5 strengthens the point that it is more efficient to find perfect square factors of a number when simplifying square roots.

Additional Samples

S3 Solve $7x^2 = 525$.

$$7x^2 = 525$$

$$\frac{7x^2}{7} = \frac{525}{7}$$

$$x^2 = 75$$

$$x = \pm\sqrt{75}$$

$$x = \pm\sqrt{25} \cdot \sqrt{3}$$

$$x = \pm5\sqrt{3}$$

S4 The perimeter of a rectangle is 44 feet. Its width is 10 feet. Find y, the length of the diagonal.

Step 1 Find the length of the rectangle. Let x = the length of the rectangle.

$$2x + 20 = 44$$

$$2x = 24$$

$$x = 12$$

Step 2 Find the length of the diagonal.

$$10^2 + 12^2 = y^2$$

$$100 + 144 = y^2$$

$$244 = y^2$$

$$\sqrt{244} = y$$

$$\sqrt{4 \cdot 61} = y$$

$$\sqrt{4} \cdot \sqrt{61} = y$$

$$2\sqrt{61} = y$$

BY THE WAY...

The first known use of the symbol $\sqrt{}$ in print was in a German book published in 1525. The symbol eventually replaced the notation R_x which had been commonly used to represent the square root. R_x came from the first and last letters of the word *radix*, a Latin translation of an Arabic word for "root."

Talk it Over

5. Explain why it is not efficient to rewrite $\sqrt{50}$ as $\sqrt{5} \cdot \sqrt{10}$ when you want to simplify $\sqrt{50}$.

6. To simplify $\sqrt{72}$, Jeremy writes:

$$\sqrt{72} = \sqrt{9 \cdot 8}$$

$$= \sqrt{9} \cdot \sqrt{8}$$

$$= \sqrt{9} \cdot \sqrt{4 \cdot 2}$$

$$= \sqrt{9} \cdot \sqrt{4} \cdot \sqrt{2}$$

$$= 3 \cdot 2\sqrt{2}$$

$$= 6\sqrt{2}$$

Martha writes:

$$\sqrt{72} = \sqrt{36 \cdot 2}$$

$$= \sqrt{36} \cdot \sqrt{2}$$

$$= 6\sqrt{2}$$

Which method do you think is easier to use? Explain.

7. **a.** Simplify $-\sqrt{40}$.

   b. What are the solutions of $x = \pm\sqrt{40}$?

   c. What are the solutions of $x^2 = 40$?

Sample 3

Solve $5x^2 = 315$.

Sample Response

$$5x^2 = 315$$

$$\frac{5x^2}{5} = \frac{315}{5} \quad \longleftarrow \quad \text{Divide both sides by 5 to get } x^2 \text{ by itself.}$$

$$x^2 = 63$$

$$x = \pm\sqrt{63} \quad \longleftarrow \quad \text{Undo the squaring.}$$

$$x = \pm\sqrt{9 \cdot 7} \quad \longleftarrow \quad \text{Write 63 as the product of 9 and 7.}$$

$$x = \pm\sqrt{9} \cdot \sqrt{7} \quad \longleftarrow \quad \text{Use the property } \sqrt{ab} = \sqrt{a} \cdot \sqrt{b}.$$

$$x = \pm 3\sqrt{7} \quad \longleftarrow \quad \text{Write } \sqrt{9} \text{ as 3.}$$

The solutions are $3\sqrt{7}$ and $-3\sqrt{7}$.

In some situations, only a positive answer to an equation like the one in Sample 3 makes sense. If the variable is the length of a segment, use only the positive solution.

488

Unit 9 Reasoning and Measurement

Answers to Talk it Over

5. It is not efficient to rewrite $\sqrt{50}$ as $\sqrt{5} \cdot \sqrt{10}$ because neither 5 nor 10 has a perfect square factor.

6. Martha's; She factored out the largest square root and avoided three steps.

7. **a.** $-2\sqrt{10}$

   b. $-2\sqrt{10}$ and $2\sqrt{10}$

   c. $-2\sqrt{10}$ and $2\sqrt{10}$

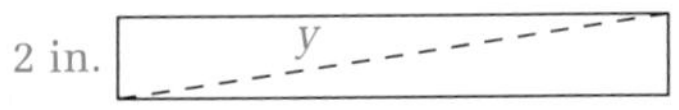

Sample 4

The perimeter of a rectangle is 32 in. Find the length of a diagonal of the rectangle. Write the answer in simplified radical form.

Sample Response

Perimeter = 32 in.

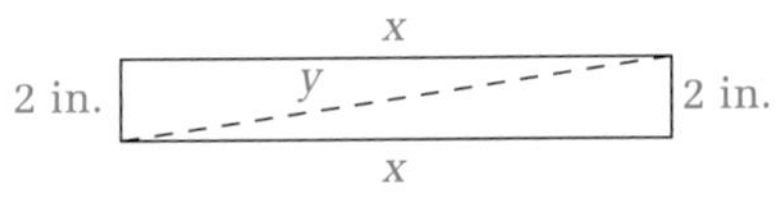

Problem Solving Strategy: Break the problem into parts

Step 1 Find the length of the rectangle.

Let x = the length of the rectangle.

$2x + 4 = 32$ ← Use the formula $P = 2l + 2w$.

$2x = 28$ ← Subtract 4 from both sides.

$x = 14$ ← Divide both sides by 2.

Step 2 Find the length of the diagonal.

$14^2 + 2^2 = y^2$ ← Use the Pythagorean theorem.

$196 + 4 = y^2$

$200 = y^2$

$\sqrt{200} = y$ ← Find the positive square root.

$\sqrt{100 \cdot 2} = y$ ← Write 200 as the product of 100 and 2.

$\sqrt{100} \cdot \sqrt{2} = y$ ← Use the property $\sqrt{ab} = \sqrt{a} \cdot \sqrt{b}$.

$10\sqrt{2} = y$ ← Write $\sqrt{100}$ as 10.

The length of the diagonal of the rectangle is $10\sqrt{2}$ in.

Look Back ←

Explain how the product properties of square roots can be used to simplify $\sqrt{24}$ and to multiply $\sqrt{3}$ and $\sqrt{11}$.

9-2 Exercises and Problems

1. a. **Reading** Put one of the properties that you learned in this lesson in your own words.
   b. Give a numerical example of the property.

Simplify.

2. $\sqrt{90}$
3. $\sqrt{68}$
4. $\sqrt{125}$
5. $3\sqrt{8}$
6. $\sqrt{25\pi}$
7. $\sqrt{3} \cdot \sqrt{7}$
8. $\sqrt{6} \cdot \sqrt{8}$
9. $\sqrt{15} \cdot \sqrt{24}$
10. $\sqrt{7} \cdot \sqrt{7}$
11. $(\sqrt{11})^2$
12. $2\sqrt{3} \cdot 4\sqrt{6}$
13. $5\sqrt{12} \cdot 3\sqrt{20}$

Look Back

Students' responses to the Look Back question provide a good opportunity to evaluate their understanding of these properties.

APPLYING

Suggested Assignment

Standard 2–21, 26–37

Extended 1–24, 26–37

Integrating the Strands

Number Exs. 1–13, 25

Algebra Exs. 14–19, 26–35

Geometry Exs. 20–24, 27–31, 36, 37

Logic and Language Exs. 1, 25, 26

Communication: Reading

To put one of the product properties of square roots in their own words in Ex. 1, students must read, think about, and interpret that property, thus offering them an opportunity to better understand the property. You may wish students to put all three properties in their own words.

Answers to Look Back

Answers may vary. Examples are given. Four is a perfect square factor of 24, so rewrite $\sqrt{24}$ as $\sqrt{4 \cdot 6}$; using the product property of square roots, $\sqrt{4 \cdot 6} = \sqrt{4} \cdot \sqrt{6} = 2\sqrt{6}$. Using the product property of square roots, $\sqrt{3} \cdot \sqrt{11} = \sqrt{3 \cdot 11} = \sqrt{33}$.

Answers to Exercises and Problems

1. a. Answers may vary. An example is given. The square root of the product of two non-negative real numbers is equal to the product of their square roots.
   b. $\sqrt{7 \cdot 11} = \sqrt{7} \cdot \sqrt{11}$
2. $3\sqrt{10}$
3. $2\sqrt{17}$
4. $5\sqrt{5}$
5. $6\sqrt{2}$
6. $5\sqrt{\pi}$
7. $\sqrt{21}$
8. $4\sqrt{3}$
9. $6\sqrt{10}$
10. 7
11. 11
12. $24\sqrt{2}$
13. $60\sqrt{15}$

Using Technology

Depending upon what type of calculator students are using, they may need to enter different key sequences for Ex. 25. For example, some students may need to enter

45 $\boxed{\sqrt{}}$ ÷ 5 $\boxed{\sqrt{}}$ $\boxed{=}$ for part (a). Have students refer to their manuals if necessary.

Solve for x.

14. $x^2 = 8$

15. $x^2 = 48$

16. $2x^2 = 36$

17. $15x^2 = 180$

18. $y^2 + y^2 = x^2$

19. $(2a)^2 + (5a)^2 = x^2$

20. a. On the map at the right, what is the distance from point A to point B?

   b. What is the distance from point B to point C?

   c. What is the distance from point C to point A? Give your answer in simplified radical form.

   d. If each unit on the grid represents five miles, what is the actual distance from point C to point A, to the nearest mile?

21. A square $ABCD$ has vertices with coordinates $A(4, -4)$, $B(4, 4)$, $C(-4, 4)$, and $D(-4, -4)$. Find the length of the diagonal $\overline{AC}$ in simplified radical form.

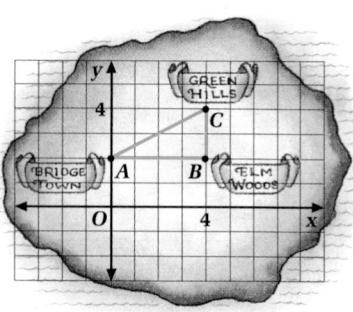

22. Find the length of the hypotenuse of this triangle in terms of x. Simplify the answer.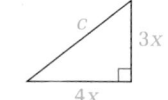

23. The maximum speed that some types of boats can travel can be estimated using the formula $s^2 = 1.44x$, where s is the speed in nautical miles per hour (or "knots") and x is the length of the boat's water line in feet.

   a. Solve the formula for s.

   b. Estimate the maximum speed of a boat with a water line of 25 ft.

24. a. The triangle shown is an isosceles right triangle with legs of length a. Write a formula in simplified radical form for the length of the hypotenuse, c, in terms of a.

   b. Use the formula you wrote in part (a). Find the length of the hypotenuse of an isosceles right triangle that has legs of length 6 cm.

25 TECHNOLOGY Enter the key sequences on your calculator to find each answer.

   a. $\dfrac{\sqrt{45}}{\sqrt{5}}$ ⟶ 45 INV x² ÷ 5 INV x² = ?

   b. $\dfrac{\sqrt{32}}{\sqrt{2}}$ ⟶ 32 INV x² ÷ 2 INV x² = ?

   c. Use your answers to parts (a) and (b) to simplify $\dfrac{\sqrt{75}}{\sqrt{3}}$ without using your calculator. Describe a shortcut.

Answers to Exercises and Problems

14. $2\sqrt{2}, -2\sqrt{2}$

15. $4\sqrt{3}, -4\sqrt{3}$

16. $3\sqrt{2}, -3\sqrt{2}$

17. $2\sqrt{3}, -2\sqrt{3}$

18. $y\sqrt{2}, -y\sqrt{2}$

19. $a\sqrt{29}, -a\sqrt{29}$

20. a. 4

   b. 2

   c. $2\sqrt{5}$

   d. about 22 mi

21. $8\sqrt{2}$

22. $5x$

23. a. $s = 1.2\sqrt{x}$

   b. about 6 knots

24. a. $c = a\sqrt{2}$

   b. $6\sqrt{2}$ cm

25. a. 3

   b. 4

c. Descriptions may vary. An example is given.

Rewrite $\dfrac{\sqrt{75}}{\sqrt{3}}$ as $\sqrt{\dfrac{75}{3}}$.

Do the division and then find the square root: $\sqrt{\dfrac{75}{3}} = \sqrt{25} = 5$.

26. Counterexamples may vary. An example is given. Suppose $a = 4$ and $b = 9$.

Then $\sqrt{a + b} = \sqrt{4 + 9} = \sqrt{13}$. Also, $\sqrt{a} + \sqrt{b} = \sqrt{4} + \sqrt{9} = 2 + 3 = 5$. Since $5 = \sqrt{25}$, 5 does not equal $\sqrt{13}$. Therefore, $\sqrt{a + b}$ does not equal $\sqrt{a} + \sqrt{b}$.

27. 20 m

28. $\sqrt{29}$ in. or about 5.4 in.

29. 15 ft   **30.** 12 cm

490

Ongoing ASSESSMENT

26. Writing Kristin wonders whether the statement $\sqrt{a+b} = \sqrt{a} + \sqrt{b}$ is true for all values of a and b. Can you find a counterexample? Write a note to Kristin to help her decide.

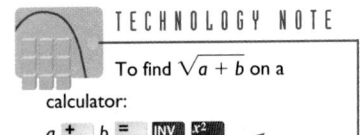

TECHNOLOGY NOTE

To find $\sqrt{a+b}$ on a calculator:

$a\;+\;b\;=\;\boxed{INV}\;\boxed{x^2}$

Review PREVIEW

For Exercises 27–30, find the missing length in each right triangle. *(Section 9-1)*

27.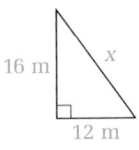
16 m x
12 m

28.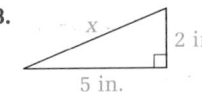
x 2 in.
5 in.

29.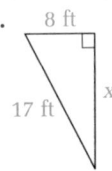
8 ft
17 ft x

30.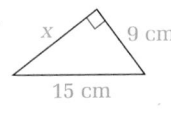
x 9 cm
15 cm

31. The vertices of a triangle are $A(3, 2)$, $B(5, 4)$, and $C(4, -1)$. Draw a dilation of $\triangle ABC$ with center $A(3, 2)$ and scale factor 2. *(Section 6-6)*

Solve. *(Sections 5-2, 5-6)*

32. $\frac{2}{3}x = 0$

33. $3x + 34.5 = 0$

34. $0 = -5x$

35. $39 - 13x = 0$

 ······ **Working on the Unit Project**

Use the empty milk carton or other container you collected.

36. Construct a boat using these steps.

1 Open the top of the carton completely.

2 Cut through two opposite side edges and along a diagonal of the base of the carton.

3 Use either half of the carton to construct a boat with triangular ends.

37. a. Measure the sides and height of a triangular end of the boat you built.

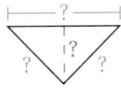

  b. Explain why you can use the formula you found in Exercise 36 on page 485 to approximate the height of a triangular end of your boat.

  c. Test the formula you wrote in Exercise 36 on page 485. How close is the height from the formula to the measured height you found in part (a)?

You will need to use tape or a stapler to keep the walls of the boat in place.

9-2 Investigating Properties of Square Roots

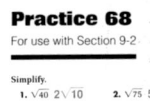

Answers to Exercises and Problems

31.

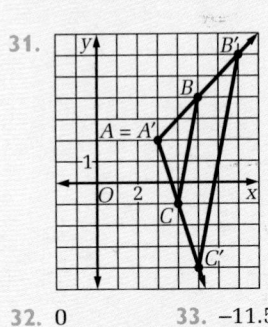

32. 0   **33.** −11.5

34. 0   **35.** 3

36. Boats may vary.

37. a. Answers may vary, depending on boats. Probable measurements in cm: sides: 9.5 cm, 9.5 cm, 13.5 cm; height: 6.5 cm

  b. Answers may vary. An example is given. From the measurements in part (a), the triangular end of the boat is an isosceles triangle, so the formula in Exercise 36 on page 485 can be used to approximate the height of the triangle.

  c. Results may vary. Using the measurements given in part (a), the height from the formula is about 6.7 cm, which is close to 6.5 cm, the measured height.

If-Then Statements and Converses

IS IT RIGHT?

What does this ad say to you? Does it lead you to believe the following statements?

1) If you are a champion, then you choose Zephyrs.

2) You are a champion if you choose Zephyrs.

Statements like these are called **if-then statements** or *conditional statements*. The *if* part is called the **hypothesis** and the *then* part is called the **conclusion.**

hypothesis   conclusion
If you are a champion, then you choose Zephyrs.

conclusion   hypothesis
You are a champion if you choose Zephyrs.

Notice that the words "if" and "then" are not parts of the hypothesis and conclusion.

If-then statements are either true or false. One way to show that an if-then statement is false is to find a counterexample.

> **Watch Out!**
> The hypothesis does not always come first. Also, the word *then* does not always appear in an if-then statement.

Sample 1

Tell whether each if-then statement is *True* or *False*.

a. If it is 11 P.M., the sun has set.

b. If a triangle is equilateral, then it is isosceles.

Unit 9 Reasoning and Measurement

This time-lapse photo shows the sun in the sky over Kavtovic, Alaska from 9 P.M. to 4 A.M. on June 24.

Sample Response

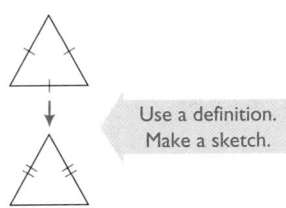

Use a definition. Make a sketch.

a. There is a counterexample.
In parts of Alaska, the sun never sets during some summer nights.

Apply what you know about geography.

The statement is false.

b. By definition, an isosceles triangle has *at least two* sides of equal length. Since every equilateral triangle has *three* sides of equal length, every equilateral triangle is also an isosceles triangle. There is no counterexample.

The statement is true.

Sample 2

Identify the hypothesis and the conclusion of each statement.

a. If concert tickets cost less than $10, I will buy one.

b. We will go to the park if it is sunny.

Sample Response

a. hypothesis: Concert tickets cost less than $10.
   conclusion: I will buy one.

b. hypothesis: It is sunny.
   conclusion: We will go to the park.

Talk it Over

1. Suppose there really is a brand of shoes called Zephyrs. Are statements (1) and (2) on page 492 *True* or *False*? Explain.

2. Discuss whether you think each statement is *True* or *False*.

   a. If the sun is shining, then there are no clouds in the sky.

   b. If a figure is a square, it is a rectangle.

3. Write at least one if-then statement about the Zephyr advertisement or about another subject. Identify the hypothesis and the conclusion of your statement. Tell whether your statement is *True* or *False*. Explain why.

9-3 If-Then Statements and Converses

493

493

Converse of an If-Then Statement

The **converse** of an if-then statement is formed by interchanging the "if" and "then" parts.

if-then statement: If you are an 18 year old U.S. citizen, **then** you can vote.
converse: **If** you can vote, **then** you are an 18 year old U.S. citizen.

The converse of a true if-then statement is not necessarily true. In the example above, the converse is false because voters may be older than 18.

Sample 3

Write the converse of each if-then statement and tell whether the converse is *True* or *False*.

a. If you are in the kitchen, then you are in the house.
b. If an integer is divisible by 2, then it is even.

Sample Response

a. converse: If you are in the house, then you are in the kitchen.

Make a sketch.

There is a counterexample. You could be in the house, but in a room that is not the kitchen.

The converse is false.

b. converse: If an integer is even, then it is divisible by 2.

Use a definition.

The definition of an even integer is that it is an integer that is divisible by 2. There is no counterexample.

The converse is true.

▸ Now you are ready for:
Exs. 1–20 on pp. 496–497

An Important Converse

Sometimes a statement and its converse are both true. You know from arithmetic that if you multiply any number by zero, the result is zero. This property can be stated as an if-then statement.

If $a = 0$ or $b = 0$, then $ab = 0$.

The *zero-product property* is the converse of this statement.

ZERO-PRODUCT PROPERTY

If a product of factors is zero, one or more of the factors must be zero.

If $ab = 0$, then $a = 0$ or $b = 0$.

Sample 4

Solve $x(x + 1) = 0$.

Sample Response

Use the zero-product property.

At least one factor must be 0. ⟶ $x(x + 1) = 0$

$x = 0$ or $x + 1 = 0$.

$x + 1 - 1 = 0 - 1$

The equation has *two* solutions. $x = -1$

The solutions are 0 and -1.

Talk it Over

4. Do both $x = 0$ and $x = -1$ make the equation $x(x + 1) = 0$ true? Explain.

5. a. Use the zero-product property to solve $-1.3x = 0$.

   b. Describe another method of solving $-1.3x = 0$.

6. Can you use the zero-product property to solve $x(x + 1) = -1$? Explain why or why not.

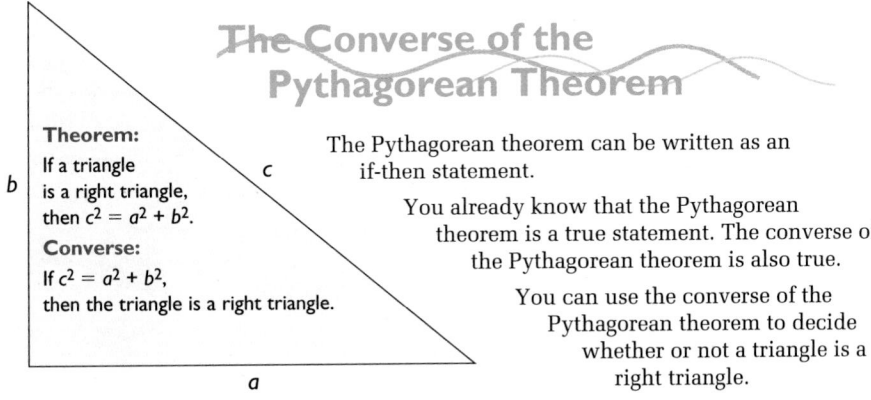

The Converse of the Pythagorean Theorem

Theorem:
If a triangle is a right triangle, then $c^2 = a^2 + b^2$.

Converse:
If $c^2 = a^2 + b^2$, then the triangle is a right triangle.

The Pythagorean theorem can be written as an if-then statement.

You already know that the Pythagorean theorem is a true statement. The converse of the Pythagorean theorem is also true.

You can use the converse of the Pythagorean theorem to decide whether or not a triangle is a right triangle.

9-3 If-Then Statements and Converses

495

Mathematical Procedures

The zero product property provides an important procedure for solving higher-order equations. In Unit 10, students will use this property to solve quadratic equations and to find the x-intercepts of their graphs.

Additional Sample

S4 Solve $(m - 2)m = 0$.

By the zero-product property, at least one of the factors must be 0. Either $(m - 2) = 0$ or $m = 0$.

$(m - 2) = 0$ or $m = 0$

$m - 2 + 2 = 0 + 2$

$m = 2$

The solutions are 2 and 0.

Answers to Talk it Over

4. Yes.

5. a. $-1.3 = 0$ or $x = 0$;
   Since $-1.3 \neq 0$, $x = 0$.

   b. Divide both sides by -1.3; $x = 0$.

6. No. To use the property, you must start with a product equal to 0. The product given is equal to -1.

Sample 5

The lengths of the sides of a triangle are given. Is the triangle a right triangle? How do you know?

a. 15 ft, 20 ft, 25 ft   **b.** 8 cm, 10 cm, 13 cm

Sample Response

If a triangle is a right triangle, the longest side will be the hypotenuse.

a.

$a^2 + b^2 \overset{?}{=} c^2$

$15^2 + 20^2 \overset{?}{=} 25^2$

$225 + 400 \overset{?}{=} 625$

$625 = 625$

The triangle is a right triangle because $a^2 + b^2 = c^2$.

> Let the two shorter sides be a and b. Let the longest side be c.

> Compare $a^2 + b^2$ and c^2.

b.

$a^2 + b^2 \overset{?}{=} c^2$

$8^2 + 10^2 \overset{?}{=} 13^2$

$64 + 100 \overset{?}{=} 169$

$164 \neq 169$

The triangle is *not* a right triangle because $a^2 + b^2 \neq c^2$.

Look Back

State the hypothesis and the conclusion of the Pythagorean theorem. Explain how to form the converse of the Pythagorean theorem.

▶ Now you are ready for:
Exs. 21–51 on pp. 497–499

9-3 Exercises and Problems

1. **Reading** What is one way to show that an if-then statement is false?

For Exercises 2–9, do these things.

a. Identify the hypothesis and the conclusion of each statement.

b. Tell whether the statement is *True* or *False*. If it is false, give a counterexample.

2. If you visit the State Aquarium, then you will see some fish.

3. If you see an advertisement for a product, you will buy the product.

4. If the measure of an angle is less than 90°, the angle is acute.

5. If $x^2 = 25$, then $x = \pm \sqrt{25}$.

6. If a figure is a rectangle, then the area of the figure is the product of the length and width of the figure.

7. If a and b are nonnegative integers, then $\sqrt{ab} = \sqrt{a} \cdot \sqrt{b}$.

8. If $\angle A = 30°$, then $\sin A = 0.5$.

9. If $2y = 0$, then $y = 0$.

BY THE WAY...

The Maya of Mexico and Central America were among the first peoples to use a symbol to represent zero. The symbol looked somewhat like a shell, as shown in this section of a Mayan manuscript.

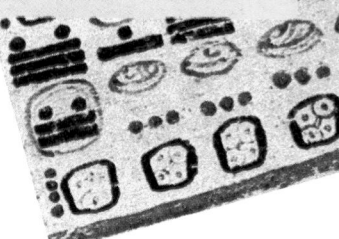

Answers to Look Back

Hypothesis: A triangle is a right triangle.

Conclusion: The square of the length of the hypotenuse of the triangle is equal to the sum of the squares of the lengths of the legs.

Explanations may vary. An example is given. To form the converse of the Pythagorean theorem, rewrite the conclusion using the word "if" and replacing "hypotenuse" with "one side" and "legs" with "other two sides." Rewrite the hypothesis using the word "then."

Answers to Exercises and Problems

1. Find a counterexample that contradicts the statement.

2–9. Counterexamples may vary. Examples are given.

2. **a.** Hypothesis: You visit the state Aquarium. Conclusion: You will see some fish.

For Exercises 10–18, do these things.

a. Write the converse of the statement.

b. Tell whether the converse is *True* or *False*. If it is false, give a counterexample.

10. If you are in Africa, then you are south of the equator.

11. If you are in Zimbabwe, then you are south of the equator.

12. If you are at Lake Tanganyika, then you are in Africa.

13. If m and n are even numbers, then $(m + n)$ is an even number.

14. A number is divisible by 3 if the number is divisible by 6.

15. If a figure is a triangle, then the area of the figure is half the area of some parallelogram with the same base and height.

16. If the coordinates of a point are $(0, -3)$, then the point is on the y-axis.

17. If the probability of an event is 1, then the event is certain to happen.

18. If $(x + 3)(x - 4) = 0$, then $x + 3 = 0$.

19. **Group Activity** Work in a group of three or four students.

    a. Write if-then statements suggested by your day at school or by an advertisement on TV or in a magazine.

    b. **Writing** Write a paragraph explaining why your statements and their converses are true or false.

20. **Open-ended** Write a true if-then statement whose converse is also true.

pulley
15 ft
8 ft

Solve.

21. $\frac{3}{5}t = 0$

22. $y(y - 4) = 0$

23. $m(m + 1) = 0$

24. $q(5q + 4) = 0$

25. $3j(j - 2) = 0$

26. $12z(3 - z) = 0$

The lengths of the sides of a triangle are given. Is it a right triangle? How do you know?

27. 7 yd, 24 yd, 25 yd

28. 8 in., 11 in., 15 in.

29. 4 in., 5 in., 6 in.

30. 24 mm, 26 mm, 10 mm

31. 2 ft, 8 ft, 7 ft

32. 34 mm, 16 mm, 30 mm

33. Use the photo and diagram at the left. A science class is doing an experiment using pulleys to raise a pole. Some students say that the pole will stand at a right angle to the ground when the length of rope between the pulley and the top of the pole is 17 ft. Are they correct?

34. Suppose the pole in Exercise 33 is 12 ft long. Let d be the distance between the pulley and the base of the pole, and let l be the length of the rope between the pulley and the top of the pole. Find two pairs of integers d and l for which the pole will stand at a right angle to the ground.

9-3 If-Then Statements and Converses   **497**

6. b. True.

7. a. Hypothesis: a and b are non-negative integers. Conclusion: $\sqrt{ab} = \sqrt{a} \cdot \sqrt{b}$

    b. True.

8. a. Hypothesis: $\angle A = 30°$ Conclusion: $\sin A = 0.5$

    b. True.

9. a. Hypothesis: $2y = 0$ Conclusion: $y = 0$

    b. True.

10–18. Counterexamples may vary. Examples are given.

10. a. If you are south of the equator, then you are in Africa.

    b. False. You could be in Australia.

11. a. If you are south of the equator, then you are in Zimbabwe.

    b. False. You could be in another country south of the equator, such as Uruguay.

12. a. If you live in Africa, then you are at Lake Tanganyika.

    b. False. You could be somewhere else in Africa, such as Lake Victoria.

13. a. If $(m + n)$ is an even number, then m and n are even numbers.

    b. False. Suppose $(m + n) = 4$. The values for m and n could be 3 and 1, which are odd numbers.

14–34. See answers in back of book.

Answers to Exercises and Problems

    b. Most likely answer: True. (If answered False, accept reasonable counterexamples.)

3. a. Hypothesis: You see an advertisement for a product. Conclusion: You will buy the product.

    b. False. I see many advertisements for products that I do not want to buy or that are too expensive for me to buy.

4. a. Hypothesis: The measure of an angle is less than 90°. Conclusion: The angle is acute.

    b. Most likely answer: True. Possible answer: False. An angle of 0° is less than 90°, but a 0° angle is not an acute angle.

5. a. Hypothesis: $x^2 = 25$ Conclusion: $x = \pm\sqrt{25}$

    b. True.

6. a. Hypothesis: A figure is a rectangle. Conclusion: The area of the figure is the product of the length and the width of the figure.

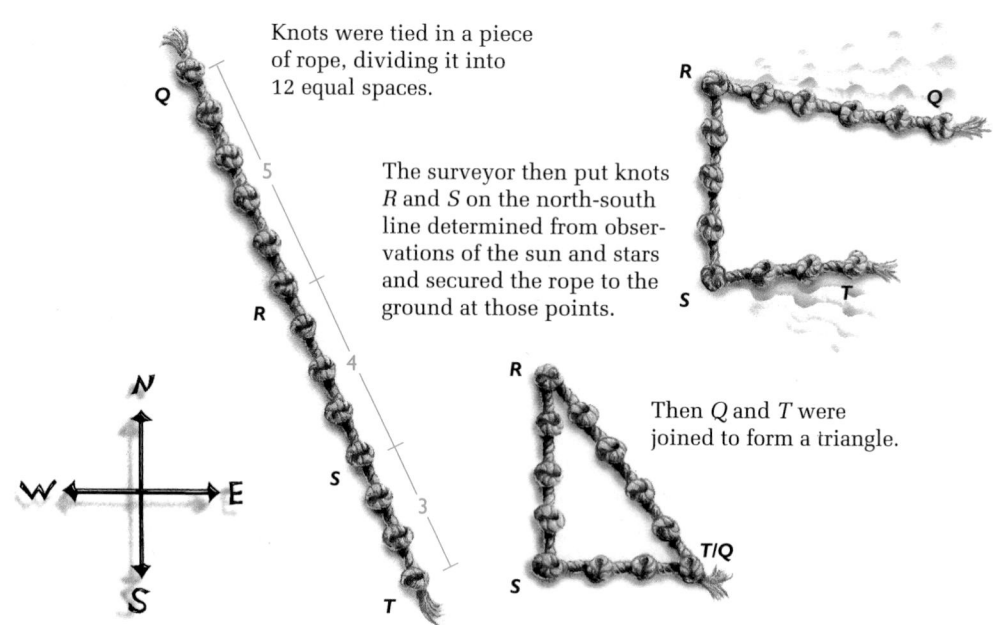

connection to HISTORY

Surveyors in ancient Egypt may have used knotted ropes to build walls at right angles.

Knots were tied in a piece of rope, dividing it into 12 equal spaces.

The surveyor then put knots R and S on the north-south line determined from observations of the sun and stars and secured the rope to the ground at those points.

Then Q and T were joined to form a triangle.

35. The knots at R and S divide the rope into what lengths?

36. How do you know that $\angle RST = 90°$?

37. Egyptian pyramids were built so that their four sides faced directly north, south, east, and west. Does $\overline{ST}$ lie on an east-west line? How do you know?

Ongoing ASSESSMENT

38. **Open-ended** Write a problem situation about school or home that can be solved using the Pythagorean theorem. Write another problem situation that can be solved using the converse of the Pythagorean theorem.

Review PREVIEW

Simplify. *(Section 9-2)*

39. $\sqrt{72}$ 40. $\sqrt{8} \cdot \sqrt{6}$ 41. $5\sqrt{3} \cdot \sqrt{12}$ 42. $10\sqrt{15} \cdot 8\sqrt{6}$

A die is rolled. Find each probability. *(Section 6-2)*

43. $P(\boxdot)$ 44. $P(\boxdot \text{ or } \boxdot)$ 45. $P(\text{number} > 6)$ 46. $P(\text{not } \boxdot)$

498   **Unit 9** Reasoning and Measurement

BY THE WAY...

Entrances to Egyptian pyramids often face north toward the North Star. The main entrances of Navajo *hogans* face east toward the sunrise.

---

35. 3, 4, and 5

36. Reasons may vary. An example is given. If the square of the length of the longest side of $\triangle RST$ is equal to the sum of the squares of the lengths of the other two sides, then the triangle is a right triangle, and $\angle RST = 90°$. Since $3^2 + 4^2 = 5^2$ (9 + 16 = 25; 25 = 25), then $\angle RST = 90°$.

37. Yes. Explanations may vary. An example is given. $\overline{RS}$ lies on a north-south line, and $\angle RST = 90°$. This means that $\overline{RS}$ is perpendicular to $\overline{ST}$. So, $\overline{ST}$ must lie on an east-west line.

38. Problems may vary. Examples are given. 1) Marcus lives across the street from a rectangular park. The park has a diagonal path that goes from corner to corner. The length of the park is 1.5 km and the width is 2 km. Marcus noticed that if he starts at a corner of the park and walks to the opposite corner, it takes less time to walk along the diagonal path than to walk along the sides of the park. Explain why. (*Answer:* You walk 3.5 m along the sides, while the diagonal path is only 2.5 m long.) 2) Yariela is building a square wooden frame with each side measuring 12 in. She estimates that when the diagonals measure about 17 in., the frame will be correctly formed. Do you agree or disagree? Explain. (*Most likely answer:* I agree; $12^2 + 12^2 \approx 17^2$, so the angles are right angles.)

Working on the Unit Project

For Exercises 47–49, tell whether you think each statement is *True* or *False*. Give a reason for your choice.

47. If a boat has a hole in it, the boat sinks.

48. If an object can float in fresh water, the object can float in salt water.

49. If an object is made of steel, the object cannot float in water.

50. Write the converse of each statement in Exercises 47–49. Tell whether you think the converse is *True* or *False*. Give a reason for your choice.

51. Gather some small objects (for example, a coin, a pencil, a toothpick, a paper clip, a piece of paper, cardboard, wood, or string). Tell whether you think the probability is 0 or 1 that each object will float in a pan of water. Then test your predictions.

Unit 9 — CHECKPOINT 1

1. **Writing** How would you teach a friend to multiply square roots? Start with $\sqrt{10} \cdot \sqrt{15}$ and write an explanation of how to get a result of $5\sqrt{6}$.

Find the missing length in each right triangle.   9-1

2.
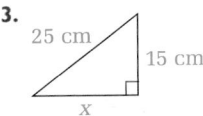
36 in.   15 in. x

3.
25 cm   15 cm x

even number: 2n
odd number: 2p+1
even × odd = 2n(2p+1)
  = 4np + 2n
  = 2(2np + n)
2(2np + n) is even, because it is a multiple of 2.

4. Jonah claims that the product of an odd number and an even number is an even number. He writes the steps shown at the left. What kind of reasoning is he using?

Simplify.   9-2

5. $\sqrt{300}$

6. $\sqrt{6} \cdot \sqrt{42}$

7. Solve $6x^2 = 300$.

8. Use the statement "If a number is odd, then it is a multiple of 3."   9-3

   a. Identify the hypothesis and the conclusion of the statement.

   b. Tell whether the statement is *True* or *False*. If it is false, give a counterexample.

   c. Write the converse of the statement.

9. Solve $x(2x + 3) = 0$.

10. The lengths of the sides of a triangle are 45 m, 50 m, and 14 m. Is it a right triangle? How do you know?

9-3 If-Then Statements and Converses   **499**

Working on the Unit Project

Students can work in small groups to share their knowledge and reasoning as they do these exercises.

Quick Quiz (9-1 through 9-3)

See page 542.

Practice 69 For use with Section 9-3

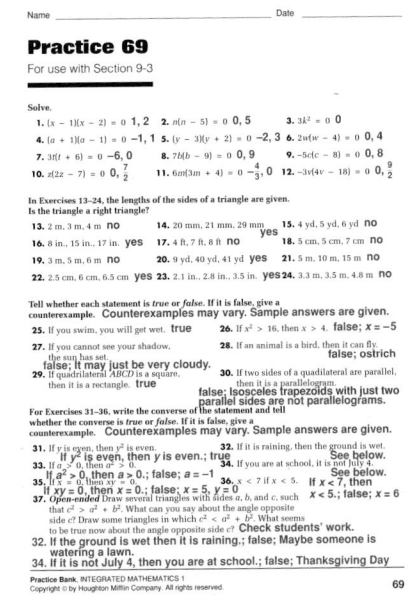

Answers to Checkpoint

1. Explanations may vary. An example is given. Suppose you start with $\sqrt{10} \cdot \sqrt{15}$. You can rewrite this product as $\sqrt{10 \cdot 15}$, or $\sqrt{150}$. See if you can find any perfect square factors for the number under the square root symbol. A perfect square factor of 150 is 25, so you can write $\sqrt{150}$ as $\sqrt{25 \cdot 6}$. Rewrite this as $\sqrt{25} \cdot \sqrt{6}$. Since $\sqrt{25} = 5$, the final answer is $5\sqrt{6}$.

2. 39 in.   3. 20 cm

4. deductive   5. $10\sqrt{3}$

6. $6\sqrt{7}$ 7. $5\sqrt{2}, -5\sqrt{2}$

8. a. Hypothesis: A number is odd. Conclusion: It is a multiple of 3.

   b. False. Counterexamples may vary. An example is given. The number 5 is odd, but it is not a multiple of 3.

   c. If a number is a multiple of 3, then it is odd.

9. $0, -\dfrac{3}{2}$

10. No; because $45^2 + 14^2 = 2221$ and $50^2 = 2500$.

Answers to Exercises and Problems

39. $6\sqrt{2}$ 40. $4\sqrt{3}$

41. 30   42. $240\sqrt{10}$

43. $\dfrac{1}{6}$ 44. $\dfrac{1}{3}$

45. 0   46. $\dfrac{5}{6}$

47–49. Choices and counterexamples may vary. Examples are given.

47. False. The hole could be above the water level.

48. True.

49. False. Submarines are made of steel and they float in water.

50. Ex. 47: If a boat sinks, the boat has a hole in it. False; the boat could have been loaded with too much cargo or a storm could have caused the boat to capsize and sink. Ex. 48: If an object can float in salt water, the object can float in fresh water. False; an egg will float in salt water but will sink in fresh water. Ex. 49: If an object cannot float in water, the object is made of steel. False; a softball is made of leather and rubber and it cannot float in water.

51. Answers may vary, depending on objects chosen.

Objectives and Strands
See pages 474A and 474B.

Spiral Learning
See page 474B.

Materials List
➤ $8\frac{1}{2} \times 11$ in. paper
➤ $\frac{1}{4}$ in. diameter disks of paper
➤ Graph paper

Recommended Pacing
Section 9-4 is a two-day lesson.

Day 1

Pages 500–501: Exploration,
Exercises 1–5

Day 2

Pages 502–504: Theoretical
Geometric Probability through
Look Back, *Exercises 6–20*

Extra Practice
See pages 635–638.

Warm-Up Exercises
Warm-Up Transparency 9-4

Support Materials
➤ Practice 70
➤ Enrichment 62 in the Activity
Bank
➤ Study Guide 9-4
➤ Problem Set 20
➤ Diagram Masters 2, 22 in the
Explorations Lab Manual
➤ McDougal Littell Mathpack
software: *Probability Constructor*
➤ Probability Constructor Activity
Book: Activities 4, 7, 10, 14, and
15
➤ Quiz 9-4
➤ Alternative Assessment 3

Section 9-4

Focus
Use lengths and areas to
determine theoretical
geometric probabilities.

Geometric Probability

A Likely Spot

EXPLORATION

How can you use geometry to find the
probability that an event will happen?

• **Materials:** $8\frac{1}{2} \times 11$ paper; several $\frac{1}{4}$ in. diameter
disks of paper (use a paper punch)

• **Work with another student.**

① Divide your sheet of paper into four regions by drawing
diagonals. Shade one region, as shown in the photograph.

② In this experiment, you are going to drop a paper disk
from a point above the center of your sheet of paper.
Before you begin, predict the probability that the disk
will land in the shaded
region. Discuss your
reasoning.

500

Answers to Exploration

1. Check drawings.
2. Predictions and explanations may vary. An exam-
ple is given. I predict that
the probability that the
disk will land on the shad-
ed region is about $\frac{1}{4}$, or

0.25, because the shaded
region appears to be about
one fourth of the sheet of
paper.

3, 4. Results may vary. An
example is given. After
dropping the disk 20
times, there were 4 wins
and 16 losses. The experimental probability should
be about 0.25.

③ Hold a disk about 4 in. above the center of the paper and then drop the disk. If the disk lands in the shaded region, record it as a *win* in a table like the one shown in the photograph on page 500. If it does not land in the shaded region, record it as a *loss*. (If the disk lands on a line or off the paper, do not count it. Drop the disk again.)

④ Drop the disk 20 times and record your results.

⑤ Calculate the experimental probability: $\frac{\text{number of wins}}{\text{20 trials}}$.

Do your results support the prediction you made in step 2?

⑥ Collect the experimental probabilities calculated by all the groups in your class. Find the mean of these values. Do the results support the prediction you made in step 2?

⑦ Find the ratio of the shaded area of your paper to the area of the whole sheet of paper. Compare this ratio to the experimental probability you found in step 6.

▶ **Now you are ready for:**
Exs. 1–5 on p. 504

501

Exploration

Making reasonable predictions is an important mathematical skill. The goal of this Exploration is to have students learn experimental techniques to support or make geometric probability predictions. Students may need to review briefly the notion of experimental probability developed in Section 6-2.

Using Technology

The *Probability Constructor* software can be used to simulate the experiment in the Exploration.

Reasoning

Students may wish to compare the Exploration, which finds experimental probability, with Sample 1 on page 502, which determines theoretical probability. As part of this activity, students may wish to create the diagram on page 502, drop their paper disks onto it, and determine an experimental probability for Sample 1.

Answers to Exploration

5. Yes.

6. Answers may vary. The mean should be about 0.25.

7. $\frac{1}{4}$, or 0.25; Comparisons may vary. The ratio should be about the same as the experimental probability.

Additional Sample

S1 A desk is 29 inches wide and 60 inches long. On it is a desk pad that is 17 inches wide and 22 inches long. If a person randomly flips a paperclip on the desk, what is the probability that it will land on the desk pad?

probability $= \dfrac{\text{area of pad}}{\text{area of desk}} =$

$\dfrac{17(22)}{29(60)} = \dfrac{374}{1740} \approx 0.21$

The probability the paperclip will land on the desk pad is about 0.21.

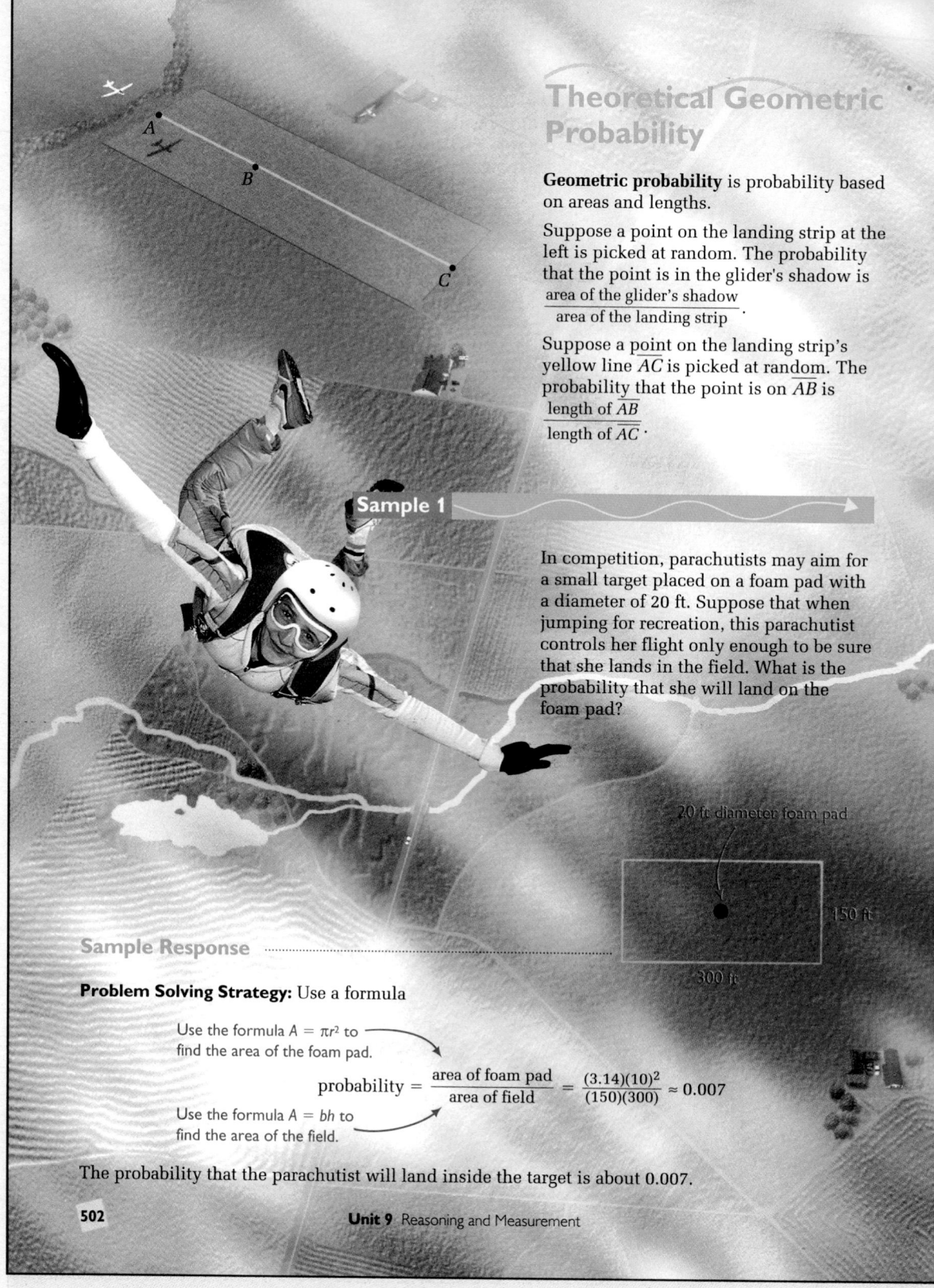

Theoretical Geometric Probability

Geometric probability is probability based on areas and lengths.

Suppose a point on the landing strip at the left is picked at random. The probability that the point is in the glider's shadow is $\dfrac{\text{area of the glider's shadow}}{\text{area of the landing strip}}$.

Suppose a point on the landing strip's yellow line $\overline{AC}$ is picked at random. The probability that the point is on $\overline{AB}$ is $\dfrac{\text{length of } \overline{AB}}{\text{length of } \overline{AC}}$.

Sample 1

In competition, parachutists may aim for a small target placed on a foam pad with a diameter of 20 ft. Suppose that when jumping for recreation, this parachutist controls her flight only enough to be sure that she lands in the field. What is the probability that she will land on the foam pad?

20 ft diameter foam pad

150 ft

300 ft

Sample Response

Problem Solving Strategy: Use a formula

Use the formula $A = \pi r^2$ to find the area of the foam pad.

Use the formula $A = bh$ to find the area of the field.

probability $= \dfrac{\text{area of foam pad}}{\text{area of field}} = \dfrac{(3.14)(10)^2}{(150)(300)} \approx 0.007$

The probability that the parachutist will land inside the target is about 0.007.

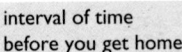

Sample 2

A delivery person will deliver a package to your home sometime between 4:00 P.M. and 6:00 P.M. What is the probability that you will miss the delivery person if you get home at 4:15 P.M.?

Sample Response

Method ❶

Problem Solving Strategy: Draw a diagram

Draw a time line divided into 15-minute intervals.

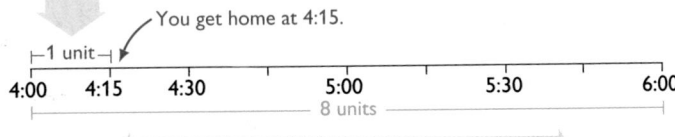

interval of time before you get home

You get home at 4:15.

├─1 unit─┤

4:00   4:15   4:30   5:00   5:30   6:00

├──────── 8 units ────────┤

interval of time when delivery person may arrive

Find the ratio of the length of the segment for the interval of time before you get home to the length of the segment for the interval of time when the delivery person may arrive.

$$\text{probability} = \frac{\text{length of segment for 4:00–4:15}}{\text{length of segment for 4:00–6:00}} = \frac{1}{8}$$

The probability that you will miss the delivery person is $\frac{1}{8}$.

Method ❷

Find the probability that the delivery person arrives before you get home.

Find the probability that a time chosen at random from 4:00 P.M. to 6:00 P.M. is between 4:00 P.M. and 4:15 P.M.

$$\text{probability} = \frac{\text{minutes from 4:00 to 4:15}}{\text{minutes from 4:00 to 6:00}} = \frac{15}{(2)(60)} = \frac{15}{120} = \frac{1}{8}$$

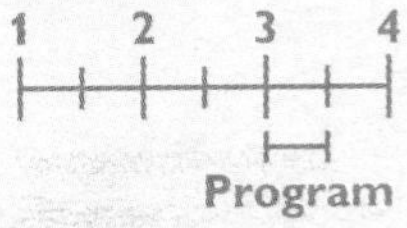

The probability that you will miss the delivery person is $\frac{1}{8}$.

Talk it Over

1. Would you use a ratio of *lengths* or a ratio of *areas* to find each probability?

   a. the probability that a point is in a particular region

   b. the probability that a point is on a particular segment

2. In Sample 2, which method do you prefer, drawing a time line or counting the number of minutes? Why?

9-4 Geometric Probability

503

Look Back

What facts do you need to know to find the theoretical probability that this spinner will land on the green sector? Describe how you could find the theoretical probability if these facts were given. Do you need these facts to find the experimental probability? Explain.

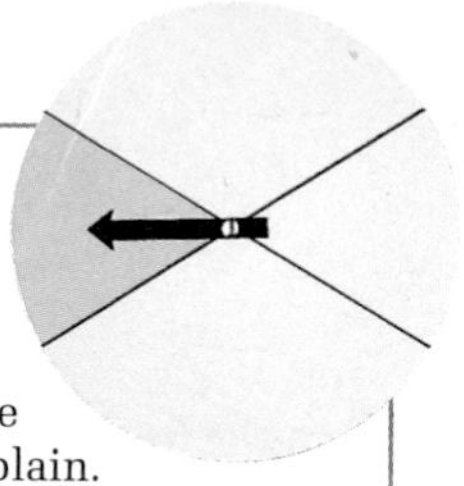

····▶ Now you are ready for:
Exs. 6–20 on pp. 504–506

9-4 Exercises and Problems

1. Suppose you perform a probability experiment like the one in the Exploration using the piece of paper shown at the right. Do you think the probability that a disk will land in the shaded region of this piece of paper is *greater than* or *less than* the probability that a disk will land in the shaded region of the paper used in the Exploration? Why?

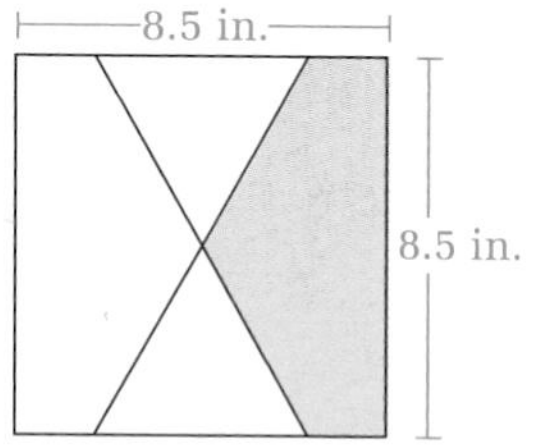

⊢—8.5 in.—⊣

8.5 in.

2. Would the results of the Exploration be different if the sheet of paper you used was a *circle* divided into four equal parts?

3. a. In step 5 of the Exploration, you found the ratio $\frac{\text{number of wins}}{20 \text{ trials}}$.

Use the table you made in step 3 to find the ratio $\frac{\text{number of losses}}{20 \text{ trials}}$.

b. What is the sum of the two ratios in part (a)?

Find the probability that a point chosen at random from each figure is in the shaded region. Then tell which probability is greater.

4. A.   B.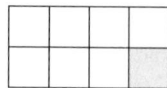

5. A. 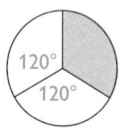   B.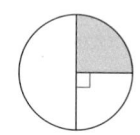

6. **Reading** Read the paragraph on theoretical geometric probability on page 502. Suppose you pick a point at random on $\overline{AC}$. How can you find the probability that the point is on $\overline{BC}$?

7. Which would be greater, the probability that a point chosen at random from *circle A* or from *circle B* is in the shaded region?

A.    B.

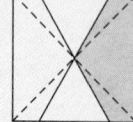

8. A 44 ft × 165 ft rectangular barge has sunk in a river. A salvage boat anchors at a random spot in the shaded section of river so that a diver can search for the barge.

   a. What is the approximate shape of the shaded section of river?

   b. What is the area of the shaded section of river?

   c. What is the area of the barge?

   d. What is the probability that the boat will anchor over the sunken barge?

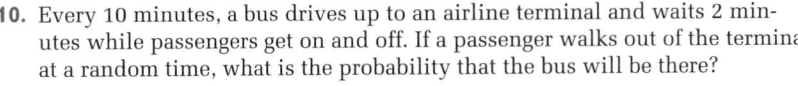

9. Jerome is expecting a call from his grandmother sometime between 12:00 noon and 6:00 P.M. During this time, he goes outside to shoot baskets for 45 min. What is the probability that his grandmother will call while he is outside?

10. Every 10 minutes, a bus drives up to an airline terminal and waits 2 minutes while passengers get on and off. If a passenger walks out of the terminal at a random time, what is the probability that the bus will be there?

11. **Writing** In a game of darts, the players aim for the bull's-eye at the center of the dart board.

    a. Does the skill of the player affect the probability that a player will hit the bull's-eye? Explain.

    b. Would you use *experimental* or *theoretical* probability to find the probability that a particular player will hit the bull's-eye? Explain why and how.

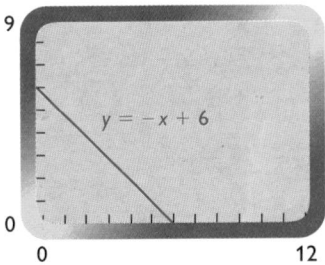

12. a. Find the area of the calculator screen at the left in the units marked on the axes.

    b. Find the area of the triangle whose sides are the x-axis, the y-axis, and the line $y = -x + 6$.

    c. Suppose a point of the calculator screen is chosen at random. What is the probability that the point lies on or below the line $y = -x + 6$?

13. Suppose a point of this calculator screen is chosen at random. Find the probability that each statement is true.

    a. The point lies on or above the line $y = x + 3$.

    b. The point lies on or below the line $y = x$.

    c. The coordinates of the point are a solution of the inequality $y \le x + 3$.

    d. The coordinates of the point are a solution of this system of inequalities: $y \le x + 3$
       $\qquad\qquad\qquad\qquad y \ge x$

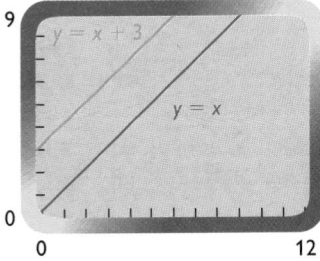

9-4 Geometric Probability   505

Communication: Writing

Ex. 11 provides an excellent opportunity for students to apply geometric probability to a real-life situation. Part (a) points out that the experimental probability of a situation may change, even though the geometric object involved does not. Part (b) requires students to make a choice between experimental or theoretical probability and then to explain their choice.

Integrating the Strands

Ex. 13 integrates the strands of algebra, geometry, and probability. It uses the concepts of inequalities, lines in a plane, equations of lines, a coordinate grid, randomness, and probability. Discuss this integration with students to reinforce the interrelated nature of mathematics.

Using Technology

Exs. 12 and 13 can be expanded by having students graph more than one or two lines. Students can create triangles and quadrilaterals on screen and then find the probabilities of a point being inside or outside each figure.

the player's goal of getting the dart to land as close to the center as possible mean that the points in the dart board are not equally likely to be hit. The experimental probability for one player will differ from the experimental probability for another player; the only way to estimate this probability is for the player to throw some darts (say, 20 darts) and for the rate of success $\left(\dfrac{\text{number of bull's eyes}}{\text{number of darts thrown}}\right)$ to be computed.

12. a. 108 units²   b. 18 units²

   c. about 0.17 or 17%

13. a. about 0.17 or 17%

   b. about 0.63 or 63%

   c. about 0.83 or 83%

   d. about 0.21 or 21%

Answers to Exercises and Problems

4. a. $\frac{1}{4}$

   b. $\frac{1}{8}$; probability in part (a)

5. a. $\frac{1}{3}$

   b. $\frac{1}{4}$; probability in part (a)

6. The probability that the point is on $\overline{BC}$ = $\dfrac{\text{length of } \overline{BC}}{\text{length of } \overline{AC}}$.

7. B

8. a. parallelogram

   b. 250,000 yd²

   c. 7260 ft²

   d. about 0.003 or 0.3%

9. $\frac{1}{8}$ or 0.125   10. $\frac{1}{5}$ or 0.2

11. a. Yes. Explanations may vary. An example is given. The more skillful a player is, the more likely the player is to hit

the bull's eye, that is, the greater the player's probability of hitting the bull's eye.

   b. experimental; Explanations may vary. An example is given. Theoretical probability is used when the points in a region are equally likely to be hit. The skill of a particular player and

505

Students' designs and written responses to Ex. 15 will demonstrate how well they have understood geometric probability. As there will be a number of different designs, you may wish to have students display them. Ex. 15 would also make an excellent group activity.

Working on the Unit Project

After students have completed their research for Exs. 19 and 20, have some students share what they found with the class and ask others to summarize their ideas about water and buoyancy.

Practice 70 For use with Section 9-4

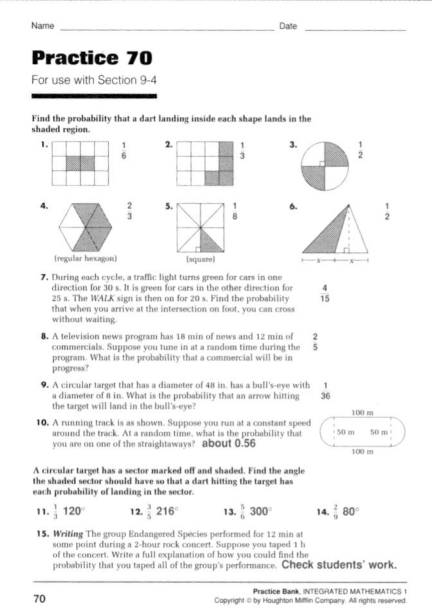

14. a. 45; 60; 30

b. $P = \frac{2x}{360}$ or $P = \frac{x}{180}$

c.

$0 < x \le 180$

d. Explanations may vary. Examples are given. The graph represents a function because it passes the vertical-line test. The graph does represent a direct variation situation because it passes through the origin.

506

---

14. a. What must the value of x be so that the probability of the spinner landing on a yellow sector is $\frac{1}{4}$? $\frac{1}{3}$? $\frac{1}{6}$?

b. Let P = the probability that the spinner will land on a yellow sector. Write an equation for P in terms of x.

c. Graph your equation from part (b). What values of x make sense?

d. Does your graph represent a function? a direct variation situation? Explain why or why not.

Ongoing **ASSESSMENT**

15. a. Open-ended Design a dart board with three possible scores: 100, 200, and 300. Use these guidelines.

➤ The probability of getting 100 should be 50%.

➤ The probability of getting 200 should be 30%.

➤ The probability of getting 300 should be 20%.

(Assume that a dart will always land on the dart board and has an equal chance of landing anywhere on the board.)

b. Writing Describe your design. How did you make sure that the probabilities of getting each score met the guidelines?

Review **PREVIEW**

16. a. Write the converse of the statement "If Raymond lives in Toronto, then he lives in Canada."

b. Tell whether the converse is *True* or *False*. If it is false, find a counterexample. *(Section 9-3)*

17. Suppose a figure is a rectangle. What are three things you know about the figure? *(Section 1-7)*

18. Write each formula. *(Sections 5-7, 7-3)*

a. area of a triangle　　**b.** circumference of a circle

Working on the Unit Project

19. Research Read the entry on *water* in an encyclopedia or a science book.

a. What is the weight in grams of 1 cubic centimeter of water?

b. What is the weight in pounds of 1 cubic foot of water?

c. What does H_2O stand for?

d. Is ice heavier than water?

20. Research Read the entry on *buoyancy* in an encyclopedia or a science book. Describe what determines the buoyancy of an object.

506　　　　　　　　**Unit 9** Reasoning and Measurement

---

15. a. Designs may vary. An example is given.

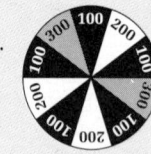

b. My design is a circular board with 10 sections of equal size. To meet the guidelines for the probabilities, I made 5 sections 100 points: $\frac{5}{10}$ = 50%; 3 sections 200 points: $\frac{3}{10}$ = 30%;

and, 2 sections 300 points: $\frac{2}{10}$ = 20%.

16. a. If Raymond lives in Canada, then he lives in Toronto.

b. False. Raymond may live somewhere else in Canada, such as Montreal.

17. Answers may vary. An example is given. The figure has four sides, opposite pairs of sides are con-

gruent and parallel and all angles measure 90°.

18. a. $A = \frac{1}{2}bh$

b. $C = \pi d$ or $C = 2\pi r$

19. a. 1 g　　**b.** 62.4 lb

c. the chemical formula for a water molecule: two atoms of hydrogen and one atom of oxygen

d. No. Ice is lighter than water.

20. See answer in back of book.

Surface Area of Space Figures

It **+** All **+** Adds **+** Up

─ Focus

Find the surface areas of prisms, cylinders, and pyramids.

Talk it Over

Suppose you cut open this popcorn box and lay it flat.

16 cm

4.5 cm

10 cm

1. What is the shape of each side of the box? What are the dimensions of each side?

2. What formula can you use to find the area of each side?

3. Find the total area of all the sides of the box.

4. Explain why your answer to question 3 is called the *surface area* of the popcorn box.

A **prism** is a space figure with two sides that are congruent, parallel polygons. These sides are called the **bases** of the prism. The other sides of the prism are called **faces**. In this book, the faces of prisms are rectangles.

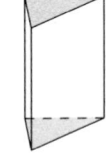
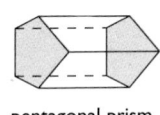

pentagonal prism

triangular prism

A prism is named by the shape of its bases. The popcorn box is a *rectangular prism* because the bases are rectangles.

$X_?\triangle ab$

FORMULA: SURFACE AREA OF A PRISM

Surface area (S.A.) = areas of two bases + areas of faces

Surface area is measured in square units.

Answers to Talk it Over

1. rectangle; top and bottom: 4.5 cm by 10 cm; sides: 4.5 cm by 16 cm; front and back: 10 cm by 16 cm

2. $A = lw$

3. 554 cm^2

4. Explanations may vary. An example is given. All the faces make up the surface of the figure, so the total area of all the faces is the area of the surface of the figure.

PLANNING

Objectives and Strands
See pages 474A and 474B.

Spiral Learning
See page 474B.

Materials List
➤ Ruler
➤ Graph paper

Recommended Pacing
Section 9-5 is a two-day lesson.
Day 1
Pages 507–509: Talk it Over through Sample 2, *Exercises 1–16*
Day 2
Pages 509–511: Surface Area of a Pyramid through Look Back, *Exercises 17–32*

Extra Practice
See pages 635–638.

Warm-Up Exercises
Warm-Up Transparency 9-5

Support Materials
➤ Practice 71
➤ Enrichment 63 in the Activity Bank
➤ Study Guide 9-5
➤ Problem Set 20
➤ Additional Exploration 9
➤ Diagram Master 2 in the Explorations Lab Manual
➤ Quiz 9-5
➤ Alternative Assessment 4

TEACHING

Using Manipulatives

You may wish to obtain a number of popcorn boxes and actually have students cut them open as described in Talk it Over questions 1–4.

Error Analysis

The intuitive sense of a base of a prism that students have is the face the prism is resting on. For example, if the triangular prism were placed on a table with a rectangular face on the table, students usually call that face the base. Stress that the orientation of the prism does not determine its bases. By definition, the bases of a prism are congruent and *parallel* sides.

Additional Sample

S1 Find the surface area of the triangular prism.

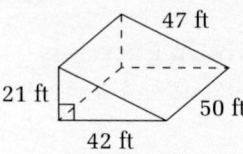

The bases are triangles.
$A = 2\left(\frac{1}{2}\right)(21)(42) = 882 \text{ ft}^2$
The faces are rectangles.
$A = (21)(50) + (42)(50) + (47)(50)$
$= 1050 + 2100 + 2350$
$= 5500 \text{ ft}^2$
S.A. $= 882 + 5500 = 6382 \text{ ft}^2$

Using Manipulatives

As before, you may wish to provide a cylindrical ice-cream or oatmeal carton for students to cut up as they discuss Talk it Over questions 5 and 6. Students can label the parts as shown in the text and then actually measure them to find the surface area.

Limited English Proficiency

It may help second-language learners avoid confusion if you tell them that the word *space* in the phrase *space figure* does not refer to outer space; rather, it refers to the three-dimensional region enclosed by the figure.

Find the surface area of this triangular prism.

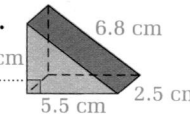

Sample Response

> Surface area (S.A.) is the sum of the areas of the bases and the faces.

Problem Solving Strategy: Make a sketch

Make sketches of each base and each face of the prism.

The bases are triangles.   The faces are rectangles.

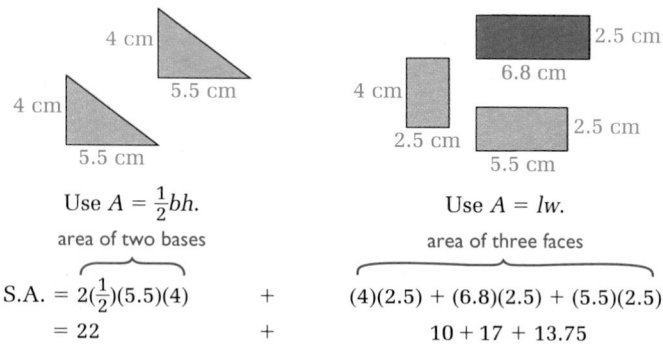

Use $A = \frac{1}{2}bh$.   Use $A = lw$.
area of two bases   area of three faces

S.A. $= 2(\frac{1}{2})(5.5)(4)$ $+$ $(4)(2.5) + (6.8)(2.5) + (5.5)(2.5)$
$= 22$ $+$ $10 + 17 + 13.75$
≈ 62.8

The surface area is about 62.8 cm².

Surface Area of a Cylinder

The shape of a tin can is an example of a *cylinder*. A **cylinder** is a space figure with a curved surface and two congruent parallel bases that are circles. You have to look straight at one end to see a circle.

cylinder

Talk it Over

5. Imagine cutting up a cylinder as shown below. Explain what is happening in each step.

6. In the last step, what does $2\pi r$ represent?

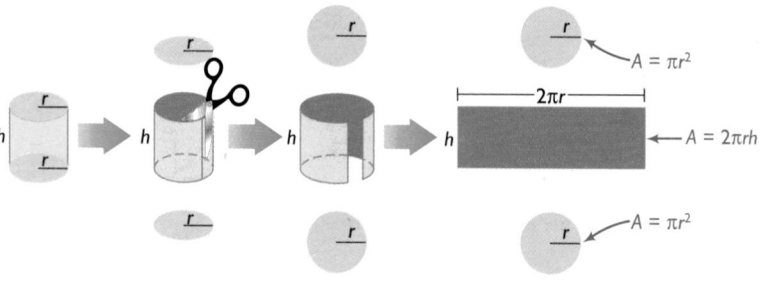

Answers to Talk it Over

5. Explanations may vary. An example is given. In step 1, the circular bases are cut out, and the curved surface is cut vertically. In step 2, the circular bases are laid flat, and the curved surface is opened. Step 3 shows the area for each part of the cylinder. The circular bases each have an area of πr^2. The curved surface

   flattens into a rectangle having an area of $2\pi rh$.

6. In the last step, $2\pi r$ represents the length of the rectangle, which is the same as the circumference of each base.

FORMULA: SURFACE AREA OF A CYLINDER

Surface area = areas of two circular bases + area of curved surface

$$S.A. = 2\pi r^2 + 2\pi rh$$

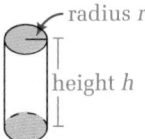

radius r

height h

Sample 2

A company packages vegetables in this can. How much material is needed to make the can? Give your answer to the nearest square inch.

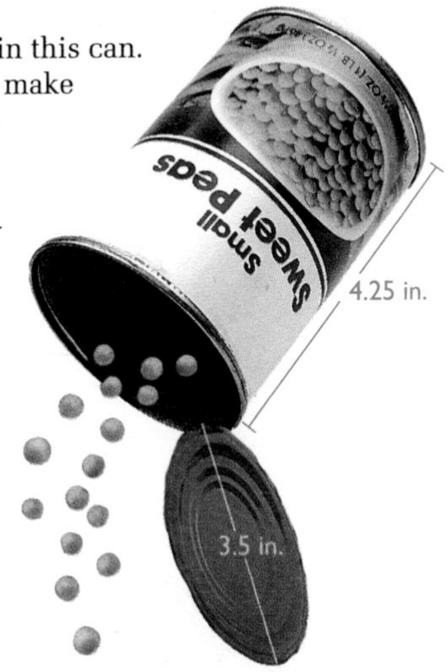

Sample Response

Problem Solving Strategy: Break the problem into parts

The can is a cylinder. Use the formula $S.A. = 2\pi r^2 + 2\pi rh$.

Step 1 Find the radius of the can.

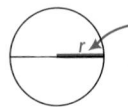

The radius of the can is half the diameter.

$$r = \frac{1}{2}(3.5) = 1.75$$

Step 2 Use the radius to find the surface area of the can.

$S.A. = 2\pi r^2 + 2\pi rh$ ⟵ Write the formula for the surface area of a cylinder.

$= 2\pi(1.75)^2 + 2\pi(1.75)(4.25)$ ⟵ Substitute 1.75 for r and 4.25 for h.

$\approx 19.24 + 46.73$

≈ 65.97

The company will need about 66 in.² of material.

4.25 in.

3.5 in.

▶ Now you are ready for:
Exs. 1–16 on pp. 512–513

Surface Area of a Pyramid

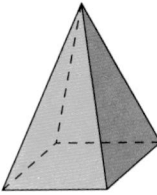

A **pyramid** is a space figure with only one base. A pyramid is identified by the shape of its base. Its faces are triangles.

In this book, you will work only with *regular pyramids*. A **regular pyramid** has these properties.

➤ The base is a regular polygon.

➤ All faces are congruent isosceles triangles.

9-5 Surface Area of Space Figures

509

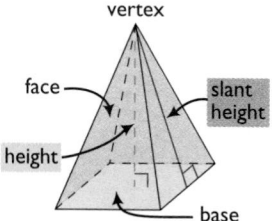

The **height** of a regular pyramid is the length of the perpendicular segment from the vertex to the center of the base.

The **slant height** of a regular pyramid is the height of one of the congruent triangular faces.

vertex · face · slant height · height · base

FORMULA: SURFACE AREA OF A PYRAMID

Surface Area = area of base + areas of faces

Talk it Over

7. These diagrams show a regular square pyramid being unfolded into a flat figure. Which dimension is given in the first figure, the *height* of the pyramid or the *slant height*?

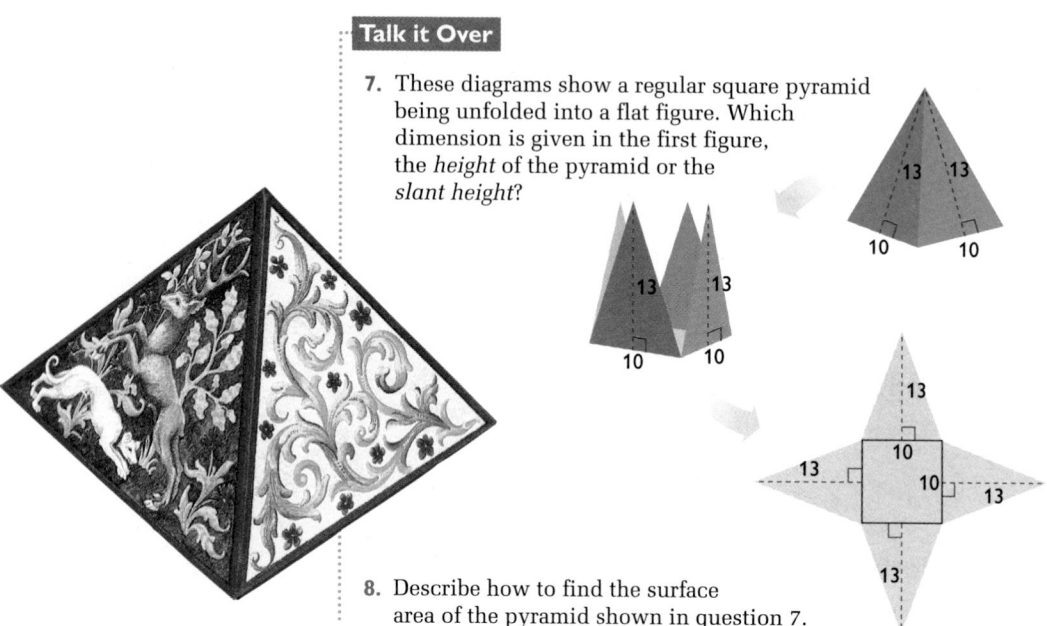

8. Describe how to find the surface area of the pyramid shown in question 7.

Sample 3

Find the surface area of this regular square pyramid.

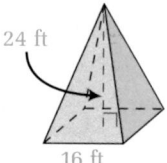

24 ft

16 ft

510   Unit 9 Reasoning and Measurement

Problem Solving Strategy: Break the problem into parts

To find the surface area, you need to know the slant height. The slant height is not given, but you can find it using the other dimensions.

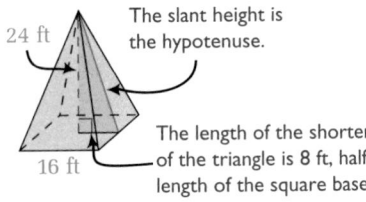

The slant height is the hypotenuse.

24 ft

16 ft

The length of the shorter leg of the triangle is 8 ft, half the length of the square base.

Step 1 Find the slant height.

Sketch the pyramid. Draw a right triangle inside the pyramid using the 24 ft height as one of the legs.

Use the Pythagorean theorem to find the slant height.

$c^2 = a^2 + b^2$

$c^2 = 24^2 + 8^2$ ⟵ Substitute 24 for a and 8 for b.

$c^2 = 640$

$c = \sqrt{640}$ ⟵ Find the positive square root.

$c \approx 25.30$

The slant height is about 25.30 ft.

Step 2 Find the surface area of the pyramid.

S.A. = area of square base + area of four triangular faces

$\text{S.A.} = s^2 + 4(\frac{1}{2}bh)$

$= 16^2 + 4(\frac{1}{2})(16)(25.30)$ ⟵ Substitute 16 for s and for b and 25.30 for h.

≈ 1065.6

The surface area of the regular square pyramid is about 1065.6 ft².

Look Back ⟵

What space figures are these? Tell how you would find the surface area of each figure.

·······▶ **Now you are ready for:**
Exs. 17–32 on pp. 513–514

9-5 Surface Area of Space Figures

511

Additional Sample

S3 Find the surface area of this regular square pyramid.

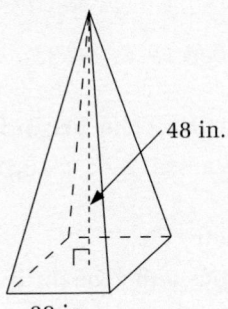

48 in.

20 in.

Step 1. **Find the slant height.**

$c^2 = a^2 + b^2$

$c^2 = 10^2 + 48^2$

$c^2 = 2404$

$c = \sqrt{2404}$

$c = 49.03$

The slant height is 49.03 in.

Step 2. **Find the surface area.**

$\text{S.A.} = 4(\frac{1}{2}bh) + s^2$

$\text{S.A.} = 4(\frac{1}{2})(20)(49.03) + 20^2$

$\text{S.A.} = 2361.2$

The surface area is 2361.2 in².

····················

Look Back

Tell students to record their responses to this activity in their journals for future reference. ···············●

Answers to Look Back ········:

cylinder: surface area = $2\pi r^2 + 2\pi rh$;

pyramid: surface area = $4(\frac{1}{2}bh) + s^2$;

prism: surface area = area of the two bases + area of the faces

511

9-5 Exercises and Problems

1. **Reading** What are two strategies for solving problems about surface area?

Identify each type of prism and find its surface area.

2.
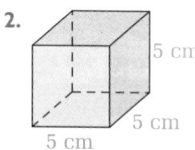
5 cm
5 cm
5 cm

3.
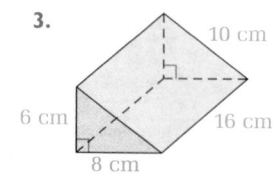
10 cm
6 cm
8 cm
16 cm

4.
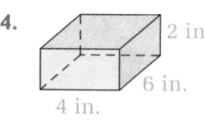
2 in.
6 in.
4 in.

5.
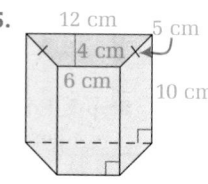
12 cm
5 cm
4 cm
6 cm
10 cm

connection to **ART**

6. Follow the steps to draw a pentagonal prism.

Draw two congruent pentagons.

Connect the corresponding vertices.

Dash the lines that cannot be seen.

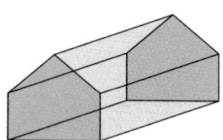

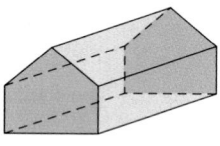

7. a. Follow the steps in Exercise 6 but draw two congruent triangles in the first step.

   b. Name the space figure you have drawn.

8. Follow the steps to draw a cylinder.

Draw two congruent ovals.

Join the ovals.

Dash the lines that cannot be seen.

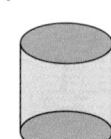

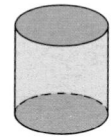

9. a. **Open-ended** Draw a building or a scene that includes at least one prism and one cylinder.

   b. Give the space figures in your drawing dimensions. Find their surface areas.

512 **Unit 9** Reasoning and Measurement

Answers to Exercises and Problems

1. make a sketch; break the problem into parts

2. cube; 150 cm^2

3. triangular prism; 432 cm^2

4. rectangular prism; 88 in.^{2}

5. trapezoidal prism; 352 cm^2

6. Drawings may vary.

7. a. Drawings may vary.

   b. triangular prism

8. Drawings may vary.

9. a, b. Answers may vary. Examples are given.

   a. toy barn and silo

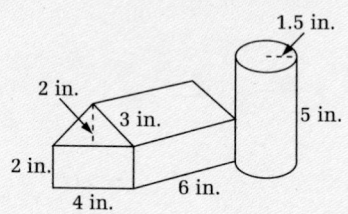
1.5 in.
2 in.
3 in.
5 in.
2 in.
4 in.
6 in.

   b. S.A. of barn = area of roof + area of building (including floor) = 44 + 64 = 108 in.^{2}; S.A. of silo = $2\pi r^2 + 2\pi rh$ = $2\pi(1.5)^2 + 2\pi(1.5)(5) \approx$ 14.14 + 47.12 ≈ 61.26 in.^{2}; S.A. of barn and silo is about 169.26 in.^{2}.

10. about 6 cans

10. Dana is painting the walls and ceiling of a rectangular classroom with dimensions 35 ft × 25 ft × 12 ft. Each can of paint covers 400 ft². About how many cans of paint will she need to paint the room?

11. **Writing** Describe how the two figures at the right are alike and how they are different.

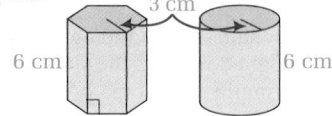
3 cm
6 cm   6 cm

Find the surface area of each cylinder or half cylinder.

12.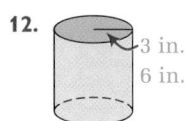
3 in.
6 in.

13.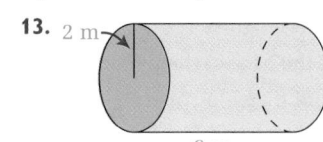
2 m
6 m

14.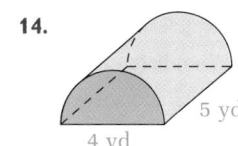
5 yd
4 yd

15. a. Draw two cylinders with the same radius. Make the height of one cylinder twice the height of the other cylinder.

  b. Do you think the surface area of the taller cylinder will be twice the surface area of the shorter cylinder? Check your conjecture by finding the surface area of each cylinder.

16. Jeremy Szabo wants to paint the sides and top of this mailbox. He has enough paint to cover about 2600 in.². Does he have enough paint for two coats?

half cylinder

CROSS RANCH

11 in.

30 in.

14 in.

Identify each space figure and find its surface area.

17.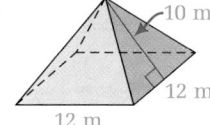
10 m
12 m
12 m

18.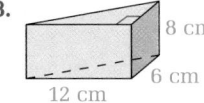
8 cm
6 cm
12 cm

19.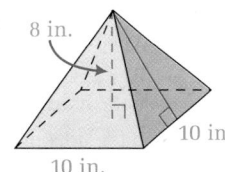
8 in.
10 in.
10 in.

Answers to Exercises and Problems

11. Descriptions may vary. An example is given. The two figures are alike because they are both space figures with congruent, parallel bases and have heights of 6 cm. They are different because the first figure is a hexagonal prism with a surface area of about 154.8 cm², while the second figure is a cylinder with a surface area of about 169.6 cm².

12. about 169.6 in.²

13. about 100.5 m²

14. about 64.0 yd²

15. a. Drawings may vary.

  b. Answers may vary. An example is given. No, the formula shows that doubling the height does not double the sur-face area: $2\pi r(2h) + 2\pi r^2 = 4\pi rh + 2\pi r^2$, not $2(2\pi rh + 2\pi r^2)$ or $4\pi rh + 4\pi r^2$.

16. No.

17. square pyramid; 384 m²

18. triangular prism; about 323.3 cm²

19. square pyramid; about 289 in.²

Communication: Writing

Ex. 11 allows students to compare and contrast a hexagonal prism with a cylinder. Students should consider what happens as the number of sides of the prism increases.

Reasoning

Ex. 15 should help students clear up a common error in reasoning, namely that doubling a dimension doubles the area or volume. You may wish to have students experiment further with the concepts presented in this exercise.

Practice 71 For use with Section 9-5

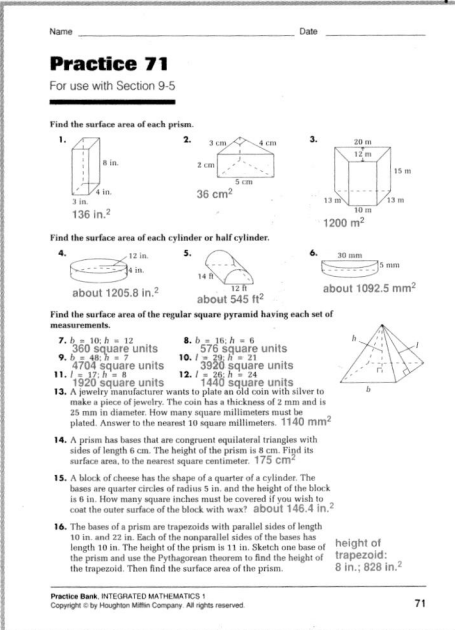

20. A *tetrahedron* is a regular pyramid whose base is a triangle. The base and faces of this tetrahedron are congruent equilateral triangles. Find its surface area.

21. **Writing** Jody draws a regular square pyramid and labels the edges of the base 6 in., the height 8 in., and the slant height 8 in. Pat claims that Jody's pyramid cannot really exist. Do you agree or disagree? Explain why.

22. The surface area of a regular square pyramid is 168 cm². The area of the base is 36 cm².

   a. Sketch the pyramid. Label the slant height x.

   b. What is the length of one edge of the base of the pyramid?

   c. Write and solve an equation to find the value of x.

23. **Using Manipulatives** This diagram is a pattern for a pyramid. Create a pattern for a regular square pyramid whose base has an area of 4 in.² and whose slant height is 3 in.

Ongoing ASSESSMENT

24. **Writing** Describe how the methods for finding the surface areas of prisms, cylinders, and pyramids are alike and how they are different.

Review PREVIEW

For Exercises 25–27, suppose a point on $\overline{XZ}$ is picked at random. Find the probability that the point is on each segment. *(Section 9-4)*

25. $\overline{XY}$ 26. $\overline{YZ}$ 27. $\overline{XZ}$

28. Sketch a trapezoid. What dimensions would you need to know to find the area of your trapezoid? *(Section 5-7)*

Find the volume of each box. *(Toolbox Skill 19)*

29. a box with length 5 in., width 7 in., and height 2 in.

30. a box with length 6 in., width 3 in., and height 3 in.

Working on the Unit Project

31. What type of space figure is your boat?

32. a. Make a sketch of your boat. Record its measurements in centimeters. (Use the conversion factor 1 in. = 2.54 cm if necessary.)

   b. Find the area of a triangular base of your boat.

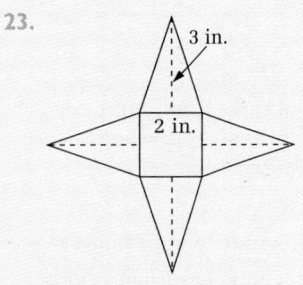
514

Volumes of Prisms and Cylinders

Focus
Find the volumes of prisms and cylinders.

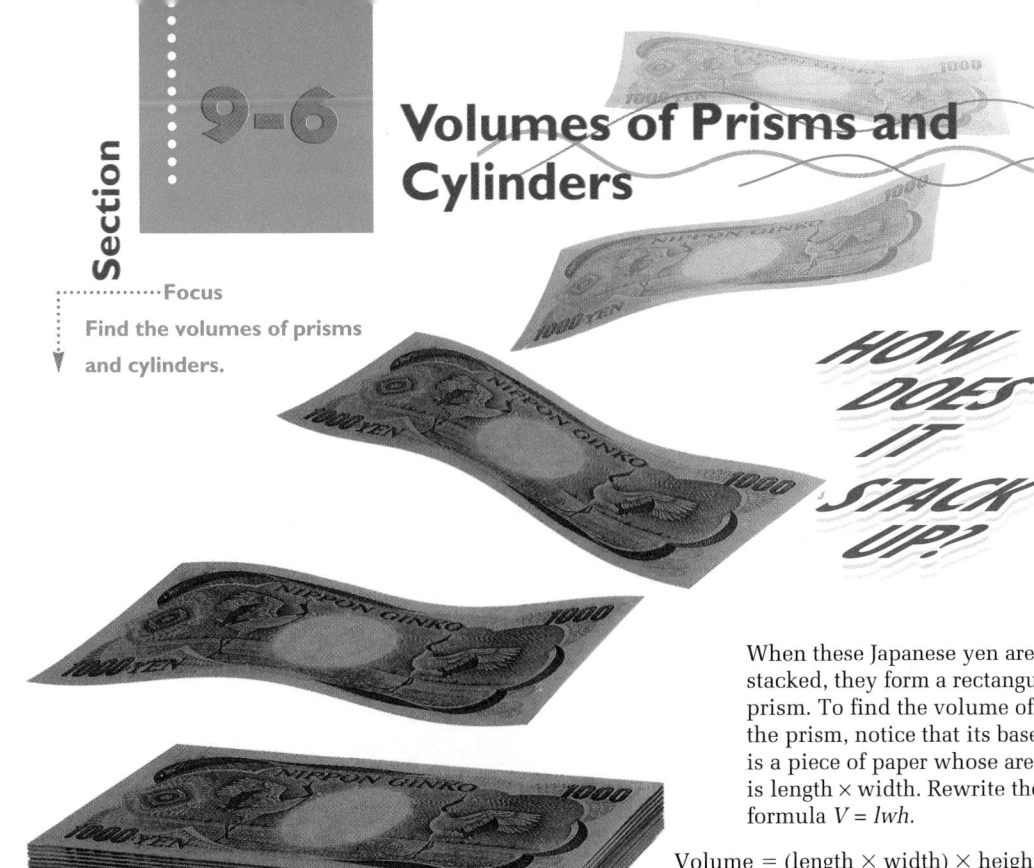

HOW DOES IT STACK UP?

When these Japanese yen are stacked, they form a rectangular prism. To find the volume of the prism, notice that its base is a piece of paper whose area is length × width. Rewrite the formula $V = lwh$.

Volume = (length × width) × height
= area of a base × height

You can think of the volume of any prism as the volume of layers of paper with the same shape as the base. For all prisms, the formula for finding the volume is the same.

FORMULA: VOLUME OF A PRISM

Volume of a prism = area of a base × height of the prism

$$V = Bh$$

Volume is measured in cubic units.

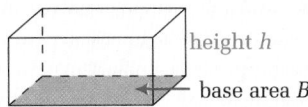

height h

base area B

Answers to Exercises and Problems

32. a. Sketches and measurements may vary. An example is given.

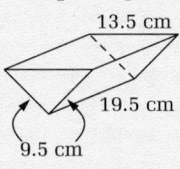

13.5 cm

19.5 cm

9.5 cm

b. Answers may vary. An example is given based on the diagram in part (a). Using $b = 13.5$ cm and $h = 6.7$ cm, the area of the triangular base is about 45.2 cm².

515

Communication: Discussion

When discussing the fact that the formula for finding the volume of a prism is the same for all prisms, remind students that the base can be any regular polygon; therefore, different formulas are used to calculate B, the area of the base.

Error Analysis

Some students may still think a prism must always rest on its base. This is not true. Remind these students that the bases of a prism are two congruent polygons lying in parallel planes.

Additional Sample

S1 Find the volume of the prism.

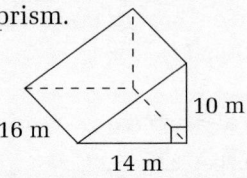
10 m
16 m
14 m

Step 1. Find the area of a base of the prism.

$B = \frac{1}{2}bh$

$B = \frac{1}{2}(14)(10)$

$B = 70$

The area of the base is 70 m².

Step 2. Find the volume of the prism.

$V = Bh$

$V = 70(16)$

$V = 1120$

The volume of the prism is 1120 m³.

Talk it Over

In question 1, students should understand that h represents the height of the trapezoid in one formula and the height of the prism in the other formula.

Sample 1

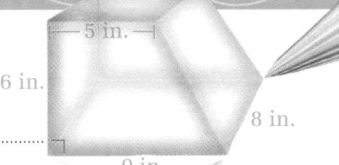
5 in.
6 in.
8 in.
9 in.

Find the volume of this prism.

Sample Response

Problem Solving Strategy: Break the problem into parts

Step 1 Find the area of a base of the prism.

A base of a prism is one of the two congruent, parallel polygons. A base of this prism is a trapezoid.

$b_1 = 5$
$h = 6$
$b_2 = 9$

$B = \frac{1}{2}(b_1 + b_2)h$ ← Write the formula for the area of a trapezoid.

$= \frac{1}{2}(5 + 9)(6)$ ← Substitute 5 for b_1, 9 for b_2, and 6 for h.

$= \frac{1}{2}(14)(6)$

$= 42$

The area of a base of the prism is 42 in.²

Step 2 Find the volume of the prism.

$V = Bh$ ← Write the formula.

$= (42)(8)$ ← Substitute 42 for B, the area of a base. Substitute 8 for h, the height of the prism.

$= 336$

The volume of the prism is 336 in.³

THE CLASSIC "A"
Capacity: 2
Floor size: 7 ft x 6 ft
Peak height: 4 ft
Weight: 8 lbs 2 oz

Unbeatable value! Our classic "A-Frame" style tent features mosquito-proof mesh doors on either end; reinforced seams; full-coverage fly for maximum storm protection. Easy assembly with five shock-corded aluminum poles. Fully waterproof. Color: loden.

Talk it Over

1. In Sample 1, the values substituted for h in the formulas $B = \frac{1}{2}(b_1 + b_2)h$ and $V = Bh$ are not the same. Explain.

2. Before Will buys the tent shown in this ad, he wants to make sure that it is as roomy as his old tent.

   a. The tent and its floor form a prism. What kind of prism?

   b. Explain how Will can find the volume of the tent.

Sample 2

The volume of this rectangular prism is 12.5 ft³. Find the missing dimension h.

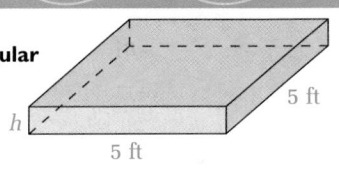
5 ft
h
5 ft

Unit 9 Reasoning and Measurement

Answers to Talk it Over

1. Explanations may vary. An example is given. In the formula $A = \frac{1}{2}(b_1 + b_2)h$, h represents the height of the trapezoidal base of the prism, which is 6 in. In the formula $V = Bh$, h represents the height of the prism, which is 8 in. Although h represents a height in each formula, it represents different heights, so the values substituted are not the same.

2. **a.** triangular prism

   b. First find the area of the triangular base using $A = \frac{1}{2}bh = \frac{1}{2}(6)(4) = 12$ ft². Next multiply its base area times the height: $V = Bh = 12 \cdot 7 = 84$ ft³.

Use what you know: the length, the width, and the volume of the prism.

$V = Bh$ ←———— Write the formula for the volume of a prism.

$V = 25h$ ←———— The base is a square. $B = 5 \cdot 5$.

$12.5 = 25h$ ←———— Substitute 12.5 for V.

$\dfrac{12.5}{25} = \dfrac{25h}{25}$ ←———— Divide both sides by 25.

$0.5 = h$

The height of the rectangular prism is 0.5 ft.

······▶ **Now you are ready for:**
Exs. 1–10 on p. 519

Volume of a Cylinder

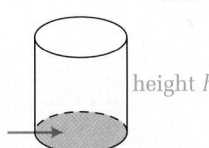

The pictures show a prism changing into a cylinder. As the number of faces increases, the prism looks more and more like a cylinder.

The formula $V = Bh$ can be used to find the volume of a cylinder.

FORMULA: VOLUME OF A CYLINDER

Volume of a cylinder = area of a base × height of the cylinder

$V = \mathbf{B}h$

$V = \pi r^2 h$

height h

base area $B = \pi r^2$ ———▶

Sample 3

Find the volume of this penny. Give your answer to the nearest hundredth.

$r = 0.374$ in.

$h = 0.062$ in.

Sample Response

Write the formula for the volume of a cylinder. ———▶

A penny is a cylinder.

$V = Bh$

$= \pi r^2 h$ ←———— The base is a circle with area πr^2.

$= (\pi)(0.374)^2(0.062)$ ←———— Substitute 0.374 for r and 0.062 for h.

≈ 0.03 ←———— $\boxed{\pi}\ \boxed{\times}\ 0.374\ \boxed{x^2}\ \boxed{\times}\ 0.062\ \boxed{=}$

The volume of a penny is about 0.03 in.³.

9-6 Volumes of Prisms and Cylinders

517

Additional Samples

S2 The volume of a rectangular prism is 86.4 m³. Find the missing dimension h.

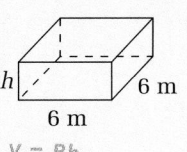

h ⟍ 6 m
6 m

$V = Bh$

$V = 36h$

$86.4 = 36h$

$\dfrac{86.4}{36} = \dfrac{36h}{36}$

$2.4 = h$

The height of the rectangular prism is 2.4 m.

S3 Find the volume of a quarter. Give your answer to the nearest hundredth.

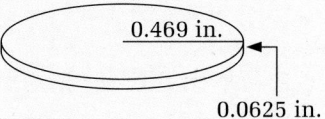

0.469 in.

0.0625 in.

$V = Bh$

$V = \pi r^2 h$

$V = \pi(0.469)^2(0.0625)$

$V \approx 0.04$

The volume of a quarter is about 0.04 in³.

518

Additional Sample

S4 Find the volume of this figure.

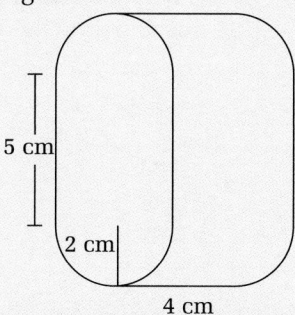

5 cm

2 cm

4 cm

The base is two half circles and a rectangle.

Step 1. Find the area of the base.

$B = \pi r^2 + lw$

$B = \pi(2)^2 + 5(4)$

$B \approx 32.6$

The area of the base is about 32.6 cm².

Step 2. Find the volume.

$V = Bh$

$V \approx 32.6(4)$

$V \approx 130.4$

The volume is about 130.4 cm³.

Talk it Over

Question 4 introduces the term *composite space figure.* You may wish to have students create some of their own composite space figures to help them better understand the concept.

Look Back

Use the Look Back for a class discussion to help students solidify their understanding of how to identify the bases of space figures.

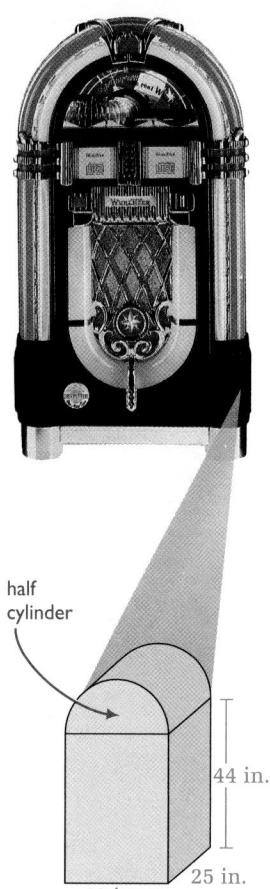

half cylinder

44 in.

32 in.

25 in.

► Now you are ready for:
Exs. 11–29 on pp. 520–522

Sample 4

The dimensions of this jukebox are shown below. Find the volume of the jukebox.

Sample Response

Problem Solving Strategy: Break the problem into parts

Step 1 Find the area of the base of the jukebox.

⊖ + ▭   The front of the jukebox is the base. It is made of a half circle and a rectangle.

$B = \frac{1}{2}$(area of the circle) + area of the rectangle

$= \frac{1}{2}\pi r^2 + lw$

$= \frac{1}{2}(\pi)(16)^2 + (44)(32)$ ◄── Substitute 16 for r, 44 for l, and 32 for w.

$\approx 402.1 + 1408$ ◄── 0.5 × π × 16 x² + 44 × 32 =

≈ 1810.1

The area of the base is about 1810.1 in.².

Step 2 Find the volume of the space figure.

$V = Bh$

$\approx (1810.1)(25)$ ◄── Substitute 1810.1 for B and 25 for h.

$\approx 45{,}252.5$ in.³

The volume of the jukebox is about 45,253 in.³.

Talk it Over

3. In Sample 4, how do you know that the 25 in. label gives the height of the space figure?

4. Figures like the one in Sample 4 are often called *composite space figures.* Why do you think this name is used?

Look Back ◄

The formula $V = Bh$ can be used to find the volume of many kinds of space figures. The shapes at the right are designs for notepads. All the pieces of paper in a pad will have the same shape. How can you decide which side of each pad to use as the base in the formula $V = Bh$?

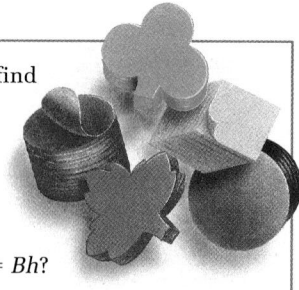

Unit 9 Reasoning and Measurement

Answers to Talk it Over

3. Answers may vary. An example is given. The bases of the space figure are the two sides which are congruent and parallel. The front and back are the bases of the jukebox. The distance between the bases is the height of the space figure. The distance between the front and back of the jukebox is 25 in. Therefore, the 25 in. label is the height.

4. Answers may vary. An example is given. The word *composite* is an adjective meaning "composed of" or "made out of." This means a space figure is made out of other space figures by putting them together. The jukebox is a space figure made out of a half cylinder and a rectangular prism.

Answers to Look Back

Answers may vary. An example is given. In order to use the formula $V = Bh$, the space figure must have two congruent, parallel sides that are the bases. Each piece of paper in a notepad has the same shape; that shape is the base of the space figure.

9-6 Exercises and Problems

1. **Reading** Is the prism in Sample 1 sitting on one of its two *bases*? How can you tell?

Find the volume of each prism.

2.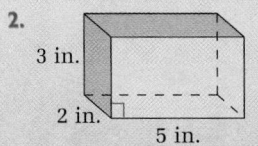
3 in.
2 in.
5 in.

3.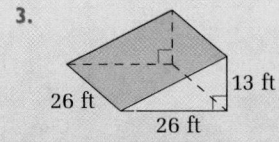
26 ft
13 ft
26 ft

4.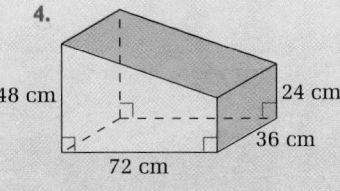
48 cm
24 cm
36 cm
72 cm

Find the missing dimension h.

5.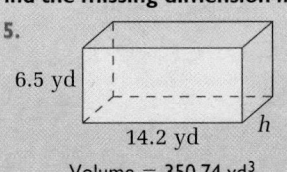
6.5 yd
14.2 yd
h
Volume = 350.74 yd³

6.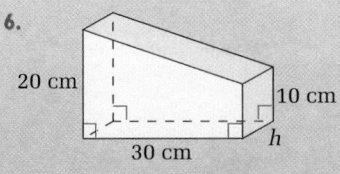
20 cm
10 cm
30 cm
h
Volume = 5400 cm³

7.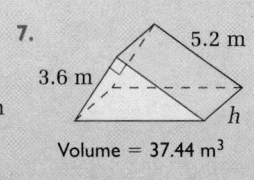
5.2 m
3.6 m
h
Volume = 37.44 m³

8. **Open-ended** Describe a situation in which a person would need to find surface area. Describe a situation in which a person would need to find volume.

9. Liz is buying fish for her aquarium. To avoid overcrowding the tank, she should buy one fish for every gallon of water. (There are 231 in.³ of water in one gallon.) How many fish can Liz have in her new aquarium if she fills it to a depth of 11 in.?

10. a. Make a table to record the volume of water in Liz's aquarium when the water is 1 in. deep, 2 in. deep, and 3 in. deep. Graph the volume of water as a function of depth.

   b. What is the slope of the line you graphed in part (a)? How is the slope related to the dimensions of the aquarium?

   c. Sam's aquarium has a 12 in. × 24 in. base. He is going to graph the volume of water as a function of depth for his aquarium. Predict whether the slope of the line he graphs will be *greater than* or *less than* the slope of the line you graphed in part (a).

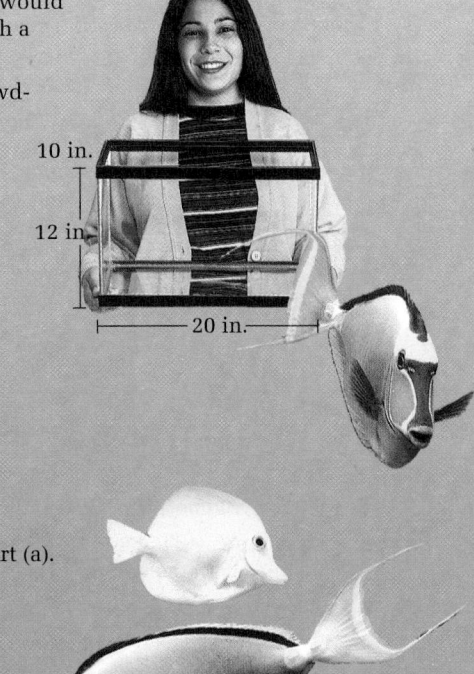
10 in.
12 in.
20 in.

9-6 Volumes of Prisms and Cylinders   519

APPLYING

Suggested Assignment

Standard 2–21, 23–29

Extended 1–21, 23–29

Integrating the Strands

Algebra Exs. 10, 28

Functions Ex. 10

Measurement Exs. 2–29

Geometry Exs. 1–29

Logic and Language Exs. 1, 8, 10, 17, 22, 29

Communication: Reading

Ex. 1 affords another opportunity for students to solidify their understanding of what the base of a prism is. For Exs. 2–7, you may wish to have students identify the bases of each figure before they actually do the calculations.

Integrating the Strands

Ex. 10 integrates the five strands of algebra, functions, measurement, geometry, and logic and language.

Answers to Exercises and Problems

1. No. The shape of the figure it is sitting on is not exactly that same shape as the side parallel to it.

2. 30 in.³

3. 4394 ft³

4. 93,312 cm³

5. 3.8 yd

6. 12 cm

7. 4 m

8. Answers may vary. Examples are given. If I want to paint a container I built to hold my tapes, I would have to find the surface area to know how much paint I need. I would have to find the volume of a sandbox to know how much sand it would hold.

9. 9 fish

10. a.

Depth of water (in.)	Volume (in.³)
1	200
2	400
3	600

See graph at right.

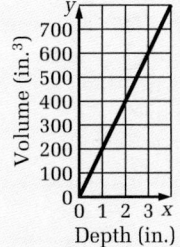
Depth (in.)

b. 200; The slope is equal to the area of the base.

c. greater; The slope of his line will be 288.

Find the volume of each space figure.

11. 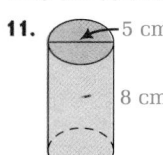 5 cm
8 cm

12. 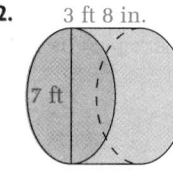 3 ft 8 in.
7 ft

13. 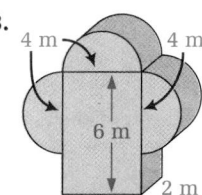 4 m   4 m
6 m
2 m

Find the missing dimension h.

14. 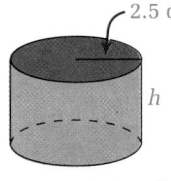 2.5 cm
h
Volume ≈ 58.9 cm³

15. 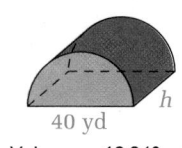 h
40 yd
Volume ≈ 18,840 yd³

16. 8 in.
h
4.3 in.
Volume ≈ 628.9 in.³

17. **Writing** Which will increase the volume of a cylinder more, doubling its *height* or doubling its *radius*? Explain.

18. A cake recipe calls for a 9 in. × 13 in. × 2 in. rectangular pan. A chef wants to use two identical cylindrical pans instead. Which of the sets shown would you recommend the chef use? Give a reason for your choice.

diameter: 9 in.
height: 1.75 in.

diameter: 8 in.
height: 2.5 in.

19. These two mugs have the same volume.

8 cm
9.25 cm
11.7 cm

a. Find the volume of the taller mug.

b. Use your answer to part (a) to help you write an equation about the volume of the shorter mug.

c. Solve the equation you wrote in part (b).

d. What is the diameter of the shorter mug?

20. The thickness of a new penny can be as small as 0.059 in. or as large as 0.067 in. Find the smallest and the largest possible volume of a new penny. Give your answers to the nearest thousandth.

> **BY THE WAY...**
>
> Bees know the importance of being math-wise! Their honeycombs are made of regular hexagonal prisms. This shape allows bees to store the most honey using the least amount of beeswax.

520

21. **Packaging** A food manufacturer needs to package 56 crackers in a box. Each cracker is about $\frac{1}{4}$ in. thick and has a diameter of about $2\frac{1}{4}$ in.

   a. If the crackers are packed in two cylindrical rows as shown, what are the approximate dimensions of the box? (Assume the crackers touch the top, bottom, and sides of the box.)

   b. Find the ratio of the surface area to the volume for the box.

   c. How else could the crackers be packaged? Experiment with other shapes and sizes of boxes. Can you design a box that has a smaller surface area to volume ratio?

Ongoing ASSESSMENT

22. **Group Activity** Work with another student.

   a. Collect data on the dimensions of different sizes and shapes of cans. Record your data in the first three rows of a table like the one shown.

   b. Complete the table by finding each can's surface area (S.A.), volume, and the $\frac{\text{surface area}}{\text{volume}}$ ratio.

   c. For which can is the surface area to volume ratio the smallest? the largest? Sketch these cans and label their dimensions.

   d. Compare your results with those of your classmates. How do size and shape affect the $\frac{\text{surface area}}{\text{volume}}$ ratio?

   e. What recommendations would you make to the manufacturers of the cans you measured?

TYPE OF CAN	?	?	?
Height	?	?	?
Diameter	?	?	?
S.A.	?	?	?
Volume	?	?	?
S.A. Volume	?	?	?

Review PREVIEW

Identify each space figure and find its surface area. (Section 9-5)

23.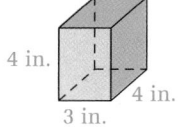
4 in.
4 in.
3 in.

24.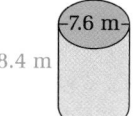
7.6 m
8.4 m

25.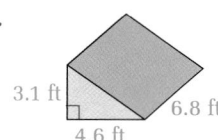
3.1 ft
6.8 ft
4.6 ft

26. Convert 72 km/h to m/s. (Section 7-5)

27. Find the height of this pyramid. (Sections 9-1, 9-5)

28. Two points on a line are (2, −3) and (−1, 5). Write an equation for the line. (Section 8-4)

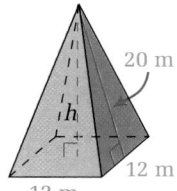
20 m
h
12 m
12 m

9-6 Volumes of Prisms and Cylinders

Answers to Exercises and Problems

22. Data and answers may vary. Examples are given.

a, b.

Type of can	Soup	Vegetable	Juice
Height	$3\frac{3}{4}$ in.	$4\frac{1}{8}$ in.	$3\frac{1}{2}$ in.
Diameter	$2\frac{1}{2}$ in.	3 in.	2 in.
S.A.	39.3 in.^{2}	53.0 in.^{2}	28.3 in.^{2}
Volume	18.4 in.^{3}	29.1 in.^{3}	11.0 in.^{3}
$\frac{\text{S.A.}}{\text{Volume}}$	2.1	1.8	2.6

c. vegetable; juice; Sketches may vary.

Tomato Juice
Peas

d. The smaller cans seem to have a greater surface area to volume ratio than the larger cans.

Assessment: Performance

Ex. 22 provides assessment of students' understanding of both volume and surface area, as well as of ratio. The performance of each group can be judged through a class discussion of results and recommendations. Allow groups to compare their conclusions and share their recommendations.

e. The results indicate that the packaging of larger quantities is less expensive per unit volume than the packaging of smaller quantities. I might recommend that manufacturers package their products in larger cans. This would lower packaging costs and increase profits.

23. rectangular prism; 80 in.^{2}

24. cylinder; about 291.3 m^2

25. triangular prism; about 104.3 ft^2

26. 20 m/s

27. about 19.1 m

28. $y = -\frac{8}{3}x + \frac{7}{3}$

521

Working on the Unit Project

Imagine your boat floating in water. Part of the boat will be below the water level. Here are two scientific facts that you may have learned in your research on buoyancy and water.

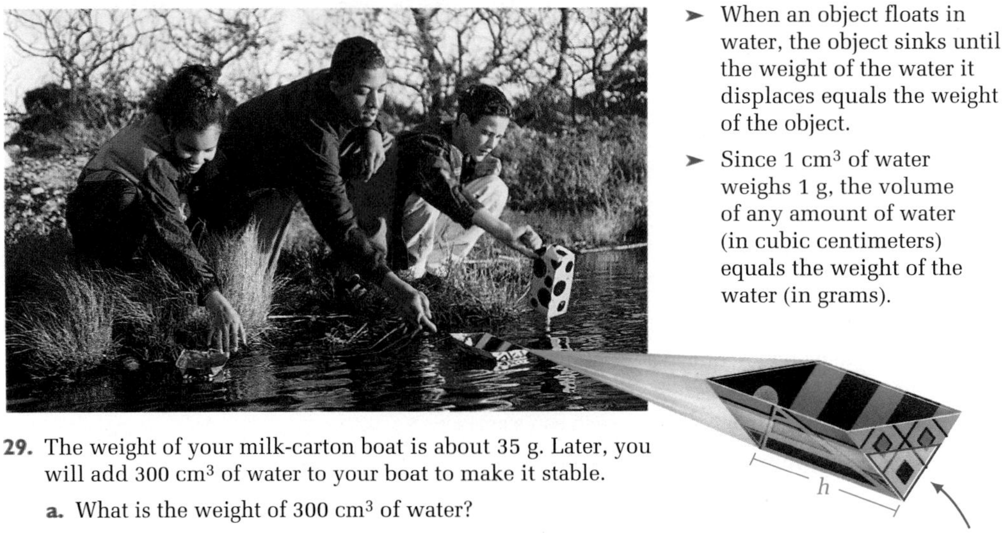

➤ When an object floats in water, the object sinks until the weight of the water it displaces equals the weight of the object.

➤ Since 1 cm³ of water weighs 1 g, the volume of any amount of water (in cubic centimeters) equals the weight of the water (in grams).

29. The weight of your milk-carton boat is about 35 g. Later, you will add 300 cm³ of water to your boat to make it stable.

a. What is the weight of 300 cm³ of water?

b. What will be the total weight of your boat with the added water?

c. What will be the weight of the water your boat will displace when it floats?

d. What volume of water will your boat displace?

e. When you float your boat, the part under water will be shaped like a triangular prism as shown above. Complete this equation: $Bh = \underline{\ ?\ }$.

f. Find B for your boat.

B = area of part of base that is under water

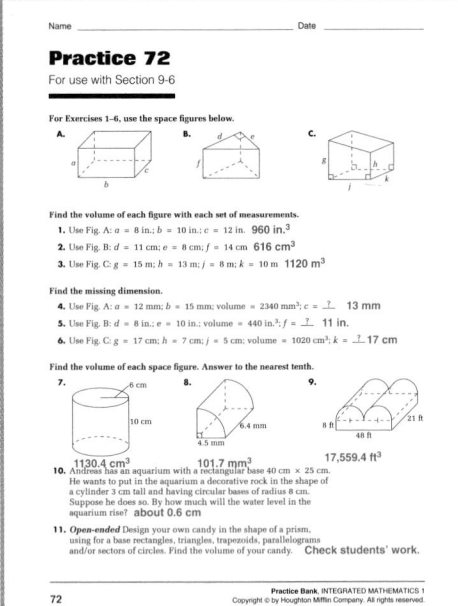

$Unit\ 9$ **CHECKPOINT 2**

1. **Writing** Describe how to find the surface area of this space figure.

14 yd   14 yd
12 yd   7.5 yd

9-4

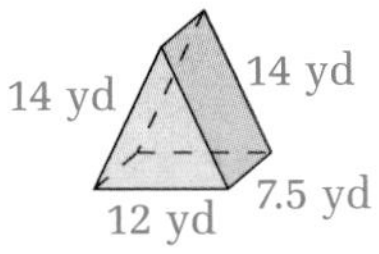

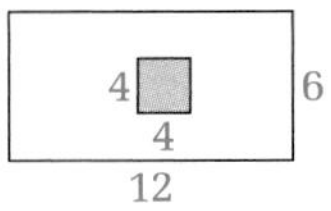

2. Suppose a point in the rectangle at the left is picked at random. What is the probability that the point is in the shaded region?

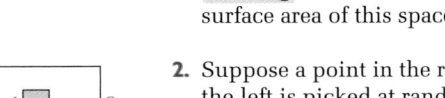
4 [ ] 6
4
12

Find the surface area and volume of each space figure. 9-5, 9-6

3. a regular square pyramid with base edge length 20 mm and height 18 mm (Find surface area only.)

4. a cylinder with diameter 3.25 in. and height 1.5 in.

Unit 9 Reasoning and Measurement

Answers to Exercises and Problems

29. a. 300 g

b. 335 g

c. 335 g

d. 335 cm³

e. 335 cm³

f. Answers may vary. Example: about 17.2 cm²

Answers to Checkpoint

1. Descriptions may vary. An example is given. The space figure is a triangular prism. To find the area of the bases you need to find the height of the triangles: $h = \sqrt{14^2 - 6^2} \approx 12.6$ yd. The area of the bases is $2\left(\frac{1}{2} \cdot 12 \cdot 12.6\right) = 151.2$.
   Find the area of the faces: $7.5 \cdot 14 = 105$;

$7.5 \cdot 14 = 105$;
$7.5 \cdot 12 = 90$. Add the areas to find the surface area: $151.2 + 105 + 105 + 90 = 451.2$ yd².

2. $\frac{2}{9}$, or about 0.2

3. about 1224 mm²

4. about 31.9 in.²; about 12.4 in.³

9-7 Volumes of Pyramids and Cones

Focus
Find the volumes
of pyramids and
cones.

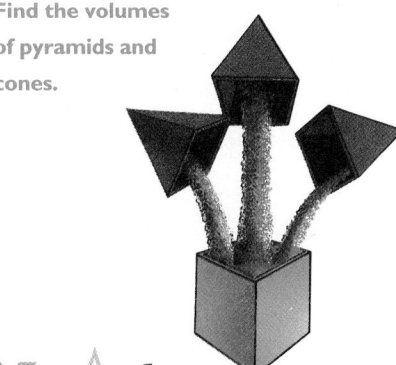

Fill 'er Up

Talk it Over

Does it surprise you to know that the total volume of three identical pyramids equals the volume of one prism with the same base area and height? You will test this idea in the exercises.

1. What is the formula for the volume of a prism?

2. Suppose you know the volume of the green prism. How can you find the volume of one of the red pyramids?

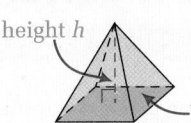

FORMULA: VOLUME OF A PYRAMID

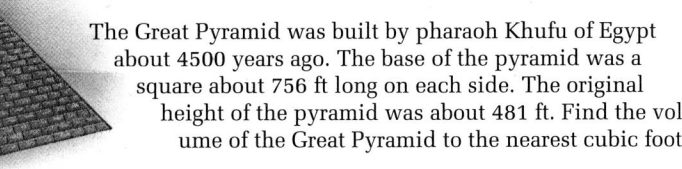

height h ... base area B

Volume of a pyramid = $\frac{1}{3}$ × area of base × height of the pyramid

$$V = \frac{1}{3}Bh$$

Sample 1

The Great Pyramid was built by pharaoh Khufu of Egypt about 4500 years ago. The base of the pyramid was a square about 756 ft long on each side. The original height of the pyramid was about 481 ft. Find the volume of the Great Pyramid to the nearest cubic foot.

Sample Response

$V = \frac{1}{3}Bh$ ← Write the formula for volume of a pyramid.

$\approx \frac{1}{3}(756^2)(481)$ ← The area of the square base is 756 · 756.
Substitute 756^2 for B and 481 for h.

$\approx 91{,}636{,}272$

The volume of the pyramid was about 91,636,272 ft³.

9-7 Volumes of Pyramids and Cones　　　**523**

Answers to Talk it Over

1. $V = Bh$

2. Answers may vary. An example is given. Divide the volume of the prism by 3.

PLANNING

Objectives and Strands
See pages 474A and 474B.

Spiral Learning
See page 474B.

Materials List
➤ Stiff paper
➤ Ruler
➤ Scissors
➤ Masking tape
➤ Sand or uncooked rice
➤ Graph paper
➤ Protractor
➤ Metric ruler

Recommended Pacing
Section 9-7 is a one-day lesson.

Extra Practice
See pages 635–638.

Warm-Up Exercises
Warm-Up Transparency 9-7

Support Materials
➤ Practice 73
➤ Enrichment 65 in the Activity Bank
➤ Study Guide 9-7
➤ Problem Set 21
➤ Additional Exploration 10
➤ Diagram Master 2 in the Explorations Lab Manual
➤ Quiz 9-7

Questions 1 and 2 provide an introduction to the formula for finding the volume of a pyramid and an intuitive conceptualization of its meaning.

Additional Sample

S1 Find the volume of this pyramid.

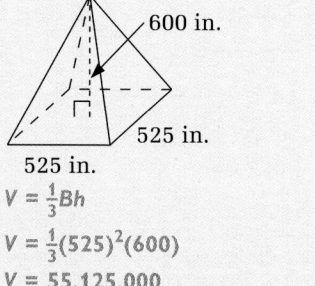

600 in.

525 in.

525 in.

$V = \frac{1}{3}Bh$

$V = \frac{1}{3}(525)^2(600)$

$V = 55,125,000$

The volume is 55,125,000 in.^{3}.

Communication: Drawing

To illustrate the hexagonal base of a pyramid changing to a circle, students can draw a circle, mark off and draw six equal sides, bisect each arc and draw the polygon by connecting the points of bisection, and then bisect each arc again and draw another polygon.

Additional Sample

S2 Find the volume of the cone below.

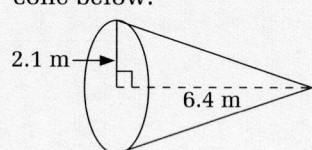

2.1 m

6.4 m

$V = \frac{1}{3}Bh$

$V = \frac{1}{3}\pi r^2 h$

$V = \frac{1}{3}\pi (2.1)^2(6.4)$

$V \approx 29.6$

The volume of the cone is about 29.6 m^3.

524

Volume of a Cone

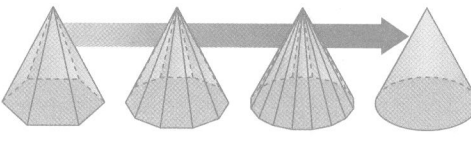

A **cone** is a space figure with one circular base and a vertex that is directly above the center of the base. The pictures show a pyramid changing into a cone. Notice that as the number of faces increases, the base of the pyramid looks more like a circle.

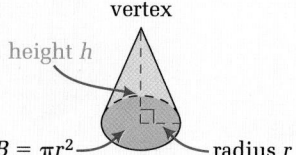

The formula $V = \frac{1}{3}Bh$ can be used to find the volume of a cone.

FORMULA: VOLUME OF A CONE

Volume of a cone $= \frac{1}{3} \times$ **area of base** $\times$ **height of the cone**

$$V = \frac{1}{3}Bh$$

$$V = \frac{1}{3}\pi r^2 h$$

vertex

height h

base area $B = \pi r^2$ — radius r

Sample 2

11 yd

12 yd

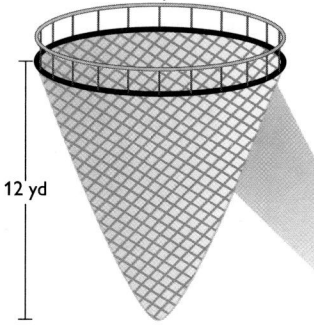

Aquaculture is the science of raising marine fish and shellfish for food. Some farmers raise salmon in large cone-shaped pens that are under water. Find the volume of the pen shown.

Unit 9 Reasoning and Measurement

Use the formula for the volume of a cone.

$$V = \frac{1}{3}Bh$$

$$= \frac{1}{3}\pi r^2 h \quad \longleftarrow \quad \text{The base is a circle with area } \pi r^2.$$

$$= (\frac{1}{3})(\pi)(5.5)^2(12) \quad \longleftarrow \quad \text{The radius is half of 11. Substitute } 5.5 \text{ for } r \text{ and } 12 \text{ for } h.$$

$$\approx 380.1 \quad \longleftarrow \quad 1 \div 3 \times \boxed{\pi} \times 5.5 \boxed{x^2} \times 12 =$$

The volume of the cone-shaped pen is about 380.1 yd³.

Talk it Over

The cylinder and cone at the right have the same base and height.

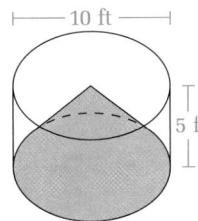

10 ft

5 ft

3. How are the volumes of the cylinder and cone related?

4. What is the volume of the cone?

Look Back ◄────

Describe how the methods for finding the volume of a pyramid and the volume of a cone are alike and how they are different.

9-7 Exercises and Problems

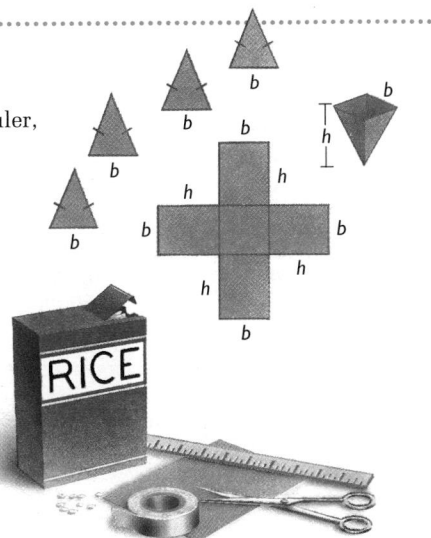

1. **Using Manipulatives** You will need stiff paper, a ruler, scissors, masking tape, and sand or uncooked rice.

   a. Cut out four identical isosceles triangles. Tape the edges of the four triangles together to build a pyramid with an open base.

   b. Make a paper pattern for an open box that is a rectangular prism. Your prism should have the same base area and height as your pyramid. Cut out your prism and tape it together.

   c. Fill your pyramid with sand or rice. Pour the contents of the pyramid into the prism. How many times do you need to fill the pyramid to fill the prism completely?

   d. What formula does your experiment support?

9-7 Volumes of Pyramids and Cones

525

After students discuss questions 3 and 4, relate the idea that the volume of a cone is one third the volume of a cylinder having the same base and height to the fact that the volume of a pyramid is one third the volume of a prism with the same base and height. These connections can help students to remember the formulas and also to understand the geometric relationships involved.

Look Back

After students describe their methods verbally, have them enter a written description in their journals.

APPLYING

Suggested Assignment

Standard 2–9, 14, 15, 17–25

Extended 2–11, 13–15, 17–25

Integrating the Strands

Algebra Exs. 8, 22

Measurement Exs. 1–6, 8–11, 13–18, 25

Geometry Exs. 1–21, 23, 24

Logic and Language Exs. 1, 2, 8, 12, 16, 23–25

Answers to Talk it Over

3. The volume of the cone is $\frac{1}{3}$ the volume of the cylinder.

4. about 130.9 ft³

Answers to Look Back

Descriptions may vary. An example is given. The methods for finding the volume of a pyramid and the volume of a cone are alike because both require multiplying $\frac{1}{3}$ times the area of the base times the height. The methods are different because the formulas used to find the areas of the bases are not the same.

Answers to Exercises and Problems

1. a, b. Constructions may vary.

   c. 3

   d. $V = \frac{1}{3}Bh$ or $V = \frac{1}{3}lwh$

2. **Reading** What does B stand for in the formula for the volume of a pyramid, $V = \frac{1}{3}Bh$?

Find the volume of each space figure.

3.
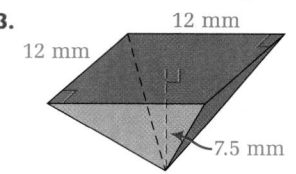

4.
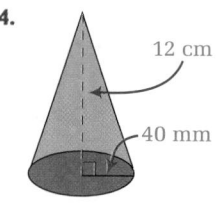

5.
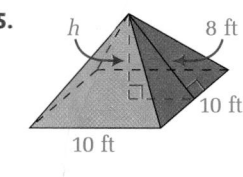

6. **a.** Predict which of these two cone-shaped paper cups will hold the most water.

   b. Find the volume of each cone to test your prediction.

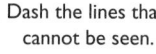

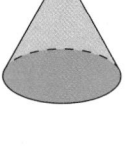

7. **Drawing** Follow the steps to draw a cone.

   Draw an oval and a point over the center.

   Join the point to the oval.

   Dash the lines that cannot be seen.

8. **a.** Draw two cones that have the same base, but one cone is twice as tall as the other. Record the diameter and height of each cone. Make a prediction about the relationship between the volumes of the cones. Then check your prediction.

   b. Draw two cones that have the same height, but the diameter of one cone is twice the diameter of the other cone. Record the diameter and height of each cone. Make a prediction about the relationship between the volumes of the cones. Then check your prediction.

   c. **Writing** Which has a greater effect on the volume of a cone, changing the *height* or changing the *diameter*? Explain your answer.

9. The Great Pyramid is the largest of the Pyramids of Giza. The second largest pyramid was built for pharoah Khafre. The base of Khafre's pyramid was a square about 708 ft long on each side. The height was about 471 ft. Find the volume of Khafre's pyramid.

> **BY THE WAY...**
>
> Ancient Egyptians made their homes out of mud but used stones to build their tombs. The pyramids were made to last forever.

Answers to Exercises and Problems ···

2. B stands for the area of the base, lw.

3. 360 mm³

4. about 201 cm³

5. about 208.17 ft³

6. **a.** the cone on the left

   b. Volume of the cone on the left is about 9.42 in.³; volume of the cone on the right is about 8.42 in.³.

7, 8. Drawings may vary.

8. **a.** Predictions may vary. The volume of the taller cone, with twice the height, is twice the volume of the shorter cone.

   b. Predictions may vary. The volume of the wider cone, with twice the diameter, is 4 times the volume of the narrower cone.

c. diameter; Explanations may vary. An example is given. The volume of a cone is $V = \frac{1}{3}\pi r^2 h$.

Doubling the height is the same as multiplying the formula by 2, which doubles the volume $[V = \frac{1}{3}\pi r^2(2h)]$.

Doubling the diameter also doubles the radius.

connection to HISTORY

For hundreds of years Native American groups of the Great Plains made their tipis from buffalo hides sewn together and stretched over wooden frames.

This tipi is one of many model tipis made by members of the Kiowa and Kiowa-Apache tribes between 1891 and 1904.

10. Although tipis are not always true cones, you can use a cone to approximate the shape of a tipi.

   a. Find the approximate volume of the model tipi.

   b. When the tipi is spread flat, it almost forms a half circle. Use this fact to estimate the surface area of the model tipi. (*Hint:* First find the tipi's slant height.)

   c. The surface area you found in part (b) does not include the floor. Find the approximate area of the floor of the model tipi.

   d. Suppose the artist who made the model tipi used a scale of 1 in. : 0.7 ft. Find the approximate height and diameter of the full-size tipi. Then find its volume, its surface area, and the area of its floor.

11. **a.** Find the volume, the surface area, and the area of the floor of each of these tents.

   b. Compare your answers from part (a) to your answers from part (d) of Exercise 10. What advantage does the tipi have over the tents?

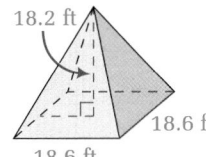

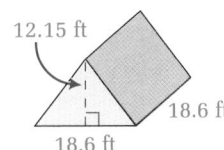

12. **Research** Find out how the Kiowa and Kiowa-Apache peoples, or another group who used the tipi, lived during the 1800s and earlier. Describe some of the advantages that the tipi had for their way of life.

9-7 Volumes of Pyramids and Cones

527

Answers to Exercises and Problems

Since the radius is squared, doubling it is the same as multiplying the formula by 2^2 or 4, which means the volume is multiplied by a factor of 4 $[V = \frac{1}{3}\pi(2r)^2 h].$

9. about 78,698,448 ft³

10. **a.** about 6123 in.³

   b. about 1413 in.²

   c. about 706.5 in.²

   d. height ≈ 18.2 ft, diameter ≈ 21 ft, $V ≈ 2100.2$ ft³, S.A. ≈ 692.4 ft², floor area ≈ 346.2 ft²

11. **a.** left tent: $V ≈ 2098.8$ ft³, S.A. ≈ 758.9 ft², floor area ≈ 346 ft²; right tent: $V ≈ 2101.7$ ft³, S.A. ≈ 795.2 ft², floor area ≈ 346 ft²

   b. A tipi has the advantage of using less material (because of smaller surface area) than a tent that will give you about the same volume and floor area.

12. Research and answers may vary.

527

Application

Exercises such as 14 and 15 can be used to make students aware of the idea that there is *wasted* space when an object is placed inside of another having a different shape. This idea can be related to the practical problem of *packing* products of any kind by a manufacturer for shipment to a retail store or an individual customer. Packing materials cost a business money, and therefore it is advantageous to reduce the amount of material used to reduce costs. You might wish to broach this topic with students and ask them in what type (shape) of container they would ship square, round, rectangular, cylindrical, or cone-shaped objects.

Assessment: Task

The activity in Ex. 16 allows students to creatively demonstrate an important concept. As a summation of the task, call upon one or more groups to carry out their demonstration for the entire class.

13. **a.** What trigonometric ratio can you use to find the height of this cone?
 b. Find the height of the cone.
 c. Find the volume of the cone.

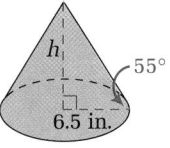

14. For a math project, a student cuts pyramid-shaped and cone-shaped blocks from cube-shaped wooden blocks.

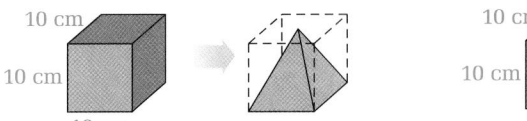

 a. How much wood will *not* be used to make the pyramid-shaped block?
 b. How much wood will *not* be used to make the cone-shaped block?

15. **a.** Find the volume of the regular square pyramid at the right.
 b. Suppose a cone is put inside the pyramid as shown. Find the volume of the cone.
 c. The cone's volume is what percent of the pyramid's volume?

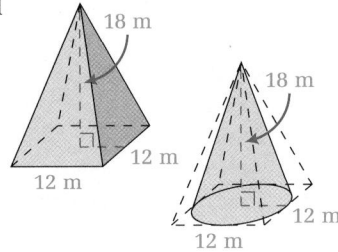

Ongoing ASSESSMENT

16. **Group Activity** Work in a group of four people. Design and carry out a demonstration to show that the volume of a cone is one third of the volume of a cylinder with the same base area and height. Describe the work done by your group.

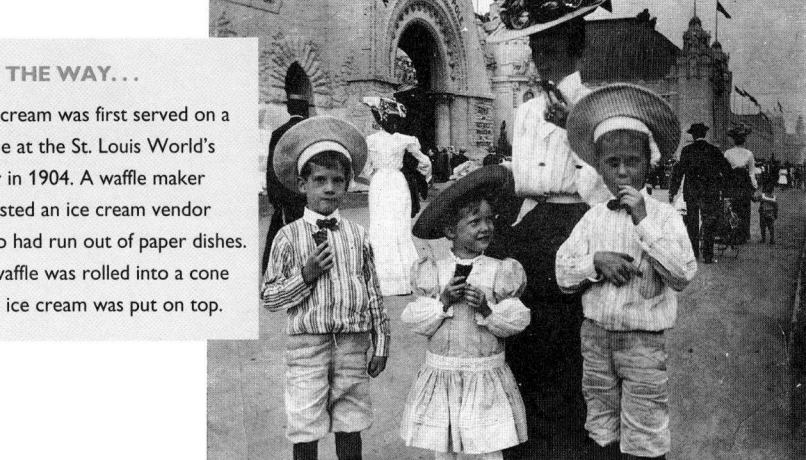

BY THE WAY...

Ice cream was first served on a cone at the St. Louis World's Fair in 1904. A waffle maker assisted an ice cream vendor who had run out of paper dishes. A waffle was rolled into a cone and ice cream was put on top.

Answers to Exercises and Problems

13. **a.** tan 55°
 b. $h \approx 9.28$ in.
 c. $V \approx 410.72$ in.^{3}
14. **a.** $\frac{2}{3}$ of the wood, or about 667 cm^3
 b. about 738 cm^3
15. **a.** 864 m^3
 b. about 678.6 m^3
 c. about 78.5%
16. Demonstrations may vary. An example is given. Use an empty can as the cylinder. Shape a cone out of paper so that it has the same base and height as the can. Tape the cone to form a container. Fill the cone with beans or rice, and then empty it into the can. Repeat three times.
17. cylinder; about 282.7 ft^3
18. triangular prism; about 17.1 m^3
19. 7.5
20. 15
21. 4:3
22. 2.5, −2.5
23. **a–c.** Drawings may vary. The measures of the angles are equal.

Identify each space figure and find its volume. *(Section 9-6)*

17.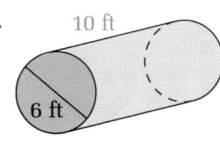
10 ft

6 ft

18.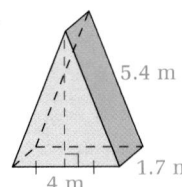
5.4 m

4 m   1.7 m

Polygon JKLMN ~ polygon PQRST. Find each measure. *(Section 6-5)*

19. *KL*   20. *SR*

21. Polygon *PQRST* is a dilation of polygon *JKLMN*. What is the scale factor? *(Section 6-6)*

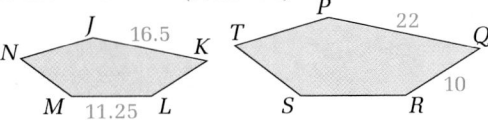

22. Solve the proportion $\frac{x^2}{4} = \frac{25}{16}$. *(Sections 6-5, 9-2)*

 Working on the Unit Project

23. **a.** Draw two parallel lines and a third line that crosses them. Label the angles as shown. Measure ∠1 and ∠2. What do you notice?

   b. Draw another pair of parallel lines with a third line that crosses them. Draw the third line at a different angle than you used above. Find and measure two angles that are in corresponding positions the way ∠1 and ∠2 are in this diagram. What do you notice?

   c. Repeat part (b) several more times.

   d. Make a conjecture about the measures of angles in corresponding positions when a line crosses two parallel lines.

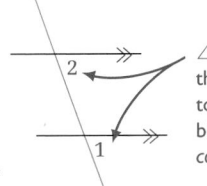
∠1 and ∠2 are *below* the parallel lines and to the *right* of the blue line. They are in *corresponding positions.*

24. The diagram below shows one triangular base of your boat. When your boat is in the water, a portion of the base (△*XYT*) will be under water. As long as the boat is level, $\overline{XY}$ will be parallel to $\overline{RS}$.

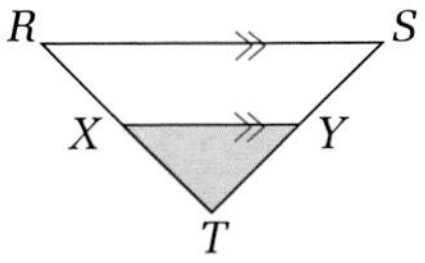

   a. How is $\overline{RT}$ like the blue line in the diagram in Exercise 23?

   b. Using the conjecture that you made in part (d) of Exercise 23, what can you say about ∠*TXY* and ∠*R*?

   c. What can you say about ∠*XTY* and ∠*RTS*? Why?

   d. What two triangles in the diagram are similar? How do you know?

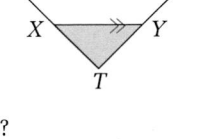
R   S
X   Y
T

25. Use a ruler to mark a centimeter scale along the height of one triangular base of your boat. Include a scale mark at least every 5 mm. Describe how you can use the scale to find the *draft* of your boat (see page 474) when it is in the water.

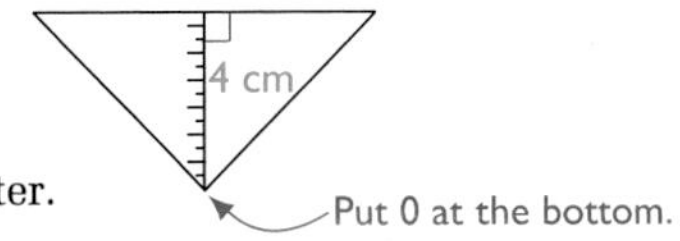

4 cm

Put 0 at the bottom.

Working on the Unit Project
Before students begin Ex. 23, ask them to take time to read it, understand what they are to do, and plan how they will record their work so they can make and explain their conjectures.

Practice 73 For use with Section 9-7

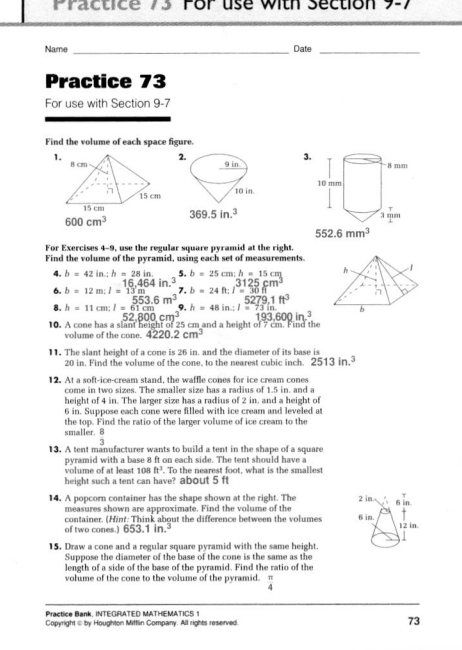

Answers to Exercises and Problems

23. **d.** Conjectures may vary. An example is given. When a line crosses two parallel lines, the measures of angles in corresponding positions will be the same.

24. **a.** It is a line that crosses two parallel lines.

   b. Their measures are equal.

   c. Their measures are equal because they are the same angle. The length of the sides of an angle does not affect its measure.

   d. △*XYT* and △*RST*; Triangles that have two pairs of congruent corresponding angles are similar.

25. The draft is how much of the boat is under water. The centimeter reading at the water line tells you how many centimeters of the boat is under water.

529

Objectives and Strands
See pages 474A and 474B.

Spiral Learning
See page 474B.

Materials List
➤ Graph paper

Recommended Pacing
Section 9-8 is a two-day lesson.
Day 1
Pages 530–531: Talk it Over 1 through Talk it Over 8, *Exercises 1–11*
Day 2
Pages 532–533: Volumes of Similar Space Figures through Look Back, *Exercises 12–33*

Extra Practice
See pages 635–638.

Warm-Up Exercises
Warm-Up Transparency 9-8

Support Materials
➤ Practice 74
➤ Enrichment 66 in the Activity Bank
➤ Study Guide 9-8
➤ Problem Set 21
➤ Diagram Master 2 in the Explorations Lab Manual
➤ Quiz 9-8
➤ Test 38
➤ Alternative Assessment 6

Section 9-8

Similar Figures: Area and Volume

Mountains
AND MOLEHILLS

Focus
Use ratios of corresponding lengths, areas, and volumes in similar figures.

Talk it Over

Do you remember this misleading graph from Unit 3? The ratio of the *number* of cable systems in 1990 to the *number* of cable systems in 1970 is about 4 to 1.

Cable Systems in the United States, 1970–1990

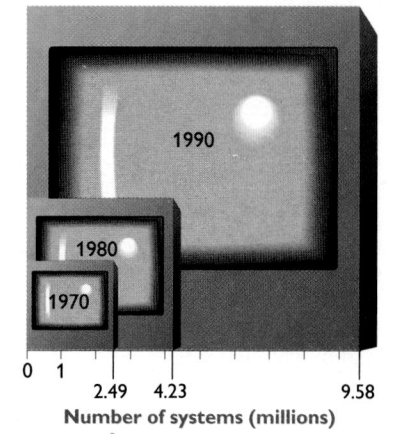

0   1   2.49   4.23   9.58
Number of systems (millions)

1. The fronts of the TV sets are similar rectangles. Do the *areas* of the fronts of the TV sets for 1990 and 1970 look as if they are in the ratio 4 to 1? Why or why not?

2. The TV sets are space figures with the same shape. Does it look as if the *volumes* of the TV sets for 1990 and 1970 are in the ratio 4 to 1? Why or why not?

3. Which part of this graph gives you the correct impression of the growth in the number of cable systems from 1970 to 1990?

The triangles at the left are similar.

4. What is the ratio of corresponding lengths of the figures?

5. What is the ratio of the areas of the figures?

6. Is the ratio you found for question 5 equivalent to $3^2:4^2$?

The ratio of corresponding lengths of two similar figures is always the same. The ratio of the areas of the similar figures is not the same as the ratio of corresponding lengths. If you look at several pairs of similar figures, you will see a pattern.

X ⌐ab

AREAS OF SIMILAR FIGURES

If the ratio of corresponding lengths of two similar figures is $a:b$, then the ratio of their areas is $a^2:b^2$.

Answers to Talk it Over

1. No. Reasons may vary. An example is given. You would need more than 4 fronts of TV sets of the size representing 1970 to cover the area of the front of the TV set representing 1990.

2. No. Reasons may vary. An example is given. You could fit more than 4 TV sets of the size representing 1970 into the TV set representing 1990.

3. Answers may vary. An example is given. The heights of the TV sets appear to be in the same ratio as the number of cable systems for each year.

4. $\frac{3}{4}$

5. $\frac{9}{16}$

6. Yes.

Sample 1

The ratio of the lengths of the corresponding sides of two similar pentagons is 8:5. What is the ratio of their areas?

Sample Response

The ratio of their areas is

$(8^2):(5^2)$, or $64:25$, or $\frac{64}{25}$. ◀ ∘ ∘ ∘ ∘ ∘ ∘

> **Watch Out!**
> The pentagons do not necessarily have areas of 64 and 25. Their areas might be 128 and 50.

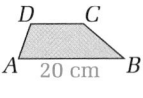

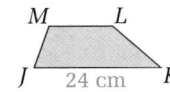

Sample 2

Trapezoid *ABCD* ~ trapezoid *JKLM*. The area of trapezoid *ABCD* is 150 cm². What is the area of trapezoid *JKLM*?

Sample Response

Problem Solving Strategy: Break the problem into parts

Step 1 Find the ratio of the lengths of corresponding sides $\overline{AB}$ and $\overline{JK}$.

$$\frac{AB}{JK} = \frac{20}{24} = \frac{5}{6} \longleftarrow \text{Write the ratio in its simplest form.}$$

Step 2 Set up a proportion.

Let x = the area of trapezoid *JKLM*.

$$\frac{\text{area of trapezoid } ABCD}{\text{area of trapezoid } JKLM} \longrightarrow \frac{150}{x} = \frac{25}{36} \longleftarrow \text{The ratio of areas is } (5^2):(6^2), \text{ or } 25:36.$$

Step 3 Solve the proportion.

$$\frac{150}{x} = \frac{25}{36}$$

$$(150)(36) = 25x \longleftarrow \text{Use cross products.}$$

$$5400 = 25x$$

$$\frac{5400}{25} = \frac{25x}{25} \longleftarrow \text{Divide both sides by 25.}$$

$$216 = x$$

The area of trapezoid *JKLM* is 216 cm².

Talk it Over

7. Suppose you know the scale factor of two similar figures. How can you find the ratio of their areas?

8. The ratio of the areas of two similar triangles is 49:36. What is the ratio of the lengths of their corresponding sides?

·····▶ **Now you are ready for:**
Exs. 1–11 on pp. 533–534

9-8 Similar Figures: Area and Volume **531**

Answers to Talk it Over

7. Since the scale factor is the ratio of corresponding lengths of similar figures, you can square the numerator and square the denominator of the scale factor to find the ratio of the areas.

8. 7:6

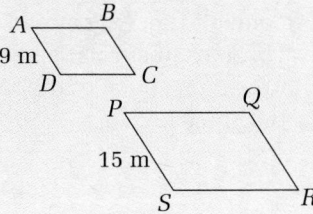

TEACHING

Talk it Over

Before students discuss questions 1–6, you may wish to review the definition of similarity. Questions 4–6 lead students to discover the relationship between the areas of similar figures.

Additional Samples

S1 The ratio of the lengths of the corresponding sides of two similar hexagons is 4:9. What is the ratio of their areas?

The ratio of their areas is $(4^2):(9^2)$, or $16:81$, or $\frac{16}{81}$.

S2 Parallelogram *ABCD* ~ parallelogram *PQRS*. The area of parallelogram *ABCD* is 180 m². What is the area of parallelogram *PQRS*?

Step 1. Find the ratio of the lengths of corresponding sides in simplest form.

$$\frac{AD}{PS} = \frac{9}{15} = \frac{3}{5}$$

Step 2. Set up a proportion. Let x = the area of parallelogram *PQRS*.

$$\frac{180}{x} = \frac{9}{25}$$

Step 3. Solve the proportion.

$$\frac{180}{x} = \frac{9}{25}$$

$$(180)(25) = 9x$$

$$4500 = 9x$$

$$500 = x$$

The area of parallelogram *PQRS* is 500 m².

532

Talk it Over

Question 8 on page 531 gives students the opportunity to use the areas of similar polygons concept in reverse. Questions 9–12 on this page allow students to discover the relationship between volumes of similar space figures.

Visual Thinking

Encourage students to make sketches of several pairs of similar space figures. Ask them to explain the relationship that the sketches suggest. This activity involves the visual skills of *generalization* and *communication*.

Using Manipulatives

Providing models of similar space figures for students to handle will help clarify the concept that the ratio of the volumes of similar space figures is the cube of the ratio of the sides.

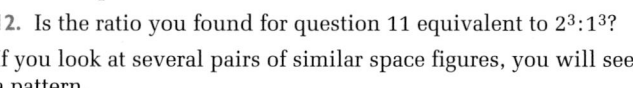

Volumes of Similar Space Figures

Similar space figures have the same shape but not necessarily the same size. To decide whether two space figures are similar, first check whether their bases are similar. Then check whether all their corresponding lengths are in proportion.

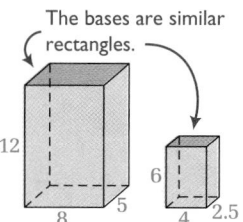

The bases are similar rectangles.

Talk it Over

9. How can you tell that these two prisms are similar?

10. What is the ratio of corresponding lengths of the prisms?

11. What is the ratio of the volumes of the prisms?

12. Is the ratio you found for question 11 equivalent to $2^3:1^3$?

If you look at several pairs of similar space figures, you will see a pattern.

VOLUMES OF SIMILAR SPACE FIGURES

If the ratio of corresponding lengths of two similar space figures is $a:b$, then the ratio of their volumes is $a^3:b^3$.

BY THE WAY...

To understand how a dinosaur ran, paleontologists need to know the dinosaur's body mass. The scientists use the volume of a plastic model to estimate the volume of the actual dinosaur. Then they multiply this value by 1000 kg/m³ to approximate the dinosaur's mass.

Talk it Over

13. Paleontologists and other scientists use plastic replicas of dinosaurs. The scientist in the photo is measuring the water a model displaces to find the volume of the model.

a. Suppose the model is made to a scale of 1:40. What is the ratio of corresponding lengths of the model and the actual dinosaur?

b. What value would the scientist multiply the volume of the model by to estimate the volume of the dinosaur?

Unit 9 Reasoning and Measurement

Answers to Talk it Over

9. The bases are similar rectangles. All corresponding lengths are in proportion:
$\frac{12}{6} = \frac{8}{4} = \frac{5}{2.5} = \frac{2}{1}$.

10. 2:1

11. 8:1

12. Yes.

13. a. 1:40

   b. 64,000

Sample 3

Two cylinders are similar if their diameters are in the same ratio as their heights. What is the ratio of the volumes of these two similar cylinders?

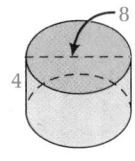

Sample Response ....

The ratio of their corresponding lengths is 3:4, so the ratio of their volumes is $(3^3):(4^3)$, or 27:64.

Sample 4

Two similar regular square pyramids have base edges with lengths 8.5 in. and 12 in. The volume of the larger pyramid is 54 in.^{3}. What is the volume of the smaller pyramid?

Sample Response ....

Problem Solving Strategy: Use a proportion

Let x = the volume of the smaller pyramid.

$$\begin{array}{c}\text{volume of smaller pyramid} \\ \hline \text{volume of larger pyramid} \end{array} \longrightarrow \quad \frac{x}{54} = \frac{(8.5)^3}{12^3} \quad \longleftarrow \begin{array}{l}\text{The ratio of corresponding lengths is 8.5:12,} \\ \text{so the ratio of the volumes is } (8.5)^3:12^3.\end{array}$$

$$\frac{x}{54} = \frac{614.125}{1728}$$

$$54 \cdot \frac{x}{54} = 54 \cdot \frac{614.125}{1728} \quad \longleftarrow \text{Multiply both sides by 54.}$$

$$x \approx 19.2$$

The volume of the smaller pyramid is about 19.2 in.^{3}.

···▶ Now you are ready for:
Exs. 12–33 on pp. 534–538

Look Back ◀───────

Describe three ideas you have learned about similar space figures.

9-8 Exercises and Problems

1. The ratio of the lengths of the corresponding sides of two similar triangles is 3:4. What is the ratio of their areas?

2. The ratio of the lengths of the corresponding sides of two similar rectangles is 6:5. What is the ratio of their areas?

3. The scale factor of two similar figures is 7:10. What is the ratio of their areas?

4. All circles are similar figures. Suppose the diameter of one circle is 12 cm and the diameter of a second circle is 8 cm. What is the ratio of their areas?

9-8 Similar Figures: Area and Volume

Additional Samples

S3 The ratio of the lengths of the edges of two similar prisms is 7:12. What is the ratio of their volumes?

The ratio of their volumes is $(7^3):(12^3)$ or 343:1728.

S4 Two similar cones have diameters of 6 in. and 10 in. The volume of the smaller cone is 56 in.^{3}. What is the volume of the larger cone?

Use a proportion. Let x = the volume of the larger cone.

$$\frac{56}{x} = \frac{6^3}{10^3}$$

$$\frac{56}{x} = \frac{216}{1000}$$

$$216x = 56,000$$

$$x \approx 259.3$$

The volume of the larger cone is about 259.3 in.^{3}.

Look Back

The Look Back would make a good journal entry. You may want to suggest that students support their ideas with examples. ············●

APPLYING

Suggested Assignment

Standard 1–7, 9–13, 18–27, 29–33

Extended 1–27, 29–33

Integrating the Strands

Geometry Exs. 1–28, 30–33

Trigonometry Ex. 10

Statistics and Probability Ex. 29

Discrete Mathematics Ex. 29

Logic and Language Exs. 8, 12, 14–17, 28, 29

Answers to Look Back ·········

Answers may vary. Examples are given. (1) Two space figures are similar if their bases are similar and all corresponding lengths are in proportion. (2) If the ratio of the corresponding lengths of two similar space figures is $a:b$, then the ratio of their volumes is $a^3:b^3$. (3) Two cylinders are similar if their diameters are in the same ratio as their heights.

Answers to Exercises and Problems ·········

1. 9:16
2. 36:25
3. 49:100
4. 9:4

Find the ratio of the areas of each pair of similar figures. Then find the missing area.

5.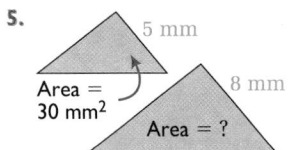

6.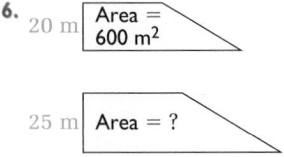

7.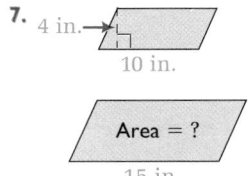

8. **Writing** Suppose you work at a store that develops film. A customer wants a photograph enlarged to "twice its original size." What questions would you ask the customer?

9. $\triangle ABC$ has vertices with coordinates $A(0, 0)$, $B(1, 2)$, and $C(3, 0)$.

   a. Sketch $\triangle ABC$ on a coordinate plane. Find its area.

   b. Draw a dilation of $\triangle ABC$ with center $(0, 0)$ and scale factor 3.

   c. What is the area of the image of $\triangle ABC$, $\triangle A'B'C'$?

   d. Suppose you enlarged $\triangle ABC$ by a scale factor of 20. Find the area of $\triangle A'B'C'$ without sketching the triangle. Describe your method.

10. a. How do you know that these two triangles are similar?

    b. Use sin 30° to find the value of x.

    c. Find the ratio of corresponding lengths of the triangles.

    d. Find the ratio of the area of the small triangle to the area of the large triangle.

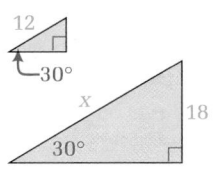

11. **Photocopying** Copy machines often have percent settings for making enlarged or reduced copies. Suppose you set a copy machine to reduce this page to 50% of its original size. The copy will be half as long and half as wide as this page.

    a. To what percent of its original area will the area of this page be reduced if the copy machine is set to reduce to 50%? 75%? 60%?

    b. Suppose the copy machine is set to enlarge to 120%. To what percent of its original area will the area of the page be enlarged?

12. **Reading** How can you decide whether two space figures are similar?

13. **Career** Exploding houses and terrifying monsters add thrills to movies. Special-effects designers often use scale models to create these scenes. Suppose a special-effects designer plans a model of a *pteranodon*, using a scale of 1 in.: 3 ft.

    a. Some pteranodons had wingspans of about 25 ft. What should the wingspan of the model be?

    b. Find the ratio of the volume of the original pteranodon to the volume of the scale model.

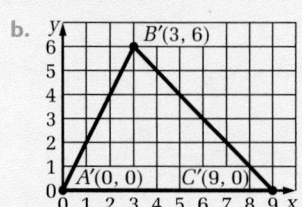

534

In *Gulliver's Travels* by Jonathan Swift, Captain Lemuel Gulliver travels to several fictional nations of the world. At one point in his travels he is held captive in Lilliput, the land of the Lilliputians, "human creature[s] not six inches high." He tells this story.

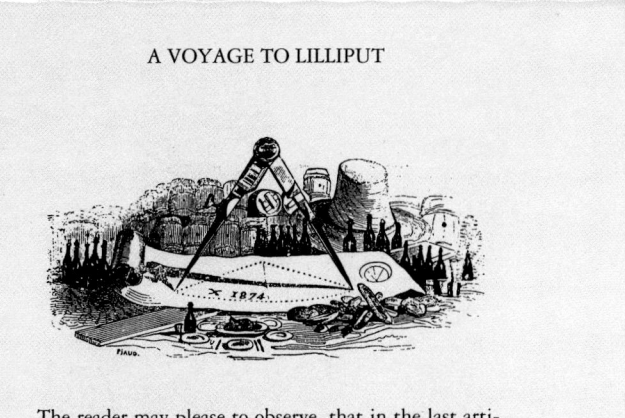

A VOYAGE TO LILLIPUT

The reader may please to observe, that in the last article for the recovery of my liberty, the Emperor stipulates to allow me a quantity of meat and drink sufficient for the support of 1728 Lilliputians. Some time after, asking a friend at Court how they came to fix on that determinate number, he told me, that his Majesty's mathematicians, having taken the height of my body by the help of a quadrant, and finding it to exceed theirs in the proportion of twelve to one, they concluded from the similarity of their bodies, that mine must contain at least 1728 of theirs, and consequently would require as much food as was necessary to support that number of Lilliputians. By which, the reader may conceive an idea of the ingenuity of that people, as well as the prudent and exact economy of so great a prince.

14. Explain why the Lilliputians might think that Gulliver is a giant.

15. What is the ratio of Gulliver's height to the height of a Lilliputian?

16. What is the ratio of the amount of food needed to support Gulliver to the amount of food needed to support one Lilliputian?

17. If you were one of the Emperor's mathematicians, how would you explain why the number 1728 is used to find the amount of food Gulliver needs?

9-8 Similar Figures: Area and Volume

535

Application

Use the answer to Ex. 15 (12 to 1) to explore this question: Would it be possible for a human being to be 12 times as tall as say a 6-foot person weighing 175 pounds? The person's height would be 72 feet (a real giant) and, by the fact that volumes of space figures are in the ratio of $\frac{a^3}{b^3}$ for lengths a and b, then the volume (and thus weight) of the giant human would be 72^3 to 6^3 or 1728 times the weight of the 6-ft person (1728×175) or 302,400 pounds! The feet of the giant human would cover an area 12^2 times greater than that of the 6-ft person, and thus the bones would have to support 12 times more weight ($12^2 \cdot 12 = 12^3$). Ask students if they think humans could be 12 times taller (No; the bones would break.) Point out that the best weightlifters in the world can lift, at most, twice their weight.

Answers to Exercises and Problems

The area of the enlargement would be 3×400, or 1200 square units.

10. **a.** Two pairs of corresponding angles are congruent, 30° and 90°.
    b. 36
    c. 3:1
    d. 1:9

11. **a.** 25%; 56.25%; 36%
    b. 144%

12. if their bases are similar and all corresponding lengths are in proportion

13. **a.** about $8\frac{1}{3}$ in.
    b. 27 ft³:1 in.³, or 46,656 in.³:1 in.³

14. Answers may vary. An example is given. The mathematicians found that Gulliver's height exceeded the Lilliputians's height in the proportion 12:1.

15. 12:1

16. 1728:1

17. Explanations may vary. An example is given. Using the ratio of the heights of Gulliver and the Lilliputians, 12:1, I computed the ratio of the volumes of their bodies, $12^3:1^3 = 1728:1$. From this, I determined that Gulliver would need as much food as 1728 Lilliputians.

At first glance, all pairs of figures in Exs. 18–21 appear similar. Students must use their knowledge of similarity to determine which pairs are really similar and then support their conclusions with valid reasons.

In this studio you can see many examples of similar space figures.

Decide whether the space figures in each pair are similar. Give reasons for your answers.

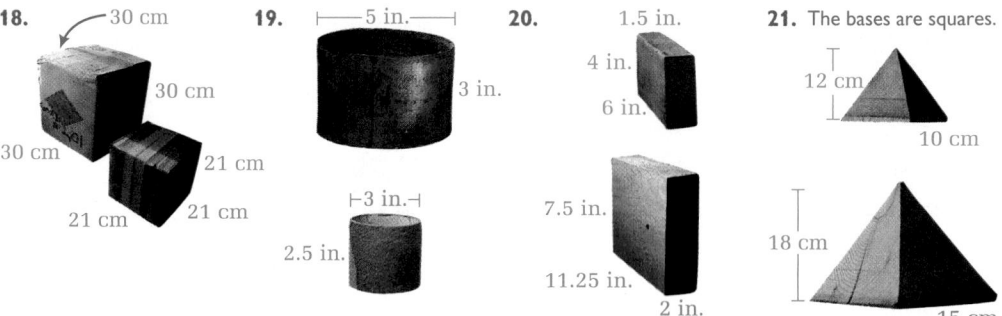

18. 30 cm, 30 cm, 30 cm, 21 cm, 21 cm, 21 cm

19. 5 in., 3 in., 3 in., 2.5 in.

20. 1.5 in., 4 in., 6 in., 7.5 in., 11.25 in., 2 in.

21. The bases are squares. 12 cm, 10 cm, 18 cm, 15 cm

22. The ratio of corresponding lengths of two similar prisms is 6:5. What is the ratio of their volumes?

23. The ratio of corresponding lengths of two similar cylinders is 4:15. What is the ratio of their volumes?

Answers to Exercises and Problems

18. Yes. The ratio of sides is 7:10.

19. No. The ratio of the diameters is not the same as the ratio of the heights.

20. No. The ratio of the widths is not the same as the ratio of the lengths and the heights.

21. Yes. The ratio of all corresponding sides is 2:3.

22. 216:125

23. 64:3375

Cooperative Learning

Ex. 28 gives students the opportunity to discover the relationship between the surface areas of similar space figures. Although all groups should reach the same conclusion, the test used on each group's conjecture may be different and should be shared with the rest of the class.

Find the ratio of the volumes of each pair of similar space figures. Then find the missing volume. In Exercises 25 and 26, leave your answers in terms of π.

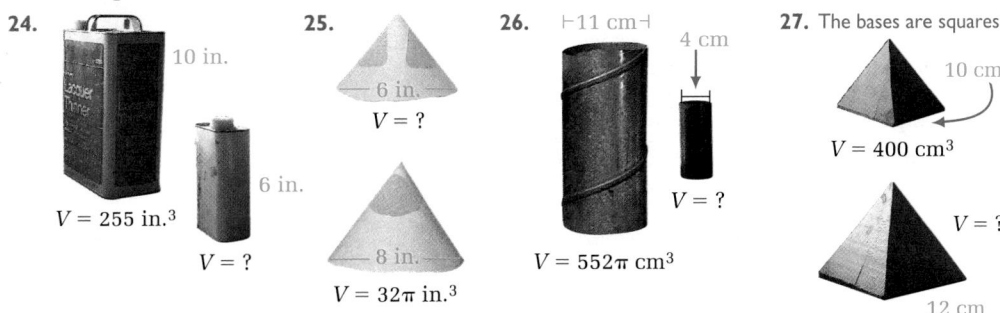

24.
10 in.
$V = 255$ in.^{3}
6 in.
$V = ?$

25.
6 in.
$V = ?$
8 in.
$V = 32\pi$ in.^{3}

26. ⊢11 cm⊣
4 cm
$V = ?$
$V = 552\pi$ cm^3

27. The bases are squares.
10 cm
$V = 400$ cm^3
$V = ?$
12 cm

28. Group Activity Work with another student. Make a conjecture about how the ratio of the surface areas of two similar space figures is related to the ratio of their corresponding lengths. Design and carry out a test of your conjecture.

9-8 Similar Figures: Area and Volume

Answers to Exercises and Problems

24. $27:125$; 55.1 in.^{3}

25. $27:64$; 13.5π in.^{3}

26. $64:1331$; 26.54π cm^3

27. $125:216$; 691.2 cm^3

28. Conjectures and tests may vary. For similar space figures with corresponding lengths a and b, the ratio of the surface areas is $a^2:b^2$.

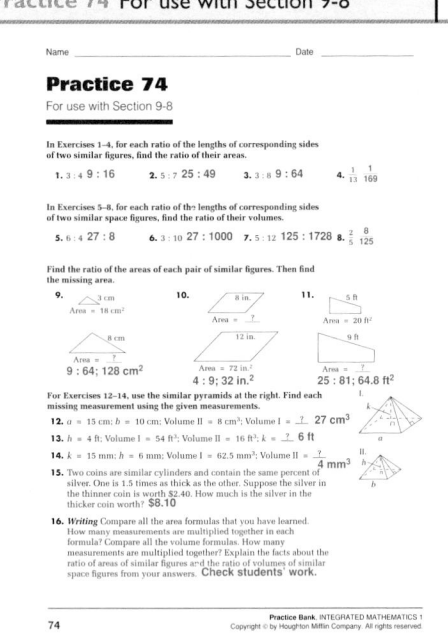

Practice 74 For use with Section 9-8

Ongoing **ASSESSMENT**

29. a. **Writing** Which of the three graphs below gives the most accurate picture of the decrease in whole milk sales? Write a paragraph explaining your reasons.

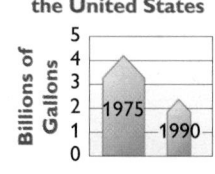

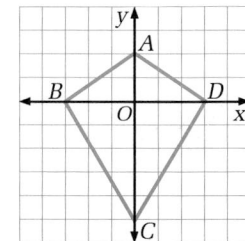

b. **Open-ended** In 1975, 1.3 billion gallons of low-fat milk were sold in the United States. In 1990, 2.8 billion gallons were sold. Draw a graph that accurately represents this increase in sales.

Review **PREVIEW**

30. Find the volume of a cone with radius 8 yd and height 20 yd. *(Section 9-7)*

31. a. What are the coordinates of the vertices of quadrilateral *ABCD*? *(Section 4-1)*

b. What is the line of symmetry? *(Section 1-7)*

c. Translate quadrilateral *ABCD* 4 units to the right. What are the coordinates of the vertices of the image *A′B′C′D′*? *(Section 4-3)*

d. Rotate quadrilateral *ABCD* 180°. What are the coordinates of the vertices of the image *A″B″C″D″*? *(Section 4-5)*

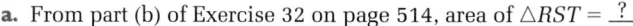

Working on the Unit Project

32. Label a sketch of one base of your boat as shown. △*XYT* is the part of the base that will be under water when you float your boat. Use the results of earlier *Working on the Unit Project* exercises to complete each statement.

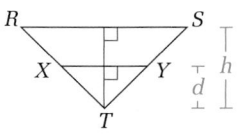

a. From part (b) of Exercise 32 on page 514, area of △*RST* = _?_.

b. From part (f) of Exercise 29 on page 522, area of △*XYT* = _?_.

c. From part (a) of Exercise 37 on page 491, height of △*RST*, *h* = _?_.

33. In Exercise 24 on page 529 you discovered that △*RST* ~ △*XYT*.

a. Use the values from Exercise 32 to write a proportion to find *d*, the height of △*XYT*.

b. Solve your proportion for *d*.

c. Explain why your answer to part (b) is the *draft* of your boat.

538 **Unit 9** Reasoning and Measurement

Answers to Exercises and Problems

29. a. Answers may vary. An example is given. The first pictograph gives the most accurate picture of the decrease in whole milk sales because it uses the same symbol for each year to represent 1 billion gallons. You can compare the symbols to see that about $1\frac{3}{4}$ billion gallons more were sold in 1975. In the middle graph, the dimensions of the symbol for 1975 are about twice as large as the dimensions of the symbol for 1990. This means that the area of the larger symbol is about four times the area of the smaller symbol, giving the impression that sales were about four times greater in 1975, which is not true. Similarly, in the graph on the right, the larger space figure has dimensions about twice as large as the smaller. This means that the volume of the larger figure is about 8 times the volume of the smaller, implying that sales in 1975 were about 8 times the sales in 1990. The actual numbers for sales show that this is not true.

b. Graphs may vary. An example is given.

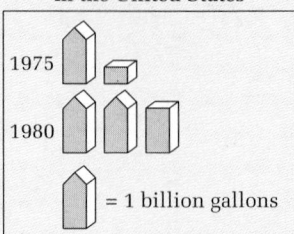

538

Completing the Unit Project

Now you are ready to find the actual draft of the boat you built.

➤ Fill your boat with 300 cm³ of water and float the boat in a large sink or container.

➤ Read the boat's draft from the scale that you marked on your boat. Record your results.

Now you can determine how much cargo your boat can carry.

➤ Select one object for the first test and predict how many of these objects will cause your boat to sink.

➤ Record your predictions and then test your guesses by filling your boat with these objects.

➤ Count how many items your boat can hold until it sinks and record these results.

When you have gathered data from at least three cargo tests, describe your observations in a report. Include the effect of the size of the object, the relative heaviness of the object, and the number of each object it took to sink your boat.

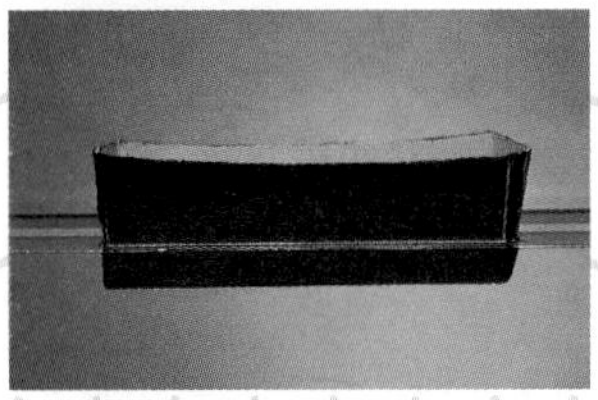

Look Back ◄

Get together with a small group of students. Share your results. Discuss these questions: How close is the actual draft to your prediction? Why might the actual draft and the predicted draft be different? What have you learned about buoyancy?

Alternative Projects

Project 1: Boating in Various Seas

Water takes up less space as it starts to cool but more space as it freezes. Write a story describing how this affects the draft of a boat passing through the Arctic, through the North Atlantic, through the Panama Canal, and across the Equator.

Project 2: Design a Container

Create a design for a container to keep things hot or cold. First decide what objects you want to carry in your container and then plan a shape to fit those objects. Keep in mind that a container with a large surface area relative to its volume will generally lose heat faster than one with a smaller surface area.

Unit 9 Completing the Unit Project

539

Answers to Exercises and Problems

30. about 1340.4 yd³

31. a. $A(0, 2)$; $B(-3, 0)$; $C(0, -5)$; $D(3, 0)$

b. $x = 0$ or y-axis

c. $A'(4, 2)$; $B'(1, 0)$; $C'(4, -5)$; $D'(7, 0)$

d. $A''(0, -2)$; $B''(3, 0)$; $C''(0, 5)$; $D''(-3, 0)$

32. a. about 45.2 cm²

b. about 17.2 cm²

c. about 6.7 cm

33. a. $\dfrac{\text{area of } \triangle RST}{\text{area of } \triangle XYT} = \dfrac{h^2}{d^2};$

$\dfrac{45.2}{17.2} = \dfrac{6.7^2}{d^2}$

b. about 4.1 cm

c. The value of d, about 4.1 cm, is the depth of the part of the boat that is below the water level, so it is the draft of the boat.

Quick Quiz (9-7 through 9-8)

Find the volume of each space figure. [9-7]

1.

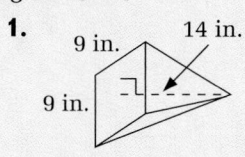

9 in. 14 in.

9 in.

378 in.³

2.

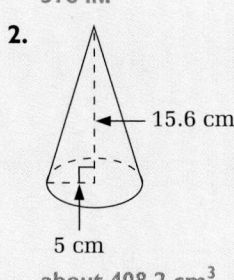

15.6 cm

5 cm

about 408.2 cm³

3. 3 m

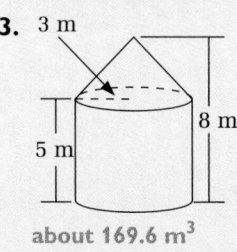

8 m

5 m

about 169.6 m³

The ratio of the lengths of corresponding sides of two similar trapezoids is 4:9. [9-8]

4. What is the ratio of their areas? **16:81**

5. If the area of the smaller trapezoid is 128 m², what is the area of the larger trapezoid? **648 m²**

The ratio of the diameters of two similar cylinders is 3:5. [9-8]

6. What is the ratio of their volumes? **27:125**

7. If the volume of the larger cylinder is 1375 ft³, what is the volume of the smaller cylinder? **297 ft³**

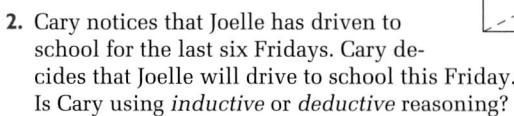

Unit 9 REVIEW AND ASSESSMENT

1. Find the length of the diagonal of this rectangle.

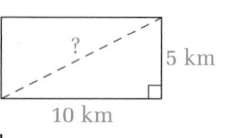

? | 5 km

10 km

9-1

2. Cary notices that Joelle has driven to school for the last six Fridays. Cary decides that Joelle will drive to school this Friday. Is Cary using *inductive* or *deductive* reasoning?

Simplify.

3. $\sqrt{96}$

4. $2\sqrt{3} \cdot 4\sqrt{8}$

9-2

5. Solve $3y^2 = 81$.

6. Use the statement "If you have a fever, then you should stay home from school."

9-3

   a. Identify the hypothesis and conclusion of the statement.

   b. Write the converse of the statement.

   c. Tell whether the converse in part (b) is *True* or *False*. If it is false, give a counterexample.

7. **Open-ended** Write a statement from mathematics or from your life at school or at home that is false, but whose converse is true.

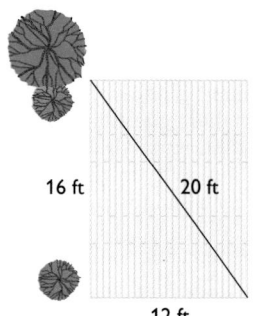

16 ft 20 ft

12 ft

Solve for x.

8. $x(x + 4) = 0$ **9.** $6x(x - 2) = 0$

10. The lengths of the sides of a deck are 12 ft and 16 ft. A carpenter measures the diagonal from one corner to the other, as shown. The diagonal is 20 ft. Is the corner of the deck a right angle? How do you know?

11. a. A paper bag has a rectangular base that is 6 in. wide and 13 in. long. A circular hole with diameter 4 in. has been cut out of the base. If you drop a small bead into the bag without aiming, what is the probability that it will fall through the hole?

9-4

   b. What would the probability that you found in part (a) be if the hole was a 4 in. by 4 in. square?

12. The side of a cube is 5 in. long. What is the surface area of the cube?

9-5

13. Find the surface area of a regular square pyramid with base edge length 8 in. and slant height 3 in.

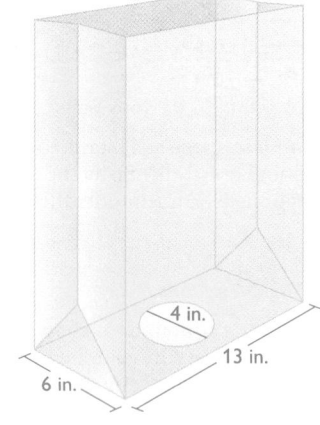

4 in.

13 in.

6 in.

Answers to Unit 9 Review and Assessment ··

1. a. about 11.2 km

2. inductive

3. $4\sqrt{6}$

4. $16\sqrt{6}$

5. $3\sqrt{3}, -3\sqrt{3}$

6. a. Hypothesis: You have a fever. Conclusion: You stay home from school.

   b. If you stay home from school, you have a fever.

   c. False. You could stay home from school because you sprained your ankle.

7. Answers may vary. An example is given. If the ground is wet in the morning, it rained last night; false; it might have snowed. If it rained last night, the ground is wet in the morning.

8. 0, −4

9. 0, 2

10. Yes; the Pythagorean theorem applies: $(12)^2 + (16)^2 = (20)^2$.

11. a. about 0.16

   b. about 0.21

12. 150 in.²

13. 112 in.²

14. 72 ft³

14. The wall of a downtown office building lobby contains a floor-to-ceiling aquarium. The aquarium is a rectangular prism with length 3 ft and width 2 ft. Its walls are 12 ft high. How much water does the aquarium hold?

9-6

15. **Writing** A manufacturer has a choice of two different cans to use in packaging a new product. Which of these two cans would you recommend the manufacturer use? Explain your choice. Consider the amount of material needed to produce the can. What other factors might influence the manufacturer's decision?

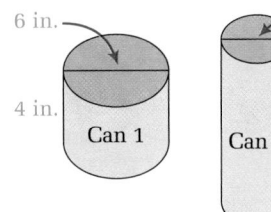

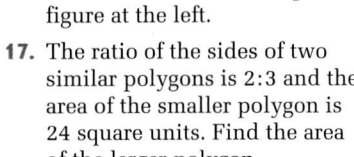

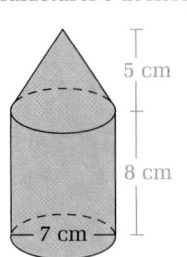

16. Find the volume of the space figure at the left.

9-7

17. The ratio of the sides of two similar polygons is 2:3 and the area of the smaller polygon is 24 square units. Find the area of the larger polygon.

9-8

18. The height of one cylinder is 5 times the height of a similar cylinder. How many times greater is the volume of the larger cylinder than the volume of the smaller cylinder?

19. **Self-evaluation** What did you learn about geometry and about space figures in this unit that you did not know before?

20. **Group Activity** Work with another student.

The two right triangles shown below are similar. Find the area of the smaller triangle using the two methods described below.

a. One student should use the Pythagorean theorem to find x. Then find the area of the smaller triangle.

b. Another student should use the fact that the two triangles are similar. Write and solve a proportion to find the area of the smaller triangle.

c. Show that your two answers are the same. Which method do you think is easier to use? Why?

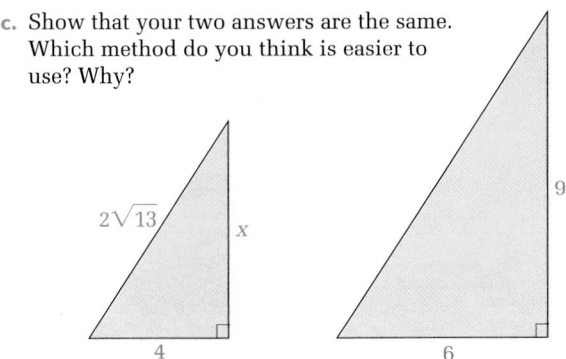

Answers to Unit 9 Review and Assessment

15. Can 1; Explanations may vary. An example is given. Can 1 and Can 2 have equal volumes, but Can 1 has less surface area than Can 2. This means that Can 1 holds as much as Can 2, but less materials are needed to make Can 1. Other factors that might influence the manufacturer's decision would be the information displayed on the label, the type of product, and the size and shape of shipping cartons.

16. about 372.02 cm^3

17. 54 square units

18. 125 times

19. Answers may vary.

20. a. 6; 12 square units

b. $\frac{4}{9} = \frac{A}{27}$, $A = 12$ square units

c. Both answers are 12 square units; choices of methods may vary. An example is given. For this problem, the method using ratios and a proportion was easier because the numbers are small whole numbers, which made the calculations simpler.

Quick Quiz (9-1 through 9-3)

Find the missing length for each triangle. [9-1]

1.

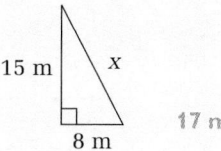

15 m x

8 m   17 m

2.

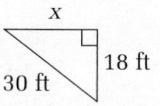

x

18 ft

30 ft   24 ft

Simplify. [9-2]

3. $\sqrt{54}$ $3\sqrt{6}$

4. $\sqrt{5} \cdot \sqrt{7}$ $\sqrt{35}$

Solve for x. [9-2]

5. $2x^2 = 24$ $x = \pm 2\sqrt{3}$

6. $z^2 + z^2 = x^2$ $x = \pm z\sqrt{2}$

Tell whether each statement is *True* or *False*. If it is false, give a counterexample. [9-3]

7. If $y = 0$, then $4y = 0$.   True.

8. If $x^2 = 25$, then $x = 5$.

False; x may equal –5.

Write the converse of each statement. Tell whether it is *True* or *False*. If it is false, give a counterexample. [9-3]

9. If $(x + 1) = 0$, then $(x + 1)(x - 2) = 0$.

If $(x + 1)(x - 2) = 0$, then $(x + 1) = 0$. False; $(x - 2)$ can equal 0.

10. If $5x = 0$, then $x = 0$.

If $x = 0$, then $5x = 0$. True.

IDEAS AND (FORMULAS) $= X^2_{sP_s}$

➤ **Reasoning** If-then statements are either true or false. For an if-then statement to be true, it must be true in all cases. *(p. 492)*

➤ **Reasoning** One way to show that an if-then statement is false is to find a counterexample. *(p. 492)*

➤ **Reasoning** The converse of a true if-then statement is not necessarily true. *(p. 494)*

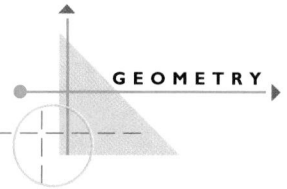

ALGEBRA $) x^2$

$(\times$

➤ To multiply and simplify square roots of nonnegative numbers, use these properties. *(p. 486)*

$$\sqrt{ab} = \sqrt{a} \cdot \sqrt{b} \qquad \sqrt{a} \cdot \sqrt{b} = \sqrt{ab} \qquad \sqrt{a} \cdot \sqrt{a} = a$$

➤ To simplify an expression in radical form, look for perfect square factors of the number under the $\sqrt{\ }$ symbol. *(p. 487)*

➤ If a product of factors is zero, one or more of the factors must be zero. (zero-product property) *(p. 495)*

If $ab = 0$, then $a = 0$ or $b = 0$.

➤ You can use the zero-product property to solve equations. These equations may have more than one solution. *(p. 495)*

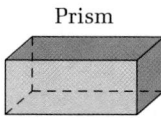

GEOMETRY

➤ In any right triangle, the square of the length of the hypotenuse is equal to the sum of the squares of the lengths of the legs. (Pythagorean theorem) *(p. 478)*

➤ The acute angles of a right triangle are complementary. *(p. 480)*

➤ If the square of the length of one side of a triangle is equal to the sum of the squares of the lengths of the other two sides, then the triangle is a right triangle. (converse of the Pythagorean theorem) *(p. 495)*

➤ You can use the converse of the Pythagorean theorem to decide whether or not a triangle is a right triangle. *(p. 495)*

➤ Prisms and pyramids are named by the shapes of their bases. *(pp. 507, 509)*

➤ You can use formulas to find the surface areas and volumes of different space figures. *(pp. 507, 509, 510, 515, 517, 523, 524)*

Cylinder

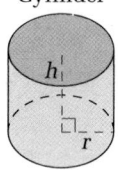

h

r

Prism

S.A. = areas of bases + areas of faces

V = area of base × height = Bh

S.A. = $2\pi r^2 + 2\pi rh$

V = area of base × height = $\pi r^2 h$

Pyramid

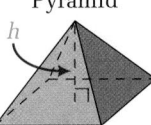

Cone

S.A. = area of base + area of faces

$V = \frac{1}{3} \times$ area of base $\times$ height $= \frac{1}{3}Bh$

$V = \frac{1}{3} \times$ area of base $\times$ height

$= \frac{1}{3}\pi r^2 h$

➤ If the ratio of corresponding lengths of two similar figures is $a:b$, then the ratio of their areas is $a^2:b^2$. *(p. 530)*

➤ If the ratio of corresponding lengths of two similar space figures is $a:b$, then the ratio of their volumes is $a^3:b^3$. *(p. 532)*

STATISTICS & PROBABILITY ▶

➤ Suppose a point in the rectangle below is picked at random. Then the probability that the point is in the shaded region is $\dfrac{\text{area of shaded region}}{\text{area of rectangle}}$. *(p. 502)*

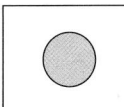

➤ Suppose a point on $\overline{AC}$ is picked at random. Then the probability that the point is on $\overline{AB}$ is $\dfrac{\text{length of } \overline{AB}}{\text{length of } \overline{AC}}$. *(p. 502)*

$A \qquad B \qquad\qquad C$

Unit 9 Review and Assessment

543

An 8-inch square target has a 2-inch square bullseye. [9-4]

1. What is the theoretical probability that a disk dropped on the target will hit the bullseye? $\frac{1}{16}$

2. How many times would you expect to land on the bullseye if you dropped a disk 80 times? **5**

3. A friend is to come to your home between 4:00 P.M. and 5:00 P.M. If you make a 5-minute telephone call during that time, what is the probability that she will arrive while you are talking on the telephone? [9-4] $\frac{1}{12}$

Find the surface area of each figure. [9-5]

4.

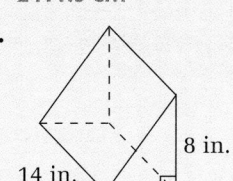

8 cm

40 cm

2411.5 cm²

5.

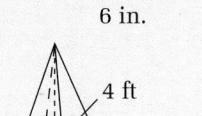

8 in.

14 in.

6 in.   384 in.²

6.

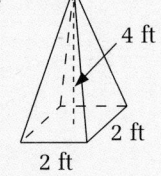

4 ft

2 ft

2 ft   20.5 ft²

Find the volume of each figure. [9-6]

7.

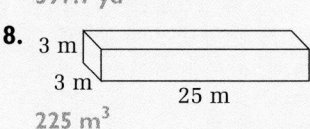

4 yd

18 yd   3 yd

597.7 yd³

8.

3 m

3 m   25 m

225 m³

543

Quadratic Equations as Models

OVERVIEW

➤ **Unit 10** introduces reflections and presents quadratic equations and their graphs. Students learn to represent quadratic relationships with tables, equations, and graphs and to relate changes in equations to transformations of the corresponding graphs. They recognize that problem situations can be modeled by quadratic equations and graphs, and they use factoring, graphing, and the quadratic formula to solve equations. In preparation for the work on factoring, students can use the review of prime factorization in the **Student Resources Toolbox** on pages 648 and 649.

➤ The theme of the Unit Project is to write a math story. Students investigate a variety of situations modeled by equations and include one in which the trajectory of a batted ball is modeled by an equation from this unit.

➤ Connections to language arts, sports, business education, art, history, and science are integrated into the teaching materials and exercises.

➤ Problem-solving strategies used in Unit 10 include *Work Backwards* (Section 10-7) and *Use a Formula* (Section 10-8).

Unit Objectives

Section	Objectives	NCTM Standards
10-1	• Recognize and draw the mirror image of an object.	1, 2, 4, 5, 8
10-2	• Translate and reflect the graph of $y = x^2$.	1, 2, 4, 5, 6, 8
10-3	• Use a table of values to sketch a parabola.	1, 2, 4, 5, 6, 8
	• Use algebra to find the coordinates of the points where a parabola crosses the axes.	
10-4	• Use shortcuts for simplifying powers.	1, 2, 3, 4, 5
10-5	• Factor and expand algebraic expressions.	1, 2, 4, 5, 6, 8
	• Use factoring to find the x-intercepts of parabolas.	
10-6	• Multiply two binomials.	1, 2, 4, 5, 6, 8
	• Use a formula to find the line of symmetry of a parabola.	
10-7	• Factor trinomials.	1, 2, 4, 5, 6, 8
	• Sketch the graph of a parabola.	
10-8	• Solve problems about situations modeled by parabolas.	1, 2, 4, 5, 6, 8

Section	Connections to Prior and Future Concepts
10-1	**Section 10-1** introduces the concepts of reflection and orientation. The idea of reflection is related to the idea of line symmetry, introduced in Section 1-7. Transformations were studied in Sections 4-3, 4-4, 6-6, and will be further investigated in Section 10-2 of this book; in Units 3, 4, and 5 of Book 2; and in Units 5, 9, and 10 of Book 3.
10-2	**Section 10-2** expands the study of transformations to include reflecting and translating parabolas on the coordinate plane. Parabolas will be explored throughout this unit; in Units 2, 4, and 9 of Book 2; and in Units 1, 2, and 6 of Book 3.
10-3	**Section 10-3** continues the investigation of parabolas begun in Section 10-2. Students graph a parabola from a table of values and locate the intercepts of the graph. Parabolas are further explored in the remainder of this unit and in Books 2 and 3, as noted above.
10-4	**Section 10-4** extends the study of exponents begun in Sections 1-3 and 2-3 to include finding products of powers and simplifying powers. Exponents are frequently used in advanced mathematics, and students will work with exponents and powers throughout Books 2 and 3.
10-5	**Section 10-5** uses algebra tiles to illustrate factoring and expanding algebraic expressions. Factors for quadratic expressions are further explored in Sections 10-6, 10-7, and 10-8 of Book 1, as well as in Unit 4 of Book 2 and Unit 2 of Book 3. A review of factoring integers is presented on page 648 of the Toolbox.
10-6	**Section 10-6** extends the concept of expanding expressions from Section 10-5 to multiplication of two binomials. Multiplication of binomials will be further studied in Section 10-7 and is a skill that will be used throughout Books 2 and 3. This section also continues the study of parabolas, investigating the features of a parabola whose equation is in expanded form. Quadratic functions and graphs are emphasized in Unit 4 of Book 2 and in Unit 2 of Book 3.
10-7	**Section 10-7** combines many skills from the Unit. Students graph a quadratic equation by locating the vertex, intercepts, and line of symmetry. They also learn to factor simple trinomials. Graphing and factoring quadratic functions will be extended and applied in Unit 4 of Book 2 and in Unit 2 of Book 3.
10-8	**Section 10-8** introduces the quadratic formula and uses quadratic equations to solve problems. Students use the skills developed throughout Unit 10 to solve quadratic equations. Quadratic equations will be further studied in Books 2 and 3.

Integrating the Strands ··

Strands	Sections
Number	10-3, 10-4, 10-6, 10-7
Algebra	10-1, 10-2, 10-3, 10-4, 10-5, 10-6, 10-7, 10-8
Functions	10-2, 10-3, 10-4, 10-5, 10-6, 10-7, 10-8
Measurement	10-6
Geometry	10-1, 10-2, 10-4, 10-5, 10-6, 10-7, 10-8
Logic and Language	10-1, 10-2, 10-5

Section Planning Guide

➤ Essential exercises and problems are indicated in boldface.
➤ Ongoing work on the Unit Project is indicated in color.
➤ Exercises and problems that require student research, group work, manipulatives, or graphing technology are indicated in the column headed "Other."

Section	Materials	Pacing	Standard Assignment	Extended Assignment	Other
10-1	protractor, ball	Day 1	**1**, 2, **3**, **4**, 5, **6–8**	**1**, 2, **3**, **4**, 5, **6–8**, 9–11	12
		Day 2	**13–22**, 23–30, 31	**13–22**, 23–30, 31	31
10-2	graphing technology	Day 1	**2–4**, **6–16**	1, **2–4**, **6–16**	5
		Day 2	**17–41**, 43–52, 53, 54	**17–41**, 43–52, 53, 54	42
10-3	graphing technology	Day 1	1–3, **4–14**, 17–24, 25	1–3, **4–14**, 15–24, 25	
10-4	graphing technology	Day 1	**1–19**	**1–19**	
		Day 2	**20–27**, **31**, **32**, 35–42, 43	**20–27**, 29, 30, **31**, **32**, 35–42, 43	28, 33, 34, 43
10-5	algebra tiles	Day 1	**1–3**, 7	**1–3**, 7	4–6
		Day 2	**8–13**, 14, **15–23**, **25–28**, 31–40, 41–43	**8–13**, 14, **15–23**, 24, **25–28**, 29, 31–40, 41–43	30
10-6	algebra tiles	Day 1	**5–12**, 13	4, **5–12**, 13	1–3
		Day 2	**14**, **16–25**, **27**, **28**, 29, 31–40, 41, 42	**14**, 15, **16–25**, 26, **27**, **28**, 29, 31–40, 41, 42	30
10-7	algebra tiles	Day 1	**1**, **2**, **6–23**	**1**, **2**, **6–23**	3–5
		Day 2	**25–27**, 29, **30–32**, 34–42, 43	24, **25–27**, 28, 29, **30–32**, 34–42, 43	33
10-8		Day 1	**2–15**	1, **2–15**	
		Day 2	**17–24**, 26–33, 34, 35	16, **17–24**, 26–33, 34, 35	25
Review		**Day 1**	**Unit Review**	**Unit Review**	
Test		**Day 2**	**Unit Test**	**Unit Test**	

Yearly Pacing	Unit 10 Total	Units 1–10 Total	Remaining	Total
	19 days (2 for Unit Project)	164 days	0 days	164 days

Support Materials

➤ See **Project Book** for notes on Unit 10 Project: Write a Math Story.
➤ "UPP" and "disk" refer to **Using Plotter Plus** booklet and **Plotter Plus** disk.
➤ Warm-up exercises for each section are available on **Warm-Up Transparencies**.
➤ "FI" and "GI" refer, respectively, to the McDougal Littell Mathpack software Activity Books for **Function Investigator** and **Geometry Inventor**.

Section	Study Guide	Practice Bank	Problem Bank	Activity Bank	Explorations Lab Manual	Assessment Book	Visuals	Technology
10-1	10-1	Practice 76	Set 22	Enrich 67	Add. Expl. 11 Master 2	Quiz 10-1	Folder 10	GI Acts. 16–18
10-2	10-2	Practice 77	Set 22	Enrich 68	Add. Expl. 12 Masters 2, 3	Quiz 10-2	Folder 10	UPP, page 36 Parabola Plotter (disk) Parabola Quiz (disk)
10-3	10-3	Practice 78	Set 22	Enrich 69	Masters 2, 3	Quiz 10-3, Test 41		
10-4	10-4	Practice 79	Set 23	Enrich 70		Quiz 10-4		
10-5	10-5	Practice 80	Set 23	Enrich 71	Masters 6, 7	Quiz 10-5	Folder 1	
10-6	10-6	Practice 81	Set 23	Enrich 72	Add. Expl. 13 Masters 6, 7	Quiz 10-6 Test 42	Folders 1, 10	UPP, pages 21–24 FI Act. 17
10-7	10-7	Practice 82	Set 24	Enrich 73	Add. Expl. 14 Masters 2, 6, 7	Quiz 10-7	Folder 1	
10-8	10-8	Practice 83	Set 24	Enrich 74	Masters 2, 3	Quiz 10-8, Test 43		UPP, page 37
Unit 10	Unit Rev.	Practice 84	Unif. Prob. 10	Fam. Inv. 10		Tests 44–47		

UNIT TESTS

Spanish versions of these tests are on pages 158–161 of the **Assessment Book.**

Name _____ Date _____ Score _____

Test 44

Test on Unit 10 (Form A)

Directions: Write the answers in the spaces provided.

1. How has the graph of $y = x^2$ been transformed to produce the graph of $y = x^2 + 6$?
translated up 6 units

In Questions 2 and 3, write the equation of the line of symmetry and the coordinates of the vertex for the graph of each equation.

2. $y = (x - 3)^2$ **3.** $y = -(x + 7)^2$

4. Graph the parabola $y = (x - 1)(x + 4)$ on the coordinate plane at the right.

5. Without graphing, find the x-intercepts and y-intercept of the graph of $y = (x + 4)(x - 2)$.

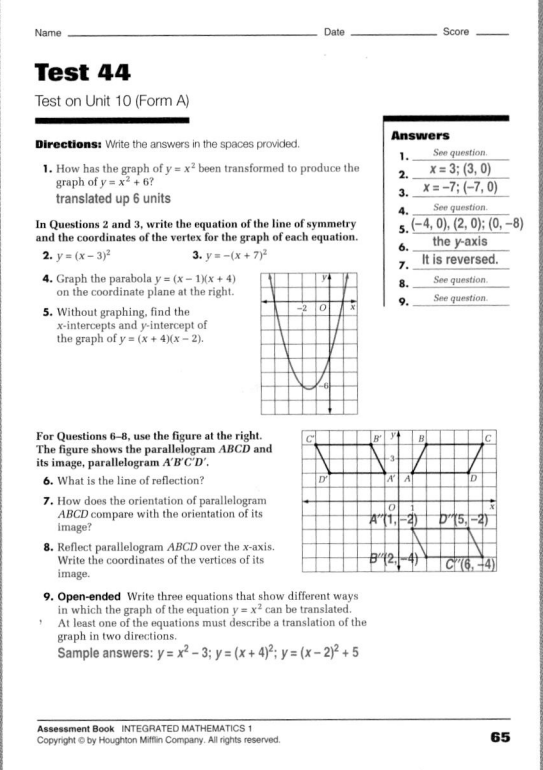

For Questions 6–8, use the figure at the right. The figure shows the parallelogram $ABCD$ and its image, parallelogram $A'B'C'D'$.

6. What is the line of reflection?

7. How does the orientation of parallelogram $ABCD$ compare with the orientation of its image?

8. Reflect parallelogram $ABCD$ over the x-axis. Write the coordinates of the vertices of its image.

9. Open-ended Write three equations that show different ways in which the graph of the equation $y = x^2$ can be translated. At least one of the equations must describe a translation of the graph in two directions.
Sample answers: $y = x^2 - 3$; $y = (x + 4)^2$; $y = (x - 2)^2 + 5$

Answers
1. _See question._
2. $x = 3$; $(3, 0)$
3. $x = -7$; $(-7, 0)$
4. _See question._
5. $(-4, 0)$, $(2, 0)$; $(0, -8)$
6. the y-axis
7. It is reversed.
8. _See question._
9. _See question._

65

Name _____ Date _____ Score _____

Test 44 (continued)

Directions: Write the answers in the spaces provided.

Simplify.

10. $m^3 \cdot m \cdot m^5$ **11.** $(-3c^3)(4c^5d^4)$

12. $(r^5)^3$ **13.** $(-2xy^4)^5$

In Questions 14 and 15, expand each product.

14. $3x(7x + 8)$ **15.** $-2x(3x - 8)$

16. Factor the right side of the equation $y = 3x^2 + 12x$ completely. Then find the x-intercepts and the y-intercept of the graph of the equation.
$y = 3x(x + 4)$; x-int: $(0, 0)$, $(-4, 0)$; y-int: $(0, 0)$

Factor each trinomial.

17. $x^2 + 3x + 2$ **18.** $x^2 - 5x - 14$ **19.** $x^2 + x - 12$

In Questions 20 and 21, use the quadratic formula to solve each equation.

20. $2x^2 - 5x + 2 = 0$ **21.** $x^2 = -3.1x - 1.3$

22. Use the line of symmetry, the vertex, and the intercepts to sketch the graph of the equation $y = x^2 + 2x - 3$.

Answers
10. m^9
11. $-12c^8d^4$
12. r^{15}
13. $-32x^5y^{20}$
14. $21x^2 + 24x$
15. $-6x^2 + 16x$
16. _See question._
17. $(x + 1)(x + 2)$
18. $(x - 7)(x + 2)$
19. $(x - 3)(x + 4)$
20. $\frac{1}{2}$ or 2
21. -0.5 or -2.6
22. _See question._
23. _See question._

23. Writing Compare the factoring method to the quadratic formula when solving a quadratic equation. Tell when it is appropriate to use each method.
Sample answer: While the quadratic formula can be used to solve any quadratic equation, the factoring method can only be used when there is a factorable expression involved.

Name _____ Date _____ Score _____

Test 45

Test on Unit 10 (Form B)

Directions: Write the answers in the spaces provided.

1. How has the graph of $y = x^2$ been transformed to produce the graph of $y = x^2 - 4$?
translated down 4 units

In Questions 2 and 3, write the equation of the line of symmetry and the coordinates of the vertex for the graph of each equation.

2. $y = (x + 7)^2$ **3.** $y = -(x - 3)^2$

4. Graph the parabola $y = (x - 4)(x + 1)$ on the coordinate plane at the right.

5. Without graphing, find the x-intercepts and y-intercept of the graph of $y = (x - 4)(x + 2)$.

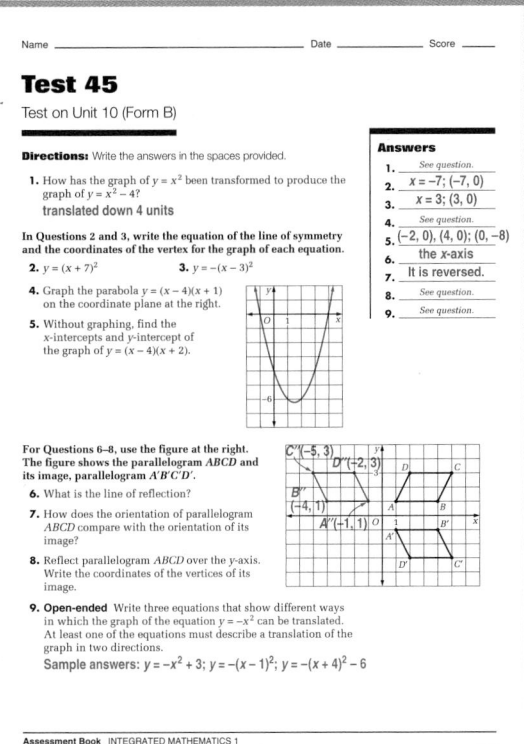

For Questions 6–8, use the figure at the right. The figure shows the parallelogram $ABCD$ and its image, parallelogram $A'B'C'D'$.

6. What is the line of reflection?

7. How does the orientation of parallelogram $ABCD$ compare with the orientation of its image?

8. Reflect parallelogram $ABCD$ over the y-axis. Write the coordinates of the vertices of its image.

9. Open-ended Write three equations that show different ways in which the graph of the equation $y = -x^2$ can be translated. At least one of the equations must describe a translation of the graph in two directions.
Sample answers: $y = -x^2 + 3$; $y = -(x - 1)^2$; $y = -(x + 4)^2 - 6$

Answers
1. _See question._
2. $x = -7$; $(-7, 0)$
3. $x = 3$; $(3, 0)$
4. _See question._
5. $(-2, 0)$, $(4, 0)$; $(0, -8)$
6. the x-axis
7. It is reversed.
8. _See question._
9. _See question._

67

Name _____ Date _____ Score _____

Test 45 (continued)

Directions: Write the answers in the spaces provided.

Simplify.

10. $m^4 \cdot m \cdot m^2$ **11.** $(-3c^2)(4c^3d^6)$

12. $(r^3)^6$ **13.** $(-2xy^7)^3$

In Questions 14 and 15, expand each product.

14. $2x(5x + 6)$ **15.** $-3x(6x - 5)$

16. Factor the right side of the equation $y = 7x^2 + 14x$ completely. Then find the x-intercepts and the y-intercept of the graph of the equation.
$y = 7x(x + 2)$; x-int: $(0, 0)$, $(-2, 0)$; y-int: $(0, 0)$

Factor each trinomial.

17. $x^2 - 5x + 6$ **18.** $x^2 - 7x + 6$ **19.** $x^2 + x - 20$

In Questions 20 and 21, use the quadratic formula to solve each equation.

20. $4x^2 - 5x + 1 = 0$ **21.** $x^2 = -2.3x - 1.2$

22. Use the line of symmetry, the vertex, and the intercepts to sketch the graph of the equation $y = x^2 - 2x - 3$.

Answers
10. m^7
11. $-12c^5d^6$
12. r^{18}
13. $-8x^3y^{21}$
14. $10x^2 + 12x$
15. $-18x^2 + 15x$
16. _See question._
17. $(x - 3)(x - 2)$
18. $(x - 6)(x - 1)$
19. $(x - 4)(x + 5)$
20. $\frac{1}{4}$ or 1
21. -1.5 or -0.8
22. _See question._
23. _See question._

23. Writing Compare the factoring method to the quadratic formula when solving a quadratic equation. Tell when it is appropriate to use each method.
Sample answer: While the quadratic formula can be used to solve any quadratic equation, the factoring method can only be used when there is a factorable expression involved.

Software Support

McDougal Littell Mathpack
Function Investigator
Geometry Inventor

Outside Resources

Books/Periodicals

Swetz, Frank and J. S. Hartzler, eds. *Mathematical Modeling in the Secondary School Curriculum: A Resource Guide of Classroom Exercises.* Reston, VA: NCTM, 1991.

Fishman, Joseph. "Analyzing Energy and Resource Problems: An Interdisciplinary Approach with Mathematical Modeling." *Mathematics Teacher* (November 1993): pp. 628–633.

Manipulatives

Perl, Terri. "Manipulatives and the Computer: A Powerful Partnership for Learners of All Ages." *Classroom Computer Learning* (March 1990): 20–22.

Algeblocks. Cincinnati, OH: Southwestern Publications.

Software

Edwards, Lois A. and Zhongyi Sun. *Data Models: A Mathematical Modeling Tool.* Scotts Valley, CA: Sunburst.

Harvey, Wayne and Judah L. Schwartz. *The Function Supposer: Explorations in Algebra.* Newton, MA: Education Development Center, 1992.

Quadratic Equations as Models

PROJECT GOALS

➤ Students write a math story that includes at least three situations that are modeled by equations or formulas from earlier units and one situation that is modeled by an equation from this unit.

➤ When writing the story, students choose the setting, characters, and plot along with equations and formulas studied in the course that can be used in the story.

➤ Students work together in a cooperative group and appreciate each other's contributions to the project.

PROJECT PLANNING

Project Teams

Before students begin working as part of a team, have individual students write a brief description of a sports experience.

Have students work on the project in groups of four. One way for the individuals in the group to distribute the work is as follows:

1. Editor: coordinates project planning and checks all aspects of the project.

2. Writer: outlines and writes the final draft of the story.

3. Illustrator: draws related illustrations or graphs for the story.

4. Grapher: determines reasonable values for the variables in the equations and formulas used in the story.

Individual students will need a few days to examine the earlier units and find appropriate equations and formulas to use in the story. The team needs to discuss the individual choices and select what will be used.

"Thar she blows!" — On the dark, choppy surface of the ocean, the first sign of a whale is often its spout. When a whale breathes out, air and water mix to form a spout that arcs out of the whale's blowhole. Water droplets in the spout follow a ⌒-shaped *trajectory*, or path, that is a parabola. The spout often rises as high as 25 ft before being pulled back down by gravity. When a whale or dolphin leaps, its center of gravity also traces a parabolic path.

544

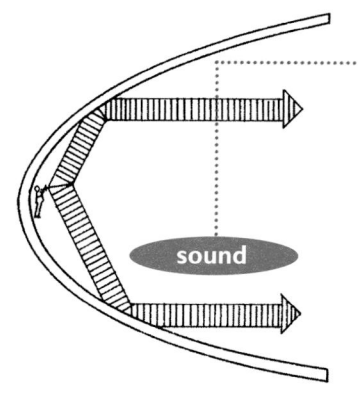

sound

Acoustical engineers use parabolas to direct sound waves. The parabolic curve of a band shell directs the music to the audience at an outdoor concert. During a TV reporter's interview, small parabolic reflectors direct the speakers' words to the microphone. Car headlights and some telescopes contain parabolic surfaces that direct light rather than sound waves.

Ballplayers and many other athletes know a lot more about parabolas than they may think. The outfielder running to catch a fly ball is actually predicting the parabolic trajectory of the batted ball. The pitcher who throws a knuckle ball or curve ball is trying to outwit the batter by making the ball change course from the expected parabola.

Suggested Rubric for Unit Project

4 The math story is creative and appealing. It contains a setting, characters, and a plot incorporating situations using equations and formulas from previous units and from this unit. The equations chosen are appropriate models for the situations. The variables in the equations and formulas have reasonable values.

3 The math story is acceptable but not particularly creative. The equations and formulas from this course have acceptable values for the variables.

Write a Math Story

Your group's project is to write a story that reveals the mathematics in the world around you.

Include at least three situations modeled by equations or formulas from earlier units.

Include a situation in which the trajectory of a batted ball is modeled by one of the equations from this unit.

By the end of the unit, you can figure out whether each of the equations below applies to a ball kicked or hit off a player's head, and how far and how high the ball travels. Include one of these trajectories in your story, too.

| Equations for the Paths of Some Soccer Balls | | |
| (x = horizontal distance in feet and y = vertical distance in feet) | | |
Strike angle	Starting speed	Equation
20°	93 ft/s	$y = x(0.364 - 0.002x)$
45°	30 ft/s	$y = -0.036x^2 + x + 6$
45°	103 ft/s	$y = -0.003x^2 + x$
60°	30 ft/s	$y = -0.071(x + 3)(x - 27.5)$

WELCOME TO PARABOLIC PARK

How could we resist the fireworks and fountains of Parabolic Park, only two miles away! Thirty minutes was all it took our group to cover the uphill distance at a pace of 15 min per mile (d=rt). The slope was an easy tan 10°—no problem. When we arrived, we found that the ads told a true story: "If you come to Parabolic Park, then the parabolas will dazzle you."

545

Support Materials

The *Project Book* contains information about the following topics for use with this Unit Project.

➤ Project Description
➤ Teaching Commentary
➤ Working on the Unit Project Exercises
➤ Completing the Unit Project
➤ Assessing the Unit Project
➤ Alternative Projects
➤ Outside Resources

Limited English Proficiency

Writing a math story is a valuable project for students acquiring English. It reviews the mathematical concepts of the course and requires written descriptions of these concepts in a story setting. Have a student fluent in English work as a partner with a student acquiring English.

ADDITIONAL BACKGROUND

Multicultural Note

The people of China probably used rockets as projectiles before their first documented use in the twelfth century. The black powder fuel of these first rockets consists of charcoal, saltpeter, and sulfur, all of which are found in China.

Around A.D. 1150, ground firework-rockets were used as projectiles during battle. The firework-rocket was attached to an arrow with weights. These weights allowed the rocket to travel a greater distance. By the year 1300, the rockets were designed so that they could fly between 500 and 1100 yards, and their use had already spread throughout Asia and Europe.

Suggested Rubric for Unit Project

2 The math story is incomplete. It may not contain a setting, characters, or a plot. The situations modeled on equations and formulas from previous units in the course are incomplete. The situations modeled on the baseball and soccer equations are incomplete. The values for the variables in the equations and formulas are not reasonable. This story should be returned with suggestions for improvements and a new deadline.

1 The story cannot be evaluated. It is illegible, incomplete, or not understandable. The story should be returned with a new deadline for completion. The group should be encouraged to speak with the teacher as soon as possible so that they understand the purpose and the format of the project.

Parabolic Surface

If a parabola is rotated about its axis of symmetry, the surface described by its rotation is a parabolic surface. Parabolic surfaces are used for such things as radar antennas, solar collectors, and auto headlights. A parabolic surface can reflect energy from a point or focus incoming energy at a point. The energy is then sent in one direction. The point is known as the focus and lies on the axis of symmetry.

ALTERNATIVE PROJECTS

Project 1, page 606

Mickey Mantle's Record Home Run

Find out the details about Mickey Mantle's longest home run in Yankee Stadium. Model the path of the ball with an equation and a graph and determine about how far from home plate the ball would have landed if it had not hit anything before it reached the ground.

Project 2, page 606

Making Parabolas from Straight Lines

Research how parabolas can be formed by straight lines using paper folding, paper and pencil, or string. Create a design that contains at least three parabolas formed by straight lines.

Getting Started

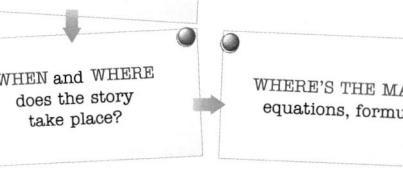

WHAT TYPE OF STORY? humor, science fiction, mystery, adventure, sports

WHO are your characters?

WHAT are they doing?

WHEN and WHERE does the story take place?

WHERE'S THE MATH? equations, formulas

WHAT KIND OF PICTURES? drawings, photos, cutouts

For this project you should work in a group of four students. Here are some ideas to help you get started.

☞ Think of a setting for your story.

☞ Outline your story.

☞ Review the equations and formulas you studied in this course. Discuss which ones you could use in your group's story.

☞ Think about where you can find reasonable values for the variables in the equations and formulas your group has decided to use in its story.

☞ Plan to meet later to discuss the baseball equations in the "Working on the Unit Project" exercises and the soccer equations in the table on page 545. For each equation, figure out all you can about the trajectory of the ball and how it was struck.

Working on the Unit Project

Your work in Unit 10 will help you write your math story.

Related Exercises:

Section 10-1, Exercise 31
Section 10-2, Exercises 53, 54
Section 10-3, Exercise 25
Section 10-4, Exercise 43
Section 10-5, Exercises 41–43
Section 10-6, Exercises 41, 42
Section 10-7, Exercise 43
Section 10-8, Exercises 34, 35

Alternative Projects p. 606

Can We Talk   TRAJECTORIES

➤ Where else have you seen parabolas in nature?

➤ Do you think you can kick a soccer ball as far as you can hit a baseball? a volleyball? Why or why not?

➤ Suppose two balls are hit at the same speed, but at different angles. Which ball do you think will land farther away? Why?

➤ Why would someone who designs baseball stadiums want to know about the trajectory of a baseball after it has been hit?

➤ How do you think wind affects the path of a kicked, thrown, batted, headed, or punched ball?

546   **Unit 10** Quadratic Equations as Models

Answers to Can We Talk?

➤ Shooting stars seem to follow a pattern similar to a parabola. Any object that is thrown or kicked into the air will follow a path similar to a parabola as it passes back to Earth.

➤ Soccer balls and volley balls are heavier and larger than softballs and baseballs. That is one reason why they do not go as far as a baseball.

➤ A ball hit at an angle closer to the ground will go further than one hit higher into the air. A ball hit at a 45° angle is the most likely to go the longest distance.

➤ Stadium designers need to know how far a ball could be hit and what are the most frequent patterns in order to make the game challenging for players and to protect the spectators and surrounding areas.

➤ Wind can carry a ball farther or can block its travel, depending on its force and direction.

Reflections

Focus

Recognize and draw the mirror image of an object.

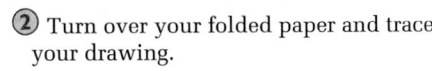

EXPLORATION

What happens when you flip a shape over a line?

- **Materials: paper, rulers, protractors**
- **Work with another student.**

Draw with a ruler.

Draw dashed lines for your tracing.

Draw the fold line.

① Fold a sheet of paper in half. Without unfolding your paper, draw any polygon you want. Be sure to use a dark pencil or pen.

② Turn over your folded paper and trace your drawing.

③ Unfold your paper and draw a line along the crease. It should look as if you have flipped your original drawing over the fold line.

④ Connect all corresponding vertices.

⑤ The fold line divides each segment you drew in step 4 into two parts. Measure the length of each part. Write the measurements on your drawing.

⑥ The fold line crosses each segment you drew in step 4. Measure each angle formed. Write the measurements on your drawing.

⑦ Look at the measurements you have written. What is the relationship between the fold line and the segments you drew?

⑧ Hold your paper so that the fold line is vertical. Pick any point on the right side of your original drawing. Is the corresponding point on the tracing located on its right side? Try other points. What do you find?

⑨ Compare your results. What do you find?

10-1 Reflections

547

Answers to Exploration

1–6. Outcomes may vary.

7. The fold line is perpendicular to each segment and divides each segment in half.

8. No. The results are the same with other points. Points on the right side of the original drawing end up on the left side of the tracing, and points on the left side of the original drawing end up on the right side of the tracing.

9. Results should be the same.

Objectives and Strands
See pages 544A and 544B.

Spiral Learning
See page 544B.

Materials List
➤ 8½ × 11 in. paper
➤ Ruler
➤ Protractor
➤ Graph paper
➤ Ball

Recommended Pacing
Section 10-1 is a two-day lesson.
Day 1
Pages 547–548: Exploration through Properties of a Reflection, *Exercises 1–12*
Day 2
Pages 549–551: Sample 1 through Look Back, *Exercises 13–31*

Extra Practice
See pages 638–639.

Warm-Up Exercises
Warm-Up Transparency 10-1

Support Materials
➤ Practice 76
➤ Enrichment 67 in the Activity Bank
➤ Study Guide 10-1
➤ Problem Set 22
➤ Additional Exploration 11
➤ Diagram Master 2 in the Explorations Lab Manual
Overhead Visual 10
➤ McDougal Littell Mathpack software: *Geometry Inventor*
➤ Geometry Inventor Activity Book: Activities 16–18
➤ Quiz 10-1
➤ Alternative Assessment 1, 2

TEACHING

Exploration

When students complete the Exploration, they will have discovered the properties of a reflection listed on this page. They should repeat steps 1–8 for another polygon, however, so as not to draw conclusions based upon the examination of a single figure.

You may wish to point out that transformations that produce congruent images are often referred to as *rigid motions*. A key property of a rigid motion is that it preserves distance; that is, if two points A and B on a figure are x units apart, then the image points A' and B' will also be x units apart.

Multicultural Note

You might want to discuss with students the photograph on this page of the Taj Mahal.

Located in Agra, India, about 110 miles south of New Delhi, the Taj Mahal was built as a tomb for Mumtaz Mahal, the wife of a Muslim ruler, Shah Jehan. With the help of 20,000 workers, including builders, jewelers, and artists from all over the world, the Taj Mahal took 18 years to build and was completed in 1648.

The octagonal building is made of white marble and extends for 186 feet on its longest side. It rests on a vast platform of red sandstone 313 feet square. The walls of the mausoleum are 70 feet high and are topped by a massive bulb-shaped dome. The platform on the tip of the dome's pinnacle soars 243 feet in the air–the height of a 20-story building.

The tomb stands in an elaborate walled garden with marble pavements, fountains, and reflecting pools, and is reflected in its entirety in the long pool at the front.

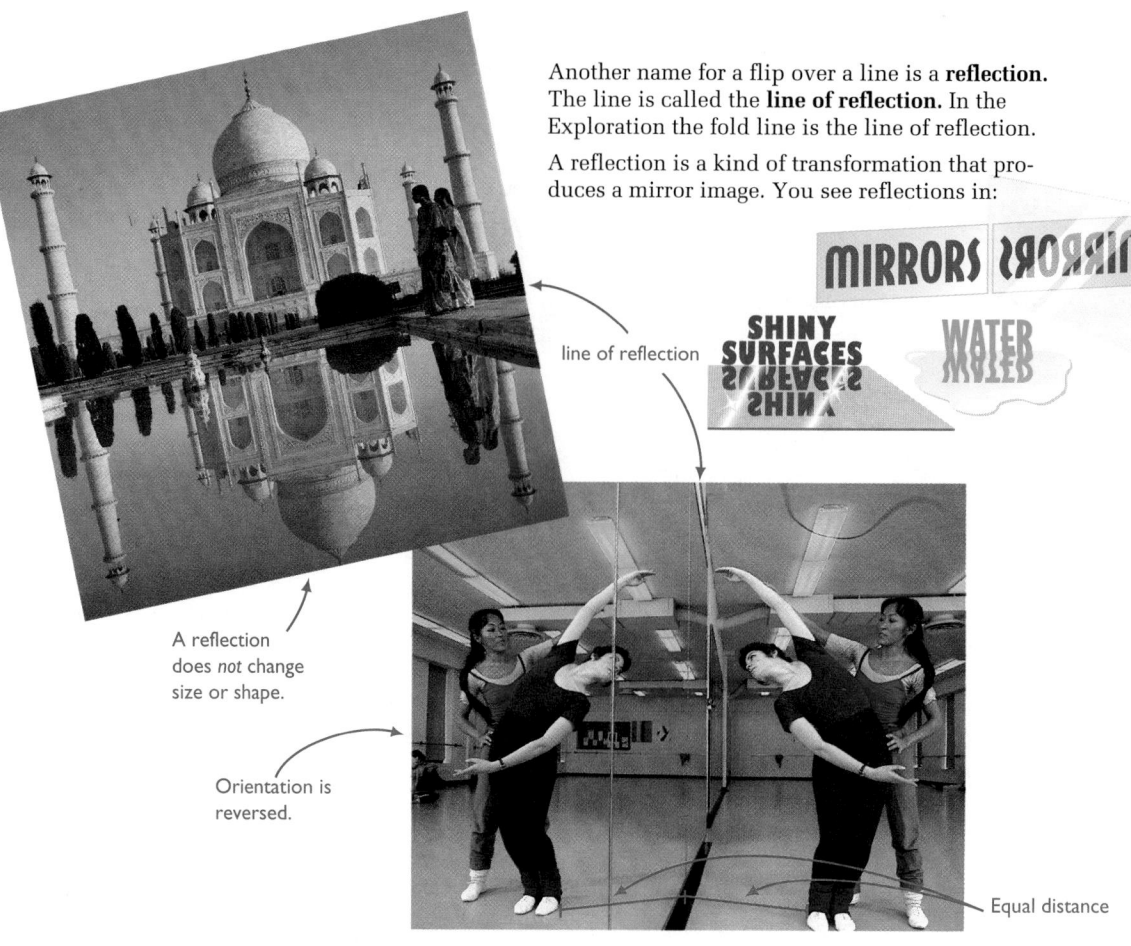

line of reflection

A reflection does *not* change size or shape.

Orientation is reversed.

Equal distance

Another name for a flip over a line is a **reflection.** The line is called the **line of reflection.** In the Exploration the fold line is the line of reflection.

A reflection is a kind of transformation that produces a mirror image. You see reflections in:

MIRRORS

SHINY SURFACES

WATER

In a mirror, right and left are switched because a reflection reverses *orientation*. **Orientation** is the direction in which the points on a figure are ordered.

PROPERTIES OF A REFLECTION

➤ The image is congruent to the original figure.

➤ The orientation of the image is reversed.

➤ The line of reflection is perpendicular to and cuts in half any segment connecting corresponding points on the image and the original figure.

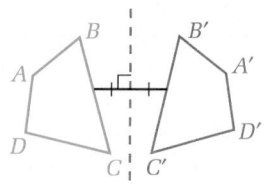

:······➤ Now you are ready for:
: **Exs. 1–12 on pp. 551–552**

Writing Tell whether each transformation is a reflection. If it is, trace the diagram and draw the line of reflection. If it is not, tell why not.

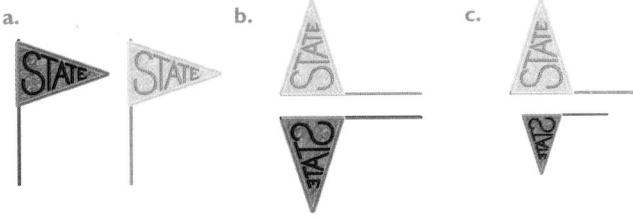

a.   b.   c.

Sample Response

a. This is *not* a reflection. The image is congruent to the original figure but the orientation is not reversed.

b. This is a reflection. The image is congruent to the original and its orientation is reversed.

The line of reflection is halfway between the original and the image.

c. This is *not* a reflection. The orientation of the image is reversed, but the image is not congruent to the original figure.

Check You can check that the diagram shows a reflection by tracing and folding or with a MIRA® transparent mirror.

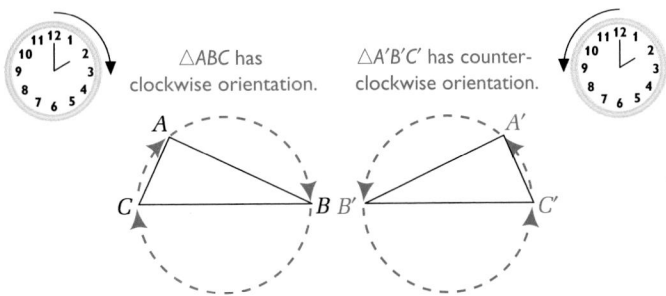

BY THE WAY...

Watch water swirling down a drain. It spins clockwise above the Equator and counterclockwise below it—and so do tornadoes and hurricanes.

The cause of this reversal of orientation, called the *Coriolis effect*, is the spin of Earth on its axis.

The orientation of a polygon may be *clockwise* or *counterclockwise*.

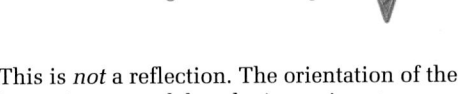

△*ABC* has clockwise orientation.

△*A'B'C'* has counterclockwise orientation.

10-1 Reflections

S1 Tell whether each transformation is a reflection. If it is, trace the diagram and draw the line of reflection.

a.

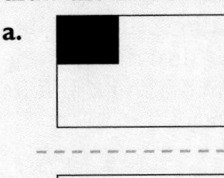

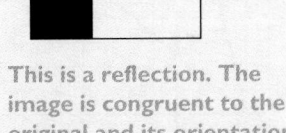

This is a reflection. The image is congruent to the original and its orientation is reversed.

b.

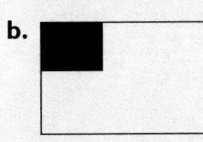

This is not a reflection. The orientation of the image is reversed, but the image is not congruent to the original figure.

c.

This is not a reflection. The image is congruent to the original figure, but the orientation is not reversed.

Additional Sample

S2 Copy △XYZ and the coordinate grid on graph paper. Reflect △XYZ over the y-axis. Write the coordinates of the vertices of its image and describe the orientation of △XYZ and its image.

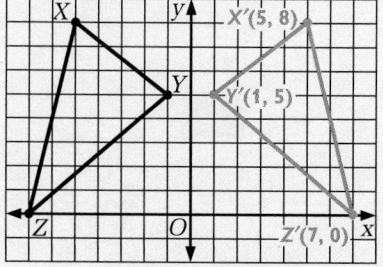

△**XYZ has clockwise direction.** △**X′Y′Z′ has counterclockwise direction.**

Talk it Over

Questions 1–4 help students to understand the properties of a reflection. Question 4(b) leads to the observation that a reflection can reflect a figure onto itself. In this case, the figure is said to be *invariant* (unchanged) for the reflection.

Reasoning

After students have studied Sample 2, you may wish to ask them to make a generalization about reflecting a point (a, b) over the x-axis and over the y-axis.

x-axis: $(a, b) \rightarrow (a, -b)$

y-axis: $(a, b) \rightarrow (-a, b)$

Mathematical Procedures

Students should understand that a figure can be reflected over any line. On graph paper, have students draw a coordinate axis, the line $y = -x$, and a triangle. Ask students how they would reflect the triangle over the line and then have them perform the reflection. Ask students to state any conclusions about the procedure.

Using Technology

The *Geometry Inventor* software can be used to find the reflection image of a polygon.

Reflections on a Coordinate Plane

When a polygon is drawn on a coordinate plane, you can reflect it over the horizontal or the vertical axis. These axes are also called the **x-axis** and the **y-axis**.

Sample 2

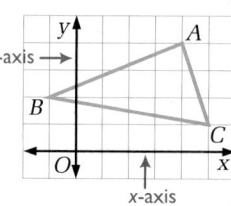

Copy △ABC and reflect it over the x-axis. Describe the orientation of △ABC and its image.

TECHNOLOGY NOTE

With geometric drawing software, you can reflect any figure you draw over any line you draw.

Sample Response

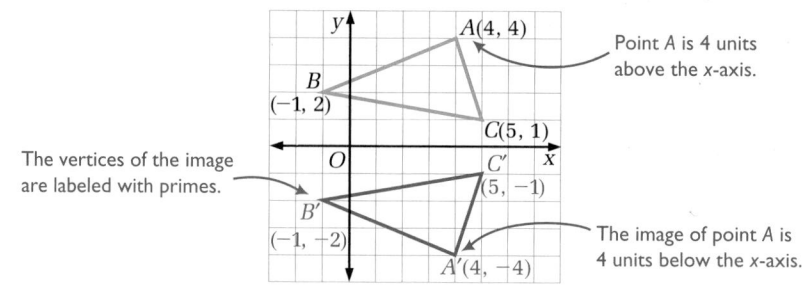

Point A is 4 units above the x-axis.

The vertices of the image are labeled with primes.

The image of point A is 4 units below the x-axis.

△ABC has counterclockwise orientation.

△A′B′C′ has clockwise orientation.

Talk it Over

1. What is the line of reflection in Sample 2? What is its equation?

2. In Sample 2, a point on △ABC is 3 units above the line of reflection. Is the corresponding point on △A′B′C′ above or below the line of reflection? How many units above or below?

3. Suppose a polygon is reflected over a vertical line. One point on the polygon is 2 units to the right of the line of reflection. Where is the image of this point located?

4. **a.** Discuss the symmetric shape in the photo at the left in terms of a reflection. Identify the line of reflection, the original, and the image.

   b. Can a figure be its own reflection image? Explain.

550   **Unit 10** Quadratic Equations as Models

Answers to Talk it Over

1. the x-axis; $y = 0$
2. below; 3 units below
3. 2 units to the left of the line of reflection
4. Answers may vary. Examples are given.

   a. The whale's tail in the photo has symmetry because one half of the tail is a reflection of the other half.

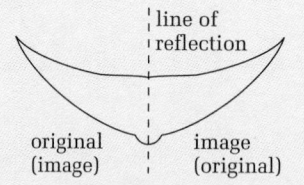

line of reflection

original (image)   image (original)

   b. Yes, if it has symmetry. An example is the whale's tail in part (a).

Answers to Look Back

Answers may vary. An example is given. A reflection is like a translation, rotation, and dilation because a reflection does not change the shape of the figure. A reflection is different from the other three

Look Back ◄

Now you are ready for:
Exs. 13–31 on pp. 553–554

How is a reflection like the other transformations you have learned about—translations, rotations, and dilations? How is it different?

10-1 Exercises and Problems

1. **Reading** To decide whether a transformation is a reflection, what three properties should you look for?

2. **a.** Repeat the Exploration using your printed first name instead of a polygon.

   b. What property of reflections makes your name hard to read?

   c. Describe what happens to your original printed name and its reflection when you look at them in a mirror.

3. Imagine you are looking at yourself in a mirror.

   a. When you raise your right arm, what does your mirror image do?

   b. When you walk toward the mirror, what does your mirror image do?

   c. Suppose you are wearing a T-shirt that says "Plant a Tree." When you look in a mirror, can you read the words on your T-shirt?

4. **a.** Print the word AMBULANCE so that someone can read it in a mirror.

   b. When is it important to be able read this word in a mirror?

5. **Writing** Which of these photos do you think show a reflection? Explain your choice.

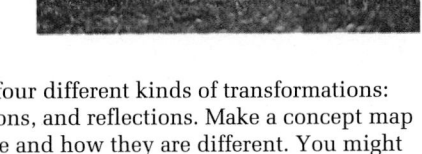

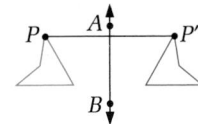

6. You have now worked with four different kinds of transformations: rotations, translations, dilations, and reflections. Make a concept map that shows how they are alike and how they are different. You might want to include a sketch of each type of transformation.

7. Point P' is the image of point P after a reflection over $\overleftrightarrow{AB}$. How is $\overleftrightarrow{AB}$ related to segment PP'?

10-1 Reflections

Limited English Proficiency

To help students with limited English proficiency work effectively on transformations, encourage them to prepare illustrated bilingual note cards for each of the following terms: *rotations, translations, dilations,* and *reflections.*

Look Back

Have students give written responses to the Look Back. These questions provide an opportunity for students to summarize and compare the characteristics of the transformations they have studied. Students should include these responses in their journals.

APPLYING

Suggested Assignment
Standard 1, 3, 4, 6–8, 13–31
Extended 1–11, 13–31

Integrating the Strands
Algebra Exs. 27–30
Geometry Exs. 1–31
Logic and Language Exs. 6, 24

Answers to Look Back

transformations because, unlike the others, a reflection reverses orientation. Unlike a dilation, but like a translation and a rotation, a reflection does not change the size of a figure.

Answers to Exercises and Problems

1. The image is congruent to the original. The orientation is reversed. The line of reflection divides in half and is perpendicular to each segment connecting corresponding points on the image and the original.

2. **a.** Answers may vary.

   b. The orientation is reversed.

   c. The reflection looks exactly like the original when held up to a mirror.

3. **a.** raises its left arm

   b. walks towards me

   c. Yes, but the words and letters are backwards.

4. **a.** AMBULANCE *(mirror-reversed)*

4. **b.** Answers may vary. An example is given. A driver must be able to recognize an approaching ambulance by looking in a rear-view mirror.

5. The first and last photos show true reflections because the animals are reflected in water. The middle photo is not a reflection, it's just two different penguins bending in opposite ways, which looks like a reflection.

6. See answer in back of book.

7. $\overleftrightarrow{AB}$ is perpendicular to $\overline{PP'}$ and divides $\overline{PP'}$ in half.

551

Communication: Drawing

For Ex. 8, have students separate the letters of the alphabet into those that are changed with reflection and those that are not changed with reflection. Remind them to consider both horizontal and vertical reflection. Ask them to illustrate their answers with drawings.

Research

There are mathematics books written in Braille. Students may find it interesting to research how number and operational symbols ($+, -, \times, \div$) are written in Braille.

8. Answers may vary. Examples are given.

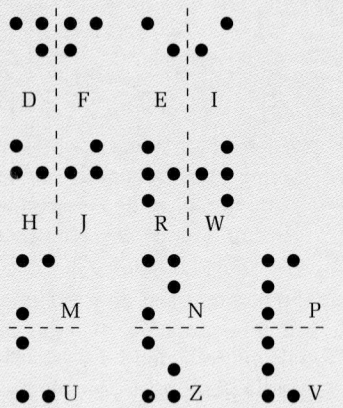

9. Answers may vary. Examples are given.

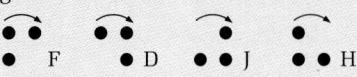

10. Answers may vary. Examples are given.

8. **Open-ended** The words MOM, COOKBOOK, and HAT HUT are unchanged when they are reflected over the dashed line shown. Write three other words that are unchanged when reflected over a vertical or horizontal line through the word, and draw each line of reflection.

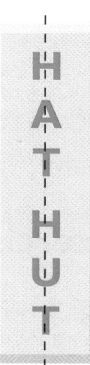

connection to LANGUAGE ARTS

Braille signs are read with the fingertips. Each sign is a pattern of raised dots on a 3×2 matrix. In the alphabet shown, the large black circles represent raised dots. The tiny circles represent positions in the matrix that do *not* contain raised dots.

Use the Braille alphabet shown.

9. Some pairs of Braille letters are reflections of each other. Draw each pair with the line of reflection between the letters.

10. Some pairs of Braille letters are rotations of each other. Draw each pair with a curved arrow to show the rotation.

11. There are *no* pairs of Braille signs that are translations of each other. Why do you think this is so?

12. **Research** There are also Braille signs for words, parts of words, and punctuation. Do some research about Braille. What other signs can you find that are reflections of each other?

> **BY THE WAY...**
>
> The Braille alphabet was invented almost 175 years ago by a fifteen-year-old French student named Louis Braille. Braille himself was visually impaired.

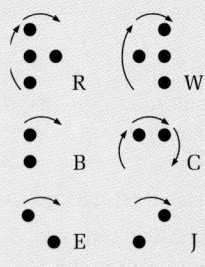

11. Answers may vary. An example is given. To a blind person, translations would be read as the same letter because a translation

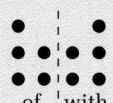

does not change the position of the dots in the matrix .

12. Answers may vary. An example is given.

of ¦ with

13. Yes.

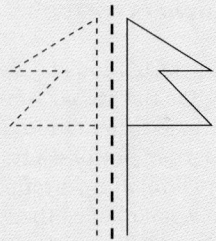

14. No. The image has the same orientation as the original. This is a rotation.

Writing Tell whether each transformation is a reflection. If it is, trace the flags and draw the line of reflection. If it is not, tell why not.

13.

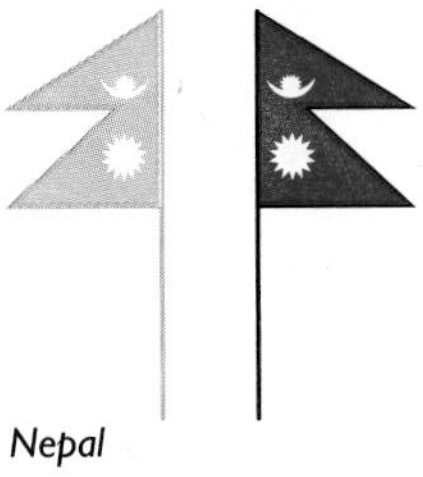

Nepal

14.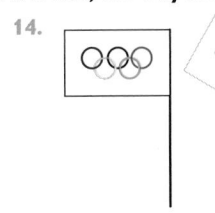

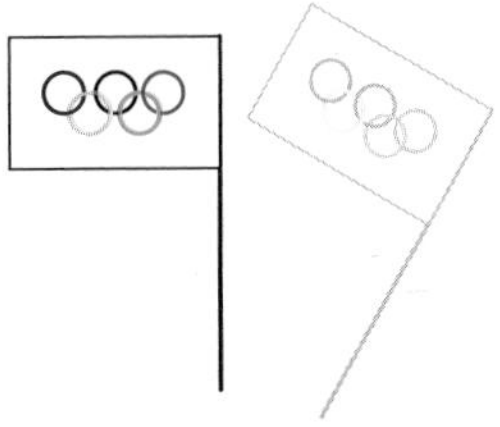

Olympics

15.

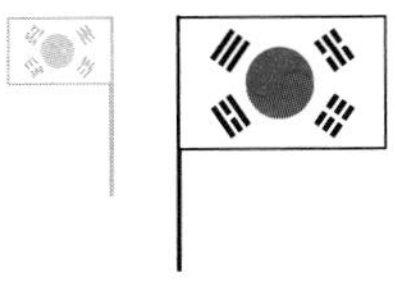

Republic of Korea

Each diagram shows a figure and its image. What is the line of reflection?

16.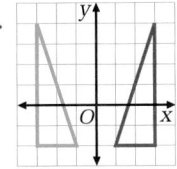

17.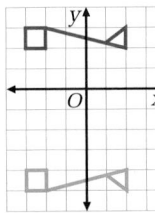

18.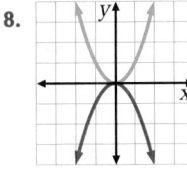

19.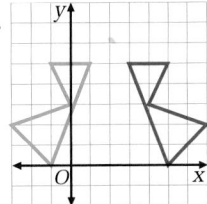

Copy each figure and reflect it over the indicated line. Describe the orientation of the original figure and its image.

20. **a.** x-axis

   b. y-axis

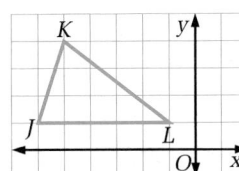

21. **a.** x-axis

   b. y-axis

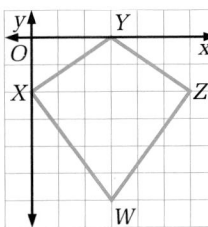

22. **a.** x-axis

   b. y-axis

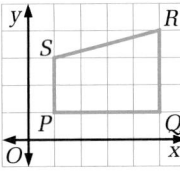

23. The x-axis is the line of symmetry for the figure shown. Suppose you reflect the figure over the x-axis.

   a. Describe the image after this transformation.

   b. What are the coordinates of the point of the figure that lies on the x-axis? What are the coordinates of its image?

   c. What are the coordinates of the image of point A? Explain how you arrived at your answer.

   d. What are the coordinates of the image of point B? Explain how you arrived at your answer.

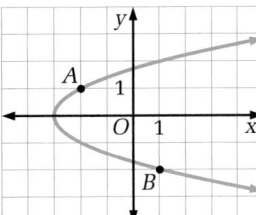

In Talk it Over question 4(b), students learned that a figure can be identical to its reflection image. Ask students if they think that a point can be its own image under a reflection. Have them suggest an example for which this is true. When a point is its own image, it is called a *fixed point* for the transformation. The figure shows segment AB bisected by the line of reflection k at point P. When $\overline{AB}$ is reflected in k, point P is a fixed point.

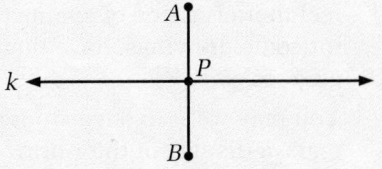

Answers to Exercises and Problems

15. No. The image is not congruent to the original figure. This is a dilation combined with a reflection.

16. the y-axis, or the line $x = 0$

17. the line $y = -1$

18. the x-axis, or the line $y = 0$

19. the line $x = 2$

20. $\triangle JKL$ has clockwise orientation and $\triangle J'K'L'$ has counterclockwise orientation for both reflections.

   a, b.

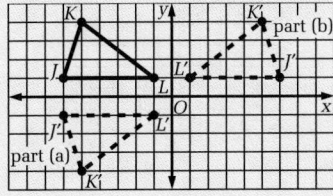

21. $WXYZ$ has clockwise orientation and $W'X'Y'Z'$ has counterclockwise orientation for both reflections.

   a, b. See graph at right.

22. $PQRS$ has counterclockwise orientation and $P'Q'R'S'$ has clockwise orientation for both reflections.

   a, b. See graph at right.

21. a, b.

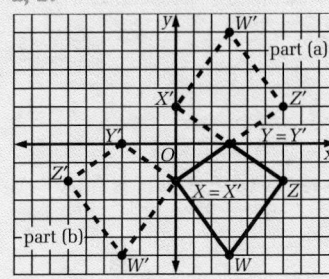

22. a, b.

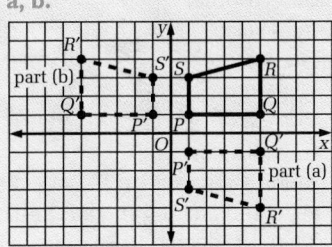

23. See answers in back of book.

553

Practice 76 For use with Section 10-1

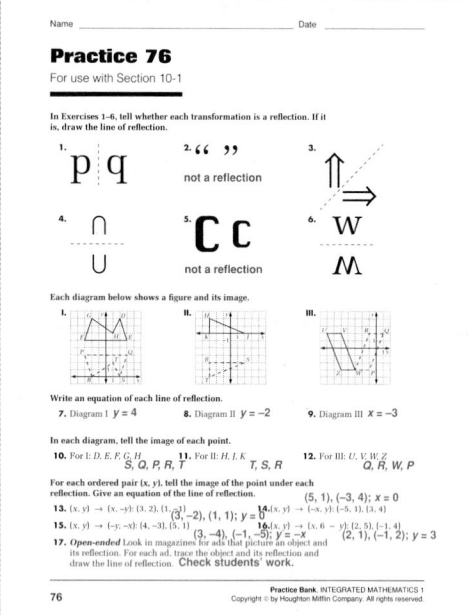

24. **Writing** Draw any picture you wish. For example, you could draw a flower, a geometric shape, a combination of the two, or something else. Then create a design by reflecting, translating, and rotating your picture. Describe how you made your design.

Review **PREVIEW**

25. Find the ratio of the surface areas of a cylinder with radius 6 in. and a similar cylinder with radius 15 in. *(Section 9-8)*

26. The ratio of the heights of two similar prisms is 3:2. The volume of the smaller prism is 720 cm³. What is the volume of the larger prism? *(Section 9-8)*

Graph each equation. *(Section 8-3)*

27. $x = 2$ 28. $x = -1$ 29. $y = 4$ 30. $y = -3$

Working on the Unit Project

31. **Group Activity** Work in a group of three students. Two of you should toss a ball back and forth in front of a wall while the third observes the path of the ball (you can use a crumpled piece of paper for a ball). Toss the ball a few different ways, and take turns until each student has had a chance to be the observer.

a. Each of you should make a sketch of one path that he or she observed.

b. Does the path you drew have symmetry? If so, sketch the line of symmetry.

c. Compare everyone's drawings. How are they alike? How are they different?

554

Answers to Exercises and Problems

24. Drawings and descriptions may vary. An example is given.

I began with the crescent moon shape ABC and shaded the left half. Then I reflected this shape over line AC. I reflected the entire oval over the horizontal line through point O and rotated the resulting figure 90° clockwise to produce the oval below the original one. I translated the second oval up and to the right to touch my first two ovals and rotated the resulting figure 90° clockwise again to produce the oval on the right.

25. 4:25

26. 2430 cm³

27.

28.

29–31. See answers in back of book.

554

Transforming Parabolas

PLANNING

Objectives and Strands
See pages 544A and 544B.

Spiral Learning
See page 544B.

Materials List
➤ Graphics calculator or graphing software
➤ Graph paper

Recommended Pacing
Section 10-2 is a two-day lesson.
Day 1
Pages 555–556: Opening paragraph through Exploration, *Exercises 1–16*
Day 2
Pages 557–558: Sample 1 through Look Back, *Exercises 17–54*

Extra Practice
See pages 638–639.

Warm-Up Exercises
Warm-Up Transparency 10-2

Support Materials
➤ Practice 77
➤ Enrichment 68 in the Activity Bank
➤ Study Guide 10-2
➤ Problem Set 22
➤ Additional Exploration 12
➤ Diagram Masters 2, 3 in the Explorations Lab Manual
Overhead Visual 10
➤ Using Plotter Plus: Parabolas
➤ Using Mac or IBM Plotter Plus Disk: Parabola Plotter and Parabola Quiz
➤ Quiz 10-2
➤ Alternative Assessments 3–7

Focus

Translate and reflect the graph of $y = x^2$.

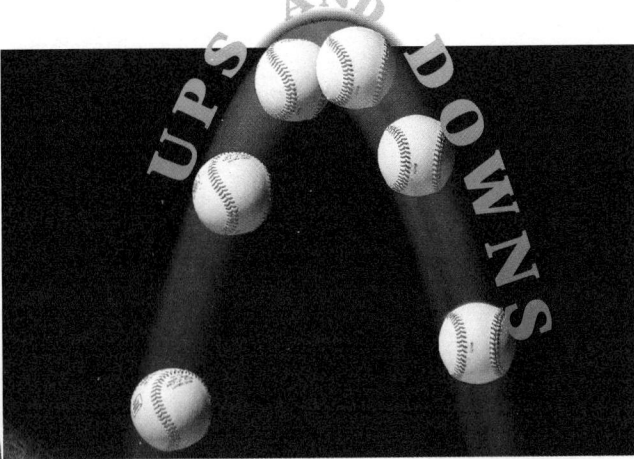

A strobe photo shows that the path of a thrown ball looks like an upside-down U.

A thown or batted ball, water rising from a fountain, a whale's spout, and the light trails of fireworks all follow a curve called a **parabola.** This curve has the same general shape as the graph of the *squaring function* $y = x^2$.

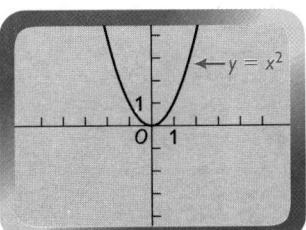

Talk it Over

1. Look at the graph of $y = x^2$.
   a. What is the line of symmetry of the graph?
   b. The point of a parabola that lies on the line of symmetry is called the **vertex** of the parabola. What are the coordinates of the vertex of the graph?

2. Look at the graph of $y = -x^2$ (read $-x^2$ as "the opposite of x^2").
   a. What is the line of symmetry of the graph?
   b. What are the coordinates of the vertex of the graph?
   c. Suppose you want to find the value of y when $x = 2$. Which do you write, $(-2)^2$ or -2^2? Does it matter?

3. How do the values of y compare for $y = -x^2$ and $y = x^2$ when $x = 1$? when $x = 0$? when $x = -2$? in general?

4. The graph of $y = -x^2$ is a reflection of the graph of $y = x^2$. What is the line of reflection?

10-2 Transforming Parabolas

555

Answers to Talk it Over

1. a. the y-axis, or the line $x = 0$
   b. (0, 0)

2. a. the y-axis, or the line $x = 0$
   b. (0, 0)
   c. -2^2; Yes, it matters because $(-2)^2 = (-2)(-2) = 4$ and $-2^2 = -(2 \cdot 2) = -4$.

3. Answers may vary. An example is given. When $x = 1$, the values of y are opposites (-1 and 1); when $x = 0$, the values of y are the same (0); when $x = -2$, they are opposites (-4 and 4). In general, for a given value $x \neq 0$, the corresponding values of y are opposites.

4. the x-axis, or the line $y = 0$

Talk it Over

Question 1(b) defines the vertex of a parabola. Students should realize that this point is the minimum (or maximum) point on the graph of a parabola. Question 2(c) points out the important difference between $(-2)^2$ and -2^2. Questions 3 and 4 allow students to compare $y = x^2$ and $y = -x^2$.

Exploration

The goal of the Exploration is to lead students to discover the relationship between the basic function $y = x^2$ and the translated parabolas $y = x^2 + a$ and $y = (x + a)^2$. To make certain all students understand the relationships presented here, you may wish to have the groups compare their summaries from step 6.

EXPLORATION

How can you find other functions whose graphs are parabolas?

* **Materials:** graphics calculators or graphing software
* **Alternative approach:** Make a table of values for each function and sketch the parabola on graph paper.
* **Work in a group of four students.**

1. Using a standard viewing window, graph the function $y = x^2$.

2. On the same axes, graph the function $y = x^2 + 4$. Describe the relationship between the graphs of the two functions.

3. Repeat step 2 comparing the graph of $y = x^2$ with the graph of $y = x^2 - 4$.

4. Repeat step 2 comparing the graph of $y = x^2$ with the graph of $y = (x + 4)^2$.

5. Repeat step 2 comparing the graph of $y = x^2$ with the graph of $y = (x - 4)^2$.

6. Summarize your findings by answering these questions.

   a. How is the graph of $y = x^2$ translated when a number is added to or subtracted from x^2?

   b. How is the graph of $y = x^2$ translated when a number is added to or subtracted from x *before* it is squared?

······► **Now you are ready for:**
  Exs. 1–16 on pp. 558–560

Unit 10 Quadratic Equations as Models

Answers to Exploration ···············

1.

$y = x^2$	
x	y
-2	4
-1	1
0	0
1	1
2	4

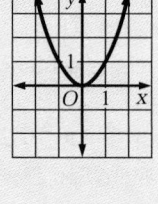

2.

$y = x^2 + 4$	
x	y
-2	8
-1	5
0	4
1	5
2	8

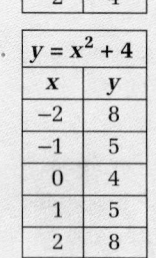

The graph of $y = x^2 + 4$ is the same as $y = x^2$, but translated 4 units up.

3.

$y = x^2 - 4$	
x	y
-2	0
-1	-3
0	-4
1	-3
2	0

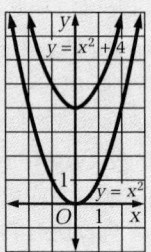

The graph of $y = x^2 - 4$ is the same as $y = x^2$, but translated 4 units down.

4.

$y = (x + 4)^2$	
x	y
-6	4
-5	1
-4	0
-3	1
-2	4

See graph at right. The graph of $y = (x + 4)^2$ is the same as $y = x^2$, but translated 4 units to the left.

5.

$y = (x - 4)^2$	
x	y
2	4
3	1
4	0
5	1
6	4

See graph at right. The graph of $y = (x - 4)^2$ is the same as $y = x^2$, but translated 4 units to the right.

4.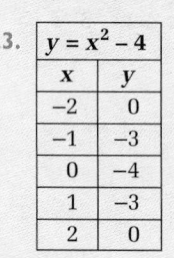

5.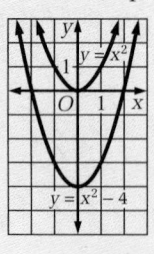

6. See answers in back of book.

Sample 1

Tell how to translate the graph of $y = x^2$ in order to produce the graph of each equation.

a. $y = x^2 + 2$ b. $y = x^2 - 2$

c. $y = (x + 2)^2$ d. $y = (x - 2)^2$

Sample Response

a. $y = \underbrace{x^2 + 2}$

2 is added to x^2

b. $y = \underbrace{x^2 - 2}$

2 is subtracted from x^2

c. $y = \underbrace{(x + 2)^2}$

2 is added to x

d. $y = \underbrace{(x - 2)^2}$

2 is subtracted from x

Translate the graph of $y = x^2$ 2 units up.

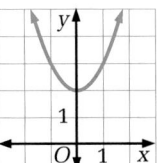

Translate the graph of $y = x^2$ 2 units down.

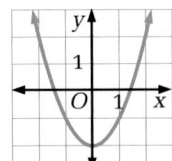

Translate the graph of $y = x^2$ 2 units to the left.

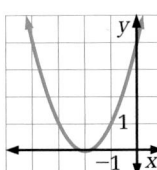

Translate the graph of $y = x^2$ 2 units to the right.

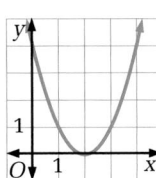

BY THE WAY...

A water jet from a fountain forms a parabola. If there are many jets, their outlines also form a parabola.

Talk it Over

5. Which of the parabolas in Sample 1 have the y-axis as their line of symmetry?

6. Which of the parabolas in Sample 1 have their vertex on the x-axis?

7. In which quadrants does each parabola in Sample 1 appear?

8. Can the value of y be a negative number for any of the parabolas in Sample 1? How do you know?

Answers to Talk it Over

5. the parabolas in parts (a) and (b)

6. the parabolas in parts (c) and (d)

7. (a) I, II; (b) I, II, III, IV; (c) I, II; (d) I, II

8. Yes. The graph of the parabola in part (b) is in the III and IV quadrants. In these quadrants, the value of the y-coordinate for each point is negative.

Additional Sample

S1 Tell how to translate the graph of $y = -x^2$ to produce the graph of each equation.

a. $y = -x^2 + 3$

3 is added to $-x^2$. Translate the graph of $y = -x^2$ 3 units up.

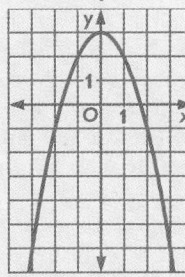

b. $y = -x^2 - 3$

3 is subtracted from $-x^2$. Translate the graph of $y = -x^2$ 3 units down.

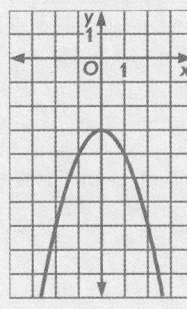

c. $y = -(x + 3)^2$

3 is added to x. Translate the graph of $y = -x^2$ 3 units to the left.

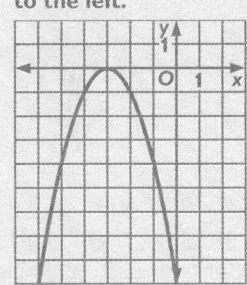

d. $y = -(x - 3)^2$

3 is subtracted from x. Translate the graph of $y = -x^2$ 3 units to the right.

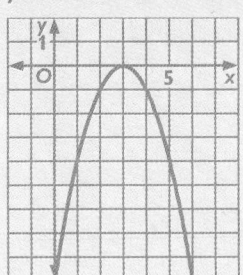

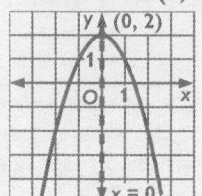
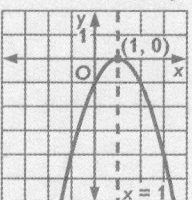

Sample 2

For the graph of each equation, write the equation of the line of symmetry and the coordinates of the vertex.

a. $y = x^2 - 5$ **b.** $y = (x + 6)^2$

Sample Response

a. $y = x^2 - 5$

Translate $y = x^2$ down 5 units.

The vertex moves down 5 units.

The line of symmetry stays the same.

The equation of the line of symmetry is $x = 0$.

Substitute 0 for x in the equation $y = x^2 - 5$ to find the value of y.

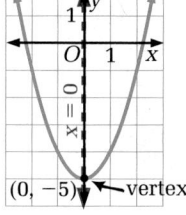

The vertex is at the point $(0, -5)$.

b. $y = (x + 6)^2$

Translate $y = x^2$ to the left 6 units.

The vertex moves to the left 6 units.

The line of symmetry moves to the left 6 units.

The equation of the line of symmetry is $x = -6$.

Substitute -6 for x in the equation $y = (x + 6)^2$ to find the value of y.

The vertex is at the point $(-6, 0)$.

▶ **Now you are ready for:**
Exs. 17–54 on pp. 560–561

Look Back

Explain how you would change the equation $y = x^2$ to produce a graph translated up, down, left, or right.

10-2 Exercises and Problems

1. **Reading** How are the graphs of $y = x^2$ and $y = -x^2$ shown on page 555 alike? How are they different?

Use the diagram showing the path of a beach ball in the air.

2. What does the vertex mean in terms of this situation?

3. What horizontal distance does the ball travel from the moment it is thrown in the air until it falls back down?

4. What horizontal distance has the ball traveled when it first begins to fall back down?

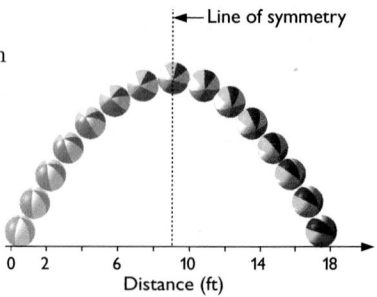

← Line of symmetry

Distance (ft)

Answers to Look Back

Answers may vary. An example is given. To produce a graph that is translated: (1) up, add a positive number to x^2; (2) down, subtract a positive number from x^2; (3) left, add a positive number to x before squaring; (4) right, subtract a positive number from x before squaring.

Answers to Exercises and Problems

1. Answers may vary. An example is given. They have the same shape, vertex, and line of symmetry. For corresponding points on each graph, the values of the x-coordinates are the same, but the values of the y-coordinates are opposites.

2. Answers may vary. An example is given. The vertex is the point where the ball stopped rising and started falling.

3. 18 ft

4. about 9 ft

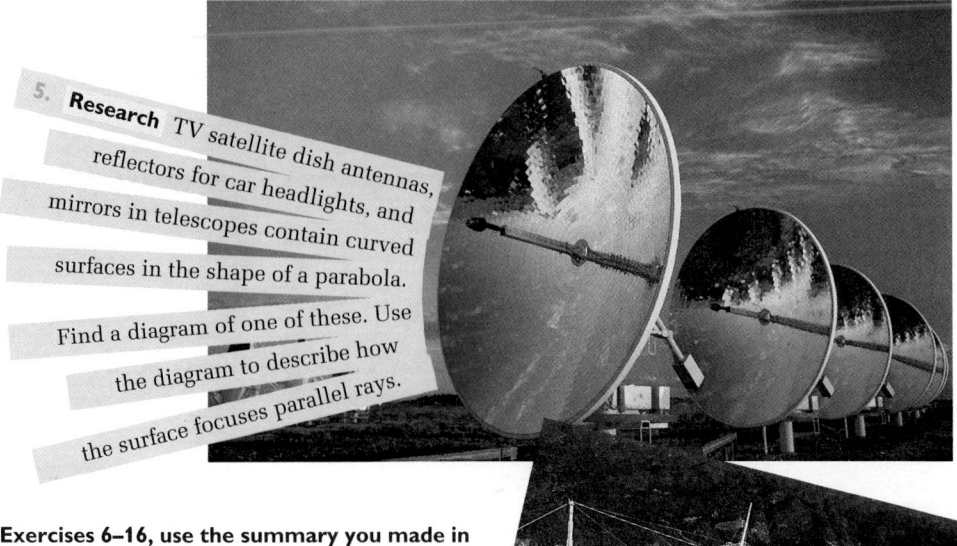

5. Research TV satellite dish antennas, reflectors for car headlights, and mirrors in telescopes contain curved surfaces in the shape of a parabola.

Find a diagram of one of these. Use the diagram to describe how the surface focuses parallel rays.

For Exercises 6–16, use the summary you made in the Exploration.

Match each equation with the translation of $y = x^2$ that produces it.

6. $y = x^2 - 9$ **A.** to the left 9 units
7. $y = x^2 + 9$ **B.** up 9 units
8. $y = (x - 9)^2$ **C.** down 9 units
9. $y = (x + 9)^2$ **D.** to the right 9 units

Match each equation with its graph.

10. $y = (x + 3)^2$ 11. $y = x^2 + 3$ 12. $y = (x - 3)^2$
13. $y = -x^2$ 14. $y = x^2 - 3$

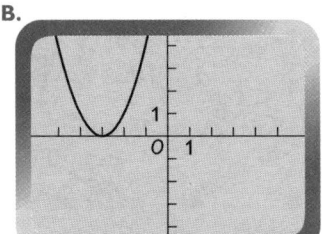

A.

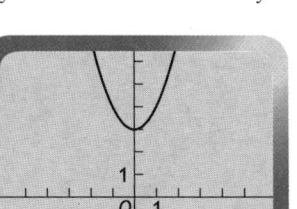

B.

C. 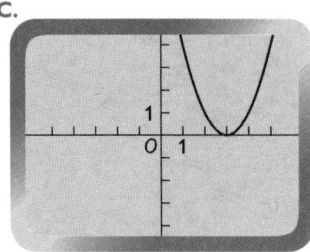   **D.** 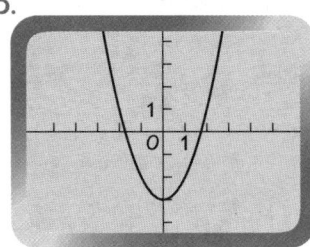   **E.**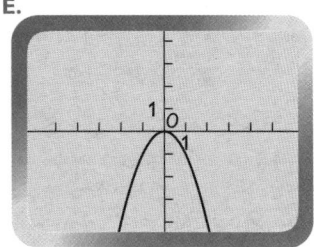

BY THE WAY...

Are ETs on other worlds calling to us from across the universe? In Arecibo, Puerto Rico, a telescope with a 1000 foot wide parabola-shaped disk is waiting to pick up their signals.

10-2 Transforming Parabolas   **559**

Answers to Exercises and Problems

5. Incoming light or sound waves are reflected to a point called the focus, concentrating the light or signal. The same diagram can represent a light bulb or transmitter located at the focus. In this case, the light or signal is reflected out in parallel rays.

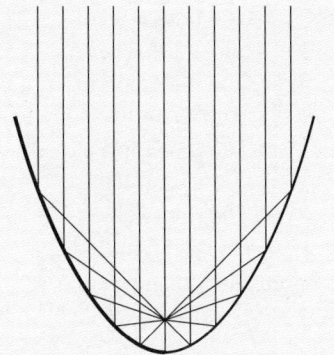

6. C
7. B
8. D
9. A
10. B
11. A
12. C
13. E
14. D

15. **Writing** Ariel says that $y = x^2 + 7$ *cannot* be the equation of the path of water from a fountain. Do you agree with her? Why or why not?

16. a. What translation of $y = -x^2$ produces each graph?

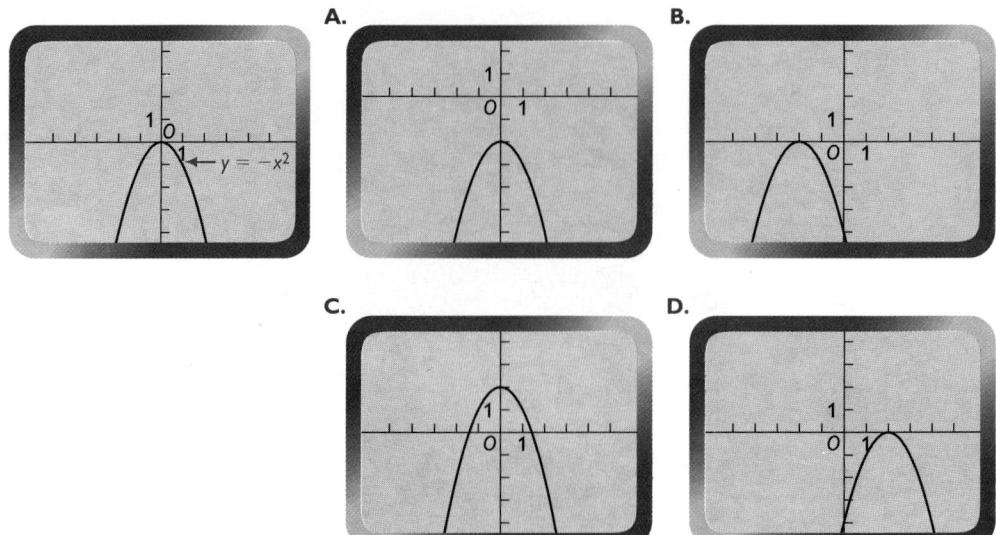

b. Using a graphics calculator or a table of values for each equation, match each graph in part (a) with one of these equations.

  I. $y = -x^2 + 2$ **II.** $y = -x^2 - 2$ **III.** $y = -(x - 2)^2$ **IV.** $y = -(x + 2)^2$

Tell how to translate the graph of $y = x^2$ or $y = -x^2$ to produce the graph of each equation.

17. $y = x^2 + 10$ **18.** $y = x^2 - 7$ **19.** $y = (x + 11)^2$

20. $y = (x - 7)^2$ **21.** $y = -x^2 + 10$ **22.** $y = -x^2 - 6$

23. $y = -(x + 6)^2$ **24.** $y = (x + 3)^2 - 4$ **25.** $y = (x - 1)^2 + 2$

Which equations in Exercises 17–25 have graphs with each characteristic?

26. The line of symmetry is the y-axis.   **27.** The line of symmetry is the x-axis.

28. The vertex is on the x-axis.   **29.** The vertex is on the y-axis.

30. There is only one point on the x-axis.   **31.** There is no point on the x-axis.

For the graph of each equation, do these things.

a. Write the equation of the line of symmetry.

b. Write the coordinates of the vertex.

32. $y = (x + 12)^2$ **33.** $y = x^2 + 8$ **34.** $y = x^2 - 12$ **35.** $y = (x - 11)^2$

For Exercises 36–41, write an equation whose graph fits each description and has the same shape as the graph of $y = x^2$.

36. translation of $y = x^2$ to the right 1 unit   **37.** translation of $y = x^2$ down 9 units

---

Answers to Exercises and Problems

15. Answers may vary. An example is given. Yes, I agree. The graph of $y = x^2 + 7$ is U-shaped. The path of water from a fountain would be shaped like an upside-down U, rising to a high point and then falling.

16. a. A: down 2 units; B: left 2 units; C: up 2 units; D: right 2 units

b.

Equation	Graph
I. $y = -x^2 + 2$	C
II. $y = -x^2 - 2$	A
III. $y = -(x - 2)^2$	D
IV. $y = -(x + 2)^2$	B

17. Translate the graph of $y = x^2$ up 10 units.

18. Translate the graph of $y = x^2$ down 7 units.

19. Translate the graph of $y = x^2$ left 11 units.

20. Translate the graph of $y = x^2$ right 7 units.

21. Translate the graph of $y = -x^2$ up 10 units.

22. Translate the graph of $y = -x^2$ down 6 units.

23. Translate the graph of $y = -x^2$ left 6 units.

24. Translate the graph of $y = x^2$ left 3 units and down 4 units.

25. Translate the graph of $y = x^2$ right 1 unit and up 2 units.

26. the equations in Exercises 17, 18, 21, and 22

27. none

28. the equations in Exercises 19, 20, and 23

38. vertex at the point $(0, 5)$

39. vertex at the point $(-10, 0)$

40. translation of $y = -x^2$ up 4 units

41. translation of $y = -x^2$ to the left 5 units

Ongoing ASSESSMENT

42. Group Activity Work in a group of three students.

   a. Have one group member describe a translation of the parabola $y = x^2$. Have another member draw a graph of the translated parabola. Have the third member of the group write an equation for the translated parabola.

   b. Exchange roles and repeat the process until each group member has given a description, drawn a graph, and written an equation. Work together to check that the graphs and equations match the translations. Correct any errors that you find.

Review PREVIEW

$\triangle A'B'C'$ is the reflection image of $\triangle ABC$. Tell whether each statement is *True* or *False*. If it is false, rewrite the statement so that it is true. *(Section 10-1)*

43. The y-axis is the line of reflection.

44. $\triangle A'B'C'$ is congruent to $\triangle ABC$.

45. $\triangle A'B'C'$ has the same orientation as $\triangle ABC$.

46. The midpoint of $\overline{CC'}$ is on the x-axis.

Solve. *(Section 2-7)*

47. $3x + 9 = 0$ **48.** $15 - 4x = 15$ **49.** $12x - 24 = 0$

Use the zero-product property to solve each equation. *(Section 9-3)*

50. $-5k = 0$ **51.** $2(t + 4) = 0$ **52.** $3x(x - 1) = 0$

 Working on the Unit Project

53. Suppose the origin on a coordinate plane represents the point where a baseball is hit. Which equation has a graph that might be the path of the baseball? Explain how you made your choice.

   A. $y = x^2$ **B.** $y = (x - 10)^2 + 100$

   C. $y = -x^2$ **D.** $y = -(x - 10)^2 + 100$

54. For the graph you chose in Exercise 53, write the equation of the line of symmetry and the coordinates of the vertex.

10-2 Transforming Parabolas

561

Integrating the Strands

Ex. 42 integrates the geometric concepts of translation and parabola with the algebraic equation for the parabola $y = x^2$, which is also a function. Students must also use mathematical language to describe their translations and write equations.

Assessment: Task

Ex. 42 requires students to give a verbal, visual, and symbolic description of a translation of $y = x^2$. Students who can do all three successfully have achieved a complete understanding of translating parabolas.

Practice 77 For use with Section 10-2

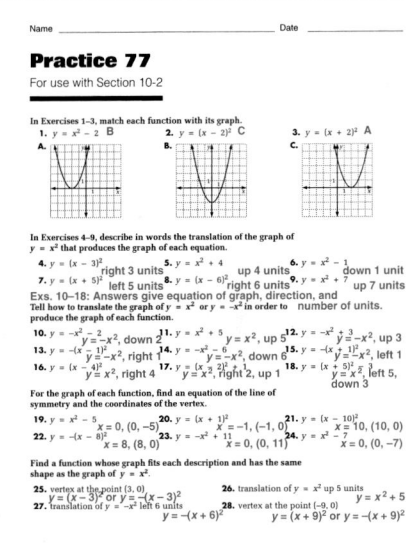

Answers to Exercises and Problems

29. the equations in Exercises 17, 18, 21, and 22

30. the equations in Exercises 19, 20, and 23

31. the equations in Exercises 17, 22, and 25

32. a. $x = -12$ **b.** $(-12, 0)$

33. a. $x = 0$ **b.** $(0, 8)$

34. a. $x = 0$ **b.** $(0, -12)$

35. a. $x = 11$ **b.** $(11, 0)$

36. $y = (x - 1)^2$

37. $y = x^2 - 9$

38. $y = x^2 + 5$

39. $y = (x + 10)^2$

40. $y = -x^2 + 4$

41. $y = -(x + 5)^2$

42. a, b. Answers may vary.

43. False. The x-axis is the line of reflection.

44. True.

45. False. The orientation of $\triangle A'B'C'$ is reversed.

46. True. **47.** -3

48. 0 **49.** 2

50. 0 **51.** -4

52. 0 or 1

53. D; Explanations may vary. An example is given. The path of a ball is shaped like an upside-down U. The graphs of the equa-

tions in (A) and (B) are shaped like U's, not like upside-down U's. The graph of the equation in (C) has its highest point at the origin, while the path of the ball must extend above the origin. The only choice left, D, has the right shape.

54. $x = 10$; $(10, 100)$

561

Objectives and Strands
See pages 544A and 544B.

Spiral Learning
See page 544B.

Materials List
➤ Graphics calculator or graphing software
➤ Graph paper

Recommended Pacing
Section 10-3 is a one-day lesson.

Extra Practice
See pages 638–639.

Warm-Up Exercises
💡 Warm-Up Transparency 10-3

Support Materials
➤ Practice 78
➤ Enrichment 69 in the Activity Bank
➤ Study Guide 10-3
➤ Problem Set 22
➤ Diagram Masters 2, 3 in the Explorations Lab Manual
➤ Quiz 10-3
➤ Test 41
➤ Alternative Assessments 8, 9

Section

10-3 Factors and Intercepts

WHAT PRICE IS RIGHT?

Focus
Use a table of values to sketch a parabola, and use algebra to find the coordinates of the points where a parabola crosses the axes.

The survey showed that for every $1 drop in price, the class will sell 5 more sweatshirts.

Predicting Income from Sweatshirt Sales

Drop in price	Price	Number of sweatshirts sold	Income (price × shirts sold)
0	$20	30	$20 × 30 = $600
$1	$20 − $1 = $19	30 + 5(1) = 35	$19 × 35 = $665
$2	$20 − $2 = $18	30 + 5(2) = 40	$18 × 40 = $720
$3	$20 − $3 = $17	30 + 5(3) = 45	$17 × 45 = $765
$4	$20 − $4 = $16	30 + 5(4) = 50	$16 × 50 = $800
$5	$20 − $5 = $15	30 + 5(5) = 55	$15 × 55 = $825
$6	$20 − $6 = $14	30 + 5(6) = 60	$14 × 60 = $840
$7	$20 − $7 = $13	30 + 5(7) = 65	$13 × 65 = $845
$8	$20 − $8 = $12	30 + 5(8) = 70	$12 × 70 = $840
$9	$20 − ? = ?	30 + ? = ?	? × ? = ?
$10	?	?	?
⋮	⋮	⋮	⋮
x	$20 − x$	$30 + 5x$	$(20 − x)(30 + 5x)$

I think that we can earn a lot of money for our class trip by selling sweatshirts. Last year the class sold 30 shirts at $20 each. Now that we've done this survey, what do *you* think? Can we raise even more money if we lower the price?

Talk it Over

1. What do you notice about the income from sweatshirt sales as the price drops from $20 to $13?

2. a. Complete the table for drops in price of $9 and $10.
   b. How is income affected when the price is below $13?

3. a. What will the income be if the drop in price is $20?
   b. What will the income be if no shirts are sold?

4. The table shows information about two variables: the drop in price and the income.
   a. Which is the control variable? Which is the dependent variable?
   b. Does the table represent a function? How can you tell?

Unit 10 Quadratic Equations as Models

Answers to Talk it Over

1. Answers may vary. An example is given. The income increases.

2. a. $9: $20 − $9 = $11;
       30 + 5(9) = 75;
       $11 × 75 = $825;
   $10: $20 − $10 = $10;
       30 + 5(10) = 80;
       $10 × 80 = $800

   b. Answers may vary. An example is given. The income decreases.

3. a. 0
   b. 0

4. a. the drop in price; the income
   b. Yes. There is only one income value for each drop in price.

Suppose you let y represent the income from selling sweatshirts. Then the table and the equation

$$y = (20 - x)(30 + 5x)$$

model the same situation. A third way to model the situation is to draw a graph.

Sample 1

Graph $y = (20 - x)(30 + 5x)$.

Sample Response

TECHNOLOGY NOTE

With a graphics calculator or graphing software, use the table of values to help you set an appropriate viewing window.

Watch Out!

Plot enough points to see the shape of the graph. Include the vertex and make sure that your graph is symmetric.

1 Make a table of values for x and y. Choose both positive and negative values for x.

$y = (20 - x)(30 + 5x)$		
x	y	(x, y)
-6	0	$(-6, 0)$
-3	345	$(-3, 345)$
0	600	$(0, 600)$
7	845	$(7, 845)$
10	800	$(10, 800)$
15	525	$(15, 525)$
20	0	$(20, 0)$

2 Plot the points and draw a smooth curve.

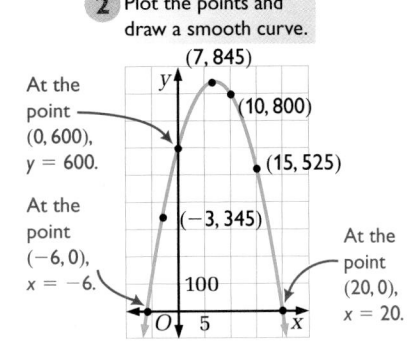

At the point $(0, 600)$, $y = 600$.

At the point $(-6, 0)$, $x = -6$.

At the point $(20, 0)$, $x = 20$.

The graph of the function $y = (20 - x)(30 + 5x)$ is a parabola.

Talk it Over

Use the graph in Sample 1.

5. For which values of x does the graph model the income from sweatshirt sales? Explain.

6. The class president predicts that the class will raise the most money by reducing the price to $13. Explain how the graph shows this.

7. Horizontal and vertical intercepts are also called **x-intercepts** and **y-intercepts**. What is the y-intercept of the graph?

8. a. An x-intercept of the graph occurs when $x = 20$. Substitute 20 for x in the factor $(20 - x)$. What do you notice?

   b. What do you think will happen when you substitute -6 for x in the factor $(30 + 5x)$? Test your guess.

10-3 Factors and Intercepts

563

Answers to Talk it Over

5. Explanations may vary. An example is given. values of x such that $0 \le x \le 20$; The price cannot drop more than $20. (You could say $-6 \le x \le 20$ if negative values of x are viewed as price increases.)

6. Answers may vary. An example is given. The highest point on the graph is the vertex at $(7, 845)$, so dropping the price $7 to $13 maximizes income at $845.

7. The y-intercept is 600.

8. a. Answers may vary. An example is given. The factor $20 - x$ equals 0 when $x = 20$, so $y = 0$.

   b. Answers may vary. An example is given. The factor $30 + 5x$ equals 0 when $x = -6$, so $y = 0$.

TEACHING

Using Technology

Using a TI-82, the data on page 562 can be plotted by editing list L1 as the Drop in Price and L2 as the Income and then running the following program:

```
PROGRAM: SETLINE
GridOff
CoordOn
Func
FnOff
Plot1 (XYLine, L1, L2, 0, …)
PlotsOn 1
ZoomStat
```

Talk it Over

For question 4(b), students need to recall the definition of a function as a relationship in which there is only one value of the dependent variable for each value of the control variable.

Additional Sample

S1 Graph
$$y = (2 + 0.5x)(4 - 2x).$$

Make a table of values for x and y.

$y = (2 + 0.5x)(4 - 2x)$		
x	y	(x, y)
-6	-16	$(-6, -16)$
-4	0	$(-4, 0)$
-1	9	$(-1, 9)$
0	8	$(0, 8)$
2	0	$(2, 0)$
4	-16	$(4, -16)$

Plot the points and draw a smooth curve.

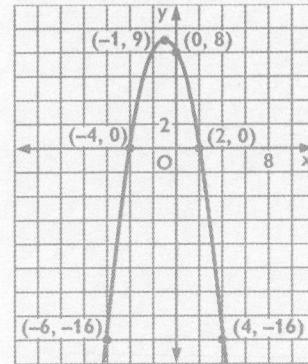

The graph is a parabola.

563

Question 7 introduces the terms *x*-intercept and *y*-intercept. Students should realize that *x*-intercepts are points having coordinates $(x, 0)$ and *y*-intercepts have coordinates $(0, y)$.

Mathematical Procedures

Review the zero-product property at this time. Students should understand that the property can be used to find the *x*-intercepts of an equation that can be factored into linear terms.

Additional Sample

S2 Without graphing, find the *x*-intercepts and the *y*-intercepts for the graph of $y = (2 + 0.5x)(4 - 2x)$.

Find the *x*-intercepts. The graph crosses the *x*-axis where $y = 0$.
$y = (2 + 0.5x)(4 - 2x)$
$0 = (2 + 0.5x)(4 - 2x)$
By the zero-product property, $(2 + 0.5x) = 0$ or $(4 - 2x) = 0$. Therefore, $x = -4$ or $x = 2$. The *x*-intercepts are -4 and 2.
Find the *y*-intercept. The graph crosses the *y*-axis where $x = 0$.
$y = (2 + 0.5x)(4 - 2x)$
$\quad = (2)(4)$
$\quad = 8$
The *y*-intercept is 8.

Look Back

Use the Look Back question for a class discussion.

Finding the x-Intercepts of a Graph

You can use the zero-product property and algebra to find the *x*-intercepts of the graph of an equation written in the form
$$y = \text{factor} \times \text{factor}.$$

Sample 2

Without graphing, find the x-intercepts and the y-intercept for the graph of $y = (20 - x)(30 + 5x)$.

Sample Response

Step 1. Find the *x*-intercepts. The graph crosses the *x*-axis where $y = 0$.
$$y = (20 - x)(30 + 5x)$$
$$0 = (20 - x)(30 + 5x) \quad \longleftarrow \text{Substitute 0 for } y.$$

By the zero-product property, when a product is zero at least one of the factors is zero. Set both factors equal to zero.

$20 - x = 0$	*or*	$30 + 5x = 0$
$20 - x + x = 0 + x$		$30 + 5x - 30 = 0 - 30$
$20 = x$		$5x = -30$
		$\dfrac{5x}{5} = \dfrac{-30}{5}$
		$x = -6$

The *x*-intercepts are 20 and -6.

Step 2. Find the *y*-intercept.

The graph crosses the *y*-axis where $x = 0$.
$$y = (20 - x)(30 + 5x)$$
$$y = (20 - 0)(30 + 5 \cdot 0) \quad \longleftarrow \text{Substitute 0 for } x.$$
$$y = (20)(30)$$
$$y = 600$$
The *y*-intercept is 600.

BY THE WAY...

Any freely hanging cable forms a curve that looks like a parabola but is really a curve called a *catenary*.

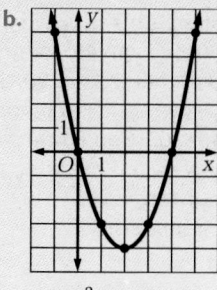

Suspending weights from the cable turns the catenary into a parabola! This happens when a suspension bridge like the Golden Gate in San Francisco is built.

Look Back ←

The graph of $y = (x + 3)(x - 2)$ is a parabola. What does its equation tell you about its intercepts?

Unit 10 Quadratic Equations as Models

Answers to Look Back

Answers may vary. An example is given. To find the *y*-intercept, set $x = 0$; this gives $y = 3(-2) = -6$. To find the *x*-intercepts, set each factor equal to zero; this gives $x = -3$ or $x = 2$, so the *x*-intercepts are -3 and 2.

Answers to Exercises and Problems

1. Answers may vary. An example is given. The zero-product property says that if a product of factors is zero, one or more of the factors must be zero. Therefore, you can set each factor, $(20 - x)$ and $(30 + 5x)$, equal to zero and solve to find the values of *x*.

2. a. 5; 0; −3; −4; −3; 0; 5

b.

c. $y = x^2$

d. 0 and 4   e. 0

10-3 Exercises and Problems

1. **Reading** Read the statement of the zero-product property on page 495. How does this statement help you solve the equation $(20 - x)(30 + 5x) = 0$ in Sample 2 on page 564?

2. **a.** Complete the table of values.

   b. Plot the ordered pairs on a coordinate plane and draw a smooth curve through the points.

   c. Does your graph look more like the graph of $y = x^2$ or of $y = -x^2$?

   d. What are the x-intercepts of your graph?

   e. What is the y-intercept of your graph?

$y = x(x - 4)$	
x	**y**
−1	?
0	?
1	?
2	?
3	?
4	?
5	?

3. **Writing** Colby is graphing a parabola. He is having trouble deciding which points to pick for his table of values. What would you suggest to help Colby?

Graph each parabola.

4. $y = (x - 2)x$

5. $y = (x - 3)(3 - x)$

6. $y = (x - 4)(x + 4)$

Without graphing, find the x-intercepts and the y-intercept of the graph of each equation.

7. $y = x(x + 7)$

8. $y = (x + 12)(x - 1)$

9. $y = (x + 7)(4 - x)$

10. $y = (x + 4)(x + 13)$

11. $y = (x - 4)(3x + 13)$

12. $y = (4x - 1)(3x + 4)$

13. **a.** Suppose you translate the graph of $y = (x + 4)(x + 13)$ to the right 4 units. What are the x-intercepts of the new graph?

   b. Suppose you translate the graph of $y = (x + 4)(x + 13)$ down 5 units. What is the y-intercept of the new graph?

14. During a golf tournament, Chet Washington hits a ball that lands 200 yd from the tee. The path of the ball is a parabola.

   a. Sketch a possible path for the ball, using the x-axis as the ground and the origin as the tee.

   b. What are the x-intercepts of the graph?

   c. What do the x-intercepts mean in terms of the situation?

   d. **Writing** Amalia wrote the equation $y = -2x(4x - 800)$ for a possible path of the ball. Elizabeth wrote the equation $y = -2x(2x - 100)$. Which student's equation *cannot* be correct? Why?

10-3 Factors and Intercepts   **565**

APPLYING

Suggested Assignment
Standard 1–14, 17–25
Extended 1–25

Integrating the Strands
Number Exs. 21–24
Algebra Exs. 1–25
Functions Exs. 2, 4–20

Mathematical Procedures
For Exs. 4–6, students should plot enough points to see the shape of the parabola.

Using Technology
Students can check their answers to Exs. 7–12 by graphing the equations on a graphics calculator and using the TRACE feature to find the x-intercepts and the y-intercepts. The TRACE feature is discussed on page 615 of the Technology Handbook.

7. x-intercepts: 0 and −7; y-intercept: 0

8. x-intercepts: −12 and 1; y-intercept: −12

9. x-intercepts: −7 and 4; y-intercept: 28

10. x-intercepts: −4 and −13; y-intercept: 52

11. x-intercepts: 4 and $-\frac{13}{3}$; y-intercept: −52

12. x-intercepts: $\frac{1}{4}$ and $-\frac{4}{3}$; y-intercept: −4

13. **a.** 0 and −9   **b.** 47

14. **a.**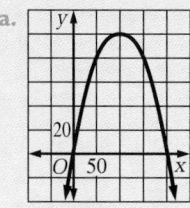

   b. 0 and 200

   c. the position of the ball when it was hit and the position when it landed

   d. Elizabeth's equation, $y = -2x(2x - 100)$, cannot be correct because this equation has x-intercepts at 0 and 50 rather than at 0 and 200.

Answers to Exercises and Problems

3. Answers may vary. An example is given. Locate the vertex. Put this number in the middle of the table. Enter some x-values that are less than the x-coordinate of the vertex above the vertex value and some x-values that are greater than the x-coordinate of the vertex below the vertex value.

4.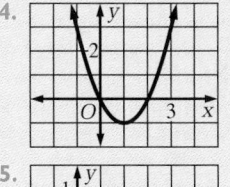

5.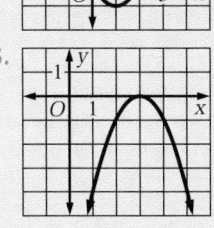

6.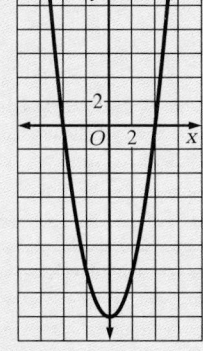

565

Exs. 15 and 16 can be used to introduce students to the economic concepts of price, supply, and demand for a product. The demand for a product (number sold) is not only a function of supply (which can also affect price). A simple model is that as demand increases, supply increases, and price can also increase. If supply overtakes demand, however, then the oversupply can affect prices adversely, driving them down in the hope of stimulating demand. The *equilibrium price* is the point at which supply and demand are balanced. In Step 2 of Sara Ortiz's presentation, the equilibrium price is the maximum point on the parabola. At a lower price, sales keep increasing because demand is increasing. At a higher price, demand and sales start to decrease.

Answers to
Exercises and Problems

15. **a.** linear decay

 b. price; number sold

 c. The number sold decreases. Answers may vary. An example is given. Fewer people are willing to buy the computer at a higher price.

 d. 3000; None will be sold when the price is $3000.

 e. Answers may vary. An example is given. The maximum *y*-value shown on the graph is 300,000.

16. **a.** price; sales dollars

 b. 0 and 3000; Answers may vary. An example is given. The price must be under $3000.

 c. 0; Answers may vary. An example is given. No money is made from laptop sales when the price is $0.

 d. $1500; Answers may vary. An example is given. The maximum yearly income of $225,000,000 occurs when the price is $1500.

17. See answer in back of book.

18. down 7 units

19. up 3 units

20. right 1 unit

---

connection to **BUSINESS ▶ EDUCATION**

Use Sara Ortiz's presentation.

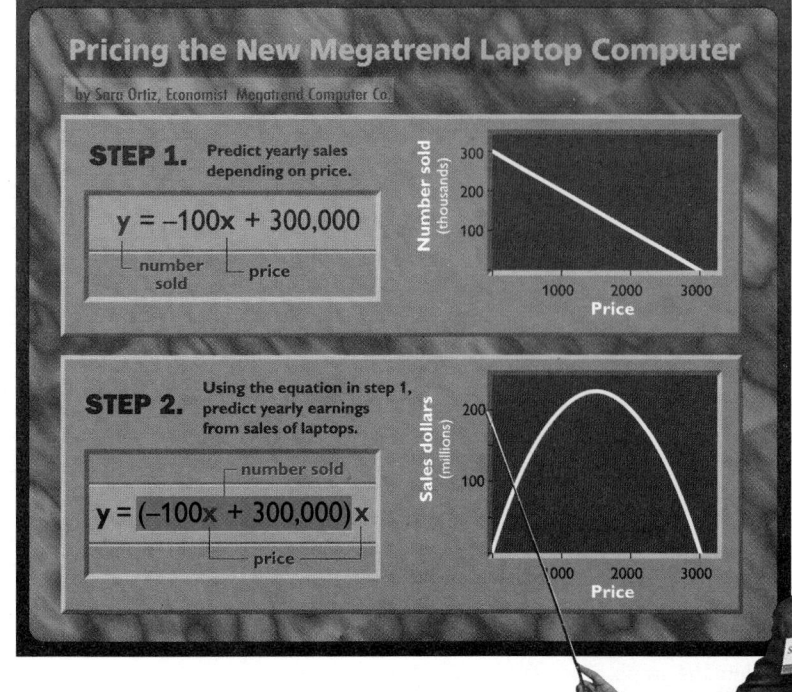

Pricing the New Megatrend Laptop Computer

by Sara Ortiz, Economist Megatrend Computer Co.

15. Look at the equation and graph in Step 1 of Sara Ortiz's presentation.

 a. Is this function an example of *linear growth* or *linear decay*?

 b. Which is the control variable? Which is the dependent variable?

 c. As the price goes up, what happens to the number sold? Why do you think this happens?

 d. What is the *x*-intercept? What does it mean in terms of the situation?

 e. When Sara Ortiz wrote the equation, she assumed that Megatrend will never sell more than 300,000 laptops in any given year. How can you tell?

16. Look at the equation and graph in Step 2 of Sara Ortiz's presentation. The graph is a parabola.

 a. Which is the control variable in the equation above? Which is the dependent variable?

 b. What are the *x*-intercepts? What do they mean in terms of the situation?

 c. What is the *y*-intercept? What does it mean in terms of the situation?

 d. What price should Megatrend set for the laptop computer? Why?

566 **Unit 10** Quadratic Equations as Models

---

21. 10^7

22. 10^2

23. 10^4

24. $\frac{1}{10^4}$, or 10^{-4}

25. **a.** Assume the ball is hit from ground level.

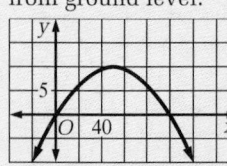

b. 0 and 100

c. Yes. Answers may vary. An example is given. We don't know how high the ball went, so the highest point could be at (50, 5), (50, 8), (50, 6.2), and so on.

d. No. Answers may vary. An example is given. The graph of the equation is a parabola that opens downward and has the proper *x*-intercepts. However, the ball could never go up 2500 ft, or almost half a mile, as the location of the vertex at (50, 2500) indicates, especially since we know that the ball was hit close to the ground.

17. **Writing** Write a summary of what you can tell about a parabola from its equation. Give at least one example for each property.

Tell how to translate the graph of $y = x^2$ in order to produce the graph of each equation. *(Section 10-2)*

18. $y = x^2 - 7$ **19.** $y = x^2 + 3$ **20.** $y = (x - 1)^2$

Write each expression as a power of ten. *(Section 1-3)*

21. $10^5 \times 10^2$ **22.** $10^{10} \times 10^{-8}$ **23.** $\dfrac{10^9}{10^5}$ **24.** $\dfrac{10^4}{10^8}$

 Working on the Unit Project

25. Suppose a batter hits a ball that is very low to the ground. The ball lands at a spot 100 ft away.

   a. Suppose the ball is hit from the ground. Sketch a graph of a possible path of the ball. Use the x-axis as the ground and the origin as the point where the ball was hit.

   b. What are the x-intercepts of your path?

   c. Is there more than one possible path with these intercepts? Explain.

   d. Could the graph of the equation $y = -x(x - 100)$ represent the ball's path? Describe your reasoning.

Unit 10   **CHECKPOINT 1**

1. **Writing** Find a picture of a parabola. Sketch it on graph paper. Tell at least three things you know about it.

2. Copy the figure and reflect it over each axis. Write the coordinates of the vertices of its image. Describe the orientation of the original figure and its image.   **10-1**

   a. x-axis   **b.** y-axis

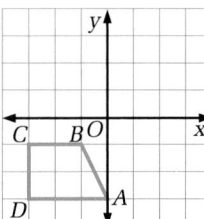

Tell how to transform the graph of $y = x^2$ in order to produce the graph of each equation.   **10-2**

3. $y = x^2 + 8$ **4.** $y = (x - 1)^2$ **5.** $y = -x^2$

6. Graph the parabola $y = (x + 2)(x - 3)$. Plot enough points to show the shape of the curve.   **10-3**

Without graphing, find the x-intercepts and the y-intercept of the graph of each equation.

7. $y = (x + 5)(x + 11)$ **8.** $y = x(x - 12)$

Answers to Checkpoint

1. Answers may vary. An example is given, using the parabola shown in Sample 1(c) on page 557. The equation of the line of symmetry is $x = -2$. The vertex is $(-2, 0)$. The x-intercept is -2. The y-intercept is 4.

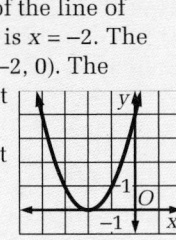

2. a.

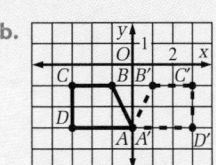

$A'(0, 3), B'(-1, 1),$
$C'(-3, 1), D'(-3, 3);$
$ABCD$ has counterclockwise orientation and $A'B'C'D'$ has clockwise orientation.

b.

$A'(0, -3), B'(1, -1),$
$C'(3, -1), D'(3, -3);$
$ABCD$ has counterclockwise orientation and $A'B'C'D'$ has clockwise orientation.

Assessment: Portfolio

For Ex. 17, encourage students to include examples to illustrate how to tell the properties of a parabola from its equation. Students should consider this summary for inclusion in their portfolios.

Quick Quiz (10-1 through 10-3)

See page 608.

Practice 78 For use with Section 10-3

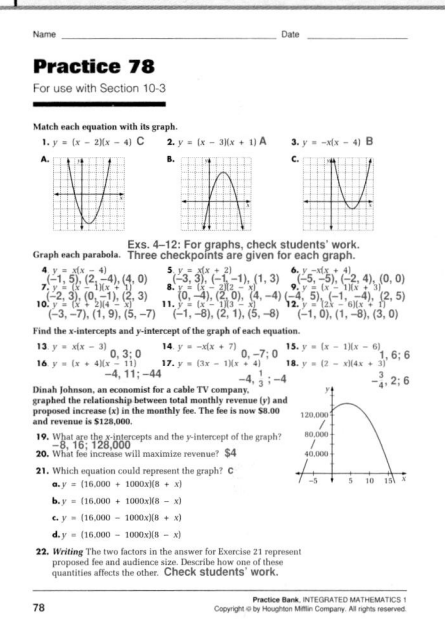

3. Translate the graph of $y = x^2$ up 8 units.

4. Translate the graph of $y = x^2$ right 1 unit.

5. Reflect the graph of $y = x^2$ over the x-axis.

6.

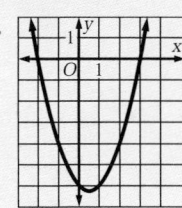

7. x-intercepts: -5 and -11; y-intercept: 55

8. x-intercepts: 0 and 12; y-intercept: 0

PLANNING

Objectives and Strands
See pages 544A and 544B.

Spiral Learning
See page 544B.

Materials List

➤ Calculator

➤ Graphics calculator or graphing software

Recommended Pacing
Section 10-4 is a two-day lesson.

Day 1
Pages 568–570: Talk it Over through Sample 2, *Exercises 1–19*

Day 2
Pages 570–572: Talk it Over through Look Back, *Exercises 20–43*

Extra Practice
See pages 638–639.

Warm-Up Exercises
Warm-Up Transparency 10-4

Support Materials

➤ Practice 79

➤ Enrichment 70 in the Activity Bank

➤ Study Guide 10-4

➤ Problem Set 23

➤ Quiz 10-4

➤ Alternative Assessment 10

Section 10-4

Working with Powers

Focus
Use shortcuts for simplifying powers.

The POWER OF SHORTCUTS

Talk it Over

1. Write each product as a power of 4.
   a. $4 \cdot 4 \cdot 4 = 4^?$
   b. $4 \cdot 4 = 4^?$
   c. $(4 \cdot 4 \cdot 4) \cdot (4 \cdot 4) = 4^? \cdot 4^?$
   d. $(4 \cdot 4 \cdot 4) \cdot (4 \cdot 4) = 4^?$

2. Write each product as a power of x.
   a. $x \cdot x \cdot x = x^?$
   b. $x \cdot x = x^?$
   c. $(x \cdot x \cdot x) \cdot (x \cdot x) = x^? \cdot x^?$
   d. $(x \cdot x \cdot x) \cdot (x \cdot x) = x^?$

3. Use your calculator to complete parts (a) and (b) below to see if $4^3 \cdot 4^2 = 4^5$.
   a.  ?
   b. ?

4. Use your answers to questions 1–3. Do you see a way to write $3^6 \cdot 3^4$ as a power of 3 without writing out the factors? Describe a shortcut.

5. Marissa drew a diagram to show how she came up with a shortcut for multiplying $5^{12} \cdot 5^9$. Explain her reasoning.

6. Draw a diagram like Marissa's to show that $(-3)^2 \cdot (-3)^3 = (-3)^{2+3}$.

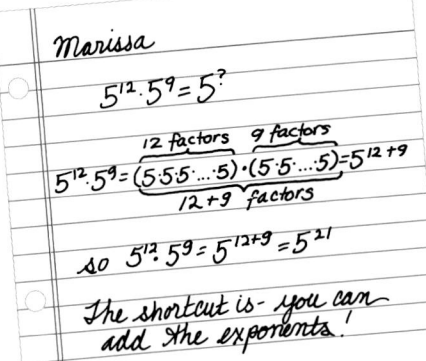

Marissa

$5^{12} \cdot 5^9 = 5^?$

12 factors   9 factors
$5^{12} \cdot 5^9 = (5 \cdot 5 \cdot 5 \cdots 5) \cdot (5 \cdot 5 \cdots 5) = 5^{12+9}$
          12+9 factors

so $5^{12} \cdot 5^9 = 5^{12+9} = 5^{21}$

The shortcut is - you can add the exponents!

568   Unit 10 Quadratic Equations as Models

Answers to Talk it Over

1. a. 4^3
   b. 4^2
   c. $4^3 \cdot 4^2$
   d. 4^5

2. a. x^3
   b. x^2
   c. $x^3 \cdot x^2$
   d. x^5

3. a. 1024
   b. 1024

4. Yes. Shortcuts may vary. An example is given. To write the product $3^6 \cdot 3^4$ as a power of 3, add the exponents: $3^{6+4} = 3^{10}$.

5. Explanations may vary. An example is given. She noticed that 5^{12} contains 12 factors of 5 and 5^9 contains 9 factors of 5, so $5^{12} \cdot 5^9$ contains 12 + 9, or 21 factors of 5 in all. So she found that instead of writing out all the factors she could add the exponents and write $5^{12} \cdot 5^9$ as $5^{12+9} = 5^{21}$.

6. $(-3)^2 \cdot (-3)^3 = (-3)^?$
   $(-3)^2 \cdot (-3)^3 =$
   2 factors   3 factors
   $\underbrace{(-3)(-3)} \cdot \underbrace{(-3)(-3)(-3)}$ =
   2 + 3 factors
   $(-3)^{2+3}$
   So $(-3)^2 \cdot (-3)^3 = (-3)^{2+3} = (-3)^5$

Marissa's diagram can be used to model a general rule:

$$a^m \cdot a^n = \underbrace{\overbrace{(a \cdot a \cdot \ldots \cdot a)}^{m \text{ factors}} \cdot \overbrace{(a \cdot a \cdot \ldots \cdot a)}^{n \text{ factors}}}_{m + n \text{ factors}} = a^{m + n}$$

PRODUCT OF POWERS RULE

$$a^m \cdot a^n = a^{m + n}$$

same base

Example:

$$2^3 \cdot 2^7 = 2^{3 + 7}$$
$$= 2^{10}$$

To multiply powers that have the same base, you add the exponents.

Sample 1

Simplify.

a. $a^6 \cdot a^5$ b. $x \cdot x^2$ c. $y^2 \cdot y^3 \cdot y^4$

Sample Response

a. $a^6 \cdot a^5 = a^{6 + 5}$ ⟵ The powers have the same base. Use the product of powers rule.

$= a^{11}$

b. $x \cdot x^2 = x^1 \cdot x^2$ ⟵ Rewrite x as x^1.

$= x^{1 + 2}$ ⟵ Use the product of powers rule.

$= x^3$

c. $y^2 \cdot y^3 \cdot y^4 = (y^2 \cdot y^3) \cdot y^4$ ⟵ Group two of the factors.

$= y^{2 + 3} \cdot y^4$ ⟵ Use the product of powers rule.

$= y^5 \cdot y^4$

$= y^{5 + 4}$ ⟵ Use the product of powers rule.

$= y^9$

Talk it Over

7. Which property of multiplication allows you to group two factors in part (c) of Sample 1?

8. In part (c) of Sample 1, can you use the shortcut $y^2 \cdot y^3 \cdot y^4 = y^{2 + 3 + 4}$? Why or why not?

10-4 Working with Powers

569

Answers to Talk it Over

7. associative property of multiplication

8. Yes. Reasons may vary. An example is given.
$y^2 \cdot y^3 \cdot y^4 = y^{2+3} \cdot y^4 = y^{(2+3)+4} = y^{2+3+4}$, since the way you group numbers being added doesn't change the sum.

Talk it Over

Questions 1–6 lead students to discover the product of powers rule.

Additional Sample

S1 Simplify.

a. $b^4 \cdot b^3$
$b^4 \cdot b^3 = b^{4+3}$
$= b^7$

b. $y^3 \cdot y$
$y^3 \cdot y = y^3 \cdot y^1$
$= y^{3+1}$
$= y^4$

c. $x^4 \cdot x^2 \cdot x^2$
$x^4 \cdot x^2 \cdot x^2 = (x^4 \cdot x^2) \cdot x^2$
$= x^{4+2} \cdot x^2$
$= x^6 \cdot x^2$
$= x^{6+2}$
$= x^8$

Talk it Over

Question 8 leads students to realize that the product of powers rule applies to more than just two powers with the same base.

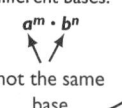

Watch Out!
The product of powers rule does not apply to powers with different bases.

$a^m \cdot b^n$

not the same base

To simplify some products you begin by regrouping the factors.

Sample 2

Simplify.

a. $2a^2 \cdot 9a^3$

b. $-x^2y \cdot 3xy^3$

Sample Response

a. $2a^2 \cdot 9a^3 = (2 \cdot 9) \cdot (a^2 \cdot a^3)$ ⟵ Group numbers and group powers with the same base.

$= 18 \cdot a^{2+3}$ ⟵ The powers have the same base. Use the product of powers rule.

$= 18a^5$

b. $-x^2y \cdot 3xy^3 = (-1x^2y) \cdot (3xy^3)$ ⟵ Rewrite $-x^2y$ as $-1x^2y$.

$= (-1 \cdot 3) \cdot (x^2 \cdot x) \cdot (y \cdot y^3)$ ⟵ Group numbers, group powers of x, and group powers of y.

$= -3 \cdot (x^2 \cdot x^1) \cdot (y^1 \cdot y^3)$ ⟵ Rewrite x as x^1 and y as y^1.

$= -3 \cdot (x^{2+1}) \cdot (y^{1+3})$ ⟵ Use the product of powers rule for powers with the same base.

$= -3x^3y^4$

⋯➤ Now you are ready for:
Exs. 1–19 on pp. 572–574

Talk it Over

9. Use the product of powers rule to complete each statement.
   a. $(3^4)^2 = 3^4 \cdot 3^4 = 3^?$
   b. $(a^2)^3 = a^2 \cdot a^2 \cdot a^2 = a^?$

10. Look at your answers to question 9. Describe a shortcut for simplifying $(a^2)^3$ without writing out the factors.

11. Dan made this diagram for his shortcut.

a^m is a factor n times n terms

$(a^m)^n = \underbrace{a^m \cdot a^m \cdot \ldots \cdot a^m} = a^{\overbrace{m+m+\ldots+m}} = a^{m \cdot n}$

Explain his reasoning.

Dan's diagram models this general rule:

POWER OF A POWER RULE

$(a^m)^n = a^{m \cdot n}$

To find a power of a power, you multiply the exponents.

Example:

$(x^3)^4 = x^{3 \cdot 4}$

$= x^{12}$

Answers to Talk it Over

9. **a.** 3^8
   b. a^6

10. Shortcuts may vary. An example is given. Multiply the exponents:
    $(a^2)^3 = a^{2 \cdot 3} = a^6$.

11. Explanations may vary. An example is given. $(a^m)^n$ means that there are n factors of a^m. Using the product of powers rule, you add the exponents to get $a^{m+\ldots+m}$. There are n terms in the sum, so $a^{m+\ldots+m} = a^{m \cdot n}$.

Simplify.

a. $(x^5)^2$

b. $(y^4)^5 \cdot y^4$

Sample Response

a. $(x^5)^2 = x^{5 \cdot 2}$ ⟵ Use the power of a power rule.

$\quad = x^{10}$

b. $(y^4)^5 \cdot y^4 = (y^{4 \cdot 5}) \cdot y^4$ ⟵ Use the power of a power rule.

$\quad = y^{20} \cdot y^4$

$\quad = y^{20 + 4}$ ⟵ Use the product of powers rule.

$\quad = y^{24}$

BY THE WAY...

The number 10^{100} is called a *googol*. The number 10 raised to the googol power is called a *googolplex*.

Talk it Over

12. Simplify.

a. $(2x)^3 = 2x \cdot 2x \cdot 2x = \underline{?}$

b. $(xy)^3 = \underline{?} \cdot \underline{?} \cdot \underline{?} = \underline{?}$

c. $(xy^2)^3 = \underline{?} \cdot \underline{?} \cdot \underline{?} = \underline{?}$

13. Look at your answers to question 12. Describe a shortcut for simplifying $(2x)^3$ without writing out the factors. Explain your reasoning.

14. Meghan made this diagram for her shortcut. Explain her reasoning.

$$\overbrace{ }^{ab \text{ is a factor } n \text{ times}} \quad \overbrace{ }^{n \text{ factors}} \quad \overbrace{ }^{n \text{ factors}}$$

$$(ab)^n = (ab) \cdot (ab) \cdot \ldots \cdot (ab) = (a \cdot a \cdot \ldots \cdot a) \cdot (b \cdot b \cdot \ldots \cdot b) = a^n \cdot b^n$$

Meghan's diagram models this general rule:

POWER OF A PRODUCT RULE

$$(ab)^n = a^n \cdot b^n$$

To find a power of a product, you find that power of each factor and then multiply.

Example:

$(2x)^4 = 2^4 \cdot x^4$

$\quad = 16x^4$

10-4 Working with Powers

571

Answers to Talk it Over

12. a. $8x^3$

b. $xy \cdot xy \cdot xy = x^3y^3$

c. $xy^2 \cdot xy^2 \cdot xy^2 = x^3y^6$

13. Shortcuts may vary. An example is given. Write each of the factors 2 and x to the third power: $2^3 \cdot x^3$, or $8x^3$.

14. Explanations may vary. An example is given. $(ab)^n$ means that there are n factors of ab. Therefore, there are n factors of a and n factors of b. We write n factors of a as a^n and n factors of b as b^n, so $(ab)^n$ is the same as $a^n \cdot b^n$.

a. $(6p)^4$

$$(6p)^4 = 6^4 \cdot p^4$$
$$= 1296p^4$$

b. $\left(-c^3\right)^4$

$$\left(-c^3\right)^4 = \left(-1c^3\right)^4$$
$$= (-1)^4 \cdot \left(c^3\right)^4$$
$$= c^{12}$$

c. $\left(-3x^3y^4\right)^2$

$$\left(-3x^3y^4\right)^2 = (-3)^2 \cdot \left(x^3\right)^2 \cdot \left(y^4\right)^2$$
$$= 9x^6y^8$$

Reasoning

Ask students to make a generalization about when −1 raised to a power is 1 and when it is −1. (−1 raised to an even power is 1; −1 raised to an odd power is −1.)

Look Back

The Look Back would make a good class discussion, which would help to clarify students' understanding of all three rules.

APPLYING

Suggested Assignment

Standard 1–27, 31–33, 35–43

Extended 1–27, 29–33, 35–43

Integrating the Strands

Number Exs. 17, 18, 35

Algebra Exs. 1–14, 17–43

Functions Exs. 30, 31, 33, 34, 36, 37

Geometry Exs. 15–16, 38

Sample 4

Simplify.

a. $(4n)^3$ **b.** $(-k^2)^3$ **c.** $(2a^5b^2)^4$

Sample Response

a. $(4n)^3 = 4^3 \cdot n^3$ ◄—— Use the power of a product rule.
$$= 64n^3$$

b. $(-k^2)^3 = (-1k^2)^3$ ◄—— Rewrite $-k^2$ as $-1k^2$.
$$= (-1)^3 \cdot (k^2)^3$$ ◄—— Use the power of a product rule.
$$= -1k^6$$ ◄—— Use the power of a power rule.
$$= -k^6$$

c. $(2a^5b^2)^4 = 2^4 \cdot (a^5)^4 \cdot (b^2)^4$ ◄—— Use the power of a product rule.
$$= 16a^{20}b^8$$ ◄—— Use the power of a power rule.

Look Back ◄——

There are three rules about powers in this section. Which rule would you use to simplify each expression?

$$(2x)^5, \quad x^{12} \cdot x^5, \quad \text{and} \quad (x^{12})^5$$

Explain how to decide which rule to use.

····▶ **Now you are ready for:**
Exs. 20–43 on pp. 574–575

10-4 Exercises and Problems

1. **Reading** Why can you use the product of powers rule to simplify $x^4 \cdot x^3$ but not to simplify $x^4 \cdot y^3$?

Simplify.

2. $x^4 \cdot x^3$ 3. $t^2 \cdot t^8$ 4. $y^5 \cdot y$ 5. $k^3 \cdot k^2 \cdot k^5$

6. $x^2 \cdot x \cdot x^3$ 7. $6a^3 \cdot 9a^3$ 8. $y \cdot 5y^2$ 9. $(-4z^2)(7z^3)$

10. $(-b^2)(3b^3)$ 11. $ab \cdot a^3b$ 12. $(-8x^3y^3)(x^2y^2)$ 13. $(4m^5)(-2mn^2)$

14. **Writing** Suppose your friend missed today's class. Write a note to convince your friend that $x^5 \cdot x^3 = x^{5+3}$. What would you show? What would you say?

Answers to Look Back ··

Answers may vary. An example is given. You would use the power of a product rule to simplify $(2x)^5$, the product of powers rule to simplify $x^{12} \cdot x^5$, and the power of a power rule to simplify $(x^{12})^5$. To determine which rule to use, look carefully at the expression to be simplified. If you are multiplying two powers with the same base, use the product of powers rule: add the exponents. If the expression is a power in parentheses raised to a power, use the power of a power rule: multiply the exponents. If the expression is a product of two or more factors raised to a power, use the power of a product rule: raise each factor to the given power.

This mylar and fabric wall hanging was designed and created by artist Patricia Malarcher. She uses a basic geometric grid design and explores ways of breaking up geometric shapes such as squares. A twelve-inch square is her basic design element.

In Exercises 15 and 16, write the area of each square and compare it to the area of the original square.

Original:

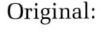

x

x

Area $= x^2$

BY THE WAY...

Patricia Malarcher's art is influenced by Islamic prayer rugs and by fabric draped over shrines in Bali and over doorways in Mexico and Italy.

15. 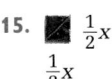 $\frac{1}{2}x$

$\frac{1}{2}x$

16. $2x$

$2x$

17. a. Complete each equation.

$$\frac{4^5}{4^3} = \frac{4 \cdot 4 \cdot 4 \cdot 4 \cdot 4}{4 \cdot 4 \cdot 4}$$

$$= \underline{?} \cdot \underline{?}$$

$$= 4^?$$

b. Complete each equation.

$$\frac{x^7}{x^2} = \frac{x \cdot x \cdot x \cdot x \cdot x \cdot x \cdot x}{x \cdot x}$$

$$= \underline{?} \cdot \underline{?} \cdot \underline{?} \cdot \underline{?} \cdot \underline{?}$$

$$= x^?$$

c. From your answers to parts (a) and (b), do you see a way to write $\frac{5^{12}}{5^9}$ as a power of 5 without writing out the factors? Describe a shortcut.

18. a. Use the shortcut you wrote in part (c) of Exercise 17 to complete the equation $\frac{6^2}{6^2} = 6^?$.

b. Look at the equation in part (a). What do you think is the value of 6^0? Explain your reasoning.

10-4 Working with Powers

573

Answers to Exercises and Problems

1. To use the product of powers rule, the bases must be the same.

2. x^7 3. t^{10}

4. y^6 5. k^{10}

6. x^6 7. $54a^6$

8. $5y^3$ 9. $-28z^5$

10. $-3b^5$ 11. a^4b^2

12. $-8x^5y^5$ 13. $-8m^6n^2$

14. Answers may vary. An example is given. I would show this to my friend:
$x^5 \cdot x^3 =$
$(x \cdot x \cdot x \cdot x \cdot x)(x \cdot x \cdot x) =$
x^8. I would say, "Simplify $x^5 \cdot x^3$ by adding the exponents."

15. $\frac{1}{4}x^2$; one fourth the area of the original square

16. $4x^2$; four times the area of the original square

17. a. $4 \cdot 4 = 4^2$

b. $x \cdot x \cdot x \cdot x \cdot x = x^5$

c. Yes. Shortcuts may vary. An example is given. Write $\frac{5^{12}}{5^9}$ as $5^{12-9} = 5^3$. The shortcut is to subtract the exponents.

18. a. $6^{2-2} = 6^0$

b. Explanations may vary. An example is given. 6^0 must equal 1 because $6^0 = \frac{6^2}{6^2}$. Any number divided by itself equals 1.

Reasoning

Ex. 19 leads students to conclude that $x^0 = 1$. You may wish to point out that this is true only if $x \neq 0$ because 0^0 has no meaning.

Communication: Discussion

You may wish to use Ex. 31 for a class discussion. Students can share their reasoning and use examples to support their thinking.

19. Use the product of powers rule to simplify $x^0 \cdot x^b = x^{? + ?} = x^?$. What do you conclude about the value of x^0?

Simplify.

20. $(k^5)^3$ **21.** $(y^3)^8$ **22.** $(2x)^3$ **23.** $(-d^5)^2$

24. $(2x^3)^4$ **25.** $(-3h^3)^5$ **26.** $(-4m^2n)^2$ **27.** $(5y^6z^3)^4$

28 TECHNOLOGY Which calculator key sequence is *not* a way to calculate $(2^5)^2$?

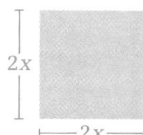

a. 2 y^x 5 = x^2

b. 2 y^x 5 y^x 2 =

c. 2 y^x 1 0 =

d. 2 y^x 7 =

29. **Writing** Use the area models to write a convincing argument that $(2x)^2$ is greater than $2x^2$ for any positive value of x.

x
x
x
x
2x
2x

30. The calculator shows the graphs of $y = 2x^2$ and $y = (2x)^2$ on the same axes. For what values of x is $2x^2$ less than $(2x)^2$?

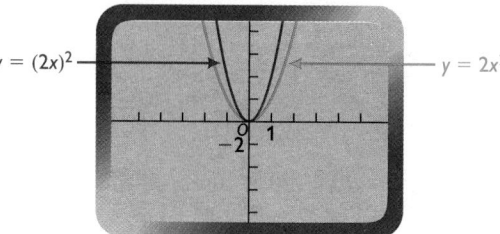

y = (2x)²
y = 2x²
O
1
−2

31. Terry and Alec disagree on an answer.

Who is right?

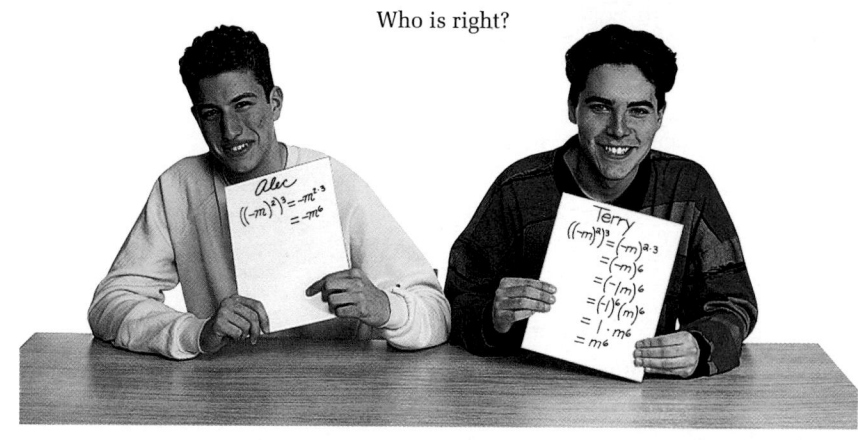

Alec
((−m)²)³ = −m^{2·3}
= −m⁶
Terry
((−m)²)³ = (−m)^{2·3}
= (−m)⁶
= (−1m)⁶
= (−1)⁶(m)⁶
= 1 · m⁶
= m⁶

32. Insert parentheses in the expression on the left side of each equation to make a true statement.

a. $2a^2b^3 = 2a^6b^3$ b. $2a^2b^3 = 4a^2b^3$ c. $2a^2b^3 = 8a^6b^3$

Answers to Exercises and Problems

19. $x^{0+b} = x^b$; x^0 is equal to 1.
20. k^{15}
21. y^{24}
22. $8x^3$
23. d^{10}
24. $16x^{12}$
25. $-243h^{15}$
26. $16m^4n^2$
27. $625y^{24}z^{12}$
28. d
29. Answers may vary. An example is given. If a square is $2x$ long on each side, then it can be divided into 4 small squares, each with side of length x, which is twice the area of 2 squares with side of length x.
30. all except $x = 0$, when they are both equal to 0.
31. Terry
32. a. $2(a^2b)^3$
    b. $(2a)^2b^3$
    c. $(2a^2b)^3$
33. a. $y = x^2$: Quadrants I and II; $y = x^3$: Quadrants I and III
    b. $y = x^2$
    c. Answers may vary. An example is given. Similarities: Both contain the points (0, 0) and (1, 1), and their shapes are similar in Quadrant I. Differences: The graph of $y = x^2$ is U-shaped (both "ends" go up), but the graph of $y = x^3$ rises from the left, levels off, then rises again toward the right. Also, the graph of $y = x^3$ is "steeper" than the graph of $y = x^2$.
34. a. $y = x^4$ and $y = x^6$: Quadrants I and II; $y = x^5$ and $y = x^7$: Quadrants I and III
    b. Answers may vary. An example is given. All the graphs contain the points (0, 0) and (1, 1) and look somewhat similar in Quadrant I. The graphs of $y = x^4$ and $y = x^6$ (even powers) are both U-shaped. The graphs of $y = x^5$ and $y = x^7$ (odd powers) both have shapes closer to the shape of the graph of $y = x^3$. Also, the greater the power, the "steeper" the graph is in most places.
    c. Predictions may vary. An example is given. Since x^8 is an even power, I predict that it will have a U-shaped graph; since x^9 is an odd power, I predict that it will have a graph that is like a U that has had its left half reflected over the x-axis. Graphing shows that these predictions are correct.

y
1
O
1
x
y
1
O
2
x

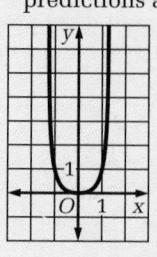

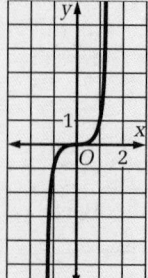

33. TECHNOLOGY Use a graphics calculator or graphing software. Graph $y = x^2$ and $y = x^3$ on the same screen.

a. In which quadrants does each graph lie?

b. For which graph is the y-axis a line of symmetry?

c. **Writing** Write about how the graphs of $y = x^2$ and $y = x^3$ are alike and how they are different.

34. TECHNOLOGY Functions of the form $y = x^n$ are called *power functions*. Use a graphics calculator or graphing software to graph $y = x^4$, $y = x^5$, $y = x^6$, and $y = x^7$ on the same axes.

a. In which quadrants does each graph lie?

b. Describe the similarities and differences of the graphs. Are some more alike than others? Which?

c. Predict what the graphs of $y = x^8$ and $y = x^9$ will look like. Graph the equations to check your predictions.

Ongoing ASSESSMENT

35. **Writing** Fred wonders whether $2^4 \cdot 2^3 = 2^{12}$ is true. Write him a note describing two or three ways he can find out.

Review PREVIEW

Find the x- and y-intercepts of the graph of each equation. *(Section 10-3)*

36. $y = (x - 5)(x - 2)$

37. $y = x(x + 3)$

38. a. Write a statement represented by the figure formed by x-tiles and 1-tiles. *(Section 1-5)*

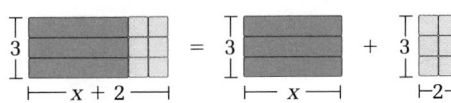

b. Draw a diagram similar to the one in part (a) to represent the statement $2(x + 3) = 2x + 2 \cdot 3$.

Use the distributive property to rewrite each expression. *(Section 1-5)*

39. $2(n - 5)$ **40.** $6(4r + 3)$ **41.** $-3(x + 12)$ **42.** $(m + n)8$

 Working on the Unit Project

43. **Research** The pull of gravity is not the same on every planet. Do some research to find out how this affects a ball thrown or dropped on other planets.

10-4 Working with Powers   **575**

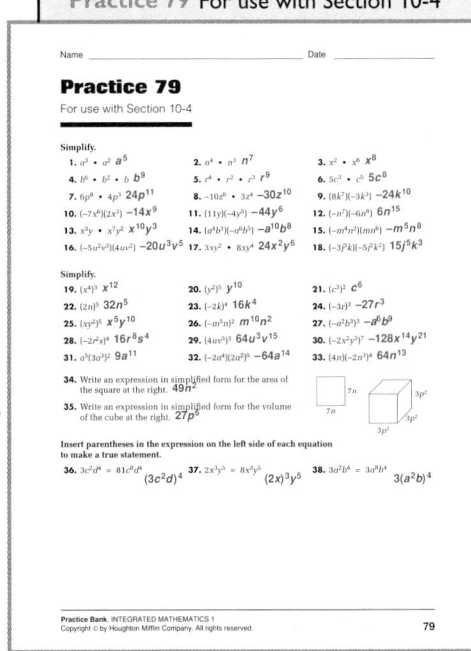
Answers to Exercises and Problems

35. Answers may vary. Examples are given. (1) Use the definition of an exponent: $2^4 = 2 \cdot 2 \cdot 2 \cdot 2$ and $2^3 = 2 \cdot 2 \cdot 2$. Write out 2^{12} as a product $(2^{12} = 2 \cdot 2 \cdot 2 \cdot 2 \cdot 2 \cdot 2 \cdot 2 \cdot 2 \cdot 2 \cdot 2 \cdot 2 \cdot 2)$ and compare it to $2^4 \cdot 2^3 = 2 \cdot 2 \cdot 2 \cdot 2 \cdot 2 \cdot 2 \cdot 2$. (2) Use a calculator with an exponent key to evaluate

2^{12} (4096) and $2^4 \cdot 2^3$ (128) and compare the answers. (3) Use the product of powers rule to simplify $2^4 \cdot 2^3 = 2^{4+3} = 2^7$.

36. x-intercepts: 5 and 2; y-intercept: 10

37. x-intercepts: 0 and -3; y-intercept: 0

38. a. $3(x + 2) = 3x + 3 \cdot 2$

b.

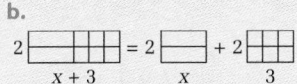

39. $2n - 10$

40. $24r + 18$

41. $-3x - 36$

42. $8m + 8n$

43. Answers may vary. An example is given. The greater the surface gravity, the faster a ball thrown or dropped will fall back to the planet's surface. For example, a ball will hit the ground faster on Jupiter than on Earth, but slower on Mars than on Earth.

575

Using Technology

Exs. 33 and 34 allow students to explore and make conjectures about power functions. Students may find it interesting to extend Ex. 34 to higher powers.

Assessment: Portfolio

Ex. 35 can be used to assess students' understanding of $a^m \cdot a^n = a^{m+n}$ and whether they are confusing the three power rules presented in this section. Students should consider including their responses in their portfolios.

Working on the Unit Project

Direct students to organize the results of their research so that it is presented in a format that can be understood by others.

Practice 79 For use with Section 10-4

Factored Form and x-Intercepts

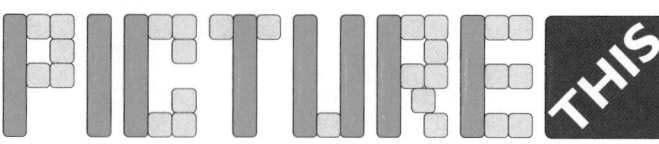

Focus
Factor and expand algebraic expressions, and use factoring to find the x-intercepts of parabolas.

EXPLORATION

How can you picture an algebraic expression?

- **Materials: algebra tiles**
- **Work with another student.**

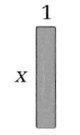

Area = ?   Area = ?

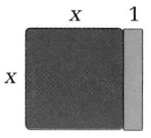

Area = ?

① Suppose your algebra tiles have the dimensions shown. What is the area of each kind of tile? What is the area of a rectangle made with the two tiles?

② **a.** Build a rectangle with these tiles. The tiles should touch but not overlap.

b. Complete: Length of your rectangle = ?
Width of your rectangle = ?

c. Complete:
Area of your rectangle = length × width
= ?

d. Complete:
Area of your rectangle = sum of the areas of the tiles
= ?

e. Write a mathematical statement that relates the expressions you wrote in parts (c) and (d).

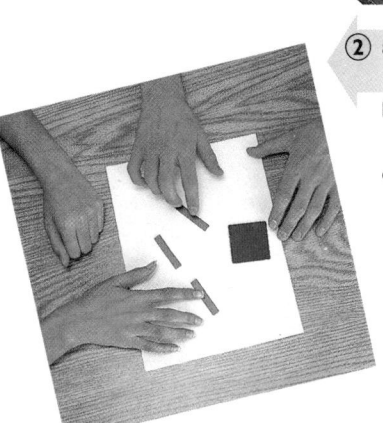

Answers to Exploration

1. x; x^2; $x + x^2$, or $x(x + 1)$
2. a.
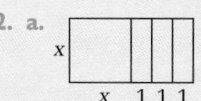

b. $x + 3$; x
c. $(x + 3)x$, or $x(x + 3)$
d. $x^2 + 3x$
e. $x(x + 3) = x^2 + 3x$

3 **a.** Build a rectangle with these tiles.

  b. Repeat parts (b)–(e) of step 2 for the rectangle you build with these tiles.

4 **a.** Build a different rectangle with the same tiles you used in step 3.

  b. How does the area of this rectangle compare to the area of the rectangle you built in step 3?

5 Complete each equation by writing two expressions for the area of the rectangle. If a tile is red, subtract the area of that tile from the others.

Example:

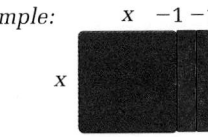

Area = $x(x - 2) = x^2 - 2x$

a.

Area = $\underline{?} \times \underline{?} = \underline{?} + \underline{?}$

b.

Area = $\underline{?} \times \underline{?} = \underline{?} - \underline{?}$

▶ Now you are ready for:
Exs. 1–7 on p. 580

Expanding and Factoring Expressions

In the Exploration you wrote pairs of equal expressions for the area of a rectangle.

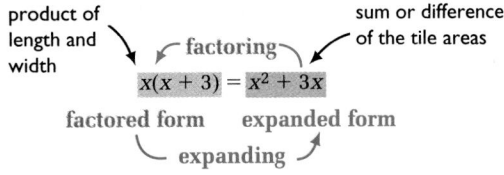

product of length and width ↘ factoring ↙ sum or difference of the tile areas

$x(x + 3) = x^2 + 3x$

factored form   expanded form

⌣ expanding ⌣

Sample 1

Expand $x(x - 3)$.

Sample Response

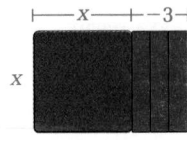

$x(x - 3) = x \cdot x - x \cdot 3$ ← Use the distributive property.

$= x^2 - 3x$

10-5 Factored Form and x-Intercepts   **577**

TEACHING

Exploration

The goal of the Exploration is to develop equal expressions for the area of a rectangle. This approach provides a visual and concrete basis for expanding an algebraic expression and for factoring the expanded form.

Mathematical Procedures

A procedure that permeates mathematics at all levels is the doing and undoing of a particular process. You can add two numbers, for example, and then undo the addition by using subtraction. You can square a number and then undo the squaring process by taking the square root. In this section, students learn to expand the factored form of an expression and then undo the expanded form by factoring it.

Additional Sample

S1 Expand $x(x + 5)$.

$x(x + 5) = x \cdot x + x \cdot 5$

$= x^2 + 5x$

Answers to Exploration

3, 4. Answers may vary. Examples are given.

3. a.

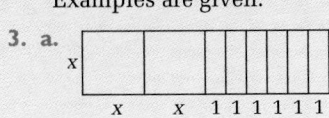

b. (b) $2x + 6$; x

  (c) $(2x + 6)x$, or $x(2x + 6)$

  (d) $2x^2 + 6x$

  (e) $x(2x + 6) = 2x^2 + 6x$

4. a.

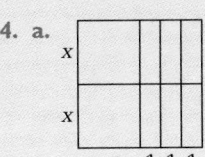

b. The area is the same, $2x^2 + 6x$.

5. a. $x(x + 4) = x^2 + 4x$

  b. $x(x - 1) = x^2 - x$

577

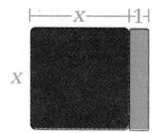

Sample 2

Factor $x^2 + x$.

Sample Response

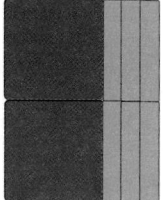

$x^2 + x = x \cdot x + x \cdot 1$ ◄— Rewrite each term as a product.

$= x \cdot x + x \cdot 1$ ◄— Look for a common factor in each term.

$= x(x + 1)$ ◄— Use the distributive property in reverse.

In the Exploration you built two rectangles with an area of $2x^2 + 6x$.

An expression like $2x^2 + 6x$ is called a *binomial*. A **binomial** is the sum of two *monomials*. A **monomial** can be any of these.

➤ a number, like 8 or -8,

➤ a variable raised to any power, like x^2,

➤ the product of numbers and variables raised to any power, like $2x^2$, $6x$, or $-8x$.

By building two rectangles with the same area, you showed that there is sometimes more than one way to factor an expression.

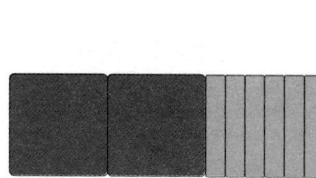

Area: $x(2x + 6) = 2x^2 + 6x$ Area: $2x(x + 3) = 2x^2 + 6x$

Look at the two different factored forms of $2x^2 + 6x$.

greatest common factor

$$x(2x + 6) = 2x(x + 3)$$

The expression $2x(x + 3)$ has been **factored completely** since x and 3 have no common factors other than 1.

Finding x-Intercepts

You can find the x-intercepts of some parabolas without graphing if you begin by factoring completely one side of the equation for the parabola.

The graph of $y = 3x^2 - 6x$ is a parabola. Find the x-intercepts of the graph.

Sample Response

Student Resources Toolbox
p. 648 *Factors*

Step 1. Factor the binomial $3x^2 - 6x$ completely.

$3x^2 - 6x$

$3 \cdot x \cdot x - 6 \cdot x$ ← Rewrite each power as a product.

$3 \cdot x \cdot x - 2 \cdot 3 \cdot x$ ← Write each number as a product of primes.

$3 \cdot x \cdot x - 2 \cdot 3 \cdot x$ ← Look for the common factors.

$3x(x - 2)$ ← Factor out $3x$, the greatest common factor.

Step 2. Write the original equation in its factored form.

$y = 3x(x - 2)$

$0 = 3x(x - 2)$ ← The x-intercepts occur when $y = 0$. Substitute 0 for y and solve.

$3x = 0$ *or* $x - 2 = 0$ ← Use the zero-product property. Set each factor equal to zero.

$x = 0$ $\quad x - 2 + 2 = 0 + 2$

$x = 2$

The x-intercepts of the parabola are 0 and 2.

BY THE WAY...

Secret codes can be made by multiplying two 100-digit numbers. To break the code, you must factor the 200-digit expanded product. Not even a computer can factor a number this big!

▶ Now you are ready for:
Exs. 8–43 on pp. 580–582

Talk it Over

1. **a.** Describe how to find the greatest common factor of 12 and 15.

   b. Find the greatest common factor of the two terms of $12x^2 - 15x$.

2. Explain how to solve $0 = 5x(x + 1)$.

Look Back ◀

The expression $4x^2 - 16x$ has the form $ax^2 + bx$. Explain how to factor completely an expression of this form.
The expression $4x(x - 4)$ has the form $ax(x + b)$. Explain how to expand an expression of this form.

Answers to Talk it Over

1. **a.** List all the factors of 12 and all of the factors of 15, then select the largest factor that is common to both numbers.

   b. $3x$

2. Answers may vary. An example is given. Use the zero-product property and set each factor equal to zero and solve for x:
   $5x = 0$, $x = 0$;
   $x + 1 = 0$, $x = -1$.

Answers to Look Back

Explanations may vary. Examples are given. To factor an expression of the form $ax^2 + bx$, find the greatest common factor, g, of a and b such that $cg = a$ and $dg = b$. There is also an x common to both terms. Therefore, $ax^2 + bx = gx(cx + d)$. To expand an expression of the form $ax(x + b)$, use the distributive property: $ax(x + b) = ax \cdot x + ax \cdot b = ax^2 + (ab)x$.

Additional Sample

S3 The graph of $y = 4x^2 + 2x$ is a parabola. Find the x-intercepts of the graph.

Step 1 Factor the binomial completely.

$4x^2 + 2x$

$4 \cdot x \cdot x + 2 \cdot x$

$2 \cdot 2 \cdot x \cdot x + 2 \cdot x$

$2 \cdot (2 \cdot x) \cdot x + (2 \cdot x)$

$2x(2x + 1)$

Step 2 Write the original equation in factored form.

$y = 2x(2x + 1)$

Substitute 0 for y and solve.

$2x = 0$ or $2x + 1 = 0$

$x = 0$ $\qquad 2x = -1$

$\qquad\qquad x = -\frac{1}{2}$

The x-intercepts of the parabola are 0 and $-\frac{1}{2}$.

Multicultural Note

Most secret codes have been broken. One code, however, that was constructed in 1942 by using the language of the Native American Navajos, has never been broken. Without the help of the Navajos and their code, military officials today state that many victories in World War II may not have been won. In 1982, President Ronald Reagan named August 14 "National Navajo Code Talkers Day."

Look Back

Use the Look Back for a class discussion to summarize the concepts of the section. Be sure students recognize that factoring and expanding are opposite processes that undo one another. Use the discussion to review the meaning of the greatest common factor and that to factor an expression always means to factor it completely.

APPLYING

Integrating the Strands

Algebra Exs. 1–43

Functions Exs. 15–23, 29, 30, 41–43

Geometry Exs. 1–7, 29

Logic and Language Exs. 29, 43

Cooperative Learning

You may want to use Exs. 1–7 as a cooperative group activity to enhance students' understanding of the equality of the algebraic expressions involved.

Answers to
Exercises and Problems

1. $x(x + 4)$

2. $x(3x - 1)$

3. $x(x + 2)$

4. a, b.

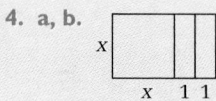

c. $x^2 + 2x = x(x + 2)$

5. a, b.

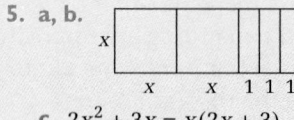

c. $2x^2 + 3x = x(2x + 3)$

6. Answers may vary. Examples are given based on the rectangle in part (a).

a.

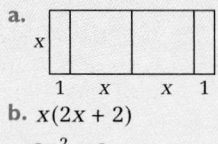

b. $x(2x + 2)$

c. $2x^2 + 2x$

7. Answers may vary. An example is given. Yes, it is possible because each tile has at least one side of length x. All the tiles can be lined up to form a rectangle with height x.

10-5 Exercises and Problems

Express the area of each rectangle built from algebra tiles as the product of its length and width.

1. 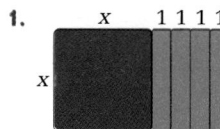 **2.** 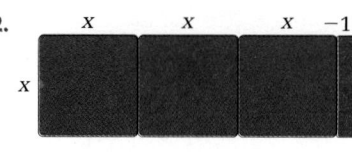 **3.**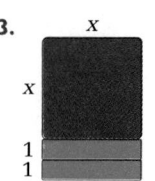

Using Manipulatives For Exercises 4–6, use algebra tiles or paper models.

4. a. Build a rectangle using the tiles at the right.

   b. Sketch your rectangle and label the length and width.

   c. Express the area of your rectangle in two ways, as a sum and as a product.

5. Repeat Exercise 4 using two x^2-tiles and three x-tiles.

6. a. **Open-ended** Use at least one x^2-tile and at least one x-tile to build any rectangle you want.

   b. Express the area of the rectangle you built in part (a) as a product.

   c. Express the area of the rectangle you built in part (a) as a sum.

7. **Writing** Do you think it is possible to build a rectangle using any number of x^2-tiles and any number of x-tiles? Give a reason for your answer.

Expand each product.

8. $x(x + 9)$ **9.** $-x(x - 15)$ **10.** $(x + 11)x$

11. $6x(x - 7)$ **12.** $3x(2x + 4)$ **13.** $-4x(5x - 9)$

14. **Reading** Give two examples of binomials and monomials that appear on page 578.

The graph of each equation is a parabola. For each equation, do these things.

a. Factor one side completely.

b. Find the x-intercepts and the y-intercept of the graph.

15. $y = x^2 - 5x$ **16.** $y = x^2 + 10x$ **17.** $y = 12x + x^2$

18. $y = -x^2 - 5x$ **19.** $y = -x^2 + 10x$ **20.** $6x - x^2 = y$

21. $y = 8x^2 - 40x$ **22.** $-4x^2 + 12x = y$ **23.** $y = 12x - 6x^2$

24. a. **Open-ended** Write an expression in the form $ax^2 + bx$ that can be factored.

   b. Write an expression in the form $ax^2 + bx$ that can be factored in two ways.

Tell whether each expression is a *monomial* or a *binomial*.

25. $x - 2$ **26.** $3x^2 + 8$ **27.** $-7a^2b^3$ **28.** $6y^3 + 4xy$

---

8. $x^2 + 9x$

9. $-x^2 + 15x$

10. $x^2 + 11x$

11. $6x^2 - 42x$

12. $6x^2 + 12x$

13. $-20x^2 + 36x$

14. monomials: $2x^2$ and $6x$; binomials: $x^2 + x$ and $2x^2 + 6x$

15. a. $x(x - 5)$

   b. x-intercepts: 0 and 5; y-intercept: 0

16. a. $x(x + 10)$

   b. x-intercepts: 0 and -10; y-intercept: 0

17. a. $x(12 + x)$

   b. x-intercepts: 0 and -12; y-intercept: 0

18. a. $-x(x + 5)$

   b. x-intercepts: 0 and -5; y-intercept: 0

19. a. $-x(x - 10)$

   b. x-intercepts: 0 and 10; y-intercept: 0

20. a. $x(6 - x)$

   b. x-intercepts: 0 and 6; y-intercept: 0

21. a. $8x(x - 5)$

   b. x-intercepts: 0 and 5; y-intercept: 0

22. a. $-4x(x - 3)$

   b. x-intercepts: 0 and 3; y-intercept: 0

23. a. $6x(2 - x)$

   b. x-intercepts: 0 and 2; y-intercept: 0

29. Urban Planning A town is planning to put a concrete walkway around a Japanese garden. The walkway should be the same width on each side. It must be at least 5 ft wide to allow two wheelchairs to pass.

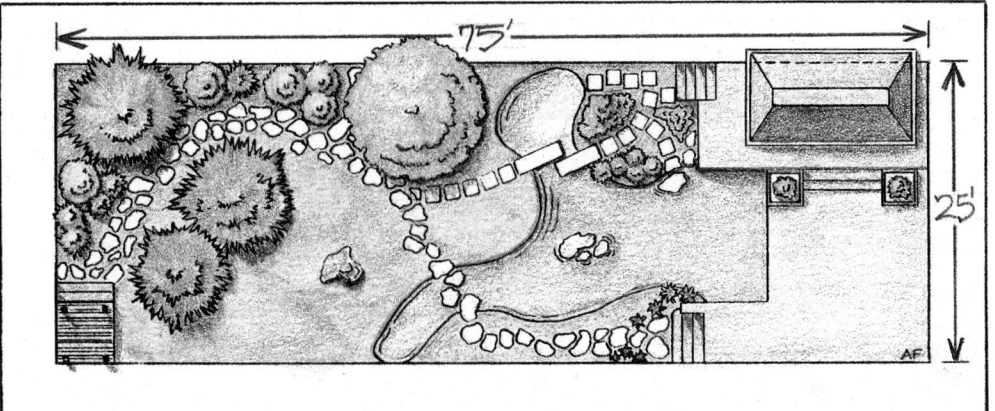

a. Let x represent the width of the walkway. Write an expression in expanded form to represent the area of the walkway. (*Hint:* Divide the walkway into squares and rectangles. Add up the areas of the pieces.)

b. Factor your expression.

c. Let y represent the area of the walkway. Write an equation for the area.

d. Find the x-intercepts of the graph of your equation.

e. Complete the table of values for your equation.

x	y
-70	?
-35	?
-25	?
-15	?
20	?

f. Use the x-intercepts and the table of values to sketch the parabola on graph paper. Mark the horizontal axis with units of 5 and the vertical axis with units of 500.

g. **Writing** In which quadrants does the graph contain ordered pairs that are not realistic solutions for the situation? Explain.

h. Suppose there is only enough concrete to cover an area of 1500 ft². Use the graph to estimate the largest possible width for the walkway. Round your answer to the nearest foot.

BY THE WAY...

The NAMES Project AIDS Memorial Quilt consists of 12 ft square sections with walkways around them. New sections are being added all the time. On August 5, 1993, the sections alone had an area of 10 acres. Including the walkways, the quilt covered an area of 16 acres on that date.

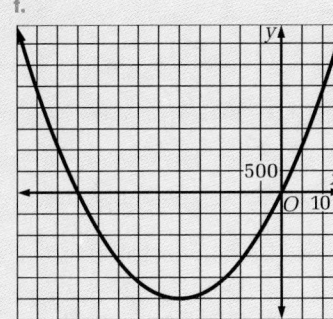

Integrating the Strands

The strands of algebra, functions, geometry, and logic and language are interwoven in Ex. 29 on urban planning. This problem first engages students in visualizing the walkway geometrically. The algebraic analysis leads to an understanding of the realistic solutions for the situation.

Communication: Writing

Ex. 29(g) demonstrates that a mathematical solution may not be an answer to a real-life situation. Students should realize that it is important to look at mathematical solutions to be sure they make sense within the context of a real-life problem.

Answers to Exercises and Problems

24. Answers may vary. Examples are given.

a. $7x^2 + 9x$

b. $7x^2 + 70x$

25. binomial

26. binomial

27. monomial

28. binomial

29. a. $4x^2 + 200x$

b. $4x(x + 50)$

c. $y = 4x(x + 50)$

d. 0 and -50

e.

x	y
-70	5600
-35	-2100
-25	-2500
-15	-2100
20	5600

f.

g. Quadrants II and III; In these quadrants, x is negative. Since x represents the width of the walkway, x cannot be negative.

h. 6 ft (Even though the actual value is closer to 7, you cannot round up or you will not have enough concrete.)

581

Practice 80 For use with Section 10-5

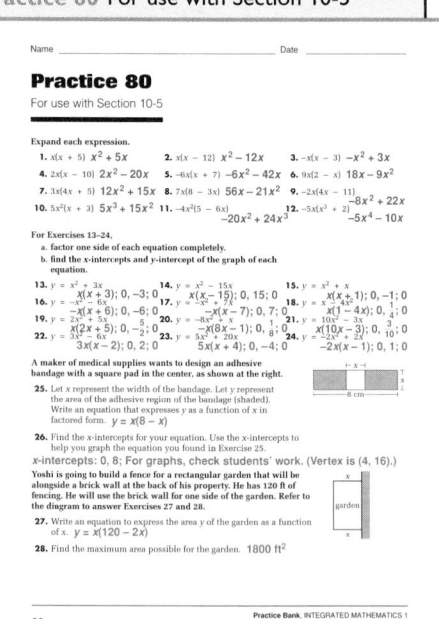

Ongoing **ASSESSMENT**

30. **Group Activity** Work in a group of three students. Take turns until each student has been first, second, and third. Have the first student write an equation in the form $y = ax^2 + bx$. Have the second student factor completely the right side of the equation. Have the third student name the x-intercepts of the graph of the equation.

 Review **PREVIEW**

Simplify. *(Sections 10-4, 10-5)*

31. $m^3 \cdot 2m^5$ 32. $(-5x^2)(-2x^3)$ 33. $(3ac)^3$ 34. $(3n^2)^2$

35. $x^2 + x + 7x + 7$ 36. $x^2 - 2x - 10x - 8$ 37. $x^2 + 4x - 5x + 14$

For the graph of each equation, write the equation of the line of symmetry and the coordinates of the vertex. *(Section 10-2)*

38. $y = x^2 - 13$ 39. $y = x^2 + 42$ 40. $y = (x - 11)^2$

Working on the Unit Project

41. One equation for the path of a ball hit at a 45° angle is $y = x(1 - 0.005x)$.

   a. Find the x-intercepts of the graph of the equation.

   b. What do the x-intercepts represent in this situation?

   c. Write the expanded form of the equation.

42. Another equation for the path of a ball hit at a 45° angle is $y = -0.003x^2 + x$.

   a. Write the factored form of the equation.

   b. Find the x-intercepts of the graph of the equation.

   c. In this situation, what is represented by the x-intercepts?

43. Which of the equations in Exercises 41 and 42 represents the path of the ball that lands farther away? What do you think made it land farther away?

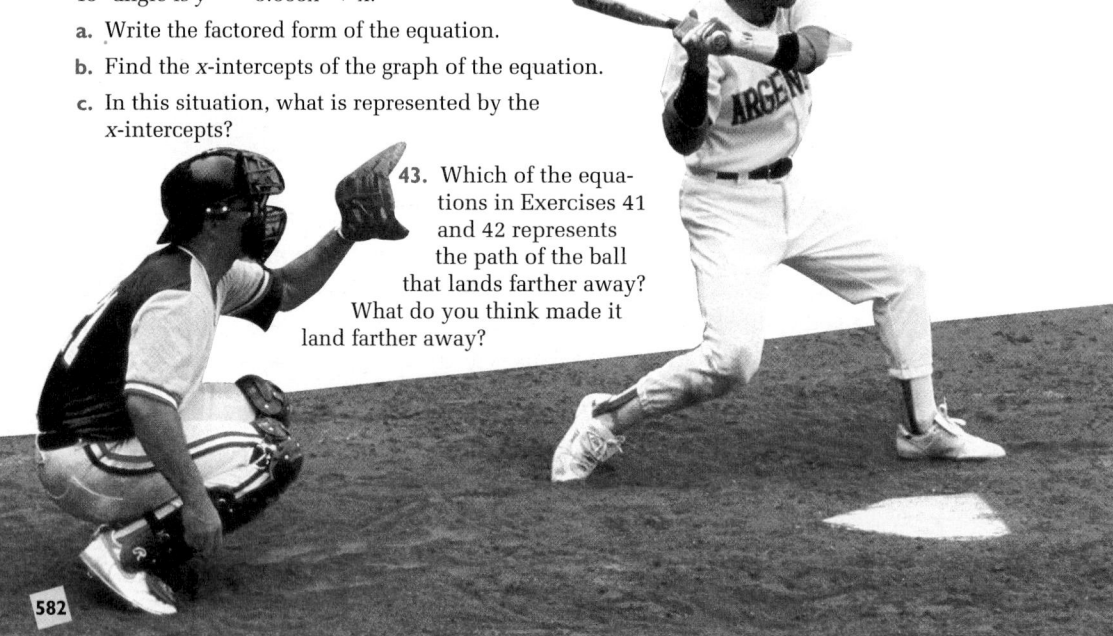

582

Answers to Exercises and Problems

30. Answers may vary. An example is given. $y = 24x^2 + 12x$; $y = 12x(2x + 1)$; x-intercepts: 0 and $-\dfrac{1}{2}$

31. $2m^8$

32. $10x^5$

33. $27a^3c^3$

34. $9n^4$

35. $x^2 + 8x + 7$

36. $x^2 - 12x - 8$

37. $x^2 - x + 14$

38. $x = 0$; $(0, -13)$

39. $x = 0$; $(0, 42)$

40. $x = 11$; $(11, 0)$

41. a. 0 and 200

   b. The origin represents the point where the ball was hit; the ball traveled 200 units horizontally before landing.

c. $y = x - 0.005x^2$

42. a. $y = -x(0.003x - 1)$

   b. 0 and $333\frac{1}{3}$

   c. The origin represents the point where the ball was hit; the ball traveled $333\frac{1}{3}$ units horizontally before landing.

43. the equation in Exercise 42; Answers may vary. An example is given. Since both balls were hit at a 45° angle, the ball in Exercise 42 must have traveled farther because it was hit harder.

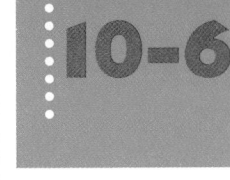

Section **10-6**

Expanded Form and Line of Symmetry

THE LONG AND THE SHORT OF IT

Focus

Multiply two binomials and use a formula to find the line of symmetry of a parabola.

Talk it Over

1. Can you build a rectangle with the dimensions

   length = $x + 1$

   width = $x + 1$

   using only x^2-tiles and x-tiles?

2. a. In the rectangle shown, what is the area of the piece shown with dashed lines?

   b. How many 1-tiles do you need to fill in the dashed area?

   c. Express the area of the rectangle as the product of its length and width.

   d. Express the area of the rectangle as the sum of the areas of the tiles.

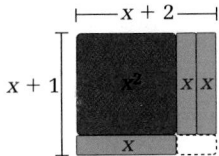

3. a. Express the area of this rectangle as the product of its length and width.

   b. Express the area of this rectangle as the sum of the areas of the tiles.

   c. Are the expressions you wrote in parts (a) and (b) equal?

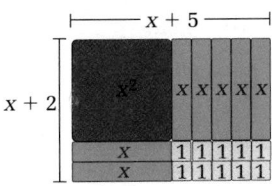

One way to find the product of two binomials is to build a rectangle from algebra tiles. Another way is to use the distributive property.

Answers to Talk it Over

1. No, you will be left with a square "hole" in your tile rectangle.
2. a. 2 square units
   b. 2
   c. $(x + 2)(x + 1)$
   d. $x^2 + 3x + 2$
3. a. $(x + 5)(x + 2)$
   b. $x^2 + 7x + 10$
   c. Yes.

PLANNING

Objectives and Strands

See pages 544A and 544B.

Spiral Learning

See page 544B.

Materials List

- Algebra tiles

Recommended Pacing

Section 10-6 is a two-day lesson.

Day 1

Pages 583–584: Talk it Over through Sample 2, *Exercises 1–13*

Day 2

Pages 585–586: Talk it Over through Look Back, *Exercises 14–42*

Extra Practice

See pages 638–639.

Warm-Up Exercises

- Warm-Up Transparency 10-6

Support Materials

- Practice 81
- Enrichment 72 in the Activity Bank
- Study Guide 10-6
- Problem Set 23
- Additional Exploration 13
- Diagram Masters 6, 7 in the Explorations Lab Manual
- Overhead Visuals 1, 10
- Using Plotter Plus: Expanded Form and the Line of Symmetry of a Parabola
- Quiz 10-6
- Test 42
- Alternative Assessments 11, 12

TEACHING

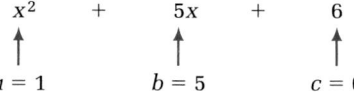

Talk it Over

Questions 1–3 give a visual meaning to finding the product of two binomials. This approach provides a foundation for finding such a product algebraically using the distributive property.

Additional Sample

S1 Expand $(x + 3)(x + 4)$.

$(x + 3)(x + 4)$
$= (x + 3)x + (x + 3)4$
$= x^2 + 3x + 4x + 12$
$= x^2 + 7x + 12$

Talk it Over

Question 4 illustrates that the distributive property may be used on either term, pointing out that there is more than one way to expand the product of two binomials.

Additional Sample

S2 Expand $(2x + 4)(x - 1)$.

$(2x + 4)(x - 1)$
$= (2x + 4)x - (2x + 4)1$
$= 2x^2 + 4x - 2x - 4$
$= 2x^2 + 2x - 4$

Error Analysis

In expansions involving subtraction, such as Sample 2, students may not distribute the subtraction sign correctly. A remedy for this error is to have students rewrite the subtraction as addition before expanding. For example, Sample 2 would be rewritten as $(x + (-3))(2x + 1)$.

Sample 1

Expand $(x + 2)(x + 3)$.

Sample Response

$(x + 2)(x + 3) = (x + 2)x + (x + 2)3$ ⟵ Use the distributive property.
$= x^2 + 2x + 3x + 6$ ⟵ Use the distributive property again.
$= x^2 + 5x + 6$ ⟵ Combine like terms.

BY THE WAY...

Biologists use trinomials to track the changes in DNA as it breaks apart and recombines. The long strands of DNA form complicated loop patterns.

Each pattern is represented by a different trinomial.

Talk it Over

4. For the first step of Sample 1, Sonia wrote
$$(x + 2)(x + 3) = x(x + 3) + 2(x + 3).$$
Will she get the same answer?

In Sample 1 you multiplied two binomials. The product is the sum of three monomials and has the form $ax^2 + bx + c$.

$$x^2 \quad + \quad 5x \quad + \quad 6$$
$$\uparrow \qquad\qquad \uparrow \qquad\qquad \uparrow$$
$$a = 1 \qquad\quad b = 5 \qquad\quad c = 6$$

An expression with three monomial terms is called a **trinomial**.

Sample 2

Expand $(x - 3)(2x + 1)$.

Sample Response

$(x - 3)(2x + 1) = (x - 3)2x + (x - 3)1$ ⟵ Use the distributive property.
$= 2x^2 - 6x + x - 3$ ⟵ Use the distributive property again.
$= 2x^2 - 5x - 3$ ⟵ Combine like terms.

▶ Now you are ready for:
Exs. 1–13 on p. 587

Answers to Talk it Over

4. Yes.

Talk it Over

Questions 6 and 7 present two different methods for finding the equation of the line of symmetry for a parabola. As students discuss these methods, you may wish to have them compare and contrast the methods with one another.

Talk it Over

5. What is the equation of the line of symmetry of the three parabolas shown?

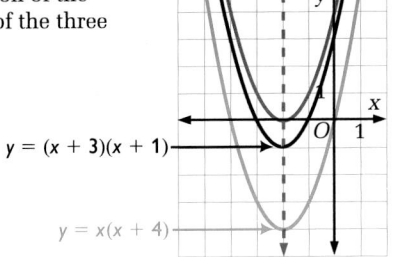

$y = (x + 2)(x + 2)$

$y = (x + 3)(x + 1)$

$y = x(x + 4)$

6. Angela says she can find the equation of the line of symmetry from the equation $y = (x + 3)(x + 1)$ without drawing the graph. Here is her thinking.

"The x-intercepts are −3 and −1."   a. How does she know?

"The line of symmetry crosses the x-axis halfway between the x-intercepts."   b. Which point is halfway between $(−3, 0)$ and $(−1, 0)$?

"So, the line of symmetry is x = −2."   c. Do you agree?

7. Celine found this formula in an algebra book.

> For the graph of $y = ax^2 + bx + c$, the equation of the line of symmetry is $x = -\dfrac{b}{2a}$.

She wanted to see how this works for the graphs of the equations in question 5. Here is what she did.

"I rewrote each equation in the form $y = ax^2 + bx + c$."   a. What did she write for each equation?

"I found that the coefficients a and b are the same in each equation."   b. Does this surprise you?

"I substituted the values of a and b in $x = -\dfrac{b}{2a}$."   c. Did she get the same answer as Angela did in question 6?

8. Why do you think the coefficient c does not appear in Celine's formula?

10-6 Expanded Form and Line of Symmetry

585

Answers to Talk it Over

5. $x = -2$

6. a. When the graph crosses the x-axis, the value of y is zero: $(x + 3)(x + 1) = 0$. By the zero-product property, $x + 3 = 0$ or $x + 1 = 0$, so $x = -3$ or $x = -1$.

  b. $(-2, 0)$

  c. Yes.

7. a. $y = x^2 + 4x + 4$,
  $y = x^2 + 4x + 3$,
  $y = x^2 + 4x$

  b. Answers may vary.

  c. Yes, since
  $$x = -\frac{b}{2a} = -\frac{4}{2(1)} = -2,$$
  she got $x = -2$

8. The line of symmetry of the given equations is a vertical line. Changing the value of c moves the vertex up or down along this line.

Additional Sample

S3 Without graphing, find each feature of the graph of the equation $y = x^2 + 2x - 15$.

a. the equation of the line of symmetry

$$x = -\frac{b}{2a} = -\frac{2}{2 \cdot 1} = -1$$

The equation of the line of symmetry is $x = -1$.

b. the coordinates of the vertex

$y = (-1)^2 + 2(-1) - 15$

$y = -16$

The coordinates of the vertex are $(-1, -16)$.

c. the y-intercept

$y = x^2 + 2x - 15$

$y = 0^2 + 2(0) - 15$

$y = -15$

The y-intercept is -15.

··················

Look Back

If students have difficulty with this activity, they should review Talk it Over questions 5–7. If the Look Back is used as a class discussion, be sure that students correct any misunderstandings they may have about any of the various methods.

·············●

Without graphing, find each feature of the graph of the equation $y = x^2 + 6x - 17$.

a. the equation of the line of symmetry

b. the coordinates of the vertex

c. the y-intercept

Sample Response ··················

a. Use the formula for the line of symmetry.

$$x = -\frac{b}{2a}$$

$$= -\frac{6}{2 \cdot 1} \qquad \longleftarrow \text{ Substitute 6 for } b \text{ and 1 for } a.$$

$$= -3$$

The equation of the line of symmetry is $x = -3$.

b. The vertex is on the line of symmetry. The equation of the line of symmetry tells the x-coordinate of the vertex.

$$y = x^2 + 6x - 17$$

$$= (-3)^2 + 6(-3) - 17 \qquad \longleftarrow \text{ Substitute } -3 \text{ for } x.$$

$$= -26$$

The coordinates of the vertex are $(-3, -26)$.

c. The value of x is 0 where the graph crosses the y-axis.

$$y = x^2 + 6x - 17$$

$$= 0^2 + 6(0) - 17 \qquad \longleftarrow \text{ Substitute 0 for } x.$$

$$= -17$$

The y-intercept is -17.

BY THE WAY...

To record a concert in stereo, sound engineers use a parabolic disk to direct the sound to two microphones placed on opposite sides of the line of symmetry.

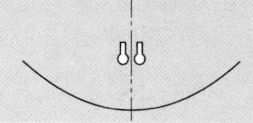

Look Back ◄

Describe as many different methods as you can for finding the equation of the line of symmetry of a parabola without graphing. If a method can be used with only certain kinds of equations, be sure to include this information.

······► Now you are ready for:
Exs. 14–42 on pp. 588–590

Unit 10 Quadratic Equations as Models

Answers to Look Back ··

Answers may vary. Three methods have been presented (a and b are integers): (1) For equations of the form $y = x^2 + a$, the line of symmetry is $x = 0$. For equations of the form $y = (x + a)^2$, the equation of the line of symmetry is $x = -a$. (2) For equations in factored form $y = (x + a)(x + b)$, find the x-intercepts. The line of symmetry is a vertical line halfway between the x-intercepts. (3) Use the formula $x = -\frac{b}{2a}$ with equations of the form $y = ax^2 + bx + c$.

10-6 Exercises and Problems

Using Manipulatives For Exercises 1–3, do these things.

a. Write each product in expanded form.

b. Check your answer to part (a) by using algebra tiles to build a rectangle. Sketch your rectangle.

1. $(x + 1)(x + 2)$ **2.** $(x + 2)(x + 3)$ **3.** $(x + 4)(x + 1)$

4. A shortcut for expanding the product of two binomials is called *FOIL*. "FOIL" is made up of the first letters of the words First, Outer, Inner, and Last.

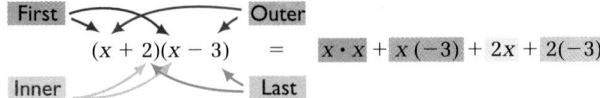

$(x + 2)(x - 3) = x \cdot x + x(-3) + 2x + 2(-3)$

Use the distributive property to show that FOIL works.

Expand each product.

5. $(x + 2)(x + 3)$ **6.** $(x + 5)(x + 4)$ **7.** $(x - 4)(x + 4)$ **8.** $(x - 7)(x + 7)$

9. $(x - 2)(x - 2)$ **10.** $(x + 4)(x + 4)$ **11.** $(x + 2)(2x - 5)$ **12.** $(4x - 6)(2x + 7)$

connection to HISTORY

13. Five years before he was elected president of the United States, James A. Garfield proved the Pythagorean theorem this way.

① Draw a rectangle and cut along the diagonal to form two triangles.

② Put the two triangles next to each other and draw a dashed line to make a third triangle.

③ Draw dashed lines to extend each side to make a square.

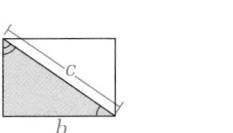 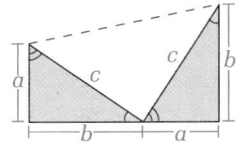 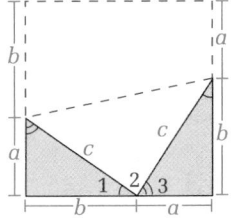

a. What is the sum of the measures of angles 1, 2, and 3?

b. Why is the sum of the measures of angles 1 and 3 equal to 90°?

c. What is the measure of angle 2?

d. What is the area of the square formed by extending the sides as shown? Write the product in expanded form.

e. Find the sum of the areas of the three triangles.

f. Study the diagram. Write an equation that shows how the total area of the three triangles is related to the area of the square.

g. Combine like terms to simplify the equation you wrote in part (f). Rewrite the equation so c^2 is alone on one side.

Answers to Exercises and Problems

1. a. $x^2 + 3x + 2$

b.
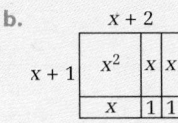

2. a. $x^2 + 5x + 6$

b.
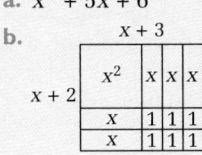

3. a. $x^2 + 5x + 4$

b.
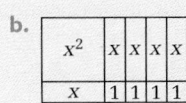

4. $(x + 2)(x - 3) =$
$(x + 2)x - (x + 2)3 =$
$x \cdot x + 2x - 3x - 2 \cdot 3 =$
$x^2 - x - 6$; You get the same answer with the FOIL method.

5. $x^2 + 5x + 6$

6. $x^2 + 9x + 20$

7. $x^2 - 16$

8. $x^2 - 49$

9. $x^2 - 4x + 4$

10. $x^2 + 8x + 16$

11. $2x^2 - x - 10$

12. $8x^2 + 16x - 42$

13. a. $180°$

APPLYING

Suggested Assignment

Standard 5–14, 16–25, 27–29, 31–42

Extended 4–29, 31–42

Integrating the Strands

Number Exs. 37–40

Algebra Exs. 1–36, 41

Functions Exs. 16–30, 35, 36, 41, 42

Measurement Exs. 13, 29

Geometry Exs. 13, 29, 30, 41

Mathematical Procedures

The use of shortcuts to do mathematical procedures, as in Ex. 4, is an efficient way to work. However, such shortcuts should only be used if students understand the reasons why they work. The FOIL method is based firmly upon the distributive property. Use Exs. 5–12 to demonstrate this fact.

Research

For Ex. 13, ask students to find out how many different proofs of the Pythagorean theorem exist. A book on the history of mathematics may be helpful in this research.

13. b. The sum of the measures of the two acute angles in each right triangle is 90°. Since $\angle 3$ is equal in measure to the other angle marked the same way, the sum of the measures of angles 1 and 3 is 90°.

c. $180° - 90° = 90°$

d. $(a + b)(a + b)$; $a^2 + 2ab + b^2$

e. $2 \cdot \frac{1}{2}ab + \frac{1}{2}(c)(c) = ab + \frac{1}{2}c^2$

f. The dashed line divides the square into two congruent parts. Therefore,
$$ab + \frac{1}{2}c^2 = \frac{1}{2}(a^2 + 2ab + b^2).$$

g. $ab + \frac{1}{2}c^2 = \frac{1}{2}a^2 + ab + \frac{1}{2}b^2$;
$$c^2 = a^2 + b^2$$

14. What is the equation of the line of symmetry of these three parabolas?

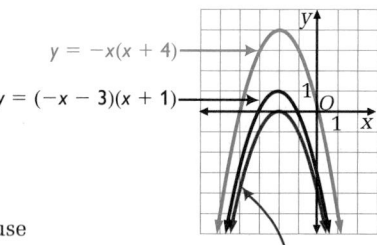
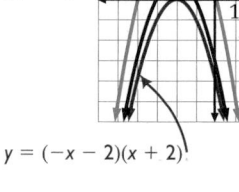

$y = -x(x + 4)$
$y = (-x - 3)(x + 1)$
$y = (-x - 2)(x + 2)$

15. **Reading** Questions 6 and 7 on page 585 describe two methods for finding the line of symmetry from an equation without drawing the graph. Describe how to use Abe's and Celine's methods on one of the equations in Exercise 14.

Writing Match each equation with its graph. Then explain how you decided which graph matches each equation.

16. $y = x^2 + 5x - 6$
17. $y = x^2 + 5x + 6$
18. $y = x^2 - 5x + 6$
19. $y = x^2 - 5x - 6$

A.
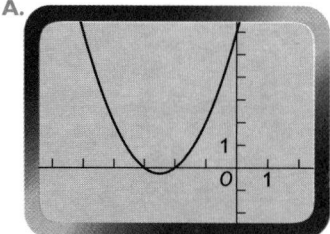

B.
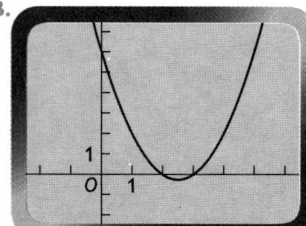

C.
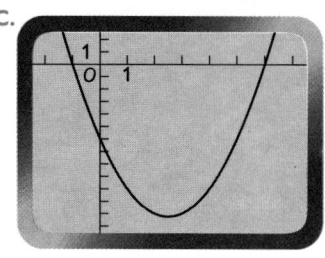

D.
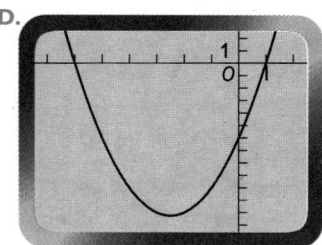

Without graphing, find each feature of the graph of each equation.

a. the equation of the line of symmetry

b. the coordinates of the vertex

c. the y-intercept

20. $y = x^2 + 4x - 11$
21. $y = x^2 + 8x + 12$
22. $y = x^2 - 4x - 15$
23. $y = x^2 + 16x + 15$
24. $y = x^2 + 6x + 18$
25. $y = x^2 - 6x + 7$

26. a. **Writing** Describe how you can use the factored form of an equation for a parabola to find some features of its graph.

b. **Writing** Describe how you can use the expanded form of an equation for a parabola to find some features of its graph.

Use equations (a)–(h). The graphs of these equations are parabolas.

a. $y = x^2 - 4$
b. $y = x(x - 8)$
c. $y = (x + 4)^2$
d. $y = -x^2 + 8x$
e. $y = -x^2 + 8x - 1$
f. $y = x^2 - 8x + 8$
g. $y = x^2 + 4$
h. $y = (x - 4)^2$

27. Which parabolas have the same line of symmetry?

28. Which parabolas have the same vertex?

from (3) as the x-coordinate and substituting that value in the equation to find the y-coordinate of the vertex.

b. If the equation of a parabola is given in expanded form, then you can find (1) the equation of the line of symmetry by using the equation $x = -\dfrac{b}{2a}$; (2) the coordinates of the vertex by

using the value of x from (1) as the x-coordinate and substituting that value in the equation to find the y-coordinate of the vertex; (3) the y-intercept by substituting 0 for x in the equation.

27. a and g; b, d, e, f, and h

28. none

29. a. Diagrams may vary. An example is given.

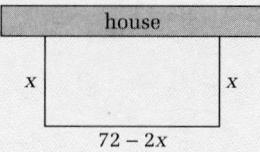

b. $72 - 2x$

c. $y = x(72 - 2x)$

d. 0 and 36; You cannot build a pen with a width of 0 ft or less, or of 36 ft or more.

29. Roxanne found 72 ft of leftover fencing in the garage. She decided to use it to make a rectangular pen for her dog Wolf. She wants to make the area of the pen as large as possible. Her mother allows her to use their house as one side of the pen.

  a. Draw a diagram to represent this situation.

  b. Let x be the width of the pen. Write an expression for the length of the pen in terms of x.

  c. Let y be the area of the pen. Use your answer to part (b) to write an equation for the area of the pen.

  d. Find the x-intercepts of the graph of the equation you wrote in part (c). What do the x-intercepts mean in this situation?

  e. Find the coordinates of the vertex of the graph. What does the vertex mean in this situation?

  f. What length and width should Roxanne use to build the pen?

Ongoing **ASSESSMENT**

30. **Group Activity** Work with another student.

  For each translation (a)–(d) of the graph of $y = x^2$, find the equation of the line of symmetry and the coordinates of the vertex. Take turns using these methods.

  Method ❶
  Sketch a graph of the translated parabola.

  Method ❷
  Write an equation in expanded form for the translated parabola. Use the formula $x = -\dfrac{b}{2a}$.

  Compare your answers. If they are different, find any errors and correct them together.

  a. 6 units left   **b.** 7 units right   **c.** 4 units left   **d.** 2 units right

Review **PREVIEW**

Expand each product. *(Section 10-5)*

31. $(x + 3)x$ **32.** $x(2x - 3)$ **33.** $3x(4x + 5)$ **34.** $-2x(x - 8)$

35. Factor the right side of the equation $y = -2x^2 - 6x$. Then find the x-intercept(s) and y-intercept of the graph of the equation.

36. Find the x-intercepts of the graph of $y = (x + 11)(x - 20)$. *(Section 10-3)*

Write all the pairs of factors for each number. *(Toolbox Skill 7)*

37. 15   **38.** 17   **39.** 24   **40.** 21

10-6 Expanded Form and Line of Symmetry   **589**

Integrating the Strands

The problem situation in Ex. 29 incorporates ideas from geometry, algebra, measurement, and functions. The interaction of mathematical concepts from different strands provides students with the tools they need to solve practical problems.

Assessment: Task

Ex. 30 allows for ongoing assessment of the concepts of Section 10-2 as well as those of this section. Students having problems with translating $y = x^2$ should review Section 10-2.

30. c. $y = x^2 + 8x + 16$; line of symmetry: $x = -\dfrac{b}{2a} = -\dfrac{8}{2 \cdot 1} = -4$; coordinates of vertex: $(-4, 0)$

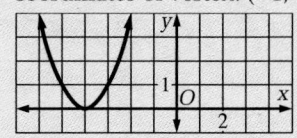

d. $y = x^2 - 4x + 4$; line of symmetry: $x = -\dfrac{b}{2a} = -\dfrac{-4}{2 \cdot 1} = 2$; coordinates of vertex: $(2, 0)$

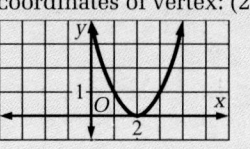

Answers to Exercises and Problems

29. e. $(18, 648)$; The y-coordinate represents the largest possible area. The x-coordinate represents the width that gives the largest area.

  f. $l = 36$ ft, $w = 18$ ft

30. a. $y = x^2 + 12x + 36$; line of symmetry: $x = -\dfrac{b}{2a} = -\dfrac{12}{2 \cdot 1} = -6$; coordinates of vertex: $(-6, 0)$

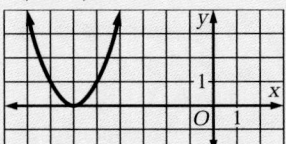

b. $y = x^2 - 14x + 49$; line of symmetry: $x = -\dfrac{b}{2a} = -\dfrac{-14}{2 \cdot 1} = 7$; coordinates of vertex: $(7, 0)$

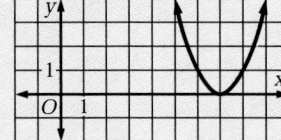

31. $x^2 + 3x$ **32.** $2x^2 - 3x$

33. $12x^2 + 15x$ **34.** $-2x^2 + 16x$

35. $-2x(x + 3)$; x-intercepts: 0 and -3; y-intercept: 0

36. -11 and 20

37. 1 and 15, -1 and -15, 3 and 5, -3 and -5

38. 1 and 17, -1 and -17

39. 1 and 24, -1 and -24, 2 and 12, -2 and -12, 3 and 8, -3 and -8, 4 and 6, -4 and -6

40. 1 and 21, -1 and -21, 3 and 7, -3 and -7

Practice 81 For use with Section 10-6

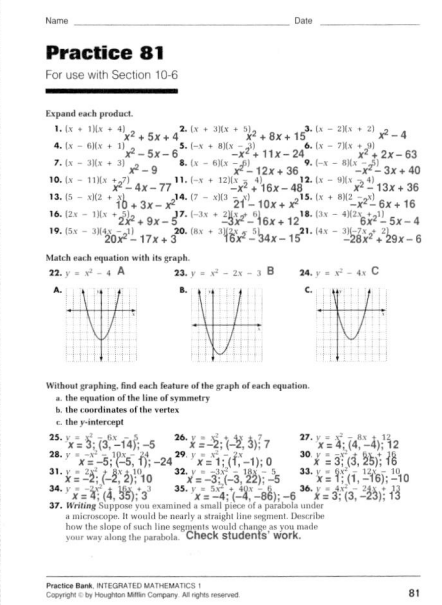

 Working on the Unit Project

41. An equation for the path of a ball hit at a 45° angle with an initial speed of 80 ft/s is $y = -0.005x^2 + x$. Find each feature of the graph of this equation.

   a. the y-intercept

   b. the equation of the line of symmetry

   c. the coordinates of the vertex

   d. the maximum height reached by the ball

42. Suppose you repeated Exercise 41 with the equation $y = -0.005x^2 + x + 4$. Predict how the results would be different. Then repeat Exercise 41 with this equation to check your prediction.

Unit 10   CHECKPOINT 2

1. **Writing** Factor the expression $6x^2 + 18x$. Explain your method.

   Simplify.   10-4

   2. $m \cdot m^2 \cdot m^4$ **3.** $(-a^2b)(3a^3b^5)$

   4. $(y^2)^5$ **5.** $(-3m^2n)^4$

   Expand each product.   10-5

   6. $x(x + 5)$ **7.** $x(x - 10)$ **8.** $-2x(x + 6)$

   The graph of each equation is a parabola. For each equation, do these things.

   a. Factor one side completely.

   b. Find the x-intercepts and the y-intercept of the graph.

   9. $y = x^2 + 3x$ **10.** $y = -5x^2 + 10x$

   11. Expand the product $(x + 3)(x - 4)$.   10-6

   12. Without graphing, find each feature of the graph of the equation $y = x^2 - 8x + 6$.

   a. the equation of the line of symmetry

   b. the coordinates of the vertex

   c. the y-intercept

590 Unit 10 Quadratic Equations as Models

Answers to Checkpoint

1. $6x(x + 3)$; Explanations may vary. An example is given. $6x^2 = 2 \cdot 3 \cdot x \cdot x$ and $18x = 2 \cdot 3 \cdot 3 \cdot x$. The greatest common factor of $6x^2$ and $18x$ is $2 \cdot 3 \cdot x$, or $6x$. Factor out $6x$.

2. m^7

3. $-3a^5b^6$

4. y^{10}

5. $81m^8n^4$

6. $x^2 + 5x$

7. $x^2 - 10x$

8. $-2x^2 - 12x$

9. **a.** $x(x + 3)$

   b. x-intercepts: 0 and -3; y-intercept: 0

10. **a.** $-5x(x - 2)$

    b. x-intercepts: 0 and 2; y-intercept: 0

11. $x^2 - x - 12$

12. **a.** $x = 4$

    b. $(4, -10)$

    c. 6

Using Factors to Sketch $y = x^2 + bx + c$

TTRIPLE play

$x^2 + 5x + 6$

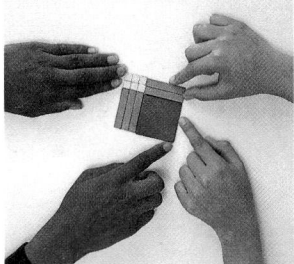

$x^2 + 3x + 2$

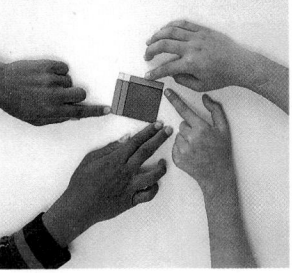

$x^2 + 6x + 9$

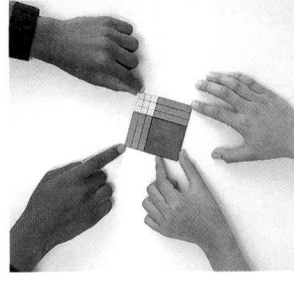

Talk it Over

1. Zoe and Sandy built rectangles to model some trinomials.

   a. What do you notice about the number of [▬] tiles and the number of [▢] tiles in each model?

   b. Build a rectangle for $x^2 + 8x + 15$. Repeat part (a) for your rectangle.

2. Try to build a rectangle for $x^2 + 2x + 3$. What do you notice?

3. Expand each product.

   a. $(x + 5)(x + 3)$ b. $(x - 2)(x + 6)$ c. $(x - 3)(x - 8)$

4. For each product in question 3, look at the numbers in your expanded form and the numbers in the factored form. How are these numbers related?

5. Make a conjecture about the relationship between the integers b and c in any trinomial of the form $x^2 + bx + c$ and the numbers in the factored form.

10-7 Using Factors to Sketch $y = x^2 + bx + c$ 591

PLANNING

Objectives and Strands
See pages 544A and 544B.

Spiral Learning
See page 544B.

Materials List
➤ Algebra tiles
➤ Graph paper

Recommended Pacing
Section 10-7 is a two-day lesson.
Day 1
Pages 591–593: Talk it Over through Sample 3, *Exercises 1–23*
Day 2
Pages 594–595: Sketching the Graph of $y = x^2 + bx + c$ through Look Back, *Exercises 24–43*

Extra Practice
See pages 638–639.

Warm-Up Exercises
Warm-Up Transparency 10-7

Support Materials
➤ Practice 82
➤ Enrichment 73 in the Activity Bank
➤ Study Guide 10-7
➤ Problem Set 24
➤ Additional Exploration 14
➤ Diagram Masters 2, 6, 7 in the Explorations Lab Manual
Overhead Visual 1
➤ Quiz 10-7
➤ Alternative Assessments 13, 14

Answers to Talk it Over

1. a. Answers may vary. An example is given. The number of small square tiles is the product of the numbers of tall thin tiles in the two groups below and to the right of the large square tile.

   b. There are 8 tall thin tiles and 15 small square tiles; $3 + 5 = 8$ and $3 \cdot 5 = 15$.

	x	1	1	1	1	1
x	x^2	x	x	x	x	x
1	x	1	1	1	1	1
1	x	1	1	1	1	1
1	x	1	1	1	1	1

2. No rectangle can be built using 1 x^2-tile, 2 x-tiles, and 3 unit tiles.

3. a. $x^2 + 8x + 15$

   b. $x^2 + 4x - 12$

   c. $x^2 - 11x + 24$

4, 5. Answers may vary. An example is given.

4. The coefficient of x in the expanded form is equal to the sum of the two numbers in the factored form, and the constant term in the expanded form is equal to the product of the two numbers in the factored form.

5. I think that the integer b is equal to the sum of the numbers in the factored form and that the integer c is equal to the product of these numbers.

591

Talk it Over

The Talk it Over questions help students see the relationship between the integers b and c in $x^2 + bx + c$ and the numbers in factored form. This relationship forms the basis of the procedure for factoring a trinomial whose coefficient of the quadratic term is one.

....

Additional Sample

S1 Factor $x^2 + 11x + 18$.

Step 1 List pairs of integers whose product is 18.
   1, 18   2, 9   3, 6

Step 2 Add the integers in each pair.
   1 + 18 = 19
   2 + 9 = 11
   3 + 6 = 9

Step 3 Substitute the pair of integers that add up to the coefficient of x in the expanded form. Since 2 + 9 = 11, use 2 and 9.
   $x^2 + 11x + 18 = (x + 2)(x + 9)$

....

Talk it Over

Question 6 leads students to the type of reasoning necessary when the cooefficient of x is negative. Students should recognize that only two negative integers have a positive product and a negative sum.

Sample 1

Factor $x^2 + 7x + 12$.

Sample Response

Problem Solving Strategy: Work backward

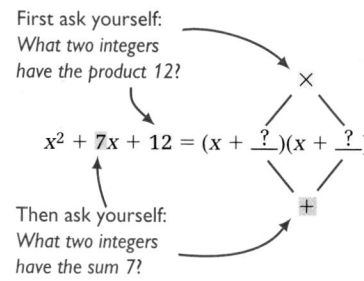

First ask yourself:
What two integers have the product 12?

$$x^2 + 7x + 12 = (x + \underline{\ ?\ })(x + \underline{\ ?\ })$$

Then ask yourself:
What two integers have the sum 7?

$$x^2 + 7x + 12 = (x + 3)(x + 4)$$

① List pairs of integers whose product is 12.
   1, 12   2, 6   3, 4

② Add the integers in each pair.
   1 + 12 = 13   2 + 6 = 8   3 + 4 = 7

③ Substitute the pair of integers that add up to the coefficient of x in the expanded form. Since 3 + 4 = 7, use 3 and 4.

Check

Method ①

Multiply using the distributive property.
$$x^2 + 7x + 12 \overset{?}{=} (x + 3)(x + 4)$$
$$x^2 + 7x + 12 \overset{?}{=} (x + 3)x + (x + 3)4$$
$$x^2 + 7x + 12 \overset{?}{=} x^2 + 3x + 4x + 12$$
$$x^2 + 7x + 12 = x^2 + 7x + 12 \ ✔$$

Method ②

Use algebra tiles to make a rectangle.
$$(x + 3)(x + 4) = x^2 + 7x + 12 \ ✔$$

$\longmapsto x + 3 \longmapsto$

$x + 4$

Talk it Over

6. Suppose you want to factor $x^2 - 7x + 12$.
   a. How is this trinomial different from the one factored in Sample 1?
   b. What questions would you ask yourself to find the factors?
   c. What are the factors of $x^2 - 7x + 12$?

7. In Sample 1, why are pairs of integers like -1 and -12 *not* listed?

Answers to Talk it Over

6. a. The coefficient of x is -7, not 7.

   b. What two integers have the sum -7? What two integers have the product 12?

   c. $x - 3$ and $x - 4$

7. Answers may vary. An example is given. The sum of two negative integers is a negative integer, but the sum of the factors must be 7, a positive integer.

Sample 2

Factor $x^2 + 4x - 12$.

Sample Response

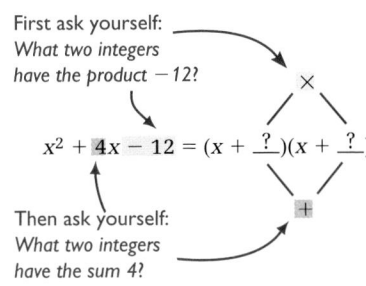

First ask yourself:
What two integers have the product −12?

$x^2 + \underline{4}x - 12 = (x + \underline{\ ?\ })(x + \underline{\ ?\ })$

Then ask yourself:
What two integers have the sum 4?

$x^2 + 4x - 12 = (x - 2)(x + 6)$

1 List pairs of integers whose product is −12.

   1, −12   2, −6   3, −4

  −1, 12   −2, 6   −3, 4

2 Add the integers in each pair.

  $1 + (-12) = -11$ $2 + (-6) = -4$ $3 + (-4) = -1$

  $(-1) + 12 = 11$ $(-2) + 6 = 4$ $(-3) + 4 = 1$

3 Substitute the pair of integers that add up to the coefficient of x in the expanded form. Since $(-2) + 6 = 4$, use −2 and 6.

> ### Talk it Over
>
> **Explain how to factor each expression.**
>
> **8.** $x^2 - 4x - 12$ **9.** $x^2 - 4x + 4$

Sample 3

Factor $x^2 + 3x + 10$.

Sample Response

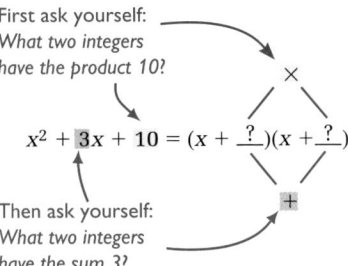

First ask yourself:
What two integers have the product 10?

$x^2 + \underline{3}x + 10 = (x + \underline{\ ?\ })(x + \underline{\ ?\ })$

Then ask yourself:
What two integers have the sum 3?

1 List pairs of integers whose product is 10.

  1, 10   2, 5

2 Add the integers in each pair.

  $1 + 10 = 10$ $2 + 5 = 7$

3 There is no pair of integers that have a product of 10 and that add up to 3.

The expression $x^2 + 3x + 10$ cannot be factored.

······▶ **Now you are ready for:**
Exs. 1–23 on pp. 595–596

Additional Samples

S2 Factor $x^2 + 7x - 18$.

Step 1 **List pairs of integers that have a product of −18.**

   1, −18   2, −9   3, −6

  −1, 18 −2, 9 −3, 6

Step 2 **Add the integers in each pair.**

  $1 + (-18) = -17$

  $(-1) + 18 = 17$

  $2 + (-9) = -7$

  $(-2) + 9 = 7$

  $3 + (-6) = -3$

  $(-3) + 6 = 3$

Step 3 **Substitute the pair of integers that add up to the coefficient of x in the expanded form. Since −2 + 9 = 7, use −2 and 9.**

  $x^2 + 7x - 18 = (x - 2)(x + 9)$

S3 Factor $x^2 - 5x + 10$.

Step 1 **List pairs of integers that have a product of 10.**

   1, 10   2, 5

  −1, −10 −2, −5

Step 2 **Add the integers in each pair.**

  $1 + 10 = 11$

  $-1 + (-10) = -11$

  $2 + 5 = 7$

  $-2 + (-5) = -7$

Step 3 **There is no pair of integers that have a product of 10 and a sum of −5. The expression cannot be factored.**

Answers to Talk it Over

8, 9. Explanations may vary. Examples are given.

8. List pairs of integers whose product is −12. Add the integers in each pair to find the pair whose sum is −4. Then write the factored form $(x + 2)(x - 6)$.

9. List pairs of integers whose product is 4. Add the integers in each pair to find the pair whose sum is −4. Then write the factored form $(x - 2)(x - 2)$.

593

Additional Sample

S4 Use the vertex, intercepts and symmetry to sketch the graph of $y = x^2 + 2x - 3$.

Step 1 Find the coordinates of the vertex.

$$x = -\frac{b}{2a}$$

$$x = -\frac{2}{2 \cdot 1} = -1$$

Use $x = -1$ to find y.

$$y = (-1)^2 + 2(-1) - 3$$

$$y = -4$$

The coordinates of the vertex are $(-1, -4)$.

Step 2 Find the y-intercept. The y-intercept is found where $x = 0$.

$$y = 0 + 2 \cdot 0 - 3$$

$$y = -3$$

Step 3 Find the x-intercepts. Factor:

$$y = (x + 3)(x - 1)$$

Use the zero-product property.

$$0 = (x + 3)$$

$$x = -3$$

or

$$0 = (x - 1)$$

$$x = 1$$

The x-intercepts are -3 and 1.

Step 4 Sketch the graph.

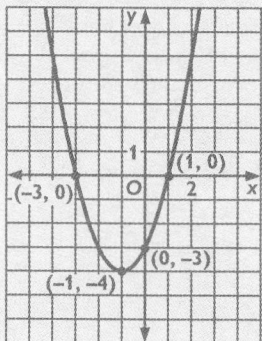

Sketching the Graph of $y = x^2 + bx + c$

You can graph a parabola from its equation by using a table of values or by using a graphics calculator or graphing software. You can also use the line of symmetry, the vertex, and the intercepts to sketch a parabola.

Sample 4

Use the line of symmetry, the vertex, and the intercepts to sketch the graph of $y = x^2 + 4x - 12$.

Sample Response

Step 1. Find the line of symmetry and the vertex.

The vertex is on the line of symmetry of the parabola. First find the equation of the line of symmetry.

$$x = -\frac{b}{2a} \quad \longleftarrow \text{ Use the formula for the line of symmetry.}$$

$$x = -\frac{4}{2 \cdot 1} \quad \longleftarrow \text{ Substitute 1 for } a \text{ and 4 for } b.$$

$$x = -2$$

The equation of the line of symmetry is $x = -2$.

The equation of the line of symmetry gives the x-coordinate of the vertex. Use this value to find the y-coordinate of the vertex.

$$y = (-2)^2 + 4(-2) - 12 \quad \longleftarrow \text{ Substitute } -2 \text{ for } x.$$

$$y = -16$$

The coordinates of the vertex are $(-2, -16)$.

Step 2. Find the y-intercept.

The y-intercept is found where $x = 0$.

$$y = 0^2 + 4 \cdot 0 - 12 \quad \longleftarrow \text{ Substitute 0 for } x.$$

$$y = -12$$

The y-intercept is -12.

Step 3. Find the x-intercepts.

Use the factored form of the equation from Sample 2:

$$y = (x - 2)(x + 6)$$

By the zero-product property,

$$0 = x - 2 \quad or \quad 0 = x + 6$$

$$2 = x \quad\quad\quad\quad -6 = x$$

The x-intercepts are 2 and -6.

BY THE WAY...

Switch on a flashlight and hold it sideways against a wall in a darkened room.

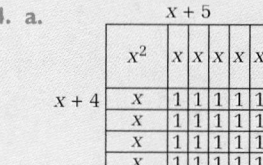

The light from the flashlight forms a parabola.

Unit 10 Quadratic Equations as Models

Answers to Look Back

Answers may vary. An example is given. To find the x-intercepts of $y = x^2 + 3x + 2$, rewrite the right side of the equation in factored form and use the zero-product property: $0 = (x + 2)(x + 1)$, so $x + 2 = 0$ or $x + 1 = 0$. This means that $x = -2$ or $x = -1$. The x-intercepts are -2 and -1.

Answers to Exercises and Problems

1. $x^2 + 4x + 3 = (x + 3)(x + 1)$

2. $x^2 + 8x + 12 = (x + 6)(x + 2)$

3. a.

	$x + 4$			
	x^2	x	x	x	x
$x + 4$	x	1	1	1	1
	x	1	1	1	1
	x	1	1	1	1
	x	1	1	1	1

b. $(x + 4)(x + 4)$

c. $(x + 4)x + (x + 4)4 = x^2 + 4x + 4x + 16 = x^2 + 8x + 16$

4. a.

	$x + 5$				
	x^2	x	x	x	x	x
$x + 4$	x	1	1	1	1	1
	x	1	1	1	1	1
	x	1	1	1	1	1
	x	1	1	1	1	1

b. $(x + 5)(x + 4)$

Step 4. Sketch the graph.

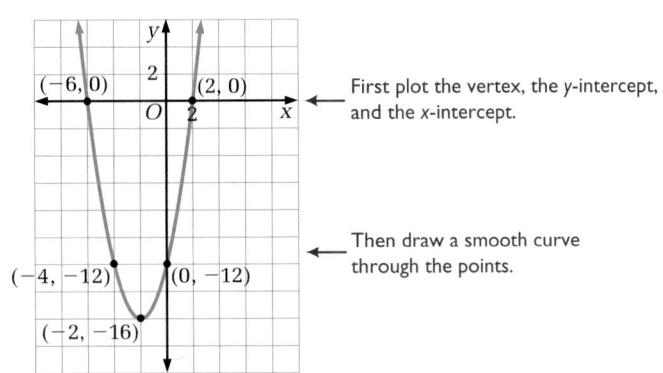

First plot the vertex, the *y*-intercept, and the *x*-intercept.

Then draw a smooth curve through the points.

The line of symmetry is $x = -2$. Plot the point opposite $(0, -12)$.

······▶ **Now you are ready for:** Exs. 24–43 on pp. 596–597

Look Back ◄

Explain how to use algebra to find the *x*-intercepts of the graph of $y = x^2 + 3x + 2$.

Express the area of each rectangle in expanded form and in factored form.

1.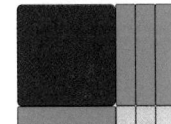

2.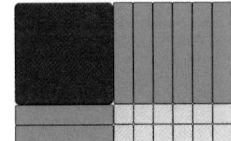

Using Manipulatives Use algebra tiles.

a. **Build a rectangle for each trinomial.**

b. **Use your rectangles to help you factor each trinomial.**

c. **Check your answers to part (b) by multiplying the factors.**

3. $x^2 + 8x + 16$ **4.** $x^2 + 9x + 20$ **5.** $x^2 + 9x + 14$

Factor each trinomial.

6. $x^2 + 6x + 5$ **7.** $x^2 + 9x + 18$ **8.** $x^2 - 11x + 18$

9. $x^2 - 5x + 6$ **10.** $x^2 + 8x + 12$ **11.** $x^2 - 13x + 12$

Writing **Factor each trinomial that can be factored using integers. For each trinomial that cannot be factored, explain why not.**

12. $x^2 - 12x + 20$ **13.** $x^2 + 9x + 10$ **14.** $x^2 - 5x + 24$

Answers to Exercises and Problems

4. c. $(x + 5)x + (x + 5)4 =$
$x^2 + 5x + 4x + 20 =$
$x^2 + 9x + 20$

5. a.

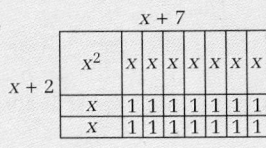

b. $(x + 7)(x + 2)$

c. $(x + 7)x + (x + 7)2 =$
$x^2 + 7x + 2x + 14 =$
$x^2 + 9x + 14$

6. $(x + 5)(x + 1)$

7. $(x + 6)(x + 3)$

8. $(x - 9)(x - 2)$

9. $(x - 3)(x - 2)$

10. $(x + 6)(x + 2)$

11. $(x - 12)(x - 1)$

12. $(x - 10)(x - 2)$

13. cannot be factored; The only pairs of positive factors of 10 are 1, 10 and 2, 5. Neither pair has the sum 9.

14. cannot be factored; The only pairs of negative factors of 24 are these: $-1, -24; -2, -12; -3, -8;$ and $-4, -6$. None of these pairs has the sum -5.

The Look Back question would make an excellent journal entry. Students may want to share their explanations with a partner. ·············●

APPLYING

Suggested Assignment

Standard 1, 2, 6–23, 25–27, 29–32, 34–43

Extended 1, 2, 6–32, 34–43

Integrating the Strands

Number Exs. 1–5

Algebra Exs. 1–43

Functions Exs. 24–28

Geometry Exs. 1–5, 15–17, 33

Using Technology

For Exs. 12–14, have students graph the equations on a graphics calculator and use the TRACE feature to observe that the equations that can be factored cross the *x*-axis at an integer. Those that cannot be factored cross between integers.

Error Analysis

For exercises in which students have to factor a trinomial, common errors involve using the wrong integers. After students write their factors, they should always expand the expression to see if the product equals the original trinomial. If it does not, then the choice of integers was incorrect and different integers have to be tried.

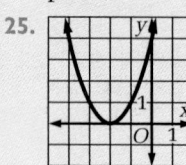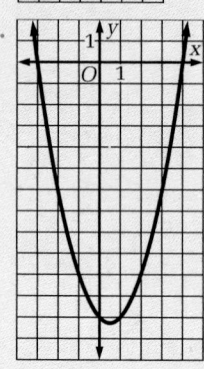
Find an expression for the height of each figure with the given base(s) and area.

15. Area = $x^2 + 7x + 10$

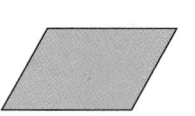

$x + 5$

16. Area = $x^2 + 13x + 42$

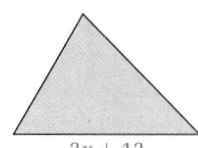

$2x + 12$

17. Area = $x^2 - 11x + 30$

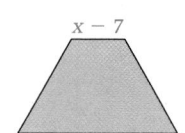

$x - 7$

$x - 3$

Factor each trinomial.

18. $x^2 + 3x - 10$

19. $x^2 - 2x - 8$

20. $x^2 - 3x - 10$

21. $x^2 - 8x + 12$

22. $x^2 - 7x + 10$

23. $x^2 + 6x - 16$

24. **Reading** Sample 4 shows how to sketch the graph of a parabola. List the steps. How do these steps help you make an accurate sketch?

Use the line of symmetry, the vertex, and the intercepts to sketch the graph of each equation.

25. $y = x^2 + 4x + 4$

26. $y = x^2 - x - 12$

27. $y = x^2 - 6x - 7$

28. Alissa is throwing her bridal bouquet from a balcony that is 14 ft high. The path of the bouquet as it flies through the air is a parabola. An equation for this parabola is $y = -x^2 + x + 20$, where y represents the height in feet of the bouquet and x represents its distance in feet from the edge of the balcony.

a. Can the equation $y = -x^2 + x + 20$ also be written $y = -(x^2 - x - 20)$? Explain.

b. Suppose Alissa's mother is standing 12 ft from the balcony, her brother is standing 8 ft from the balcony, and her best friend is standing 5 ft from the balcony. Which of them is in the best position to catch the bouquet? Explain your reasoning.

29. a. Copy and complete the table.

b. Describe at least two patterns that you see in your table.

c. Use the patterns in the table to help you write this product in expanded form: $(x - a)(x + a) = $ _?_

Factored Form	Expanded Form
$(x - 2)(x + 2)$	?
$(x - 3)(x + 3)$	?
$(x - 5)(x + 5)$	?

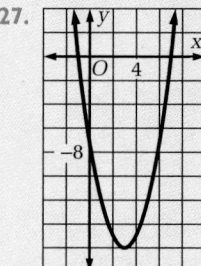
596

Describe how these trinomials are different from the other trinomials in this section. Then factor each trinomial.

30. $2x^2 - x - 6$ **31.** $3x^2 - 9x + 6$ **32.** $2x^2 + 6x + 6$

 Ongoing **ASSESSMENT**

33. **Group Activity** Work with another student.

| $x^2 + 4x + 4$ |
| $x^2 + 6x + 9$ |
| $x^2 + 8x + 16$ |
| $x^2 + 10x + 25$ |

a. Discuss any patterns you see in the four trinomials shown at the right. Describe the patterns using mathematical terms such as *coefficient* and *constant*.

b. Use algebra tiles to build a rectangle for each of the four trinomials.

c. Describe anything special you notice about the rectangles you built.

d. Express the area of each rectangle as a product.

e. Why do you think the trinomials are called *perfect square trinomials*?

f. Write two more perfect square trinomials and factor them.

Review **PREVIEW**

Expand each product. *(Section 10-6)*

34. $(x + 2)(x + 6)$ **35.** $(x - 7)(x + 4)$ **36.** $(x + 8)(x - 5)$

Solve. *(Sections 2-7, 2-9)*

37. $9x + 12 = 4$ **38.** $x^2 = 144$ **39.** $x^2 = 28$

Evaluate each expression when $a = 3$, $b = 2$, $c = -5$, and $d = 9$.
(Section 1-2)

40. $b^2 - 4ac$ **41.** $-\dfrac{24}{2a}$ **42.** $-b + \sqrt{d}$

Working on the Unit Project

43. **Baseball** An equation for the path of a baseball hit at a 45° angle with an initial speed of about 103 ft/s is $y = -0.003x^2 + x + 4$. In factored form, this equation can be approximated by $y = -0.003(x + 4)(x - 337)$.

a. At what height off the ground is the ball hit? Which form of the equation helps you answer this question most easily?

b. How far from home plate will the ball hit the ground? Which form of the equation helps you answer this question most easily?

c. Suppose the left field fence is 325 ft from home plate and stands 10 ft high. Will a ball hit down the left field line go over the fence? Why or why not?

10-7 Using Factors to Sketch $y = x^2 + bx + c$ **597**

Practice 82 For use with Section 10-7

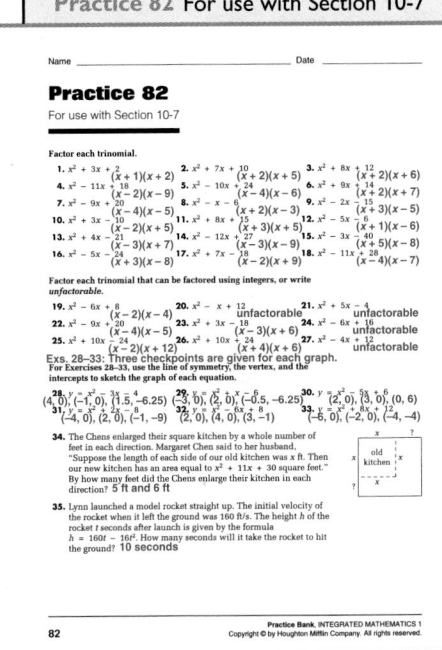

Answers to Exercises and Problems

30–32. Answers may vary. An example is given. The x^2-term has a coefficient other than 1 or –1.

30. $(2x + 3)(x - 2)$

31. $3(x - 2)(x - 1)$

32. $2(x^2 + 3x + 3)$

33. See answers in back of book.

34. $x^2 + 8x + 12$

35. $x^2 - 3x - 28$

36. $x^2 + 3x - 40$

37. $-\dfrac{8}{9}$

38. 12, –12

39. $2\sqrt{7}, -2\sqrt{7}$

40. 64

41. –4

42. 1

43. a. 4 ft; Answers may vary. An example is given. The expanded form helps answer this question most easily because the constant term is the y-intercept, which is the height when the ball is hit.

b. 337 ft; Answers may vary. An example is given. The factored form helps answer this question most easily because the ball hits the ground when $y = 0$, that is, at the x-intercepts, which can be found using the zero-product property.

c. Yes. If you substitute $x = 325$ into the original equation, you will get $y = 12.125$, which is greater than 10.

597

Objectives and Strands
See pages 544A and 544B.

Spiral Learning
See page 544B.

Materials List
➤ Graphics calculator
➤ Graph paper

Recommended Pacing
Section 10-8 is a two-day lesson.
Day 1
Pages 598–601: Opening para-
graph through Talk it Over,
Exercises 1–15
Day 2
Pages 601–603: Using the
Quadratic Formula through Look
Back, *Exercises 16–35*

Extra Practice
See pages 638–639.

Warm-Up Exercises
Warm-Up Transparency 10-8

Support Materials
➤ Practice 83
➤ Enrichment 74 in the Activity
Bank
➤ Study Guide 10-8
➤ Problem Set 24
➤ Diagram Masters 2, 3 in the
Explorations Lab Manual
➤ Using Plotter Plus: Approximating
Solutions of Polynomial Equations
➤ Quiz 10-8
➤ Test 43
➤ Alternative Assessments 15, 16

Section **10-8** **Solving Quadratic
Equations**

Focus
Solve problems about
situations modeled by
parabolas.

FORMULA
For $\sqrt{\text{Success}}$

Talk it Over

1. a. Model the student-driver's rule of thumb with an equation.
One car length is about 15 ft. Use d and s for the variables.

   b. Is your equation for this rule of thumb a linear function?
How do you know?

   c. What is the control variable?

   d. What is the dependent variable?

2. a. Is the formula $d = 0.05s^2 + s$ a linear function? How do
you know?

   b. What is the control variable?

   c. What is the dependent
variable?

"My driver's education textbook
contains this formula for the distance
a car travels once the driver has
decided to stop the car."
d = stopping distance in feet
s = speed in miles per hour
$d = 0.05s^2 + s$

"I use this rule of thumb for keeping
far enough in back of the car ahead:
"Leave one car length between
your car and the car in front for
every 10 miles per hour of speed."

Driving School

598   **Unit 10** Quadratic Equations as Models

Answers to Talk it Over

1. a. $d = \frac{15s}{10}$, or $d = 1.5s$, or
$d = \frac{3s}{2}$

   b. Yes. Reasons may vary.
An example is given.
The equation has the
form $y = kx$, so it is an
example of direct varia-

tion, which is a special
kind of linear function.

   c. s

   d. d

2. a. No. Reasons may vary.
An example is given. A
linear function has the
variable raised to the

first power and can be
written in the form
$y = mx + b$; since this
equation does not have
that form, it is not linear.

   b. s

   c. d

TECHNOLOGY NOTE

When you graph a quadratic equation on a graphics calculator, you enter the equation in the form $y = ax^2 + bx + c$. Use the TRACE feature to find the coordinates of points on the graph.

Suppose the stopping distance of a car is 60 ft. When you substitute 60 for d in the textbook formula $d = 0.05s^2 + s$, you produce the equation

$$60 = 0.05s^2 + s.$$

This new equation is a *quadratic equation*. A **quadratic equation** is an equation that can be written in the form

$$0 = ax^2 + bx + c$$

where a, b, and c are numbers and a is not equal to zero. A function that leads to a quadratic equation is called a *quadratic function*. The graph of a quadratic function is a parabola.

Talk it Over

3. **a.** Change $60 = 0.05s^2 + s$ to the form $0 = ax^2 + bx + c$.

   b. What are the values of a, b, and c in the equation you wrote?

4. How is a quadratic equation different from a linear equation?

Finding the Solutions of $0 = ax^2 + bx + c$

The solutions of a quadratic equation of the form $0 = ax^2 + bx + c$ are found where the graph of $y = ax^2 + bx + c$ crosses the x-axis.

SOLVING A QUADRATIC EQUATION

➤ First rewrite the equation in the form $0 = ax^2 + bx + c$.

➤ Then find the x-intercepts of the graph of $y = ax^2 + bx + c$.

The x-intercepts are the solutions.

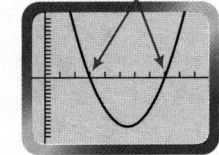

Section 10-3 includes a way to use the zero-product property to find the x-intercepts of a parabola without graphing. To use this method, you must first write the equation in factored form. Many equations cannot be factored. To find the solutions of equations like these, you can draw the graph of the equation.

10-8 Solving Quadratic Equations

599

Talk it Over

Questions 1 and 2 draw the distinction between a linear function and a quadratic function.

Reasoning

Ask students to explain why the coefficient a cannot be zero in a quadratic equation. (If a is zero, the equation is a linear equation.)

Communication: Discussion

In discussing the solutions of $0 = ax^2 + bx + c$, point out that the x-intercepts provide the values of x that make the right side of the equation $(ax^2 + bx + c)$ equal to zero. Thus, $0 = 0$ is a true statement and the x-intercepts are solutions. At more advanced levels, these solutions are the *zeros* of the function.

Visual Thinking

Check students' understanding by asking them to draw concept maps of a linear equation and a quadratic equation. Have them discuss the differences between their maps with the class. This activity involves the visual skills of *recall* and *communication*.

Answers to Talk it Over

3. **a.** $0 = 0.05s^2 + s - 60$

   b. $a = 0.05$, $b = 1$, $c = -60$

4. Answers may vary. An example is given. A linear equation can be written in the form $0 = mx + b$ and a quadratic equation can be written in the form $0 = ax^2 + bx + c$.

600

Additional Sample

S1 Using the textbook formula $d = 0.05s^2 + s$, what is the fastest speed at which eight car lengths is a safe following distance?

Repeat the steps shown in the Sample 1 Response with 8 car lengths substituted for *d*. The *x*-intercepts are −60 and 40. Therefore, the fastest speed at which it is safe to drive eight car lengths behind another car is 40 mi/h.

Reasoning

In Sample 1, the fastest speed for four car lengths is 25 mi/h. In Additional Sample S1, the fastest speed for eight car lengths is 40 mi/h. Ask students why the fastest speed for eight car lengths is not twice the fastest speed for four car lengths. (Stopping distance is not a linear function of speed; it is a quadratic function. As a car begins to move faster, the distance needed to stop it increases faster than that of a linear function.)

Visual Thinking

Ask students to pick points on the graph and explain what they mean. Ask them to suggest how they would use the graph as a visual display in a driver's education class. This activity involves the visual skills of *generalization* and *communication*.

Sample 1

Use the textbook formula $d = 0.05s^2 + s$. What is the fastest speed at which four car lengths is a safe following distance? One car length is 15 ft.

car in front

my car

Sample Response

Problem Solving Strategy: Use a formula

Step 1. Write an equation in the form $0 = ax^2 + bx + c$.

$$d = 0.05s^2 + s \qquad \longleftarrow \text{Use the textbook formula.}$$
$$4 \cdot 15 = 0.05s^2 + s \qquad \longleftarrow \text{Substitute four car lengths for } d.$$
$$60 = 0.05s^2 + s$$
$$60 - 60 = 0.05s^2 + s - 60 \qquad \longleftarrow \text{Subtract 60 from both sides.}$$
$$0 = 0.05s^2 + s - 60$$

Step 2. Replace *s*, the control variable, with *x*. Graph the function $y = 0.05x^2 + x - 60$. You might want to use a graphics calculator.

1 Make a table of values. Use enough values to show where the graph crosses the *x*-axis.

The sign of *y* changes when the graph crosses the *x*-axis.

$y = 0.05x^2 + x - 60$	
x	*(x, y)*
−50	(−50, 15)
−40	(−40, −20)
−30	(−30, −45)
−20	(−20, −60)
−10	(−10, −65)
0	(0, −60)
10	(10, −45)
20	(20, −20)
30	(30, 15)
40	(40, 60)

2 Then plot the points and draw a smooth curve.

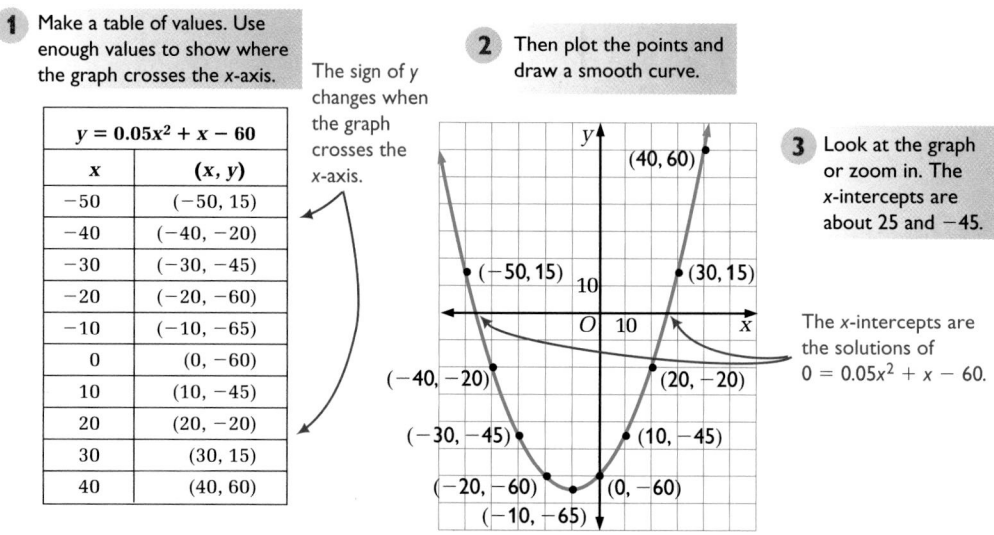

3 Look at the graph or zoom in. The *x*-intercepts are about 25 and −45.

The *x*-intercepts are the solutions of $0 = 0.05x^2 + x - 60$.

The *x*-intercepts are about 25 and −45, but you cannot drive −45 mi/h. The fastest speed at which it is safe to drive four car lengths behind another car is about 25 mi/h.

Unit 10 Quadratic Equations as Models

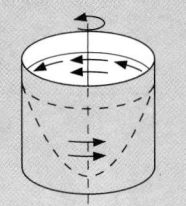

BY THE WAY...

Carefully spinning any liquid in a pan forms a parabola-shaped surface. Try it yourself with water!

Talk it Over

5. Is it safe to use the student-driver's rule of thumb at the speed found in the response to Sample 1? Why or why not?

6. Answer these questions about the equation and graph in Sample 1.

   a. Use the formula $x = -\dfrac{b}{2a}$ to write the equation of the line of symmetry of the graph.

   b. About how far is each of the x-intercepts from the point where the line of symmetry crosses the x-axis?

   c. Express each x-intercept in terms of the answers to parts (a) and (b).

▶ Now you are ready for:
Exs. 1–15 on pp. 603–604

Using the Quadratic Formula

Unless you have a graphics calculator or graphing software, solving a quadratic equation by graphing may take a long time. Also, it is often difficult to read the solutions from a hand-drawn graph accurately.

Another way to solve any quadratic equation is to use the *quadratic formula*. The **quadratic formula** uses the values a, b, and c in the equation $0 = ax^2 + bx + c$ to find the values of x that are solutions of the equation.

QUADRATIC FORMULA

For an equation in the form $0 = ax^2 + bx + c$, there are two solutions.

x-coordinate of the vertex, which is on the line of symmetry ⟶   ⟵ distance from each x-intercept to the line of symmetry

$$x = -\frac{b}{2a} \pm \frac{\sqrt{b^2 - 4ac}}{2a}$$

"$\pm$" means there are two solutions.

The solutions are:

$$x = -\frac{b}{2a} + \frac{\sqrt{b^2 - 4ac}}{2a} \quad \text{and} \quad x = -\frac{b}{2a} - \frac{\sqrt{b^2 - 4ac}}{2a}$$

10-8 Solving Quadratic Equations

601

Answers to Talk it Over

5. No. Explanations may vary. An example is given. For a speed of 25 mi/h, Jason's rule of thumb gives $\dfrac{25 \text{ mi/h}}{10 \text{ mi/h}} \cdot 15 \text{ ft} = 37.5 \text{ ft}$, which underestimates stopping distance by more than a car length.

6. a. $x = -10$

   b. about 35 units

   c. $-45 = -10 - 35$ and $25 = -10 + 35$

S2 Use the quadratic formula to solve $0 = 2x^2 + x - 25$.

$0 = 2x^2 + x - 25$

$0 = ax^2 + bx + c$

$a = 2, b = 1, c = -25$

$x = -\dfrac{b}{2a} \pm \dfrac{\sqrt{b^2 - 4ac}}{2a}$

$x = -\dfrac{1}{2(2)} \pm \dfrac{\sqrt{1^2 - 4(2)(-25)}}{2(2)}$

$x = -\dfrac{1}{4} \pm \dfrac{\sqrt{201}}{4}$

$x \approx -\dfrac{1}{4} \pm \dfrac{14.2}{4}$

$x \approx -0.25 \pm 3.55$

$x \approx -0.25 + 3.55 \approx 3.3$ or

$x \approx -0.25 - 3.55 \approx -3.8$

Using Technology

Challenge students to write a program on a graphics calculator or a computer to solve a quadratic equation using the quadratic formula. The input should be *a, b,* and *c.*

Multicultural Note

Neither the authors of the *Nine Chapters on the Mathematical Art* nor the date of its composition is exactly known. It was formed by the collective effort and wisdom of mathematicians of several centuries. The book was compiled in question-and-answer form and was used as a textbook for many centuries. It contained a total of 246 problems that covered important principles of arithmetic, geometry, and algebra. The book also circulated in Japan and Korea at one time and had a large influence on the development of mathematics in those countries.

Sample 2

Use the quadratic formula to solve $0 = 0.05x^2 + x - 60$. Round decimal answers to the nearest tenth.

Sample Response

Step 1. Decide what values to use for *a, b,* and *c.*

$0 = 0.05x^2 + x - 60$

$0 = 0.05x^2 + 1x + (-60)$

$a = 0.05, b = 1,$ and $c = -60$

Step 2. Write the quadratic formula.

$x = -\dfrac{b}{2a} \pm \dfrac{\sqrt{b^2 - 4ac}}{2a}$

$x = -\dfrac{1}{2(0.05)} \pm \dfrac{\sqrt{1^2 - 4(0.05)(-60)}}{2(0.05)}$ ← Substitute 0.05 for *a*, 1 for *b*, and -60 for *c*.

$x = -10 \pm \dfrac{\sqrt{13}}{0.10}$

$x \approx -10 \pm \dfrac{3.61}{0.10}$ ← Use a calculator to find $\sqrt{13}$.

$x \approx -10 \pm 36.1$

$x \approx -10 + 36.1$ *or* $x \approx -10 - 36.1$

$x \approx 26.1$ *or* $x \approx -46.1$

The solutions are about 26.1 and about -46.1

Check

$60 = 0.05x^2 + x$ ← Use the *original* equation. → $60 = 0.05x^2 + x$

$60 \stackrel{?}{=} 0.05(26.1)^2 + 26.1$ ← Substitute for *x.* → $60 \stackrel{?}{=} 0.05(-46.1)^2 + (-46.1)$

$60 \approx 60.16$ ✔ $60 \approx 60.16$ ✔

Talk it Over

7. Why is the expression $0.05x^2 + x$ not exactly equal to 60 when 26.1 and -46.1 are substituted for *x*?

8. Which answer in Sample 2 is a solution to the problem situation in Sample 1? Did the graph in Sample 1 produce a good enough estimate for the problem situation? Why or why not?

Answers to Talk it Over

7. because 26.1 and -46.1 are approximations obtained by using a rounded square root of a number that is not a perfect square

8. Answers may vary. An example is given. 26.1; Yes, I think the estimate of 25 mi/h is good enough because it is very close to the more accurate answer of 26.1 mi/h. Also, the estimate is slightly low, which actually increases the margin of safety.

Answers to Look Back

Answers may vary. Examples are given. I would solve the first equation by factoring because I see how to factor the right side: $0 = (x + 9)(x - 4)$, so the solutions are -9 and 4. I would solve the second equa-

tion by graphing if I have a graphics calculator or graphing software, or by using the quadratic formula if I have a scientific calculator, because I cannot factor the right side.

X·Lab

THREE METHODS FOR SOLVING QUADRATIC EQUATIONS

➤ **Factoring**
Fastest method for equations that are easy to factor.

➤ **Graphing**
Works well if you are using a graphics calculator or graphing software.

➤ **Quadratic Formula**
Useful when factoring and graphing are not practical.
You will need a scientific calculator.

Look Back ←
Which method would you use to solve each of these quadratic equations? Why?
$$0 = x^2 + 5x - 36 \qquad 0 = 3.4x^2 - 2.3x - 8.6$$

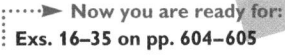
:····➤ **Now you are ready for:**
: Exs. 16–35 on pp. 604–605

10-8 Exercises and Problems

1. **Open-ended** Different cars and road conditions may produce different stopping distance formulas. Describe how each of these conditions could affect the stopping distance.
   a. how quickly the driver begins to brake (reaction time)
   b. the condition of the road
   c. the weight of the car

Use the stopping distance formula $d = 0.05s^2 + s$.

2. What is the stopping distance for a speed of 15 mi/h?

3. What is the stopping distance for a speed of 55 mi/h?

4. What is the fastest speed at which 175 ft is a safe following distance?

5. What is the fastest speed at which five car lengths is a safe following distance? (A car length is about 15 ft.)

Use the graph of $y = x^2 + 6x + 8$ at the right.

6. a. What are the x-intercepts of the graph?
   b. What are the solutions of the equation $0 = x^2 + 6x + 8$?

7. a. Where does the graph cross the line $y = 3$?
   b. What are the solutions of the equation $3 = x^2 + 6x + 8$?

Make a graph to solve each quadratic equation.

8. $0 = x^2 + 4x + 4$ 9. $-4 = -4.8x^2 - 6x$ 10. $10 = 3x^2 - 4x + 8$

BY THE WAY...

NASA uses a plane flying in a parabolic path to create a weightless environment for astronaut training and some scientific experiments.

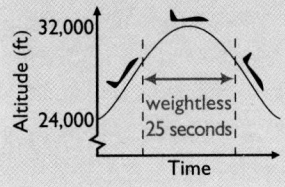

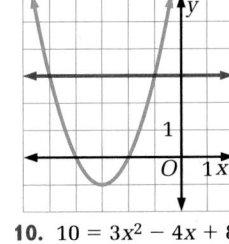

10-8 Solving Quadratic Equations

603

Answers to Exercises and Problems ·····························

1. Answers may vary. Examples are given.
   a. The faster the reaction time, the shorter the stopping distance.
   b. The drier the road, the shorter the stopping distance.
   c. The heavier the car, the longer the stopping distance.

2. 26.25 ft   3. 206.25 ft
4. 50 mi/h   5. 30 mi/h

6. a. –2 and –4
   b. –2 and –4

7. a. at $x = -1$ and at $x = -5$
   b. –1 and –5

8. –2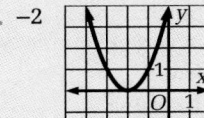

9. Estimates may vary. An example is given: about 0.5 and about –1.75.
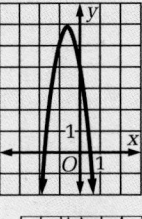

10. Estimates may vary. An example is given: about 1.75 and about –0.4.
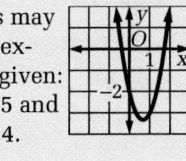

Talk it Over
Be sure that students recognize that the answers in Sample 2 are approximate answers.

Research
Students who enjoy mathematics will find it interesting to research the history of developing formulas to solve quadratic and higher degree equations. This project has the potential to open up many new and fascinating insights into the development of algebra during the past four hundred or more years.

Look Back
Use the Look Back for a class discussion. Students should choose the method that is the fastest and most accurate to solve each equation.   ·············•

APPLYING

Suggested Assignment
Standard 2–15, 17–24, 26–35
Extended 1–24, 26–35

Integrating the Strands
Algebra Exs. 1–35
Functions Exs. 1–15, 17–26, 30–35
Geometry Exs. 6–10, 12–14, 30–34

Research
For Ex. 1, students may find it interesting to research stopping distances for various vehicles under various conditions. Make bulletin board space available for them to display their findings.

Interdisciplinary Problems

Quadratic equations serve as models for many diverse real-world situations. In this section, Exs. 1–5, 11, 24, 26, 34, and 35 provide a few examples of such situations. Students can also find other examples in previous sections of this unit. Quadratic equations appear frequently whenever scientists write mathematical descriptions of physical events. For example, problem situations involving motion, electricity, work and energy, chemical reactions, economic forecasting models, engineering, design work, and construction of machines, ships, airplanes, bridges, buildings, trains, and automobiles all involve the solution of quadratic equations.

11. **Marine Biology** When dolphins are traveling very fast, they use less energy if they leap. The length and height of a dolphin's leap are related to the dolphin's speed. The formula for the height of one dolphin's leap was found to be $y = -0.4x^2 + x$.

The dolphin's height (in meters) above the water level is y.

The dolphin's horizontal distance (in meters) from its starting point is x.

a. How high above the water is this dolphin when $x = 2$?

b. How far is this dolphin from its starting point when it is 0.6 m high?

For Exercises 12–14, use the graph.

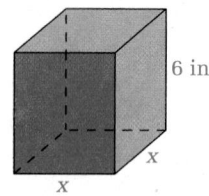

12. What are the x-intercepts of the graph?

13. What is the equation of the line of symmetry?

14. Which equation matches the graph?

   A. $y = (x + 1)(x + 5)$ **B.** $y = (x - 1)(x + 5)$ **C.** $y = (x - 1)(x - 5)$

15. a. Rewrite the equation $12 = x^2 + x - 8$ in the form $0 = ax^2 + bx + c$.

   b. Factor the equation you wrote in part (a).

   c. Use your answer to part (b) to find the solutions of the equation.

16. **Reading** In the quadratic formula, what does the $\pm$ symbol mean?

17. a. Solve the equation $0 = x^2 - 2x - 8$ by factoring.

   b. Solve the equation $0 = x^2 - 2x - 8$ using the quadratic formula.

   c. **Writing** Compare your answers to part (a) and part (b). Which method do you think is better for this equation?

Solve each quadratic equation. Tell which method you used.

18. $0 = x^2 - 4$ 19. $0 = -5x^2 + 3x + 2$ 20. $8 = x^2 + 6x$

21. $x^2 - 10x = -21$ 22. $2 = -0.036x^2 + x + 9$ 23. $0.5x^2 - 0.5x - 4 = 6$

24. The height of a square prism is 6 in.

   a. Write an expression for the surface area of the prism.

   b. The surface area of the prism is 176 in.^{2}. Write an equation in terms of x for the surface area.

   c. Solve the equation you wrote in part (b). What does your solution represent in terms of the prism?

6 in.

x

x

Unit 10 Quadratic Equations as Models

Answers to Exercises and Problems

11. a. 0.4 m

   b. The dolphin will reach a height of 0.6 m twice: at 1 m on the way up and at 1.5 m on the way down.

12. 1 and 5   13. $x = 3$

14. C

15. a. $0 = x^2 + x - 20$

   b. $0 = (x + 5)(x - 4)$

   c. −5, 4

16. Answers may vary. An example is given. $\pm$ means that the two roots are on either side of the line of symmetry $x = -\dfrac{b}{2a}$; one root is $\dfrac{\sqrt{b^2 - 4ac}}{2a}$ units to the right; one root is $\dfrac{\sqrt{b^2 - 4ac}}{2a}$ units to the left.

17. a, b. 4, −2

   c. Answers may vary. An example is given. I think factoring is much easier than using the quadratic formula because factoring takes me less time.

18–23. Estimates and methods may vary. Examples are given.

18. 2, −2; factoring

19. −0.4, 1; graphing or quadratic formula

20. about −7.1, about 1.1; graphing or quadratic formula

21. 3, 7; factoring

22. about 33.6, about −5.8; graphing or quadratic formula

23. −4, 5; graphing or quadratic formula

25. Group Activity Work in a group of six students. Divide the group into two teams.

➤ Students on Team 1 solve equations (a)–(c) by graphing.

➤ Students on Team 2 solve the equations with the quadratic formula.

➤ When both teams are finished, compare solutions. Look for any errors and correct them.

Quadratic Equations
a. $6 = 5x^2 + 8x + 2$
b. $15 = 3x^2 - 4x + 10$
c. $-3 = 4x^2 + 4x - 2$

Ongoing **ASSESSMENT**

26. Open-ended The equation $y = -4.9x^2 + 3.03x + 0.22$ gives the height in meters of a table tennis ball above the table. The height is represented by y and the number of seconds the ball is in the air is represented by x.

a. Suppose you want to know how long it takes for the ball to reach a height of 0.3 m. Do you think that factoring, graphing, or the quadratic formula is the best method for finding the answer? Explain your choice.

b. Use the method you chose to find the answer.

Review **PREVIEW**

Factor each trinomial. *(Section 10-7)*

27. $x^2 - 8x + 7$ **28.** $x^2 + x - 20$ **29.** $x^2 - 2x - 24$

30. Use the vertex and the intercepts to sketch the graph of the equation $y = x^2 + 7x - 18$.

Find the x-intercepts and the y-intercept of the graph of each equation. *(Sections 8-1, 10-5)*

31. $y = -2x + 1$ **32.** $y = 3x$ **33.** $y = x^2 - 3x$

 Working on the Unit Project

34. An equation for the path of a baseball hit at a 50° angle with an initial speed of 80 ft/s is $y = -0.006x^2 + 1.192x + 4$. How far from home plate will the ball first hit the ground?

35. When a baseball is hit at a 45° angle, the equation $r = \dfrac{v^2}{32}$ gives the *range*, how far away the ball hits the ground. In the equation, v represents the initial speed of the ball in feet per second. Suppose the range is 300 ft. What is the initial speed?

605

Assessment: Standard

For Ex. 26, you may wish to collect students' papers and assign a grade to their work.

Working on the Unit Project

Exs. 34 and 35 can be used for class discussion. Students' responses will provide a summary of quadratic equations as models.

Quick Quiz (10-7 through 10-8)

See page 607.

Practice 83 **For use with Section 10-8**

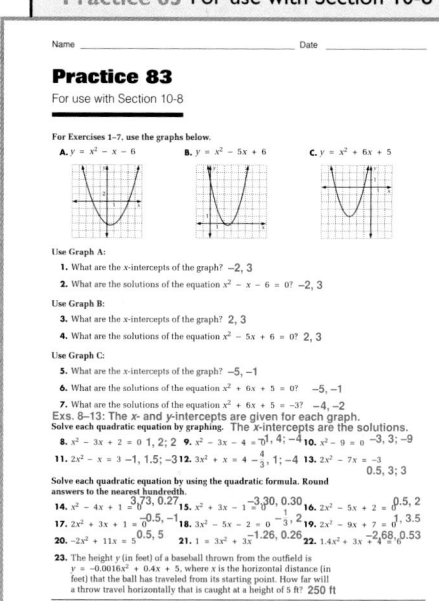

Answers to Exercises and Problems

24. a. $2x^2 + 24x$

b. $176 = 2x^2 + 24x$

c. about 5.1 and about -17.1; -17.1 cannot be the length and width of the prism, so the approximate length and width must be 5.1 in.

25. a. -2, 0.4

b. about 2.1, about -0.8

c. -0.5

26. a. Answers may vary. An example is given. I think that the quadratic formula is the best method because it gives an accurate solution quickly.

b. about 0.03 s

27. $(x - 7)(x - 1)$

28. $(x + 5)(x - 4)$

29. $(x - 6)(x + 4)$

30.

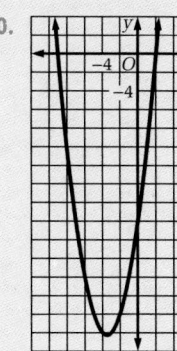

31. x-intercept: 0.5; y-intercept: 1

32. x-intercept: 0; y-intercept: 0

33. x-intercepts: 0 and 3; y-intercept: 0

34. about 202 ft

35. about 98 ft/s

605

Completing the Unit Project

Now you are ready to write your math story. It should include these things.

➤ a setting, characters, and a plot

➤ at least three situations modeled by equations or formulas you studied earlier in this course

➤ at least one situation modeled by a baseball equation given in the "Working on the Unit Project" exercises

➤ at least one situation modeled by a soccer equation in the table on page 545

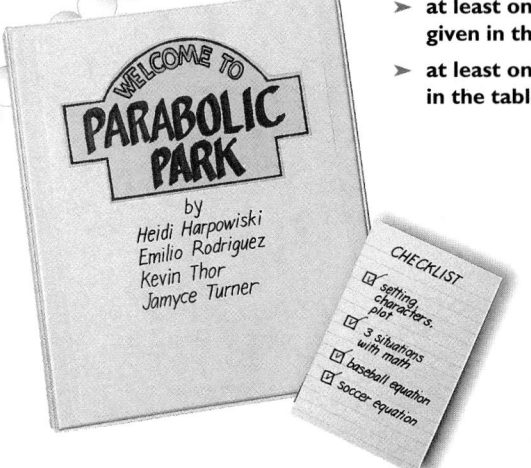

Look Back

Which form of the baseball and soccer path equations was most helpful in finding how far the ball traveled or how high it traveled? What difference did you find between the soccer equations when the ball was kicked and when it was struck with the head?

Alternative Projects

Project 1: **Mickey Mantle's Record Home Run**

On May 22, 1963, Mickey Mantle hit a home run in Yankee Stadium that nearly went out of the park. This is reportedly the closest anyone has ever come to hitting an out-of-the-park home run in this stadium.

Find out the details about this home run. Model the path of the ball with an equation and a graph. About how far from home plate would the ball have landed if it had not hit anything before it reached the ground?

Project 2: **Making Parabolas from Straight Lines**

Parabolas can be formed by straight lines. Research how this is done—with paper folding, paper and pencil, or string. Then create a design that contains at least three parabolas formed by straight lines.

1. The diagram shows a figure in blue and its image in red. 10-1

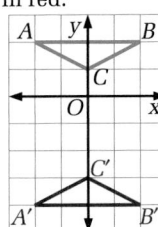

  a. What is the line of reflection?

  b. What is the relationship between $\overline{BB'}$ and the line of reflection?

  c. How far from the line of reflection is point C? point C'?

  d. How does the orientation of $\triangle ABC$ compare with the orientation of its image?

Tell whether each transformation is a reflection. If it is, trace the figure and draw the line of reflection. If it is not, tell why not.

2.

3.

4. Copy $\triangle ABC$ from Exercise 1. Reflect it over the x-axis. Write the coordinates of the vertices of its image.

5. Tell how to transform the graph of $y = x^2$ in order to produce the graph of $y = x^2 - 15$. 10-2

6. Find the equation of the line of symmetry and the coordinates of the vertex for the graph of $y = (x + 8)^2$.

7. Graph the parabola $y = (x + 3)(2x - 1)$. Plot enough points to show the shape of the curve. 10-3

8. Without graphing, find the x-intercepts and the y-intercept of the graph of $y = (x - 2)(x + 6)$.

Simplify. 10-4

9. $b^2 \cdot b \cdot b^7$ **10.** $-6x^2 \cdot 5x^2y^3$ **11.** $(m^4)^7$ **12.** $(-5ab^2)^3$

13. Expand the expression $2x(3x - 5)$. 10-5

14. Factor the right side of the equation $y = 7x^2 + 21x$ completely. Then find the x-intercepts and y-intercept of the graph of the equation.

15. **Open-ended** Write the equation of a parabola that has x-intercepts at -3 and 0.

16. Expand the product $(x - 3)(x - 6)$. 10-6

17. Without graphing, find each feature of the graph of the equation $y = x^2 - 10x - 4$.

  a. the equation of the line of symmetry

  b. the coordinates of the vertex

  c. the y-intercept

Unit Support Materials

➤ Unit 10 Cumulative Practice 84

➤ Unit 10 Study Guide Review

➤ Unifying Problem 10

➤ Unit Tests 44 and 45

➤ Cumulative Tests 46 and 47

Quick Quiz (10-7 through 10-8)

Factor. [10-7]

1. $x^2 + 9x + 14$ $(x + 2)(x + 7)$

2. $x^2 - 5x + 6$ $(x - 2)(x - 3)$

3. $x^2 + 4x - 12$ $(x + 6)(x - 2)$

4. Use the vertex, intercepts and symmetry to sketch the graph of $y = x^2 - x - 6$. [10-7]

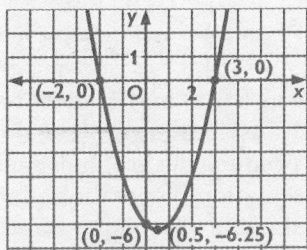

Use the quadratic formula to solve each equation. [10-8]

5. $0 = x^2 + 2x - 6$ $-3.65, 1.65$

6. $0 = 6x^2 - 5x + 1$ $0.5, 0.33$

7. $0 = x^2 - 8$ $-2.83, 2.83$

7.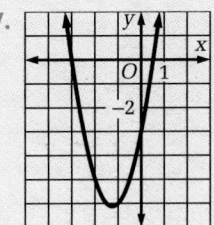

8. x-intercepts: 2 and -6; y-intercept: -12

9. b^{10}

10. $-30x^4y^3$

11. m^{28}

12. $-125a^3b^6$

13. $y = 6x^2 - 10x$

14. $y = 7x(x + 3)$; x-intercepts: 0 and -3; y-intercept: 0

15. $y = x(x + 3)$

16. $x^2 - 9x + 18$

17. a. $x = 5$

  b. $(5, -29)$

  c. -4

Answers to Unit 10 Review and Assessment

1. a. $y = -1$

  b. The line of reflection divides $\overline{BB'}$ in half and is perpendicular to $\overline{BB'}$.

  c. 2 units; 2 units

  d. The orientation of $\triangle ABC$ is clockwise. The orientation of $\triangle A'B'C'$ is reversed; it is counter-clockwise.

2. No. The image is not congruent to the original figure and the orientation is not reversed.

3. Yes.

4.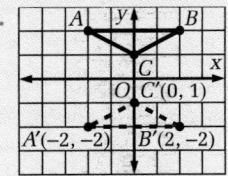

5. Translate the graph of $y = x^2$ down 15 units.

6. a. $x = -8$

  b. $(-8, 0)$

1. Reflect the figure over the x-axis. [10-1]

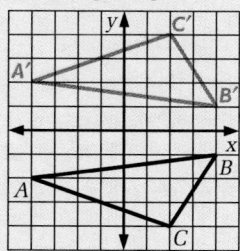

2. Reflect the figure over the y-axis. [10-1]

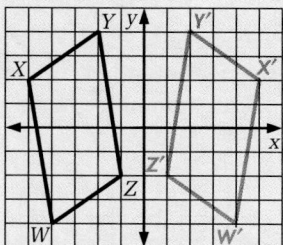

For Exs. 3 and 4, show how to translate the graph of $y = x^2$ to produce the graph of each equation. Write the equation of the line of symmetry and the vertex for the transformed equation. [10-2]

3. $y = x^2 - 3$

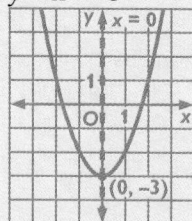

4. $y = (x + 1)^2$

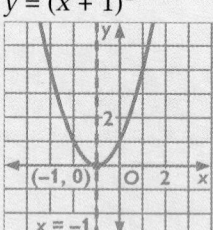

5. Graph the parabola $y = (x - 3)(2 - x)$. [10-3]

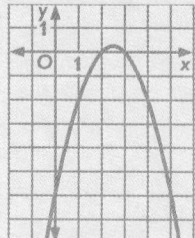

6. Without graphing, find the x-intercepts and the y-intercept for the equation $y = (x + 5)(x - 3)$. [10-3]

x-intercepts: –5 and 3;
y-intercept: –15

Factor each trinomial that can be factored using integers. For each trinomial that cannot be factored, explain why not.   10-7

18. $x^2 - 10x + 16$ 19. $x^2 - 3x - 18$ 20. $x^2 + 6x - 8$

21. Use the vertex, the x-intercepts, and the y-intercept to sketch the graph of the equation $y = x^2 + 2x - 15$.

22. **Writing** Write about the similarities and differences between solving quadratic equations and solving linear equations.   10-8

23. Solve the quadratic equation $0.4 = 2.4x^2 - 0.06x - 1$. Tell which method you used.

24. **Self-evaluation** You have learned several methods for solving a quadratic equation.

 a. Describe the method you like the most. What do you like about it?

 b. Describe the method you like the least. Why is it your least favorite?

25. **Group Activity** Work in a group of three students. Sit in a circle.

 a. Each person should write an expression that is a product of two binomials. On another piece of paper, write the expression in expanded form.

 b. Pass your expanded version to the person on your left, and the factored version to the person on your right. You will receive two pieces of paper yourself.

 c. Factor the expanded expression you receive. Write the factored version below the expanded expression.

 d. Expand the factored expression you receive. Write the expanded version above the factored expression.

 e. As a group, look at all your pieces of paper and compare them to see if your work is correct. There should be three pairs of matching papers.

 f. Choose one of the expressions. Suppose it is the rule for a function. Draw a graph of the function. On your graph, label the coordinates of the x-intercepts, the y-intercept, and the vertex.

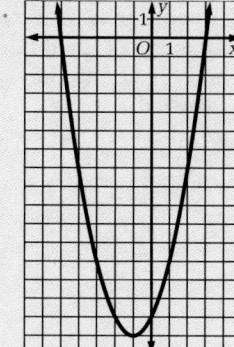

Answers to Unit 10 Review and Assessment

18. $(x - 8)(x - 2)$

19. $(x - 6)(x + 3)$

20. cannot be factored; The only pairs of factors of –8 are these: 1, –8; –1, 8; 2, –4; and –2, 4. None of these pairs has the sum 6.

21.

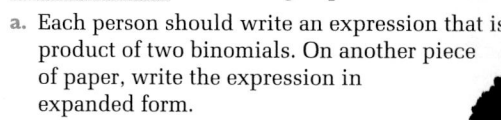

22. Answers may vary. An example is given. With linear equations, we get the variable alone and produce one solution. With quadratic equations, we make one side of the equation zero and solve by factoring, graphing, or the quadratic formula. In either case, the solutions are the x-intercepts of the graph.

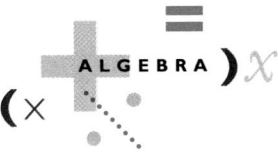

IDEAS AND (FORMULAS)=x^2

ALGEBRA

➤ You can use the zero-product property and algebra to find the x- and y-intercepts of a parabola from its equation. *(p. 564)*

➤ The product of powers rule says that to multiply powers that have the same base, you add the exponents *(p. 569)*:

$$a^m \cdot a^n = a^{m+n}$$

➤ The power of a power rule says that to find a power of a power, you multiply the exponents *(p. 570)*:

$$(a^m)^n = a^{m \cdot n}$$

➤ The power of a product rule says that to find a power of a product, you find that power of each factor and then multiply *(p. 571)*:

$$(ab)^n = a^n \cdot b^n$$

➤ You can use the quadratic formula

$$x = -\frac{b}{2a} \pm \frac{\sqrt{b^2 - 4ac}}{2a}$$

to solve a quadratic equation of the form $0 = ax^2 + bx + c$. *(p. 601)*

GEOMETRY

➤ A reflection produces an image that is congruent to the original, but reversed in orientation. Any segment that connects corresponding points on the original figure and its image is perpendicular to the line of reflection and is intersected at its midpoint by the line of reflection. *(p. 548)*

➤ The graph of $y = -x^2$ is a reflection of the graph of $y = x^2$ over the x-axis. *(p. 555)*

➤ You can draw some parabolas by translating the graph of $y = x^2$ up, down, left, or right. The numbers in the equation tell you the length and the direction of the translation. *(p. 556)*

➤ The line of symmetry of the graph of $y = ax^2 + bx + c$ is the line $x = -\frac{b}{2a}$. *(p. 585)*

➤ You can use the line of symmetry, the vertex, and the x- and y-intercepts to sketch the graph of $y = ax^2 + bx + c$. *(p. 594)*

Key Terms

- **reflection** (p. 548)
- **x-axis** (p. 550)
- **vertex** (p. 555)
- **factored form** (p. 577)
- **monomial** (p. 578)
- **quadratic equation** (p. 599)

- **line of reflection** (p. 548)
- **y-axis** (p. 550)
- **x-intercept** (p. 563)
- **expanded form** (p. 577)
- **factored completely** (p. 578)
- **quadratic formula** (p. 601)

- **orientation** (p. 548)
- **parabola** (p. 555)
- **y-intercept** (p. 563)
- **binomial** (p. 578)
- **trinomial** (p. 584)

Unit 10 Review and Assessment

609

Answers to Unit 10 Review and Assessment

23. about 0.8, about −0.8; Methods may vary. An example is given. I used the quadratic formula.

24. a, b. Answers may vary.

25. a–f. Answers may vary.

Contents of Student Resources

Student Resources

Technology Handbook

Using a Graphics Calculator

This handbook introduces you to the basic features of most graphics calculators. Because there are so many different kinds of calculators, this handbook does not always give specific keystrokes. Check your calculator's instruction manual for any details not provided here.

Performing Calculations

➤ **The Keyboard**

Look closely at your calculator's keyboard. Notice that most keys serve more than one purpose. Each key is labeled with its primary purpose, and labels for any secondary purposes appear somewhere near the key. You may need to press 2nd, SHIFT, or ALPHA to use a key for a secondary purpose.

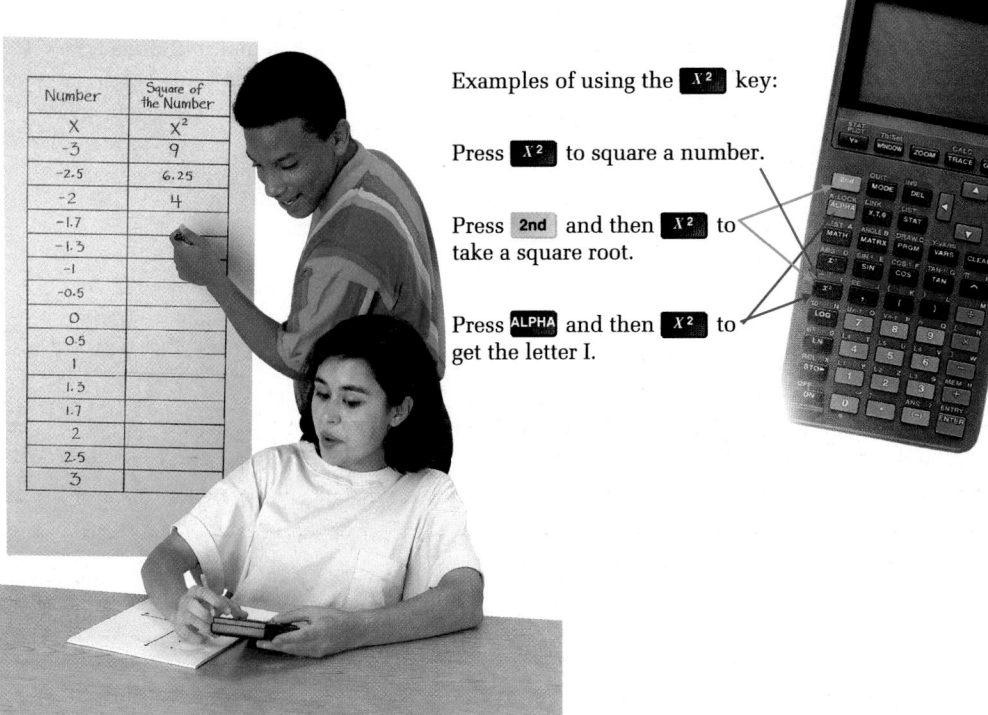

Number	Square of the Number
X	X²
-3	9
-2.5	6.25
-2	4
-1.7	
-1.3	
-1	
-0.5	
0	
0.5	
1	
1.3	
1.7	
2	
2.5	
3	

Examples of using the X^2 key:

Press X^2 to square a number.

Press 2nd and then X^2 to take a square root.

Press ALPHA and then X^2 to get the letter I.

Technology Handbook   611

➤ **The Home Screen**

Your calculator has a "home screen" where you can do calculations. You can usually enter a calculation on a graphics calculator just as you would write it on a piece of paper. This is unlike a scientific calculator, where you often must work backwards. For example, compare the steps for calculating $\sqrt{2}$:

Graphics Calculator

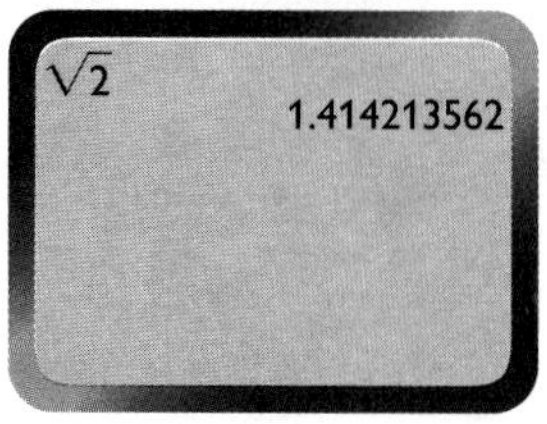

$\sqrt{2}$
                     1.414213562

1. Enter $\sqrt{}$.
2. Enter 2.
3. Press ENTER or EXE.

Scientific Calculator

1. Enter 2.

                     2

2. Enter $\sqrt{}$.

                1.414213562

Shown below are other things to remember as you enter calculations on your graphics calculator.

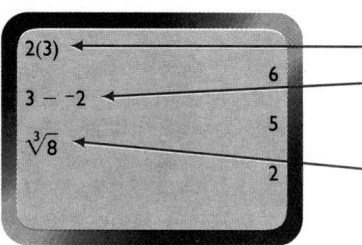

2(3)   6

3 − −2   5

$\sqrt[3]{8}$ 2

The calculator may recognize implied multiplication.

The calculator has a subtraction key, ▬, and a negation key, (−). If you use these incorrectly, you will get an error message.

You may need to get a cube root (or other operation that is not often used) from a MATH menu.

Try This

1. Use your calculator to find the value of each expression.
   a. $(-1.5)^2$ **b.** $-2(33 - 47)$ **c.** $\tan 45°$

2. **a.** Which calculation, $-15 + 29 \div 2$ or $(-15 + 29) \div 2$, gives the correct value for the average of −15 and 29?

   b. Find the average of −15, 29, and 43.

3. Find $\sqrt{ab}$ when $a = 32$ and $b = 128$.

Student Resources

Answers to Try This ············

1. **a.** 2.25
   b. 28
   c. 1
2. **a.** $(-15 + 29) \div 2$
   b. 19
3. 64

Displaying Graphs

➤ ### The Viewing Window

When you use a graphics calculator to display graphs, think of the screen as a "viewing window" that lets you look at a portion of the coordinate plane.

On many calculators, the standard viewing window uses values from −10 to 10 on both the *x*- and *y*-axes. You can adjust the viewing window by pressing the **RANGE** or **WINDOW** key and entering new values for the window variables.

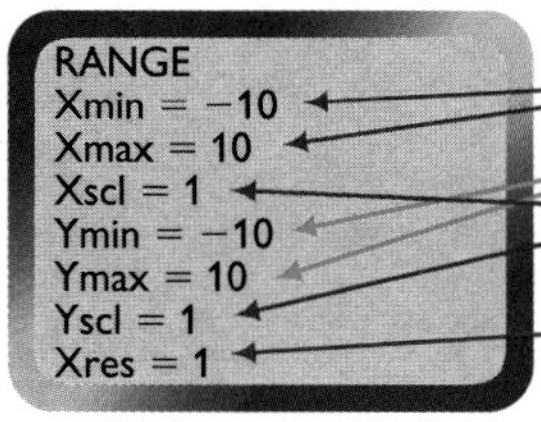

RANGE
Xmin = −10
Xmax = 10
Xscl = 1
Ymin = −10
Ymax = 10
Yscl = 1
Xres = 1

The *x*-axis will be shown for $-10 \le x \le 10$.

The *y*-axis will be shown for $-10 \le y \le 10$.

With scale variables set to equal 1, tick marks will be 1 unit apart on both axes.

Some calculators have a resolution variable. This controls how "smooth" the graph will look.

➤ ### Entering and Graphing a Function

To graph a function, enter its equation and set the variables for an appropriate viewing window. (You may need to experiment to find a good viewing window.) The graph of $y = \frac{1}{2}x + 3$ is shown using the standard viewing window.

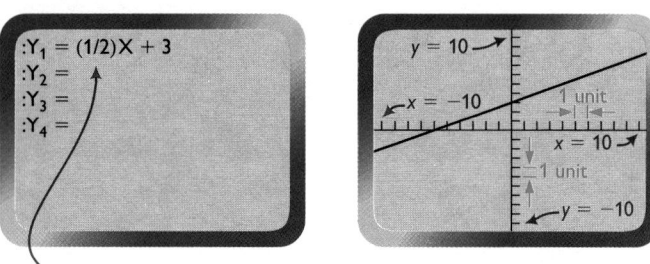

:Y₁ = (1/2)X + 3
:Y₂ =
:Y₃ =
:Y₄ =

Use parentheses when entering the function. If you enter $y = 1/2x + 3$ instead, the calculator may interpret the equation as $y = \frac{1}{2x} + 3$.

➤ **Function Form**

To graph an equation, you must enter it in the form $y = \dots$, which is called *function form* because y is expressed as a function of x. For example, before graphing $3x - 4y = 8$, first solve the equation for y. Since $y = \frac{3}{4}x - 2$, enter either $y = (3/4)x - 2$ or $y = 0.75x - 2$.

Try This

4. Enter and graph each equation separately. Use the standard viewing window. You may need to put the equation in function form first.

   a. $y = 3x + 1$ **b.** $x + 2y = 4$ **c.** $y = |x|$ **d.** $y = \frac{5}{x}$

5. Find a good viewing window for the graph of $y = 65 - 3x$. Be sure your window shows where the graph crosses both axes.

➤ **Squaring the Screen**

A "square screen" is a viewing window with equal unit spacing on the two axes. For example, the graph of $y = x$ is shown for two different windows.

Standard Viewing Window   **Square Screen Window**

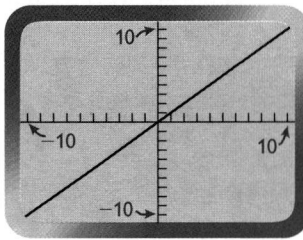 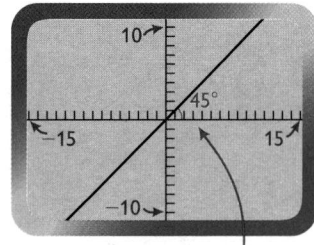

On a square screen, the line $y = x$ makes a 45° angle with the x-axis.

On many graphics calculators, the ratio of the screen's height to its width is about 2 to 3. To get a square screen, choose values for the window variables that make the "length" of the y-axis about two-thirds the "length" of the x-axis:

$$(\text{Ymax} - \text{Ymin}) \approx \frac{2}{3}(\text{Xmax} - \text{Xmin})$$

Student Resources

Answers to Try This

4. a.
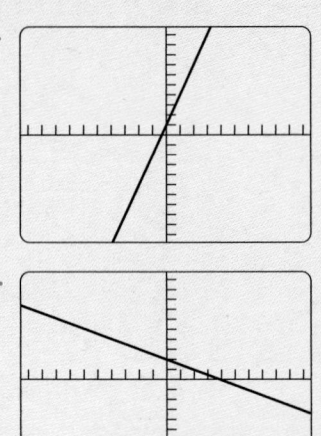

   b.

c.
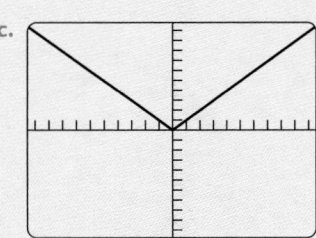

d.
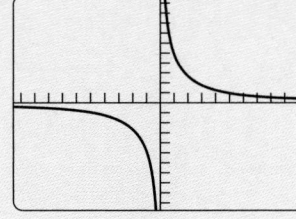

5. Graphs may vary. An example is given.

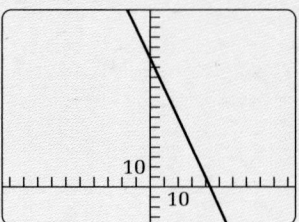

Reading a Graph

➤ The TRACE Feature

After a graph is displayed, you can use the calculator's TRACE feature. When you press **TRACE**, a flashing cursor appears on the graph. The x- and y-coordinates of the cursor's location are shown at the bottom of the screen. Press the left- and right-arrow keys to move the TRACE cursor along the graph.

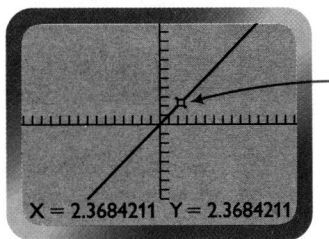

X = 2.3684211   Y = 2.3684211

The TRACE cursor is at the point (2.3684211, 2.3684211) on the graph of $y = x$.

➤ Friendly Windows

As you press the right-arrow key while tracing a graph, you may notice that the x-coordinate increases by "unfriendly" increments. Your calculator may allow you to control the x-increment, Δx, directly. If not, you can control it indirectly by choosing an appropriate Xmax for a given Xmin. For example, on a TI-81 graphics calculator, choose Xmax so that

$$\text{Xmax} = \text{Xmin} + 95\Delta x.$$

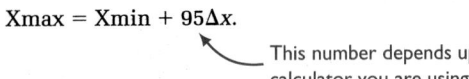

This number depends upon the calculator you are using.

Suppose you want Δx to equal 0.1. If Xmin $= -5$, then set Xmax equal to $-5 + 95(0.1)$, or 4.5. This gives a "friendly window" where the TRACE cursor's x-coordinate will increase by 0.1 each time you press the right-arrow key.

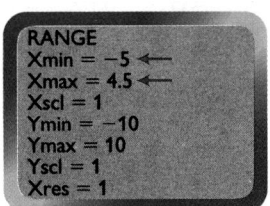

RANGE
Xmin = −5 ←
Xmax = 4.5 ←
Xscl = 1
Ymin = −10
Ymax = 10
Yscl = 1
Xres = 1

Technology Handbook

6. Graph the equation $y = x^2 + 2x - 1$. Choose a friendly window where $\Delta x = 0.1$. Use the TRACE feature to determine, to the nearest tenth, the x-coordinate of each of the two points where the graph crosses the x-axis.

➤ The TABLE Feature

Instead of tracing the graph of an equation, you may wish to examine a table of values. Not all calculators have a TABLE feature. Check to see if yours does.

The screen shows a table of values for $y = x^2 + 2x - 1$.

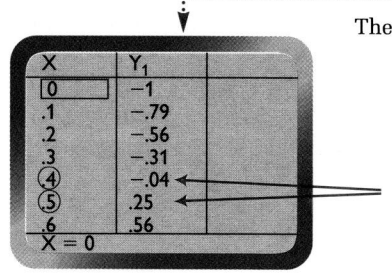

Notice that the y-values change sign between $x = 0.4$ and $x = 0.5$

Taking a Closer Look at a Graph
➤ The ZOOM Feature

Suppose you are interested in the point where the graph of the equation $y = x^2 + 2x - 1$ crosses the positive x-axis. Tracing the graph shows that the x-coordinate of the point is between 0.4 and 0.5.

Move the TRACE cursor to a point just below the x-axis. The y-coordinate of this point is negative but close to 0.

Move the TRACE cursor to a point just above the x-axis. The y-coordinate of this point is positive but close to 0.

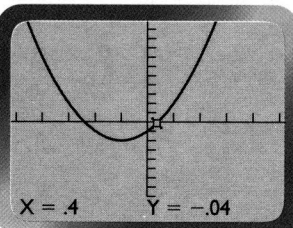

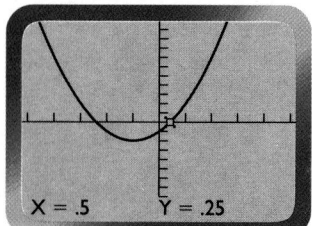

To get a closer look at the point of interest, you can use the ZOOM feature. Your calculator may have more than one way to zoom. A common way is to put a "zoom box" around the point. The calculator will then draw what's inside the box at full-screen size.

Student Resources

---

6. approximate x-intercepts:
−2.4 and 0.4

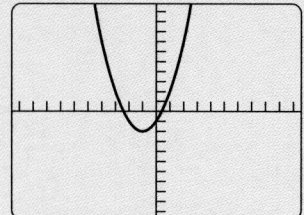

Define a "zoom box" . . .   **and then zoom.**

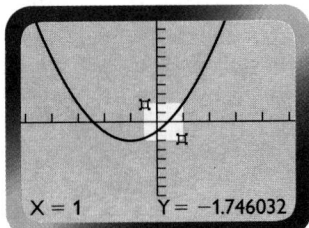

X = 1   Y = −1.746032

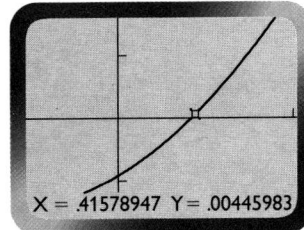

X = .41578947   Y = .00445983

You create a "zoom box" by fixing first one corner and then the opposite corner of the box. (See your calculator's instruction manual for details.)

Tracing the new graph reveals that the graph crosses the x-axis between $x = 0.41$ and $x = 0.42$. Repeated zooming and tracing give better estimates.

Try This

7. Try zooming in on the point where the graph of $y = x^2 + 2x - 1$ crosses the negative x-axis. Between what two values, to the nearest hundredth, does the x-coordinate of the point lie?

➤ **Solving Problems with a Graphics Calculator**

Suppose a school's student council sponsors a dance that costs $248 for a band, decorations, and so on. The council plans to charge $3.50 per couple attending the dance. How many couples must attend the dance for the student council to recover its costs?

One way to solve this problem is to graph the equation $y = 3.5x - 248$, since the council's profit or loss, y, is equal to its income ($3.50 per couple times x couples) minus its costs ($248). Where the graph crosses the x-axis tells you how many couples must attend for the council to *break even*.

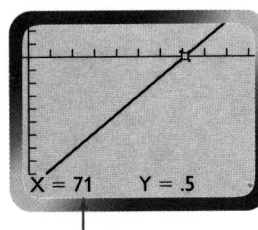

X = 71   Y = .5

Tracing the graph shows that 71 couples must attend for the council to break even.

Try This

8. Suppose the student council wants the dance to make a profit of $100. Use your calculator to find how many couples must attend.

Technology Handbook

Answers to Try This

7. −2.42 and −2.41
8. 100 couples

Displaying Statistical Graphs

➤ ### Histograms, Line Graphs, and Box-and-Whisker Plots

Many graphics calculators can display histograms, line graphs, and sometimes even box-and-whisker plots of data that you enter. For example, the histogram below displays the data about the readers of *Galaxy* magazine (see pages 150–151).

Readers of **galaxy** Magazine

Age group	Frequency
10–14	1110
15–19	3398
20–24	4344
25–29	3215
30–34	332
35–39	112

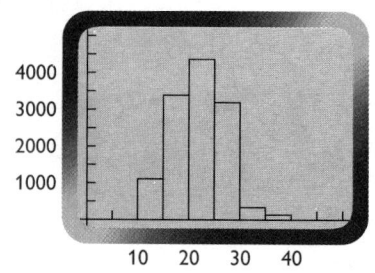

➤ ### Scatter Plots and Curve Fitting

Curve fitting is the process of finding an equation that describes a set of ordered pairs. Often, the first step is to graph the paired data in a scatter plot. For example, the scatter plot below displays the data for an Olympic event (see page 416). It also shows a *regression line* that the calculator fit to the data.

Men's Winning Times in Olympic 400 m Freestyle Swimming

Year	Country of winner	Years after 1960	Time (seconds)
1960	AUS	0	258.3
1964	USA	4	252.2
1968	USA	8	249.0
1972	USA	12	240.27
1976	USA	16	231.93
1980	USSR	20	231.31
1984	USA	24	231.23
1988	E.GER	28	226.95
1992	UT	32	225.00

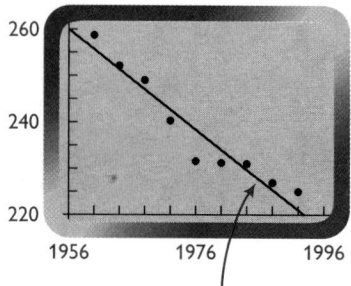

The equation of the regression line is $y = 2325 - 1.056x$ where x is the year and y is the winning time.

Student Resources

Using a Spreadsheet

In addition to using a graphics calculator, you may want to use a computer with a spreadsheet program. A spreadsheet can help you solve a problem like this one: Suppose you want to buy a CD player that costs $195, including tax. You already have $37 and can save $9 per week. After how many weeks can you buy the player? (See Sample 1 in Section 5-1.)

A spreadsheet is made up of cells named by a column letter and a row number, like A3 or B4. You can enter a label, a number, or a formula into a cell.

	A	B
	CD Savings	
1	Week number	Total saved
2	0	37
3	= + A2 + 1	= + B2 + 9
4	= + A3 + 1	= + B3 + 9
5	= + A4 + 1	= + B4 + 9
6	= + A5 + 1	= + B5 + 9
7	= + A6 + 1	= + B6 + 9
8	= + A7 + 1	= + B7 + 9
9	= + A8 + 1	= + B8 + 9
10	= + A9 + 1	= + B9 + 9
11	= + A10 + 1	= + B10 + 9
12	= + A11 + 1	= + B11 + 9
13	= + A12 + 1	= + B12 + 9
14	= + A13 + 1	= + B13 + 9
15	= + A14 + 1	= + B14 + 9
16	= + A15 + 1	= + B15 + 9
17	= + A16 + 1	= + B16 + 9
18	= + A17 + 1	= + B17 + 9
19	= + A18 + 1	= + B18 + 9
20	= + A19 + 1	= + B19 + 9

Cell B1 contains the label "Total saved."

Cell B2 contains the number 37.

Cell B3 contains the formula "=+B2+9." This formula tells the computer to take the number in cell B2, add 9 to it, and put the result in cell B3. (Likewise, the formula in cell A3 tells the computer to take the number in cell A2, add 1 to it, and put the result in cell A3.)

Instead of typing a formula into each cell individually, you can use the spreadsheet's copy and fill commands.

In this spreadsheet, the computer has replaced all the formulas with calculated values. You can have the computer draw a scatter plot with a line connecting the plotted points. As you can see, you will have enough money to buy the CD player after 18 weeks.

	A	B
	CD Savings	
1	Week number	Total saved
2	0	37
3	1	46
4	2	55
5	3	64
6	4	73
7	5	82
8	6	91
9	7	100
10	8	109
11	9	118
12	10	127
13	11	136
14	12	145
15	13	154
16	14	163
17	15	172
18	16	181
19	17	190
20	18	199

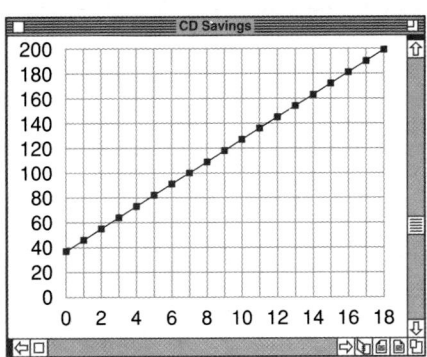

Technology Handbook

Extra Practice

Unit 1

Evaluate each variable expression for $a = 5$, $b = 2$, and $c = 15$. `1-2`

1. $9 - a$ **2.** $7b$

3. $\frac{1}{3}c$ **4.** $8 \div b$

5. bc **6.** $a + c$

7. $\frac{c}{a}$ **8.** $c - b$

9. $2c - a$ **10.** $3a - c$

11. $4a + 3b$ **12.** $ab + c$

Write as a power of ten. `1-3`

13. $10 \cdot 10 \cdot 10$ **14.** $10 \cdot 10$ **15.** $10 \cdot 10 \cdot 10 \cdot 10$

16. $10^8 \cdot 10^5$ **17.** $10^7 \cdot 10$ **18.** $10^3 \cdot 10^3$

19. $\frac{10^{12}}{10^9}$ **20.** $\frac{10^{50}}{10}$ **21.** $\frac{10^4}{10^4}$

22. $10^2 \cdot 10^3 \cdot 10^4$ **23.** $10^8 \cdot 10^{15} \cdot 10^{11}$ **24.** $\frac{10^{14}}{10^{10}} \cdot 10^5$

Evaluate each variable expression for $x = 4$, $y = 8$, and $z = 3$. `1-4`

25. $x + yz$ **26.** $(x + y)z$ **27.** $9y - xz$

28. $xy - z^2$ **29.** $(xy - z)^2$ **30.** $z + 10y \div x$

31. $x^2 + 4x + 3$ **32.** $z^2 + z + 7$ **33.** $3y^2 + 9y + 5$

34. $2z^3 + 5z + 6$ **35.** $y^3 + 5y^2$ **36.** $x^3 - 3x + 8$

Rewrite each expression without parentheses. `1-5`

37. $4(a - b)$ **38.** $9(x + \text{?})$ **39.** $2(2r + s)$

40. $3(5x - y)$ **41.** $\frac{1}{2}(4x^2 - 8)$ **42.** $\frac{1}{6}(12b^2 + 36b)$

Combine like terms if possible. If not, explain why. `1-5`

43. $7a + 5a$ **44.** $k + 3k$ **45.** $9d - d$

46. $x^2 + 8x$ **47.** $5x + 3y$ **48.** $3k^2 + 2k + 1$

49. $3r + r^3 + 4r + 6$ **50.** $m + 2m^2 + 7m + 3$ **51.** $x^3 + 5x^2 + 7x + 1$

52. $2b + 4(b + 5) + 7$ **53.** $8x + 6(x + 3) - 9$ **54.** $2(a + 1) + 5(a + 2)$

Answers to Extra Practice Unit 1

1. 4
2. 14
3. 5
4. 4
5. 30
6. 20
7. 3
8. 13
9. 25
10. 0
11. 26
12. 25
13. 10^3
14. 10^2
15. 10^4
16. 10^{13}
17. 10^8
18. 10^6
19. 10^3
20. 10^{49}
21. 10^0
22. 10^9
23. 10^{34}
24. 10^9
25. 28
26. 36
27. 60
28. 23
29. 841
30. 23
31. 35
32. 19
33. 269
34. 75
35. 832
36. 60
37. $4a - 4b$
38. $9x + 27$
39. $4r + 2s$
40. $15x - 3y$
41. $2x^2 - 4$
42. $2b^2 + 6b$
43. $12a$
44. $4k$
45. $8d$
46. no like terms
47. no like terms
48. no like terms
49. $r^3 + 7r + 6$
50. $2m^2 + 8m + 3$
51. no like terms
52. $6b + 27$
53. $14x + 9$
54. $7a + 12$

Name the congruent polygons. Then list the pairs of congruent sides. 1-6

55.

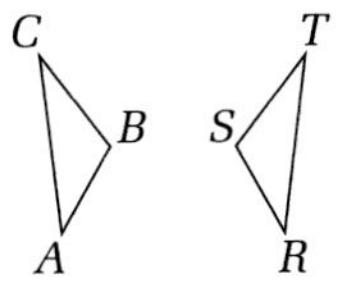

56.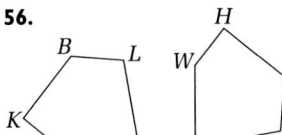

57.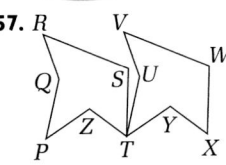

58.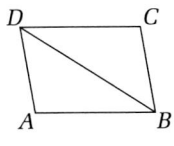

59.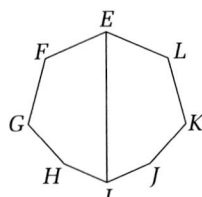

60.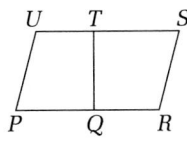

What name best describes the quadrilateral? 1-7

61.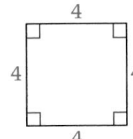

62.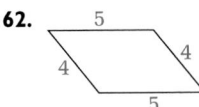

63.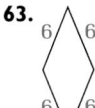

64.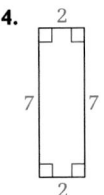

65.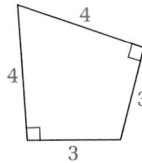

66.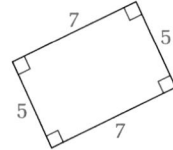

67.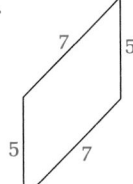

68.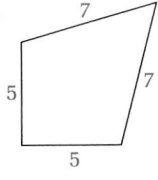

Draw all lines of symmetry for each figure or write *no symmetry*. 1-7

69.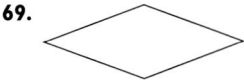

70.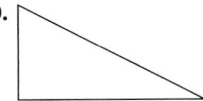

71.

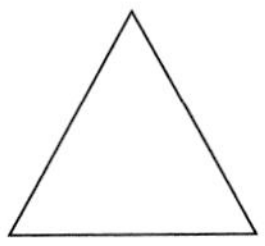

72.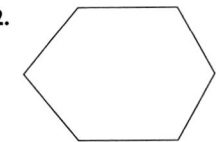

73.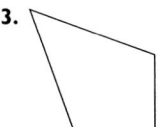

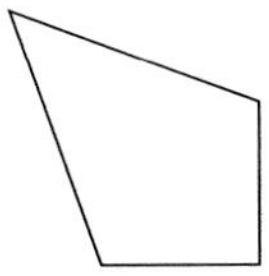

74.

Extra Practice

621

60. quadrilateral $PQTU \cong$
quadrilateral $STQR$;
side $PQ \cong$ side ST,
side $QT \cong$ side TQ,
side $TU \cong$ side QR,
side $UP \cong$ side RS

61. square

62. parallelogram

63. rhombus

64. rectangle

65. kite

66. rectangle

67. parallelogram

68. kite

69.

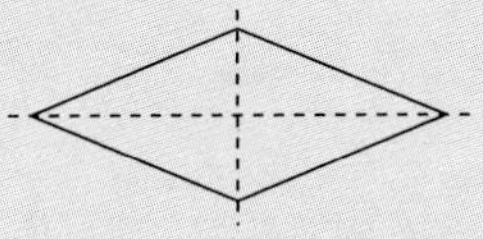

70. no symmetry

71.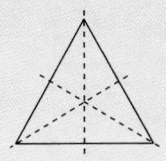

72.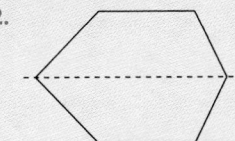

73.

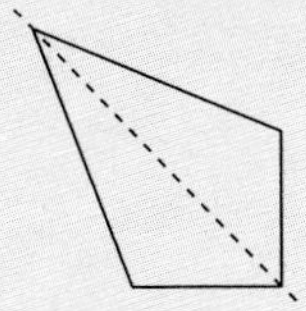

74. no symmetry

Answers to Extra Practice Unit 1

55. triangle $ABC \cong$
triangle RST;
side $AB \cong$ side RS,
side $BC \cong$ side ST,
side $CA \cong$ side TR

56. pentagon $BLACK \cong$
pentagon $ITEWH$;
side $BL \cong$ side IT,
side $LA \cong$ side TE,
side $AC \cong$ side EW,
side $CK \cong$ side WH,
side $KB \cong$ side HI

57. hexagon $PQRSTZ \cong$
hexagon $TUVWXY$;
side $PQ \cong$ side TU,
side $QR \cong$ side UV,
side $RS \cong$ side VW,
side $ST \cong$ side WX,
side $TZ \cong$ side XY,
side $ZP \cong$ side YT

58. triangle $ABD \cong$
triangle CDB;
side $AB \cong$ side CD,
side $BD \cong$ side DB,
side $AD \cong$ side CB

59. pentagon $EFGHI \cong$
pentagon $ELKJI$;
side $EF \cong$ side EL,
side $FG \cong$ side LK,
side $GH \cong$ side KJ,
side $HI \cong$ side JI,
side $IE \cong$ side IE

621

Unit 2

Evaluate each expression for the given values of the variables. `2-2`

1. **a.** $7 - x$ when $x = -8$
   b. $7x$ when $x = -8$
2. **a.** $y^2 + 3y$ when $y = 5$
   b. $y^2 + 3y$ when $y = -5$
3. **a.** $-k$ when $k = -\frac{7}{8}$
   b. $|k|$ when $k = -\frac{7}{8}$
4. **a.** $rs - s$ when $r = -4$ and $s = 6$
   b. $rs - s$ when $r = -4$ and $s = -6$
5. **a.** $m^2 \div n$ when $m = -8$ and $n = -2$
   b. $m^2 + n$ when $m = -8$ and $n = -2$
6. **a.** $at - 16t^2$ when $a = 10$ and $t = 5$
   b. $-4.9t^2 + at$ when $a = 10$ and $t = 5$

Write each number in scientific notation. `2-3`

7. 0.057
8. 2,350,000
9. 64,970
10. 0.000091
11. 48 billion
12. 8 ten thousandths

Simplify. Write each number in decimal notation. `2-3`

13. $(9 \times 10^5)(4 \times 10^2)$
14. $10^{-6} \cdot 10^6$
15. 10^{-3}
16. $\dfrac{1.8 \times 10^7}{3.0 \times 10^5}$
17. $\dfrac{1.2 \times 10^4}{2.5 \times 10^8}$
18. $\dfrac{3 \times 10^4}{6 \times 10^4}$

In the diagram, $\angle COD$ is a right angle, $\overrightarrow{OB}$ bisects $\angle GOC$, and $\angle AOG = 60°$. Use angle relationships to find the measure of each angle. `2-5`

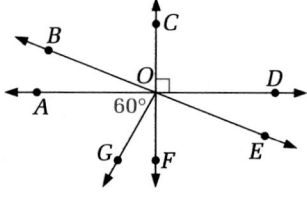

19. $\angle FOG$
20. $\angle GOC$
21. $\angle BOG$
22. $\angle DOE$
23. $\angle EOF$
24. $\angle AOB$
25. $\angle COE$
26. $\angle BOF$

Find the unknown angle measures. `2-5`

27.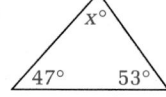
28.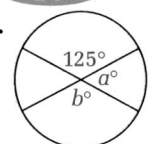
29.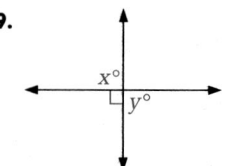

Student Resources

Answers to Extra Practice Unit 2 ···

1. a. 15   b. −56
2. a. 40   b. 10
3. a. $\frac{7}{8}$ b. $\frac{7}{8}$
4. a. −30   b. 30
5. a. −32   b. 62
6. a. −350   b. −72.5
7. 5.7×10^{-2}
8. 2.35×10^6
9. 6.497×10^4

10. 9.1×10^{-5}
11. 4.8×10^{10}
12. 8×10^{-4}
13. 360,000,000
14. 1
15. 0.001
16. 60
17. 0.000048
18. 0.5

19. 30°   20. 150°
21. 75°   22. 15°
23. 75°   24. 15°
25. 105°   26. 105°
27. $x = 80°$
28. $a = 55°$, $b = 125°$
29. $x = y = 90°$

Simplify if possible. If not, write *not possible*.   2-6

30. $(7a)(-2b)$ **31.** $\frac{1}{2}(8r^2)$ **32.** $x \cdot 2x \cdot 3x$

33. $3s - 4st - 2t + 7s$ **34.** $2a^2 + 4ab + b^2 - 7a$

35. $4x^3 - x^2 + 7x + x^2 - 3x - 1$ **36.** $y^2 + 2y - y^3 + 7y^2$

37. $-4w + 5k - 7k + w + 3wk$ **38.** $s^2t^2 - 8st + 7 + 7s$

Solve.   2-7

39. $d + 3 = 10$ **40.** $-2x = -2$

41. $m - 8 = 0$ **42.** $7n = -21$

43. $8 = r + 18$ **44.** $3 - y = 5$

45. $2k + 19 = 19$ **46.** $5a - 9 = 11$

47. $1 + 8t = 57$ **48.** $x + x + 9 = 17$

49. $35 + 90 + b = 180$ **50.** $v + v + 70 = 180$

Solve.   2-8

51. $\frac{x}{3} - 1 = 5$ **52.** $3 = 0.04j + 1$

53. $-7 = 2 - \frac{y}{5}$ **54.** $2.5 - 7.5m = 3.1$

55. $\frac{z}{8} + 7 = 2$ **56.** $24 = 18 + 15r$

57. In a recent season, the Houston Astros had as many wins as losses. They played 162 games in all. How many wins did they have?

58. Rory bought a 9-volt battery for $2.79 and a three-pack of video tapes on sale. The total cost excluding tax was $8.76. Find the cost of each tape.

Estimate each number within a range of two integers. Then use a calculator to find each number to the nearest hundredth.   2-9

59. $\sqrt{30}$ **60.** $\sqrt{105}$

61. $\sqrt[3]{70}$ **62.** $\sqrt[3]{200}$

Find the square roots of each number.   2-9

63. $\frac{64}{49}$ **64.** 0.81

65. 625 **66.** 0.09

67. Write the numbers in order from least to greatest:
$\sqrt{87}$, 9.4, and $\sqrt[3]{800}$.

Extra Practice   623

Answers to Extra Practice Unit 2

30. $-14ab$

31. $4r^2$

32. $6x^3$

33. $10s - 4st - 2t$

34. not possible

35. $4x^3 + 4x - 1$

36. $-y^3 + 8y^2 + 2y$

37. $-3w - 2k + 3wk$

38. not possible

39. 7

41. 8

43. −10

45. 0

47. 7

49. 55

51. 18

53. 45

55. −40

40. 1

42. −3

44. −2

46. 4

48. 4

50. 55

52. 50

54. −0.08

56. 0.4

57. 81   **58.** $1.99 each

59. $\sqrt{30}$ is a number between 5 and 6; 5.48

60. $\sqrt{105}$ is a number between 10 and 11; 10.25

61. $\sqrt[3]{70}$ is a number between 4 and 5; 4.12

62. $\sqrt[3]{200}$ is a number between 5 and 6; 5.85

63. $\frac{8}{7}, -\frac{8}{7}$

64. 0.9, −0.9

65. 25, −25

66. 0.3, −0.3

67. $\sqrt[3]{800}$, $\sqrt{87}$, 9.4

For each data set, find the range, the mean, the median, the mode(s), and the outlier(s) when they exist. `3-2`

1. Number of touchdowns in a season by leading football scorers:
   11, 11, 11, 11, 13, 14, 15, 19

2. Per season points scored by Smythe Division NHL teams:
   39, 74, 81, 82, 84, 96

3. Number of football passes caught in a season by leading receivers:
   64, 67, 68, 68, 69, 77, 78, 84, 93, 108

4. Wins in a season by National League baseball teams
   (consider the data as a single set):
   Eastern Division—73, 75, 76, 76, 89, 92, 96
   Western Division—64, 72, 72, 77, 86, 90, 96

Solve. `3-2`

5. $\frac{x-8}{6} = 9$

6. $\frac{x+93}{2} = 96$

7. $\frac{275+x}{4} = 89$

8. $\frac{x+25+56}{3} = 30$

9. $\frac{2x+1}{4} = 1.5$

10. $\frac{3x-7}{5} = 4$

Graph each inequality on a number line. `3-3`

11. $x \geq -2$

12. $x < 5$

13. $-3.5 < x \leq -2$

14. $x \neq -1$

15. $x \leq 1\frac{1}{2}$

16. $-3 < x < 3$

Write an inequality for the graph. `3-3`

17. —2   0   2

18. —1 0 1 2 3 4

19. 100   104   108

Make a stem-and-leaf plot for the data in the specified exercise (above). `3-4`

20. Exercise 2

21. Exercise 3

22. Exercise 4

Make a box-and-whisker plot to compare the two sets of data. `3-5`

23. Number of touchdowns in a season by leading football scorers
    American Conference: 8, 8, 8, 8, 9, 9, 10, 11, 12
    National Conference: 11, 11, 11, 11, 13, 14, 15, 19

24. Wins in a season by Campbell Conference NHL teams
    Norris Division: 30, 32, 36, 36, 43
    Smythe Division: 17, 31, 33, 35, 36, 42

25. Wins in a season by National League baseball teams:
    Eastern Division—73, 75, 76, 76, 89, 92, 96
    Western Division—64, 72, 72, 77, 86, 90, 96

624   **Student Resources**

Answers to
Extra Practice Unit 3

1. mean = 13.125; median = 12;
   mode = 11; outlier = 19; range = 8

2. mean = 76; median = 81.5;
   mode = none; outlier = 39;
   range = 57

3. mean = 77.6; median = 73;
   mode = 68; outlier = 108;
   range = 44

4. mean = 81; median = 76.5;
   mode = 72, 76, 96; outlier = none;
   range = 32

5. 62

6. 99

7. 81

8. 9

9. 2.5

10. 9

11. —4 —2 0 2 4

12. —2 0 2 4 6

13. —5 —4 —3 —2 —1 0

14. —4 —2 0 2 4

15. —2 1 0 1 2

16. —4 —2 0 2 4

17. $-1 \leq x < 1$

18. $x < 1\frac{1}{3}$

19. $x \geq 106$

20. 3 | 9
     4 |
     5 |
     6 |
     7 | 4
     8 | 1 2 4
     9 | 6

21. 6 | 4 7 8 8 9
     7 | 7 8
     8 | 4
     9 | 3
    10 | 8

22. 6 | 4
     7 | 2 2 3 5 6 6 7
     8 | 6 9
     9 | 0 2 6 6

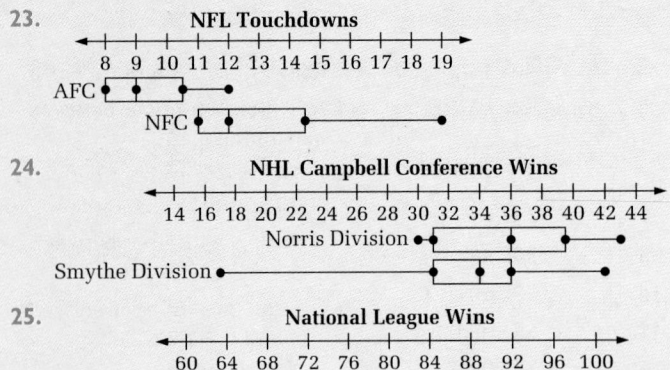

23. **NFL Touchdowns**
    8 9 10 11 12 13 14 15 16 17 18 19
    AFC
    NFC

24. **NHL Campbell Conference Wins**
    14 16 18 20 22 24 26 28 30 32 34 36 38 40 42 44
    Norris Division
    Smythe Division

25. **National League Wins**
    60 64 68 72 76 80 84 88 92 96 100
    Eastern Division
    Western Division

26–28. Answers may vary.
Examples are given.

26. I chose a stem-and-leaf
plot. It divides the data
into intervals for visual
interpretation.

**Medals by Country, 1992
Winter Olympics**

0 | 7 7 7 9
1 | 1 4
2 | 0 1 3 6

Choose an appropriate type of graph to display the data and draw the graph. Explain your choice. **3-6**

26. Total number of medals earned by the top ten countries in the 1992 Winter Olympics: 7, 7, 7, 9, 11, 14, 20, 21, 23, 26

27. U.S. jewelry purchases in 1992

gemstones, 46%
plain gold, 35%
costume, 13%
silver, 5%
platinum, 1%

28. Percent of U.S. homes with electric heat

Year	1981	1983	1985	1987	1989
Percent	18.6	18.5	20.8	22.7	24.6

Unit 4

For Exercises 1–4, plot the points and connect them in order. Then write the specific name of the polygon formed and find its area. **4-2**

1. $R(3, 3)$, $S(-2, 3)$, $T(-3, -2)$

2. $W(-2, 0)$, $X(0, 4)$, $Y(2, 0)$, $Z(0, -3)$

3. $J(1, 3)$, $K(1, -2)$, $L(5, -1)$, $M(5, 1)$

4. $A(-4, -1)$, $B(1, -1)$, $C(1, -4)$, $D(-2, -6)$, $E(-4, -4)$

Find the coordinates of the vertices of the polygon in the specified exercise above after the given translation. **4-3**

5. Exercise 1; 2 units up

6. Exercise 2; 7 units left

7. Exercise 3; 3 units right and 4 units down

8. Exercise 4; 5 units left and 1 unit down

Write the coordinates of P' after each translation of $P(-3, 2)$. **4-3**

9. $(x, y) \rightarrow (x + 3, y - 1)$

10. $(x, y) \rightarrow (x - 1, y - 4)$

Extra Practice   **625**

27. I chose a circle graph. This is a good way to display data that represents various percentages of a whole.

U.S. Jewelry Purchases

28. These data seemed like the type that is often presented in a misleading manner. I drew two line graphs to demonstrate this. Notice how the second graph, with the broken axes, gives the impression of a large increase in use of electric heat.

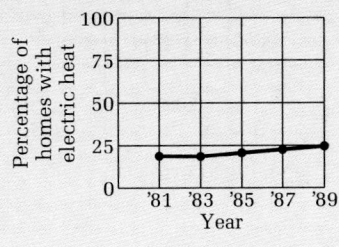

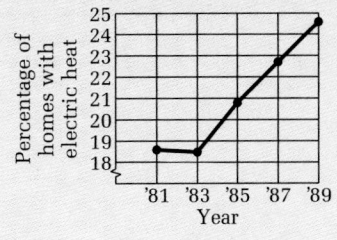

1.

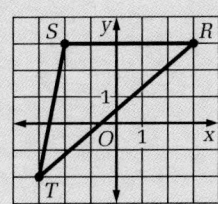

obtuse triangle; 12.5 sq. units

2.

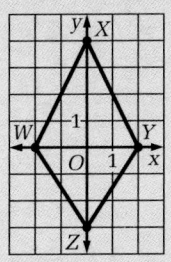

kite; 14 sq. units

3.

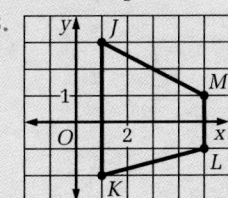

trapezoid; 14 sq. units

4.

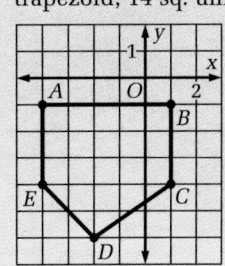

pentagon; 20 sq. units

5. $R'(3, 5)$, $S'(-2, 5)$, $T'(-3, 0)$

6. $W'(-9, 0)$, $X'(-7, 4)$, $Y'(-5, 0)$, $Z'(-7, -3)$

7. $J'(4, -1)$, $K'(4, -6)$, $L'(8, -5)$, $M'(8, -3)$

8. $A'(-9, -2)$, $B'(-4, -2)$, $C'(-4, -5)$, $D'(-7, -7)$, $E'(-9, -5)$

9. $(0, 1)$

10. $(-4, -2)$

625

The following graphs show rotations around the origin. Describe the direction and amount of the rotation of each graph. 4-4

11.

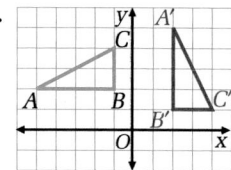

12.

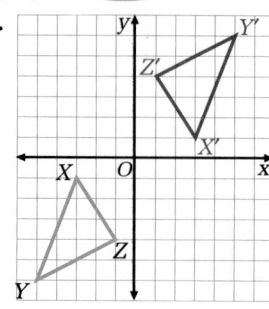

13.

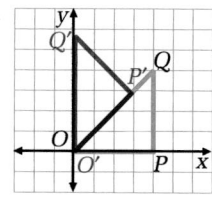

Copy the figure on polar graph paper. Draw each indicated rotation of the figure around the origin. 4-4

14. 70° clockwise

15. 140° counterclockwise

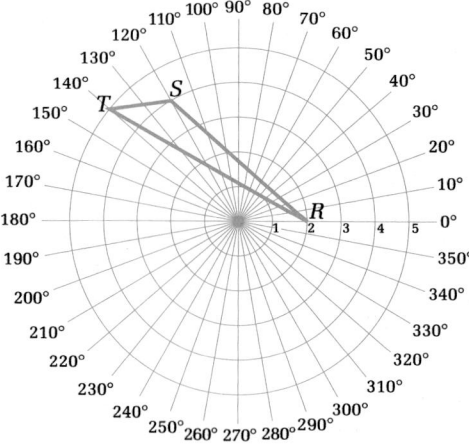

Make a scatter plot for each relationship. State whether the scatter plot shows a *positive correlation*, a *negative correlation*, or *no correlation*. 4-5

16.

Amount financed at 8% for a 25-year mortgage	Monthly payment
$70,000	$540.27
$80,000	$617.45
$90,000	$694.63
$100,000	$771.82

17.

New England States	Area (mi²)	1990 population
Connecticut	4,845	3,287,116
Maine	30,865	1,227,928
Massachusetts	7,838	6,016,425
New Hampshire	8,969	1,109,252
Rhode Island	1,045	1,003,464
Vermont	9,249	562,758

626   **Student Resources**

14.

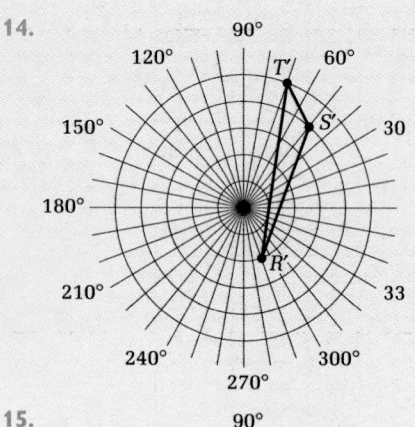

15.

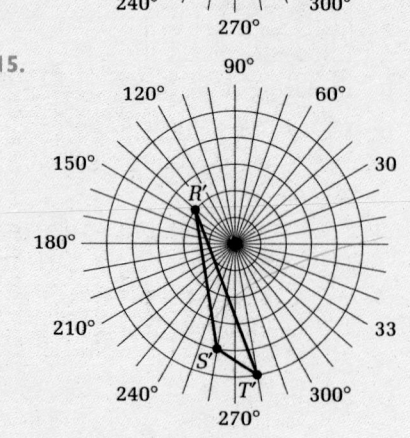

Answers to Extra Practice Unit 4 ···

11. 90° clockwise

12. 180° in either direction

13. 45° counterclockwise

14–15. See graphs at left.

16.

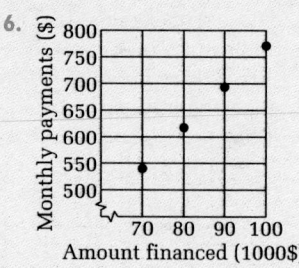

positive correlation

17.

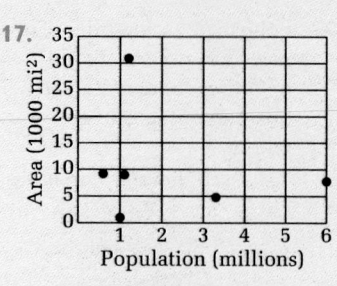

no correlation

Tell whether each graph represents a function. Explain your answer. 4-6

18.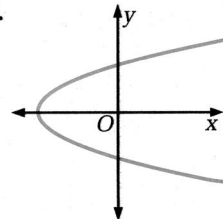

19.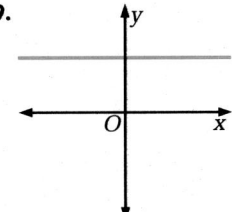

20.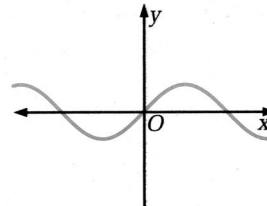

Choose the most reasonable graph for each situation. 4-6

21. The amount of gas in the tank of Felicia's car is a function of the length of time Felicia has been driving.

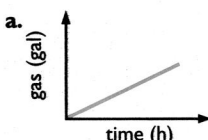

22. The cost of a mushroom pizza is a function of its diameter.

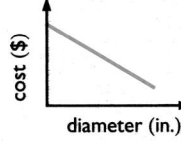

Write an equation to represent each function. 4-7

23.

x	y
−6	2
−3	1
0	0
3	−1
6	−2

24.

x	y
−2	−5
−1	−4
0	−3
1	−2
2	−1

25.

x	y
−2	3
−1	2
0	1
1	0
2	−1

Write each function as an equation. Then graph the function. 4-7

26. The area of a square depends on the length of each side.

27. The Fashion Shoppe is offering 20% off on all its merchandise during its spring sale. The sale price of an item depends on the original price of the item.

28. To find a Fahrenheit temperature if you know the Celsius temperature, multiply the Celsius temperature by 1.8 and then add 32.

Answers to Extra Practice Unit 4

18. No; fails vertical-line test.

19. Yes; passes vertical-line test.

20. Yes; passes vertical-line test.

21. b   22. c

23. $y = -\frac{1}{3}x$

24. $y = x - 3$

25. $y = -x + 1$

26. $A = s^2$, where A is the area, and s is side length.

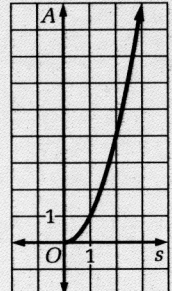

27. $s = 0.8P$, where s is sale price, and P is original price.

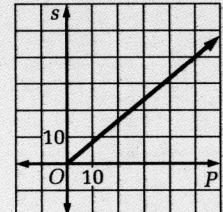

28. $F = 1.8C + 32$, where F is Fahrenheit, and C is Celsius.

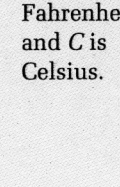

Unit 5

Simplify. 5-2

1. $-(2x - 3)$
2. $-3(-2a + 5)$
3. $-(-x - 3y)$
4. $-2t - 4(2t - t^2)$
5. $8 - (-9 - 7m)$
6. $7y - (8y + z)$

Solve. 5-2

7. $7 - x = 15$
8. $-2d - 3 = 9$
9. $-(r + 2) = 5$
10. $-5(1 - 2b) = 20$
11. $8 = 7y - 4(y - 1)$
12. $3a + 2 + a = 34$
13. $5(x + 3) - 6x = -12$
14. $1.9k - 3 - 7.5k = 4$
15. $-2 = -6(n - 1) + 4$

Solve. 5-3

16. $3x = 12 - x$
17. $5c - 3 = 6c$
18. $2r + 15 = -r$
19. $0.6(b - 5) = b$
20. $4j = 6(j - 9)$
21. $-8(1 - b) = 6b$
22. $9 - 4(1 - w) = -w$
23. $3k + 8(k + 6) = 4$
24. $2(3x - 1) + 7 = x$
25. $-6(4d + 5) + 7d = -2d$

Solve and graph each inequality.

26. $-3x > -9$
27. $8 - y \le 15$
28. $15 > 7z - 6$
29. $-4(a + 1) \ge 2$
30. $8 > 5 - 9b$
31. $3(8 - j) < -12$
32. $-(5 - 2p) \le -1$
33. $2k < 8k - 24$
34. $-m \ge 3(m - 16)$

Solve each equation for the variable indicated. 5-5

35. $P = 4s$, for s
36. $2k = x - 3y$, for x
37. $I = Ed^2$, for E
38. $2x + y = 4$, for y
39. $h = at - 16t^2$, for a
40. $s = \pi r^2 + \pi rl$, for l

Solve. 5-6

41. $35 = -14k$
42. $-\frac{2}{3}b = -6$
43. $\frac{3}{4}c = 18$
44. $\frac{1}{6}z - 7 = -1$
45. $2 - \frac{3}{2}t = 11$
46. $-\frac{1}{3}r + 5 = 1$
47. $\frac{5}{8}x + 3 = -2$
48. $11 - \frac{1}{7}v = 18$
49. $\frac{5}{6}s + 1 = -9$

Solve each equation for y. 5-6

50. $3x - 2y = 6$
51. $ax - by = c$
52. $c = ay + b$
53. $x - \frac{3}{4}y = 12$
54. $2x - \frac{1}{2}y = -4$
55. $\frac{1}{2}x + \frac{1}{3}y = 6$

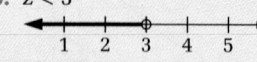

628

Student Resources

Answers to
Extra Practice Unit 5

1. $-2x + 3$
2. $6a - 15$
3. $x + 3y$
4. $-10t + 4t^2$
5. $17 + 7m$
6. $-y - z$
7. -8
8. -6
9. -7
10. 2.5
11. $\frac{4}{3}$
12. 8
13. 27
14. -1.25
15. 2
16. 3
17. -3
18. -5
19. -7.5
20. 27
21. 4
22. -1
23. -4
24. -1
25. -2

26. $x < 3$

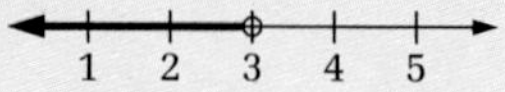

27. $y \ge -7$

28. $z < 3$

29. $a \le -\frac{3}{2}$

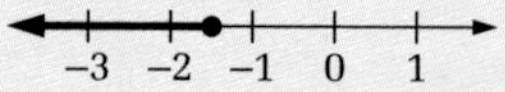

30. $b > -\frac{1}{3}$

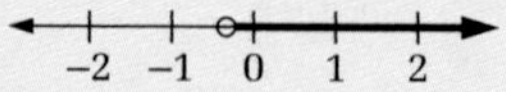

31. $j > 12$

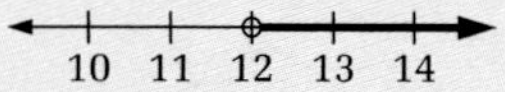

32. $p \le 2$

33. $k > 4$

34. $m \le 12$

35. $s = \frac{P}{4}$

36. $x = 2k + 3y$
37. $E = \frac{I}{d^2}$
38. $y = 4 - 2x$
39. $a = \frac{h + 16t^2}{t}$
40. $l = \frac{s - \pi r^2}{\pi r}$
41. $-\frac{5}{2}$
42. 9
43. 24
44. 36
45. -6
46. 12
47. -8
48. -49
49. -12
50. $y = \frac{3}{2}x - 3$
51. $y = \frac{ax - c}{b}$
52. $y = \frac{c - b}{a}$
53. $y = \frac{4}{3}x - 16$
54. $y = 4x + 8$
55. $y = -\frac{3}{2}x + 18$

628

Find the area of each parallelogram, triangle, and trapezoid. `5-7`

56.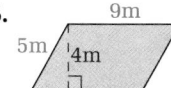

57.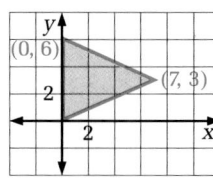

58.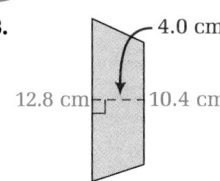

59. A right triangle with perpendicular sides of lengths 9 m and 10 m

60. A parallelogram with base 7.5 cm and height 4 cm

61. A trapezoid with bases 7 ft and 10 ft, and with height 8 ft

Find the specified dimension. `5-7`

62. The height of a parallelogram with area 9.5 in.^{2} and base 2.5 in.

63. The base of a triangle with area $12(y - 3)$, base $6y$, and height 2

64. The height of a trapezoid with bases 7 and 15, height $(x - 4)$, and area 55

Solve each system of equations. `5-8`

65. $y = 3x$
$2x + y = 20$

66. $b = 7a$
$a + 2b = 21$

67. $2r + 3s = -5$
$6s = 2r$

68. $d = 5c$
$3c - d = 6$

69. $7a + b = 28$
$b = 2a + 1$

70. $y = x - 2$
$5x - y = 10$

71. $b = a + 2$
$4a + 5b = 16$

72. $t = s + 5$
$6s - t = 30$

73. $y = 3x - 6$
$7x + 2y = 66$

Unit 6

A survey of 500 registered voters from Carverville found that 257 favored increased funding for schools, 185 oppposed it, and the rest were undecided. `6-2`

1. Suppose more than 50% of the voters must vote in favor of increased funding for schools for the funding to be increased. Is it likely that the funding will be increased? Explain.

2. a. Find the probability that a voter is undecided about the funding issue.

   b. Is the probability in part (a) theoretical or experimental probability? Explain.

Answers to Extra Practice Unit 5 ··························

56. 36 m^2

57. 21 sq. units

58. 46.4 cm^2

59. 45 m^2

60. 30 cm^2

61. 68 ft^2

62. 3.8 in.

63. 36

64. 5

65. $x = 4, y = 12$

66. $a = \frac{7}{5}, b = \frac{49}{5}$

67. $r = -\frac{5}{3}, s = -\frac{5}{9}$

68. $c = -3, d = -15$

69. $a = 3, b = 7$

70. $x = 2, y = 0$

71. $a = \frac{2}{3}, b = 2\frac{2}{3}$

72. $s = 7, t = 12$

73. $x = 6, y = 12$

Answers to Extra Practice Unit 6 ··························

1. Yes. The experimental probability that a voter will be in favor is 0.514, which is more than 50%.

2. a. 0.116

   b. experimental; The probability is based on an experiment.

A card is drawn at random from a deck of ten cards numbered from 1 to 10. Find the probability of getting the indicated card. 6-2

3. an even or an odd number

4. a multiple of 3

5. a factor of 20

6. a number that is greater than 20

7. not a 7

8. a number that is less than 8

Solve each proportion. 6-3

9. $\dfrac{x}{18} = \dfrac{15}{27}$ **10.** $\dfrac{32}{m} = \dfrac{48}{36}$ **11.** $\dfrac{35}{64} = \dfrac{5.6}{t}$ **12.** $\dfrac{250}{12} = \dfrac{k}{15}$

13. $\dfrac{2.5}{a} = \dfrac{8}{28}$ **14.** $\dfrac{2.4}{2} = \dfrac{3}{y}$ **15.** $\dfrac{9}{0.02} = \dfrac{b}{5}$ **16.** $\dfrac{r}{1.5} = \dfrac{1.28}{4}$

17. $\dfrac{0.4}{8} = \dfrac{x}{6}$ **18.** $\dfrac{0.4}{0.14} = \dfrac{z}{7}$ **19.** $\dfrac{3}{k} = \dfrac{25}{7}$ **20.** $\dfrac{40}{g} = \dfrac{48}{105}$

Quadrilateral ABCD ~ quadrilateral WXYZ.

Complete each statement. 6-5

21. $\angle X = $? °

22. $\angle W \cong \angle$?

23. $YZ = $? units

24. $AB = $? units

25. perimeter of $ABCD = $? units

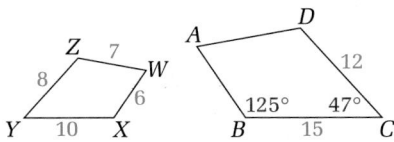

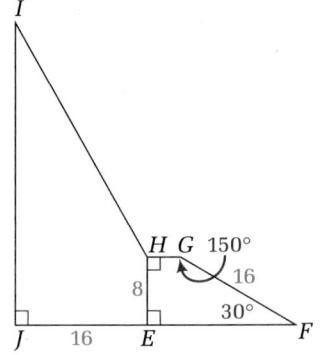

Quadrilateral EFGH ~ quadrilateral JIHE.

Complete each statement. 6-5

26. $\angle HEF$ corresponds to $\angle$? .

27. $\angle EHI = $? °

28. $\angle I = $? °

29. $HI = $? units

30. $GH = $? units

31. Name two similar triangles in the diagram. Explain why they must be similar. 6-5

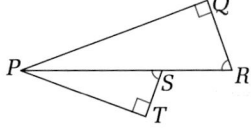

32. $\triangle A' B' C'$ is the image of $\triangle ABC$ under a dilation. Find the coordinates of the center of the dilation and the scale factor. 6-6

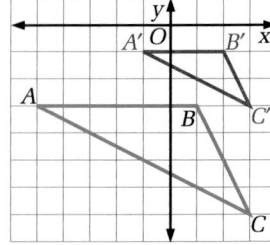

Answers to Extra Practice Unit 6

3. 1 **4.** 0.3 **21.** 125° **22.** A **33.**

5. 0.5 **6.** 0 **23.** 8 **24.** 9

7. 0.9 **8.** 0.7 **25.** 46.5 **26.** J

9. $x = 10$ **10.** $m = 24$ **27.** 150° **28.** 30°

11. $t = 10.24$ **12.** $k = 312.5$ **29.** 32 **30.** 4

13. $a = 8.75$ **14.** $y = 2.5$ **31.** $\triangle PQR \sim \triangle PTS$;

15. $b = 2250$ **16.** $r = 0.48$ $\angle Q \cong \angle T$ and $\angle S \cong \angle R$,

17. $x = 0.3$ **18.** $z = 20$ so the triangles are similar.

19. $k = 0.84$ **20.** $g = 87.5$

32. $(3, 1); \dfrac{1}{2}$

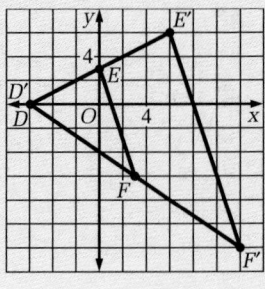

Draw a dilation of △DEF with vertices D(−6, 0), E(0, 3), and F(3, −6) using the specified center of dilation and scale factor. 6-6

33. center *D,* scale factor 2

34. center (0, 0), scale factor $\frac{1}{3}$

35. center (6, 0), scale factor $\frac{4}{3}$

Use a calculator to write each ratio as a decimal rounded to hundredths. 6-7

36. sin 7°   **37.** sin 75°

38. cos 58°   **39.** cos 60°

Write each ratio as a fraction and as a decimal rounded to hundredths. 6-7

40. sin *A*   **41.** cos *A*

42. sin *C*   **43.** cos *C*

44. Use your answer to Exercise 40 and a calculator to find the measure of ∠*A* the nearest degree. 6-7

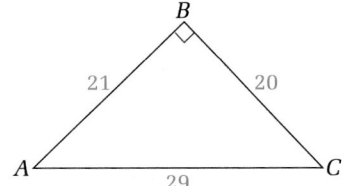

Unit 7

Use a calculator to write each ratio as a decimal rounded to hundredths. 7-1

1. tan 58°   **2.** tan 9°

3. tan 81°   **4.** tan 45°

Refer to right △ABC. 7-1

5. Find tan *A* and tan *B.*

6. Use the inverse tangent key on a calculator to find the measure of ∠*A* to the nearest degree.

7. Use right △DEF. Find the length of *EF* to the nearest 0.1 cm. 7-1

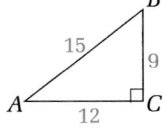

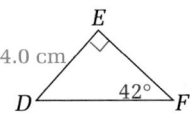

Answers to Extra Practice Unit 6 ··

34. 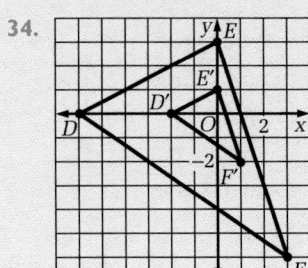   **35.**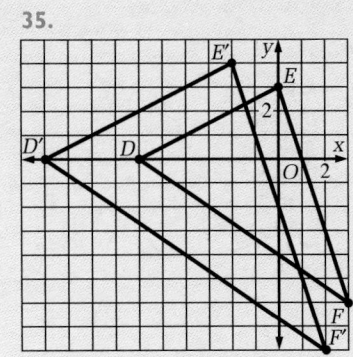

36. 0.12   **37.** 0.97

38. 0.53   **39.** 0.50

40. $\frac{20}{29}$; 0.69

41. $\frac{21}{29}$; 0.72

42. $\frac{21}{29}$; 0.72

43. $\frac{20}{29}$; 0.69

44. 44°

Answers to
Extra Practice Unit 7 ···················

1. 1.60

2. 0.16

3. 6.31

4. 1.00

5. $\frac{3}{4}$; $\frac{4}{3}$

6. 37°

7. 4.4 cm

8. about 157.1 ft

9. about 9.4 mm

10. about 4.7 in.

11. about 267 m

12. about 3.1 in.

13. about 12.6 m

14. about 125.7 ft

15. about 8.7 cm

16.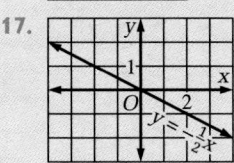

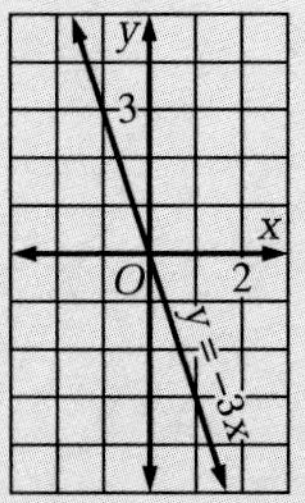

17.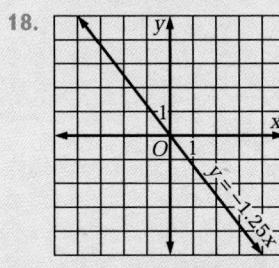

18.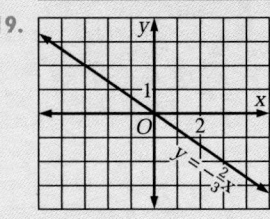

19.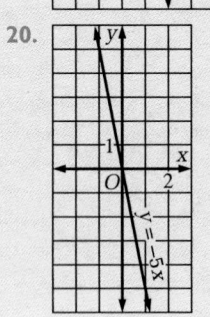

20.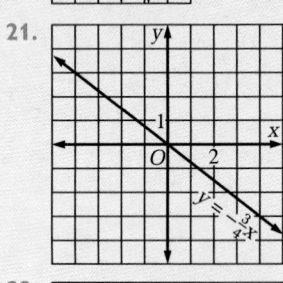

21.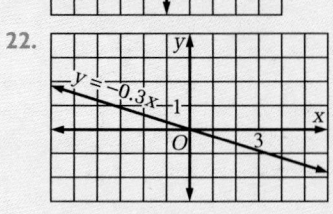

22.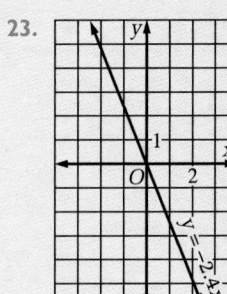

632

Find the circumference of each circle. 7-3

8. The diameter is 50 ft.

9. The radius is 1.5 mm.

10. The radius is $\frac{3}{4}$ in.

11. The diameter is 85 m.

Find each arc length. 7-3

12.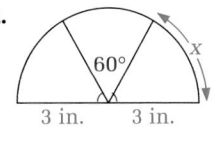

13.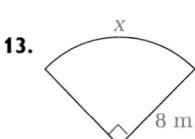

14.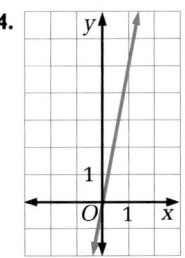

15.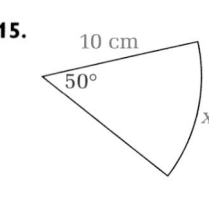

Graph each equation. 7-4

16. $y = -3x$

17. $y = -\frac{1}{2}x$

18. $y = -1.25x$

19. $y = -\frac{2}{3}x$

20. $y = -5x$

21. $y = -\frac{3}{4}x$

22. $y = -0.3x$

23. $y = -2.4x$

Write an equation of the form $y = kx$ to describe each line. 7-4

24.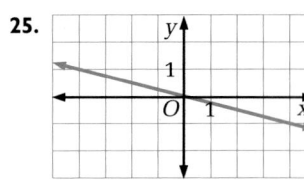

25.

26.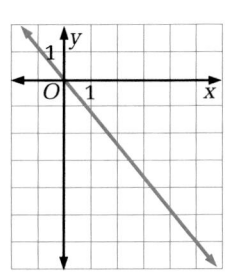

Each expression shows the units of a conversion problem. Use dimensional analysis to find the unit(s) of the answer. 7-5

27. $\frac{km}{mi} \times mi = \underline{\ ?\ }$

28. $\frac{dollars}{h} \times \frac{h}{day} = \underline{\ ?\ }$

29. $\frac{cycles}{min} \times \frac{min}{s} = \underline{\ ?\ }$

30. $\frac{cents}{kWh} \times kWh = \underline{\ ?\ }$

31. $\frac{min}{page} \times pages \times \frac{h}{min} = \underline{\ ?\ }$

32. $\frac{dollars}{lb} \times \frac{lb}{oz} \times oz = \underline{\ ?\ }$

632   **Student Resources**

23.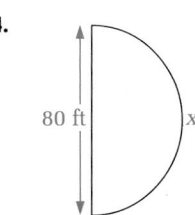

24. $y = 5x$

25. $y = -\frac{1}{4}x$

26. $y = -1.2x$

27. km

28. $\frac{dollars}{day}$

29. $\frac{cycles}{s}$

30. cents

31. h

32. dollars

33. about 154 in.²

34. about 314 m²

35. about 2.5 mm²

36. about 0.4 in.²

37. about 25.1 m²

38. about 117.8 in.²

39. $\frac{9\pi}{8}$ cm², or about 3.5 cm²

Find the area of the circle described. `7-6`

33. The radius is 7 in.

34. The diameter is 20 m.

35. The diameter is 1.8 mm.

36. The radius is $\frac{3}{8}$ in.

Find the area of each sector. In Exercise 39, the area of the entire circle is 9π cm². `7-6`

37.
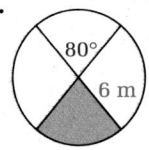

38.
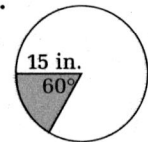

39.
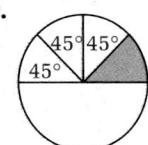

Unit 8

Without graphing, find the slope and the vertical intercept of the line modeled by each equation. `8-1`

1. $y = 2x + 7$

2. $y = \frac{1}{3} - \frac{2}{3}x$

3. $y = 0.7x - 3.5$

4. $y = 7x$

5. $y = 5 - x$

6. $y = -\frac{1}{8}x + \frac{7}{8}$

Find the slope and vertical intercept of each graph. Then write an equation for the line. `8-1`

7.
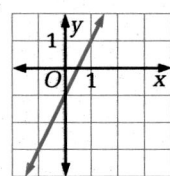

8.
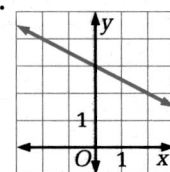

9.
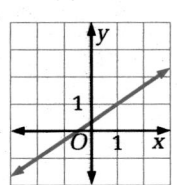

10.
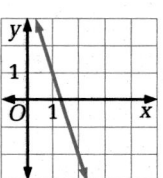

Rewrite each equation in slope-intercept form. `8-2`

11. $5x - y = 7$

12. $2x + 5y = -10$

13. $\frac{1}{2}x + y = -4$

14. $8x - 5y = -15$

15. $2x + \frac{1}{3}y = 6$

16. $8x - \frac{2}{3}y = 2$

Find the intercepts of the graph of each equation. Then use the intercepts to graph the equation. `8-2`

17. $x + 4y = -8$

18. $4x - 9y = -18$

19. $\frac{3}{2}x - 5y = 15$

20. $-0.6x + 0.5y = 3$

21. $2.5x + 2y = -5$

22. $\frac{3}{4}x - \frac{4}{3}y = 6$

Extra Practice

633

Answers to Extra Practice Unit 8

1. 2; 7

2. $-\frac{2}{3}; \frac{1}{3}$

3. 0.7; −3.5

4. 7; 0

5. −1; 5

6. $-\frac{1}{8}; \frac{7}{8}$

7. 2; −1; $y = 2x - 1$

8. $-\frac{1}{2}; 3; y = -\frac{1}{2}x + 3$

9. $\frac{2}{3}; \frac{1}{3}; y = \frac{2}{3}x + \frac{1}{3}$

10. −3; 4; $y = -3x + 4$

11. $y = 5x - 7$

12. $y = -\frac{2}{5}x - 2$

13. $y = -\frac{1}{2}x - 4$

14. $y = \frac{8}{5}x + 3$

15. $y = -6x + 18$

16. $y = 12x - 3$

17. $(0, -2), (-8, 0)$
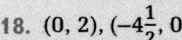

18. $(0, 2), (-4\frac{1}{2}, 0)$
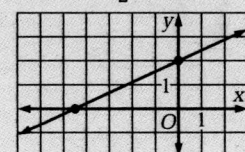

19. $(0, -3), (10, 0)$
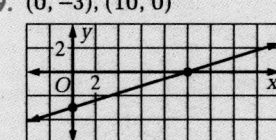

20. $(0, 6), (-5, 0)$
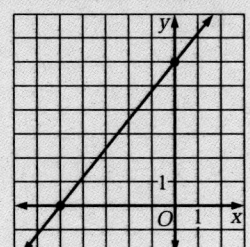

21. $(0, -\frac{5}{2}), (-2, 0)$
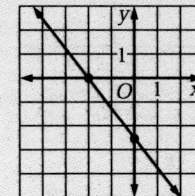

22. $(0, -\frac{9}{2}), (8, 0)$
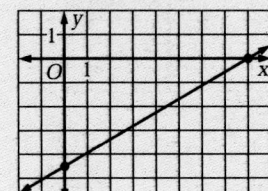

23. undefined; $x = -3$

24. 0; $y = 4$

25. 0; $y = -2.5$

26. undefined; $x = 3.5$

27. Slope is undefined.

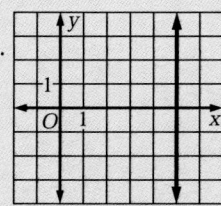

28. Slope is 0.

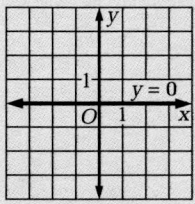

29. Slope is undefined.

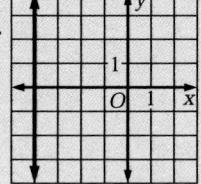

30. Slope is 0.

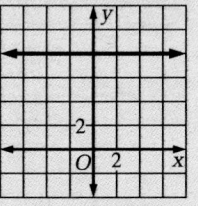

31. Slope is undefined.

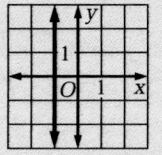

32. Slope is 0.

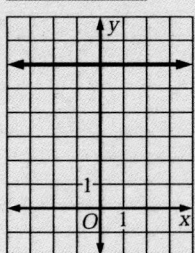

33. Slope is undefined.

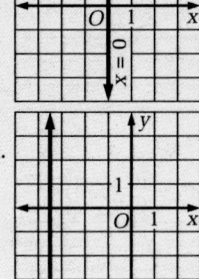

34. Slope is undefined.

35. $y = \frac{3}{5}x - 2$

36. $y = 3x$

Find the slope of each line and write an equation for the line. 8-3

23.
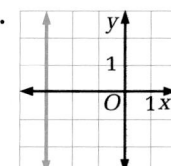

24.
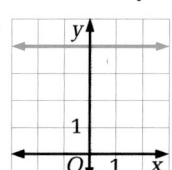

25.
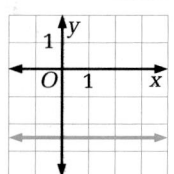

26.
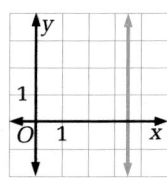

Graph each equation and find the slope of the line. 8-3

27. $x = 5$

28. $y = 0$

29. $5x = -20$

30. $-\frac{1}{2}y = -4$

31. $1.2x = -1.2$

32. $-3y = -18$

33. $\frac{3}{4}x = 0$

34. $-2x = 7$

Write an equation for the line. If there is too little information, state this. 8-4

35. The slope is $\frac{3}{5}$ and the vertical intercept is -2.

36. The slope is 3 and the origin is on the line.

37. The points (7, 5) and (5, 9) are on the line.

38. The slope is undefined.

39. The horizontal intercept is -6 and the vertical intercept is 4.

40. The slope is $-\frac{1}{4}$ and the point (4, 3) is on the line.

41. The slope is -5 and the horizontal intercept is 1.

42. The vertical intercept is -8.

43. The points (8, -1) and (3, -1) are on the line.

44. The points (2, 3) and (3, 2) are on the line.

Graph the system of equations. Estimate the solution of the system or write _no solution_ if there is none. 8-5

45. $y = 3x + 4$
    $y = -3x - 6$

46. $y = 2x + 3$
    $y = 2x - 3$

47. $y = -\frac{1}{2}x + 2$
    $y = -2x - 2$

48. $y = \frac{2}{3}x$
    $y = \frac{2}{3}x + 3$

49. $y = 0.4x$
    $y = 4x - 2$

50. $y = -2.5x + 1$
    $y = 0.5x + 5$

Graph each inequality. 8-6

51. $x \leq 0$

52. $y > -3$

53. $y \leq \frac{1}{2}x$

54. $y > -3x$

55. $y < 2x + 3$

56. $y \geq -x - 1$

57. $\frac{1}{3}x + 2 > y$

58. $-\frac{2}{5}x - 4 \leq y$

59. $3x + 2y \leq -10$

60. $5y - x > -5$

61. $4x - 3y > 6$

62. $3x + 4y \geq 12$

634   **Student Resources**

37. $y = -2x + 19$

38. too little information

39. $y = \frac{2}{3}x + 4$

40. $y = -\frac{1}{4}x + 4$

41. $y = -5x + 5$

42. too little information

43. $y = -1$

44. $y = -x + 5$

45–50. Estimates may vary. Examples are given.

45. $\left(-1\frac{2}{3}, -1\right)$

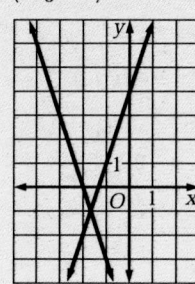

46. no solution

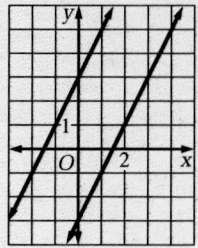

Graph each system of inequalities. 8-7

63. $y \le 4$
$x > -2$

64. $y < 2x + 5$
$y > -3x + 2$

65. $y \le -2x + 4$
$y \le 3x + 4$

66. $y > -x$
$x \le 3$

67. $2x - y > 4$
$3x + 5y \ge -15$

68. $x + 3y > -3$
$4x - 5y < -10$

Unit 9

Find the missing length in each triangle. 9-1

1.
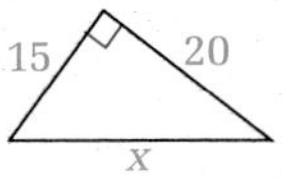

2.
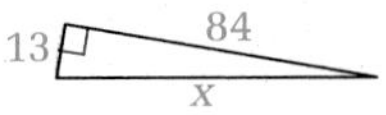

3.
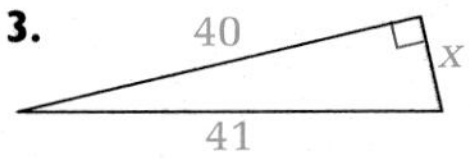

4.
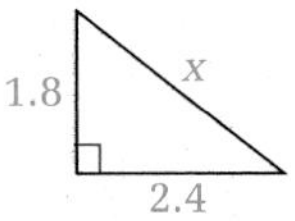

5.
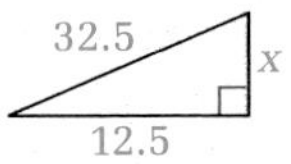

6.
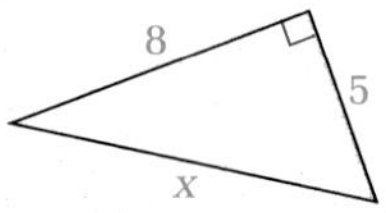

7.
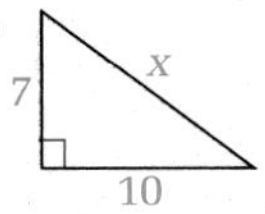

8.
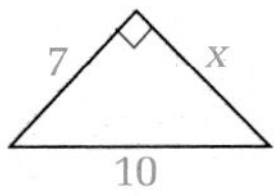

Simplify. 9-2

9. $\sqrt{48}$ **10.** $2\sqrt{27}$ **11.** $\sqrt{90}$ **12.** $5\sqrt{175}$

13. $\sqrt{7} \cdot \sqrt{7}$ **14.** $\sqrt{6} \cdot \sqrt{14}$ **15.** $\sqrt{24} \cdot \sqrt{3}$ **16.** $\sqrt{28} \cdot 3\sqrt{45}$

17. $(-\sqrt{6})^2$ **18.** $(-7\sqrt{2})^2$ **19.** $3\sqrt{5} \cdot 2\sqrt{15}$ **20.** $6\sqrt{18} \cdot \frac{1}{3}\sqrt{6}$

Solve for *x*. 9-2

21. $x^2 = 56$ **22.** $5x^2 = 140$ **23.** $x^2 + (2x)^2 = 100$

24. $4x^2 = 25$ **25.** $\frac{1}{2}x^2 = 81$ **26.** $x^2 + 3x^2 = y^2$

Write the converse of the statement and tell whether the converse is *true* or *false*. If it is false, give a counterexample. 9-3

27. If m and n are even numbers, then $m \times n$ is an even number.

28. If $x > -2$, then $-2x > 4$.

29. A pair of angles are congruent if they are vertical angles.

30. If $k = 6$, then $|k| = 6$.

Answers to Extra Practice Unit 9

1. 25
2. 85
3. 9
4. 3
5. 30
6. $\sqrt{89} \approx 9.4$
7. $\sqrt{149} \approx 12.2$
8. $\sqrt{51} \approx 7.1$
9. $4\sqrt{3}$
10. $6\sqrt{3}$
11. $3\sqrt{10}$
12. $25\sqrt{7}$
13. 7
14. $2\sqrt{21}$
15. $6\sqrt{2}$
16. $18\sqrt{35}$
17. 6
18. 98
19. $30\sqrt{3}$
20. $12\sqrt{3}$
21. $x = \pm 2\sqrt{14}$
22. $x = \pm 2\sqrt{7}$
23. $x = \pm 2\sqrt{5}$
24. $x = \pm\frac{5}{2}$
25. $x = \pm 9\sqrt{2}$
26. $x = \pm\frac{1}{2}y$
27. If $m \times n$ is an even number, then m and n are even numbers. False; $3 \times 4 = 12$.
28. If $-2x > 4$, then $x > -2$. False; If $x = -3$, then $-2x > 4$, but $x < -2$.
29. If a pair of angles are congruent, then they are vertical angles. False; Two congruent angles in two different triangles are not vertical angles.
30. If $|k| = 6$, then $k = 6$. False; $k = -6$.

Answers to Extra Practice Unit 8

47. $\left(-2\frac{2}{3}, 3\frac{1}{3}\right)$

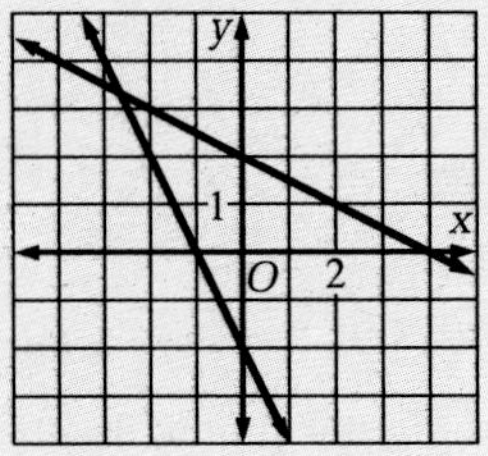

48. no solution

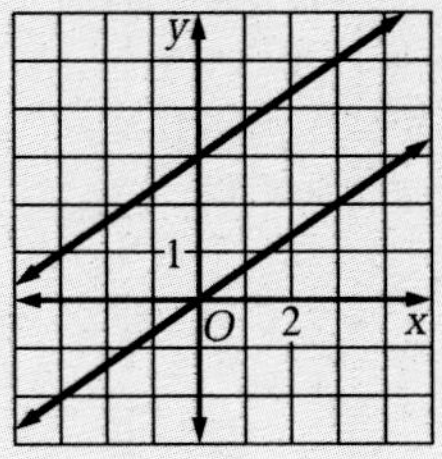

49. about $\left(\frac{1}{2}, \frac{1}{4}\right)$

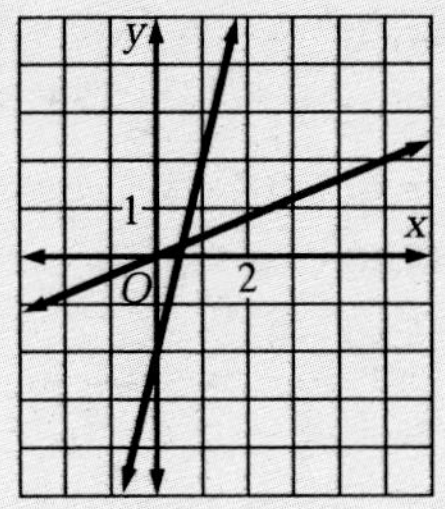

50–68. See answers in back of book.

Solve. 9-3

31. $-1.6r = 0$.

32. $z(z - 4) = 0$

33. $2m(m + 3) = 0$

34. $4j^2 = 0$

35. $x(9 + x) = 0$

36. $5y(1 - 4y) = 0$

The lengths of the sides of a triangle are given. Is the triangle a right triangle? How do you know? 9-3

37. 8 cm, 10cm, 12cm

38. 15 in., 8 in., 17 in.

39. 29 m, 20 m, 21 m

40. 6 yd, 8 yd, 7 yd

41. 40 ft, 9 ft, 41 ft

42. 16 mm, 12 mm, 8 mm

A point P on $\overline{AG}$ is chosen at random. Find the probability that P is on the specified segment. 9-4

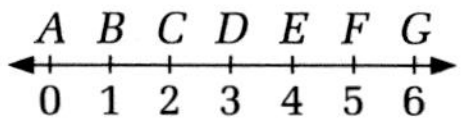

$$\begin{array}{ccccccc} A & B & C & D & E & F & G \\ \hline 0 & 1 & 2 & 3 & 4 & 5 & 6 \end{array}$$

43. $\overline{AC}$ **44.** $\overline{BE}$ **45.** $\overline{EF}$ **46.** $\overline{BF}$ **47.** $\overline{GA}$ **48.** $\overline{BG}$

Find the probability that a point chosen at random from the region shown is in the shaded region. 9-4

49. 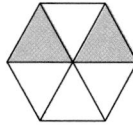 **50.** 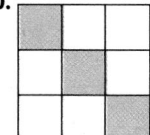 **51.** 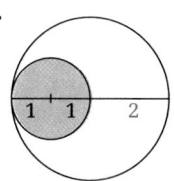 **52.**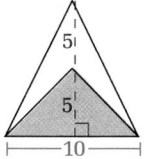

Identify each space figure and find its surface area. 9-5

53. 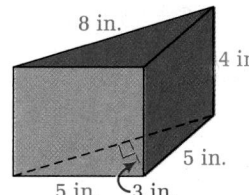 **54.** 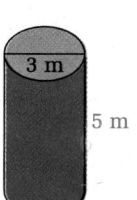 **55.**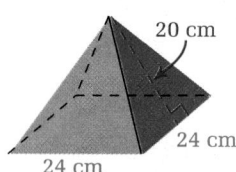

Student Resources

Answers to Extra Practice Unit 9 ···

31. 0

32. 0, 4

33. 0, −3

34. 0

35. 0, −9

36. 0, $\frac{1}{4}$

37. No; $8^2 + 10^2 \neq 12^2$.

38. Yes; $8^2 + 15^2 = 17^2$.

39. Yes; $20^2 + 21^2 = 29^2$.

40. No; $6^2 + 7^2 \neq 8^2$.

41. Yes; $9^2 + 40^2 = 41^2$.

42. No; $8^2 + 12^2 \neq 16^2$.

43. $\frac{1}{3}$

44. $\frac{1}{2}$

45. $\frac{1}{6}$

46. $\frac{2}{3}$

47. 1

48. $\frac{5}{6}$

49. $\frac{1}{3}$

50. $\frac{1}{3}$

51. $\frac{1}{4}$

52. $\frac{1}{2}$

53. triangular prism; 96 in.²

54. cylinder; about 61.3 m²

55. pyramid; 1536 cm²

Identify each space figure and find its surface area. 9-5

56.
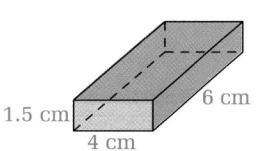
1.5 cm 6 cm
4 cm

57.
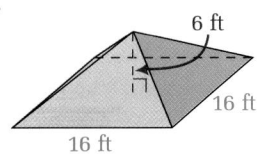
6 ft
16 ft
16 ft

58.
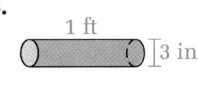
1 ft 3 in.

59.
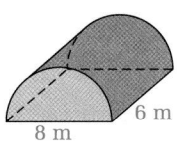
6 m
8 m

60.
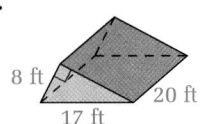
8 ft 20 ft
17 ft

61.
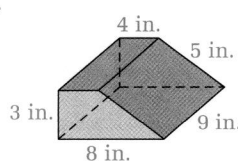
4 in.
5 in.
3 in. 9 in.
8 in.

Find the volume of the space figure above. 9-6

62. The figure in Exercise 53

63. The figure in Exercise 54

64. The figure in Exercise 56

65. The figure in Exercise 58

66. The figure in Exercise 59

67. The figure in Exercise 60

68. The figure in Exercise 61

Find the volume of the space figure. 9-7

69. The figure in Exercise 55

70. The figure in Exercise 57

71. A cone with radius 12 mm and height 15 mm

72.
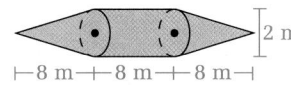
2 m
8 m 8 m 8 m

Find the ratio of the areas of each pair of similar figures. Then find the unknown area. 9-8

73.
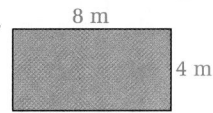
8 m
4 m
6 m
area = ?

74.
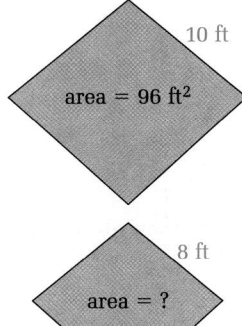
10 ft
area = 96 ft²
8 ft
area = ?

75.
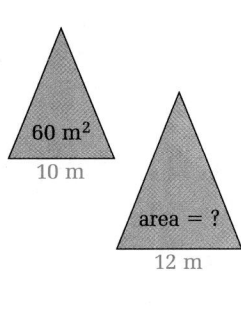
60 m²
10 m
area = ?
12 m

Extra Practice

637

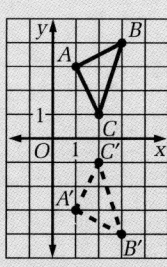

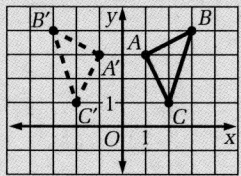

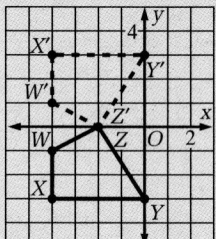

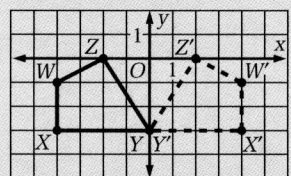

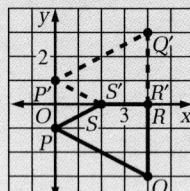

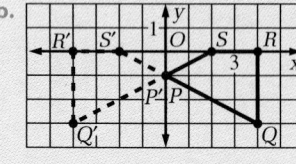

Find the ratio of the volumes of each pair of similar space figures. Then find the unknown volume. 9-8

76. Two rectangular prisms with heights 9 cm and 12 cm, if the volume of the smaller one is 180 cm^3

77. Two cylinders with diameters 12 m and 8 m, if the volume of the larger one is 180 πm^3 (Give the volume of the smaller cylinder in terms of π.)

Unit 10

Copy each figure and reflect it over (a) the *x*-axis and then (b) the *y*-axis. Describe the orientation of the original figure and its image. 10-1

1.

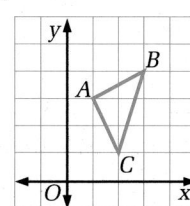

2.

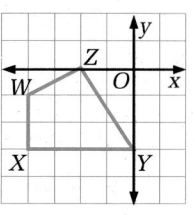

3.

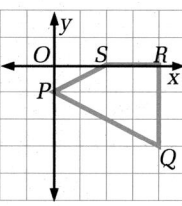

Match each function with its graph. 10-2

4. $y = -(x + 2)^2$ **5.** $y = (x - 2)^2$ **6.** $y = -x^2 + 2$ **7.** $y = x^2 - 2$

A.

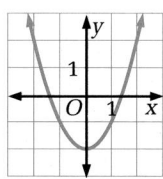

B.

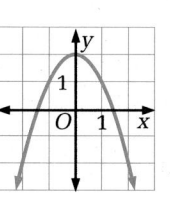

C.

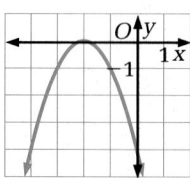

D.

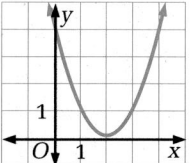

Graph each parabola. 10-3

8. $y = -x(x + 4)$ **9.** $y = (x + 4)(x - 2)$ **10.** $y = (2x - 6)(x - 1)$

Without graphing, find the *x*-intercepts and the *y*-intercept of the graph of each equation. 10-3

11. $y = x(x - 4)$ **12.** $y = -x(x + 2)$ **13.** $y = (x + 8)(x - 3)$

14. $y = (2x - 1)(x + 7)$ **15.** $y = (3x - 2)(4x + 1)$

638 Student Resources

Answers to Extra Practice Unit 10

1. a.

$\triangle ABC$ has clockwise orientation, and $\triangle A'B'C'$ has counterclockwise orientation.

b.

$\triangle ABC$ has clockwise orientation, and $\triangle A'B'C'$ has counterclockwise orientation.

2. a.

Quadrilateral *WXYZ* has counterclockwise orientation, and quadrilateral *W′X′Y′Z′* has clockwise orientation.

b.

Quadrilateral *WXYZ* has counterclockwise orientation, and quadrilateral *W′X′Y′Z′* has clockwise orientation.

3. a.

Quadrilateral *PQRS* has counterclockwise orientation, and quadrilateral *P′Q′R′S′* has clockwise orientation.

b.

Quadrilateral *PQRS* has counterclockwise orientation, and quadrilateral *P′Q′R′S′* has clockwise orientation.

4. C
5. D
6. B
7. A
8.

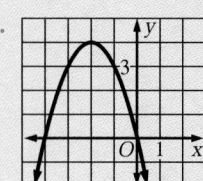

9.

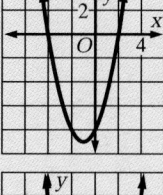

10.

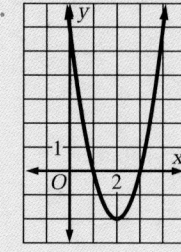

11. 0 and 4; 0
12. 0 and –2; 0
13. –8 and 3; –24
14. $\frac{1}{2}$ and –7; –7
15. $\frac{2}{3}$ and $-\frac{1}{4}$; –2
16. x^7
17. $6b^6$
18. $-6m^8$
19. $35x^4y^3$
20. $-2a^{12}b^5$
21. t^{12}
22. j^{14}
23. $3125a^{15}$
24. $16a^4$
25. $-64a^3b^9$
26. $81x^8y^{16}$
27. $64r^{30}s^{12}$

Simplify. `10-4`

16. $x^6 \cdot x$

17. $2b^2 \cdot 3b^4$

18. $(-3m^3)(2m^5)$

19. $5xy^2 \cdot 7x^3y$

20. $-a^5b^2 \cdot 2a^7 b^3$

21. $(t^4)^3$

22. $(-j^7)^2$

23. $(5a^3)^5$

24. $(-2a)^4$

25. $(-4ab^3)^3$

26. $(3x^2y^4)^4$

27. $(-2r^5 s^2)^6$

Expand each expression. `10-5`

28. $x(x - 6)$

29. $-a(a + 4)$

30. $2x(x - 2)$

31. $3x(2x - 5)$

32. $-4x(1 - 3x)$

33. $7x(4x + 9)$

Factor the binomial side of the equation. Then find the x-intercepts and the y-intercept of the graph of each equation. `10-5`

34. $y = x^2 + 8x$

35. $y = 3x - x^2$

36. $y = -x^2 - 7x$

37. $x^2 - 4x = y$

38. $-x^2 + 9x = y$

39. $y = 2x^2 - 6x$

Expand each product. `10-6`

40. $(x - 5)(x - 2)$

41. $(x - 5)(x + 2)$

42. $(x + 5)(x - 2)$

43. $(x - 6)(x - 6)$

44. $(x - 6)(x + 6)$

45. $(x + 6)(x + 6)$

46. $(3x + 1)(x - 2)$

47. $(4x - 3)(x - 4)$

48. $(5x - 2)(x + 2)$

49. $(2x + 7)(5x + 4)$

50. $(8x + 3)(3x - 2)$

51. $(7x - 1)(3x + 4)$

Without graphing, find (a) the equation of the line of symmetry, (b) the coordinates of the vertex, and (c) the y-intercept of the graph of each equation. `10-6`

52. $y = x^2 - 2x - 3$

53. $y = x^2 + 2x + 1$

54. $y = x^2 - 2x + 3$

55. $y = x^2 - 10x - 5$

Factor each trinomial that can be factored using integers. If a trinomial cannot be factored, write _not possible_. `10-7`

56. $x^2 - 11x + 28$

57. $x^2 + 12x - 28$

58. $x^2 + 27x - 28$

59. $x^2 - 13x + 30$

60. $x^2 - 17x - 30$

61. $x^2 - x - 30$

Use the line of symmetry, the vertex, and the intercepts to sketch the graph of each equation. `10-7`

62. $y = x^2 - 6x + 9$

63. $y = x^2 - 6x + 5$

64. $y = x^2 + 3x - 10$

65. $y = x^2 - 2x - 3$

66. $y = x^2 + 8x + 15$

67. $y = x^2 - x - 6$

Solve each quadratic equation. Tell which method you used. `10-8`

68. $0 = 0.5x^2 - 8$

69. $x^2 - 14x = -45$

70. $7x - x^2 = 3.75$

71. $0 = x^2 - 6x + 8$

72. $x^2 - 3x + 1 = 0$

73. $1 = 2x^2 - 4x - 5$

74. $0.2x^2 - 1.5x = 4$

75. $3x^2 - x = 14$

76. $5 = 2x^2 - 6x + 3$

77. $2x^2 - x = 1$

Extra Practice

Answers to Extra Practice Unit 10 ·······················

28. $x^2 - 6x$ **29.** $-a^2 - 4a$

30. $2x^2 - 4x$ **31.** $6x^2 - 15x$

32. $-4x + 12x^2$

33. $28x^2 + 63x$

34. $x(x + 8)$; 0 and -8; 0

35. $x(3 - x)$; 0 and 3; 0

36. $-x(x + 7)$; 0 and -7; 0

37. $x(x - 4)$; 0 and 4; 0

38. $x(-x + 9)$ or $-x(x - 9)$; 0 and 9; 0

39. $2x(x - 3)$; 0 and 3; 0

40. $x^2 - 7x + 10$

41. $x^2 - 3x - 10$

42. $x^2 + 3x - 10$

43. $x^2 - 12x + 36$

44. $x^2 - 36$

45. $x^2 + 12x + 36$

46. $3x^2 - 5x - 2$

47. $4x^2 - 19x + 12$

48. $5x^2 + 8x - 4$

49. $10x^2 + 43x + 28$

50. $24x^2 - 7x - 6$

51. $21x^2 + 25x - 4$

52. a. $x = 1$ b. $(1, -4)$ c. -3

53. a. $x = -1$ b. $(-1, 0)$ c. 1

54. a. $x = 1$ b. $(1, 2)$ c. 3

55. a. $x = 5$ b. $(5, -30)$ c. -5

56. $(x - 7)(x - 4)$

57. $(x + 14)(x - 2)$

58. $(x + 28)(x - 1)$

59. $(x - 10)(x - 3)$

60. not possible

61. $(x - 6)(x + 5)$

62.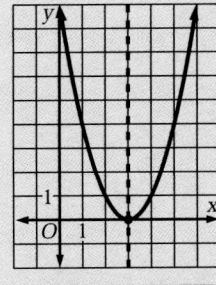

63.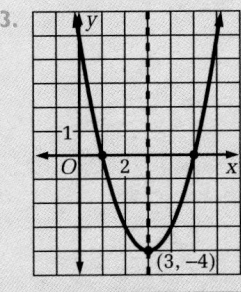

(3, -4)

64.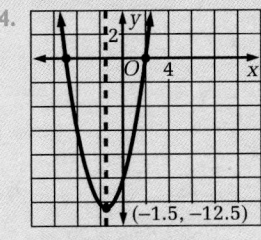

($-1.5, -12.5$)

65.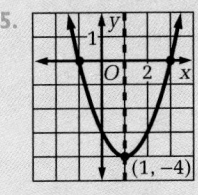

(1, -4)

66.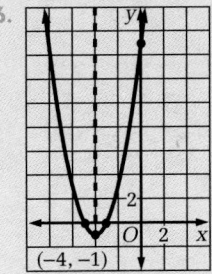

($-4, -1$)

67.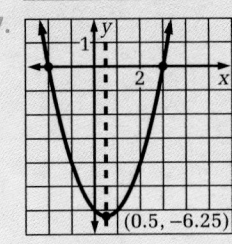

(0.5, -6.25)

68–77. Methods and estimates may vary. Answers are given.

68. 4 and -4 **69.** 9 and 5

70. about 6.4 and 0.6

71. 2 and 4

72. about 2.6 and 0.4

73. 3 and -1

74. about 9.6 and -2.1

75. -2 and $\frac{7}{3}$

76. about 3.3 and -0.3

77. 1 and $-\frac{1}{2}$

Toolbox

➤Numbers and Operations

Integers

| Skill 1 | Using Properties of Addition and Multiplication |

➤ **Commutative Property** You can add or multiply numbers in any order.

$$7 + 5 = 5 + 7 \qquad 7(5) = 5(7)$$

➤ **Associative Property** When you add or multiply three or more numbers, you can group the numbers as you like, without changing the sum or product.

$$(8 + 3) + 2 = 8 + (3 + 2) \qquad 8 \times (3 \times 2) = (8 \times 3) \times 2$$

➤ **Identity Property of Addition** Adding 0 to a number does not change the number.

$$46 + 0 = 46$$

➤ **Identity Property of Multiplication** Multiplying a number by 1 does not change the number.

$$46 \times 1 = 46$$

You can use these properties to help you compute in your head. Just change the order of the numbers and regroup.

Example $27 + 18 + 13 = (27 + 13) + 18$ ◄——— Reorder and regroup.
$$= \quad 40 \quad + 18 \quad ◄——— \text{Add within parentheses first.}$$
$$= 58$$

Example $50 \times 7 \times 2 \quad = (50 \times 2) \times 7$ ◄——— Reorder and regroup.
$$= \quad 100 \quad \times 7 \quad ◄——— \text{Multiply within parentheses first.}$$
$$= 700$$

Use the properties of addition and multiplication to find each answer mentally.

1. $36 + 15 + 25$ **2.** $27 + 0 + 9$ **3.** $12 + 19 + 18$

4. $7 \times 1 \times 7$ **5.** $25 \times 3 \times 4$ **6.** $9 \times 2 \times 5$

7. $17 + 9 + 21 + 13$ **8.** $34 + 0 + 16 + 51$ **9.** $2 \times 1 \times 50 \times 11$

10. $9 \times 5 \times 7 \times 20$ **11.** $11 + 15 + 29$ **12.** $4 \times 6 \times 50$

640 **Student Resources**

Answers to Skill 1

1. 76
2. 36
3. 49
4. 49
5. 300
6. 90

7. 60
8. 101
9. 1100
10. 6300
11. 55
12. 1200

Skill 2 Comparing Integers

The **integers** are the numbers

..., −3, −2, −1, 0, 1, 2, 3, ...

3 dots mean *and so on.*

To compare two integers on a *number line,* remember that the *greater* number is to the *right* of the lesser number on a horizontal number line.

$-3 < 3$	$3 > -3$
−3 is less than 3.	3 is greater than −3.

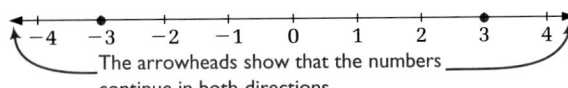

−4 −3 −2 −1 0 1 2 3 4
The arrowheads show that the numbers continue in both directions.

Example To write the numbers −3, 3, and −1 in order from least to greatest, first locate them on a number line.

−4 −3 −2 −1 0 1 2 3 4

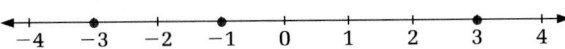

Look at the number line above. Write the numbers in order as they appear from left to right: −3, −1, 3. Using symbols, write the *inequality*

$-3 < -1 < 3$.

You read this inequality as "−1 is between −3 and 3."

Show each pair of integers on a number line. Then compare the integers using > or <.

1. 7, 10 **2.** −4, 0 **3.** 5, −7 **4.** −5, −7

Write each set of numbers in order from least to greatest. Then complete the inequality ? < ? < ?.

5. 0, 2, −1 **6.** −3, 0, −5 **7.** 3, −7, 1 **8.** −2, −4, −6

9. 12, −9, −19 **10.** −17, −27, −7 **11.** 56, −100, 0 **12.** −1, −10, 10, 1

Answers to Skill 2

1.

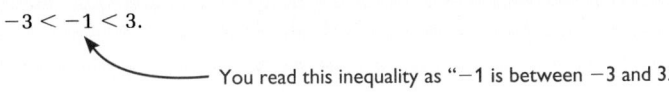

5 6 7 8 9 10 11 12

$7 < 10$

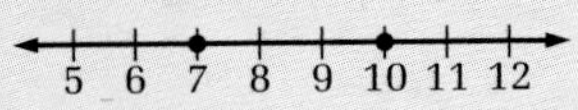

2.

−6 −5 −4 −3 −2 −1 0 1 2

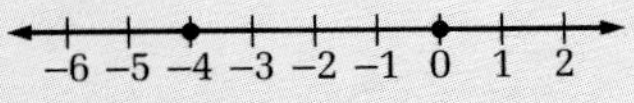

$-4 < 0$

3.

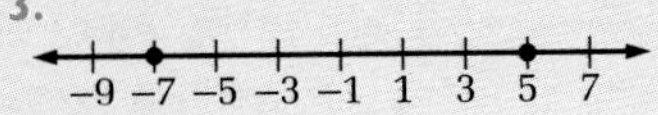

−9 −7 −5 −3 −1 1 3 5 7

$5 > -7$

4.

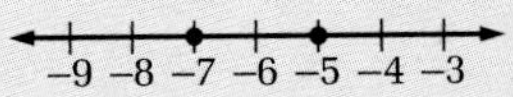

−9 −8 −7 −6 −5 −4 −3

$-5 > -7$

5. −1, 0, 2; $-1 < 0 < 2$
6. −5, −3, 0; $-5 < -3 < 0$
7. −7, 1, 3; $-7 < 1 < 3$
8. −6, −4, −2; $-6 < -4 < -2$
9. −19, −9, 12; $-19 < -9 < 12$
10. −27, −17, −7; $-27 < -17 < -7$
11. −100, 0, 56; $-100 < 0 < 56$
12. −10, −1, 1, 10; $-10 < -1 < 1 < 10$

Think of positive integers as arrows pointing to the right and negative integers as arrows pointing to the left.

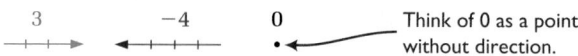

Think of 0 as a point without direction.

Begin each sum at zero on a number line.

Examples   3 + 2

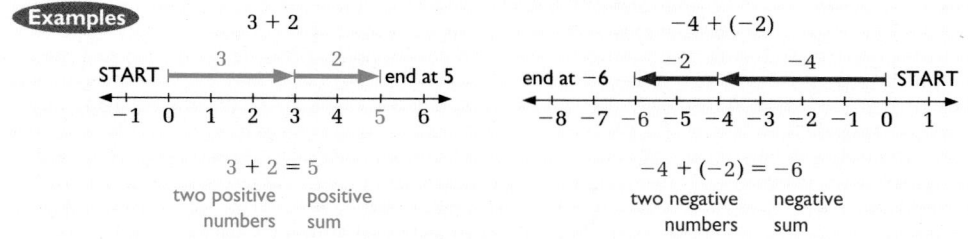

3 + 2 = 5

two positive   positive
numbers   sum

−4 + (−2)

−4 + (−2) = −6

two negative   negative
numbers   sum

The sum of a positive and a negative integer can be positive, negative, or zero. The direction of the longer arrow gives the sign of the sum.

Examples   3 + (−2)   2 + (−4)   −2 + 2

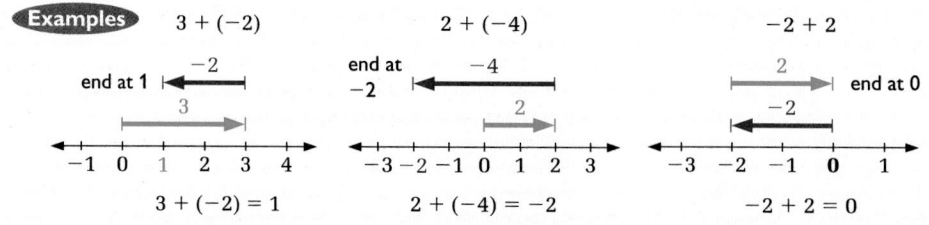

3 + (−2) = 1   2 + (−4) = −2   −2 + 2 = 0

Arrow for 3 is longer; sum is positive.   Arrow for −4 is longer; sum is negative.   Arrow lengths are equal; sum is 0.

When the integers to be added have *opposite* signs, subtract and choose the appropriate sign if the result is not 0.

Find each sum. Think of a number line.

1. 7 + 2   **2.** −7 + (−2)   **3.** 7 + (−2)   **4.** −7 + 2

5. 3 + (−3)   **6.** 0 + (−8)   **7.** −4 + 9   **8.** 3 + (−9)

9. −8 + (−10)   **10.** 6 + (−5)   **11.** −5 + 5   **12.** −5 + (−5)

13. −50 + 40   **14.** 15 + (−6)   **15.** −30 + (−3)   **16.** −50 + 50

Answers to Skill 3 ···

1. 9	9. −18
2. −9	10. 1
3. 5	11. 0
4. −5	12. −10
5. 0	13. −10
6. −8	14. 9
7. 5	15. −33
8. −6	16. 0

You can use integer chips to model integers.

+1, or 1   −1   0   −2   0 + 2, or 2

Since 1 + (−1) = 0, the plus and minus chips form a *zero pair*, with the value 0.

Examples To add with integer chips, you model the first number. Add chips for the second number. Count the chips.

3 + 2 = 5   −3 + (−2) = −5

Examples Find each difference.   **a.** −2 − (−1)   **b.** −2 − 1   **c.** −2 − (−5)

Method ❶ Use integer chips.

a.  Subtract 1 negative chip.

1 negative chip remains.

−2 − (−1) = −1

b. Add a zero pair to give you a positive chip to subtract.

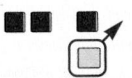 Subtract 1 positive chip. 3 negative chips remain.

−2 − 1 = −3

c. Add zero pairs until you have 5 negative chips to subtract.

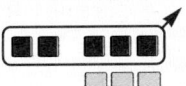 Subtract 5 negative chips. 3 positive chips remain.

−2 − (−5) = 3

Method ❷ To *subtract* an integer, *add its opposite.*

a. −2 − (−1) = −2 + 1   **b.** −2 − 1 = −2 + (−1)   **c.** −2 − (−5) = −2 + 5
= −1   = −3   = 3

Subtract using integer chips. Then subtract by adding an opposite.

1. 7 − 5   **2.** 3 − 4   **3.** 6 − (−2)

Subtract.

4. −3 − 2   **5.** −2 − (−3)   **6.** 0 − 6   **7.** 0 − (−1)   **8.** 5 − (−5)
9. 9 − 0   **10.** −4 − 4   **11.** −4 − (−4)   **12.** 25 − 35   **13.** 10 − (−30)

Toolbox   643

Answers to Skill 4

1. 2   8. 10
2. −1   9. 9
3. 8   10. −8
4. −5   11. 0
5. 1   12. −10
6. −6   13. 40
7. 1

Example Multiply $2 \times (-4)$.

Method ❶ Two groups of -4 is the same as 1 group of -8.

 $2 \times (-4) = -8$

Method ❷ Multiplying by a negative number changes the sign.

$$2 \times (-4) = 2 \times (-1 \times 4)$$
$$= -1 \times (2 \times 4) \longleftarrow \text{Reorder and regroup}$$
$$= -1 \times 8$$
$$= -8$$

When you multiply or divide nonzero integers, be careful to choose the correct sign (positive or negative) for the product or quotient. (See the table.)

× or ÷	+	−
+	+	−
−	−	+

Examples Multiply or divide.

a. $-7 \times (-4) = 28$ **b.** $\dfrac{-15}{3} = -5$ **c.** $-18 \div (-6) = 3$

Remember: When you multiply any number by 0, the result is 0. When you divide 0 by a nonzero number, the result is 0, but you cannot divide by 0.

Examples Multiply or divide.

a. $0 \times 9 = 0$ **b.** $0 \times (-12) = 0$ **c.** $0 \div 7 = 0$

Multiply or divide.

14. $8 \times (-1)$ **15.** $-8 \times (-1)$ **16.** -8×0

17. $20 \div (-5)$ **18.** $-72 \div 8$ **19.** $-27 \div (-9)$

20. -5×5 **21.** 5×5 **22.** $-5 \times (-5)$

23. $18 \div 6$ **24.** $0 \div (-4)$ **25.** $30 \div (-3)$

26. $-1 \div (-1)$ **27.** $6 \times (-8)$ **28.** $-1 \times (-2) \times (-3)$

29. $-3 \times 2 \times (-5)$ **30.** $9 \times (-4) \div 2$ **31.** $(6 \div 2) \times (-7)$

Student Resources

Answers to Skill 4

14. -8 23. 3

15. 8   24. 0

16. 0   25. -10

17. -4 26. 1

18. -9 27. -48

19. 3   28. -6

20. -25 29. 30

21. 25   30. -18

22. 25   31. -21

One-Step Equations

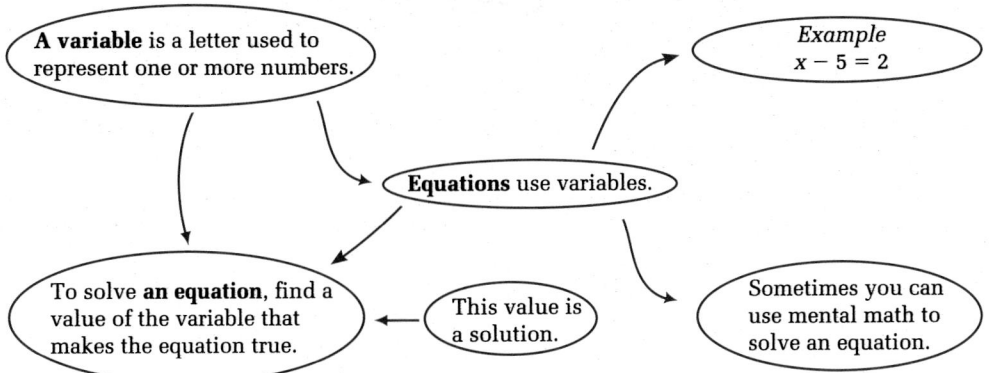

Skill 5 Solving One-Step Equations with Mental Math

A **variable** is a letter used to represent one or more numbers.

Equations use variables.

Example
$x - 5 = 2$

To solve **an equation**, find a value of the variable that makes the equation true.

This value is a solution.

Sometimes you can use mental math to solve an equation.

To use mental math to solve an equation, think about what the equation says.

Example Solve each equation using mental math.

a. $3x = -12$

3 times a number is -12.

$3(-4) = -12$, so $x = -4$.

b. $n - 5 = 2$

When 5 is subtracted from a number, the result is 2.

$7 - 5 = 2$, so $n = 7$.

Is the given number a solution of the equation? Write Yes or No.

1. $9 - x = 2$; 7
2. $y + 5 = 8$; -13
3. $-5n = 45$; -9

4. $z - 2 = 12$; -10
5. $6m = 3$; 18
6. $6 + x = 15$; 9

Solve each equation using mental math.

7. $x + 2 = 8$
8. $y - 4 = 0$
9. $-4z = 20$

10. $\frac{n}{5} = 2$
11. $-5 + t = -1$
12. $5 - a = 6$

13. $-k = 9$
14. $d \div 6 = -4$
15. $b + 7 = 3$

16. $m - 6 = -1$
17. $-3x = -3$
18. $\frac{y}{-4} = -8$

19. $8r = 56$
20. $8 + r = 56$
21. $m - 2 = -4$

22. $\frac{m}{3} = -4$
23. $m + 2 = -4$
24. $2m = -4$

Toolbox

645

Answers to Skill 5

1. Yes.
2. No.
3. Yes.
4. No.
5. No.
6. Yes.
7. 6
8. 4

9. -5
10. 10
11. 4
12. -1
13. -9
14. -24
15. -4
16. 5

17. 1
18. 32
19. 7
20. 48
21. -2
22. -12
23. -6
24. -2

You can use algebra tiles to model and solve equations. Make the same changes to both sides of the equation until there is one x-tile ( ) left on one side.

Example　Solve $x + (-4) = -1$.

First, model the equation.

$$x + (-4) = -1$$

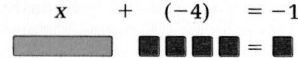

Then add 4 "plus" chips to *both* sides.

$$x + (-4) + 4 = -1 + 4$$

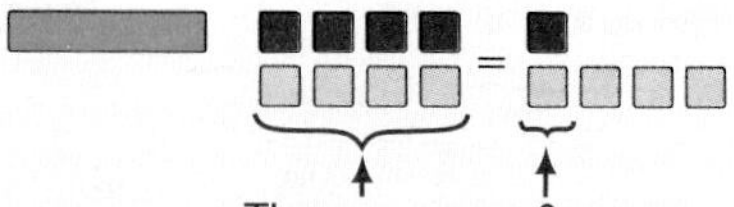

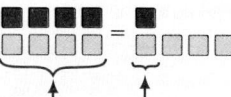

$$x + 0 = 3$$

These represent 0.

$$x = 3$$

Notice: Since $x + (-4) = x - 4$, you can model and solve

$$x - 4 = -1$$

the same way as

$$x + (-4) = -1.$$

Use algebra tiles or sketch the model.

1. a. What equation does the model represent?

b. Solve the equation you wrote in part (a) by making the same change to both sides of the model.

Solve.

2. $x - 3 = 2$　　　　　　　　**3.** $1 + x = 5$

4. $-2 + x = -5$　　　　　　**5.** $x - 1 = 1$

6. $x + 6 = 2$　　　　　　　**7.** $-3 + x = -3$

646　　　　　　　**Student Resources**

Answers to Skill 6

1. a. $x + 2 = -1$

   b. $x = -3$

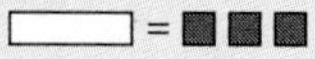

2. 5

3. 4

4. −3

5. 2

6. −4

7. 0

Example Solve $3x = -6$.

Method ❶ Use a model.

To find what x equals (modeled by one x-tile), divide each side into three identical groups.

$3x = -6$

Each x-tile represents -2, so $x = -2$.

Method ❷ Use algebra to solve $3x = -6$.

$$3x = -6$$

$$\frac{3x}{3} = \frac{-6}{3} \quad \longleftarrow \text{ Divide both sides by 3.}$$

$$x = -2$$

Use algebra tiles or sketch the model.

8. a. What equation does the model below represent?

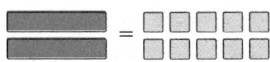

   b. Use two methods to solve the equation. Include a sketch of what you did with the tiles to solve the equation.

Solve.

9. $5x = 20$ **10.** $2x = -8$

11. $-3x = 18$ **12.** $-4x = -16$

13. $7x = 42$ **14.** $6x = -6$

15. $\frac{x}{2} = 4$

   (*Hint:* Multiply both sides by 2.)

Toolbox

647

Answers to Skill 6

8. **a.** $2x = 10$

   b. (1) [model: $x = 5$]

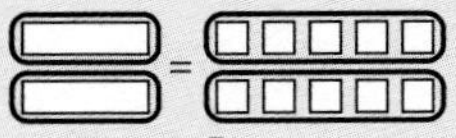

   $x = 5$

   (2) $2x = 10$

   $\frac{2x}{2} = \frac{10}{2}$

   $x = 5$

9. 4

10. -4

11. -6

12. 4

13. 6

14. -1

15. 8

Factors

Since $7 \cdot 5 = 35$, 7 and 5 are **factors** of 35.

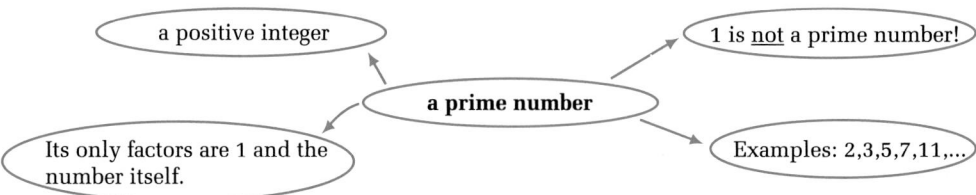

To find the **prime factorization** of a whole number, write it as a product of prime numbers.

Example Write the prime factorization of 135.

Method ❶ See if you can identify a factor at sight. Here, since 135 has 5 as the ones' digit, 135 is divisible by 5.

$$135 = 5 \cdot 27$$
$$\underset{3 \cdot 3 \cdot 3}{\underbrace{\hphantom{27}}}$$

$135 = 3 \cdot 3 \cdot 3 \cdot 5$ ◀——— Write the primes in increasing order.

Method ❷ Test the prime numbers in order as divisors to see which are factors.
Remember: A prime number can be a factor more than once.

135 ÷ 2 is not a whole number; 2 is *not* a factor.

135 ÷ 3 = 45; make a *factor tree* as shown.

Using the factor tree, $135 = 3 \cdot 3 \cdot 3 \cdot 5$.

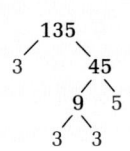

Write the prime factorization of each number that is not a prime number. If the number is prime, write *prime*.

1. 40	**2.** 39	**3.** 41	**4.** 84	**5.** 104
6. 102	**7.** 110	**8.** 125	**9.** 97	**10.** 147
11. 252	**12.** 144	**13.** 133	**14.** 58	**15.** 176

Student Resources

Answers to Skill 7 ···

1. 2 × 2 × 2 × 5
2. 3 × 13
3. prime
4. 2 × 2 × 3 × 7
5. 2 × 2 × 2 × 13
6. 2 × 3 × 17
7. 2 × 5 × 11
8. 5 × 5 × 5

9. prime
10. 3 × 7 × 7
11. 2 × 2 × 3 × 3 × 7
12. 2 × 2 × 2 × 2 × 3 × 3
13. 7 × 19
14. 2 × 29
15. 2 × 2 × 2 × 2 × 11

A **common factor** of two numbers is a number that is a factor of both. For example:

$$3 \cdot 8 = 24 \ \text{ and } \ 3 \cdot 12 = \mathbf{36}$$

3 is a common factor of **24** and **36**.

The **greatest common factor** (**GCF**) of two numbers is the *largest* number that is a common factor of both.

Example Find the GCF of 24 and 36.

First write the prime factorization of each number.

$24 = 2 \cdot 2 \cdot 2 \cdot 3$ ←——— prime factorization

$36 = 2 \cdot 2 \cdot 3 \cdot 3$ ←——— prime factorization

Circle each number that is in *both* lists to find the common factors:

$24 = ②\cdot ②\cdot 2 \cdot ③$

$36 = ②\cdot ②\cdot ③\cdot 3$

$②\cdot ②\cdot ③ = 12$ is the GCF.

A **common multiple** of two numbers is a number that is a multiple of both. For example:

$$8 \cdot 10 = 80 \ \text{and } 20 \cdot 4 = 80$$

80 is a common multiple of 8 and 20.

The **least common multiple** (**LCM**) of two numbers is the *smallest* number that is a common multiple of both.

Example Find the LCM of 8 and 20.

First find the greatest common factor of the two numbers.

$8 = ②\cdot ②\cdot 2$

$20 = ②\cdot ②\cdot 5$

$GCF = ②\cdot ②= 4$

To find the LCM, multiply the GCF by the factors that are not circled:

$LCM = 4 \cdot 2 \cdot 5 = 40$.

Find the GCF and the LCM of each pair of numbers.

1. 6, 9   **2.** 10, 55   **3.** 35, 42   **4.** 30, 45

5. 9, 27   **6.** 18, 48   **7.** 25, 50   **8.** 12, 20

9. 21, 30   **10.** 4, 32   **11.** 22, 55   **12.** 60, 100

Toolbox

Answers to Skill 8

1. 3; 18

2. 5; 110

3. 7; 210

4. 15; 90

5. 9; 27

6. 6; 144

7. 25; 50

8. 4; 60

9. 3; 210

10. 4; 32

11. 11; 110

12. 20; 300

Fractions

Reciprocals are two numbers whose product is 1. Every number except 0 has exactly one reciprocal. Zero has no reciprocal.

$$2 \text{ and } \frac{1}{2} \text{ are reciprocals: } 2 \times \frac{1}{2} = 1$$

$$\frac{3}{4} \text{ and } \frac{4}{3} \text{ are reciprocals: } \frac{3}{4} \times \frac{4}{3} = 1$$

$$-\frac{3}{4} \text{ and } -\frac{4}{3} \text{ are reciprocals: } \left(-\frac{3}{4}\right) \times \left(-\frac{4}{3}\right) = 1$$

Notice that a positive number has a positive reciprocal, and a negative number has a negative reciprocal.

Example Find the reciprocal of $\frac{2}{7}$.

Find the reciprocal of a fraction by exchanging the *numerator* and the *denominator*.

$$\frac{2}{7} \quad \begin{matrix} \leftarrow \text{numerator} \\ \leftarrow \text{denominator} \end{matrix}$$

The reciprocal of $\frac{2}{7}$ is $\frac{7}{2}$.

Example Find the reciprocal of -3.

First write -3 as a fraction whose denominator is 1: $-\frac{3}{1}$.

Then find the reciprocal.

The reciprocal of -3 is $-\frac{1}{3}$.

Example Find the reciprocal of $1\frac{1}{3}$.

First write the mixed number $1\frac{1}{3}$ as a fraction: $\frac{4}{3}$.

Then find the reciprocal.

The reciprocal of $1\frac{1}{3}$ is $\frac{3}{4}$.

Find the reciprocal of each number.

1. $\frac{3}{5}$ 2. $-\frac{7}{4}$ 3. -6 4. 1 5. 6

6. $-2\frac{1}{4}$ 7. $1\frac{5}{6}$ 8. $-\frac{1}{9}$ 9. $-4\frac{2}{3}$ 10. 0.7

650   **Student Resources**

Answers to Skill 9

1. $\frac{5}{3}$ 6. $-\frac{4}{9}$

2. $-\frac{4}{7}$ 7. $\frac{6}{11}$

3. $-\frac{1}{6}$ 8. -9

4. 1 9. $-\frac{3}{14}$

5. $\frac{1}{6}$ 10. $\frac{10}{7}$

To add (or subtract) fractions with the same denominator, you add (or subtract) the numerators.

Example $\frac{7}{8} + \frac{5}{8} = \frac{7+5}{8}$ ←——— Add the numerators.

$$= \frac{12}{8}$$

$$= \frac{3}{2}, \text{ or } 1\frac{1}{2}$$ ←——— Write the answer in simplest form.

Example $\frac{5}{6} - \frac{1}{6} = \frac{5-1}{6}$ ←——— Subtract the numerators.

$$= \frac{4}{6}$$

$$= \frac{2}{3}$$ ←——— Write the answer in simplest form.

To add or subtract fractions with *different* denominators, you rewrite them as equivalent fractions with the same denominator. Use the LCM of the denominators as the new denominator. Then add or subtract.

Example $\overset{\times \frac{2}{2}}{\frac{11}{12} + \frac{1}{18}} = \frac{33}{36} + \frac{2}{36}$ ←——— 36 is the LCM of 12 and 18.

$\times \frac{3}{3}$

$$= \frac{33 + 2}{36}$$ ←——— Add the numerators.

$$= \frac{35}{36}$$

Add or subtract. Write the answer in lowest terms.

1. $\frac{11}{18} + \frac{3}{18}$ 2. $\frac{7}{12} - \frac{1}{12}$ 3. $\frac{8}{9} - \frac{1}{9}$ 4. $\frac{3}{4} + \frac{1}{4}$

5. $\frac{5}{6} + \frac{13}{18}$ 6. $\frac{1}{6} + \frac{3}{10}$ 7. $\frac{1}{3} - \frac{1}{12}$ 8. $\frac{5}{6} - \frac{4}{9}$

9. $-\frac{3}{5} + \frac{1}{3}$ 10. $-\frac{2}{7} - \frac{3}{8}$ 11. $\frac{5}{18} - \frac{11}{22}$ 12. $\frac{47}{50} + \left(-\frac{3}{10}\right)$

13. $-\frac{2}{3} + \left(-\frac{1}{9}\right)$ 14. $-\frac{2}{11} - \left(-\frac{3}{11}\right)$ 15. $\frac{7}{8} - \left(-\frac{1}{6}\right)$ 16. $\frac{3}{5} + \frac{5}{7}$

Toolbox

651

Answers to Skill 10 ···

1. $\frac{7}{9}$ 7. $\frac{1}{4}$ 12. $\frac{16}{25}$

2. $\frac{1}{2}$ 8. $\frac{7}{18}$ 13. $-\frac{7}{9}$

3. $\frac{7}{9}$ 9. $-\frac{4}{15}$ 14. $\frac{1}{11}$

4. 1   10. $-\frac{37}{56}$ 15. $1\frac{1}{24}$

5. $1\frac{5}{9}$ 11. $-\frac{2}{9}$ 16. $1\frac{11}{35}$

6. $\frac{7}{15}$

To multiply two fractions, you multiply the numerators and the denominators. Remember to simplify before you multiply if possible.

Example

$$7 \times \frac{3}{4} = \frac{7}{1} \times \frac{3}{4}$$ ◀—— Write 7 as $\frac{7}{1}$.

$$= \frac{7 \times 3}{1 \times 4}$$ ◀—— Multiply numerators.
◀—— Multiply denominators.

$$= \frac{21}{4}, \text{ or } 5\frac{1}{4}$$

Example

$$\frac{5}{6} \times \frac{9}{10} = \frac{5 \times 9}{6 \times 10}$$ ◀—— Multiply numerators.
◀—— Multiply denominators.

$$= \frac{\overset{1}{\cancel{5}} \times \overset{3}{\cancel{9}}}{\underset{2}{\cancel{6}} \times \underset{2}{\cancel{10}}}$$ ◀—— Simplify.

$$= \frac{3}{4}$$

To divide by a number, multiply by its reciprocal.

Example

$$\frac{3}{4} \div \frac{2}{3} = \frac{3}{4} \times \frac{3}{2}$$ ◀—— The reciprocal of $\frac{2}{3}$ is $\frac{3}{2}$.

$$= \frac{3 \times 3}{4 \times 2}$$

$$= \frac{9}{8}, \text{ or } 1\frac{1}{8}$$

Example

$$\frac{4}{9} \div 6 = \frac{4}{9} \times \frac{1}{6}$$ ◀—— The reciprocal of 6 is $\frac{1}{6}$.

$$= \frac{4 \times 1}{9 \times 6}$$

$$= \frac{2}{27}$$

Multiply or divide. Write the answer in lowest terms.

17. $\frac{3}{4} \times \frac{2}{7}$ 18. $\frac{2}{3}\left(-\frac{3}{5}\right)$ 19. $\left(-\frac{1}{3}\right)\left(-\frac{5}{6}\right)$ 20. $\frac{4}{15} \times \frac{9}{20}$

21. $\frac{3}{5} \div 9$ 22. $4 \div \frac{1}{2}$ 23. $-\frac{15}{16} \div \frac{3}{8}$ 24. $-\frac{1}{2} \div \left(-\frac{2}{3}\right)$

25. $\left(-\frac{9}{10}\right)\left(-\frac{3}{5}\right)$ 26. $\frac{7}{12} \times \frac{10}{21}$ 27. $\frac{11}{18}\left(-\frac{27}{44}\right)$ 28. $\frac{5}{7}\left(-\frac{7}{5}\right)$

29. $-\frac{9}{10} \div \left(-\frac{3}{5}\right)$ 30. $\frac{3}{8} \div \frac{5}{6}$ 31. $\frac{5}{9} \div \frac{3}{4}$ 32. $\frac{5}{7} \div \left(-\frac{7}{5}\right)$

Answers to Skill 10 ···

17. $\frac{3}{14}$

18. $-\frac{2}{5}$

19. $\frac{5}{18}$

20. $\frac{3}{25}$

21. $\frac{1}{15}$

22. 8

23. $-2\frac{1}{2}$

24. $\frac{3}{4}$

25. $\frac{27}{50}$

26. $\frac{5}{18}$

27. $-\frac{3}{8}$

28. -1

29. $1\frac{1}{2}$

30. $\frac{9}{20}$

31. $\frac{20}{27}$

32. $-\frac{25}{49}$

Percent

Skill 11   Writing Percents

Percent means *divided by 100.*

Example   To write a percent as a decimal, move the decimal point two places to the *left* and remove the percent symbol.

$$24\% = 0.24 \longleftarrow \text{percent to decimal}$$

$$24\% = \frac{24}{100} = \frac{6}{25} \longleftarrow \text{percent to fraction in lowest terms}$$

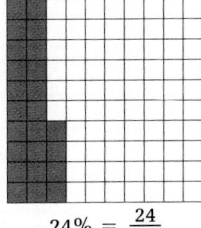

$$24\% = \frac{24}{100}$$

Example   To write a decimal as a percent, move the decimal point two places to the *right* and write the percent symbol.

$$0.75 = 75\% \longleftarrow \text{decimal to percent}$$

$$0.75 = \frac{75}{100} = \frac{3}{4} \longleftarrow \text{decimal to fraction in lowest terms.}$$

To write a fraction as a percent, write the fraction as a decimal and then write the decimal as a percent.

Example $\dfrac{6}{5} = \dfrac{6 \cdot 20}{5 \cdot 20} = \dfrac{120}{100} = 1.20 = 120\%$

Example $\dfrac{2}{3} = 2 \div 3 = 0.6666\ldots = 66.\overline{6}\% \approx 66.7\%$

You should memorize the chart below.

Equivalent Percents, Decimals, and Fractions

$1\% = 0.01 = \frac{1}{100}$	$33\frac{1}{3}\% = 0.\overline{3} = \frac{1}{3}$	$66\frac{2}{3}\% = 0.\overline{6} = \frac{2}{3}$
$10\% = 0.1 = \frac{10}{100}$	$40\% = 0.4 = \frac{2}{5}$	$75\% = 0.75 = \frac{3}{4}$
$20\% = 0.2 = \frac{1}{5}$	$50\% = 0.5 = \frac{1}{2}$	$80\% = 0.8 = \frac{4}{5}$
$25\% = 0.25 = \frac{1}{4}$	$60\% = 0.6 = \frac{3}{5}$	$100\% = 1$

Toolbox

653

Write each percent as a decimal and as a fraction in lowest terms.

1. 70% **2.** 32% **3.** 125% **4.** $\frac{3}{4}$% **5.** 7.5%

Write as a percent and as a fraction in lowest terms.

6. 0.66 **7.** 0.333... **8.** 0.009 **9.** 0.375 **10.** 1.6

Write as a decimal and as a percent.

11. $\frac{18}{25}$ **12.** $\frac{29}{1000}$ **13.** $\frac{72}{125}$ **14.** $\frac{27}{20}$ **15.** $\frac{1}{6}$

Skill 12 The Percent One Number is of Another

To find the percent one number is of another, begin by writing an equation. Then solve and write your answer as a percent.

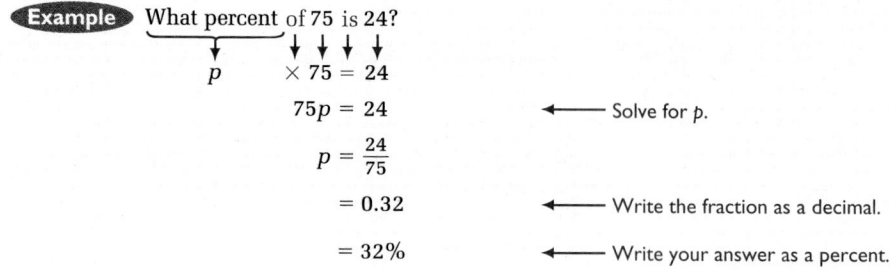

Example What percent of 75 is 24?

$$p \times 75 = 24$$

$$75p = 24 \qquad \longleftarrow \text{Solve for } p.$$

$$p = \frac{24}{75}$$

$$= 0.32 \qquad \longleftarrow \text{Write the fraction as a decimal.}$$

$$= 32\% \qquad \longleftarrow \text{Write your answer as a percent.}$$

Remember: The word *of* implies *multiply*.

Find each answer.

1. What percent of 60 is 24?

2. What percent of 24 is 60?

3. What percent of 125 is 5?

4. What percent of 96 is 36?

5. 6 is what percent of 80?

6. 20 is what percent of 16?

7. 7.5 is what percent of 25?

8. 13.5 is what percent of 72?

9. 2 is what percent of 400?

10. What percent of 57 is 19?

Answers to Skill 11

1. 0.7; $\frac{7}{10}$ **2.** 0.32; $\frac{8}{25}$

3. 1.25; $\frac{5}{4}$ **4.** 0.0075; $\frac{3}{400}$

5. 0.075; $\frac{3}{40}$ **6.** 66%; $\frac{33}{50}$

7. $33\frac{1}{3}$%; $\frac{1}{3}$ **8.** 0.9%; $\frac{9}{1000}$

9. 37.5%; $\frac{3}{8}$ **10.** 160%; $\frac{8}{5}$

11. 0.72; 72%

12. 0.029; 2.9%

13. 0.576; 57.6%

14. 1.35; 135%

15. $0.1\overline{6}$; $16.\overline{6}$%, or $16\frac{2}{3}$%

Answers to Skill 12

1. 40% **2.** 250%

3. 4% **4.** 37.5%

5. 7.5% **6.** 125%

7. 30% **8.** 18.75%

9. 0.5%

10. $33\frac{1}{3}$%

Skill 13 Percent of Increase or Decrease

To find a percent of increase or decrease, use these equations:

$$\text{percent of increase} = \frac{\text{amount of increase}}{\text{original amount}}$$

$$\text{percent of decrease} = \frac{\text{amount of decrease}}{\text{original amount}}$$

Example A price increased from \$120 to \$135. Find the percent of increase.

$$\text{percent of increase} = \frac{\text{amount of increase}}{\text{original amount}}$$

$$= \frac{135 - 120}{120}$$ ← Amount of increase equals new amount minus original amount.

$$= \frac{15}{120} = 0.125 = 12.5\%$$

Find the percent of increase or decrease.

1. original price, \$25 new price, \$23

2. original price, \$.50 new price, \$.55

3. original price, \$6.25 new price, \$6.40

4. original price, \$12.50 new price, \$9.25

Skill 14 Finding a Percent of a Number

To find a percent of a number, first rewrite the percent as a decimal or as a fraction. Then multiply by the given number.

Example Find: **a.** 25.6% of 375 **b.** $5\frac{1}{2}\%$ of \$40 **c.** $33\frac{1}{3}\%$ of 60

a. Rewrite 25.6% as a decimal: 0.256

Multiply by the given number: $0.256(375) = 96$

25.6% of 375 is 96.

b. Rewrite $5\frac{1}{2}\%$ as a fraction: $\frac{5.5}{100}$

Multiply by the given number: $\frac{5.5}{100}(40) = \frac{220}{100} = 2.2$

$5\frac{1}{2}\%$ of \$40 is \$2.20

c. $33\frac{1}{3}\%$ of $60 = \frac{1}{3} \cdot 60 = 20$

Answers to Skill 13

1. 8% decrease
2. 10% increase
3. 2.4% increase
4. 26% decrease

Find each number.

1. 6% of 30

2. 75% of 28

3. $66\frac{2}{3}$% of 39

4. 2.5% of 148

5. 105% of 94

6. $9\frac{1}{4}$% of 80

7. 40% of 275

8. 0.4% of 20

9. $62\frac{1}{2}$% of 72

10. A blouse that sells for $26.50 is on sale at a 30% discount. Find the amount of discount and the new price.

11. Estimate 79% of $50.

12. If $\frac{1}{2}$% of 500 manufactured items are defective, is it reasonable to say that about 250 items are defective? Explain.

Skill 15   When a Percent of a Number is Known

To find a number when a percent of it is known, write an equation where n is the number you want to find. Write the percent as a decimal or fraction and solve the equation.

Example 8% of what number is 30?

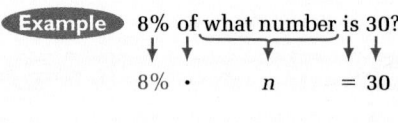

$$8\% \cdot n = 30$$

$$0.08 \cdot n = 30$$

$$n = \frac{30}{0.08} = 375$$

The number is 375.

Example 9 is $33\frac{1}{3}$% of what number?

$$9 = 33\frac{1}{3}\% \cdot n$$

$$9 = \frac{1}{3} \cdot n$$

$$27 = n$$

The number is 27.

---

1. 40% of what number is 32?

2. 32% of what number is 40?

3. 6% of what number is 57?

4. 25% of what number is 162?

5. 14 is $66\frac{2}{3}$% of what number?

6. 9 is 2.4% of what number?

7. 2 is $\frac{1}{4}$% of what number?

8. 12 is 160% of what number?

9. 0.7% of what number is 3.5?

10. 62.5% of what number is 8?

11. 80 is 128% of what number?

12. 128 is 80% of what number?

13. Kim earned a score of 92% on a math final and got 12 questions wrong. How many questions were on the final?

Student Resources

Answers to Skill 14

1. 1.8
2. 21
3. 26
4. 3.7
5. 98.7
6. 7.4
7. 110
8. 0.08
9. 45
10. $7.95; $18.55
11. Answers may vary. An example is given. 79% is almost 80%, so, if 80% of $50 is $40 (0.8($50) = $40),

then 79% of $50 is a little less than $40.

12. Answers may vary. An example is given. No; 250 items are half the items, not $\frac{1}{2}$%; 1% of the items is 5 items, so a $\frac{1}{2}$% defective rate would yield about two or three defective items per 500 items.

Answers to Skill 15

1. 80
2. 125
3. 950
4. 648
5. 21
6. 375
7. 800
8. 7.5
9. 500
10. 12.8
11. 62.5
12. 160
13. 150

➤Measurement

Angles

Skill 16 Using a Protractor

Angle S is formed by two rays (called *sides*) with the same endpoint (called the *vertex*).

side
S
side

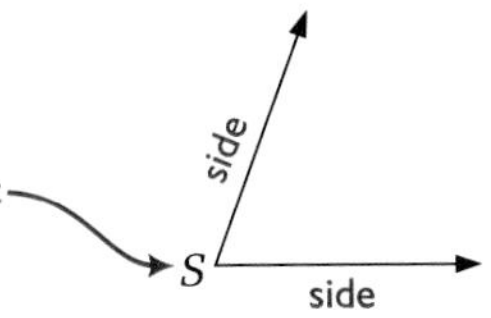

An angle is measured in *degrees*.

Example **Use a protractor to draw an angle that measures 132°.**

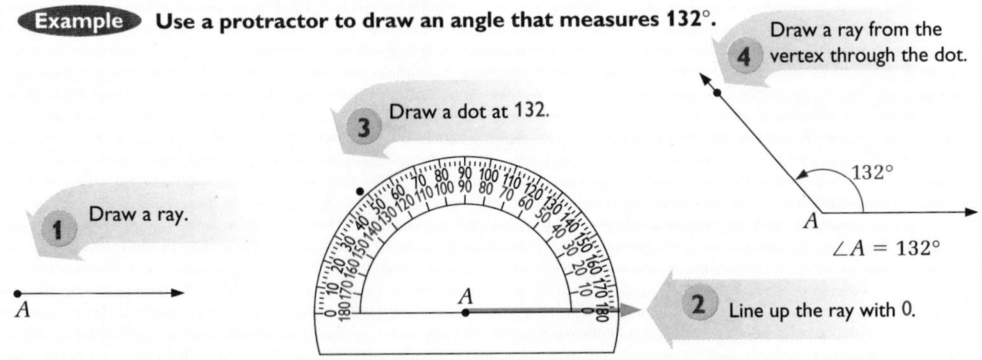

1 Draw a ray.
A
2 Line up the ray with 0.
3 Draw a dot at 132.
A
4 Draw a ray from the vertex through the dot.
132°
A
∠A = 132°

When you measure or draw an angle, be careful to choose the scale on the protractor that has 0 on one side of the angle.

Use a protractor to find the measure of each angle.

1.

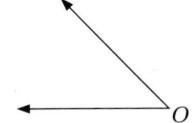

O

2.

E

3.

P

4.

T

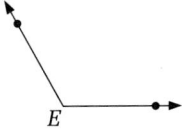

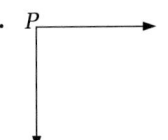

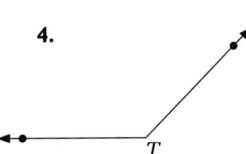

Draw an angle with the specified measure.

5. 60° **6.** 25° **7.** 108° **8.** 180°

9. The sum of the measures of $\angle 1$, $\angle 2$, and $\angle 3$ is 180°.

a. If $\angle 1 = 55°$ and $\angle 3 = 37°$, find the measure of $\angle 2$.

b. If $\angle 1 = 55°$ and $\angle 2 = x°$, write the measure of $\angle 3$ in terms of x.

1
2
3

Answers to Skill 16

1. 45°
2. 120°
3. 90°
4. 135°
5.

6.

7.

8.

9. a. 88°
   b. $(125 - x)°$

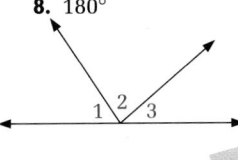

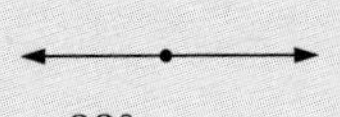

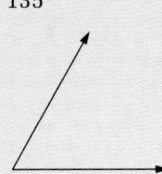

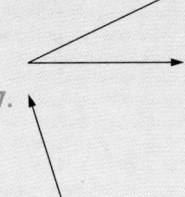

Perimeter

The **perimeter** (*P*) of a figure is the distance around it.

Examples

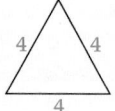

$P = 4 + 4 + 4$
$= 3(4)$

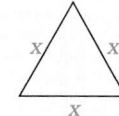

$P = x + x + x$
$= 3x$

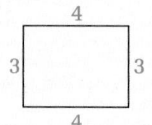

$P = 3 + 4 + 3 + 4$
$= (2 \times 3) + (2 \times 4)$

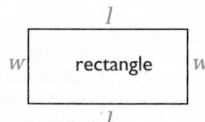

$P = l + w + l + w$
$= 2l + 2w$

Find the perimeter.

1.

2.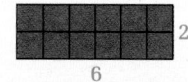

3.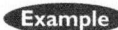

Area

The **area** (*A*) of a figure is the number of square units enclosed.

Examples

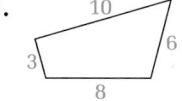

$A = 6 \times 2$
$= 12$ square units

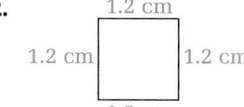

$A = $ length $\times$ width
$= l \times w$

$A = $ side $\times$ side
$= s \times s$

Example   Find the area of a rectangle with length 10 m and width 3.5 m.

$A = l \times w = 10(3.5) = 35$ m² ⟵ meters $\times$ *meters* = square meters (m²)

Answers to Skill 17

1. 27
2. 4.8 cm
3. 2x + 12

Find the area of each figure.

1. a square with sides of length 8 cm

2. a rectangle with length 11 ft and width 9 ft

3. a square with sides of length 1.5 mm

4. a rectangle with length 8 yd and width t yd

5. a rectangle with length 2.4 in. and width 1.2 in.

Volume

Skill 19   Finding Volume

The **volume** (V) of a space figure is the amount of space it encloses. An amount of space is measured in *cubic units*, such as cubic inches (in.^{3}), cubic centimeters (cm^3), and so on.

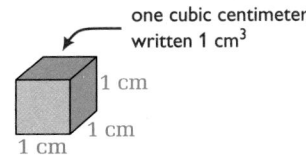
one cubic centimeter, written 1 cm^3

Example The box below has two layers. Each layer encloses $4 \times 3 = 12$ cubic units, so the entire box encloses $12 \times 2 = 24$ cubic units.

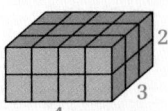
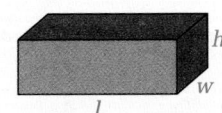

$V = 4 \times 3 \times 2$

$= 24$ cubic units

$V = \text{length} \times \text{width} \times \text{height}$

$= l \times w \times h$

Example Find the volume of a box with length 8.6 mm, width 4 mm, and height 3 mm.

$l \times w \times h = (8.6)(4)(3) = 103.2$ mm^2

Find the volume of each box.

1. a box with length 20 cm, width 15 cm, and height 8 cm

2. a box with length 12 in., width 2.5 in., and height 0.75 in.

3. a box with length 4 ft, width $\frac{1}{2}$ ft, and height 2 ft

4. A cube is a box whose length and width and height are all the same length. Find the volume of each cube.

   a. a cube with sides of length 7 yd

   b. a cube with sides of length 0.5 ft

Toolbox

659

Answers to Skill 18

1. 64 cm^2

2. 99 ft^2

3. 2.25 mm^2

4. $8t$ yd^2

5. 2.88 in.^{2}

Answers to Skill 19

1. 2400 cm^3

2. 22.5 in.^{3}

3. 4 ft^3

4. a. 343 yd^3

   b. 0.125 ft^3

➤ Graphing

A Coordinate System

Skill 20   Locating Points on a Coordinate Plane

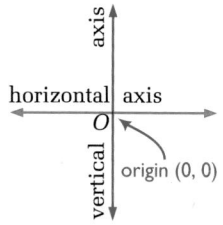

A **coordinate system** is a grid of horizontal and vertical lines.

A horizontal line and a vertical line are chosen as *axes*. The **origin** is the point where the axes cross.

The positive direction is to the right of the origin on the horizontal axis and upward on the vertical axis.

The **coordinates** of a point in a coordinate plane are an **ordered pair** of numbers in the form (**horizontal coordinate**, **vertical coordinate**) or (*x*, *y*).

Example   Point *A* on the graph at the right is 4 units to the left of the origin and 1 unit up in a vertical direction. The coordinates of *A* are $(-4, 1)$.

Example   Point *B* is 4 units to the right of the origin and 0 units in a vertical direction. The coordinates of *B* are $(4, 0)$.

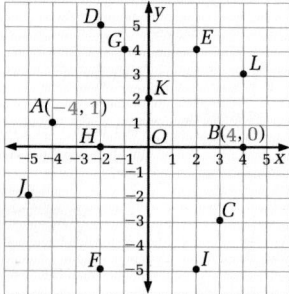

Use the graph. Name the coordinates of each point.

1. *C*   2. *D*   3. *E*   4. *F*   5. *G*

6. *H*   7. *I*   8. *J*   9. *K*   10. *L*

Skill 21   Graphing Points on a Coordinate Plane

To **graph** or **plot** a point, you start at the origin.

Example   Graph $M(-2, -3)$.

The horizontal coordinate -2 tells you to move left 2 units from the origin. The vertical coordinate -3 tells you to move down 3 units from there.

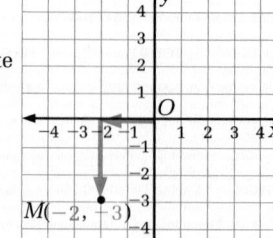

Graph each point on a coordinate plane.

1. $P(3, 2)$ 2. $Q(5, -1)$ 3. $R(-4, 0)$

4. $S(0, 2)$ 5. $T(-5, -3)$ 6. $U(-2, 2)$

Answers to Skill 20

1. $(3, -3)$ 6. $(-2, 0)$

2. $(-2, 5)$ 7. $(2, -5)$

3. $(2, 4)$ 8. $(-5, -2)$

4. $(-2, -5)$ 9. $(0, 2)$

5. $(-1, 4)$ 10. $(4, 3)$

Answers to Skill 21

1–6.

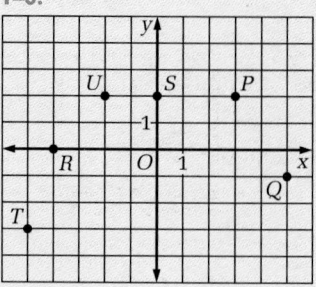

Data Displays

Skill 22 | Making a Bar Graph

Nutritional Information about One Brand of Turkey Meats			
	Percent Fat	Calories per slice	Fat per slice in grams
Turkey bologna	21	70	5
Turkey ham	4	40	2
Turkey salami	12	50	3

A bar graph is a visual display of data that fall into distinct categories, such as the data in this table.

Example You can use a bar graph to display the percent of fat in the turkey meats.

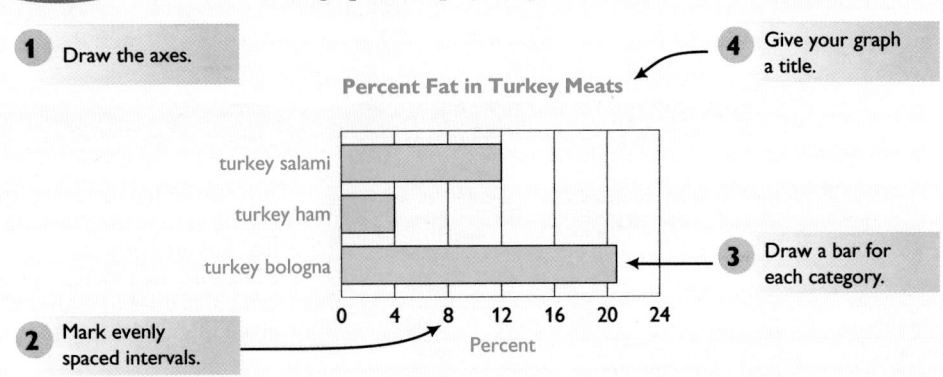

1 Draw the axes.

4 Give your graph a title.

Percent Fat in Turkey Meats

3 Draw a bar for each category.

2 Mark evenly spaced intervals.

Draw a bar graph for each set of data from the table.

1. grams of fat per slice of turkey meats

2. calories per slice of turkey meats

Skill 23 | Making a Line Graph

Average Monthly Temperatures (°F)				
	Jan.	Apr.	Jul.	Oct.
Juneau, Alaska	24.2	39.7	56.0	42.2
Miami, Florida	67.2	75.2	82.6	78.3
Peoria, Illinois	21.6	51.4	75.5	54.0

The table shows the average temperature in each of four months for three cities. You can use a line graph to show how data change over time.

Toolbox

661

Answers to Skill 22

1.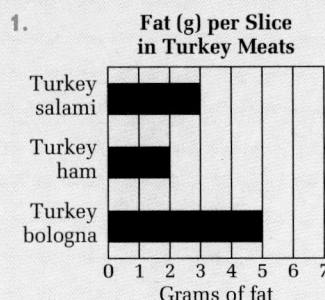

2.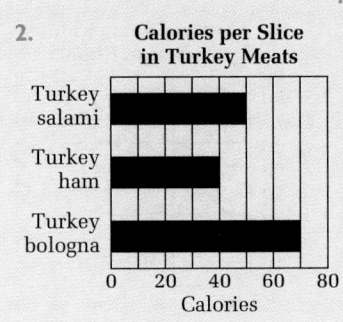

Example Use the temperature table at the bottom of page 661. Display the average monthly temperature in Juneau, Alaska, on a line graph.

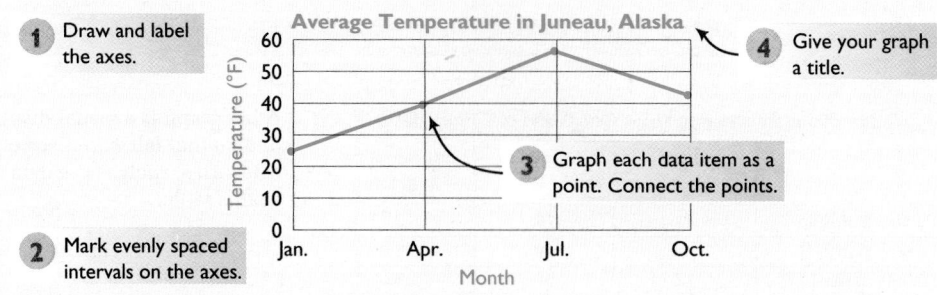

1 Draw and label the axes.

2 Mark evenly spaced intervals on the axes.

3 Graph each data item as a point. Connect the points.

4 Give your graph a title.

Draw a line graph of the average monthly temperature for each city.

1. Miami, Florida

2. Peoria, Illinois

Skill 24 Making a Pictograph

National Conference Leading Punters 1991-1992			
	Number of punts	Longest in yards	Total yards
Newsome	72	84	3243
Barnhardt	67	62	2947
Landeta	53	71	2317
Arnold	60	62	2609

A pictograph and a bar graph can be used to display data that fall into distinct categories. In a pictograph a symbol is used to represent a given number of items.

Example Draw a pictograph to show the number of punts kicked by each player.

Let a football represent 10 punts. Newsome's 72 punts are about 7 symbols.

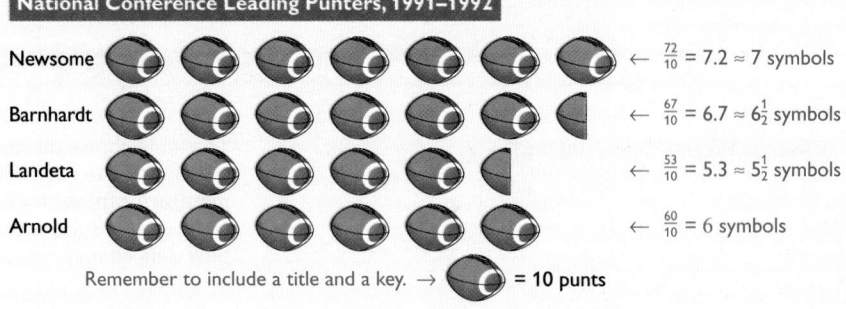

National Conference Leading Punters, 1991–1992

Newsome ← $\frac{72}{10} = 7.2 \approx 7$ symbols

Barnhardt ← $\frac{67}{10} = 6.7 \approx 6\frac{1}{2}$ symbols

Landeta ← $\frac{53}{10} = 5.3 \approx 5\frac{1}{2}$ symbols

Arnold ← $\frac{60}{10} = 6$ symbols

Remember to include a title and a key. → = 10 punts

Student Resources

Answers to Skill 23 ··

1.

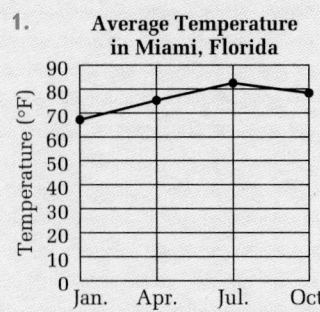

Average Temperature in Miami, Florida

2.

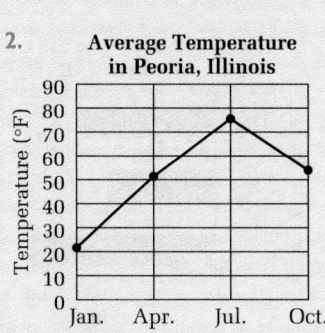

Average Temperature in Peoria, Illinois

Draw a pictograph for each set of data from the table on page 662.

1. the longest punt for each player

2. the total number of yards kicked by each player

Skill 25 **Making a Circle Graph**

A circle graph shows how parts relate to a whole and to each other.

Example Draw a circle graph to show the wins, losses, and ties for the Canadiens.

Standings of Three Teams in the National Hockey League, 1991-1992			
Team	Wins	Losses	Ties
Canadiens	41	28	11
Bruins	36	32	12
Sabres	31	37	12

Follow these steps to figure out how big to make the "wins" sector of the circle. Do the same for the other two sectors.

Write the wins as a fraction of the whole.

$$\frac{41}{80} \cdot 360° = 184.5°$$

Find the total number of games:
$41 + 28 + 11 = 80$

Multiply by 360° to find the number of degrees for the sector.

Follow these steps to draw the circle graph.

1. Draw a circle.

2. Use a protractor to draw the angle for each sector.

3. Label each sector and give your graph a title.

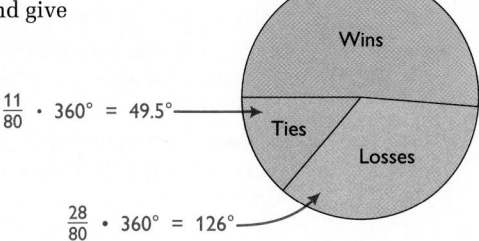

Canadiens' Record 1991–1992

$$\frac{11}{80} \cdot 360° = 49.5°$$

$$\frac{28}{80} \cdot 360° = 126°$$

Use the table above. Draw a circle graph to show the wins, losses, and ties for each team.

1. Bruins

2. Sabres

Toolbox 663

Answers to Skill 24

1.
National Conference
Longest Punts 1991–1992

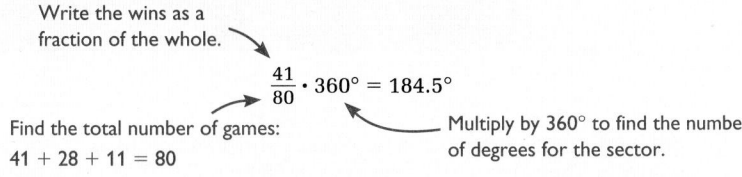

Newsome
Barnhardt
Landeta
Arnold

= 10 yards

2.
National Conference
Punting Yardage 1991–1992

Newsome
Barnhardt
Landeta
Arnold

= 500 yards

Answers to Skill 25

1. Bruins' Record
1991–1992

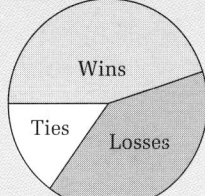

Wins
Ties
Losses

2. Sabres' Record
1991–1992

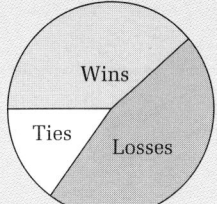

Wins
Ties
Losses

➤Problem Solving

One step toward being a good problem solver is to use an organized plan for approaching a problem. Here is a four-step plan.

Understand	• Read the problem. • Decide what facts are given and what you need to find. • There may be more information than you need or not enough. (The Table of Measures on page 666 can supply a missing fact.)
Plan	• Choose one or more strategies for solving the problem. (Get ideas from the list of strategies on page 665.) • Choose the correct operations for calculations.
Work	• Use the strategies you have chosen. • Estimate before you calculate, if you can. • Do any necessary calculations. (For tips on using a graphics calculator or a spreadsheet, see the Technology Handbook.)
Answer	• Check whether your answer is reasonable. • Read the problem again and give your answer using the words of the problem.

Example Students at the Booster Club Car Wash earned $41 between 7:30 A.M. and 9:30 A.M. They charged $7 per truck and $4 per car. How many of each kind of vehicle did they wash in those two hours?

Understand Facts given: prices charged, $7 per truck and $4 per car

total earned, $41

time period, 2 hours ◀— Unnecessary information

Facts to find: the number of trucks and of cars washed

Plan If you knew the number of trucks, ◀ It seems as if there is not you could calculate the number   enough information.
of cars, and vice versa.

Guess an equal number of trucks ◀— Try the strategy of "guess and cars. Check the total earned   and check."
using multiplication and addition.
Continue to guess and check until
you find a correct total of $41.

Work Guess 5 trucks and 5 cars: (5 × $7) + (5 × $4) = $55 ◀— Too high
Guess 3 trucks: (3 × $7) + (5 × $4) = $41 ◀— The correct total

Answer The students washed 3 trucks and 5 cars.

Answers to Skill 26·············

1. 7 calculators

2. 2562 mi

3. 209 student tickets

Use the four-step plan to solve each problem.

1. The Computer Club has saved $520 to buy graphics calculators. Each calculator costs $72 including tax. How many calculators can the club buy?

2. Anika Miller participates in a frequent flyer plan. She needs 20,000 mi to earn a free airline ticket. Her January statement showed that she had earned 12,102 frequent flyer miles. Trips in February and April earned her 3534 mi and 1802 mi. How many more miles does Anika need to get a free ticket?

3. Student tickets for the basketball playoffs cost $2 each. All other tickets cost $5 each. Suppose 347 tickets worth $1108 were sold. How many of them were student tickets?

Skill 27   Choosing Strategies

Good problem solvers have a checklist of strategies in their toolbox of skills. Here are examples of some strategies you may have used.

 **Examples**

- Make a table   ←——— When you have data to organize
- Draw a diagram   ←——— When words describe a picture
- Guess and check   ←——— When you do not seem to have enough information
- Use an equation   ←——— When you know a relationship between quantities
- Use a formula   ←——— When you recognize a known relationship
- Identify a pattern   ←——— When you can examine several cases
- Solve a simpler problem ←— When easier numbers help you understand the problem
- Work backward   ←——— When you are looking for a fact leading to a known result
- Use a proportion   ←——— When you know that two ratios are equal

Solve each problem.

1. One week Winona Jones earned $306.25 for 35 hours of work at the library. The next week she worked 28 hours. How much did she earn that week?

2. The seniors sold 287 boxes of stationery to earn money for their gift to the school. Each box of stationery included 18 sheets of paper and 12 envelopes. The students collected $1363.25 in sales. What did each box cost?

3. A machine manufactures an item every 12 seconds. Three people are needed to operate the machine. How many items are manufactured in an 8 h work day?

4. A rectangular vegetable garden has a perimeter of 46 ft and a length of 10 ft. What is its area?

Toolbox **665**

Answers to Skill 27

1. $245
2. $4.75
3. 2400 items
4. 130 ft^2

Table of Measures

Time

60 seconds (s) = 1 minute (min)
60 minutes = 1 hour (h)
24 hours = 1 day
7 days = 1 week
4 weeks (approx.) = 1 month

$\left.\begin{array}{l}\text{365 days}\\\text{52 weeks (approx.)}\\\text{12 months}\end{array}\right\}$ = 1 year

10 years = 1 decade
100 years = 1 century

Metric	**United States Customary**

Length (Metric)

10 millimeters (mm) = 1 centimeter (cm)

$\left.\begin{array}{l}\text{100 cm}\\\text{1000 mm}\end{array}\right\}$ = 1 meter (m)

1000 m = 1 kilometer (km)

Length (U.S.)

12 inches (in.) = 1 foot (ft)

$\left.\begin{array}{l}\text{36 in.}\\\text{3 ft}\end{array}\right\}$ = 1 yard (yd)

$\left.\begin{array}{l}\text{5280 ft}\\\text{1760 yd}\end{array}\right\}$ = 1 mile (mi)

Area (Metric)

100 square millimeters = 1 square centimeter
(mm²)   (cm²)

10,000 cm² = 1 square meter (m²)

10,000 m² = 1 hectare (ha)

Area (U.S.)

144 square inches (in.²) = 1 square foot (ft²)

9 ft² = 1 square yard (yd²)

$\left.\begin{array}{l}\text{43,560 ft}^2\\\text{4840 yd}^2\end{array}\right\}$ = 1 acre (A)

Volume (Metric)

1000 cubic millimeters = 1 cubic centimeter
(mm³)   (cm³)

1,000,000 cm³ = 1 cubic meter (m³)

Volume (U.S.)

1728 cubic inches (in.³) = 1 cubic foot (ft³)

27 ft³ = 1 cubic yard (yd³)

Liquid Capacity (Metric)

1000 milliliters (mL) = 1 liter (L)

1000 L = 1 kiloliter (kL)

Liquid Capacity (U.S.)

8 fluid ounces (fl oz) = 1 cup (c)

2 c = 1 pint (pt)

2 pt = 1 quart (qt)

4 qt = 1 gallon (gal)

Mass

1000 milligrams (mg) = 1 gram (g)

1000 g = 1 kilogram (kg)

1000 kg = 1 metric ton (t)

Weight

16 ounces (oz) = 1 pound (lb)

2000 lb = 1 ton (t)

Temperature — Degrees Celsius (°C)

0°C = freezing point of water

37°C = normal body temperature

100°C = boiling point of water

Temperature — Degrees Fahrenheit (°F)

32°F = freezing point of water

98.6°F = normal body temperature

212°F = boiling point of water

Table of Symbols

Symbol		Page	Symbol		Page
•	× (times)	10	±	plus-or-minus sign	112
a^n	nth power of a	19	$\sqrt[3]{a}$	cube root of a	114
( )	parentheses—a grouping symbol	27	>	is greater than	144
[ ]	brackets—a grouping symbol	28	<	is less than	144
			≥	is greater than or equal to	144
≅	congruent, is congruent to	40	≤	is less than or equal to	144
−3	negative 3	64	(x, y)	ordered pair of numbers	184
$-a$	opposite of a	64	$\triangle ABC$	triangle ABC	191
$\lvert a \rvert$	absolute value of a	64	$A \rightarrow A'$	point A goes to point A' after a transformation	198
≈	is approximately equal to	73			
$\overleftrightarrow{AB}$	line AB	82	$\dfrac{1}{a}$	reciprocal of a	230
$\overline{AB}$	segment AB	82	$a : b$	ratio of a to b	301
AB	the length of AB	82	$P(E)$	probability of event E	308
$\overrightarrow{AB}$	ray AB	85	~	similar, is similar to	329
∠A	angle A	85	$\sin A$	sine of A	344
°	degree(s)	86	$\cos A$	cosine of A	344
=	equals, is equal to	99	$\tan A$	tangent of A	361
$\stackrel{?}{=}$	is this statement true?	100	π	pi, a number approximately equal to 3.14	376
$\sqrt{a}$	nonnegative square root of a	112	m	slope	418

Table of Squares and Square Roots

No.	Square	Sq. Root	No.	Square	Sq. Root	No.	Square	Sq. Root
1	1	1.000	51	2,601	7.141	101	10,201	10.050
2	4	1.414	52	2,704	7.211	102	10,404	10.100
3	9	1.732	53	2,809	7.280	103	10,609	10.149
4	16	2.000	54	2,916	7.348	104	10,816	10.198
5	25	2.236	55	3,025	7.416	105	11,025	10.247
6	36	2.449	56	3,136	7.483	106	11,236	10.296
7	49	2.646	57	3,249	7.550	107	11,449	10.344
8	64	2.828	58	3,364	7.616	108	11,664	10.392
9	81	3.000	59	3,481	7.681	109	11,881	10.440
10	100	3.162	60	3,600	7.746	110	12,100	10.488
11	121	3.317	61	3,721	7.810	111	12,321	10.536
12	144	3.464	62	3,844	7.874	112	12,544	10.583
13	169	3.606	63	3,969	7.937	113	12,769	10.630
14	196	3.742	64	4,096	8.000	114	12,996	10.677
15	225	3.873	65	4,225	8.062	115	13,225	10.724
16	256	4.000	66	4,356	8.124	116	13,456	10.770
17	289	4.123	67	4,489	8.185	117	13,689	10.817
18	324	4.243	68	4,624	8.246	118	13,924	10.863
19	361	4.359	69	4,761	8.307	119	14,161	10.909
20	400	4.472	70	4,900	8.367	120	14,400	10.954
21	441	4.583	71	5,041	8.426	121	14,641	11.000
22	484	4.690	72	5,184	8.485	122	14,884	11.045
23	529	4.796	73	5,329	8.544	123	15,129	11.091
24	576	4.899	74	5,476	8.602	124	15,376	11.136
25	625	5.000	75	5,625	8.660	125	15,625	11.180
26	676	5.099	76	5,776	8.718	126	15,876	11.225
27	729	5.196	77	5,929	8.775	127	16,129	11.269
28	784	5.292	78	6,084	8.832	128	16,384	11.314
29	841	5.385	79	6,241	8.888	129	16,641	11.358
30	900	5.477	80	6,400	8.944	130	16,900	11.402
31	961	5.568	81	6,561	9.000	131	17,161	11.446
32	1,024	5.657	82	6,724	9.055	132	17,424	11.489
33	1,089	5.745	83	6,889	9.110	133	17,689	11.533
34	1,156	5.831	84	7,056	9.165	134	17,956	11.576
35	1,225	5.916	85	7,225	9.220	135	18,225	11.619
36	1,296	6.000	86	7,396	9.274	136	18,496	11.662
37	1,369	6.083	87	7,569	9.327	137	18,769	11.705
38	1,444	6.164	88	7,744	9.381	138	19,044	11.747
39	1,521	6.245	89	7,921	9.434	139	19,321	11.790
40	1,600	6.325	90	8,100	9.487	140	19,600	11.832
41	1,681	6.403	91	8,281	9.539	141	19,881	11.874
42	1,764	6.481	92	8,464	9.592	142	20,164	11.916
43	1,849	6.557	93	8,649	9.644	143	20,449	11.958
44	1,936	6.633	94	8,836	9.695	144	20,736	12.000
45	2,025	6.708	95	9,025	9.747	145	21,025	12.042
46	2,116	6.782	96	9,216	9.798	146	21,316	12.083
47	2,209	6.856	97	9,409	9.849	147	21,609	12.124
48	2,304	6.928	98	9,604	9.899	148	21,904	12.166
49	2,401	7.000	99	9,801	9.950	149	22,201	12.207
50	2,500	7.071	100	10,000	10.000	150	22,500	12.247

Student Resources

Table of Trigonometric Ratios

Angle	Sine	Cosine	Tangent	Angle	Sine	Cosine	Tangent
1°	.0175	.9998	.0175	46°	.7193	.6947	1.0355
2°	.0349	.9994	.0349	47°	.7314	.6820	1.0724
3°	.0523	.9986	.0524	48°	.7431	.6691	1.1106
4°	.0698	.9976	.0699	49°	.7547	.6561	1.1504
5°	.0872	.9962	.0875	50°	.7660	.6428	1.1918
6°	.1045	.9945	.1051	51°	.7771	.6293	1.2349
7°	.1219	.9925	.1228	52°	.7880	.6157	1.2799
8°	.1392	.9903	.1405	53°	.7986	.6018	1.3270
9°	.1564	.9877	.1584	54°	.8090	.5878	1.3764
10°	.1736	.9848	.1763	55°	.8192	.5736	1.4281
11°	.1908	.9816	.1944	56°	.8290	.5592	1.4826
12°	.2079	.9781	.2126	57°	.8387	.5446	1.5399
13°	.2250	.9744	.2309	58°	.8480	.5299	1.6003
14°	.2419	.9703	.2493	59°	.8572	.5150	1.6643
15°	.2588	.9659	.2679	60°	.8660	.5000	1.7321
16°	.2756	.9613	.2867	61°	.8746	.4848	1.8040
17°	.2924	.9563	.3057	62°	.8829	.4695	1.8807
18°	.3090	.9511	.3249	63°	.8910	.4540	1.9626
19°	.3256	.9455	.3443	64°	.8988	.4384	2.0503
20°	.3420	.9397	.3640	65°	.9063	.4226	2.1445
21°	.3584	.9336	.3839	66°	.9135	.4067	2.2460
22°	.3746	.9272	.4040	67°	.9205	.3907	2.3559
23°	.3907	.9205	.4245	68°	.9272	.3746	2.4751
24°	.4067	.9135	.4452	69°	.9336	.3584	2.6051
25°	.4226	.9063	.4663	70°	.9397	.3420	2.7475
26°	.4384	.8988	.4877	71°	.9455	.3256	2.9042
27°	.4540	.8910	.5095	72°	.9511	.3090	3.0777
28°	.4695	.8829	.5317	73°	.9563	.2924	3.2709
29°	.4848	.8746	.5543	74°	.9613	.2756	3.4874
30°	.5000	.8660	.5774	75°	.9659	.2588	3.7321
31°	.5150	.8572	.6009	76°	.9703	.2419	4.0108
32°	.5299	.8480	.6249	77°	.9744	.2250	4.3315
33°	.5446	.8387	.6494	78°	.9781	.2079	4.7046
34°	.5592	.8290	.6745	79°	.9816	.1908	5.1446
35°	.5736	.8192	.7002	80°	.9848	.1736	5.6713
36°	.5878	.8090	.7265	81°	.9877	.1564	6.3138
37°	.6018	.7986	.7536	82°	.9903	.1392	7.1154
38°	.6157	.7880	.7813	83°	.9925	.1219	8.1443
39°	.6293	.7771	.8098	84°	.9945	.1045	9.5144
40°	.6428	.7660	.8391	85°	.9962	.0872	11.4301
41°	.6561	.7547	.8693	86°	.9976	.0698	14.3007
42°	.6691	.7431	.9004	87°	.9986	.0523	19.0811
43°	.6820	.7314	.9325	88°	.9994	.0349	28.6363
44°	.6947	.7193	.9657	89°	.9998	.0175	57.2900
45°	.7071	.7071	1.0000				

Tables

A-1 Domain and Range

For use after Section 4-7

Focus
Describe the domain and
the range of a function.

Talk it Over

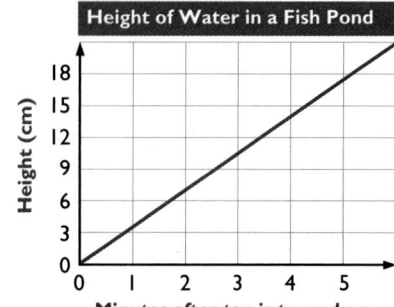

1. The graph above represents a function. Identify the control variable and the dependent variable.

2. Describe all possible values of the control variable. Then describe all possible values of the dependent variable.

3. Did your descriptions include the values between the numbers shown on the scales of the axes? Why or why not?

In Question 2, you described the *domain* and the *range* of the water height function. For any function, the **domain** is all the values of the control variable. The **range** is all the values of the dependent variable. The range depends on the domain.

Sample

Describe the domain and the range of the function represented by the graph.

Sample Response

The control variable is the number of days. The domain is the numbers 0, 1, 2, 3, 4, ...

The dependent variable is the amounts earned. The range is the dollar amounts 0, 50, 100, 150, ...

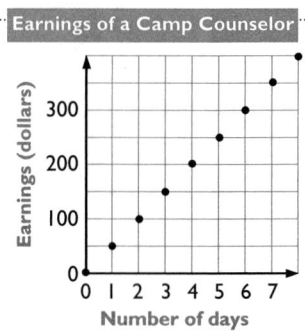

The points on the graph are not connected. The domain and range values are discrete.

Answers to Talk it Over

1. Control variable: minutes after tap is turned on; dependent variable: height of water

2. Control variable values: all amounts of time from 0 through 6 minutes; dependent variable values: all heights from 0 through 21 cm

3. Yes; both the control variable and the dependent variable are continuous quantities.

4. Descriptions may vary. An example is given. The domain is the number of days: 0, 1, 2, 3.... The range is the dollar amounts 0, 0.10, 0.20, 0.30.... There may be an upper limit to the domain and the range.

For example, once you owe more in fines than a book is worth, the library may charge you the cost of replacing the book.

Choosing Reasonable Values

Various conditions can limit or restrict the domain and range values of a function. Thinking about such restrictions can help you choose domain and range values that make sense for a particular function.

Talk it Over

4. A library charges $.10 for each day a book is overdue. Your total fine depends on the number of days a book is overdue. Describe a reasonable domain and range for the function. Explain your thinking.

A-1 Exercises and Problems

Describe the domain and the range of the function represented by each graph.

1.
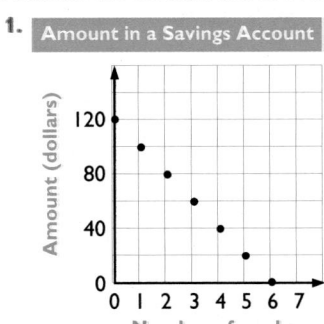

2.
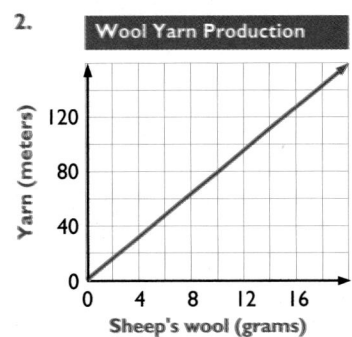

For Exercises 3–7, describe a reasonable domain and range for each function. Explain your thinking.

3. John Ramos is buying gasoline for his car. The price per gallon is $1.29. His total cost depends on the amount of gasoline he buys.

4. The height of a burning candle depends on how long it burns. Lisa found that the height of one candle decreased by about 1 in. per hour. The original height of the candle was 10 in.

5. Tariq has 12 tickets to sell to a play. The tickets cost $3.50 each. The amount of money he collects depends on the number of tickets he sells.

6. Jan Amdahl pays for some groceries with a $50 bill. The amount of change she receives depends on the cost of the groceries.

7. Customers at a music store can use a coupon for $1.50 off any purchase costing at least $15. The store allows only one coupon per customer. The amount that a customer using a coupon pays depends on the amount of his or her purchase.

Answers to Exercises and Problems

1. The domain is the number of weeks: 0, 1, 2, 3, 4, 5, 6. The range is the dollar amounts 120, 100, 80, 60, 40, 20, 0.

2. The domain is any number of grams, 0 or greater. The range is any number of meters, 0 or greater.

3–7. Answers may vary. Examples are given.

3. The domain is the number of gallons from 0 through the total number of gallons his tank holds. If g is the total number of gallons his tank holds, then the range is the dollar amounts to the nearest penny from 0 through $1.29g$.

4. Let t = the number of hours that pass and h = the candle height. Then $h = 10 - t$. So the domain is all numbers of hours from 0 through 10. The range is all heights from 10 in. through 0 in.

5. Let t = the number of tickets sold and A = the amount collected. Then $A = 3.5t$. So the domain is the number of tickets sold: 0, 1, 2, ..., 12. The range is the dollar amounts 0, 3.50, 7.00, ..., 42.

6. Let g = the cost of Jan's groceries and m = the change she receives. Then $m = 50 - g$. So the domain is the costs from $0 through $50. The range is the amount of change from $50 through $0.

7. Let a = the amount of a customer's purchase and p = the amount the customer pays after the $1.50 discount. Then $p = a - 1.50$. So the domain is the dollar amounts to the nearest penny ≥ $15. The range is the dollar amounts to the nearest penny ≥ $13.50.

A-2 Slope and y-Intercept

For use after Section 8-1

Ilona Rostov wants to sell some hand-made pins. She first draws a line to predict how much money she can make if she charges $25 for each pin. Then she realizes that it is going to cost her $300 to buy materials for the pins. She draws a second line to show her expected profit minus her start-up costs.

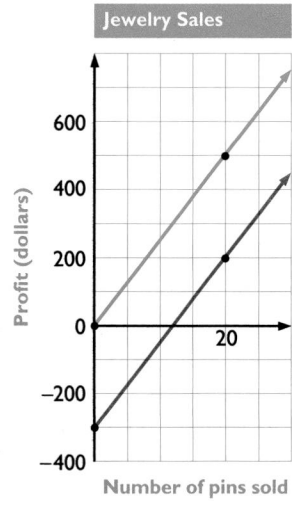

Jewelry Sales

-----Focus

Analyze and interpret the effects of changing the slope and the y-intercept of a graph.

Talk it Over

1. What do you notice about the slopes of the two lines? What real-world information do the slopes give you?

2. Give the vertical intercept of each graph. What real-world information do the vertical intercepts give you?

You can use what you know about the slope-intercept form of an equation to predict how the graph of one equation will compare with another.

Sample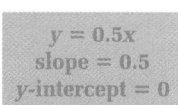

Describe how the graphs of y = 2x and y = 0.5x will compare with the graph of y = x. Then graph all three equations.

Sample Response

Each equation is in slope-intercept form, $y = mx + b$, where m = slope and b = y-intercept.

$y = x$	$y = 2x$	$y = 0.5x$
slope = 1	slope = 2	slope = 0.5
y-intercept = 0	y-intercept = 0	y-intercept = 0

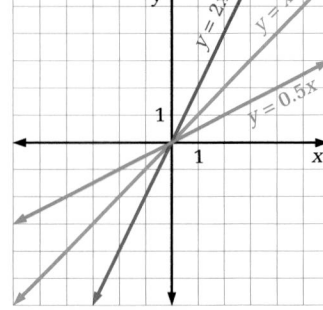

The graph of $y = 2x$ will be steeper than the graph of $y = x$ because its slope is greater. The graph of $y = 0.5x$ will be less steep because its slope is less. All three graphs will pass through the origin.

Answers to Talk it Over

1. The slopes are the same; the slope is the amount Ilona charges per pin ($25).

2. The vertical intercept of the first line is 0; this the amount of money made before any pins are sold, disregarding start-up costs. The vertical intercept of the second line is −300; before she makes a profit, Ilona needs to make $300 to pay off her start-up costs.

3. A–m; B–k; C–n; D–l; The graphs have the same slope, but different y-intercepts. You can match the graphs by checking the y-intercepts.

4. Descriptions may vary. An example is given. Changing the value of m changes the slope. When $m > 0$, the graph slopes up from left to right, and becomes steeper as m increases. When $m < 0$, the graph slopes down from left to right. It becomes steeper as $|m|$ increases. Changing the value of b changes the y-intercept. As b increases or decreases, the graph moves up or down the y-axis, respectively. The graph crosses the y-axis at the origin when $b = 0$, above the origin when $b > 0$, and below the origin when $b < 0$.

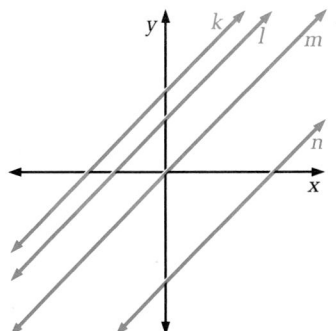

Talk it Over

3. Match each line at the left with one of the equations below. Explain your thinking.

A. $y = x$ B. $y = x + 3$

C. $y = x - 4$ D. $y = x + 2$

4. Describe how changing the values of m and b in an equation in the form $y = mx + b$ changes the graph of the equation. Be sure to talk about both positive and negative values of m and b.

 A-2 Exercises and Problems

1. Describe how the graphs of $y = -3x$ and $y = 3x$ compare with the graph of $y = x$. Then graph all three equations.

2. Describe how the graphs of $y = x - 20$ and $y = x + 20$ compare with the graph of $y = x$. Then graph all three equations.

3. In Section 8-1, you used the equation $h = 110t + 6000$ to model the height h above sea level of a balloon t minutes after it takes off from a location that is 6000 ft above sea level. Height is measured in feet.

   a. Suppose a balloon takes off from Leadville, Colorado (elevation 10,152 ft above sea level), and rises at a rate of 55 ft/min. Write an equation to describe the height above sea level of this balloon t minutes after it takes off.

   b. Describe how the graph of the equation you wrote in part (a) will compare with the graph of the equation $h = 110t + 6000$. Then graph both equations on the same set of axes.

4. Does doubling the value of m in an equation of the form $y = mx$ double the angle between the graph and the x-axis? Explain your reasoning.

5. Look back at the jewelry-sales graph on page 672. Suppose Ilona decides to lower her start-up costs by using cheaper materials. She also decides to charge less for each pin.

   a. Write an equation to model this new situation.

   b. Describe how the graph of your equation will compare with the lines on the jewelry-sales graph on page 672. Then graph your equation.

6. One way to graph an equation is to make a table of values, plot points, and draw a line through the points. Based on what you know about slope-intercept form, describe another method for graphing an equation.

Answers to Exercises and Problems

1. The graph of $y = 3x$ will slope up from left to right and will be steeper than the graph of $y = x$. The graph of $y = -3x$ will slope down from left to right. All three graphs will pass through the origin.

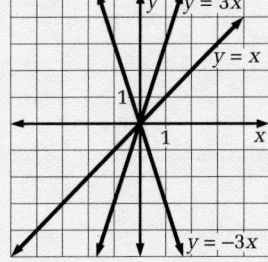

2. All three graphs will have the same slope. The graph of $y = x$ will have a y-intercept of 0. The graph of $y = x - 20$ will have a y-intercept of -20. The graph of $y = x + 20$ will have a y-intercept of 20. (See graph at top of next column.)

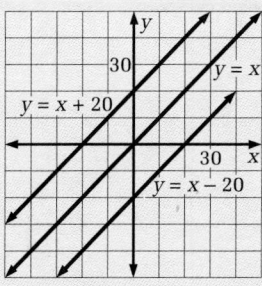

3. a. $h = 55t + 10{,}152$

   b. The graph of $h = 55t + 10{,}152$ will be less steep than the graph of $h = 110t + 6000$ because its slope is smaller. Its vertical intercept is greater, so it will cross the vertical axis at a higher point.

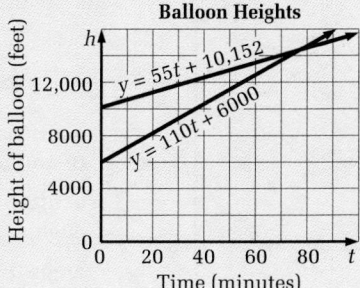

Balloon Heights

4. No; assuming the scales on the axes are the same, the graph of $y = x$ forms a 45° angle with the x-axis. If doubling the slope doubled the angle, then the graph of $y = 2x$ would form a 90° angle with the x-axis. It does not.

5. a. Equations may vary. An example is given. Let p = profit and n = number of pins sold; $p = 20n - 200$. (The vertical intercept should be > -300 and < 0. The slope should be > 0 and < 25.)

   b. Answers may vary. An example is given. In general, answers should note that the graph of the equation from part (a) will not be as steep as the lines on page 672 and will cross the vertical axis between 0 and -300.

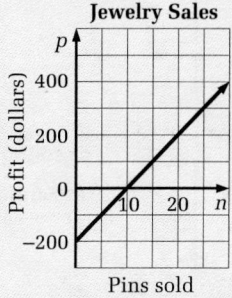

Jewelry Sales

6. Answers may vary. An example is given. First rewrite the equation in slope-intercept form. Plot the y-intercept first. Then use the fact that slope = $\dfrac{\text{rise}}{\text{run}}$ to count over to a second and a third point on the graph. Connect the points with a line.

Glossary

absolute value (**p. 64**) The distance that a number is from zero on a number line.

absolute value function (**p. 230**) The function $y = |x|$.

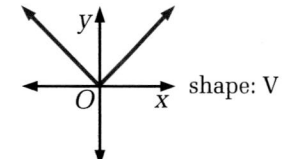

shape: V

acute angle (**p. 86**) An angle that measures between 0° and 90°.

acute triangle (**p. 88**) A triangle with three acute angles.

angle (**p. 85**) A figure formed by two rays that have the same endpoint, called the *vertex*.

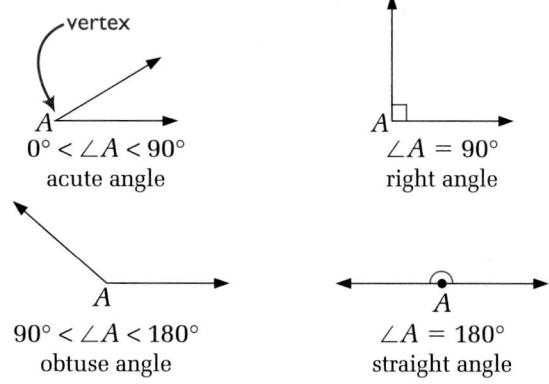

vertex

$0° < \angle A < 90°$
acute angle

$\angle A = 90°$
right angle

$90° < \angle A < 180°$
obtuse angle

$\angle A = 180°$
straight angle

arc (**p. 377**) Part of a circle. Any angle with its vertex at the center of a circle is a *central angle of the circle*. The region formed by a central angle and its arc is a *sector*.

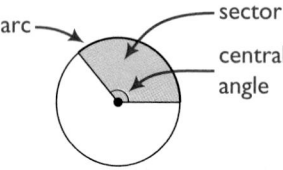

arc

sector

central angle

arc length (**p. 377**) The measurement of an arc.

at random (**p. 308**) *See* sampling.

base (**pp. 281, 282, 283**) *See* parallelogram, trapezoid, triangle.

base of a power (**p. 19**) *See* power.

bases of a prism (**p. 507**) *See* prism.

binomial (**p. 578**) The sum of two monomials.

boundary line (**p. 456**) *See* linear inequality.

box–and–whisker plot (**p. 158**) A method for displaying the median, quartiles, and extremes of a data set.

Test Scores

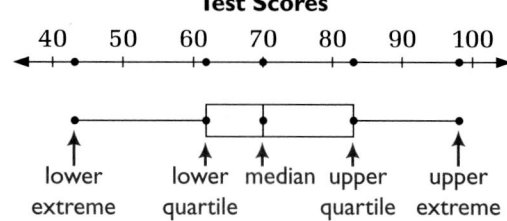

40   50   60   70   80   90   100

lower extreme   lower quartile   median   upper quartile   upper extreme

cell (**p. 129**) *See* spreadsheet.

center of dilation (**p. 337**) *See* dilation.

center of rotation (**p. 202**) *See* rotation.

central angle of a circle (**p. 85**) An angle with its vertex at the center of a circle. *See also* arc.

circumference (**p. 375**) The perimeter of a circle. A segment that joins two points of the circumference with the center of the circle is a *diameter*.

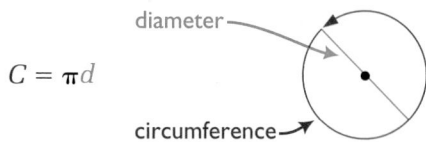

diameter

$C = \pi d$

circumference

coefficient (**p. 33**) A number multiplied by a variable in a term of an expression. *See also* term.

combined inequality (**p. 144**) An inequality with two inequality signs such that a value of a variable is between two quantities. A combined inequality like $-3 < x \le 5$ is graphed on a number line as a segment or *interval*.

complementary angles (**p. 86**) Two angles whose measures have the sum 90°.

complementary events (**p. 310**) Two events such that only one or the other is possible. For example, the event "E happens" is the complement of the event "E does not happen."

concept map (p. 5) A visual summary that helps you remember the connections between ideas.

conclusion (p. 492) *See* if-then statement.

conditional statements (p. 492) *See* if-then statement.

cone (p. 524) A space figure with one circular base and a vertex.

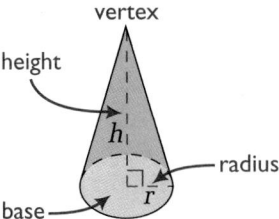

congruent (p. 38) Having the same size and shape.

congruent angles (p. 87) Angles with equal measures.

congruent segments (p. 82) Segments with equal measures.

conjecture (p. 21) A statement, opinion, or conclusion based on observation.

constant term (p. 33) Any term of an expression which contains only a number. *See also* term.

continuous (p. 60) Quantities that are measured.

control variable (p. 218) *See* function.

converse (p. 494) A statement in "if-then" form obtained by interchanging the "if" and "then" parts of an original if-then statement.

converse of the Pythagorean theorem (p. 495) If the square of the length of one side of a triangle is equal to the sum of the squares of the lengths of the other two sides, then the triangle is a right triangle.

conversion factor (p. 393) A conversion factor is a ratio of two equal (or approximately equal) quantities that are measured in different units.

coordinate geometry (p. 190) Geometry involving the drawing, analyzing, finding area, and comparison of figures on a coordinate plane.

coordinate plane (p. 184) A grid formed by two perpendicular number lines or *axes* intersecting at the *origin*. The axes split the plane into four *quadrants*.

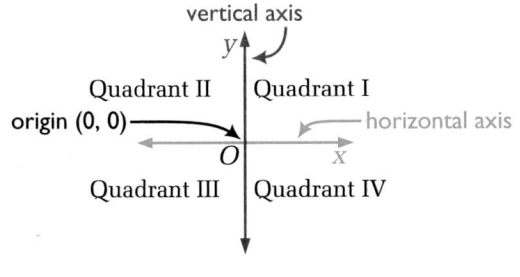

coordinates (p. 184) The unique *ordered pair* of real numbers associated with each point in a coordinate plane. The first number of the ordered pair is the *x-coordinate*, the second number is the *y-coordinate*.

correlation (p. 212) The relationship between two data sets. Two data sets can have positive correlation if they increase or decrease together, negative correlation if one set increases as the other set decreases, or no correlation.

corresponding vertices and sides (p. 40) In congruent polygons, matching vertices are corresponding vertices and matching sides are corresponding sides.

cosine (p. 344) *See* trigonometric ratios.

counterexample (p. 22) An example that shows that a statement is false.

cross products (p. 315) Equal products formed by multiplying the numerator of each of a pair of equal fractions by the denominator of the other.

cube root (p. 114) One of three equal factors of a number.

cylinder (p. 508) A space figure with a curved surface and two parallel, congruent circular bases.

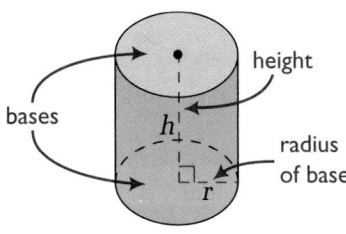

deductive reasoning (p. 479) Using facts, definitions, and accepted principles and properties to prove general statements.

dependent variable (p. 218) *See* function.

diagonal (p. 424) A segment joining two nonconsecutive vertices of a polygon.

diameter (p. 375) *See* circumference.

dilation (p. 337) A transformation in which the original figure and its image are similar. Lines drawn through the corresponding points on the original figure and its image meet in a point called the *center of dilation*.

dimensional analysis (p. 392) A problem solving strategy where you cancel units of measurement as if they are numbers.

dimensions (p. 128) *See* matrix.

direct variation (p. 360) A linear function defined by an equation of the form $y = kx$, $k \neq 0$. You say that y is directly proportional to x.

discrete (p. 60) Quantities that are counted.

distance from a point to a line (p. 281) The length of the perpendicular segment from a point to a line.

distributive property (p. 32) Each term inside a set of parentheses can be multiplied by a factor outside the parentheses. For example, $3(x + 2) = 3x + 6$.

endpoint (p. 82) A point that marks the first point or the last point of a line segment. *See also* ray.

equation (p. 99) A statement formed by placing an equals sign between two numerical or variable expressions.

equilateral triangle (p. 44) A triangle with all three sides congruent.

equivalent equations (p. 100) Equations that have the same solution set.

equivalent inequalities (p. 263) Inequalities that have the same solution set.

evaluate (p. 12) To find the value of a variable expression when a number is substituted for the variable.

event (p. 308) *See* outcome.

expanded form (p. 577) The form an expression is in when it has no parenthesis.

$$x(x + 3) = x^2 + 3x$$
factored form   expanded form

experimental probability (p. 308) In an experiment, the ratio of the number of times an event occurs to the number of times the experiment is run.

exponent (p. 19) *See* power.

extremes (p. 158) The lowest and highest numbers of a data set.

faces of a prism (p. 507) *See* prism.

factor (p. 19) When two or more numbers or variables are multiplied, each of the numbers or variables is a factor of the product.

factored completely (p. 578) When the greatest common factor of all the terms in an expression is 1.

factored form (p. 577) The form an expression has when it is factored completely.

$$x^2 + 3x = x(x + 3)$$
expanded form   factored form

fitted line (p. 212) *See* scatter plot.

frequency (p. 150) The number of times an event or data item occurs within an interval.

frequency table (p. 151) A table that displays the exact number of data items in an interval.

function (p. 220) A relationship between two variables in which the value of the *dependent variable* is dependent on the value of the *control variable*. There can only be one value of the dependent variable for each value of the control variable.

geometric probability (p. 502) Probability based on areas and lengths.

graph of an equation (p. 426) The points whose coordinates are the solutions of the equation.

height (pp. 281, 282) *See* parallelogram, triangle.

height of a regular pyramid (p. 510) *See* regular pyramid.

heptagon (p. 39) A polygon with seven sides.

hexagon (p. 39) A polygon with six sides.

histogram (p. 150) A type of bar graph that displays frequencies.

horizontal axis (p. 184) *See* coordinate plane.

horizontal intercept (p. 428) *See* x-intercept.

hyperbola (p. 230) The graph of $xy = k$, $k \neq 0$. *See also* reciprocal function.

hypotenuse (p. 344) *See* right triangle.

hypothesis (p. 492) *See* if-then statement.

identity function (p. 230) The function $y = x$.

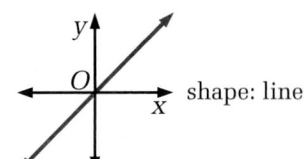

shape: line

if-then statement (p. 492) A statement with an *if* part and a *then* part. The *if* part is the *hypothesis* and the *then* part is the *conclusion*. Also called a *conditional statement*.

image (p. 337) The result of a transformation.

inductive reasoning (p. 479) A method of reasoning in which a conjecture is made based on several observations.

inequality (p. 144) A statement formed by placing an inequality sign between two numerical or variable expressions.

integer (p. 113) Any number that is a positive or negative whole number or zero.

Student Resources

interval (p. 145) *See* combined inequality.

irrational number (p. 113) A real number that cannot be written as the quotient of two integers.

isosceles triangle (p. 44) A triangle with at least two sides congruent.

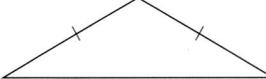

kite (p. 45) A quadrilateral with two pairs of consecutive sides equal in measure. These pairs do not have a side in common.

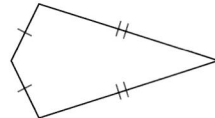

legs of a right triangle (p. 344) *See* right triangle.

length of a segment (p. 82) The measure of the distance between and including the two endpoints of a segment.

like terms (p. 33) Terms with the same variables or variable powers.

line (p. 82) A straight arrangement of points that extends forever in opposite directions.

line plot (p. 137) A method of displaying a data set on a number line. It is helpful in showing the outliers and the range of a data set.

line of reflection (p. 548) *See* reflection.

line of symmetry (p. 45) *See* symmetry.

linear combination (p. 426) The result of adding two or more linear expressions.

linear decay (p. 420) A decreasing linear function that can be defined by $y = mx + b, m < 0$.

linear equation (p. 427) An equation of a line.

linear function (p. 420) A function that can be defined by $y = mx + b$.

linear growth (p. 420) An increasing linear function that can be defined by $y = mx + b, m > 0$.

linear inequality (p. 456) An inequality whose graph on a coordinate plane is a region bounded by a line, called the *boundary line*.

lower quartile (p. 158) The median of the data in the lower half of a data set.

margin of error (p. 323) In an experiment or poll, the interval most likely to include the exact result.

mathematical model (p. 241) An equation or graph that represents a real-life problem. Using such equations or graphs is *modeling*.

matrix (p. 128) An arrangement of numbers in rows and columns. The number of rows by the number of columns gives you the *dimensions* of the matrix.

mean (p. 136) The sum of the data in a data set divided by the number of items.

median (p. 136) In a data set, the middle number or the average of the two middle numbers when the data are arranged in numerical order.

midpoint (p. 82) The point that divides a segment into two congruent parts.

mode (p. 136) The most frequently occurring item, or items, in a data set.

modeling (p. 241) *See* mathematical model.

monomial (p. 578) A number, a variable, or the product of a number and one or more variables.

obtuse angle (p. 86) An angle that measures between 90° and 180°.

obtuse triangle (p. 88) A triangle with one obtuse angle.

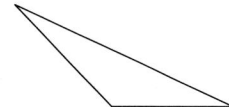

octagon (p. 39) A polygon with eight sides.

opposites (p. 64) A number and its opposite are the same distance from 0 on the number line but on opposite sides. The opposite of 3 is −3.

order of operations (p. 26) A set of rules that orders the way you simplify an expression: simplify inside parentheses, calculate any powers, multiply or divide left to right, and finally add or subtract left to right.

ordered pair (p. 184) *See* coordinates.

orientation (p. 548) The direction, clockwise or counterclockwise, in which the points on a figure are ordered.

origin (p. 184) *See* coordinate plane.

outcome (p. 308) One possible result in an experiment in a probability problem. A set of outcomes is an *event*.

outliers (p. 137) Data values that are much larger or much smaller than the other values in a data set, and therefore not typical of the data set.

parabola (pp. 229, 555) The graph of $y = ax^2 + bx + c, a \neq 0$. The point where the curve turns is either the maximum or the minimum and is called the *vertex*. *See also* squaring function.

parallel lines (p. 44) Two lines in the same plane that do not intersect.

parallelogram (p. 45) A quadrilateral with both pairs of opposite sides parallel.

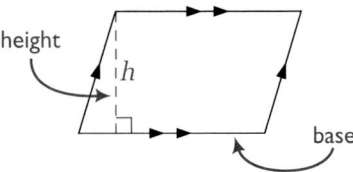

pentagon (p. 39) A polygon with five sides.

perfect cube (p. 114) A number whose cube root is an integer.

perfect square (p. 114) A number whose square root is an integer.

perpendicular (p. 44) Two lines, segments, or rays that intersect to form right angles.

polygon (p. 39) A plane figure formed by line segments, called *sides*. Each side intersects exactly two other sides, one at each endpoint or *vertex*. No two sides with a common vertex are on the same line.

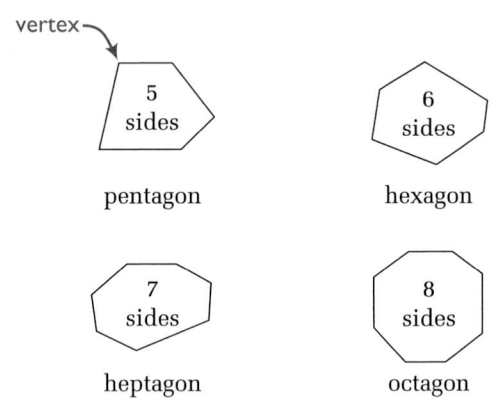

population (p. 321) The entire set of objects being studied.

power (p. 19) A number that is used as a factor a given number of times. In the power 5^2, 5 is the *base* and 2 is the *exponent*.

prism (p. 507) A space figure with two parallel congruent *bases*. The other sides of the prism are *faces*.

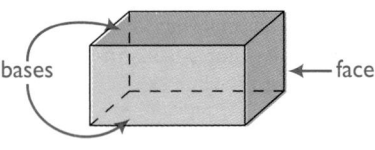

rectangular prism

probability (p. 308) The ratio of the number of favorable outcomes to the total number of outcomes.

proportion (p. 314) An equation that shows two equal ratios.

pyramid (p. 509) A space figure with one base and triangular faces.

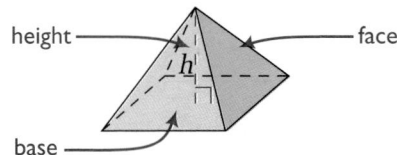

Pythagorean theorem (p. 478) If the length of the hypotenuse of a right triangle is c and the lengths of the legs are a and b, then $c^2 = a^2 + b^2$.

quadrant (p. 184) *See* coordinate plane.

quadratic equation (p. 599) Any equation that can be written in the form $ax^2 + bx + c = 0$, $a \neq 0$.

quadratic formula (p. 601) The formula

$$x = -\frac{b}{2a} \pm \frac{\sqrt{b^2 - 4ac}}{2a},$$

given $0 = ax^2 + bx + c$, $a \neq 0$.

quadratic function (p. 599) Any function of the form $y = ax^2 + bx + c$, $a \neq 0$.

quadrilateral (p. 39) A polygon with four sides.

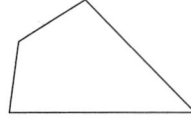

radical form (p. 487) An expression that uses the radical symbol ($\sqrt{}$). For example, $\sqrt{11}$.

radius (p. 376) One half of a diameter of a circle.

range (p. 137) The difference between the extremes in a data set.

rate (p. 302) A ratio that compares the amounts of two different kinds of measurements, for example, meters per second.

ratio (p. 301) The quotient you get when one number is divided by a second number not equal to zero.

rational number (p. 113) A real number that can be written as a quotient of two integers $\frac{a}{b}$, $b \neq 0$.

ray (p. 85) A part of a line that starts at a point, the *endpoint*, and extends forever in one direction.

real number (p. 113) Any number that is either rational or irrational.

reciprocals (p. 275) Two numbers whose product is 1.

reciprocal function (p. 230) The function $y = \frac{1}{x}$.

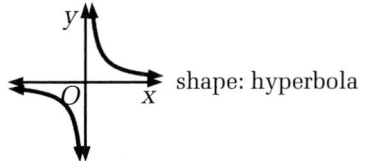 shape: hyperbola

rectangle (p. 45) A quadrilateral with four right angles.

rectangular prism (p. 507) *See* prism.

reflection (p. 548) The image you get when you flip a figure over a line. This line is the *line of reflection*.

regular pyramid (p. 509) A pyramid in which the base is a regular polygon, and all other faces are congruent isosceles triangles. The *height* is the perpendicular length from the vertex to the center of the base. The height of a face is the *slant height*.

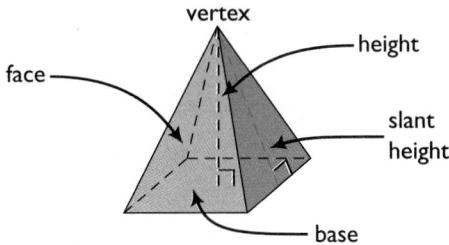

rhombus (p. 45) A quadrilateral with four congruent sides.

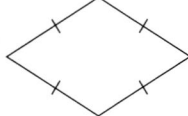

right angle (p. 86) An angle that measures 90°.

right triangle (p. 44) A triangle with one right angle. The side opposite the right angle is the *hypotenuse*, the other sides are the *legs*.

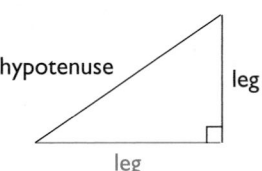

rise (p. 361) The vertical change between two points on a line.

rotation (p. 202) A turn around a point called the *center of rotation*. A figure that looks the same after a rotation of less than 360° has *rotational symmetry*.

rotational symmetry (p. 205) *See* rotation.

run (p. 361) The horizontal change between two points on a line.

sample (p. 321) A subset of the population on which a study or an experiment is being done.

sampling (p. 323) Choosing a sample from a population for an experiment or study. The sample is chosen *at random*, meaning each member of the population has an equal chance to be chosen.

scale (p. 330) The ratio of the size of a representation of an object to the actual object.

scale drawing (p. 330) A drawing representing and similar to an actual object.

scale factor (p. 337) The ratio of a length on an image to the corresponding length on the original figure of a dilation.

scalene triangle (p. 44) A triangle with no sides congruent.

scatter plot (p. 212) A graph that shows the relationship between two sets of data. A line that passes close to most of the data points is called a *fitted line*.

scientific notation (p. 72) A number written as a number that is at least one but less than ten, multiplied by a power of ten.

sector (p. 402) *See* arc.

segment (p. 82) Two points on a line and all points between them.

side (p. 40) *See* polygon.

similar figures (p. 329) Two figures with the same shape, but not necessarily the same size.

similar space figures (p. 532) Space figures that have the same shape, but not necessarily the same size.

sine (p. 344) *See* trigonometric ratios.

slant height of a regular pyramid (p. 510) *See* regular pyramid.

slope (p. 361) The measure of the steepness of a line given by the ratio of rise to run for any two points on the line.

slope-intercept form of an equation (p. 418) An equation of the form $y = mx + b$, where m represents the slope and b represents the y-intercept of a line.

solution (p. 99) Values for variables that make an equation true.

solution of an equation with two variables (p. 426)
An ordered pair of values that makes an equation with two variables true.

solution of a system (p. 291) Values for variables that make a system of equations true.

solution of a system of inequalities (p. 464) Values for variables that make a system of inequalities true.

solution region (p. 456) *See* linear inequality.

solving an equation (p. 99) Finding all values of a variable that make an equation true.

spreadsheet (p. 129) A computerized version of a matrix. Each position in a spreadsheet is a *cell*.

square (p. 45) A quadrilateral with four right angles and four congruent sides.

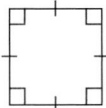

square root (p. 112) One of two equal factors of a number.

squaring function (p. 230) The function $y = x^2$.

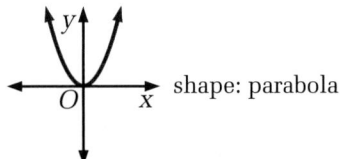

shape: parabola

standard form of a linear equation (p. 427) An equation in the form $ax + by = c$, where a, b, and c are integers and a and b are not both zero.

stem-and-leaf plot (p. 152) A display of data where each number is represented by a *stem* and a *leaf*.

straight angle (p. 86) An angle that measures 180°.

substitute (p. 12) To replace variables with given values.

supplementary angles (p. 86) Two angles whose measures have the sum 180°.

symmetry (p. 45) When a polygon can be folded so that one half fits exactly over the other half, the polygon has symmetry. The fold line is called the *line of symmetry*.

system of equations (p. 291) Two or more equations in the same variables.

system of inequalities (p. 464) Two or more inequalities in the same variables.

tangent ratio (p. 361) *See* trigonometric ratios.

term (p. 33) Each expression in a sum.

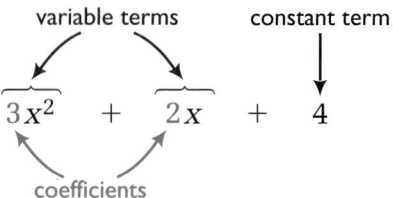

terms of a proportion (p. 314) The numbers or variables in a proportion.

theoretical probability (p. 309) When all outcomes of an experiment are equally likely, the probability of an event is the ratio of favorable outcomes to the number of possible outcomes.

transformation (p. 202) A change made to an object or its position.

translation (p. 197) Sliding a figure without changing its size or shape and without turning or flipping it over. When a figure can translate within a pattern, the pattern has *translational symmetry*.

translational symmetry (p. 199) *See* translation.

trapezoid (p. 193) A quadrilateral with one pair of parallel sides, called bases. The other sides are legs.

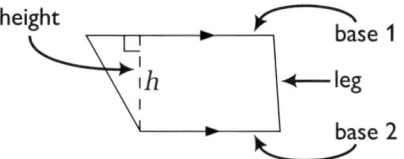

triangle (p. 39) A polygon with three sides.

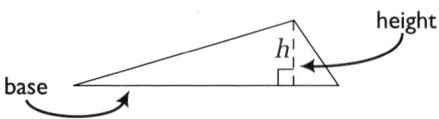

trigonometric ratios (p. 344) The *cosine*, *sine*, and *tangent ratios*.

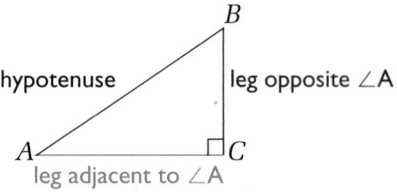

$$\cos A = \frac{\text{adjacent}}{\text{hypotenuse}} = \frac{AC}{AB}$$

$$\sin A = \frac{\text{opposite}}{\text{hypotenuse}} = \frac{BC}{AB}$$

$$\tan A = \frac{\text{opposite}}{\text{adjacent}} = \frac{BC}{AC}$$

trinomial (p. 584) The sum of three monomials.

unit rate (p. 302) A rate for one unit of a given quantity.

upper quartile (p. 158) The median of the data in the upper half of a data set.

variable (p. 10) A symbol, usually a letter, that represents a number.

variable expression (p. 10) An expression that contains a variable.

variable terms (p. 256) Terms of an expression that contain a variable. *See also* term.

variation constant (p. 360) The nonzero constant k in a direct variation defined by $y = kx$.

vertex of an angle (p. 85) *See* angle.

vertex of a parabola (p. 555) *See* parabola.

vertex of a polygon (p. 40) *See* polygon.

vertical angles (p. 87) Two congruent angles formed by intersecting lines and facing in opposite directions.

vertical axis (p. 184) *See* coordinate plane.

vertical intercept (p. 418) *See* y-intercept.

vertical-line test (p. 221) When two or more points of a graph lie in the same vertical line, the graph is not a function.

x-axis (pp. 226, 550) The horizontal axis in the coordinate plane.

x-intercept (p. 563) The x-coordinate of a point where a graph intersects the x-axis (where $y = 0$). Also called a *horizontal intercept*.

y-axis (pp. 226, 550) The vertical axis in the coordinate plane.

y-intercept (p. 563) The y-coordinate of a point where a graph intersects the y-axis (where $x = 0$). Also called a *vertical intercept*.

zero-product property (p. 495) When a product of factors is zero, one or more of the factors must be zero. If $ab = 0$, then $a = 0$ or $b = 0$.

Index

Student Resources

Student Resources

Stock a Music Store (Unit 3), 125, 126, 134, 142, 149, 157, 163, 169, 175, 176

Theme Poster Contest (Unit 2), 55, 56, 63, 70, 78, 84, 92, 98, 104, 110, 118, 119

Write a Math Story (Unit 10), 545, 546, 554, 561, 567, 575, 582, 590, 597, 605, 606

See also Group Activities.

Property(ies)
associative, 640
commutative, 640
distributive, 32, 53
of equality, 99-100
of exponents, 75, 568-572
identity, 640
of inequality, 264
of opposites, 248
of a proportion, 315
of square roots, 486, 542
of transformations, 197, 202, 237, 337, 354, 547, 548, 609
zero-product, 495, 542, 564

Proportion(s), 314-316
a problem solving strategy, 316, 323, 324, 331, 378, 384, 402, 533
cross products, 315
direct variation, 384
terms of, 314

Pyramid
height of, 510
regular, 509
slant height of, 510
surface area, 509
volume of, 523

Pythagorean theorem, 477-479
converse of, 495

Quadrant, 184

Quadratic equation, 599
solution of, 599, 603

Quadratic formula, 601

Quadratic function, 599

Quadrilateral, 39, 44-45

Quartiles, 158

Radical form, 487

Radicals, 112-115, 478-479, 486-489

Radius, 376

Random, 308

Range, 137
line plot, 137

Range of a function, 670-671

Rate, 302-303
converting units of, 394-396
unit price, 302
unit rate, 302

Ratio, 301-303, 530-533

Rational number, 113

Ray, 85

Reading, in exercises, 6, 7, 14, 23, 28, 42, 61, 75, 82, 89, 96, 103, 108, 116, 131, 139, 147, 154, 160, 166, 172, 185, 194, 199, 206, 215, 222, 231, 244, 250, 258, 259, 266, 267, 272, 277, 285, 292, 311, 316, 325, 332, 340, 348, 370, 378, 388, 396, 405, 423, 429, 435, 443, 451, 459, 465, 481, 489, 496, 504, 512, 519, 526, 534, 551, 558, 565, 572, 580, 588, 596, 604

Real number, 113

Reasoning, 542
conclusion, 492
conditional statement, 492
conjecture, 21
counterexample, 22
deductive, 479-480, 486, 492-494
hypothesis, 492
if-then statement, 492
inductive, 11-12, 14 (Exs. 5, 6), 21-22, 71, 92 (Ex. 30), 424 (Ex. 33), 477, 479-480
interpreting data and graphs, 4-5, 130, 135, 170, 212, 214, 367, 420, 530
visual thinking. *See* Manipulatives, Modeling, *and* Multiple representations.

See also Algorithmic thinking, Patterns, *and* Writing. *Also, Talk it Over discussion questions in each section promote critical thinking skills.*

Reciprocal(s), 275-277, 650

Reciprocal function, 230

Rectangle, 45
area of, 192, 193, 658
perimeter of, 13, 658

Reduction, 338

Remote interior angle, 92

Reflection, 547-550
line of, 548
orientation of, 548
properties of, 548

Regular pyramid, 509

Research (out of textbook), in exercises, 8, 17, 63, 78, 82, 97, 109, 131, 140, 142, 155, 157, 169, 187, 201, 209, 217, 254, 259, 267, 274, 279, 304, 305, 320, 326, 327, 333, 335, 348, 350, 363, 364, 374, 379, 381, 397, 399, 423, 432, 460, 462, 481, 506, 527, 552, 575

Review
Checkpoints, 30, 78, 98, 149, 210, 269, 320, 382, 438, 454, 522, 567, 590

Extra Practice, 620-639

Review and Assessment, 51-53, 120-123, 177-179, 235-237, 295-297, 352-354, 409-413, 470-473, 540-543, 607-609

Review/Preview. *The exercise set in each section of each unit contains Review/Preview exercises.*

Toolbox (examples and practice of prealgebra skills), 640-665

Rhombus, 45

Right angle, 44-45, 86

Right triangle, 44, 88
area of, 191-192
hypotenuse of, 344

Rise, 361

Rotation(s), 202-206
amount of turn, 203
center of, 202
drawing, 204-205
symmetry, 205

Rotational symmetry, 205, 206

Row
of a matrix, 128
of a spreadsheet, 129

Rules
of exponents, 75
of powers, 569-572

Run, 361

Sample, 321

Sampling, 321-325

Scale, 330, 334

Scale drawings, 330-331

Scale factor, 337

Scalene triangle, 44

Scatter plot(s), 212-214, 460, 472
fitted line of, 212, 369, 439-440

Credits

DESIGN

Book Design: Two Twelve Associates, Inc.
Cover Design: Two Twelve Associates, Inc. Photographic collage by Susan Wides; student photography by Ken Karp.
Electronic Technical Art: American Composition & Graphics, Inc.

ACKNOWLEDGMENTS

17 From "Animalimericks" in *Jamboree: Rhymes for All Times,* by Eve Merriam. Copyright © 1962, 1964, 1966, 1973, 1984 by Eve Merriam. Reprinted by permission of Marian Reiner.

69, 259 From "To Build a Fire," by Jack London. First appeared in *The Century Magazine,* by The Century Company, August, 1908. Subsequently published in *Lost Face,* by Jack London. Copyright © 1910 by Macmillan Publishing Company.

132 Data for Exercises 11–14 from The Recording Industry Association of America's 1992 Year-end Statistics.

188 From *The Hunt for Red October,* by Tom Clancy. Annapolis, Maryland: United States Naval Institute, 1984.

245 From *Rules of Thumb 2,* by Tom Parker. Boston: Houghton Mifflin Company, 1987.

335 From "The Adventure of the Musgrave Ritual," by Arthur Conan Doyle. Originally appeared in Arthur Conan Doyle's *The Memoirs of Sherlock Holmes,* published in 1893 by George Newnes, Limited. First U.S. edition, Copyright © 1894 by Harper & Brothers.

336 From *A More Perfect Union* in Houghton Mifflin Social Studies, by Armento et al. Copyright © 1991 by Houghton Mifflin Company. Reprinted by permission of Houghton Mifflin Company.

535 From *Gulliver's Travels,* by Jonathan Swift. Originally published in London by B. Motte, Publishers, 1726.

STOCK PHOTOGRAPHY

v Michael Newman/Photo Edit (t); **viii** © Allsport USA/ Michael Hans; **ix** P.L.I./Westlight; **xi** Photo Tal Streeter, *The Art of the Japanese Kite,* Weatherhill Inc. (t); **xi** Superstock (c); **xi** Dr. Jeremy Burgess/Science Photo Library/Photo Researchers (inset); **xii** © 1993 Mark Zemnick (t); **xiii** Dan Paul (b); **xxvi** © James Hart (r); Kevin Morris/Allstock (l); (logo sketches) Courtesy Larry DesJarlais, (finished logo) The Smithsonian Institution National Museum of the American Indian National Campaign Office; **1** (Seattle, Colorado, Florida, Milwaukee, Toronto, New York) The Major League logos pictured in this book are reproduced with the permission of Major League Properties, Inc. and are the exclusive property of Major League Baseball Clubs. The logos may not be reproduced without the consent of Major League Properties, Inc. or the appropriate individual Club.; (Utah) Courtesy

NBA Properties, Inc.; **3** (Vietnam) Scribner/The Picture Cube; (USA) Frank Siteman/The Picture Cube; (Alaska) Soames Summerhays/Photo Researchers, Inc.; (Africa) Owen Franken/Stock Boston; (S. America) Juan Pablo Lira/The Image Bank; (India) Porterfield/Chickering/Photo Researchers, Inc.; **9** Travelpix 1991/FPG International (l); Dave Wilhelm/The Stock Market (tr); Robert Frerck/ Tony Stone Images/Chicago Inc. (br); **10-11** John Elk, III/ Stock Boston; **11** © Bob Daemmrich; **12** John Colwell/ Grant Heilman Photography; **15** Thomas Del Brase/ The Stock Market; **17** Photo by Bachrach; **19** Fred Conrad/© 1985 *Discover* Magazine; **30** Betsy Pillsbury; **37** Michael Newman/Photo Edit; **48** Blair Seitz/Photo Researchers, Inc.; **54** Curt Teich Postcard Archives, Lake County Museum; **55** Stephen Derr/The Image Bank (b); Chuck Carter/© National Geographic Society (t); **57** © Allsport USA/Darrell Ingham (t); © Bob Daemmrich (b); **58** Tom Van Sant, Geosphere Project, Santa Monica/Photo Researchers, Inc.; **60** Richard Megna/Fundamental Photo (tl & r); © 1990 Bruce Iverson (b); **61** Fred Ward/Black Star; **62** Nantucket Nectars Wood Pressed Apple Juice Label; **64** Sonya Hagopian/Woods Hole Oceanographic Institution (t); Rod Catanagh/ Woods Hole Oceanographic Institution (b); **67** © Kindra Clineff (t); Terence Turner/ FPG International (b); **70** National Baseball Hall of Fame; **73** Dr. Jeremy Burgess/Science Photo Library/Photo Researchers (inset); Superstock; **76** © Bruce Iverson (tl); Photo Researchers, Inc. (bl); © David Liebman (bc); Doc White/Images Unlimited, Inc. (tc); © Allsport USA/ Vandystadt (br); Warren Morgan/Westlight (tr); **77** Stamp design © 1993 United States Postal Service. All rights reserved; **78** Tony Stone Images/Chicago Inc.; **81** EROS Data Center/USGS; **83** Courtesy Belize Tourist Board (r); Courtesy Indonesian Consulate (l); **85** Tom Croke/ Lightwave; **91** Vince Streano/Tony Stone Images/Chicago Inc. (l); Alex MacLean/Landslides (c); Guido Alberto/The Image Bank (r); **92** István Bodóczky; **93** Chris Mooney/ FPG International; **94** L. Hafencher/H. Armstrong Roberts; **96** Alex MacLean/Landslides (l); © Kindra Clineff (r); **97** Theodores Wores (1859-1939), Dai Nichi Do Garden, Nikko, Japan, 1887. Oil on canvas. Courtesy Jim Lyons.; **99** Photo Tal Streeter, *The Art of the Japanese Kite,* Weatherhill Inc.; **103** Courtesy of Jose Sainz (l); Alex MacLean/Landslides (c); Jonathan L. Barkin/The Picture Cube (r); **104** Excavations of the Metropolitan Museum of Art, 1936, and Rogers Fund, 1937. (37.40. 26) (t); America Hurrah, New York, NY (l); Myron/Tony Stone Images/ Chicago Inc. (br); **110** © 1992 Jack Vartoogian; **115** © 1993 Terry Wild Studio; **116** © Jerry Sinkovec (t); Philip & Karen Smith/Tony Stone Images/Chicago Inc. (b); **124** (1900) Northwind Picture Archives; (1910) Courtesy Antique Automobile Club of America ; (1920 & 1930) Collection Lester Glassner; (1940) FPG International; **124-125** (1950) Courtesy Antique Automobile Club of America; (1960) FPG International; (1970) From the Collection of the Museum at F.I.T., NY; (1980)

Rob Lang/FPG International; (1990) General Motors; (1990 woman) Mark Scott/FPG International; Reprinted courtesy *The Boston Globe* (tl & c); **126** Reprinted courtesy *The Boston Globe* (tl & r); **127** © Bob Daemmrich; **128** Courtesy City Year; **133** Philip Bailey/The Stock Market; **139** Doug Armond/Tony Stone Images/Chicago Inc.; **143** © Allsport USA/Mike Powell; **144** Leo Mason/ Sports Illustrated; **146** John Biever/Sports Illustrated; **147** © Allsport USA/Mike Powell; **162** Ted Mahieu/The Stock Market; **164** Andrew Lichtenstein/Impact Visuals; **168** Jeff Greenberg/Photo Researchers, Inc.; **172** National Audubon Society/Photo Researchers, Inc.; **173** NASA JSC/ Starlight; **177** Randall Hyman/Stock Boston; **180** Superstock (t); Copyright © by Universal City Studios, Inc. Courtesy of MCA Publishing Rights, a Division of MCA Inc. (bl to r); **180-181** From *The Animation Book* by Kit Laybourne. Copyright © 1979 by Kit Laybourne. Reprinted by permission of Crown Publishers, Inc.; **181** Ian Jackson/ Ace/Picture Perfect USA, Inc. (b); **183** Institute of Nautical Archaeology; **186** Amadeo Vergani/The Stock Market (l); L. Villota/The Stock Market (c); Ben Simmons/The Stock Market (r); **187** Chuck Close, ERIC, 1990 (in progress), in artist's studio 5/90, #20919. Photo by Bill Jacobson, courtesy The Pace Gallery; **188** Sovfoto (t); © David Muench (c, bl & br); **190** Werner Forman/Art Resource, New York (t); **197** Courtesy Schindler Elevator Corporation; **199** Caroline Wood/Allstock (l); **200** Collection of Clare & Joseph Fischer (bl); Elkus Collection (#370-911), California Academy of Sciences, San Francisco, CA (tl); Piet Mondrian, Composition with Red, Yellow, Blue "Ex-Mart Stam," 1928, Giraudon/Art Resource, New York/© Estate of P. Mondrian/E.M. Holtzman Trust, New York, NY (tc); Wooden carved cup, Kuba people of Zaire. Smithsonian Institution, Museum of African Art. Tufino/Art Resource, New York (br); **202** © Allsport USA/Michel Hans; **204** F. Gohier/Photo Researchers, Inc.; **205** © Animals Animals (r); **206** © Animals Animals; **208** © Doranne Jacobson (tl, tr, c); Superstock (b); © John Running (inset top); © Stephen Trimble (inset bottom); **210** © Doranne Jacobson (c); Dan McCoy/Rainbow (r); David Crossley/ Pueblo to People 1991 (l); **222** Robert A. Jureit/The Stock Market; **228** Chris Noble/Allstock; **231** Tom Ives/The Stock Market; **235** © Doranne Jacobson (c & r); Lacquer screen with dragon design. Private Collection. Werner Forman/Art Resource, New York, NY; **238** Richard Hutchin/Photo Edit (t); **238-239** Jerry Howard/Stock Boston; **239** Larry Lawfer/The Picture Cube (t); Aneal Vohra/Unicorn Stock Photo (b); **243-244** © Kindra Clineff; **245** © Tom Pantages (t); Michael Newman/Photo Edit (b); **251** Aneal Vohra/The Picture Cube; **252** Henry Kaiser/ Leo de Wys; **253** © Duomo/David Madison; **258** Chuck Keeler/Tony Stone Images/Chicago Inc. (l); R. Paul Gerda/ Leo de Wys (r); **267** NASA; **272** © 1977 by Sidney Harris, *What's So Funny About Science?*, William Kaufman Inc.; **275** Courtesy Patchwork Quilt Tsushin, Tokyo; **278** Bob Daemmrich/Stock Boston; **283** James A. Sugar/ Black Star; **286** Don Smetzer/Tony Stone Images/Chicago Inc. (tl); Superstock (br); The Western Reserve Historical Society, Cleveland, Ohio (b); **287** Paul A. Hein/Unicorn Stock Photo; **289** Bill Gallery/Stock Boston; **298** J. Kugler/FPG International (l); Aldo Mastrocola/ Lightwave (c); Travelpix/FPG International (r); **301** Courtesy of Sony Electronics Inc.; **303** Courtesy Theater Imax Keong Emas, Jakarta, Indonesia; **305** NASA;

309 Ed Bishop; **310** Superstock; **314** © Allsport USA/ Bud Symes; **317** Soames Summerhay/Photo Researchers, Inc.; **318** David Young-Wolff/Photo Edit; **319** © Kindra Clineff; **321** Tom & Pat Leeson/Photo Researchers, Inc.; **326** Andy Sacks/Tony Stone Images/Chicago Inc.; **328** Department of Special Collections, The University of Chicago Library; **333** © Ed Braverman; **334** © Stephen Feld (b); Courtesy *Nutshell News*, photo by Jim Cook (t); **335** The Bettmann Archive (r); **346-347** © Peter Chapman; **356** © Fred Zahradnik (r); From *The World Book Encyclopedia*. © 1994 World Book, Inc. By permission of the publisher. (l); **357** Greg Ballinger/*Thrasher* Magazine (tl to r); © Duomo/William R. Sauaz (b); **359** Dennis Hallinan/FPG International; **361** Rheinland Pfalz/ Westlight (b); Paul Chesley/Tony Stone Images/Chicago Inc. (t); **363** Superstock; **364** Terry Qing/FPG International (t); Courtesy Profit by Stair Systems, Inc. (b); **366** David Stoeklein/The Stock Market (b); Telegraph Colour Library/FPG International (t); **369** Doug Handel/ The Stock Market; **371** Ben Rose/The Image Bank; **372** Bob Abraham/The Stock Market; **373** Ellis Herwig/ Stock Boston (t); Dr. David Jones/Photo Researchers, Inc. (b); **376** J. Taposchaner/FPG International; **378** Roces Roadskate; **379** Bill Parsons/DDB Stock Photography; **380** Copyright (1990) Hansen Planetarium, Salt Lake City, Utah; Reproduced with Permission (l); **381** Alex MacLean/Landslides; **382** © 1993 Mark Zemnick (t); **386** © James W. Kay; **388** © Lander-Goldberg; **391** Dennis O'Clair/Tony Stone Images/Chicago Inc.; **392** © Bob Daemmrich; **394** Telegraph Colour Library/FPG International; **398** Fred Espenak/Science Photo Library/ Photo Researchers, Inc.; **401** F. Gohier/ Photo Researchers, Inc.; **402** America Hurrah, New York, NY; **403** Steve Leonard/Tony Stone Images/Chicago Inc.; **409** NASA; **411** © Bob Daemmrich (t); **414** Wide World Photos; **414-415** Manny Millan/*Sports Illustrated*; **415** © Steve Finberg; **417** P.L.I./Westlight; **418** Paul Murphy/Unicorn Stock Photo (on ground); **418-419** Mauritius/Westlight (in air); **420** © Bob Daemmrich; **427** L.L.T. Rhodes/Devaney Stock Photos; **429** Alex MacLean/Landslides; **430** Guy Marche/FPG International (r); Barry L. Runk/Grant Heilman Photography (l); **432** © Bob Daemmrich; **438** Oscar Palmquist/Lightwave (b); The Bettmann Archive (tl); Neil Leifer/*Sports Illustrated* (tc); © Duomo (tr); **439** Superstock; **440-441** © Duomo/ Steven E. Sulton; **444** Dallas & John Heaton/Westlight; **446** Lori Adamski Peek/ Tony Stone Images/Chicago Inc.; **449** Chris Boylan/ Unicorn Stock Photo; **453** Ray Scioscia/Habitat for Humanity; **461** Ancient Art & Architecture (r); Ted Clutter/Photo Researchers, Inc. (l); **462** Aldo Mastrocola/Lightwave; **463** Mary Kate Denny/Photo Edit; **465** Ben Edwards/Tony Stone Images/Chicago Inc. (t); Anthony Cassidy/Tony Stone Images/Chicago Inc. (b); **467** Tom Pantages (l); **470** © Allsport USA/Gary Newkirk; **471** © Kindra Clineff; **474** Master Builders, Inc. (t); Telegraph Color Library/FPG International (b); **475** (clipper) Luis Posenda/FPG International; (Viking) Wide World Photos; (junk) R. Ian Lloyd/The Stock Market; (felucca) K. Herring/One World Photographic; (kayak) © Wolfgang Kaehler; (dhow) Hugh Sitton/Tony Stone Images/Chicago Inc.; (canoes) Leonard Harris/Stock Boston; **476** Kevin Schafer/Allstock; **478-479** Courtesy Sony Electronics Inc.; **481** Metropolitan Museum of Art, Fletcher Fund, 1956 (56.171.38); **482** © Duomo/

Al Tielemans (t); **484** George A. Plimpton Collection, Rare Book & Manuscript Library, Columbia University; **490** Leo Mason/The Image Bank; **492** Globus Brothers/The Stock Market; **493** Wiancko Braasch/Allstock; **496** Barbara & Justin Kerr; **498** Nancy Dudley/Stock Boston; **502** © Allsport USA/Vandystadt/Didier Klein; **505** David Blum/Pixel Studios/The Stock Market; **510** Copyright © 1992 Dorling Kindersley Limited, London. From *Press-out Romantic Gift Boxes*, published in the United States by Dorling Kindersley Publishing Inc., NY (l); **512** Donovan Reese/Tony Stone Images/Chicago Inc.; **513** Richard Olsenius/© National Geographic Society; **514** Courtesy Sony Electronics Inc.; **517** Sam Gray (bl); **518** Wurlitzer Jukeboxes; **519** Morris Lane/The Stock Market (fish); **520** Grant Heilman Photography (br); **524** Courtesy T. Skretting AS; **527** Catalogue No. 245023, Department of Anthropology, Smithsonian Institution (r); Photography by Edward S. Curtis, courtesy San Diego Museum of Man (l); **528** Missouri Historical Society; **532** © 1993 Louis Psihoyos/Matrix; **534** John Sibbick © National Geographic Society; **535** From Great Ocean Publishers Edition of *Gulliver's Travels* by Jonathan Swift with illustrations by J.J. Grandville. (t); **539** Richard Megna/Fundamental Photo; **544** Doug Perrine/Innerspace Vision (t); **544-545** Sportschrome East/West (b); **545** Tim Murphy/The Image Bank (r); **548** Bill Stanton/International Stock Photo (r); Suzanne & Nick Geary/Tony Stone Images/Chicago Inc. (l); **550** Flip Nicklin/Minden Pictures; **551** © Stan Osolinski (b); Mitch Reardon/Tony Stone Images/Chicago Inc. (br); Darrell Gulin/ Allstock; **552** Willie L. Hill, Jr./Stock Boston; **555** © Doranne Jacobson (r); Bert Blokhuis/Tony Stone Images/Chicago Inc. (l); Dick Luria/FPG International (t); **559** Philippe Psaila/Gamma-Liaison (r); Nadia Mackenzie/Tony Stone Images/Chicago Inc.; **561** Ralph Cowan/Tony Stone Images/Chicago Inc.; **565** © Duomo (b); © Kindra Clineff; **573** © Rhoda Sidney; Courtesy Patricia Malarcher; **582** © Duomo/David Madison; **587** The Bettmann Archive; **590** © Henry Horenstein; **597** Chuck Rydlewski/*Sports Illustrated*; **598** © Bob Daemmrich; **600** © 1993 Ron Kimball; **604** Stuart West-Moreland/Allstock; **605** © Henry Horenstein; **607** Robert Caputo/Stock Boston

ASSIGNMENT PHOTOGRAPHERS

Kindra Clineff **v** (b), **14, 31, 39** (r), **42** (l), **45, 50, 56, 60, 79** (c), **79** (b), **108, 119, 125** (t), **134, 135, 137, 140, 142, 155** (tr), **179, 240, 282, 294, 299** (tl), **299** (bc), **299** (tr), **299** (bl), **300, 351, 374, 382** (b), **408, 436, 447, 455, 469, 476, 491, 494, 504, 506, 507, 509, 510, 511, 515, 517, 520, 521, 545** (c), **556** (t), **606**

David Conover **102, 314** (r), **478, 482** (b), **497, 522, 536-537, 537**

Steven Greenberg **227** (r)

Richard Haynes/RM International **vi, x, xiv, xv, 1, 3** (b), **13, 16, 23, 35, 39** (t), **39** (c), **39** (l), **42** (r), **44, 65, 71, 88, 122, 155** (b), **160-161, 190** (b), **193, 199** (c), **199** (r), **211, 214, 219, 225, 227** (c), **261, 264, 269, 270, 274, 277, 279, 280-281, 285, 288, 299** (br), **306, 307, 308, 322, 343, 367, 368, 375, 380** (r), **400, 411** (b), **416, 425, 448, 475** (t), **485, 491, 500, 501, 547, 549, 554, 556** (b), **562, 562-563** (b), **576, 577, 591, 608**

Ken Karp **22, 26, 27, 28, 36, 48, 63, 79** (t), **181** (t), **189, 192, 247, 433, 467** (t), **487, 519, 566, 574, 585**

David Shopper **vii, 185**

ILLUSTRATIONS

Mary Azarian **69, 255, 256, 259**
Annie Bissett **77, 90, 118, 148** (b), **167, 169** (b), **313, 325, 350** (b)
Roger Boehm **298-299**
Jana Brenning **74, 121, 188** (l)
John Carrozza **490, 505**
Chris Costello **183, 356** (t)
Christine Czernota **200, 202, 208, 224, 235, 298** (l), **545**
Dartmouth Publishing, Inc. **59**
DECODE, Inc. **211, 333** (b)
DLF Group **7, 18, 83, 98, 120, 139, 169** (t), **184, 185, 186, 188** (r), **234** (b), **303, 393, 483, 497**
Bob Doucet **523**(t&b), **527**
Nancy Chandler Edwards **5, 25** (b), **126** (c), **165** (b), **178** (b), **370, 390, 437, 480, 484, 499**
Arn Franzen **581** (t)
Walt Fournier **6, 20, 23**
George Guzzi **349**
Piotr Kaczmarek **68, 148** (c), **154, 156, 164, 177** (t), **178** (br), **189** (b), **198, 216-217, 273, 502, 549** (t), **558** (b), **566**
Joe Klim **249, 267, 312** (b), **415, 477, 498, 518, 522, 525**
Jason P. Lee **258**
Andrew Meyer **54, 317, 435, 575** (b), **597**
Morgan Cain & Associates **132** (b), **133** (b), **143, 150, 151, 152, 163, 172, 174, 201, 321, 332, 397** (tl), **409, 410** (br), **452, 453, 455, 538, 564, 586, 594, 601**
Linda Phinney-Crehan **25** (t)
Neil Pinchin **125** (l&r), **126** (tl), **126** (r)
Lisa Rahon **xii, 516**
Photo manipulations by Lisa Rahon **xxvi, 10, 11, 183, 189, 195, 197, 199, 206, 214, 219, 228, 252-253** (b), **356, 357, 364** (cl), **372, 418, 419, 475, 482, 492, 513, 515, 519, 544, 600**
Patrice Rossi **8, 15**
Fred Schrier **272**
George Ulrich **358** (t), **378** (tr), **382** (tl), **405** (bl), **406** (tr), **407** (c), **544**

TYPOGRAPHIC TITLES

Frank Loose Design **3, 10, 19, 26, 31, 37, 44, 301, 306, 314, 321, 328, 337, 343, 359, 367, 375, 383, 392, 400, 417, 425, 433, 439, 447, 455, 463, 477, 486, 492, 500, 507, 515, 523, 530, 555, 562, 568, 583, 591, 598**

Midnight Oil Studios **57, 64, 71, 79, 85, 93, 99, 105, 111, 127, 135, 143, 150, 158, 164, 170, 183, 190, 218, 225, 241, 247, 255, 261, 270, 275, 280, 289**

DECODE, Inc. **211**

Scott Kim **547**

Selected Answers

 Unit 1

Pages 3 and 5 Talk it Over
1–2. Answers may vary. Examples are given. **1. a.** slippery when wet **b.** steep hill **c.** merge left **d.** railroad crossing
2. a. food/restaurant **b.** boarding area **c.** waiting area **d.** emergency phone/security **3.** Answers may vary. An example is given. The person has learned about issues involved with the subject of communication. **4.** Answers may vary. An example is given. One might include ideas about shapes of signs, colors of signs, messages, proper placement, and so on.

Pages 6–9 Exercises and Problems
9, 11, 13, 15. Answers may vary. Examples are given.
9. restrooms **11.** elevators **13.** two-way traffic
15. intersection **17.** Answers may vary. An example is given. Immigration is resulting in an increase in the number (and percentage) of American homes where a language other than English is used for conversation; article **19.** Wyoming; table **21.** Georgia; table **23.** Answers may vary. An example is given. The number of people speaking a language other than English at home in 1990, and the percent of change in this category for each state from 1980 to 1990
27. Answers may vary. An example is given. Native and Total People Who Speak Each Language **32.** 16
33. 16 **34.** 16 **35, 36.** Answers may vary.

Pages 11–12 Talk it Over
1. The variable expressions are alike because they both multiply a variable by 12. They are different because the first expression has a d for days and the second one has an h for hours. **2.** Answers may vary. An example is given. How many quarters are in x dollars? **3.** 2,000,001 toothpicks **4. a.** 15 **b.** $\frac{1}{2}$
c. 21 **d.** 25

Pages 14–16, 18 Exercises and Problems
1. A variable is a letter used to represent one or more numbers. **3. a.** 168; 8760 **b.** $24d$

5.

shape	perimeter
1	16
2	32
3	48
4	64
5	80

perimeter = $16n$

7. $p + t$ **9.** $p + t$ **11.** Answers may vary. An example is given. They use the same variables. The variables represent different things. **13.** D **15.** A

27. $x = \frac{1}{2}$

Perimeter

Number of x-tiles	Long rectangles	Tall rectangles
1	3	3
2	4	5
3	5	7
4	6	9
5	7	11

31. $2x + 6$ **42.** B **43.** Choices may vary. **44.** 2401
45. 59,049 **46.** 0.125

Pages 20–22 Talk it Over
1. $2^8 = 256$ **2.** $2^{12} = 4096$ **3.** Answers may vary. An example is given. I prefer the exponent key because it is faster. **4.** about 3.9 mi **5. a.** 9 zeros
b. $10^9 = 1,000,000,000$ **6. a.** n zeros **b.** 100 zeros
7. Answers may vary. An example is given. The sum of the exponents of the two powers of 10 on the left is equal to the power of ten on the right. **8. a.** 10^8
b. 10^8 **c.** 10^9 **9.** Answers may vary. An example is given. To multiply powers of ten, add the exponents.
10. a. 10^2 **b.** 10^4 **c.** 10^7 **11.** Answers may vary. An example is given. To divide powers of ten, subtract the exponents.

Pages 23–25 Exercises and Problems
1. Answers may vary. An example is given.
$2 \cdot 2 \cdot 2 \cdot 2 \cdot 2$; $2^3 \cdot 2^2$; $2^4 \cdot 2^1$ **3.** 4^3; four cubed
5. x^3; x cubed **7.** z^6; z to the sixth power **9.** 5^2
11. 7^3 **13.** x^6 **15, 17, 19.** Explanations may vary. An example is given. **15.** 4^3, 4^4, 4^5, 4^6; Four is used as a factor an increasing number of times. Since four is a

positive number greater than one, the larger the exponent, the larger the number. **17.** $\left(\frac{1}{2}\right)^5, \left(\frac{1}{2}\right)^4, \left(\frac{1}{2}\right)^3, \left(\frac{1}{2}\right)^2$;

Since $\frac{1}{2} < 1$, multiplying it by itself makes it smaller (the numerator stays the same, but the denominator gets larger). The larger the exponent, the smaller the number. **19.** $(0.1)^2, (0.3)^2, (0.4)^2, (0.8)^2$; Each base is multiplied by itself the same number of times (2). Since the bases are all positive, the larger the base, the larger the number. **21.** $x^2 + 1$; 50 **23.** $x^2 + x + 1$; 57 **25.** $x^2 + 4x + 4$ **29.** 10^{12} **31.** 10^{500} **33.** 10^0 **35.** 10^{446} **37.** 10^{a-b} **39. a.** 10^{15} **b.** 10^{15} **c.** 10^{15} **d.** 10^{a+b+c} **42. a.** $6.59s$ **b.** $16.99l$ **c.** $32.95; $50.97 **43.** Answers may vary. Examples are given. The word polygon means many angles. A triangle is a 3-gon. A quadrilateral is a 4-gon. **44.** about 88

Page 26 Talk it Over
1. Answers may vary, but should be near 78. **2.** The calculator added 84 and 2, multiplied this result by 75, and then divided by 3. **3.** The graphics calculator did the multiplication first, then the division, and did the addition last. **4.** Enter $\boxed{(}\,\boxed{8}\boxed{4}\,\boxed{+}\,\boxed{2}\,\boxed{\times}\,\boxed{7}\boxed{5}\,\boxed{)}$ $\boxed{\div}\,\boxed{3}\,\boxed{\text{ENTER}}$; 78 **5.** Enter $\boxed{2}\,\boxed{\times}\,\boxed{7}\boxed{5}\,\boxed{+}\,\boxed{8}\boxed{4}\,\boxed{=}\,\boxed{\div}\,\boxed{3}$ $\boxed{=}$

Pages 28–29 Exercises and Problems
3. 23 **5.** 13 **7.** 3 **9.** 12 **11.** 1009 **13.** 43 **15.** 9 **17.** 77 **19.** 70 **23.** $48 \div 4 + 8 - (4 + 2) = 14$ **25.** $48 \div (4 + 8) - 4 + 2 = 2$ **36–37.** Counterexamples may vary. Examples are given. **36.** When $x = 2$, $2^2 + 2 \neq 2^3 (4 + 2 \neq 8)$. **37.** When $x = 2$, $2^2 \cdot 2^4 \neq 2^8$ $(4 \cdot 16 \neq 256)$. **38.** Yes. **39.** No. **40.** Yes. **41. a.** Answers may vary. Examples are given: $x = 4$: $2x + 3 = 11, 2(x + 3) = 14$; $x = 3$: $2x + 3 = 9, 2(x + 3) = 12$; $x = 2$: $2x + 3 = 7$, $2(x + 3) = 10$ **b.** addition and multiplication; For $2x + 3$, multiply first, then add. For $2(x + 3)$, add first, then multiply. **c.** $2(x + 3)$; The expression $2(x + 3)$ is equivalent to $2x + 6$, which is 3 more than $2x + 3$.

Page 30 Checkpoint
1. Answers may vary. Examples are given. The biggest increase from 1980 to 1990 is in the number of people speaking Vietnamese at home. The number of people speaking Thai at home has more than doubled from 1980 to 1990. Among languages listed, the fewest number of people still speak Mon-Khmer. **2.** The languages are listed in order of highest percent change to lowest percent change from 1980 to 1990. **3.** 12 **4.** 24 **5.** 31 **6.** 10^8 **7.** 10^{18} **8.** 10^7 **9.** 82 **10.** 90 **11.** 6571 **12.** 40

Pages 31, 33 Talk it Over
1. Alicia adds the cost of the sandwich and the drink first and then multiplies by 4. Ben multiplies the cost of the sandwich by 4 and the cost of the drinks by 4,

and then adds these totals together. **2.** $15.36; They are the same. **3.** Answers may vary. An example is given. Alicia's; $2.25 + $.75 = $3.00, and 4(3) is easier to solve than $(4 \cdot 2.25) + (4 \cdot 0.75)$. **4.** $12 - 0.2 = 11.80$ **5.** Answers may vary. An example is given. Think of $.89 as $.90 - $.01: $4(0.9 - 0.01) = 4(0.9) - 4(0.01) = 3.56$. **6.** Answers may vary. An example is given. When combining like terms, use the distributive property in reverse to group the coefficients of the like terms. Add the coefficients to get the new coefficient of the variable. **7.** unlike terms

Pages 34–36 Exercises and Problems
1. $600 + 100$ **3.** $200 + 40$ **5.** 1025 **7.** 34 **9.** 4790 **11.** 2500 **13. a.** Answers may vary. Examples are given. Add together the area of each wall, or find the perimeter of the floor and multiply by the height. **b.** $ab + ac + ad + ae$ **15.** $x(x + 3); x^2 + 3x$ **17.** $6a - 3b$ **19.** $4t^2 + 1$ **25.** They cannot be combined because there are no like terms. **27.** $9y + 19$ **29.** $9y + 54$ **33.** 100 **34.** 361 **35.** 841 **36.** 10^2 **37.** 10^{10} **38.** 10^4 **39.** $2(v + u); 2v + 2u$ **40.** $3(s + t); 3s + 3t$

Pages 38–39 Talk it Over
1–2. Answers may vary. **3.** Answers may vary. An example is given. I think they should be considered the same because if either one is flipped over, it will match the other one exactly. Also, the pieces are the same shape and size. **4. a.** turn **b.** flip **c.** slide **5.** Yes. **6–8.** Answers may vary.

Pages 41–43 Exercises and Problems
1. slide **3.** turn **5.** No. **7.** Answers may vary. An example is given. Someone in a group may suggest an idea, a method, or a solution that you would not have suggested. Someone in a group might correct a mistaken idea of yours and set you back on a correct solution path. Working in a group allows you to get more answers to a problem situation. **9.** Answers may vary. **13.** Answers may vary. **17.**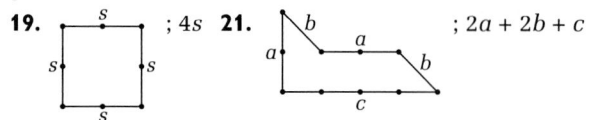

19, 21. Choices of variables may vary. Examples are given.

19. ; $4s$ **21.** ; $2a + 2b + c$

23. pentagon $HIJKL \cong$ pentagon $ONMLK$; side HI corresponds to side ON, side IJ corresponds to side NM, side JK corresponds to side ML **26.** $10x + 15$ **27.** $6y^2 - 18y + 24$ **28.** 0 **29.** 0 **30.** $4a$ **31.** $2b + 2c$

Integrated Mathematics

1. Answers may vary. An example is given. Parallelograms have opposite sides that are parallel and congruent. **2.** Answers may vary. An example is given. Kites have two pairs of adjacent sides that are congruent. **3.** Yes. Yes. **4.** parallelograms that are not rhombuses **5.** kite, isosceles triangle **6.** rhombus, rectangle; The square has four lines of symmetry. **7–10.** Answers may vary Examples are given.
7. Several words ("triangel," "inchs," and "perimters") are misspelled, and "maybe 'cause" is not standard grammar. **8.** The writer uses the terms "quadrilateral," "congruent," "parallelogram," "kites," "perimeters," "triangles," and "rectangles" correctly. **9.** The first report mentions the fact that when working with the triangle that had side lengths which were consecutive numbers (3, 4, 5), the perimeters of the figures made were consecutive even numbers (14, 16, 18). This idea could be developed inductively by testing other triangles with consecutive numbers as side lengths. It could be developed deductively by testing a general triangle with consecutive sides a, $a + 1$, and $a + 2$. **10.** The first report simply shows the two different triangles with no pictures of the shapes formed from these triangles. In the second report, the writer shows the triangles forming quadrilaterals, and labels the sides to show which congruent sides were joined. This gives the reader a visual summary of the shapes formed in the triangle exercise.

Pages 47–49 Exercises and Problems
1. Jeanine: equilateral (also isosceles); Alain: isosceles; Helen: scalene; Duane: right, scalene **3.** 22, 24, 26
5. perimeter = 4b perimeter = 2a + 2b

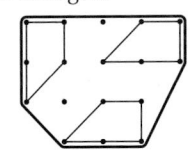

7. Trapezoids and perimeters may vary. Examples are given. **a.**   **b.** 20; 18

9. rhombus **11.** parallelogram

13.   **15.** no symmetry

19. Quadrilateral $EFGH$ has two pairs of opposite sides that are congruent, so it is a parallelogram.
27. Explanations may vary. An example is given. From Exercise 6, we know that for congruent scalene triangles with sides a, b, and c, the perimeters of all possible quadrilaterals are 2a + 2b, 2a + 2c, and 2b + 2c. These perimeters will hold true for any type of

triangle, for if the triangles are not scalene then we know that if the triangles are isosceles then $a = b$ or $a = c$ or $b = c$ or if the triangles are equilateral we know $a = b = c$. **29.** Answers may vary.
30. $a + b + 3c$ **31.** 2a + 2b **32.** 100; ten squared or one hundred **33.** 1000; ten cubed or one thousand **35.** 1,000,000,000; ten to the ninth power or one billion

Pages 51–52 Unit 1 Review and Assessment
1. German, Italian, Polish **2.** Explanations may vary. An example is given. A bar graph might be hard to read because the numbers are close. **3.** Answers may vary. An example is given. Spanish is the language spoken most at home by United States residents 5 years old and over who speak a language other than English; Portuguese is spoken the least; speaking Polish has declined from 1980 to 1990. **4.** $n + 4$
5. $x \div 4$ **6.** 4x **7. a.** 252 mi **b.** 21x

8.

Rectangle	Perimeter
1	2x + 4
2	4x + 4
3	6x + 4
4	8x + 4
5	10x + 4

9. 10^{11} **10.** 10^3 **11.** 10^{12} **12. a.** Answers may vary. An example is given. $10^2 + 10^3 \neq 10^{2+3}$, because $100 + 1000 \neq 100,000$. **b.** No. **c.** $10^a \cdot 10^b = 10^{a+b}$
13. 1041 **14.** 6 **15.** 3740 **16.** 1386 **17.** $12x^2 + 3x$
18. There are no like terms. **19.** $5x^2 + 10x$
20. Amy's mistake is that she added the 4 and 6 together and then used the distributive property: $10(3x + 12) = 30x + 120$. She should have used the distributive property first, $6(3x + 12) = 18x + 72$, and then combined like terms: $4 + 18x + 72 = 18x + 76$.
21. Answers may vary. Examples are given. $ABGK \cong JEDH$; $ABGK \cong SNMQ$; $JEDH \cong SNMQ$
22. A *flip* across the line that passes through D and H will show that $ABGK \cong SNMQ$. A *slide* up two units and to the right one unit will show that $SNMQ \cong JEDH$. A *slide* up two units and to the right one unit and a *flip* will show that $ABGK \cong JEDH$.
23. Answers may vary. An example is given. Using $ABGK$ and $JEDH$, side $AB \cong$ side JE, side $BG \cong$ side ED, side $GK \cong$ side DH.
24. hexagon   **25.** Choices of variables may vary. An example is given. $a + b + c + 2d + e$

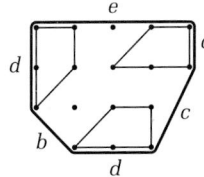

26. a. no perpendicular sides **b.** sides *AC* and *MO*; sides *AM* and *CO* **c.** parallelogram **27. a.** sides *LR* and *RJ*; sides *RJ* and *JD*; sides *JD* and *DL*; sides *DL* and *LR* **b.** sides *LR* and *DJ*; sides *LD* and *RJ* **c.** rectangle **28. a.** sides *KW* and *WO*; sides *WO* and *OC*; sides *OC* and *CK*; sides *CK* and *KW* **b.** sides *KC* and *WO*; sides *KW* and *CO* **c.** square **29. a.** no perpendicular sides **b.** sides *LC* and *WN*; sides *LW* and *CN* **c.** rhombus

Unit 2

Pages 57–60 Talk it Over
1. Answers may vary. An example is given. thousands **2.** Answers may vary. An example is given. Split the crowd into equal-size smaller sections. Count or estimate the number of people in one of these smaller sections, then multiply by the number of sections. **3.** Answers may vary. **4.** Answers may vary. **5.** Answers may vary. Examples are given. Yes, hair could be separated into equal parts and a smaller group counted. Yes, you could choose a section of the movie theater and count the number of people seated in that section. No, you cannot see the entire population of a city in order to divide it into equal groups. You would have to conduct a survey.
6. Answers may vary. Examples are given. A fraction might be used to indicate yardage of fabric; a decimal might be used to indicate a price; a negative number might be used to indicate a business loss.
7–9. Answers may vary. Examples are given.
7. exact; Each student has only one school identification number, and each school identification number identifies only one student. **8.** estimated; The population of a state capital is constantly changing. People move in and out of the city and there are births and deaths. Generally, it is not possible to determine an exact population count and an estimate is used instead. **9.** estimated; People often read books over several days and experience many interruptions.
10. Answers may vary. An example is given. about 1 million groups **11.** Answers and methods may vary. Examples are given. thousands; Count the number of math books laid flat, end to end, that will fit along a longer wall of the classroom; then count the number of math books, laid flat, side by side, that will fit along a shorter wall of the classroom. Count or estimate the number of books that would fit in a pile from the floor to the ceiling of the classroom. Multiply these three counts to estimate the number of books that could be stacked in the classroom. **12.** the population of the United States in 1990 and the population of the world in 1990; Yes. **13.** Estimates may vary. An example is given. thousands **14.** Answers may vary. **15.** 0.8 **16.** continuous **17.** continuous **18.** discrete **19.** (16) measure; (17) measure; (18) count

Pages 60–63 Exercises and Problems
1. a–d. Answers may vary. Examples are given.
a. hundreds **b.** Split the dish into 4 parts, count the bacteria in one part and multiply by 4. **c.** about 220 **d.** My guess was right. **3.** Guesses, estimates, and methods may vary. An example is given. hundreds; about 130 photographs; Suppose there are 13 photographs in Unit 2 of your textbook and there are 10 units in the textbook; 13 · 10 = 130 photographs.
5. Answers may vary. Examples are given. identify: school ID number; order: periods of the day; count: number of students in each class; measure: length of playing field **7, 9.** Estimates may vary. Examples are given. **7.** thousands **9.** thousands **11.** more than **13.** unreasonable; Los Angeles has fewer residents than New York City, which had only about 7.3 million residents in 1990. **15.** about 1.1 billion **17.** No, the average weight of an adult is much less than 400 lb, so it is very unlikely that the combined weight of the 10 adults is near the capacity weight of the elevator. **19, 21.** Answers may vary. **23.** 1
25. discrete **27.** continuous
36.

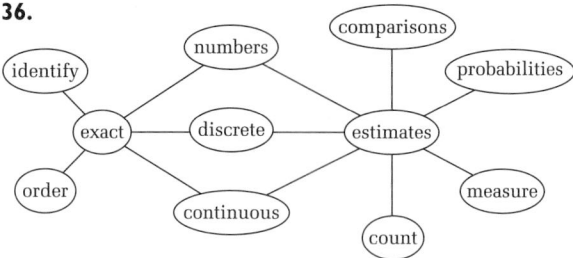

37. 2 · 3 · 7 **38.** 2 · 2 · 3 · 3 **39.** 2 · 7 · 7 **40.** 5 · 5 · 5
41.
$$\xleftarrow{\quad\bullet\;\;\bullet\;\;|\;\;\bullet\;|\;|\;|\;|\;\bullet\;|\;\bullet\;\bullet\;|\;\bullet\quad}\rightarrow$$
  −7 −5   0   3 4   7
7 and −7 are the same distance from 0.

Pages 64–66 Talk it Over
1. They are used to show measurement below sea level. Answers may vary. Examples are given: temperature below zero, overdrawn bank balance
2. between −6 and −9, closer to −9 **3. a.** the absolute value of twelve; 12 **b.** the absolute value of negative four; 4 **4.** 10 and −10 **5.** −7; 2; 0 **6.** *negative 12* or *the opposite of 12* **7.** −15; 27 **8.** 0 **9.** 0 **10.** 0 **11.** 0
12. −1.3, 5, 0; The results are the same as those in Sample 1. The opposite of a number is equal to the product of −1 and the number: −*x* = −1 · *x*. **13.** 88
14. Answers may vary. An example is given. I can do Sample 1 by mental math. I prefer to use a calculator for Sample 2 because it involves more than one calculation at a time.

1. absolute value: 17; opposite: 17 **3.** absolute value: $\frac{1}{2}$; opposite: $\frac{1}{2}$ **5.** absolute value: 3,785,200; opposite: 3,785,200 **7.** False. When $x = -2$, $-x = -(-2) = 2$, which is positive. **9.** True. **11.** False. For example, when $x = -3$, $|-3| \neq -3$. **13.** -53 **15.** 18 **17.** 24 **19.** -19 **21.** -180 **23.** -30 **25.** -4.5 **27.** Answers may vary. An example is given. Exercises 18, 20, 21, and 26 can be easily done by mental math. In Exercise 21, just add 120 and 60 together in your head and keep the negative sign; you get -180. In Exercise 26, multiplying 21 by 100 is easy and you know that the product of two negatives is a positive; the product is 2100. **31.** $300
33. $3 + 18 \div 2 \cdot (-3) = 3 + 9 \cdot (-3) = 3 + (-27) = -24$
35. $-9 \div 2 \cdot 10 \div 3 = (-4.5) \cdot 10 \div 3 = -45 \div 3 = -15$
37. $\frac{-20 - 5}{2} = \frac{-25}{2} = -12.5$ **39.** -12.5 **41.** -16 **43.** -3
49. $37°C$ **51.** $-243.4°F$ **55–57.** Estimates may vary. Examples are given. **55.** millions **56.** thousands **57.** thousands **58.** base, exponent or power; factor **59.** 10^{14} **60.** 10^{40} **61.** 10^{15} **62.** 10^{27}

Pages 71, 73, 74 Talk it Over
1. a. 6 **b.** 5 **c.** 6 **2.** $\frac{1}{10^{12}}$, $\frac{1}{1,000,000,000,000}$
3, 4. Descriptions may vary. Examples are given.
3. 3.4×10^4; You can think of 34,000 as 3.4 ten thousands. Since ten thousand is the fourth power of ten, 34,000 is the same as 3.4×10^4. **4.** 2.19×10^{-6}; The first digit not equal to zero, 2, is in the millionths' place. Since millionths is the sixth decimal place, 0.00000219 is the same as 2.19×10^{-6}. **5.** 9,300,000; nine million, three hundred thousand **6.** 0.00048; forty-eight hundred-thousandths **7.** No, it should be 4×10^{14}. **8.** No, it should be 1.27×10^{-5}.
9. You need to know how long the recording time is.
10. 10^5 **11.** 10^{-1}; 10^{-2} **12.** 10,000; 100,000; 10 EXP 4 means $10 \cdot 10^4$ or 10^5.

Pages 75–77 Exercises and Problems
1. Scientific notation writes a number as the product of a number greater than or equal to one and less than 10, and a power of 10. **3. a.** Descriptions may vary. Examples are given. Use a calculator, rewrite 10^{-7} as $\frac{1}{10^7}$ and find its decimal form, or use the rule that 10^{-7} has seven decimal places (six zeros to the right of the decimal point, followed by a one).
b. 0.0000001 **5.** $=$ **7.** $>$ **9.** $<$ **11.** 10^{10} **13.** 10^0
17. 2.39×10^{-8} Cal **19.** 5.036×10^5 kg **21.** 0.04 m
23. 2.2 m **25.** 26 m **27.** C **29.** G **31.** 1.44×10^{10}
33. 3.78×10^{-7} **35.** 2×10^3 **37.** Explanations may vary. An example is given. Rewrite the product as $(3.4)(1.2)(10^{15} \cdot 10^{23})$. This can be simplified to give the answer 4.08×10^{38}; [3.4 EE 15 × 1.2 EE 23 =]
43. Estimates may vary. Examples are given. United

States: about $3.4 \times 10^8 = 340,000,000$; United Kingdom: about $6 \times 10^7 = 60,000,000$; Japan: about $3 \times 10^8 = 300,000,000$ **45.** about $4 \times 10^7 = 40,000,000$
48. 1.8 **49.** -48 **50.** -3.2 **51.** 4 **52.** 24 **53.** 0.8
54. 600

Page 78 Checkpoint
1. Answers may vary. An example is given. You could estimate how many kernels fit in a handful or a measuring cup and then estimate how many of these handfuls or cupfuls would fit in a shoebox.
2. Estimates may vary. An example is given. hundreds **3.** millions **4.** discrete **5.** Answer may vary with location of city. **6.** 7 **7.** 14 **8.** 3.45×10^{-4}
9. 2.671×10^7 **10.** 800,000,000,000

Pages 79–82 Talk it Over
1. 1 in. **2.** 1 in. **3.** 1 m **4.** 1 mi **5.** 1 m; Answers may vary. An example is given. Look in a dictionary or encyclopedia under *measurement* or *metric system*. **6. a.** one hundredth; 10^{-2} **b.** one thousand; 10^3
7–8. Estimates may vary. Examples are given.
7. about 90 ft or about 30 m **8.** about 7 ft or about 2 m **9.** Answers may vary. An example is given. A hairdresser is likely to be good at estimating inches or centimeters; a carpenter, inches, feet, and yards, or millimeters, centimeters, and meters; and a runner, yards and miles, or meters and kilometers.
10. Estimates may vary. An example is given: about 1700 mi north to south; about 2800 mi east to west **11.** Answers may vary. An example is given. Find the total area of the large rectangle that contains both A and B and subtract the area of the small rectangle in the upper right-hand corner. **12. a.** about 7 squares
b. 2800 mi^2 **13.** $\overleftrightarrow{AB}$ represents the line shown. $\overline{AB}$ is the segment with endpoints A and B. AB is the distance from A to B, which is 6 units. **14.** Estimates may vary. An example is given. about 1.5 in.

Pages 82–84 Exercises and Problems
3. QR is about 1 in. or 2.5 cm; QT is about 0.5 in. or 1.5 cm; TS is about 0.5 in. or 1.5 cm; SR is about $\frac{3}{8}$ in. or 1 cm **7.** No. The distance is approximately 1425 km. **9.** about 7,600,000 km^2 **15.** True.
17. False. **19.** True. **24.** 2.5×10^{-7}; 0.00000025
25. parallelogram **26.** rectangle **27.** kite
28. Definitions may vary. Examples are given. An acute angle is an angle that measures between 0° and 90°. A right angle is an angle that measures 90°. An obtuse angle is an angle that measures between 90° and 180°.

Pages 85–87 Talk it Over
1. Answers may vary but should include three of the following: $\overrightarrow{OW}$, $\overrightarrow{OX}$, $\overrightarrow{OY}$, $\overrightarrow{OZ}$. **2.** Answers may vary but should include three different angles from among the following: $\angle 1$, $\angle 2$, $\angle 3$, $\angle WOX$, $\angle WOY$, $\angle WOZ$,

∠XOW, ∠XOY, ∠XOZ, ∠YOW, ∠YOX, ∠YOZ, ∠ZOW, ∠ZOX, ∠ZOY. **3.** ∠WOX, ∠XOW **4.** vertex **5.** No. Explanations may vary. An example is given. There are many angles with vertex *O*, so it isn't clear which angle "∠*O*" would identify. **6.** Answers may vary but should include two different angles from among the following: ∠1, ∠2, ∠3, ∠WOX, ∠WOY, ∠WOZ, ∠XOW, ∠XOY, ∠XOZ, ∠YOW, ∠YOX, ∠ZOW, ∠ZOX, ∠ZOY. **7.** "under 15" and "15" **8. a.** "15" and "15.5" **b.** 90° **c.** They are complementary. **9–11.** Answers may vary. Examples are given. **9.** Method 1: Draw a straight angle. Sketch two rays to divide the angle into three equal parts. Each part is close to a 60° angle. Method 2: Divide a 90° angle into thirds. Two thirds that are side by side are close to a 60° angle. **10.** Sketch a 45° angle as described in Sample 1, part (b). Sketch a 30° angle adjacent to the 45° angle using Method 2 of question 9. The result is close to a 75° angle. **11.** Draw a straight angle. From its vertex, draw a ray not perpendicular to the sides of the straight angle. The larger angle will be an obtuse angle.

Pages 89–92 Exercises and Problems
1. Answers may vary. An example is given. An angle can be named one way by a number or two ways by three letters. The middle letter must always be the vertex, but the other two letters, which represent a point on each side of the angle, can be either first or last. **3.** $\overrightarrow{OP}$, $\overrightarrow{OL}$, $\overrightarrow{OM}$, $\overrightarrow{ON}$ **5.** ∠*LOM* (∠*MOL*), ∠*PON* (∠*NOP*) **7.** ∠*LON* **9. a.** supplementary **b.** Yes. ∠*AOB* = 90° and ∠*LOP* = 90° **11.** 60° **13. a.** Answers may vary. **b.** Answers may vary. An example is given. ∠4 = ∠2 ≈ 70°; ∠1 = ∠3 ≈ 110° **c.** Answers may vary. **15, 17.** Sketches, methods, and descriptions may vary. Examples are given. **15.** Sketch a 90° angle and divide it in half to estimate a 45° angle. Then divide in half one of the 45° angles to estimate a 22° angle.

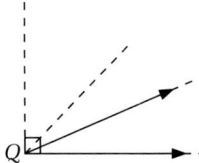

17. Sketch a straight angle. Sketch two rays to divide the angle into three angles of equal measure. Combine two of the angles, as shown, to estimate a 120° angle.

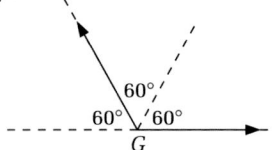

19. ∠*AKB*, ∠*GKD* **21.** 45° **23.** 22.5° **25.** *y* + 102 + 32 = 180; 46° **27.** *c* + 6 + 14 = 90; 70° **32.** segment *AB* **33.** length of segment *JK* **34.** line *RS* **35.** length of segment *PQ* **36.** commutative property of multiplication **37.** associative property of addition **38.** 4*x* + 8 **39.** 8*a* + 6

Pages 93, 95 Talk it Over
1. 90 + 90 + 3*x* + 2*x* + 3*x* = 180 + 8*x*
2. *y* + 12 + 12 + *y* + 2*y* = 24 + 4*y* **3.** 18y^2
4. $x^3 + 8x^3 + x^3 + 3x^3 = 13x^3$ **5.** No. **6.** Yes. **7.** No.
8. No. It cannot be simplified, because there are no like terms.

Pages 96–98 Exercises and Problems
3. 2(3*a* + 3*b*) + 2(7*a* + 7*b* − 1) = 20*a* + 20*b* − 2
5. 2(*x* · *z*) + 2(*y* · *z*) + 2(*x* · *y*) = 2*xz* + 2*yz* + 2*xy*
7. 30m^2 **9.** 40*ac* **11.** 14*h* + 7*hk* + 11*k* **13.** 11m^3 + 4*n*
15. 2*xy* **17.** 8*cf* − 6*f* + 3 **19.** 9*y* · 5*z* = 45*yz*
21. 10*c* · 5*c* · 3*c* = 150c^3 **23.** C **25.** D

35. 40° is a little less than half of 90°.

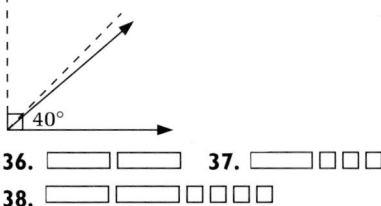

36. ▭▭ ▯▯▯ **37.** ▭▯▯▯
38. ▭▭▭ ▯▯▯▯

Page 98 Checkpoint
1. Explanations may vary. An example is given. To estimate distances on a map, you use the given scale as a ruler to determine lengths by finding out how many of these scale units are in a given distance and then multiply the value of the scale by the number of units. To estimate areas on a map, you determine the dimensions of a given area using the method described above and then use the appropriate area formula. To estimate angle measures, you compare a given angle to an angle of known measure, usually a 90° or 180° angle, and estimate depending on whether the given angle is smaller or larger than the angle it is being compared to. **2.** about 500 km **3.** No, it measures approximately 800 km. **4.** about 720,000 km² **5.** *AB* ≈ $\frac{3}{4}$ in. ≈ 2 cm; *AC* ≈ 1$\frac{1}{4}$ in. ≈ 4.5 cm; *BC* ≈ 1 in. ≈ 2.5 cm **6.** False. *AX* = *XC* if *X* is the midpoint of $\overline{AC}$. **7.** 22 + 19 + *x* = 180, *x* = 139°

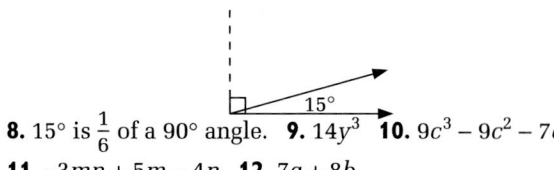

8. 15° is $\frac{1}{6}$ of a 90° angle. **9.** 14y^3 **10.** 9c^3 − 9c^2 − 7*c*
11. −3*mn* + 5*m* − 4*n* **12.** 7*a* + 8*b*

Page 101 Talk it Over

1. No. Reasons may vary. An example is given. $46 + 18 = 64$, not 28, so 23 is not a solution. **2.** Yes. Reasons may vary. An example is given. Both equations have the solution 2. Equations with the same solution are equivalent. **3.** Subtract 9 from both sides. 45 **4.** Divide each side by 6. 6.5 **5.** Add 10 to both sides, then divide both sides by 6. 37

Pages 103–104 Exercises and Problems

5. 7 **7.** 3 **9.** 5 **11.** 4 **13.** The correct solution is 11. The other solver added 6 to both sides instead of subtracting 6 from both sides. **15.** The correct solution is 6. The other solver divided by 5 first instead of adding 10 first to both sides of the equation. In doing so, the solver forgot to divide 10 by 5. **19.** B **21.** D **23.** $n + 69 + 98 + 54 = 360$; 139° **25.** $x + x + 40 = 180$; 70°; $y + y + 30 = 116$; 43 cm **30.** $-32x^3$ **31.** The expression cannot be simplified because there are no like terms. **32.** $-5a^3 + 3ab + 4a$ **33.** -1 **34.** -16 **35.** 4.8 **36.** -50

Page 106 Talk it Over

1. a. $-49 = 4n + 7$

Undo addition of 7. Subtract 7 from both sides.

$-49 - 7 = 4n + 7 - 7$

$-56 = 4n$

Undo multiplication by 4. Divide by 4.

$-\dfrac{56}{4} = \dfrac{4n}{4}$

$-14 = n$

b. You would add 7 to both sides instead of subtracting 7. **2.** 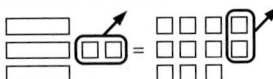; -14

Page 108, 110 Exercises and Problems

3. 16 **5.** -2 **7.** 10.5 **9.** 1540 **11.** -108 **13.** $-0.08\overline{3}$
15. $255.50 = 150.50 + 35x$; 3 h
17. $11.97 = 1.29 + 12x$; $0.89
19. $33.88 = 3.99x + 4.48x$; 4 yd
27. a. $3x + 2 = 11$ **b.**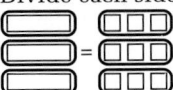

Subtract two unit tiles from each side.

Divide each side into three identical groups.

Look at one of the groups.

$x = 3$

28. 63,000,000 **29.** 0.000018 **30.** 11, 11, 121
31. 7, 7, 7, 343

Pages 111, 112, 114 Talk it Over

1. 3; 4; 5 **2.** 6 units; 7 units; 8 units; 10 units; 20 units **3.** greater than; The diagonal of a square with side 1 is greater than 1. **4.** less than; One way is to measure two units on the grid and compare this to the length of the diagonal. **5.** BD is greater than AB and BC; BD is greater than 3; BD is less than 4. **6.** PQ is greater than AB and BC; PQ is less than BD; PQ is greater than 2; PQ is less than 3. **7.** Answers may vary. Examples are given: 1.5; 2.25; 2.75; 3.25; 3.5 **8.** $\sqrt{2} \approx 1.41$; $\sqrt{5} \approx 2.24$; $\sqrt{8} \approx 2.83$; $\sqrt{10} \approx 3.16$; $\sqrt{13} \approx 3.61$ **9.** Yes, the results should be about the same. Yes, the square roots and the estimated lengths of the sides are about the same. Explanations may vary. An example is given. In the formula for the area of a square, $A = s^2$, the side, s, is the positive square root of the area, A. **10.** 27 cm³ **11.** 2 cm

Pages 116–118 Exercises and Problems

3. $\dfrac{1}{9}$ and $-\dfrac{1}{9}$ (or $0.\overline{1}$ and $-0.\overline{1}$) **5.** $\dfrac{10}{3}$ and $-\dfrac{10}{3}$ (or $3.\overline{3}$ and $-3.\overline{3}$) **7.** rational **9.** rational **11.** rational **13.** $\sqrt{21}$ is a number between 4 and 5. 4.58 **15.** $\sqrt{38}$ is a number between 6 and 7. 6.16 **19.** $\sqrt[3]{90}$ is a number between 4 and 5. 4.48 **21.** $10 < \sqrt[3]{1300} < 11$; $\sqrt[3]{1300} \approx 10.91$ **23.** F **25.** C **27.** E **29.** $\sqrt[3]{14}$, 3, $\sqrt[3]{48}$, $\sqrt{48}$, $\sqrt[3]{1100}$, $\sqrt{169}$, 18, $\sqrt{700}$ **33.** 15 cm **35.** about 10.86 ft **37.** 4 cm **39.** about 7.05 in. **41.** 8 units, 9 units **48.** 0.9 **49.** 64 **50.** -2 **51.** 25% **52.** 36 **53.** 420

54.

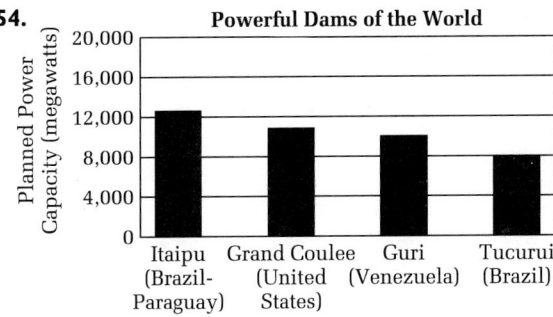

Powerful Dams of the World

Pages 120–122 Unit 2 Review and Assessment

1. Answers may vary. hundreds; 340; Count the number of spaces between the words in the first line (10) and multiply by the number of lines (34). **2.** millions **3.** Answers may vary. **4.** continuous **5.** 4 **6.** 7 **7.** 1 **8.** 28 **9.** 0.00000001 **10.** 1 **11.** 7.151×10^8 **12.** 9.06×10^{-3} **13.** 3.065×10^{-3}; 0.003065; 3065 millionths **14. a.** 4.8×10^{12} m (rounding 5.79 to 6 and 78 to 80) **b.** 4,516,200,000,000 m **15.** about 1250 km **16.** No. It is closer to 750 km. **17.** about 562,500 km² **18.** about 5,625,000 km² **19.** True. **20.** True. **21.** False. **22.** Answers may vary. Examples are given. $\angle DCE$ and $\angle DCA$, $\angle ECB$ and $\angle BCA$ **23.** $\angle ACB$ **24.** Answers may vary.

Examples are given. ∠*DCE* and ∠*ACB*, ∠*DCA* and ∠*ECB*   **25. a.** 52° **b.** 128° **26.** 100° is slightly larger than a right angle.

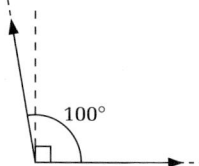

100°

27. Paragraphs may vary. An example is given. Measurements can be estimated or exact. You need to estimate when the number of objects to be counted is very large or if you do not have precise measuring instruments. You can estimate a number of objects by estimating a small portion of the whole and then multiplying this estimate by the number of portions of that size in the whole. Also, you can estimate an unknown quantity by relating it to a known quantity, such as the population of a large city. You can estimate distances by using a map scale or by using a familiar object, like a foot, whose estimated length is known. When the measurement that you need is limited in some way, such as the dimensions of a door through which you must carry a piece of furniture or the width of a shade that you have to order for a window, then an exact measurement is needed. Otherwise, an estimate is fine.   **28.** $10mn + 5m$
29. $-2k^3 - k^2 - 14$ **30.** $96a^2$ **31.** $(3a)(12a) = 36a^2$
32. $40 + 90 + x = 180$ or $40 + x = 90$; $x = 50$
33. $x + 50 + x = 90$; $2x + 50 = 90$; $x = 20$ **34.** 105
35. −50 **36.** 4 **37.** $45.12 = 1.39n + 0.49n$; 24 people
38. $8.51 = 8y + 2.19$; $.79 **39.** rational
40. 11 and −11 **41.** $4 < \sqrt[3]{72} < 5$ **42.** about 12.25 ft
43. about 4.33 mm

Unit 3

Pages 128–130 Talk it Over
1. 0.17; dropout rate in 1991 **2.** No. **3.** Yes. Answers may vary depending on survey used. **4.** Answers may vary. An example is given. The dropout rate shows that the percentage of dropouts per year declines, so the line graph is decreasing. The line graph for dropouts increases because the number of dropouts increases by year, but this is probably due to the increase in participants. **5.** Answers may vary. An example is given. The trends are easier to see by looking at the graphs, because they show at a glance whether there has been an increase or decrease.
6. row 1 and column A **7.** 29,036; B4; C5 represents interest income in 1991 and B4 represents donations of goods and services in 1990. **8.** B6 and C6; Formulas can use information from other cells of the spreadsheet directly to compute values and save time. **9.** rows 2, 3, 4, and 5 **10.** a column

Pages 131–134 Exercises and Problems
1. 3×5 **3.** tax, in dollars, per gallon of gas in Japan
5. Answers may vary. An example is given. A stacked bar graph is better because it enables the reader to compare how the total cost of gas varies by country.
7. a. Answers may vary. An example is given. The tides occur later every day. High and low tides stay roughly the same distance apart. **b.** Answers may vary. An example is given: high tide—around 5 P.M., low tide—around 11 P.M. **11. a.** compact discs
b. Answers may vary. Using the graph seems faster.
13. a. Cassettes had the highest value of shipments, followed by compact discs. Sales of albums and singles were much lower. **b.** The value of cassette shipments increased from 1986 to 1988, and declined in the following year. **c.** In part (a) you read the values vertically on the line representing 1988, while in part (b) you read the values horizontally, following the line representing cassette shipments. **15.** 301.8; C3; D4 represents net income or loss in 1989 and C3 represents expenses in 1988. **17.** row 4
19, 21. Answers may vary. Examples are given.
19. pictograph; shows the slight difference in numbers better **21.** circle graph; shows half the total much better **25.** 14 m **26.** Answers may vary. Example: 0.5 **27.** Answers may vary. Example: 0.75
28. 7 **29.** −30 **30.** 8

Pages 135–137 Talk it Over
1. Answers may vary. An example is given. The numbers are decimals that range from 0.3 to 5.1. **2.** Yes, 1.1 appears three times. **3.** Yes, the two types of muffins have much higher fat contents than the other breads and crackers. **4.** Estimates may vary.
5. The amounts 5.1 and 4.0 are very different from the other amounts. These amounts increase the total sum, which is used to find the mean, but their sizes do not affect the median or the mode. **6.** No. The mean is not necessarily greater than the median or the mode(s) in a data set. For example, in the data set 2, 5, 9, 12, 12, the median, 9, and the mode, 12, are both greater than the mean, 8; in the data set 5, 5, 5, 5, the mean, the median, and the mode all equal 5. **7.** Answers may vary. An example is given. The median appears most typical because it is close in value to 9 of the 11 values. **8.** corn and bran muffins; avoid eating these muffins **9. a.** No. 1.1 **b.** mean: about 0.85; median: 0.9; mode: 1.1; the mean

Pages 139–142 Exercises and Problems
1. Methods may vary. An example is given. Arrange the values in order from smallest to largest: 12, 15, 17, 23, 43, 56. There are 6 numbers in the list, so the median is the average of the third and fourth numbers, 17 and 23. The median is 20. **3.** Yes, if each entry appears the same number of times. **5.** Choices for most typical numbers and reasons may vary.

Examples are given. mean: about 134.4; median: 134.5; modes: 122, 144; The modes are most typical since they represent 6 of the 14 data values. The mean and median are 8 units away from the closest data value. **7.** Answers may vary. Examples are given. There are no outliers in Exercises 4 and 5; in Exercise 6 the outlier is 593. **9.** It is not possible to find the mean because the data are not numerical. If there were an odd number of sizes, then the middle size when the sizes are listed from smallest to largest would be the median. Since there are an even number of sizes and you cannot find the "average" of two sizes, there is no median. The mode is XL. **11.** about 75.7 **13. a–c.** Answers may vary. Examples are given. **a.** The mean gives equal importance to all her scores because it is the average. **b.** The median is in the middle and is not influenced by a single score that may not be representative of her work. **c.** The median is more typical of her scores. She has one poor grade which affects her average. **15. a.** Ad Pros: mean: $23,800; median: $25,000; mode: $25,000. Lunch Fare: mean: about $24,167; median: $19,000; modes: $12,000 and $25,000 **b.** Lunch Fare; Ad Pros; Ad Pros **19.** −33 **21.** 918; since $\frac{\text{sum}}{6} = 153$, sum = 6 • 153 = 918 **23.** No; reasons may vary. An example is given. She would need to get 108 on the last test to have a mean score of 90, and a score of 108 is impossible. **27.** No. The median age has nothing to do with the current year, but rather with population changes and life expectancy. **30.** Answers may vary. An example is given. A matrix would be a more useful tool for displaying this data, because I could easily use the data to estimate monthly fuel costs. **31.** 3^7 **32.** x^3

33. ◄━●━┼━┼━┼━┼━┼━┼━●━► −5 < 3
−5 −4 −3 −2 −1 0 1 2 3

34. ◄━┼━●━┼━●━┼━┼━┼━► −4 < −2
−5 −4 −3 −2 −1 0 1 2

Pages 143, 145 Talk it Over

1. distance **2.** about 8.13 m **3.** 1928, 1961, 1964; 1962 and 1964, 1965 and 1967 **4.** The record jumps would have to be greater than or equal to 8.96 m. **5.** Long jumps before this date were less than 7.98 m. **6.** greatest increase: 1968; smallest increase: 1965; The biggest interval represents the greatest increase and the smallest interval represents the smallest increase. **7.** to include 8.96 m and show the record could be tied at a future date; The symbol > means "greater than," while the symbol ≥ means "greater than or equal to." The symbol < means "less than," while the symbol ≤ means "less than or equal to." **8.** If j is greater than 0, then 0 must be less than j.

9. Answers may vary. An example is given. Yes. No. In Sample 1, any and all numbers greater than or equal to 8.96 do represent a possible long jump record, and this is what the graph shows. However, realistically speaking, the length of a long jump must be limited and cannot go into infinity. In Sample 2, there are an infinite number of points between 0 and 1.91, yet there are not an infinite number of past high jump records.

Pages 146–148 Exercises and Problems

3. B **5.** ◄━┼━●━┼━┼━┼━⊕━┼━►
−2 0 2 4 6 8 10

x is greater than or equal to zero and less than 8.

7.
−1.6   1.1
◄━┼━●━┼━┼━┼━┼━●━┼━►
−2 −1.5 −1 −0.5 0 0.5 1 1.5

x is greater than or equal to −1.6 and less than or equal to 1.1.

9. $-2 \le t \le 30$, t = surface temperature
◄━●━━━━━━●━►
−2   30

11. $90 < d < 180$, d = degree measure of an obtuse angle ◄━┼━━⊕━━⊕━►
0   90   180

13. $40 \le s \le 55$, s = speed limit
◄━┼━━━━●━━●━►
15   40 55

25. mean: about 32.93; median: 32; mode: 37
26. 15 and −15; about 6.08
27.

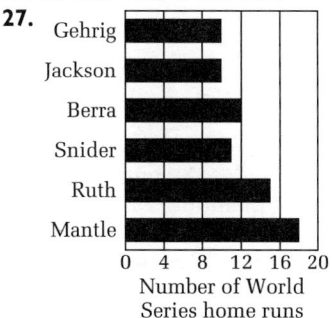

Number of World
Series home runs

Page 149 Checkpoint

1. Answers may vary. An example is given. Hi! I decided to write to tell you how my grades have improved. Although the median (middle) score is theoretically 78, it was not one of my actual scores. My first four scores were 75 or lower, my last four scores were all higher than 80. Although my mean (average) score is about 79, my mode (most common) score is 84! The line graph shows the upward trend of my quiz grades.

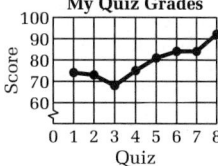

My Quiz Grades

2. a.

	1968–1969	1978–1979	1988–1989
Men	56%	51%	47%
Women	44%	49%	53%

b. Women's share of awarded bachelor's degrees is increasing, while men's share is decreasing.
c. A series of circle graphs would best show the varying percentage of the whole for each gender, over time. One could see the trend quicker than by reading a table. **3. a.** mean: about 12.4; median: 12; modes: 12, 13; Choices and explanations may vary. An example is given. Since the data are closely grouped except for one value, the mean, median, and modes are almost equivalent. I think all of them are typical for the data set. **b.** range: 9 months, outlier: 18 months **c.** Answers may vary. Examples are given. (1) The mean is most affected, decreasing from about 12.4 to about 11.7. (2) The mode is most affected because there are now three modes, 11, 12, 13, instead of two modes. **4.** Answers may vary. An example is given. She needs a combined score of 196 or more on the two tests. For instance, scores of 98 and 98, 99 and 97, or 96 and 100 will give her a mean of at least 90.
5. $8 \le d \le 40$, where d = diameter in feet;

```
◄─┼─●─┼──┼──┼──┼──┼►
  0  8  16  24  32  40  48
```

6. $t \le 32$, where t = temperature at which water will freeze;

```
◄────────┼──────┼────►
         0      32
```

Pages 150–153 Talk it Over
1. the graph on the left; It includes persons of age 20, and all persons up to but not including 25.
2. $20 \le a < 25$, a = age **3.** about 4500 **4.** Answers may vary. An example is given. Histograms and bar graphs are both ways to compare data in different categories. Each type of graph uses bars to show amounts. Each category in a histogram is an interval, the height of the bar shows the frequency in the interval, and there are no spaces between bars. A bar graph shows a specific value for a specific category, and has a space between bars to separate the categories. **5.** There are no spaces between bars of a histogram because they represent parts of a continuous interval on the number line. **6.** No. Both show data for intervals only, not for specific ages. **7.** Answers may vary. An example is given. The exact number was 8 over my estimate. **8. a.** 108 **b.** Answers may vary. Example: 225–249, 250–274, 275–299, 300–324, 325–349 **c.** Answers may vary . An example is given.

Number of *Galaxy* Copies Sold Each Week

Number sold	Frequency
225–249	4
250–274	5
275–299	7
300–324	6
325–349	3

9. Yes. The stem, "15," appears only once, but the leaf, "6," appears twice. **10.** Answers may vary. An example is given. Count from the first leaf and the last leaf at the same time until you get to the middle number (median) or numbers (median is average of two numbers). 155.5 **11.** No. 181 is an outlier, since on most days there are more than enough spaces.
12. Answers may vary. An example is given. They are alike in that they each divide the data into intervals, show the frequency in each interval, and the number of "leaves" in any entry corresponds to the height of the histogram bar. They are different in that the histogram shows only the frequency of values, while the stem-and-leaf plot also shows the actual values.

Pages 154–157 Exercises and Problems
3. Answers may vary. An example is given. This group covers the largest interval, 25 years, except for the "over 74" interval. It is expected that there are more people living that are under 25 than there are over 74. **5.** Estimates may vary by region. Examples are given. Northeast: about 50 million; West: about 50 million; Midwest: about 60 million; South: about 85 million **7.**

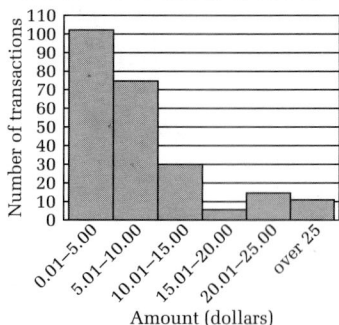

11. a.

Average Number of Days of Precipitation per Year

```
 3 | 5 6
 4 |
 5 | 1
 6 | 1
 7 |
 8 | 2 6 9
 9 | 0 0 6 7 8 9
10 | 1 4 4 4 7 9
11 | 2 3 4 5 5 6 7 9
12 | 2 4 4 5 5 5 6 6 7 8 9
13 | 4 4 5
14 |
15 | 2 4 4 6 6
16 | 9
17 |
18 |
19 |
20 | 9
```

Integrated Mathematics

b. 115.5 days

19. $v \geq 18$, v is voting age.

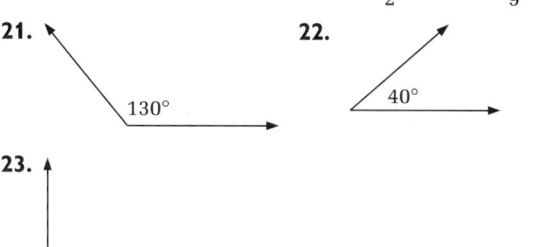

20. $2 \leq a \leq 9$, a is the increase in average global temperature from now until 2050.

21. [angle diagram] 130°

22. [angle diagram] 40°

23. 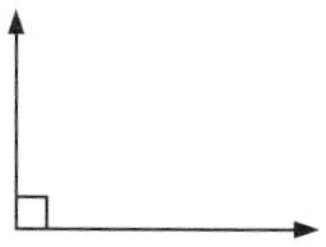 [right angle diagram]

24. median: 184; range: 197; outlier: 279

Pages 159–160 Talk it Over
1. a. 28.5; 12 **b.** These ranges are shown by the lengths of the whiskers in the plot. **2. a.** 25%; 25% **b.** No. Explanations may vary. An example is given. The percents will remain about the same because the lower and upper quartiles are the medians of the bottom 50% and top 50%, so it does not matter how many scores are in the data set. **3. a.** 50% **b.** These scores are where the box is drawn in the plot. **c.** No. Explanations may vary. An example is given. The quartiles and the median divide the data set into 4 parts with each part representing about 25% of the data, so the middle two parts represent about 50% for any number of scores in the data set. **4.** Answers may vary. An example is given. *Quartile* sounds like "quarter" or "quart," which refer to $\frac{1}{4}$ or 25% of a dollar and $\frac{1}{4}$ or 25% of a gallon, respectively. The quartiles help to divide the data into four parts, each part representing 25% of the data. **5.** No. Comparing where the boxes and whiskers start and end is enough to make those statements. **6.** Class I has a smaller box indicating a closer clustering of scores. **7.** Estimates and statements may vary. An example is given. Class I: about 73; Class II: about 60; 75% of Class I scored above 73, whereas 75% of Class II scored above 60.

Pages 160–163 Exercises and Problems
3. Descriptions may vary. An example is given. When the 32 data values are listed from smallest to largest, the lower quartile is the average of the eighth and ninth numbers in the list, or 11.
5. a.

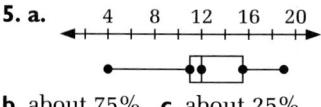

b. about 75% **c.** about 25%

7. a.

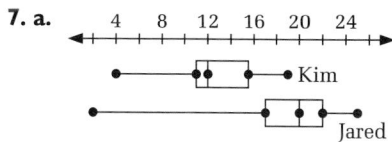

b. Statements may vary. Examples are given. The plot for Jared's class extends farther in both directions than the plot for Kim's class, so Jared's class has a wider range of ages than Kim's class. The median of Jared's data is greater than the upper extreme of Kim's data, so about 50% of the brothers and sisters of the students in Jared's class are older than all of the brothers and sisters of the students in Kim's class. **c.** Answers may vary. An example is given. More of the brothers and sisters of students in Jared's class are at least 17 years old than are less than 17 years old. **d.** Answers may vary. An example is given. Yes, agree. The box portion of the plot for Jared's data lies completely to the right of the box portion of the plot for Kim's data. This shows that 75% of the brothers and sisters of Jared's classmates are older than 75% of the brothers and sisters of Kim's classmates. Also, if you choose an age at random from Jared's data, it is more than 50% likely that the age is greater than the upper extreme of Kim's data. **9. a.** 7 **b.** 15 **c.** Answers may vary. Examples are given. 15 > 7 (or 15 is about twice 7); the left whisker in Jared's plot is longer than (about twice as long as) the left whisker in Kim's plot. **11.** about $350; about $200 **13.** about 25% **15.** No. The median cost of repairs for the Facet is $350, so $300 is lower. **21.** about 59 **22.** 0–1.9. This means that most volunteers work 0–1.9 hours per week. **23.** You need to know the exact amount of hours that each volunteer has worked each week. **24.** $9a - 12b$ **25.** $3x^2$ **26.** Answers may vary. An example is given. A line graph would show specifically the change in tide over a 24-hour period.

Page 165 Talk it Over
1. Answers may vary. An example is given. You could use a table to display each state and percent of trash recycled by that state. **2.** Answers may vary. An example is given. A bar graph could be used to display these data. **3.** Answers may vary. An example is given. You could use a histogram or a stem-and-leaf plot.

Pages 166–169 Exercises and Problems
1. Answers may vary. Examples are given. The bar graph shows that the percentage of tires recycled is about the same as the percentage of paper recycled. The circle graph shows that almost half the landfill space is taken up by paper. The box-and-whisker plot shows that about 50% of Southerners contribute more tons of trash per person, per year, than 75% of Westerners. The histogram shows that about 30 states recycle 10% or less of their trash. **3.** a, c, d **5.** a, d

7. Answers may vary. Examples are given. A line graph can be used to show trends over time. A line graph can be used to compare two sets of data, especially over time. A line graph can be used to estimate an unknown data value between two known data values. **9, 11.** Answers may vary. Examples are given. **9. a.** Canada produces almost twice as much ice cream as Japan. **b.** This graph gives a quick comparison of different nations' ice cream production. The ice cream cone makes the subject matter clear. **11. a.** Only 3 out of 17 Running Club members ran 20 or more times in October. **b.** The plot quickly displays this small data set, and allows each data value to be seen. **13, 15, 17.** Choices may vary. Examples are given. **13.** circle graph; shows parts of a whole **15.** box-and-whisker plot **17.** Answers may vary. Examples are given. **a.** A line plot shows outliers best. It shows exactly where the data fall and how they cluster so outliers can easily be seen. **b.** A histogram conceals outliers because it only shows an interval of data. How exact data values are grouped in the interval cannot be seen.

20.

```
0 100            1000            1900
```

21. about 8400 km^2 **22.** Record Place; about 20 more cassettes **23.** False.

Pages 170–171 Talk it Over

1. Answers may vary. An example is given. Yes. The steepness of the various graphs makes it appear that income increases either slowly or quite dramatically. **2. a.** $1000; $10,000; $1000 **b.** 1 year; 1 year; 4 years **c.** The bigger the interval between scale marks on the horizontal axis, the more vertical the graph; the bigger the interval between scale marks on the vertical axis, the more horizontal the graph. **3.** about 33.7%; Answers may vary. An example is given. Graph A; Graph B suggests no change, and Graph C suggests too dramatic a change for 33.7%. **4.** There were about twice as many calculators in 1991 as there were in 1990. **5.** The area of the 1991 calculator is about four times the area of the 1990 calculator. **6.** No. Answers may vary. An example is given. The 1991 calculator should have twice the area (same width, twice the height) of the 1990 calculator.

Pages 172–175 Exercises and Problems

5. False. **7.** True. **11.** Answers may vary. An example is given. The graph does not show the one hurricane in March. The table tells us exactly how many hurricanes occurred, but you would have to estimate this on the graph. **13.** No. Reasons may vary. For example, the data in the table are given per month, so one month is the smallest interval of time that could be used for a histogram. **15. a.** Graph B **b.** Graph A

c. Graph A **17.** The actual number of cable systems in 1970 and 1990, 2.49 million and 9.58 million, are indicated along the horizontal axis; 9.58 > 2.49, so there was an increase from 1970 to 1990. Also, the television symbol for 1990 is much larger than the symbol for 1970, indicating a big increase from 1970 to 1990. **19.** b **23–26.** Choices may vary. Examples are given. **23.** bar graph; Each jumper's bar would show the length of the jump and could be compared easily. **24.** histogram; It shows the frequency (number of visitors) in an interval (each day of a month). **25.** circle graph; It shows the percentage of total graduates for each age. **26.** 7×10^5 **27.** 2.34×10^7 **28.** 6×10^{-3} **29.** 4.61×10^{-5}

30–33.

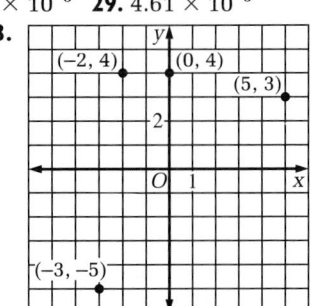

Pages 177–178 Unit 3 Review and Assessment

1. 4×2 **2.** plastic rope collected at state beaches in 1989 **3. a, b.** Answers may vary. Examples are given. **3. a.** More glass beverage bottles and plastic rope were collected in 1989 than in 1990. More metal beverage cans and cigarette filters were collected in 1990 than in 1989. **b.** My statements are more easily seen by the graph. **4.** mean: 24; median: 26; mode: 18; The mean and median are most typical because both are near the middle of the data set. **5.** 12 **6.** No. It is only 2 units from the next score. **7.** 28 or higher **8.** $35 \leq s \leq 55$, s = speed of Laurie Chin;

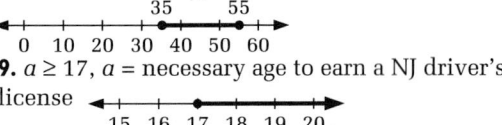

9. $a \geq 17$, a = necessary age to earn a NJ driver's license

```
    15  16  17  18  19  20
```

10. a, b. Answers may vary. Examples are given. **a.** Most VCR's sold cost between $200.00 and $299.00; the number of VCR's sold in the $100.00–$199.00 price range is the same as the number of VCR's sold in the $300.00–399.00 and the $500.00–$599.00 ranges. **b.** exact cost of VCR's and how many different models there are in each price range

11. a. Homeroom Class Sizes

```
1 | 2 4 5 7 7
2 | 1 3 3 3 3 4 6 6 6 8 9
3 | 0 1 4 4 8
4 | 3
5 | 1 6
6 | 7
```

 12

Integrated Mathematics

b. 26 **c.** Statements may vary. An example is given. About half the class sizes were in the 20–30 student range. **12. a.**

Homeroom Class Sizes

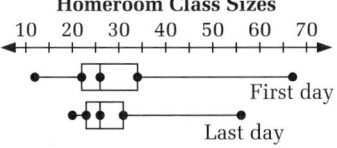

b. Answers may vary. Examples are given. There was a wider range of homeroom class sizes on the first day than on the last day. The median class size on the first day was the same as the median class size on the last day. **13–17.** Answers may vary. Examples are given. **13.** A box-and-whisker plot is an excellent way to compare two large sets of data. The plot allows you to compare the medians, quartiles, and ranges. **14.** A circle graph shows best what portion of Sara's total monthly income was spent on various items. **15.** line graph; It is best for showing change over time. **16.** The new graph is misleading because of the scale gap of 10,000 units. The new graph shows that $\frac{1}{3}$ fewer bottles were collected from 1989–1990, when it was actually $\frac{1}{4}$ fewer. The new graph makes it seem that 4 times more filters were collected from 1989–1990, when it was really 3 times more. Also, it appears from the new graph that in 1989 twice as many bottles were collected as filters which is false. **17.** Circle graphs show parts of a whole. Circle graphs are based on percentages. Angle measures are used to divide up the circle.

Unit 4

Pages 183–184 Talk it Over

1. Answers may vary but should include: Each square is labeled (N or S), (E or W), (1, 2, 3, or 4). The numbers increase as you move away from the common vertex of N1 W1, N1 E1, S1 E1, S1 W1. **2.** N = North, S = South, E = East, W = West **3.** crossbow **4.** N1 E1 **5.** (1, 4) **6.** point E **7.** point C or B **8.** quadrant II **9.** Start at the origin. Move 4 units to the right, then 3 units down and place a dot at that location. **10. a.** the point two units to the right and two units below the upper left-hand corner **b.** plates

Pages 185–186, 189 Exercises and Problems

1. No; reasons may vary. An example is given. (1, 3) is the point one unit to the right and three units above the origin. This is different from (3, 1) which is three units to the right and one unit above the origin. **3.** the U.S. Capitol **5.** Dept. of State **7.** C1, C2 **9.** C5, C6 **11.** northeast **13.** quadrant III **15.** about 0°, 78 W **34. a.** Answers may vary. For example, the highest salaries are earned for language programming in New York, the lowest for data entry in Dallas. **b.** Answers may vary. For example, a

combined histogram using a single set of axes and color coding for the cities would make many comparisons easier.

35. a.

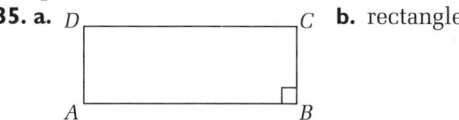

b. rectangle

Pages 191, 192, 194 Talk it Over

1. Answers may vary. Examples are given. An equilateral triangle has all 60° angles; $\triangle ABC$ has one 90° angle. Also, $\overline{AC}$ is longer than $\overline{AB}$ or $\overline{BC}$, so $\triangle ABC$ is not equilateral. **2.** Yes; $\overline{FE}$ and $\overline{HJ}$; $\overline{GH}$ and $\overline{ED}$; $\overline{GF}$ and $\overline{JD}$. **3.** It has 4 right angles and the opposite sides are equal in length. **4.** The width 4 is the same as the difference between the first coordinates of X and Y. **5.** It is the difference of their second coordinates. **6. a.**

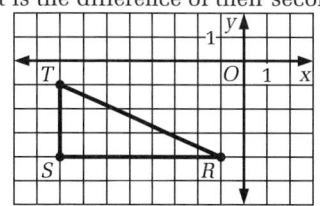

b. area of $\triangle RST$ = 10.5 square units **7.** Answers may vary. For example, another method would be to count the number of squares in rectangle *NSMR*. **8.** Add the areas of rectangle *LSNP* and $\triangle NSM$ to get the area of trapezoid *LMNP*.

Pages 194–196 Exercises and Problems

1, 3. Answers may vary. Examples are given.
1. description; Describing a square is simple: a polygon with 4 right angles and 4 equal sides. **3.** coordinates; There are different types of parallelograms, so coordinates would probably be simpler than a description. **7.** pentagon; The figure has 5 sides.

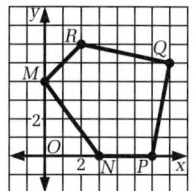

9. square; The figure has 4 equal sides and 4 right angles.

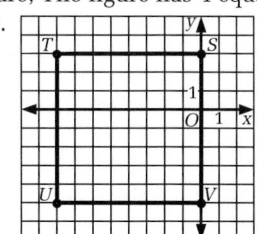

11, 13, 15. Methods may vary. Examples are given.
11. 22.75 square units; I found the area of a rectangle with length 7 and width 6.5. Half this area is the area of the triangle. **13.** 17.5 square units; I found the

area of a rectangle with length 5 and width 7. Half the area of the rectangle is the area of the triangle.
15. 16.5 square units; I found the area of a rectangle with length 7 and width 3. Then I subtracted the areas of the two triangles that are not part of the trapezoid. **17.** Calvin **22.** 49 **23.** 13 **24.** $9x - 12$
25. $9a + 3ab$ **26. a.** Check graphs. **b.** $(4, -2)$
c. $(-2, -2)$ **d.** right triangle

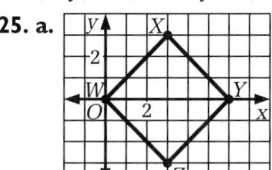

Pages 197–198 Talk it Over
1. 4 units; right **2.** Yes. **3.** $B(-3, 1)$ and $B'(1, 1)$; They have the same y-coordinate. **4.** left or right
5. up or down

Pages 199–201 Exercises and Problems
3. $T'(-3, -1)$, $E'(-2, 2)$, $D'(1, -2)$ **5.** $T'(-6, 3)$, $E'(-5, 6)$, $D'(-2, 2)$ **7.** not a translation **9.** $J'(-6, 1)$, $K'(0, 3)$, $L'(-1, 6)$ **17.** $P'(1, 7)$ **19.** $P'(2, 5)$
21. $(x, y) \rightarrow (x + 4, y - 6)$
25. a.

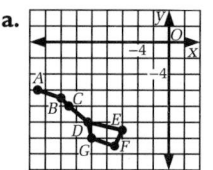

b. square; All angles are right angles and all 4 sides are congruent. **26.** 11 **27.** $-15\frac{2}{3}$ **28.** $-8\frac{1}{2}$ **29.** 90°
30. 180° **31.** 270°

Pages 203–204 Talk it Over
1. No. **2.** Answers may vary. An example is given. In order for someone to copy your rotation, you need to specify the center of the rotation and the amount and direction of the turn. Directions: Put your finger on a corner of the paper and turn it 30° clockwise.
3. A turn 180° clockwise moves points to the same place as a turn 180° counterclockwise. **4. a.** 270° clockwise **b.** 90° clockwise

Pages 206–209 Exercises and Problems
1. a–c. Answers may vary. Examples are given.
a. about a 90° rotation from a vertical to a horizontal position, with the center of rotation at the base of the trunk **b.** about a 40° rotation clockwise and counterclockwise, with the center of rotation at the hinge of the scissors **c.** The sun appears to rotate about 180° from east to west each day, with the center of rotation at my location. **3.** 90° counterclockwise or 270° clockwise **5.** 90° clockwise or 270° counterclockwise

7. a–b.

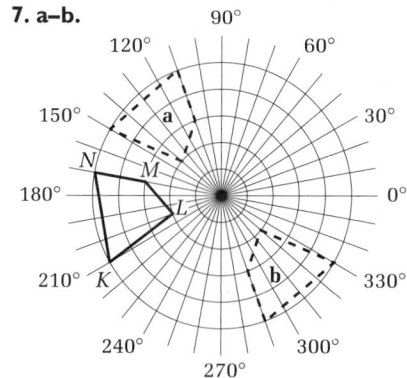

13. Yes. The design has symmetry at 60°, 120°, 180°, 240°, and 300°. **15.** Yes. The design has symmetry at about 26°, 52°, 78°, 104°, 130°, 156°, 182°, 208°, 234°, 260°, 286°, 312°, and 338°. **23.** $A'(5, -4)$, $B'(-2, -6)$
24.

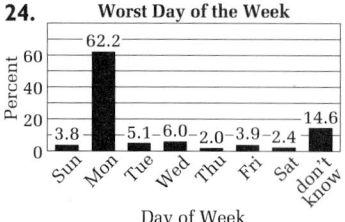

Page 210 Checkpoint
1. Answers may vary. Two uses of coordinate systems are maps and computer screens. Both use horizontal and vertical distances for locations. Coordinates of a location on a map describe a square region, while the coordinates of a location on a computer screen indicate a pixel of light.
2. a. **b.** quadrant III

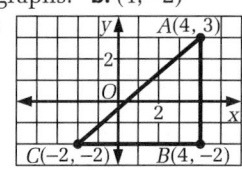

c. The bowl of the Big Dipper is a quadrilateral. The lengths of opposite sides are not the same, so the bowl is not a parallelogram. **3.** origin or $(0, 0)$ **4.** 4.5 square units **5.**

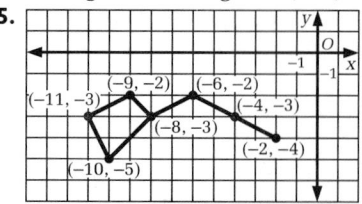

6. 180°, either counterclockwise or clockwise
7. translational symmetry **8.** no symmetry
9. rotational symmetry

Pages 212–213 Talk it Over
1. Answers may vary. A positive correlation is expected. **2.** Answers may vary. One might reason there is no correlation or that there is a negative cor-

Integrated Mathematics

relation—girls, who tend to be shorter, tend to have longer fingernails than boys. **3.** No. 270 horsepower is a good prediction based on the data, but the actual car may have a different horsepower. **4.** about 2600 cc **5.** Answers may vary. Examples are given. **a.** No. All of the points lie below the line, so the line does not lie as close as it could to many of the points.

Pages 214–217 Exercises and Problems
1. a. One point would be directly above the other.
b. One point would lie to the left of the other.
3. a. Predictions may vary. **b.** Graphs may vary.
c. Answers may vary, based on graphs in part (b).
5. negative correlation **7.** positive correlation
13. no correlation

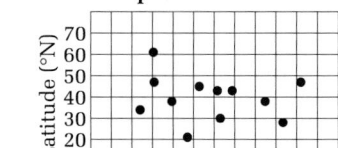
Precipitation and Latitude

15. positive correlation

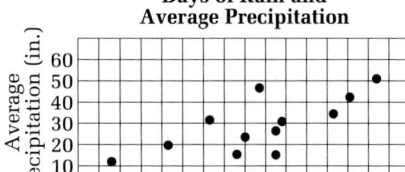
Days of Rain and Average Precipitation

17, 19, 21. Answers may vary. Examples are given.
17. about 45°F **19.** about 15 in. **21.** c **24.** 180°
25. 90° **26.** 360°

27. ![number line from -4 to 3 with filled dot at -2 and arrow right]
$-4\ -3\ -2\ -1\ 0\ 1\ 2\ 3$

28. ![number line from -3 to 6 with arrow left]
$-3\ -2\ -1\ 0\ 1\ 2\ 3\ 4\ 5\ 6$

29. ![number line from -3 to 8 with open circle at -1 and filled dot at 6]
$-3\ -2\ -1\ 0\ 1\ 2\ 3\ 4\ 5\ 6\ 7\ 8$

30. a. $17.25 **b.** $40.25 **c.** $5.75x$ dollars

Pages 218–219 Talk it Over
1. Answers may vary; for example, the change in temperature on a January morning in St. Paul, MN.
2. highest: 6°F; 12:00 noon; lowest: −5°F; 2:00 or 3:00 A.M. **3.** Temperature increased from 3:00 A.M. to 9:00 A.M. and from 11:00 A.M. to 12:00 noon. Temperature decreased from 12:00 midnight to 2:00 A.M. and from 10:00 A.M. to 11:00 A.M. **4.** Yes. **5.** No. **6.** box size; The control variable is usually put on the horizontal axis. **7.** dependent variable: height from the ground; control variable: time during the ride

Pages 222–224 Exercises and Problems
1. horizontal; vertical **3.** The time it takes Michael to run the 100 m dash is a function of his speed. dependent: Michael's time; control: Michael's speed
5. a. dependent: length; control: age
b.

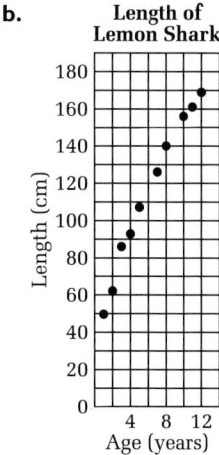
Length of Lemon Shark

c. Yes; passes the vertical-line test **d.** about 115 cm
e. about 9 years **7.** b **9.** function; passes the vertical-line test

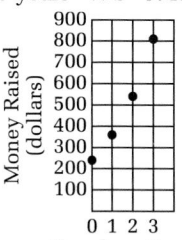

11. a. three **b.** one **c.** Graph B; There is only one y for each x. **17.** Answers may vary. The data points should be close to an upward-slanting line.

18. $\frac{1}{4}$ **19.** $-\frac{1}{4}$ **20.** 2 **21.** $-\frac{8}{3}$ **22.** −20; 0; −45 **23.** 10; 8; 11 **24.** 2; 0; −3 **25.** $-\frac{1}{2}$; undefined; $\frac{1}{3}$

Pages 227, 229, 230 Talk it Over
1, 2. Choices of variables may vary. **1.** control: the number of minutes you have already watched, w; dependent: the number of minutes left on the tape, t; $t = 108 - w$ **2.** control: number of tickets sold, t; dependent: profit earned by senior class, p; $p = 5t - 500$ **3.** Carmella cannot burn negative calories in negative minutes. Both calories and minutes have to be positive quantities and this is true only in quadrant I. **4.** Answers may vary, but the point should be of the form $(x, 5x)$. **5.** vertical; control **6.** Each y-value is the same as its x-value. **7.** y is the reciprocal of x.

Pages 231–233 Exercises and Problems

1. table of values, equation, graph, verbal description
3. $y = \frac{x}{2}$ or $y = 0.5x$ **5, 7, 9.** Choices of variables may vary. **5. a.** control variable: total sales, s; dependent variable: total pay, p **b.** $p = 0.045s + 350$ **c.** $590.66
7. Let t = time (seconds); let d = distance (feet); $t = \frac{d}{1000}$ **9.** Let n = number of people; let c = amount each person pays; $c = \frac{800}{n}$

17. $y = 5.3x$ **19.** $y = 30x$

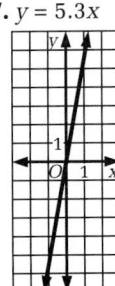

21. a. $5.24 **b.** $x + 0.05x = 1.05x$ **c.** Let x = total price before tax; let y = total cost; $y = x + 0.05x$ or $y = 1.05x$ **d.** **e.** Yes.

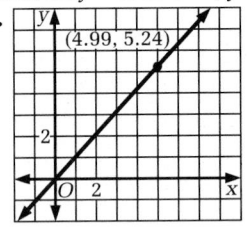

(4.99, 5.24)

23. absolute value function or $y = |x|$ **25.** reciprocal function or $y = \frac{1}{x}$ **27.** No. **29.** Yes. **31.** quadrants I, II **33.** quadrants I, II **35, 37.** Answers may vary.

35. conjecture: line **37.** conjecture: V-shaped

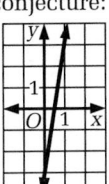 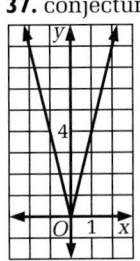

41. Answers may vary. There should be an x-value with two y-values. **42.** 21 **43.** 4 **44.** $\frac{-27}{7}$

Pages 235–236 Unit 4 Review and Assessment
1. Answers may vary. **2. a.** $A(-1, 1)$; $B(-1, -1)$; $C(1, -1)$ **b.** quadrant III **3. a.**

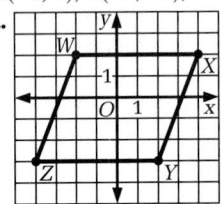

b. parallelogram **c.** 30 square units

4. $P'(-6, 6)$, $Q'(-1, 6)$, $R'(-1, 2)$, $S'(-6, 4)$ **5.** 180°
6. 90° clockwise **7.** 90° counterclockwise **8.** no symmetry **9.** translational **10.** rotational
11.

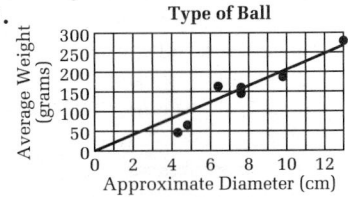

12. positive correlation **13–14.** Estimates may vary. Examples are given. **13.** about 110 g **14.** about 8.5 cm **15.** No. **16.** Yes. **17.** No. **18. a.** The amount of money left is the dependent variable. The number of tapes bought is the control variable. **b.** The amount of money left is a function of the number of tapes bought. **c.** $a = 30 - 3.5t$, where a = amount of money left (in dollars) and t = number of tapes bought. **d.**

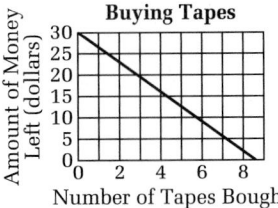

19. $y = x + 2$

20.

x	y
−2	7
−1	5
0	3
1	1
2	−1
3	−3

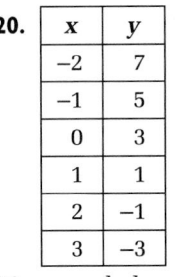

21. a parabola

x	y
−2	1
−1.5	−0.75
−1	−2
−0.5	−2.75
0	−3
0.5	−2.75
1	−2
1.5	−0.75
2	1

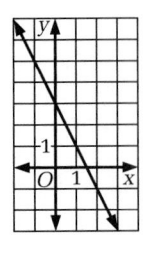

22. Answers may vary. An example is given. Coordinates can be used to show movement. By changing the coordinates you can move shapes up or down, to the left or to the right, and diagonally. Coordinates can be used to show locations on maps by identifying squares where buildings and other places can be found. Coordinates can be used to make graphs that show relationships between two quantities.

16

Integrated Mathematics

Pages 243–244

1. Answers may vary. An example is given. A table is good, because you can make comparisons week by week until the exact answer is found.

2. $37 + 10w = 195$; $y = 37 + 10x$ **3, 4.** Situations may vary. Examples are given. **3.** Four copies of Shakespeare's *The Tempest* cost $15. How much does one copy cost? **4.** A 100-point quiz consists of a 50-point essay and two short-answer questions. If the short-answer questions are worth an equal number of points, how many points is each short-answer question worth?

Pages 244–246 Exercises and Problems

3. Twenty tickets must be sold.

Number of Tickets Sold	Value of Tickets	
16	$108.00	
17	$114.75	
18	$121.50	
19	$128.25	← < $131
20	$135.00	← > $131

5. The population was less than 30,000,000 in 1984.

	A	B
1	Year	Population
2	1991	32,664,000
3	1990	B2 − 419,783
4	1989	B3 − 419,783
5	1988	B4 − 419,783
6	1987	B5 − 419,783
7	1986	B6 − 419,783
8	1985	B7 − 419,783
9	1984	B8 − 419,783
⋮	⋮	⋮

7. Solve $2(800) + 2x = 1950$. The lengths of the other sides are 800 ft, 175 ft, and 175 ft. **11, 13, 15.** Choice of variables may vary. **11.** $t = 7p$, where t = time in minutes and p = number of pounds of vegetables **13.** $e = \frac{1}{20}f$ or $e = \frac{f}{20}$, where e = number of English words in a situation using everyday French and f = total number of words used **15.** $m = \frac{2.75p}{t}$, where m = number of minutes you wait in line, p = number of people ahead of you, and t = number of tellers **17.** Ex. 14 (finding daily high temperatures)

19.

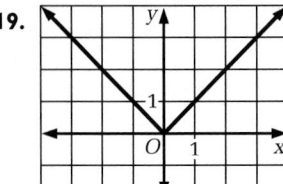

20.

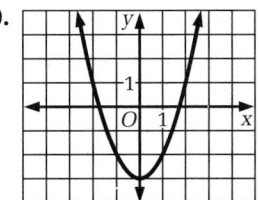

21.

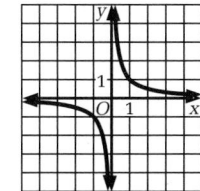

22. −7 **23.** 7 **24.** $-\frac{3}{5}$ **25.** 0 **26.** 0.75 **27.** 6.3 **28.** −3 **29.** −3 **30.** −2 **31.** −2 **32.** $2x + 8$ **33.** $-4k - 21$ **34.** $\frac{2}{3}t + 10$ **35.** $15x - 46$

Pages 247, 250 Talk it Over

1. a–d. oral exercises **2.** Answers may vary. An example is given. I think Alison's method makes more sense because Keith's method suggests that all opposites are negative. **3. a.** Yes. **b.** Yes. **c.** Yes. **d.** Yes. **4.** Answers may vary. An example is given. Find the opposite of the first term. Next change the subtraction signs to addition signs and the addition signs to subtraction signs. Then evaluate the new expression.

5.
$$6 - x = -2$$
$$6 - x + x = -2 + x$$
$$6 = -2 + x$$
$$6 + 2 = -2 + x + 2$$
$$8 = x$$

6.
$$(n - 2)180 = 450$$
$$\frac{(n - 2)180}{180} = \frac{450}{180}$$
$$n - 2 = 2.5$$
$$n - 2 + 2 = 2.5 + 2$$
$$n = 4.5$$

Pages 250–251, 254 Exercises and Problems

3. $-x - 6$ **5.** $-5k - 3$ **7.** $-2x - 5$ **9.** 17 **11.** 10 **13.** 7 **15.** 11 **17.** −2 **19.** 5.7 **21.** Yes. If $n = 3$, then the polygon is a triangle. No. Since no polygon can have less than three sides. No. A polygon cannot have partial sides. Yes. n can be any whole number greater than 2. **23.** Yes. **32.** Choice of variables may vary. $T = 45 + 38h$, where T = total earnings, h = number of hours. If $h = 4$, then $T = \$197$. **33.** Answers may vary. Player 1 is more consistent in his or her scores. Player 2 received better scores than Player 1 on 5 of the 9 holes. **34.** −10 **35.** 13 **36.** −1

Pages 255–257 Talk it Over

1. tortoise: 2 mi; hare: 350 mi **2.** 17.5 mi **3. a.** $(t - 12)$ h **b.** t h **c.** $35(t - 12)$ mi **d.** $0.2t$ mi **4.** The distances are equal. **5.** $35(t - 12) = 0.2t$ **6.** You can find out how many hours the race lasted. **7.** Use $t \approx 12.1$ and $r = 0.2$ in $D = rt$ to get $D = r \cdot t \approx 2.42$ mi. **8.** 6 min

Pages 258–260 Exercises and Problems

3. rate **5.** time **7.** tortoise: 0.2 mi/h, hare: 35 mi/h **9.** −3 **11.** −36 **13.** 10 **15.** $\frac{2}{3}$ **17.** about 6 h 2 min **19. a.** $25,000 + 1000y$ **b.** $20,000 + 1500y$ **c.** 10 years **21. a.** 2 mi/h **b.** $10 = 3.75r + 1.25(4)$; $r = 1\frac{1}{3}$ mi/h **26.** 3 **27.** −4 **28.** −6 **29.** 8^3 **30.** 5^2 **31.** 3^4 **32.** c **33.** Let a = age in months of a dollar bill; $13 \le a \le 18$

Selected Answers

1. a and d   **2.** Answers may vary. An example is given. Adding –2 is the same as subtracting 2, subtracting –2 is the same as adding 2.   **3.** No. –1 is not less than –1. Yes. –8(–5) ≤ 40 is true.   **4.** There are an infinite number of points on the graph.   **5.** Answers may vary. An example is given. $x < 11$ **6.** Answers may vary. An example is given. $25 < -5x$ **7.** Answers may vary. An example is given. Choose any x less than or equal to 16, for example, $x = 0$. Substitute this value into the original inequality. Simplify to see if the result is true.   **8. a.** Divide by 5.   **b.** Add 4.   **c.** Divide by –3.   **9.** In the third step, dividing both sides by –6 reversed the inequality symbol.
10. a. $t \leq 12{,}000$; t = no. of tickets   **b.** $w < 2$; w = weeks to complete report   **c.** $8 + w \leq s$; w = minimum wage, s = Claire's hourly wage   **d.** $s > 10 + a$; a = average test score, s = Kwai's score   **11. a.** over: >; under: <; less than: <; more than: >; maximum: ≤; minimum: ≥   **b.** Answers may vary. An example is given. The maximum amount that she can spend is $100.

3. C   **5.** $q < -\frac{3}{2}$

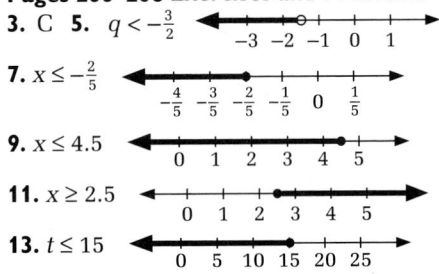

7. $x \leq -\frac{2}{5}$

9. $x \leq 4.5$

11. $x \geq 2.5$

13. $t \leq 15$

15. c   **17.** b   **19.** a   **21.** $50{,}000 > 4.5t$; $t < 11{,}111\frac{1}{9}$; maximum temperature is about 11,111°F.   **25.** up to 296 newsletters   **27.** 100 ft or less   **29.** at least 15 lawns   **31.** no solution

33. 5   **34.** 9   **35.** –19   **36.** –5   **37.** –8   **38.** –6   **39.** D: distance; r: rate; t: time   **40.** P: perimeter; l: length; w: width   **41.** V: volume; l: length; w: width; h: height

1. Answers may vary. An example is given. First use the distributive property to get $5x - 25 + 2x = 17$. Then combine the like terms $5x$ and $2x$ to get $7x - 25 = 17$. Next, add 25 to both sides to undo the subtraction of 25; this gives $7x - 25 + 25 = 17 + 25$ or $7x = 42$. To undo the multiplication by 7, divide both sides by 7; this gives $\frac{7x}{7} = \frac{42}{7}$; $x = 6$.

2.

Number of weeks	Total saved ($)
0	500
1	500 + 22(1) = 522
2	500 + 22(2) = 544
⋮	⋮
31	500 + 22(31) = 1182
32	500 + 22(32) = 1204

Alma needs 32 weeks.
3. Answers may vary. An example is given. On Saturday I bought a $29 jacket and three small plants and spent $47 in all. How much did each plant cost?
4. $3x + 2(x - 5) = 90$; $x = 20$; 60° and 30°
5. $2x + 4(x + 3) = 180$; $x = 28$; 56° and 124°   **6.** –13
7. 3   **8.** 1.5   **9.** 2   **10.** $y \leq 4$

11. $m < -3$

12. $c < -2$

13. $7(40) + 10x > 380$; $x > 10$; 11 h

1. $\frac{D}{t} = r$; Distance divided by time equals rate.
2. Answers may vary. An example is given. To solve $I = prt$ for r, divide each side by pt: $\frac{I}{pt} = r$.

1. $128 = r \cdot 32$ $D = rt$
Divide both sides by the coefficient of r.
$$\frac{128}{32} = \frac{r \cdot 32}{32} \qquad \frac{D}{t} = \frac{rt}{t}$$
$$4 = r \qquad\qquad \frac{D}{t} = r$$
r is a number. r is an algebraic expression.

3. $d = \frac{C}{\pi}$ **5.** $\frac{E}{c^2} = m$ **7.** $\frac{-r + 220p}{p} = a$ **9.** $s = b + c$

11. $y = 100 - x$ **13.** $y = 75 - 5x$ **15.** $y = \frac{c - ax}{b}$

17. a. $s = 200 + 0.12d$ **b.** $d = \frac{s - 200}{0.12}$ **c.** $3333.33
d. $1238.00   **24.** $x < -2$

25. $y \leq -4$

26. $x > -2$

27. 80.6%   **28.** $\frac{1}{3}$ **29.** $\frac{5}{2}$ **30.** $-\frac{6}{11}$ **31.** $\frac{7}{9}$

1. 12   **2.** Answers may vary. Examples are given. $\frac{2}{5}$ and $\frac{5}{2}$, –3 and $-\frac{1}{3}$, –1 and –1   **3.** 1   **4.** Division by zero is undefined.   **5.** multiplying, reciprocal   **6.** Explanations may vary. An example is given. Since the factors in a product can be grouped in any way,

$\frac{1}{6} \cdot 6h = \left(\frac{1}{6} \cdot 6\right)h$. Since $\frac{1}{6} \cdot 6 = 1$ and multiplying by 1 does not change a number, $\left(\frac{1}{6} \cdot 6\right)h = 1 \cdot h = h$.
7. Yes. $4(0.25) = 1$ **8.** No. The product is not 1, it is -1. **9.** Add 19 to both sides of the equation. Then multiply both sides by $-\frac{4}{3}$, the reciprocal of $-\frac{3}{4}$.

10. Methods may vary. Examples are given. (1) Subtract 12 from both sides and divide both sides by 0.4, or (2) subtract 12 from both sides, rewrite 0.4 as $\frac{4}{10}$ or $\frac{2}{5}$, and multiply both sides by the reciprocal $\frac{5}{2}$. I prefer (2), using the reciprocal, because dividing by decimals takes too long.

Pages 277–279 Exercises and Problems
1. The goal is to get the variable alone. **3.** 32 **5.** 16
7. 30 **9.** $y = \dfrac{18 - 9x}{-4}$ **11.** $y = \dfrac{15 - 7x}{2}$ **13.** $x = \frac{2}{5}y - 1$
15. Enter "1" and then press the $\boxed{\div}$ key. Then enter "2.5" and press the $\boxed{=}$ key. This gives 0.4, the reciprocal of 2.5. **22.** $\dfrac{P - b}{2} = s$ **23.** $\dfrac{46 - 5y}{-6} = x$
24. $2x - 14 = y$ **25.** $y = 2x$ **26.** 250 **27.** 25 **28.** 3 **29.** 14

Pages 281, 283 Talk it Over
1. a. $\dfrac{A}{b} = h$ **b.** Replace A with 25 and b with 10, then divide. **2.** Answers may vary. An example is given. I agree because the width of a rectangle is the perpendicular distance between two opposite sides. This is the same as the height of the parallelogram (the distance to the base from a point on the opposite side). **3. a.** A height drawn from a vertex must be perpendicular to the base opposite the vertex. Sometimes the base must be extended in order to meet the height at a right angle. **b.** Yes. **c.** No.

Pages 285–288 Exercises and Problems
1. a. the segment drawn from P perpendicular to $\overline{AB}$ **b.** same as (a) **3.** 28 ft^2 **5.** 5.2 cm **7.** 9.5 cm **9.** 69,882 cm^2 or 6.9882 m^2 **15. a.** 1912 ft^2 **b.** The perimeter of the square bandstand would be about 15 ft greater. **c.** Recommendations may vary. An example is given. The perimeter of the octagonal bandstand is less than that of a square bandstand. The octagonal shape will probably cost less. **17.** height: 10 units; area: 80 square units **19.** Explanations may vary. An example is given. Multiply both sides by 2, then divide both sides by $(b_1 + b_2)$ to get $h = \dfrac{2A}{b_1 + b_2}$
21. 75 **22.** -16 **23.** -16 **24.** $(-3, 4)$, $(-8, 1)$, $(-4, -6)$, $(-1, 4)$ **25.** $y = 20 - 2x$ **26.** $y = -6 - \dfrac{x}{6}$
27. $y = \frac{5}{4}(16 + 4x)$ or $y = 20 + 5x$

Pages 289, 291 Talk it Over
1. a. Their sum is 90°. **b.** $n + 30 = 90$ **c.** 60°
2. a. $w + z = 90$ **b.** No; reasons may vary. An example is given. The equation has two variables, and to find the unknown measures you need to write the equation in terms of only one variable. **c.** Answers may vary. An example is given. $w + w = 90$ or $2w = 90$; 45°, 45° **3.** The conditions cannot all be met at the same time. **4. a.** Cliff labeled the acute angles by using the definition of complementary. His equation relates the complement of the angle with four times the angle. **b.** Emma used x for one angle and the condition that the second angle is 4 times the first to label the angles. Her equation shows the sum of these two to be 90°. **c.** Answers may vary. An example is given. Emma's method; it seems more straightforward.

Pages 292–293 Exercises and Problems
3. correct **5.** $x = -9$, $y = -72$ **7.** $p = 6$, $m = 15$ **9.** 30° and 150° **11. a.**

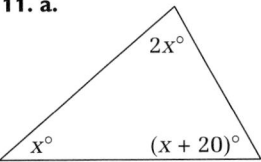

b. 40°, 60°, and 80° **17.** Answers may vary. An example is given. $b = 2 - 3a$; $a = 1$, $b = -1$ **20.** The rectangle's area equals the parallelogram's area (24 units^2). The triangle's area is 12 units^2.
21. $q < 4$

22. $a \le -5$

23. $m \ge 9$

24. 78% **25.** 156% **26.** $43\frac{1}{3}\%$ **27.** $87\frac{1}{2}\%$

Pages 295–296 Unit 5 Review and Assessment
1. a.

Number of Pages	Total Charges
1	$6 + 2(1) = 8$
2	$6 + 2(2) = 10$
3	$6 + 2(3) = 12$
⋮	⋮
n	$6 + 2(n)$

b.

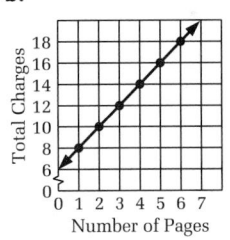

c. $6 + 2n = t$, where n = number of pages, t = total charges **d.** 12 pages **2.** Yes. The polygon would have 12 sides. **3.** 135 s **4.** Answers may vary.
5. -26 **6.** -5 **7.** 8 **8.** 12 **9.** -2 **10.** $\frac{1}{2}$
11. a. Explanations may vary. An example is given. John forgot to put a negative sign in front of the 5. He should have had $-5 = 4x$ and then $-\frac{5}{4} = x$.
b. Dwayne forgot to reverse the inequality sign when he multiplied by -1. The result should be $-5 < y$.

12. $x < -3$

13. $a \geq -4$

14. $x > 4$

15. 19 h **16.** $\dfrac{I}{pr} = t$ **17.** $y = 180 - 4x$ **18.** $\dfrac{P}{T} + 2 = n$ or

$\dfrac{P + 2T}{T} = n$ **19.** $x = \dfrac{21}{4}y + \dfrac{35}{4}$ **20.** $x = \dfrac{8}{3}(y - 6)$ or

$x = \dfrac{8}{3}y - 16$ **21.** $y = \dfrac{4}{5}(1 - 5x)$ or $y = \dfrac{4}{5} - 4x$ **22.** $\dfrac{y}{4} = x$

23. Area of a parallelogram = bh **24.** Area of a
trapezoid = $\frac{1}{2}(b_1 + b_2)h$ **25.** about 105,000 mi^2
26. $x = 4, y = -1$ **27.** $y = 3x, y + 2x = 180$; two angles
of 36°, 1 angle of 108°

Unit 6

Pages 301–302 Talk it Over

1. 20 in. **2.** $\dfrac{16}{12}$ **3.** Yes. The aspect ratio, $\dfrac{16}{12}$, can be

simplified to $\dfrac{4}{3}$. **4.** about 15% **5.** Answers may vary.

Examples are given: discounts at stores, newspaper
survey results, test grades.

Pages 303–305 Exercises and Problems

3. 1:4 **5.** 4:1 **13.** $a = 1, b = 3$ **14.** $r = -5, s = 5$
15. $d = 13, e = 2$ **16.** Answers may vary.

Pages 309–311 Talk it Over

1. Since the probability is a little more than 50%, it is
just slightly more likely the person is a female than

not. **2.** $P(E) = \dfrac{\text{male residents at least 25 yr}}{\text{old with 1–3 years of college}}{\text{total residents at least 25 yr old}} =$

$\dfrac{13,720}{158,694} \approx 0.086$. The probability is about 0.09 or 9%.
3. three **4.** 0; Explanations may vary. An example is
given. When a die is rolled, the side facing up indicates the number rolled. It is not possible to have two
sides of a die facing up. **5.** The largest value that a
probability of an event can have is 1, which represents 100% probability. Most events have a probability less than one. The complement would then be 1
minus the probability of this event.
6. $P(\text{not } E) + P(E) = 1$

Pages 311–313 Exercises and Problems

1. impossible **3.** certain **7. a.** Answers and explanations may vary. An example is given. The probability
of choosing a female is about 58%, and the probability of choosing a male is about 42%. Therefore, there
is a slightly better probability of a female student
being chosen. **b.** about 0.42 or 42% **9.** Answers
may vary. An example is given. No, the probability
for the event will remain at 50%. Tossing a fair coin
in an example of theoretical probability and the prob-

ability of a tail is equally as likely as the probability
of a head regardless of any previous outcomes.
11. a. about 0.24 or 24% **b.** about 0.76 or 76%
13. thin crusts **15.** $\dfrac{12}{52} = \dfrac{3}{13} \approx 0.23$ or 23%

17. $\dfrac{39}{52} = \dfrac{3}{4} = 0.75 = 75\%$ **19.** $\dfrac{26}{52} = \dfrac{1}{2} = 0.5 = 50\%$

21. $\dfrac{1}{52} \approx 0.02$ or 2% **23.** red: 0.2; yellow: 0.2;

green: 0.1; orange: 0.1; tan: 0.1 **25.** $\dfrac{8}{5}$ **26.** $\dfrac{37}{3}$ **27.** $\dfrac{3}{5}$

28. square roots ± 5; cube root 2.92
29. square roots ± 7.62; cube root 3.87
30. square roots ± 8.49; cube root 4.16 **31.** 721.5
32. -162 **33.** -5.74

Page 315 Talk it Over

1. Descriptions may vary. An example is given. The
first ratio must equal the second ratio. Multiplying
the denominator of the second ratio, 4, by 3 gives the
denominator of the first ratio, 12. So multiply the
numerator of the second ratio, 3, by 3 to get 9 as the
numerator of the first ratio. Then $x = 9$. **2. a.** 60
b. $4x = 45$ **c.** The factors of the left side of the equation in part (b), 4 and x, are the numerator of the first
ratio and the denominator of the second ratio. The
factors of the right side, 15 and 3 ($45 = 3 \cdot 15$), are the
denominator of the first ratio and the numerator of
the second ratio. **3.** Descriptions may vary. An
example is given. Multiply both sides by the LCD of
the fractions, $6y$. This gives $5y = 72$. Divide both
sides by 5; $y = 14.4$. **4.** Yes. Explanations may vary.
An example is given. Rewrite 5 as $\dfrac{5}{1}$ to get $\dfrac{5}{1} = \dfrac{12}{y}$.

Then use cross products to solve.

Pages 316–317, 319 Exercises and Problems

1. 2, 3, 8, 12; $2 \cdot 12 = 3 \cdot 8$, $24 = 24$ **3.** 7.2 **5.** 1250
7. about 2.3 **9.** D **11.** Answers may vary. Examples
are given. $\dfrac{x}{7} = \dfrac{6}{84}, \dfrac{7}{x} = \dfrac{84}{6}$ **13.** $\dfrac{n}{36} = \dfrac{4}{9}, \dfrac{36}{n} = \dfrac{9}{4}, \dfrac{n}{4} = \dfrac{36}{9},$

$\dfrac{4}{n} = \dfrac{9}{36}, \dfrac{4}{9} = \dfrac{n}{36}, \dfrac{9}{4} = \dfrac{36}{n}, \dfrac{36}{9} = \dfrac{n}{4}, \dfrac{9}{36} = \dfrac{4}{n}$

15, 17, 19. Explanations may vary. Examples are
given. **15.** Proportional reasoning is not appropriate
because people do not grow at a constant rate.
17. Proportional reasoning is not appropriate because
many factors determine gasoline mileage, such as
road surface, tire wear, and weather conditions.

19. Proportional reasoning is appropriate but the
wrong proportion is used in the problem. The number of seconds depends on the number of cuts, not
the number of pieces. Two cuts are needed to get
3 pieces. So $\dfrac{12 \text{ s}}{2 \text{ cuts}} = 6$ s/cut. To get 4 pieces, 3 cuts
are needed. At 6 s/cut, 3 cuts take $3(6) = 18$ s. So, it
would take 18 s (not 16 s) to saw a log into 4 pieces.

30. $\frac{2}{12} = \frac{1}{6} \approx 0.17$ or 17% **31.** $\frac{7}{12} \approx 0.58$ or 58%

32. $\frac{9}{12} = \frac{3}{4} = 0.75 = 75\%$ **33.** control variable: time; dependent variable: distance **34.** control variable: number of lawns cut; dependent variable: earnings **35.** 90 **36.** 1175 **37.** 62.5

Pages 320 Checkpoint
1. Answers may vary. An example is given. Two types of ratios I have studied in this unit are experimental probability and theoretical probability. For each ratio, the possible values are any number on a scale from 0 to 1, including 0 and 1. The ratios are alike because they describe the probability of an event. They are different because experimental probability describes events based on actual observations, and theoretical probability describes events based on outcomes that are likely to occur without doing an experiment. **2. a.** 65°, 25° **b.** 13:18 **3.** about 0.48 or 48% **4.** $\frac{1}{2} = 0.5 = 50\%$ **5.** $\frac{1}{5} = 0.2 = 20\%$

6. $\frac{4}{5} = 0.8 = 80\%$ **7.** 0.9 **8.** 4 **9.** 16 **10.** 450 mi

Pages 321, 325 Talk it Over
1. Answers may vary. An example is given. An ocean can be very large, which would make it difficult to be sure all the whales in every part are counted. Since whales are constantly in motion, it would be hard to tell if a whale had already been counted or not. Usually a city park has a very large squirrel population. Squirrels can move rapidly and easily climb trees, hiding above eye level. Also, the squirrel population is constantly changing as squirrels run in and out of the park. **2.** Answers may vary. Examples are given. **a.** Estimate the area that the crowd takes up. Count the number of people in one part of the area. Estimate how many of these parts are in the total area and multiply this number by the number of people who can fit in this part. **b.** It would be difficult to use this method. A deer moves among the trees, whereas crowds are usually still. **3.** Multiply $3 \cdot 2000$ and $50 \cdot x$. Then divide by 50. **4.** The estimate would be less certain due to the increased overall range of numbers of faulty bulbs. A larger margin gives more room for error.

Pages 325–327 Exercises and Problems
1. a. the second handful of dried beans **b.** all the dried beans **5. a.** about 3135 residents **b.** about 2610 to 3920 residents **7. a.** about 207,690 to 265,650 votes **b.** Answers may vary. An example is given. Johnson's and Rubinski's intervals overlap, so it is possible either could win, but the results of the exit poll show Johnson is favored. **c.** Reasons may vary. Examples are given. The margin of error may be larger than originally anticipated. The sample of 400 is so small compared to the population of 1,295,000 as to be insignificant. Only 483,000 or about 37% of reg-

istered voters voted, so perhaps the pre-election poll included people who did not end up voting at all. The pre-election poll counts changeable opinions, whereas the exit poll counts unchangeable votes. **11.** 36 **12.** about 15.7 **13.** 30 **14.** R corresponds to X, S corresponds to Y, T corresponds to Z **15.** 14 **16.** 28

Pages 330–331 Talk it Over
1. pentagon **2.** $\angle J$ corresponds to $\angle P$, $\angle K$ corresponds to $\angle Q$, $\angle L$ corresponds to $\angle R$, $\angle M$ corresponds to $\angle S$, $\angle N$ corresponds to $\angle T$; $\overline{JK}$ corresponds to $\overline{PQ}$, $\overline{KL}$ corresponds to $\overline{QR}$, $\overline{LM}$ corresponds to $\overline{RS}$, $\overline{MN}$ corresponds to $\overline{ST}$, $\overline{NJ}$ corresponds to $\overline{TP}$ **3. a.** 1:1; the measures of the corresponding angles are equal. **b.** Yes. Since the corresponding angles are congruent and the corresponding sides are in proportion, the triangles are similar. **4.** Descriptions may vary. An example is given. Use a ruler to find the dimensions of store 18 on the drawing. Use the scale and the drawing dimensions to write proportions that can be solved to find the actual dimensions. Multiply the actual dimensions to find the area in square feet. **5.** The proportion would have 1 in. replaced with $\frac{3}{4}$ in.

$$\frac{\frac{3}{4}}{40} = \frac{1\frac{15}{16}}{a}; \ 77.5 = \frac{3}{4}a; \ a \approx 103 \text{ ft}$$

Pages 332–336 Exercises and Problems
1. a. ~ **b.** Answers may vary. An example is given. Congruent figures have corresponding sides of equal length and corresponding angles of equal measure. Similar figure have corresponding angles of equal measure and corresponding sides in proportion, not necessarily equal. Congruent figures are always similar. Therefore, the symbol for congruence contains an equals sign and the symbol for similarity. Similar figures are not always congruent, so the symbol for similarity does not contain the equals sign. **3.** 74° **5.** 6 **7.** No. Explanations may vary. An example is given. Solving the proportion $\frac{1.5}{2} = \frac{x}{4}$, you get $x = 3$. The ball would reach the player 3 m above the floor. The player might catch the pass, but not easily. **11.** A distance of 2 units on the drawing represents a distance of 1 unit on an actual grasshopper. The drawing is twice as large as an actual grasshopper. **13.** Answers may vary. An example is given: about $\frac{1}{4}$ in. **15. a.** 2

b. 4 **19.** Explanations may vary. An example is given. Each triangle contains a 90° angle. The vertical angles are congruent. Since two pairs of corresponding angles are congruent, the triangles are similar. 50 m **29.** about 597,000 people **30.** mean: about 79; median: 81; mode: 84

31. parallelogram

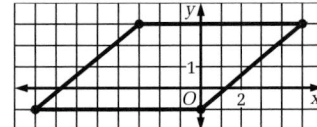

Pages 337, 339 Talk it Over

1. Answers may vary. An example is given. The top right globe has the same height but twice the width. The bottom left globe has the same width but twice the height. The bottom globe has twice the height and twice the width. **2.** No. Yes, the bottom right graphic is similar to the original. **3.** Answers may vary. An example is given. $\frac{A'C'}{AC} = \frac{3}{1}$, $\frac{A'B'}{AB} = \frac{3}{1}$, $\frac{B'C'}{BC} = \frac{3}{1}$; $\angle A \cong \angle A'$, $\angle B \cong \angle B'$, $\angle C \cong \angle C'$; the triangles are similar. $\frac{J'K'}{JK} = \frac{1}{2}$, $\frac{K'L'}{KL} = \frac{1}{2}$, $\frac{L'M'}{LM} = \frac{1}{2}$, $\frac{M'J'}{MJ} = \frac{1}{2}$; $\angle J \cong \angle J'$, $\angle K \cong \angle L'K'J'$, $\angle L \cong \angle L'$, $\angle M \cong \angle L'M'J'$; the quadrilaterals are similar. **4. a.** The dilation in Sample 1 is an enlargement. **b.** The dilation in Sample 2 is a reduction. **c.** A scale factor that is greater than 1 results in an enlargement. **d.** A scale factor between 0 and 1 results in a reduction. **5.** Drawings may vary. An example is given. The center of dilation is at P and the scale factor is $\frac{1}{2}$.

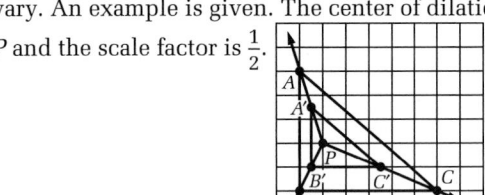

Pages 340–342 Exercises and Problems

3. dilation **5. a.** $(-1, 0)$ **b.** 2 **7. a.** $(-3, 6)$ **b.** 2

9.

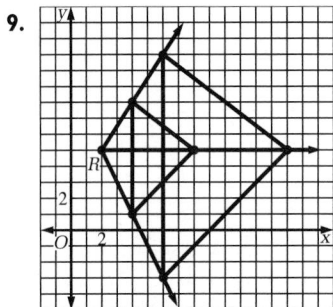

11.

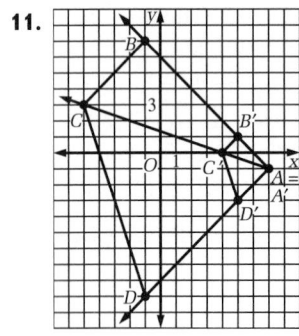

17. $\angle A = \angle D = 63°$; $\angle B = 42°$; $\angle F = 75°$; $AB = 39$; $FE = 24$ **18.** 24 **19.** 11 **20.** -9 **21.** $\frac{1}{6}$ **22.** 2 to 3 **23.** 3:4

Pages 344–347 Talk it Over

1. a. $\sin 53° \approx 0.79863551$ **b.** Answers may vary. An example is given. If the calculator answer were rounded to the nearest hundredth, it would be the same as the ratio I found in the Exploration. The values may be slightly different due to small measuring errors in measuring the angle and rounding the lengths to the nearest half unit.
2. a. $\cos 53° \approx 0.6018150232$ **b.** Answers may vary. An example is given. If the calculator answer were rounded to the nearest hundredth, it would be the same as the ratio I found in the Exploration. The values may be slightly different due to small measuring errors in measuring the angle and rounding the lengths to the nearest half unit. **3.** $\overline{AC}$; $\overline{BC}$
4. $\cos K \approx 0.89$, $\sin K \approx 0.44$, $\cos L \approx 0.44$, $\sin L \approx 0.89$ Statements may vary. An example is given. The cosine of $\angle K$ is the same as the sine of $\angle L$, and the sine of $\angle K$ is the same as the cosine of $\angle L$.
5. Explanations may vary. An example is given. Question 4 suggests that the sine of one acute angle of a right triangle is equal to the cosine of the other acute angle. **6.** $\sin 37° = \cos 53° \approx 0.6018150232$; $\cos 37° = \sin 53° \approx 0.79863551$ **7.** No, the hypotenuse will always be the longest side and its value will always be in the denominator of the ratio.
8. closer to **9.** Answers may vary. An example is given. Let d be the distance between the foot of the ladder and the base of the building. You can find d by using the equation $\cos 75° = \frac{d}{12}$. **10.** Answers may vary. An example is given. I would use the sine ratio since it is the ratio of the leg opposite the angle over the hypotenuse. The leg opposite has a length equal to the height of the ramp, and the hypotenuse has a value equal to the length of the ramp.

Pages 348–350 Exercises and Problems

3. 0.53 **5.** 0.98 **7.** $\frac{5}{13}$, 0.38 **9.** $\frac{5}{13}$, 0.38

11. a–d. Answers may vary.
13. Drawings may vary. An examples is given.

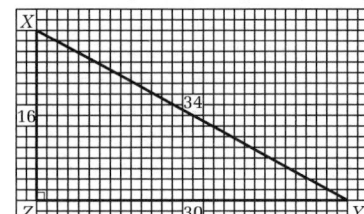

17. $\sin 50° = \frac{x}{12}$, $\cos 40° = \frac{x}{12}$; about 9.2

19, 21. Sketches may vary. Examples are given.
19. maximum height: **21.** about 96.4 m
about 14.5 ft

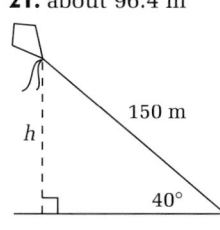

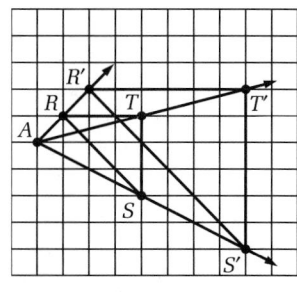

23. about 119.4 mi   **26.**

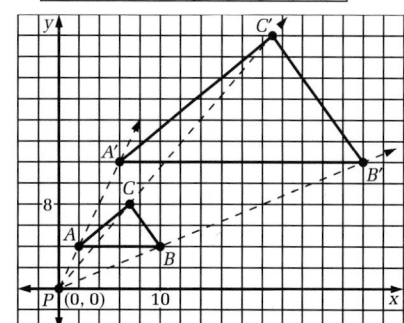

27. Yes. The broken scale gives the impression that station WRRR is much more popular than WSSS or WTTT when it actually only has about 3000 more listeners than WSSS and about 2000 more listeners than WTTT. **28.** $\frac{4}{12} = \frac{1}{3} \approx 0.33$

Pages 352–354 Unit 6 Review and Assessment
1. 6:1 **2.** \$.16/oz **3.** There are about 19,200 women attending the game. **4.** 0.2 or 20% **5.** 0.3 or 30% **6.** 0.65 or 65% **7.** Since 35 of 45 hits were against a right-handed pitcher, the probability of the next hit being against a right-handed pitcher is about 0.78 or 78%. **8.** 8 **9.** Yes. **10.** 3 **11.** No. **12.** $\frac{1}{6} \approx 0.17$ or 17% **13.** $\frac{1}{2} = 0.5 = 50\%$ **14.** $\frac{5}{6} \approx 0.83$ or 83%

15. $\frac{1}{2} = 0.5 = 50\%$ **16.** Answers may vary. An example is given. Probability has many important uses, from interpreting the results of polls to telling the fairness of games. There are two types of probability, experimental probability and theoretical probability. Both tell the chances that a certain event is likely to happen. However, experimental probability is based on observations and recorded results, while theoretical probability is based on the knowledge that the possible outcomes of an event are equally likely to happen without doing an experiment. You can use data from a table to find the probability of an event by using the formula for experimental probability:

$$P(E) = \frac{\text{number of times } E \text{ is observed to happen}}{\text{total number of observations}}.$$ **17.** 15

18. $\frac{4}{3}$ or about 1.33 **19.** 20 stacks **20.** Candidate A:

about 1,380,000 votes; Candidate B: about 1,120,000 votes; Candidate C: about 500,000 votes **21.** No. **22.** Candidate A: 1,230,000 to 1,530,000 votes; Candidate B: 970,000 to 1,270,000 votes; Candidate C: 350,000 to 650,000 votes **23.** False; any two squares are similar, but one square may be smaller, so they would not be congruent. **24.** True. **25.** Answers may vary. An example is given. Christie's father wants to measure the ropes for a tree swing from a branch about halfway up a tree in the backyard. He needs to know the height of the tree. He is 6 ft tall and his shadow is 4 ft long. The tree's shadow is 12 ft long. How tall is the tree? **26. a.** Explanations may vary. An example is given. Both triangles have $\angle C$ in common, and each triangle has a right angle. Since two pairs of corresponding angles are congruent, the triangles are similar. **b.** 20 **c.** 10 **27.** $8\frac{1}{3}$ in.

28.

Vertices of Original	Vertices of Image
$A(2, 4)$	$A'(6, 12)$
$B(10, 4)$	$B'(30, 12)$
$C(7, 8)$	$C'(21, 24)$

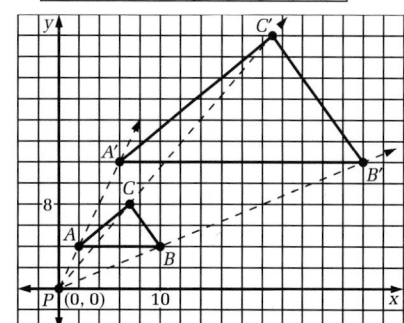

29.

Vertices of Original	Vertices of Image
$A(2, 4)$	$A'(2, 4)$
$B(10, 4)$	$B'(6, 4)$
$C(7, 8)$	$C'(4.5, 6)$

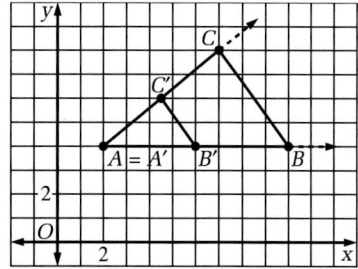

30. $\frac{1}{3}$ **31.** 60 **32.** $\frac{4}{5}$ **33.** $\frac{4}{5}$ **34.** $\frac{3}{5}$ **35.** $\frac{3}{5}$ **36. a.** The wires are about 2020 ft each. **b.** The distance from the base is about 917 ft.

Pages 359, 360, 362 Talk it Over

1. The heights of the posts increase by 1.5 ft as the distance from the eaves increases by 2 ft. **2.** For each triangle, the ratio $\frac{h}{d}$ equals 0.75.

Post	Horizontal distance d of posts from eaves (ft)	Height h of post (ft)	$\frac{h}{d}$
1	2	1.5	0.75
2	4	3	0.75
3	6	4.5	0.75
4	8	6	0.75
5	10	7.5	0.75

3. Explanations may vary. An example is given. Two triangles are similar when two pairs of corresponding angles are congruent. The triangles formed by the posts all have right angles and all have the angle formed with the eaves in common. Since the triangles are similar, corresponding sides are proportional, so the ratios $\frac{h}{d}$ are the same. **4.** Multiply both sides of the equation by d. **5.** Substitute 5 for h in the equation $h = 0.75d$; $5 = 0.75d$; divide both sides by 0.75; $\frac{5}{0.75} = \frac{0.75}{0.75}d$; $d = 6\frac{2}{3}$. **6.** larger **7.** $\frac{4}{3}$; The tangent ratio of $\angle B$ is the reciprocal of the tangent ratio of $\angle A$. **8. a.** about 0.7536 **b.** Answers may vary. An example is given. It is difficult with a protractor to find the exact angle measure. While 37° is close, the actual angle measure is slightly smaller. **9.** about 36.9°; This measure is smaller by one tenth of a degree. **10.** larger; Explanations may vary. An example is given. For a steeper roof, the height of the roof would be greater. This would make the tangent ratio, $\frac{\text{height}}{\text{horizontal distance}}$, larger.

Pages 363–366 Exercises and Problems

1. 1.25 **3.** 10 **5. a.** $h = \frac{4}{3}d$ **b.** $\frac{4}{3}$ **7. a.** rise = 3 ft, run = 4 ft **b.** slope = $\frac{3}{4}$ = 0.75; Yes. **c.** $\frac{1.5}{2}$ = 0.75 **d.** $\frac{3}{4}$ = 0.75 **13.** $\overline{RT}$ **15. a.** 2 **b.** about 1.963 **c.** No. Explanations may vary. An example is given. If a triangle is formed using the given lengths, the measure of $\angle T$ would be close to, but not exactly equal to, 63°.

17. a.

Angle	10°	20°	30°	40°	50°	60°	70°	80°
Tangent	0.1763	0.3640	0.5774	0.8391	1.1918	1.7321	2.7475	5.6713
Sine	0.1736	0.3420	0.5000	0.6428	0.7660	0.8660	0.9397	0.9848
Cosine	0.9848	0.9397	0.8660	0.7660	0.6428	0.5000	0.3420	0.1736

17. b. Answers may vary. Examples are given. The sine of an angle is equal to the cosine of the complement of the angle; sine and tangent values increase as the angle increases, whereas cosine values decrease;

sine and cosine values are numbers less than 1, whereas tangent values may be greater than 1; the tangent value is equal to the sine value divided by the cosine value.

19. a.

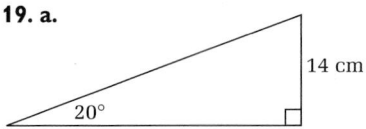

b. Explanations may vary. An example is given. To find the length of the adjacent leg, use the equation $\tan 20° = \frac{14}{\text{adjacent leg}}$. To find the hypotenuse, use the equation $\sin 20° = \frac{14}{\text{hypotenuse}}$. **c.** 38.5 cm, 40.9 cm **22.** $\sin A \approx 0.9231$; $\cos A \approx 0.3846$ **23.** 2.4 **24.** 240 **25.** 0.41$\overline{6}$, or $\frac{5}{12}$ **26.** $l = \frac{1}{3}h$

Pages 369–370 Talk it Over

1. about 33.6 in.; This answer is close to the prediction in Sample 2. **2.** Yes. Explanations may vary. An example is given. When you multiply each side of the equation $\frac{B}{D} = 0.8$ by D, you obtain the equivalent equation $B = 0.8D$. **3.** From the 12 in. mark on the vertical axis, draw a horizontal line across until it reaches the fitted line. Then draw a vertical line down until it reaches the horizontal axis. Estimate the drop height at the crossing point. **4.** Answers may vary. An example is given. Rashid could use the equation $\frac{B}{D} = 0.8$ or $B = 0.8D$; substitute 12 for B and then solve for D; $\frac{12}{D} = 0.8$; $12 = 0.8D$; $D = \frac{12}{0.8} = 15$; He should use a drop height of about 15 in. **5.** The variation constant, 0.8, and the slope, 0.81$\overline{6}$, are close. **6.** $\frac{4}{3} \approx 1.3$

Pages 370–374 Exercises and Problems

1. B and D; B represents the bounce height, D represents the drop height.

3. ➤ No. No.
   ➤ Answers may vary. Most balls display direct variation.
   ➤ Answers may vary. The variation constants should be nearly the same for the same type of ball. Differences among groups' results could be due to measurement errors or rounding differences.
   ➤ Variation constants depend on the balls used. The ball with the most bounce has the largest variation constant. The ball with the least bounce has the smallest variation constant.

5. Yes. For each data point, the ratio $\frac{\text{height}}{\text{width}}$ is close to 0.07. **7.** Yes. For each data point, the ratio $\frac{\text{cost}}{\text{errors}}$ is close to 21. **9.** Reasons may vary. An example is given. **9. a.** No. The water level does not change steadily over time. **b.** continuous

Integrated Mathematics

11. a. control: speed; dependent: stopping distance

b. $\frac{15}{10} = 1.5$, $\frac{40}{20} = 2$, $\frac{75}{30} = 2.5$, $\frac{120}{40} = 3$, $\frac{175}{50} = 3.5$, $\frac{240}{60} = 4$,

$\frac{315}{70} = 4.5$ **c.** No. Explanations may vary. An example

is given. The graph is a curve, not a straight line, and

the ratio $\frac{\text{dependent variable}}{\text{control variable}}$ increases steadily.

13. about $\frac{1}{33}$, or about 0.03 **15.** about 3.9 atm

17. a. Each ratio $\frac{D}{W}$ is equal to about 0.017. **b.** The

dosage varies directly with the weight. about 0.017;

$D = \frac{1}{60}W$ or $D \approx 0.017\,W$. **c.** 1.25 tablets; Methods

may vary. For example, substitute 75 for W in the

equation in part (b) and solve to find the value of D,

or continue the pattern established in the table, not-

ing that the values of W are increasing by 15 and the

corresponding values of D are increasing by 0.25.

17. d.

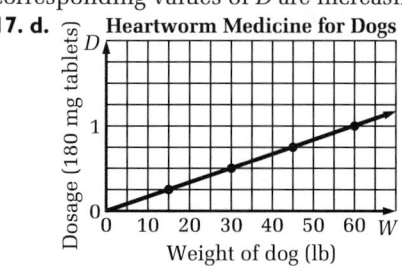

Heartworm Medicine for Dogs

e. $\frac{1}{60}$ or about 0.017 **23. a.** 1 **b.** 45° **25.** about 1.22

26.

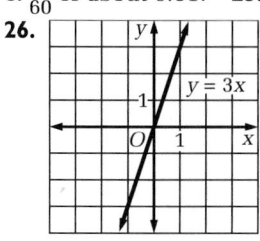

$y = 3x$

27.

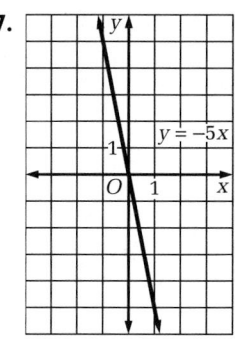

$y = -5x$

28.

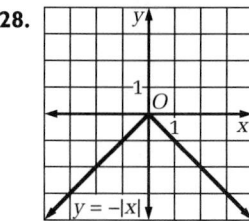

$y = -|x|$

29. $P = 2l + 2w$ **30.** $P = 4s$ **31.** $P = 3s$

Page 377 Talk it Over

1.

$\frac{45°}{360°}$	$\frac{135°}{360°}$	$\frac{60°}{360°}$	$\frac{30°}{360°}$	$\frac{90°}{360°}$
$\frac{1}{8}$	$\frac{3}{8}$	$\frac{1}{6}$	$\frac{1}{12}$	$\frac{1}{4}$

2. 72°

Page 379, 381 Exercises and Problems

5. about 50.2 ft (using $\pi \approx 3.14$) **7. a.** about 37.3 ft

9. about 9 in. **11.** 4 $P = 4s$

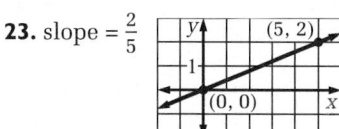

13. a, b, and c

15. about 3.1 ft

23. slope $= \frac{2}{5}$

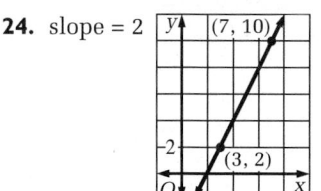

24. slope $= 2$

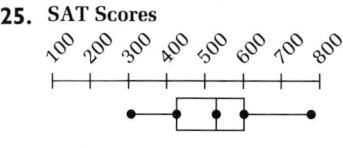

25. SAT Scores

26. 18 **27.** −1 **28.** 3

Page 382 Checkpoint

1. Answers may vary. An example is given. For one
variable to vary directly with another, the two quantities must have a constant ratio. Find the ratio for
each data pair. If the ratio is constant or almost constant, one variable varies directly with the other.

2. a. $\frac{7}{4}$ or 1.75 **b.** $\frac{4}{7}$ or about 0.571 **3.** about 13.9 ft

4. Estimates may vary. Examples are given. **a.** about
13 in. **b.** about 37 in. **5.** Estimates may vary; for
example, about 0.7. **6. a.** about 1256 cm **b.** about
19.1 ft **c.** $C = \pi d$ **7. a.** about 44 in. (using $\pi \approx 3.14$)
b. about 7.3 in. **c.** about 14.7 in.

Pages 383–387 Talk it Over

1. Answers may vary. An example is given. The equations are different because they involve different
quantities, variables, and constants, and they have
different solution pairs. They all have the form variable = constant × another variable, so each is a direct
variation equation.

2.

equation	$h = 0.75d$	$B = 0.8D$	$C = \pi d$
control variable	d	D	d
dependent variable	h	B	C
variation constant	0.75	0.8	π

3. Method 1; Explanations may vary. An example is given. Method 1 is easier because you just need to multiply each time by the variation constant, 8.4, to find the amount earned, rather than setting up a proportion. **4. a.** 0 **b.** 3.14159; 0.75; 0.8
c. Point (2, 6.28318) is on the graph of $C = \pi d$; point (6, 4.5) is on the graph of $h = 0.75d$; point (4, 3.2) is on the graph of $B = 0.8D$. **5.** Answers may vary. An example is given. The three lines are similar because they all have the general form for direct variation, $y = kx$, and all pass through the origin. They are different because the three lines in question 4 have positive slopes or variation constants, while these three have negative slopes or variation constants. **6.** Lines with negative slope slant downward from left to right. Lines with positive slopes slant upward from left to right. **7.** -1 **8. a, b.** Explanations may vary. Examples are given. **a.** To go from $(-1, 2)$ to $(2, -4)$, you move 3 units to the right (a run of 3) and 6 units down (a rise of -6). **b.** To go from $(2, -4)$ to $(-1, 2)$, you move 3 units to the left (a run of -3) and 6 units up (a rise of 6). **c.** Regan: $\dfrac{\text{rise}}{\text{run}} = \dfrac{-6}{3} = -2$;

Lianon: $\dfrac{\text{rise}}{\text{run}} = \dfrac{6}{-3} = -2$ **9.** Yes. The rise has a change of -2, which is down, for every run change of 1 to the right.

Pages 388–391 Exercises and Problems
1. 8.4 represents the hourly rate of pay, $8.40.
3. $12.93 **5. a.** 1875 words **b.** 320 min, or 5 h 20 min
7. a. 57.6 min **b.** Yes, the skater covers about 2.8 mi.
9. $y = \dfrac{1}{3}x$ **11. a.** Yes. All the points lie close to a fitted line passing through the origin. **b.** $y = \dfrac{1}{2}x$ **13. a.** No. The line connecting the points does not pass through the origin.

17. 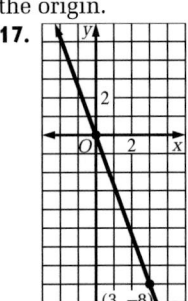 **19.**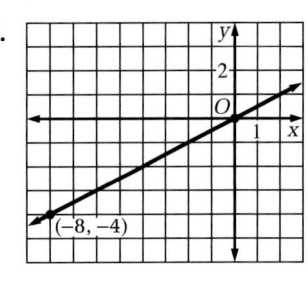

23. C **25.** B **27.** Explanations may vary. An example is given. A: $y = \dfrac{5}{3}x$; B: $y = -\dfrac{3}{5}x$; Graph A shows positive slope and Graph B shows negative slope.

29. 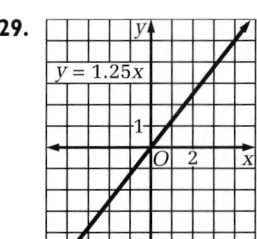 **31.**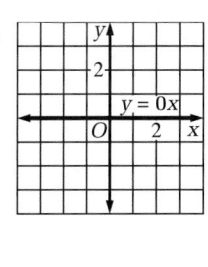

35. about 44.0 in. **36.** about 69.1 cm **37.** $-6x - 4y$
38. $-20a + 8b$ **39.** $7c - 35d$ **40.** $m + n$ **41.** $\dfrac{3}{25}$
42. $2\dfrac{1}{2}$

Pages 393, 394, 396 Talk it Over
1. Estimates may vary. An example is given. $3 \times 1.2 = 3.6$, which is about equal to 3.57; $5 \times 1.2 = 6$, which is about equal to 5.95. **2.** number of gallons of gasoline; the cost of the gasoline; 1.189 **3.** No. Reasons may vary. An example is given. Because the amounts in a conversion factor are equal, the ratio represents the number 1. So when you multiply by a conversion factor, you are changing the units of the expression but not the value. **4.** $\dfrac{1 \text{ in.}}{2.54 \text{ cm}}$; about 4.7 in. **5.** Answers may vary. Examples are given. $\dfrac{60 \text{ s}}{1 \text{ h}}, \dfrac{7 \text{ days}}{1 \text{ week}}, \dfrac{3 \text{ ft}}{1 \text{ yd}}, \dfrac{100 \text{ cm}}{1 \text{ m}}, \dfrac{36 \text{ in.}}{1 \text{ yd}}$ **6.** Answers may vary. An example is given. The conversion factor $\dfrac{1 \text{ mi}}{1.61 \text{ km}}$ is the same as the conversion factor $\dfrac{0.621 \text{ mi}}{1 \text{ km}}$ because $\dfrac{1}{1.61} \approx \dfrac{0.621}{1}$. **7. a.** $d = 44t$
b. 1320 ft

Pages 396–399 Exercises and Problems
3. $\dfrac{\text{h}}{\text{s}}$ **5.** $\dfrac{\text{ft}}{\text{min}}$ **7. a.** distance traveled, number of gallons **b.** 22 mi/gal **9.** about 80 km **11.** 38,016 ft
13. 3875 pesos **15. a.** 1.3 pesos **b.** $C = 1.3L$
c. 32.5 pesos **17. a.** The green line represents super unleaded gasoline. **b.** the cost of 10 gallons of each type of gasoline **c.** the number of gallons of each type of gasoline that can be bought for $8 **d.** In the graph on the left, the gray vertical segment represents the amount of money you save by buying 10 gal of regular unleaded rather than 10 gal of super unleaded gasoline. In the graph on the right, the gray horizontal segment represents the additional amount of gasoline you can afford by buying $8 worth of regular unleaded rather than $8 worth of super unleaded gasoline. **21.** about 0.513 mi **23.** about 47.2 mi/h

28. 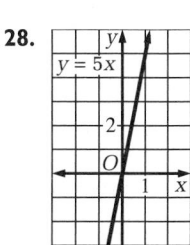 **29.**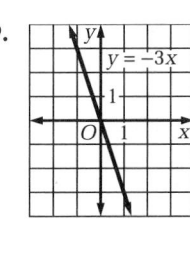

30. 80 **31.** $5\frac{5}{9}$ or about 5.6 **32.** 180 **33.** 135°
34. 120° **35.** 65°

Page 401 Talk it Over
1. The variable r represents the length of an irrigation arm; it cannot be a negative number. **2.** If you multiply both sides of $\frac{A}{\pi} = r^2$ by π, you get the equivalent equation $A = \pi r^2$.

Pages 404–407 Exercises and Problems
3. about 113.0 ft² (using $\pi \approx 3.14$) **5.** about 28.3 square units **7.** Answers may vary. An example is given. I agree. The circumference is 4π units and the area is 4π square units. The kinds of units are different, but the number of units is the same.

9. a.

r (cm)	r^2 (cm²)	A (cm²)	$\frac{A}{r^2}$
1	1	3.14	3.14
2	4	12.56	3.14
3	9	28.26	3.14
4	16	50.24	3.14
5	25	78.50	3.14
6	36	113.04	3.14

b. There is a constant ratio, 3.14. **c.** 3.14.
d. Answers may vary. An example is given. I predict a straight line through the origin with a slope equal to 3.14.

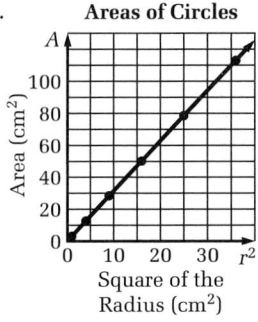

Areas of Circles

11. about 19.6 in.² **13.** about 20.4 cm² **15.** about 17.4 ft² **17. a.** about 18.8 in.² **b.** about 14.1 in.²
c. A slice of Tom's pizza is about $1\frac{1}{3}$ times as large as a slice of Vera's pizza. **19.** $1836.90 (using $\pi \approx 3.14$)

21. a. about 16.75 ft² (using $\pi \approx 3.14$) **b.** 17 marigolds **28.** 750 mi/h **29.** $198/yd² **30.** 90° clockwise or 270° counterclockwise **31.** −20 **32.** 63 **33.** −1

Pages 409–411 Unit 7 Review and Assessment
1. Answers may vary. An example is given. The slope, tangent ratio, and direct variation ratio are different ways to express the ratio of the length of a vertical leg to the length of a horizontal leg in a right triangle. The tangent can be thought of as the slope of the hypotenuse and the slope as the "rise" of the opposite leg over the "run" of the adjacent leg. A straight line passing through the origin can be expressed by an equation with a variation constant. This variation constant is the same as the slope. While the tangent ratio involves an angle, the slope is a rate and does not necessarily involve an angle. You might find the slope of the roof to determine its steepness, the tangent of an angle in order to find the length of a side or an angle measure for a triangle, and the variation constant to write an equation relating two variables. **2. a.** $h = s \tan 35°$ **b.** 0.700
c. about 14 ft **d.** about 60 ft **e.** tan 35° or 0.700
3. a. For each central angle, $\frac{M}{W} = 0.025$.
b. $\frac{M}{W} = 0.025$, or $M = 0.025W$ **c.** 7 m
4. a.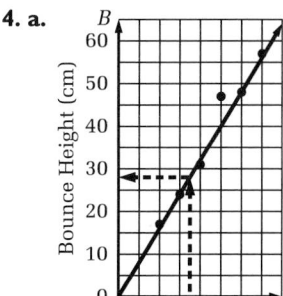

b. bounce height of 47 cm from a drop height of 100 cm **c.** Yes. Explanations may vary. An example is given. The remaining points are close to a straight line that passes through the origin. Also, for the remaining data each ratio $\frac{B}{D}$ is close to 0.4.
d. about 0.4 **e.** about 28 cm **5.** about 2,716,100 mi (using $\pi \approx 3.14$) **6.** 6 complete turns **7.** about 41.1 in. **8. a.** about 0.657; inches per pound
b. about 3.3 in. **c.** about 4.6 lb **9. a.** C; The line slants down from left to right. **b.** B; Lines A and B have positive slope because they slant up from left to right. B has the smallest slope because it has a small rise compared to the run. **c.** A; The slope of line A is positive and greater than that of line B.

10. a.

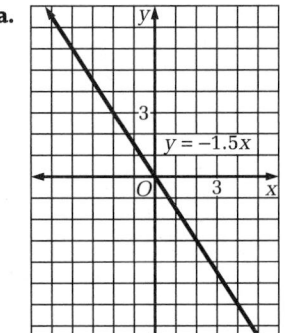

$y = -1.5x$

b. −1.5

c. quadrants II and IV   **d.** 6

11. about 5840 days (4 to 5 more if you include an extra day for each leap year)   **12.** $y = \frac{1}{365}d$; $\frac{1}{365}$ or about 0.0027   **13.** about 42,190 m   **14.** about 4.17 h   **15.** about 95.3 ft/s   **16. a.** $51\frac{3}{7}°$, or about 51.4°   **b.**

b. about 16.1 in.^{2} (using $\pi \approx 3.14$)   **17. a.** about 42.1 in.^{2} **b.** about 0.27   **c.** No. Explanations may vary. For example, multiplying both parts of a ratio by 2 does not change the ratio.

Unit 8

Pages 417–419 Talk it Over

1.

Time in minutes (t)	Height of balloon in feet (h)
0	6000
1	6110
2	6220
3	6330
4	6440
5	6550
6	6660

2. $h = 6000 + 110t$ **3.** control: t; dependent: h
4. Graphs may vary. An example is given. The points lie on a straight line.

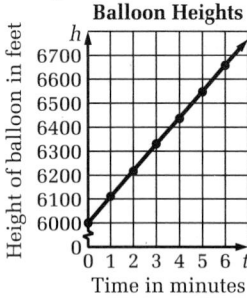

Balloon Heights

5. Answers may vary. Examples are given. *Linear* means in a straight line. In this situation, *growing* means increasing in height above the ground.   **6.** 110; The slope means that for every minute after the balloon takes off, the balloon rises 110 ft more above sea level.

7. (0, 6000); This point represents the height of the balloon above sea level before it begins rising.
8. Answers may vary. Examples are given. Both equations have the variables h for height and t for time and model rise of a balloon. The slope in the equation for question 2 is 110 and the vertical intercept is 6000, while the slope in the equation $h = 60t + 8200$ is 60 and the vertical intercept is 8200.   **9.** Explanations may vary. An example is given. If you redraw the graph so there is no gap in the scale, but still let two units on the scale equal 100 ft, the graph of the line will look the same (only higher up). If you decide to let one unit on the scale equal 100 ft, then the graph of the line rises more steeply. Either way, the slope will not change.   **10.** Yes. The graph passes the vertical-line test.   **11. a.** No. Reasons may vary. An example is given. The graph does not pass through the origin.   **b.** Explanations may vary. An example is given. The vertical intercept in the equation $h = 60t + 8200$ is 8200. The vertical intercept in a direct variation equation is always zero.   **12.** Yes; Yes.   **13.** y; x

Pages 421–424 Exercises and Problems
1. C   **3.** A

5. a.

Weight in tons (w)	Total cost in dollars (c)
0	25
1	44
2	63
3	82

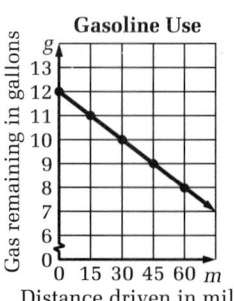

Sand Sales at Better Building Supplies

b. slope: 19; vertical intercept: 25; Explanations may vary. An example is given. The slope shows that the cost increases $19 for each additional ton of sand sold. The vertical intercept is the delivery charge.
c. $c = 19w + 25$

7. a.

Distance driven in miles (m)	Gas remaining in gallons (g)
0	12
15	11
30	10
45	9
60	8

Gasoline Use

b. slope: $-\frac{1}{15}$; vertical intercept: 12; Explanations may vary. An example is given. The slope shows that for each 15 miles driven, 1 gallon of gasoline is used. The vertical intercept represents the amount of gasoline in the car, 12 gallons, before driving any distance.

c. $g = -\frac{1}{15}m + 12$ **9.** Answers and explanations may vary. An example is given. I agree with Nancy that the slope of the line is always the rate given in the situation because the rate shows how the dependent variable changes based on the control variable. This is the same as the slope or the ratio $\frac{\text{rise}}{\text{run}}$. I also agree that the vertical intercept is always the starting value. It is the point where the value of the control variable is zero. **11.** $3, -5$ **13.** $-1, -\frac{1}{2}$ **19, 21.** Explanations may vary. Examples are given. **19.** Yes; Yes. The situation can be represented by the equation $c = 2.5m$, where c is the total cost and m is the number of movies attended. The equation has the general form of direct variation. The slope is positive, which indicates growth, and the graph is a line through the origin. **21.** No. The cost will not increase or decrease, so the the rate of growth is 0. Since the line does not pass through the origin, this is not an example of direct variation. **25. a.** $\frac{2}{3}, 0$ **b.** $y = \frac{2}{3}x$ **c.** linear growth **d.** Yes. The line goes through the origin. **27. a.** $4, -8$ **b.** $y = 4x - 8$ **c.** linear growth **d.** No. The line does not go through the origin. **29, 31.** Explanations may vary. Examples are given. **29.** neither; The graph is not a straight line. **31.** neither; The graph is not a line. **34.** about 26.2 in.^{2} **35.** $-\frac{3}{2}$ or -1.5 **36.** -11 **37.** 26 **38.** 26 **39.** -6 **40.** $y = \frac{c}{b}$ **41.** $y = 60 - x$ **42.** $y = b - ax$ **43.** $y = \frac{1}{3}x + 50$

Pages 426, 428, 429 Talk it Over
1. 4; by the circle around $48 at the point (0, 4); by the point (0, 4) **2.** Yes; No. Explanations may vary. An example is given. It is not possible to buy a fraction of a tape. **3.** No. Explanations may vary. An example is given. The order of the coordinates is important because each coordinate represents a specific variable. If you change the order, the coordinates no longer represent the correct variable. **4. a, b.** Answers may vary. Examples are given. **a.** All the equations have the same slope, $-\frac{2}{3}$. However, they all have different vertical intercepts. **b.** All the equations have the same variable expression, $8t + 12d$, but they each equal different amounts of money. The equations in step 3 are in slope-intercept form, while the equations in step 5 (b) are not. **5.** $-\frac{3}{2}$; linear decay **6.** 1; If $r = 0$, then Dynah does not spend any time running. She would spend 1 h walking the 4 mi. **7. a, b.** Explanations may vary. Examples are given. **a.** No. You cannot have a negative amount of time. **b.** Yes. It is possible to run or walk for part of an hour. **c.** $0 \le r \le \frac{2}{3}$

8. a. Substitute 0 for x and solve to find the value for y. **b.** Substitute 0 for y and solve to find the value for x. **c.** Substitute any value for x (or y) other than those for the intercepts and solve to find the value for y (or x).

Pages 429–432 Exercises and Problems
1. a. the cost of 1 tape **b.** the cost of 1 compact disc **c.** the number of compact discs bought **d.** the number of tapes bought **e.** the total cost of t tapes **f.** the total cost of d compact discs **3. a.** $r - \frac{1}{4}w + 0b$ or $r - \frac{1}{4}w$ **b.** $r - \frac{1}{4}w = 32$

5. a. Tables may vary. An example is given.

w	l
10	120
20	110
30	100
40	90
50	80
60	70

b. $2l + 2w = 260$ **c.** Graphs may vary. They should represent part of the line below.

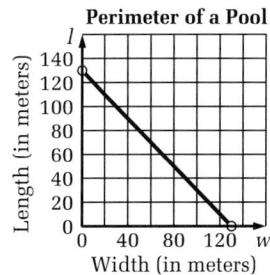

Perimeter of a Pool

d. Explanations may vary. An example is given. The graph should not cross either axis because if $l = 0$ or $w = 0$, then you would have a line segment instead of a rectangle and a pool cannot be a line segment.
9, 11. Answers may vary. Examples are given.
9. $(11, 0)$, $\left(0, \frac{11}{3}\right)$, $(2, 3)$ **11.** $\left(-\frac{2}{5}, 0\right)$, $(0, -1)$, $(2, -6)$
13. $y = -\frac{2}{3}x - 2$ **15.** $y = 12x - \frac{4}{3}$
17. horizontal intercept: 18; vertical intercept: 12
19. a. $0.054s + 0.072c = 18$
24. a. $c = 1.25m + 32$

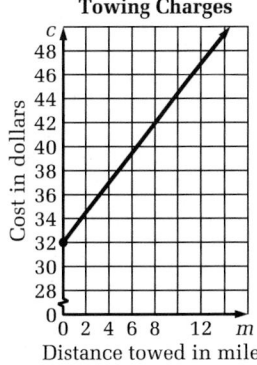

Towing Charges

b. 1.25, 32; The slope indicates that the cost increases by $1.25 for each mile towed. The vertical intercept is the cost of hooking up the car.

Selected Answers

25. a.

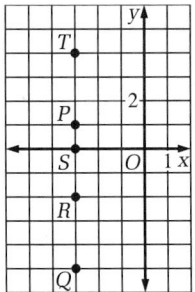

b. All the points lie on a horizontal line and have *y*-coordinate 4. **c.** Yes. Explanations may vary. An example is given. For each value of the control variable, there is only one value for the dependent variable.

26. a.

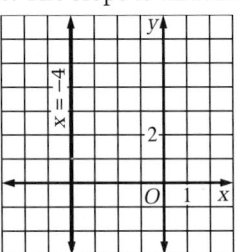

b. All the points lie on a vertical line and have *x*-coordinate –3. **c.** No. Explanations may vary. An example is given. There is more than one value for the dependent variable when the control variable equals –3.

Page 434 Talk it Over
1. The slope is undefined.

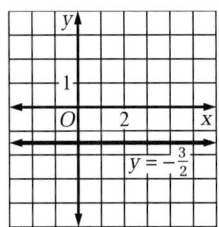

2. The slope is 0.

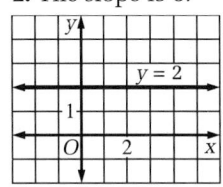

3. The slope is 0.

4. Descriptions and explanations may vary. An example is given. The equation in question 3 is different from the other two because the variable, *y*, has a coefficient of 2 instead of a coefficient of 1 as the others do. This difference does not change the fact that the graph is a horizontal line. If you divide both sides by 2, you get $y = -\frac{3}{2}$.

Page 435, 437 Exercises and Problems
1, 3. Estimates may vary. Examples are given.
1. about 0 **3.** Since the line appears to be vertical, the slope is undefined. **5.** 0; $y = -2$ **7.** 0; $y = 4$

9. The slope is undefined.

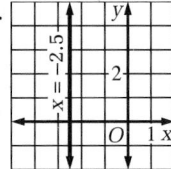

11. The slope is 0.

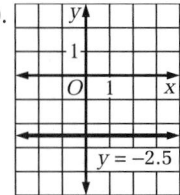

13. $x = 4$ **15.** $y = 6$ **17.** Choices of coordinates may vary. An example is given: (0, –4), (0, –2), (0, 1), (0, 6). They each have *x*-coordinate 0. $x = 0$ **19.** True.

21. False. **28.** $y = -\frac{2}{3}x + 4$ **29.** $y = \frac{4}{9}x - 8$

30. $y = 0x + 5$ **31.** not possible; The slope is undefined, and the equation has no vertical intercept.

32. $12\frac{2}{3}$ **33.** –8 **34.** 378

35. –2; –3; $y = -2x - 3$

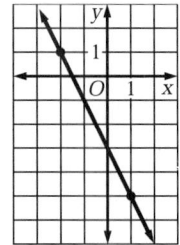

Page 438 Checkpoint
1. Answers may vary. An example is given. Linear growth and direct variation situations are alike because each can be represented by an equation in slope-intercept form. The graphs of both are straight lines. Linear growth can have a vertical intercept other than 0, while direct variation always passes through the origin. Linear growth never has negative slope, while direct variation can have a positive or negative slope. **2.** A, B **3.** A **4.** Equations and graphs may vary. Examples are given. $y = 0.5x + 2.5$

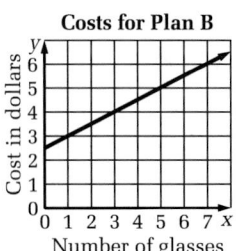

Costs for Plan B
Cost in dollars / Number of glasses

5. 0 **6.** $15j + 10s$

7–9. Graphs, intercepts, and equations may vary depending on choices of control and dependent variables. Examples are given.

7.

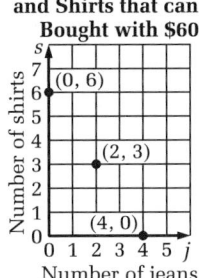

Combinations of Jeans and Shirts that can be Bought with $60

8. Yes. The vertical intercept represents buying only shirts, and the horizontal intercept represents buying only jeans.
9. a. $15j + 10s = 60$

b. $-\frac{3}{2}$; The number of shirts that Mark can buy decreases by 3 for every 2 pairs of jeans he buys.

10. vertical intercept: −4; horizontal intercept: 6   **11.** vertical intercept: 3; horizontal intercept: 6
12. The slope is 0.   **13.** $x = -3$

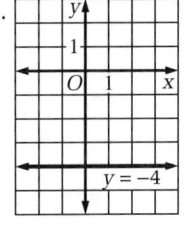

Pages 440, 441, 443 Talk it Over
1. Yes.   **2.** No. Answers may vary. An example is given. Covering up the scales, it might seem as if each line on the graph represents 1 unit, making the slope appear to be $-\frac{1}{6}$. The unequal scales make the rate of change appear to be more gradual than it would appear with equal scales.   **3. a.** 241 s; The prediction is 9.07 s more than the actual time.   **b.** 217 s
4. No, but the results should be close.   **5.** substituting the values for the slope and the vertical intercept in the equation $y = mx + b$ **6.** $y = -\frac{3}{5}x + 8$ **7. a.** dependent variable   **b.** (12, 41)   **c.** a decrease of 7 dollars per week; negative; m; Explanations may vary. An example is given. The rate of change is the same as the slope. Since Kemitra is spending money every week, the amount she has decreases, so the slope is negative.   **d.** $y = -7x + 125$

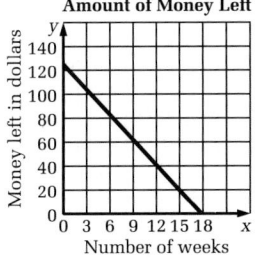

Amount of Money Left

e. The vertical intercept represents the amount of money Kemitra is given for her birthday. The horizontal intercept represents the number of weeks it takes for her to spend all her money.   **f.** $104

Pages 443–446 Exercises and Problems
3. $y = -x + 7$ **5.** $y = 2$ **7.** $y = 2x + 1$ **9. a.** $F = \frac{9}{5}C + 32$
or $C = \frac{5}{9}F - \frac{160}{9}$ **b.** 20°C   **11. a–c.** Graphs, equations, and predictions may vary. Examples are given.

a.

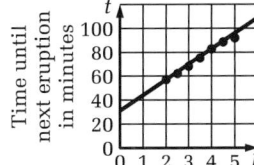

Activity at Old Faithful

b. $t = 13l + 31$ **c.** about 85.6 min   **13. a.** $\overline{AB}$: $\frac{5}{2}$; $\overline{BC}$: 0;

$\overline{CD}$: $\frac{5}{2}$; $\overline{AD}$: 0   **b.** $\overline{AB}$ and $\overline{CD}$; $\overline{BC}$ and $\overline{AD}$

c. $\overleftrightarrow{AB}$: $y = \frac{5}{2}x + 3$; $\overleftrightarrow{BC}$: $y = 3$; $\overleftrightarrow{CD}$: $y = \frac{5}{2}x - \frac{19}{2}$;

$\overleftrightarrow{AD}$: $y = -2$ **15.** $y = \frac{4}{5}x + 11$ **17.** $y = -\frac{3}{2}x + 8$

19. $y = \frac{3}{4}x + 3$ **21.** $x = -3$ **23. a.** control variable

b. (10, 8.7)   **c.** $.75 per mile   **d.** $c = 0.75m + 1.2$

e.

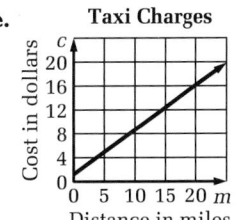

Taxi Charges

f. $4.95; $13.95; by the points (5, 4.95) and (17, 13.95)
g. The vertical intercept represents the initial charge before driving any distance.   **26.** $y = 5$ **27.** $x = 1$
28. $x = 5.5$, $y = 27.5$ **29.** $x = 4$, $y = 15$ **30.** $x = -2$, $y = -8$

Pages 449–450 Talk it Over
1. For any value of n that is greater than 20, the line representing Videobusters is below the line representing Movies Plus. This means that the cost at Videobusters is below or less than the cost at Movies Plus.   **2.** For $c = 2.50n$, $50 = 2.5(20)$, and $50 = 50$. For $c = 2.00n + 10$, $50 = 2.00(20) + 10$, $50 = 40 + 10$, and $50 = 50$.   **3.** No. Explanations may vary. An example is given. The graph shows diagonal lines because the distance increases as the time increases. This comparison of distance and time does not involve the direction in which they are running.   **4.** Estimates may vary. An example is given. Rita; about 11 s; Sandy: about 12.5 s; Rita wins.   **5.** Both equations have the same slope, 8.   **6.** If the lines intersect, then the system has a solution. If the lines are parallel, then the system has no solution.

Pages 451–452 Exercises and Problems

1. a. Group I **b.** No. **3.** Estimates may vary. Examples are given. **a.** (2, 3) **b.** (2, 3)

7. Estimates may vary. An example is given: about $\left(2\frac{1}{3}, \frac{1}{3}\right)$.

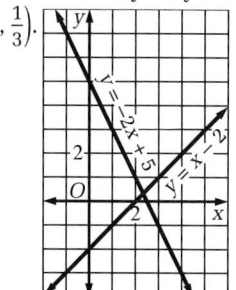

9. (−6, −6)

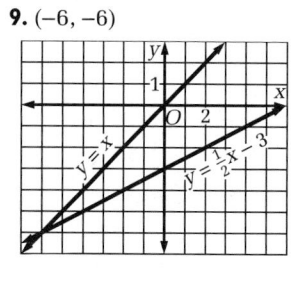

11. Descriptions may vary. An examples is given. The dog chases the squirrel but never catches up with it because they are both running at the same rate, and the squirrel had a 5 m head start. **13.** solution

15. no solution **19.** $y = \frac{1}{3}x - \frac{2}{3}$, $y = \frac{1}{4}x - \frac{1}{2}$; solution

Page 454 Exercises and Problems

25. $y = \frac{5}{6}x - 3$ **26.** $y = \frac{5}{3}x - 11$

27. Yes. $x < 5.5$

28. Yes. $x \geq 5$

29. Yes. $x \geq 4$

Page 454 Checkpoint

1. Write the equations in slope-intercept form. If the values for m are equal and the values for b are not, then there is no solution. Graph the equations. If the lines are parallel, then there are no solutions.

2. $y = 5x - 19$ **3.** $y = -\frac{1}{4}x + \frac{1}{4}$ **4.** solution **5.** no solution **6.** solution **7.** Choices of variables may vary. An example is given. Let c be the total cost in dollars and let g be the number of gallons bought. $c = g + 10$, $c = 1.5g$; $g = 20$, $c = 30$ (At 20 gallons, the cost for either is $30.)

Pages 456–458 Talk it Over

1. (a) $y \geq x$ **2.** First, graph the boundary line $y = 3$ as a solid line. Then shade the region above the line. **3. a, b.** Descriptions may vary. Examples are given. **a.** The graph of $x = 2$ is a vertical line, parallel to the y-axis, with all points having x-coordinate 2. **b.** The graph of $x \leq 2$ includes the graph of $x = 2$, and the region to the left of the line is shaded. **4.** solid; below **5.** dashed; above **6.** solid; above **7.** dashed; below **8.** Answers may vary. An example is given. The coordinates of all the points in the shaded region make the inequality $y \leq 4 - \frac{2}{3}x$ true. Since this inequality is the slope-intercept form of the original inequality, $8x + 12y \leq 48$, they both have the same

solutions. **9.** all points with whole-number coordinates in the region bounded by the x-axis, the y-axis, and the line $y = 4 - \frac{2}{3}x$ **10.** No. Explanations may vary. An example is given. If the inequality is in slope-intercept form, then the symbol $<$ means to graph below the boundary line. However, if the inequality is in standard form, you need to check a point on one side of the boundary line to determine which region to shade. **11.** If one point in the region makes the inequality true, then all the points in the region make the inequality true and this is the correct region to shade. **12.** The boundary line is a boundary for the solution region. A point on this line may be a solution to the inequality, but it will not help you determine which region is the solution region.

13. $y > \frac{2}{3}x - 4$; Yes.

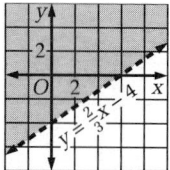

Page 459–462 Exercises and Problems

3. C **5.** D

7.

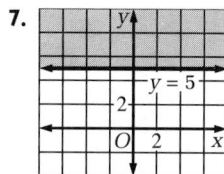

9.

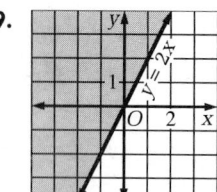

11.

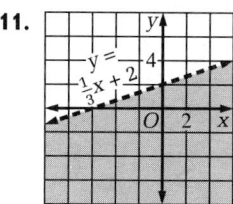

13.

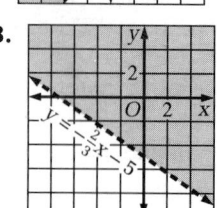

17. a. (8, 7), (7, 4), (7.5, 6), (5.5, 4)
b, c. The slope is 1. **d.** below **e.** $>$; $<$
f. Explanations may vary. An example is given. If lines were drawn from the origin to each point on the graph, the lines would lie below the line $y = x$. This means that each line must have a slope less than 1. Since the slopes of these lines are represented by the ratio $\frac{BC}{AB}$, which is also the cosine ratio for $\angle B$, $\cos B$ must be less than 1.

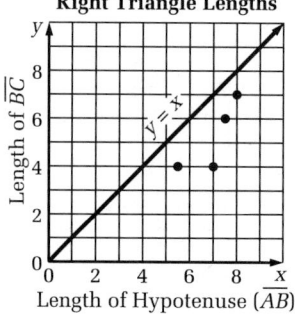

Right Triangle Lengths

Length of $\overline{BC}$ vs. Length of Hypotenuse ($\overline{AB}$)

19.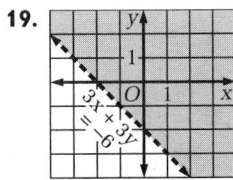

21.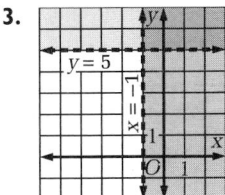

23.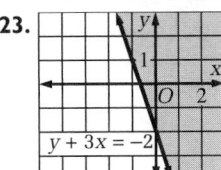

25.

$x + 2y \le 6$

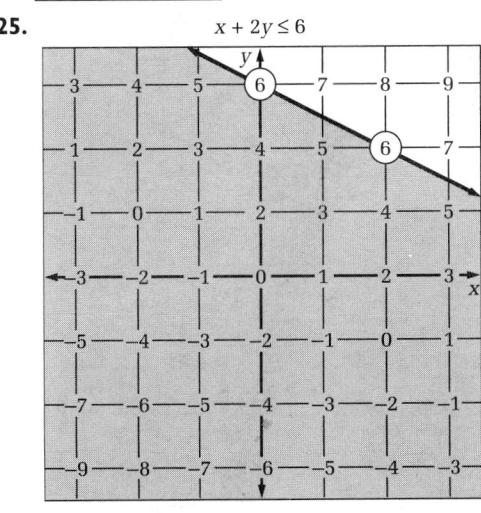

13.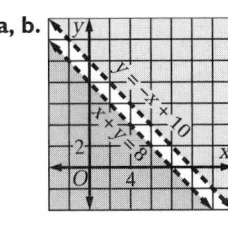

15. a, b.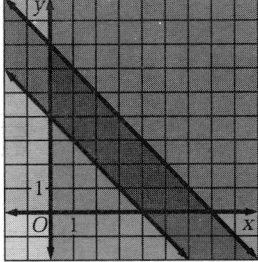

c. The lines are parallel.   **d.** The regions do not overlap. Explanations may vary. An example is given. The boundary lines do not overlap since they are parallel, so the only way the regions could overlap is if the region between the two parallel lines contained solution points and this does not happen.

23. Descriptions may vary. An examples is given. trapezoid

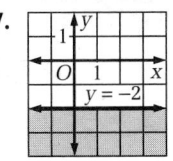

32. solution   **33.** no solution   **34.** no solution

35. about 188 books   **36.** $\frac{5}{6} \approx 83\%$

37.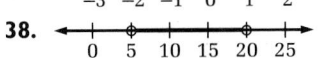

38.

26.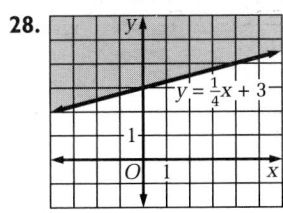

27.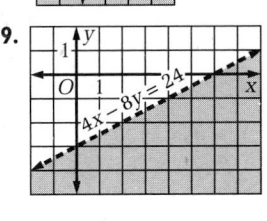

28.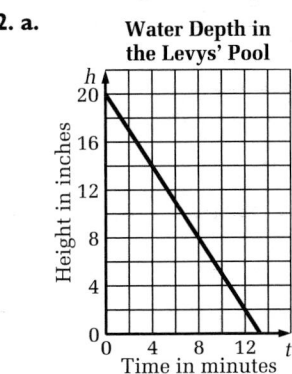

29.

30. about 6 mm   **31.** about 17.5 cm   **32.** $a = 35°$; $b \approx 5.7$ m   **33.** angle a and 55°; Their sum is 90°.
34. $7 < \sqrt{50} < 8$ **35.** $7 < \sqrt{58} < 8$ **36.** $13 < \sqrt{193} < 14$
37. $0 < \sqrt{0.18} < 1$ **38.** $0 < \sqrt{0.0136} < 1$

Pages 470–471 Unit 8 Review and Assessment
1. a. $\frac{2}{3}$; 2   **b.** $y = \frac{2}{3}x + 2$ **c.** linear growth   **d.** No. The line does not pass through the origin.

2. a.

Water Depth in the Levys' Pool

Pages 464 Talk it Over
1. The boundary line is dashed because the graph represents only failing grades, and 60 is a passing grade. Also, 60 is not a solution of $x < 60$.   **2.** No.
3. at least 67 (assuming scores must be whole numbers)   **4.** Answers may vary. An example is given. (80, 10); none; 80 is the mid-year test score, and 10 is the year-end score. A student with these grades will not have a passing weighted average.

Pages 465–468 Exercises and Problems
3. No.   **5.** No.   **7.** Yes.

9.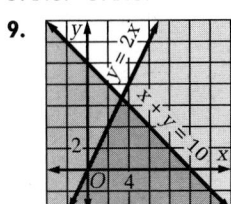

11.

b. slope: –1.5; vertical intercept: 20; The slope means that the water level decreases 1.5 in. per minute. The vertical intercept is the height of the water before the pool is drained, 20 in. **c.** $h = -1.5t + 20$ **d.** $13\frac{1}{3}$ min; horizontal intercept **3. a–e.** Answers may vary depending on the choice of control variable. Examples are given. Let r = running distance and b = biking distance. **a.** $7.5r + 20b = 30$ **b.** $r = -2\frac{2}{3}b + 4$

c.

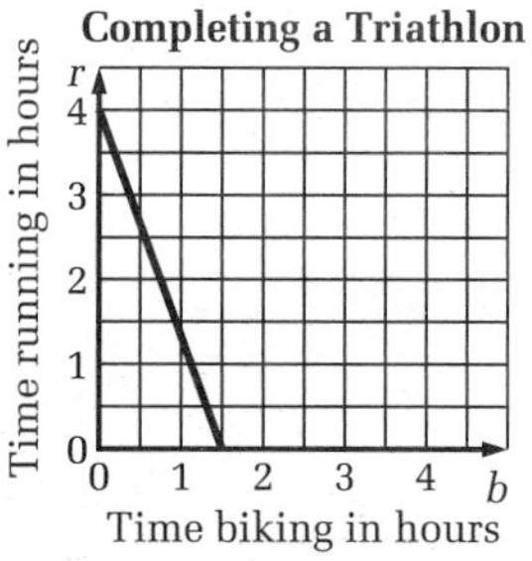
Completing a Triathlon
Time running in hours
r
4
3
2
1
0
0 1 2 3 4 b
Time biking in hours

d. horizontal intercept: 1.5; vertical intercept: 4; The horizontal intercept means that Jim will spend 1.5 h biking the 30 mile trail if he does not run. The vertical intercept means that Jim will spend 4 h running the 30 mile trail if he does not ride his bicycle.

4. The slope is undefined.

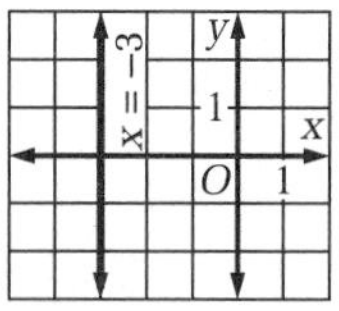
x = –3
y
1
x
O 1

5. The slope is 0.

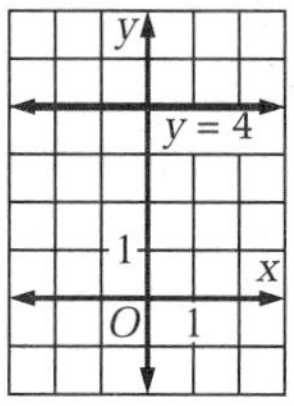
y
y = 4
1
x
O 1

6. a.

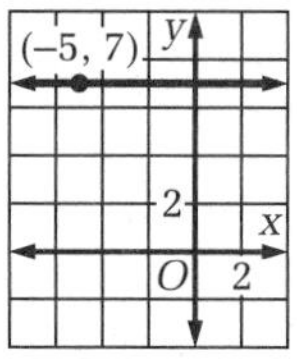
(–5, 7)
y
2
x
O 2

b. $y = 7$

7. Descriptions and explanations may vary. An example is given. To graph $y = 2x - 3$, I would graph the vertical intercept, –3, which is obvious in the equation because it is in slope-intercept form. I would then use the slope and move up 2 and right 1 to graph another point, (1, –1). I'd check this point in the equation before drawing the line. To graph $5x - 4y = 12$, I would substitute 0 for x to find the vertical intercept, (0, –3), and then substitute 0 for y to find the horizontal intercept, $\left(2\frac{2}{5}, 0\right)$. I would then draw a line through the two intercepts. I like the method for using intercepts because they are usually easy points to calculate.

8. $y = \frac{1}{2}x - 2$ **9.** $y = 2x + 11$ **10.** Equations may vary. An example is given. $y = -x - 11$

11. a. $y = 0.75x + 16.25$ **b.** slope: 0.75; vertical intercept: 16.25; The slope means that the fraction of the hourly pay (one hour per month) that goes towards union dues is 0.75. The vertical intercept is the fixed amount each member must pay toward union dues, \$16.25, regardless of the rate of hourly pay.

c. \$27.13 **12. a.** Estimates may vary. An example is given. $\left(2\frac{1}{3}, 3\frac{2}{3}\right)$ **b.** $\left(2\frac{1}{4}, 3\frac{3}{4}\right)$ **13.** no solution

14. solution **15.** solution **16. a.** Let c = total cost and w = weight of package in ounces. $c = w + 5$ and $c = 1.25w + 3.50$ **b.**

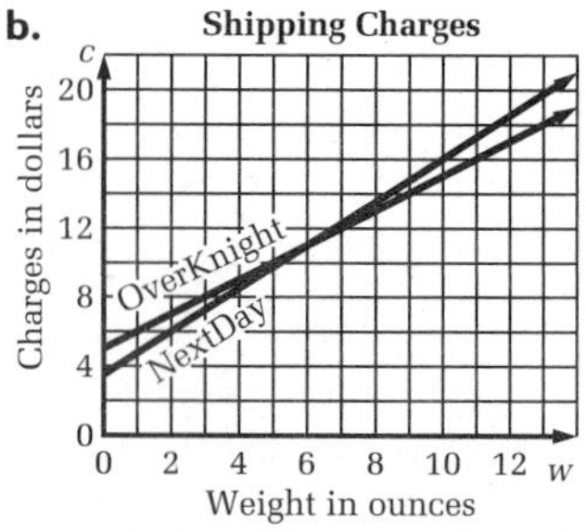
Shipping Charges
c
Charges in dollars
20
16
12
8
4
0
OverKnight
NextDay
0 2 4 6 8 10 12 w
Weight in ounces

c. 6 oz **d.** for packages that weigh more than 6 oz

17.

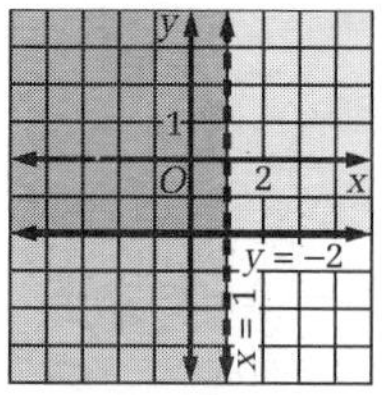
y
1
O 2 x
y = –2
x = 1

18.

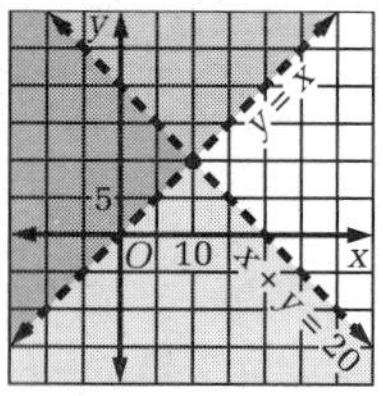
y
y = x
5
O 10 x
x + y = 20

19.

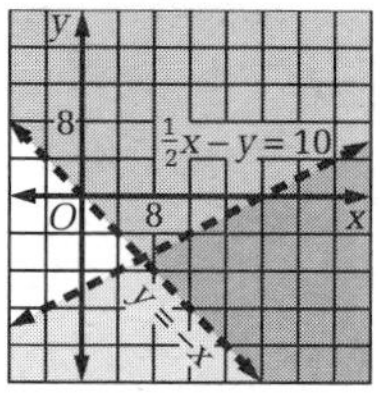
y
8
$\frac{1}{2}x - y = 10$
O 8 x
y = –x

20.

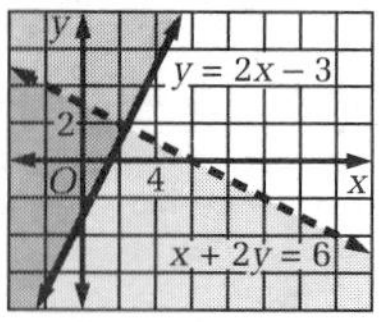
y
y = 2x – 3
2
O 4 x
x + 2y = 6

21. a. $x + y \leq 65$ **b.** $20x + 25y \leq 1200$

c.

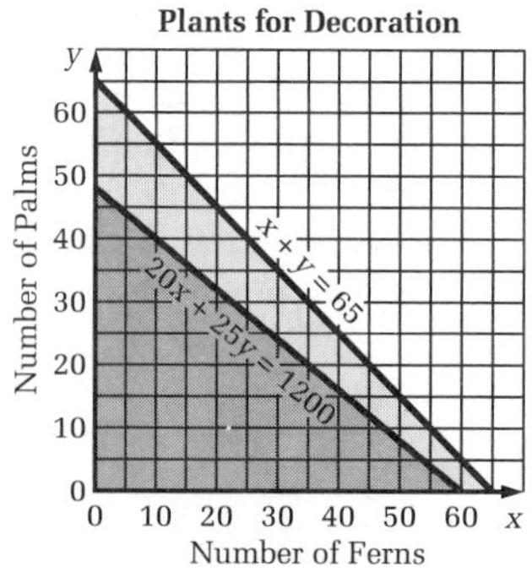
Plants for Decoration
y
Number of Palms
60
50
40
30
20
10
0
x + y = 65
20x + 25y = 1200
0 10 20 30 40 50 60 x
Number of Ferns

d. 48; This number is the vertical intercept of the boundary line $20x + 25y \leq 1200$.

Page 480 Talk it Over

1. the definition of complementary angles; two angles whose measures add up to 90° **2.** No. Yes. Explanations may vary. An example is given. The inductive method shows that for only 6 right triangles, the acute angles are complementary. Maybe if Charles had tried a seventh triangle, he would have had a different result. The deductive method uses a general right triangle, not a specific one. Barbara uses known definitions and properties to show and prove that the acute angles of a right triangle are complementary.

Pages 481–484 Exercises and Problems

1. a–c. Drawings and tables may vary. The results should support the conjecture made in the Exploration. **3.** b and c **5.** making a sketch; Answers may vary. An example is given. Each problem gives measurements of figures. Making a sketch of each figure helps you understand the problem and makes it easier to solve. **7.** 12 in. **9.** about 21.8 mi **11.** $h = 12$ m; area = 60 m² **13.** about 127.3 ft **23.** deductive **25.** inductive

29. 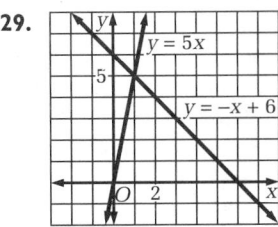 **30.**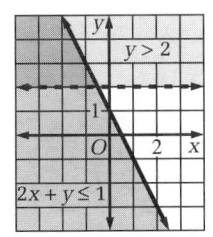

31. 0.2 **32.** $\frac{1}{3}$ **33.** 8 **34.** 14 **35.** 20 ft

Pages 486, 488 Talk it Over

1. Arguments may vary. An example is given. Since the triangle is a right triangle, you can use the Pythagorean theorem to find the value of x. The hypotenuse is x, so $x^2 = 3^2 + 3^2 = 9 + 9 = 18$. $x^2 = 18$, so $x = \sqrt{18}$. **2.** Corresponding sides of similar triangles are in proportion; arguments may vary. An example is given. Since the triangles are similar, the corresponding sides have the same ratio. Therefore, $\frac{1}{3} = \frac{\sqrt{2}}{x}$. Using cross products, $x = 3\sqrt{2}$. **3. a.** They are equal. **b.** Yes; explanations may vary. An example is given. Since $18 = 9 \cdot 2$, $\sqrt{18} = \sqrt{9 \cdot 2}$. Also, $\sqrt{9} = 3$, so $\sqrt{9} \cdot \sqrt{2} = 3\sqrt{2}$. From part (a), $\sqrt{18} = 3\sqrt{2}$. Therefore, $\sqrt{9 \cdot 2} = \sqrt{9} \cdot \sqrt{2}$. **c.** They both equal about 4.24. **4.** The result is the number a. **5.** It is not efficient to rewrite $\sqrt{50}$ as $\sqrt{5} \cdot \sqrt{10}$ because neither 5 nor 10 has a perfect square factor. **6.** Martha's; She factored out the largest square root and avoided three steps.

Pages 489–491 Exercises and Problems

3. $2\sqrt{17}$ **5.** $6\sqrt{2}$ **7.** $\sqrt{21}$ **9.** $6\sqrt{10}$ **11.** 11 **13.** $60\sqrt{15}$ **15.** $4\sqrt{3}, -4\sqrt{3}$ **17.** $2\sqrt{3}, -2\sqrt{3}$ **19.** $a\sqrt{29}, -a\sqrt{29}$ **21.** $8\sqrt{2}$ **27.** 20 m **28.** $\sqrt{29}$ in. or about 5.4 in. **29.** 15 ft **30.** 12 cm

31.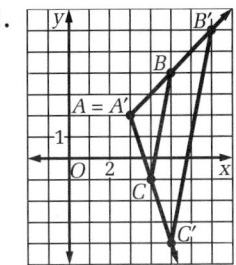

32. 0 **33.** −11.5 **34.** 0 **35.** 3

Pages 493, 495 Talk it Over

1. Explanations may vary. An example is given. Statement (1) is false because all champions do not buy the same brand of shoes. Statement (2) is false because choosing a particular brand of shoes does not make someone a champion. **2. a, b.** Discussions may vary. Examples are given. **a.** The statement is false. It is possible for the sky to be partly cloudy with the sun shining through the clouds. **b.** The statement is true. A square is a particular kind of rectangle. **3.** Answers may vary. An example is given. Statement: If you can ice skate, then you are a hockey player. Hypothesis: You can ice skate. Conclusion: You are a hockey player. The if-then statement is false because someone who can ice skate may not skate well enough to be a hockey player. Also, the person may be a figure skater, a speed skater, or an entertainer in an ice show. **4.** Yes. **5. a.** −1.3 = 0 or $x = 0$; Since $−1.3 \neq 0$, $x = 0$. **b.** Divide both sides by −1.3; $x = 0$. **6.** No. To use the property, you must start with a product equal to 0. The product given is equal to −1.

Pages 496–498 Exercises and Problems

1. Find a counterexample that contradicts the statement. **3, 5, 7, 9.** Counterexamples may vary. Examples are given. **3. a.** Hypothesis: You see an advertisement for a product. Conclusion: You will buy the product. **b.** False. I see many advertisements for products that I do not want to buy or that are too expensive for me to buy. **5. a.** Hypothesis: $x^2 = 25$ Conclusion: $x = \pm\sqrt{25}$ **b.** True. **7. a.** Hypothesis: a and b are non-negative integers. Conclusion: $\sqrt{ab} = \sqrt{a} \cdot \sqrt{b}$ **b.** True. **9. a.** Hypothesis: $2y = 0$ Conclusion: $y = 0$ **b.** True. **11, 13, 15, 17.** Counterexamples may vary. Examples are given.

11. a. If you are south of the equator, then you are in Zimbabwe. **b.** False. You could be in another country south of the equator, such as Uruguay. **13. a.** If $(m + n)$ is an even number, then m and n are even numbers. **b.** False. Suppose $(m + n) = 4$. The values for m and n could be 3 and 1, which are odd numbers. **15. a.** If the area of a figure is half the area of some parallelogram with the same base and height, then the figure is a triangle. **b.** True. **17. a.** If an event is certain to happen, then the probability of the event is 1. **b.** True. **21.** 0 **23.** 0, −1 **25.** 0, 2 **27.** Yes; because $7^2 + 24^2 = 625$ and $25^2 = 625$ **29.** No; because $4^2 + 5^2 = 41$ and $6^2 = 36$. **31.** No; because $2^2 + 7^2 = 53$ and $8^2 = 64$. **33.** Yes. **39.** $6\sqrt{2}$ **40.** $4\sqrt{3}$ **41.** 30 **42.** $240\sqrt{10}$ **43.** $\frac{1}{6}$ **44.** $\frac{1}{3}$ **45.** 0 **46.** $\frac{5}{6}$

Page 499 Checkpoint
1. Explanations may vary. An example is given. Suppose you start with $\sqrt{10} \cdot \sqrt{15}$. You can rewrite this product as $\sqrt{10 \cdot 15}$, or $\sqrt{150}$. See if you can find any perfect square factors for the number under the square root symbol. A perfect square factor of 150 is 25, so you can write $\sqrt{150}$ as $\sqrt{25 \cdot 6}$. Rewrite this as $\sqrt{25} \cdot \sqrt{6}$. Since $\sqrt{25} = 5$, the final answer is $5\sqrt{6}$. **2.** 39 in. **3.** 20 cm **4.** deductive **5.** $10\sqrt{3}$ **6.** $6\sqrt{7}$ **7.** $5\sqrt{2}, -5\sqrt{2}$ **8. a.** Hypothesis: A number is odd. Conclusion: It is a multiple of 3. **b.** False. Counterexamples may vary. An example is given. The number 5 is odd, but it is not a multiple of 3. **c.** If a number is a multiple of 3, then it is odd. **9.** $0, -\frac{3}{2}$
10. No; because $45^2 + 14^2 = 2221$ and $50^2 = 2500$.

Page 503 Talk it Over
1. a. ratio of areas **b.** ratio of lengths **2.** Answers may vary. An example is given. I prefer Method 2, counting the number of minutes, not only because it is less work, but also because it makes more sense to me to compare minutes instead of line segments when dealing with time.

Page 504–506 Exercises and Problems
1. greater than; Explanations may vary. An example is given. If you draw the lines across the square, as shown at right, you can see that the area of the shaded region is greater than one fourth of the total area. Therefore, the probability that a disk will land in the shaded region is greater than $\frac{1}{4}$, or 0.25.

3. a. Answers may vary. The ratio should be about 0.75. **b.** Answers may vary. The sum should be close to 1. **5. a.** $\frac{1}{3}$ **b.** $\frac{1}{4}$; probability in part (a) **7.** B **9.** $\frac{1}{8}$ or 0.125 **16. a.** If Raymond lives in Canada, then he lives in Toronto. **b.** False. Raymond may live somewhere else in Canada, such as Montreal.
17. Answers may vary. An example is given. The figure has four sides, opposite pairs of sides are congru-

ent and parallel and all angles measure 90°.
18. a. $A = \frac{1}{2}bh$ **b.** $C = \pi d$ or $C = 2\pi r$

Pages 507, 508, 510 Talk it Over
1. rectangle; top and bottom: 4.5 cm by 10 cm; sides: 4.5 cm by 16 cm; front and back: 10 cm by 16 cm **2.** $A = lw$ **3.** 554 cm^2 **4.** Explanations may vary. An example is given. All the faces make up the surface of the figure, so the total area of all the faces is the area of the surface of the figure. **5.** Explanations may vary. An example is given. In step 1, the circular bases are cut out, and the curved surface is cut vertically. In step 2, the circular bases are laid flat, and the curved surface is opened. Step 3 shows the area for each part of the cylinder. The circular bases each have an area of πr^2. The curved surface flattens into a rectangle having an area of $2\pi rh$. **6.** In the last step, $2\pi r$ represents the length of the rectangle, which is the same as the circumference of each base. **7.** the slant height **8.** Descriptions may vary. An example is given. Find the area of the base; $10 \times 10 = 100$ square units. Find the area of one face; $\frac{1}{2}(10)(13) =$ 65 square units. Since there are four faces, multiply this area by 4; $4(65) = 260$ square units. Add the area of the base and the area of the faces to find the total surface area; $100 + 260 = 360$ square units.

Pages 512–514 Exercises and Problems
3. triangular prism; 432 cm^2 **5.** trapezoidal prism; 352 cm^2 **7. a.** Drawings may vary. **b.** triangular prism **11.** Descriptions may vary. An example is given. The two figures are alike because they are both space figures with congruent, parallel bases and have heights of 6 cm. They are different because the first figure is a hexagonal prism with a surface area of about 154.8 cm^2, while the second figure is a cylinder with a surface area of about 169.6 cm^2. **13.** about 100.5 m^2 **17.** square pyramid; 384 m^2 **19.** square pyramid; about 289 in.^{2} **25.** about 0.29 or 29% **26.** about 0.71 or 71% **27.** 1 or 100% **28.** Sketches may vary. You need to know the length of each base and the height. **29.** 70 in.^{3} **30.** 54 in.^{3}

Pages 516, 518 Talk it Over
1. Explanations may vary. An example is given. In the formula $A = \frac{1}{2}(b_1 + b_2)h$, h represents the height of the trapezoidal base of the prism, which is 6 in. In the formula $V = Bh$, h represents the height of the prism, which is 8 in. Although h represents a height in each formula, it represents different heights, so the values substituted are not the same. **2. a.** triangular prism **b.** First find the area of the triangular base using $A = \frac{1}{2}bh = \frac{1}{2}(6)(4) = 12$ ft^2. Next multiply its base area times the height: $V = Bh = 12 \cdot 7 = 84$ ft^3.

3. Answers may vary. An example is given. The bases of the space figure are the two sides which are congruent and parallel. The front and back are the bases of the jukebox. The distance between the bases is the height of the space figure. The distance between the front and back of the jukebox is 25 in. Therefore, the 25 in. label is the height.   **4.** Answers may vary. An example is given. The word *composite* is an adjective meaning "composed of" or "made out of." This means a space figure is made out of other space figures by putting them together. The jukebox is a space figure made out of a half cylinder and a rectangular prism.

Pages 519–521 Exercises and Problems
3. 4394 ft^3 **5.** 3.8 yd   **7.** 4 m   **9.** 9 fish   **11.** about 157.1 cm^3 **13.** about 85.7 m^3 **15.** about 30.0 yd
23. rectangular prism; 80 in.^{2} **24.** cylinder; about 291.3 m^2 **25.** triangular prism; about 104.3 ft^2
26. 20 m/s   **27.** about 19.1 m   **28.** $y = -\frac{8}{3}x + \frac{7}{3}$

Page 522 Checkpoint
1. Descriptions may vary. An example is given. The space figure is a triangular prism. To find the area of the bases you need to find the height of the triangles: $h = \sqrt{14^2 - 6^2} \approx 12.6$ yd. The area of the bases is $2\left(\frac{1}{2} \cdot 12 \cdot 12.6\right) = 151.2$. Find the area of the faces: $7.5 \cdot 14 = 105; 7.5 \cdot 14 = 105; 7.5 \cdot 12 = 90$. Add the areas to find the surface area: $151.2 + 105 + 105 + 90 = 451.2$ yd^2.   **2.** $\frac{2}{9}$, or about 0.2   **3.** about 1224 mm^2 **4.** about 31.9 in.^{2}; about 12.4 in.^{3}

Pages 523, 525 Talk it Over
1. $V = Bh$ **2.** Answers may vary. An example is given. Divide the volume of the prism by 3.   **3.** The volume of the cone is $\frac{1}{3}$ the volume of the cylinder.
4. about 130.9 ft^3

Pages 526–529 Exercises and Problems
3. 360 mm^3 **5.** about 208.17 ft^3 **7.** Drawings may vary.   **9.** about 78,698,448 ft^3 **15. a.** 864 m^3 **b.** about 678.6 m^3 **c.** about 78.5%   **17.** cylinder; about 282.7 ft^3 **18.** triangular prism; about 17.1 m^3 **19.** 7.5   **20.** 15   **21.** 4:3   **22.** 2.5, −2.5

Pages 530–532 Talk it Over
1. No. Reasons may vary. An example is given. You would need more than 4 fronts of TV sets of the size representing 1970 to cover the area of the front of the TV set representing 1990.   **2.** No. Reasons may vary. An example is given. You could fit more than 4 TV sets of the size representing 1970 into the TV set rep-

resenting 1990.   **3.** Answers may vary. An example is given. The heights of the TV sets appear to be in the same ratio as the number of cable systems for each year.   **4.** $\frac{3}{4}$ **5.** $\frac{9}{16}$ **6.** Yes.   **7.** Since the scale factor is the ratio of corresponding lengths of similar figures, you can square the numerator and square the denominator of the scale factor to find the ratio of the areas.   **8.** 7:6   **9.** The bases are similar rectangles. All corresponding lengths are in proportion: $\frac{12}{6} = \frac{8}{4} = \frac{5}{2.5} = \frac{2}{1}$.   **10.** 2:1   **11.** 8:1   **12.** Yes.   **13. a.** 1:40   **b.** 64,000

Pages 533–538 Exercises and Problems
1. 9:16   **3.** 49:100   **5.** 25:64; 76.8 mm^2 **7.** 4:9; 90 in.^{2} **11. a.** 25%; 56.25%; 36%   **b.** 144%
13. a. about $8\frac{1}{3}$ in.   **b.** 27 ft^3:1 in.^{3}, or 46,656 in.^{3}:1 in.^{3} **19.** No. The ratio of the diameters is not the same as the ratio of the heights.   **21.** Yes. The ratio of all corresponding sides is 2:3.
23. 64:3375   **25.** 27:64; 13.5π in.^{3} **27.** 125:216; 691.2 cm^3 **30.** about 1340.4 yd^3 **31. a.** $A(0, 2)$; $B(-3, 0)$; $C(0, -5)$; $D(3, 0)$ **b.** $x = 0$ or y-axis   **c.** $A'(4, 2)$; $B'(1, 0)$; $C'(4, -5)$; $D'(7, 0)$ **d.** $A''(0, -2)$; $B''(3, 0)$; $C''(0, 5)$; $D''(-3, 0)$

Pages 540–541 Unit 9 Review and Assessment
1. a. about 11.2 km   **2.** inductive   **3.** $4\sqrt{6}$ **4.** $16\sqrt{6}$ **5.** $3\sqrt{3}, -3\sqrt{3}$ **6. a.** Hypothesis: You have a fever. Conclusion: You stay home from school.   **b.** If you stay home from school, you have a fever.   **c.** False. You could stay home from school because you sprained your ankle.   **7.** Answers may vary. An example is given. If the ground is wet in the morning, it rained last night; false; it might have snowed. If it rained last night, the ground is wet in the morning.   **8.** 0, −4   **9.** 0, 2   **10.** Yes; the Pythagorean theorem applies: $(12)^2 + (16)^2 = (20)^2$.   **11. a.** about 0.16   **b.** about 0.21   **12.** 150 in.^{2} **13.** 112 in.^{2} **14.** 72 ft^3 **15.** Can 1; Explanations may vary. An example is given. Can 1 and Can 2 have equal volumes, but Can 1 has less surface area than Can 2. This means that Can 1 holds as much as Can 2, but less materials are needed to make Can 1. Other factors that might influence the manufacturer's decision would be the information displayed on the label, the type of product, and the size and shape of shipping cartons.   **16.** about 372.02 cm^3 **17.** 54 square units   **18.** 125 times

Unit 10

Page 550 Talk it Over

1. the x-axis; $y = 0$ **2.** below; 3 units below
3. 2 units to the left of the line of reflection
4. Answers may vary. Examples are given. **a.** The whale's tail in the photo has symmetry because one half of the tail is a reflection of the other half.

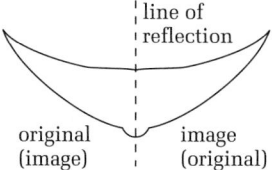

b. Yes, if it has symmetry. An example is the whale's tail in part (a).

Pages 551–554 Exercises and Problems

1. The image is congruent to the original. The orientation is reversed. The line of reflection divides in half and is perpendicular to each segment connecting corresponding points on the image and the original.
3. a. raises its left arm **b.** walks towards me **c.** Yes, but the words and letters are backwards. **7.** $\overleftrightarrow{AB}$ is perpendicular to $\overline{PP'}$ and divides $\overline{PP'}$ in half.
13. Yes.

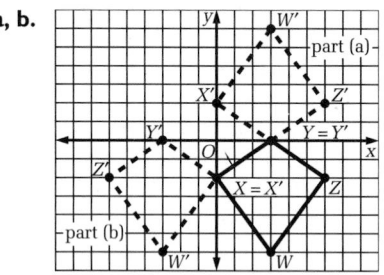

15. No. The image is not congruent to the original figure. This is a dilation combined with a reflection.
17. the line $y = -1$ **19.** the line $x = 2$ **21.** WXYZ has clockwise orientation and W'X'Y'Z' has counterclockwise orientation for both reflections.

a, b.

25. $4:25$ **26.** 2430 cm^3

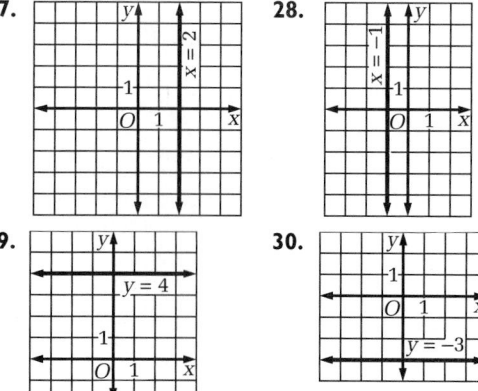

27. **28.**

29. **30.**

Pages 555, 557 Talk it Over

1. a. the y-axis, or the line $x = 0$ **b.** (0, 0)
2. a. the y-axis, or the line $x = 0$ **b.** (0, 0) **c.** -2^2; Yes, it matters because $(-2)^2 = (-2)(-2) = 4$ and $-2^2 = -(2 \cdot 2) = -4$. **3.** Answers may vary. An example is given. When $x = 1$, the values of y are opposites (-1 and 1); when $x = 0$, the values of y are the same (0); when $x = -2$, they are opposites (-4 and 4). In general, for a given value $x \neq 0$ of x, the corresponding values of y are opposites. **4.** the x-axis, or the line $y = 0$ **5.** the parabolas in parts (a) and (b)
6. the parabolas in parts (c) and (d) **7.** (a) I, II; (b) I, II, III, IV; (c) I, II; (d) I, II **8.** Yes. The graph of the parabola in part (b) is in the III and IV quadrants. In these quadrants, the value of the y-coordinate for each point is negative.

Pages 558–561 Exercises and Problems

3. 18 ft **7.** B **9.** A **11.** A **13.** E **15.** Answers may vary. An example is given. Yes, I agree. The graph of $y = x^2 + 7$ is U-shaped. The path of water from a fountain would be shaped like an upside-down U, rising to a high point and then falling. **17.** Translate the graph of $y = x^2$ up 10 units. **19.** Translate the graph of $y = x^2$ left 11 units. **21.** Translate the graph of $y = -x^2$ up 10 units. **23.** Translate the graph of $y = -x^2$ left 6 units. **25.** Translate the graph of $y = x^2$ right 1 unit and up 2 units. **27.** none
29. the equations in Exercises 17, 18, 21, and 22
31. the equations in Exercises 17, 22, and 25
33. a. $x = 0$ **b.** (0, 8) **35. a.** $x = 11$ **b.** (11, 0)
37. $y = x^2 - 9$ **39.** $y = (x + 10)^2$ **41.** $y = -(x + 5)^2$
43. False. The x-axis is the line of reflection.
44. True. **45.** False. The orientation of $\triangle A'B'C'$ is reversed. **46.** True. **47.** -3 **48.** 0 **49.** 2 **50.** 0
51. -4 **52.** 0 or 1

Pages 562–563 Talk it Over

1. Answers may vary. An example is given. The income increases. **2. a.** $9: \$20 - \$9 = \$11$; $30 + 5(9) = 75$; $\$11 \times 75 = \$825$; $\$10$: $\$20 - \$10 = \$10$; $30 + 5(10) = 80$; $\$10 \times 80 = \$800$
b. Answers may vary. An example is given. The

Integrated Mathematics

income decreases. **3. a.** 0 **b.** 0 **4. a.** the drop in price; the income **b.** Yes. There is only one income value for each drop in price. **5.** Explanations may vary. An example is given. values of x such that $0 \le x \le 20$; The price cannot drop more than \$20. (You could say $-6 \le x \le 20$ if negative values of x are viewed as price increases.) **6.** Answers may vary. An example is given. The highest point on the graph is the vertex at (7, 845), so dropping the price \$7 to \$13 maximizes income at \$845. **7.** The y-intercept is 600. **8. a.** Answers may vary. An example is given. The factor $20 - x$ equals 0 when $x = 20$, so $y = 0$. **b.** Answers may vary. An example is given. The factor $30 + 5x$ equals 0 when $x = -6$, so $y = 0$.

Pages 565, 567 Exercises and Problems

5.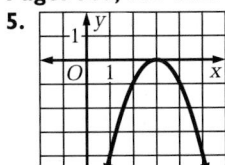

7. x-intercepts: 0 and -7; y-intercept: 0
9. x-intercepts: -7 and 4; y-intercept: 28
11. x-intercepts: 4 and $-\dfrac{13}{3}$; y-intercept: -52
13. a. 0 and -9 **b.** 47 **18.** down 7 units
19. up 3 units **20.** right 1 unit **21.** 10^7 **22.** 10^2
23. 10^4 **24.** $\dfrac{1}{10^4}$, or 10^{-4}

Page 567 Checkpoint

1. Answers may vary. An example is given, using the parabola shown in Sample 1(c) on page 557. The equation of the line of symmetry is $x = -2$. The vertex is $(-2, 0)$. The x-intercept is -2. The y-intercept is 4.

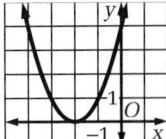

2. a. 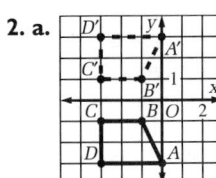 $A'(0, 3)$, $B'(-1, 1)$, $C'(-3, 1)$, $D'(-3, 3)$; $ABCD$ has counter-clockwise orientation and $A'B'C'D'$ has clockwise orientation.

b. 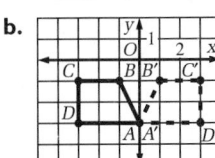 $A'(0, -3)$, $B'(1, -1)$, $C'(3, -1)$, $D'(3, -3)$; $ABCD$ has counter-clockwise orientation and $A'B'C'D'$ has clockwise orientation.

3. Translate the graph of $y = x^2$ up 8 units.
4. Translate the graph of $y = x^2$ right 1 unit.
5. Reflect the graph of $y = x^2$ over the x-axis.

6.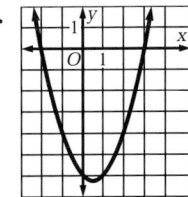

7. x-intercepts: -5 and -11; y-intercept: 55
8. x-intercepts: 0 and 12; y-intercept: 0

Pages 568–571 Talk it Over

1. a. 4^3 **b.** 4^2 **c.** $4^3 \cdot 4^2$ **d.** 4^5 **2. a.** x^3 **b.** x^2 **c.** $x^3 \cdot x^2$ **d.** x^5 **3. a.** 1024 **b.** 1024 **4.** Yes. Shortcuts may vary. An example is given. To write the product $3^6 \cdot 3^4$ as a power of 3, add the exponents: $3^{6+4} = 3^{10}$. **5.** Explanations may vary. An example is given. She noticed that 5^{12} contains 12 factors of 5 and 5^9 contains 9 factors of 5, so $5^{12} \cdot 5^9$ contains 12 + 9, or 21 factors of 5 in all. So she found that instead of writing out all the factors she could add the exponents and write $5^{12} \cdot 5^9$ as $5^{12+9} = 5^{21}$.

6. $(-3)^2 \cdot (-3)^3 = (-3)^?$
$$(-3)^2 \cdot (-3)^3 =$$
$$\underbrace{2 \text{ factors}}_{} \quad \underbrace{3 \text{ factors}}_{}$$
$$\underbrace{(-3)(-3) \cdot (-3)(-3)(-3)}_{2 + 3 \text{ factors}} = (-3)^{2+3}$$
So $(-3)^2 \cdot (-3)^3 = (-3)^{2+3} = (-3)^5$

7. associative property of multiplication **8.** Yes. Reasons may vary. An example is given. $y^2 \cdot y^3 \cdot y^4 = y^{2+3} \cdot y^4 = y^{(2+3)+4} = y^{2+3+4}$, since the way you group numbers being added doesn't change the sum. **9. a.** 3^8 **b.** a^6 **10.** Shortcuts may vary. An example is given. Multiply the exponents: $(a^2)^3 = a^{2 \cdot 3} = a^6$. **11.** Explanations may vary. An example is given. $(a^m)^n$ means that there are n factors of a^m. Using the product of powers rule, you add the exponents to get $a^{m+\ldots+m}$. There are n terms in the sum, so $a^{m+\ldots+m} = a^{m \cdot n}$. **12. a.** $8x^3$ **b.** $xy \cdot xy \cdot xy = x^3y^3$ **c.** $xy^2 \cdot xy^2 \cdot xy^2 = x^3y^6$ **13.** Shortcuts may vary. An example is given. Write each of the factors 2 and x to the third power: $2^3 \cdot x^3$, or $8x^3$. **14.** Explanations may vary. An example is given. $(ab)^n$ means that there are n factors of ab. Therefore, there are n factors of a and n factors of b. We write n factors of a as a^n and n factors of b as b^n, so $(ab)^n$ is the same as $a^n \cdot b^n$.

Pages 572–575 Exercises and Problems

1. To use the product of powers rule, the bases must be the same. **3.** t^{10} **5.** k^{10} **7.** $54a^6$ **9.** $-28z^5$ **11.** a^4b^2 **13.** $-8m^6n^2$ **15.** $\dfrac{1}{4}x^2$; one fourth the area of the original square **17. a.** $4 \cdot 4 = 4^2$ **b.** $x \cdot x \cdot x \cdot x \cdot x = x^5$ **c.** Yes. Shortcuts may vary. An example is given. Write $\dfrac{5^{12}}{5^9}$ as $5^{12-9} = 5^3$. The shortcut is to subtract the exponents.

Selected Answers

19. $x^{0+b} = x^b$; x^0 is equal to 1. **21.** y^{24} **23.** d^{10}
25. $-243h^{15}$ **27.** $625y^{24}z^{12}$ **31.** Terry
36. x-intercepts: 5 and 2; y-intercept: 10
37. x-intercepts: 0 and -3; y-intercept: 0
38. a. $3(x + 2) = 3x + 3 \cdot 2$
b. $2\ \boxed{} = 2\ \boxed{} + 2\ \boxed{}$
$\quad\ \ x + 3 \qquad\qquad x \qquad\quad 3$
39. $2n - 10$ **40.** $24r + 18$ **41.** $-3x - 36$ **42.** $8m + 8n$

Page 579 Talk it Over
1. a. List all the factors of 12 and all of the factors of 15, then select the largest factor that is common to both numbers. **b.** $3x$ **2.** Answers may vary. An example is given. Use the zero-product property and set each factor equal to zero and solve for x: $5x = 0$, $x = 0$; $x + 1 = 0$, $x = -1$.

Pages 580, 582 Exercises and Problems
1. $x(x + 4)$ **3.** $x(x + 2)$ **9.** $-x^2 + 15x$ **11.** $6x^2 - 42x$
13. $-20x^2 + 36x$ **15. a.** $x(x - 5)$ **b.** x-intercepts: 0 and 5; y-intercept: 0 **17. a.** $x(12 + x)$ **b.** x-intercepts: 0 and -12; y-intercept: 0 **19. a.** $-x(x - 10)$
b. x-intercepts: 0 and 10; y-intercept: 0
21. a. $8x(x - 5)$ **b.** x-intercepts: 0 and 5; y-intercept: 0 **23. a.** $6x(2 - x)$ **b.** x-intercepts: 0 and 2; y-intercept: 0 **25.** binomial **27.** monomial
31. $2m^8$ **32.** $10x^5$ **33.** $27a^3c^3$ **34.** $9n^4$
35. $x^2 + 8x + 7$ **36.** $x^2 - 12x - 8$ **37.** $x^2 - x + 14$
38. $x = 0$; $(0, -13)$ **39.** $x = 0$; $(0, 42)$ **40.** $x = 11$; $(11, 0)$

Pages 583–585 Talk it Over
1. No, you will be left with a square "hole" in your tile rectangle. **2. a.** 2 square units **b.** 2
c. $(x + 2)(x + 1)$ **d.** $x^2 + 3x + 2$ **3. a.** $(x + 5)(x + 2)$
b. $x^2 + 7x + 10$ **c.** Yes. **4.** Yes. **5.** $x = -2$
6. a. When the graph crosses the x-axis, the value of y is zero: $(x + 3)(x + 1) = 0$. By the zero-product property, $x + 3 = 0$ or $x + 1 = 0$, so $x = -3$ or $x = -1$.
b. $(-2, 0)$ **c.** Yes.
7. a. $y = x^2 + 4x + 4$, $y = x^2 + 4x + 3$, $y = x^2 + 4x$
b. Answers may vary. **c.** Yes, since
$x = -\dfrac{b}{2a} = -\dfrac{4}{2(1)} = -2$, she got $x = -2$ **8.** The line of symmetry of the given equations is a vertical line. Changing the value of c moves the vertex up or down along this line.

Pages 587–589 Exercises and Problems
5. $x^2 + 5x + 6$ **7.** $x^2 - 16$ **9.** $x^2 - 4x + 4$
11. $2x^2 - x - 10$ **17.** A; y-intercept is 6 and line of symmetry is $x = -2.5$. **19.** C; y-intercept is -6 and line of symmetry is $x = 2.5$. **21. a.** $x = -4$
b. $(-4, -4)$ **c.** 12 **23. a.** $x = -8$ **b.** $(-8, -49)$ **c.** 15
25. a. $x = 3$ **b.** $(3, -2)$ **c.** 7 **27.** a and g; b, d, e, f, and h **31.** $x^2 + 3x$ **32.** $2x^2 - 3x$ **33.** $12x^2 + 15x$

34. $-2x^2 + 16x$ **35.** $-2x(x + 3)$; x-intercepts: 0 and -3; y-intercept: 0 **36.** -11 and 20 **37.** 1 and 15, -1 and -15, 3 and 5, -3 and -5 **38.** 1 and 17, -1 and -17
39. 1 and 24, -1 and -24, 2 and 12, -2 and -12, 3 and 8, -3 and -8, 4 and 6, -4 and -6 **40.** 1 and 21, -1 and -21, 3 and 7, -3 and -7

Page 590 Checkpoint
1. $6x(x + 3)$; Explanations may vary. An example is given. $6x^2 = 2 \cdot 3 \cdot x \cdot x$ and $18x = 2 \cdot 3 \cdot 3 \cdot x$. The greatest common factor of $6x^2$ and $18x$ is $2 \cdot 3 \cdot x$, or $6x$. Factor out $6x$. **2.** m^7 **3.** $-3a^5b^6$ **4.** y^{10}
5. $81m^8n^4$ **6.** $x^2 + 5x$ **7.** $x^2 - 10x$ **8.** $-2x^2 - 12x$
9. a. $x(x + 3)$ **b.** x-intercepts: 0 and -3; y-intercept: 0
10. a. $-5x(x - 2)$ **b.** x-intercepts: 0 and 2; y-intercept: 0 **11.** $x^2 - x - 12$ **12. a.** $x = 4$ **b.** $(4, -10)$
c. 6

Pages 591–593 Talk it Over
1. a. Answers may vary. An example is given. The number of small square tiles is the product of the numbers of tall thin tiles in the two groups along two sides of the large square tile. **b.** There are 8 tall thin tiles and 15 small square tiles; $3 + 5 = 8$ and $3 \cdot 5 = 15$.

	x	1	1	1	1	1
x	x^2	x	x	x	x	x
1	x	1	1	1	1	1
1	x	1	1	1	1	1
1	x	1	1	1	1	1

2. No rectangle can be built using 1 x^2-tile, 2 x-tiles, and 3 unit tiles. **3. a.** $x^2 + 8x + 15$ **b.** $x^2 + 4x - 12$
c. $x^2 - 11x + 24$ **4, 5.** Answers may vary. An example is given. **4.** The coefficient of x in the expanded form is equal to the sum of the two numbers in the factored form, and the constant term in the expanded form is equal to the product of the two numbers in the factored form. **5.** I think that the integer b is equal to the sum of the numbers in the factored form and that the integer c is equal to the product of these numbers. **6. a.** The coefficient of x is -7, not 7.
b. What two integers have the sum -7? What two integers have the product 12? **c.** $x - 3$ and $x - 4$
7. Answers may vary. An example is given. The sum of two negative integers is a negative integer, but the sum of the factors must be 7, a positive integer.
8. Explanations may vary. An example is given. List pairs of integers whose product is -12. Add the integers in each pair to find the pair whose sum is -4. Then write the factored form $(x + 2)(x - 6)$.

Pages 595–597 Exercises and Problems
1. $x^2 + 4x + 3 = (x + 3)(x + 1)$ **7.** $(x + 6)(x + 3)$
9. $(x - 3)(x - 2)$ **11.** $(x - 12)(x - 1)$ **13.** cannot be factored; The only pairs of positive factors of 10 are 1, 10 and 2, 5. Neither pair has the sum 9. **15.** $x + 2$
17. $x - 6$ **19.** $(x - 4)(x + 2)$ **21.** $(x - 6)(x - 2)$

Integrated Mathematics

23. $(x + 8)(x - 2)$

25. 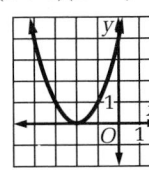   **27.**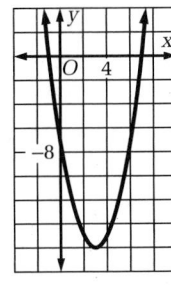

31. Answers may vary. An example is given. The x^2-term has a coefficient other than 1 or −1. $3(x - 2)(x - 1)$ **34.** $x^2 + 8x + 12$ **35.** $x^2 - 3x - 28$ **36.** $x^2 + 3x - 40$ **37.** $-\frac{8}{9}$ **38.** 12, −12 **39.** $2\sqrt{7}, -2\sqrt{7}$ **40.** 64 **41.** −4 **42.** 1

Pages 598, 599, 601, 602 Talk it Over

1. a. $d = \frac{15s}{10}$, or $d = 1.5s$, or $d = \frac{3s}{2}$ **b.** Yes. Reasons may vary. An example is given. The equation has the form $y = kx$, so it is an example of direct variation, which is a special kind of linear function. **c.** s **d.** d
2. a. No. Reasons may vary. An example is given. A linear function has the variable raised to the first power and can be written in the form $y = mx + b$; since this equation does not have that form, it is not linear. **b.** s **c.** d **3. a.** $0 = 0.05s^2 + s - 60$ **b.** $a = 0.05, b = 1, c = -60$ **4.** Answers may vary. An example is given. A linear equation can be written in the form $ax + by = c$ and a quadratic equation can be written in the form $0 = ax^2 + bx + c$. **5.** No. Explanations may vary. An example is given. For a speed of 25 mi/h, the student-driver's rule of thumb gives $\frac{25 \text{ mi/h}}{10 \text{ mi/h}} \cdot 15 \text{ ft} = 37.5 \text{ ft}$, which underestimates stopping distance by more than a car length.
6. a. $x = -10$ **b.** about 35 units **c.** $-45 = -10 - 35$ and $25 = -10 + 35$ **7.** because 26.1 and −46.1 are approximations obtained by using a rounded square root of a number that is not a perfect square
8. Answers may vary. An example is given. 26.1;Yes, I think the estimate of 25 mi/h is good enough because it is very close to the more accurate answer of 26.1 mi/h. Also, the estimate is slightly low, which actually increases the margin of safety.

Pages 603–605 Exercises and Problems

3. 206.25 ft **5.** 30 mi/h **7. a.** at $x = -1$ and at $x = -5$
b. −1 and −5 **9.** Estimates may vary. An example is given: about 0.5 and about −1.75.

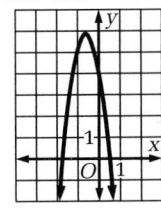

11. a. 0.4 m **b.** The dolphin will reach a height of 0.6 m twice: at 1 m on the way up and at 1.5 m on the way down. **13.** $x = 3$ **15. a.** $0 = x^2 + x - 20$ **b.** $0 = (x + 5)(x - 4)$ **c.** −5, 4 **17. a, b.** 4, −2 **c.** Answers may vary. An example is given. I think factoring is much easier than using the quadratic formula because factoring takes me less time.
19, 21, 23. Estimates and methods may vary. Examples are given. **19.** −0.4, 1; graphing or quadratic formula **21.** 3, 7; factoring **23.** −4, 5; graphing or quadratic formula **27.** $(x - 7)(x - 1)$
28. $(x + 5)(x - 4)$ **29.** $(x - 6)(x + 4)$

30.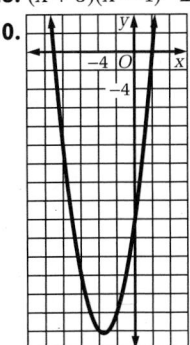

31. x-intercept: 0.5; y-intercept: 1 **32.** x-intercept: 0; y-intercept: 0 **33.** x-intercepts: 0 and 3; y-intercept: 0

Pages 607–608 Unit 10 Review and Assessment

1. a. $y = -1$ **b.** The line of reflection divides $\overline{BB'}$ in half and is perpendicular to $\overline{BB'}$. **c.** 2 units; 2 units
d. The orientation of $\triangle ABC$ is clockwise. The orientation of $\triangle A'B'C'$ is reversed; it is counterclockwise.
2. No. The image is not congruent to the original figure and the orientation is not reversed.
3. Yes.   **4.**

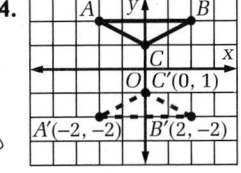 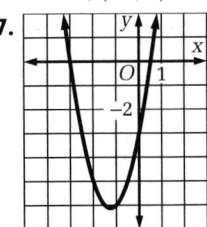

5. Translate the graph of $y = x^2$ down 15 units.
6. $x = -8$; $(-8, 0)$
7.

8. x-intercepts: 2 and −6; y-intercept: −12 **9.** b^{10}
10. $-30x^4y^3$ **11.** m^{28} **12.** $-125a^3b^6$ **13.** $6x^2 - 10x$
14. $y = 7x(x + 3)$; x-intercepts: 0 and −3; y-intercept: 0
15. Answers may vary. $y = x(x + 3)$
16. $x^2 - 9x + 18$ **17. a.** $x = 5$ **b.** $(5, -29)$ **c.** −4
18. $(x - 8)(x - 2)$ **19.** $(x - 6)(x + 3)$ **20.** cannot be

factored; The only pairs of factors of -8 are these: 1, -8; -1, 8; 2, -4; and -2, 4. None of these pairs has the sum 6.

21.
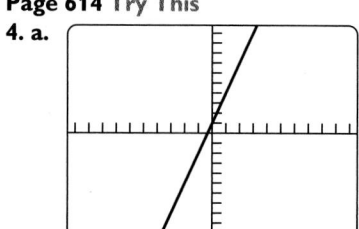

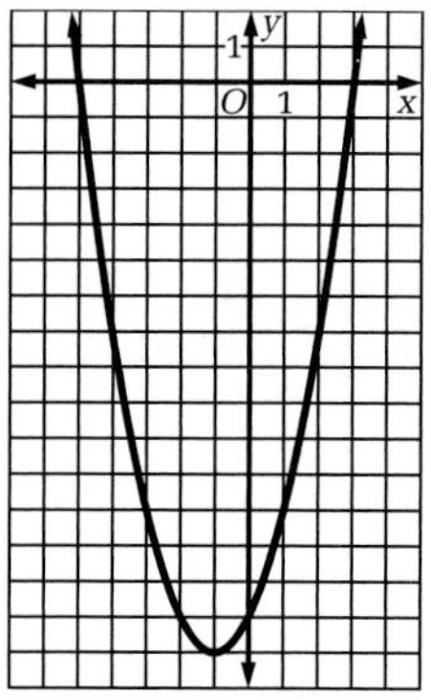

d.
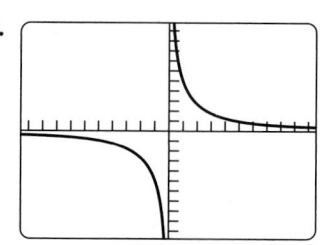

5. Graphs may vary. An example is given.
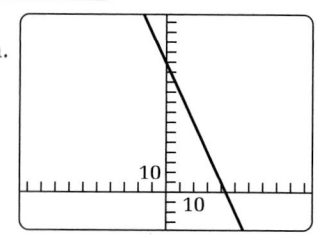

22. Answers may vary. An example is given. With linear equations, we get the variable alone and produce one solution. With quadratic equations, we make one side of the equation zero and solve by factoring, graphing, or the quadratic formula. In either case, the solutions are the x-intercepts of the graph.
23. about 0.8, about -0.8; Methods may vary. An example is given. I used the quadratic formula.

Page 616 Try This
6. approximate x-intercepts: -2.4 and 0.4
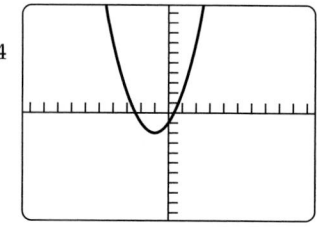

Technology Handbook

Page 612 Try This
1. a. 2.25 **b.** 28 **c.** 1 **2. a.** $(-15 + 29) \div 2$ **b.** 19
3. 64

Page 614 Try This
4. a.
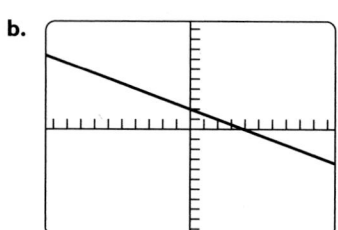

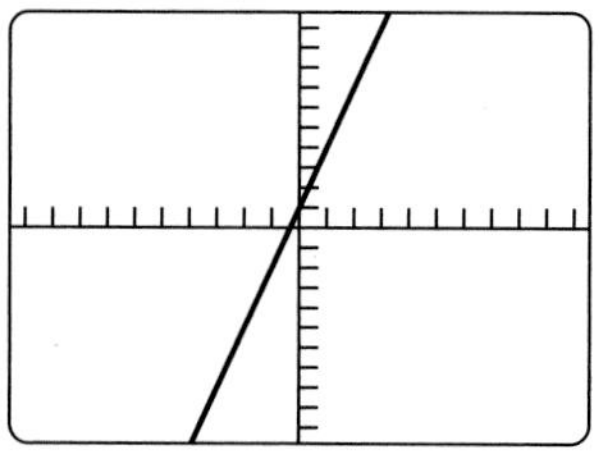

b.
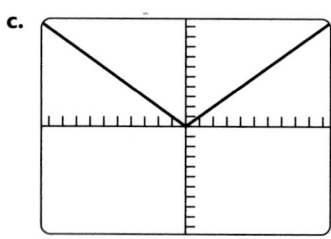

c.

Page 616 Try This
7. -2.42 and -2.41 **8.** 100 couples

Extra Practice

Pages 620–621 Extra Practice Unit 1
1. 4 **3.** 5 **5.** 30 **7.** 3 **9.** 25 **11.** 26 **13.** 10^3 **15.** 10^4
17. 10^8 **19.** 10^3 **21.** 10^0 **23.** 10^{34} **25.** 28 **27.** 60
29. 841 **31.** 35 **33.** 269 **35.** 832 **37.** $4a - 4b$
39. $4r + 2s$ **41.** $2x^2 - 4$ **43.** $12a$ **45.** $8d$ **47.** no like terms **49.** $r^3 + 7r + 6$ **51.** no like terms **53.** $14x + 9$
55. $\triangle ABC \cong \triangle RST$; $\overline{AB} \cong \overline{RS}, \overline{BC} \cong \overline{ST}, \overline{CA} \cong \overline{TR}$
57. hexagon $PQRSTZ \cong$ hexagon $TUVWXY$;
$\overline{PQ} \cong \overline{TU}, \overline{QR} \cong \overline{UV}, \overline{RS} \cong \overline{VW}, \overline{ST} \cong \overline{WX}, \overline{TZ} \cong \overline{XY}$,
$\overline{ZP} \cong \overline{YT}$ **59.** pentagon $EFGHI \cong$ pentagon $ELKJI$;
$\overline{EF} \cong \overline{EL}, \overline{FG} \cong \overline{LK}, \overline{GH} \cong \overline{KJ}, \overline{HI} \cong \overline{JI}, \overline{IE} \cong \overline{IE}$
61. square **63.** rhombus **65.** kite **67.** parallelogram
69.
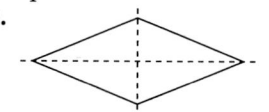
71.
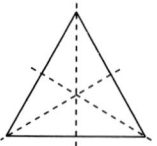

73.
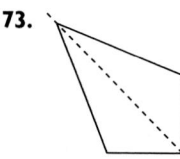

42

Integrated Mathematics

Pages 622–623 Extra Practice Unit 2

1. a. 15 **b.** −56 **3. a.** $\frac{7}{8}$ **b.** $\frac{7}{8}$ **5. a.** −32 **b.** 62

7. 5.7×10^{-2} **9.** 6.497×10^4 **11.** 4.8×10^{10}
13. 360,000,000 **15.** 0.001 **17.** 0.000048 **19.** 30°
21. 75° **23.** 75° **25.** 105° **27.** $x = 80°$ **29.** $x = y = 90°$
31. $4r^2$ **33.** $10s - 4st - 2t$ **35.** $4x^3 + 4x - 1$
37. $-3w - 2k + 3wk$ **39.** 7 **41.** 8 **43.** −10 **45.** 0
47. 7 **49.** 55 **51.** 18 **53.** 45 **55.** −40 **57.** 81
59. $\sqrt{30}$ is a number between 5 and 6; 5.48
61. $\sqrt[3]{70}$ is a number between 4 and 5; 4.12
63. $\frac{8}{7}, -\frac{8}{7}$ **65.** 25, −25 **67.** $\sqrt[3]{800}, \sqrt{87}, 9.4$

Pages 624–625 Extra Practice Unit 3

1. mean = 13.125; median = 12; mode = 11;
outlier = 19; range = 8 **3.** mean = 77.6; median = 73;
mode = 68; outlier = 108; range = 44 **5.** 62 **7.** 81
9. 2.5

11.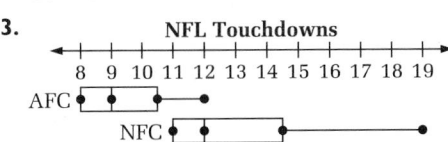

13.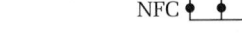

15.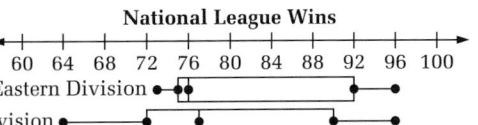

17. $-1 \le x < 1$ **19.** $x \ge 106$

21.

6	4 7 8 8 9
7	7 8
8	4
9	3
10	8

23.

NFL Touchdowns

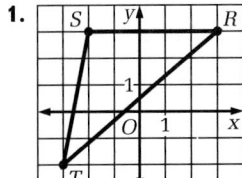

25.

National League Wins

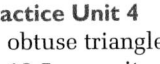

27. Answers may vary.
An examples is given.
I chose a circle graph.
This is a good way to
display data that
represents various
percentages of a whole.

U.S. Jewelry Purchases

Pages 625–627 Extra Practice Unit 4

1. obtuse triangle; 12.5 sq. units

3. trapezoid; 14 sq. units

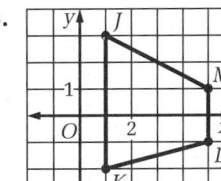

5. $R'(3, 5)$, $S'(-2, 5)$, $T'(-3, 0)$ **7.** $J'(4, -1)$, $K'(4, -6)$,
$L'(8, -5)$, $M'(8, -3)$ **9.** (0, 1) **11.** 90° clockwise
13. 45° counterclockwise
15.

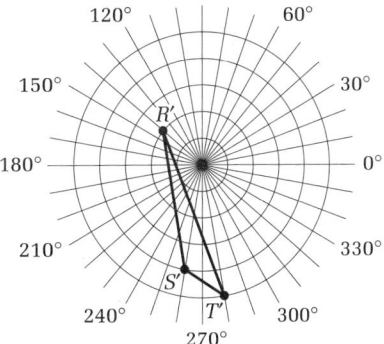

17.

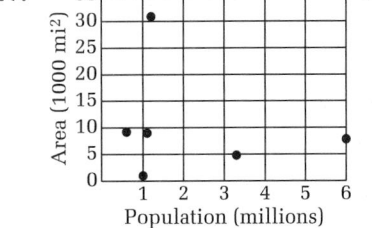

no correlation

19. Yes; passes vertical-line test. **21.** b **23.** $y = -\frac{1}{3}x$
25. $y = -x + 1$ **27.** $s = 0.8P$, where s is sale price, and
P is original price.

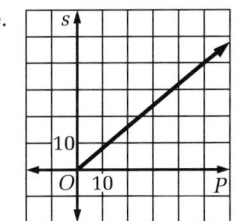

Pages 628–629 Extra Practice Unit 5

1. $-2x + 3$ **3.** $x + 3y$ **5.** $17 + 7m$ **7.** −8 **9.** −7 **11.** $\frac{4}{3}$
13. 27 **15.** 2 **17.** −3 **19.** −7.5 **21.** 4 **23.** −4 **25.** −2
27. $y \ge -7$

29. $a \le -\frac{3}{2}$

31. $j > 12$

33. $k > 4$

35. $s = \frac{P}{4}$ **37.** $E = \frac{I}{d^2}$ **39.** $a = \frac{h + 16t^2}{t}$ **41.** $-\frac{5}{2}$ **43.** 24

Selected Answers

45. −6 **47.** −8 **49.** −12 **51.** $y = \dfrac{ax - c}{b}$ **53.** $y = \dfrac{4}{3}x - 16$

55. $y = -\dfrac{3}{2}x + 18$ **57.** 21 sq. units **59.** 45 m^2

61. 68 ft^2 **63.** 36 **65.** $x = 4$, $y = 12$ **67.** $r = -\dfrac{5}{3}$, $s = -\dfrac{5}{9}$

69. $a = 3$, $b = 7$ **71.** $a = \dfrac{2}{3}$, $b = 2\dfrac{2}{3}$ **73.** $x = 6$, $y = 12$

Pages 629–631 Extra Practice Unit 6

1. Yes. The experimental probability that a voter will be in favor is 0.514, which is more than 50%. **3.** 1
5. 0.5 **7.** 0.9 **9.** $x = 10$ **11.** $t = 10.24$ **13.** $a = 8.75$
15. $b = 2250$ **17.** $x = 0.3$ **19.** $k = 0.84$ **21.** 125°
23. 8 **25.** 46.5 **27.** 150° **29.** 32 **31.** $\triangle PQR \sim \triangle PTS$; $\angle Q \cong \angle T$ and $\angle S \cong \angle R$, so the triangles are similar.

33.
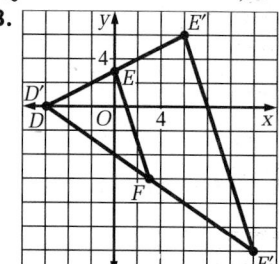

35.
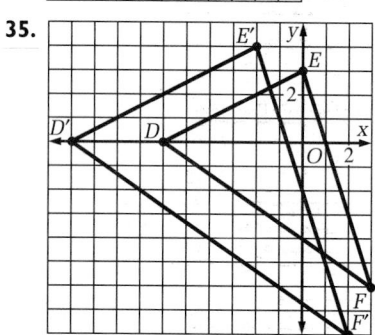

37. 0.97 **39.** 0.50 **41.** $\dfrac{21}{29}$; 0.72 **43.** $\dfrac{20}{29}$; 0.69

Pages 631–632 Extra Practice Unit 7

1. 1.60 **3.** 6.31 **5.** $\dfrac{3}{4}$; $\dfrac{4}{3}$ **7.** 4.4 cm **9.** about 9.4 mm
11. about 267 m **13.** about 12.6 m **15.** about 8.7 cm

17.
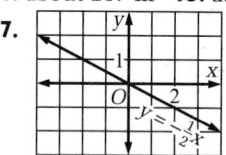

19.
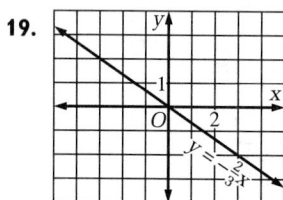

21.
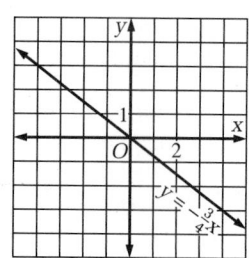

23.
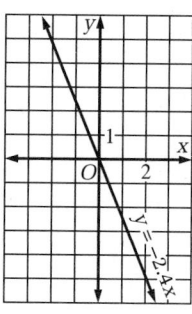

25. $y = -\dfrac{1}{4}x$ **27.** km **29.** $\dfrac{\text{cycles}}{\text{s}}$ **31.** h **33.** 154 in.^{2}

35. 2.5 mm^2 **37.** 25.1 m^2 **39.** $\dfrac{9\pi}{8}$ cm^2, or about 3.5 cm^2

Pages 633–635 Extra Practice Unit 8

1. 2; 7 **3.** 0.7; −3.5 **5.** −1; 5 **7.** 2; −1; $y = 2x - 1$
9. $\dfrac{2}{3}$; $\dfrac{1}{3}$; $y = \dfrac{2}{3}x + \dfrac{1}{3}$ **11.** $y = 5x - 7$ **13.** $y = -\dfrac{1}{2}x - 4$
15. $y = -6x + 18$

17. $(0, -2)$, $(-8, 0)$
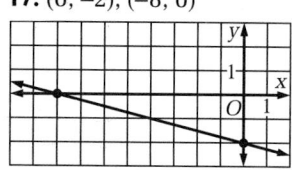
19. $(0, -3)$, $(10, 0)$
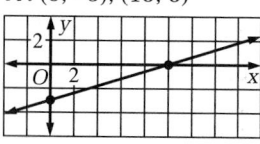

21. $\left(0, -\dfrac{5}{2}\right)$, $(-2, 0)$
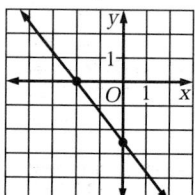

23. undefined; $x = -3$ **25.** 0; $y = -2.5$

27. Slope is undefined.
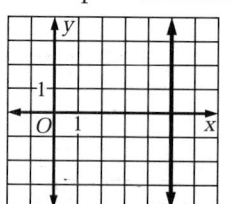
29. Slope is undefined.
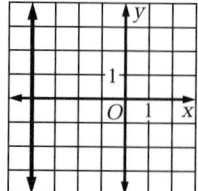

31. Slope is undefined.
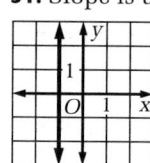
33. Slope is undefined.
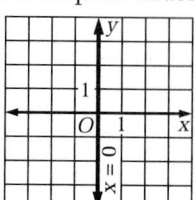

35. $y = \dfrac{3}{5}x - 2$ **37.** $y = -2x + 19$ **39.** $y = \dfrac{2}{3}x + 4$

44

Integrated Mathematics

41. $y = -5x + 5$ **43.** $y = -1$

45, 47, 49. Estimates may vary. Examples are given.

45. $\left(-1\frac{2}{3}, -1\right)$

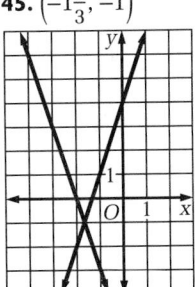

47. $\left(-2\frac{2}{3}, 3\frac{1}{3}\right)$

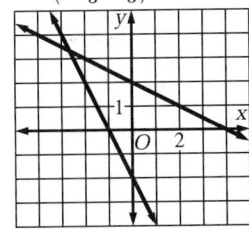

49. about $\left(\frac{1}{2}, \frac{1}{4}\right)$

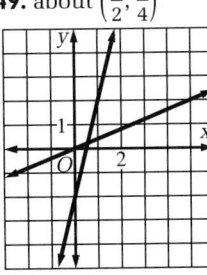

51.

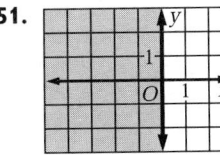

53.

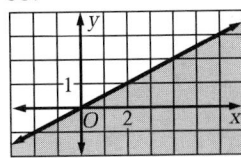

55.

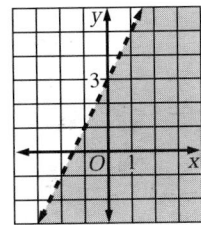

57.

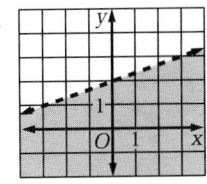

59.

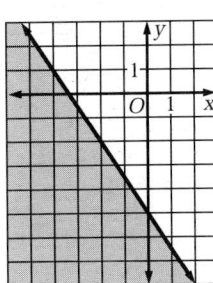

61.

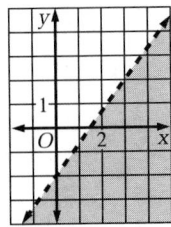

63.

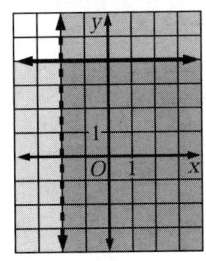

65.

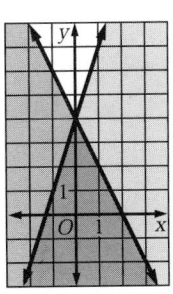

67.

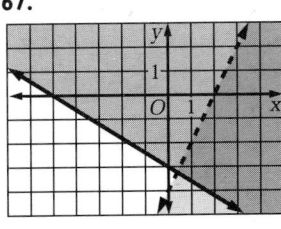

Pages 635–637 Extra Practice Unit 9

1. 25 **3.** 9 **5.** 30 **7.** $\sqrt{149} \approx 12.2$ **9.** $4\sqrt{3}$ **11.** $3\sqrt{10}$
13. 7 **15.** $6\sqrt{2}$ **17.** 6 **19.** $30\sqrt{3}$ **21.** $x = \pm2\sqrt{14}$
23. $x = \pm2\sqrt{5}$ **25.** $x = \pm9\sqrt{2}$ **27.** If $m \times n$ is an even number, then m and n are even numbers. False; $3 \times 4 = 12$. **29.** If a pair of angles are congruent, then they are vertical angles. False; Two congruent angles in two different triangles are not vertical angles.
31. 0 **33.** 0, −3 **35.** 0, −9 **37.** No; $8^2 + 10^2 \neq 12^2$.
39. Yes; $20^2 + 21^2 = 29^2$. **41.** Yes; $9^2 + 40^2 = 41^2$.
43. $\frac{1}{3}$ **45.** $\frac{1}{6}$ **47.** 1 **49.** $\frac{1}{3}$ **51.** $\frac{1}{4}$ **53.** triangular prism;
96 in.^{2} **55.** pyramid; 1536 cm^2 **57.** pyramid; 576 ft^2
59. half-cylinder; about 173.7 m^2
61. trapezoidal prism; 216 in.^{2} **63.** about 35.3 m^3
65. about 84.8 in.^{3} **67.** 1200 ft^3 **69.** 3072 cm^3
71. about 2261.9 mm^3 **73.** 16:9; 18 m^2
75. 25:36; 86.4 m^2 **77.** 27:8; $53\frac{1}{3}\pi$ m^3

Pages 638–639 Extra Practice Unit 10

1. a.

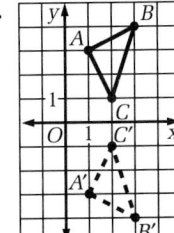

$\triangle ABC$ has clockwise orientation, and $\triangle A'B'C'$ has counterclockwise orientation.

b.

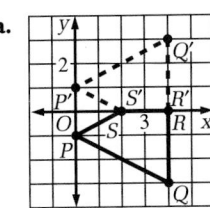

$\triangle ABC$ has clockwise orientation, and $\triangle A'B'C'$ has counterclockwise orientation.

3. a.

Quadrilateral $PQRS$ has counterclockwise orientation, and quadrilateral $P'Q'R'S'$ has clockwise orientation.

3. b.

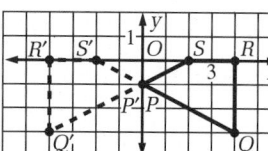

Quadrilateral *PQRS* has counterclockwise orientation, and quadrilateral *P'Q'R'S'* has clockwise orientation.

5. D **7.** A **9.**

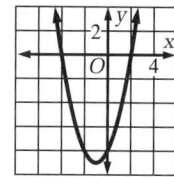

11. 0 and 4; 0 **13.** −8 and 3; −24 **15.** $\frac{2}{3}$ and $-\frac{1}{4}$; −2

17. $6b^6$ **19.** $35x^4y^3$ **21.** t^{12} **23.** $3125a^{15}$ **25.** $-64a^3b^9$

27. $64r^{30}s^{12}$ **29.** $-a^2 - 4a$ **31.** $6x^2 - 15x$

33. $28x^2 + 63x$ **35.** $x(3 - x)$; 0 and 3; 0 **37.** $x(x - 4)$; 0 and 4; 0 **39.** $2x(x - 3)$; 0 and 3; 0 **41.** $x^2 - 3x - 10$

43. $x^2 - 12x + 36$ **45.** $x^2 + 12x + 36$

47. $4x^2 - 19x + 12$ **49.** $10x^2 + 43x + 28$

51. $21x^2 + 25x - 4$ **53. a.** $x = -1$ **b.** $(-1, 0)$ **c.** 1

55. a. $x = 5$ **b.** $(5, -30)$ **c.** −5 **57.** $(x + 14)(x - 2)$

59. $(x - 10)(x - 3)$ **61.** $(x - 6)(x + 5)$

63.

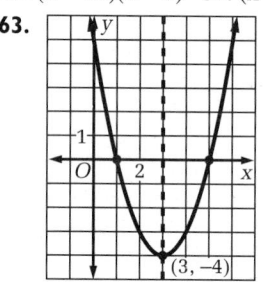

65.

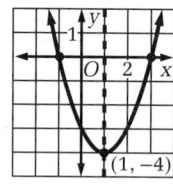

67.

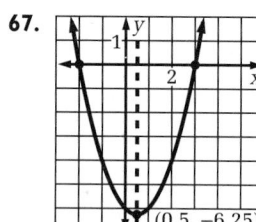

69, 71, 73, 75, 77. Methods and estimates may vary. Answers are given. **69.** 9 and 5 **71.** 2 and 4

73. 3 and −1 **75.** −2 and $\frac{7}{3}$ **77.** 1 and $-\frac{1}{2}$

Toolbox Skills

Page 640 Skill 1

1. 76 **2.** 36 **3.** 49 **4.** 49 **5.** 300 **6.** 90 **7.** 60 **8.** 101 **9.** 1100 **10.** 6300 **11.** 55 **12.** 1200

Page 641 Skill 2

1. $7 < 10$

2. $-4 < 0$

3. $5 > -7$

4. $-5 > -7$

5. −1, 0, 2; $-1 < 0 < 2$ **6.** −5, −3, 0; $-5 < -3 < 0$

7. −7, 1, 3; $-7 < 1 < 3$ **8.** −6, −4, −2; $-6 < -4 < -2$

9. −19, −9, 12; $-19 < -9 < 12$ **10.** −27, −17, −7; $-27 < -17 < -7$ **11.** −100, 0, 56; $-100 < 0 < 56$

12. −10, −1, 1, 10; $-10 < -1 < 1 < 10$

Page 642 Skill 3

1. 9 **2.** −9 **3.** 5 **4.** −5 **5.** 0 **6.** −8 **7.** 5 **8.** −6 **9.** −18

10. 1 **11.** 0 **12.** −10 **13.** −10 **14.** 9 **15.** −33 **16.** 0

Pages 643–644 Skill 4

1. 2 **2.** −1 **3.** 8 **4.** −5 **5.** 1 **6.** −6 **7.** 1 **8.** 10 **9.** 9

10. −8 **11.** 0 **12.** −10 **13.** 40 **14.** −8 **15.** 8 **16.** 0

17. −4 **18.** −9 **19.** 3 **20.** −25 **21.** 25 **22.** 25 **23.** 3

24. 0 **25.** −10 **26.** 1 **27.** −48 **28.** −6 **29.** 30 **30.** −18

31. −21

Page 645 Skill 5

1. Yes. **2.** No. **3.** Yes. **4.** No. **5.** No. **6.** Yes. **7.** 6

8. 4 **9.** −5 **10.** 10 **11.** 4 **12.** −1 **13.** −9 **14.** −24

15. −4 **16.** 5 **17.** 1 **18.** 32 **19.** 7 **20.** 48 **21.** −2

22. −12 **23.** −6 **24.** −2

Pages 646–647 Skill 6

1. a. $x + 2 = -1$ **b.** $x = -3$ □ = ■ ■ ■

2. 5 **3.** 4 **4.** −3 **5.** 2 **6.** −4 **7.** 0

8. a. $2x = 10$

b. (1) ▭▭ = ⬭⬭ (2) $2x = 10$
$\frac{2x}{2} = \frac{10}{2}$
$x = 5$ $x = 5$

9. 4 **10.** −4 **11.** −6 **12.** 4 **13.** 6 **14.** −1 **15.** 8

Page 648 Skill 7

1. $2 \times 2 \times 2 \times 5$ **2.** 3×13 **3.** prime

4. $2 \times 2 \times 3 \times 7$ **5.** $2 \times 2 \times 2 \times 13$ **6.** $2 \times 3 \times 17$

7. $2 \times 5 \times 11$ **8.** $5 \times 5 \times 5$ **9.** prime **10.** $3 \times 7 \times 7$

11. $2 \times 2 \times 3 \times 3 \times 7$ **12.** $2 \times 2 \times 2 \times 2 \times 3 \times 3$

13. 7×19 **14.** 2×29 **15.** $2 \times 2 \times 2 \times 2 \times 11$

Page 649 Skill 8

1. 3; 18 **2.** 5; 110 **3.** 7; 210 **4.** 15; 90 **5.** 9; 27

6. 6; 144 **7.** 25; 50 **8.** 4; 60 **9.** 3; 210 **10.** 4; 32

11. 11; 110 **12.** 20; 300

Page 650 Skill 9

1. $\frac{5}{3}$ **2.** $-\frac{4}{7}$ **3.** $-\frac{1}{6}$ **4.** 1 **5.** $\frac{1}{6}$ **6.** $-\frac{4}{9}$ **7.** $\frac{6}{11}$ **8.** −9

9. $-\frac{3}{14}$ **10.** $\frac{10}{7}$

Pages 651–652 Skill 10

1. $\frac{7}{9}$ **2.** $\frac{1}{2}$ **3.** $\frac{7}{9}$ **4.** 1 **5.** $1\frac{5}{9}$ **6.** $\frac{7}{15}$ **7.** $\frac{1}{4}$ **8.** $\frac{7}{18}$ **9.** $-\frac{4}{15}$

10. $-\frac{37}{56}$ **11.** $-\frac{2}{9}$ **12.** $\frac{16}{25}$ **13.** $-\frac{7}{9}$ **14.** $\frac{1}{11}$ **15.** $1\frac{1}{24}$

16. $1\frac{11}{35}$ **17.** $\frac{3}{14}$ **18.** $-\frac{2}{5}$ **19.** $\frac{5}{18}$ **20.** $\frac{3}{25}$ **21.** $\frac{1}{15}$ **22.** 8

23. $-2\frac{1}{2}$ **24.** $\frac{3}{4}$ **25.** $\frac{27}{50}$ **26.** $\frac{5}{18}$ **27.** $-\frac{3}{8}$ **28.** -1 **29.** $1\frac{1}{2}$

30. $\frac{9}{20}$ **31.** $\frac{20}{27}$ **32.** $-\frac{25}{49}$

Pages 654 Skill 11

1. $0.7; \frac{7}{10}$ **2.** $0.32; \frac{8}{25}$ **3.** $1.25; \frac{5}{4}$ **4.** $0.0075; \frac{3}{400}$

5. $0.075; \frac{3}{40}$ **6.** $66\%; \frac{33}{50}$ **7.** $33\frac{1}{3}\%; \frac{1}{3}$ **8.** $0.9\%; \frac{9}{1000}$

9. $37.5\%; \frac{3}{8}$ **10.** $160\%; \frac{8}{5}$ **11.** $0.72; 72\%$

12. $0.029; 2.9\%$ **13.** $0.576; 57.6\%$ **14.** $1.35; 135\%$

15. $0.1\overline{6}; 16.\overline{6}\%$, or $16\frac{2}{3}\%$

Page 654 Skill 12

1. 40% **2.** 250% **3.** 4% **4.** 37.5% **5.** 7.5%

6. 125% **7.** 30% **8.** 18.75% **9.** 0.5% **10.** $33\frac{1}{3}\%$

Page 655 Skill 13

1. 8% decrease **2.** 10% increase **3.** 2.4% increase
4. 26% decrease

Page 656 Skill 14

1. 1.8 **2.** 21 **3.** 26 **4.** 3.7 **5.** 98.7 **6.** 7.4 **7.** 110
8. 0.08 **9.** 45 **10.** $7.95; $18.55 **11.** Answers may
vary. An example is given. 79% is almost 80%, so, if
80% of $50 is $40 (0.8($50) = $40), then 79% of $50
is a little less than $40. **12.** Answers may vary. An
example is given. No; 250 items are half the items,
not $\frac{1}{2}\%$; 1% of the items is 5 items, so a $\frac{1}{2}\%$ defective
rate would yield about two or three defective items
per 500 items.

Page 656 Skill 15

1. 80 **2.** 125 **3.** 950 **4.** 648 **5.** 21 **6.** 375 **7.** 800
8. 7.5 **9.** 500 **10.** 12.8 **11.** 62.5 **12.** 160 **13.** 150

Page 657 Skill 16

1. 45° **2.** 120° **3.** 90° **4.** 135°

5.   **6.**   **7.**

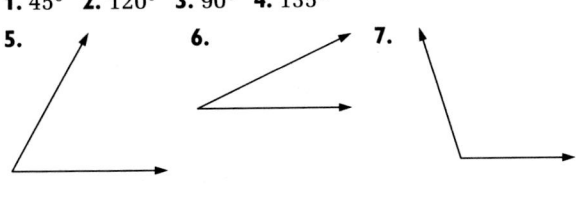

8.

9. a. 88° **b.** $(125 - x)°$

Page 658 Skill 17

1. 27 **2.** 4.8 cm **3.** $2x + 12$

Page 659 Skill 18

1. 64 cm² **2.** 99 ft² **3.** 2.25 mm² **4.** $8t$ yd²
5. 2.88 in.²

Page 659 Skill 19

1. 2400 cm³ **2.** 22.5 in.³ **3.** 4 ft³ **4. a.** 343 yd³
b. 0.125 ft³

Page 660 Skill 20

1. $(3, -3)$ **2.** $(-2, 5)$ **3.** $(2, 4)$ **4.** $(-2, -5)$ **5.** $(-1, 4)$
6. $(-2, 0)$ **7.** $(2, -5)$ **8.** $(-5, -2)$ **9.** $(0, 2)$ **10.** $(4, 3)$

Page 660 Skill 21

1–6.

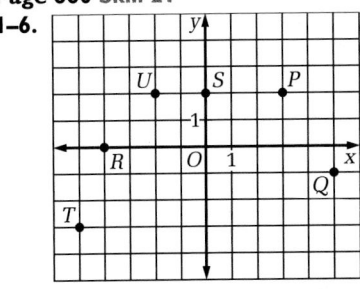

Page 661 Skill 22

1.

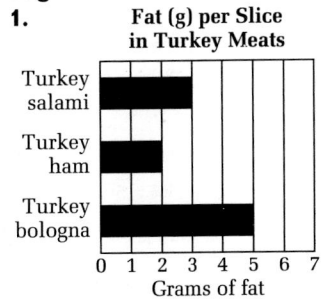

Fat (g) per Slice in Turkey Meats

2.

Calories per Slice in Turkey Meats

Page 662 Skill 23

1.

**Average Temperature
in Miami, Florida**

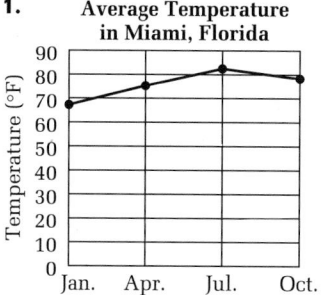

2.

**Average Temperature
in Peoria, Illinois**

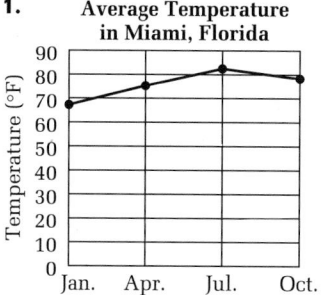

Page 663 Skill 24

1. **National Conference
Longest Punts 1991–1992**

Newsome 🌰🌰🌰🌰🌰🌰🌰🌰🌰🌰   🌰 = 10 yards

Barnhardt 🌰🌰🌰🌰🌰🌰🌰

Landeta 🌰🌰🌰🌰🌰🌰🌰

Arnold 🌰🌰🌰🌰🌰🌰🌰

2. **National Conference
Punting Yardage 1991–1992**

Newsome 🌰🌰🌰🌰🌰🌰🌰   🌰 = 500 yards

Barnhardt 🌰🌰🌰🌰🌰🌰🌰

Landeta 🌰🌰🌰🌰🌰

Arnold 🌰🌰🌰🌰🌰

Page 663 Skill 25

1.   **2.**

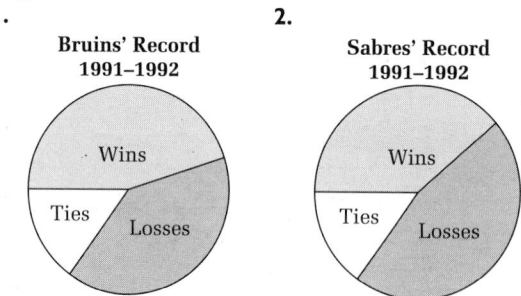

**Bruins' Record
1991–1992**   **Sabres' Record
1991–1992**

Page 664 Skill 26

1. 7 calculators **2.** 2562 mi **3.** 209 student tickets

Page 665 Skill 27

1. $245 **2.** $4.75 **3.** 2400 items **4.** 130 ft^2

Appendix 1

Pages 670, 671 Talk it Over
1. Control variable: minutes after tap is turned on; dependent variable: height of water **2.** Control variable values: all amounts of time from 0 through 6 minutes; dependent variable values: all heights from 0 through 21 cm **3.** Yes; both the control variable and the dependent variable are continuous quantities.
4. Descriptions may vary. An example is given. The domain is the number of days: 0, 1, 2, 3…. The range is the dollar amounts 0, 0.10, 0.20, 0.30…. There may be an upper limit to the domain and the range. For example, once you owe more in fines than a book is worth, the library may charge you the cost of replacing the book.

Page 671 Exercises and Problems
1. The domain is the number of weeks: 0, 1, 2, 3, 4, 5, 6. The range is the dollar amounts 120, 100, 80, 60, 40, 20, 0. **3.** The domain is the number of gallons from 0 through the total number of gallons his tank holds. If g is the total number of gallons his tank holds, then the range is the dollar amounts to the nearest penny from 0 through 1.29g. **5.** The domain is the number of tickets sold: 0, 1, 2, …, 12. The range is the dollar amounts 0, 3.50, 7.00, …, 42.

Appendix 2

Pages 672, 673 Talk it Over
1. The slopes are the same; the slope is the amount Ilona charges per pin ($25). **2.** The vertical intercept of the first line is 0; this is the amount of money made before any pins are sold, disregarding start-up costs. The vertical intercept of the second line is −300; Ilona needs to make $300 to pay off her start-up costs. **3.** A−m; B−k; C−n; D−l; You can match the graphs by checking the y-intercepts. **4.** Descriptions may vary. An example is given. Changing the value of m changes the slope. When $m > 0$, the graph slopes up from left to right, and becomes steeper as m increases. When $m < 0$, the graph slopes down from left to right. It becomes steeper as $|m|$ increases. Changing the value of b changes the y-intercept. The graph crosses the y-axis at the origin when $b = 0$, above the origin when $b > 0$, and below the origin when $b < 0$.

Page 673 Exercises and Problems
1. The graph of $y = 3x$ will slope up from left to right and will be steeper than the graph of $y = x$. The graph of $y = -3x$ will slope down from left to right. All three graphs will pass through the origin. **3. a.** $h = 55t + 10,152$ **b.** The graph of $h = 55t + 10,152$ will be less steep than the graph of $h = 110t + 6000$ because its slope is smaller. Its vertical intercept is greater, so it will cross the vertical axis at a higher point.

Additional Answers

Unit 1

Page 23 Exercises and Problems
20. a.

Round	1	2	3	4	5	6	7	8	9	10
Number of letters	5	25	125	625	3125	15,625	78,125	390,625	1,953,125	9,765,625

b. 5^n **c.** round 13

Page 47 Exercises and Problems
5. perimeter = $4b$ perimeter = $2a + 2b$

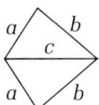 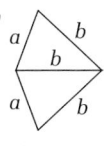

6. perimeter = $2a + 2b$ perimeter = $2a + 2c$ perimeter = $2b + 2c$

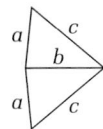 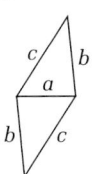

Page 49 Exercises and Problems
34. 1,000,000; ten to the sixth power or one million
35. 1,000,000,000; ten to the ninth power or one billion
36. Answers may vary.

Unit 2

Page 82 Exercises and Problems
5. a–c. Answers may vary. A sample table of values is given. Examples are given, based on the table.
a. Lengths of Upper Sections of Thumbs

Person	1	2	3	4	5
Length	$1\frac{1}{8}$ in.	1 in.	$1\frac{1}{8}$ in.	$1\frac{1}{2}$ in.	$1\frac{1}{4}$ in.

b. $1\frac{1}{8}$ in. **c.** Person 2

d. Methods may vary. An example is given. If your measurement is more than an inch, say $1\frac{1}{4}$ in., you could count each 4 thumb lengths as 5 in. A similar procedure can be used if your measurement is less than an inch.

Unit 3

Page 149 Checkpoint
5. $8 \le d \le 40$, where d = diameter in feet;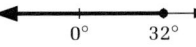

6. $t \le 32°$, where t = temperature at which water will freeze;

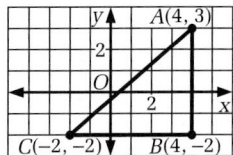

Unit 4

Page 196 Exercises and Problems
21. Answers may vary. An example is given. The area of *ABCDEFGHI* is 32 square units.

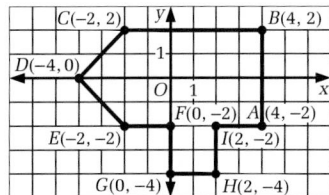

22. 49 **23.** 13 **24.** $9x - 12$ **25.** $9a + 3ab$ **26. a.** Check graphs.
b. $(4, -2)$ **c.** $(-2, -2)$ **d.** right triangle

27. a. The parallelogram is flopping down to the right. The coordinates of the two top vertices change; the bottom two do not. **b.** 10; 8; 6; 0; yes; the rectangle **c.** False.

Page 201 Exercises and Problems
33. Sketches and descriptions may vary. Examples are given. Suppose a car travels up a hill and becomes horizontal again at the top of the hill. As the car moves to the right, the first coordinates increase. As the car 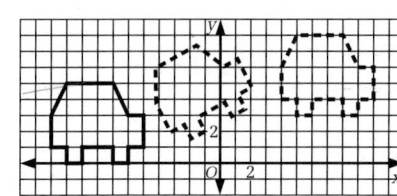 moves up the hill, the second coordinates increase. As the car levels off at the top of the hill, some of the second coordinates increase and some decrease. The third figure is a translation of the original figure 15 units right and 3 units up.

A-1

Page 233 Exercises and Problems

40. Analyses may vary. An example is given.The function $y = x^3$ could be named the cubing function. The graph is a curve that lies in quadrant I and quadrant III. Each part of the curve looks like half of a parabola, although the cubing function and squaring function pass through different points (the part of the $y = x^3$ curve in quadrant III looks as if the left half of the parabola $y = x^2$ has been turned down).

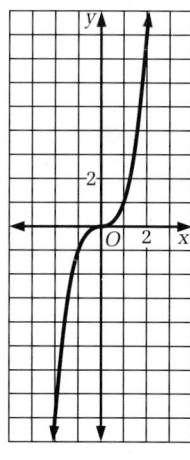

Unit 5

Page 246 Exercises and Problems

33. $-4k - 21$ **34.** $\frac{2}{3}t + 10$ **35.** $15x - 46$ **36. a, d.** Answers may vary.

b.

Number of Cars	Total Earnings ($)
1	3
2	6
⋮	⋮
x	$3x$

c. need to wash fewer cars; need to wash more cars

Unit 6

Page 342 Exercises and Problems

16. a–d. Polygons and answers may vary. Examples are given based on the dilations in part (a). **a.** The center of dilation is at $P(1, 1)$. The scale factors are $\frac{1}{2}$ and 2.

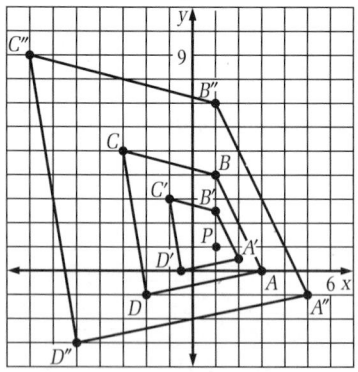

b. $ABCD \sim A'B'C'D' \sim A''B''C''D''$ **c.** $\frac{A'B'}{AB} = \frac{C'D'}{CD} = \frac{1}{2}, \frac{B''C''}{BC} = \frac{A''D''}{AD} = \frac{2}{1}$

Page 349 Exercises and Problems

19–22. Sketches may vary. Examples are given.

19. maximum height: about 14.5 ft

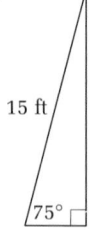

20. 7.5 m

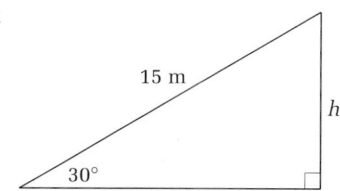

21. about 96.4 m

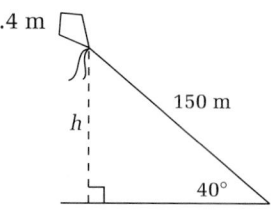

22. a.

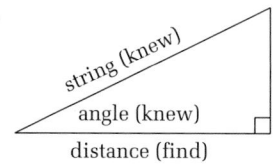

b. He may have used the cosine ratio. **c.** $\cos 28° = \frac{d}{1800}$; $d \approx 1589.3$ ft

Unit 7

Page 373 Exercises and Problems

21. a. Predictions may vary, depending on ball chosen and drop heights used. For example, if a ball with a variation constant of 0.6 is chosen and a drop height of 50 cm is chosen, students might predict that the heights of the first three bounces will be 0.6(50) = 30 cm, 0.6(30) = 18 cm, and 0.6(18) = 10.8 cm, respectively. **b.** Drop heights and data may vary. **c.** The first bounce height varies directly with the drop height, and each bounce height after the first varies directly with the previous bounce height. Consider the model $\frac{B}{D} = 0.6$. If D is the drop height, and B_1, B_2, and B_3 are the first three bounce heights, then $B_1 = 0.6D$, $B_2 = 0.6B_1$, and $B_3 = 0.6B_2$. Alternatively, $B_1 = 0.6D$, $B_2 = (0.6)(0.6)D$, and $B_3 = (0.6)(0.6)(0.6)D$. **d, e.** Results may vary.

Page 374 Exercises and Problems

26.

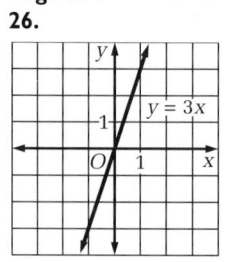

27.

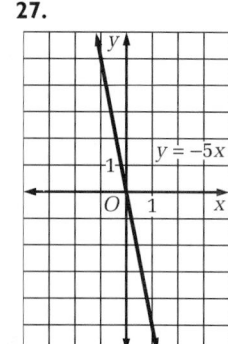

28.

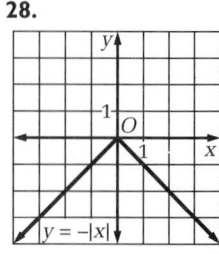

Page 390 Exercises and Problems

32. b.

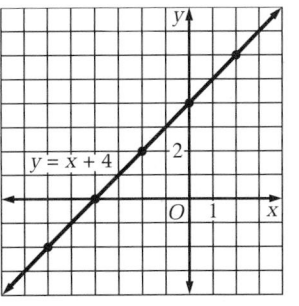 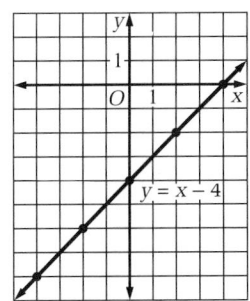

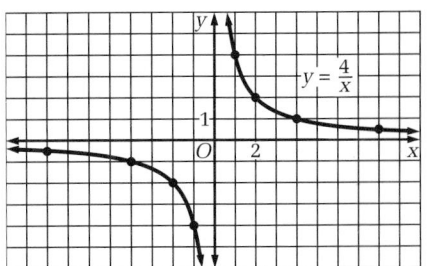

Unit 8

Page 421 Exercises and Problems

6. a.

Distance beneath surface in km (d)	Temperature in °C (t)
0	20
1	45
2	70
3	95
4	120

Temperature Beneath Earth's Surface

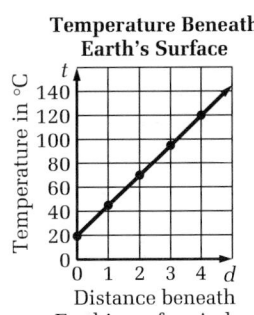

Distance beneath Earth's surface in km

b. slope: 25; vertical intercept: 20; Explanations may vary. An example is given. The slope shows that for each kilometer beneath Earth's surface, the temperature increases by 25°C. The vertical intercept represents the temperature at Earth's surface, 20°C. **c.** $t = 25d + 20$

7. a.

Distance driven in miles (m)	Gas remaining in gallons (g)
0	12
15	11
30	10
45	9
60	8

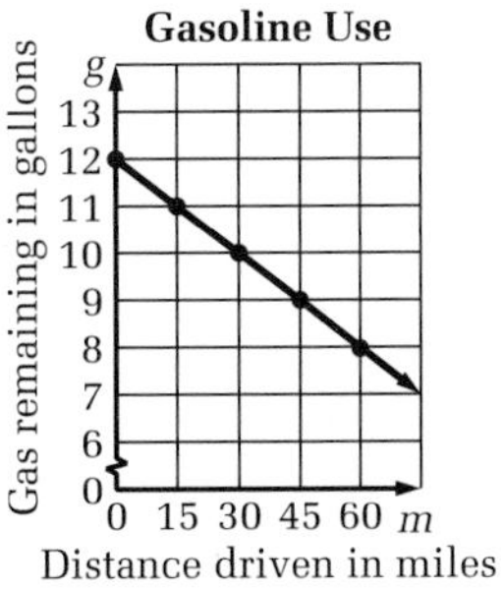

Gasoline Use

Distance driven in miles

b. slope: $-\frac{1}{15}$; vertical intercept: 12; Explanations may vary. An example is given. The slope shows that for each 15 miles driven, 1 gallon of gasoline is used. The vertical intercept represents the amount of gasoline in the car, 12 gallons, before driving any distance.

c. $g = -\frac{1}{15}m + 12$

8. a.

Distance driven in miles (m)	Cost in dollars per day (c)
0	45.00
10	48.90
20	52.80
30	56.70

Cost of a Truck Rental for One Day

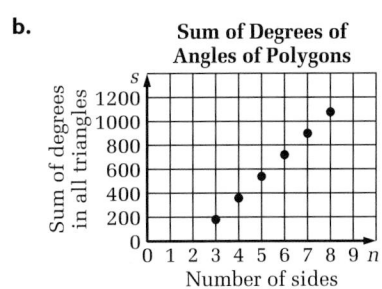

Distance driven in miles

b. slope: 0.39; vertical intercept: 45; Explanations may vary. An example is given. The slope shows that for each mile, the cost of the rental increases $.39. The vertical intercept is the daily rental charge before any driving is done. **c.** $c = 0.39m + 45$

Page 422 Exercises and Problems

18. a.

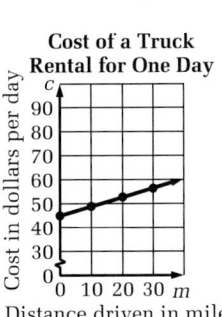

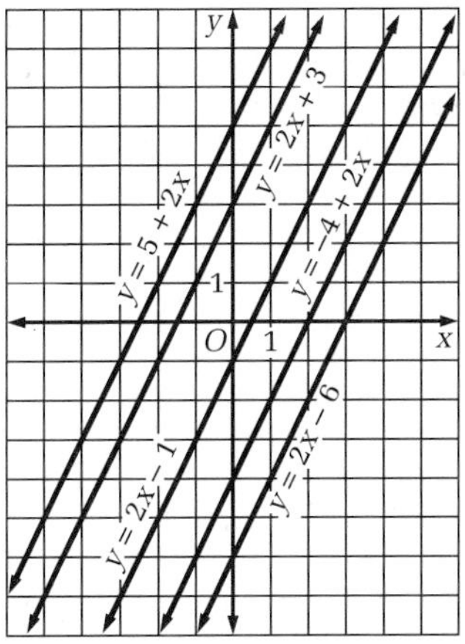

b. Answers may vary. Examples are given. They each have slope 2 because the value of m in each equation is 2. $y = 2x - 6$

Page 424 Exercises and Problems

33. a.

POLYGON	Number of sides	Number of triangles formed	Sum of degrees in all triangles
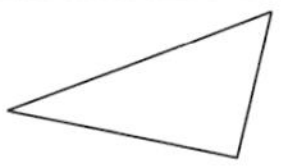	3	1	180
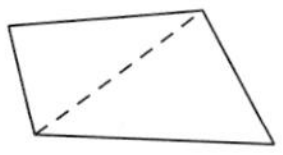	4	2	360
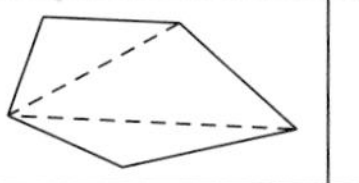	5	3	540
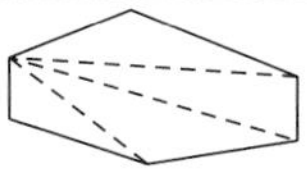	6	4	720
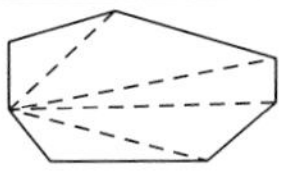	7	5	900
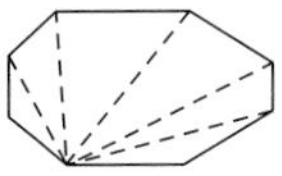	8	6	1080

b.

Sum of Degrees of Angles of Polygons

Number of sides

A-3

c. 1440° **d.** 12 **e.** Yes. slope: 180; vertical intercept: −360; $s = 180n − 360$ **f.** Descriptions and explanations may vary. Examples are given. The number of triangles formed is 2 less than the number of sides. Yes. No. It is linear because the points fall along a line, but it is not direct variation because the line does not contain the origin.

Page 431 Exercises and Problems
20. a.

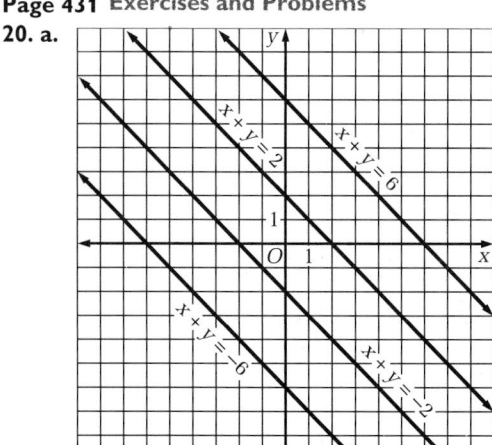

b. 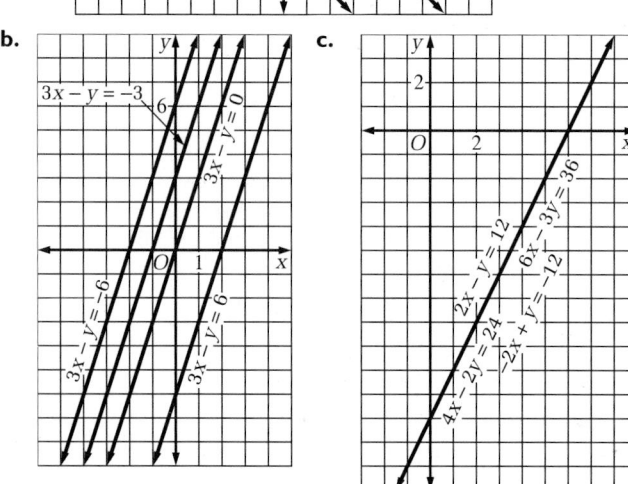 **c.**

d. For parts (a) and (b), the equations in each are all in standard form, with the coefficients of the variables the same and the graphs of the lines parallel. In part (c), the equations are equivalent equations in standard form and all the equations have the same line graph.

Page 432 Exercises and Problems
24. a. $c = 1.25m + 32$

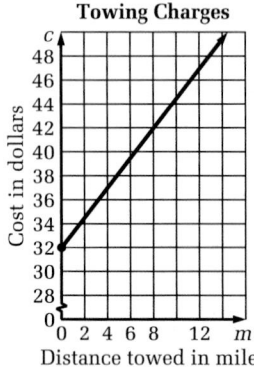

b. 1.25, 32; The slope indicates that the cost increases by $1.25 for each mile towed. The vertical intercept is the cost of hooking up the car.

25. a.

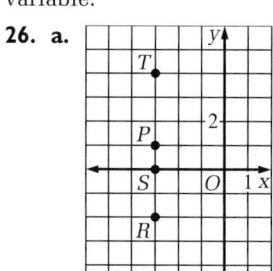

b. All the points lie on a horizontal line and have y-coordinate 4.
c. Yes. Explanations may vary. An example is given. For each value of the control variable, there is only one value for the dependent variable.

26. a.

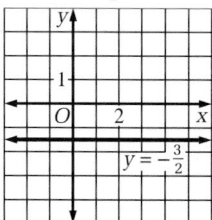

b. All the points lie on a vertical line and have x-coordinate −3.
c. No. Explanations may vary. An example is given. There is more than one value for the dependent variable when the control variable equals −3.

Page 434 Talk it Over
3. The slope is 0.

4. Descriptions and explanations may vary. An example is given. The equation in question 3 is different from the other two because the variable, y, has a coefficient of 2 instead of a coefficient of 1 as the others do. This difference does not change the fact that the graph is a horizontal line. If you divide both sides by 2, you get $y = -\frac{3}{2}$.

Page 446 Exercises and Problems
25. Answers may vary. An example is given.

How to Write Equations for Lines
I.   Using a Graph
    A. Find the slope using two points.
    B. Locate vertical intercept.
    C. Substitute values in $y = mx + b$.
    D. Example: show graph of $y = 2x + 1$.
II. Using Two Points
    A. Find slope.
    B. Use slope, one point, and $y = mx + b$ to find b.
    C. Substitute values in $y = mx + b$.
    D. Example: (−1, 2) and (3, 5).
III. Using One Point and the Slope
    A. Substitute the slope and coordinates of the point in $y = mx + b$ to find b.
    B. Substitute values in $y = mx + b$.
    C. Example: slope = −2, (1, 3).

In this chapter, you will learn three ways to write equations for lines depending on the information you are given. You may have a graph, two points, or one point and the slope. Each way is demonstrated by an example.

Page 454 Checkpoint
4. solution **5.** no solution **6.** solution **7.** Choices of variables may vary. An example is given. Let c be the total cost in dollars and let g be the number of gallons bought. $c = g + 10$, $c = 1.5g$; $g = 20$, $c = 30$ (At 20 gallons, the cost for either is $30.)

Page 460 Exercises and Problems
15. a. Flakies: $.0048/g; Nutri-Flakes: $.0046/g; Wheat Delight: $.0052/g; Treat-O-Wheat: $.0051/g **b.** Nutri-Flakes
c, d.

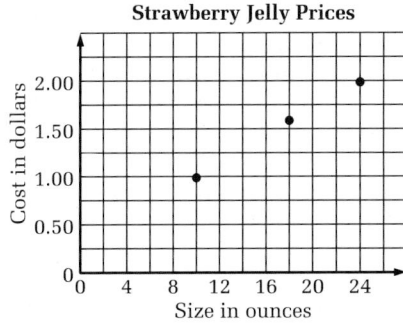

e. above: Wheat Delight at (380, 1.99); below: Flakies at (300, 1.45), and Nutri-Flakes at (355, 1.65) **f.** Wheat Delight is more expensive per gram than Treat-O-Wheat. Nutri-Flakes and Flakies are less expensive per gram than Treat-O-Wheat.
16. Product choices may vary. An example is given.

a. Strawberry Jelly

Size	Cost
10 oz	$.99
18 oz	$1.59
24 oz	$1.99

b. No, the unit price is not constant.

c. Descriptions may vary. An example is given. The three points do not lie on a straight line. However, if I draw a line from the origin through the point (24, 1.99), the other two points will lie above the line, showing that the cost per ounce for these sizes of jelly is greater.
d. Answers may vary. An example is given. Unit price is usually less for larger sizes because it does not cost twice as much to package a product that is twice the size. That is, the unit packaging cost decreases as the size increases. This causes the unit price to decrease as the size of the product increases. **17. a.** (8, 7), (7, 4), (7.5, 6), (5.5, 4)
b, c. The slope is 1.

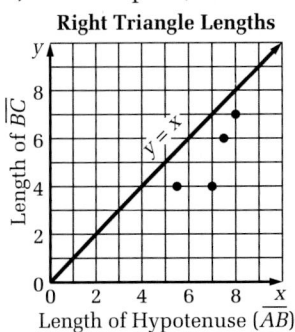

d. below **e.** >; < **f.** Explanations may vary. An example is given. If lines were drawn from the origin to each point on the graph, the lines would lie below the line $y = x$. This means that each line must have a slope less than 1. Since the slopes of these lines are represented by the ratio $\frac{BC}{AB}$, which is also the cosine ratio for $\angle B$, cos B must be less than 1.

Page 461 Exercises and Problems
24.

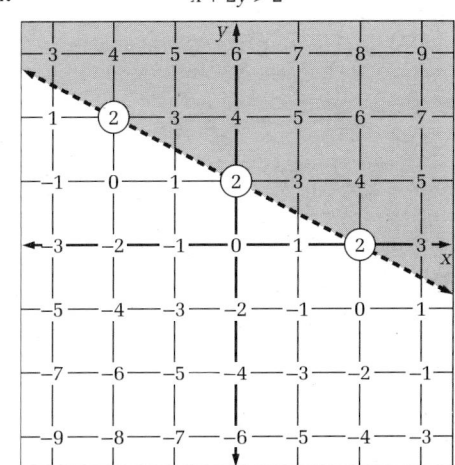

25.

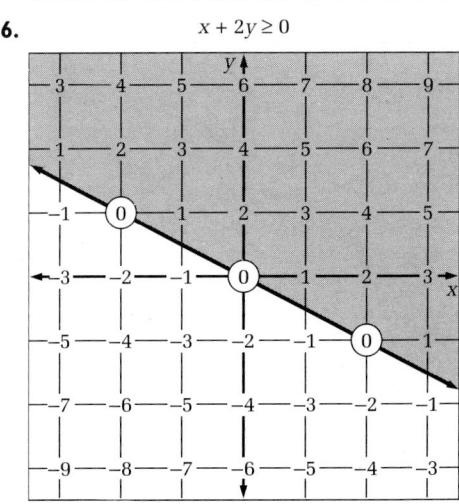

26.

31. Quizzes may vary. An example is given.
Graph each inequality.
1) $y > -2x$ 2) $y \le 3x + 1$ 3) $y < \frac{1}{2}x - 2$ 4) $2x + 4y \ge -8$

5) Geneva has \$20 to buy pencils and notebooks at the store. Pencils cost \$1, and notebooks cost \$2. Write and graph an inequality to represent the amount Geneva can spend at the store. Put the number of notebooks on the vertical axis.

I think my questions are good because they tell if you know how to graph inequalities in both slope-intercept form and standard form. They also tell if you understand the inequality symbols and the use of a solid or dashed line. Question 5 will show if you understand how to model a practical situation using an inequality.

Page 467 Exercises and Problems
19. a. Let n = number of hours worked for neighbor and p = number of hours worked for photographer. $n \le 4$ and $6n + 5p > 90$

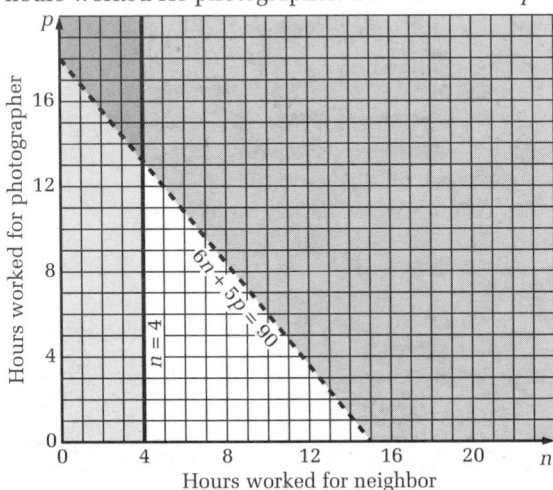

b. Answers may vary. Any values (n, p) in the solution region will enable Nadine to earn more than \$90. For example, Nadine could work 1 hour for her neighbor and 17 hours for the photographer to earn \$91. **20. a.** Let b = number of bowls and p = number of plates. $b \ge 4$ and $6b + 8p \le 50$

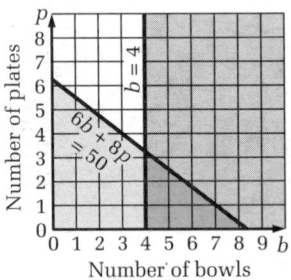

b. 7 bowls and 1 plate

21. a, b.

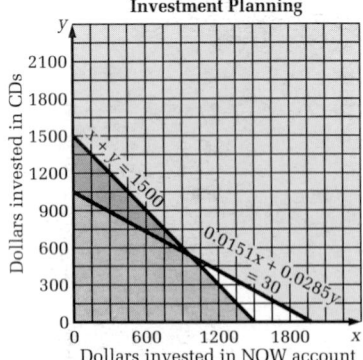

Investment Planning

c. a triangle; vertices may vary due to rounding. An example is given. (0, 1500), (0, 1052.63), (951.49, 548.51)

d. Answers may vary due to rounding. Examples are given.

Vertex	Amount in NOW	Amount in CD	Amount in cash	Interest
(0, 1500)	\$0	\$1500	\$0	\$42.75
(0, 1052.63)	\$0	\$1052.63	\$447.37	\$30.00
(951.49, 548.51)	\$951.49	\$548.51	\$0	\$30.00

e. invest \$951.49 in the NOW account and \$548.51 in the CD

Unit 9

Page 480 Look Back
Answers may vary. An example is given. Inductive reasoning involves using observations to determine a pattern and make a conjecture. The conjecture is based on the observations and cannot be proven for all situations. Deductive reasoning involves using facts, definitions, and accepted principals and properties to show that a situation is always true. It can be used to prove general statements.

Sample situation: I observed that when I did not do my homework, the teacher seemed to call on me more because I did not raise my hand to answer questions. So I decided to raise my hand even when I had not done my homework. In this situation, I used inductive reasoning. I realized that by not doing my homework, I was not prepared for class, I did not understand the material very well, and I did poorly on tests and quizzes. After examining these facts, I used deductive reasoning and decided to start doing my homework so that I could do better in school.

Page 497 Exercises and Problems
14. a. If a number is divisible by 3, then it is divisible by 6. **b.** False. If a number is divisible by 3 and is an odd number, it is not divisible by 6. **15. a.** If the area of a figure is half the area of some parallelogram with the same base and height, then the figure is a triangle. **b.** True. **16. a.** If a point is on the y-axis, then the coordinates of the point are (0, –3). **b.** False. There are an endless number of points on the y-axis. **17. a.** If an event is certain to happen, then the probability of the event is 1. **b.** True. **18. a.** If $x + 3 = 0$, then $(x + 3)(x - 4) = 0$. **b.** True. **19.** Answers may vary. Examples are given. **a.** Statement: If I go to soccer practice after school, then I don't get home until 5:30. Converse: If I don't get home until 5:30, then I went to soccer practice after school. **b.** My statement is true because soccer practice ends at 5:00. Then I take a shower and change, and ride my bicycle home. The converse of my statement is false. I don't have soccer practice every day. Sometimes I go to the library or visit with my friends, so I might not get home until 5:30.

20. Answers may vary. An example is given. Statement: If a shirt originally priced at \$20 is discounted 10%, then the discount is \$2. Converse: If the discount is \$2, then a shirt originally priced at \$20 is discounted 10%. **21.** 0 **22.** 0, 4 **23.** 0, –1 **24.** 0, $-\frac{4}{5}$ **25.** 0, 2 **26.** 0, 3

27. Yes; because $7^2 + 24^2 = 625$ and $25^2 = 625$. **28.** No; because $8^2 + 11^2 = 185$ and $15^2 = 225$. **29.** No; because $4^2 + 5^2 = 41$ and $6^2 = 36$. **30.** Yes; because $24^2 + 10^2 = 676$ and $26^2 = 676$. **31.** No; because $2^2 + 7^2 = 53$ and $8^2 = 64$. **32.** Yes; because $16^2 + 30^2 = 1156$ and $34^2 = 1156$. **33.** Yes. **34.** $l = 15$, $d = 9$; $l = 13$, $d = 5$

Page 506 Exercises and Problems
20. Descriptions may vary. An example is given. Buoyancy is the ability or tendency of an object to float in liquid or to rise in air or gas. Buoyancy also refers to the lifting effect of a fluid. The buoyant force acting on an object is equal to the weight of the fluid displaced by the object (this force acts vertically upward through the center of gravity). The more dense a fluid is, the more buoyant it is said to be (because it will support more weight).

Page 551 Exercises and Problems

6. Concept maps may vary. An example is given.

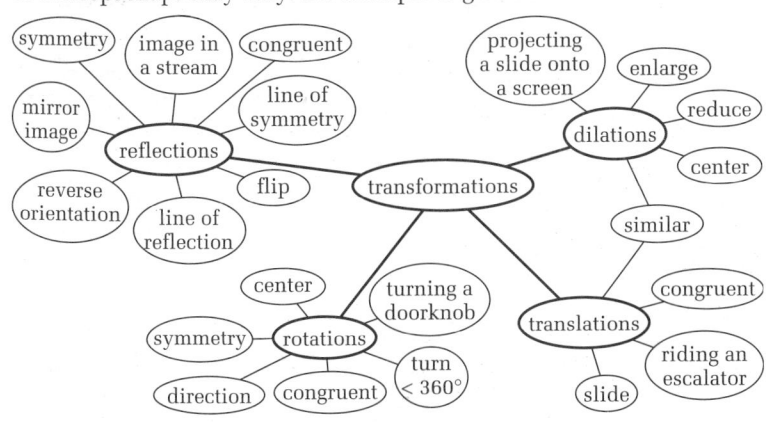

Page 553 Exercises and Problems

23. a. The image is the same as the original figure. b. (−3, 0); (−3, 0) c. (−2, −1); Explanations may vary. An example is given. Point A is 1 unit above the x-axis, so its image is in the same position relative to the y-axis but 1 unit below the x-axis. d. (1, 2); Explanations may vary. An example is given. Point B is 2 units below the x-axis, so its image is in the same position relative to the y-axis but 2 units above the x-axis.

Page 554 Exercises and Problems

29. 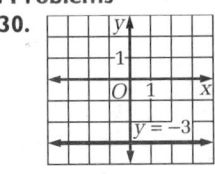   30.

31. a. Answers may vary. The sketches should show an arch-shaped path. b. Yes, each path has a vertical line of symmetry through the highest point. c. Answers may vary. An example is given. All are curved and have a vertical line of symmetry through the highest point, but some are flattened and low while others are steep and high.

Page 556 Exploration

6. a. The graph of $y = x^2$ is translated up if a positive number is added to x^2 and is translated down if a positive number is subtracted from x^2. b. The graph of $y = x^2$ is translated to the left if a positive number is added to x before it is squared, and is translated to the right if a positive number is subtracted from x before it is squared.

Page 567 Exercises and Problems

17. Summaries may vary. An example is given. (1) If the x-squared term has a negative sign in front, the graph opens downward. Example: $y = -x^2$. (2) If a number is added to or subtracted from the x^2- or $-x^2$-term, the graph is translated up or down. This changes the vertex but not the line of symmetry. Example: $y = x^2 - 4$ is symmetric about the y-axis and its vertex is at (0, −4). (3) If a number is added to or subtracted from x before squaring, the graph is translated to the left or the right. This shifts both the vertex and the line of symmetry. Example: $y = (x - 2)^2$ is symmetric about the line $x = 2$ and its vertex is at (2, 0). (4) If the equation of a parabola is written in the form $y =$ factor × factor, you can find the x-intercepts by setting y equal to 0 and using the zero-product property. Example: $y = (x - 4)(x + 3)$ has x-intercepts at 4 and −3 since $x - 4 = 0$ or $x + 3 = 0$. (5) You can always find the y-intercept of a parabola by substituting 0 for x in the equation. Example: $y = x^2 + 3x + 5$ has a y-intercept of 5 since $y = 0^2 + 3 \cdot 0 + 5 = 5$.

Page 597 Exercises and Problems

33. a. Answers may vary. An example is given. For each trinomial, if you find half the coefficient of x and square the result, the product is the constant term.

b.
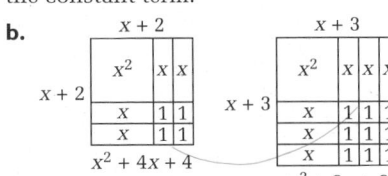
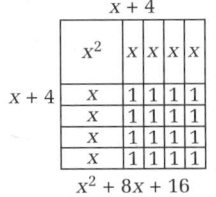
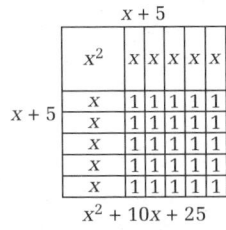

c. Each rectangle is a square. d. (x + 2)(x + 2), (x + 3)(x + 3), (x + 4)(x + 4), (x + 5)(x + 5) e. because each one is the square of a binomial, that is, each trinomial has two identical binomial factors f. Answers may vary. An example is given. $x^2 + 2x + 1 = (x + 1)(x + 1)$ and $x^2 + 12x + 36 = (x + 6)(x + 6)$

Extra Practice

Pages 634–635 Exercises and Problems

50. $\left(-1\frac{1}{3}, 4\frac{1}{3}\right)$ 51. 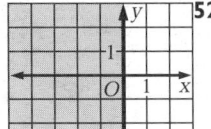   52.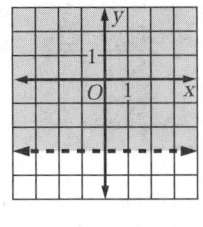

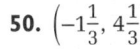
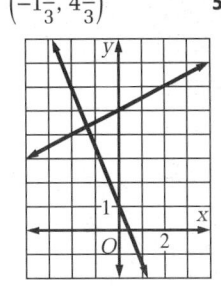

53. 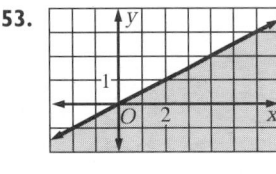   54. 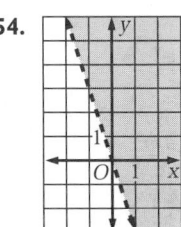   55.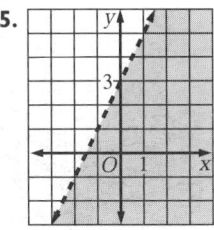

56. 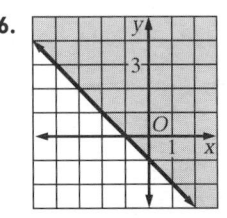   57.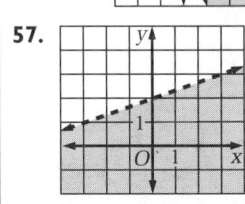

58. 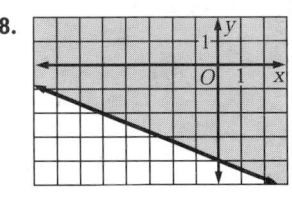   59.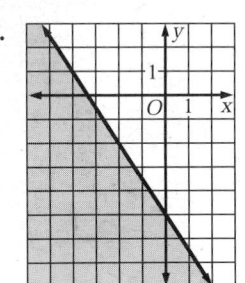

A-7

60.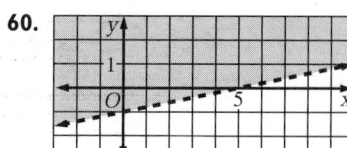

61.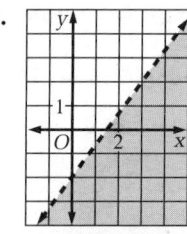

62.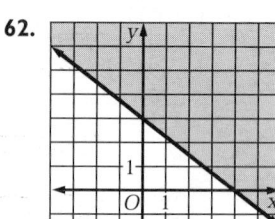

63.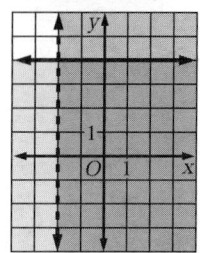

64.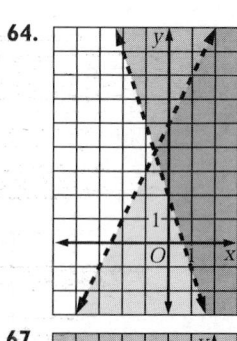

65.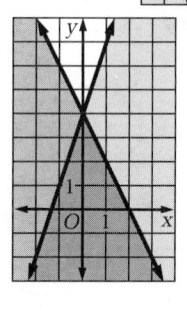

66.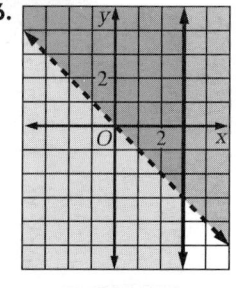

67.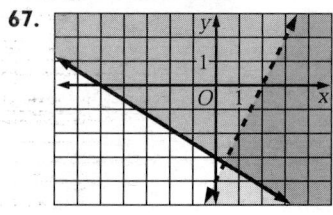

68.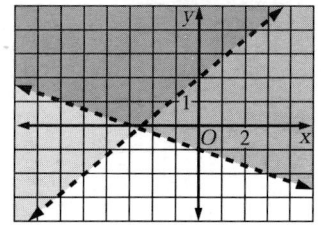